THE PROBOSCIDEA

Evolution and Palaeoecology of Elephants and Their Relatives

FRONTISPIECE

This rendition depicts how the Proboscidea may have undergone three major radiation events, starting from the bottom. Depicted at the centre front of each stage are species that form the spine of these radiation events: *Palaeomastodon* at the centre of the first, *Gomphotherium* at the centre of the second, and *Primelephas* at centre of the third. *Moeritherium*, in the first radiation, is depicted standing in water; gomphotheres and stegodons in the second radiation are shown in varied habitats; *Mammuthus*, in the third radiation, stands in snow-covered tundra, *Elephas* is in front of Elephant Rock in Sri Lanka at the centre back, and *Loxodonta* is in front of an acacia trees in East Africa. *Anthracobune*, known only from teeth, skull fragments, and a few isolated postcranial bones, is not shown, but is now classified within the Proboscidea. This South Asian genus is of early to middle Eocene age, and belongs in the first radiation.

Artwork by Gary H. Marchant, with suggestions from Jann S. Grimes and Dominique Visset.

For simplicity, whenever possible, we used the monograph of Osborn (1936, 1942, details below; see also Osborn 1942, pp. 1604–5) as a primary source; other references are indicated (some illustrations are modified). The sequence in which these taxa appear follows the classification (Table 16.1 and Appendix B) and Figure 16.5. Special thanks to V. Cannella, N. Court, R. W. Graham, J. E. Kalb, W. D. Lambert, H. Saegusa, and D. Wilson for their comments and suggestions.

Key to proboscideans illustrated

1. *Moeritherium trigodon* (after Osborn 1936, pp. 18, 38, 66, 79).
2. *Numidotherium koholense* (after Mahboubi *et al.* 1986, pp. 26–39; N. Court, personal communication 1992).
3. *Barytherium grave* (after Savage and Long 1986, p. 148; N. Court, personal communication 1992).
4. *Prodeinotherium bavaricum* (after Osborn 1936, p. 101).
5. *Palaeomastodon beadnelli* (after Osborn 1936, pp. 66, 149).
6. *Mammut americanum* (after Osborn 1936, frontispiece and pp. 164, 190).
7. *Gomphotherium angustidens* (after Osborn 1936, p. 287).
8. *Platybelodon grangeri* (after Osborn 1936, pp. 118, 460).
9. *Rhynchotherium tlascalae* (after Osborn 1936, pp. 118, 503).
10. *Cuvieronius hyodon* (after Osborn 1936, pp. 118, 542, 543; Simpson and Paula Couto 1957, Plate 1).
11. *Tetralophodon longirostris* (after Osborn 1936, p. 342).
12. *Anancus arvernensis* (after Osborn 1936, pp. 118, 630).
13. *Stegolophodon cautleyi* (after Osborn 1936, pp. 849, 850; H. Saegusa, personal communication 1992).
14. *Stegodon ganesa* (after Osborn 1936, pp. 855, 872).
15. *Primelephas gomphotheroides* (after Maglio 1973, frontispiece; J. E. Kalb, personal communication 1992).
16. *Loxodonta africana* (after Osborn 1942, pp. 1147, 1189, 1190, 1194, 1196).
17. *Elephas maximus* (after Osborn 1942, pp. 1147, 1306, 1311, 1314, 1322, 1326).
18. *Mammuthus primigenius* (after Osborn 1942, pp. 1116, 1126, 1147).

G H Marchant '94

THE PROBOSCIDEA

Evolution and Palaeoecology of Elephants and Their Relatives

Edited by

JEHESKEL SHOSHANI

Wayne State University, Detroit, Michigan USA and

Elephant Research Foundation, Bloomfield Hills, Michigan, USA

and

PASCAL TASSY

URA 12, CNRS, Université Paris VI, Laboratoire de Paléontologie des Vertébrés et de Paléontologie Humaine, Paris, France

OXFORD UNIVERSITY PRESS

This book has been printed digitally and produced in a standard specification in order to ensure its continuing availability

OXFORD
UNIVERSITY PRESS

Great Clarendon Street, Oxford OX2 6DP

Oxford University Press is a department of the University of Oxford.
It furthers the University's objective of excellence in research, scholarship, and education by publishing world-wide in

Oxford New York

Auckland Bangkok Buenos Aires Cape Town Chennai
Dar es Salaam Delhi Hong Kong Istanbul Karachi Kolkata
Kuala Lumpur Madrid Melbourne Mexico City Mumbai Nairobi
São Paulo Shanghai Taipei Tokyo Toronto

Oxford is a registered trade mark of Oxford University Press
in the UK and in certain other countries

Published in the United States
by Oxford University Press Inc., New York

Reprinted 2004

ISBN 0-19-854652-1

Antony Rowe Ltd., Eastbourne

Dedication

This book is dedicated to those who devoted much of their careers to studying fossil and living proboscideans. Most notable among these individuals was Henry Fairfield Osborn (1857–1935). The pinnacle of his career was his monumental two volume publication entitled *Proboscidea*, published posthumously, one volume in 1936, the second in 1942, the inspiration of this book. Included in this present volume is a contribution by the late Heinz Tobien (1911–93) who died in March 1993; Tobien's many publications and scientific insight during the past four decades on the evolution of the Proboscidea will continue to enrich us. Others who devoted much of their lives to the study of proboscideans were: Charles William Andrews (1866–1924), Erwin Hinckley Barbour (1856–1947), Michel Beden (1941–84), Paulus Edward Pieris Deraniyagala (1900–73), Abdel Wahid Gaziry (1941–89), Dirk Albert Hooijer (1919–93), and Hikoshichiro Matsumoto (1887–1975).

Henry Fairfield Osborn
1857–1935

Heinz Tobien
1911–1993

FOREWORD

VINCENT J. MAGLIO

More than 200 years ago the first fossil proboscideans were described in the world's literature, nearly a century before Charles Darwin formalized and documented the theory of evolution. During the 150 years culminating in the publication of Henry Fairfield Osborn's classic treatise on the order, an overwhelming proliferation of taxa was proposed. By 1939, 90 fossil genera and over 550 extinct species of Proboscidea had been named. Since that time, modern concepts of evolution, population variability, sexual dimorphism, palaeogeography, and systematics have reversed this trend, so that today only about one-third of that former taxonomic diversity is recognized as valid. Nevertheless, the Proboscidea remains as one of the great mammalian orders of the Cenozoic, one that achieved enormous morphological variety, and wide ecological adaptation.

It has been 50 years since the second volume of Osborn's treatise was published posthumously. During these years a great deal of new fossil material has been amassed, and a large number of palaeontologists, biologists, and geologists have been drawn to these wondrous animals. The present volume brings together these ideas, and summarizes our current knowledge of the origin and evolution of this remarkable group of mammals. Studies on dental homologies and attention to skeletal morphology have opened renewed interest in the position of the Proboscidea among eutherians, and the relationships of several other early mammalian orders. Immunological studies on residual proteins, whilst still in their infancy and not yet applicable to Pliocene–Miocene fossils, are none the less very exciting developments that may eventually provide supporting evidence for our changing systematic concepts. Fossil proboscideans from China, the Middle East, and the North Sea have been added to our knowledge, and re-evaluations of several groups, such as the Elephantidae and Stegodontidae, have shed new light on major evolutionary processes. Correlation of fossil remains with palaeoecological and palaeobotanical evidence is allowing us for the first time to visualize proboscideans as living populations, rather than just mineralized dental characteristics. The mechanisms controlling zoogeographical distribution and insular dwarfism are beginning to allow a dynamic evolutionary history to be assembled for this group. Modern concepts of plate tectonics and transcontinental connections, combined with improved understanding of palaeoclimatic and palaeobotanical conditions resulting in 'filter bridges' between major geographical regions are providing us with a better understanding of the distribution of fossil proboscidean taxa. Finally, the extraordinary extinctions that devastated the order in the late Pleistocene are coming under increasing scrutiny as we investigate life history data deduced from dental lamination studies, seasonal mortality, and growth rates.

The evolutionary success of the Proboscidea is reflected by their morphological and taxonomic diversity, their nearly world-wide distribution on every continent except Australia and Antarctica, and their longevity through nearly 50 million years. Their great past ability to migrate and adapt to changing climatic and interspecific competitive conditions provides a unique laboratory for the testing of evolutionary theories and the development of new concepts.

The impetus for these studies on fossil proboscideans, while interesting from an evolutionary point of view, is not of academic interest alone. The human race seems determined to drive all other living species inexorably to the brink of extinction. Our shameful inability to control world environments, the dwindling and fragmentation of elephant natural ranges, and the scandalous poaching fed by the world's senseless hunger for ivory trinkets will ultimately doom the two remaining species to memory and to lifeless museum exhibits. It is only with a better understanding of their past and present that we can hope to plan for their future.

The knowledge summarized in this book is only a beginning, not an end. There remains much to be done, and time is growing short. We can only hope that the continued efforts of dedicated workers in numerous disciplines, combined with public education and governmental co-operation may avert yet another great evolutionary disaster ... the final extinction of the order Proboscidea.

THE EDITORS

Jeheskel (Hezy) Shoshani teaches Comparative Vertebrate Morphology and Evolution at Wayne State University, Detroit, Michigan, USA. He is interested in cladistic relationships among mammalian orders, with emphasis on proboscideans, sirenians, and hyracoideans. Hezy has conducted conservation and palaeontological related fieldwork in Africa, Asia, and North America. He edited the popular book *Elephants: majestic creatures of the wild* (Rodale Press, Emmaus, Pennsylvania).

Pascal Tassy teaches palaeontology and systematics at the University Pierre-et-Marie-Curie, Paris, France. He is interested in mammalian evolution and phylogenies—especially of proboscideans and allies. Pascal has conducted palaeontological fieldwork in various parts of the world, including France, Bulgaria, Pakistan, and Kenya. He has written popular books such as *Le message des fossiles* (Hachette, Paris) (translated as *The message of fossils* (McGraw-Hill, New York)).

PREFACE

This volume was first conceived in June 1979 at a symposium on extant elephants, organized by J. Shoshani for the annual meeting of the American Society of Mammalogists in Corvallis, Oregon, USA. But it was not until August 1989, when a symposium entitled 'Evolution and Paleoecology of Proboscidea' was held in Rome, Italy, during the Fifth International Theriological Congress, that these ideas crystallized. There were 22 papers and posters presented at the symposium organized by J. Shoshani and P. Tassy. These works were updated and expanded, and invited papers were added to enlarge the scope of the studies, so that this volume comprises a panorama of the many aspects of proboscidean evolution.

The Proboscidea: evolution and palaeoecology of elephants and their relatives offers a distillation of the current scholarly research on some of the major aspects of proboscidean evolution since the 1936 and 1942 publications of *Proboscidea* by Henry Fairfield Osborn. It is hoped that this volume will serve as an introduction to the present knowledge of the group, not only for proboscideanologists, but also for zoologists, palaeontologists, archaeologists, and students of evolution in general.

A historical overview of the Proboscidea, basic anatomy, and nomenclature introduces the subject (Part I). Diversity of proboscideans covering a span of nearly 50 million years gives a palaeontological key for understanding trends in their evolution. Comparison with other extant mammals sheds some light on the early differentiation of the group. Most of the literature (including morphological and molecular data) converges on the conclusion that sea cows, order Sirenia, are the extant sister group of proboscideans, and together with Desmostylia they constitute the monophyletic taxon Tethytheria. Yet, the relation of tethytheres to other mammalian orders, especially hyracoids and perissodactyls is a much debated question (Part II).

In a general sense, the Proboscidea may be viewed as having undergone three major radiation events — a scenario which is portrayed in the Frontispiece. The early proboscideans, including mammutids which became extinct only 10 000 years ago, constitute the first radiation (Part III). A traditional phylogenetic image of elephants rooted in a large ancestral group named the Gomphotheriidae has long been accepted. Introduction of cladistic methods about three decades ago has helped elucidate this picture, but also yielded new problems, such as those of relationships among 'gomphotheres', stegodontids, and mammutids. Conflicting views on the definition of amebelodons and the monophyly of stegodontids exist. The paraphyletic gomphotheres, amebelodons, platybelodons (shovel tuskers), and those closer to true elephants, including Stegodontidae, form the second radiation (Part IV). The family Elephantidae, differentiated more than 5 million years ago, embodies the third radiation (Part V). This section is devoted to the evolution and palaeobiology of mammoths, as well as detailed studies of dwarfism by insularity, which occurred several times in elephantids and in stegodontids.

Elephants, as the largest living terrestrial animals, have always fascinated people. The two extant proboscidean species, *Loxodonta africana* from Africa, and *Elephas maximus* from Asia, are endangered species. As such, they need a determined policy for conservation. Once a formidable group, found on all continents except Australia and Antarctica, the proboscideans now are restricted to these two species. Hence, elephants offer a unique model for understanding related mammalian forms of the past. Extinction of the living elephants could be a scientific disaster, as well as a social one. The core of a major division of this book (Part VI) consists of the present status of living elephants and patterns of proboscidean extinctions. The summary chapter has been written in a 'user friendly' style to enable a wide range of students to view this complex order in as simple a presentation as possible (Part VII).

We are fortunate to have had the opportunity to edit this volume and to be in company with some of the most notable students of this order. We thank all the authors who contributed to the knowledge of one of the most remarkable examples of mammalian evolution, the Proboscidea.

January 1993
Bloomfield Hills, Michigan
Paris

J. S.
P. T.

PREFACE

Ce livre a d'abord été conçu en juin 1979 à l'occasion d'un symposium sur les éléphants actuels, organisé par J. Shoshani dans le cadre de la réunion annuelle de l'American Society of Mammalogists à Corvallis, Oregon, USA. Mais c'est en Italie, à Rome, en août 1989, que l'idée se cristallisa lors du 5ème Congrès Thériologique International, avec l'organisation, par J. Shoshani et P. Tassy, d'un symposium intitulé « Evolution and Paleoecology of Proboscidea ». Vingt-deux communications orales et posters y furent présentés. Ces travaux ont été mis à jour et approfondis, et des articles invités sont venus s'ajouter afin d'élargir le champ des études, de telle sorte que le livre puisse être un panorama des aspects nombreux de l'évolution des proboscidiens.

The Proboscidea: evolution and palaeoecology of elephants and their relatives offre un concentré des recherches actuelles sur quelques-uns des aspects les plus importants de l'évolution des proboscidiens, menées depuis 1936 et 1942, dates de la publication de *Proboscidea* par Henry Fairfield Osborn. Nous espérons que ce volume servira d'introduction à l'état des connaissances sur le groupe, non seulement auprès des « proboscideologistes » mais aussi des zoologistes, des archéologues, des paléontologistes et des évolutionnistes en général.

Une revue historique des Proboscidea, de leur anatomie et de la nomenclature, introduit le sujet (Ière Partie). La diversité des proboscidiens couvre à peu près les derniers 50 millions d'années et donne une clé paléontologique pour comprendre leurs tendances évolutives. La comparaison avec les autres mammifères actuels éclaire la différenciation initiale du groupe. Dans sa grande majorité, la littérature (y compris les données morphologiques et moléculaires) converge vers une conclusion: les vaches-marines (ordre des Sirenia) sont le groupe frère actuel des proboscidiens, et, avec les Desmostylia, ils constituent le taxon monophylétique des Tethytheria. II reste que la parenté des téthythères avec les autres ordres de mammifères, notamment les hyracoïdes et les périssodactyles, est une question vivement débattue (IIème Partie).

De façon générale, on peut considérer que les Proboscidea ont connu trois radiations majeures – un scénario qui est illustré par le Frontispice. Les premiers proboscidiens, inclus les mammutidés qui se sont éteints il n'y a que 10 000 ans, forment la première radiation (IIIème Partie). On a depuis longtemps proposé une image phylogénétique devenue traditionnelle où les éléphants s'enracinent dans un vaste groupe ancestral nommé les Gomphotheriidae. On a commencé à améliorer cette image grâce à l'introduction des méthodes cladistiques depuis une trentaine d'années mais de nouveaux problèmes ont alors surgi, comme ceux touchant aux parentés entre « gomphothères », stégodontidés et mammutidés. Des points de vue contradictoires persistent sur la définition des amébélodontes et sur la monophylie des stégodontidés. Le groupe paraphylétique des gomphothères, amébélodontes, platybélodontes (à défenses-en-pelle) et les formes plus proches des vrais éléphants, Stegodontidae inclus, forment la deuxième radiation (IVème Partie). La famille des Elephantidae s'est différenciée il y a plus de cinq millions d'années et concrétise la troisième radiation (Vème Partie). Cette section est consacrée à l'évolution et à la paléobiologie des mammouths et au nanisme insulaire qui s'est manifesté plusieurs fois chez les éléphants et stégodontidés.

Les éléphants sont les plus grands animaux terrestes: ils ont toujours été objet de fascination. Les deux espèces actuelles, *Loxodonta africana* en Afrique et *Elephas maximus* en Asie, sont en danger. Elles nécessitent une politique de conservation ferme et déterminée. D'un groupe extraordinaire, connu sur tous les continents, à l'exception de l'Australie et de l'Antarctique, il ne subsiste plus aujourd'hui que ces deux espèces. C'est pourquoi les éléphants offrent un modèle unique pour comprendre les mammifères du passé qui leur sont apparentés. L'extinction des éléphants actuels serait un désastre à la fois scientifique et social. Le présent statut des éléphants et les modalités des différentes extinctions des proboscidiens forment le centre d'une importante partie de ce livre (Vlème Partie). Le chapitre final, en forme de résumé, a été écrit dans un style « facile » pour permettre au plus grand nombre de lecteurs de saisir rapidement les aspects majeurs de cet exceptionnel ordre de mammifères (VIIème Partie).

Nous avons eu la bonne fortune de pouvoir diriger un tel volume, et de côtoyer ainsi quelques-uns des plus éminents spécialistes de cet ordre. Nous remercions tous les auteurs qui ont contribué à la connaissance de l'un des plus remarquables exemples d'évolution mammalienne: les Proboscidea.

Janvier 1993
Bloomfield Hills, Michigan
Paris

J.S.
P.T.

ACKNOWLEDGEMENTS

We take this opportunity to acknowledge the efforts by Ernesto Capana and Luigi Boitani of the Organizing Committee for the Fifth International Theriological Congress (Rome, Italy) which was the original impetus for the manuscripts included in this volume. Heartfelt thanks are extended to all our contributors and to the Scientific Advisory Committee of this volume for their assistance. Many of the contributors also served as reviewers, and we are greatly indebted to them (specific acknowledgements are given at the end of each chapter). Jon E. Kalb's suggestions for the layout of the book and his efforts in renewing contact with Vincent J. Maglio are much appreciated. We also extend our appreciation for services by the libraries at Wayne State University, Michigan (especially to Jacqueline Beaty and Barbara Overbeck), the Elephant Research Foundation, Michigan (especially to M. Pamela Bedore, Estelle R. Davidson, and Ronald Orenstein), and Cranbrook Institute of Science, Michigan (CIS, especially to Gretchen Young-Weiner), and to Lawrence Hutchinson and Stewart Bailey (of CIS) for the use of a laser printer.

The staff at the American Museum of Natural History, New York, especially Malcolm C. McKenna, Susan K. Bell, Joan Davis, and the library personnel (Sarah F. Granato in particular), are recognized for their continuous help with various aspects of this publication. J. Shoshani is indebted to his best friend, his wife Sandra, who has been extremely tolerant and supportive in very difficult times. P. Tassy wishes to acknowledge INSERM, U.155 (Director J. Feingold), URA 12, CNRS (Director P. Taquet) for logistic support and Dominique Visset and Véronique Barriel for help with illustrations and typing.

Understandably, the compilation of this volume has been a slow process, involving the co-ordination and mobilization of sources from around the world. Helping with the 'donkey' work of this volume was a small staff in Michigan, including Jann S. Grimes and Gary H. Marchant (both skilfully drew and/or corrected some illustrations) and Eleanor C. Marsac, the late Jerry G. Marsac, Jules L. Pierce, and Susan Wolak (who helped with typing, proofreading, editing, and translations from other languages). Computer assistance was provided by George Overbeck. Special gratitude is extended to Oxford University Press, for without their extended support, the publication of this book might have been further delayed.

Cover illustration: *Gomphotherium angustidens*, artwork by Gary H, Marchant, modified after Osborn (1936) and Tassy (1985).

J. S.
P. T.

CONTENTS

PART IV THE SECOND RADIATION: GOMPHOTHERES AND STEGODONTIDS

PART V THE THIRD RADIATION: ELEPHANTIDAE

PART VI PALAEOECOLOGY, TAPHONOMY, EXTINCTION, AND CONSERVATION

PART VII SUMMARY AND CONCLUSIONS

APPENDICES

CONTRIBUTORS

LARRY D. AGENBROAD
Department of Geology, Quaternary Studies Program, Northern Arizona University, Flagstaff, Arizona 86011, USA

ALEXANDER O. AVERIANOV
Zoological Institute Russia, Academy of Sciences, Universitetskaya nab. 1, 199034 St. Petersburg, Russia

FACHROEL AZIZ
Quaternary Geology Laboratorium, Jalan Dr. Junjunan 236, Bandung, Indonesia

GORDEN L. BELL, Jr.
Museum of Geology, South Dakota School of Mines and Technology, Rapid City, South Dakota 57701, USA

LUCIA CALOI
Dipartimento di Scienze della Terra, Universita degli Studi, 'La Sapienza', Piazzale Aldo Moro, 5-00100 Roma, Italy

KENNETH E. CAMPBELL, Jr.
Natural History Museum of Los Angeles County, 900 Exposition Boulevard, Los Angeles, California 90007, USA

RODOLFO M. CASAMIQUELA
Fundación Ameghino, Casilla 375, (8500) Viedma, Rio Negro, Argentina

EDWIN H. COLBERT
Museum of Northern Arizona, Route 4, Box 720, Flagstaff, Arizona 86001, USA

NICHOLAS COURT
Sedgwick Museum, University of Cambridge, Downing Street, Cambridge CB2 3EQ, England

JOHN DE VOS
National Museum of Natural History, P. O. Box 9517, 2300 RA, Leiden, The Netherlands

THOMAS D. DILLEHAY
Department of Anthropology, University of Kentucky, 211 Lafferty Hall, Lexington, Kentucky 40504, USA

IAIN DOUGLAS-HAMILTON
Co-ordinator EEC African Elephant Survey and Conservation Programme, P. O. Box 54667, Nairobi, Kenya

JOSEPH P. DUDLEY
Department of Biology and Wildlife, University of Alaska, Fairbanks, Alaska 99775-0180, USA

MARTIN S. FISCHER
Institut für Spezielle Zoologie und Evolutionsbiologie, Friedrich-Schiller Universität, Erbeststr. 1, 07743 Jena, Germany

DANIEL C. FISHER
Museum of Paleontology and Department of Geological Sciences, University of Michigan, Ann Arbor, Michigan 48109-1079, USA

CARL D. FRAILEY
Johnson County Community College, Overland Park, Kansas 66210, and Museum of Natural History, University of Kansas, Kansas 66045, USA

DAVID J. FROEHLICH
Vertebrate Paleontology Laboratory (Texas Memorial Museum), J. J. Pickle Research Campus, and Department of Geological Sciences, University of Texas, Austin, Texas 78712, USA

JIAN GUAN
Department of Paleontology, Beijing Natural History Museum, 126 Tian Qiao Street, Beijing, 100050, People's Republic of China

JOHN M. HARRIS
Earth Science Division, Natural History Museum, 900 Exposition Blvd., Los Angeles, California 90007, USA

JON E. KALB
Vertebrate Paleontology Laboratory (Texas Memorial Museum), J. J. Pickle Research Campus, University of Texas, Austin, Texas 78712, USA

TASSOS KOTSAKIS
Centro di Studio per la Geologia dell' Italia Centrale, C.N.R., c/o Dipartimento di Scienze della Terra, Universita degli Studi 'La Sapienza', Piazzale Aldo Moro, 5-00100 Roma, Italy

W. DAVID LAMBERT
Florida Museum of Natural History, Department of Natural Sciences, Museum Road, University of Florida, Gainesville, Florida 32611-2035, USA

ADRIAN LISTER
Department of Biology, University College London, Gower Street, London WC1E 6BT, England

JEROLD M. LOWENSTEIN
Box 0568 MR122, University of California, San Francisco, California 94143, USA

W. PATRICK LUCKETT
Department of Anatomy, University of Puerto Rico, Medical Sciences Campus, GPO Box 365067, San Juan, Puerto Rico 00936, USA

VINCENT J. MAGLIO
c/o Jeheskel Shoshani, Department of Biological Sciences, Wayne State University, Detroit, Michigan 48202, USA

ANA V. MAZO
Museo Nacional de Ciencias Naturales, Departamento de Paleontologia, José Gutiérrez Abascal, 2,28006 Madrid, Spain

JIM I. MEAD
Department of Geology, Quaternary Studies Program, Northern Arizona University, Flagstaff, Arizona 86011, USA

FRANCES MICHELMORE
Consultant Ecologist/GIS Specialist, EEC African Elephant Survey and Conservation Programme, United Nations Environment Programme/Global Environment Monitoring System, P. O. Box 47074, Nairobi, Kenya

DICK MOL
Gudumholm 41, 2133 HG Hoofddorp, The Netherlands

MARIA R. PALOMBO
Dipartimento di Scienze della Terra, Universita degli Studi 'La Sapienza', Piazzale le Aldo Moro, 5-00100 Roma, Italy

CARMELO PETRONIO
Dipartimento di Scienze della Terra, Universita degli Studi 'La Sapienza', Piazzale le Aldo Moro, 5-00100 Roma, Italy

LIDIA ROMEO-PITTMAN
Instituto Geológico, Minéro y Metalúrgico, Av. Canadá, San Borja, Apartado 889, Lima 41, Péru

V. LOUISE ROTH
Zoology Department, Duke University, Durham, North Carolina 27708-0325, USA

HARUO SAEGUSA*
Department of Geology and Mineralogy, Faculty of Science, Kyoto University, Kyoto, Japan. *Current address: Museum of Nature and Human Activities, Hyogo, Yayoigaoka, Sanda City 669-13, Japan

CHARLES SANTIAPILLAI
Department of Zoology, University of Peradeniya, Peradeniya, Sri Lanka

JEFFREY J. SAUNDERS
Illinois State Museum, Research and Collections Center, 1011 E. Ash Street, Springfield, Illinois 62703, USA

ROBERT J. G. SAVAGE
Department of Geology, The University of Bristol, Bristol BS8 1RJ, England

JEHESKEL SHOSHANI
Department of Biological Sciences, Wayne State University, Detroit, Michigan 48202, USA
and
Elephant Research Foundation
106 East Hickory Grove Road
Bloomfield Hills, Michigan 48304, USA

PAUL Y. SONDAAR
Natuurmuseum Rotterdam, Westzeedijk 345, P.O. Box 23452, 3001 KL Rotterdam, The Netherlands

RAMAN SUKUMAR
Centre for Ecological Sciences, Indian Institute of Science, Bangalore — 560 012, India

PASCAL TASSY
Université Paris VI, Laboratoire de Paléontologie des Vertébrés et de Paléontologie Humaine (URA 12, CNRS), 4, Place Jussieu, 75252, Paris Cedex 05, France

EITAN TCHERNOV
Department of Evolution, Systematics and Ecology. The Institute of Life Sciences, The Hebrew University of Jerusalem, 91904, Jerusalem, Israel

HEINZ TOBIEN
Johannes-Gutenberg Universität, Institüt für Geowissenschaften, Postfach 3980, Saarstrasse 21, D 6500 Mainz 1, Germany

NANCY E. TODD
Geobiology Program, The George Washington University, Washington, D.C. 20052, USA

GERT D. VAN DEN BERGH
c/o National Museum of Natural History, P.O. Box 9517, 2300 RA Leiden, The Netherlands

HANS VAN ESSEN
Burgemeester Bloemersstraat 62, 6952 BB Dieren, The Netherlands

ROBERT M. WEST*
Cranbrook Institute of Science, Box 801, Bloomfield Hills, Michigan 48304, USA. *Current address: RMW Science Action, 2900 Mckinley Street, NW, Washington DC 20015, USA

Members of the Scientific Advisory Committee

JOHN EISENBERG
Department of Natural Sciences, Florida Museum of Natural History, Museum Road, University of Florida, Gainesville, Florida 32611-20358, USA

RUSSELL W. GRAHAM
Illinois State Museum, Research and Collections Center, 1011 E. Ash Street, Springfield, Illinois 62703, USA

RICHARD M. LAWS
No. 3 The Footpath, Cambridge CB3 7PX, England

MALCOLM C. MCKENNA
Department of Vertebrate Paleontology, American Museum of Natural History, Central Park West at 79th Street, New York 10024, USA

Papers and posters presented at the Fifth International Theriological Congress, Rome, Italy, 22–29 August 1989

Listed are names of authors and the titles (only for those who are not contributing to this volume) as they appear in the *Abstracts of papers and posters*, Vol. I, pp. 139–67.

LARRY D. AGENBROAD and JIM I. MEAD
LUCIA CALOI, TASSOS KOTSAKIS, MARIA RITA PALOMBO, AND CARMELO PETRONIO
C. S. CHURCHER
Mammuthus as an indicator of the age of deposits in the Old Crow Basin, Yukon Territory, Canada.

MARTIN S. FISCHER
GARY HAYNES
When the mountains fall down: African elephants and bone accumulations in the wild.

YUKISHIGE KOZAWA, MASAHIKO AKIYAMA, HIDETOSHI KAMIYA, and ELEPHANT RESEARCH GROUP OF JAPAN
Feeding habit estimation by carbon isotope composition and teeth structure in fossil elephant.

HENRYK KUBIAK
Paleoecology of middle-European Quaternary proboscideans.

ADRIAN M. LISTER
JEROLD M. LOWENSTEIN and JEHESKEL SHOSHANI
ANA V. MAZO
DICK MOL
V. LOUISE ROTH
HARUO SAEGUSA
JEFFREY SAUNDERS
ROBERT J. G. SAVAGE and NICHOLAS COURT
Skeletal anatomy of *Barytherium.*

JEHESKEL SHOSHANI, ROBERT J. G. SAVAGE, and ROBERT M. WEST
PAUL Y. SONDAAR, FACHROEL AZIZ, GERT D. VAN DEN BERGH, and S. TASEER HUSAIN
PASCAL TASSY
EITAN TCHERNOV and JEHESKEL SHOSHANI
HERBERT THOMAS, PASCAL TASSY, and SEVKET SEN
Paleogene proboscidean remains from the southern Dhofar (sultanate of Oman).

S. DAVID WEBB
Early mammoths in North America.

HENRY FAIRFIELD OSBORN AND THE PROBOSCIDEA

BY EDWIN H. COLBERT

Henry Fairfield Osborn was a large man who had large thoughts that inspired him to undertake large projects, which, on the whole, led to large results. It was Osborn, perhaps more than anyone, who made the American Museum of Natural History in New York the largest museum of its kind in the world. He visualized and put into practice the extensive collecting programme and the varied research projects in vertebrate palaeontology that made the Museum pre-eminent as a world centre of this scientific discipline. He was the author of an extended series of palaeontological papers and monographs, culminating in the massive two-volume treatise on the Proboscidea, in its own way, a final expression of Osborn's predilection for bigness.

Although in his early years he was very much concerned with Mesozoic mammals — among the tiniest of fossils — his later research interests were involved with extinct animals of large and even gigantic size, resulting in large, and even gigantic, publications. Such, for example, was the American Museum Memoir on sauropod dinosaurs (Osborn and Mook 1921) the Memoir on fossil horses (largely the work of W. D. Matthew) (Osborn 1918), the two-volume monograph on titanotheres, published by the United States Geological Survey (Osborn 1929*c*, *d*), and the Proboscidea monograph, a special publication of the American Museum (Osborn 1936, 1942).

Osborn was not a person given to working alone; almost always, his manuscripts were prepared with the help of collaborators (Matthew, for example) or assistants (generally anonymous). Indeed, the preparation of so massive a publication as the Proboscidea monograph required the efforts of a considerable number of people. During the last 5 years of Osborn's life, I was his part-time research assistant, having succeeded C. C. Mook, Mook in turn having followed W. K. Gregory, who was Osborn's assistant for many years.

Osborn was born on 8 August 1857 in Fairfield, Connecticut, of wealthy parents, and had all of the advantages that might be wished for during his early years. In 1873 he entered Princeton University, where he became a very close friend of William Berryman Scott. Osborn had every intention of going 'into the business of railroads and finance' — as related by Scott — while Scott, in turn, 'had determined to study medicine' (Scott 1939).

Scott tells in his autobiography how he, Osborn, and a friend, Francis Spier, decided in 1876 that it would be fun to go west and look for fossils. In the summer of 1877 they explored the Bridger Basin in Wyoming, and thus began the palaeontological careers of Scott and Osborn.

After graduating from Princeton in 1877, Scott and Osborn went to England for graduate study. Osborn first studied embryology at Cambridge University under Francis Balfour. He then studied comparative anatomy under Huxley in London, where he had the pleasure of being introduced to Darwin, who had come to visit Huxley at his laboratory.

In 1880 he returned to Princeton, and was appointed to the faculty in 1881. Ten years were spent at Princeton, where he continued his close association with Scott, also a member of the faculty. In 1891 Osborn was appointed Da Costa Professor at Columbia University (then Columbia College) in New York, and was instrumental in founding the Department of Biology. He was simultaneously appointed to a Curatorship at the American Museum of Natural History, where he founded the Department of Vertebrate Paleontology.

Thus Osborn had a dual career — at Columbia, where he built up an active research and teaching biological department, and where he trained such outstanding palaeontologists as William Diller Matthew, William King Gregory, and Clive Forster Cooper, and at the American Museum of Natural History, where he closely superintended the growth and the activities of the Department of Vertebrate Paleontology and gathered together through the years such eminent palaeontological scholars as Matthew, Gregory, Walter Granger, Jacob Wortman, William Gidley, Oliver Perry Hay, Barnum Brown, George Gaylord Simpson, and others. Achievements such as these might be thought sufficient for a single lifetime, but Osborn served for many years as President of the Museum Board of Trustees, in which capacity he was a leading force in building the Museum into the great institution it is today.

In addition, Osborn was long outstanding in the activities and the development of the New York Zoological Society — the famous 'Bronx Zoo'.

Much of Osborn's success in his role as administrator and promoter at the Museum, at Columbia, and at the Zoological Society was (in addition to his natural aggressive energy) due to his position in New York society. He moved in influential circles; the financial doors of Wall Street were open to him, and he was the personal friend of such giants as J. P. Morgan (his maternal uncle) and President Theodore Roosevelt. Alternatively, much of his success in the scientific world was due to his undoubted ability as a scholar. In prosecuting his research he depended upon the help of assistants, but it was his background of rigorous training, insight, and original thinking that established him as an authoritative palaeontologist and evolutionist.

With the resources of the American Museum of Natural History, combined with his own funds, it was perhaps inevitable that Osborn, who had no personal experience with parsimony, would think of research in large terms, for his day. (The reader should remember that he lived and worked before the days of National Science Foundation grants.) Consequently, he envisioned several large research projects to be carried forward under his name almost from the beginning of his association with the American Museum. There were five such projects, four of them already mentioned, namely three monographic studies on perissodactyl ungulates, the horses, the rhinoceroses (Osborn 1898), and the titanotheres; one on sauropod dinosaurs; and the one with which we are concerned, namely the Proboscidea monograph.

It should be mentioned that very early in their careers, Scott and Osborn decided to divide certain groups of fossil mammals between themselves so that their research interests would not conflict. Osborn elected the perissodactyls or odd-toed ungulates, while Scott chose the artiodactyls or even-toed ungulates. In accordance with his decision to work on perissodactyls, Osborn became involved with three of the research projects mentioned above.

Osborn, in a sense, inherited the project on the sauropod dinosaurs in 1900, when he was appointed as vertebrate palaeontologist for the United States Geological Survey, to succeed O. C. Marsh, who had died in 1897 (Osborn 1930).

Osborn wrote in 1930 that 'The fifth monograph — that on the proboscideans — was of my own choice; it practically began in 1900 when I predicted that Africa would prove to have been the home of the proboscideans, and it has filled up all my spare time and energy since June 30, 1920, when I transmitted the monograph on the titanotheres to the Director of the Geological Survey' (Osborn 1930, p. 70).

Osborn's first paper involving Proboscidea was his brief study of the angulation of the limbs as adaptations to weight, published in 1900 (Osborn 1900*a*). In the same year, he published in *Science* an exposition on the geological and faunal relations between Europe and North America, in which he set forth 'the theory of the successive invasions of an African fauna'. (Osborn 1900*b*). It was in this paper that he predicted an African centre of origin for Proboscidea; a similar speculation had been put forward by Hans Stehlin of Basle. An African origin for Proboscidea was soon corroborated, owing to explorations in the Fayûm region of Egypt around the turn of the century by H. L. Beadnell of the Egyptian geological survey, and also to the researches of C. W. Andrews of the British Museum, who studied the fossils Beadnell had collected. It was Andrews who, in 1901, named and described *Moeritherium*, the late Eocene ultimate proboscidean ancestor, as well as *Palaeomastodon*, the early Oligocene proboscidean showing the strongly retracted nasal bones (indication of a proboscis) and the enlarged incisor tusks that were to be the definitive features of proboscideans through 35 million years of evolutionary history (Andrews 1901*b*).

The discoveries in Egypt inspired a prompt reaction on the part of Osborn; in 1907, 6 years after Andrews' publication of *Moeritherium* and *Palaeomastodon*, Osborn mounted an expedition for further exploration of the Fayûm sediments. As a result, a considerable collection of Eocene–Oligocene mammals was obtained, including excellent fossils of the ancestral and early proboscideans (Osborn 1907*b*).

The Fayûm expedition of 1906–1907 was a good example of Osborn's field methods. He was an active organizer but not a practitioner of palaeontological field work; it is difficult to imagine him going into the heat and dust and becoming involved in the hard, sweaty, everyday, nitty gritty labour of intensive field collecting. He conceived of what had to be done, procured the necessary funds and personnel, then turned over the prosecution of the campaign to one of his trusted lieutenants — Matthew, Granger, or Brown. Frequently, he would visit the collecting party in the field, but his presence was that of an overseeing

Fig. 1 Camel caravan of V. P. Expedition to the Fayûm, Egypt, 1907, as it passed by the Ferrar Butte. (Photograph by W. Granger, courtesy of the American Museum of Natural History, New York.)

visitor. He persuaded Morris K. Jessup, President of the Board of Trustees of the American Museum of Natural History, to provide the funds necessary for the Fayûm expedition. He went to Egypt with his family, and Granger and George Olsen of the Museum's Department of Vertebrate Paleontology. He carried with him a characteristically enthusiastic letter from President Theodore Roosevelt to Lord Cromer (Egypt was then under British rule) and so, of course, all doors were opened. He and his party went to the Fayûm site with members of the Egyptian Geological Survey, numerous 'native' workers who would do the heavy digging, and a large camel caravan (Fig. 1). After preliminary observations of the region, Osborn delegated supervision of the project to Granger, and went on his way.

The fossils collected by the Fayûm expedition may have been among the strong stimuli that turned Osborn's thoughts toward the problem of proboscidean evolution as a long-range project. From the time of his first paper in 1900 until 1933, less than 2 years before his death, he published more than fifty contributions having to do with proboscidean evolution, the forerunners of the immense Proboscidea monograph.

Osborn's proboscidean contributions were, in part, descriptive and in part philosophical. He had his own ideas concerning evolutionary mechanisms, and coined his own particular terms for the various evolutionary processes as he envisaged them. (Osborn, like most scholars of his generation, had a good grounding in the classics, so he was never at a loss when it came to the creation of new words.)

One of his first creations was *adaptive radiation*, a useful term that is widely used today (Gregory 1935). Some of his other inventions have not fared as well. *Aristogenesis*, to indicate the origin of new parts, and *rectigradations*, to indicate what might be called 'determinate variations' are forgotten favourites that appear in many of his proboscidean contributions (Osborn 1934*a*, 1938).

A concept prevalent in his writings on proboscideans is that of *polyphyletic evolution*. Early in his scientific career, Osborn seemed impressed by parallel evolution, and he applied this concept to his depiction of proboscidean evolution. This is apparent in his early papers on proboscideans, but it becomes especially notable in the two volumes of his Proboscidea monograph, where he shows the various phylogenetic lines of Mastodontoidea, for example, evolving separately from a 'Lower Eocene, Primary stock of Proboscidea' (Osborn 1936), or the several lines of Elephantoidea developing from a pre-Pliocene, unspecified ancestor (Osborn, 1936). For him, it would seem that nothing was ancestral to anything.

It should be noted, however, that the numerous independent lines of evolution, as he saw them, were constrained within the four primary proboscidean subdivisions that he recognized in 1921; the Moeritherioidea, Dinotherioidea, Mastodontoidea, and Elephantoidea (Osborn 1921*a*).

Osborn was prone to issue definitive, dated statements and published in 1925, in the *Proceedings of the American Philosophical Society*, a paper entitled 'Final conclusions on the evolution, phylogeny and classification of the Proboscidea' (Osborn 1925). They were not final, because he lived for 10 years after 1925, and had many changes of mind concerning those conclusions.

Since circumstances dictated that I should frequently be closely associated with Osborn, I got to know him rather well. I may be one of the very few people alive today who was personally acquainted with Professor Osborn, and for this reason, I am taking the liberty of writing about Osborn, not only in relation to his work on the Proboscidea monograph, but as I saw him as a person.

I saw him as a large, rather portly man, with strong features accentuated by an ample, bulbous nose. He had a full head of somewhat curly, iron-grey hair and a clipped moustache. He was always formally dressed, even when the weather was hot (indeed it is not possible for me to imagine him in his shirt sleeves), and he behaved in an aloof manner. All in all, he was an imposing figure.

He was incredibly pompous, vain, arrogant, conceited, remarkably self-centred, and almost entirely lacking in humility. He was a gifted person who (when I knew him) could look back on a long lifetime of great accomplishments, but his pomposity and his conceit spoiled the impression he made on the casual beholder. For those who knew him well, he was often hard to put up with. Yet, there was an underlying streak of good fellowship and kindness; I experienced it from time to time. He was sincerely interested in my accomplishments and success — as long as my own efforts did not get in the way of what I was supposed to be doing for him.

His attitude grated on the sensibilities of his colleagues and contemporaries. It led, in part, to the departure of W. D. Matthew from the Museum (after a tenure of more than 30 years), and it frequently led to feelings of frustration on the part of those who worked with him or under him. W. K. Gregory was one person who knew how to get along with Osborn; he could differ with the great man without raising hackles.

It was interesting to see the contrast in personalities between Osborn and his lifelong friend, Scott, who was a man of human dimensions, with a down-to-earth vision of life, as is apparent from his autobiography.

I felt a bit overwhelmed when I began my association with Osborn in 1930. He was formidable and impressive; I was a mere graduate student. I decided that if I was to be his research assistant, it was up to me to report my findings as I saw them, respectfully but firmly. What I had to say was not always to his liking; he did not wish to be contradicted. But when I felt that the evidence was on my side I stuck to my guns, and I think he respected me for holding to my views. At any rate we had a good working relationship.

The Proboscidea monograph was then in its final state of preparation. The Osbornian method of compiling and writing that great work was unique; I am sure that no scientific publication has ever been put together as were the two volumes of the monograph. Much of the text was done by assistants, especially observations on specimens and the long quotations from other authors. Osborn would revise text submitted to him and of course he would write substantial blocks of text — all of which would be integrated by an editorial assistant. After the text was reviewed and retyped, it would go to the Museum print shop, which occupied the basement of one of the buildings facing Seventy-Seventh Street. There, a dedicated compositor, Nick Caggana, would set the text into type and a galley proof would go upstairs to Professor Osborn. He would make liberal corrections and changes on the galley in red pencil and it would go down to be reworked; up it would come again for more changes, and so on, several times.

In the meantime, a corps of artists and a photographer would prepare illustrations. Page proof would then be prepared and sent up to Osborn time and again for corrections and changes. Hawthorne Daniel, then Editor of Museum publications, told me that there had been occasions when page proof had been corrected and changed as many as 17 times by Osborn. Those were indeed expansive and expensive days. Those were the days of metal type and letter press. Daniel told me that there were many tons of type on racks in the basement, waiting to be put on the presses.

I came into the picture when all of this was going on, so, of course, I was involved in it to some extent. I was more particularly involved in new projects, especially Osborn's work on the length of enamel in molar teeth of elephantoids. He thought that the progress of Pleistocene time could be measured by the increase in the enamel length of proboscidean molars from early to late Pleistocene deposits. I felt that the process was not as orderly and evenly paced as he imagined, and we had some lively arguments on that score (Osborn and Colbert 1931).

Work on the Proboscidea monograph continued, and in 1935, Osborn died quietly at his desk in his home overlooking the Hudson River. He died without seeing the Proboscidea monograph in print. There was work to be done on both volumes, so the completion of monograph was carried out by those of us who remained.

The Proboscidea monograph, the culmination of Osborn's long interest in this order of mammals, is truly a monument to the man. Like another monument, the American Museum of Natural History, it is large and impressive, and as such, it is a fitting memorial. Whatever may be one's judgement of Osborn, it must be admitted that he was an impressive palaeontologist.

ABBREVIATIONS, SYMBOLS, AND TERMS

DENTAL RELATED

Symbols for permanent cheek teeth are written with upper case letters, e.g. 'P' for premolars, 'M' for molars. 'P3' and 'M3' refer to either maxillary or mandibular third premolar and molar, whereas 'M^3' and 'M_3' refer to upper and lower third molars, respectively. Deciduous teeth are designated with a lower case 'd' (e.g. 'dp2' = deciduous second premolar). Intermediate teeth are dp4, M1, and M2.

NON-DENTAL RELATED

aff. Affinity (e.g. *Archaeobelodon* aff. *filholi*, species examined shows close affinity to *A. filholi*). This and other abbreviations are explained in Schenk *et al.* (1956, pp. 27–9).

BC Before Christ.

BP Before present.

Carbon dating The symbol '^{14}C' is used.

ex gr. Of the group of (used similarly to 'aff.' above).

'**Fig.**' v. '**fig.**', and '**Table**' v. '**table**' Figures and tables which appear in this volume are written with an upper case 'F' or 'T' ('Fig.', 'Table'), whereas those from other sources are written with lower case 'f' or 't' ('fig.', 'table').

Geological time periods and stratigraphic settings For geological time periods, we use 'early', 'middle', or 'late' (lower case letters, e.g. late Eocene, early Miocene). For stratigraphic settings and rock units or formations, we use 'Lower', 'Middle', and 'Upper' (upper case letters, e.g. Upper Fossil Wood Zone, Fayûm, Egypt).

Ma *Mega annum*, i.e. million years before present (Van Couvering and Berggren 1977, p. 285).

'**Mastodon**' v. '**mastodont**' The common name 'mastodon' is used only for the American mastodon (*Mammut americanum*), spelled without a 't' to differentiate it from past spelling which implied a more inclusive and phylogenetically heterogeneous grouping of taxa. Members of the family Mammutidae are called mammutids; all non-mammutid proboscideans are referred to either by their scientific names or by specific common names. (Note: the late H. Tobien who has been, for many years, using the common name 'mastodont' for gomphotheres and mammutids, agreed that it is important to define and differentiate clearly among so many 'mastodonts'.)

Time range Oldest proceeding to most recent, e.g. '50 000–10 000 years BP'.

NOTES FOR LITERATURE CITED

Numbers in { } Numbers in { } at the end of each reference are the numbers of the chapters in which that reference is cited.

Not all references in the book appear in the reference list If a contributor refers to the authorship of a taxon (e.g. *Gomphotherium angustidens* (Cuvier, 1817)), that reference (Cuvier 1817) may not appear if it was not used for a reason other than the authorship. Note: if the combined name of a genus and species is currently being used as it was originally described, then the authorship is written without parentheses, e.g. *Elephas maximus* Linnaeus, 1758. If, however, the authorship is written in parentheses, e.g. *Mammut americanum* (Kerr, 1792), it means that the species has been transferred to a genus other than the one for which it was proposed (Kerr used *Elephas americanus*).

Sequence of entries Except for citations for 'de Blainville', which are listed under 'B', all citations, such as 'de Vos, J.' and 'van der Maarel, F. H.', are listed in alphabetical order under 'd' and 'v' (not under 'Vos, J. de' and 'Maarel, F. H. van der').

PART I

Introduction

ONE

Historical overview of classification and phylogeny of the Proboscidea

PASCAL TASSY AND JEHESKEL SHOSHANI

Extant elephants were first named Proboscidea by Illiger in 1811. At that time, they were known as 'pachyderms'. According to Cuvier's *Le règne animal* (1817), the pachyderms also included tapirs, pigs, hippos, and hyraxes. In 1834, Blainville clustered elephants and sirenians together in the 'gravigrades', a monophyletic group which was named Tethytheria by McKenna in 1975 (which includes the orders Proboscidea, Sirenia, and Desmostylia).

The classification and phylogeny of proboscideans is a never-ending academic game since — except for the two living species *Loxodonta africana* and *Elephas maximus* — the group includes only fossil forms, many still to be discovered. Nevertheless, it is possible to extract some basic tendencies in the manner systematists used to classify proboscideans in pre- and post-Darwinian times. One topic, soon debated, was the major divisions between and among 'mastodons' and elephants, and the way to recognize which taxa to include or exclude from these divisions. Another problematic issue was the position of 'less elephant-like' proboscideans, such as *Deinotherium* — without upper tusks but with two big downturned lower tusks — described by Kaup in 1829, or *Moeritherium* — a pig-like mammal — described by Andrews in 1901*c–d*.

In 1821, Gray subdivided the Proboscidea into two families: Elephantidae and 'Mastodonadae'. For the latter name, 'Mastodontidae', coined by Girard in 1852, prevailed. Both familial names were based on 'mastodonte', coined by Cuvier in 1806, when the differences between the extinct '*Elephas americanus*' Kerr, 1792 (now called *Mammut americanum*, the American mastodon) and the living elephants were demonstrated.

The first important hierarchical classification of proboscideans is pre-Darwinian (1859) and is due to Falconer who recognized *Deinotherium* as a proboscidean and subdivided the other proboscideans into two genera, '*Mastodon*' and '*Elephas*' (Fig. 1.1). The system was primarily based on the number of lophs on the molars. '*Mastodon*' was subdivided into two subgenera, '*Trilophodon*' (a junior synonym of *Gomphotherium*) having dp4, M1, M2 with three lophs, and '*Tetralophodon*' having dp4, M1, M2 with four lophs, or even five. Molars with six lophs or more characterized the genus *Elephas*. The shape of lophs was employed as a characteristic for two groups inside '*Trilophodon*': the true mastodons (today's mammutids) with interlophs opened, on the one hand, and the other mastodons (today's gomphotheres), with interlophs blocked by supplementary cusps, on the other. Three subgenera were recognized inside '*Elephas*', depending on the number of lophs: six to eight (*Stegodon*), and more than eight ('*Euelephas*' and '*Loxodon*').

In 1866, Gaudry had produced the first phylogeny ever (Fig. 1.2), including extinct and extant species in the geological frame. This phylogeny showed that one Miocene European species, '*Mastodon pyrenaicus*', was ancestral, both to those mastodons called '*Bunolophodon*' by Vacek, and to a group formed by true mastodons and elephants.

It was Vacek who, in 1877, built a system based first on the shape of the lophs of molars (Fig. 1.3) '*Mastodon*' was subdivided into two subgenera: *Zygolophodon* (true mastodons) with opened interlophs, and '*Bunolophodon*' (a junior synonym of *Gomphotherium*) with supplementary cusps in the interlophs. The second subgenus included trilophodont and tetralophodont species.

Cope, in 1889, proposed a system based primarily on the number and shape of tusks (Fig. 1.4). The proboscideans were divided into two families: 'Dinotheriidae' and Elephantidae. The latter comprised five genera which were placed in three groups: Cope's group I, '*Tetrabelodon*' (a junior synonym of *Gomphotherium*) with two upper and two lower tusks; Cope's group II, '*Dibelodon*', '*Mastodon*', and '*Emmenodon*' (a junior synonym of *Stegodon*) with only two upper tusks; and Cope's group III, *Elephas*. Cope held *Emmenodon* as the ancestor of *Elephas*.

In the beginning of the twentieth century, the discovery by Andrews (1903, 1906) in the Fayûm of Egypt, of Eocene and Oligocene proboscideans, *Moeritherium*, *Palaeomastodon*, and *Phiomia*, altered the preceding schemes. *Moeritherium*, first found in the Eocene, was an odd proboscidean with enlarged incisors but without true tusks, no retracted nares (interpreted as having no trunk), an elongated skull, and very simple molars. It was first considered by Andrews as the ancestor of other proboscideans, particularly of the more elephant-like genera of the Oligocene, *Palaeomastodon* and *Phiomia*. Andrews ultimately discovered *Moeritherium* also in the Oligocene layers and abandoned this

Genera, Subgenera, and Species.				
MASTODON	Subgen. 1. TRILOPHODON....	(*a.*) Colliculi acuti valliculæque transversi	Spec. 1. M. (Trilophodon)	Borsoni (*I. Hays*)
			„ 2. M. (Triloph.)	tapiroides (*Cuv.*)
			„ 3. M. (Triloph.)	Ohioticus (*Blumb.*)
		(*b.*) Colliculi obtusi alternatim mammillati, valliculæ interruptæ.	„ 4. M. (Triloph.)	angustidens (*Cuv.*)
			„ 5. M. (Triloph.)	Pyrenaicus (*Lart. MSS.*)
			„ 6. M. (Triloph.)	Humboldtii (*Cuv.*)
			„ 7. M. (Triloph.)	Pandionis
	Subgen. 2. TETRALOPHODON.	(*c.*) Colliculi obtusi valliculæque transversi	„ 8. M. (Tetralophodon)	longirostris (*Kaup*)
			„ 9. M. (Tetraloph.)	latidens (*Clift*)
		(*d.*) Colliculi obtusi alternatim mammillati, valliculæ interruptæ.	„ 10. M. (Tetraloph. ?)	Andium (*Cuv.*)
			„ 11. M. (Tetraloph.)	Perimensis
		(*e.*) Colliculi numero 5, obtusi alternatim mammillati, valliculæ interruptæ.	„ 12. M. (Tetraloph.)	Arvernensis (*Croizet & Jobert*)
			„ 13. M. (Tetraloph.)	Sivalensis
ELEPHAS..	Subgen. 1. STEGODON.......	(*a.*) Colliculi circiter 6, cæmento in valliculis parco	Spec. 1. E. (Stegodon)	Cliftii
		(*b.*) Colliculi 7–8, cæmento in valliculis copiosissimo	„ 2. E. (Stegod.)	bombifrons
			„ ? 3. E. (Stegod.)	? Ganesa
			„ 4. E. (Stegod.)	insignis
	Subgen. 2. LOXODON........	(*c.*) Colliculi grossè digitati, adamante crasso......	„ 5. E. (Loxodon)	planifrons
			„ 6. E. (Loxod.)	meridionalis (*Nesti*)
		(*d.*) Colliculi medio angulatim dilatati, machæridibus per detritionem rhomboideis.	„ 7. E. (Loxod.)	priscus (*Goldf.*)
			„ 8. E. (Loxod.)	Africanus (*Blumb.*)
	Subgen. 3. EUELEPHAS......	(*e.*) Colliculi subremoti, adamante crassiusculo	„ 9. E. (Euelephas)	Hysudricus
		(*f.*) Colliculi approximati medio leviter dilatati, machæridibus undulatis.	„ 10. E. (Eueleph.)	antiquus
			„ 11. E. (Eueleph.)	Namadicus
			„ 12. E. (Eueleph.)	Columbi
		(*g.*) Colliculi approximati, machæridibus valde undulatis.	„ 13. E. (Eueleph.)	Indicus (*Linn.*)
			„ 14. E. (Eueleph.)	Armeniacus
		(*h.*) Colliculi confertissimi, adamante valde attenuato, machæridibus vix undulatis.	„ 15. E. (Eueleph.)	primigenius (*Blumb.*)

Fig. 1.1 Classification of proboscideans (after Falconer 1857).

idea. Andrews included in a new family, the Palaeomastodontidae, the genera *Palaeomastodon*, *Phiomia*, and *Tetrabelodon*. These taxa were tetrabelodont, trilophodont, and shared a long mandibular symphysis. Andrews also described from Egypt the enigmatic genus *Barytherium*, the affinities of which were already much debated.

In 1924, Matsumoto proposed the hypothesis of the diphyletic origin of 'mastodons': *Palaeomastodon* was ancestral to true mastodons and *Phiomia* to bunolophodont mastodons (Fig. 1.5). Three years later, he abandoned this idea and held *Phiomia* as the ancestor of all mastodon lineages.

Osborn's studies of fossil proboscideans will remain a unique and enormous work. Osborn's classification, built in 1933, included four superfamilies (sometimes considered as suborders; Table 1.1) in the Proboscidea: 'Moeritherioidea', 'Deinotherioidea', 'Mastodontoidea', and 'Elephantoidea'. Five families of mastodontoids were recognized: 'Mastodontidae' (true mastodons = mammutids), 'Serridentidae' (serrate toothed gomphotheres), 'Bunomastodontidae' (bunolophodont mastodons = most gomphotheres), and 'Humboldtidae' (New World gomphotheres). Osborn's system was based on the concept of adaptive radiation. Basic adaptive traits were, first, the morphology of the cheek teeth, and secondly, that of the tusks. The phylogenetic tree of the Proboscidea at the genus level was called by Osborn an 'adaptive radiation'. It has the shape of a bush with numerous lineages, 'phyla' (= genera), rooted in a large stem group in the Oligocene, and even Eocene, with few interrelations (Fig. 1.6).

Simpson's classification of mammals published in 1945 included a system for proboscideans close to that of Osborn, though with much less splitting. The order included moeritherioids, elephantoids, deinotherioids, and barytherioids. The Elephantoidea included the Gomphotheriidae ('bunolophodont mastodons'), the Mammutidae ('true mastodons'), and the Elephantidae (with both Stegodontinae and Elephantinae).

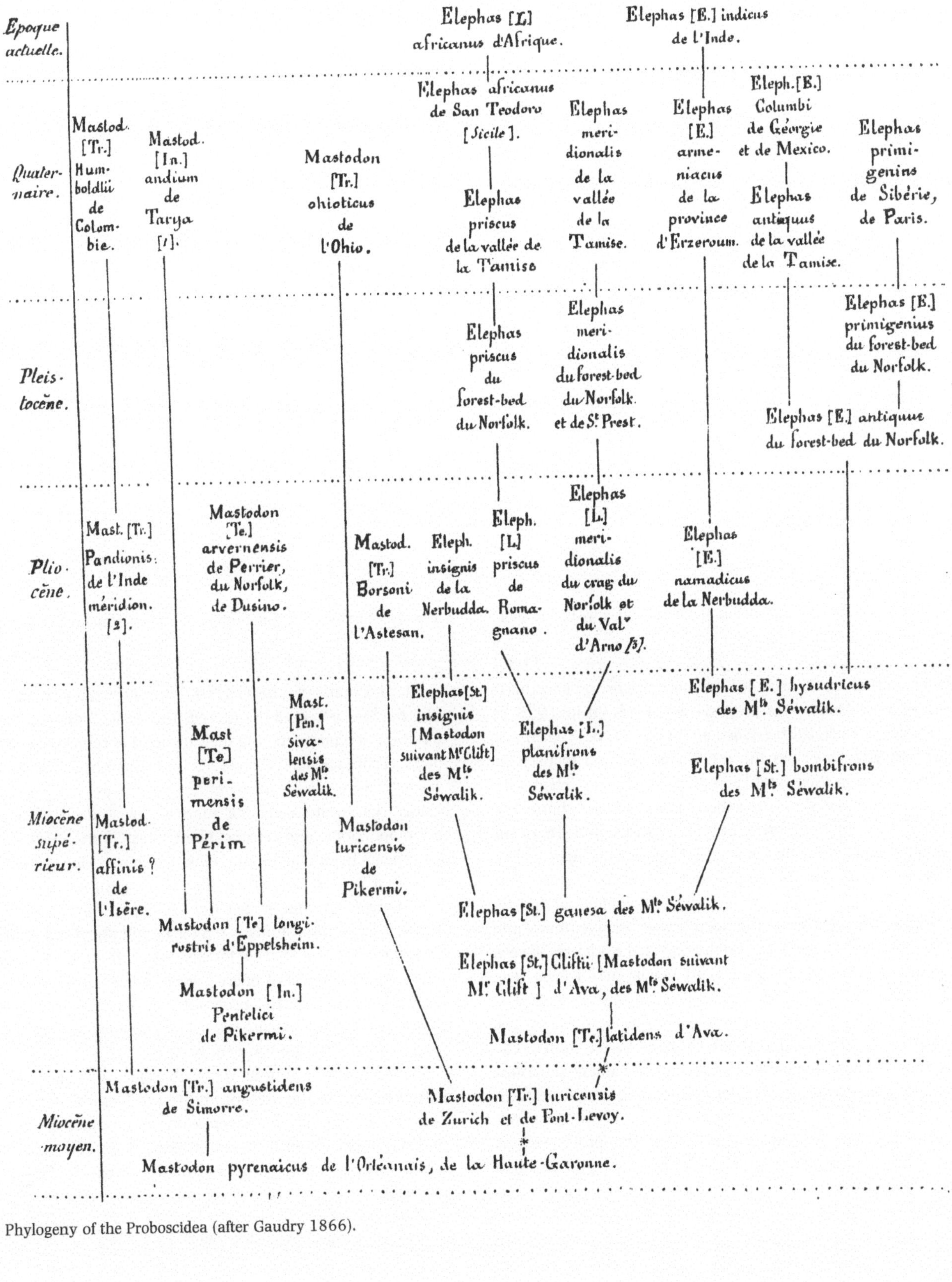

Fig. 1.2 Phylogeny of the Proboscidea (after Gaudry 1866).

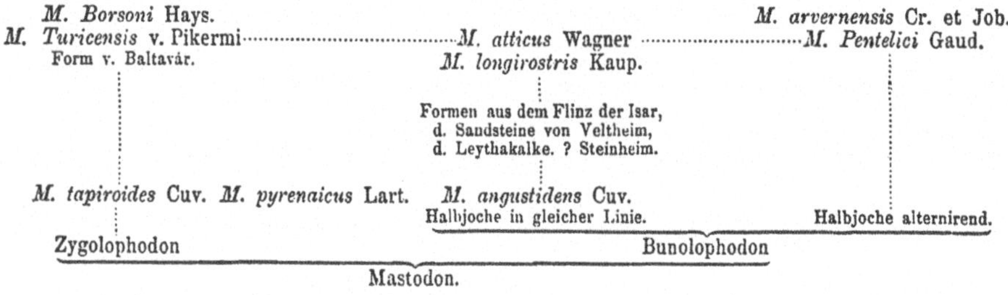

Fig. 1.3 'Classification of European mastodonts' (after Vacek 1877).

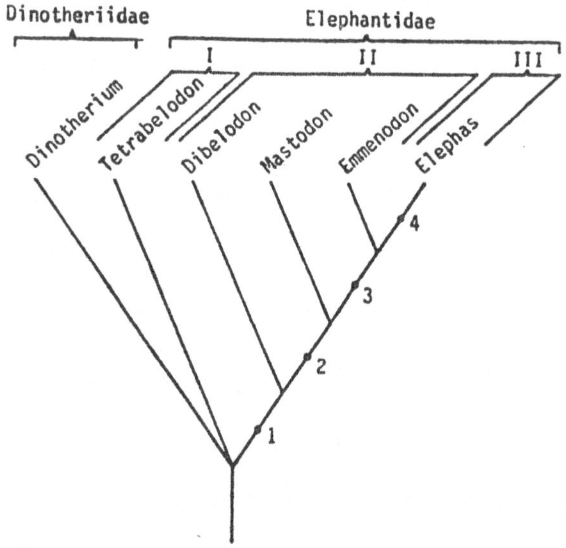

Fig. 1.4 Phylogeny and classification of the Proboscidea (redrawn after Cope 1889).

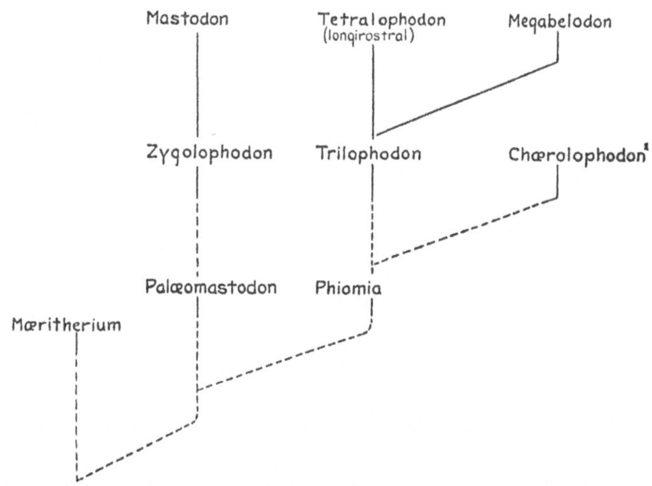

Fig. 1.5 Phylogeny of the Proboscidea (after Matsumoto 1924*b*).

In 1955, Deraniyagala excluded the moeritheres and barytheres from the proboscideans. This proposal was renewed in 1973 by Maglio who even excluded *Deinotherium*. In the 1970s, Tobien also excluded *Moeritherium* from the proboscideans, and focused on mastodons (mammutids and gomphotheres) and elephants. Tobien's system is based on the subordination of four characters: shape of lophs, number of lophs, number of tusks, and size of the mandibular symphysis, a kind of synthesis of previous approaches. The system contrasts bunolophodont mastodons and true mastodons (zygolophodons). Among the bunolophodont mastodons, Tobien demonstrated the identity of serridentines and gomphotheres, and showed the originality of the *Choerolophodon* group (pig-toothed gomphotheres). Both Maglio and Tobien (e.g. fig. 1 in Maglio 1973, p. 6; fig. 1 in Tobien 1988, p. 115) as well as Thenius in 1969, considered the stegodons as members of true mastodons (zygolophodons) without affinities to elephants. The latter were rooted among gomphotheres. In 1970 Maglio described the late Miocene African genus *Primelephas*, an intermediate form between gomphotheres and true elephants. In 1980, a considerable shift on the geographical origin of the proboscideans resulted from the description by R. M. West of the proboscidean affinities of *Anthracobune*, an Eocene genus from Pakistan.

At the end of the 1970s, the introduction of cladistic approaches perceptibly altered the basic concepts of subordination of characters and identification of taxa (see classification of the Proboscidea, Tables 1.1, 16.1 and Appendix B). Previous systems were based on evolutionary grades, not necessarily on clades. Large stem groups such as 'mastodons', trilophodont, bunolophodont, and tetrabelodont 'mastodons', were doomed to disappear in phylogenetic classifications. Hence, cladograms appeared in the literature with many terminal taxa but rather few named internal nodes (for instance, Tassy and Shoshani 1988; Tassy, this volume, Chapter 6, and Shoshani, this volume, Chapter 16). The search for sister-groups and shared derived characters of any kind offered, among the results, the return to older ideas, namely the proboscidean affinities of *Moeritherium*, *Deinotherium*, and *Barytherium* (see classification of the Proboscidea, Table 16.1 and Appendix B).

Taxonomic discussions on proboscidean clades and relationships belong to the present and not to history. A more detailed analysis of historical concepts on proboscidean classification and phylogeny (with the influence, not discussed here, of Depéret, Schlesinger, Vaufrey, and others) can be found in Tassy (1990*a*).

Table 1.1 Comparative classifications of the Proboscidea (abridged) according to six schemes

Osborn (1936,1942)	Simpson (1945)	Maglio (1973) and Coppens *et al.* (1978)	Tassy (1985, 1988*a* and this volume)	This volume (Table 16.1 and Appendix B)
				Anthracobune
• Moeritherioidea	Moeritherioidea			
• Moeritheriidae	Moeritheriidae		*Moeritherium*	Moeritheriidae
•			*Numidotherium*	Numidotheriidae
	Barytherioidea			
•	Barytheriidae		*Barytherium*	Barytheriidae
• Deinotherioidea	Deinotherioidea			
• Curtognathidae	Deinotheriidae		Deinotheriidae	Deinotheriidae
			Elephantiformes	Elephantiformes
			Palaeomastodon	*Palaeomastodon*
			Phiomia	*Phiomia*
			Hemimastodon	*Hemimastodon*
• Mastodontoidea	Elephantoidea	Mammutoidea	Elephantoidea	Elephantoidea
• Mastodontidae	Mammutidae	Mammutidae	Mammutidae	Mammutidae
Palaeomastodon				
			Eozygodon	*Eozygodon*
• *Zygolophodon*		*Zygolophodon*	*Zygolophodon*	*Zygolophodon*
• *Mastodon*	*Mammut*	*Mammut*	*Mammut*	*Mammut*
Stegolophodontinae		Stegodontidae		
Stegolophodon		*Stegolophodon, Stegodon*		
•			*Choerolophodon*	*Choerolophodon*
		Gomphotherioidea	Amebelodontidae	
• Bunomastodontidae	Gomphotheriidae	Gomphotheriidae	gomphotheres (1 and 2)	Gomphotheriidae
• Longirostrinae	Gomphotheriinae	Gomphotheriinae		
• *Trilophodon*	*Gomphotherium*	*Gomphotherium*	*Gomphotherium*	*Gomphotherium*
Amebelodontinae, *Phiomia*	*Palaeomastodon, Phiomia*	*Palaeomastodon*		
•			*Tetralophodon*	*Tetralophodon*
•			*Anancus*	*Anancus*
Stegodontoidea			*Paratetralophodon*	*Paratetralophodon*
• Stegodontidae			Stegodontidae	Stegodontidae
Stegodontinae			*Stegolophodon*	*Stegolophodon*
• *Stegodon*			*Stegodon*	*Stegodon*
Elephantoidea				
• Elephantidae	Elephantidae	Elephantidae	Elephantidae	Elephantidae
Loxodontinae	Stegodontinae	Stegotetrabelodontinae		
Loxodonta	*Stegolophodon*	*Stegotetrabelodon*	*Stegotetrabelodon*	*Stegotetrabelodon*
Palaeoloxodon	*Stegodon*	*Stegodibelodon*		*Stegodibelodon*
Hesperoloxodon				
• Elephantinae	Elephantinae	Elephantinae	Elephantinae	Elephantinae
Elephas	*Loxodonta*	*Primelephas*	*Stegodibelodon*	*Primelephas*
Hespelephas	*Elephas*	*Loxodonta*	*Primelephas*	*Loxodonta*
Platelephas		*Elephas*	*Loxodonta*	*Elephas*
Mammuthinae			*Elephas*	
Archidiskodon				
Metarchiskodon				
Parelephas				
• *Mammonteus*	*Mammuthus*	*Mammuthus*	*Mammuthus*	*Mammuthus*
Totals[a]: 5 S, 8 F, 44 G	4 S, 6 F, 24 G	2 S, 4 F, 15 G	2+ S, 5+ F, 37 G	2+ S, 8+ F, 38 G

• Comparison can be made horizontally; otherwise, refer to original sources.

[a] S, Suborders or superfamilies; F, families; G, genera (not all S, F, and G are listed; see Figs 6.1, 16.5; Tables 16.1, 34.1; and Appendix B)

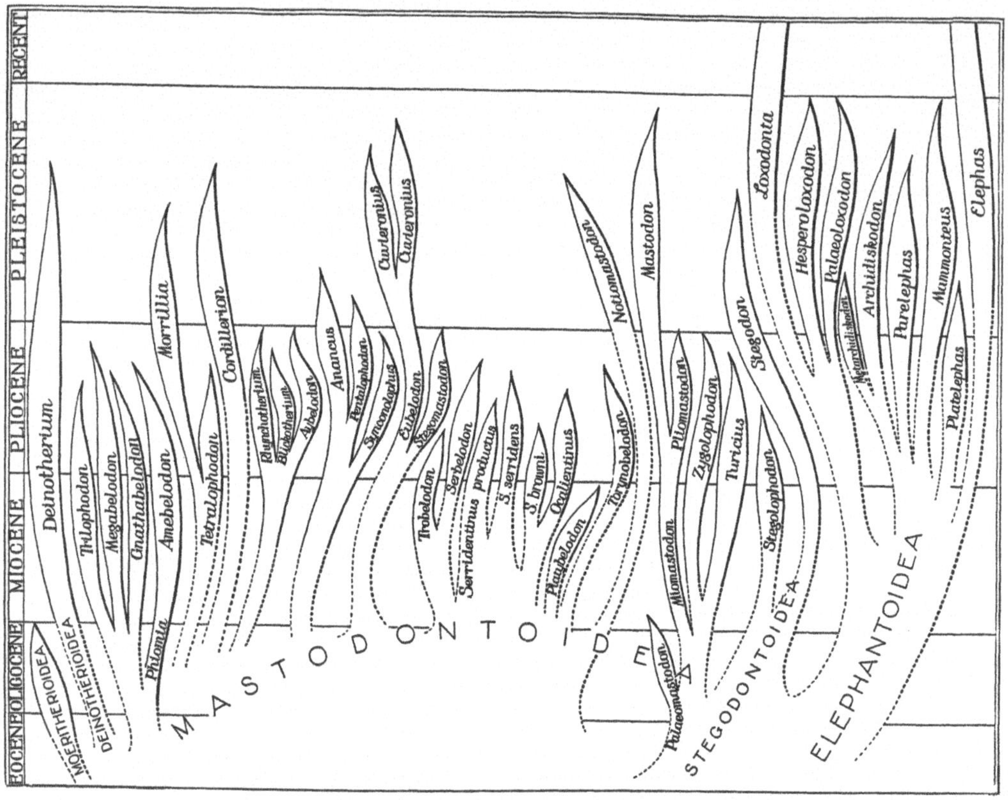

(a)

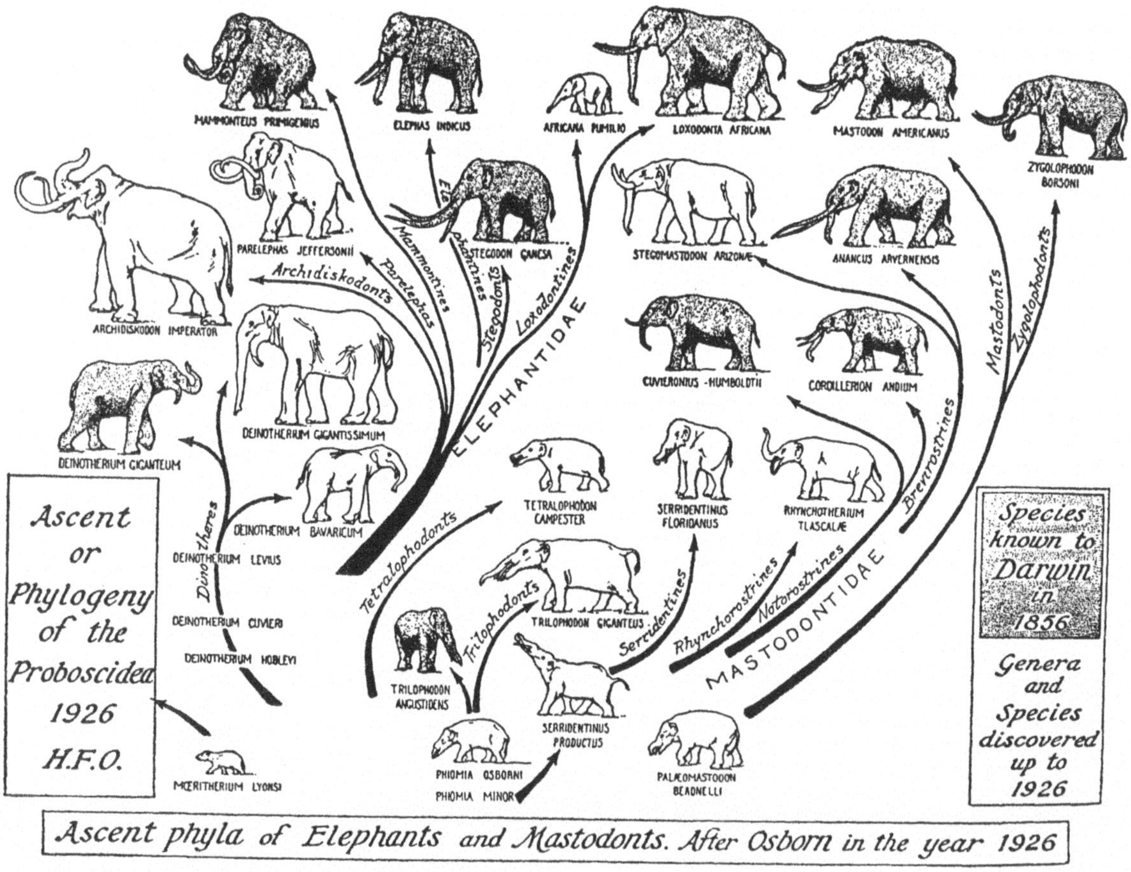

(b)

Fig. 1.6
(a) 'Adaptive radiation' of the Proboscidea (after Osborn 1936, plate XI, between pages 684–5).
(b) 'Proboscidean phyla or lines of descent, 1926' (after Osborn 1934b, fig. 1, p. 178).

TWO

Skeletal and other basic anatomical features of elephants

JEHESKEL SHOSHANI

INTRODUCTION

The purpose of this chapter is to augment the dental terminology (Tassy, this volume, Chapter 3) and ontogenetic data on incisors (Luckett, this volume, Chapter 4), and to provide brief but basic information on the anatomical (excluding functional) features found in elephants, so that non-proboscideanologists can be introduced to elephant anatomy. Living representatives of Proboscidea, *Loxodonta africana* (the African elephant) and *Elephas maximus* (the Asian elephant), were used as models for hard and soft anatomies (Fig. 2.1).

HARD ANATOMY

THE SKELETON

As in all mammals, the skeleton of an elephant is composed of four major divisions: the skull (cranium and mandible), vertebral column, appendages, and ribs and sternum (Fig. 2.2). Five sections make the vertebral column: cervical, thoracic, lumbar, sacral, and caudal vertebrae. The number of vertebrae in a region may vary according to the species.

Unlike other mammals, e.g. *Canis* or *Felis*, where the legs are in an angular position, elephants' legs are in an almost vertical position under the body. This arrangement provides a strong support for the vertebral column, thoracic and abdominal contents, and, of course, the great weight of the animal. In a cross-section of long bones there is no marrow cavity; instead the space is occupied by a network of dense cancellous bone through which vital haemopoiesis is possible, yet the bones are stronger and better able to withstand pressure than if they contained marrow cavities. A summary of the number of bones in the skeleton is given in Table 2.1. This total is similar to that of many mammals.

THE SKULL

Much of the volume of the cranium consists of 'inflated' bones which are compartmentalized to form diploë which makes the skull lighter, yet provides the strength needed. In addition, greater surface area on the cranium yields more room for muscle attachments towards the back; this support is needed when we

Fig. 2.1 Left: bull Asian elephant. *Elephas maximus maximus*, Yala National Park, Sri Lanka. Right: bull African elephant. *Loxodonta africana africana*, Amboseli National Park, Kenya. (Photographs by J. Shoshani.)

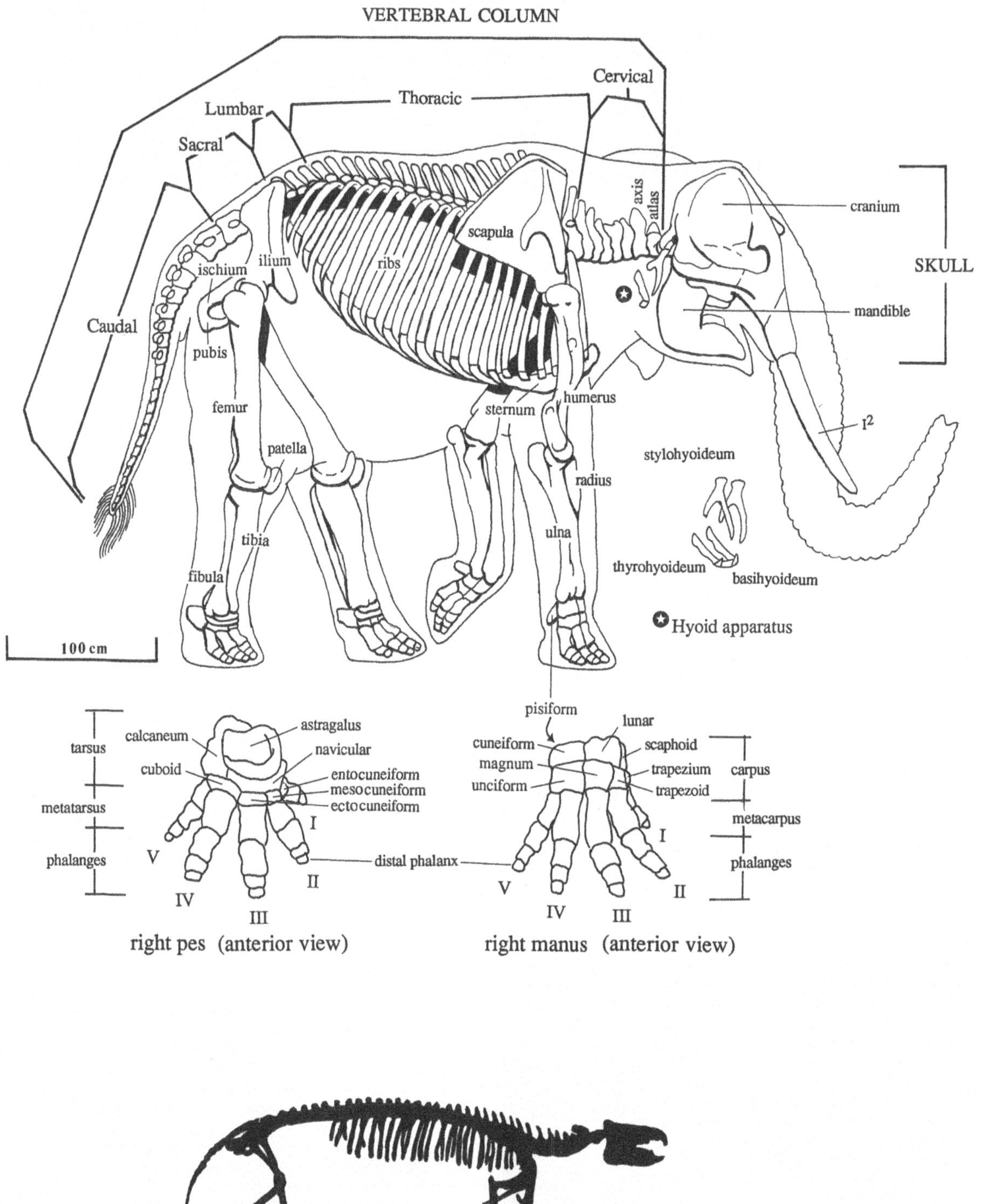

Fig. 2.2 Top: skeleton of a female African elephant (*Loxodonta africana*) in side view, showing major bones, including hyoid apparatus (structures apply to *Elephas maximus* as well; details of skull are given in Figs 2.3a and b; artwork by G. H. Marchant; data by J. Shoshani after Blair 1710; Sikes 1971; Shoshani, personal observations, 1980–92). Bottom: silhouette of skeleton of *Moeritherium*, one of the earliest proboscideans (late Eocene, Qasr el Sagha Formation, Fayûm, Egypt; after Simons 1964), for comparison; note that the skeleton of this *Moeritherium* was constructed from at least two individuals so the proportions may be misleading (G. Meyer, personal communication to P. Tassy 1977; see also caption to Plate 16.1, this volume).

Table 2.1 A summary of bones in the skeletons of living elephants, compared to that of a mammoth[a]

	Number of bones		
Name of bone, or complex of bones	*Loxodonta africana*	*Elephas maximus*	*Mammuthus primigenius*
Skull (cranium and mandible)	53–55	53–55	53–55
Hyoid apparatus	5	5	5
Vertebral column			
cervical vertebrae	7	7	7
thoracic vertebrae	20–21	19–20	18–19
lumbar vertebrae	3–4	3–5	3–5
sacral vertebrae	4–6	3–5	3–5
caudal vertebrae	18–33	24–34	21+
Sternum (4) + ribs (40–42, or 38–40, or 36–38)	44–46	42–44	40–42
Forelimbs (including scapulae and sesamoids)	82	82	82
Hind limbs (including pelves and sesamoids)	88	88	88
Total (for both sides of body)	324–347	326–345	320–329

[a]Note that the clavicle, baculum (os penis), baubellum (os clitoris), and os cordis are absent in the living elephants; the pelvis includes the acetabulum bone, visible in a fetus (based on Blair 1710, p. 143; Osborn 1942, p. 1095; Shoshani *et al.* 1982, pp. 42–43; Mariappa 1986, pp. 3–28; and personal observations).

consider that the head of an adult male African elephant may weigh 300 kg. The mandible, on the other hand, does not contain diploë; it is solid and relatively very heavy (Fig 2.3b). References consulted for the skeleton and skull include: Eales (1931), Osborn (1942), Frade (1955), Sikes (1971), Laursen and Bekoff (1978), Kingdon (1979), Shoshani *et al.* (1982), Shoshani and Eisenberg (1982), and personal observations by the author.

PREMOLARS AND MOLARS

As explained by Tassy (this volume, Chapter 3), dental characters of proboscidean taxa have changed dramatically during the past 50–45 million years. All changes are related to an increase in body size and the type of food consumed. The number of cheek teeth in adults did not change, canines were lost, and the incisors were reduced in number. In the two living species, over the lifetime of an individual, the number of teeth is the same, with this dental formula for adults; incisors 1/0, canines 0/0, decidual premolars 3/3, molars 3/3, a total of 26 teeth. The jaws of an elephant, despite their size, cannot accommodate all six cheek teeth at one time; one tooth of an elephant can weigh over 5 kg. As a result, elephants do not replace their cheek teeth in a vertical manner as most mammals do, rather in a horizontal progression. A newborn elephant has two or three small teeth in one jaw quadrant; as it becomes older, new and bigger teeth are developing from behind and slowly moving forward, as in a slow conveyor belt, to replace the old ones. In its lifetime, about 50–70 years, an elephant has six sets of teeth (Figs 2.4 and 2.5). Worn teeth move forward and fragment; pieces of teeth can be found where they fell out of the mouth, or even in faeces if they are swallowed.

The shape of the enamel loops differs in *Loxodonta* from that of *Elephas* (Fig. 2.6). In the former, the loops are lozenge shaped — hence the generic name — and in the latter they are simplified closed loops. An average number of loops for each of the cheek teeth, upper and lower, for both elephant species (African/Asian) is: 5/5, 7/8, 10/13, 10/14, 12/19, 13/24 (Laws 1966; Sikes 1966, 1971; Lark 1984; Roth and Shoshani 1988; Fig. 2.5).

Fig. 2.3 (a) Ventral (left) and dorsal (right) views of the cranium of a juvenile *Elephas maximus*, showing bones, foramina, and other features (about 1/5 natural size; after Gregory 1903, figs 1 and 2, pp. 390–391; Osborn 1942, fig. 796, p. 916; with minor changes, abbreviations are exactly as they appear in Gregory and in Osborn). Key: A na., anterior nares; a.p.f., anterior palatine foramina (canals); Bo., basioccipital; Bs., Basisphenoid; c.f., notch, a vestige of condylar foramen (?) (confluent with f.l.p.); e.a.m., (secondary) external auditory meatus; eu. eustachian opening of tympanic; Ex.o., exoccipital; f.l.m., foramen lacerum medium; f.l.p., foramen lacerum posterius; f.ov., foramen ovale (confluent with f.l.m.); Fr., frontal; f.st.m., stylomastoid foramen; i.2., tusk; i.c.c., canal for internal carotid artery; i.o.f., infraorbital foramen; L., lachrymal; Ma., malar (= jugal); Mx., maxilla; Mx.p., maxillary pouch for molars; Na., nasal; Na.ss., nareal sinus; o.c., occipital condyle; p3 (dm.2) [*sic*]*, third premolar (or second deciduous molar of authors) [*sic*]*; p4 (dm.3) [*sic*]*, fourth premolar (or third deciduous molar of authors) [*sic*]*; Pa., parietal; pal.Mx., palatine ledge of maxilla; p.a.s., alisphenoid canal; p.As., pterygoid wing of alisphenoid; pg., postglenoid ledge of squamosal; pl., palatine; P.mx., premaxilla; P.na. and p.n., posterior nares; Po.f., postorbital ridge of frontal; Ps., presphenoid; Pt., pterygoid; p.ty., post-tympanic ledge of squamosal, which with pg. forms a secondary external auditory meatus (= e.a.m.); So., supraoccipital; Sq., squamosal; tp.h., tympanohyal; ty., tympanic bulla; ty.p., anterior process of tympanic; Vo., vomer. *Note: It appears as though Osborn's (1942) 'first', 'second', and 'third' teeth in mammals are not homologous with the loci for extant elephants ▶

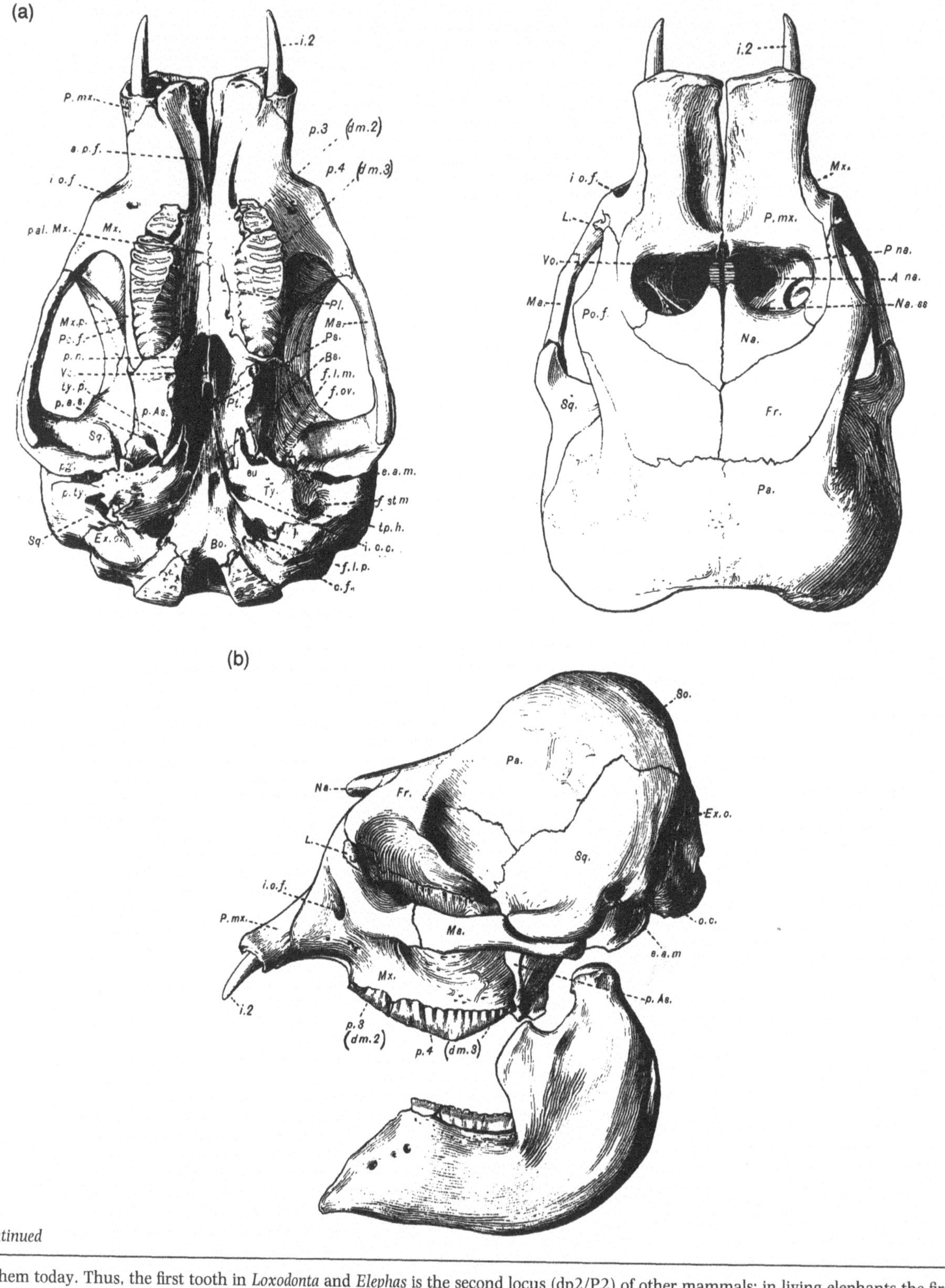

Fig. 2.3—*continued*

as we know them today. Thus, the first tooth in *Loxodonta* and *Elephas* is the second locus (dp2/P2) of other mammals; in living elephants the first locus (dp1/P1) is lost. As a result, the label 'p3 (dm2), third premolar (or second deciduous molar of authors)' should read 'dp^3, third upper deciduous premolar', and the label 'p4 (dm3), fourth premolar (or third deciduous molar of authors)' should read 'dp^4, fourth upper deciduous premolar' (P. Tassy, personal communication 1993). (b) Lateral view of skull of *Elephas maximus*, showing bones, foramina, and other features (after Gregory 1903, fig. 3, p. 392; Osborn 1942, fig. 797, p. 917; key as in (a)). The structures 'crista orbitalis ventralis' and 'crista orbitotemporalis' of Saegusa (1987), mentioned by Guan (this volume, Chapter 13), are not labelled in these figures.

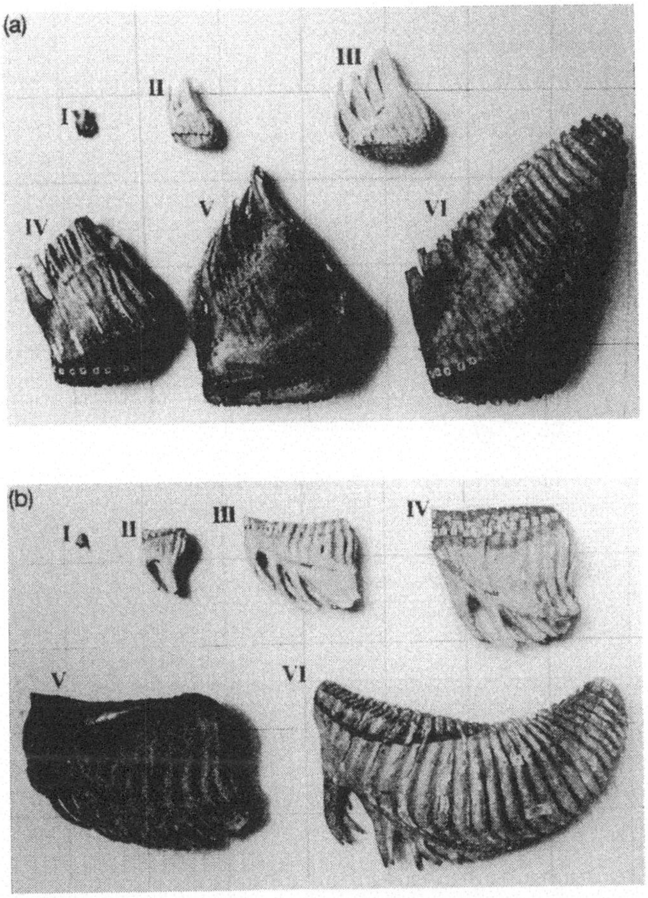

Fig. 2.4 (a) Upper and (b) lower isolated teeth of *Elephas maximus* (after Roth and Shoshani 1988). A sample of weights of teeth for the Asian species are: tooth I, 9.0 g; tooth II; 125.0 g; tooth III, 568.0 g; tooth IV, 1600.0 g; tooth V, 3685.4 g; and tooth VI, 5159.6 g. Timing and rate of tooth replacement is, for all practical purposes, equal for both species: tooth I is shed (i.e. tooth II begins to wear) at the age of 2–3 years, tooth II is replaced at 4–6 years, tooth III at 10–14 years, tooth IV at 20–25, tooth V at 30–35, and tooth VI, or part of it, usually remains in the mouth until the animal dies at 50–70 years.

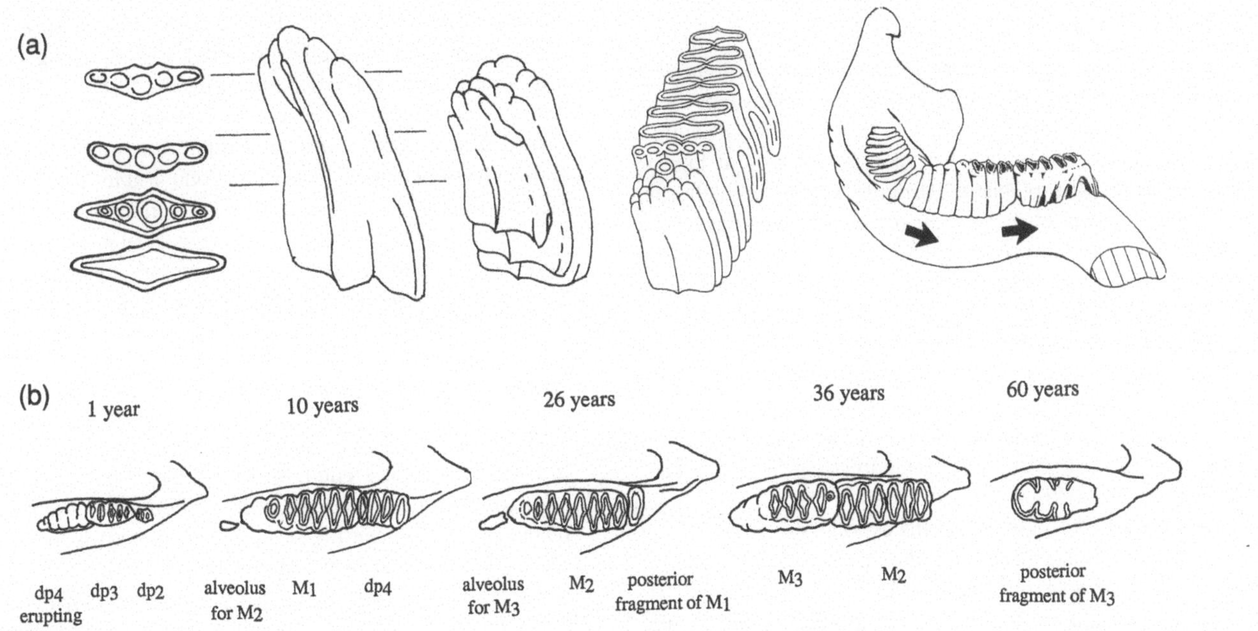

Fig. 2.5 Simplified diagrams showing: (a) cross-sections of isolated lamellae at different locations to reveal patterns on occlusal surfaces, a tooth, and a left dentary in medial view (arrows indicate direction of horizontal tooth displacement); (b) mandibulae of *Loxodonta africana* depicting teeth which are present at different ages. (a) drawn from specimens; (b) after Laws 1966; artwork by G. H. Marchant.

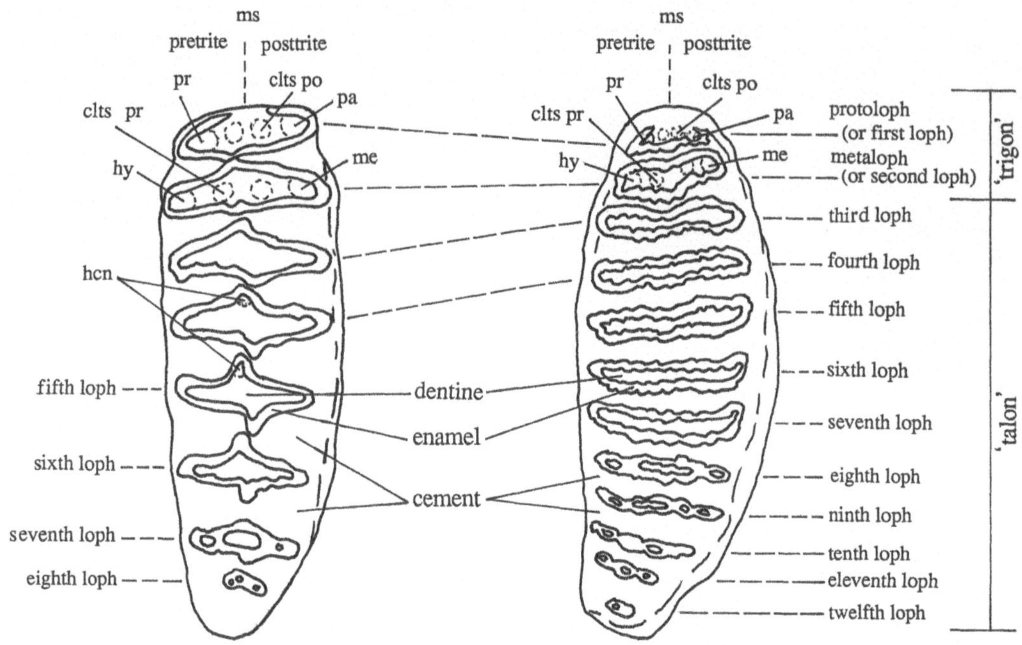

Fig. 2.6 Upper left third molars of *Loxodonta africana* (left) and *Elephas maximus* (right) in occlusal views. Key: clts po, conelets, posttrite (approximate loci of adaxial or secondary conelets of protoloph); clts pr, conelets, pretrite (approximate loci of adaxial or secondary conelets of metaloph); hcn, homologous remnant of central conule seen in gomphotheres; hy, hypocone (approximate locus of abaxial or main conelet of metaloph); me = metacone (approximate locus of abaxial or main conelet of metaloph); ms, median sulcus (approximate locus); pa, paracone (approximate locus of abaxial or main conelet of protoloph); pr, protocone (approximate locus of abaxial or main conelet of protoloph). Note: trigon and talon are in quotation marks because their position is only approximate; the original mammalian trigon includes me, pa, and pr, whereas the original mammalian talon also includes hy. Terminology after Tobien (1975, p. 196) and also partly modified after Osborn (1942, pp. 1544–52); see also Figs 9.1, 17.1, and 27.1.

INCISORS OR TUSKS

The proximal end of the tusk (I^2 in modern elephants) is occupied by a pulp cavity; in life this cavity contains blood vessels, nerve and lymphatic tissues. In elephants, deciduous incisors, or tusks, are replaced by permanent second incisors within 6–12 months of birth. Permanent tusks, which then develop, grow continuously at a rate of about 17 cm/year (Colyer and Miles 1957) and are composed mostly of dentine. The newly developing tusk bears a conical cap of smooth enamel which wears off at a later stage.

Bull elephants usually have longer, heavier tusks than cows; in the African elephant, both sexes may have tusks of equal length. In the Asian species, the males usually carry tusks; in females, tusks are small, vestigial, or absent. A bull Asian which is tuskless is called a 'Mukhna' (sometimes spelled 'Muckna'; Deraniyagala 1955). Contrary to common belief, some Asian bulls have huge tusks, bigger than the average size of the African species; one such male Asian was the 'giant tusker of Udiapur' whose tusks grew so long that they had to be trimmed at the tips (Osborn 1942, p. 1326, fig. 1179). Another Asian bull with huge tusks is 'Tommy' (currently with the Ringling Brothers and Barnum & Bailey Circus, called 'King Tusk'); even though his tusks are trimmed at their tips, in 1985 his left tusk measured 145 cm, and his right one measured 157 cm along the outside curvature. For comparative purposes, the tusks of the legendary 'Ahmed' (*L. africana*) measured 285 cm and 297 cm. The longest recorded tusks for an African elephant measured 326.4 cm, and the heaviest weighed 102.7 kg. Comparable figures for Asian elephants are: 302 cm and 39.0 kg (references to this section include: Osborn 1936, 1942; Deraniyagala 1955, 1963; Laursen and Bekoff 1978; Shoshani, personal observation 1982; Shoshani and Eisenberg 1982; Shoshani *et al.* 1987).

DENTAL STRUCTURE IN TUSKS AND MOLARS

Dentine

In transverse cross-section, a tusk exhibits a pattern of lines that criss-cross to form small diamond-shaped areas visible with the naked eye on the dentine surface. This pattern is formed by two sets of lines which begin at the centre of the tusk and curve towards the periphery. One set of lines curves clockwise and the other counterclockwise, crossing each other and dividing the surface into rhomboid-shaped areas which become progressively smaller towards the centre of the tusk (Fig. 2.7). Known as

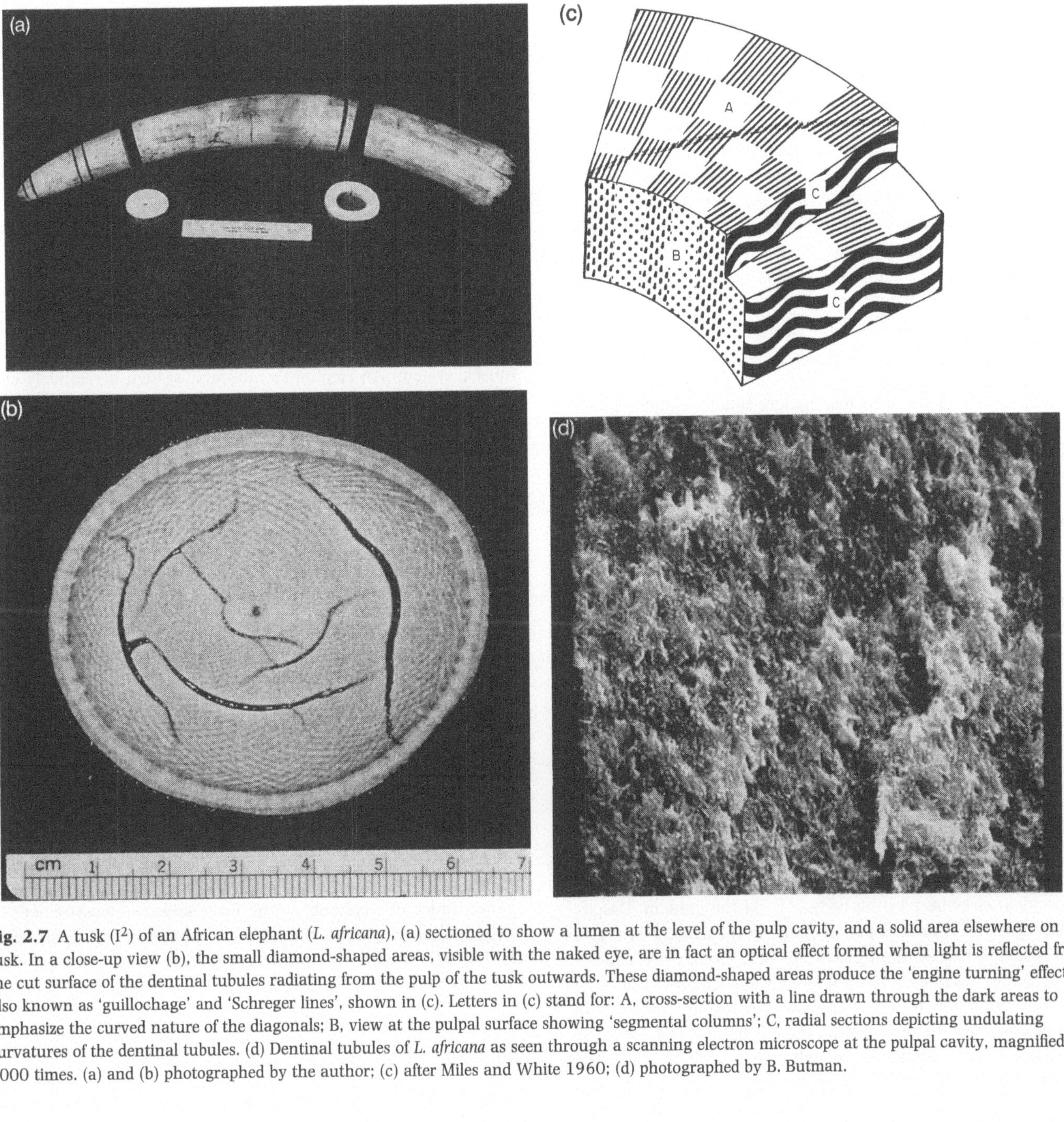

Fig. 2.7 A tusk (I^2) of an African elephant (*L. africana*), (a) sectioned to show a lumen at the level of the pulp cavity, and a solid area elsewhere on the tusk. In a close-up view (b), the small diamond-shaped areas, visible with the naked eye, are in fact an optical effect formed when light is reflected from the cut surface of the dentinal tubules radiating from the pulp of the tusk outwards. These diamond-shaped areas produce the 'engine turning' effect, also known as 'guillochage' and 'Schreger lines', shown in (c). Letters in (c) stand for: A, cross-section with a line drawn through the dark areas to emphasize the curved nature of the diagonals; B, view at the pulpal surface showing 'segmental columns'; C, radial sections depicting undulating curvatures of the dentinal tubules. (d) Dentinal tubules of *L. africana* as seen through a scanning electron microscope at the pulpal cavity, magnified 1000 times. (a) and (b) photographed by the author; (c) after Miles and White 1960; (d) photographed by B. Butman.

'engine turning', this pattern is also present in the dentine of molars and is unique within Proboscidea; it is not present in the incisors of hippopotamus, walrus, and wild boar or in the canines of narwhal (Miles and White 1960; Sikes 1971). Tobien (1978*b*, pp. 179, 184) referred to this engine turning as 'guillochage', whereas Espinoza *et al.* (1990) and Espinoza and Mann (1993) named the engine turning 'Schreger Pattern' (described by Bernhard Gottlob Schreger in 1800 (Obermayer 1881) and not to be confused with Hunter–Schreger bands in enamel). The outer angles formed by the Schreger lines are acute (<90°) in *Mammuthus* and obtuse (>115°) in *Elephas* and *Loxodonta*, although there is overlap among these taxa (Espinoza *et al.* 1990; Espinoza and Mann 1993). J. Shoshani and D. C. Fisher examined tusks of stegodontids, mammutids, and gomphotheres and noted that there are also overlapping variations among these extinct taxa and the two extant species. A thorough collection of data is needed to establish species characters and how these variations relate to density of dentinal tubules and to phylogeny. Saunders (1979) observed that more tubules (i.e. more 'elasticity') were present in mammoth tusks than in those of the American mastodon, a hypothesis also made by J. Shoshani in a proposal to the US Fish and Wildlife Service (proposal FWS/LE SPA 1, 1979 and 1980). In addition to the 'Schreger lines' ('engine turning' or 'guillochage'), one can sometimes observe concentric patterns, 'lines of Owen', in a transverse cross-section — these are the interfaces between the 'cones' discussed by Fisher (this volume, Chapter 30). On a finer scale, there are daily increments in the dentine known as 'von Ebner lines'; these increments are substructures within the bi-weekly pattern discussed by Fisher in Chapter 30.

Enamel

Under the scanning electron microscope, enamel exhibits prisms which vary in their densities and patterns among mammals and within Proboscidea (Cring 1986; Kamiya 1991, and references within). These prisms may be arranged in a geometrical pattern to give the impression of lines or bands, known as 'Hunter– Schreger bands' (Poole 1967, p. 139; von Koenigswald *et al.* 1993, pp. 304–6). Enamel also exhibits incremental concentric lines known as 'Lines of Retzius' (Retzius 1837; Peyer 1968, pl. 69a and 69b). Externally, these lines can be observed as compacted closed loops, the perikymata (Simmelink 1988), whose density and degree of curvature may be useful taxonomic characters.

SOFT ANATOMY

THE SKIN

Thicknesses of the skin vary; it is paper thin on the inside of the ears, around the mouth, and the anus and can be as much as 2.5 cm (hence 'Pachydermata', the old name of the order Proboscidea) elsewhere in the body, e.g. on the dorsum. Despite its thickness, the skin's rich nerve supply makes it a sensitive organ system. The skins of *Loxodonta* and *Elephas* are usually grey in colour; elephants in Africa often appear brown or even reddish, a condition resulting from wallowing in mudholes of coloured soil which is plastered on their bodies. In Asia wallowing usually results in darker or lighter grey colours than the original body colour. It is of interest to systematists that between the two living elephant species, *Loxodonta* is less hairy than *Elephas*, especially true of juveniles, and it is the latter genus which is more closely related to *Mammuthus*: one species in particular — *M. primigenius* — is known to have had long hair (references to this section include: Smith 1890; Neuville 1921; Eales 1925; Deraniyagala 1951*a*, *b*, 1952, 1955; Augusta and Burian 1962; Douglas-Hamilton and Douglas-Hamilton 1975; Shoshani *et al.* 1982, 1985*b*; Mariappa 1986; Moss 1988). Sikes (1971, p. 10) noted that 'Sweat glands occur, but are difficult to locate'; Shoshani *et al.* (1982, p. 60) searched for these glands and found none.

EARS

Differences in the sizes of ears among *L. africana*, *E. maximus* and *M. primigenius* can be explained based on their geographical distribution: *Loxodonta*, having originated close to the equator in Africa and having remained within the continent, has the largest ears, and *Mammuthus*, having adapted to northern climates as it spread through Europe, Asia, and North America, had the smallest; the ears of *Elephas* are intermediate in size between those of *Loxodonta* and *Mammuthus* (Lydekker 1907; Deraniyagala 1951*a*, 1955; Augusta and Burian 1962).

THE MUSCULAR SYSTEM

Like all mammals, elephants have a bilaterally symmetrical muscle system. An elephant's muscular system is relatively easy to understand, mostly because the structures are large. The vast majority of muscles referred to in this chapter are voluntary (Fig. 2.8 and references therein). Table 2.2 gives a rough estimate of the number of skeletal muscles in an elephant, not including muscles of internal organs such as the heart, intestine, and diaphragm.

Unlike humans and many other mammals, the elephant cannot rotate its forearms; the manus ('hand') stays in a fixed pronation position, with the 'palm of the hand' facing backwards. This condition possibly co-evolved with the reduction or absence of two muscles deep in the forearm — the pronator quadratus and pronator radii teres. This character of fixed pronation has been important to students of the evolution of Proboscidea, and has been used as one of the features to define extinct and extant proboscideans.

THE PROBOSCIS

An elephant's trunk, or *proboscis*, is the single most important feature of an elephant. It is the character after which the mammalian order of elephants — the Proboscidea — was named. In early fetal development, the nose and upper lip are separate; in the late fetal stages and onward the trunk is a fusion of the nose and upper lip (Fischer and Trautmann 1987). It has no bones or

Table 2.2 Estimated number of skeletal muscles in elephants

Names of organs, or complex of organs	Skeletal muscles
Head (including ears, excluding trunk)	70
Proboscis (major muscles, see details below)	7[a]
Cervical and dorsal	8
Forelimbs (including scapulae)	41
Hind limbs (including pelves)	48
Body wall	8
Sacral, caudal, and genitals	15
Estimated total	197× 2 = 394

[a]These are shown in cross-section in Fig. 2.8. An estimate of approximately 150 000 muscle fascicles (parts of muscles) was obtained for one proboscis.

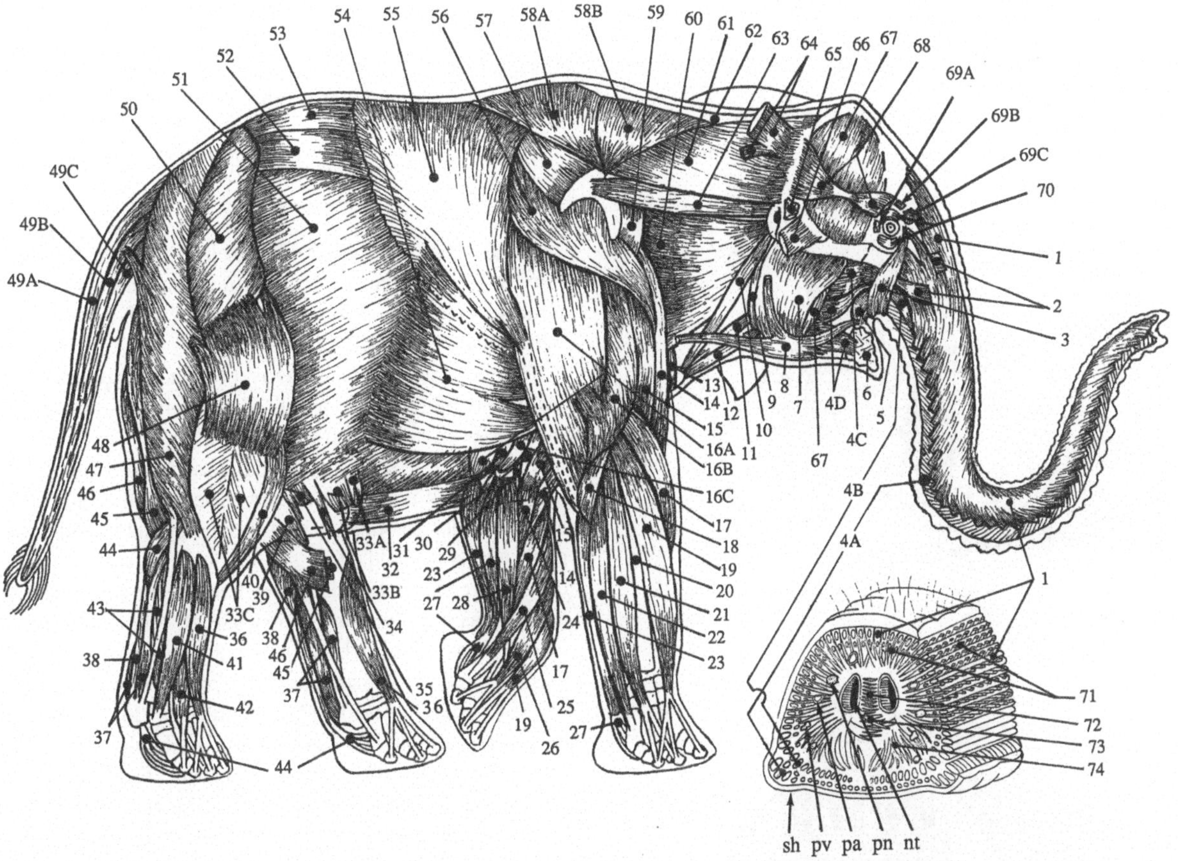

Fig. 2.8 Muscles beneath the skin of an elephant (depicted is a female *Loxodonta africana*, but structures apply to *Elephas maximus* as well). (Drawing by G. H. Marchant; data by J. Shoshani, slightly modified after Cuvier 1849; Miall and Greenwood 1878; Boas and Paulli 1908, 1925; Eales 1926, 1928; Shindo and Mori 1956; Stöcker 1957; Nielsen 1965; Shoshani, personal observations, 1980–1992; Mariappa 1986.

Key to muscles:

1 Maxillo-labialis
2 Nasalis
3 Naso-labialis
4A Pars rimana of Buccinator
4B Pars supra-labialis of Buccinator
4C Buccinator
4D Longitudinal portion of Buccinator
5 Orbicularis oris
6 Mentalis
7 Massetericus
8 Platysma
9 Digastricus
10 Sterno-mastoideus
11 Sterno-thyroideus
12 Sterno-mandibularis
13 Pectoralis
14 Biceps brachii
15 Brachialis
16A Triceps: caput magnum
16B Triceps: caput medium
16C Triceps: caput parvum
17 Extensor carpi radialis
18 Anconeus
19 Extensor communis digitorum
20 Central tendon
21 Extensor digiti quinti
22 Extensor carpi ulnaris
23 Flexor carpi ulnaris
24 Supinator longus
25 Extensor pollicis
26 Extensor indicis et pollicis
27 Palmaris longus
28 Flexor carpi radialis
29 Pronator radii teres
30 Coraco-brachialis
31 Tensor faciae antibrachii
32 Cutaneus abdominis
33A Vastus intermedius
33B Vastus medialis
33C Vastus lateralis
34 Sartorius
35 Adductor
36 Extensor digitorum longus
37 Soleus
38 Gastrocnemius
39 Gracilis
40 Rectus femoris
41 Peroneus longus
42 Peroneus anticus
43 Peroneus brevis
44 Plantaris
45 Semitendinosus
46 Semimembranosus
47 Biceps femoris
48 Tensor Faciae latae
49A Sacro-coccygeus dorsalis
49B Sacro-coccygeus lateralis
49C Sacro-coccygeus ventralis
50 Gluteus medius
51 Obliquus externeus
52 Iliocostalis
53 Longissimus dorsi
54 Cutaneus trunci
55 Latissimus dorsi
56 Deltoideus
57 Infraspinatus
58A Trapezius: thoracic portion
58B Trapezius: cervical portion
59 Supraspinatus
60 Masto-humeralis
61 Splenius capitis
62 Ligamentum nuchae
63 Omotrachelian
64 Auriculo-occipitalis

65 Adductus inferioris
66 Zygomatic auricularis
67 Temporalis
68 Occipito-frontalis
69A Post-orbicularis
69B Supra-orbicularis
69C Pre-orbicularis
70 Orbicularis oculi
71 Rectus nasi (or 'radiating muscles')
72 Septum nasi
73 Transversus nasi (? new name)
74 Rectus nasi

Other structures:
nt nasal tube
pa proboscidean artery
pn proboscidean nerve
pv proboscidean vein
sh skin with hair

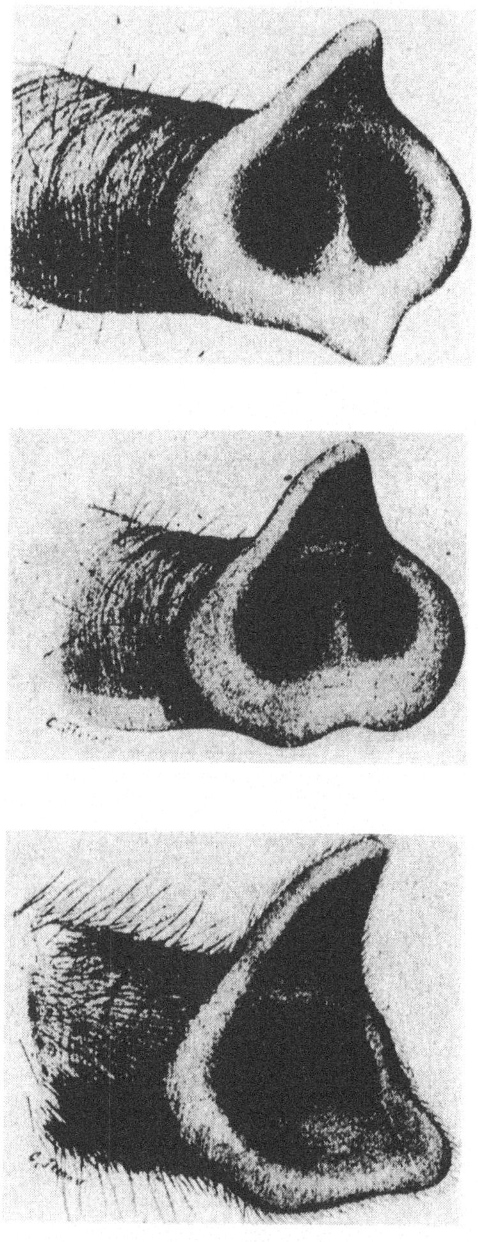

Fig. 2.9 Tips of trunks of an African elephant (*Loxodonta africana*, with two finger-like processes, top), an Asian elephant (*Elephas maximus*, with one finger-like process, centre) and a woolly mammoth (*Mammuthus primigenius*, with one finger-like process dorsally and a flap ventrally, bottom), after Flerov (1931, pp. 865–7) (see also Augusta and Burian 1962; a summary of differences among *Loxodonta*, *Elephas*, and *Mammuthus* appears in Shoshani 1986b, pp. 455–67). Note that in Osborn (1942, p. 1128) there is an error in the labelling of fig. 995.

cartilage; it is composed of muscles, blood and lymph vessels, nerves, little fat, connective tissues, skin, hair, and bristles. The alinasal cartilage is found at the external nareal opening and helps divide the nostrils.

Georges Cuvier, the famous French anatomist, and his colleagues examined the trunk of an elephant and estimated the number of muscles in it at about 40 000 (Cuvier 1849). Others have noted that some people have made an effort to count them '...but such an attempt is totally useless' (Harrison 1847, p. 394).

Two major groups of muscles (superficial and internal) are generally described. The superficial are further subdivided as dorsals, ventrals, and laterals. The deeper internal muscles are grouped as radiating and transverse. Results of investigations of a few elephant trunks (by C. Gans, F. Hensen-Smith, G. Overbeck, J. Shoshani, K. Watson, and C. Zajac) show that many of the myriad units of the so-called 'muscles' in the proboscis are, based on histological examinations, probably parts of muscles or muscle fascicles. The proboscis of one elephant (*E. maximus*, named 'Iki'; Shoshani *et al.* 1982; Riha *et al.* 1991) was studied in detail, and based on extrapolation from an extensive count, the number of muscle fascicles in that trunk was about 150 000, four times more than Cuvier's muscle estimate.

The external morphology of the distal end of the proboscis differs among *Loxodonta*, *Elephas*, and *Mammuthus*. In *L. africana*, there are two finger-like processes, one on the dorsal and one on the ventral side. In *E. maximus*, there is only one 'finger', on the dorsal side; the ventral portion is thickened and 'stumpy'.The trunk of the woolly mammoth (*M. primigenius*) possesses a broad 'flap' which extends the width of the trunk on the ventral side, and a projection on the dorsal end (Flerov 1931; Fig. 2.9). The trunk of the African elephant seems to have more folds of skin in the form of 'rings' or annulations than that of the Asian elephant. That of the African species appears to be 'floppy', while that of the Asian is more rigid. This feature can best be seen when they raise their trunks towards their foreheads; that of *Loxodonta* flattens against the forehead, while that of *Elephas* does not (Shoshani, personal observation, 1990).

INTERNAL ORGANS

The internal organs of an elephant are not proportionately larger than those of other mammals (Fig. 2.10 and references therein); in many cases, they are relatively small. For example, the brain of an adult elephant weights 4.5–5.5 kg (ratio of 1/500) and that of a human is about 1.6 kg (ratio of 1/50). Negative allometry of brain/body weight ratio has been demonstrated for many mammalian taxa (Jerison 1973; Eisenberg 1981). The heart of an

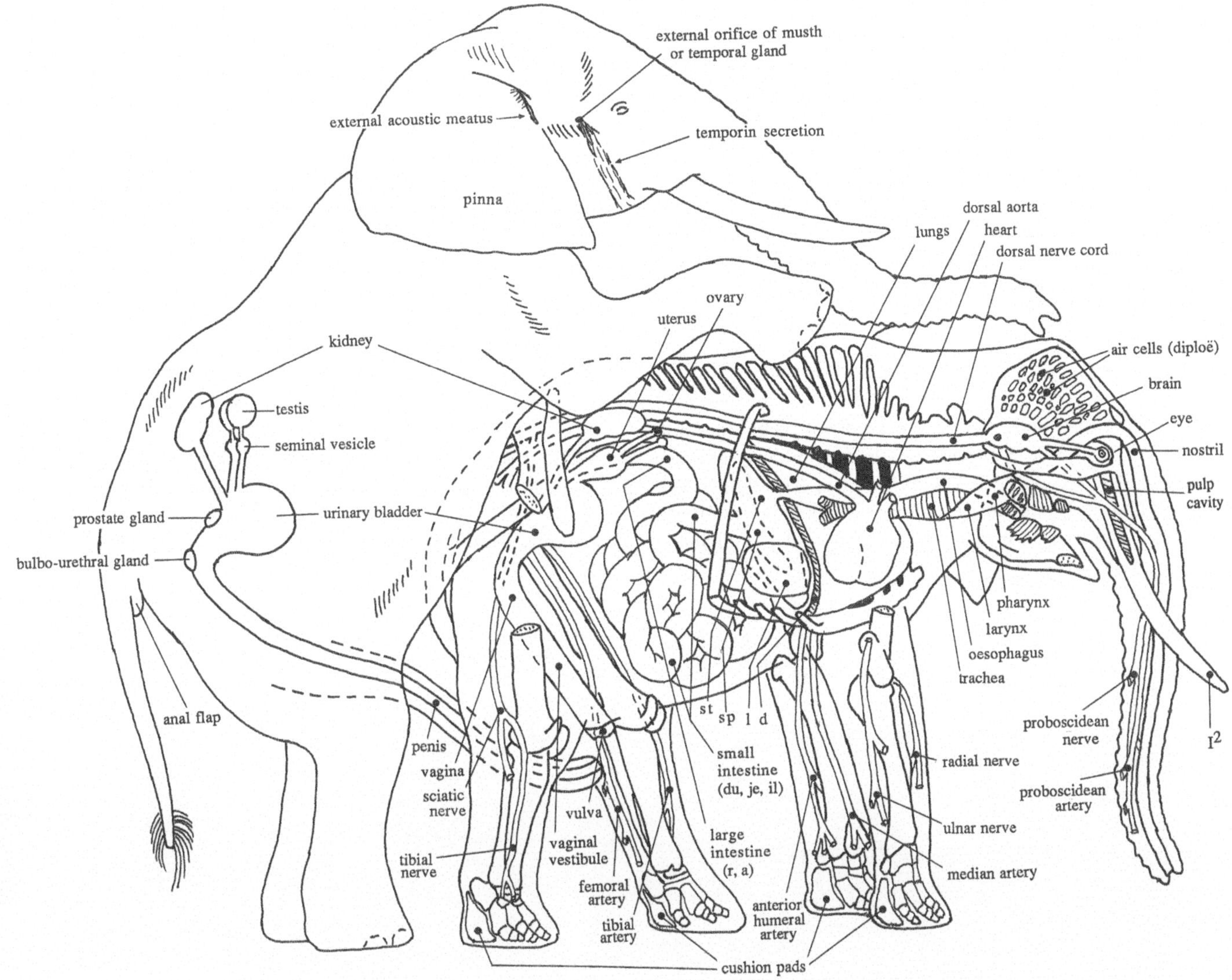

Fig. 2.10 A simplified diagram showing the internal organs of an elephant (depicted is a female, *Loxodonta africana*, but structures apply to *Elephas maximus* as well). (Drawing by G. H. Marchant; data by J. Shoshani, modified after Blair 1710; Hill 1953; Perry 1953; Shimizu *et al.*, 1960; Sikes 1971; Shoshani, personal observations, 1980–1992; Shoshani *et al.* 1982.) Key for digestive system and other structures: d, diaphragm; l, liver; sp, spleen; st, stomach. The small intestine includes: du, duodenum; il, ilium; je, jejunum. The large intestine includes: r, rectum; and ends with a, anus. The pancreas is attached to the small intestine; the caecum is the section between the small and large intestines.

elephant may weigh 12.0–21.0 kg, depending on the size and age. In relative terms, however, it weighs approximately 0.5 per cent of the total body weight. The shape of the heart is unique to elephants and sirenians; it has a bifid (double-pointed) apex instead of the typically single-pointed heart-shape (cf. heart of a mammoth, Suzuki *et al.* 1992). The lungs adhere to the diaphragm and breathing is performed with the aid of the diaphragm more than with the expansion of the rib cage. The liver is composed of 2–3 lobes, there is no gall-bladder, but the main hepatic duct is enlarged. The digestive system is of a simple type among mammals; elephants are hind-gut fermenters. The combined length of the small and large intestines may reach 35 m.

An important aspect of the elephant's physiology is the variability in its body/skin temperature which often ranges several degrees on either side of the average (35.9 °C; Benedict 1936). By doing so, elephants can withstand extreme environmental conditions without losing much energy. The ability to tolerate changes in habitat has been an extremely important feature in the evolution of this order and in the survival of the two modern species.

Reproductive organs

The male and female reproductive systems consist of typical parts found in other mammals. The testes in male elephants are permanently located inside the body, near the kidneys, a primitive condition. The penis, in a mature male, is long, muscular, and controlled by a voluntary muscle, the *levator penis*; it can reach 100 cm in length and has a diameter of 16 cm at its base.

Unlike bovids and equids, where the vulval opening is located under the tail, in elephants it is between the female's hind legs (Fig. 2.10 and references therein). The cervix, in a non-pregnant individual, may not be a distinct structure as found in most mammals. The clitoris is a well-developed organ; it can reach 40 cm long and is manipulated by the *levator clitoris* muscle. The mammary glands are located between the forelegs, a condition which enables the mother to be in touch with her calf while it nurses. (In *Elephas*, the male differs from the female in having larger 'knobs' on its head. *Loxodonta* females may be distinguished from males as they have an angular head profile, whereas that of the male is rounder).

MUSTH

Unique among mammals, the musth (temporal) gland found in *Loxodonta*, *Elephas*, and *Mammuthus* is located midway between the eye and the ear on each side of the head. The skin in this region is about 2.0 cm thick. The gland is composed of sub-units, lobes and lobules, separated by connective tissues, and has a central chamber with a single orifice, which varies from 1.5 to 2.5 cm (Eales 1925; Fernando *et al.* 1963; Shoshani 1992*a*). Secretions produced by the musth gland can be sensed by the Jacobson organ (vomero-nasal organ; Rasmussen and Hulgren 1990), located on the roof of the mouth close to the front. It functions to process chemical cues (detected by males) and transfers the information to the brain. The most recent accounts on musth are those of Poole and Moss (1981) and of Schmidt (1992). The latter noted '...in captivity at least, musth is a means for bull elephants to determine their dominance hierarchy' (Schmidt 1992, p. 96).

ACKNOWLEDGEMENTS

Without the crisp and accurate anatomical illustrations of G. H. Marchant, this chapter would be 'naked'; I thank him very much. D. C. Fisher, E. C. Marsac, J. L. Pierce, and P. Tassy critically read this chapter.

THREE

Dental homologies and nomenclature in the Proboscidea

PASCAL TASSY

INTRODUCTION

Proboscidean molars and premolars are basically bunolophodont teeth. They diverged from the plesiomorphic tethythere condition where the bilophodont/bunolophodont pattern is outlined on molars (Domning *et al.* 1986; Tassy and Shoshani 1988). The main evolutionary trend is towards the augmentation of cusps due to subdivision of the main mammalian cusps and by the augmentation of loph(id)s.

The molars of the Eocene tethythere species *Anthracobune pinfoldi* from Asia are rather close to the generalized mammalian condition, and are used to show the plesiomorphous condition for proboscideans (Figs 3.1a, 3.2a); bilophodonty is outlined and the paraconid is already reduced to form a paracristid. The dentition of the Eocene genus *Moeritherium* from Africa is close to the ancestral condition for proboscideans (Figs 3.1b, 3.2b). Central conules develop on the pretrite side from paraconule, metaconule, paraconid (= paracristid), and cristid obliqua. The third loph(id) is outlined with the occurrence of a distocrista, a distocristid, and the inflation of the posterior cingulum.

In deinotheres (Figs 3.1c, 3.2c) the number of loph(id)s is enlarged only for dp4 and M1 (which are trilophodont). The lower molars of *Palaeomastodon* have three lophids (Fig. 3.2d), whereas the upper and lower molars of *Phiomia* have three loph(id)s (Figs 3.1e, 3.2e); both are Oligocene genera. The molars of all early members of the Elephantoidea (Figs 3.1f, 3.2f) share the trilophodont grade (plesiomorphic condition): three loph(id)s on dp4, M1, M2; three or four loph(id)s on M3. The tetralophodont grade occurred only among amebelodontids and among gomphotheres (four loph(id)s on dp4, M1, M2 and five or more loph(id)s on M3). Stegodontid and elephantid molars with numerous laminae were derived from gomphothere tetralophodont molars, and reached the pentalophodont grade or more. Laminae resulted from the antero-posterior compression of the numerous loph(id)s in both stegodontid and elephantid molars. As a result, only the first and second laminae of the molars of living elephants correspond to the generalized mammalian molar pattern (see also Saegusa, this volume, Chapter 17).

The molarization of premolars only concerns P3 and P4 and is never complete. The posterior cusps of P3 are only outlined in the plesiomorphic proboscidean condition. In the Oligocene genus *Phiomia*, the ectoloph of P^4 is opened labially by the interloph (Fig. 3.1e) which makes the P^4 fully bilophodont (Fig. 3.2e). In Elephantoidea, the P^4 and P_4 are fully bilophodont (trilophodont grade) or trilophodont (tetralophodont grade). The P^3 and P_3 of the tetralophodont grade are bilophodont.

Morphological trends which do not alter the crown pattern are trends towards more lophodont teeth or more bunodont teeth. True lophodont teeth occurred early in proboscideans, e.g. the African Eocene species *Numidotherium koholense* (Mahboubi *et al.* 1986), and *Barytherium grave* (Andrews 1906) have fully lophodont molars (two lophs only). True lophodont molars are also seen in deinotheriids (Miocene up to Pleistocene). Less strongly lophodont molars are seen among Elephantoidea in the Mammutidae (trilophodont grade) and are called zygodont or zygolophodont molars. Bunodont teeth with numerous extracusps — particularly posttrite central conules in evolved species — occurred independently in various elephantoid lineages, e.g. the Old World Miocene genus *Choerolophodon*, the Old World genus *Anancus* (Miocene up to Pleistocene), and the New World genus *Stegomastodon* (Pliocene up to Pleistocene).

NOMENCLATURE

Dental nomenclature applied to the rather complicated proboscidean molars varies greatly between authors. Figure 3.3 shows one commonly used scheme of proboscidean dental terminology based on both Osborn's and Vacek's terminologies. Nevertheless, comparison with less derived teeth, such as perissodactyl teeth, is made easy if *Anthracobune* and *Moeritherium* are taken as morphological intermediates (Figs 3.1, 3.2). Hence the general mammalian nomenclature can be applied with few changes. Moreover, many extra-cusps have also been described previously in other mammalian groups. As dental nomenclature implies only topological identity, these names can be used. Since different terminologies for proboscidean molars have been used by different specialists, a lexicon with synonyms is given here for the main terms.

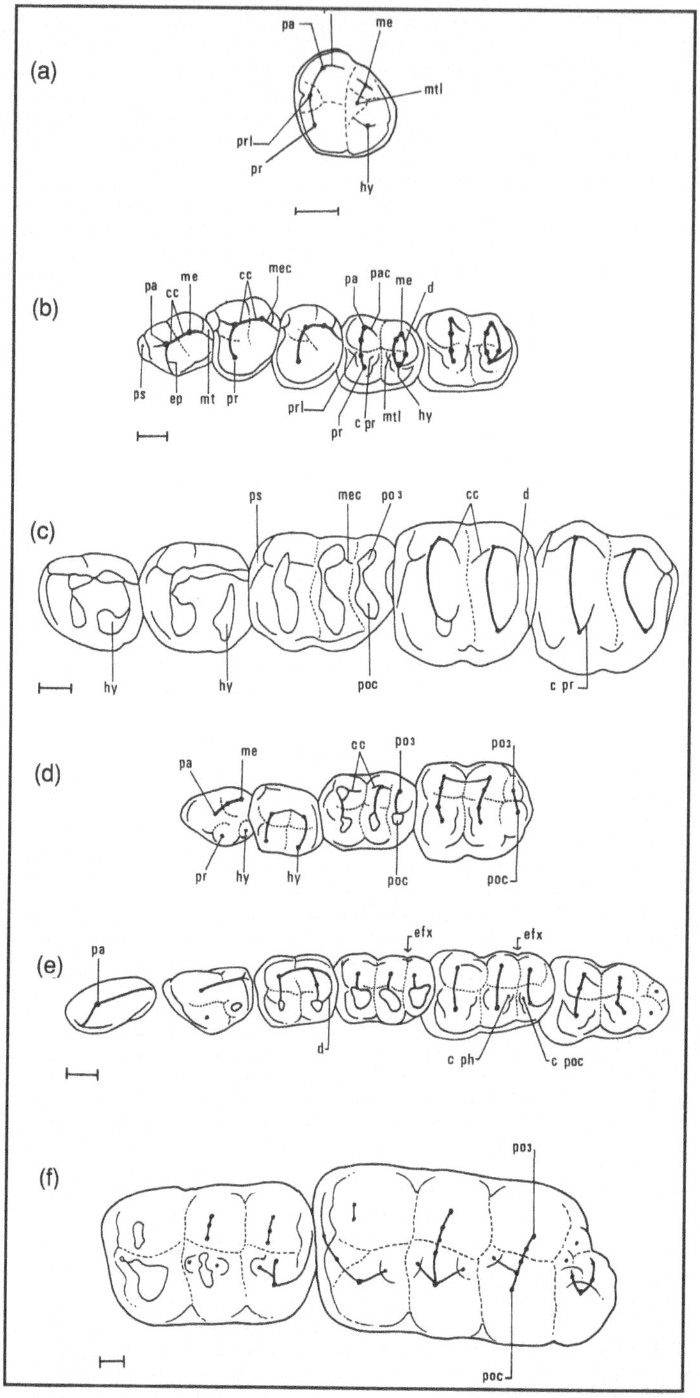

Fig. 3.1 Left upper tooth rows. (a) *Anthracobune pinfoldi*, M^3, middle Eocene. (b) *Moeritherium lyonsi*, P^2–M^2, late Eocene. (c) Deinotheriidae (*Prodeinotherium hobleyi*, early Miocene), P^3–M^3. (d) *Palaeomastodon beadnelli*, P^3–M^2, early Oligocene. (e) *Phiomia serridens*, P^2–M^3, early Oligocene. (f) *Gomphotherium sylvaticum*, M^2–M^3, early Miocene. ((b)–(e) modified from Tassy 1982). Scale = 1 cm. Key: cc, centrocrista (postparacrista + premetacrista); c ph, post hypocone crest (posterior central conule 2); c poc, pre-postentoconule crest (anterior pretrite central conule 3); c pr, postprotocone crest (posterior pretrite central conule 1); d, distocrista ; efx, ectoflexus; ep, epicrista; hy, hypocone; me, metacone; mec, postmetacrista; mt, metastyle; mtl, metaconule; pa, paracone; pac, postparacrista; poc, postentoconule; po3, posttrite cusp of third loph; pr, protocone; prl, paraconule; ps, parastyle.

LEXICON

Key to authors

AF	Anthony and Friant (1941)
C	Cuvier (1806)
DRM	Domning *et al.* (1986)
G	Graf (1957)
H	Harris (1973)
He	Hershkovitz (1971)
L	Lartet (1859)
O	Osborn (1888, 1907*a*, 1921*a*, 1936)
R	Radinsky (1969)
Re	Remane (1960)
S	Schlesinger (1917)
Sz	Szalay (1968, 1969)
T	Tassy (1977*a*, 1982)
To	Tobien (1973*b*, 1975)
V	Vacek (1877)
Vb	Vandebroek (1961)
Vv	Van Valen (1966)

abaxial cone(id) (To) On each loph(id) the conelet that is most distant from the median sulcus (= pretrite and posttrite main cusps).
adaxial cone(id) (To) On each loph(id) the conelet that is close to the median sulcus (= mesoconelet).
central conule (O) On the anterior and posterior walls of each loph(id) a conule either serrated (crest) or blunt (conule *sensu stricto*). Can occur on both pretrite and posttrite half-loph(id)s but formerly defined for the pretrite half-lophs only: see pretrite central conule, and posttrite central conule.
centrocrista (Vb) Postparacrista + premetacrista.
colline (French, L) Loph, lophid.
cone (C) Each main cusp.
conelet (O) A cusp which results from the subdivision of each main cusp.
conule (O) A cusp or cuspule on the walls of any loph(id).
convolute (G) Distocrista (He).
crescentoid (To) Central conule.
crête récurrente antérieure (French, AF) Paraconule, paraconid.
crista, German ***Cristen*** (S) = German ***Sperrleisten*** (S) The central conules of the zygodont molars.
cristid obliqua (Sz) = ***crista obliqua*** (Re) = **metalophid** (R) In proboscideans subdivided into mesoconid and protoconulid.
distocrista (He) Postero-labial crest of the hypocone.
distocristid (He) Postero-lingual crest of the entoconid.
ectocone (O) Labial main cusp of each loph(id).
ectoconelet (O) Labial cusp which results from the subdivision of each main labial cusp.

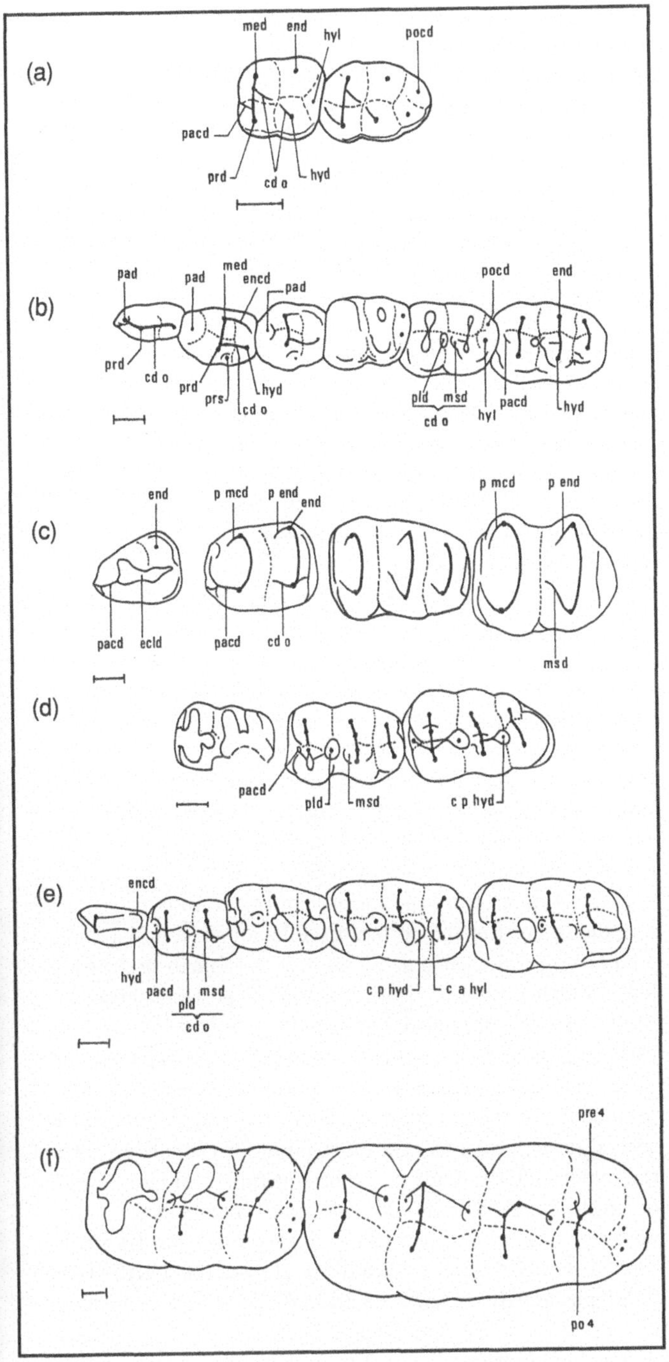

Fig. 3.2 Left lower tooth rows (a–e), right lower tooth row (f). (a) *Anthracobune pinfoldi*, M_2–M_3, middle Eocene. (b) *Moeritherium lyonsi*, P_2–M_3, late Eocene. (c) Deinotheriidae (*Prodeinotherium bavaricum*, early Miocene), P_3–M_2. (d) *Palaeomastodon beadnelli*, M_1–M_3, early Oligocene. (e) *Phiomia serridens*, P_3–M_3, early Oligocene. (f) *Gomphotherium sylvaticum*, M_2–M_3, early Miocene. ((b)–(e) modified from Tassy 1982). Scale = 1 cm. Key: cd o, cristid obliqua; c p hyd, conule or crest posthypoconid (posterior pretrite central conule 2); c a hyl, pre-hypoconulid conule or crest (anterior pretrite central conule 3); ecld, ectolophid; encd, entocristid; end, entoconid; hyd, hypoconid; hyl, hypoconulid; med, metaconid; msd, mesoconid (anterior pretrite central conule 2); pacd, paracristid; pad, paraconid; p end, pre-entoconid crest (entocristid); pld, protoconulid (posterior pretrite central conule 1); p mcd, pre-metaconid crest (entocristid); pocd, postentoconulid; po4, posttrite cusp of fourth lophid; prd, protoconid pre4, pretrite cusp of fourth lophid; prs, protostylid.

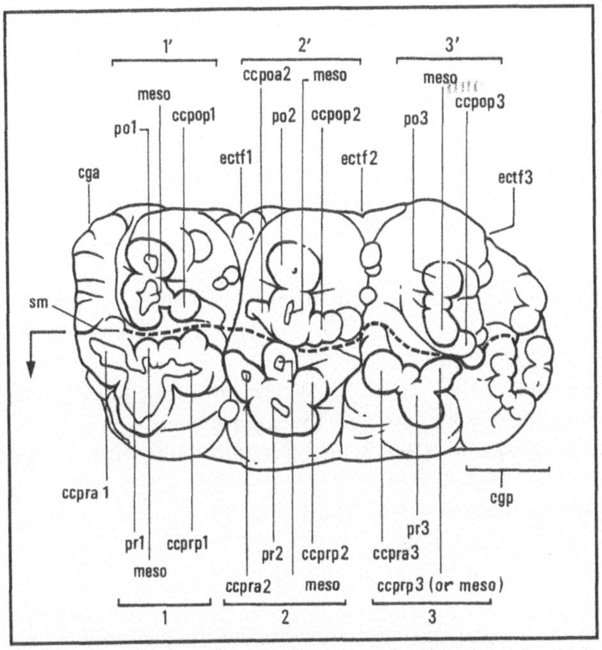

Fig. 3.3 Proboscidean dental nomenclature (left M^2 of the amebelodontine *Protanancus chinjiensis*, middle and late Miocene; after Tassy 1983). This tooth belongs to the trilophodont grade with an outlined fourth loph. The arrow is pointing to the anterior end and then towards the lingual side. Key: 1, 2, 3, pretrite halves of first, second, and third lophs; 1', 2', 3', posttrite halves of first, second and third lophs; ccpoa2, anterior posttrite central conule of 2nd loph; ccpop1, 2, 3, posterior posttrite central conules of 1st, 2nd, and 3rd lophs; ccpra1, 2, 3, anterior pretrite central conule of 1st, 2nd, and 3rd lophs; ccprp1, 2, 3, posterior pretrite central conule of 1st 2nd, and 3rd lophs; cga, anterior cingulum; cgp, posterior cingulum (here it forms an outline of the fourth loph); ectf1, 2, 3, ectoflexus of the 1st, 2nd, and 3rd interlophs; meso, mesoconelet of each half-loph, on both sides of the median sulcus (adaxial cone); po1, main posttrite cusp of the first loph (paracone); po2, main posttrite cusp of the second loph (metacone); po3, main posttrite cusp of the third loph; pr1, main pretrite cusp of the first loph (protocone); pr2, main pretrite cusp of the second loph (hypocone); pr3, main pretrite cusp of the third loph (postentoconule); sm, median sulcus.

ectoflexus (Vv) Labial notch of the interloph(id)s.
ectoloph (O) Labial loph made by pre- and postparacrista and pre- and postmetacrista.
ectoloph(id) (To) Labial half-loph(id) (not synonymous with ectoloph (O)).
entocone (O) Lingual main cusp of each loph(id).
entoconelet (O) Lingual cusp which results from the subdivision of each main lingual cusp.
entoconid (O) Second posttrite main cusp on lower molars.
entoconid II (DRM) Postentoconulid = third posttrite main cusp.

entocristid (Vv) Crest which unites metaconid and entoconid.
entoloph(id) (To) Lingual half-loph(id) = **entocone** (O).
entostyle (He) Style on the lingual cingulum.
epicrista (Vb) Lingual crest of the paracone.
geschnittenen Cristen (German, S) Posttrite zygodont crests.
hypocone (O) Second pretrite of upper molars.
hypoconid (O) Second pretrite of lower molars.
hypoconulid (O) Third pretrite of lower molars.
interloph(id) (T) = **valley** (L) Space between loph(id)s.
lobe (C) Loph or lophid, crest, ridge, *colline*.
median sulcus (O) Sulcus between pretrite and posttrite half-loph(id)s.
mesoconelet (O) The cusp close to the median sulcus (adaxial) when the main cusp of each half-loph(id) is subdivided. On the first loph, part of the paraconule which belongs to the loph; on the second loph, part of the metaconule which belongs to the loph.
mesoconid (Vv) Cusp on the anterior side of the hypocone which develops from the cristid obliqua (= 2nd anterior central conule of lower molars).
mesostyle (H) Entostyle (He).
metacone (O) Second posttrite main cusp of upper molars.
metaconid (O) First posttrite main cusp of lower molars.
metaconule (O) Usually subdivided into two cusps, the second anterior central conule and the mesoconelet 2.
metaloph (O) Loph joining metacone and hypocone (= hypoloph).
paracone (O) First posttrite main cusp of upper molars.
paraconid (O) First anterior central conule of lower molars = *crête récurrente antérieure*.
paraconule (O) Usually subdivided into first anterior central conule and mesoconelet 1 on the upper molars.
postentoconule (He) Third pretrite main cusp of upper molars.
postentoconulid (He) Third posttrite main cusp of lower molars = entoconid II.
postmetaconulid (He) First posterior posttrite conule of lower molars.
postmetacrista (Sz) Posterior crest of the metacone = second posterior posttrite zygodont crest of zygodont molars.
postparacrista (Sz) Posterior crest of the paracone = first posterior posttrite zygodont crest of zygodont molars.
postprotocone crest (T) = posterior central conule of the first pretrite half-loph.
posttrite (V) On each loph(id) the main cusp [or half-loph(id)] which is less worn (labial on upper molars and lingual on lower molars); posttrite 1 = paracone, metaconid; posttrite 2 = metacone, entoconid. As an adjective, any cusp or cuspule or crest which belongs to the less worn half-loph(id)s.
posttrite central conule (T) = **posttrite enamel pillar** (To) Conule on the anterior and posterior walls of each labial half-loph (upper molars) and each lingual half-lophid (lower molars). First posterior posttrite conule of lower molars = postmetaconulid.
posttrite enamel pillar (To) Posttrite conule.
posttrite half-loph(id) (To) Posttrite main cusps, part of the loph(id) which is less worn.
posttrite zygodont crest (T) Zygodont crest or zygodont crista, or *geschnittenen Cristen*. On upper zygodont molars: first anterior posttrite zygodont crest = preparacrista; first posterior posttrite zygodont crest = postparacrista; second anterior posttrite zygodont crest = premetacrista; second posterior posttrite zygodont crest = postmetacrista; third anterior posttrite zygodont crest = anterior crista of the third posttrite half-loph(id); third posterior posttrite zygodont crest = posterior crista of the third posttrite half-loph(id).
premetacrista (Sz) Anterior crest of the metacone.
preparacrista (Sz) Anterior crest of the paracone.
pretrite (V) On each loph(id) the main cusp [or half-loph(id)] which is more worn (lingual on upper molars, labial on lower molars); pretrite 1 = protocone, protoconid; pretrite 2 = hypocone, hypoconid. As an adjective, any cusp or cuspule or crest which belongs to the more worn half-loph(id)s.
pretrite central conule (T) = **central conule** (O) = **crescentoid** (To) = **pretrite zygodont crest** (T) Pretrite conule on each loph(id), i.e. lingual on upper molars and labial on lower molars, either serrated (crest) or blunt.

On upper molars: first anterior central conule = paraconule, *crête récurrente antérieure*; first posterior central conule = postprotocone crest = posterior cuspule or crest of the protocone; second anterior central conule = front part of the metaconule; second posterior central conule = posthypocone crest = posterior cuspule or crest of the hypocone; third anterior central conule = anterior cuspule or crest of the postentoconule; third posterior central conule = posterior cuspule or crest of the postentoconule; fourth anterior central conule = anterior cuspule or crest of the fourth pretrite half-loph; fourth posterior central conule = posterior cuspule or crest of the fourth pretrite half-loph.

On lower molars: first anterior central conule = paraconid = *crête récurrente antérieure*; first posterior central conule = protoconulid; second anterior central conule = mesoconid; second posterior central conule = posterior cuspule or crest of the hypoconid; third anterior central conule = anterior cuspule or crest of the hypoconulid; third posterior central conule = posterior cuspule or crest of the hypoconulid; fourth anterior central conule = anterior cuspule or crest of the fourth pretrite half-loph; fourth posterior central conule = posterior cuspule or crest of the fourth pretrite half-loph.

pretrite half-loph(id) (To) The part of the loph(id) which is worn first.
protocone (O) First pretrite main cusp of upper molars.
protoconid (O) First pretrite main cusp of lower molars.
protoconulid (T) Cuspid or crest which develops from the cristid obliqua on the posterior wall of the protoconid.
protoloph (O) Loph which joins protocone and paracone.
protolophid (O) Lophid which joins protoconid and metaconid.
protostylid (Vv) Labial cuspid posterior to the protoconid.
Sperrleisten (German, S) Central conule of zygodont teeth.
talon (C) Cuspules which develop from posterior cingulum.

trefoil (C) Part of each half loph(id) made of the main cusp and conules. Pretrite trefoil = pretrite half-loph(id); posttrite trefoil = posttrite half-loph.
tritoloph(id) (O) Third loph(id).
tubercule intermédiaire de néoformation (French, AF) Central conule
valley (French ***vallon***, ***vallée***, L) Interloph(id).
zygodont crest (To) Posttrite zygodont crest = crista or *geschnittenen Cristen*.

ACKNOWLEDGEMENTS

Thanks to N. Court, E. C. Marsac, J. L. Pierce, and J. Shoshani for their comments and suggestions. The figures were drawn by D. Visset.

FOUR

Ontogenetic evidence for incisor homologies in proboscideans

W. PATRICK LUCKETT

INTRODUCTION

Living elephants are characterized by the presence of greatly elongated, ever-growing tusks in the upper jaws; these modified incisors may attain a length of up to 345 cm (11.5 ft) in male *Loxodonta africana* (Sikes 1971). Following Andrews' (1906) comparison of the large tusks of Miocene through Recent proboscideans with the moderately enlarged second incisors (I2) in both jaws of the Eocene genus *Moeritherium*, most authors have accepted his proposed homologies of the tusks as specialized I2 in all fossil and extant proboscideans (e.g. Matsumoto 1923; Osborn 1936; Sikes 1971). The occurrence of hypertrophied I2 as a shared, derived attribute for all proboscideans was adopted recently by Mahboubi *et al.* (1986) in their analysis of the affinities of the Eocene genus *Numidotherium* from Algeria. Most authors have also considered the small, transitory incisor that occurs during the fetal and early postnatal period in extant elephants to be the deciduous predecessor (dI^2) of the permanent tusks in the upper jaw (Corse 1799; Rose 1894; Schaub 1948; Sikes 1971).

In contrast, however, two investigators have suggested a different homology for the permanent tusks in both fossil and living proboscideans. Anthony (1933) analysed several stages of incisor ontogeny in *Elephas* and *Loxodonta*, and he postulated that both the small transitory and the large permanent tusks belong to the same (deciduous) dentition. If true, the transitory and permanent tusks of extant elephantids would correspond to dI^2 and dI^3 of *Moeritherium*, respectively, rather than to the traditional interpretation of dI^2 and I^2. This hypothesis of incisor homologies was reiterated subsequently by Anthony and Friant (1941).

More recently, a modification of Anthony's (1933) hypothesis was proposed by Tassy (1985, 1987), based on his assessment of incisor ontogeny in fossil proboscideans. He also believed that both the transitory and permanent tusks in the upper jaw of Miocene–Recent elephantoids belonged to the primary or deciduous dentition, although he suggested that these were probably dI^3 and dI^2, respectively, a reversal of the homologies postulated by Anthony (1933). Tassy emphasized the precocity of development and eruption of the incisors in fossil and extant elephantoids as support for this hypothesis, as well as the fact that both incisors can project simultaneously from the jaws for a short period of time. Despite his excellent discussion of the problems involved in assessing incisor homologies of fossil and extant proboscideans, Tassy (1987, p. 16) admitted that the evidence to support his hypothesis 'is less than decisive', and he acknowledged that an analysis of dental ontogeny in living elephants would be useful to test this hypothesis.

Although hundreds of elephant fetuses have been collected during culling operations in East Africa (Laws *et al.* 1975; Hanks 1979), there have been only a few histological studies of dental development in either genus of living elephants. Bolk (1919) examined the lower jaw in a 205 mm crown–rump length (CR) fetus of *Loxodonta*, in which he was unable to detect any trace of a rudimentary lower incisor. He noted the presence of a developing incisor in the upper jaw, but no description or illustration was presented. Subsequently, Eales (1925, 1926) radiographed and dissected the head of a 210 mm CR *Loxodonta* fetus, in which she identified a pair of calcified deciduous incisors in the upper jaw (no measurements provided). Anthony (1933) demonstrated the presence of dentine and enamel on the transitory incisors in *Elephas* and *Loxodonta* fetuses during the first half of the pregnancy. He also claimed that the most mature of these exhibited calcification on both the transitory and permanent tusks, based on X-ray and histological studies. Later, however, Driak (1935) studied the deciduous premolars and incisors in a larger, 565 mm CR *Elephas* fetus. Dentine and enamel were well developed on the single pair of upper incisors, but no trace of a second pair of incisors could be found.

Each of these prenatal studies on elephants was very limited in scope, and only Anthony (1933) discussed problems of dental homologies. In addition, contradictory evidence was presented by Anthony (1933) and by Driak (1935), concerning the presence or absence of the permanent tusk anlage (primordium) during the first half of pregnancy; this is a key point in assessing the deciduous or successional nature of the permanent tusks. In a recent evaluation of the problems surrounding the identification of deciduous or successional tooth homologies during ontogeny in therians (Luckett 1993), it was emphasized that the epithelial connections among the oral epithelium, primary dental lamina, deciduous tooth germ, and lingual successional lamina must be evaluated in order to distinguish such homologies. Little or no attention was devoted to these epithelial relationships in previous studies of proboscidean development, and the present investigation was undertaken to provide additional criteria for assessing incisor homologies in elephants, and to re-examine the previously published data used to evaluate these homologies in extant and fossil proboscideans.

MATERIAL AND METHODS

A well-preserved 147 mm CR fetus of the African elephant, *Loxodonta africana* (Fig. 4.1a), was obtained through the courtesy of John S. Perry (England). The head was embedded in celloidin, sectioned serially at 30 μm, and the sections were stained with Azan-trichrome in the laboratory of Wolfgang Maier (Germany). A photograph of this specimen was published by Fischer and Trautmann (1987). The age of the fetus is unknown, but it is probably in the first third of pregnancy. In addition, the skull of an older *Loxodonta* fetus (AMNH 119620) was examined at the American Museum of Natural History, New York, USA (AMNH). These two fetuses were compared with other previously studied prenatal specimens of both African and Asian elephants for which histological or radiographic data were available (Table 4.1), as well as with four juvenile skulls in the collections of the AMNH, the Laboratoire d'Anatomie Comparée, Museum National d'Histoire Naturelle, Paris, France, and the Forschungsinstitut und Naturmuseum Senckenberg, Frankfurt am Main, Germany. Although the number of fetal specimens available for histological analysis is limited, this study was facilitated by comparing dental development in proboscideans with more extensive ontogenetic series of other therian mammals, including a closely graded series of *Procavia* fetuses from the first half of pregnancy.

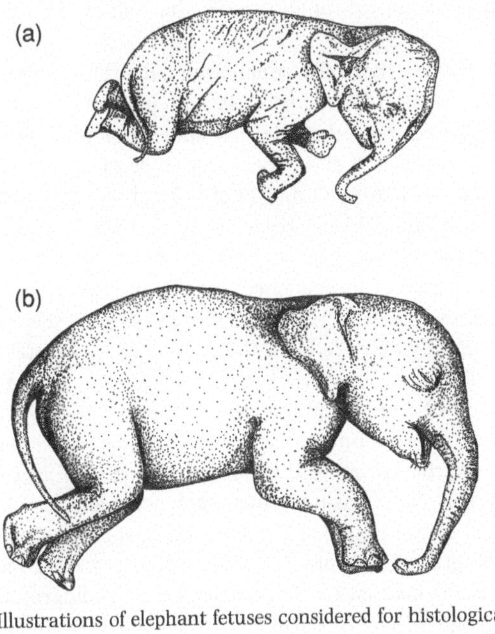

Fig. 4.1 Illustrations of elephant fetuses considered for histological study during this investigation. (a) *Loxodonta africana*, 147 mm CR fetus (drawn from the photograph in Fischer and Trautmann 1987). (b) *Elephas maximus*, 565 mm CR fetus, about 10.3 months gestational age (redrawn from Toldt 1913). Not to scale.

RESULTS

Serial sections of both jaws of the 147 mm CR *Loxodonta* fetus were examined in detail, although this investigation will be concerned primarily with the developmental state of the premaxillary region and the incisors. In the anterior region of the upper jaw, a single deciduous incisor occurs bilaterally, associated with the developing premaxillary ossification. Before considering the developmental stage of this tooth, it is useful to review the major phases of dental ontogeny in eutherian mammals, in order to understand the relationships of the developing incisor to the surrounding tissues. The earliest evidence for eutherian dental development is indicated by differentiation of a lenticular swelling of the oral epithelium, extending antero-posteriorly in each jaw quadrant. This primary dental lamina subsequently gives rise to localized thickenings along its longitudinal extent; these form the tooth buds for the primary or deciduous dentition. Each bud will undergo further differentiation to form the classic cap and bell stages of development for each primary tooth germ. Because these early ontogenetic stages are fundamentally identical in all eutherians that exhibit normal dental development, the hypothetical appearance of the three developing deciduous incisors in the Eocene proboscidean *Moeritherium* can be reconstructed to demonstrate the epithelial relationships of these teeth to the dental lamina and oral epithelium (Fig. 4.2). In a longitudinal view, each incisor is attached by a slender, primary dental lamina stalk to the oral epithelium. In later ontogenetic stages of all eutherians, the primary dental lamina fragments and disappears, so that the primary lamina stalk may appear as isolated, epithelial islets, or be completely absent. In the middle

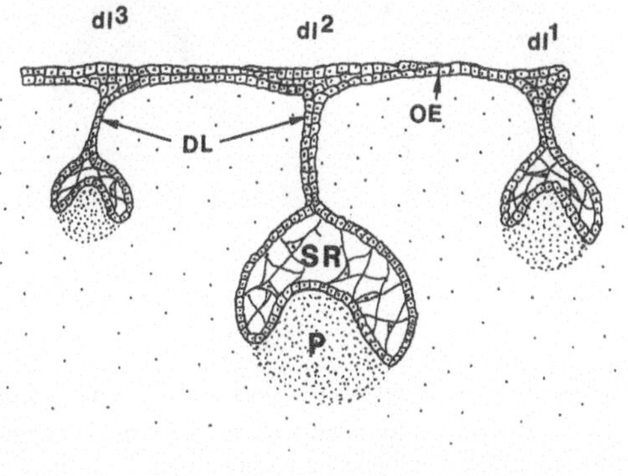

Fig. 4.2 Hypothetical appearance of developmental relationships among the oral epithelium (OE), primary dental lamina stalks (DL), and bell stage tooth germs of dI^{1-3} in the Eocene proboscideans *Moeritherium* and *Numidotherium*. Longitudinal view. P, dental papilla; SR, stellate reticulum.

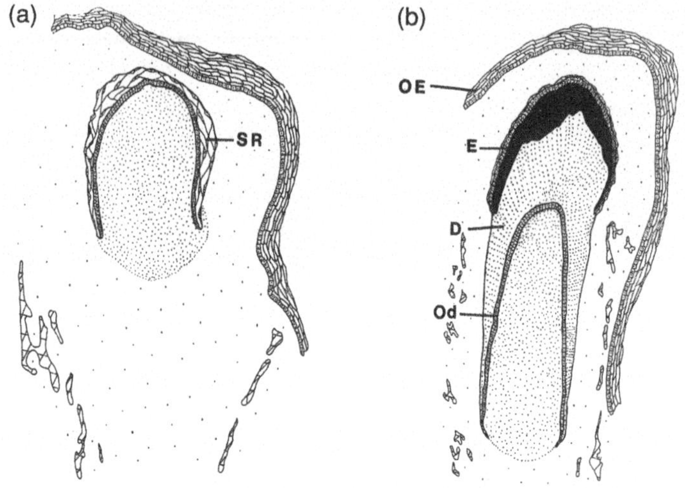

Fig. 4.3 Histological sections through two ontogenetic stages of dI^2 in elephants. (a) *Loxodonta africana* fetus, 147 mm CR, with dI^2 in late bell stage, lacking dentine or odontoblasts. Note that the primary dental lamina stalk is already disrupted from the oral epithelium. (b) *Elephas maximus* fetus, 565 mm CR, with dentine and enamel developed on the apex of the elongated dI^2 (drawn from a photograph in Driak, 1935). D, dentine; E, enamel; Od, odontoblasts; OE, oral epithelium; SR, stellate reticulum. Camera lucida drawings, transverse sections.

bell stage illustrated for *Moeritherium*, these epithelial connections are still intact, as they would be in extant eutherians at this stage.

In the 147 mm CR *Loxodonta* fetus, the deciduous incisor is in the late bell stage, with well-developed stellate reticulum, but it lacks dentine or odontoblasts (Fig. 4.3a). The primary dental lamina stalk is already disrupted, so that only isolated epithelial islands extend between the tooth and the adjacent oral epithelium (Fig. 4.4a). No trace of a primary dental lamina or abnormal deciduous incisors is detected rostral or distal to the

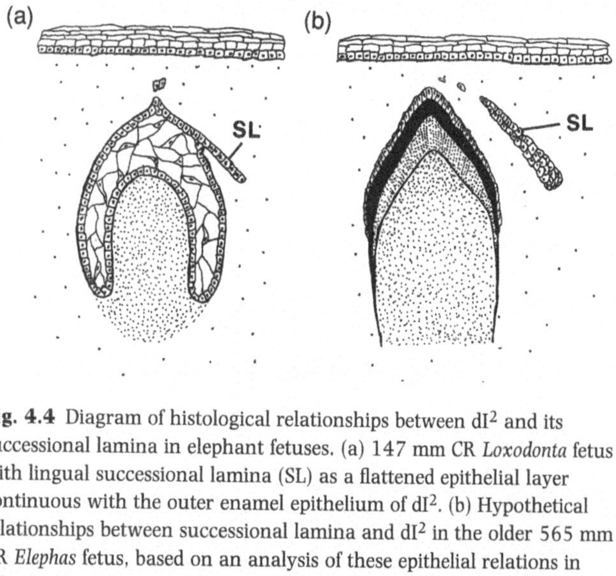

Fig. 4.4 Diagram of histological relationships between dI^2 and its successional lamina in elephant fetuses. (a) 147 mm CR *Loxodonta* fetus with lingual successional lamina (SL) as a flattened epithelial layer continuous with the outer enamel epithelium of dI^2. (b) Hypothetical relationships between successional lamina and dI^2 in the older 565 mm CR *Elephas* fetus, based on an analysis of these epithelial relations in other eutherian mammals, as well as on the absence of a recognizable successor I^2 in this fetus. Transverse sections.

deciduous incisor. The suggested homology of this tooth as dI^2 is further corroborated by the fact that the premaxillary ossification extends both anteriorly and posteriorly to the tooth germ. Examination of these relations in earlier ontogenetic stages would provide a further test for this hypothesis. A short, flat successional lamina extends on the lingual side of dI^2, in continuity with its outer enamel epithelium. The successional lamina is present along only the distal third of the tooth, so that it is not evident in sections through the anterior half of dI^2 (cf. Figs 4.3a, 4.4a).

There is no trace of a rudimentary deciduous incisor in the lower jaw of the 147 mm fetus. In both jaws, three pairs of cheek teeth have attained the late bell stage, but lack odontoblasts; these represent dP2–dP4. The appearance of the cheek teeth in later stages of *Loxodonta* and *Elephas* has been described by Bolk (1919) and Driak (1935), respectively, and will not be considered further here.

A later stage of incisor development in *Elephas maximus* was described by Driak (1935) in a 565 mm CR fetus (Fig. 4.1b); this was accurately dated to be in the 11th month of gestation (Toldt 1913). The gestation period for both elephant genera averages about 21–22 months (Toldt 1913; Hanks 1979), so that the fetus was from mid-pregnancy. Both dentine and enamel were present on the crown of the elongate dI^2 in the 565 mm fetus (Fig. 4.3b). No measurements were provided for this tooth; however, examination of Driak's radiograph suggests that dI^2 had a length of ±26 mm. No evidence for a second incisor was found by Driak, either histologically or radiographically. A similar stage of incisor development is evident in a fetal skull of *Loxodonta* (AMNH 119620), measuring 177 mm in total length. The calcified dI^2 is 27 mm long (Table 4.1), and no evidence of an additional incisor was detected by radiography.

The two elephant fetuses studied by Anthony (1933) were intermediate in size between the specimens described above; his *Loxodonta* fetus measured 350 mm CR, but no length was given for the *Elephas* specimen. The 'transitory' incisors in both fetuses had dentine and enamel on their crowns, and the lengths of these teeth were similar in both (Table 4.1). In addition, Anthony described the presence of a calcified primordium for the 'permanent' tusk, placed somewhat distolingual to the larger transitory incisor, in the 350 mm fetus. Nevertheless, the histological evidence that he presented for this 'calcified' tusk (Anthony 1933, plate I, figs 11, 12) appears to me to be only the slightly swollen successional lamina of dI^2; there is clearly no evidence for a distinct enamel organ or odontoblasts for the permanent tusk in Anthony's figures. It is unfortunate that Driak (1935) made no mention of the successional lamina for dI^2 in his older fetus, because it must have been swollen to a similar or greater degree than that of Anthony's 350 mm fetus; its hypothetical appearance along the distolingual extent of dI^2 is indicated in Fig. 4.4b.

The hypothesis that a calcified 'permanent' tusk primordium is lacking during at least the first half of gestation in elephant fetuses is corroborated by Anthony's (1933) own inability to detect this tooth in near-term fetuses or neonatal infants. The earliest distinct evidence for the presence of permanent tusks was presented by Anthony (1933) in a juvenile *Elephas*, in which a small, calcified tooth (4 mm long) lay distolingual to the elongate dI^2 (65 mm long). Without doubt, uncalcified bud, cap, and bell stages of the permanent tusk could be identified histologically in late fetal and early infant stages of elephant ontogeny, if such an enormous task were to be undertaken. The further development of the calcified permanent tusks postnatally occurs concomitantly with resorption of the root of the adjacent dI^2; this was well documented by Anthony (1933) and will not be discussed further.

DISCUSSION

SIGNIFICANCE OF HISTOLOGICAL OBSERVATIONS FOR ASSESSING INCISOR HOMOLOGIES

The observation that the primary dental lamina had already detached from the oral epithelium in the region of dI^2, and that it had disappeared both anteriorly and posteriorly to this tooth in the premaxillary region of the 147 mm fetus, renders it improbable that another deciduous incisor primordium (either dI^1 or dI^3) could be initiated at a later stage from the oral epithelium. Once the primary dental lamina becomes detached from the oral epithelium in therians, there is no evidence that an additional antemolar deciduous tooth can arise. The only likely source for the permanent tusk in later ontogenetic stages is the lingual successional lamina of dI^2. Unfortunately, Driak (1935) made no mention of the nature of the successional lamina in his 565 mm fetus, even though he found no trace of the successor tusk. Although direct evidence for the origin of the permanent tusk as the successor to dI^2 is still lacking, histological examination of the premaxillary region during the second half of pregnancy in an elephant fetus would be the best way to test this hypothesis.

A successor tooth can develop in eutherians at varying times after formation of the late bell stage by its deciduous predecessor; in many chiropterans, it is initiated before the onset of dentine formation, whereas in the hyracoid *Procavia*, the successor I^1 does not appear until a thick layer of dentine and enamel occurs on the crown of dI^1 (Luckett 1993, and unpublished observations). The developmental relationships between the large dI^1 of *Procavia* and its successor serve as a useful model for the probable late development of the permanent tusk (=I^2) in elephants from the successor lamina of dI^2. In both groups, the large anterior

Table 4.1 Prenatal stages of elephants analysed during this study

Specimen	Fetal length	Head or basicondylar length	dI^2	I^2
Loxodonta (Tübingen)	147 mm CR	58 mm HL	Late bell; no odontoblasts	Not developed
Loxodonta (Eales 1926)	210 mm CR	86 mm BCL	Calcified tooth (X-ray only)	Not detected
Elephas (Anthony 1933)	–	±86 mm BCL	Dentine and enamel (12.8 mm long)	Successional lamina only
Loxodonta (Anthony 1933)	350 mm CR	± 100 mm BCL	Dentine and enamel (9.3 mm long)	Successional lamina only
Elephas (Driak 1935)	565 mm CR	± 225 mm HL	Dentine and enamel (± 26 mm long)	No distinct tooth present
Loxodonta (AMNH 119620)	–	158 mm BCL	Calcified tooth (27 mm long)	Not detected on radiograph
Loxodonta (Anthony 1933)	–	211 mm BCL	Dentine and enamel (38 mm long)	Not detected in dried cranium

deciduous incisor possesses well-developed dentine and enamel on its crown at the time when its lingual successional lamina extends distal to the tooth as an unswollen epithelium. Consequently, the large successor incisor erupts postnatally somewhat distal to its deciduous predecessor in both taxa (Luckett, unpublished findings).

The available data on the position and eruption of the permanent tusks in both jaws of fossil proboscideans were reviewed ably by Tassy (1987), and his analysis demonstrated the same basic topographical relationships between the transitory and permanent tusks in both fossil and extant proboscideans. My preliminary examination (with J. Shoshani) of these relationships in the fossil genera *Platybelodon* and *Amebelodon* at the AMNH is in agreement with Tassy's findings. This suggests that the developmental homologies of the transitory and permanent tusks are identical in Miocene through Recent elephantoids. The present study supports the 'traditional' interpretation of these teeth as dI^2 and I^2, similar to the condition in *Moeritherium* (and probably in *Numidotherium*).

Most arguments offered by Anthony (1933) and Tassy (1987) to corroborate their hypothesis that the 'permanent' tusks in elephantoids are unreplaced deciduous incisors were based on features that do not provide reliable criteria for assessing dental homologies. The major attribute emphasized by both authors is the relative precocity of development and eruption of the permanent tusks. As noted above, this was based, in part, on the mistaken belief that a calcified permanent tusk occurs in the 350 mm *Loxodonta* fetus. In addition, Tassy (1987) demonstrated that the definitive tusks erupt relatively earlier during ontogeny in Miocene–Recent elephantoids (before dP4 erupt) than they did in *Moeritherium* (at about the same time as M3). Tassy (1987, p. 15) further suggested that this 'precocity of eruption of the permanent tusk possibly falsifies the hypothesis of dental replacement' and that the 'development of the permanent elephantoid tusk is coincident with that of dP4'. It is clear, however, that 'development' and 'eruption' of the permanent tusk in extant elephantids proceed at differing rates. In 147–565 mm CR fetuses studied by histological or radiographic methods, dI^2 and dP2–dP4 are all well-developed, whereas there is no distinct trace of the permanent tusk at this time. Clearly, development of the permanent tusk is accelerated during late fetal–early postnatal life in elephantids; in contrast, the further development and subsequent eruption of dP4 are relatively retarded during this same period. Thus, precocity of eruption of the permanent tusk provides no clue to its deciduous or successional nature. In other eutherians with greatly enlarged anterior incisors, such as soricids, procaviids, and rodents, there are considerable ontogenetic differences in the rates of development and eruption of the 'permanent' incisors, as well as differences in their deciduous or successional homologies (for further discussion of developmental heterochronies in the mammalian dentition, see Luckett 1993).

A final point concerns the fact that both the transitory and permanent incisors may protrude from the jaw for a short period of time in some fossil elephantoids, such as in the *Mammut* mandible described by Peterson (1926). Tassy (1987) considered this as further corroboration for his hypothesis that both teeth are deciduous incisors, noting that this would be a relatively uncommon condition for a deciduous tooth and its successor. None the less, both dP3 and P3 frequently occur as erupted teeth temporarily in some marsupial families, and it is the 'normal' condition for most antemolar tooth families in several genera of chiropterans (Luckett 1993). Moreover, in juvenile skulls of all three genera of extant hyracoids, the erupting tips of I^1 and I_2 appear temporarily distal to their fully erupted deciduous predecessors, similar to the relations in *Mammut* (Luckett, unpublished results).

PHYLOGENETIC IMPLICATIONS

The hypothesis that an enlarged I^2 represents a shared, derived character of unquestioned fossil and extant proboscideans (Osborn 1936; Mahboubi *et al.* 1986) is corroborated by the present study, although upper incisors are unknown for deinotheriids. It is unclear at present whether the large lower tusks are homologous in all proboscideans that possess them, especially when *Numidotherium* is compared with other genera. A knowledge of the deciduous incisors *in situ* within the mandible of *Numidotherium* would greatly facilitate an analysis of this question. The question of whether early–middle Eocene anthracobunids should be included in the Proboscidea is beyond the scope of the present study (see Shoshani and Tassy, this volume, Chapter 34). It is of interest, however, that at least one anthracobunid possessed three lower incisors, of which the alveolus for I_2 is notably larger than that for I_3 (Wells and Gingerich 1983, West 1983). Unfortunately, it is impossible to determine the relative size of I_1 in this jaw, because of breakage, and upper incisors are apparently unknown for anthracobunids. Discovery of complete anterior dentitions of anthracobunids may provide additional features for assessing their phylogenetic relationships. If anthracobunids exhibit a moderately enlarged I2 in both jaws, when compared to I1 and I3, it would be less likely that they could have been 'broadly ancestral' to both proboscideans and sirenians, *contra* Wells and Gingerich (1983).

SUMMARY AND CONCLUSIONS

Analysis of incisor development in the upper jaws of prenatal elephants, by both histological and radiographic methods, indicates that only a single deciduous incisor locus, probably dI^2, develops in *Loxodonta* and *Elephas*. Consideration of the epithelial relationships between dI^2 and the oral epithelium in a 147 mm CR fetus of *Loxodonta* renders it highly improbable that any additional deciduous incisor could develop during later stages of pregnancy. The only likely source for the permanent tusk is the lingual successional lamina of dI^2, which occurs as a flattened epithelium in the 147 mm fetus. Contrary to an earlier suggestion by Anthony (1933), there is no distinct evidence for the calcified permanent tusk during the first half of pregnancy. The temporal and spatial relationships between the transitory and permanent tusks in extant and fossil proboscideans during ontogeny provide no positive evidence that both teeth are deciduous incisors, *contra* Tassy (1987); instead, these are consistent with the assessment of fetal development, which indicates the homologies of these teeth as dI^2 and I^2 respectively. Thus, ontogenetic analysis supports the proposal by Andrews (1906) and Osborn (1936) that the enlarged tusks of Eocene to Recent proboscideans are homologous, at least in the upper jaw.

ACKNOWLEDGEMENTS

J. Perry, Cambridge, England, and W. Maier, Tübingen, Germany deserve special gratitude for making this study possible by obtaining and preparing the 147 mm *Loxodonta* fetus for histological study. Fetal and juvenile skulls of elephants were examined through the courtesy and help of G. Storch, Frankfurt, Germany; F. Jouffroy, Paris, France; and G. Musser, New York, USA. Special thanks go to J. Shoshani for introducing me to the study of fossil proboscideans, and for encouraging all aspects of this investigation. Juvenile skulls and dentitions of fossil proboscideans were examined at the AMNH through the courtesy of R. Tedford and M. C. McKenna, and J.-J. Jaeger, Université Montpellier II, Montpellier, France, permitted me to examine his collection of *Numidotherium*. Finally, appreciation and thanks go to P. Tassy, Université Paris VI, Paris, France, for his patience in discussing with me various aspects of dental eruption and morphology in fossil proboscideans, even though my study contradicted his 1987 hypothesis. Figures 4.1–4.4 were drawn by N. Hong.

PART II

Proboscidean relatives

FIVE

On the position of Proboscidea in the phylogenetic system of Eutheria: a systematic review

MARTIN S. FISCHER

INTRODUCTION

The systematic position of Proboscidea is an example of continuity in mammalian classification. Disregarding any systematic schools, Proboscidea is usually classified as belonging to an independent unit within placental mammals, and almost since early systematic studies, its proximity to sirenians has been postulated. Even in the intuitive system of Linnaeus (1758), the 'Ordo Bruta' comprises among others (e.g. *Bradypus*, *Myrmecophaga*, or *Manis*) *Elephas* and *Trichechus*.

The prevailing conception in the late eighteenth and nineteenth centuries relates to the obvious thick-skinnedness of different mammals, expressed in the term 'Pachydermes' (Geoffroy and Cuvier 1795) including 'éléphant, tapir, cochon, hippopotame, rhinocéros'. Cuvier (1817) placed sirenians as herbivore cetaceans in all of his classifications. Storr's (1780) 'Belluae' and Illiger's (1811) 'Multungula' were almost identical with Cuvier's 'Pachydermes'. At first Blainville (1816) classified 'Éléphans' as 'Ordre Gravigrades' next, but unrelated, to the 'Ongulogrades' which included 'Les Lamantins' (*Trichechus*), but later he united 'Proboscidei' and 'Sirenei' as 'Gravigrades' in one taxon (Blainville, in Gervais 1836). This is the first appearance of what we now call 'Tethytheria'.

Concerning the specific association of Proboscidea and Sirenia, later classifications (e.g. Huxley 1869, Gill 1872, Cope 1882) were actually inferior to that of Blainville. Stressing the importance of placentation as a key for the classification of mammals, Milne-Edwards (1844) associated sirenians with pachyderms. Flower and Lydekker (1891) coined the term 'Subungulata', mostly by using Cope's (1882) classification. Subungulata pointed to a relationship between hyraxes and elephants. Sirenia were not included in Flower and Lydekker's Subungulata. In some older classifications, elephants and hyraxes have been depicted next to one another, but without intended direct relationship (e.g. Huxley 1869, p. 98):

> ...both *Elephas* and *Hyrax* must go into the same primary division of the Monodelphia as the Rodentia and Carnivora. ... So far as the Elephants are concerned, I must confess that I see no difficulty in the way of an arrangement which unites the Proboscidea more closely with the Rodentia than with the Artiodactyla and Perissodactyla.

Gregory (1910) defined Ungulata differently from previous and subsequent workers. He excluded the Paraxonia from the ungulates, which he allies with carnivores, and included among others Sirenia, Proboscidea, Hyracoidea, Embrithopoda, and Mesaxonia.

The prevailing concept in our century is that Proboscidea are closely related to Sirenia. The association of these two taxa together with Hyracoidea is reflected by the later usage of the term Subungulata (e.g. Weber 1928) which Simpson (1945) renamed Paenungulata, enclosing these groups and several fossil taxa. As Simpson wrote, this grouping is developed from Blainville's 'Gravigrades', Cope's 'Taxeopoda', and Flower and Lydekker's Subungulata. Simpson (1945) stresses the 'possibly polyphyletic origin of the superorder', superorders being theoretical or even hypothetical groupings, not necessarily natural units. As will be discussed below, the relationship of Hyracoidea to Proboscidea and Sirenia is under discussion.

In 1975 McKenna coined the term 'Tethytheria' for the sister-groups Proboscidea and Sirenia (and Desmostylia). During recent years this concept has been substantiated with morphological arguments by different workers (Tassy 1979, 1981; Domning *et al.* 1986; Fischer 1986, 1989, 1990; Novacek 1986, 1989; Novacek and Wyss 1986; Shoshani 1986*a*, 1992*b*; Prothero *et al.* 1988; Tassy and Shoshani 1988; Court 1990; Thewissen and Domning 1992; Fischer and Tassy 1993). Court (1990) and Fischer and Tassy (1993) provided arguments not only for the inclusion of extinct Embrithopoda in the Tethytheria, but possibly also in the Proboscidea.

The present morphological, molecular, and parasitological evidence for proboscidean relationships will now be presented.

MORPHOLOGICAL EVIDENCE FOR TETHYTHERIA

Tethytheria as used in this paper include Proboscidea, Sirenia, and Embrithopoda. They share several derived cranial and postcranial characters.

Morphogenetic studies on fetal elephant middle-ear structures revealed that the development of the round window (fenestra rotunda) does not take place in the usual therian way in elephants but that the perilymphatic foramen persists (Fischer 1990). Subsequent dissections of the same region in adult sirenians showed that, in sirenians also, the hitherto called round window is actually a foramen perilymphaticum, and that the perilymphatic duct leaves through this opening. This unique condition in therian mammals is found also in the periotic of *Arsinoitherium* (Court 1990; Fischer and Tassy 1993). Court's recent preparation of skull material of *Numidotherium* has yielded an almost complete periotic. According to Court and Jaeger (1991), as well as our own observations (Tassy and Fischer, unpublished), *Numidotherium* could well have possessed the typical therian pattern of perilymphatic foramina (round and oval window, as well as a cochlear canaliculus). X-ray investigations have not resulted in any further evidence. The presence of the therian pattern in an early member of the proboscidean stem lineage suggests — independently of the systematic position of Embrithopoda — either a threefold parallelism in Embrithopoda, Sirenia, and Proboscidea, or a reversal of the character in *Numidotherium*.

A second tethytherian character is the shape of the tegmen tympani. In sirenians the tegmen tympani form a barrel-shaped vault over the anterior tympanic cavity (Fisher 1990). Whereas the tegmen tympani are flat in skulls of elephants, they possess the same vault-shape during fetal stages (Fischer and Tassy 1993). In *Arsinoitherium* the character state is unknown, but in other therian mammals the tegmen tympani are flat during morphogenesis.

As Tassy (1981) has pointed out, the zygomatic arch and the external auditory meatus are set very high in tethytherians, and the orbits are also shifted so far anteriorly that they are not lying above M^1. The infraorbital foramen is situated below the orbit. In sirenians and proboscideans, but not in *Arsinoitherium*, the zygomatic arches are extended laterally.

The 'taxeopod' arrangement of carpals and tarsals has long been discussed in the systematic literature. Carpal taxeopody is defined by means of an exclusive contact of the lunatum (lunar) with the capitatum (os magnum), giving the impression of serially arranged bricks. It is one of the most important characters in favour of paenungulates. But the carpus of Sirenia, Embrithopoda, and Proboscidea is not serially arranged in a strict sense, as in hyrax, because the lunatum has a medial contact with the trapezoid (Fischer and Tassy 1993). As the usual therian condition is represented by a lateral contact of the lunatum, the derived condition of tethytherians may well have passed through a strict taxeopod arrangement, as shown by hyracoids. This would favour the paenungulate concept.

MOLECULAR EVIDENCE FOR THE TETHYTHERIA

The sister-group relationships between Sirenia and Proboscidea are not supported by molecular evidence (however, see Shoshani 1992*b* for evidence). The analysis of seven protein sequences at the ordinal level by Miyamoto and Goodman (1986) supports the monophyly of the Paenungulata (with Tubulidentata included) with an unresolved basal multifurcation.

A former parsimonious tree by Kleinschmidt *et al.* (1986), based on haemoglobin only, groups Sirenia and Proboscidea. But this tree was modified (Czelusniak *et al.* 1990): the hyrax now joins first the elephants and then the manatee clusters to both (the aardvark is not represented).

The analysis of α-crystallin data supports the monophyly of Paenungulata (with Tubulidentata and Macroscelidea included), but not of Tethytheria (de Jong and Goodman 1982; McKenna 1992; de Jong *et al.* 1993). The macroscelidid *Elephantulus* has an identical sequence with *Procavia* and the two are grouped in an unresolved branching with *Trichechus* and *Orycteropus*. (See also Irwin and Wilson (1993) on limitations of molecular methods.)

PARASITOLOGICAL EVIDENCE FOR PROBOSCIDEAN AFFINITIES

The elephant louse (*Haematomyzus elephantis*) is the only representative of the taxon Rhynchophthinrina. It is not a louse in the strict sense, and it does not belong in the Anoplura. Lyal (1985) has considered a sister-group relationship of Rhynchophthinrina and Anoplura, but the recent work of Tröster (1990) on the aardvark louse (*Hybophthirus notophallus*), the most primitive Anoplura, leaves some doubt about Lyal's hypothesis. According to Tröster (1990), the systematic position of the elephant louse between Anoplura and Ischnocera is unresolved.

Leaving all problems with the 'parasitological method' out of consideration, it is interesting to note that elephants possess an ectoparasitic 'louse' that is more primitive in several characters as compared to Anoplura, and that the basic anoplurans are living on aardvarks. But no conclusions on the relationship of Proboscidea can be drawn. The only temptation lies within the idea of a basal position of Proboscidea (Tethytheria?) and Tubulidentata within Eutheria, as in the α-crystallin sequence. Parasites seem to contradict an association of these groups with hyracoids.

THE STEM-GROUP OF PROBOSCIDEA

Recent discoveries of Asian fossils, as well as reconsideration of known fossils, are of great importance for the discussion of the origin of Proboscidea and of Tethytheria. Wells and Gingerich (1983) placed these taxa in the family Anthracobunidae. According to these authors, anthracobunids could represent basal tethytherians, but they classified these fossils in the Proboscidea. The cladistic analysis of Fischer and Tassy (1993) partly confirms this view. *Jozaria* (middle Eocene, Pakistan), *Pilgrimella* (middle Eocene, Pakistan), and *Anthracobune* (middle Eocene, Pakistan) cluster together at the base of the proboscidean stem lineage. In that paper the position of *Minchenella* (late Palaeocene, China) is unresolved: it may either be even more basal than the foregoing groups or on the same branch as Desmostylia. Wells and Gingerich (1983) accepted *Minchenella* as 'the most plausible ancestor of Anthracobunidae'. If we assume Proboscidea and Desmostylia to be sister-taxa, and *Minchenella* as their 'possible ancestor' (Domning *et al.* 1986), then the geological age of Proboscidea as given by the oldest fossils may be between early Eocene and late Palaeocene.

THE STEM-GROUP OF TETHYTHERIA

The systematic position of Desmostylia is in dispute. It is either the sister-group of Proboscidea (Domning *et al.* 1986), or of Sirenia (Shoshani 1986*a*) or of Tethytheria (Tassy 1981; Fischer and Tassy 1993). Two other fossil taxa, *Phenacolophus* (Eocene, Mongolia), once thought to be an embrithopod (McKenna and Manning 1977), and *Lammidhania* (early to middle Eocene, Pakistan) could well represent stem-groups of tethytheres (Fischer and Tassy 1993).

The recent identification of stem proboscideans and tethytheres from Asia influences ideas on their origin. The long-held assumption that proboscideans are an endemic African group is now less convincing than their origin in Asia.

THE SISTER-GROUP OF TETHYTHERIA

Proboscidea and Sirenia are sister-groups. Their last common ancestor dates back at least to the middle or late Palaeocene, and probably lived in Asia. Considering that one of the groups is aquatic, that the primitive proboscidean genus *Moeritherium* very likely had semi-aquatic habits, and that Desmostylia was also aquatic, it cannot be ruled out that the last common ancestor of Tethytheria was semi-aquatic (Janis 1988).

Two different hypotheses on the sister-group of Tethytheria are discussed (Fig. 5.1). The paenungulate concept advocates a closer relationship of Hyracoidea with Tethytheria, and of this clade to Mesaxonia (Novacek 1986, 1989; Shoshani 1986*a*, 1992*b*; Wyss *et al.* 1987; Novacek *et al.* 1988). The second hypothesis prefers a closer relationship of Hyracoidea with Mesaxonia, and of both together with Tethytheria (Fischer 1986, 1989; Prothero *et al.* 1988; Fischer and Tassy 1993; cf. Thewissen and Domning 1992).

The morphological evidence for each hypothesis has been tested, and there remains evidence for both. Besides new myological data (Shoshani 1993*c*), the assumed synapomorphies of paenungulates are: serial arrangement of carpals and tarsals (see above, critically discussed by Fischer and Tassy 1993), amastoidy (Novacek 1986), and zonary placenta and quadrilobular allantois (discussed by Fischer 1989).

The primary derived character for living Perissodactyla (Hyracoidea, Mesaxonia) is the eustachian sac (Fischer 1986). Morphogenesis shows that it develops not from the broad

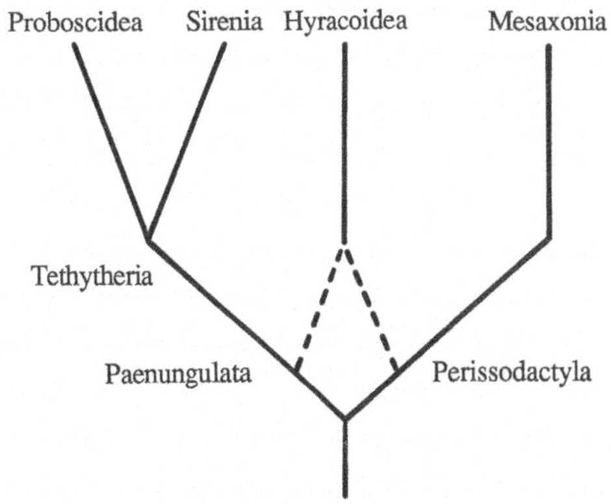

Fig. 5.1 Suggested relationships among Tethytheria (Proboscidea and Sirenia), Hyracoidea, and Perissodactyla (the latter two groups are considered together as Mesaxonia). Dashed lines imply uncertain relationships.

tubotympanal space, but after the formation of the eustachian tube (Fischer 1989). It is present in hyraxes, horses, and tapirs, but not in the rhinoceros (Fischer and Tassy 1993). Nevertheless, in all these taxa and in rhinoceros the internal carotid artery has an extrabullar course. Postcranial synapomorphies for Perissodactyla as used here, proposed by Fischer (1986), have been partly refuted, e.g. loss of acromion (Wyss *et al.* 1987), caudal origin of *semimembranosus* muscle (Shoshani 1992*b*, 1993*c*), but others (hoof structures, flattened talonavicular joint, course of *sternoscapularis* muscle) remain valid. The validity of another character, the tuber maxillae, has been questioned by Novacek and Wyss (1986), but the eutherian plesiomorphic condition (reduction of tuber maxillae after eruption of last molar, maxillae reaches only to front edge of orbit) occurs in early proboscideans (*Moeritherium*, *Numidotherium*, *Phiomia*) and in the genus *Arsinoitherium*. Therefore the persistence of the tuber maxillae has been acquired independently in Proboscidea, in Sirenia, and in Perissodactyla.

The decision between the two hypotheses on the sister-group of Tethytheria depends to a high degree on the systematic position of the entire monophyletic group consisting of the recent taxa, Proboscidea, Sirenia, Mesaxonia, and Hyracoidea, as well as several fossil taxa. Unfortunately, affinities of this group to other eutherian taxa are not yet established. Although Prothero *et al.* (1988) and Prothero (1993) hold to the view of a monophyletic Ungulata, it should be admitted that the basis for this assumption is weak. Ungulata is a group that is merely based on conventions in the systematic community and not on many synapomorphies. It is very often overlooked that mammalian systematists, e.g. Gregory (1910), Simpson (1945), or Thenius (1969), either avoided using this term, or had different opinions as to the content of this group. The most restrictive use is to include only even- and odd-toed ungulates. In a vast sense Ungulata comprises up to seven recent orders of Eutheria, even Tubulidentata, and numerous fossil taxa. The lack of clear constitutive characters for the Ungulata sometimes leads to strange definitions, e.g. those groups belong to ungulates which share a common ancestry among condylarths, i.e. ancestral ungulates (Carroll 1988).

ACKNOWLEDGEMENTS

I thank Pascal Tassy for many discussions and critical review of the text. Many thanks to M. C. McKenna, J. Shoshani, and P. Tassy for editing this chapter.

SIX

Who is who among the Proboscidea?

PASCAL TASSY

INTRODUCTION

Phylogenetic reconstruction of the Proboscidea is almost purely a palaeontological game since only two species survive from what was once a formidable group. This game is made difficult, even hazardous, because of the lacunar nature of the fossil record, the usual overemphasis on dental characters in past literature, the relatively few global studies, and fewer cladistic analyses. Nevertheless, it is possible to summarize the 'state of the art', i.e. the distribution of morphological characters observed on fossils, through parsimony analysis. The resulting cladogram (Fig. 6.1) sheds some light on the numerous unresolved problems.

MATERIAL AND METHOD

The cladogram of the Proboscidea (Fig. 6.1) is a consensus tree resulting from a parsimony analysis which was run through the Hennig86 program designed by Farris (1988). The data matrix includes 25 terminal taxa (22 proboscideans and 3 outgroups: Sirenia, Desmostylia, anthracobunids) and 138 unweighted characters, numbered 0–137 (Tables 6.1 and 6.2). Polarity was determined by the outgroup criterion. All multistate characters are additive. This study is a modification of a previous analysis (Tassy 1990*a*) based on 136 characters. Two characters are added following the works of Shoshani *et al.* (1989*a*) and Gingerich *et al.* (1990): the presence or absence of an astragalar foramen (136), and the morphology of the coracoid process of the scapula (137). Consequently, the state of the characters 9, 10, 16, 27 (astragalus) and 29 (scapula) previously coded '?', is presently known for two taxa, anthracobunids and *Numidotherium* (see also Shoshani et al. Chapter 8 this volume). A 'bug' in the former data matrix concerning characters 109–110 is corrected. Characters are not selected: any character that can be checked with confidence for each terminal taxon (personal examination or sufficient description in the literature) is kept. Lack of information due to bad preservation of fossils is coded: '?'

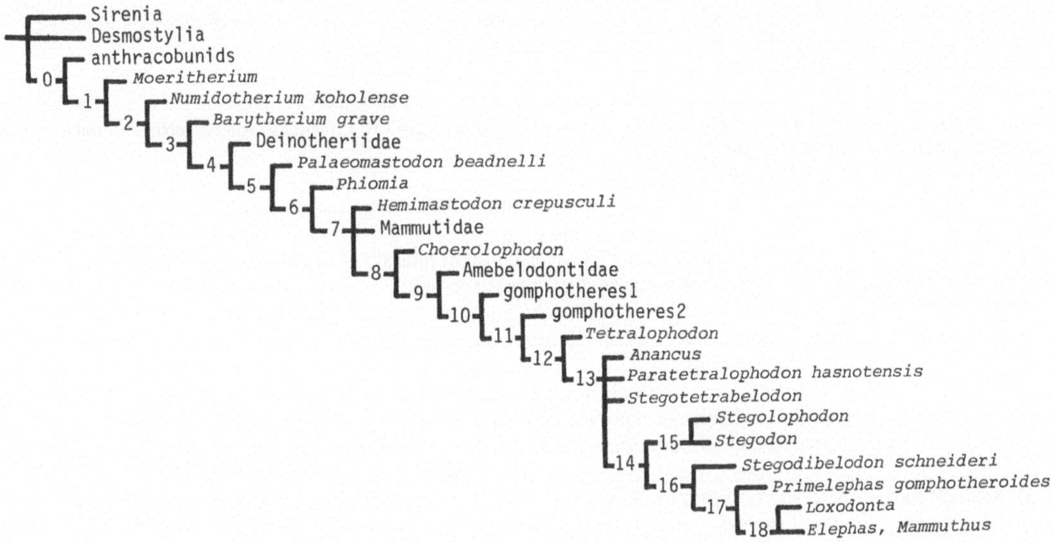

Fig. 6.1 Cladogram of the Proboscidea (strict consensus tree: command 'nelsen' of Hennig86 v.1.5) based on 138 unweighted additive characters (see Tables 6.1 and 6.2). Length, 226 steps; consistency index, 0.73; retention index, 0.89. Rooting by outgroups (Sirenia, Desmostylia and anthracobunids; the anthracobunids appear to be closer to Proboscidea).

Table 6.1 Description of the transformation states of 138 characters (numbered 0–137) used to analyse the interrelationships of the Proboscidea. The character matrix is shown in Table 6.2. The distribution of characters supporting the cladogram (strict consensus tree) obtained by Hennig86 (Fig. 6.1) is depicted in Table 6.3

0	orbit: opens into the jugal (0), or into the maxilla (1)
1	jugal: extends to the anterior border of the glenoid fossa (0), or to the rear border of the glenoid fossa (1)
2	processus ascendens of palatine: present (0), or absent in the orbitotemporal fossa (1)
3	external auditory meatus: low (0), or high (1)
4	external auditory meatus: open (0), or closed ventrally by post-tympanic and postglenoid parts of the squamosal (1)
5	squamosal: cerebral part not inflated (0), or inflated dorsally (1)
6	condylar foramen: present (0), or absent (1)
7	optic foramen and fissura sphenorbitale: more anterior position in the orbitosphenoid (0), or more posterior position (1)
8	mastoid: absent (0), or present externally (1)
9	astragalus: articulates distally with cuboid and navicular (0), or with navicular only (=foot serial) (1)
10	astragalus: lateral calcaneal facet small (0), or enlarged (1)
11	scapula: infraspinous fossa narrow (0), or enlarged (1)
12	humerus: entepicondylar foramen present (0), or absent (1)
13	humerus: epicondylar crest low (0), or high (1)
14	ulna: proximal epiphysis narrow (0), or widened with thick olecranon (1)
15	femur: shorter (0), or longer than tibia (1)
16	astragalus: tuberculum mediale small or absent (0), or protruding (1)
17	lunar: articulates with unciform (0), or not (= serial carpus) (1)
18	frontal: contacts the nasal (0), or the premaxilla (1)
19	occipital nuchal fossa: absent (0), or present (1) (due to the enlargement of the ligamentum nuchae; correlation with the presence of a deep sulcus on the neural apophysis of the axis vertebra)
20	pneumatization of cranium: reduced (0), or important (1)
21	ethmoid foramen: in the middle of the orbitotemporal fossa (0), or posteriorly displaced (1)
22	astragalus: fibular facet wide (0), or reduced (1)
23	calcaneum: fibular facet not laterally expanded (0), or laterally expanded (1)
24	femur: transverse section of diaphysis circular (0), or anteroposteriorly compressed (1)
25	tibia: tibial crest narrow (0), or enlarged with cranial depression (1)
26	tibial plate: symmetric (0), or asymmetric (1) (correlated with the asymmetry of the femoral condyles)
27	astragalus: neck short (0), or long (1)
28	ilium: narrow (0), or laterally widened (1)
29	scapula: acromion reduced (0), or protruding (1)
30	orbits: above premolars (0), or retracted above M^1 (1)
31	palatine: with a spina nasalis posterior (1), or without (0)
32	nasal fossa: at the base of external opening, sagittal process of the premaxilla absent (0), present (1), or expanded (2)
33	maxilla: supra-orbital foramen absent (0), sometimes present (1), always present (2)
34	orbitotemporal fossa: foramina visible in lateral view (0), or all compressed in the posterior notch of the fossa (1)
35	premaxilla: narrow (0), or widened with facial concavity accentuated (1)
36	paroccipital process: present (0), or absent (1)
37	cranium: cerebral part not inflated (0), inflated with sagittal crest lost (1), or inflated vertically (2)
38	premaxilla: fossa incisiva short (0), or lengthened with a correlated shortening of the insertion for the ethmoid cartilage (1)
39	canalis temporalis system: present (0), or absent (1)
40	auditory bulla: narrow (0), or inflated postero-laterally (1)
41	foramen ovale: separate from foramen lacerum medium (0), or confluent (= foramen metoticum) (1)
42	lacrimal foramen: present (0), or absent (1)
43	external nasal opening: narrow (0), wide (1), or enlarged transversely with lateral fossae (2)
44	humerus: bicipital groove shallow (0), or deep with enlarged greater tuberosity (1)
45	carpus and tarsus: high and narrow (=more 'cursorial') (0), or low and wide (=more 'graviportal') (1)
46	tympanic: does not contribute to the cerebral floor (0), contributes to the cerebral floor (1)
47	stylohyoid: posterior ramus absent (0), or present (1)
48	mandibular symphysis: short (0), or lengthened, ahead of tip of premaxillae (1)
49	angular process of dentary (=angulus mandibulae or gonion): protruding (0), or retracted (1)
50	mandibular diastema: short (0), lengthened with anterior grinding tooth close to the symphyseal border (1), or posterior to the symphyseal border (2)
51	mandibular symphysis: parallel with the horizontal ramus (0), slightly deflected downward (1), or strongly deflected (2)
52	mandibular symphysis: long (0), or shortened (=brevirostry) (1)
53	mandibular foramen: anterior border process absent (0), or present (1)
54	pelvis: distal obturator notch poorly delineated (0), well delineated (1)
55	ulna: olecranon process weakly tilted (0), or more strongly tilted medially (1)
56	maxilla: maxillary ridges (=interalveolar cristae) smooth (0), or sharp (1)

Table 6.1 *(cont.)*

57 orbit: rim subcircular (0), or with angular corners (1)
58 fronto-parietal area: flat (0), or concave with large dorsal parietal bulges (1)
59 postglenoid fossa: flat (0), or deep (1)
60 basicranium: not elevated (0), or elevated (1)
61 postorbital (zygomatic) process of frontal: sharp (0), or reduced (=tubular skull) (1)
62 lacrimal: present (0), or absent (1)
63 zygomatic process of maxilla: masseteric fossa and ventral branch absent (0), or present (1)
64 mandible: ventral border thin (0), or transversely widened (1)
65 maxilla: processus frontalis narrow (0), or inflated with processus zygomaticus of frontal (1)
66 cranium: facial part relatively short (0), or much lengthened (1) (i.e. bony palate lengthened posterior to the last tooth in function, external nasal opening and orbits much retracted, orbitotemporalis crista nearly vertical)
67 symphyseal groove of mandibula: narrow and shallow (0), or wide and deep (1)
68 magnum: medial face without (0), or with a protruding antero-dorsal facet for the trapezoid (1)
69 dI^2 and dI_1 (or dI_2): not evergrowing (0), or forming evergrowing tusks (1)
70 lower tusks: present (0), or absent (1)
71 upper tusks: short (0), or lengthened (1)
72 upper tusks: with enamel (0), or without enamel at the adult growth stage (1)
73 lower tusks: with enamel (0), or without enamel at the adult growth stage (1)
74 lower tusks: transverse section flat (0), piriform (1), or circular (2)
75 lower tusks: medial border round (0), or vertical (1)
76 upper tusks: bent downward (0), or straight or bent upward (1)
77 lower tusks: concave dorsally (0), or straight (1)
78 upper incisors: all present (0), or at least one pair is missing (1)
79 postcanine loci: eight (0), or seven or less (1)
80 pattern of dental eruption: all permanent postcanine teeth in function (=complete toothrow) (0), or 'horizontal tooth displacement' (= during life, resorption of anterior grinding teeth when posterior ones are erupting) (1)
81 lower canine: present (0), or absent (1)
82 lower incisors: all present (0), or at least one pair is missing (1)
83 upper canine: present (0), or absent (1)
84 upper incisors: present (0), or all missing (1)
85 premolars: P^1_1 present (0), or absent (1)
86 P^2_2, P^3_3, P^4_4: all present (0), or P^2_2 absent (1), or P^{2-3}_{2-3} absent (2), or P^{2-3-4}_{2-3-4} absent (3)
87 P^4: tricuspid (0), with a small metacone (1), with an enlarged metacone (2), with a hypocone (3)
88 P_4: posterior lophid absent (0), or present (1)
89 P^3: hypocone absent (0), or present (1)
90 P^{3-4}_{3-4}: supplementary posterior loph(id) absent (0), or present (1)
91 P^2: reduced (0), or inflated and caniniform (1)
92 lower molars: oblique cristid obliqua (contacts metaconid) (0), or entirely labial (contacts protoconid) (1)
93 molars: bunolophodonty incipient (metaconule close to hypocone, the four main cusps of lower molars arranged opposite each other) (0), or achieved (1) (i.e. paracone and metacone transversely enlarged and subdivided, paraconule contacts precingulum, postprotocone crest present, metaconid and entoconid enlarged transversely)
94 molars: pretrite cusps round (0), or transversely enlarged (1)
95 M_3: small hypoconulid and no postentoconulid (0), or hypoconulid transversely enlarged with postentoconulid present (1)
96 M_{1-2}: hypoconulid and postentoconulid absent (0), or present (1)
97 M^{1-2}: distocrista absent (0), or present (1)
98 lower molars: hypolophid lower or as high as protolophid (0), or higher than protolophid (1)
99 M^3: postentoconule small (0), large (1), posterolabial extra cusps enlarged (2), trilophodonty achieved (3), tetralophodonty achieved (4), five lophs or more (5)
100 M_{1-2}: second interloph without central conule (0), or with central conule (1)
101 M^1: bilophodont (0), third loph outlined (=posterior extra-cusps) (1), trilophodonty achieved (2), or 4 lophs (3)
102 M_1: bilophodont (0), trilophodont (1), tetralophodont (2)
103 M^2: bilophodont (0), third loph outlined (1), trilophodonty achieved (2), or 4 lophs (3)
104 M_2: bilophodont (0), trilophodont (1), or tetralophodont (1)
105 M_3: bilophodont (0), trilophodont (1), or four lophids with central conules in the third interlophid (2)
106 molars: with five (or more) not modified loph(id)s (0), loph(id)s modified into columnar laminae (=elephantine molars) (1), elephantine molars with height superior to width (2)
107 molars: with five (or more) not modified loph(id)s (0), loph(id)s modified into laminae inflated at their base (= stegodontid molar) (1)
108 molars: crown high or moderately brachyodont (0), or accentuated brachyodonty (1)
109 M_3: eight, or fewer, elephantine laminae (0), or more than eight (1)
110 M^3: six, or fewer, elephantine laminae (0), or more than six (1)
111 lower molars: protoconulid small (0), or large (1)

Table 6.1 *(cont.)*

112 upper molars: in the second interloph, central conules absent (0), or present (1)
113 molars: posttrite conules absent (0), or present, at least outlined (1)
114 lower molars: second and third lophids at right angle with dentary axis (0), or orientated posterobuccal (1)
115 M^3 of trilophodont grade (or more): large (0), narrow with robustness index (I = 100 width/length) <70 (1), or I < 60 (2)
116 coronary cement: absent or weak (0), or thick (1)
117 lower molars: antero-lingual cingulum inflated (0), or reduced (1)
118 dp^2: metacone absent (0), or present (1)
119 dp^2: protocone absent (0), or present (1)
120 dp^2: parastyle large (0), or small (1)
121 dp^3: distocrista (= two posterior cusps) absent (0), or present (1)
122 dp^3: metacone and hypocone on the same line (0), or metacone displaced posterior to hypocone (1)
123 dp^4_4: bilophodont (0), trilophodont (1), or tetralophodont (2)
124 dp^3: contact between pre-hypocone cusp (=anterior pretrite central conule) and post-paracone cusp (=first posterior posttrite central conule) absent (0), or present (1)
125 dp_3: contact between postprotoconid (=first posterior pretrite central (conule) and pre-entoconid cusp (=second posttrite anterior central conule) absent (0), or present (1)
126 P_{3-4}: protostylid absent (0), or present (1)
127 P^4: cusp in the middle of second loph absent (0) or present (1)
128 molars: zygolophodonty absent (0) present (1) (i.e. loph(id)s compressed antero-posteriorly with thin summital subdivision)
129 molars: pretrite conules simple and labial cristae simple (0), or subdivided into crenulated crests (1)
130 dp_2: present (0), or absent (1)
131 molars: choerolophodonty absent (0), or present (with oblique pretrite and posttrite half-loph(id)s) (1)
132 dp_3: postmetaconid cusp absent (0), or present (1)
133 molars: 'anancoidy' absent (0), or sometimes present (1), or always present (2) (i.e. alternate contact between successive pretrite and posttrite half-loph(id)s)
134 molars: anterior pretrite central conule and pretrite mesoconelet separate (0), or fused (1)
135 dp, P, and M: true lophodonty absent (0), or present (1)
136 astragalus: astragalar foramen present (0), or absent (1)
137 scapula: coracoid process protruding anteriorly (0), reduced (1), or protruding ventrally (2)

Table 6.2 Matrix of the 138 characters (nos 0–137) distributed among the twenty-five terminal taxa analysed by the parsimony program Hennig86 (v.1.5). Missing data are scored '?'. All characters are unweighted and additive. The character polarity is inferred by tne outgroup criterion, Sirenia, Desmostylia, and anthracobunids are introduced as outgroups.

	0 10 20 30 40 50 60 68
Sirenia	000000000??0100??11000?????000000000000000000000000000000000000000000
Desmostylia	000110001000100000000000100001000000000000000000??000000000000000000000
anthracobunids	?????????10?????1?????????0????????????????????0?000??????????????0?
Numidotherium	11?111111010101110011?001100?1000000100000?000??000000?000001001 1100?
Barytherium	?????????10111??1???1?0????0?1???????????????????0?000???????????1?000
Moeritherium	111111111100100 01?0000001?0000000000000000010000?00000000000001100000?
Deinotheriidae	111111111111111111111111111111100000001000?0000??0002000000000000001000
Palaeomastodon	1?111??1???111?1????11?1111???11??????????????????102000??0??????00?00?
Phiomia	11111111111111?11111111?111??111211100000000000?101000?0000000000000
Hemimastodon	???
Mammutidae	1111111111111111111111111111111112211111111110000?0101000000000000000000
Amebelodontidae	1111111111111111111111111111111102211111111211?111100000000000000000
Choerolophodon	1111111111?1111111111111111111?10221111111111111??111000??000000000111
gomphotheres1	11111111111111111111111111111111022111211111211111110000000000000000000
gomphotheres2	111111111111111111111111111111110221112111112 11?111100000000000000000
Tetralophodon	1111111111111111111111111111111102211121111211??11110000000000000000000
Anancus	11111111111111111111111111111111002211121111211 1?0110100000001 00000000
Paratetralophodon	11111111?1???????1111????????00211112111112??????1?????00001000?00??
Stegolophodon	11111111?1???????1111????????002211121111?2????01?01????00110000000?
Stegodon	111111111111111111111111111111100221112111112?1??01101000000110000000 0
Stegotetrabelodon	111111?11?111111???11???111?1?00??1?????????1???11120???????10?00?00?
Primelephas	??0??11????????1??????0?
Stegodibelodon	???01101????????????0??0?
Loxodonta	111111111111111111111111111111100201112111112111011110110001100000000
Elephas,Mammuthus	111111111111111111111111111111100201112111112111011111111111100000000

	69 80 90 100 110 120 130 137
Sirenia	00000000000000000000000000000000000000000000000000000000000000000000?1
Desmostylia	0000000001000000000000001000000000000000000000000?????1??0?00?0000010
anthracobunids	000000000100000000000000100000000000000000000000????????0000?0?00002
Numidotherium	0000000000101100101000011000100000000000000000000010?000000000000000112
Barytherium	0000000001101110100000011000100000000000000000000????????0000?0000112
Moeritherium	000000000101000102000011012110000000000000000000001000000?10000000012
Deinotheriidae	?00?120?011011111031100110011100210000000000000000110101000000000000112
Palaeomastodon	100?10??0?10111?104110111112111111100000010000000????????0000?00000??
Phiomia	10001010011011101051101111211212121100000110100010010100010000000001?
Hemimastodon	?????????????????????????11????3??????000?0?10?10????????????00?00000??
Mammutidae	1010100001111110115110?111121131212120000011011011111111000011000000 12
Amebelodontidae	10101010011111101051100111121141212120000011112011111110000000000012
Choerolophodon	11111??1?111111013???0?111121131212120000011012011111110000001110001?
gomphotheres1	1010110001111110115110011112113121212000001101201111111000000000000012
gomphotheres2	1010120001111110115110?111121131212120000011012011111110000000000012
Tetralophodon	1011120011111101151 11?111121151323220000011112011111211000000010012
Anancus	1111???1?111111 0125??1?111121151323220000011112011111211?0000002 0012
Paratetralophodon	1?11???0?111??101???????11??1?513?3?2000?0?11?21???????????00?00000??
Stegolophodon	10101101111111101251?1?111121151323220010011012 01?11112??0?00000010??
Stegodon	1111???1?11111 1013????111121151323220110011002111111 12???000001012
Stegotetrabelodon	10111201111111101251?1?11112115132322100001101211?????2?10000?00000?2
Primelephas	10??120?1?11?1??1251?1?1111211513232210011110021111????1??0000000??
Stegodibelodon	11????????1111??1?????1111?115132322100001?00211??????????00?00000??
Loxodonta	11111?01?111111013????11112115132322200111100211111112???000000012
Elephas,Mammuthus	11111?01?111111 0115??1?1111211513232220011110021111112??0000000012

The different proboscidean groups are summarized by 22 terminal taxa, mostly monophyletic, though some are paraphyletic (*Hemimastodon crepusculi*, *Paratetralophodon hasnotensis*, *Primelephas gomphotheroides*, *Stegodibelodon schneideri*, the gomphotheres, and, probably, the genus *Stegolophodon*). Only the interrelationships of elephantids are detailed here. The other main elephantoid groups: Mammutidae, *Choerolophodon*, Amebelodontidae and the paraphyletic gomphotheres are introduced as terminal taxa. I will detail in this chapter aspects of the extent of these groups.

RESULTS

Six equally parsimonious trees were found (exhaustive research by implicit enumeration, command 'ie' of Hennig86) (222 steps; consistency index (CI), 0.75; retention index (RI), 0.90). Three trees are the result of the position of *Hemimastodon*, a paraphyletic taxon which has only plesiomorphic elephantoid traits. Two trees express the conflicting evidence of the interrelationships of elephantids, stegodontids, and tetralophodont gomphotheres. Their pattern is: (*Stegotetrabelodon* (*Paratetralophodon*, *Anancus*) (Stegodontidae (*Stegodibelodon* (*Primelephas* (*Loxodonta* (*Elephas*, *Mammuthus*)))))), and (*Paratetralophodon* (*Stegotetrabelodon* (*Anancus* (Stegodontidae (*Stegodibelodon* (*Primelephas* (*Loxodonta* (*Elephas*, *Mammuthus*)))))))). The consensus tree (command 'nelsen' of Hennig86: 226 steps; CI, 0.73; RI 0.89) (Fig. 6.1) summarizes these conflicts (nodes 7 and 13). The list of diagnostic traits for each node and each monophyletic terminal taxon is given in Table 6.3. Due to lack of data, many character states are optimized and can appear on different nodes: these characters are marked with an asterisk (*). Characters with non-flexible positions in the cladogram are referred to as 'unambiguous characters' (even if they appear to be homoplasic). There are few of these characters, mainly due to the nature of palaeontological data: missing data are not balanced! The reading of these lists gives the phylogenetic information. In the following sections I will discuss only some characters of the various proboscidean clades and comment briefly upon the main results of this analysis, and also the weakest parts of it. Much work is still needed if one wants to understand the phylogeny of the Proboscidea (see Shoshani *et al.*, this volume, Chapter 8; Kalb *et al.*, this volume, Chapter 12; Shoshani, this volume, Chapter 16).

ORIGIN OF THE PROBOSCIDEA (NODE 0)

Anthracobunids, Eocene taxa from the Indian subcontinent (West 1980; Wells and Gingerich 1983), form a paraphyletic grouping (Fischer and Tassy 1993). The characters of anthracobunids are those of the species *Anthracobune pinfoldi* Pilgrim, 1940. In Fig. 6.1. this taxon appears to be the sister-group of Proboscidea (node 1). This hypothesis is based on the presence of a ventrally protruding coracoid process on the scapula in anthracobunids (137), according to Shoshani *et al.* (1989*a*), and of a tuberculum mediale on the astragalus (16), according to Gingerich *et al.* (1990). However, the astragalar trait is equivocal (possibly present at the root) since it is lacking in known Sirenia and possibly in an autapomorphic state in Desmostylia. The scapular character is unequivocal although its polarity is perhaps ambiguous, since a comparable coracoid process is known in the 'condylarth' *Periptychus*, as illustrated by Cope (1884*a*). The dichotomy between anthracobunids and Proboscidea *sensu stricto* is coherent with a vicariance event on both sides of the Tethys Sea.

MONOPHYLY OF THE PROBOSCIDEA (NODE 1)

The monophyly of the Proboscidea *sensu* Tassy and Shoshani (1988): (*Moeritherium* (*Numidotherium* (*Barytherium* (Deinotheriidae, Elephantiformes)))), has never been seriously questioned. Maglio (1973) excluded moeritheres, barytheres and even deinotheres, but only for classificatory convenience, and without arguing for the interrelationships of these groups with other mammalian orders. Nevertheless, the pattern of differentiation of these groups is not so robust, as a detailed analysis of the characters shows.

The monophyly of the Proboscidea (node 1) is supported by numerous synapomorphies discussed elsewhere (e.g. Tassy and Shoshani 1988; Tassy 1990*a*), that also could be present at node 0: none is known in anthracobunids.

The genus, *Moeritherium* Andrews, 1901*d* (described from the Eocene and early Oligocene of Egypt, Libya, Algeria, Senegal, and Mali), includes two species revised by Tobien (1971): *M. lyonsi* and *M. trigodon*. The monophyly of *Moeritherium* is supported by four characters (42, 61, 62, 126) such as the reduced zygomatic process of the frontal (61), or a protostylid on P_{3-4} (126).

The species *Numidotherium koholense* Jaeger, 1986 (in Mahboubi *et al.* 1986) is known only from the late early Eocene of Algeria. It appears to be closer to [*Barytherium* (Deinotheriidae, Elephantiformes)] due to several unambiguous cranial and postcranial features (such as the enlarged infraspinous fossa of the scapula (11), or length of the femur compared to this tibia (15)).

The genus *Barytherium* from Egypt and Libya includes *B. grave* Andrews and another small, unnamed species (Shoshani *et al.* 1989*a*). Only two unambiguous characters (78, loss of one upper incisor; 83, loss of upper canine) bring *Barytherium* closer to (Deinotheriidae, Elephantiformes). Other traits (Table 6.3) can occur either at the preceding or following nodes. This genus shares with *Numidotherium* and Deinotheriidae the lophodont condition of the molars. This trait does not appear as a synapomorphy for these groups but as an apomorphy at node 2 with a reversal to the bunodont condition in Elephantiformes. Three independent occurrences are less parsimonious, but still plausible.

SISTER-GROUP RELATIONSHIPS OF NODE 4 (DEINOTHERIIDAE, ELEPHANTIFORMES)

The Deinotheriidae Bonaparte, 1841 include two genera, *Prodeinotherium* Ehik, 1920 and *Deinotherium* Kaup, 1829 (Harris

Table 6.3 Distribution of characters listed in Table 6.1. The corresponding state is 1, except for multistate characters. In listing multistate characters, the state applying to a node is placed in parentheses. Reversals are preceded by a negative sign, e.g. –64. Equivocal character states (uncertain level of generality due to missing data and/or uncertain polarity due to homoplasy) are marked with an asterisk (*) and followed by the indication of the other preceding node where they can appear.

Node 0: 0*, 1*, 2*, 5*, 6*, 7*, 9* (root), 16* (root), 25* (root), 29* (root), 118*, 137(2)

Proboscidea (node 1): 0* (node 0), 1* (node 0), 2* (node 0), 5* (node 0), 6* (node 0), 7* (node 0), 25* (root), 29* (root), 81, 85, 87(1), 92, 93, 97, 118* (node 0)

Node 2: 11, 13, 15, 20, 21*, 25* (root), 28*, 29* (root), 36*, 64*, 65*, 82, 95, 135

Node 3: 14*, 18*, 19*, 21* (node 2), 23*, 26*, 28* (node 2), 29* (root), 36* (node 2), 37*, 64* (node 2), 65* (node 2), 78, 83, 119*, 121*, 123(1)*

Node 4: 10, 14* (node 3), 18* (node 3), 19* (node 3), 22, 23* (node 3), 26* (node 3), 27, 28* (node 2), 36* (node 2), 37* (node 3), –64 (node 2), 65* (node 2), 69*, 73, 87(3), 89, 96(1), 98, 101(1 or 2)*, 102(1), 119* (node 3), 121* (node 3), 123(1)* (node 3)

Elephantiformes (node 5): 30, 31, 32(1)*, 33 (2)*, 34*, 35*, 36* (node 2), 37* (node 3), 48, 50(1), 65* (node 2), 69* (node 4), 87 (4), 91*, 94, –95, 96(2), 99(1), 100, 101* (node 4), 103(1), 104(1), 111, 119* (node 3), –135

Node 6: 32(1)* (Elephantiformes), 33(2)* (idem), 34* (idem), 35* (idem), 36* (node 2), 37* (node 3), 65* (node 2), 87(5), 91* (Elephantiformes), 99(2), 101(2)* (node 4), 103(2), 105(1), 112, 114, 119* (node 3)

Elephantoidea (node 7): 32(2), 37* (node 3), 38, 39, 40, 41, 46*, 71, 80, 86(1), 91* (Elephantiformes), 99(3), 105(2), 115(1), 117, 120, 122

Node 8: –31, 42, 43(1), 44, 45, 46* (Elephantoidea), 47*, 49, 91* (Elephantiformes), 115(2)

Node 9: 43(2), 46* (Elephantiformes), 47* (node 8)

Node 10: 37(2), 74(1)

Node 11: 74(2)

Node 12: 51*, 72, 77, 90, 99(5), 101(3), 102(2), 103(3), 104(2), 113*, 123(2), 124, 125

Node 13: –30, –48*, 52(1)*, 60, 70*, 76, 86(2), 113* (node 12), 116

Node 14: –48* (node 13), 52* (node 13), 59, 70* (node 13), –113* (nodes 12 or 13), –114*

Stegodontidae (node 15): 70* (node 13), –74(1), 108, –114* (node 14), 134

Elephantinae (node 16): –33(0 or 2)*, 54(0 or 1)*, 55*, 70* (node 13), 106(1), –114* (node 14)

Node 17: –33 (0 or 2)* (Elephantinae), 51(1), 54* (Elephantinae), 55* (idem), 70* (node 13), 109, 110, –114* (node 14)

(*Loxodonta* (*Elephas, Mammuthus*)) (node 18): –33(0)* (Elephantinae), 54* (Elephantinae), 55* (idem), 70* (node 13), 106(2), –114* (node 14)

Loxodonta: 86(3)

(*Elephas, Mammuthus*): 53, 56, 57, 58

Stegodon: 107

Stegolophodon: –74(1)* (Stegodontidae)

Stegotetrabelodon: 51(2), 106(1), –113* (node 14)

Paratetralophodon: 113* (node 12)

Anancus: 113* (node 12), 133(2)

Tetralophodon: 113* (node 12), 133 (1)

Choerolophodon: 66, 67, 68, 86(3), 130, 131, 132

Amebelodontidae: 75, 113

Mammutidae: 128, 129

Phiomia: 75, 127

Palaeomastodon: 50

Deinotheriidae: 37* (nodes 3–6), 51(2), 74, 84, 102(2), 119* (nodes 3–6)

Barytherium: 29* (nodes 2–4)

Numidotherium: (?12), 63

Moeritherium: 42, 61, 62, 126

1978). Earliest known deinotheres occur in the early Miocene (Africa, Indian subcontinent, Europe). The Elephantiformes Tassy, 1988*a* includes the Oligocene African genera *Phiomia* and *Palaeomastodon*, and Miocene and post-Miocene Elephantoidea. The monophyly of (Deinotheriidae, Elephantiformes) implies a considerable gap in the fossil record. Many synapomorphies at node 4 are postcranial traits: the skeleton of deinotheriids is closer to that of elephants than are the skeletons of *Numidotherium* and of *Moeritherium* (e.g. the characters of the astragalus: 10, 22, 27). Many non-dental traits are unknown in *Barytherium* and could well be present at node 3 (such as the morphology of the ulna (14), calcaneum (23), and tibia (26)). Still, parallelisms can affect postcranial traits, as has been demonstrated for the unciform and unciform-triquetrum contact (Tassy 1994*c*).

Deinotheriidae were a very conservative group known from Africa, Europe, western and South-East Asia (but not China). Harris (1978) recognized six species with few distinctive characters. Latest deinotheriids are known from East Africa at around 1 Ma (*Deinotherium bozasi* Arambourg, 1934).

MONOPHYLY OF ELEPHANTIFORMES (NODE 5)

Andrews (1906) demonstrated that *Palaeomastodon beadnelli* Andrews, 1901*a* and *Phiomia serridens* Andrews and Beadnell, 1902 (= *Phiomia wintoni* Andrews, 1905) from the early Oligocene of the Fayûm (Egypt) were closer to elephants than were moeritheres or deinotheres. Sister-group relationships between *Palaeomastodon beadnelli* and (*Phiomia*, Elephantoidea) are supported by 13 unambiguous traits, mostly dental. Four cranial and mandibular traits (30, 31, 48, 50) are probably correlated with the development of a trunk, though not fully flexible as in elephants. The detailed structure of the nasal area, checked in *P. serridens* but not in *P. beadnelli* (32: premaxilla with outlined process (pillar or crest) for insertion of the ethmoid cartilage) could have evolved at node 5 but, as far as we know, not at node 4. Though deinotheres probably had a trunk, it was not of the elephantine type.

The monophyly of (*Phiomia*, Elephantoidea) is based on six dental traits, notwithstanding the traits not checked in *Palaeomastodon*. The monophyly of *P. beadnelli* is based on the lengthened mandibular diastema, a trait that occurred in parallel within differently evolved Neogene elephantoids.

All known P^4s of *Phiomia* (including small specimens possibly belonging to a third species: *Phiomia* cf. *minor* Andrews, 1904*a*) possess an extra cusp on the posterior loph, considered to be an autapomorphic trait.

MONOPHYLY OF THE ELEPHANTOIDEA (NODE 7)

Thirteen unambiguous synapomorphies diagnose the Elephantoidea. In the present state of knowledge, this evidence falsifies former hypotheses concerning the diphyly of elephantoids (Matsumoto 1924*c*; Tobien 1971, 1976). The affinities of *Hemimastodon crepusculi* (Pilgrim, 1908) from the early Miocene of the Bugti Hills, Pakistan (only two M^3s) are controversial. Characters of the trilophodont M^3s with a distinct postcingulum are those that would be expected in an ancestral morphotype for Elephantoidea (Tassy 1988*b*). However, Pickford (1987*b*) allocates the species to kubanochoerine suids, while Shoshani and McKenna (1991) treat it as an indeterminate ungulate.

Elephantoid synapomorphies (Table 6.3) include 'horizontal tooth displacement' (80) formerly called 'horizontal tooth succession' or 'horizontal tooth replacement'. The two latter expressions can be misleading. Horizontal displacement means that in elephantoids, during the life of an individual, the anterior premolars (or molars in evolved species) are lost when M3 erupts, so that the teeth immediately behind those lost move anteriorly. The tooth row is therefore never complete. Nevertheless, dp2–3–4 are replaced by premolars in primitive elephantoids. The functional succession in primitive mammutids, amebelodons and gomphotheres is: (1) dp2, dp3; then (2) dp4; then (3) P3 (which replaces dp3), M1; then (4) P4 (which replaces dp4); then (5) M2; then (6) M3; (the timing of eruption of P^2 — when present — is unknown, and is not listed).

THE MAMMUTIDAE

Mammutids or 'true mastodons' are defined by the zygolophodonty of the molars (Tobien 1975) observed in every recognized species. Two derived traits are used (128, 129: sharper loph(id)s due to antero-posterior compression, transformation of the central conules into sharp pretrite and posttrite crests).[1]

Mammutids are not a diverse group and are morphologically conservative elephantoids. They include the genera *Mammut* Blumenbach, 1799, *Zygolophodon* Vacek, 1877 (= *Turicius* Osborn, 1926*a*), and *Eozygodon* Tassy and Pickford, 1983. According to Madden (1980*b*), *Miomastodon* Osborn, 1922*a*, is a junior synonym of *Zygolophodon*. Species-level taxonomy is very problematic since most dental traits are stable and most type specimens are isolated teeth. The primitive *Eozygodon morotoensis* (Pickford and Tassy, 1980) from Uganda and Kenya is the earliest known species (22 Ma in Kenya). The widely spread '*Zygolophodon turicensis* group' includes four Eurasian species: the type species *Z. turicensis* (Schinz, 1824) in Europe, *Z. atavus* from Kazakhstan, *Z. metachinjiensis* from the Indian subcontinent, and *Z. gobiensis* from Mongolia and China. A better understanding of cranial and postcranial remains is needed in order to better elucidate the evolution of *Z. turicensis* known from early to late Miocene; only then may the synonymy of these geographically widely separated species be established.

Two competitive hypotheses imply different scenarios for the origin of North American mammutids. Based on cranial characters, a sister-group relationship between *M. borsoni* (Hay, 1834) and *Mammut americanum* (Kerr, 1792) (the latter became extinct 10 000 years ago; King and Saunders 1984) is the most parsimonious. This hypothesis implies a Pliocene migration of *Mammut* toward North America via the Bering Bridge. The most widely accepted scenario, however, is the derivation of the North American *Mammut* lineage from the middle Miocene immigrant, *Zygolophodon merriami* (Osborn, 1921*b*).

MONOPHYLY OF NODE 8: (*CHOEROLOPHODON* (AMEBELODONTIDAE (GOMPHOTHERES, STEGODONTIDAE, ELEPHANTIDS)))

This unnamed group (node 8) is *partly* comprised of Tobien's (1973*a*) 'bunodont group'. Among seven diagnostic characters are the loss of the lacrimal foramen in adults (42), and character (43 (state 1)): widened external nasal aperture with lateral fossae in the premaxillae.[2] The sister-group relationship between Amebelodontidae and (gomphotheres (Stegodontidae, elephantids)) (node 9) is based only on character 43 (state 2): nasal area with

[1] Morphological variations between 'thin' and 'rugose' morphotypes of Pleistocene *Mammut americanum* (Saunders 1977) and between 'gracile' and 'robust' morphotypes of Miocene *Zygolophodon turicensis* (Tassy 1985) were described. These descriptions were based on germ (unerupted) and unworn teeth, and are not to be confused with the detrition of molars which makes wear figures slightly different on both morphotypes.

[2] A cast of the holotype of *Gomphotherium annectens* (Matsumoto, 1925) from the Department of Earth Sciences, University of Nagoya, allows new observations. This species is thought to be the stem species of other gomphotheres (Tassy 1985, this volume, Chapter 11). Contrary to other known gomphotheres, its cranium shows a primitive, narrow nasal aperture, not expanded laterally in the premaxillae (state 0 of character 43). Hence, character 43 is homoplasic and the derived condition (state 2) does not define unambiguously node 9. The tree seen in Fig. 6.1 is only one of the six most parsimonious trees relevant to nodes 8–11. From these trees, the tree obtained by the successive weighting option has the shape: (gomphotheres 1 (*Choerolophodon* (gomphotheres 2 (Amebelodontidae (node 12))))); see Tassy (1994*a*) for details.

wider lateral fossae.[2] Due to lack of material, other traits have not been observed in *Choerolophodon* (e.g. 47: posterior branch of the stylohyoid bone (Shoshani 1986*b*)).

THE GENUS *CHOEROLOPHODON*

The monophyly of *Choerolophodon* Schlesinger, 1917 is based on numerous dental and non-dental traits, though only dental traits are checked in every recognized species (mainly 'choerolophodonty' (131): mesoconelets and central conules fused, and oblique orientation of pre- and posttrite half loph(id)s. Autapomorphies (cranial and postcranial traits, such as the low and long skull with extended bony palate and retracted orbits (66)) render this genus extremely divergent among elephantoids, though taxic diversity is rather low: six species (from early to late Miocene of Asia, eastern Europe, Africa) are allocated to the genus (for a cladistic analysis of this genus, see Tassy 1986). Nevertheless, Tobien *et al.* (1986) retain the genus *Synconolophus* Osborn, 1929 for Asian species.

The earliest species is *C. palaeindicus* (Lydekker, 1884) from the early Miocene of the Bugti Hills, Pakistan (two molars only), with the probable synonym *C. kisumuensis* (MacInnes, 1942) from the middle Miocene of Maboko, Kenya. The latest choerolophodons are dated at around 7 Ma in Europe: *C. pentelici* (Gaudry and Lartet, 1856) at Upper Maragheh, Iran, and around 6.8–6.5 Ma in Pakistan: *C. corrugatus* (Pilgrim, 1913) (= *Synconolophus dhokpathanensis* Osborn, 1929*a*) (from Khaur and Hasnot sites; Tassy 1983).

THE AMEBELODONTIDAE

The monophyly of the Amebelodontidae Barbour, 1927, is based on two dental traits (75: straight medial border of the lower tusk (Tobien 1973*a*) and 113: posttrite conules on molars; homoplasic features among elephantoids); it is one of the least robust among the main elephantoid groups (see Shoshani, this volume, Chapter 16, for a hypothesis of paraphyly of the amebelodons). The first character is homoplasic within the Amebelodontidae and occurs in most lower tusks of *Phiomia*. The second is absent in the late Miocene North American species *Serbelodon barbourensis* Frick, 1933 (a reversal?), and evolved in parallel in some gomphotheres. Hence, species included in the Amebelodontidae could form a valid stem-group for [gomphotheres (elephantids, Stegodontidae)]. Moreover, the most distinctive feature, the flat lower tusk, is a plesiomorphic trait seen in *Phiomia* and only slightly derived in the early mammutid *Eozygodon morotoensis* (tusks are lacking in all known choerolophodons).

The earliest amebelodontid species is *Archaeobelodon filholi* (Frick, 1933) from the early and middle Miocene of western Europe and Africa. More evolved taxa included *Protanancus* Arambourg, 1945 (middle Miocene and late Miocene of Africa and western and southern Asia), *Amebelodon* Barbour, 1927 (late Miocene of North America), and *Platybelodon* Borissiak, 1928 (early Miocene of Africa, middle Miocene of eastern Europe and Asia). The latest amebelodontid is *Amebelodon fricki* Barbour, 1927, from the late Hemphillian (between 5 and 3.5 Ma). The species *Platybelodon danovi* and *Platybelodon grangeri* share the presence of dentinal tubules in the lower tusks, a trait also present in the Clarendonian (late Miocene) North American species *Torynobelodon loomisi* Barbour, 1929, and in grandincisivoid specimens from the Old World (*Mastodon grandincisivus* Schlesinger, 1917). These specimens are allocated to the genus *Stegotetrabelodon* by Tobien (1978*b*), although no shared derived traits are known, except the tetralophodont grade. According to Tobien (1972) the short-jawed *Serbelodon* with simple molars is a junior synonym of the long-jawed *Amebelodon* with complicated molars.

MONOPHYLY OF NODE 10: (GOMPHOTHERES, STEGODONTIDAE, ELEPHANTIDAE)

The monophyly of this group is supported by only two traits. The first is character 37 (state 2): higher cranial vault; and the second is character 74 (state 1): the cross-section of lower tusks is piriform (that is, pear-shaped, also spelled 'pyriform'). According to Tobien (1973*a*) the latter character is the best for the group. Yet it also applies to the most primitive members of stegodontids and elephantids.

THE GOMPHOTHERES

The gomphotheres (family-name: Gomphotheriidae Hay, 1922) form a paraphyletic group and are a stem-group for both stegodontids and elephantids. They are the more diversified elephantoids and include most species of 'bunodont mastodonts'. Yet, no cladogram has emerged from previous studies (Shoshani, this volume, Chapter 16, provides evidence, at the genus level, of paraphyly of gomphotheres).

One can summarize the evolutionary trend within gomphotheres in the same way as it was done in the nineteenth century (Falconer 1857): a primitive trilophodont grade (dp4, M1, M2 with three loph(id)s and an evolved tetralophodont grade (dp4, M1, M2 with four loph(id)s giving rise to stegodontids and to elephantids.

In Fig. 6.1, gomphotheres are subdivided in two groups, each of which is paraphyletic: (1) Old World trilophodont gomphotheres, and (2) New World gomphotheres. Old World trilophodont gomphotheres belong to one genus, *Gomphotherium* Burmeister, 1837. New World gomphotheres include: *Rhynchotherium* Falconer, 1868; *Eubelodon* Barbour, 1924; *Notiomastodon* Cabrera, 1929; *Cuvieronius* Osborn, 1923; *Haplomastodon* Hoffstetter, 1950; and *Stegomastodon* Pohlig, 1912. Stem Old World gomphotheres with simple molars appear in the early Miocene of Eurasia and Africa and are called the '*Gomphotherium annectens* group' (Tassy 1985). This group includes *G. annectens* (Matsumoto, 1925) from Japan, *G. cooperi* (Osborn 1932*b*) from Pakistan, *G. sylvaticum* Tassy, 1985, from western Europe, and *G. sp.* from East Africa. From this group have emerged other, more evolved trilophodont species, such as the type species *G. angustidens* (Cuvier 1817) from Europe, as well as tetralophodont gomphotheres, and New World gomphotheres.

New World gomphotheres appear to be more closely related to tetralophodont gomphotheres (node 11) than to trilophodont gomphotheres from Eurasia, due to the presence of lower tusks with rounder cross-sections (74 (state 2)). Although this is the most parsimonious distribution of this character, I suspect it to be a weak hypothesis. The palaeogeographical implications notwithstanding, intraspecific variation in *Tetralophodon longirostris* and, mostly, the occurrence of lower tusks with a piriform cross-section in primitive *Stegolophodon* (Tassy *et al.* 1992), are strong falsifiers of this hypothesis. New World gomphotheres, in fact, include an Old World species: *Sinomastodon intermedius* de Chardin (Teilhard and Trassaert, 1937) from the Pliocene of China, which is related to brevirostrine South American gomphotheres (Tobien *et al.* 1988).

TETRALOPHODONT GOMPHOTHERES AND THE ORIGINS OF ELEPHANTIDS AND STEGODONTIDS

The hypothesis of a common origin for both elephantids and stegodontids among tetralophodont gomphotheres is warranted (see also Kalb *et al.*, this volume, Chapter 12). The rooting of Stegodontidae among Mammutidae (Maglio 1973; Tobien 1975, 1976) can be dismissed: seven dichotomies separate the two groups.

Node 12 is supported by 10 unambiguous synapomorphies, all dental characters. Six of these are correlated with the tetralophodont grade (90, 99, 101, 102, 103, 104). The dissociation of these characters is made legitimate by the observation of their independent occurrence in other taxa; for example, in Amebelodontidae (*Platybelodon grangeri*), association of trilophodont dp^4 and M^1 and tetralophodont M^2_2 (see also Guan, this volume, Chapter 13). On the contrary, no variation either intra- or intertaxa has been observed among the descendants of node 12. The genus *Tetralophodon* Falconer, 1857, includes the Vallesian species *T. longirostris* (Kaup, 1832*a*) and, perhaps, the obscure Turolian species *T. atticus* (Wagner, 1857) which could also be a primitive *Stegotetrabelodon*. Tobien *et al.* (1988) recognize the rare *Tetralophodon xiaolongtanensis* (Zhou and Zhang, 1974) from the late Miocene of China. No *Tetralophodon* is known from North America. Species allocated to this genus (such as *T. campester* Cope, 1878) belong instead to Amebelodontidae.

Node 13 is a multifurcation summarizing two trees with different groupings of the tetralophodont gomphotheres *Paratetralophodon* Tassy, 1983, *Anancus* Aymard, 1855, and the stem elephantid genus *Stegotetrabelodon* Petrocchi, 1941. Four unambiguous synapomorphies support node 13 (Table 6.3). Two cranial characters, possibly correlated (–30: orbits anteriorly placed — a reversal non-homologous with the primitive tethytherian condition; and 60: high basicranium), and two dental features that could easily be homoplasic (86 (state 2): loss of P3, and 116: thicker deposit of cement).

Anancus is a widespread genus from the late Miocene to early Pliocene of Africa, and from the late Miocene to early Pleistocene of Eurasia. *Paratetralophodon* is monotypic: *P. hasnotensis* (Osborn, 1929*a*) is known only from the Middle Siwaliks of Pakistan, and, according to Tobien *et al.* (1988), in the late Miocene of China. *Stegotetrabelodon* was first described in the late Miocene of Africa. Later, Tobien (1978*b*) and Tobien *et al.* (1988), respectively, described this genus from eastern Europe and China, though the derived features shared by the African species *S. syrticus* Petrocchi, 1941, from Sahabi (Libya) and *S. orbus* Maglio, 1970*b*, from Lothagam (Kenya) are lacking (i.e. M^2_2 with 5 laminae or more).

Because of the insertion of Stegodontidae (node 15), *Stegotetrabelodon* does not cluster with other elephantids (node 16). This is an awkward result. On the molars, the modification of 'mastodontine' ridges to form 'elephantine' lamellae appears twice: in *Stegotetrabelodon* and in Elephantidae (node 16). Hence, contrary to former results (Tassy 1990*a*), Elephantidae (*sensu* Maglio 1973) with the addition of *Stegodibelodon schneideri* Coppens, 1972, do not appear as a monophyletic unit in any of the two trees. Nevertheless, the characters at node 16 are not so robust and it is probable that fully resolved dichotomies 14, 16, 17 are somewhat misleading (see Kalb *et al.*, this volume, Chapter 12, on this topic).

The sister-group relationship between Stegodontidae Osborn, 1918, and Elephantinae Gray, 1821, (node 14) is supported by only one unambiguous trait (59: deep postglenoid fossa); however, though not described in the literature for the genus *Stegotetrabelodon* (and herein coded '?'), this trait is probably present in *S. syrticus* (L. Ginsburg, personal communication, 1991). Other traits can appear at node 13 (–48, 52, –113) or at nodes 15–18 (character –114).

THE MONOPHYLY OF STEGODONTIDAE

Two characters support the monophyly of Stegodontidae Osborn, 1918 (node 15). These are dental traits (108: strong brachyodonty, and 134: loph(id)s with fused pretrite conule and pretrite mesoconelet). The paraphyletic genus *Stegolophodon* Schlesinger, 1917, is a stem genus for *Stegodon* Falconer, 1857. The recent discovery of a primitive early Miocene *Stegolophodon* in Thailand (Tassy *et al.* 1992) demonstrates that character 134 is homoplasic. As far as dental traits are concerned, *Stegolophodon* (at least early Miocene specimens) could also be a stem-group for taxa derived from node 12. This hypothesis would imply that traits diagnosing nodes 13–14 appeared twice (the cranium of early Miocene stegolophodons is unknown). Saegusa (this volume, Chapter 17) provides additional evidence for the monophyly of Stegodontidae, whereas Kalb *et al.* (this volume, Chapter 12) consider this family paraphyletic.

THE ELEPHANTINAE (NODE 16)

Although the monophyly of the Elephantinae: (*Stegodibelodon* (*Primelephas* (*Loxodonta* (*Elephas, Mammuthus*)))) has never been questioned, only one unambiguous character supports the taxon (node 16; 106 (state 1)): molars with elephantine laminae; a trait also present in *Stegotetrabelodon*.

The sister-group relationship between *Primelephas* and (*Loxodonta* (*Elephas, Mammuthus*)) as proposed by Maglio (1973), is based on three characters (node 17). One is a homoplasic feature that occurs in *Stegotetrabelodon* (51: downward orientation of the mandibular symphysis, already outlined in *Tetralophodon* and reversed in (*Elephas, Mammuthus*)). The other two are the number of loph(id)s of M3s, higher in *Primelephas* than in *Stegodibelodon*. However, only one M^3 and two M_3s of *Stegodibelodon schneideri* are known (the only species of the genus) (Coppens 1972), so that variation for this trait may be considered. Nevertheless, *Primelephas*, with one late Miocene species from Kenya (*P. gomphotheroides* Maglio, 1970*b*) is a good ancestral morphotype for elephants.*

The monophyly of (*Loxodonta* (*Elephas, Mammuthus*)) is supported by one unambiguous trait (106 (state 2): crown height of the molars superior to the width). Other traits can occur at other nodes: the loss of the supra-orbital foramen is a reversal (−33), not homologous with the primitive proboscidean condition; it can occur at node 17, and is outlined in *P. hasnotensis*. The loss of lower tusks (70) is a very homoplasic feature (CI, 0.20). For instance, only one mandible of *S. schneideri* is known and it lacks tusks. We must, however, keep in mind the possibility of sexual dimorphism in such a simple character (see also Kalb *et al.*, this volume, Chapter 12, for the lack of relevant data for this trait in *P. gomphotheroides*).

In conclusion, the interrelationships of elephantids are less well known than the fully resolved dichotomies 14–18 would imply. Nevertheless, the sister-group relationship between *Elephas* and *Mammuthus* is well supported on morphological grounds — here four synapomorphies — (Shoshani 1986*b*).

SUMMARY AND CONCLUSIONS

This rapid overview shows that we do possess a general framework for better understanding proboscidean interrelationships. More cladistic data can be retrieved from this volume (e.g., Chapters 12, 16, 17, by Kalb *et al.*, Shoshani, and Saegusa, respectively), with somewhat conflicting results. However, what we now need, as ever, are new discoveries and new descriptions to provide new data. Nevertheless, if cladograms are to be preferred to scenarios, it must be emphasized that these descriptions and taxonomic treatments should be based on cladistic principles, even if the information is inevitably lacunar. Basic studies of the 'jungle of facts' should yield some clarification.

ACKNOWLEDGEMENTS

I thank N. Court, J. E. Kalb, E. C. Marsac, J. L. Pierce, and J. Shoshani for reviews and comments which helped to improve the text. Still, idiosyncrasies are mine.

* Note added in proof

Primelephas gomphotheroides can be a morphotype, though not necessarily a real ancestor. Kalb *et al.* (this volume, Chapter 12A) hypothesize that *P. gomphotheroides* could be related to (*Elephas, Mammuthus*). An early and primitive member of the *Loxodonta* lineage, called '*Loxodonta* sp. "Lukeino stage"' is described in the Nkondo Formation, late Miocene of Uganda and in the Lukeino Formation from Kenya (Tassy 1994*d*). Hence, the differentiation of Elephantinae should be older than six million years.

SEVEN

Proboscidean relationships based on immunological data

JEROLD M. LOWENSTEIN AND JEHESKEL SHOSHANI

INTRODUCTION

Morphological reconstructions of proboscidean evolution are now being supplemented by the immunological analysis of fossil elephant proteins. Microcomplement fixation, immunodiffusion, and radioimmunoassay (RIA) methods have been applied to examine the genetic relationships of the extinct woolly mammoth (*Mammuthus primigenius*, hereafter, *M. primigenius*), the American mastodon (*Mammut americanum*), and the living elephants, African (*Loxodonta africana*) and Asian (*Elephas maximus*). Here we review the principal finding emanating from the new field of palaeoimmunology, particularly from the very quantitative and sensitive RIA technique. In addition, we discuss new applications of RIA to identify fossils of unknown or uncertain provenance and to confirm the hybrid status of the only known *Elephas– Loxodonta* hybrid.

MATERIALS AND METHODS

Muscle specimens from three woolly mammoths, *M. primigenius* — Yuribey at ^{14}C age 10 000 years (Vereshchagin 1982), Magadan ('Dima', Plate 7.1) at 40 000 years (Vereshchagin and Mikhelson 1981), and Khatanga at 53 000 years (Arslanov *et al.* 1982) — were obtained from the Soviet Union. The tissues were extracted in 12 per cent NaCl for 48 hours and dialysed against 0.02 M phosphate buffer pH 7.1 (Prager *et al.* 1980; Shoshani *et al.* 1981). Bone from the partial skeleton of an American mastodon, *Mammut americanum* ('Elmer'), ^{14}C age 10 200 years, was extracted in ethylenediaminetetraacetic acid (EDTA) 0.2 M, pH 7.4 for 7 days at room temperature (Lowenstein *et al.* 1981) and then lyophilized. Bones, nails, and hoofs of other animals were similarly extracted in EDTA but not lyophilized. In addition, bone and tusk samples of proboscidean and non-proboscidean taxa whose geological ages ranged from the Eocene to the Pleistocene were also tested with RIA (Table 7.1). Antisera to the fossil extracts, as well as to purified albumin and collagen of the living elephant species, were raised in rabbits and chickens by subcutaneous injections (Lowenstein *et al.* 1981; Shoshani *et al.* 1985*a*, *b*).

The RIA procedure is a solid-phase double-antibody technique (Lowenstein 1980, 1985) that employs rabbit antiserum as the first, and ^{125}I-labelled goat anti-rabbit γ-globulin (GARGG) as the second antibody. For quantitative discrimination between closely related species such as the mammoth and the two extant elephant species, RIA is used as a competitive inhibition technique in which a solid-phase and liquid-phase protein compete with each other for available binding sites (Lowenstein *et al.* 1981). For use with chicken antisera, this is modified to a triple-antibody method, introducing sequentially chicken anti-proboscidean proteins, rabbit anti-chicken γ-globulin, and ^{125}I-labelled goat anti-rabbit γ-globulin (Shoshani *et al.* 1985*a,b*). In this technique, nanogram amounts of albumin and microgram amounts of collagen can be detected by RIA.

RESULTS AND DISCUSSION

RESULTS WITH EOCENE TO PLEISTOCENE SAMPLES

Extraction of proteins from dinosaur bones had been reported by Miller and Wyckoff (1968), from gomphotheres by Tuross (1989), and from the American mastodon and the woolly mammoth by Shoshani *et al.* (1985*b*). Thus, when we obtained Eocene (*c.* 55–50 Ma) bone samples, it was not a question of whether or not we would be able to extract proteins, but whether or not they would be intact enough to produce a specific antigen–antibody reaction with anti-elephant antibody. As noted in Table 7.1, when these samples were used as heterologous (not the original species against which the antibody was made) taxa in the RIA experiments, only the most recent proboscidean sample (no. 16) — late Pleistocene ?*Mammuthus* or *Elephas* tusk sample — produced twice the value of non-specific binding (NSB), which implied that there was enough protein to be detected. Other Elephantidae samples (nos 14 and 15) did not approach twice the NSB, although some protein was detected. It is of interest that the Pliocene bone samples of *Deinotherium giganteum* (no. 13) and late Miocene *Platybelodon tongxinensis* (no. 11) had small amounts of detectable protein; in *Deinotherium* it was collagen, and in *Platybelodon* it was albumin. Unfortunately, none of these reac-

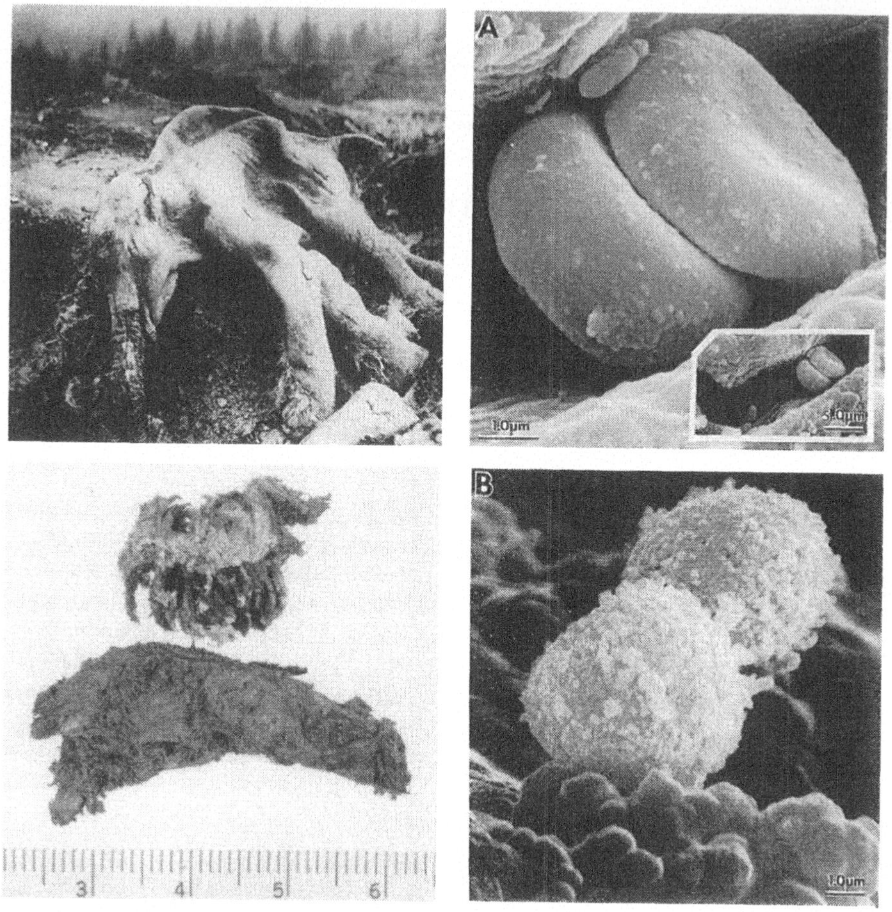

Plate 7.1. Top Left: *Mammuthus primigenius*, male calf ('Dima', discovered in Siberia in 1977), approximately 40 000 years old (photographed *in situ* by A. V. Lozhkin). Top right: Electron micrographs of blood cells observed in Dima's tissues: (A) erythrocytes (RBC), still in a biconcave disc shape inside a capillary; (B) leucocytes with microvilli on the surfaces within a blood vessel. Bottom left: Pieces of *psoas major* muscle from Dima, used in the immunological experiments. (After Barnhart *et al.* 1980; Shoshani *et al.* 1981.)

tions had elephant-specificity (either to *Loxodonta* or to *Elephas*). Future researchers may be interested in preparing extracts from some of the samples listed in Table 7.1 and employing them to produce antibodies in rabbits for testing relationships within Proboscidea further, or in chickens for testing relationships among paenungulate taxa.

During the excavation at the Shelton Mastodon Site (Shoshani *et al.* 1989*b*), a 12 000-year-old, small fragment of tissue was suspected of being a mastodon (*Mammut americanum*) toenail. To test this hypothesis, we made extracts of an array of specimens that included elephant and human toenails, cow, horse and moose hoofs, and elephant and mastodon bones. When analysed by RIA, the putative mastodon toenail turned out to be immunologically similar to cervid tissue, most likely a fragment of antler. This was later confirmed by morphological comparisons (Table 7.2).

RESULTS WITH PLEISTOCENE TO RECENT, MAMMUTIDAE AND ELEPHANTIDAE SAMPLES

Because of resemblances in the skull and teeth, *Mammuthus* and *Elephas* have been thought to be more closely related to each other than either of them is to *Loxodonta* (Shoshani *et al.* 1985*b*). Some workers have included *Mammuthus* within the genus *Elephas* (Blumenbach 1779; Krause 1978). Competitive inhibition of serum albumins by RIA, which gives the best quantitative discrimination between closely related taxa, shows the three species to be equally distant immunologically, with no pair having a measurably greater affinity (Lowenstein *et al.* 1981; Shoshani *et al.* 1985*a,b*) (Fig. 7.1).

Muscle from 'Dima', the 40 000-year-old mammoth, was found by RIA to have about 1 per cent of the albumin concentration

Table 7.1 Radioimmunoassay results with tissues of extant and extinct proboscideans and non-proboscideans dating as far back as the Eocene (*c.* 50 Ma)[a–c]

Taxa and samples tested[d,e]	Antibody to[f]					
	A *Homo* alb.	B *Lox.* alb.	C *Homo* coll.	D *Ele.* coll.	E Mean	F Mean /nsb[b]
Non-proboscidean taxa						
1. *Crocodylus* vertebra (no CGM cat. no.), early Oligocene, Fayûm	−0.21	0.49	0.03	0.99	0.32	0.25
2. *Eosiren* bone fragment (BF) (BMNH M8930), Eocene Fayûm	1.14	0.23	−0.27	1.13	0.56	0.43
3. *Arsinoitherium zitteli* femur (BMNH M8466), Oligocene, El Kemm	−0.15	−0.34	0.73	−0.25	0.00	0.00
4. *Arsinoitherium zitteli* BF (no CGM cat. no.), Oligocene, Fayûm	0.42	0.17	0.75	0.58	0.48	0.37
Proboscidean taxa						
5. *Moeritherum* femur (CGM 29869), late Eocene, Fayûm	0.78	−0.07	0.05	0.02	0.19	0.14
6. *Moeritherium 'lyonsi'* BF (BMNH M10206), Eocene, Fayûm	−0.53	0.20	1.10	0.38	0.29	0.22
7. *Barytherium grave*, humerus (CGM 10012), late Eocene, Fayûm	−0.44	−0.04	−0.07	0.13	−0.11	−0.08
8. *Palaeomastodon 'wintoni'*, BF (BMNH M9450), early Oligocene, Fayûm	0.10	0.82	0.39	0.19	0.37	0.29
9. *Palaemastodon* femur (CGM 8188), early Oligocene, Fayûm	−0.02	−0.36	0.51	0.24	0.09	0.07
10. *Gomphotherium angustidens* BF (CGM 30925), early Miocene, Europe	0.94	0.00	0.81	0.47	0.09	0.07
11. *'Platybelodon tongxinensis'* BF (no cat. no.), late Miocene, China	1.12	0.65	0.08	0.08	0.48	0.37
12. *Amebelodon floridanus* BF (no cat. no.), late Miocene, USA	0.10	0.08	−0.17	0.44	0.11	0.09
13. *Deinotherium giganteum* molar (BMNH M21270), Pliocene, Europe	−0.24	0.02	2.15	0.21	0.53	0.41
14. ? *Mammuthus* tusk (no cat. no.), early Pleistocene, 'Ubeidi'ya, Israel	1.15	0.17	0.37	0.01	0.42	0.33
15. ? *Mammuthus* or *Elephas* tusk (no BMNH cat. no.), Pleistocene, England	−0.55	0.02	1.57	0.11	0.29	0.22
16. ? *Mammuthus* or *Elephas* tusk (no cat. no.), late Pleistocene, Israel	2.88	1.07	0.70	0.75	1.35	1.05
Non-specific binding (nsb) value[g]	1.92	1.39	1.23	0.62	1.29	

[a] Numbers in columns A–D are percentage binding of the second (radioactive) antibody minus the non-specific binding (nsb). These values are not considered to be significant unless they are greater than the nsb, i.e. >1 in column F.

[b] Non-specific binding (nsb) value is obtained by applying antibody to a plastic well containing no fossil material. Based on other experiments, binding to fossil material is not considered significant unless the value is greater than twice that of the non-specific binding.

[c] Examples. For sample no. 1: the value of '−0.21' under column A was obtained from this formula: 1.71 −1.92 = −0.21 (where 1.71 (not given) is the percentage binding of the second antibody and 1.92 is the nsb value given at the bottom of the table). For sample no. 16 under column A: 4.80 −1.92 = 2.88. Note that 2.88 is greater than twice the nsb value for albumin (1.92); the result obtained for sample no. 16 is therefore significant, whereas all other values are not.

[d] Arrangement of taxa within the non-proboscidean taxa is non-mammalian, then approximately from oldest to youngest. Within the proboscidean taxa it is approximately from oldest to youngest.

[e] All the samples were given to J. Shoshani, courtesy of the following individuals: samples from the Cairo Geological Museum (CGM), Egypt, by Bahar el Chashab and Ragi Essia; samples from the British Museum, Natural History (BMNH), by W. Alan Gentry; sample no. 11 by J. Guan; no. 12 by F. Daniel Cring; no. 14 by E. Tchernov; and no. 16 by R. Geron.

[f] *Lox.*, *Loxodonta africana*; *Ele.*, *Elephas maximus*; alb., albumin; coll., collagen.

[g] nsb, Non-specific binding of albumin (see also note b above).

Summary of results from this table: Only one reaction, no. 16, is more than twice background (nsb), and none has elephant-specific reactions, but it does seem that most are above non-specific binding. The stronger reaction of no. 16 with antibody to *Homo* albumin is due to the difference in the strength of that particular antibody to *Homo* albumin as opposed to that of the antibody to *Loxodonta* albumin; additional tests are required to resolve this ambiguity (other details are given in the text).

present in muscle of a living elephant (Prager *et al.* 1980; Lowenstein *et al.* 1981). Stated otherwise, 99 per cent of the albumin was lost. Even though these mammoths are 'frozen', they did not remain consistently in a frozen state for millennia but almost certainly thawed and refroze repeatedly, which resulted in loss of proteins and electrolytes as well as tissue destruction. Eighty per cent of the albumin was found by molecular filtration to be in an aggregated state, while 20 per cent remained in a 'pristine' monomeric form (Prager *et al.* 1980). Immunologically, however, the albumin was more than 99 per cent identical with the albumin of the two living elephant species, which is more similar than the albumins of human, chimpanzees, and gorillas (Lowenstein *et al.* 1981).

The detectable levels of both albumin and collagen in the tissue of three mammoths declined rapidly as the age increased from 10 000 to 50 000 years (Fig. 7.2). The apparent disappearance in about half the time, based on these three points, is about 18 000 years. Even at the lowest concentrations, these

Table 7.2 Immunological identification of hoofs, nails and bones

	Anti-albumin to				
Sample tested	*Elephas*	*Bison*	*Equus*	*Cervus*	*Homo*
Elephas 'toenail'	**10.5**	0.9	5.8	5.4	7.8
Bos hoof	2.3	**5.7**	6.9	4.2	1.7
Equus hoof	10.4	1.6	**13.9**	3.6	8.1
Alces hoof	0.9	0	0	**1.2**	0.5
Homo toenail	1.3	0.5	1.6	1.8	**2.5**
Elephas bone	**8.4**	0.5	4.5	3.9	3.9
Mammut bone	**1.9**	0.5	1.3	0.7	1.1
Unknown	3.1	2.6	3.5	**6.8**	2.8

The numbers are percentage binding of radioactive second antibody in a solid-phase radioimmunoassay.

Identifying reactions are shown in bold print. All but the last specimen had been identified by source or anatomy prior to being tested. The unknown specimen, thought to be a mastodon toenail, turned out to be a fragment of deer antler.

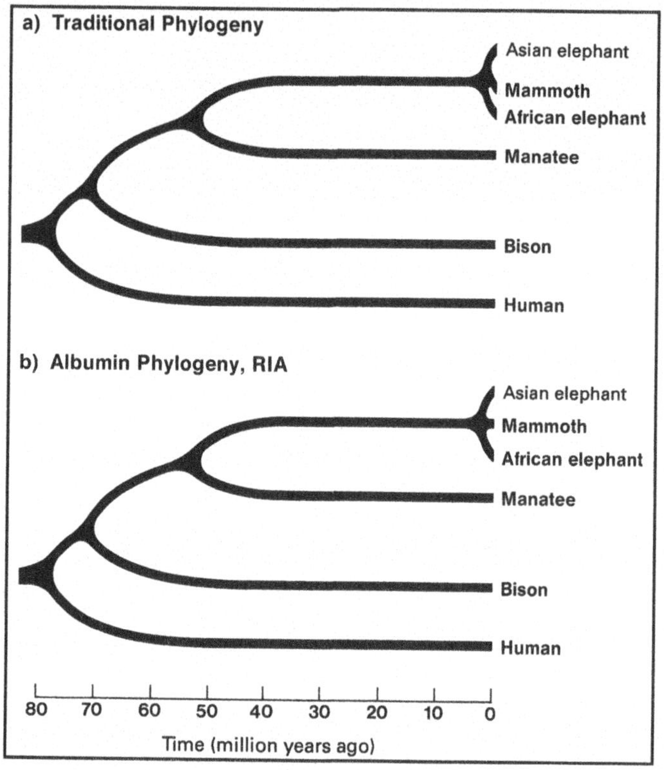

Fig. 7.1 The traditional phylogeny, based on tooth and skull similarities, places the mammoth closer to the Asian than to the African elephant. Mammoth albumin, however, is immunologically equidistant from that of the two living elephant species (after Lowenstein *et al.* 1981, p. 409). Note that in this figure and in Fig. 7.3 names of terminal species do not necessarily imply that their fossil record extended to the splitting points; rather they are representatives of mammalian orders (e.g. the line of primates, not human, of Artiodactyla, not bison, and so on).

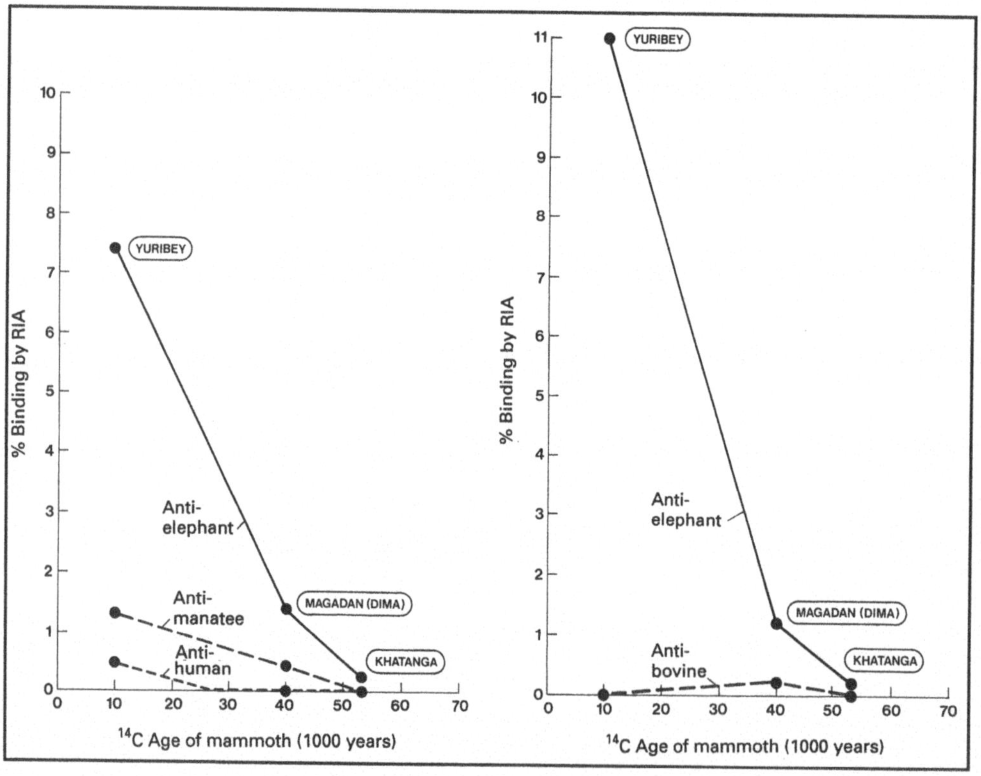

Fig. 7.2 The intensity and specificity of the reactions of tissues from three Siberian mammoths decline rapidly with increasing age. The graph on the left displays radioimmunoassay binding of antiserum to elephant albumin, that on the right antiserum to elephant collagen. The height of the reactions is proportional both to amount and specificity of protein in the fossil tissues. Note that even the weakest reaction remains species-specific — that is, the reaction with anti-elephant protein is greater than with the other antisera (after Shoshani *et al.* 1985*a*, p. 435).

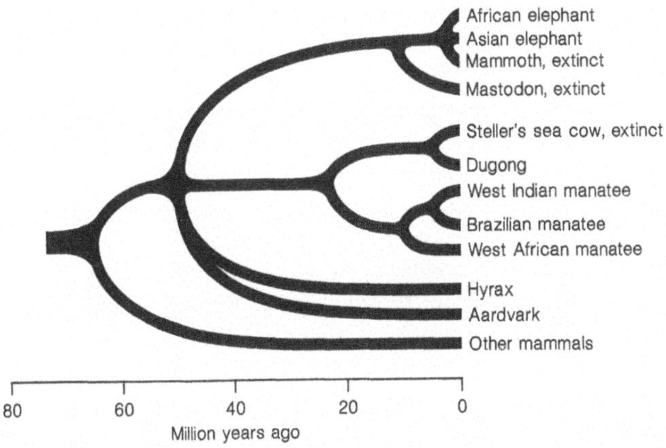

Fig. 7.3 An immunological phylogeny of paenungulate taxa with three extinct forms — *Mammuthus primigenius* (woolly mammoth), *Mammut americanum* (American mastodon), and *Hydrodamalis gigas* (Steller's sea cow). From geological evidence, it is assumed that this group diverged from other mammals about 60–65 million years ago. Branch lengths are derived from the ratio of homologous (A) and heterologous (B) reactions, by the formula: distance = 100 log A/B (i.e. 100 times the logarithm to the base of 10 of homologous/heterologous reactions). (After Lowenstein 1985, p. 543.)

proteins retain the species-specificity, reacting more strongly with antisera to elephant proteins than with antisera to manatee, human, or bovine proteins (Shoshani *et al.* 1985*b*).

Unlike the woolly mammoth (*M. primigenius*), the American mastodon (*Mammut americanum*) only weakly bound antisera to elephant albumin or collagen. Antisera to mastodon bone protein did, however, bind specifically to elephant albumin and collagen, which demonstrated that enough of the original determinants survived to stimulate rabbit antibodies that recognize these same determinants in the proteins of living elephants (Shoshani *et al.* 1985*a*).

It thus became possible to construct an immunological family tree that included two extinct as well as two extant members of Mammutidae and Elephantidae. With the addition of the extinct Steller's sea cow, *Hydrodamalis gigas* (Rainey *et al.* 1984), we were able to suggest a phylogeny of paenungulate taxa based on immunological results (Fig. 7.3).

'MOTTY': THE ONLY KNOWN HYBRID BETWEEN *L. AFRICANA* AND *E. MAXIMUS*

Animals that may not have the opportunity or inclination to react with each other in the wild may do so in zoological parks. The African and Asian elephants, classified in two distinct genera, dwell in regions thousands of kilometres apart, thus interbreeding between them has never been observed in natural circumstances. However, in the Chester Zoo, England, a male African elephant ('Jumbolino') did mate with a female Asian elephant ('Sheba'), who gave birth to a hybrid male, 'Motty', that died after 10 days of 'necrotic enterocolitis' (Howard 1979).

Motty's physical characters were described as being mixed, with the ears definitely more 'African'. Motty's mounted skin is now displayed at the British Museum (Natural History) in London. We obtained a specimen of skin from one ear and tested it by RIA for the presence of species-specific albumin. A hybrid exhibits a particular pattern of reactions, in that its albumin is immunologically more similar to each of the parent albumins than they are to each other. This pattern was previously observed in mule serum and tissue (Lowenstein, unpublished observation). Motty's albumin gave a similar reaction (Table 7.3, Figs 7.4, 7.5), results that confirm his hybrid status.

SUMMARY AND CONCLUSIONS

Residual proteins of several fossil proboscideans dating from the Eocene to the Pleistocene have been studied with a sensitive radioimmunoassay (RIA). Fossil extracts were tested with antisera to the albumins and collagens of the two living elephant species, and antisera made to fossil material were tested with elephant proteins. The albumin and collagen of three Siberian mammoths (*M. primigenius*), with ^{14}C ages 10 000, 40 000, and 50 000 years, were nearly identical immunologically with those of the living Asian and African elephants (*E. maximus* and *L. africana*). The amount of protein preserved decreased with the age of the fossil; specimens which dated to about 1–3 Ma and

Table 7.3 Immunological distances obtained by competitive inhibition experiments[a]

	Loxodonta	*Elephas*
Elephas	23	
Motty	13	10

[a] In this experiment the antibodies were made in rabbits against purified albumins of *Loxodonta africana* and *Elephas maximus*; the antigens were serum samples of *L. africana* and *E. maximus*, and an extract of Motty's skin. In competitive inhibition, a soluble and a solid-phase antigen compete for the same antibody. The soluble antigen is present in high concentration and completely inhibits binding to the solid-phase when both antigens are the same, e.g. *Loxodonta* v. *Loxodonta* (in which case the distance between them will be zero). But inhibition is incomplete when the antigens are different and, as such, reflects the degree of molecular difference between the two species. Motty's skin extract inhibited reactions of the two elephant species better than they inhibited each other. This result can only occur with a hybrid.

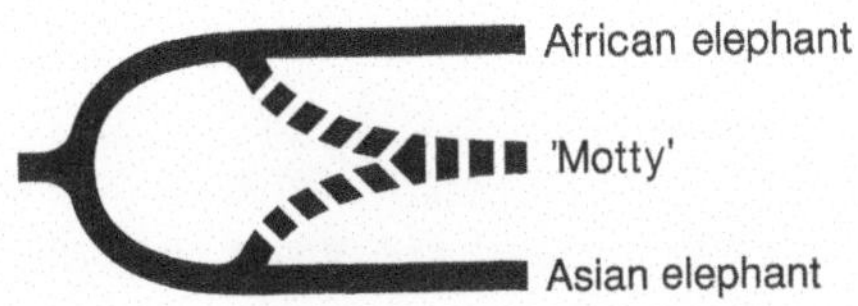

Fig. 7.4 'Motty' was a hybrid, with an African elephant father and an Asian elephant mother. The pattern of albumin reactions is unique for hybrids, in being more like each parent species than they are with each other (details in Table 7.3).

Fig. 7.5 'Motty' — shown here with his Asian elephant mother, 'Sheba' — is the only known hybrid between the African and the Asian elephants; born 11 July, 1978, died 21 July, 1978, Chester Zoo, England. (Photo credit: Kenneth W. Green, Chester Zoo.)

older showed little or no protein when tested as heterologous taxa. Although mammoth teeth have been interpreted as being more like those of *Elephas* than of *Loxodonta*, the albumins of the three species were immunologically equidistant. Since the American mastodon (*Mammut americanum*) is the only sample which showed detectable amounts of protein, it turned out to be closely related (a sister-group) to Elephantidae. RIA can be used to establish the taxon (perhaps on the family level) of fragmentary fossil specimens. One such fragment, thought to be a mastodon toenail, turned out to be a piece of deer antler (Cervidae). Tissues from 'Motty', a reported *Elephas–Loxodonta* hybrid, reacted on RIA in a manner to confirm the hybrid status.

In conclusion, molecular evolutionary studies have profoundly influenced our views of the phylogeny of living organisms, but only within the past decade have molecular techniques been applied to the study of the evolution of fossil species. Using a sensitive radioimmunoassay method, we have demonstrated that useful genetic information can be derived from fossil mammoths and mastodons; that unknown fragments of bone and tissue can be identified by their characteristic proteins; and that the hybrid status of the only known cross between an African and an Asian elephant can be confirmed post-mortem.

ACKNOWLEDGEMENTS

We thank I. Bishop, J. Clutton-Brock, and K. Bryan of the British Museum (Natural History) for providing us with a piece of Motty's skin, and Gary Scheuenstuhl for technical assistance. The Director of Chester Zoo, M. R. Brambel, helped with details on Motty and provided his photograph (Fig. 7.5). The sources of proboscidean bone and tusk samples are acknowledged in Table 7.1.

PART III

The first radiation: early proboscideans

EIGHT

The earliest proboscideans: general plan, taxonomy, and palaeoecology

JEHESKEL SHOSHANI, ROBERT M. WEST, NICHOLAS COURT, ROBERT J. G. SAVAGE, AND JOHN M. HARRIS

INTRODUCTION

This chapter includes updated synthesized information on the earliest proboscideans — *Anthracobune*, moeritheres, numidotheres, barytheres, and deinotheres. *Palaeomastodon* Andrews, 1901*a*, and *Phiomia* Andrews and Beadnell, 1902, are excluded from the description (they are included in Figs 8.1 and 8.2) because, based on the available evidence, they are classified within a well-defined suborder Elephantiformes Tassy, 1988*a*, whereas *Anthracobune*, Moeritheriidae, *Numidotherium*, Barytheriidae, and Deinotheriidae each represent a unique clade at the beginning of this cladogram prior to the appearance of Elephantiformes. In addition, anthracobunids, Moeritheriidae, Barytheriidae, and Deinotheriidae have been included or excluded from the Proboscidea (details in Tassy 1985, 1990*a*; Shoshani 1986*b*). A better understanding of these five lineages is therefore needed, for they may shed light on the relationships of more advanced proboscidean taxa, especially if there was a closer association among *Numidotherium*, Barytheriidae, and Deinotheriidae, or between Barytheriidae and Deinotheriidae, as some authors have suggested (Harris 1978; Mahboubi *et al.* 1986; Savage and Long 1986; Savage and Court 1989; Tassy 1990*a*, this volume, Chapter 6).

Geochronologically and zoogeographically, anthracobunid remains were restricted to early and middle Eocene assemblages from southern Asia, whereas moeritheres, numidotheres and barytheres were restricted to Africa, where they have been recovered from localities of Palaeogene age (Fig. 8.1, Tables 8.1 and 8.2). Deinotheres evidently originated in Africa, where they are known from early Miocene through early Pleistocene localities, but they were also distinctive components of Miocene and Pliocene assemblages in Europe and Asia.

Even though barytheres have been known for about nine decades, relatively little has been published on this group. We hope to rectify this by devoting more space to barytheres than to other early proboscideans. Figure 8.2, top, is based on character analyses by Tassy and Shoshani (1988), Savage and Court (1989), Shoshani *et al.* (1989*a*), Shoshani (this volume, Chapter 16), and Tassy (this volume, Chapter 6). Plates 8.1–8.5 depict localities as well as osteological materials of the earliest proboscideans.

Abbreviations

AMNH, American Museum of Natural History (New York, New York, USA); BMNH, British Museum, Natural History (London, England; new name, Natural History Museum); BUGM, Bristol University, Geology Museum (Bristol, England); CGM, Cairo Geological Museum (Cairo, Egypt); H-GSP, Howard University-Geological Survey of Pakistan Project (Washington, DC, USA); KA, El Kohol, Algeria; MNHN, Muséum National d'Histoire Naturelle (Paris, France); UM, University of Michigan (MP, Museum of Paleontology) (Ann Arbor, Michigan, USA); YPM, Yale Peabody Museum, Yale University (New Haven, Connecticut, USA).

DESCRIPTIONS OF TAXA

The sequence of descriptions follows the appearance of these taxa in Fig. 8.2, top (and also in Fig. 16.5), from left to right. Unless stated otherwise, diagnostic characters for *Anthracobune* are after Wells and Gingerich (1983), West (1983, 1984), Shoshani *et al.* (1989*a*), Gingerich *et al.* (1990), and Kumar (1991); for *Moeritherium* after Andrews (1906), Matsumoto (1922, 1923), Osborn (1936), Coppens and Beden (1978), and Tassy (1981); for *Numidotherium* after Mahboubi *et al.* (1984, 1986) and Court and Jaeger (1991); for *Barytherium* after Andrews (1901*c*, 1906), Savage (1969), Harris (1978), Savage and Court (1989), and Shoshani *et al.* (1989*a*); and for *Prodeinotherium* and *Deinotherium*, after Harris (1978).

ANTHRACOBUNES, FAMILY UNCERTAIN (by R. M. West)

Type species

Anthracobune pinfoldi Pilgrim, 1940.

Historical accounts and general introduction

Anthracobunid taxa were described as artiodactyls (Pilgrim 1940; Gingerich 1977; West 1980), perissodactyls (Coombs and Coombs 1979), sirenians (Sahni and Kumar 1980; Kumar 1991), and

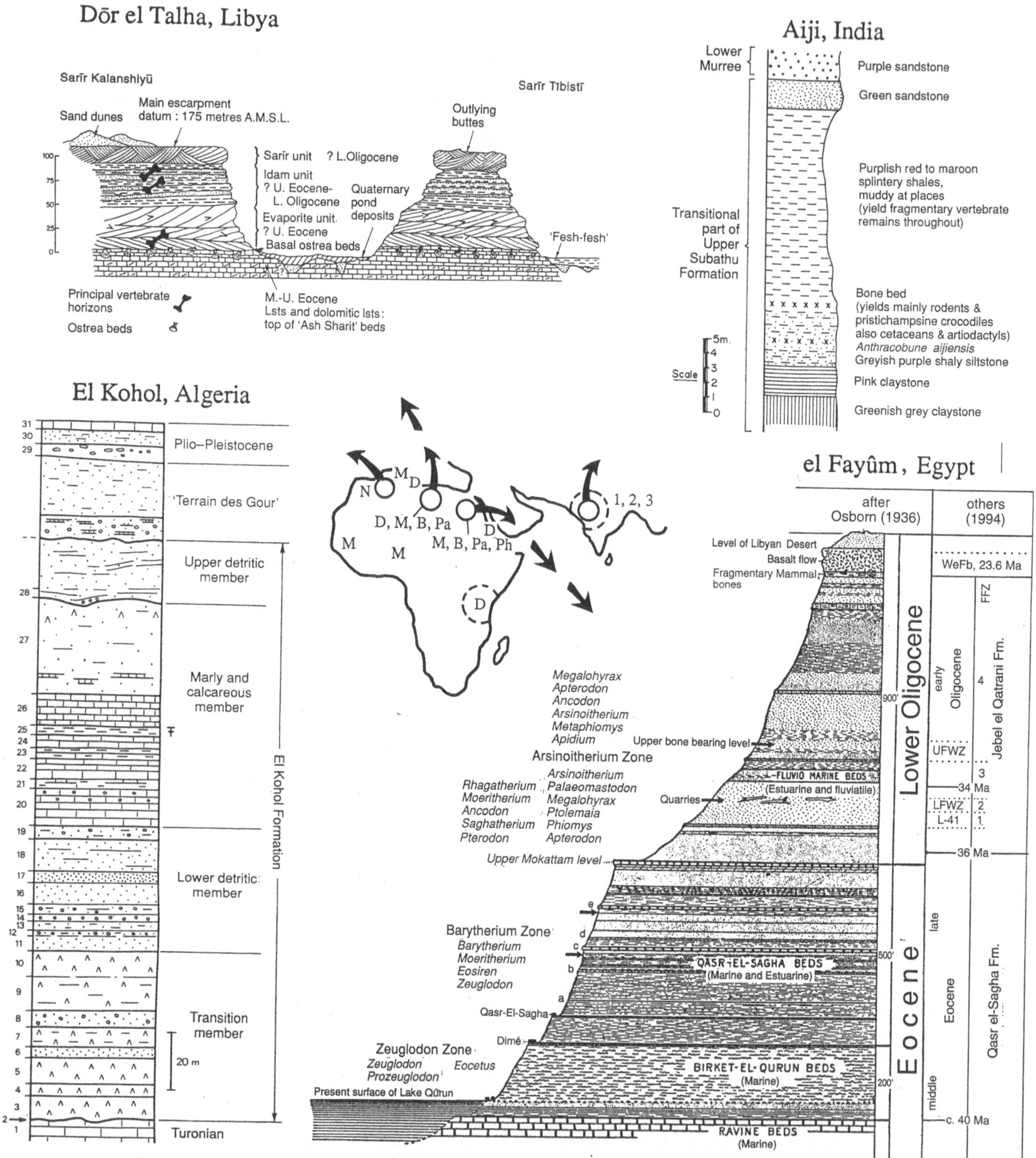
Dōr el Talha, Libya
Sarīr Kalanshiyū
Sand dunes
Main escarpment datum : 175 metres A.M.S.L.
Sarīr Tībistī
Outlying buttes
Sarīr unit ? L.Oligocene
Idam unit ? U. Eocene- L. Oligocene
Quaternary pond deposits
Evaporite unit ? U. Eocene
Basal ostrea beds
'Fesh-fesh'
100
75
50
25
0
Principal vertebrate horizons
Ostrea beds
M.-U. Eocene Lsts and dolomitic lsts: top of 'Ash Sharit' beds
Aiji, India
Lower Murree
Purple sandstone
Green sandstone
Purplish red to maroon splintery shales, muddy at places (yield fragmentary vertebrate remains throughout)
Transitional part of Upper Subathu Formation
Bone bed (yields mainly rodents & pristichampsine crocodiles also cetaceans & artiodactyls)
Anthracobune aijiensis
Greyish purple shaly siltstone
Pink claystone
Greenish grey claystone
Scale
5m.
4
3
2
1
0
El Kohol, Algeria
Plio–Pleistocene
'Terrain des Gour'
Upper detritic member
Marly and calcareous member
Lower detritic member
Transition member
El Kohol Formation
Turonian
20 m
31
30
29
28
27
26
25
24
23
22
21
20
19
18
17
16
15
14
13
12
11
10
9
8
7
6
5
4
3
2
1
N
M
D
D, M, B, Pa
M, B, Pa, Ph
D
M
M
D
1, 2, 3
el Fayûm, Egypt
after Osborn (1936)
others (1994)
Level of Libyan Desert
Basalt flow
Fragmentary Mammal bones
WeFb, 23.6 Ma
FFZ
4
UFWZ
3
34 Ma
LFWZ
2
L-41
1
36 Ma
c. 40 Ma
Lower Oligocene
early Oligocene
Jebel el Qatrani Fm.
Eocene
late Eocene
middle
Qasr el-Sagha Fm.
900'
500'
200'
Megalohyrax
Apterodon
Ancodon
Arsinoitherium
Metaphiomys
Apidium
Upper bone bearing level
Arsinoitherium Zone
Arsinoitherium
Rhagatherium
Palaeomastodon
Moeritherium
Megalohyrax
Ancodon
Ptolemaia
Saghatherium
Phiomys
Pterodon
Apterodon
FLUVIO MARINE BEDS (Estuarine and fluviatile)
Quarries
Upper Mokattam level
Barytherium Zone
Barytherium
Moeritherium
Eosiren
Zeuglodon
e
d
c
b
a
QASR-EL-SAGHA BEDS (Marine and Estuarine)
Qasr-El-Sagha
Dimé
Zeuglodon Zone
Zeuglodon
Eocetus
Prozeuglodon
BIRKET-EL-QURUN BEDS (Marine)
Present surface of Lake Qûrun
RAVINE BEDS (Marine)

Table 8.1 Proboscidean faunal comparison between those found in Dor el Talha, Libya, and el Fayûm, Egypt (data for Dor el Talha from Savage 1969; Savage and Court 1989; Wight 1980; and for el Fayûm from Andrews 1906; Simons 1968; Bown *et al.* 1982)

	Dor el Talha, Libya		el Fayûm	
Geological age	Unit or Formation	Sedimentology and mammals found	Unit or Formation	Sedimentology and mammals found
middle Oligocene	Sarir	Cross-bedded sets without fossils	no equivalent	not applicable
early Oligocene	Idam	Sets and calcilutites in mudstones fluvial channels; bioturbated, rootlet beds, many fossils Vertebrates found included fishes, reptiles, and mammals (listed only a few) Mammals: large *Barytherium* (at 18 sites) moeritheres (at 5 sites) *Palaeomastodon* (at 1 site) *Phiomia* *Arsinoitherium* (at 2 sites) sirenians, hyracoids, rodents	Jebel el Qatrani	Fluvio-marine deposits of large meandering streams and fluctuating floodbasin ponds Vertebrates found included fishes, reptiles, and mammals (listed only a few) Mammals: *Moeritherium* *Palaeomastodon* *Phiomia* *Arsinoitherium* primates, carnivores, hyracoids, rodents[a]
late Eocene	Evaporite	Sandstones, calcilutites, gypsum beds manganese nodules pisolites, limonites, oyster beds Mammals found: sirenians, cetaceans, small *Barytherium* (at 5 sites) moeritheres (at 3 sites)	Qasr el-Sagha	Alternating limestones, marls, clays, and sandstones; interdeltaic shallow marine and littoral, deltaic with channel Mammals found: sirenians, cetaceans, *Barytherium*, small and large *Moeritherium*[a]

[a] Non-proboscidean animals include a long list of vertebrates; details in Andrews (1906), Bown *et al.* (1982).

◀ **Fig. 8.1** Map of Africa and south Asia showing Eocene and Oligocene localities where proboscidean materials were found. In Asia, the dotted area includes those middle Eocene sites where anthracobunids were excavated: 1, Chorlakki and Kohat area, Northwest Frontier Province, Pakistan; 2, Basal and Ganda Kas area, Kala Chitta Hills, Punjab Province, Pakistan; 3, Subathu Formation, Jammu and Kashmir State, India (details in West 1984, p. 184; Kumar 1991, p. 224). The stratigraphic column shown at top right, is near the town of Aiji, India (Subathu Formation, India; close to site number 3), where *Anthracobune aijiensis* was found (after Kumar 1991, p. 223). In Africa, late Eocene to early Oligocene sites are at approximate areas shown on map: *Numidotherium* (N), *Moeritherium* (M), *Barytherium* (B), *Palaeomastodon* (Pa), *Phiomia* (Ph), and Miocene localities for Deinotheriidae taxa (D). The dotted area in East Africa represents a general region where deinothere remains were discovered. The stratigraphy of three sites are: El Kohol, Algeria (left, after Mahboubi *et al.* 1986, p. 19); Dor el Talha, Libya (top left, after Wight 1980, p. 314), and el Fayûm (bottom right, main section, after Osborn 1936; p. 51; the genus *Ancodon* listed twice for the 'Lower Oligocene' was, according to P. Holroyd, personal communication 1995, assigned erroneously — because it is a Eurasian taxon — *Bothriogenys* should have been used instead). The box labelled 'others (1994)' under el Fayûm is a recent, simplified composite reconstruction after Simons (1968, p. 21), Van Couvering and Harris (1991, p. 243), and Rasmussen *et al.* (1992, p. 551); key: WeFb, Widan el Faras basalt, dated 23.6 Ma; FFZ, Fayum Faunal Zone (after Rasmussen *et al.* 1992, numbered 1, to include Quarry L-41; 2, to include Quarries of LFWZ; 3, to include Quarries G and V; 4, to include Quarries of UFWZ and the rest of the upper sequence toward WeFb); UFWZ, Upper Fossil Wood Zone (containing Quarries I, J, M, O, P, and R); LFWZ, Lower Fossil Wood Zone (containing Quarries A, B, C, D, E, and F); Quarries G and V are between UFWZ and LFWZ, and Quarry L-41 is below LFWZ, but within Jebel el Qatrani Formation. The Eocene–Oligocene boundary date of 34 Ma is after Kappelman *et al.* (1992). Gingerich (1992, p. 61) places this boundary at about 36 Ma (which includes the Jebel el Qatrani Formation within the early Oligocene, and the Qasr el-Sagha Formation within the late middle and late Eocene). Gingerich's reconstruction of the Fayûm stratigraphy is similar to that of Osborn (1936). Birket-el-Qurun (at the bottom of the main section of el Fayûm) is a remnant of the much larger, ancient Lake Moeris, for which *Moeritherium* was named. Data for proboscidean materials from Africa after Tobien (1971), Wight (1971), Harris (1978), Tassy (1981), and Mahboubi *et al.* (1986). Note that a line connecting the dots through Egypt, Libya, Mali, and Senegal is the approximate coastline of the Tethys Sea/Ocean (Press and Siever 1974; Tarling 1980; Windley 1984).

Table 8.2 A summary of information (after Mahboubi *et al.* 1984, 1986) for comparison between the El Kohol formation (Algeria) and sediments and taxa found in Dor el Talha, Libya and el Fayûm, Egypt (given in Table 8.1)

	El Kohol, Algeria	
Geological age	Unit or Formation	Sedimentology and mammals found
middle Eocene	El Kohol	Sections overlying the marine Turonian: a lower red detritic member, a middle marly-calcareous member, and an upper red detritic member Vertebrates found included fishes, reptiles, and mammals (listed only a few) Mammals found: marsupials, insectivorans, carnivores, a hyracoid, and a proboscidean — *Numidotherium koholense*

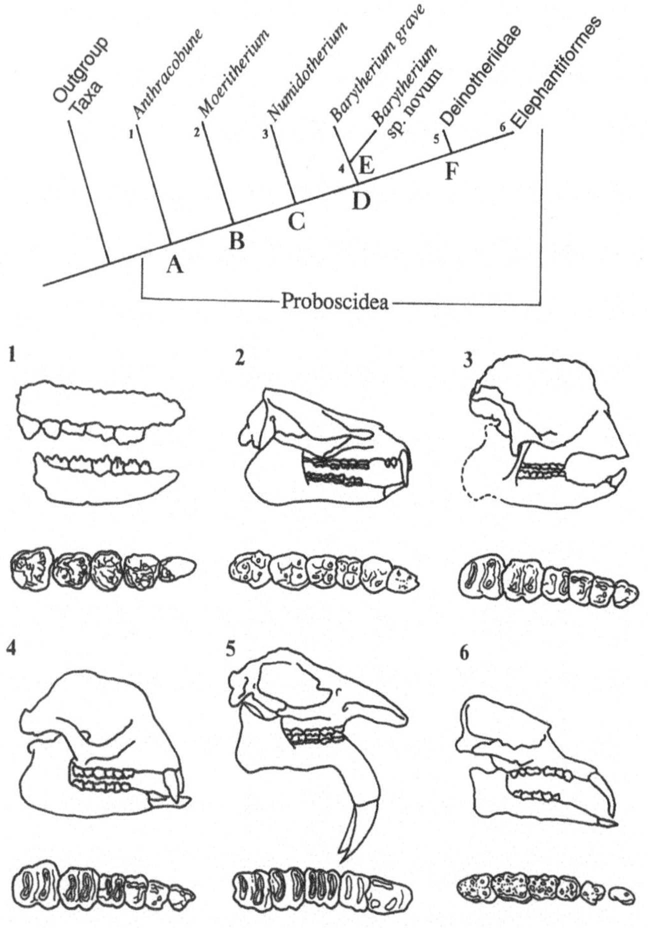

Fig. 8.2 Top: a cladogram of proboscidean taxa compared in this study (after Shoshani *et al.* 1989*a*, p. 160; Shoshani, this volume, Chapter 16; Tassy, this volume, Chapter 6). Synapomorphies are discussed in the text. Middle and Bottom: simplified portions of skulls and upper cheek teeth (not to scale) are as follows: 1, *Anthracobune* (after West 1984, p. 186; teeth (P^2–M^2) are of '*Pilgrimella pilgrimi*' after Wells and Gingerich 1983, p. 122); 2, *Moeritherium* (after Andrews 1903, p. 114; Matsumoto 1923, fig. 10; Osborn 1936, pp. 50, 61, 68, 77, 75; teeth are P^2–M^3); 3, *Numidotherium* (after Mahboubi *et al.* 1986, pp. 26–30; teeth are P^2–M^3); 4, *Barytherium* (after Savage and Long 1986; p. 148; teeth are P^2–M^3) 5, *Deinotherium* (after Blainville 1845, plate I; Osborn 1936, pp. 88–9; Savage and Long 1986, p. 148; teeth are P^3–M^3 — an illustration of *Prodeinotherium hobleyi* or *P. bavaricum* would have been a better choice than *Deinotherium giganteum* since *Prodeinotherium* represent an earlier offshoot of Deinotheriidae than *Deinotherium*); 6, *Phiomia* (after Matsumoto 1923, p. 138, fig. 10; Osborn 1936, pp. 50, 77, 243; Osborn 1942, p. 1551; Savage and Long 1986, p. 149; teeth are P^2–M^3). Approximate or average length (in cm) of skulls/teeth: for no. 1 (bone fragments), 17/9.6; for 2, 50/17; for 3, 55/15; for 4, 85/32; for 5, 100/32; and for 6, 52/26.

proboscideans (Wells and Gingerich 1983; West 1983, 1984; Shoshani *et al.* 1989a; Gingerich *et al.* 1990; Kumar 1991; Shoshani, this volume, Chapter 16). Anthracobunidae (*fide* Wells and Gingerich 1983), include *Anthracobune, Ishatherium, Jozaria, Lammidhania,* and *Pilgrimella.* West (1984), on the other hand, placed *Anthracobune* (with *Pilgrimella* and *Ishatherium* as synonyms), *Lammidhania*, and *Jozaria* in the family Moeritheriidae, order Proboscidea; he also considered '*Lammidhania* as closely related to and doubtfully separate from *Anthracobune*'. The proboscidean affinity of *Anthracobune* was also supported by Shoshani *et al.* (1989*a*), Gingerich *et al.* (1990), and Kumar (1991). Cladistic analysis by Fischer and Tassy (1993) shows that *Jozaria* is closer to Proboscidea *sensu stricto* than are *Anthracobune* and *Pilgrimella*. The species-level synonymy is changing with the discovery of new specimens and/or reinterpretation of 'old' material (West 1984; Domning *et al.* 1986; Kumar 1991).

It is evident that the taxonomic status of anthracobunid genera and species is continuously changing and not necessarily in one direction; it is becoming confusing. In addition, *Anthracobune* is the only genus for which we have both dental and non-dental elements with diagnostic characters. For these reasons, Shoshani (this volume, Chapter 16) employs only *Anthracobune* in his analysis and concludes that it is a proboscidean; this format is also followed in this chapter.

Geological settings and zoogeography

Anthracobune remains have been recovered from rocks dated to late early or early middle Eocene in northern South Asia, *c.* 50 Ma (Fig. 8.1, Plate 8.1). Specifically, *A. pinfoldi* was found near Ganda Kas, north of Basal, Attock District, Punjab Province, Pakistan (West 1984), and *A. aijiensis* was found in the Subathu Formation of Kalakot, Jammu and Kashmir, India (Kumar 1991).

(a)

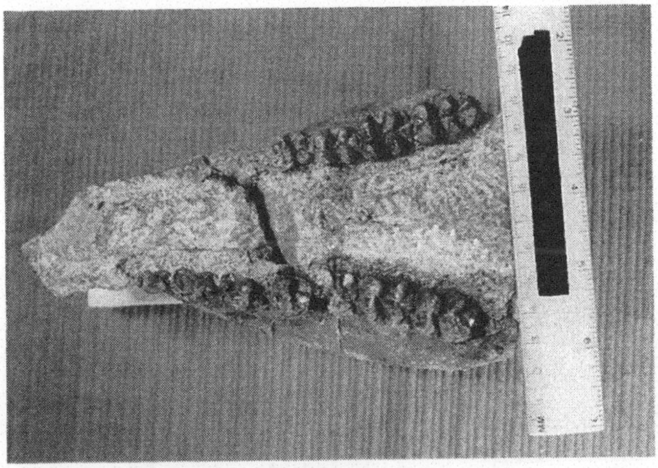

(b)

Plate 8.1. (a) Excavation site, location H-GSP 62, in Ganda Kas region, believed to be the most productive Eocene locality in Pakistan (cf. Fig. 8.1). (b) Mandible, occlusal view, of '*Anthracobune pilgrimi*', H-GSP 1981, excavated at Ganda Kas area, Pakistan. (Photographs by R. M. West; see also West 1980, 1984.)

Diagnostic characters

Dental formula

I 3/3, C 1/1, P 4/4, M 3/3 — the primitive eutherian condition.

Primitive (possibly tethytherian) characters of Anthracobune

These are 'forward placement of the (mandibular) symphysis, retention of all the incisors, only moderately enlarged canines, absence of diastema, posterior position (adjacent to the posterior end of M_3) of the ascending ramus, more molariform P_4, smaller M_3 heel, and interruption in the lophs on the lower molars. In the maxilla, primitive features of *Anthracobune* are the lack of or incomplete separation of the paracone and metacone of P^3 and P^4, the degree of independence of the molar conules, the relatively great anterior breadth of the molars, and the posterior position of the downturned and enlarged maxillary root of the zygomatic arch' (after West 1984, p. 189; see also under 'Emended diagnosis' of *Anthracobune* in West 1984, pp. 186–7, and Kumar 1991, pp. 223–4).

Other primitive (tethytherian) characters of *Anthracobune* are: scapular infraspinous fossa is larger than the supraspinous fossa (inferred from the available fragment); absence of astragalar cuboid facet; short, truncated, and semicircular head of astragalus; and presence of astragalar foramen (probably a primitive 'condylarthran' character).

Derived (proboscidean) characters of Anthracobune

A large and hooked scapular coracoid process and the presence of astragalar tuberculum mediale. A third, but homoplasic character — third lower molars with two definite cuspids at the posterior end — may be assigned to this clade (see Appendix 16.4).

Discussion

The prevailing thought is that anthracobunids are either proboscideans or very close to them. Kumar (1991, p. 238) concluded that 'The Family Anthracobunidae thus includes only four genera and five species, viz., *Anthracobune pinfoldi, A. aijiensis, Jozaria palustris, Pilgrimella pilgrimi* and *Lammidhania wardi*.' He also noted that Anthracobunidae should be included in the Proboscidea 'but still with some reservations' (Kumar 1991, p. 236). Whether anthracobunids (*fide* Wells and Gingerich 1983, or *fide* Kumar 1991; see also note c in Table 16.1) are proboscideans or not, their presence in southern Asia during the middle Eocene broadens the distribution of primitive tethytheres known from northern Africa and corroborates McKenna's (1975) Tethytheria concept.

MOERITHERES, FAMILY MOERITHERIIDAE ANDREWS, 1906 (by J. Shoshani)

Type species

Moeritherium lyonsi Andrews, 1901*d*.

Historical accounts and general introduction

Moeritherium was named after the ancient Lake Moeris in Egypt where it was first discovered in the Fayûm depression, some 60 km south-west of Cairo, in what is now the valley of the Nile River, Egypt (Andrews 1901*a*–*d*). It was thoroughly described by Andrews (1906), Matsumoto (1922, 1923), Osborn (1936), Coppens and Beden (1978), and reviewed by Tassy (1981). Seven species of moerithere have been described, as follows: *M. lyonsi* Andrews, 1901*d*; *M. gracile* Andrews, 1902; *M. ancestrale* Petronievics, 1923; *M. latidens* Deraniyagala, 1955; *M. pharaoensis* Deraniyagala, 1955; *M. trigodon* Andrews, 1904*a*; and *M. andrewsi* Schlosser, 1911. The first five species have been synonymized under the name *M. lyonsi* Andrews, 1901*d*, and the last two under *M. trigodon* Andrews, 1904*a* by Tobien (1971); more details in Deraniyagala 1955, pp. 15–16; Coppens and Beden 1978, pp. 334–5; Tobien 1971, pp. 146, 149, 1978*c*, pp. 193–4.

Whether or not *M. lyonsi* and *M. trigodon* are proboscideans has been discussed by Osborn (1936), Coppens and Beden (1978), and Tassy (1979, 1981, this volume, Chapter 6). In this volume, *Moeritherium* is considered unequivocally a proboscidean.

Geological settings and zoogeography

Moerithere remains have been recovered from late Eocene through early Oligocene sediments, dating to about 40–30 Ma, in el Fayûm, Egypt (Plate 8.2). Recently, they were recovered from Eocene localities on the fringes of the Sahara from Algeria and Libya to Mali and Senegal (Fig. 8.1, cf. Andrews 1906; Osborn 1936; Tobien 1971; Tassy 1981). True moeritheres were discovered only within the boundaries of Africa, along what is believed to be the ancient southern shore of the Tethys Sea.

Diagnostic characters

Dental formula

I 3/2, C 1/0, P 3/3, M 3/3 — loss of a lower incisor, lower canine, and two premolars (upper and lower P1) from the primitive eutherian condition.

Primitive (tethytherian) characters of Moeritheriidae

The elongated skull; long sagittal crest; external naris situated in front of cranium (not elevated, and there is no contact between the frontal and the nasal bones); infraorbital foramen short, situated below orbit; occiput broad and sloped anteriorly, giving a large area of attachment for strong neck muscles; origin of mandibular ascending ramus is at about the level of M_2; scapular infraspinous and supraspinous fossae are about the same size;

Plate 8.2 (a) Overview of the Elwyn L. Simons' camp, Quarry I, Upper Fossil Wood Zone, Jebel el Qatrani, el Fayûm, Egypt, dated to early Oligocene (c. 32 Ma; see Simons 1968, p. 21, and Gingerich 1992, p. 61 for stratigraphic position within the Fayûm depression). (b) Collecting methods at this site rely on wind erosion and subsequent sweeping in search of specimens on the surface. (c) Ventral view of '*Moeritherium andrewsi*', now *M. trigodon* (AMNH 9983 cast of CGM C-7867; this specimen is figured in Andrews 1906, plates VIII–IX, and in Osborn 1936, fig. 36, p. 61) collected in the Fayûm. (d) *Barytherium* specimens collected c. 1900 from the fluvio-marine beds, Fayûm (early Oligocene), presently in the Cairo Geological Museum, Cairo, Egypt. (Photographs by J. Shoshani, October 1982.)

absence of astragalus-cuboid contact (the last character may not be a tethytherian character).

Derived (proboscidean) characters of Moeritheriidae

I2 is enlarged; incipient development of air cells (diploë) in the cranium; eyes set far forward, above or anterior to the premolars (probably a moerithere autapomorphy); ear region of *Moeritherium* is more derived than *Numidotherium* (Court 1994*b*); mandible deep (high), massive, with short solid symphysis (both dentaries are well fused); wide ascending ramus that slopes anteriorly; mandibular condyle is high above the level of the teeth; external auditory meatus opens high on the cranium; external auditory meatus almost forms a tube with the enfolding of the squamosal and postglenoid process; zygomatic arches extend far laterally; scapular coracoid process large, hooked; radius positioned as though it may have been pronated; femur flattened in anterior-posterior view; well-developed astragalar medial process (tuberculum mediale).

Discussion

Moeritheres were comparatively small proboscideans, about the size of a large pig or a small tapir, heavily built, with long bodies, moderately short tails, and stout legs. Based on the position of the external naris, it is believed that moeritheres did not possess the characteristic trunk or proboscis, although *Moeritherium lyonsi* may have had a mobile upper lip. Yet, in the structure of the skull, particularly the incisors, the proboscidean trend is revealed. Even though *Moeritherium* was small, its limbs already showed indications of increased size, and some of the bones (e.g. the femur) were of distinctly proboscidean type.

Some students of Proboscidea considered moeritheres as at the root of the proboscidean evolutionary tree. This is true only if we exclude anthracobunids (found in Asia in older rocks than moeritheres) from discussion. Unless a rigorous cladistic analysis is conducted, caution must be exercised when employing moerithere characters as representatives of proboscideans, because in certain features, moeritheres were too specialized and are believed to have constituted a side branch rather than 'an ancestral' condition.

NUMIDOTHERES, FAMILY NUMIDOTHERIIDAE (by N. Court)

Familial designation is provisional; description of anatomical features is forthcoming.

Type species

Numidotherium koholense Mahboubi *et al.*, 1986.

Historical accounts and general introduction

Aptly called *Numidotherium koholense* (by J.-J. Jaeger, in Mahboubi *et al.* 1986), or the beast Numid from El Kohol, this proboscidean was named after 'Numidia', an ancient kingdom and Roman province in north-eastern Africa and the El Kohol locality in Algeria, where its remains were discovered in 1983. This animal represents the most recent discovery of a new bona fide proboscidean, almost 200 years after the first representatives of this order were found.

Geological settings and zoogeography

Numidotherium was discovered in sediments of about the middle Eocene from Algeria, north-west Africa (Mahboubi *et al.* 1984, 1986). The primitive nature of a hyrax (*Seggeurius amourensis*) from El Kohol suggests that this site is older than all other Eocene localities of North Africa (Court and Mahboubi 1993). At present, no other localities have yielded skeletal remains with features such as those of *Numidotherium*.

Diagnostic characters

Familial (Numidotheriidae) diagnosis is the same as for the genus *Numidotherium*.

Dental formula

I 3/2, C 1/0, P 3/3, M 3/3 — identical to that of *Moeritherium*.

Primitive (proboscidean and/or eutherian) characters of numidotheres

Periotic bone features; origin of mandibular ascending ramus is at about the level of M_2; scapular infraspinous fossa is larger than the supraspinous fossa; trochanters on femur well developed; inferred plantigrade feet; presence of os centrale in carpus; absence of astragalus-cuboid contact; and presence of the humeral entepicondylar foramen. In addition, *Numidotherium* exhibits this primitive therian feature: separation of perilymphatic foramen into fenestra cochlea rotunda and cochlea canaliculus (this separation does not occur in adult living elephants or moeritheres; Court 1992, in press; Court and Jaeger 1991; Fischer 1990).

Derived (proboscidean) characters of numidotheres

Elevated skull; pneumatized bones in the cranium; elevated external naris; modified tooth structure (e.g. I2 enlarged; transverse lophs on cheek teeth); radius fixed in pronation position: antero-posterior flattening of femur; femur lengthening and distal position of knee; well-developed astragalar medial process (tuberculum mediale); lateral (ectal) facet of astragalus usually larger than sustentacular.

Discussion

Numidotherium exhibits definite proboscidean characters although it also possesses some primitive characters not found in any known proboscideans, e.g. the presence of the entepicondylar foramen, also found in humeri of extant opossum, squirrel, and cat. *N. koholense* stood about 1–1.5 m at the shoulder and, based on the position of the external naris, it may have had a short proboscis, perhaps the length of that of a tapir. Skeletal remains of numidotheres are, by far, more abundant than those of moeritheres and those published for barytheres; such a large amount of comparative material has enabled us to better understand the basal characters of proboscideans. For example, unlike living elephants which are capable of communicating via low-frequency calls (Payne *et al.* 1986; Langbauer *et al.* 1991), *N. koholense* is believed to have been tuned to higher frequencies (Court 1992). Feeding behaviour of *N. koholense* was discussed by Court (1993*a*).

In this study, *N. koholense* is provisionally placed in the family Numidotheriidae for these reasons. Primitive characters such as the presence of an os centrale in the carpus and the presence of the humeral entepicondylar foramen have not yet been found in other early proboscideans. In addition, the astragalus is unlike those of other proboscideans (see Tassy 1982); the ulna is about the same size as the radius, and there is a large concave facet for the scaphoid on the distal end of the radius (Court, 1994*a*). The closest family into which *N. koholense* may be assigned is Barytheriidae, but due to the peculiar pedal anatomy of *Barytherium grave*, it would be difficult to include numidotheres within Barytheriidae. These observations were corroborated by independent tests conducted by J. Shoshani (see for example test no. 1 in Appendix 16.3) where the joining of *Numidotherium* and *Barytherium* raises the score of the tree by four steps (Shoshani's tests do not include Court's data).

BARYTHERES, FAMILY BARYTHERIIDAE ANDREWS, 1906 (by R. J. G. Savage)

Type species

Barytherium grave Andrews, 1901*d*.

Historical accounts and general introduction

Barytheriids have been considered enigmatic and have been both included in, and excluded from, the order Proboscidea. Andrews in his initial systematic study (1904b) placed *Barytherium* in the order Barypoda (later Barytheria), a subdivision of Amblypoda. Two years later, he (Andrews 1906) considered Barytheriidae as *incertae sedis* in the suborder Proboscidea (together with Hyracoidea and Barypoda (Arsinoitheriidae) in the same rank in the order Ungulata). Simpson (1945) classified Barytherioidea as one of the four suborders within the Proboscidea; the other three were: Moeritherioidea, Elephantoidea, and Deinotherioidea. Exclusion of barytheres from Proboscidea was noted by Osborn (1936, 1942), Deraniyagala (1955), and Harris (1978). It is perplexing that the only reference to *Barytherium* in Osborn's monograph on the Proboscidea is in the second volume (Osborn 1942, p. 1424), where that genus is mentioned in passing, in a list of mammals found in the Fayûm; Osborn considered *Barytherium* distinct enough to be placed in the order 'Barytheria' created by Andrews (1904b). Deraniyagala (1955) placed the Barytheriidae and Moeritheriidae in an order he called Moeritheria. More recent evidence, on the other hand, unequivocally places *Barytherium* within the Proboscidea (e.g. Tassy 1985; Domning *et al.* 1986; Shoshani 1986*b*; Tassy and Shoshani 1988; Tassy, this volume, Chapter 6; Shoshani and Tassy, this volume, Chapter 34).

Andrews (1901*c*, 1906) described one genus with a single species, *Barytherium grave*, from the Fayûm, Egypt. Savage (1969) reported on a second species of *Barytherium* from Dor el Talha (also spelled 'Dúr el Talhah'), Libya; no detailed description of the material of this second species was made, but it was reported to be smaller and less abundant than *B. grave*. A review of this genus and notes on additional material from Dor el Talha appeared in Harris (1978). Shoshani *et al.* (1989*a*) reported on a small species of *Barytherium* from the Fayûm, Egypt.

Geological settings and zoogeography

Fossil remains of Barytheriidae have been discovered in North Africa specifically in the Fayûm Basin, Egypt (Andrews 1901*a–d*, 1903, 1904*a–b*, 1906; Simons 1968) and in Dor el Talha, Libya (Arambourg and Magnier 1961; Savage 1969; Wight 1980). All barytheriid specimens were found in sediments of late Eocene to early Oligocene ages. In the Fayûm, the late Eocene unit is that of Qasr el-Sagha which is comparable to the Evaporite Unit in Dor el Talha, Libya, whereas the Jebel (also 'Gebel' and 'Djebel') Qatrani Formation in the Fayûm is comparable to the Idam Unit in Dor el Talha (Andrews 1906; Savage 1969; Harris 1978; Wight 1980; Savage and Court 1989).

Diagnostic characters

Dental formula

I 2/2, C 0/0, P 3/3, M 3/3 — loss of one incisor and one canine from the *Moeritherium* condition.

Primitive (tethytherian) proboscidean characters of barytheres

The origin of the mandibular ascending ramus is at about the level of M_2; scapular infraspinous fossa is larger than the supraspinous fossa; and the absence of astragalus-cuboid contact.

Derived (proboscidean) characters of barytheres

Pneumatized bones in the cranium; elevated external naris; mandible massive with well-fused symphysis extending posteriorly to about M_1; loss of teeth, see under dental formula; I2 enlarged; radius fixed in pronation position; about one-half of the lunar articulates with the ulna; a well-developed medial process (tuberculum mediale) on the astragalus; large size; and graviportal stance.

Barytherium *dental features (After Harris 1978, p. 329; Savage and Long 1986, p. 148.)*

Upper incisors vertical, lower horizontal; in the maxilla, I^2 is larger than I^1, and I^2 is chisel edged; reverse condition occurs in the mandible (I_1 is larger than I_2); diastema between incisors and premolars; P_2 triangular; P_3 with two anterior cusps and posterior loph; P_4 with a single loph, M_{1-3} bilophodont, M_3 with hypoconulid; M^{1-3} rectangular and bilophodont.

Discussion

The scapular coracoid process of *B. grave* was described by Andrews (1906, p. 174) as 'larger than in any other mammal, forming a massive hook-like process overhanging the glenoid cavity, the articular surface of which appears to have been continued for a short distance on its posterior face'; and 'The large glenoid cavity proper is nearly circular in outline'.

Based on the material described here and on that discovered in Dor el Talha, Libya (Arambourg and Magnier, 1961; Savage, 1969), it appears that there were two species of *Barytherium*; one was of elephantine size (it may have weighed about 3–4 metric tons and stood about 2.5–3 m at the shoulder) and the other was of *Moeritherium* or of *Tapirus* size (it may have weighed about 200 kg and stood about 1–1.5 m at the shoulder) and was also found in late Eocene deposits in both areas. During the late Eocene, the shores of the Tethys Sea extended beyond the present-day Mediterranean Sea. Areas north of the Sahara, including modern Egypt and Libya, were partly or wholly covered by water during the Eocene. Marine sediments of this geological period have been described by Andrews (1906), Simons (1968), Bown *et al.* (1982), and Mahboubi *et al.* (1986).

Additional descriptions of the Libyan locality and specimens of ***Barytherium***

Below are descriptions of the Libyan locality and specimens of *Barytherium* that have not been described in previous publications, or have been mentioned briefly. In either case, we centralize the information for easy reference. Skeletal remains of *Barytherium* from Dor el Talha, Libya, are currently stored at the Veterinary School of Bristol University, Langford (outside Bristol, England) and at the BUGM. The material described by Andrews (1906) is housed at the CGM. Other barythere specimens are at the BMNH, MNHN, and at the YPM. N. Court is currently undertaking a full study of all *Barytherium* specimens.

Information from Savage and Court (1989) (including excerpts transcribed from presentation by R. J. G. Savage at the 5th International Theriological Congress)

Charles W. Andrews' (1901*c–d*, 1904*b*, 1906) description of *Barytherium grave* was based on a maxillary fragment, a mandible, a humerus, a scapula, and a radius; that was all the available material. For the next 80 years or so, nothing more was discovered about barytheres.

In the 1970s, working in Libya, Savage came across more *Barytherium* material in the area of a jebel (Arabic for a mountain) called Dor el Talha which means 'the long ridge', about 400 km from the shores of the Mediterranean Sea. The geology of this part of Libya, and other parts of northern Africa can be summarized as a series of retreating shore lines in a northward direction into the Mediterranean. Thus, one can follow lacustrine and estuarine deposits with fossils in them from the Eocene through the Oligocene in Dor el Talha (central Libya), Miocene in Gebel Zelten (also spelled Jebel Zelten, north central Libya), and Pliocene in northern Libya.

In one site where the exposure is about 150 km long, one can clearly observe the geology as horizontal strata, with the older rocks at the bottom. The average thickness of the exposure is 70 m and it is divided into two formations, Evaporite Unit and Idam Unit (Wight 1980 table 1, figs 1 and 2). The Evaporite Unit was dated to the late Eocene and is equivalent to the Qasr el Sagha Formation in el Fayûm, Egypt. The Idam Unit is of early Oligocene age and is equivalent, at least in part, to the Jebel Qatrani Formation in el Fayûm. Barythere material was found in the Evaporite (small *Barytherium*) and the Idam (large *Barytherium*) units, together with other animals, in both marine and non-marine deposits (Table 8.1).

The assemblage included aquatic plants, fishes (e.g. salurids, catfish, lungfish, and some sharks), turtles, and crocodiles. The preservation of some of the animals and plants was very good. For example, there were seed cases of nymphacean or water-lily and a termite nest with many insects. We also found brain cases of reptiles and fishes, but few remains of mammals. Mammals in the lower unit included sirenians, cetaceans, barytheres, and moeritheres. The bulk of the mammal specimens was found in the upper unit, including rodents, hyracoids, arsinoitheres, moeritheres, barytheres, palaeomastodontids, and hyaenodontids.

Concentrating on the discovery of the barytheres, we were fortunate to find in the Idam beds a partial skeleton of a large *Barytherium*. The skeleton was discovered, weathering in the sands. Unfortunately, we reached it about 10 years too late. It was weathering rather badly but it was possible to retrieve much of it. The cranium of that individual was crushed and the postero-ventral part was missing; particularly, we do not have the ear region, an area with important characters. Virtually nothing behind the dentition was well preserved.

The skull

Although an old individual, the molars were clearly bilophodont (Plate 8.3), the premolars were in poor condition; it had a long diastema and extraordinarily large and interesting incisors, or tusks. In the upper jaw, the premaxilla, we noted two sockets on each side: one medial, small, and the second lateral incisor, very much larger. In the cranium of another individual, we observed bones with extensive pneumatization, and I^1 and I^2: the I^1 was a small tooth; the I^2 had a chiselled edge, and its enamel was restricted to the anterior cutting face. Both upper incisors were directed almost vertically downwards.

In the mandible there was also a long diastema between the premolars and the incisors. The arrangement of the incisors in the lower jaw was opposite from that observed in the upper jaw — the middle I_1 were the large pair, and the lateral I_2 were very much smaller, and both incisors were directed horizontally. This tusk architecture, which is similar to that of a hippopotamus, allows shearing action of the inner side of I^2 against the outer side of I_1 (Fig. 8.2).

The cheek dentition in the mandible of a mature individual showed the P_2 with the development of an incipient metaconid and a well-developed metaconid on P_3 and P_4. P_4 and all teeth (M_{1-3}) posterior to it were fully bilophodont.

From the lower, Evaporite Unit, we collected a small species of barytheriid. There were no maxillary teeth, but in the mandibular dentition we observed, on the lingual side, the development of a metaconid on P_3. In M_1–M_3, the metaconid and the protoconid cusps on the labial side joined into a complete loph.

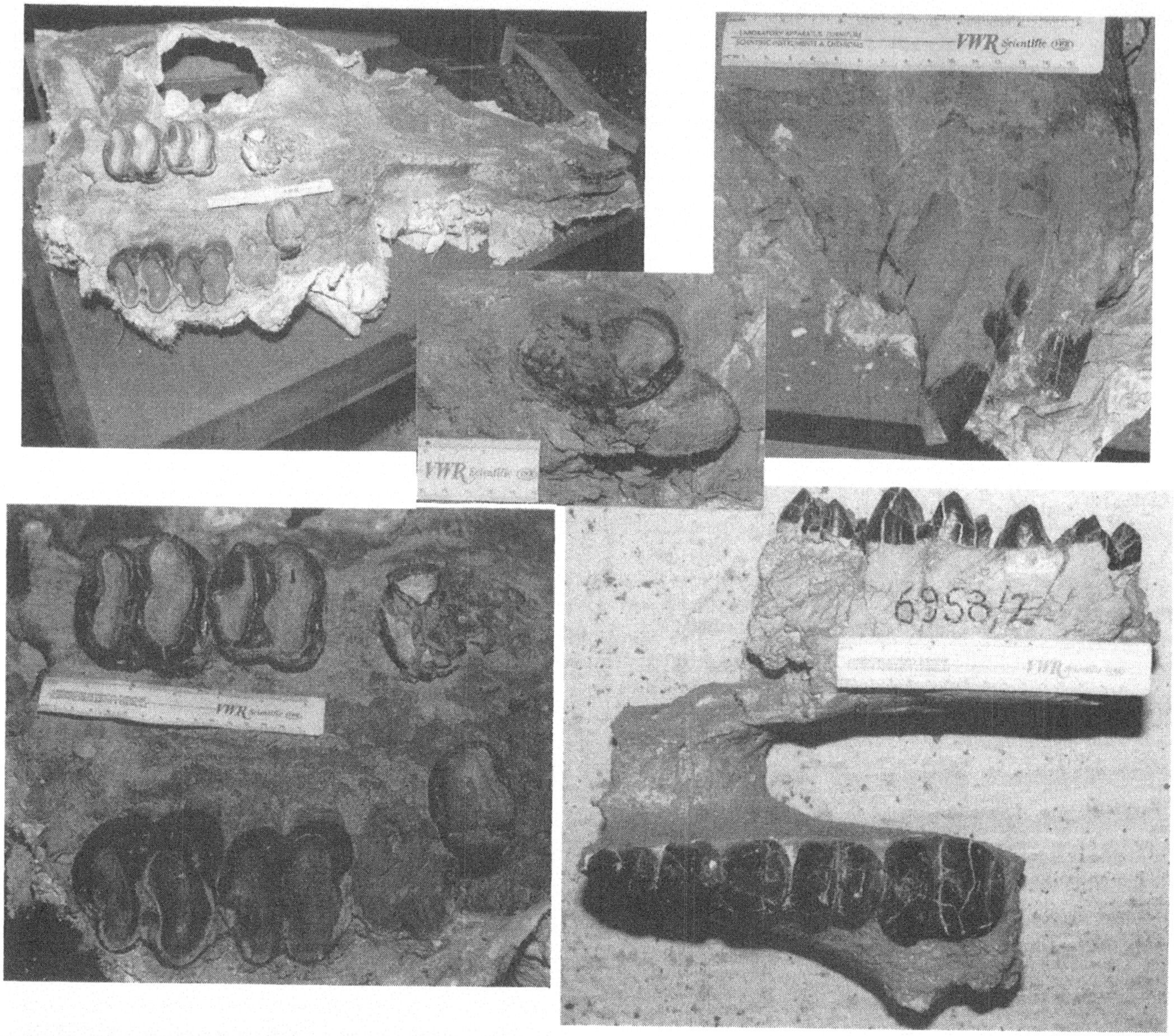

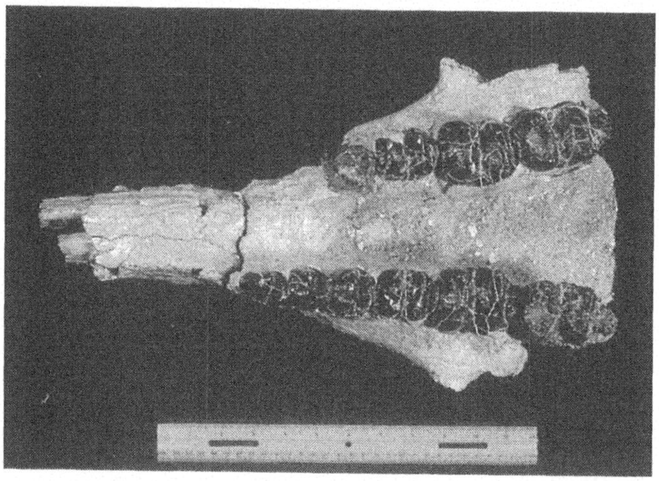

Plate 8.3 Dentition of *Barytherium grave*, collected at the Idam Unit, Dor el Talha, Libya, an early Oligocene locality. Top row: cranium of an old individual in ventral view; close-up of incisors of same animal in ventral view; close-up of incisors in lateral view. Middle row: close-up of upper palate of same individual as in upper row, depicting extensively worn P^4–M^3; mandibular fragment (BUGM 6958/2) with occlusal and lingual views of dental battery showing P_2–M_2 in the occlusal section. Bottom: mandibular fragment (same as in middle row, right) depicting a complete set of teeth on the left ramus — from P_2 through M_3. (The photograph at the bottom was taken by R. J. G. Savage; other photographs were taken by J. Shoshani.)

Pectoral girdle and appendages

The scapula was of proboscidean proportion. The most outstanding feature was a strong, downturned coracoid process for the origin of muscles which insert onto the deltoid ridge of the humerus and other areas in the foreleg (further description below). In the humerus, there was a large bicipital groove, a very prominent deltoid ridge, and it had flattened, broad ecto- and endo-supracondylar crests. On the posterior and distal end of the humerus, there was a deep olecranon fossa. Graviportal animals tend to have relatively long humeri compared to the length of the radii; in *Barytherium*, the radius is about 70 per cent of the humeral length.

Pelvic girdle and appendages

The pelvic girdle was large, broad, and heavy, with an almost elephant-like appearance, as would be expected of a graviform animal. The femur is distinctive in *Barytherium* as compared with elephants; it is flattened in the shaft, and possesses a large projecting third trochanter (observations by R. J. G. Savage). In proximal and anterior views, the head and the greater trochanter of the femur seemed to be, proportionally, about the size and shape as those of living elephants (*Loxodonta africana* and *Elephas maximus*). The patellar articulating surface was not as deep as in *L. africana* and *E. maximus*, and the patella was broader, more elongate, and flatter than that of the living elephants. The fibular distal end was wide and soon thereafter, the shaft became narrow, unlike that of *L. africana* and *E. maximus* (Plate 8.4, observations by J. Shoshani).

In the pes there was a large astragalus which was different from that of a deinothere, a gomphothere, and a modern elephant. The major difference was the large size of the sustentacular facet and the relatively small size of the ectal facet (cf. Fig. 8.3) in the astragalus of *Barytherium*. There seems to be a strong contrast among the astragali of these proboscideans, a feature noted by Tassy (1982 (where an astragalus of *Barytherium* was figured for the first time), 1985) and Shoshani (1986*b*).

Concluding remarks on the material from Dor el Talha

Based on the available material, we probably have two forms, one which was collected from the Idam Unit and was as large as *B. grave* from the Fayûm, and another which was collected from the Evaporite Unit and was of smaller stature, about one-third the size of the larger. We have not observed any significant differences between the barythere specimens from Dor El Talha (Idam Unit) and those from the Fayûm equivalent (Jebel Qatrani) and thus we consider them to belong in the same species — *B. grave*. The dental and scapular characters of the smaller taxon are almost identical to those of the larger species, *B. grave*; however, certain carpal bones of the smaller barythere species more closely resemble those of *Numidotherium* than those of *B. grave* (observations by N. Court). The smaller barythere has not been named; Shoshani *et al.* (1989*a*) referred to it as *Barytherium* species novum.

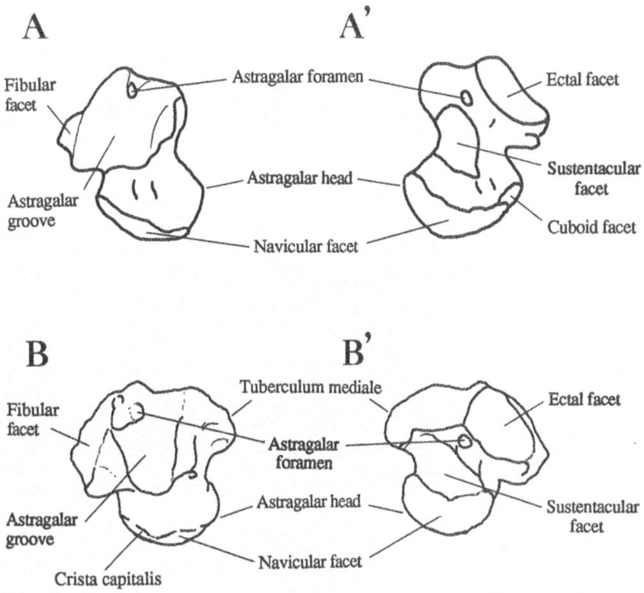

Fig. 8.3 Astragali of a 'condylarthran', *Arctocyon ferox* (top) and a generalized proboscidean, *Anthracobune* sp. (bottom) depicting terminology used in this chapter. A and B are dorsal views, A' and B' plantar views. (A and A' are simplified after Cifelli 1983, p. 3; B and B' are simplified after Gingerich *et al.* 1990, p. 75; drawings are not to scale.)

Scapulae of ***Barytherium*** *(by J. Shoshani)*

Overall, the scapula of *Barytherium* is similar to that of modern elephants in size and general architecture. The infraspinous fossa is larger than the supraspinous fossa and, based on the roots of the proximal portions of the acromion and metacromion processes, they might have been similar to those of more advanced proboscideans, including the living elephants *L. africana* and *E. maximus*. Undoubtedly, the coracoid process in *Barytherium* is the most prominent character — it is enormous and downcurved as a huge hook. This process in *Barytherium* is larger and more hooked (curved straight ventrally rather than antero-ventrally) than that of *Moeritherium* (specimen CGM 10020 is more hooked than BMNH M33343). On the anterior end of this process there is a ridge which, according to Andrews (1906, p. 117), is also present on the coracoid process of *Moeritherium*'s scapula. When viewed from the ventral side, the glenoid cavity of *Barytherium* is deeper and rounder than that of *Moeritherium*. The roundness of the glenoid cavity is accentuated by the flared and thickened rim (which acts as a supporting buttress) at the distal end of the scapula. The combination of all these characters (size and curvature of the coracoid process, and the round shape and flared, thickened rim of the glenoid cavity) distinguishes *Barytherium* scapulae from those of other closely related taxa. A comparison of scapular characters of *Barytherium* to those of other taxa is presented in Table 8.3; cf. Fig. 8.4.

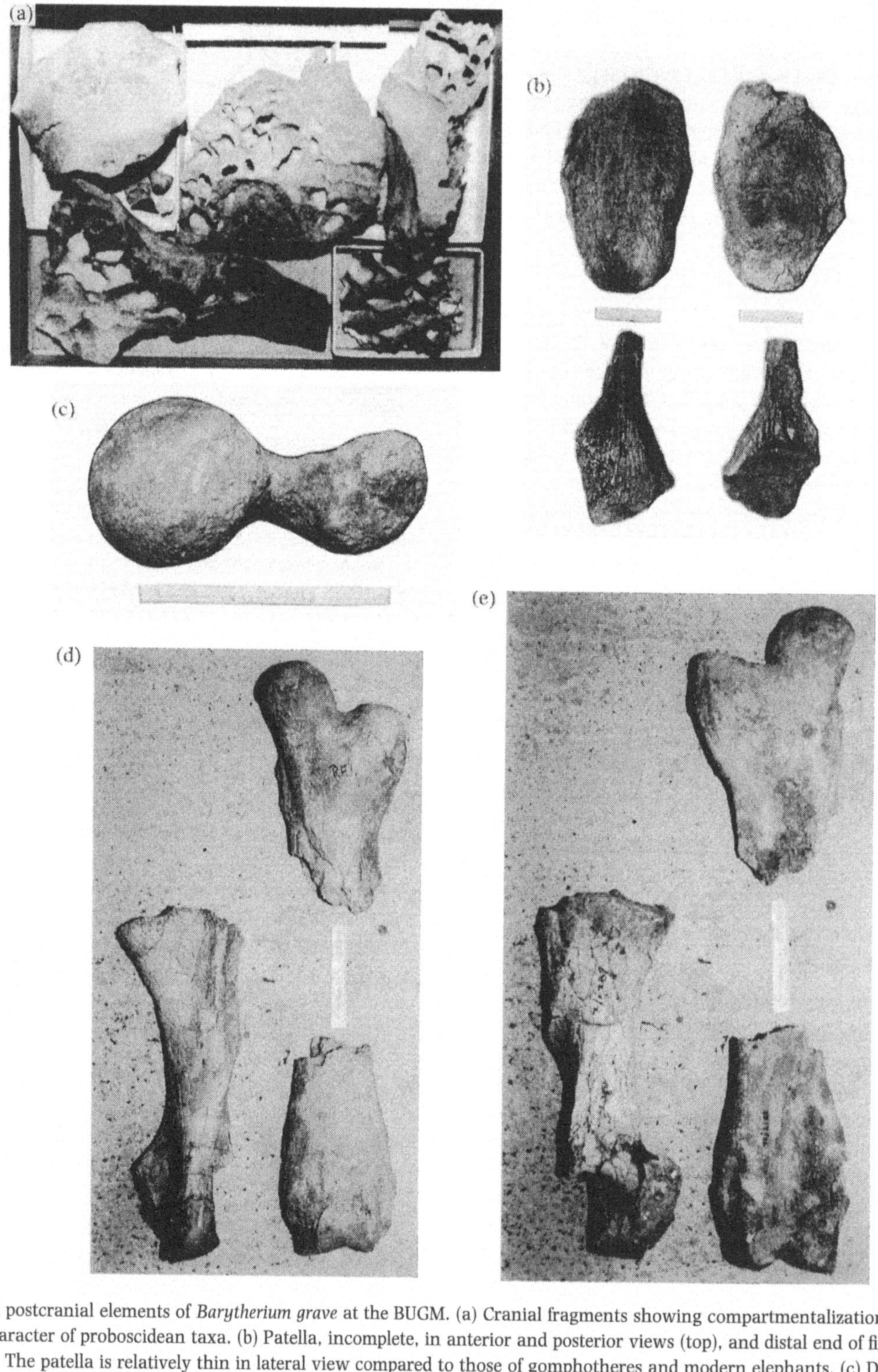

Plate 8.4 Cranial and postcranial elements of *Barytherium grave* at the BUGM. (a) Cranial fragments showing compartmentalization, i.e. diploë (air cells), an important character of proboscidean taxa. (b) Patella, incomplete, in anterior and posterior views (top), and distal end of fibula in lateral and medial views (bottom). The patella is relatively thin in lateral view compared to those of gomphotheres and modern elephants. (c) Dorsal view of left femur, showing almost round head and the greater trochanter, the shape is similar to that of modern elephants. (d) Anterior views of a major fragment of the left humerus (left) and proximal and distal ends of the left femur (right) shown in (c). Note the relatively flat patellar groove compared to those of living elephants. (e) Posterior views of the humerus and femur shown in (d). (Photographs by J. Shoshani.)

Table 8.3 Scapular characters of proboscideans and other taxa[a]

Taxon/Taxa	Coracoid process, character number					Glenoid cavity, character number				Infraspinous fossa, character number
	1	2	3	4	5	6a	6b	7	8	9
Outgroup taxa[b]	no/yes	no	no	no	?no/yes	yes	no	no	yes	no/yes
Anthracobune pinfoldi H-GSP 82-31P[c]	yes	yes	?yes	yes	yes	[yes]	[no]	[no]	[yes]	yes
Moeritherium lyonsi CGM 10020	yes	no	yes	yes	no	yes	no	no	yes	no
Numidotherium koholense KA 1-11	yes	?no	?no	???	yes	no	yes	no	no	yes
Barytherium grave CGM 10012	yes	yes	yes	yes	yes	no	no	yes	no	yes
Barytherium sp. novum BMNH M33344[d]	yes	yes	yes	yes	?no	no	no	yes	no	???
Other proboscideans[e]	yes/no	no	no	no	no	yes	yes	no	no	yes

Characters for the coracoid process are: 1, large (yes), small (no); 2, massive (yes), not massive (no); 3, hooked (yes), not hooked (no); 4, ridge present (yes), absent (no) — this ridge begins at about the tip of the coracoid process and continues dorsally towards the coracoid or anterior border of the scapula where it stops; 5, groove present (yes), absent (no) — this groove separates the glenoid cavity from a small process (the 'sus-glenoidal tubercle' of Mahboubi *et al.* 1986, p. 34) which is connected to the coracoid process.

Characters for the glenoid cavity are: 6a, shape is oval (yes), not oval (no); 6b, shape is rectangle (yes), not rectangle (no); 7, shape is round (yes), not round (no); 8, it is shallow (yes), not shallow (no).

The character for the infraspinous fossa is: 9, the infraspinous fossa is larger and/or wider than the supraspinous fossa (yes), infraspinous fossa is about the size of or smaller than the supraspinous fossa (no). In the outgroup taxa, this character is variable, often exhibiting about equal sizes of these fossae. For some extinct taxa, when the scapulae are incomplete, estimation of these character states is based on the margins.

Other notes:

a Only representative samples are tabulated.

b Desmostylia, Sirenia, Hyracoidea, Perissodactyla, and members of the so called 'Condylarthra' (see Domning *et al.* 1986; Novacek *et al.*, 1988; Prothero *et al.* 1988; Tassy and Shoshani 1988).

c Character states in brackets are based on an incomplete glenoid cavity.

d *Barytherium* sp. novum is about one-third the size of *Barytherium grave*.

e Deinotheriidae and Elephantoidea (see Tassy and Shoshani 1988).

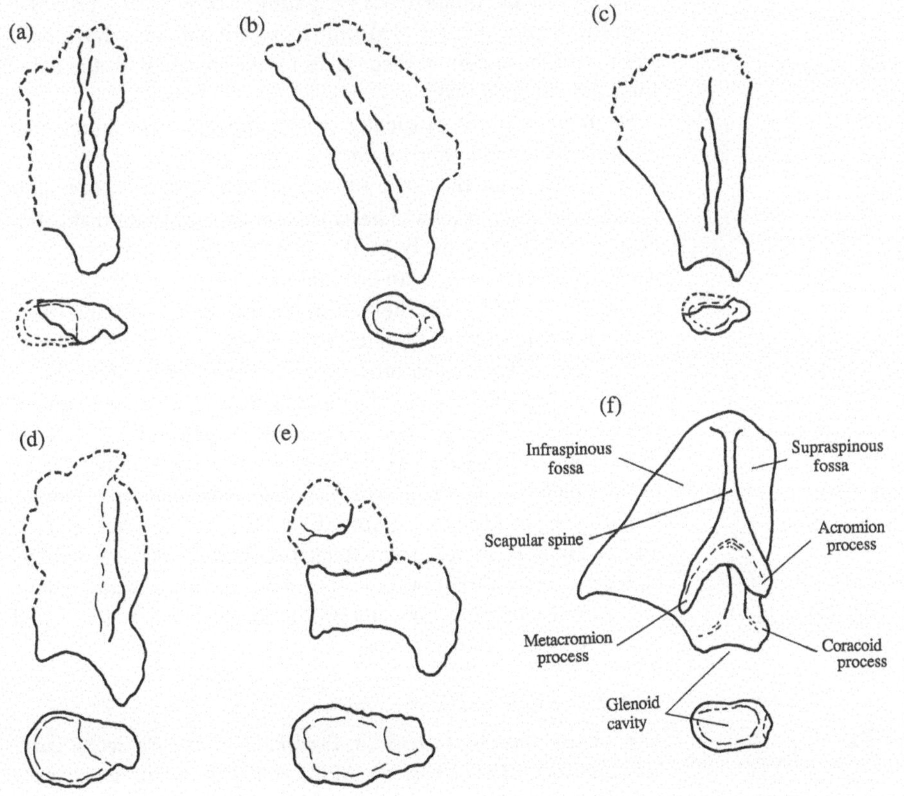

Fig. 8.4 Comparative lateral and ventral views of scapular fragments of: (a) *Anthracobune pinfoldi* (H-GSP 82-31P, right side); (b) *Moeritherium lyonsi* (CGM 10020, left side, reversed, after Andrews 1906, plate XI, fig. 5; glenoid cavity after BMNH M33343); (c) *Numidotherium koholense* (KA field number 'No. KB1'); (d) *Barytherium grave* (CGM 10012, right side, after Andrews 1906, XVII, figs 9 and 9a); (e) *Barytherium* species novum (BMNH M33344, right side); (f) A generalized Elephantiformes taxon, *Mammut americanum* (right side, after Olsen 1972).

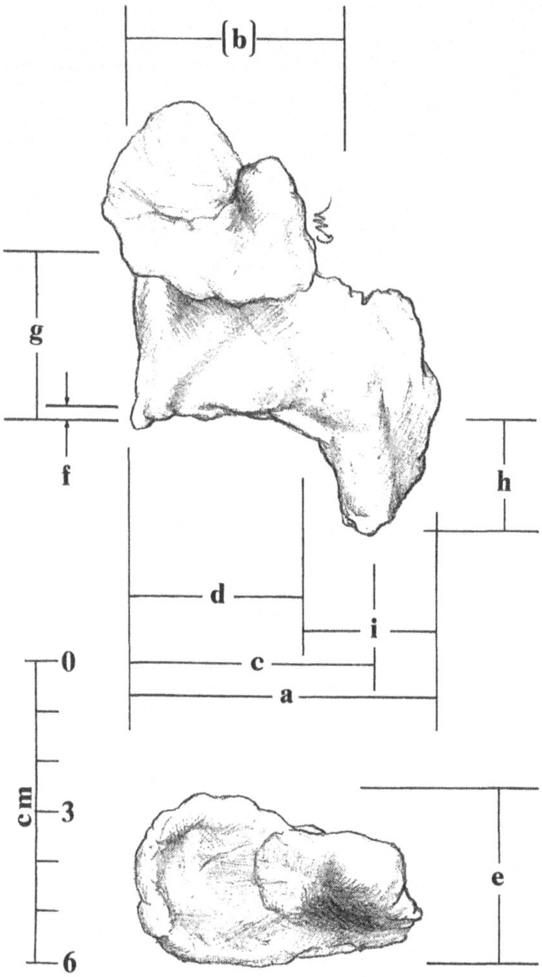

Fig. 8.5 Lateral and ventral views of the holotype scapular fragment of *Barytherium* species novum (BMNH M33344), showing locations of measurements taken (measurement a = 6.3 cm).

Observations on scapular fragments, referred to Barytherium species novum

Holotype

BMNH M33344 (Fig. 8.5), distal end of a right scapula of an adult individual in excellent condition; glenoid cavity and coracoid process complete, only a small ridge of the scapular spine, the distal ends of the infraspinous fossae and the caudal border, part of the subscapular fossae, and part of the neck; probably collected by Andrews and Beadnell in 1901 (A. W. Gentry and N. Court *In litt.*; see also Andrews 1906, p. xi).

Hypodigm (See Simpson 1945, p. 30 for definition)

The holotype, BMNH M33345, CGM 10012, BUGM 6817/24/6, 6819/43, and 6821/1.

Type locality

On the way to 'Basalt Hills', 'west of Qasr el-Sagha', Fayûm (north of Birket-el-Qurun), Egypt.

Age and distribution

Late Eocene, Qasr el-Sagha formation.

Diagnosis

Characters similar to those of *B. grave* but this specimen (an adult) is only one-third the size of *B. grave*. Specimen BMNH M33344 is about the size of *Moeritherium* but it differs from *Moeritherium* specimens (e.g. CGM 10020 described and illustrated by Andrews (1906) and BMNH M33343) in having a more massive and robust coracoid process when viewed laterally and anteriorly.

DEINOTHERES, FAMILY DEINOTHERIIDAE BONAPARTE, 1845 (by J. M. Harris)

Type species

Deinotherium giganteum Kaup, 1829.

Historical accounts and general introduction

Deinotheres were originally discovered in Europe (Plate 8.5). Known initially only from incomplete remains, their large but simple lophodont teeth misled early workers into classifying them with other bilophodont mammals such as rhinos, tapirs, and ground sloths (Osborn 1936, p. 767; Harris 1978, p. 325). A broken mandible, wrongly restored with the tusks pointing upward, led Kaup (1829) to postulate that deinotheres were related to hippopotami. A misidentified 'epipubic bone' associated with deinothere remains, convinced Solario (1864) that deinotheres were marsupials. None the less, the recovery of a partial skeleton, unearthed during the collapse of an embankment on the Prague–Brunn railroad in 1853, established that deinotheres were terrestrial quadrupeds, negating a previously proposed hypothesis (Buckland 1835) that these animals may have been sirenians that anchored themselves to river banks with the aid of their downturned tusks.

Dentally and osteologically, there can be little doubt that deinotheres are closely related to other members of the order Proboscidea even though all graviportal mammals share some postcranial adaptations. The closeness of the relationship has been debated extensively in the past but recent workers have accepted deinotheres as undisputed members of the Proboscidea (Tassy 1982, 1985; Shoshani 1986*b*; Tassy and Shoshani 1988; Shoshani, this volume, Chapter 16).

A dozen different deinothere species have been proposed at various times in the past. Harris (1978, pp. 320–3) recognized three species of *Deinotherium*: *D. giganteum* from Europe, *D. indicum* from Asia, and *D. bozasi* from Africa. He attributed other species to the smaller genus *Prodeinotherium* — *P. bavaricum* from Europe, *P. hobleyi* from Africa (e.g. Andrews 1911), and *P. pentapotamiae* and perhaps *P. orlovii*, from Asia. Recognition of geographical entities serves a useful purpose, but more work needs to be done to clarify the relationships among the species of each genus and among the genera themselves.

Geological settings and zoogeography

The oldest representatives of *Prodeinotherium* are from early Miocene localities in eastern Africa, but this genus was repre-

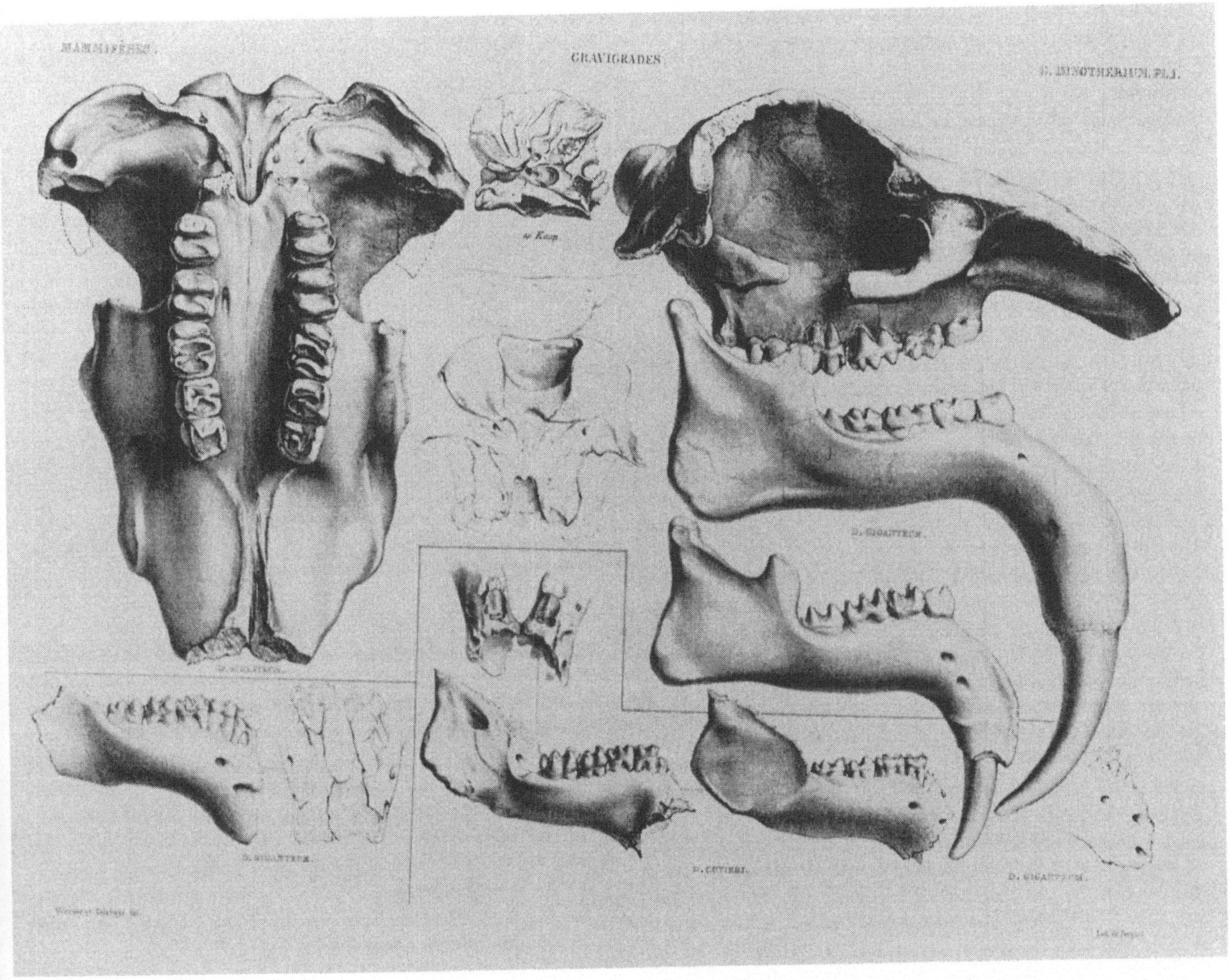

Plate 8.5 One of many extraordinary plates for that time (from Blainville 1845): plate I, '*Dinotherium giganteum*'. (A mounted skeleton of '*Deinotherium bavaricum*' at the Vienna Museum (Vienna, Austria) is depicted in Osborn 1936 (figs 63–64, pp. 99–100).)

sented in both Europe and Asia by the end of the early Miocene. The larger *Deinotherium* appeared on all three continents during the middle Miocene, initially coexisting with but ultimately outlasting the smaller genus. The last Eurasian representatives of the family are of Pliocene age, but *Deinotherium* persisted in eastern Africa until just over 1 million years ago.

Diagnostic characters

Dental formula for deciduous teeth

dI 0/1, dC 0/0, dP 3/3.

Dental formula for permanent dentition

I 0/1, C 0/0, P 2/2, M 3/3 — loss of four incisors, one canine, and two premolars from the *Moeritherium* condition. Lack of tusks in the upper jaw is an autapomorphy for Deinotheriidae.

Primitive proboscidean characters of Deinotheriidae

A flat cranium on its dorsum (most advanced proboscideans have had high, domed crania); premolars and molars of *Prodeinotherium* and *Deinotherium* had only two or three ridges, with very simple cusps; the enamel on the teeth of deinotheres was about 5–8 mm thick (compared to 1–5 mm in elephantids); cheek teeth were replaced in a vertical fashion as in other primitive proboscideans, including *Palaeomastodon* and *Phiomia*.

Derived (proboscidean) characters of Deinotheriidae

Slightly elevated skull; pneumatized bones in the cranium; elevated external naris; loss of teeth (see dental formulae); long, downrecurved incisors (I2), or tusks in the mandible; radius fixed in pronation position; a well-developed medial process (tuberculum mediale) on the astragalus; large size and graviportal stance —

Deinotherium giganteum reached about 4 m tall. Skeletal characters of Deinotheriidae have been summarized by Harris (1978).

Discussion

Despite their preservation in fluviatile, lacustrine, and even marine sediments, deinotheres were large terrestrial mammals with cranial adaptations that indicate a fully functional proboscis. They differ from other proboscideans by having downward recurving tusks in the lower jaw. The location of the most complex cheek tooth in deinotheres is the middle of the dental battery (M1). All permanent cheek teeth were present and erupted in mature individuals, although Harris (1973, 1978) interpreted a tripartite dental eruption sequence. Only size and relatively minor morphological characters distinguish the earliest deinotheres from the last known representatives of the family, whose evidently successful adaptation persisted virtually unchanged for an interval of more than 20 million years. It has yet to be established which, if any, of the other Palaeogene proboscideans were directly ancestral to deinotheres, but our knowledge of such potential ancestors is currently limited to animals that frequented coastal habitats.

The brachydont shearing cheek teeth of deinotheres were ideally suited for processing soft foliage. Loss of the upper tusks and the downturned nature of the lower tusks would have permitted more direct access of food to the mouth, and could justify an interpretation of a short, tapir-like proboscis in these animals. Many deinothere tusks show evidence of wear, although no consistent wear pattern or location has been identified. The fact that certain deinotheres had long legs but short necks enables one to argue against frequent use of their tusks for digging, and Harris (1978) suggested that these tusks could have been used as sources of purchase for manipulation of a short proboscis, or even for recognition of individual animals by other members of the same species. Deinotheres are only rare constituents in early Pleistocene assemblages from East Africa and it is likely that both the spread of grasslands and competition from the elephantids at the onset of the Pliocene contributed to the demise of this uniquely specialized but temporally persistent group.

RESULTS AND GENERAL DISCUSSION

(by J. Shoshani, R. M. West, N. Court, R. J. G. Savage, and J. M. Harris)

EARLY PROBOSCIDEAN PHYLOGENY

The cladogram at the top of Fig. 8.2 depicts a synthesis of several thoughts (those of Tassy 1982, 1985, this volume, Chapter 6; Tassy and Shoshani 1988; Savage and Court 1989; Shoshani *et al.* 1989*a*; and Shoshani this volume, Chapter 16). Note that the branching pattern of the six taxa (excluding the outgroups) is identical to that shown in Fig. 16.5, and the corresponding classification given in Shoshani, this volume, Chapter 16. The cladogram is reproduced in this chapter (modified after Shoshani *et al.* 1989*a*, p. 160) for easy reference. Given below are the synapomorphies to corroborate the bifurcations at nodes A–F in Fig. 8.2.

CHARACTERS IN SUPPORT OF THE RELATIONSHIPS SHOWN IN FIG. 8.2

Characters for *Anthracobune* (excluding those of astragalus) are after West (1984); for *Anthracobune* (astragalus only), after Gingerich *et al.* (1990); for *Moeritherium*, after Andrews (1906), Coppens and Beden (1978), and Tassy (1981); for *Numidotherium*, after Mahboubi *et al.* (1984, 1986); for *Barytherium*, after Andrews (1906), Savage (1969), Harris (1978), Savage and Court (1989), and Shoshani *et al.* (1989*a*); for *Prodeinotherium* and *Deinotherium*, after Harris (1978); and for Deinotheriidae and Elephantiformes (node F), after Tassy and Shoshani (1988).

Synapomorphies for node A in Fig. 8.2

Large and hooked scapular coracoid process; presence of astragalar tuberculum mediale; and possibly third lower molars with two definite cuspids at the posterior end (see Appendix 16.4).

Synapomorphies for node B in Fig. 8.2

Elevated skull; near closure of the external auditory meatus by the enfolding of the postglenoid process and the squamosal; confluence of the condylar foramen with the foramen lacerum posterius; anterior position of orbit; enlarged second incisors; loss of I_3 (uncertain for *Numidotherium*); loss of C_1; loss of P^1_1; M^1 and M^3 with distocrista; radius positioned or fixed in pronation position; antero-posterior flattening of femur; loss of astragalar foramen; loss of crista capitalis on astragalar head; astragalus with semicircular head and short neck (these are only a sample of the synapomorphies obtained by the parsimony analysis conducted by Shoshani (this volume, Chapter 16); see also Tassy 1990*a*, this volume, Chapter 6).

Synapomorphies for node C in Fig 8.2

Pneumatized bones in the cranium; elevated external naris with contact between frontal and premaxilla; ulna longer than radius; femur lengthening and distal position of knee; lateral (ectal) facet of astragalus larger than sustentacular (only a sample of synapomorphies is given here).

Synapomorphies for node D in Fig. 8.2

Loss of I^3; loss of C^1; I_2 only partly surrounded by enamel; large size and stance graviportal.

Synapomorphies for node E in Fig. 8.2

Straight lower tusks; massive, robust, and hooked coracoid process; glenoid cavity round, with flared, thickened rim.

Synapomorphies for node F in Fig. 8.2

Cranial foramina situated posteriorly in the orbitotemporal fossa; development of a strong ligamentum nuchae; loss of I^1 and I_1;

presence of a hypocone on P^{3-4}; bilophodont P_4 with a posthypoconid crest; M^1 and M_1 with a third loph(id); elephant-like skeleton (other details in Tassy 1990*a*, this volume, Chapter 6).

A comparison among the first five taxa given on the cladogram in Fig. 8.2 (top — from *Anthracobune* to Deinotheriidae) and Elephantiformes (number 6 on the cladogram in Fig. 8.2) can be made by evaluating the list of synapomorphies for nodes A–F given in this chapter and comparing them to the synapomorphies for Elephantiformes (node H in Appendix 16.4).

Analysis of the 24 characters (12 dental and 12 non-dental) from five taxa (*Moeritherium* (M), *Numidotherium* (N), *Barytherium* (B), *Deinotherium* (D), and *Palaeomastodon* (P)) by Savage and Court (1989) revealed that the three equally most parsimonious cladograms had these topologies:

1. (M, (N, (B, (D, P))));
2. ((B, N) (M, (D, P))); and
3. (N, (B, (M, (D, P)))).

In other words, using those specific 24 characters and five taxa, it appears that the phylogenetic position of M, N, and B may vary without changing the score of the tree. It is noted that Shoshani's results (this volume, Chapter 16; based on 123 characters from 39 taxa) corroborate the topology in (1), but topologies (2) and (3) increase the score of the tree significantly (details in Appendix 16.3; see also results by Tassy 1988*a*). The lesson from these comparisons is that a larger data set provides a more solid base than a smaller data matrix; the latter can be used to obtain a quick rough idea of the direction the results may take.

COMPARISON OF PROBOSCIDEAN SCAPULAE TO THOSE OF OTHER MAMMALS

The scapular coracoid process in mammals (Hildebrand 1974; Walker 1980) and elephants (Eales 1928; Mariappa 1986; Shoshani, personal observation) functions as the origin of the biceps and coracobrachialis muscles which flex the elbow. A large coracoid process thus provides a broad surface area for muscle attachment and therefore a strong force to flex the arm. Mammals with long and/or large coracoid processes usually possess a clavicle (Lessertisseur and Saban 1967). The barythere material from Dor el Talha, Libya includes an unidentified specimen which resembles a clavicle (large enough to match *Barytherium* proportion) but it may belong to a different taxon, perhaps not even a mammal.

A review of mammalian, and specifically proboscidean, scapulae, as described by Andrews (1906), Lessertisseur and Saban (1967), Tassy (1985), and Shoshani (1986*b*) reveals that, indeed, no other mammalian taxon possesses as large a coracoid process as that of *Barytherium* (Fig. 8.4). The closest resemblance of a coracoid process to that of *Barytherium* is that of *Moeritherium*; it is, however, smaller, more gracile, and less hooked than that of *Barytherium*. The scapular coracoid process of *Anthracobune* (specimen H-GSP 82-31P collected in Pakistan and briefly described by West 1984) seems to be almost complete. This process is also large — about the size and robustness of that of *Barytherium*. The glenoid cavities of *Anthracobune* and *Moeritherium* are also more like each other than either is to *Barytherium*; those of the latter have rounder and deeper cavities than those of the former. The scapulae of multituberculates, according to Krause and Jenkins (1983), also possessed unusually long coracoid processes. At the suggestion of J. G. M. Thewissen, J. Shoshani also examined scapulae of *Coryphodon* (specimen no. UM 74-409), *Phenacodus* (MP 80-272), and *Pleuraspidotherium* (R590 and R573) at the University of Michigan and concluded that, although these extinct ungulate-related taxa had relatively large coracoid processes, none approached the size, massiveness, and 'hook' of the coracoid process of *Barytherium* (see Addendum, however).

PALAEOECOLOGY AND DISTRIBUTION OF EARLY PROBOSCIDEANS

With the exception of the inferred habitat for *Anthracobune aijiensis* (which was a 'land mammal' from the early middle Eocene transitional part of the Subathu Formation, India; Kumar 1991, p. 237), palaeoecological and environmental conditions of the Asian anthracobunids (*fide* Wells and Gingerich 1983) were very similar to those described for the African localities. All sites were near-shore shallow-water environments. Habitats of *Anthracobune*, *Lammidhania*, and *Jozaria* were described by Wells and Gingerich (1983) and West (1984). Wells and Gingerich (1983, p. 131) noted that Kuldana Formation in Pakistan was interpreted as 'a quiet, muddy, and marshy shoreline transitional between well-drained alluvial redbeds below and marine foraminiferal limestones above'. West (1984, p. 189) indicated: 'The rock unit in the Basal area [Pakistan] is composed primarily of freshwater calcareous mudstone, fine-grained sandstone and granular calclithite. Foraminifera-bearing limestone lenses are present, indicative of the proximity of the fluctuating Tethys shoreline'. It is noted that both African and Asian localities contained numerous marine invertebrates and vertebrates, including archaeocete whales (Andrews 1906; West 1980; Gingerich and Russell 1981; Gingerich 1992).

On the habitat of *Moeritherium*, Andrews (1906, p. xvii) noted that it 'was probably an amphibious, shore, or swamp living animal'. Andrews (1906, p. 117) also observed that the scapular blade of *Moeritherium* presents 'some approach to the Sirenia type'. The environment in which *Barytherium* lived was similar to that of *Moeritherium*, but Andrews (1906, p. 176) included these intriguing comments:

> They [barytheres] occur in the same beds as *Moeritherium lyonsi* and *Eosiren libyca*, with remains of reptiles and fishes. Probably this animal lived in swampy country and perhaps was semi-aquatic. In any case the form of the fore limb is such that it was almost certainly used for something besides mere progression on firm ground, possibly, as already suggested, for digging or scraping away the surface of the soil or for forcing the animal through soft swampy ground.

It was not precisely stated on what specific characters of the forelimbs Andrews (1906) based the above comments. Presumably, the size and robustness of the limb bones and the associated processes, as well as the size of the scapular coracoid processes, could have contributed to Andrews' interpretation.

Andrews (1906, p. ix) described the Eocene fossiliferous unit of the Qasr el-Sagha Formation in the Fayûm, Egypt, where *Barytherium* and *Moeritherium* were found, as 'Alternating limestones, marls, clays, and sandstones.' Simons (1968, p. 14) expanded on this Qasr el-Sagha Formation and noted that it ranged from 'interdeltaic shallow marine and littoral' (in the lower portion of this formation) to 'deltaic and interfinger with distributary channel sands' (in the upper beds of this formation). The El Kohol Formation in Algeria, where *Numidotherium* was found (Mahboubi *et al.* 1986, p. 16) was described as including four sections overlying the marine Turonian as follows: 'transitional gypseferous member, a lower red detritic member, a middle marly-calcareous member, and an upper red detritic member'.

Since the early 1900s when Eocene proboscidean taxa were first discovered in the Fayûm, there have been numerous discoveries of Eocene proboscideans (Coppens *et al.* 1978; Bown *et al.* 1982; Wells and Gingerich 1983; West 1984; Mahboubi *et al.* 1986). The proboscidean discoveries expanded our views in two planes: geographically and phylogenetically. We learned that the geographical distribution of early proboscideans not only expanded within the African continent but, since we include *Anthracobune* in the Proboscidea, it means that the origin of Proboscidea may have been in Asia. Within Africa, fossil remains were found along the north-eastern coast (el Fayûm, Egypt; Andrews 1906), to the north-west (El Kohol, Algeria; Mahboubi *et al.* 1986), and to western Africa (Kaolack, Senegal; Coppens and Beden 1978), some 5000 km apart. A molar of *Moeritherium* from Algeria was figured by Gaudry (1893) and Tassy (1981); it was first allocated to '*Mastodon turicensis*' by Gaudry (1893). The finding of proboscidean remains, or the nearest relatives of Proboscidea, in Asia (Kohat, Pakistan (Wells and Gingerich, 1983); Basal and Ganda Kas, Pakistan, and Subathu Formation, India (West 1980, 1984; Kumar 1991); Harudi Formation, Kachchh, western India (Bajpai *et al.* 1989)) was surprising since, for decades, we believed that Eocene proboscideans were confined to Africa (Osborn 1936, 1942; Coppens *et al.* 1978).

ECOLOGICAL OBSERVATIONS

As more discoveries of Eocene–Oligocene proboscideans are made, it becomes evident that earliest members of this order were more diversified than was previously thought. With the description of *N. koholense* (Mahboubi *et al.* 1986), *A. aijiensis* (Kumar 1991), and notes on new species of *Barytherium* in this chapter, nine Eocene–Oligocene proboscidean species are recognized. They are: (1) *Anthracobune pinfoldi* Pilgrim, 1940; (2) *A. aijiensis* Kumar, 1991 (note: Kumar, 1991 placed *A. pilgrimi* (Dehm and Oettingen-Spielberg, 1958) in *Pilgrimella pilgrimi*); (3) *Moeritherium lyonsi* Andrews, 1901*d*; (4) *M. trigodon* Andrews, 1904*a*; (5) *Numidotherium koholense* Mahboubi *et al.*, 1986; (6) *Barytherium grave* Andrews, 1901*d*; (7) *B.* species novum, Shoshani *et al.*, 1989*a*; this chapter; (8) *Palaeomastodon beadnelli* Andrews, 1901*a*; and (9) *Phiomia serridens* Andrews and Beadnell, 1902 (authorship references for species (1–2) are to be found in West 1983, 1984; Kumar 1991; for (3–4) in Coppens and Beden 1978; for (5) in Mahboubi *et al.* 1986; for (6) in Harris 1978; for (7) in Shoshani *et al.* 1989*a*; and for (8–9) in Coppens *et al.* 1978). Other proposed anthracobunid taxa are: *Lammidhania wardi* (Pilgrim, 1940), *Pilgrimella pilgrimi* Dehm and Oettingen-Spielberg, 1958, and *Jozaria palustris* Wells and Gingerich, 1983 (authorship references for these species are in Wells and Gingerich 1983; West 1983, 1984; Kumar 1991).

Should we accept the hypothesis that *Phiomia* is a distinct genus from *Palaeomastodon* (see, for example, the discussion in Coppens *et al.* 1978), as was done here and elsewhere, then the total number of middle Eocene to early Oligocene proboscidean species is nine (excluding species of *Lammidhania*, *Pilgrimella*, and *Jozaria*) — four more than the number known 80 years ago. It is of interest that the diversity of proboscidean species seems to have declined during the Oligocene of Africa while that of the hyracoid species appears to have increased during the same general period (Rasmussen and Simons 1988). It is also noted that the largest member of the Hyracoidea (*Titanohyrax ultimus*) was about the same size as the smallest member of the Proboscidea (*Moeritherium trigodon*), both of which approached the size of a tapir, *Tapirus* sp. (Andrews 1906; Coppens and Beden 1978; Meyer 1978; Rasmussen and Simons 1988). This comparison accentuates the changes in niche availabilities for the Proboscidea whose members, on the whole, had larger body sizes than members of the Hyracoidea. Following the decline in the number of proboscidean species during the Oligocene, an increase occurred during the Miocene and (?)Pliocene, and then a sharp decline took place during the Pleistocene and Holocene (Maglio 1978; papers in Martin and Klein 1984; Shoshani and Tassy, this volume, Chapter 34).

SUMMARY AND CONCLUSIONS

The earliest proboscideans are classified in three families — Moeritheriidae, Barytheriidae, and Deinotheriidae — and two genera, *Anthracobune* and *Numidotherium* (no familial assignment for *Anthracobune*; *Numidotherium*, on the other hand, is placed in the family Numidotheriidae) (Table 16.1). Each of these lineages represents a unique clade, although some authors suggested closer affinities among or between any pair (e.g. between *Numidotherium* and *Barytherium*, or between *Barytherium* and *Deinotherium*) or a triplet.

The inclusion and exclusion of anthracobunids, Moeritheriidae, Barytheriidae, and Deinotheriidae in the Proboscidea has become a classic case of the fluctuating hypotheses according to the prevailing thoughts. In this volume, the above four families and *Anthracobune* are considered bona fide proboscideans. Until additional evidence becomes available, we have excluded other Asian anthracobunids from the Proboscidea and placed them in Tethytheria or possibly Proboscidea *incertae sedis* (Table 16.1; see however, Fischer and Tassy 1993).

Fossil remains of Moeritheriidae, Barytheriidae, and *Numidotherium* in Africa are limited to the Eocene–Oligocene boundary, and they appear to be localized along the shores of the ancient Tethys Sea. Skeletal remains of the family Deinotheriidae (*Prodeinotherium* and *Deinotherium*) were found in Africa, Asia, and Europe in sediments of the Miocene, Pliocene, and Pleistocene. It is probable that Deinotheriidae had its origin at an earlier date, a hypothesis yet to be corroborated.

Most of the material upon which taxa were described came from field excavations; museum collections, however, still provide excellent sources of material waiting to be 'rediscovered', as was the case with scapulae of the proposed new species of *Barytherium* that were stored at the BMNH for almost 80 years, together with *Moeritherium* specimens of approximately the same size.

ADDENDUM

The scapula of *Arsinoitherium* has a large coracoid process which may or may not be homologous to that of proboscideans (cf. Fig. 8.4 to illustrations in Andrews 1906, p. 33, and Court 1993*b*, p. 141 and plate I, 1b). Fischer and Tassy (1993) noted that the characters of the scapular coracoid process of *Arsinoitherium* are homologous with those of proboscideans; one of us (JS) believes they are not homologous. Although there are variations, proboscideans' scapular coracoid process is large, hooked, and protrudes ventrally, that of *Arsinoitherium* is large, knobby and protrudes diagonally. Microscopic studies of teeth (Pfretzschner 1992, and von Koenigswald *et al.* 1993) also indicate that *Arsinoitherium* is not a proboscidean.

ACKNOWLEDGEMENTS

Much appreciation is extended to the staff of the museums mentioned on pp. 57, 364–8 for their help in studying specimens. Special thanks to M. J. Benton for access of *Barytherium* specimens at the BUGM, and to A. W. Gentry of the BMNH for allowing J. Shoshani to borrow scapulae specimens of *Barytherium* and *Moeritherium*, and for pertinent information. The scapular fragment of *Anthracobune* was collected by a team led by R. M. West from the Howard University (HU)–Geological Survey Of Pakistan (GSP) project, directed by S. Taseer Hussain (HU) and S. M. Ibrahim Shah (GSP). The field work was supported by the GSP and by grants from the Smithsonian Institution to S. T. Hussain. Comments on versions of this chapter were made by E. C. Marsac, M. C. McKenna, J. L. Pierce, P. Tassy, and J. G. M. Thewissen. Casts of scapula specimens BMNH M33343 and 33344 were made by R. Smith, and the cast of KA field number 'No. KB1' by N. Court. The artwork is by J. S. Grimes and G. H. Marchant.

NINE

Evolution of zygodons with emphasis on dentition[1]

HEINZ TOBIEN

INTRODUCTION

Because of the structure of cheek dentition and the mandible, the zygodont mammutids represent a rather well-defined group among mammutids. (In this chapter and elsewhere in this volume, the common names 'mastodon', 'zygodon', and 'mammutid' are used only for members of the family Mammutidae, and the names 'gomphothere' and 'bunodont gomphothere' refer to members of the family Gomphotheriidae *sensu lato*, e.g. *Gomphotherium*. 'Zygodont' and 'bunodont', spelled with a 't', are used as adjectives.) Two genera represent the zygodont lineage in the Old World (*Zygolophodon* Vacek, 1877 of the middle Miocene, and *Mammut* Blumenbach, 1799 of the late Miocene, Pliocene, and Pleistocene epochs), and one genus in Africa (*Eozygodon* Tassy and Pickford, 1983 of the early Miocene). These three genera have been classified in the family Mammutidae (e.g. Tassy 1990*a*; Shoshani and Tassy, this volume, Chapter 34). The basic structural molar characters are already visible in the earliest members of *Zygolophodon*. In this chapter, I will focus on the European forms (*Zygolophodon* and *Mammut*); *Eozygodon* has been described and discussed by Tassy and Pickford (1983).

THE ZYGODONT MOLAR PATTERN

ZYGOLOPHODON AND *MAMMUT*

Genus

Zygolophodon Vacek, 1877.

Type species

Zygolophodon turicensis (Schinz, 1824) (for more details, including the non-validity of '*Zygolophodon tapiroides*', sometimes used by earlier authors, see Tassy 1985, p. 477).

Diagnosis for the genus

A mammutid with intermediate molars and M3 in which the pretrite and posttrite elements of bunodont character are arranged in clearly yoke-like transverse crests. After some wear the loph(id)s become sharpened at their summits and receive a more zygodont character. The pretrite conules are more or less developed, almost blocking the open valleys. The lophs are normal to the long axis of the crown, the lophids slightly oblique (i.e. postero-external to antero-internal). There are three lophs plus a strong cingulum or four lophs plus a weak cingulum in the M^3, and four lophids plus a more or less strong cingulum in the M_3 (Figs 9.1–9.3).

The mandible has a long symphysis with a pair of well-shaped, straight, protruding incisors of oval section. The upper tusks are curved downward, enlarged, diverging exteriorly with an enamel band at the outer side.

Genus

Mammut Blumenbach, 1799.

Type species

Mammut americanum (Kerr, 1799).

Diagnosis for the genus

A mammutid with strongly developed zygodont cheek teeth without conules. The lophs transverse to the long axis of the teeth, the lophids slightly oblique to the long axis of the teeth. The intermediate molars with no more than three loph(id)s. M3 with four loph(id)s plus cingulum. Elephantoid mandible with shortened symphysis, sometimes with rudimentary peg-like incisors. The angular process is protruding (Figs 9.4–9.7). Basicranium shortened, cranium high-domed, upper incisors without enamel, long, straight, or upturned.

DISCUSSION

In *Zygolophodon*, as in *Gomphotherium* and other gomphotheres, there are transverse crests or ridges (lophs) in the upper molars and dp^4, subdivided by a median sulcus in a labial and lingual half-ridge (=half-loph(id), Fig. 9.1). In the genus *Zygolophodon*, as

[1]The final version of this chapter was not reviewed by the author who died shortly after the last changes were sent to the editors.

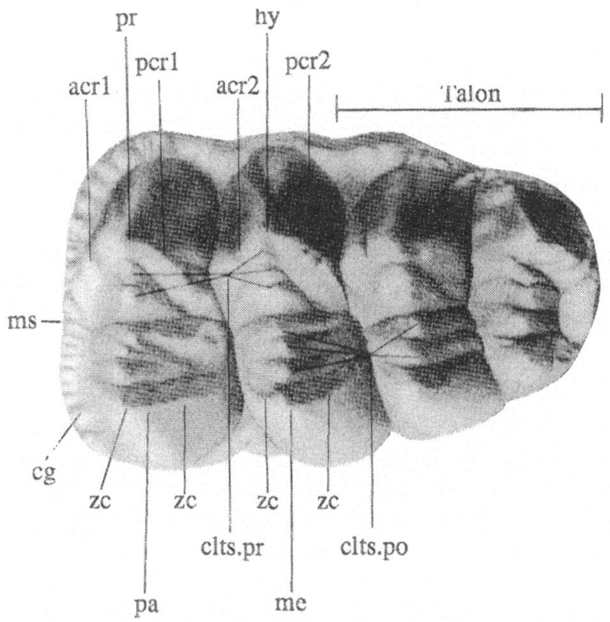

Fig. 9.1 *Zygolophodon turicensis* (Schinz, 1824), right M^3, in occlusal view. Locality: Neudorf an der March/CSSR (Spalte). Age: middle Miocene, MN6. (Reproduced from Zapfe 1954, plate 1; and Tobien 1975, p. 196, fig. 1.) Key: acr1, anterior crescentoid of the first loph; acr2, anterior crescentoid of the second loph; cg, cingulum; clts.po, conelets on the posttrite (secondary or adaxial conelets); clts.pr, conelets on the pretrite (secondary or adaxial conelets); hy, hypocone (main abaxial conelet of the second pretrite); me, metacone (main abaxial conelet of the second posttrite); ms, median sulcus; pa, = paracone (main abaxial conelet of the first posttrite); pcr1, posterior crescentoid of the first loph; pcr2, posterior crescentoid of the second loph; pr, protocone (the main or abaxial conelet of the first pretrite); zc, zygodont crest (abbreviation list in Tobien 1973*b*, pp. 144–5).

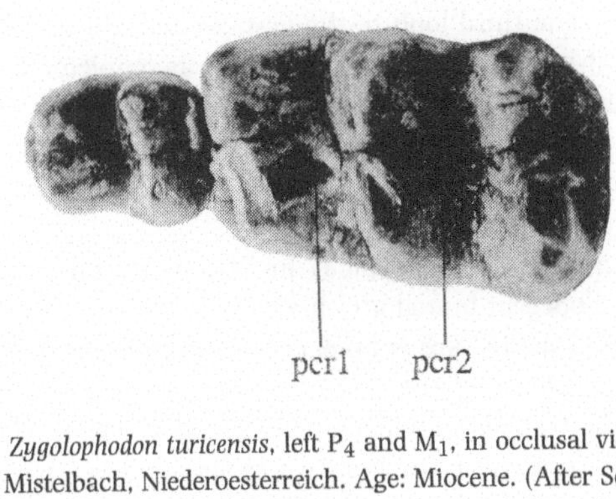

Fig. 9.2 *Zygolophodon turicensis*, left P_4 and M_1, in occlusal view. Locality: Mistelbach, Niederoesterreich. Age: Miocene. (After Schlesinger 1917, pl. 21, fig. 6.) Key: pcr1, posterior crescentoid of the first loph; pcr2, posterior crescentoid of the second loph.

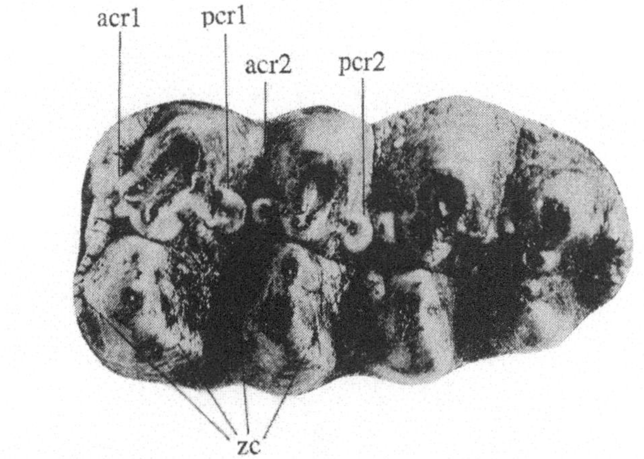

Fig. 9.3 *Zygolophodon turicensis*, left M_3, in occlusal view. Locality: Klein-Hadersdorf bei Poysdorf, Niederoesterreich. Age: middle Miocene (MN6). (After Schlesinger 1917, pl. 21, fig. 8; reproduced from Tobien 1975, p. 208, Fig. 12.) Key: acr1, anterior crescentoid of the first loph; acr2, anterior crescentoid of the second loph; pcr1, posterior crescentoid of the first loph; pcr2, posterior crescentoid of the second loph; zc, zygodont crest.

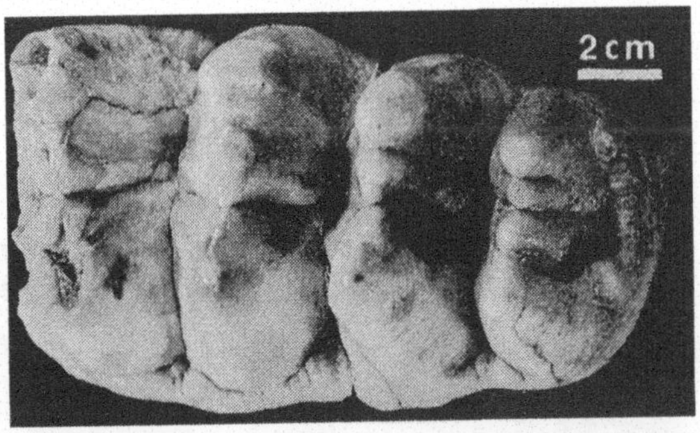

Fig. 9.4 *Mammut borsoni*, left M^3, in occlusal view (specimen no. THP–10031; THP, Natural History Museum, Tianjin (Museum of Huangho and Paiho, China). Locality 1, Yushe Basin, Shaanxi Province. Age: Jinglean (Ruscinian European Land Mammal Age, MN14b). (After Tobien *et al.* 1988, p. 163, fig. 52).

in *Mammut*, the number of the lophs in dp^4, M^1, and M^2 (= intermediate molars) is invariably three, in contrast to the bunodons, where the loph numbers vary from three to five and sometimes more (Tobien 1973*b*, p. 129). The presence of three ridges in the intermediate molars is a plesiomorphic character of all zygodons. The M^3 have three lophs plus a more or less strong posterior cingulum reaching the size of a fourth loph.

In zygodont molars, as in bunodont gomphotheres, every half loph is composed of at least two or more incompletely separated small cones (Fig. 9.1). These elements were named conelets by Osborn (1942, p. 1545). As in bunodons, the most labial conelets in the lower molars and the most lingual 'conelets' in the upper ones are stronger and larger than the conelets situated

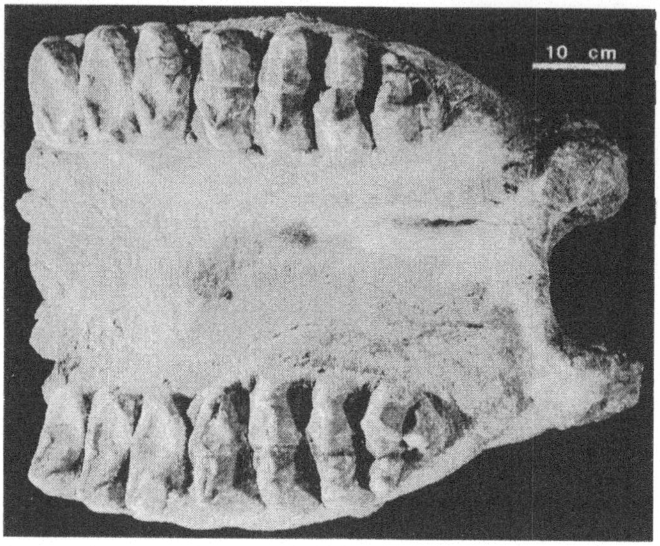

Fig. 9.5 *Mammut borsoni*, palate with both M^2 and M^3 (THP-00079). Locality: unknown, Yushe Basin, Shaanxi Province. Age: probably Jinglean. (After Tobien *et al.* 1988, p. 161, fig. 50.)

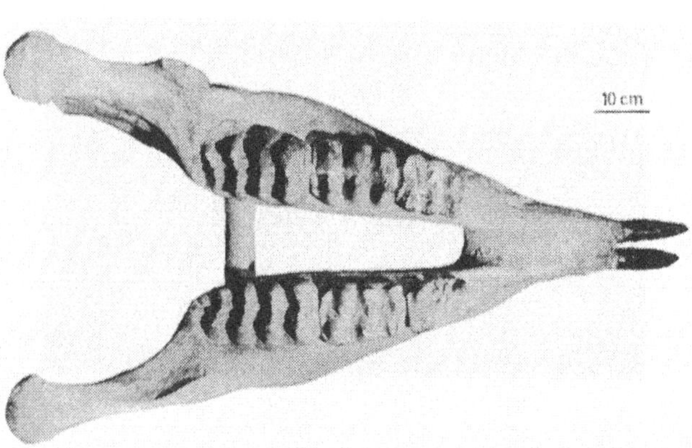

Fig. 9.6 *Mammut borsoni*, complete mandible with both M_1–M_3 and reconstructed incisors in occlusal view (THP-10000). Locality 25, Yushe Basin, Shaanxi Province. Age: Jinglean. (After Tobien *et al.* 1988, p. 172, fig. 63b).

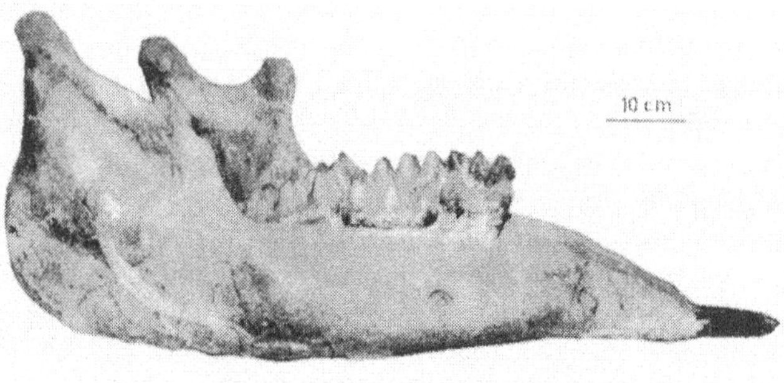

Fig. 9.7 *Mammut borsoni*, same specimen as in Fig. 9.6; labial view. (After Tobien *et al.* 1988, p. 171, fig. 63a; cf. Bakalov and Nikolov 1962, pl. 53, fig. 1a.)

nearer the sulcus, i.e. in an adaxial position, which are usually weaker and smaller (the sulcus marks the longitudinal axis of the crown, Fig. 9.1). As already pointed out by Osborn (1934*a*, p. 220), the conelets originated by binary (or ternary) fission of the primary cones, to which belong the stronger and larger abaxial conelets. There are paracone and protocone in the first ridge (protoloph) and metacone and hypocone in the second ridge (metaloph), whereas the third ridge (tritoloph) and the following ones in the M3 are derivations of the cingulum (Figs 9.1–9.3). This multiplication into conelets starts from the adaxial slopes of the primary main cones towards the sulcus.

Thus, similar to *Gomphotherium* and other bunodont gomphotheres, those conelets which are closer to the longitudinal sulcus (adaxial conelets) are of secondary origin and additions to the primary (abaxial) structures (Tobien 1973*b*). This is a unique functional construction to diversify the wear surface by multiplying the number of accessory elements, thus increasing the efficiency of the chewing process in all gomphotheres and mammutids. This phenomenon does not exist in the deinotheres.

The number of conelets in the upper molars is different in the half-lophs, being smaller in the pretrites (2–3) and somewhat larger in the posttrites (3–4) in unworn upper molars of *Zygolophodon* (Fig. 9.1), as is also observed in bunodont gomphotheres (Tobien 1973*b*, figs 1, 3, and 4). In advanced zygodons such as *Mammut borsoni*, the number of conelets increases only slightly, if at all, and more in the posttrites than in the pretrites (Figs 9.4–9.6).

Other important elements of the upper molar crowns in zygodons are the enamel crests (acr1, acr2, pcr1, and pcr2 in Fig. 9.1), running from the tips of the abaxial main conelets of the pretrite (i.e. lingual) half-lophs down to the bottom of the transverse valleys, and ending near the middle axis of the crown. These crests give the pretrite half-lophs a certain selenodont 'crescentoid' habitus, somewhat similar to the anterior and posterior arms of a selenodont or bunoselenodont molar of an artiodactyl (Fig. 9.1).

As in bunodont (Tobien 1973*b*, pp. 123–4) and zygodont molars, the crescentoids in the first two lophs are well developed. This is obviously due to the fact that this part of the molar crown is its primary, tribosphenic one (with proto-, para-, meta-, and hypocone) in which the typical tooth structure is completely represented. The third loph of the first, second, third, and fourth molar in M3 often has weakly developed, incomplete, or missing crescentoids. These parts of the molar crown are products of the talon, i.e. after all derivates of the cingulum which evidently produced only incomplete and mostly inaccurate copies of the primary proto- and metaloph (Fig. 9.1).

A further peculiarity, typical for zygodont and absent or weakly developed in bunodont cheek teeth (Tobien 1975, figs 2, 4; Tassy 1974, p. 18) are enamel crests, running on the anterior and posterior slopes of the main posttrite conelets down to the bottom of the transverse valleys. The 'zygodont crests' (Fig. 9.1), however, are present in the first two molars and M^3, particularly on the proto- and metalophs in unworn specimens in both the *Zygolophodon* and *Mammut*, but disappear or are blurred after slight wear.

Another basic character of the zygodont molars is the relative width of the crowns, compared with those of bunodont molars of the genus *Gomphotherium* (Fig. 9.8). Tobien (1971, p. 156, fig. 10; 1972, p. 172, fig. 8) pointed out this difference, using scatter diagrams of M^3 of the late Miocene *Zygolophodon turicensis* and *Gomphotherium angustidens* from the Austro-Bavarian Upper Freshwater Molasse. The greater width of the zygodont molars can also be demonstrated by using upper molars, where it becomes still more pronounced in comparing Pliocene and early Pleistocene (Villafranchian) members of the genus *Mammut* (as *M. borsoni* (Hays, 1834) and *M. americanum* (Kerr, 1792) with contemporary bunodont members (as *Tetralophodon longirostris* (Kaup, 1832) and *Anancus arvernensis* (Croizet and Jobert, 1828)).

The inclination of the labial and lingual walls of the transverse crests in the upper molars of *Zygolophodon* is as in many mammalian groups with brachyodont molars: the labial (posttrite) wall is vertical, the lingual (pretrite) wall is inclined from dorsolaterally to ventro-medially as in the case in the bunodont molars (Fig. 9.9). As in *Gomphotherium*'s M^3 and also in *Zygolophodon*, the different inclination is well developed in the first and second lophs, which correspond to the primary part of the crown, whereas the walls of the third and the fourth lophs do not follow this rule, but are more or less equally inclined (Tobien 1976, figs 27–28). These differences may be related to the fact that the third and fourth lophs are derivates of the cingulum and obviously do not produce exact copies of the primary anterior part of the crown (Fig. 9.9).

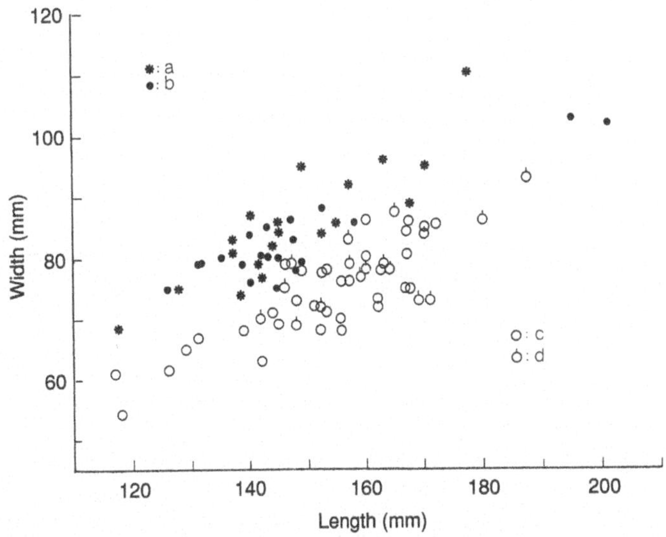

Fig. 9.8 Scatter diagram of length and width of M^3 in some bunodont gomphotheres and zygodons. Key: a, *Zygolophodon turicensis*; b, 'subtapiroid' (Schlesinger 1917) and 'zygolophodontoid' (Tassy 1974); c and d, *Gomphotherium angustidens*. A zygodont cluster (a, b) is evidently separated from a bunodont cluster (c, d). 'Subtapiroid' and 'zygolophodontid' specimens (b) are not significantly different from typical *Z. turicensis* specimens (a), but are mostly relatively broader than bunodont *G. angustidens* M^3. (After Tobien 1975, p. 217, fig. 25.)

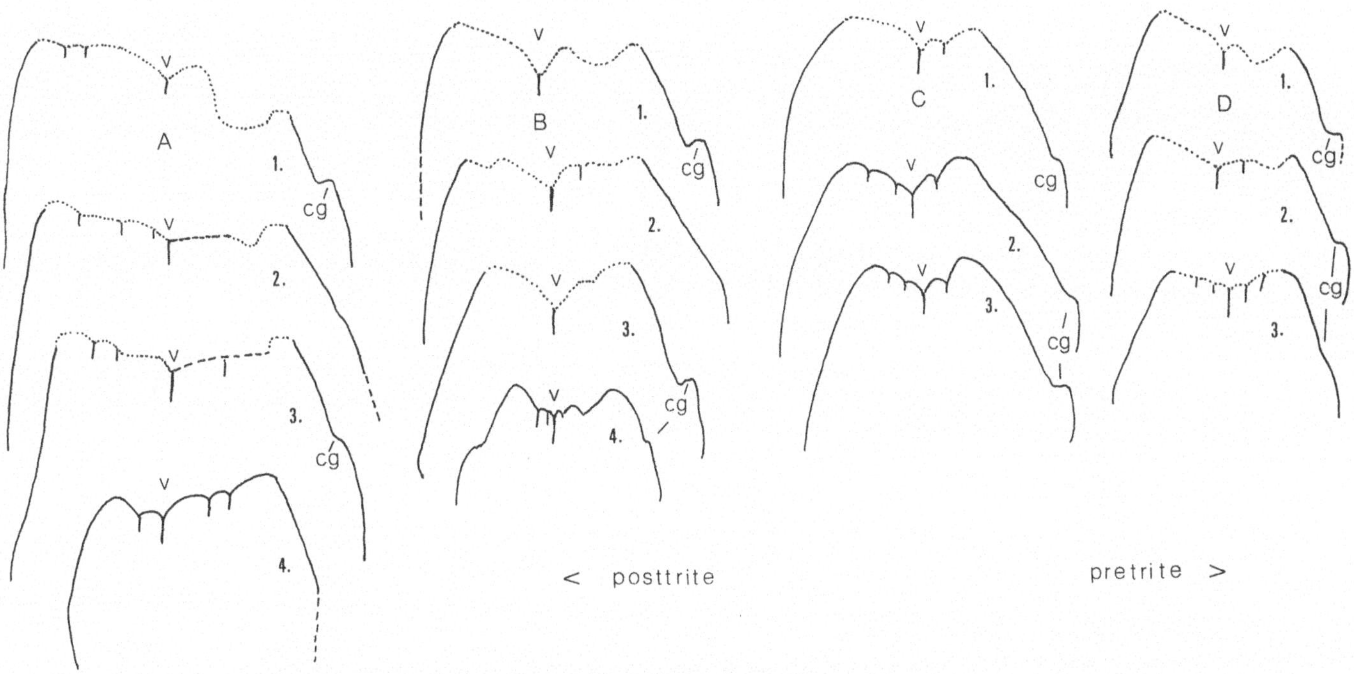

Fig. 9.9 Transverse loph profiles of upper right molars of *Zygolophodon turicensis* from the middle Miocene (MN5) of Pontlevoy-Thenay, France. The profiles cover the enamel cap of the lophs or at least its larger part: unbroken lines, unworn or slightly worn lophs; dotted lines, worn lophs; broken lines, damaged parts (for profile measuring see Tobien 1973*b*, p. 122); Key: A, M^3 left, reverted; B, left M^3, reverted; C, right M^3; D, right M^2; V, sulcus; cg, cingulum; 1., 2., 3., 4., first, second, third, fourth loph. (After Tobien 1975, p. 219, fig. 27.)

The lower molars of *Zygolophodon* are very similar to the upper ones; in the intermediate ones (dp_4 or P_4, M_1, M_2), there are three lophids. As in gomphotheriids (Tobien 1973*b*, p. 129), the lophid number in the M_3 is somewhat higher (four lophids plus a somewhat strong posterior cingulum; Figs 9.2–9.3). These molars possess a median sulcus which divides the transverse crests into half-lophids: a labial (pretrite) half-lophid, with strong crest-like crescentoids, and a lingual (posttrite) half-lophid provided with lingually situated 'zygodont crests'. The conelets of the lophids are arranged in a straight line. Visible only in unworn or slightly worn molars, the number of conelets is higher in the posttrites (3–4) than in the pretrites (2, seldom 3). Thus, the posttrites are slightly broader in anterior or posterior view, and the median sulcus is shifted labially (Tobien 1975, figs 9–16).

In unworn molars, the abaxial conelets (those away from the sulcus) are stronger than the adaxial ones (those close to the sulcus). These structures correspond in the first and second lophids to the main elements of the primary structure; protoconid and metaconid in the first lophid (=protolophid), hypoconid and entoconid in the second lophid (=metalophid). Subsequent lophids (tritolophid, tetralophid) are derivates of the cingulum. The lophids and their conelets (which are arranged in a straight line, not perpendicular to the main axis of the tooth, incompletely fused, and compressed antero-posteriorly) produce the yoke-like character of the lower molars.

In contrast to the upper molars, the lophids are obliquely positioned relative to the longitudinal axis of the crown, running from antero-lingually to postero-labially (Fig. 9.2). This oblique position of the lophids, which is the main feature in discriminating lower and upper molars in zygodons, is already developed in the early members of *Zygolophodon* of middle Miocene age. Although weakly evolved, compared with later *Zygolophodon* and *Mammut* species, the oblique position of the lophids can help to make the difficult distinction between molars of early *Zygolophodon* and *Gomphotherium*.

As in upper molars, the zygodont characters are fully developed only in the first and second lophids, whereas the subsequent lophids are incomplete. The third lophids in the M_1 and M_2, however, are more similar to the two anterior ones, than in M_3 (Figs 9.2–9.3).

The inclination of the labial and lingual walls of the transverse ridges is also well developed in the lower molars, as in gomphotheres, but in reverted sense compared with M^1–M^3, i.e. the lingual walls are vertical, the labial walls are inclined from dorsomedially to ventro-laterally. The third and fourth lophids in the M_3 usually do not follow this rule, because of their talonid-derived origin (Fig. 9.10).

In lower molars of *Zygolophodon*, the greater width compared with *Gomphotherium* molars has been demonstrated previously (Tobien 1971, fig. 10, p. 156; Tassy 1974, diagram 15, 16, pp. 53–4).

The fossil record of *Mammut* spans from the middle Miocene through the Pleistocene and it is believed to be a descendant of *Zygolophodon* (Tobien *et al.* 1988, p. 204). The typical morpho-

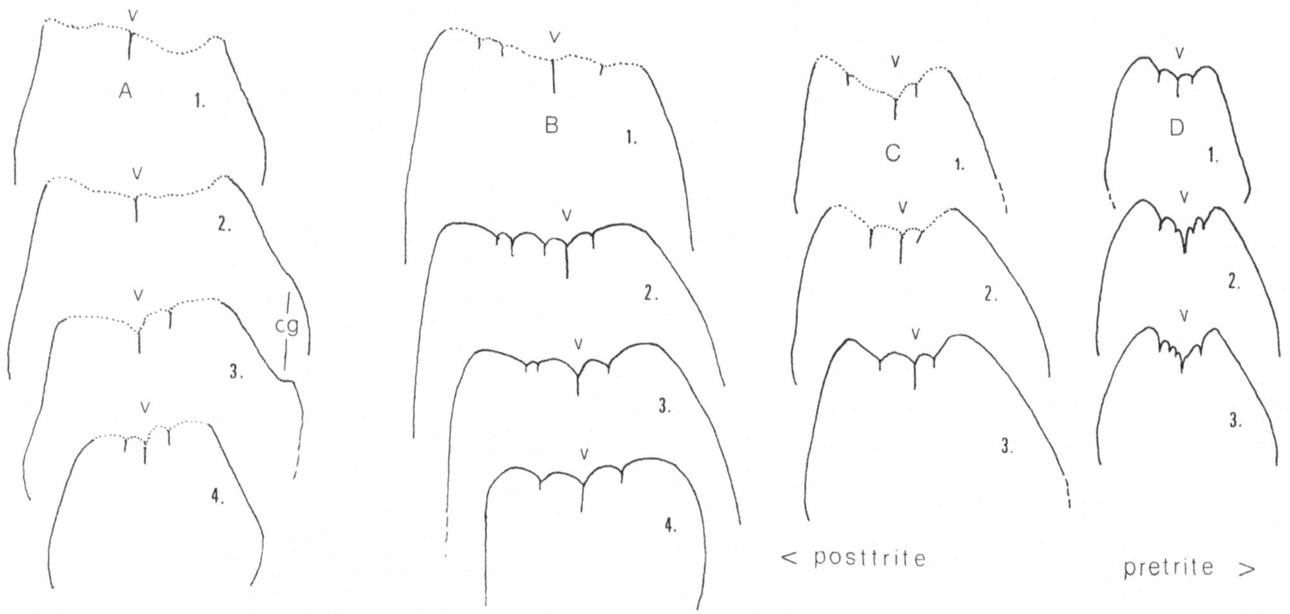

Fig. 9.10 Transverse lophid profiles of lower right molars of *Zygolophodon turicensis*. The profiles cover the enamel cap of the lophid, or at least its larger parts. Unbroken lines, unworn or slightly worn lophids; dotted lines, worn lophids; broken lines, damaged parts (for profile measuring see Tobien 1973*b*, p. 122). Key: A, right M_3, middle Miocene (MN5), Pontlevoy-Thenay, France; B, left M_3, reverted, upper Miocene ('Tortonian' = Badenian), Kalksburg, Niederosterreich; C, left M_2, reverted middle Miocene, Elgg, Kanton Zürich, Switzerland (type specimen); D, M_1 right, middle Miocene Pontlevoy-Thenay, France; V, sulcus; cg, cingulum; 1., 2., 3., 4., first, second, third, fourth lophid. (After Tobien 1975, p. 219, fig. 28.)

logical features of the zygodont molar are still more elaborated in the genus *Mammut*, compared with *Zygolophodon* and particularly with gomphotheriids.

Based on broad materials, Schlesinger (1917, p. 175; 1922, pp. 78, 82, 135) summarized the differences between the molars of *Zygolophodon* ('*Mastodon tapiroides*') and *Mammut*, represented by the Old World *M. borsoni* and the North American *M. americanum*, as follows:

(1) a slight increase of the number of transverse ridges in the M^3 (four lophs) and in the M_3 (four lophids plus talonid);

(2) the broadening of the molar crowns;

(3) the reduction of the crescentoids ('Sperrleisten' in Schlesinger 1917, 1922);

(4) the strengthening of the yoke character of the upper and lower molars, in the genus *Mammut*.

It should be emphasized that the increase of the number of transverse ridges in *Mammut* is valid only in a general manner, as *Zygolophodon* specimens are known to have a well-developed fourth loph in the M^3, and a fourth lophid plus cingulum in the M_3. Normally the fourth loph(id)s and the cingula in *Zygolophodon*, particularly in the stratigraphically earlier populations, are, in fact, in a weak 'rudimentary' condition (Tassy 1974, p. 12, figs 5–8; pl. 14, figs 1–6), whereas the M_3 in Tassy's pl. 13, figs 1, 3–5, which are of later age, are more advanced in the posterior parts of the crowns.

The broader molar crowns of *Mammut* are mainly conspicuous in comparison with contemporaneous molars of gomphotheres such as *Tetralophodon longirostris* and *Anancus arvernensis*. The variation field of the *Mammut* M_3 is more separated from the bunodont cluster, as in the M_3 distribution of *Zygolophodon* and *Gomphotherium angustidens* (Tobien 1975, fig. 6). The *M. borsoni* M_3 is not relatively broader than the *Z. turicensis* M_3, but is, on the average longer, therefore only seemingly broader. The North American *Miomastodon* and *Pliomastodon* M_3 are relatively narrow compared with those of the other zygodons (Fig. 9.11).

Comparing length and width of M_3 of *Zygolophodon* and *Mammut* (Old and New World specimens), the difference in relative width is obviously insignificant (Fig. 9.11). In the relation between length and width, there seems to be no evolutionary progress. The reinforcement of the yoke character is mainly caused by the reduction of the pretrite crescentoids which are transformed into veritable but rather weak enamel crests (Tobien 1975, figs 17–24). They contribute to the open character of the transverse valleys, and in many cases disappear after some wear, without leaving a trefoil pattern, as is known from many *Zygolophodon* molars.

Furthermore, a certain mesio-distal compression of the conelets, together with their more vague separation (compared with *Zygolophodon*), contributes to the pronounced crest-like habitus of the loph(id)s in *Mammut*. The oblique arrangement of the lophids in the lower molars is more remarkable in the genus *Mammut* than in *Zygolophodon* and is a rather distinctive character (Osborn 1936, figs 150–151). This is likewise valid for the American Mio-Pliocene genus '*Miomastodon*' Osborn, 1922 which is close to, if not identical with, the Old World *Zygolophodon* (Tobien 1964, p. 64; 1976, p. 161) and '*Pliomastodon*' Osborn, 1926*a* which is close to, if not identical with, *Mammut* (Osborn 1936, figs 98 A 1, B; 100 A 3; 102, 104).

The inclination of the lingual and labial walls in the upper molars and the reverted inclination in the lower ones is as in *Zygolophodon*. There is little progress in this character in the zygodont molar evolution, in contrast to the bunodont group where in several genera (*Choerolophodon, Synconolophus, Platybelodon, Amebelodon*), the inclination becomes subequal or equal, mainly in connection with subhypsodonty (Tobien 1973*b*, pp. 124–5).

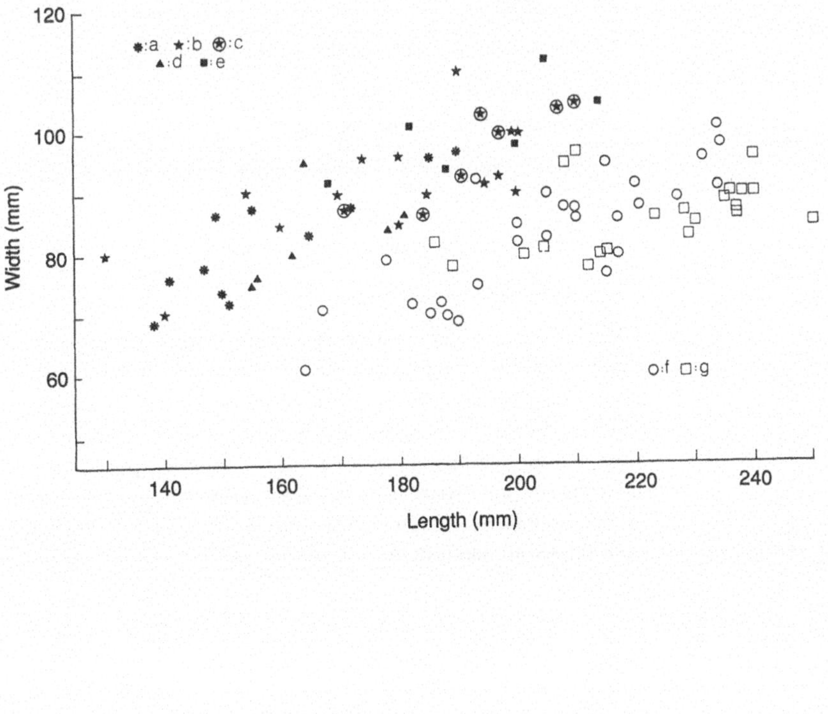

Fig. 9.11 Scatter diagram of late Neogene and Pleistocene zygodont and bunodont M_3. Key: a, Miocene *Zygolophodon turicensis*; b, Pliocene and early Pleistocene *Mammut borsoni*; c, Asiatic late Neogene zygodons, mainly *M. borsoni*; d, North American Mio/Pliocene *Miomastodon* and *Pliomastodon*; e, American and European Plio/Pleistocene *Mammut americanum*; f, European late Miocene *Tetralophodon longirostris*; g, European late Pliocene and early Pleistocene *Anancus arvernensis*. (After Tobien 1975, p. 218, fig. 26.)

The number of conelets in the transverse ridges of the upper and lower molars is substantially as in *Zygolophodon*. Figured unworn specimens of *M. borsoni* and *M. americanum* and my own observations indicate in the pretrites 2, seldom 3–4 conelets, and in the posttrites 3–4, seldom more (Figs 9.9–9.10 Tobien 1975, figs 17, 18, 20, and 22).

A peculiarity of some *Mammut* molars, not present or only vestigial in *Zygolophodon*, is a certain crenellation of the tips of the transverse ridges in unworn molars (crenellation is seen in some tooth germs of *Z. turicensis*). A striking example was depicted by von Meyer (1867, pl. 4, figs 1–5) as '*Mastodon virgatidens*' from the late Pliocene near Fulda, to the north-east of Frankfurt/ Main. The points of the conelets bear small pearl-like enamel knobs, producing a minute serration which is also developed on the crescentoids, zygodont crests, cingula, and which obviously will disappear immediately after slight wear. A similar delicate crenellation is developed on an unworn M_3 of *M. americanum* from Missouri (USA) depicted by Schlesinger (1922, pl. 22, fig. 4). Other unworn molars of *Mammut* show something of this property. Whether the crenellation is a species character or an individual variant in a population with non-crenellated molars can be decided only by a statistical evaluation of a larger sample, which is not presently at hand.

The cingula are, on the average, more developed and stronger in *Zygolophodon* on the pretrite sides of the molar crowns than in *Mammut*. Here, they are weaker or missing and found mainly in the pretrite outlets of the transverse valleys.

Occasionally, traces of cement occur on the bottom of the transverse valleys, mainly in upper and lower M3 which are rare in *Zygolophodon* (present in some *Z. turicensis* specimens from the Gers area (locality of Rajégats; Tassy 1985). This is known from the North American *Mammut* (Osborn 1936, fig. 122, p. 175; and my observations of the materials in the American Museum of Natural History, New York), as well as from the Old World *M. borsoni* (Vacek 1877, pl. 6, fig. 1).

The amount of cement is small (only a few millimetres thick), comparable to the similar incipient deposits in *Anancus*, *Tetralophodon*, and *Gomphotherium*, and quite in contrast to the thick cement covering in *Choerolophodon* and *Synconolophus* (Tobien 1973*b*, p. 130).

EVOLUTIONARY TENDENCIES IN THE ZYGODONT TOOTH PATTERN

The evolutionary changes in the long-ranging *Zygolophodon–Mammut* lineage (from later Burdigalian to the Pleistocene) are rather limited compared with those in the bunodont lineages. Conservative characters in the zygodont molar evolution are as follows.

1. The constant number of three loph(id)s in the upper and lower intermediate molars. In bunodons, an increase to four (*Tetralophodon*, *Anancus*), five (*Pentalophodon*), and more (*Stegotetrabelodon*) has been realized during evolution.
2. In the upper and lower M3, the ridge number does not exceed four; a complete fifth loph(id) has never been developed. In several bunodont groups, M3 are known with five and more ridges.
3. No tendency to hypsodonty ('subhypsodonty') has been observed in the *Mammut* molars, which would be comparable to the subhypsodont molars of, for instance, *Platybelodon grangeri*. In this context, it should be kept in mind that molars of large size are relatively high crowned, compared with smaller specimens.
4. In connection with the brachyodonty, the inclination of the side walls of the zygodont molar crowns (vertical on the posttrite sides, inclined on the pretrite) is still preserved in the latest *Mammut* populations in North America and in the Old World as well.
5. There is no significant increase in the number of conelets in the loph(id)s between *Zygolophodon* and *Mammut*.
6. There is no tendency to cement deposition (as in some bunodon lineages), or if any, only vestigial.
7. Envisioning the general tendency in zygodont molar evolution with the trend to produce sharp homogeneous molar crests, there is no trend to choerodonty which reinforces the isolation and multiplication of the conelets and conules (as in the bunodon *Choerolophodon–Synconolophus* group).
8. The formation of fully developed 'secondary trefoils' (i.e. enamel pillars at the adaxial sides of the posttrites), as in some advanced members of gomphotheres, has never been reported, particularly not from the genus *Mammut* (for comparisons, see Tobien 1973*b*, p. 137, fig. 21).

Some progressive characters, however, can be recognized in the evolution from *Zygolophodon* to *Mammut*:

(1) the reinforcement of the lophodont molar character
  (a) by a certain antero-posterior compression of the conelets in the half loph(id)s,
  (b) by an attenuation of the individualization of the conelets;

(2) the diminution and reduction of the crescentoids;

(3) the more distinctive obliqueness of the lophids in lower molars, particularly in the M_3;

(4) an occasional increase of crenellation of the crests in unworn molars;

(5) a certain reduction of the cingula; and

(6) vestigial cement deposition.

Obviously, stronger evolutionary changes from *Zygolophodon* to *Mammut* can be recognized, so far known in the incisor dentition (changing of curvature and loss of enamel band in the upper incisors, reduction and loss of the lower incisors with abbreviation of the symphysis) and in the locomotor apparatus (e.g. different carpal structure in *Zygolophodon* compared with *M. americanum* and probably also in *M. borsoni*), which cannot be discussed in detail here.

Also typical for *Zygolophodon* and *Mammut* is the contour of the mandible in lateral view (Fig. 9.7). The processus angularis mandibulae is well developed and prominent in the disto-ventral direction and the anterior border of the ramus ascendens is vertical (see, for example, Bakalov and Nikolov 1962, pl. 53). The lower borders of the ramus horizontalis and the symphysis are straight or only slightly deflected, and the incisors are horizontally implanted (Tobien 1988, p. 149, figs 39, 64a, b). This is in contrast to *Anancus* which had a weak processus angularis, the ramus ascendens was inclined, and lower border of the ramus horizontalis and the symphysis were deflected (Weithofer 1891, 1893).

SOME DATA ON THE EVOLUTIONARY HISTORY OF THE ZYGODONS

Fossil remains of zygodons are present in Africa, Asia, Europe, and North America, and are often accompanied by other bunodont species or groups. However, their repartition is somewhat restricted in certain parts.

ZYGODONS IN AFRICA

Elephantoids are obviously of African origin (e.g. in the Oligocene Djebel el Qatrani Formation of Egypt (Tassy 1982; Tobien 1988)); a rich representation can be assumed to have been there in later times. This holds true for some bunodont groups. But zygodon findings are rare. On the one hand, latest Oligocene/earliest Miocene local faunas are rare in Africa, but the rich early Miocene to latest Pliocene in East Africa (Kenya, Uganda) is poor in zygodons, except for *Eozygodon morotoensis* from the early Miocene of western Kenya. It is the smallest zygodon known (Tassy and Pickford 1983).

Despite the early Miocene age, the molar structure appears advanced. As often happens in small mammutid individuals or species, the degree of structural differentiation of the molars is low, more so as the species shows clear allometry to related, larger and later zygodons (Tassy and Pickford 1983, figs 11–13).

A fragment of a molar with zygodont habitus has been published from the early Miocene of the Lothidok-Hill in Kenya (Madden 1980*b*); it is identical with *Eozygodon morotoensis* (Tassy and Pickford 1983). No other zygodons are known from sub-Saharan Africa nor from the Pliocene of South Africa (Langebaan locality). The Pliocene of Algeria and Morocco has produced some remnants of zygodons.

ZYGODONS IN MIDDLE AND EASTERN ASIA

The genus *Zygolophodon* is likewise well represented in middle and eastern Asia. Whether the genus arrived as an immigrant from Europe or directly from Africa is an open question. It should be mentioned, however, that as of 1990, Asia Minor has not produced any remains of zygodont proboscideans (Gaziry 1976, 1980; Tobien 1990).

A local species is *Z. gobiensis*, originally described as '*Serridentinus gobiensis*' n. sp. (Osborn and Granger 1932; Osborn 1936). The species was later allocated to *Zygolophodon* (Chow and Chang 1961; Tobien *et al.* 1988). By the description and the figures, the lophids are still somewhat bunodont, and their obliquity not yet extreme. Furthermore, the straight lower border of the ramus and the pronounced processus angularis (Tobien *et al.* 1988) point to *Zygolophodon*. Compared with the European *Zygolophodon turicensis*, the Chinese species is differentiated by the smaller, sometimes absent adaxial conelets in the pretrites of the upper and lower molars.

This character has been changed in the dentition of *Mammut borsoni* from China, with its sometimes larger pretrite conelet (see p. 76). Nevertheless, a phylogenetic relationship between *Zygolophodon gobiensis* and *Mammut borsoni*, which indicates a local evolution of the populations in China, can be assumed (Tobien *et al.* 1984, table 2). In this table, an intermediate stage '*Zygolophodon* (advanced)' was inserted between *Mammut* and *Zygolophodon*. By the present information, the *Mammut* documentation is restricted to the Jinglean/sol Ruscinian and Baodean sol/Turolian.

The *Zygolophodon* occurrences are known only from Tunggurian/Astaracian sites. It might be possible that the M^3 and the two M^1 fragments of the Bahean Xiaolongtan coal mine represented an intermediate stage. The above specimens, however, need closer inspection, and we particularly need more material. A similar problem occurs with the M3 fragment from the Vallesian Ling Yanshan quarry, Kiangsu Province, determined as '?*Mammut borsoni*'. Tassy (1985, p. 511) compares *Zygolophodon gobiensis* with the European *Zygolophodon turicensis*. A certain separating character is the strong pretrite conelets in the latter.

When discussing *Zygolophodon gromovae* from Tung Gur (formerly Tunggur), Tassy (1985, p. 511) points to differences in the molar characters of the M_2–M_3 in the type mandible of *Zygolophodon gobiensis* and in the M^2+M^3 and M^3 (Dubrovo 1970, figs 1, 2) of the referred specimens of *Zygolophodon gromovae*. There are indeed differences in the habitus of the M^2+M^3 (Dubrovo 1970, fig. 1) and the M_2+M_3 of the type specimen. Be aware, however, that fully worn molars (the *Z. gromovae* M^2+M^3) are being compared with completely unworn M_2+M_3 in the *Z. gobiensis* type mandible. The former appear lightly built and simpler in structure (='*morphotype gracile*' of Tassy 1985, p. 513). The latter are more complicated and massive ('*morphotype robuste*' of Tassy 1985, p. 511; see footnote 1 in Tassy, this volume, Chapter 6).

The functional effect during attribution and occlusion of the upper and lower molar rows has a shearing or cutting component in zygodont dentitions. Tobien (1973*b*, p. 134) states: 'a considerable part of wear affects their anterior and posterior slopes, caused by partial interlocking of the upper and lower lophs and lophids during the process of mastication (see also Lehmann, 1944: 392, 397)'.

Under these circumstances, it becomes obvious that the '*morphotype gracile*' of the worn *Z. gromovae* M^2 and M^3 compared with the only slightly abraded, isolated *Z. gromovae* M^3 (Dubrovo

1970, fig. 2) and with the unworn *Z. gobiensis* M_3 of '*morphotype robuste*' character (crown more massive, lamellae thicker, and pretrites larger (Tassy 1985, p. 488)), is mainly a result of advanced trituration of the former. They are thus nearer to, if not identical with, the M_3 in the type mandible of *Zygolophodon gobiensis*. Apparently, there are different degrees of wear, which in this particular case accounted for the morphological differences. Thus, *Zygolophodon gromovae* falls into the synonymy of *Zygolophodon gobiensis* (see note Z1 in Appendix C).

Outside the People's Republic of China and further to the west, a fragmentary M_3 left from the middle Miocene of Ulan Tologoj, Western Mongolian People's Republic has lost most of the outer pretrite half of the crown and the whole first posttrite (Dubrovo 1974, p. 68, fig. 3). The length of the specimen, determined as *Zygolophodon gromovae*, is 197 mm, equal to the type M_3 of *Zygolophodon gobiensis*, as are the preserved structural elements of the crown. The lateral view of the pretrite broken side (Dubrovo 1974, fig. 3a) shows only the adaxial section; the usually higher main cones of the abaxial section are broken away. Therefore, a comparison with the crown heights of other zygodons is not feasible (Tassy 1985, p. 513). We refer the specimen to *Zygolophodon gobiensis* from Tung Gur which has (in the old stratigraphy) about the same stratigraphic age.

Moving westward, the next specimen that should be mentioned is '*Mastodon*' *atavus* (Borissiak, 1936) from the early Miocene Djilancik beds of Kazakhstan. This animal is a zygodon (Tobien 1973*a*, p. 213, footnote) and Tassy (1985, pp. 480, 503, 511) allocated it correctly to *Zygolophodon* '*turicensis* group'. However, a closer inspection of the molars is necessary to determine whether this specimen is the *Z. gobiensis* or *Z. turicensis* type. The symphysis is longer and its incisors are stronger than those of *Zygolophodon gobiensis* (Tassy 1985, fig. 210). This may be in context with the somewhat earlier age (Shanwangian/Orleanian) of the Djilancik site. Thus, the shorter symphysis and smaller incisors of *Zygolophodon gobiensis* are a step towards the reduced symphysis and incisors of *Mammut*. The skeleton from the Djilancik beds is the most complete of *Zygolophodon* specimens discovered in the whole Old World dispersal area of the genus.

Gabunia (1981, p. 196) mentions *Zygolophodon* sp. from the Sakaraulian and Tarkhanian stages (corresponding to the Orleanian; Steininger *et al.* 1985, fig. 26.1) of the Aralo-Caspian region. Specimens of *Zygolophodon turicensis* are being discovered in eastern Europe and further to the central and western parts of the continent.

As in Europe, *Zygolophodon* in central and eastern Asia probably evolved into a specific *Mammut* species, in this case, *Mammut borsoni*. Although the stratigraphic position of many Chinese findings is poor or arbitrary, it seems that the earlier levels of the Shanwangian (quasi-Orleanian), Tunggurian (quasi-Astaracian), and Bahean (quasi-Vallesian) have produced *Zygolophodon gobiensis* or related taxa, whereas in the following Baodean (quasi-Turolian) and Jinglian (quasi-Ruscinian), *Mammut borsoni* prevails (Tobien *et al.* 1988, table 4).

It should be emphasized that the sample of *Mammut* teeth from China (58 d (deciduous milk tooth), I, M), whether isolated or in a palate, in two mandibles and two ramus fragments, is the largest now known in the Palearctic. The specimens from the Saone-Rhône area (Lortet and Chantre, 1878) come closest, with about 30 dental elements, isolated or in two mandible fragments.

ZYGODONS IN SOUTHERN ASIA

Although the proboscideans in southern Asia, particularly in the Indian subcontinent, are rather diversified (Tassy 1983), zygodont taxa are rare: one, *Zygolophodon metachinjiensis* (Osborn 1929*a*, p. 4), is known from the Chinji Formation, India, and some specimens have been mentioned by Sarwar (1974, p. 246; 1977). This scarcity is similar to that in sub-Saharan Africa.

The absence of populations of *Mammut borsoni* in southern Asia may be due to the fact that this species was adapted to more moderate or cooler temperatures during the later Plio-Pleistocene. Two occurrences are known from higher latitudes: *M. borsoni* is known from the western slopes of the Ural Mountains at 52°E — east of the city of Orenburg (near the Ural River) — and 55°N — near the city of Ufa (Belyaeva 1962, p. 63). The latter occurrence was determined erroneously by Eichwald (1835) as '*Deinotherium uralense*' (see Osborn 1936, pp. 85, 87, 111). Furthermore, a '*Mastodon zaddachi*' (=*Mammut borsoni*) was described by Jentzsch (1883) from Torun (formerly Thorn, western Prussia), south of Gdansk at 53° 30′N, 18° 30′E. Currently, these are the northernmost known sites of *Mammut borsoni*.

ZYGODONS IN EUROPE

The situation in Europe is quite different: the earliest members of the genus *Zygolophodon* appear during the MN4 zone (Mein 1990), which means early Aragonian/Ramblian, together with *Gomphotherium*. They were probably immigrants from Africa. The typical species is *Zygolophodon turicensis* (Schinz, 1824) from the middle Miocene lignite of Elgg near Zurich (the genus *Turicius*, proposed by Osborn (1926*a*), falls into the synonymy of *Zygolophodon*). The external habitus of the animal is characterized by the relatively small, downcurving upper incisors, provided with an enamel band. The lower incisors are well developed and protruding (see p. 78).

The wide horizontal extension of *Zygolophodon* in central, western, and eastern Europe is emphasized by the considerable horizontal biostratigraphic occurrence from MN3b/MN4 (Ginsburg 1990, Tassy 1990*b*) through the Astaracian into the Vallesian with MN9 (Eppelsheim (Mainz Basin); Tobien 1980*a*) and the typical so-called 'Hipparion-faunas' of MN12–13 age.

Obviously, the typical *Mammut* appears during the Ruscinian (e.g. Wolfersheim: MN15), represented by the species *M. borsoni* (Hays, 1834), with the features listed above. Often accompanied by the bunodont *Anancus arvernensis* (Croizet and Jobert, 1828), *M. borsoni* is a common element in the European macrofauna, including that of the northern Mediterranean area (Spain, Italy, the Balkan peninsula, but not Asia Minor), existing into the early Pleistocene in some places.

A structurally intermediate skull fragment with long, straight, enamelless upper incisors, with a mandible having very strong, elongated lower incisors in a large and not deflected symphysis, continuing in the straight lower border of the ramus of the mandible, was described by Kubiak (1972) as *Mammut praetypicum*. Unfortunately, not much is known about the locality and age. Kubiak assumes a site near Odessa in southern Russia.

ZYGODONS IN THE AMERICAS

The first mammutids appeared in North America at a later date than in Europe, i.e. around the Orleanian/Astaracian turnover (15 Ma), that is, in the Barstovian (Woodburne 1987; Tassy 1990*b*). Among them must have been marginal populations of *Zygolophodon*, which migrated at that time via Beringia into North America. Palaeobotanical evidence for the middle Miocene shows temperate to warm– temperate zones in Beringia (Biske and Baranova 1986). The earliest North American zygodon is the genus '*Miomastodon*' (Osborn, 1922), closely related, if not identical, with the Palearctic *Zygolophodon*. More complete specimens have been discovered, mainly in north-west Oregon, but also in Nevada and Colorado.

During the Pliocene, *Miomastodon* evolved to the genus *Pliomastodon* (Osborn 1926*a–b*), obviously an intermediate form and the forerunner of the late Pliocene, Pleistocene and Holocene *Mammut americanum* (Kerr, 1792), the famous 'American mastodon' (Tobien 1964). Populations of *M. americanum* were very common in the Pleistocene and post-Pleistocene of North America. The American mastodons reached as far south as Mexico (Central America) but their remains have not been found in South America (Shoshani 1990*b*), possibly due to climatic reasons (Tobien 1976). The populations died out at the end of the Pleistocene, close to the Pleistocene–Holocene transition (details in Saunders, this volume, Chapter 27).

CONCLUDING REMARKS

Among proboscidean taxa, those with zygodont molars represent a well-defined group with a basic pattern of a 'zyg'-like tooth (i.e. a tooth with a yoke-like shape of its ridges in anterior or posterior views). The three recognized genera are: *Eozygodon* (early Miocene of Africa), *Zygolophodon* (middle Miocene of Eurasia and North America), and *Mammut* (late Miocene through the Pleistocene in Eurasia and North America). On the whole, this group was conservative throughout its 20 million years of evolution — that is, small dental changes can be observed between specimens from the early Miocene and those of the late Pleistocene. *Eozygodon*, *Zygolophodon*, and *Mammut* have been classified in the family Mammutidae (details in Tassy, this volume, Chapter 6; Shoshani and Tassy, this volume, Chapter 34).

ACKNOWLEDGEMENTS

The author is deeply indebted to J. Shoshani, P. Tassy, and E. C. Marsac for their most helpful criticism, advice, and improvements. This paper was typed by E. C. Marsac and S. Wolak and reviewed by the editors.

PART IV

The second radiation: gomphotheres and stegodontids

TEN

The earliest gomphotheres

PASCAL TASSY

INTRODUCTION

The name 'gomphotheres' is used here in a restrictive sense; it is not entirely synonymous with the 'bunodont mastodonts' (*sensu* Tobien 1973*a*) or the family name Gomphotheriidae Hay, 1922, based on the genus *Gomphotherium* Burmeister, 1837. Rather, the term gomphotheres, is used here to refer to the stem-group of both stegodontids and elephants, and thus designates a paraphyletic group (Tassy 1988*a*). Taxa included in the gomphotheres share bunodont teeth, a cranium with a transversely enlarged nasal aperture (however, see footnote 2, p. 45), and lower tusks with a piriform to oval, or even round cross-section. Hence, the gomphotheres, as defined here, do not include mammutids (zygodont teeth), nor the genus *Choerolophodon* Schlesinger, 1917 (with bunodont teeth but a narrow nasal aperture), and the amebelodontids (with bunodont teeth and a large nasal aperture but lower tusks with flat cross-section). Mammutids, choerolophodons, and amebelodons are considered as earlier offshoots of the Elephantoidea.

The earliest known gomphotheres belong to the most primitive species yet described. They can be grouped in the *Gomphotherium* '*annectens* group', which is considered as the root of all other gomphotheres.

Abbreviations

Ar, label of specimens from Artenay (Loire Basin) housed in the Museum national d'Histoire naturelle, Paris; BMNH, British Museum Natural History (now Natural History Museum); KNM, Kenya National Museum (now National Museum of Kenya); MI, label of specimens from Mwiti (Kenya), housed in the National Museum of Kenya, Nairobi; MNHN, Museum national d'Histoire naturelle, Paris.

STRATIGRAPHICAL SETTING

The fossil record demonstrates that the first representatives of the group are already cosmopolitan. The earliest gomphotheres are known from the early Miocene of Africa and Eurasia.

AFRICA

The gomphotheres are not the earliest known elephantoids from Africa. The mammutids with *Eozygodon morotoensis* at Meswa Bridge, *c.* 22 Ma (Tassy and Pickford 1983) precede all other elephantoid taxa. The earliest known African gomphotheres (*Gomphotherium* sp.) occur in East Africa at Mfwangano and Mwiti (east Turkana, Kenya), both early Miocene localities. Nevertheless, these remains are very rare: two partial lower tusks and two partial molars at Mfwangano, and one palate at Mwiti (Tassy 1986). According to Pickford (1981), two faunas are present at Mfwangano, corresponding to 'Set I' and 'Set II' of the East African biostratigraphical setting, between 20 and 17 Ma. Mwiti is probably closer to 17 Ma, which brings this locality close to the more recent early Miocene localities from northern Africa (such as Moghara in Egypt and Gebel Zelten in Libya).

EUROPE, NORTHERN ASIA

In Europe, the earliest proboscideans (Orleanian, MN4 zone, estimated between 18 and 17 Ma) are used to define the 'Proboscidean Datum Event' (Madden and Van Couvering 1976; Tassy 1990*b*). They consist of gomphotheres (*Gomphotherium*) and mammutids (*Zygolophodon*). In the locality of Artenay (Loire Basin, France) both taxa are associated (Ginsburg and Antunes 1966; Tassy 1977*a,c*), whereas only *Gomphotherium* is present in contemporaneous layers of Portugal (Fauna R2 of Lisbon) (Antunes 1990; Mazo, this volume, Chapter 14). The gomphothere formerly described as *G. angustidens* was distinguished as a more primitive species, *G. sylvaticum*, by Tassy (1985). This species is well documented with dental and postcranial remains.

The Burdigalian fauna of the Hiramaki Formation (Mino, Japan) is the most eastern record and shows the association of a gomphothere (the primitive species *Gomphotherium annectens*), with *Anchitherium*, and *Brachyodus* (Takai 1954); and association known from the MN4 zone (*sensu* Mein 1990) in western Europe (Ginsburg 1990). *G. annectens* is known from dental and skull remains (Matsumoto 1926; Makiyama 1938; Kamei *et al.* 1977). The oldest formations bearing elephantoid remains in China belong to the Lower Shanwangian, correlated with European MN4-zone faunas, where *Gomphotherium* is represented only by a tusk fragment at Jiaozigou, Linxia Basin (Qiu 1990). There are no radiometric data for the earliest proboscideans from these areas.

SOUTHERN ASIA

The earliest gomphotheres from southern Asia come from the Bugti Hills (Baluchistan, Pakistan). They precede the earliest radiometrically dated gomphotheres (genus and species undetermined) from the Kamlial Formation, which are 18.3 Ma (Barry *et al.* 1985; Johnson *et al.* 1985; Jacobs *et al.* 1990). The gomphothere from the Bugti Hills is *Gomphotherium cooperi*, known from a palate, a portion of a mandible, and isolated cheek teeth (Cooper 1922; Osborn 1932*b*).

THE LEVANT, ARABIAN PENINSULA

Northern localities in Yeroham and Rotem basins (Negev, Israel) and southern localities of Ad Dabtiyah and As Sarrar (Saudi Arabia) yielded early Miocene faunas which have been compared to mid-Orleanian faunas (Thomas *et al.* 1982; Thomas 1985; Tchernov *et al.* 1987; Whybrow 1987), probably more recent than the earlier known elephantoid faunas from Africa, Europe, and southern Asia. The remains of gomphotheres are isolated teeth. The species *G. cooperi* first described from Bugti, is cited at Ad Dabtiyah (Gentry 1987), though the age of the locality was thought to be more recent. The primitive bunodont pattern of molars from Ad Dabtiyah fits well with that of the *Gomphotherium 'annectens* group'.

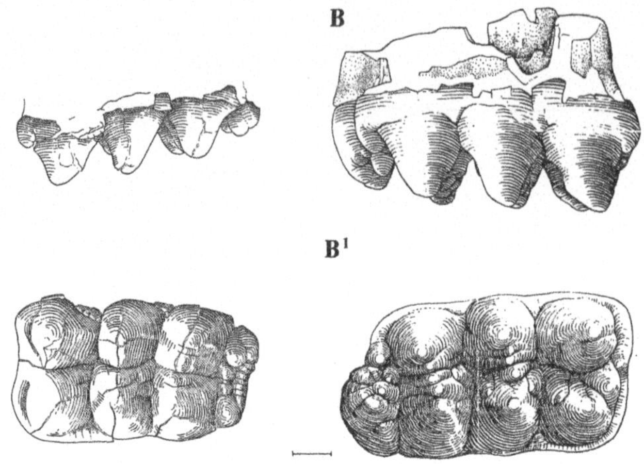

Fig. 10.1 Molars of *Gomphotherium 'annectens* group'. (A) left M^3 KNM MI1 of *Gomphotherium* sp. from Mwiti, Kenya (early Miocene); (B) right M^3 MNHN Ar 1489 of *Gomphotherium sylvaticum*, from Artenay, France (early Miocene). (A, from Tassy 1986; B, from Tassy 1977*b*.) Scale bar 2 cm.

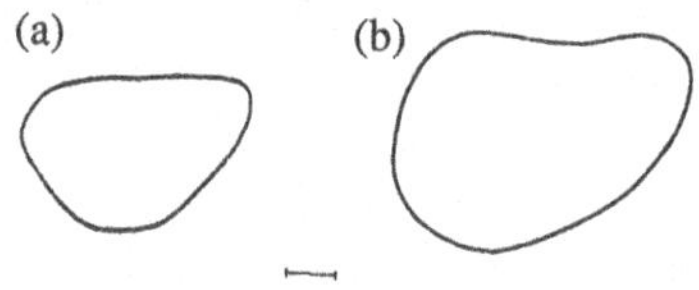

Fig. 10.2 Cross-section of right lower tusks (posterior view) of *Gomphotherium 'annectens* group'. (a) *Gomphotherium sylvaticum*, MNHN Ar 1546 from Artenay, France (early Miocene); (b) *Gomphotherium* sp. BMNH M32949 (reversed), from Mfwangano, Kenya (early Miocene). Scale bar 1 cm.

TAXONOMY

Species included in the *Gomphotherium 'annectens* group' are *G. annectens* (Matsumoto, 1925), *G. cooperi* (Osborn, 1932*b*), *G. sylvaticum* Tassy, 1985, and *Gomphotherium* sp. from East Africa. This assemblage is a 'grade group' diagnosed only by plesiomorphic traits. The cheek teeth are the only comparable features in every species. They share simple bunolophodont molars belonging to the simple pattern of the trilophodont grade. The pretrite central conules of upper molars are not inflated. Posttrite central conules are absent. The interloph(id)s are rather wide. The fourth loph of M^3 is not complete, but still fused with the posterior cingulum (Fig. 10.1). The fourth lophid of M_3 is small, even fused with the posterior cingulum in *Gomphotherium annectens*. The upper molars are large. The crown of P^4 is square in occlusal view. All these traits are plesiomorphic compared to other gomphotheres.

The lower tusks (preserved in *G. annectens, G. sylvaticum*, and *Gomphotherium* sp. from Mfwangano) have an oval or piriform cross-section without a ventral longitudinal sulcus (Fig. 10.2). The upper tusks (known in *Gomphotherium sylvaticum*) are not twisted; the enamel band is entirely situated on the lateral side and does not trend ventrally toward the tip. Unfortunately no tusk is known from *G. cooperi* from the Bugti Beds. An isolated flat lower tusk (primitive state) from Ad Dabtiyah, allocated to *G. cooperi* (Gentry 1987), would render this species a proto-gomphothere rather than a gomphothere, or could well belong to an amebelodontid.

The mandible of *G. annectens*, though not truly brevirostrine, has a relatively short symphysis compared to the alveolar distance (Makiyama 1938; Kamei *et al.* 1977). These proportions are the same as in the Oligocene species *Phiomia serridens* and the symphysis is not ventrally inclined, another trait also seen in *P. serridens*. Mandibulae of other species of *Gomphotherium 'annectens* group' are not known. Postcranial remains have only been described in *G. sylvaticum* (Tassy 1977*c*).

The relationships between primitive gomphotheres are still unclear since these species belong to a 'grade group'. Nevertheless, in Europe this grade clearly antedates the occurrence of *Gomphotherium angustidens* (Cuvier 1817) (MN5–MN9 zones, i.e. late early Miocene through late Miocene).

It is very plausible that all these taxa belong to one widespread primitive species. Yet, because of the bad preservation, many key features cannot be observed in every taxon. The evolutionary stage of the Bugti species can only be compared to other species on the basis of cheek teeth. Moreover, the allocation of *G. cooperi* to the gomphotheres is based only on the evolutionary level of the molars (molars of early Miocene choerolophodons and amebelodontids are already more derived, have more and bigger conules, and narrower upper molars). The proportions of the mandible are known only in the species from Japan.

As noted earlier, the '*annectens* group' includes primitive rare species but they are cosmopolitan (Fig. 10.3). The early gompho-

Fig. 10.3 Cosmopolitan distribution of the earliest and most primitive gomphotheres, *Gomphotherium 'annectens* group' (early Miocene). Localities are : 1, Bugti, Baluchistan, Pakistan; 2, Mfwangano, Kenya; 3, Mwiti, Kenya; 4, Hiramaki Formation, Japan; 5, Artenay (MN4 zone), France; 6, Quinta das Pedreiras (MN4 zone), Portugal; 7, Ad Dabtiyah, Saudi Arabia.

theres from the Bugti beds and from East Africa are poorly known. Because of the meagre sample size we have no idea of the variability of the characters used to separate these species from more evolved gomphotheres. Yet, it is known that in populations of the type species *G. angustidens*, intraspecific variability does not include the primitive traits shared by early Miocene gomphotheres (Tassy 1985). As a working hypothesis, we can use the *Gomphotherium 'annectens* group' as a group close to the ancestral morphotype of gomphotheres in general.

SUMMARY AND CONCLUSIONS

Though much is still to be learned in the future of the characters displayed by the earliest gomphotheres, the *Gomphotherium 'annectens* group' is the key to understanding the differentiation not only of the gomphotheres, but also of the various elephantoid lineages. Simple bunodont teeth are archetypal for any early Miocene elephantoid lineage. The association of simple bunodont molars (primitive) with lower tusks with piriform cross-section (derived) remains the best criterion to identify a primitive gomphothere. Flat lower tusks (primitive), inherited from the Oligocene genus *Phiomia*, are archetypal for early Miocene elephantoid lineages. If proto-gomphotheres are expected to be found, they will be indistinguishable from proto-amebelodontids, on the basis of characters presently used. How to recognize the stem of the stem of a stem-group? That is the question as it stands.

ACKNOWLEDGEMENTS

I thank N. Court, E. C. Marsac, J. L. Pierce, and J. Shoshani for their comments. Figures were drawn by D. Visset.

ELEVEN

Growth and sexual dimorphism among Miocene elephantoids: the example of *Gomphotherium angustidens*

PASCAL TASSY

INTRODUCTION

It has long been known that elephants are ever-growing mammals. The lifespan of the African elephant, *Loxodonta africana*, is estimated at about 60 years and growth lasts until the last molar comes into wear, i.e. 40–50 years, although the rate of growth is slowed after 40 years in males and around 30 years in females (Laws *et al.* 1975; Hanks 1979). Daubenton, in 1764, first understood that the molars of the elephant succeeded during life without vertical replacement. Identification of grinding teeth and their state of wear was then used to give an individual age to living elephants. Current dental ages are those proposed by Laws (1966) and modified by Sikes (1971), Hanks (1972), and Beden (1979) for *Loxodonta africana*; and by Roth and Shoshani (1988) for *Elephas maximus*. Such an approach was rarely applied to fossil proboscideans, due to scarcity of material. One notable exception is found in Saunders (1977) who defined five age-groups for the American mastodon, *Mammut americanum*.

The expressions 'horizontal replacement' and 'horizontal succession' have been used extensively to describe dental development in elephants. This can be misleading. I will use here the expression 'horizontal displacement'. In all known extant and extinct Elephantoidea, the eruption of teeth is comparable to that of mammals in general. The main difference is the resorption of anterior teeth (decidual premolars and premolars) before the eruption of M2 and M3. Then, newly erupted posterior teeth take the place of anterior teeth in the tooth row: horizontal displacement is due to remodelling and drift. As pointed out by Lartet (1859), premolars replaced decidual teeth in most Miocene elephantoid species. Yet, they are resorbed before M3 fully erupts, so that during life, the toothrow is never complete. In extant elephants, decidual premolars are not replaced, so that the functional succession of the six grinding teeth, dp2 to M3, is apparently 'truly horizontal'.

An attempt to estimate the growth pattern of the fossil species *Gomphotherium angustidens* (Cuvier, 1817) is presented here, made possible by the discovery of many specimens during field work (1974–84) in the Simorre area of south-western France. This broad area actually yielded Cuvier's type specimen of the species (a lower M2). In the middle Miocene (= Astaracian) locality of En Péjouan, 10 crania and 3 palates, and 18 mandibles and hemimandibles belonging to one population of *Gomphotherium angustidens* were excavated (Table 11.1).

A comparable study was made for the amebelodontid species *Archaeobelodon filholi* (Frick, 1933), based mainly on mandibles from the nearby middle Miocene locality of Sansan, housed in the Muséum national d'Histoire naturelle, Paris (see detailed measurements in Tassy 1985).

The two species share the same evolutionary stage in the dentition (trilophodont grade with replacement of dp^3/dp_3 and dp^4/dp_4 by P^3/P_3 and P^4/P_4). The same dental age scale defined in this paper (see also Tassy 1985) can be applied to both species, and to any species sharing the same grade, that is trilophodont grade. Similar growth pattern and aspects of sexual dimorphism in the shape and proportions of skull and mandible were found for both *G. angustidens* and *A. filholi*.

The dental age scale proposed here is a tool for understanding the pattern of growth and is necessary for useful metric comparisons between different species of the same grade.

Abbreviations

MNHN, Muséum National d'Histoire Naturelle, Paris; SEP, label of specimens from the locality of En Péjouan housed in the Muséum National d'Histoire Naturelle, Paris.

THE DENTAL AGE SCALE

The scale is based on (1) identification of functional grinding teeth; (2) the number of loph(id)s in wear, and their stage of wear; (3) the opening and resorption of the alveoli.

Table 11.2 summarizes the criteria for the recognition of 23 dental ages (I–XXIII), from the eruption of dp_2–dp_3 to the complete detrition of M_3. This scale is first based on the study of the mandibles, but is applied easily to the upper tooth row, even if some differences can be seen in the relative wear of corresponding teeth. The main differential wear is simply due to the occlusion, i.e. the anterior lophid of one lower molar occludes with the space between the anterior loph of the corresponding upper molar and the posterior loph of the preceding upper tooth. Hence, because of the 'horizontal displacement' of grinding teeth,

Table 11.1. Cranial and mandibular specimens of *Gomphotherium angustidens* from En Péjouan, Gers (middle Miocene, MN7 zone) according to their dental ages, as defined by Tassy (1985) and in Table 11.2

Dental ages	Crania	Palates	Maxilla right	Maxilla left	Mandibles	Dentary right	Dentary left
I	none	none	none	none	none	none	SEP 211
II	none	none	none	none	none	none	none
III	none	none	none	none	none	none	none
IV	none	none	none	none	none	none	SEP 210, SEP 253
V	none	none	none	none	none	none	none
VI	none	none	none	none	none	none	none
VII	none	none	none	none	none	none	none
VIII	none	none	none	none	none	none	none
IX	none	none	none	none	none	none	SEP 230
X	none	none	none	none	none	none	none
XI	none	none	none	none	none	none	none
XII	none	SEP 218	none	none	SEP 218	none	none
XIII	none	none	none	none	none	SEP 240	SEP 239
XIV	SEP 30	none	none	none	none	none	none
XV	SEP 18, SEP 254	none	none	none	none	SEP 280	none
XVI	none	none	none	none	SEP 249		SEP 261
XVII	SEP 186	SEP 238	none	none	none	SEP 303	none
XVIII	none	none	none	SEP 301	SEP 301	none	none
XIX	SEP 25	none	none	none	none	none	none
XX	SEP 38, SEP 251	SEP 250	none	none	SEP 187	none	SEP 216, SEP 265
XXI	SEP 185	SEP 260	none	none	SEP 302	SEP 256	none
XXII	none	none	none	none	none	none	SEP 262
XXIII	none	none	none	none	none	none	none

Table11.2. Definition of the 23 dental ages (I - XXIII) for *G. angustidens* based on lower dentition.
Key: cgs=posterior cingular cusps; L1, L2, L3, L4 = first, second, third, and fourth lophids; A=opening of alveolus; B=eruption of a tooth; C=tooth in function; C1=wear facets visible; C2=wear figures present (dentin is visible); C3=wear figures coalescent (excavated tooth); D=resorption of alveolus.

Dental	dp2	dp3			dp4			P3	P4			M1					M2					M3				Dental
ages				L1	L2	L3	cgp				L1	L2	L3	cgp		L1	L2	L3	cgp		L1	L2	L3	L4	cgp	ages
I	B	B	A																							
II	C1	C1	A																							
III	C2	C2	B																							
IV	D	C2	B (C)	C1																						
V		C2 (C3)	C	C2	(C1)																					
VI		C3	C	C2	C2																					
VII		C3 (D)	C	C2	C2	C1		(B)		A																
VIII			C	C2	C2	C2		B		B																
IX			C	C2	C2	C2	C2	C1		C	C1															
X			C	C3	C3	C3	C3	C1 (C2)		C	C2															
XI								C2	B	C	C2	C2														
XII								C2 (C3)	C1 (C2)	C	C2	C2	C1 (C2)		A											XII
XIII								C3 (D)	C2	C	C2	C2	C2	C1	B											XIII
XIV									C2	C	C2	C2	C2	C2	C	C1 (C2)										XIV
XV									C3 (D)	C	C3	C3	C3	C3	C	C2	C2	C1								XV
XVI										C3 (D)					C	C2	C2	C2		A						XVI
XVII										C3 (D)					C	C2	C2	C2	C1	B	(C1)					XVII
XVIII										D					C	C2	C2	C2	C2	C	C1	(C1)				XVIII
XIX															C	C2	C2	C2	C2	C	C2	C1 (C2)				XIX
XX															C3 (D)	C3	C3	C3	C3	C	C2	C2	C2	C1		XX
XXI															D (C3)	(C3)	(C3)	(C3)	(C3)	C	C2	C2	C2	C2		XXI
XXII															D					C	C2	C2	C2	C2	C1 (C2)	XXII
XXIII																				C	C3	C3	C3	C3	C3	XXIII

in two corresponding upper and lower molars the anterior lophid is usually more worn than the anterior loph. Also, the lower molars are often lost before the corresponding uppers. In taking into account the different molars in function, it is easy to minimize these differences and apply the scale to crania and palates. Crania and mandibles were measured as shown in Figs 11.1 and 11.2.

GROWTH PATTERN AND SEXUAL DIMORPHISM

To estimate the growth pattern of *Gomphotherium angustidens*, measurements were plotted according to the dental age of the specimens measured (i.e. Fig. 11.3). The methodology is derived from that defined by Beden (1979) for *Loxodonta africana*, but it must be kept in mind that the dental ages defined here characterize only elephantoid species which belong to the same evolutionary grade (trilophodont elephantoids with eruption of P3 and P4). In the diagrams, each dental age has the same 'length'. This methodological artefact does not alter comparisons but prevents calculation of the resulting 'curve'.

Important variations in size are first due to continuous growth up until the wear of M_3, i.e. the lengthening of the horizontal ramus (Fig. 11.3). Secondly, strong differences observed on specimens which belong to the same dental age can be attributed to sexual dimorphism. The growth pattern and secondary sexual dimorphism displayed by the fossil species *Gomphotherium angustidens* (Fig. 11.4) can be closely compared to that of *Loxodonta africana africana*, as studied by Beden (1979), although the shape of the cranium and mandible of the longirostrine gomphothere and the brevirostrine elephant are somehow different. This close similarity in development proves the antiquity

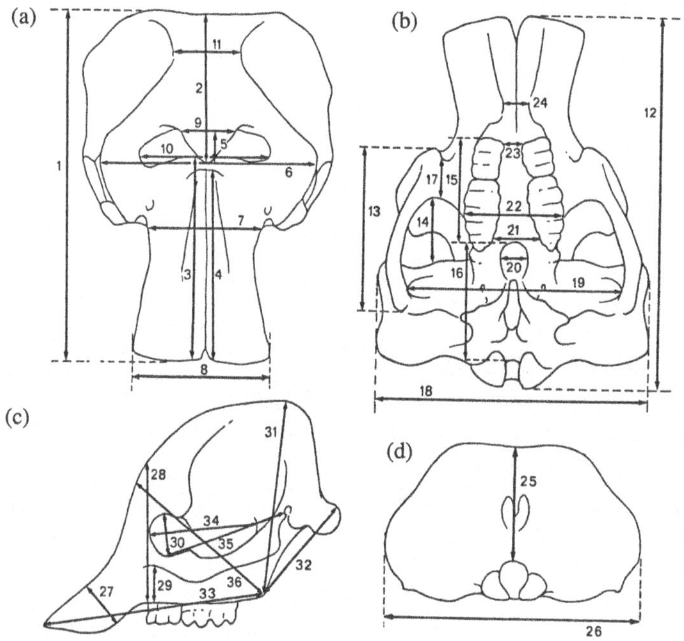

Fig. 11.1 Cranial measurements (C) taken on the elephantoid cranium. Views are: (a) antero-dorsal; (b) ventral; (c) lateral; (d) posterior. A composite cranium is shown here, mainly based on *Paratetralophodon hasnotensis* (taken from Tassy 1983). Key: 1, maximal length taken from the occipital border; 2, length of cerebral part; 3, length of premaxilla; 4, length of the fossa incisiva; 5, length of nasal bones upon the nasal fossa; 6, maximal supra-orbital width; 7, posterior rostral width (taken between the infra-orbital foramina); 8, anterior rostral width; 9, width of nasal bones upon the nasal fossa; 10, width of nasal fossa; 11, minimal cerebral width between temporal lines; 12, maximal length taken from the condyles; 13, length of the zygomatic arch taken from the processus zygomaticus of the maxilla to the posterior border of the glenoid fossa; 14, length of the orbitotemporal fossa taken at the level of the zygomatic arch; 15, palatal length from the anterior grinding tooth to the choanae; 16, length of basicranium from the choanae to the foramen magnum; 17, thickness of processus zygomaticus of the maxilla; 18, maximal cranial width taken at the zygomatic arches; 19, width of the basicranium taken at the lateral borders of the glenoid fossae; 20, maximal width of choanae; 21, internal maximal width of the palate; 22, external maximal width of the palate; 23, internal width of the palate taken at the anterior grinding teeth; 24, minimal palatal width between the interalveolar cristae (maxillary ridges); 25, sagittal height of the occipital; 26, occipital width; 27, height of premaxilla; 28, facial height taken at the anterior grinding tooth; 29, height of the maxilla ventral to the processus zygomaticus; 30, height of the orbit; 31, cranial height taken from top of cranium to the pterygoid process; 32, length of basicranium from the condyles to the pterygoid process; 33, facial length taken from tip of the rostrum to the pterygoid process; 34, length of the orbitotemporal fossa taken from the squamosal to the anterior border of the orbit; 35, mid-cranial length taken from the external auditory meatus to the ventral border of the orbit; 36, mid-cranial height taken from the pterygoid process to the dorsal border of the orbit.

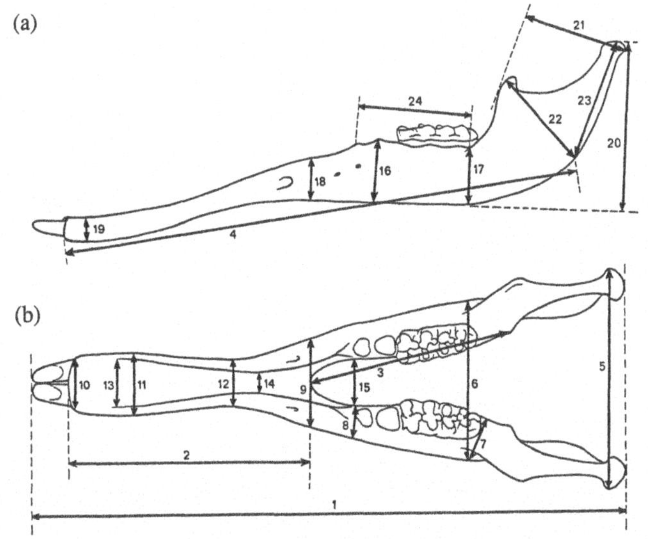

Fig. 11.2 Mandibular measurements (M) taken on the elephantoid skull. Views are: (a) lateral; (b) occlusal (a mandible of *G. angustidens*, female, is shown here). Key: 1, maximal length; 2, symphyseal length; 3, alveolar distance (*sensu* Frick 1926: taken from the most salient point of the trigonum retromolare to the symphyseal border of horizontal ramus); 4, ventral length taken from the gonion (angulus mandibulae) to the tip of the symphysis; 5, maximal width; 6, mandibular width taken at the root of ascending rami; 7, width of the horizontal ramus taken at the root of the ascending branch; 8, width of the horizontal ramus taken at the anterior of the alveolus (or of grinding tooth if the alveolus is entirely resorbed); 9, posterior symphyseal width; 10, anterior symphyseal width; 11, maximal symphyseal width; 12, minimal symphyseal width; 13, maximal width of the rostral trough; 14, minimal width of the rostral trough; 15, internal width between the anterior alveoli (or grinding teeth if the alveoli are resorbed); 16, maximal height of the horizontal ramus (measurement taken perpendicular to the ventral border of the ramus); 17, height of the horizontal ramus taken at the root of the ascending branch (measurement as in no. 16); 18, rostral height taken at the symphyseal border (measurement taken perpendicular to the ventral border of the symphyseal rostrum); 19, rostral height taken at the tip of rostrum (measured as in no. 18); 20, maximal mandibular height taken at the condyle perpendicular to the ventral border of the horizontal ramus; 21, maximal depth of the ascending ramus; 22, depth between gonion and the coronoid process; 23, height between gonion and condyle; 24, mid-alveolar length taken on the buccal side between the anterior alveolus (or grinding tooth if the alveolus is resorbed) and the root of the ascending ramus.

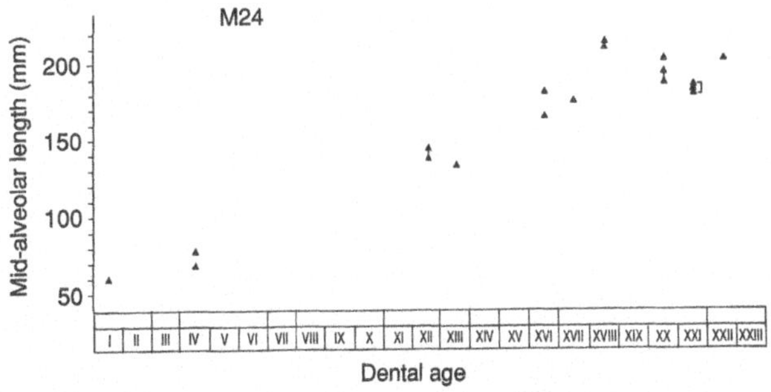

Fig. 11.3 Plots of mid-alveolar length of horizontal ramus (mandibular measurement no. 24 in Fig. 11.2 = M24) against dental age, for *Gomphotherium angustidens* from En Péjouan (Miocene).

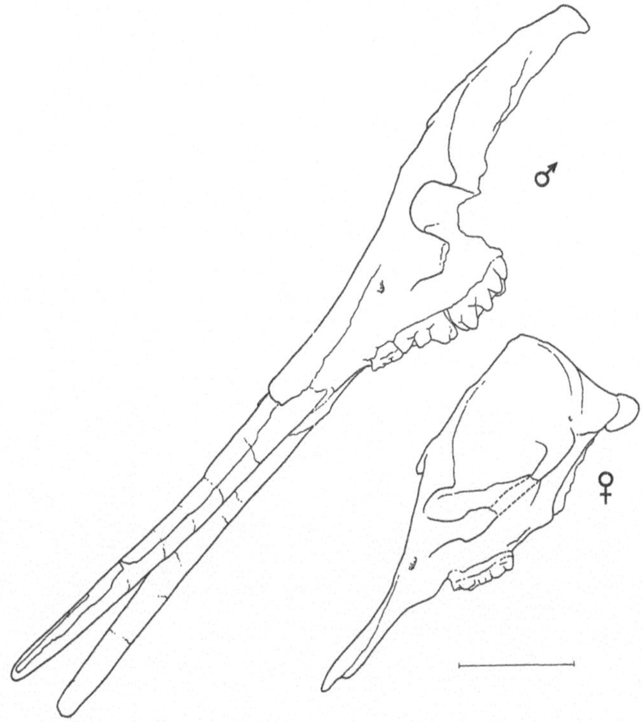

Fig. 11.4 Sexual dimorphism in *Gomphotherium angustidens* (En Péjouan, Miocene). Male (SEP 186, MNHN) is a young adult (dental age XVII). Female (SEP 185, MNHN) is an old adult (dental age XXI). Scale = 18 cm.

of the mode of growth of the Elephantoidea, which greatly antedates the morphological transformation of the elephantine molars.

Several parameters taken on the cranium and mandible demonstrate that sexual dimorphism guides the growth pattern of *Gomphotherium angustidens*. For these parameters, measurements clearly separate two groups interpreted as males (larger) and females, when the dental age is taken into account.

The height of the horizontal ramus (Fig. 11.5) is affected by sexual dimorphism after dental age XV (i.e. eruption of M_2). This applies either to the maximal height, usually situated at the anterior grinding tooth, or at the root of the ascending ramus. Measurements of the width of the ramus (Fig. 11.6) give the same result. The higher mandibles are also the thickest. At the same dental age (XX), mandibles attributed to males (such as SEP 216) are higher and thicker than those attributed to females (SEP 187), even those of older females (SEP 302 and SEP 262 of dental ages XXI and XXII).

Facial measurements are strongly affected by sexual dimorphism. For example, facial width (Fig. 11.7) is taken either at the tip of the premaxillae (cranial measurement no. 8 in Fig. 11.1

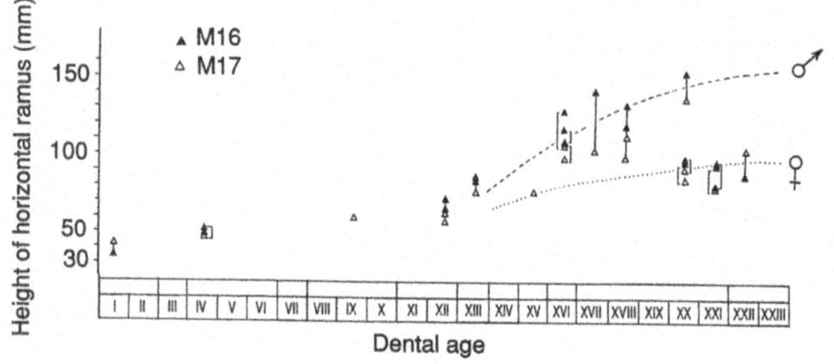

Fig. 11.5 Plots of height of horizontal ramus (mandibular measurements nos. 16 and 17 in Fig. 11.2 = M16 and M17) against dental age, for *G. angustidens* from En Péjouan (Miocene).

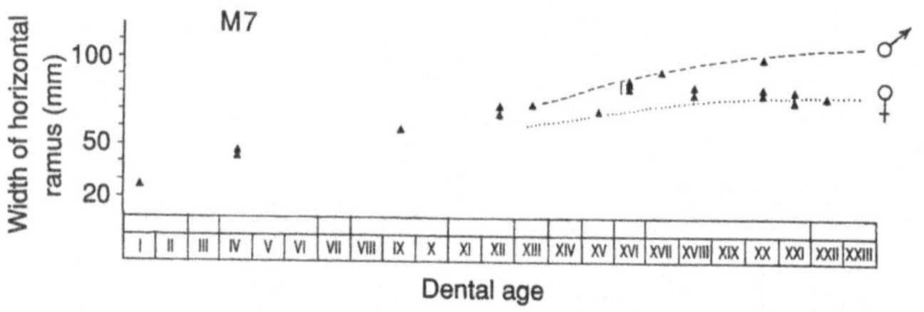

Fig. 11.6 Plots of width of horizontal ramus (mandibular measurement no. 7 in Fig. 11.2 = M7) against dental age, for *G. angustidens* from En Péjouan (Miocene).

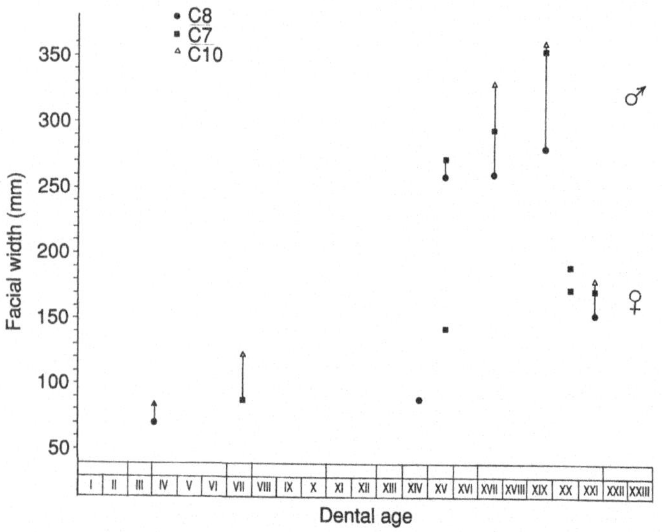

Fig. 11.7 Plots of facial width (cranial measurements nos. 7, 8, 10 in Fig. 11.1 = C7, C8, and C10) against dental age, for *G. angustidens* from En Péjouan (Miocene).

= C8), at the external nasal aperture (C10), or at the foremen infraorbitalis (C7, Fig. 11.7). The same result is obtained when the width is taken at the processus zygomaticus of the frontal bones. The length of the rostrum (C4) shows the same pattern (Fig. 11.8). Two groups are clearly separated. Specimens with a long rostrum

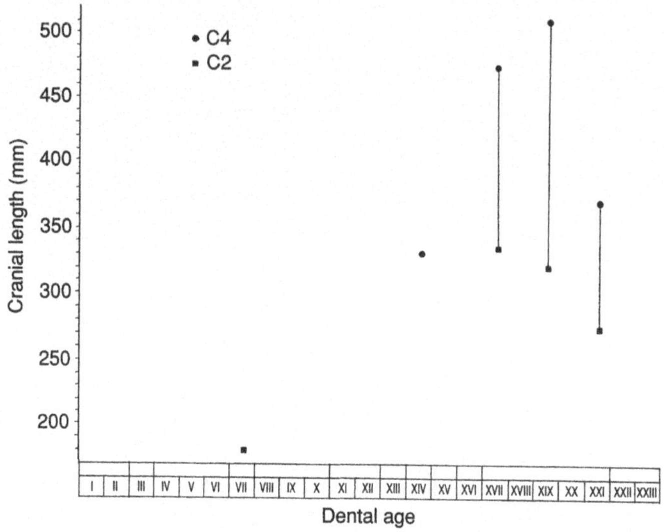

Fig. 11.8 Plots of cranial length (cranial measurements nos. 4 and 2 in Fig. 11.1 = C4 and C2) against dental age, for *G. angustidens* from En Péjouan (Miocene).

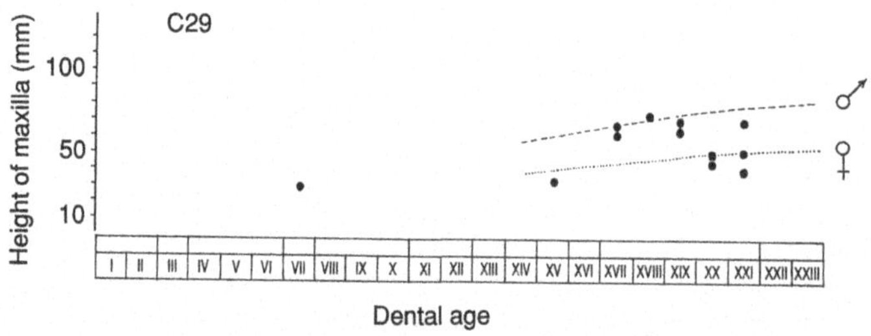

Fig. 11.9 Plots of height of maxilla (cranial measurement no. 29 in Fig. 11.1 = C29) against dental age, for *G. angustidens* from En Péjouan (Miocene).

also have a wide face and can be allocated to males (SEP 186, SEP 25). The same distribution is seen for the height of the maxilla ventral to the processus zygomaticus (Fig. 11.9).

Cerebral measurements are less clearly separated into two groups, either the length (C2 in Fig. 11.8) or the width taken at the linea temporalis, an expected result if these variations are due to sexual dimorphism and not to taxonomic reasons. The same result is found for the internal width of the palate and the thickening of the zygomatic process of the maxilla.

Size variations seen in *G. angustidens* are most easily attributed to sexual dimorphism than to the association of two species. All parameters that separate the specimens found at En Péjouan into two groups are those that separate male and female *Loxodonta a. africana*. The same correlation between parameters is found in the fossil species and in the African elephant (large face, long rostrum, and high maxilla). Beden (1979) has shown the existence of a linear relation ($y = ax + b$) between several parameters in *L. a. africana*. The fossil material studied here is partly damaged and the number of

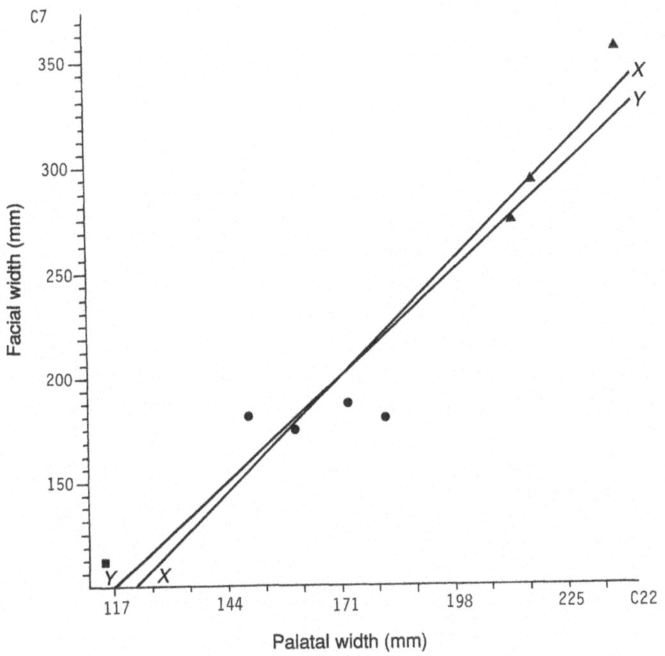

Fig. 11.10 Plots of facial width (cranial measurement no. 7 in Fig. 11.1 = C7) against palatal width (cranial measurement no. 22 in Fig. 11.1. = C22), for *G. angustidens* from En Péjouan (Miocene). For X, statistical mean = 181.00, standard deviation = 40.07, regression line $X = 0.48Y + 74.14$. For Y, statistical mean = 220.13, standard deviation = 79.09, regression line $Y = 1.89X–122.29$. Triangles, males; closed circles, females; square, juvenile from indeterminate locality in the Gers area (Musée d'Histoire Naturelle de Toulouse, collection E. and L. Lartet).

measurements that can be taken on one specimen is reduced. It appears, however, that the relation between growth of the rostrum and of the palate found for *G. angustidens* (Fig. 11.10) is of the same kind as that seen in *L. a. africana*. In Fig. 11.10 a juvenile cranium of *G. angustidens* from an unknown Miocene locality in the Gers area (south-western France) fits perfectly into the distribution of parameters measured on mature skulls from En Péjouan.

Moreover, the few anatomical associations in the locality of En Péjouan confirm sexing based only on size and proportions. On the one hand, both a maxilla and a mandible (SEP 301A–B), found associated, can be independently attributed to a male (high maxilla and thick ramus). On the other, a cranium (SEP 38) and a mandible (SEP 187), though not found articulated, were supposed to belong to the same individual because of perfect occlusion. For all significant parameters both cranium and mandible belong to the smaller group (female); their association is therefore confirmed.

DENTAL TRAITS

Enlargement of the premaxillae is naturally tied to growth of the tusks. In the population of *G. angustidens*, crania with a short and narrow rostrum possess small tusks (SEP 30) or no tusks at all (SEP 38, SEP 186, SEP 251). This extreme variation, previously unnoticed in fossil proboscideans (except dwarf elephants), is not only due to the individual age (small tusks associated with decidual grinding teeth) but also to sexual dimorphism (Fig.

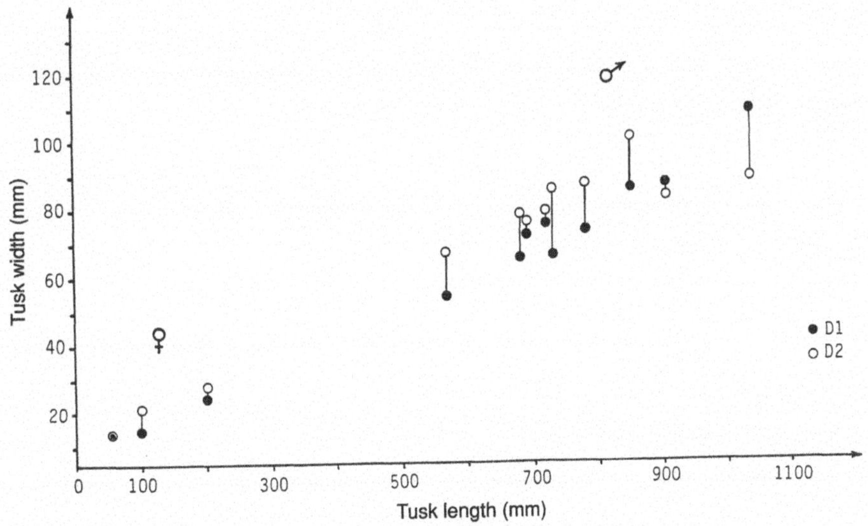

Fig. 11.11 Plots of tusk width (transverse section) against length, for *G. angustidens* from En Péjouan (Miocene). Two widths are taken: D1, the diameter perpendicular to enamel band; and D2, the diameter in parallel to the enamel band. The length is taken from the section to the tip of tusks.

11.11). Such a dimorphism is known in the extant species *E. maximus*. Females of the Asian elephant lack tusks or have very small tusks, called 'tushes' (note: this is a confusing term, frequently applied to both transitory and permanent incisors, and Shoshani, personal communication, 1992, prefers to discontinue its use); in some populations both males and females lack tusks (Deraniyagala 1955).

Variation displayed by the permanent molars of the population of *G. angustidens* found at En Péjouan can also be explained on the basis of sexual dimorphism (Fig. 11.12). Two size groups are separated (with a slight overlapping for M^3s: Figs 11.12–11.14). All big M3s which were found *in situ* within maxillae and mandibles belong to bigger crania and mandibles attributed to males: hence, longer M^3_3 belong to males. A simple taxonomic consequence is that in most cases small gomphothere teeth found in Miocene layers belong to females rather than to small species.

TAXONOMIC IMPLICATIONS

Sexual dimorphism affects the size and proportions of the cranium and the mandible only after dental age XII, and, most markedly, after dental age XIV. Hence, differences in size and proportions of young specimens should be attributed firstly to taxonomic differences. The amebelodontid species *Archaeobelodon filholi* is morphologically close to *Gomphotherium angustidens* since they both belong to the primitive trilophodont grade. Both species

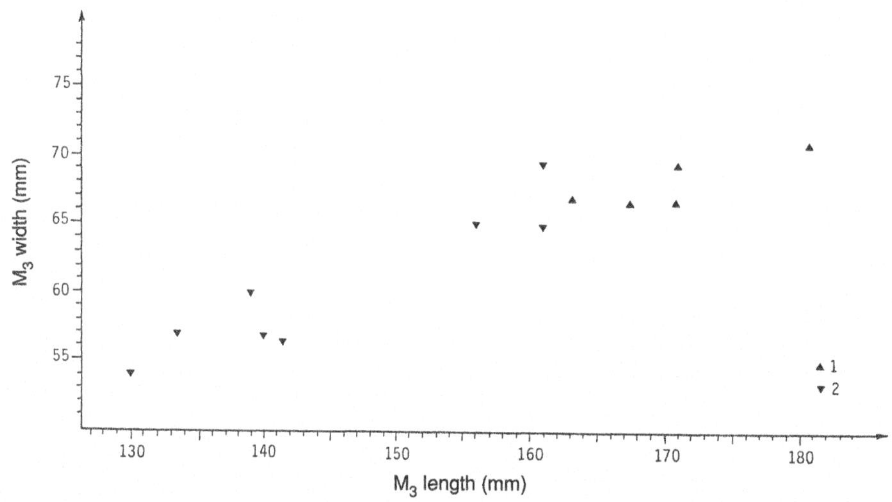

Fig. 11.12 Scatter diagram of M_3 of *G. angustidens* from En Péjouan (Miocene): plots of width against length. 1, unworn teeth; 2, worn teeth.

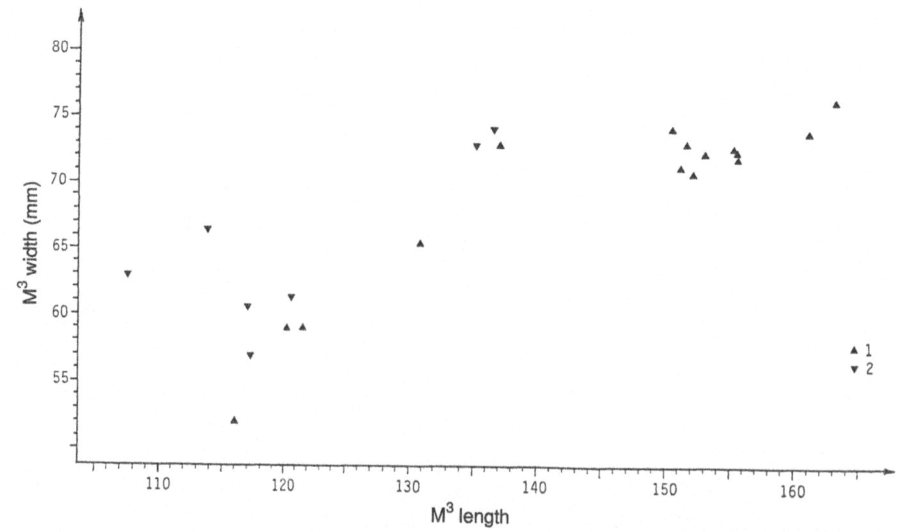

Fig. 11.13 Scatter diagram of M^3 of *G. angustidens* from En Péjouan (Miocene): plots of width against length. 1, unworn teeth; 2, worn teeth.

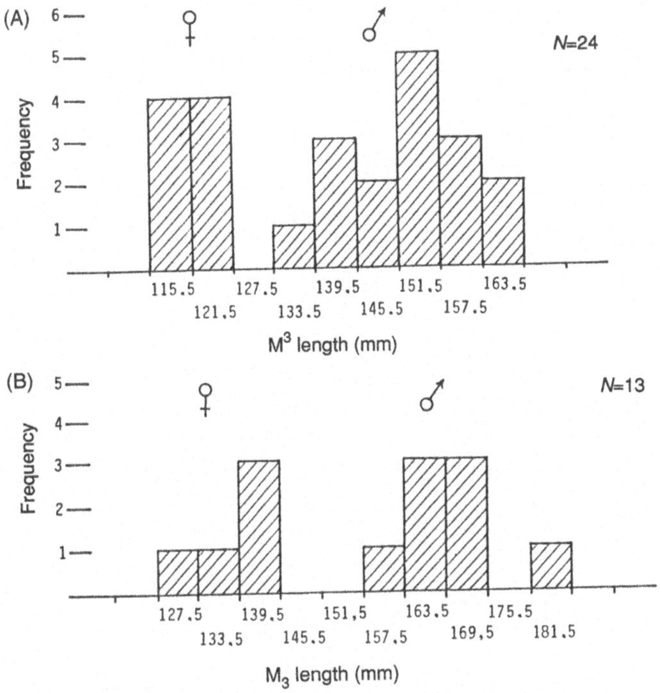

Fig. 11.14 Histograms of M^3s of *G. angustidens* from En Péjouan (Miocene). (A) length (mm) of M^3; (B) length (mm) of M_3. N = total number of specimens.

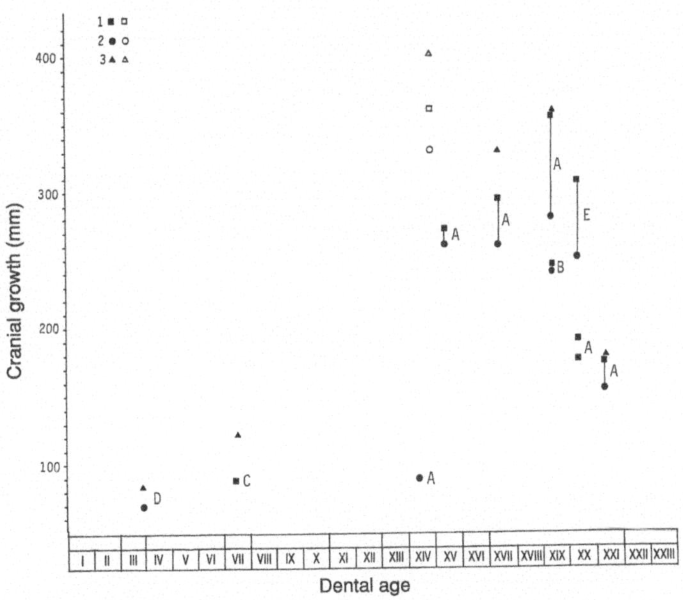

Fig. 11.15 Comparison of the mandibular growth of *Gomphotherium angustidens* and *Archaeobelodon filholi* (height of horizontal ramus M16, see explanation in Fig. 11.5). Key: 1, *G. angustidens* from En Péjouan (middle Miocene); 2, *A. filholi* from Sansan (middle Miocene); 3, *A.* aff. *filholi* from Buluk (= West Stephanie), Kenya (early Miocene).

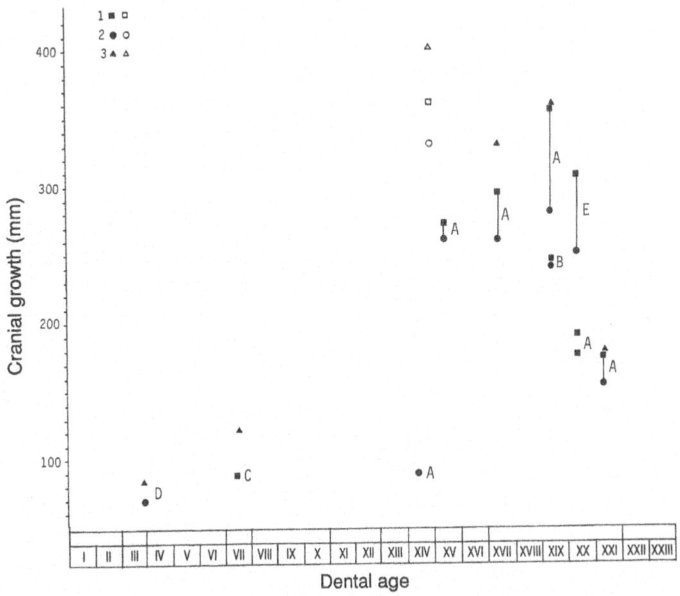

Fig. 11.16 Comparison of the cranial growth of *G. angustidens* (solid symbols) and *A. filholi* (undarkened symbols). 1, facial width (cranial measurement no. 7 in Fig. 11.1 = C7); 2, rostral width (cranial measurement no. 8 in Fig. 11.1 = C8); 3, nasal area width (cranial measurement no. 10 in Fig. 11.1 = C10). Key to localities which yielded *G. angustidens*: A, En Péjouan; B, Ceramica Mirasierra (from Mazo 1976); C, Gers, indeterminate locality; D, Castelnau Barbarens; E, Le Seignou (all middle Miocene localities). The cranium of *A.* aff. *filholi* is from Buluk, Kenya (early Miocene).

were lumped together until Tobien (1973*a*) and Tassy (1984) recognized their affinities. The growth pattern of the mandible supports this view. The mandibles of *A. filholi* show exactly the same pattern of growth and the same kind of sexual dimorphism as in *G. angustidens* (Fig. 11.15). The main difference is that mandibles of *A. filholi* are larger than those of *G. angustidens* for any dental age, including young stages. Males of *A. filholi* with M_3 in function (dental ages XX and XXI) are much larger than male *G. angustidens* of the same age, while females of *A. filholi* (dental age XVIII) fit with male *G. angustidens*.

The differences seen between young individuals are particularly relevant in making taxonomic decisions. Mandibles of young *A. filholi* of dental ages V and IX are of the same size or larger than mandibles of *G. angustidens* of dental ages XII and XIII. The skull of *A. filholi* is poorly known. But one specimen from the early Miocene of Kenya, described by Tassy (1986) as *A.* aff. *filholi*, of dental age XIV or XV, is much bigger than any skull of *G. angustidens* of the same individual age or even older (Fig. 11.16).

SUMMARY AND CONCLUSIONS

Metric comparisons between elephantoid taxa can only be relevant if the individual age of the specimens is examined. The use of a dental age scale allows for such comparisons. Elephantoid species that belong to the primitive trilophodont grade, i.e. that share the same pattern of growth and tooth eruption, can be compared relatively to the 23 dental ages defined in this chapter. The series of *Gomphotherium angustidens* from En Péjouan can be used as a reference for ageing and sexing specimens, after which metric data can be used for taxonomic purposes.

Nevertheless, environmental factors considerably alter the growth rates of living elephants. For example, though much less well studied than *Loxodonta africana africana*, the forest African elephant *L. a. cyclotis* is known to be smaller than the savanna subspecies. Among the Elephantoidea, series of fossil populations are very rare discoveries. Metric data for taxonomic purposes must be used very cautiously. On the one hand, sympatry of two size groups should first be attributed to sexual dimorphism, although the taphonomic association of two species can occur without there having been true sympatry. On the other hand, variations between distant populations of the same species can be due to local environmental factors or, precisely, to differences at the population level. Still, even taking into account these possible factors, the generalized use of the dental age scale proposed here might shed light on the growth pattern of different trilophodont species and potentially upon their taxonomy.

ACKNOWLEDGEMENTS

Many thanks to W. P. Luckett who pointed out to me the ambiguous meaning of the usual expression 'horizontal replacement'. N. Court's, J. Kalb's, and J. Shoshani's comments helped to improve the text. The drawings were made by D. Visset.

TWELVE A

Phylogeny of African and Eurasian Elephantoidea of the late Neogene

JON E. KALB, DAVID J. FROEHLICH, AND GORDEN L. BELL

INTRODUCTION

It has been nearly 200 years since Johann Blumenbach gave us the first systematic description of an African proboscidean, a right lower second molar, which in 1797 he referred to as '*Elephas africanus*' (Osborn 1942, pp. 1192, 1197). In 1825, Étienne Geoffroy Saint-Hilaire and Frédéric Cuvier published their classic work, *Histoire Naturelle des Mammifères*, in which they created a new genus, '*Loxodonte*', for this 'éléphant d'Afrique', based on the unique lozenge-shaped enamel on the plates of its molars. Two years later, in 1827, an article appeared in the *Zoological Journal* of London by an anonymous writer who referred to this redesignation of names for '*Loxodonta africanus*' as the 'dismemberment of the genus *Elephas*' (Anonymous, 1827, p. 140; see also Domning and Shoshani, 1982). Ever since, palaeontologists have been describing, redescribing, and 'dismembering' African proboscideans.

Today, under the aegis of phylogenetic systematics, this rendering continues in earnest. Pascal Tassy and colleagues have led the way in this field over the past decade, using an increasingly expanded data matrix of taxa and characters (Tassy 1982, 1983, 1985, 1986, 1988a, 1990a; Tassy and Darlu 1986, 1987; Tassy and Shoshani 1988). Saegusa (1987) has also used cladistics to elucidate the far-flung species of the genus *Stegodon*. As always, these workers have built on the pioneering efforts of others who, over a 130-year period, constructed successively more detailed 'conventional' phylogenetic models of proboscidean evolution. These workers include: Falconer (1857), Gaudry (1866), Vacek (1877), Adams (1881), Weithofer (1888), Cope (1889), Depéret (1907*a*), Schlesinger (1912), Soergel (1912), Boule and Thevenin (1920), Matsumoto (1924*a*,*b*, 1926), Osborn (1936), Petrocchi (1954), Aguirre (1969*a*), Thenius (1969), Maglio (1970*a*, 1973), Coppens *et al.* (1978), Beden (1985), Mebrate and Kalb (1985) and Tobien (1986*a*, *b*).

In this chapter, we add to this collective work by examining the systematic phylogeny of the African Elephantoidea of the late Neogene, as the superfamily is used by Tassy (1988*a*) (Table 12A.1; see Tassy and Shoshani 1988, fig. 11.5, Node D). Pending our own cladistic evaluation of the systematics of this group given below, we use Tassy's classification scheme (Table 12A.1) to refer to the taxa under study in this chapter. In this respect, the term 'gomphothere' is an informal, operative term used only to facilitate discussion of certain taxa as they are listed in Table 12A.1 that broadly share a bunodont, trifoliate enamel pattern on their molars.The term does not imply a monophyletic relationship.

We are particularly interested in amplifying the work of Tassy (1986) on the genus *Anancus*, a late-surviving 'gomphothere'. We do so by examining relationships between African species of *Anancus* recently described by Kalb and Mebrate (1993). As given in Appendix 12A.1, we examine:

(1) an early form of *Anancus* from Member A of the Lukeino Formation (Lukeino-A), Kenya and the Adu-Asa and lower Sagantole formations, Ethiopia, which is generally similar to the tetralophodont, *A. kenyensis*, from Kanam East, Kenya;

(2) a provisionally described progressive tetralophodont *Anancus* from Langebaanweg, South Africa;

(3) a pentalophodont, *Anancus petrocchii*, from Sahabi, Libya; and

(4) a newly described progressive pentalophodont *Anancus* from the Sagantole Formation, Ethiopia (Appendix 12A.1). Inclusion of other known species of *Anancus* in our analysis is beyond the scope of this chapter, including diverse fossils assigned to the North African species, *Anancus osiris* (Arambourg 1970; Tassy 1986).

Within our analysis, we include *Tetralophodon longirostris*, the putative ancestor of *Anancus*, as proposed by Tobien (1978*c*, p. 188), as well as its close relative, *Paratetralophodon hasnotensis* (Tassy 1983). We also include the large elephantoid originally described as '*Mastodon*' *grandincisivus* by Schlesinger (1917), and later referred to as *Tetralophodon* by Osborn (1936), *Stegotetrabelodon* by Tobien (1978*c*) and Gaziry (1987*b*), and an amebelodont by Tassy (1986). Tassy and Darlu (1986, 1987) have demonstrated the closer (but still ambiguous) relationship of '*M.*' *grandincisivus* to progressive 'gomphotheres' and its distant

Table 12A.1 Classifications of late Neogene Proboscidea. Taxa are limited to those studied in this chapter.

Coppens *et al.* (1978)

- Order Proboscidea
  - Suborder Gomphotherioidea
    - Family Gomphotheiridae
      - Subfamily Gomphotheriinae
        - Genus *Phiomia*
        - Genus *Tetralophodon*
      - Subfamily Anancinae
        - Genus *Anancus*
    - Family Elephantidae
      - Subfamily Stegotetrabelodontinae
        - Genus *Stegotetrabelodon*
        - Genus *Stegodibelodon*
      - Subfamily Elephantinae
        - Genus *Primelephas*
        - Genus *Loxodonta*
        - Genus *Mammuthus*
        - Genus *Elephas*
  - Suborder Mammutoidea
    - Family Stegodontidae
      - Genus *Stegolophodon*
      - Genus *Stegodon*

Tassy (1988*a*)

- Order Proboscidea
  - Suborder Elephantiformes
    - Plesion *Phiomia*
    - Superfamily Elephantoidea
      - Plesion Gomphotheres
      - Plesion Stegodontidae
      - Family Elephantidae
        - Plesion *Stegotetrabelodon*
        - Subfamily Elephantinae
          - Plesion *Stegodibelodon*
          - Plesion *Primelephas*
          - Genus *Loxodonta*
          - Plesion *Mammuthus*
          - Genus *Elephas*

This chapter

- Order Proboscidea
  - Suborder Elephantiformes
    - Superfamily Elephantoidea
      - 'Gomphotheres'
      - Plesion *'Mastodon' grandincisivus*
      - Plesion *Tetralophodon longirostris*
      - Plesion *Paratetralophodon hasnotensis*
      - Genus *Anancus*
        - *Anancus 'kenyensis'*
        - *Anancus* sp. (Langebaanweg-type)
        - Pentalophodont *Anancus* Group
          - *Anancus petrocchii*
          - *Anancus* sp. (Sagantole-type)
    - Family Elephantidae
      - 'Stegomorphs'
        - Plesion *Stegolophodon*
        - Genus *Stegotetrabelodon*
          - *Stegotetrabelodon syrticus*
          - *Stegotetrabelodon orbus*
        - Plesion *Stegodibelodon schneideri*
        - Plesion *Stegodon*
      - Subfamily Elephantinae
        - Plesion *Primelephas gomphotheroides*
        - Genus *Loxodonta*
          - *Loxodonta* Group
            - 'cf. *Loxodonta*' (Lukeino-A type)
            - *L. exoptata*
            - *L. atlantica*
            - *L. africana*
          - *Loxodonta adaurora*
      - Unnamed Group
        - Genus *Mammuthus*
        - Genus *Elephas*

relationship to *Stegotetrabelodon*. In several studies since then, Tassy (1988*a*, 1990*a*) and Tassy and Shoshani (1988) have asserted the ancestral positions of *Tetralophodon*, *Paratetralophodon*, and *Anancus* relative to *Stegotetrabelodon*, but left unresolved specific relationships of these genera to each other.

We also look at relationships of *Stegolophodon* and *Stegodon*. Osborn (1942, p. 1539), Maglio (1973, p. 15), Coppens *et al.* (1978), Tassy (1988*a*, 1990*a*), and Tassy and Shoshani (1988) all excluded these genera from the Elephantidae. Simpson (1945), however, included *Stegolophodon* and *Stegodon* within the Elephantidae as the subfamily Stegodontinae. Because of certain similarities between these two predominantly Asian genera, *Stegolophodon* and *Stegodon* have long been regarded as monophyletic (see Simpson 1945, pp. 245–9; Maglio 1973,

p. 15, table 1), most recently by Tassy (1988*a*, 1990*a*, table 4, fig. 19). However, Taruno (1985) has placed *Stegolophodon* in the family Gomphotheriidae and retained *Stegodon* in the family Stegodontidae, along with '*Eostegodon*' (Yabe 1950), a tetralophodont Asian genus Yabe and Taruno recognize as being ancestral to *Stegodon*.

Also, we look at the alleged monophyly of the four-tusked *Stegotetrabelodon* and the two-tusked *Stegodibelodon* within the subfamily Stegotetrabelodontinae, as proposed by Coppens (1972) and Coppens *et al.* (1978) (Table 12A.1), and the status of these genera within the Elephantidae and with respect to the elephantines. This includes examining the relationship of the North African species, *Stegotetrabelodon syrticus*, with the sub-Saharan species of the same genus, *S. orbus*.

Finally, we examine the subfamily Elephantinae (the 'true elephants'). In particular, we analyse the putative ancestral status of *Primelephas* in relation to *Loxodonta*, and to *Mammuthus* and *Elephas*, and the alleged ancestral relationship of *Loxodonta adaurora* to all other loxodonts, including the early loxodont from Lukeino-A described by Tassy (1986) as 'cf. *Loxodonta*'.

METHODS

GENERAL

In our study, we have adopted the parsimony methods formulated by Hennig (1966). As these methods have since been developed, organisms are grouped by shared derived characters and outgroup comparison, which is used to determine the polarity of character transformation (Swofford and Olsen 1990). We have used the computer program, 'Phylogenetic Analysis Using Parsimony' (PAUP), of Swofford (1989, Version 3.0A) using unordered, rooted, branch-and-bound search settings. PAUP's default options were utilized except where noted. For character-state optimization, we relied on the default option, accelerated transformations (ACCTRAN), which places equivocal character transformations at the first available position on the tree, that is, down tree branches. We also report on the alternative optimization mode, delayed transformations (DELTRAN), which places equivocal character transformations up tree branches. These options have the effect of either increasing reversals (with ACCTRAN) or increasing convergences (with DELTRAN). Hence, our placement of equivocal characters should be viewed as tentative pending the recovery of more complete fossil remains (and our ability to substitute information for missing data in our matrix).

Characters used in our analysis are described in Appendix 12A.2 and summarized in Table 12A.2. No assumptions have been made in scoring character states in our data matrix with regard to incomplete or missing data; we have never assumed what a character should be in the absence of actual fossil evidence. However, several tests of our results were made as described below using questionably referred or hypothetical material. Most taxa examined in our analysis, and described in detail in Appendix 12A.1, are intentionally limited to the specific level, often to particular specimens. The monophyly of these taxa is assumed; our purpose is to analyse relationships between groups, not to broaden or explore existing groups. This being the case, we have omitted autapomorphic characters from our analysis, since they would be consistent with any tree and would only serve to inflate tree statistics, such as the consistency index (CI). In the case of phylogenetic change within a taxon, characters have been polarized and scored generally from the earliest representative of that taxon, in order to further eliminate autapomorphies.

TERMINOLOGY AND IDENTIFICATION

Terms used to distinguish features of gomphothere dentition are adopted from Tassy (1986, fig. 2 and pl. 8) and Tobien (1973*b*, pp. 116–19); those used to distinguish elephantid dentition are adopted from Maglio (1973, pp. 8–13). The terms 'tetralophodont' and 'pentalophodont' used below are descriptive only and refer to elephantoids that have a maximum of four or five loph(id)s, respectively, on the intermediate molars (dp4, M1, M2).

As outlined in Kalb and Mebrate (1993), identification of particular molars of elephantoids is indicated in the following manner. For 'gomphotheres', inner (lingual) half-loph(id)s always occupy a more anterior position on upper and lower molars. Although not as accentuated, this same condition is distinguished in elephantids by the anteriormost position of the lingual half-plate with respect to the buccal half-plate, on both upper and lower molars. Secondly, lower molars of Elephantinae can always be distinguished by the concave-forward (anteriorly) shape of the posterior face of the enamel loop of worn plates; for upper molars, the posterior face is always convex-forward. The anterior and posterior faces of enamel loops of both upper and lower molars of all other elephantoids are convex, except *Stegodon*, whose enamel faces may be straight, subparallel or a mixture of shallow convex or shallow concave. Also convex are the adjoining upper half-lophs and lower half-lophids of worn molars of gomphotheres. In all cases, convexity and concavity is best expressed at the lateral margins of the enamel loops of moderately worn molars. Finally, for gomphotheres, posterior conules are dominant on the buccal (pretrite) side of lower molars, and the lingual (also pretrite) side of upper molars. Posterior columns, as relic conules, are positioned similarly on the lower and upper molars of elephantids.

SAMPLE

We have used skulls and dentition exclusively for character selection since invariably these parts are the most documented in the fossil record. Postcranial remains are beyond the scope of this analysis; also, they are often rare and limited to isolated specimens, or restricted to a few taxa. Since teeth are the most common fossils documented, all but 10 characters in our data matrix are based on dentition, including tusks. Most taxa used in

Table 12A.2. Data matrix for the cladogram in Fig 12A.1 and 12A.2. See Appendices 12A.1 and 12A.2 for a detailed description of outgroup and ingroup taxa and characters

Character		*Phiomia*	*"M." grandincisivus*	*Tetralophodon*	*Paratetralophodon*	*A. "kenyensis"*	*A.* Langebaanweg	*A. petrocchii*	*A.* Sagantole	*Stegolophodon*	*Stegot. syrticus*	*Stegot. orbus*	*Stegodibelodon*	*Stegodon*	*Primelephas*	*Loxodonta* Group	*Lox. adaurora*	*Mammuthus*	*Elephas*
CRANIUM/MANDIBLE																			
1 Cranium: flattened (0), rounded (1).	1	0	0	1	1	1	9	9	9	9	1	9	9	1	9	1	1	1	1
2 Occipital condyles: protruding (0), recessed (1).	2	0	0	0	0	0	9	9	9	9	1	9	9	0	9	0	0	1	1
3 Basicranium: not erect (0), erect (1).	3	0	0	0	1	1	9	9	9	9	1	9	9	1	9	1	1	1	1
4 Cranium: non-globular (0), globular (1).	4	0	0	0	0	0	9	9	9	9	0	9	9	0	9	1	1	0	0
5 Fronto-parietal region: not concave (0), concave (1).	5	0	0	0	0	9	9	9	9	9	0	9	9	0	9	0	0	1	1
6 Mandible width: narrow (0), wide (1).	6	0	1	0	9	1	9	0	9	1	1	9	1	1	9	1	1	1	1
TUSKS/MANDIBULAR SYMPHYSIS																			
7 Premaxillary tusk sheaths: narrow (0), flaring (1).	7	0	9	0	0	9	9	9	9	9	9	9	9	0	9	1	1	1	0
8 Upper tusks: downturned (0), not downturned (1).	8	0	1	0	0	1	9	9	9	9	1	9	9	1	9	1	1	1	1
9 Upper tusks: not upturned (0), upturned (1).	9	0	0	0	0	0	9	9	9	9	1	9	9	1	9	1	1	1	1
10 Extended mandibular symphysis: not downturned (0), downturned(1).	10	0	0	1	1	9	9	9	9	1	1	1	1	9	9	9	9	9	9
11 Mandibular symphysis: extended (0), shortened or fully reduced (1).	11	0	0	0	0	1	9	1	9	1	0	0	1	1	9	1	1	1	1
12 Mandibular symphysis: extended or shortened (0), fully reduced (1)	12	0	0	0	0	1	9	1	9	0	0	0	0	1	9	1	1	1	1
13 Lower tusks: present (0), absent or vestigial (1).	13	0	0	0	0	1	9	1	9	1	0	0	1	1	9	1	1	1	1
14 Lower tusks: flat or pyriform (0), circular (1).	14	0	0	1	9	9	9	9	9	9	1	1	9	9	9	9	9	9	9
MOLARS																			
15 Curvilinear alignment of apices: absent (0), present (1).	15	0	0	0	0	0	0	0	0	1	1	1	1	1	1	1	1	1	1
16 Distinctly brachydont molars: absent (0), present (1).	16	0	0	0	0	0	0	0	0	1	0	0	0	1	0	0	0	0	0
17 Anancoidy: absent (0), present (1).	17	0	0	0	0	1	1	1	1	0	0	0	0	0	0	0	0	0	0
18 Posterior face of lophids or lower plates: very convex (0), shallow convex or absent (1).	18	0	0	0	0	0	0	0	0	1	1	1	1	1	1	1	1	1	1
19 Posterior face of lophids or lower plates: not concave (0), shallow concave or very concave (1).	19	0	0	0	0	0	0	0	0	0	0	0	0	1	1	1	1	1	1
20 Posterior face of lophids or lower plates: convex or mixed (0), concave (1)	20	0	0	0	0	0	0	0	0	0	0	0	0	0	1	1	1	1	1
21 Anterior face of lophids or lower plates: convex (0), concave or mixed (1).	21	0	0	0	0	0	0	0	0	0	0	0	0	1	1	0	1	1	1
22 Anterior face of lophids or lower plates: convex or mixed (0), concave (1).	22	0	0	0	0	0	0	0	0	0	0	0	0	0	1	0	1	1	1
23 Plates: absent (0), present (1).	23	0	0	0	0	0	0	0	0	0	1	1	1	1	1	1	1	1	1
24 Loph(id)s/plates: expanded (0), distinctly compressed (1).	24	0	0	0	0	0	0	0	0	0	0	0	1	1	1	1	1	1	1
MOLAR ACCESSORIES																			
25 Median sulcus: present (0), absent (1).	25	0	0	0	0	0	0	0	0	0	0	0	0	1	1	1	1	1	1
26 Median sulcus: complete (0), incomplete (1).	26	0	0	0	0	0	0	0	0	1	1	1	1	9	9	9	9	9	9
27 Four apices (cones, conelets): present (0), five or more (1).	27	0	0	0	0	0	0	0	0	0	0	0	1	1	1	1	1	1	1
28 On M_2-M_3, posterior posttrite conules/columns: anteriorly (0), posteriorly (1).	28	0	0	0	0	0	1	1	1	0	0	0	0	0	0	0	1	0	0
29 On M^2-M^3 with lophs, three posterior posttrite conules: absent (0), present (1).	29	0	0	0	0	0	1	9	1	0	9	9	9	9	9	9	9	9	9
30 On M_2 and M_3, posterior pretrite conules/columns posteriorly: present (0), reduced/fused (1).	30	0	0	0	0	0	0	0	0	0	0	1	1	1	0	1	1	1	1
31 On M^3, posterior pretrite conules/columns posteriorly: absent (0), present (1).	31	0	1	1	1	1	0	0	0	1	1	1	9	9	9	1	1	9	9
LOPH(ID)/PLATE NUMBER																			
32 On M_2 or M^2, *minimum* of: three (0), four (1), five (2), six (3).	32	0	1	1	1	1	1	2	2	1	2	2	9	3	2	3	9	3	3
33 On M_3, *minimum* of: three (0), five (1), six (2), seven (3), eight (4).	33	0	1	1	1	1	2	2	2	2	3	3	3	3	4	4	4	4	4
34 On M^3, *minimum* of: three (0), five (1), six (2), seven (3), nine (4).	34	0	1	1	1	1	2	9	3	1	1	1	3	3	3	4	4	4	4

our study, including primitive species of isolated genera named in our data matrix (e.g. *Stegolophodon*, *Elephas*), are represented by only a handful of molars for each taxon, and fewer specimens still of individual teeth in the jaw. Hence, our sample number is very limited and may not distinguish between normal, or as the case may be, 'abnormal', variation in plate number or morphology (Maglio 1972; Roth 1989). Taxa with crania are even more limited, since six of the taxa have no associated crania at all and five are represented by one cranium each. These limitations may cause much of the equivocation of character placement in our cladistic analysis.

Characters given in Appendix 12A.2 from, or modified from, Tassy (1990*a*, tables 6–7) are characters 3, 5, 8–14, 16, 17, 23, 30–34. Remaining characters used by Tassy, but not used here, are those that are autapomorphic for the taxa used in our analysis, or are characters that do not vary in our ingroup. We chose not to use character 116 from Tassy (1990*a*, table 6), concerning the lack or amount of cement in valleys between loph(id)s or plates, pending further study, because of the considerable variation of this feature temporarily and geographically in the Neogene (e.g. Tobien 1973*b*, p. 130).

Abbreviations

Egy, Egypt; KNM, Kenya National Museum (Nairobi, Kenya); LT, Lothagam; LU, Lukeino; Lukeino-A, Member A, Lukeino Formation, Kenya; MCZ, Museum of Comparative Zoology, Harvard University (Cambridge, Massachusetts); NMB, Naturhistorisches Museum, Osteologische Abteilung (Basel, Switzerland).

RESULTS OF ANALYSIS

GENERAL

The PAUP analysis of our data matrix (Table 12A.2) resulted in a single most parsimonious tree (Fig. 12A.1) with 61 transformations and a consistency index (CI) of 0.689. The tree is constructed from 18 taxa using 34 characters. Three characters (32–34) are multistate; these concern loph(id) or plate number on molars. Multistate characters were used when the characters could not be broken down into binary states. Of the total characters, six (2, 8, 12, 16, 28, 32) show unequivocal convergence; six characters (11, 13, 21, 22, 30, 31) show unequivocal reversals. Those characters showing equivocation, either as convergences or reversals (indicated by DELTRAN or ACCTRAN), are characters 2, 6, 7, 9, 30, 32 (2–3), 33 (1–3), and 34 (2–3) (Fig. 12A.2). Of these, over half relate to multistate characters, which inherently allow for more possible changes, and equivocation; also, unordered multistate characters inherently allow for more equivocation. Because we omitted autapomorphies from our analysis, the apomorphies on the terminal branches of our tree are nearly all homoplasic (Fig. 12A.1). Hence, some convergences and reversals are an artefact of our analysis and should not be interpreted to necessarily indicate anything unusual about the pattern of elephantoid evolution within our taxa.

Repeated analyses were made assigning hypothetical character states to characters that were not preserved (those lacking data), in order to test the limits of existing data and the stability of our final tree. For instance, the assignment of a hypothetical lower tusk to *Primelephas* (character 13) did not change its position in the cladogram; rather, it produced a less parsimonious tree (that was one step longer). This test was made because previous cladistic analyses (e.g. Tassy 1990*a*, p. 42, character 70) had been made assuming the presence of lower tusks based on evidence from Maglio (1970*b*, 1973). Our evaluation of the fossil material, however, suggests that no data are known for the anterior lower dentition of this genus (see Appendix 12A.1: *Primelephas*). Another test was made assigning the same hypothetical cranium character states to *Stegotetrabelodon orbus* and *Stegodibelodon schneideri* that we assigned to *Stegotetrabelodon syrticus*. In this case, the position of these taxa in our cladogram remained unchanged, although elephantine relationships changed. This occured apparently because the distribution of character 2 created equivocation with respect to the hypothetical outgroup for the elephants, which resulted in the loss of some resolution within this group. We also assigned cranium characters to *Stegolophodon* based on Tassy's (1983) assessment that the juvenile cranium '*Stegolophodon*' *cautleyi progressus* originally described by Osborn (1929*a*, 1942) is indeed *Stegolophodon*, and not a zygodont mammutid as proposed by Tobien (1978*a*). By assuming the former, a longer most parsimonious tree (tree length = 63) was obtained that included two reversals (on characters 1 and 9) and a lower consistency index (0.667), suggesting that Tobien is correct that the skull is at least more primitive than *Stegolophodon*. We also note that even when assuming that the cranium was *Stegolophodon*, this did not change the position of the genus in our tree. Finally, we note that *Stegolophodon* and *Stegodon* were never linked in a monophyletic group (e.g. Tassy 1990*a*, fig. 17), or as successive sister taxa, despite multiple permutations of our cladogram. Forcing a linkage between *Stegolophodon* and *Stegodon* required a tree with a minimum of eight extra steps in our analysis. Hence, even if we were to assume seven hypothetical characters linking *Stegolophodon* and *Stegodon*, this would not change the typology of our tree.

CLASSIFICATION

The classification and terminology used here (Table 12A.1) for the elephantoids included in our study are based generally on the classificatory guidelines set forth by Tassy (1988*a*, p. 47). Within our classification scheme, the superfamily Elephantoidea (*sensu* Tassy 1988*a*, fig. 1, table 1) is divided into '*M.*' *grandincisivus*, the informal 'gomphotheres' and the monophyletic family Elephantidae. As previously mentioned, the name 'gomphotheres' is operative only and does not imply a monophyletic group; as used here, the term refers the paraphyletic series consisting of *Tetralophodon longirostris*, *Paratetralophodon hasnotensis*, and *Anancus*. The genus *Anancus* is comprised of *A.* '*kenyensis*' (as defined in Appendix 12A.1), the Langebaanweg-type *Anancus*, and the 'Pentalophodont *Anancus* Group', *A. petrocchii* and the Sagantole-type *Anancus*. As previously indicated, the term 'pentalophodont' refers to loph(id) number on intermediate molars; in this instance, the term is used to distinguish the derived *Anancus* group with a maximum of five loph(id)s on the

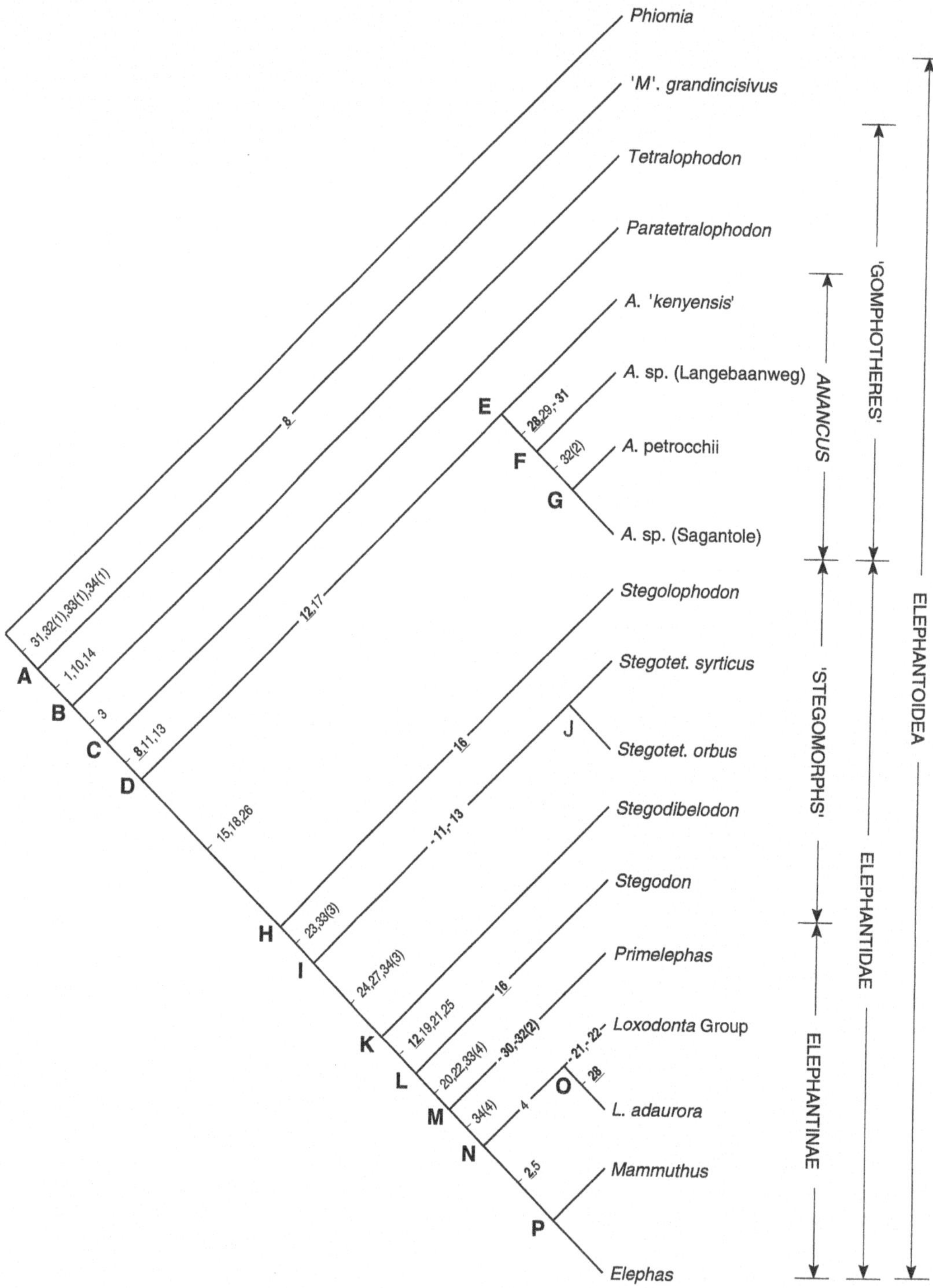

Fig. 12A.1 Cladogram of late Neogene Proboscidea. All apomorphies are unequivocal; based on the data matrix in Table 12A.2. Key: regular numbers show no homoplasy; bold, underlined numbers show convergence; bold numbers with minus signs show reversals; numbers in parentheses indicate multistate characters; bold letters refer to internal nodes (Editors' note: See Saegusa (this volume, Chapter 17) for discussion of the monophyly of Stegodontidae (*Stegolophodon* and *Stegodon*), and results from cladistic analysis by Shoshani (this volume, Chapter 16)).

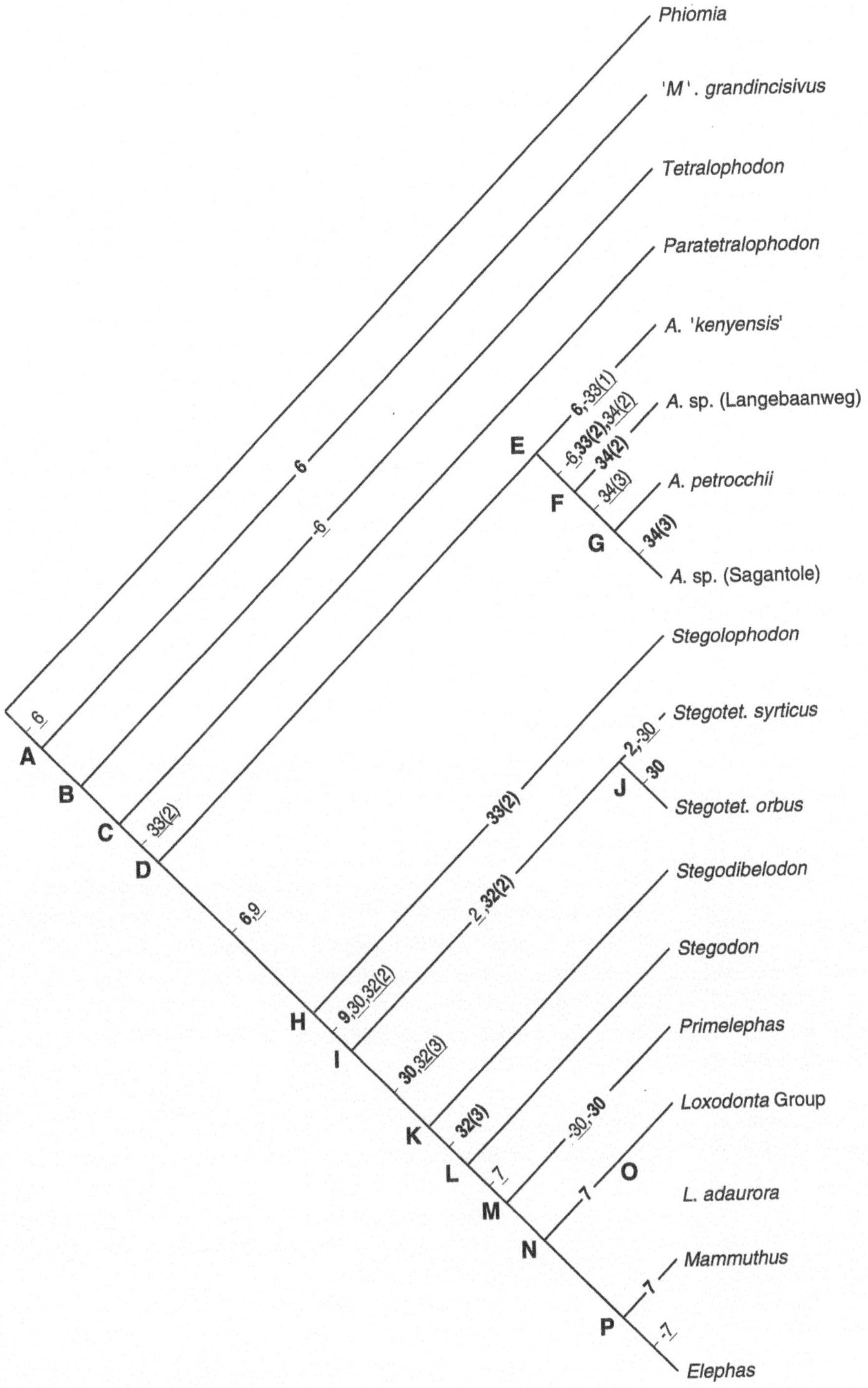

Fig. 12A.2 Equivocal characters placed on the cladogram of Figure 12A.1. Character-state optimization: ACCTRAN is indicated by regular underlined numbers, DELTRAN by bold numbers. Reversals are noted by minus signs, and multistate characters by parentheses.

dp4, M1, M2 from the plesiomorphic and paraphyletic 'tetralophodont' *Anancus*, *A. 'kenyensis'* and the Langebaanweg *Anancus*.

As defined in this chapter, the Elephantidae consist of two groups, the informal 'stegomorphs' and the subfamily Elephantinae. The former is an operative name only, which we introduce to refer to the 'stego'-prefixed paraphyletic series of taxa that are morphologically intermediate between the gomphotheres and the Elephantinae. The term is used simply to facilitate discussion of these early elephantids. Hence, the stegomorphs — in reference to 'roof-like' transverse loph(id)s or plates on the molars — consist of the genera *Stegolophodon*,

Stegotetrabelodon, Stegodibelodon, and *Stegodon*. These same taxa were formerly included in the Elephantidae as the subfamily Stegodontinae (*Stegolophodon, Stegodon*) by Simpson (1945) and the subfamily Stegotetrabelodontinae (*Stegotetrabelodon, Stegodibelodon*) by Coppens (1972) and Coppens *et al.* (1978) (Table 12A.1). The subfamily Elephantinae is retained as intended by Maglio (1973, fig. 15) with *Primelephas* occupying an ancestral position with respect to (or sharing a common ancestor with) the loxodonts, *Mammuthus* and *Elephas*. The latter two genera are sister taxa and together are a sister-group with respect to the loxodonts. In turn, the genus *Loxodonta* is comprised of the species *L. adaurora* and its sister-group. The latter includes all remaining loxodonts, referred to as the '*Loxodonta* Group'. As described in Appendix 12A.1, this group consists of 'cf. *Loxodonta*' from Lukeino-A, *L. exoptata, L. atlantica*, and the living African elephant, *L. africana*.

SYNAPOMORPHIES

In the following pages we describe the synapomorphies by referring to the internal nodes in Figures 12A.1 and 12A.2 as regards both unequivocal and equivocal character placements; also, we refer to sister-group relationships and to individual clades. Shared derived characters are summarized below with individual character numbers used in Table 12A.2 set in brackets; multistate characters are in parentheses. Reversals are noted by minus (–) signs, and equivocal characters are noted with an asterisk (*). Additional features distinguishing taxa, but not included in our cladistic analysis, are supplemented from the literature where useful.

Node A: Elephantoidea

This node is supported by four unequivocal characters. The M2 of the ingroup have four or more loph(id)s/plates [32 (1)], the M_3 have five or more lophids/plates [33 (1)], and the M^3 have five or more lophs/plates [34 (1)]. On the M^3, posterior pretrite columns are present posteriorly [31].

Much of the ingroup consists of elephantoids with relatively wide mandibles [6]* compared with the very narrow plesiomorphic condition present in *Phiomia* and other early proboscideans (see Appendix 12A.2, character 6). Two possible transformations exist for character 6: either it can be an early elephantoid character reversed in *Tetralophodon* and derived *Anancus* (ACCTRAN); or, it can be a basal elephantid character that is convergently derived in '*M.*' *grandincisivus* and *A.* '*kenyensis*' (DELTRAN) (Fig. 12A.2). We note, however, that the flaring width of the mandible of '*M.*' *grandincisivus* from Hungary is based on a reconstruction (Tobien 1978*b*, pl. 18, fig. 14b), which may warrant re-examination. A mandible from Bulgaria referred to by Tobien (1978*b*, pl. 19, fig. 15) as '*M.*' *grandincisivus* is also flaring; however, the reconstruction of this specimen suggests that the mandible may have been prepared wider than it should be (Bakalov and Nikolov 1962, pl. 66). Hence, it is possible that the '*M.*' *grandincisivus* mandible warrants the plesiomorphic state, which may help resolve the equivocation present in this character (and thus character 6 for *T. longirostris* may not be reversed after all).

'Mastodon' grandincisivus

This taxon is distinguished by having straight upper tusks [8]. The lower tusks are pyriform in cross-section (Tobien 1978b, figs. 4 and 5), an autapomorphic character (see character 14, Appendix 12A.2) omitted from our analysis. The pyriform lower tusks are also plesiomorphic when compared with our outgroup (Osborn 1936, pl. V, pp. 236/237). Concerning mandible width [6]*, see Node A for a discussion of character 6 (Fig. 12A.2).

Node B

Based on a single cranium each reported for '*M.*' *grandincisivus* and *T. longirostris* (Tobien 1978*b*, pl. 18, fig. 14a; Mottl 1969, fig. 22b), that of the latter compares more with the 'rounded' derived condition, which characterizes this group [1]. Of the extended mandibular symphyses, those that are downturned characterize the derived state [10]. Only the subhorizontal mandibular symphysis of '*M.*' *grandincisivus* (Tobien 1978*b*, pl. 18, fig. 14b) more closely reflects the plesiomorphic condition of our outgroup [10]. Taxa with lower tusks are circular in cross-section [14].

Tetralophodon longirostris

This taxon is characterized by having a narrow mandible, a reversed condition [–6]*; see discussion under Node A.

Node C

This group features a single synapomorphy in our analysis, an erect cranium first seen in *P. hasnotensis* [3]. According to Tassy's cladistic analysis (1990*a*, table 6, character 60), this character alone distinguishes this species from *Tetralophodon*. Other features of *P. hasnotensis*, however, can be found in Tassy (1983). We note that Tobien *et al.* (1988, p. 108) refer to *P. hasnotensis* as belonging to the genus *Tetralophodon*.

Node D

Straight or upturned upper tusks characterize this group [8]. The upper tusks of *A.* '*kenyensis*' and of the genus *Anancus* generally are characteristically straight (Kalb and Mebrate 1993; Osborn 1936, p. 630) (as are those of '*M.*' *grandincisivus* (Tobien 1978*b*, pl. 18, fig. 14a)). The mandibular symphyses are shortened or fully reduced [11] and the lower tusks are vestigial or missing (except for the reversal of these characters in *Stegotetrabelodon*) [13]. The ACCTRAN option places character 33(2) at node D. However, because a continuous increase in loph(id) (and plate) number in increasingly more progressive species of gomphotheres and other elephantoids has been well documented (e.g. Tobien 1973*b*, p. 129; Maglio 1972, p. 642), our preferred placement of character 33(2) is to have it derived convergently within *Anancus* (node F) and the *Stegolophodon* clade, using the DELTRAN option. As such, character 33(1) would not be reversed in the *A.* '*kenyensis*' clade; also, character 33(3) becomes further derived, and unequivocal, at node I (Fig. 12A.1). The elephantoids are thus shown to have a minimum of five lophids in nodes A–D and in the *A. 'kenyensis'* clade, but a minimum of six lophids in the three derived *Anancus* and in the *Stegolophodon* clade. At node I, a minimum of seven lophids are present.

Node E: *Anancus*

This genus is distinguished by alternating and interlocking pretrite and posttrite half-loph(id)s ('anancoidy') [17], also a fully reduced mandibular symphysis [12] that is convergent at node L.

Anancus 'kenyensis'

This species differs from other *Anancus* in our study by having a minimum of five lophids on the M_3, a reversed character [–33 (1)]* (Fig. 12A.2). Five lophids are present in the 'kenyensis-morph' described by Tassy from Lukeino-A (1986, table 17). See discussion of character 6 under Node A.

Node F: *Anancus* sp. (Langebaanweg) — *A. petrocchii*/*A.* sp. (Sagantole)

This group constitutes the derived species of *Anancus* in our study (Fig. 12A.1). Whereas *A. 'kenyensis'* has a derived 'wide' mandible (Tassy 1986, pl. XI, fig. 1a), the 'narrow' mandible of *A. petrocchii* (Petrocchi 1943, fig. 72) reflects the plesiomorphic condition [–6]* (see discussion under Node A). We note that a 'progressive' species of *Anancus*, *A. sinensis* from China, also has a narrow mandible (Tobien *et al.* 1988, fig. 20). The presence of posterior posttrite conules extending to the posterior end of the M_2–M_3 characterizes the three derived *Anancus* [28], as does the presence of two additional (or a total of three) posterior posttrite conules at the anterior end of the M^2–M^3 [29]. The absence of posterior pretrite conules posteriorly on the M^3 of the derived *Anancus* constitutes a reversal (Table 12A.2) since these conules are present posteriorly in *A. 'kenyensis'* [- 31]. We note that posterior pretrite conules are also absent on the M^3 of *A. sinensis* (Tobien *et al.* 1988, e.g. figs 13–14). As noted under Node D, because a linear increase in loph(id) number is an inherent pattern in elephantoid evolution, our preferred interpretation of characters 34(2) is to treat it as a synapomorphy at node F.

Anancus sp. (Langebaanweg)

This *Anancus* has a minimum of six lophs on the M^3 [34(2)]* and a minimum of six lophids on the M_3 (see discussions under Nodes D and F). *A. 'kenyensis'* has a minimum of five lophs on both the M^3 and the M_3. The Langebaanweg *Anancus* and *A. 'kenyensis'* differ from the Pentalophodont *Anancus* Group (node G) by being tetralophodont on the M2. The South African *Anancus*, however, is clearly more progressive than *A. 'kenyensis'* by having supplementary posterior posttrite conules on both the M^2 and M^3.

Node G: Pentalophodont *Anancus* Group

This clade is characterized by a minimum of five loph(id)s on the M2 [32(2)]. Concerning the equivocal placement of character 34(3), we note that only a single M^3 is documented for this clade, an M^3 of the Sagantole-type *Anancus* from Ethiopia [34(3)]*.

Anancus sp. (Sagantole)

The few specimens and the poor illustrations of the type molars of *A. petrocchii* from Sahabi, Libya (Petrocchi 1943, 1954) limit comparison of this pentalophodont with the Sagantole pentalophodont. Based on the evidence available, however, it is apparent that the M_2 from Sahabi has single conules on the posttrite side of the molar; an M_2 from Kanapoi, Kenya, which Kalb and Mebrate (1993) include with the Sagantole-type *Anancus*, has double conules on the posttrite side of the molar (Coppens *et al.* 1978, fig. 17.4B; Tassy 1986, pl. XIII, fig. 3). The Kanapoi M_2 also has double conules on the pretrite side of the molar, which are not apparent on the Sahabi M_2. Concerning character 34(3), see discussion above under Node G.

Node H: *Elephantidae*

Apices of the molars are aligned [15] — this character does not imply the formation of plates (in the case of *Stegolophodon*) unless the apices fully disappear with moderate wear to form a fully connecting enamel envelope. In this respect, the median sulcus of this group is 'incomplete', that is, the sulcus does not extend to the posterior end of the molar [26]. The posterior face of enamel loops of lower plates of worn molars may have a shallow convex configuration (the 'stegomorphs') or the posterior face may be concave (elephants) [18]. In the most conservative interpretation, a wide mandible (character 6) is a synapomorphy at this position [6]*. Because of the absence in our data pertaining to upper dentition for *Stegolophodon*, character 9 may be placed at either node H or node I [9]*. The upper tusks of the elephantids are upturned [9]*, although *Stegolophodon* does not preserve this character and, consequently, placement of this character is equivocal.

Stegolophodon

Early representatives of this genus reportedly differ from other elephantoids by having particularly brachyodont molars (Tassy 1990*a*, table 6, character 108) [16]. This genus stands out from all others in our ingroup by having an enamel band on the upper tusks of adults (Osborn 1942, p. 839), an autapomorphic feature omitted from our analysis. Concerning character 33(2), see discussion under Node D [33(2)]*.

Node I

Concerning character 9 (DELTRAN), see discussion under Node H [9]*. This group features fully formed plates on worn molars [23]. The M2 have a minimum of five plates either at this node (ACCTRAN) or on the *Stegotetrabelodon* clade (DELTRAN) [33(2)]*. Character 30 is an equivocal character that pertains to the presence of posterior pretrite conules/columns posteriorly on the M_2–M_3 [30]*. ACCTRAN places the derived state of this character at this node, but shows it to be reversed with *S. syrticus* and *Primelephas*: DELTRAN shows this character to be convergently derived with *S. orbus* and at node K, but reversed with *Primelephas*. This equivocation may not be resolved with the present data, although we note that the higher placement of this character (DELTRAN) conforms more with the general trend whereby the reduction of pretrite columns on the M_2–M_3 favour higher elephantoid (elephantid) taxa, and the presence of pretrite conules posteriorly characterize lower elephantoid taxa.

Node J: *Stegotetrabelodon*

The genus *Stegotetrabelodon* is noted for having 'recessed' occipital condyles [2]* (Maglio 1973, fig. 7B) (see Appendix 12A.2, character 2). The character is equivocal because only the specimen of *S. syrticus* has a skull. The extended mandibular symphysis [–11] and the lower tusks [–13] reflect the plesiomorphic condition. The DELTRAN option places character 32(2) at this node, as indicated under Node I [32(2)]*.

Stegotetrabelodon syrticus

This species is distinguished by having posterior pretrite columns that extend to the posterior end of the molar; these columns are reduced or fused posteriorly in *S. orbus*, a distinction first noted by Maglio (1970a). See discussion of character 30 under Node I [30]*. See discussion of character 2 under Node J [2]*.

Stegotetrabelodon orbus

See Node I for discussion of character 30 [30]*.

Node K

Plates are distinctly compressed compared to the expanded plates of more plesiomorphic taxa [24]. Five or more apices are present on unworn molars [27]. The M^3 have a minimum of seven or more plates [34 (3)]. The M2 typically have a minimum of six or more plates [32 (3)]*. The equivocal placement of character 32 at this node or node L is because no M2 are available for *Stegodibelodon*. See Node I for discussion of the equivocation of character 30.

Node L: *Stegodon* — Elephantinae

The mandibular symphyses of this group are fully reduced (as are those in *Anancus* — see Node E) [12]. The posterior face of the enamel loops of plates of lower molars is shallow concave or very concave [19]. The anterior face of enamel loops of plates may also be shallow concave, or 'mixed', that is, they may display a shallow concave configuration in combination with straight, or shallow convex enamel loops [21]. Lower molars usually reveal some concavity, a feature unique to the *Stegodon*/Elephantinae monophyly. A median sulcus is absent in this group or confined to the anteriormost end of the molar [25]. See Node K concerning the equivocation of character 32(3).

Stegodon

Like *Stegolophodon*, early representatives of this genus reportedly stand out by having particularly brachyodont molars (Tassy 1990a, table 6, character 108) [16]. The presence of abundant, tiny apices, or mammillae, on the plates of the molars is a feature unique to this genus of the taxa examined here. The plates of the genus *Stegodon* are unique in that the lower molars may display varied enamel loop curvature. The tendency is for the enamel loops of *Stegodon* to be straight or subparallel, but the loops may be curvilinear either in a convex or a concave fashion, or both, on the same molar (Osborn 1942, p. 853) (see character 18 note, Appendix 12A.2).

Node M: *Elephantinae*

The premaxillary tusk sheaths in the loxodonts and *Mammuthus* are flaring [7]*. This character could be placed at node N or within *Mammuthus* because the derived character is not preserved in *Primelephas* and is not present in *Elephas*. The posterior face of the enamel loops of plates of lower molars of the 'true elephants' are all concave [20], which is an easily recognizable, single feature that distinguishes the elephantines from all other elephantoids. This concavity may be apparent across the width of the entire enamel loop, or may be seen just at the lateral margins of the molar. The anterior face of the enamel loops of lower plates are also concave, except in the *Loxodonta* Group [22]. On the M_3, the elephants have eight or more plates [33 (4)].

Primelephas

This 'ancestral elephant' features posterior pretrite columns extending to the posterior end of the M_3, a plesiomorphic condition (e.g. Maglio and Ricca 1978, pl. 2, fig. 3) [–30]*. Both ACCTRAN and DELTRAN place this character in the same position; see Node I for further discussion of character 30. A second plesiomorphic feature that is present within *Primelephas* are M2 with a minimum of five plates [–32(2)]. As noted in Appendix 12A.1 ('*Primelephas*'), we do not believe that present evidence indicates that *Primelephas* had lower tusks. The absence of any verifiable cranium or mandible material for this genus makes it the least represented taxon in our analysis.

Node N: *Loxodonta* — *Mammuthus*/*Elephas*

The 'progressive elephants' are noted by having a minimum of nine plates on the M^3 [34 (3)].

Node O: *Loxodonta*

According to Maglio (1973, p. 22), Beden (1987*a*, p. 261) and others (Appendix 12A.2, character 4), a single feature that characterizes this genus is the 'globular'-shaped skull with a recurved face [4]. See Node M for discussion of character 7.

Loxodonta Group

This clade is distinguished by the anterior face of the enamel loops of worn lower molars being convex-shaped, a plesiomorphic condition [–21,–22]. The anterior face of enamel loops of worn lower molars of *L. adaurora* is concave-shaped. See Appendix 12A.1 for further distinctions between the *Loxodonta* Group and *L. adaurora*.

Loxodonta adaurora

This species is distinguished by having posterior posttrite columns that extend to the posterior end of the M_2–M_3 [28]. When combined with the posterior pretrite columns, 'twin' posterior columns are uniquely present. These twin columns are partially visible on the type M_3 for the species from Kanapoi (Maglio 1970*a*, pl. V, fig. 15). A well-worn M_2 from the Sagantole Formation, Ethiopia shows these twin posterior columns as being more fully developed and isolated posteriorly (Kalb and Mebrate 1993).

Node P: Unnamed Group: *Mammuthus* and *Elephas*

The skulls of these sister taxa are noted by having recessed occipital condyles [2]; also, the fronto-parietal region of the skulls are uniquely concave [5].

Mammuthus

See Node M for discussion of the equivocation of character 7.

Elephas

The narrow premaxillary tusk sheaths of *Elephas*, a plesiomorphic condition [−7]*, distinguishes this genus from *Mammuthus* (see discussion under Node M). The tusk sheaths of *Mammuthus* are flaring; also, the upper tusks are twisted or spiralled (Maglio 1973), an autapomorphic character (not included in our data matrix). There are no clear differences yet reported in molar morphology distinguishing *Elephas* from *Mammuthus*. Coppens *et al.* (1978, p. 357), however, note that the enamel in *Mammuthus* molars (at least in the later species) is 'very finely wrinkled or wavy'.

PHYLOGENETIC SYSTEMATICS

As seen above, our study has reached an overall resolution of the phylogenetic relationships of late Neogene proboscideans. It is evident that '*M.*' *grandincisivus* is not directly related to African *Stegotetrabelodon* as proposed by Tobien (1978*b*, p. 188); instead, the latter is a distantly related elephantoid. On existing evidence, '*M.*' *grandincisivus* is more closely related to *T. longirostris*, as was demonstrated by Tassy and Darlu (1986, 1987); however, as Osborn (1936, p. 360) proposed, '*M.*' *grandincisivus* is more primitive, as indicated by its flattened cranium, downturned mandibular symphysis, and pyriform lower tusks. We note, however, that the cranium of '*M.*' *grandincisivus* is entirely reconstructed (Schlesinger 1922, pl. 1). In respect to the lower tusks, Tassy (1986, figs. 13, 14, 18) has compared '*M.*' *grandincisivus* to *Amebelodon* (also see Tassy and Darlu 1987), although the wide mandible and straight upper tusks (characters 6 and 8) of '*M.*' *grandincisivus* are features associated with more derived taxa (Table 12A.2). Concerning *T. longirostris*, Tobien (1978*b*) compares a wide array of fossils in Europe with this species. Of these, those fossils most like the type material of *T. longirostris* (referenced in Appendix 12.1) compare closely with *P. hasnotensis* from Asia (Tassy 1983), according to our findings and inferred by Tobien *et al.* (1988, p. 108). Tassy (1985), however, notes that the facial morphology of both species (known only from one cranium each) is different.

Whatever refined analyses may tell us about the relationships of *T. longirostris* and *P. hasnotensis* to one another, our cladogram indicates that a longirostrine tetralophodont 'gomphothere' of one form or another is the sister taxon of a brevirostrine tetralophodont gomphothere, the genus *Anancus*, and of its sister group, the Elephantidae (Fig. 12A.1). In turn, the elephantids are comprised of both longirostrine and brevirostrine species, for primitive and progressive species respectively. Although an analysis of all known species of *Anancus* is beyond our purview (see Tassy 1986, figs 39–41), our results demonstrate that a simple tetralophodont *Anancus* (*A. 'kenyensis'*) shares a common ancestor with more complex, derived forms of tetralophodont (the Langebaanweg *Anancus*) and pentalophodont *Anancus* (*A. petrocchii* and the Sagantole *Anancus*). We also see that the pentalophodonts are sister taxa, one known from North Africa (*A. petrocchii*) and the other from eastern Africa (Sagantole *Anancus*) (Appendix 12A.1).

Stegolophodon and *Stegotetrabelodon* retain the plesiomorphic condition of tetralophodonty on one or more of their intermediate molars (dp4–M1–M2), and together with *Stegodibelodon*, retain convex–convex opposing enamel loops on lower molars (characters 18 and 19). All three genera have elongated and subhorizontal mandibular symphyses (Coppens 1972, pl. 1); however, *Stegotetrabelodon* has a greatly extended mandibular symphysis and fully developed lower incisors, which are plesiomorphic features (characters 11 and 12). *Stegodibelodon schneideri*, with its less elongated yet tuskless mandibular symphysis, can be characterized as a form morphologically intermediate between the fully longirostrine *Stegotetrabelodon* and the fully brevirostrine *Stegodon*.

Our analysis does not provide evidence for a *Stegolophodon*–*Stegodon* clade, and places the former close to the gomphotheres and *Stegodon* proximal to the elephants. As our cladogram and classification illustrate, however, we place both *Stegolophodon* and *Stegodon* within the Elephantidae (Fig. 12A.1, Table 12A.1). Our analysis shows *Stegolophodon* to be the most primitive elephantid and *Stegodon* to be the sister taxon of the elephantines. *Stegodon* has long been recognized as having developed changes in skull and dental morphology similar to the elephants.

As a paraphyletic series, the 'stegomorphs' represent a transitional morphological series of 'grades' between *Anancus* and the monophyletic Elephantinae. With the elephants, we see the 'completion' of various morphological trends that began in the stegomorphs. Among these: mandibles become fully brevirostrine; plates become fully formed, and fully compressed, as apices become fully aligned; the posterior face of enamel loops of lower molars become fully concave; intermediate molars lose their tetralophodonty; medial sulci disappear; and posterior columns become fully reduced or fused to molar plates. Overall, these changes represent a major and dramatic transformation in elephantid dentition.

When more is known about *Primelephas*, it may prove to be more closely related to *Mammuthus* and *Elephas*, as the sister taxon of their putative common ancestor, rather than being the sister taxon of all elephantines. If this is the case, then the loxodont clade would be the sister-group of a *Primelephas/Mammuthus/Elephas* clade. Unlike the *Loxodonta* Group, *Primelephas* does not possess a convex anterior face on the enamel loops of its lower molars; and, unlike *L. adaurora*, it does not possess twin (posttrite and pretrite) posterior columns on lower molars (characters 19 and 24) (Table 12A.2).

SUMMARY AND CONCLUSIONS

For our analysis, we used the PAUP program (1989, Version 3.0A) using unordered, rooted, branch-and-bound settings. We also used a detailed methodology with respect to construction of our data matrix, the description of taxa (Appendix 12A.1), and the selection and identification of characters (Appendix 12A.2). Our results produced a single most parsimonious tree with 61 steps and a consistency index of 0.689. No autapomorphic characters were used, nor were assumptions made in scoring our data matrix with regard to missing or incomplete data. Various tests were made as to the stability of our tree using hypothetical characters, particularly as regards incomplete data and disputed taxonomic assignments. The conclusions of our study are as follows:

1. *'M.' grandincisivus* is the most distantly related taxon in our ingroup (Fig. 12A.1). Based on the characters used in our analysis, very little (character 3) separates *T. longirostris* from *P. hasnotensis*. The latter may be regarded as a derived species of *Tetralophodon*, as viewed by Tobien *et al.* (1988), or *P. hasnotensis* may best be regarded as a separate genus, as described by Tassy (1983). Either way, further analysis may prove *P. hasnotensis* to be the sister taxon of *T. longirostris*.
2. A longirostrine tetralophodont gomphothere is the sister taxon of a monophyletic group comprised of a brevirostrine gomphothere, *Anancus*, on the one hand, and the Elephantidae.
3. The tetralophodont *A. 'kenyensis'* (as defined in Appendix 12A.1) is the sister taxon of all other *Anancus* in our analysis, including derived tetralophodont (the Langebaanweg *Anancus*) and pentalophodont *Anancus* (*A. petrocchii* and the Sagantole *Anancus*). The North African pentalophodont (*A. petrocchii*) is the sister taxon of the eastern African pentalophodont (the Sagantole *Anancus*).
4. Inclusion of *Stegolophodon* and *Stegodon* into a monophyletic Stegodontidae or Stegodontinae is not valid on the basis of our study; rather, our analysis shows *Stegolophodon* to be the sister taxon of all other (African) elephantids, and *Stegodon* to be the sister taxon of the Elephantinae.
5. The North African *Stegotetrabelodon syrticus* is the sister taxon of the sub-Saharan *Stegotetrabelodon orbus*. Inclusion of *Stegodibelodon* into a monophyletic group (Stegotetrabelodontinae) with *Stegotetrabelodon* is not supported, nor is it valid to include *Stegodibelodon* within the Elephantinae.
6. The paraphyletic taxa we refer to as 'stegomorphs' represent a morphologically intermediate series between *Anancus* and elephants. *Stegolophodon* is the most primitive stegomorph, while *Stegodon* is the most advanced. Relic gomphothere features present in early stegomorphs disappear or nearly disappear in *Stegodon* and in early elephants.
7. On present evidence, *Primelephas* appears to share a common ancestor with all other elephantines, but with more data may prove to share a common ancestor with just *Mammuthus* and *Elephas*. If this is the case, the loxodonts would be the sister-group of a *Primelephas/Mammuthus/Elephas* clade.
8. *Loxodonta adaurora* is the sister taxon of the *Loxodonta* Group (as defined in Appendix 12A.1). Because *L. adaurora* appears to be derived independently, it may not be the ancestor of all loxodonts as proposed by Maglio (1973).

ACKNOWLEDGEMENTS

For helpful comments and discussions, we particularly thank Tim Rowe, also R. Toomey III, C. Brochu, and A. Czebieniak. For additional assistance we thank Sharon Robertson, L. Froehlich, C. Hallock, and C. Kehoe. Finally, we are grateful to the editors of this volume, J. Shoshani and P. Tassy, for their remarkable endurance.

Appendix 12A.1: Taxa

The following describes the outgroup and ingroup taxa specifically used in our analysis, with the particular references we relied upon provided accordingly. The outgroup taxon, *Phiomia serridens*, is described from middle Palaeogene deposits from the Fayûm of Egypt; the remaining taxa, comprising our ingroup, are all of late Neogene age (Coppens *et al.* 1978).

PHIOMIA SERRIDENS **ANDREWS AND BEADNELL, 1902**

We follow the usage of Tobien (1978*c*), who includes within *P. serridens* the species, *P. minor*, *P. wintoni*, *P. barroisi*, and *P. osborni*. An exception is specimen NMB Egy 32 (Tobien 1978*c*, fig. 13), which Tassy (1982, p. 235) indicates may belong to a separate, undesignated species of *Phiomia*. Skull features of *P. serridens* (as the taxon is redefined by Tobien) are illustrated in Osborn (1936, fig. 183; also, see Tobien 1976, fig. 5). The genus *Phiomia* is used as the outgroup rather than *Palaeomastodon* because of the completeness of the fossil material for *Phiomia* reported, and the greater loph(id) number of its molars, which allows for a more reasonable comparison of molar characters of this genus with those in the ingroup.

'MASTODON' GRANDINCISIVUS SCHLESINGER, 1917

Refers specifically to the material originally described as '*Mastodon* (*Bunolophodon*) *grandincisivum* nova species' from Maragheh (Iran), Mithridate (near Kertsch, Russia), Pestszentlörinez (Hungary), and Mannersdorf (Austria) (Schlesinger 1917, 1922). We prefer to refer provisionally to this taxon using the original generic name (although no longer in use), until a thorough assessment is made of its taxonomic status. In this regard, Tobien (1978*b*), Osborn (1936), and Tassy (1986, fig. 18; 1990*a*, fig. 4; also see Tassy and Darlu 1987) have compared '*M*'. *grandincisivus* to *Stegotetrabelodon*, *Tetralophodon*, and an amebelodont, respectively.

TETRALOPHODON LONGIROSTRIS KAUP, 1832*a*

The specimens are from Eppelsheim, Esselborn, Breitenfeld, and Kornberg, Germany (Klähn 1931; Osborn 1936; Lehmann 1950; Bergounioux and Crouzel 1960; Mottl 1969, 1970; Tobien 1978*b*). Even though aspects of the Kornberg fossils are 'grandincisivoid' as noted by Tobien (1978*b*, p. 185), we agree with Tassy and Darlu (1986, p. 599, characters 25–26) that they more closely resemble *T. longirostris*.

PARATETRALOPHODON HASNOTENSIS (OSBORN, 1929)

The taxon as defined by Tassy (1983), and referred specimens therein described by Osborn (1929*a*, 1936) and Tobien (1972). For this analysis, we include palatal, mandibular, and dental material described and referred to by Tobien *et al.* (1988) as *Tetralophodon* (*Paratetralophodon*) cf. *hasnotensis*.

ANANCUS 'KENYENSIS' (MACINNES, 1942)

Specifically, the specimens discussed in Kalb and Mebrate (1993) and referred to as *A.* cf. *kenyensis* from Member A, the Lukeino Formation (Lukeino-A), Kenya (Tassy 1986) and *A. kenyensis* from the Adu-Asa and lower Sagantole formations, Ethiopia (Kalb *et al.* 1982). Since the type molars of *A. kenyensis* from Kanam East, Kenya, as figured by MacInnes (1942), are badly worn or only partially preserved, any detailed comparison to these specimens is limited. Thus, the assignment of the Lukeino-A fossils to *A.* cf. *kenyensis* and the Adu-Asa and lower Sagantole fossils to *A. kenyensis* are somewhat arbitrary and are based on both morphological and age considerations.

ANANCUS SP. (LANGEBAANWEG-TYPE)

The South African *Anancus* discussed in Kalb and Mebrate (1993) and referred to by Coppens *et al.* (1978, p. 348). This is a progressive tetralophodont *Anancus*, well represented from the Pelletal Phosphorite and Quartzose Sand Members of the Varswater Formation, 'E' Quarry, and by a single M_2 from the Lower Beds, Baard's Quarry (Hendey 1970*b*, pl. 3-A; 1978, fig. 1; 1982, p. 39).

ANANCUS PETROCCHII COPPENS, 1965

The specimens of the pentalophodont *Anancus* from Sahabi, Libya described by Petrocchi (1943, 1954) (Tassy 1986; Kalb and Mebrate 1993).

ANANCUS SP. (SAGANTOLE-TYPE)

The pentalophodont *Anancus* from the Sagantole Formation, Ethiopia (Kalb and Mebrate 1993), as well as an M_2 from the Kanapoi beds, Kenya (Coppens *et al.* 1978, fig. 17.4-B; Tassy 1986, p. XIII-3), which is included in this same taxon.

STEGOLOPHODON SCHLESINGER, 1917

Represented by '*Stegotetrabelodon*' *maluvalensis* (Sarwar 1977), a mandible from the Dhok Pathan Formation, the Siwalik Group, Maluval, Pakistan, which Tassy (1983) and Tassy and Darlu (1986) regard as *Stegolophodon*. Dental characters for the genus are based on figured specimens referred to as *Stegolophodon* in Osborn (1942), Tassy (1983), Tobien (1975, p. 222), and Taruno (1985). As discussed on p. 105, the juvenile skull '*Stegolophodon*' *cautleyi progressus* is not included among the material for this genus. We also exclude from present analysis the sparse material assigned to a new species of *Stegolophodon*, *S. hueiheensis*, by Chow (1959), and to a new genus, '*Rulengchia*', by Zhou and Zhang (1983). These fossils represent a single individual with very poorly preserved M3 and a brevirostrine mandible (see Tobien *et al.* 1988).

STEGOTETRABELODON SYRTICUS PETROCCHII, 1941

The genus and species from Sahabi, Libya (Petrocchi 1941, 1943, 1954; Maglio 1970*a*, 1973; Gaziry 1982, 1987*b*). Cranium morphology is based on the reconstruction in Maglio (1973, fig. 7).

STEGOTETRABELODON ORBUS MAGLIO, 1970*b*

The species from Lothagam, Kenya, described by Maglio (1970*b*, 1973) and Maglio and Ricca (1978). We note that specimen KNM LT-354 is a right M_2, not a right M^2 (Maglio and Ricca 1978, pl. 1, fig. 6), as indicated by the anteriormost position of the inner half-lophids and the prominent posterior pretrite columns extending to the last plate. Included in this species are molars from the Adu-Asa Formation, Ethiopia (Kalb and Mebrate 1993). For further discussion of *S. orbus* see *Primelephas gomphotheroides* below.

STEGODIBELODON SCHNEIDERI COPPENS, 1972

The mandible and the isolated M^3 described by Coppens (1972) from Menalla and Kolinga-1, Chad. A similar mandible is described from the Adu-Asa Formation (Kalb and Mebrate 1993).

STEGODON FALCONER AND CAUTLEY, 1847

Sources for the cranial characters of this predominantly Asian genus are from Saegusa (1987) and Osborn (1942); dental characters come from Osborn (1942) and Taruno (1985). Isolated molars of *Stegodon* from Africa are also generally considered in this analysis, as figured in MacInnes (1942), Petrocchi (1954) Coppens (1967), Cooke and Coryndon (1970), Beden (1976) and Kalb and Mebrate (1993).

PRIMELEPHAS GOMPHOTHEROIDES MAGLIO, 1970*b*

Refers to the type and hypodigm specimens from Lothagam which we identify (and partly re-identify) as follows: right and left M^3 (both KNM LT-351) (Maglio 1970*b*, pl. 3, figs 7 and 9, respectively); right and left M^2 (both KNM LT-358) Maglio 1973, pl. 3, fig. 3 and Maglio and Ricca 1978, pl. 2, fig. 6, respectively); and an incomplete left M_3 (KNM LT-358) (Maglio and Ricca 1978, pl. 2, fig. 3). We include with the Lothagam material a partial right M_2 (MCZ 54-67K) (number and photo provided by V. Maglio), also a complete right M_2 within a jaw fragment (KNM LT-342) misidentified as *S. orbus* in Maglio (1970*b*, pl. 2, fig. 4), as first noted by Gaziry (1987*b*, pp. 197–8). The latter M_2 has concave–concave opposing faces of enamel loops, a character unique to plates of lower molars of all elephantines.

We exclude from the Lothagam specimens identified as *P. gomphotheroides* a right and left M_1, a left lower incisor fragment, and associated mandibular symphysis fragments, which we believe are *S. orbus*. All but the left M_1 are figured in Maglio (1970*b*, pl. 14, figs 11 and 12) and Maglio and Ricca (1978 pl. 1, fig. 3, pl. 2, fig. 8) under the specimen numbers KNM LT-358 and MCZ 290–67K (as inscribed in fig. 12, pl. 4, Maglio 1970*b*; also see Maglio and Ricca 1978, pp. 24–5). An additional isolated incisor fragment referred to by Maglio and Ricca (1978, p. 25) as *Primelephas* is evidently associated with these same specimens. The right M_1 figured in Maglio and Ricca (1978, pl. 1, fig. 3) has distinctive convex–convex opposing faces of enamel loops, a 'gomphothere' feature characteristic of plates of lower molars of *Stegotetrabelodon*, like M_1 from Sahabi (Gaziry 1982, fig. 2; 1987*b*, fig. 11). It is apparent that these associated dental and mandibular fragments represent a single pre-adult individual (compared with the subadult type mandible, KNM LT-354 of *S. orbus*; Maglio 1973, pl. 2, figs 1–3). Hence, we know of no M_1, extended mandibular symphysis, or lower incisors that are, as yet, reported for *Primelephas*.

We also include with *P. gomphotheroides* a right M^2 (KNM LU-718) from Lukeino-A (Tassy 1986, pl. 14, fig. 1) and additional M^2 from the Adu-Asa Formation (Kalb and Mebrate 1993).

LOXODONTA ANONYMOUS, 1827

Referred to in the text as the '*Loxodonta* Group' — in our analysis refers to the lineage of loxodonts including: 'cf. *Loxodonta*' from Lukeino-A (Tassy 1986, pl. 14, fig. 6); *L. exoptata* from Laetoli, Tanzania, East Turkana (Koobi Fora) and Kanjera, Kenya, the Omo Valley, Ethiopia (MacInnes 1942, pl. 8, figs 4 and 5; Maglio 1969*b*; Beden 1983, fig. 3.9, 1987*a*, 1987*b*, fig. 7), and the Matabaietu Formation, Ethiopia (Kalb and Mebrate 1993); *L. atlantica* (Beden, 1987*a*, pl. 8.5, 1987*b*; Coppens 1965, pls. 11–12; Maglio, 1973); and *L. africana* from various localities in Africa (Osborn 1942, pp. 910, 1172–6, 1197–200, 1273; Coppens 1965, pls. 15–16; Laws 1966; Sikes 1966, 1971; Maglio 1970*b*, 1973; Beden 1979). Note: *Loxodonta adaurora* (see below) is distinguished from this group of loxodonts by having a concave–forward anterior face on enamel loops of worn lower molars (characters 21 and 22). Conversely, the *Loxodonta* Group differs from *L. adaurora* by having convex–forward anterior faces (a relic 'gomphothere' feature) on enamel loops of worn lower molars. Hence, the combination of medially expanded concave–convex enamel loops on lower molars is unique to all species of the *Loxodonta* Group and distinguishes this group from *L. adaurora*, as well as from all other elephantoids. The convex–concave ('lozenge'-shaped) enamel loop is formed due to longitudinal offset between the ('pretrite' and 'posttrite') half-plates, which causes the anterior face of the total enamel loop of lower molars to be wholly or partially expanded (depending on the degree of wear) in a convex fashion; the posterior face is expanded in a concave fashion due to incorporation into the total loop of the posterior pretrite column and to the forward rotation of the outer portion of the half-plates. The lateral margins of the plates on the lower molars, however, may reflect the concave–concave condition, particularly with little wear, a condition unique to all elephantines. A pretrite column posterior to the enamel loop of the lower molar is present in *all* loxodonts, but only *L. adaurora* has twin posterior columns, that is, posttrite as well as pretrite posterior columns (see characters 28 and 30, Appendix 12A.2). In respect to the monophyly of the *Loxodonta* Group, Beden (1985, fig. 2) regarded *L. exoptata–L. atlantica–L. africana* to be of the same lineage and placed *L. adaurora* on a separate lineage. We agree with Beden's assessment, and add 'cf. *Loxodonta*' from Lukeino-A as the basal member of the lineage (see Tassy 1986, fig. 45).

LOXODONTA ADAURORA MAGLIO, 1970*b*

Refers specifically to this species, apart from the *Loxodonta* Group described above. Type and hypodigm specimens of *L. adaurora* from Kanapoi, Ekora, and Lothagam (Unit–3), Kenya (Maglio 1970*a*, 1973), two specimens from the 'Kubi Algi' beds, Kenya (FS 160/17–17A and FS 160/18–18a) (photos and specimen numbers provided by V. Maglio), and specimen L151–1 from the Sagantole Formation, Ethiopia (Kalb and Mebrate 1993). See note under *Loxodonta* above. The informal term 'loxodont' used in the test often refers to both the *Loxodonta* Group and to *L. adaurora*.

***MAMMUTHUS* BROOKES, 1828**

The various referred species for the genus from: Osborn (1942, pp. 910, 976), Coppens (1965 — see Maglio 1973, p. 53), Arambourg (1970), Maglio (1970*b*, p. 18; 1973), and Maglio and Hendey (1970); also, *Mammuthus* from Hadar, Ethiopia (Johanson and Edey 1981, pp. 96/97, fig. 4; Beden 1985, p. 44).

***ELEPHAS* LINNAEUS, 1758**

Type, hypodigm and referred material of *E. ekorensis* from Ekora and Kanapoi, Kenya (Maglio 1970*b*) and other African material of the genus in Maglio (1973), Beden (1983), and Kalb and Mebrate (1993). Skull and mandible morphology for the genus *Elephas* from Osborn (1942, pp. 910, 956–9, 976).

Appendix 12A.2: Characters

Figured specimens illustrating the morphological characters used in our phylogenetic analysis can be found in the references cited in Appendix 12A.1. Supplementary references that illustrate or highlight specific features are cited in this appendix. Key to character states: plesiomorphic (0), derived (1, 2, 3, 4), missing data (9).

SKULL

1. Cranium: flattened (0), raised (1). Flattened relates to the subparallel relationship of the dorsal surface of the cranium with respect to the ventral surface of the maxilla or the occlusal surface of the upper molars. The plesiomorphic condition of this character, and of characters 2 and 3, can be seen clearly in the earliest proboscideans (see Tassy and Shoshani 1988, fig. 11.4).
2. Occipital condyles: protruding (0), recessed (1). 'Protruding' pertains to whether the occipital condyles extend beyond the posterior side of the cranium (as seen in profile) when it is in the erect position, or whether it is 'recessed' because the posterior cranium distinctly overlaps, or rises above the occipital condyles.
3. Basicranium: not erect (0), erect (1).
4. Cranium: non-globular (0), globular (1). This character refers to the loxodontoid 'globular' cranium with a recurved face, as characterized by Maglio (1972, p. 22), Beden (1987*a*, p. 261), and others (e.g. Anonymous 1827, p. 141; Osborn 1942, fig. 1042).
5. Fronto-parietal region of cranium: not concave (0), concave (1). This condition is illustrated in Maglio (1970*b*, figs 2 and 3; 1973, fig. 33).
6. Maximum width of the mandible from the lateral outer margin of the right ramus to the outer margin of the left ramus: narrow (0), wide (1). The plesiomorphic condition can be seen by comparing figures in Osborn (1936, pl. V, pp. 236/237, also figs 316, 591; 1942, fig. 794).

TUSKS

7. Premaxillary tusk sheaths distally: narrow (0), flaring (1). See Maglio (1970*b*, figs 2 and 4, p. 18), Sikes (1971, pl. 18), and Beden (1987*a*, p. 261).
8. Upper tusks: downturned (0), not downturned (1). Characters 8 and 9 distinguish between upper tusks that are downturned, upturned, and straight.
9. Upper tusks: not upturned (0), upturned (1). See note under character 8.
10. Extended mandibular symphysis: not downturned (0), downturned (1).
11. Mandibular symphysis: extended (0) shortened, or fully reduced (1). Characters 11 and 12 distinguish between mandibular symphyses that are fully longirostrine, fully brevirostrine, and those intermediate in length, such as *Stegodibelodon*.
12. Mandibular symphysis: extended or shortened (0), fully reduced (1). See note under character 11.
13. Lower tusks: present (0), absent or vestigial (1). Vestigial lower tusks or incisor cavities are often present in the earliest species of brevirostrine elephantoids, including elephantines (e.g. Hooijer 1958, pl. 32; Maglio 1970*b*, 1973; Sarwar 1977, fig. 56; Tobien *et al.* 1988, fig. 86).
14. Lower tusks in cross-section: flattened or pyriform (0), circular (1). The pyriform shape of '*M.*' *grandincisivus* is illustrated in Tobien (1978*b*, p. 178).

MOLARS

15. Distinct alignment of apices (cones, conelets, columns) in curvilinear manner, as seen particularly on unworn molars: absent (0), present (1). This character does not necessarily imply the formation of full plates, unless the apices disappear with moderate wear and form a fully connecting enamel envelope. See character 23.
16. Molar brachyodonty: not accentuated (0), accentuated (1). From Tassy (1990*a*, p. 43, character 108).
17. Alternating and interlocking pretrite and posttrite half-loph(id)s ('anancoidy'), as seen particularly on worn molars: absent (0), present (1). See Tobien (1973*b*, p. 130).

18. Posterior face of enamel loops of worn lophids or lower plates: very convex (0), shallow convex or absent (1). Note: *Stegodon* may show a mixed combination of parallel, subparallel, shallow convex–convex or shallow concave–concave opposing faces of enamel loops on plates of the same molar, as an intermediate condition with respect to other elephantids (for illustrations, see Osborn 1942, chapter XIV).

19. Posterior face of enamel loops of worn lophids or lower plates: not concave (0), shallow, very concave, or mixed (1). See note under character 18.

20. Posterior face of enamel loops of worn lophids or lower plates: convex or mixed (0), concave (1). See note under character 18.

21. Anterior face of enamel loops of worn lophids or lower plates: convex (0), concave or mixed (1). See note under character 18.

22. Anterior face of enamel loops of worn lophids or lower plates: concave or mixed (0), concave (1). See note under character 18.

23. Plates: absent (0), present (1). A plate is referred to here as a fully connecting enamel envelope with moderate wear, that is compressed antero-posteriorly.

24. Loph(id)s or plates: expanded (0), distinctly compressed (1). This is a generalized evaluation based on lamellar frequency (Maglio 1973, p. 12).

MOLAR ACCESSORIES

25. Median sulcus: present (0), completely absent or present only at the anteriormost end of the molar (1).

26. Median sulcus: extends the length of the molar (0), extends to the mid-region of the molar (1).

27. Number of apices (cones, conelets) as part of the main loph(id)/plate alignment: four (0), five or more (1). Apices refer to fully developed apical features, not those that are superficially divided by longitudinal grooves.

28. On M_2 and M_3, posterior central posttrite conules or columns: confined to anterior end of molar only or absent (0), extends to posterior end of molar (1).

29. On M^2 and M^3 with lophs (not plates), three posterior posttrite conules (at anterior end of molar): absent (0), present (1).

30. On M_2 and M_3, posterior central pretrite conules or columns on posterior part of molar: present (0), reduced, fused, or incorporated into lophids/plates (1). Refers to those taxa that have the derived condition relative to those taxa whose posterior pretrite conules are distinct and isolated when the molar is unworn, and create a distinct posterior fold when worn. The plesiomorphic condition must extend to the posterior end of the molar.

31. On M^3, posterior central pretrite conules or columns on the posterior half of the molar: absent (0), present (1).

PLATE NUMBER

32. On M_2 and M^2, minimum number of loph(id)s or plates: three (0), four (1), five (2), or six (3). (Note: by 'minimum' we mean that if the plate number for a given molar varies from 7 to 9, the number used in our analysis is seven).

33. On M_3, minimum number of lophids or plates: four (0), five (1), six (2), seven (3), or eight (4). See note under character 32.

34. On M^3, minimum number of lophs or plates: three (0), five (1), seven (2), or nine (3). See note under character 32.

TWELVE B

Palaeobiogeography of late Neogene African and Eurasian Elephantoidea

JON E. KALB, DAVID J. FROEHLICH, AND GORDEN L. BELL

INTRODUCTION

Our cladistic analysis of late Miocene to Pleistocene proboscideans presented in this volume (Kalb *et al.*, this volume, Chapter 12A) allows us new insights on the temporal and palaeogeographical distribution of late Neogene elephantoids, as summarized in Figs 12B.1 and 12B.2 (cf. Kalb 1995). The terminology and taxa discussed in this chapter are described in detail in the text and in Appendix 12A.1 of Chapter 12A. This report is not intended to be an exhaustive account of the palaeobiogeography of late Neogene elephantoids, rather we will address only dates and faunal patterns that bear on our particular cladistic analysis. In our discussion, Neogene epoch and European stage boundaries are based on Harland *et al.* (1990, fig. 3.15) (Fig. 12B.1); supplementary references are noted as necessary.

CHRONOSTRATIGRAPHY

EARLY ELEPHANTOIDS

'*Mastodon*' *grandincisivus* from western Eurasia is known from sites as old as 'Sarmatian' (12.8–11.6 Ma) (late Serravallian) to 'upper Pannonian' (*c.* 8.5–7.0 Ma) (late Tortonian; Tobien 1978*b*; Steininger *et al.* 1990, fig. 1). A 'grandincisivoid' specimen from Djebel Semenne, Tunisia, described by Tobien (1978*b*, p. 184), is believed to be 'middle Vallesian' (*c.* 11–10 Ma; Steininger *et al.* 1990, fig. 1). *Tetralophodon longirostris* is known from western Eurasia from widespread localities of 'lower Pannonian' or 'Vallesian' (early Tortonian) age (*c.* 11.5–8.8 Ma; Tobien 1978*b*; Mein 1990; Steininger *et al.* 1990, fig. 1). Incomplete material comparable to *T. longirostris* is reported from a dozen or so Tortonian-aged localities in northern Africa (Bergounioux and Crouzel 1956, p. 547; Coppens *et al.* 1978; Tobien 1978*b*; Geraads 1989); also, from near Chorora, Ethiopia (10.7–10.5 Ma; Tiercelin *et al.* 1979), and the Samburu Hills, Kenya (~ 9 Ma; Nakaya *et al.* 1984; Pickford *et al.* 1984). *Tetralophodon* of uncertain age and identification has recently been recovered from Uganda (M. Pickford, personal communication, 1992). *Paratetralophodon hasnotensis* is described from the Dhok Pathan Formation (Hasnot region, 7.4–4.6 Ma) of the Middle Siwaliks of Pakistan and from deposits from the stratigraphically younger Tatrot zone (3.4–2.5 Ma) of the Upper Siwaliks (Sarwar 1977, p. 101; Azzaroli and Napoleone 1982; Barry *et al.* 1982; Tassy 1983, p. 276; Flynn *et al.* 1990, p. 596). Tobien *et al.* (1988, pp. 108 and 200) describe a related form from Tortonian-aged deposits (10.4–6.7 Ma) in China.

ANANCUS

The earliest documented and datable occurrence of this genus in Eurasia may be the primitive *A. perimensis* (Tassy 1986, fig. 40) from the Dhok Pathan Formation (7.4–4.6 Ma) in the Siwaliks, although the range of this species in this formation is uncertain (Tassy 1983, fig. 38; Flynn *et al.* 1990, p. 596). *Anancus* is also reported from the lower Gaozhuang Formation (5.35–4.5 Ma), China, which appears to bracket the Mio-Pliocene boundary (5.2 Ma; Tedford *et al.* 1991). The earliest occurrences of the African tetralophodont, *A. 'kenyensis'* (as defined in Appendix 12A.1) is from the Mpesida beds, Kenya (6.4–6.0 Ma) and from Member A of the Lukeino Formation (Lukeino-A), Kenya (*c.* 6.2–5.6 Ma; Hill *et al.* 1985, 1986; Tassy 1986). This species is also present in the Adu-Asa Formation, Ethiopia and probably Unit 1 of the Lothagam Group (Lothagam-1), which on faunal grounds are believed to be late Miocene in age (*c.* 6.0–5.2 Ma) (Smart 1976; Kalb and Mebrate 1993; Kalb 1993). The derived tetralophodont *Anancus* from Langebaanweg, South Africa comes from deposits of apparent early Pliocene age (*c.* 4.5–4.0 Ma) (Hendey, 1970*a*, 1970*b*, 1978). The Sahabi Formation, which yielded the pentalophodont *A. petrocchii*, is widely believed to be basal Pliocene (see Boaz 1987; Boaz *et al.* 1979), although Geraads (1989, fig. 4) and Tassy (1990*b*, fig. 4) suggested the Sahabi fauna is latest Miocene. The Sagantole-type pentalophodont *Anancus* from the upper Sagantole Formation, Ethiopia is of similar age (*c.* 3.75–3.5 Ma; Haileab and Brown 1992; Kalb and Mebrate 1993) and type as a pentalophodont *Anancus* from Kanapoi, Kenya (4.0–3.75 Ma; Powers 1980; Brown *et al.* 1985). The Warwire Formation, Uganda (3.6–3.4 Ma; Pickford *et al.* 1991)

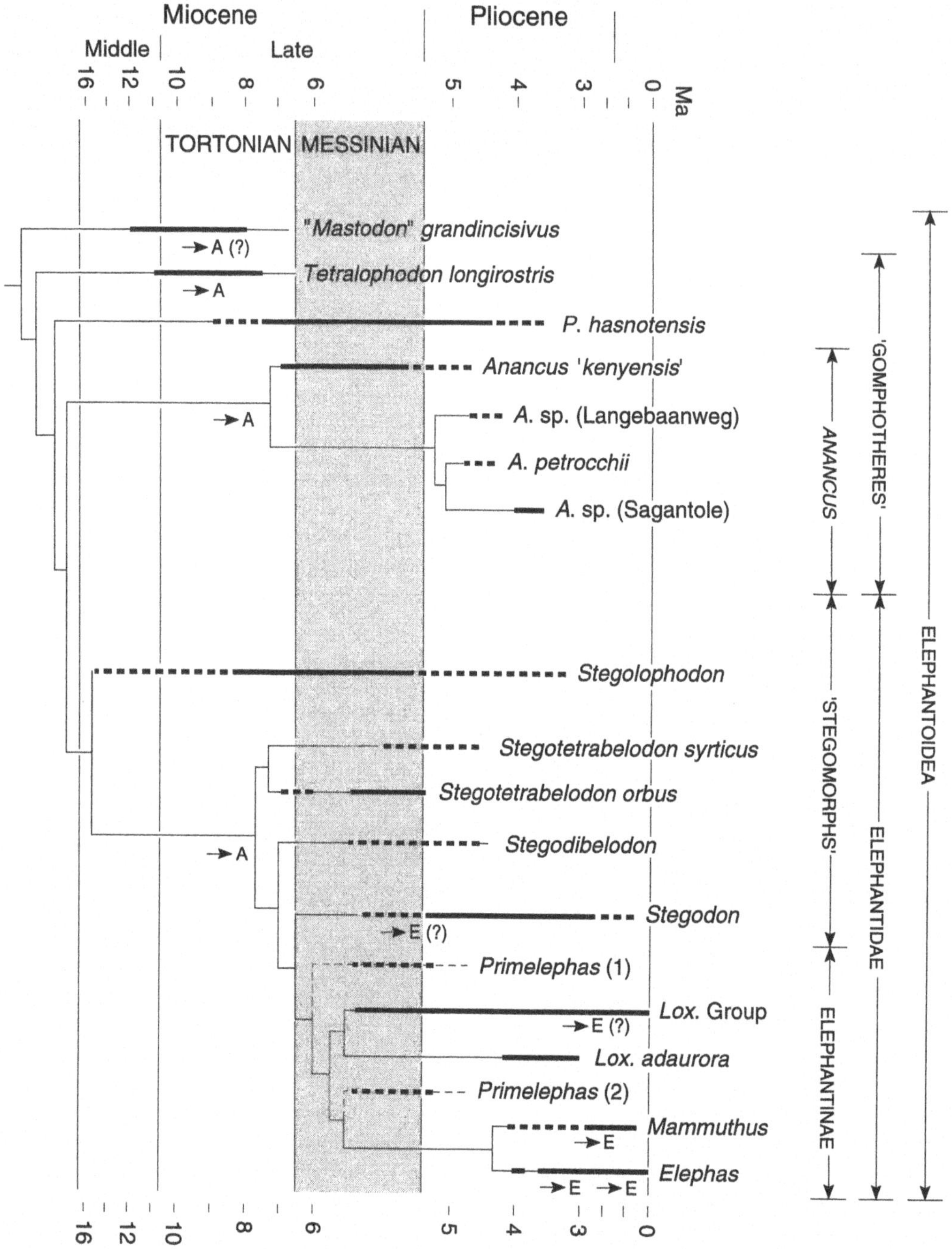

Fig. 12B.1 Phylogenetic relationships of late Neogene Proboscidea. Neogene epoch and European stage boundaries are from Harland *et al.* (1990, fig. 3.15). Bold lines refer to known or uncertain (dashed) time ranges of taxa. Thin lines refer to proposed phylogenetic relationships. Arrow symbols refer to migrations to **A**(frica) or **E**(urasia). Based on Fig. 12A.1, this volume.

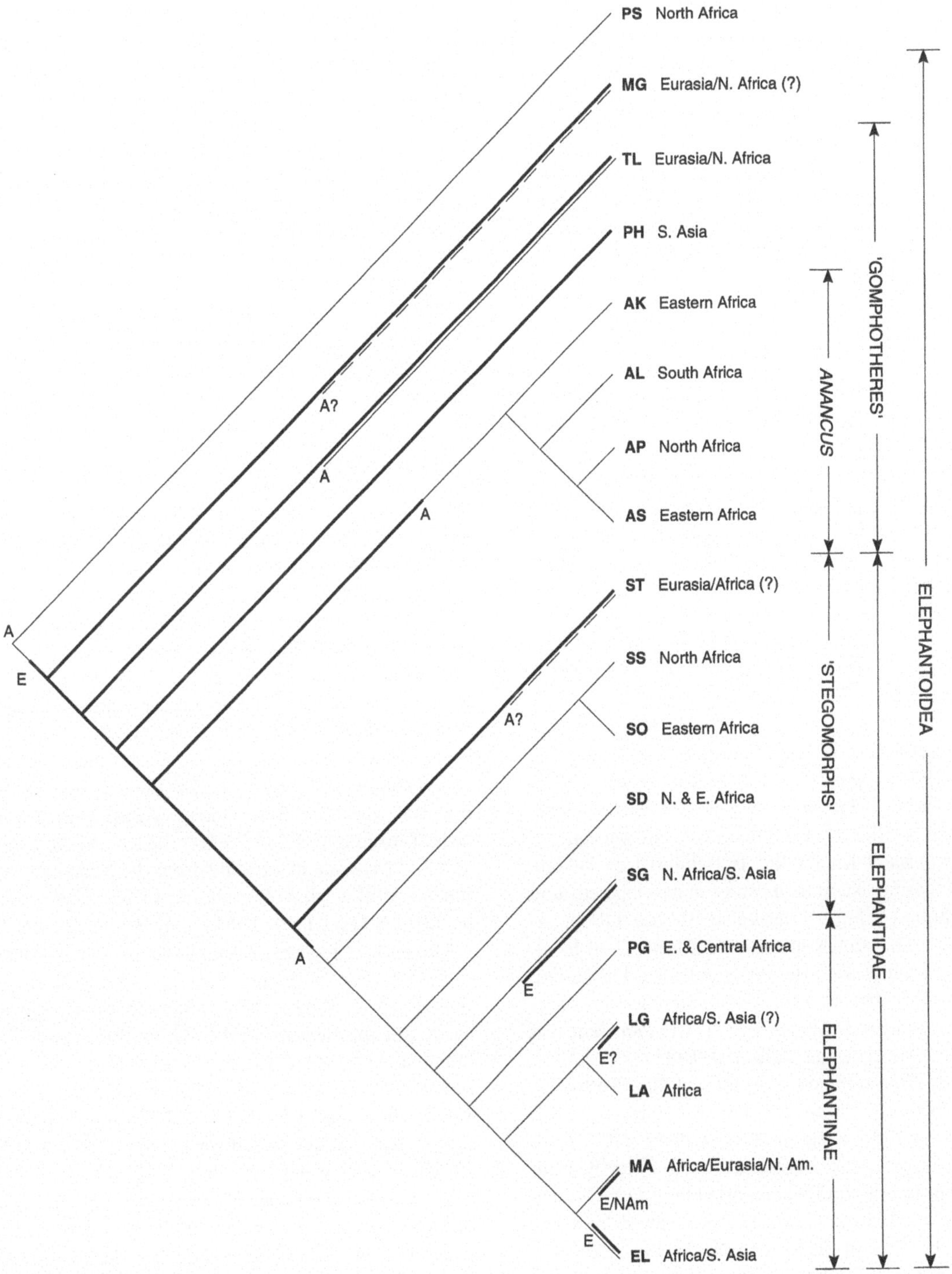

Fig. 12B.2 Geographical cladogram of late Neogene Proboscidea. Bold lines refer to known or possible (dashed) distribution of taxa in **E**(urasia), thin lines in **A**(frica). See Fig. 12B.1 for sequence in which taxa appear; abbreviations are of the same taxa, except for PS which stands for *Phiomia serridens*. Based on Fig. 12A.1, this volume.

has yielded an *Anancus* of an undescribed type; however, its age and provenance suggest comparison with a very derived (Sagantole-type?) *Anancus* from Nyawiega, Uganda (Cooke and Coryndon 1970, pl. 1, fig. B–C).

Overall, evidence suggests that the Eurasian *Anancus*, *A. perimensis* may have preceded the African *Anancus* species, although the age range of *A. perimensis* requires confirmation. Otherwise, it is apparent that the genus persisted in sub-Saharan Africa for at least 2.5 million years. In northern Africa, the tetralophodont *A. osiris* is reported to have survived well into the Pliocene (Tassy 1986, fig. 41).

EARLY ELEPHANTIDS

The earliest elephantid, *Stegolophodon*, is known in Asia from deposits of Pliocene and late Miocene age (e.g. Osborn 1942, p. 838; Sarwar 1977; Tassy 1983, fig. 38, p. 323; Johnson *et al.* 1985; Taruno 1985; Tobien *et al.* 1988, table 4; Flynn *et al.* 1990, p. 596; Qiu 1990, table 2). *Stegolophodon* has been reported to be present in middle Miocene deposits as well (Raza *et al.* 1984; Ginsburg and Tassy 1985; Tobien *et al.* 1988, p. 194) (Fig. 12B.1), although these occurrences await further confirmation (P. Tassy, personal communication, 1992; Kalb *et al.*, this volume, Chapter 12A, Appendix 12A.1). The earliest *Stegotetrabelodon* fossils in Africa are those referred by Tassy (1986, fig. 6) to *S. orbus* from the Mpesida beds (6.4–6.0 Ma) (Hill *et al.* 1986). The type locality for this species is Lothagam-1 (Maglio 1973), believed by many authors to be late Miocene in age (*c.* 6.0–5.2 Ma; e.g. Smart 1976; Powers 1980; Brown *et al.* 1985). *Stegotetrabelodon syrticus* is known only from Sahabi (Petrocchi 1943, 1954), which is either earliest Pliocene or late Miocene.

Stegodibelodon schneideri from Chad is estimated to be late Miocene to early Pliocene, based on its stage of development and on its associated fauna (*A. osiris, Primelephas, Stegodon*) (Coppens 1965, 1967, 1972). The proboscidean genera from Chad localities are the same as those from the lower Adu-Asa Formation, Ethiopia, which has yielded an edentulous mandible similar to that of *S. schneideri* (Kalb and Mebrate 1993). The total Adu-Asa fauna suggests a late Miocene age (Kalb *et al.* 1982).

Stegodon is known from numerous localities of Plio-Pleistocene age in southern Asia and China (e.g. Osborn 1942, fig. 730; Sarwar 1977; Taruno 1985; Saegusa 1987; Tedford *et al.* 1991). The earliest reported occurrence of *Stegodon* in Asia may be from the upper Mahui Formation in China, reported to be late Miocene in age (*c.* 5.5–5.3 Ma; Tedford *et al.* 1991; R. Tedford, personal communication, 1992). From the Upper Siwaliks, the Tatrot and Pinjor zones, Sarwar (1977) has identified several species of *Stegodon*, including the early species, *S. bombifrons* and *S. insignis*. The Tatrot zone is reportedly middle Pliocene in age (3.4–2.5 Ma; Azzaroli and Napoleone 1982; Barry *et al.* 1982). From eastern Asia, Saegusa (1987) refers to *Stegodon* as being Plio-Pleistocene in age.

From Africa, *Stegodon* has been described from Uganda (Cooke and Coryndon 1970; Pickford *et al.* 1991; Zaire (Sanders 1990), the Omo Valley, Ethiopia (Beden 1976), Chad (Coppens 1967), and Sahabi, Libya (Petrocchi 1943, 1954). Boaz (1987) suggests an early Pliocene age for Sahabi, although Geraads (1989) and Tassy (1986, fig. 18) propose a late Miocene age. Kalb and Mebrate (1993) describe a juvenile mandible from the Middle Awash, Ethiopia, tentatively assigned to the lower Adu-Asa Formation, that may date to the latest Miocene (*c.* 6.0– 5.2 Ma). *Stegodon* from the Warwire Formation, Uganda has a confirmed middle Pliocene age (3.6–3.4 Ma) (Pickford *et al.* 1991). *Stegodon* from upper Member B, Shungura Formation, Ethiopia may be the latest known evidence (~ 3.0 Ma) of the genus in Africa (Beden 1976; Brown *et al.* 1985, fig. 2).

Overall, evidence suggests that *Stegodon* existed in Africa from the late Miocene to ~ 3.0 Ma, or for a period of at least 2 million years. From China, *Stegodon* may also have first appeared in Asia during the late Miocene, and lasted well into the Pleistocene.

ELEPHANTINAE

The recovery of *Primelephas* and 'cf. *Loxodonta*' from Lukeino-A (Tassy 1986, fig. 6) shows that these two early elephants were contemporaneous at the end of the Miocene (6.2–5.6 Ma) (Hill *et al.* 1985, 1986). Likewise, the recovery of *L. adaurora* from Lothagam-3 (Maglio 1973) and of *L. exoptata* (of the *Loxodonta* Group) from Laetoli shows that the two *Loxodonta* lineages were contemporaneous at least to *c.* 3.75–3.5 Ma. *Loxodonta*, similar to 'cf. *Loxodonta*' from Lukeino-A, may also be present at Kanam East (MacInnes 1942, pl. VII, figs 8–10), which appears to be basal Pliocene in age (*c.* 5.2–4.75 Ma; Pickford 1986, 1987a). *Loxodonta exoptata*, also of the *Loxodonta* Group, but more derived than 'cf. *Loxodonta*' from Lukeino-A, is known from the Laetolil beds, Tanzania (*c.* 4.5–3.75 Ma) (Beden 1987*a*; Drake and Curtis 1987). '*Loxodonta* cf. *exoptata*' from the Ndolanya beds, Tanzania (Beden 1987*a*, pl. 8.7), believed to be middle Pliocene in age (*c.* 3.0–2.5 Ma) (Harris 1987), may be a very early *L. atlantica*.

The earliest reported *Mammuthus* is *M. subplanifrons* from the early Pliocene of Langebaanweg, South Africa (Maglio and Hendey 1970; Maglio 1973), although an M_3 recovered from the basal tuff at the site of Wee-ee in the Sagantole Formation, Ethiopia is very similar to the South African M_3 (Kalb and Mebrate 1993). Apparently, the Wee-ee tuff is early Pliocene (~ 4.1 Ma) (Haileab and Brown 1992; Kalb and Mebrate 1993), of a similar reported age as Langebaanweg (*c.* 4.5–4.0 Ma) (Hendey 1978). A more progressive, but as yet undescribed, *Mammuthus* is known from the middle Hadar Formation, Ethiopia (~ 3.0 Ma) (Beden 1985) and perhaps from the Kalaloo beds of the Sagantole Formation (*c.* 3.7–3.5 Ma) (Haileab and Brown 1992; Kalb and Mebrate 1993). Thus, whereas the earliest occurrence of *M. subplanifrons* is of early Pliocene age, the Hadar-type *Mammuthus* is confined more to the middle Pliocene.

The earliest known species of *Elephas* may be *E. ekorensis* from Ekora and Kanapoi, Kenya (~ 3.75 Ma) (Powers 1980; Brown *et al.* 1985). *Elephas recki brumpti* is reported from just below and above the Cindery Tuff (3.73 Ma) in the upper Sagantole Formation (Clark *et al.* 1984; Haileab and Brown 1992; Kalb and Mebrate 1993).

PALAEOBIOGEOGRAPHY

FIRST MIGRATIONS

The earliest proboscideans are confined to the African Eocene and Oligocene (Coppens *et al.* 1978; Tassy and Shoshani 1988). The earliest entry of proboscideans into Eurasia may have been in the earliest Miocene (~ 22 Ma or more) with the eastward migration from Africa into Asia Minor and southern Asia of primitive elephantoids (Savage 1990; *sensu* Tassy 1990*a*, fig. 17; 1990*b*, p. 249). Among the latter were amebelodons, and 'trilophodont gomphotheres', notably the genus *Gomphotherium*. The time of this proposed migration corresponded to one of lowered sea-levels and lowered temperatures (Haq *et al.* 1988, fig. 14). The subsequent northern expansion of these early elephantoid migrants into western Europe has previously been referred to as the 'Proboscidean Datum Event' (Madden and Van Couvering 1976). This event occurred during the middle Burdigalian (*c.* 19–17 Ma) (Tassy 1990*b*, p. 249), a time that corresponds to one of increased temperatures and elevated sea-levels (Haq *et al.* 1988, fig. 14).

EURASIAN ORIGINS

Throughout the middle Miocene, the Langhian and Serravallian stages (16.3–10.4 Ma; Harland *et al.* 1990, fig. 3.15), amebelodons and trilophodont gomphotheres dispersed widely across Eurasia (Tassy 1990*b*, figs 4 and 5), during a relatively stable period of high sea-levels and warm temperatures (Barry *et al.* 1985, p. 639; Haq *et al.* 1988, fig. 14). If '*M.*' *grandincisivus* is an amebelodon, as proposed by Tassy and Darlu (1986) and Tassy (1990*b*, fig. 4), it could well be a descendant of the initial radiation of amebelodons into western Europe and elsewhere. Likewise, it is possible that some of these same descendants may have inhabited the Middle East and Arabia and crossed back into Africa in limited numbers in the late Miocene. Such an event could have occurred at the base of the Tortonian (10.5 Ma), that coincides with a sharp drop in global sea-level (Haq *et al.* 1988, fig. 14). This scenario would explain the presence of Tortonian-aged '"grandincisivoid" amebelodons' in Tunisia, Zaire, and Arabia (Tobien 1978*b*, p. 184; Madden 1982; Madden *et al.* 1982) and the presence of the progressive amebelodon, *Platybelodon* (= *Amebelodon*) *cyrenaicus*, at Sahabi, Libya (Gaziry 1987*b*; Tassy 1990*b*, fig. 4). *In situ* evolution of such late Miocene African amebelodons in Africa, however, is also possible because this group is well represented in sub-Saharan Africa in middle Miocene deposits (Tassy 1986, 1990*b*, fig. 4).

The alleged direct descendants of the 'trilophodont gomphotheres', the 'tetralophodont gomphotheres', including *Tetralophodon longirostris* and *Paratetralophodon hasnotensis* (Tobien 1978*b*, p. 188; Tassy 1990*a*, fig. 17), are widely represented across Eurasia in late Miocene deposits of Tortonian age (Tobien 1978*b*, Tassy 1983, fig. 38; Flynn *et al.* 1990). The scattering of Tortonian-aged localities in northern Africa bearing *T. longirostris* suggests that this elephantoid may have also crossed into Africa during the low sea-levels at the base of the Tortonian (10.5 Ma). More derived populations of longirostrine tetralophodont gomphotheres reported at Chorora, Ethiopia (10.7–10.5 Ma) and the Samburu Hills, Kenya (~ 9 Ma) may be a result of this same migration. These fossils are the only other longirostrine tetralophodont gomphotheres known from sub-Saharan Africa.

Anancus was present throughout Eurasia during the late Miocene (e.g. Osborn 1936; Tobien 1978*b*; Tassy 1986; Tobien *et al.* 1988; Tedford *et al.* 1991). As proposed by Tassy (1986, pp. 102, 126), *A. perimensis* from southern Asia (*c.* 7.4–4.6 Ma), or its immediate descendant, may have crossed into Africa during the middle Tortonian (*c.* 8.0–7.5 Ma), a time of low sea-levels and documented faunal interchange (Barry *et al.* 1985; Haq *et al.* 1988, fig. 14). *Anancus* is unknown in Africa prior to Mpesida times (6.4–6.0 Ma), and pre-Messinian brevirostrine tetralophodont gomphotheres reported in Africa are also rare (Tassy 1986, pp. 82–7).

As documented from Lukeino-A, Kanam, the Middle Awash, and Lothagam-1, *A. 'kenyensis'* was present in eastern Africa throughout Messinian times (6.4–5.2 Ma). Whether this early tetralophodont was restricted to eastern Africa is unknown, but it is clear that by the middle Pliocene (~ 4.0 Ma), more derived forms of tetralophodont and pentalophodont *Anancus* covered the length of the continent, as indicated by the South African (Langebaanweg-type), north African (*A. petrocchii*), and eastern African (Sagantole-type) *Anancus* populations. At this same time, more derived populations of *Anancus* also covered the breadth of Eurasia (Tassy 1986; Tobien *et al.* 1988).

The genus *Stegolophodon* appears to have been endemic to Eurasia at least during the Tortonian and Messinian (e.g. Osborn 1942, p. 814; Tassy 1983; Taruno 1985). It is possible that small numbers of the genus made their way to Africa, perhaps in the earliest Tortonian or, more likely, in the Messinian, as indicated by isolated (and very questionable) fossils from the Congo (Hooijer 1963, 1970) and Sahabi (Petrocchi 1943, 1954). Whatever the case, it is clear that by the earliest Messinian, the sister-group of *Stegolophodon* (the remaining elephantids) had begun its rapid diversification and expansion in Africa.

AFRICAN ORIGINS

The genus *Stegotetrabelodon* first appeared in Africa during the latest Tortonian or the earliest Messinian, while *Stegodibelodon* and *Stegodon* first appeared apparently in the early Messinian. All known fossils of these early elephantids are confined to north of the Equator, which may relate to their limited southern expansion in Africa, palaeoenvironmental reasons, or to limited fossiliferous exposures of that age in sub-Saharan Africa.

The plesiomorphic *Stegotetrabelodon syrticus* is known only at Sahabi. If this site is in fact basal Pliocene (Boaz 1987), then the absence of early elephantines at Sahabi, and the presence of *S. syrticus* and a panoply of progressive proboscideans [*Platybelodon* (= *Amebelodon*) *cyrenaicus*, '*Stegolophodon sahabianus*', *Stegodon syrticus, Anancus petrocchii*] (Petrocchi 1954; Gaziry 1987*b*), suggests that northern Africa may have become isolated during the earliest Pliocene by conditions similar to those today (Kalb and Jolly 1982). A simpler explanation, however, is that Sahabi is of a late Miocene, pre-elephant age.

Stegotetrabelodon orbus is, thus far, known exclusively in eastern Africa (Kenya, Ethiopia, possibly Uganda) (Cooke and Coryndon 1970, Pl. 3, figs A–B; Tassy 1986; Kalb and Mebrate 1993). Its distance from Sahabi and its derived nature could suggest geographical expansion of the genus from a northern African origin. The type locality of *Stegodibelodon schneideri* is northern Chad (Coppens 1972; Coppens and Koeniguer 1976, fig. 1). The only other known fossil that may belong to this genus is from the Middle Awash, Ethiopia (Kalb and Mebrate 1993).

Traditionally, *Stegodon* has been regarded as endemic to southern Asia (Osborn 1942). However, our cladistic analysis (Kalb *et al.*, this volume, Chapter 12A) suggests that both the immediate ancestors of *Stegodon* and its sister-group, the Elephantinae, may be of African origin (Fig. 12B.1). Based on this finding and on the occurrences and available dates for the genus discussed above, we may hypothesize that *Stegodon* or its immediate ancestor originated in Africa during the early Messinian, then shortly crossed into the Middle East (Hooijer 1958, 1961) and southern Asia, most likely at the end of the Miocene, during well-documented periods of low sea-level in the Mediterranean and elsewhere (Haq *et al.* 1988, fig. 14). The limited numbers of *Stegodon* in Africa, none found later than the middle Pliocene (~ 3 Ma), and all restricted to the northern half of the continent, further suggest that expansion of the genus was constrained, perhaps by the rapidly diversifying elephants. In Asia, however, where abundant numbers of *Stegodon* are found in Plio-Pleistocene deposits across the continent, the genus flourished, initially alongside the more plesiomorphic *Stegolophodon*, then later in association with more derived populations of elephant migrants.

Primelephas and the earliest loxodonts inhabited Africa in the late Miocene and the early Pliocene, while more derived loxodonts, early *Mammuthus*, and early *Elephas* were present in Africa in the early Pliocene (Maglio 1973; Beden 1985; Tassy 1986; Kalb and Mebrate 1993). The first migration of *Elephas* into the southern Levant (Hooijer 1958) and Eurasia apparently began very soon after the first appearance of the genus in Africa (Maglio 1973). A second major influx of *Elephas*, accompanied by *Mammuthus*, and probably *Loxodonta* (e.g. Hooijer 1954*a*, plts. 11–12), began in the middle Pliocene (Maglio 1973, figs 48–50; Coppens *et al.* 1978, figs. 17.1 and 17.2).

CONCLUSIONS

The cladistic analysis presented in Chapter 12A conforms well with the chronostratigraphical and palaeobiogeographical distribution of late Neogene elephantoids described here. The lower taxa in our cladogram ('*Mastodon*' *grandincisivus, Tetralophodon longirostris, Paratetralophodon hasnotensis*, 'ancestral' *Anancus*, and *Stegolophodon*) appear earliest in the fossil record; also, they are probably of Eurasian origin. The higher taxa in our cladogram ('derived' *Anancus, Stegotetrabelodon, Stegodibelodon, Stegodon*, Elephantinae) appear later in the fossil record; they are probably of African origin. Our conclusions are as follows:

1. Evidence indicates that '*M.*' *grandincisivus, T. longirostris, P. hasnotensis*, 'ancestral' *Anancus, Stegolophodon*, and possibly *Stegotetrabelodon*, originated in Eurasia during the middle Neogene (*c.* 12–8 Ma). All appear to have been descendants of early elephantoids that migrated from Africa to Eurasia during the early Neogene.

2. A period of much reduced sea-levels and temperature reduction at the base of the Tortonian (10.5 Ma) is likely to have triggered the southern migration of *T. longirostris*, and perhaps 'grandincisivoid' amebelodons, into northern Africa. Another sea-level reduction during the middle Tortonian (8.0–7.5 Ma) may have coincided with the earliest migration of *Anancus* into Africa.

3. The Elephantidae, the sister-group of *Anancus*, appears to have originated in Eurasia in the early to middle Tortonian, with the appearance of *Stegolophodon* and the putative ancestor of the remaining elephantids. This ancestor may have crossed into Africa at the same time as early *Anancus*. *Stegolophodon* itself may have crossed into Africa during the Messinian, although this is questionable.

4. The rapid diversification and expansion of the African elephantids in the late Neogene coincided with the decline and eventual extinction of African gomphothere lineages. This began with the extinction of African *Tetralophodon* in the middle or late Tortonian and concluded with the eventual extinction of derived *Anancus* in the Pliocene.

5. *Stegotetrabelodon, Stegodibelodon*, and *Stegodon* first appeared in Africa by the early Messinian. The decline of these elephantids coincided with the rapid spread in the early Pliocene of the Elephantinae, the sister-group of *Stegodon*.

6. The elephantines are of African origin. *Primelephas* and the two loxodont lineages appear nearly contemporaneously in the fossil record during the middle Messinian, while the earliest reported *Mammuthus* and *Elephas* are early Pliocene in age.

7. The earliest migration of African elephantids into Eurasia may have begun with *Stegodon* (or its immediate ancestor) in the late Messinian, accompanying well-documented low sea-levels in the Mediterranean and elsewhere. For a period, *Stegodon* and *Stegolophodon* co-existed in southern Asia, al-

though *Stegodon* survived well into the Pleistocene. The elephants migrated into Eurasia, beginning first, apparently, with *Elephas* in the middle Pliocene, followed shortly by more migrations of *Elephas*, also *Mammuthus*, and probably *Loxodonta*.

ACKNOWLEDGEMENTS

We thank J. Shoshani, P. Tassy, and John Harris for their useful comments. We also thank the editors and Oxford University Press for their good will and consideration.

THIRTEEN

On the shovel-tusked elephantoids from China

JIAN GUAN

INTRODUCTION

Fossil remains of Proboscidea from the People's Republic of China (hereafter China) indicate a rich and diverse fauna which was an important component of proboscidean evolution. Historical reviews of these fossils have been given by Osborn and Granger (1932), Osborn (1936, 1942), Hopwood (1935), and Tobien *et al.* (1984, 1986, 1988). Tables 13.1 and 13.2 include some of the most productive localities of Neogene to Pleistocene elephantoids (especially gomphotheres; 'Elephantoidea' after Tassy 1988*a*), and a list of taxa found in China, respectively. This chapter focuses on a few gomphothere genera and species from China, the gomphotheres *sensu lato* include amebelodontids.

Shovel-tusked gomphotheres (genus *Platybelodon*) were so named because of their most striking features — shovel-like inferior incisors. They stood about 1.5–2.8 m tall at the shoulder and were assigned to more than 15 species. These taxa flourished during the middle Miocene in the Old World and extended into North America in the late Miocene. Central Asia is considered an important centre of radiation of shovel-tusked gomphotheres, based on *P. grangeri* and *P. danovi* from Inner Mongolia (Nei Mong-gol) and the former Soviet Union, which were discovered during the 1920s. The Central Asiatic Expeditions, sponsored by the American Museum of Natural History (AMNH, New York) collected numerous specimens, many of which were proboscideans. Based on the studies of the gomphothere fossils, four new species were established by Osborn and Granger (e.g. Osborn 1924*c*, 1929*b*; Osborn and Granger 1931, 1932). By far, the most typical proboscidean from the late Miocene of Inner Mongolia, China was *P. grangeri* Osborn, 1929*b*. More recent field work in Mongolia was carried out by a team from the AMNH (McKenna, personal communication 1992) and IVPP (Qiu *et al.* 1987*b*).

Abbreviations

AMNH, American Museum of Natural History (New York, USA); BNHM, Beijing Natural History Museum (Beijing, China); BPV, Beijing, Palaeontology, Vertebrate (abbreviations for specimens at BNHM); IVPP, Institute of Vertebrate Palaeontology and Palaeoanthropology (Beijing, China); OAT, Office of Archaeology of Tongxin, China.

SHOVEL-TUSKED ELEPHANTOIDS FROM CHINA

Platybelodon Borissiak, 1928 and *Amebelodon* Barbour, 1927 are typical members of this group, having intermediate trilophodont molars and a long, well-developed shovel-like symphysis with two strong plate-like lower incisors. The main differences from *Gomphotherium* are: (1) reduced upper tusk (only in *Platybelodon*; shorter than the symphysis with lower incisor); (2) compression on both dorsal and ventral sides of the inferior incisors; and (3) shovel-like symphysis with oversized and elongated plate-like lower incisors.

RESEARCH HISTORY

Tobien *et al.* (1986, 1988) reviewed proboscidean findings in China with emphasis on gomphotheres. Their studies expanded our knowledge of the classification and systematics of these groups.

Osborn (1936) summarized the knowledge on platybelodons; about five species of *Platybelodon* (including *Torynobelodon*) and six species of *Amebelodon* were discovered in the sediments of the Miocene and Pliocene. These groups share the shovel-tusks and gomphotheriid molars. Despite the accumulated materials, little systematic work on *Amebelodon* and *Platybelodon* has been conducted. Among the more important phyletic reconstructions were those of Barbour (1927, 1929, 1931) and Osborn (1936). Barbour placed *Amebelodon* and *Platybelodon* (including *Torynobelodon*) into the family Amebelodontidae. Osborn (1936) considered Amebelodontinae a branch of Gomphotheriidae, with the subdivision of *Platybelodon* in the Old World, and *Amebelodon* in the New World (cf. Gregory 1945). Tassy (1985), however, placed *Platybelodon* and *Amebelodon* in Amebelodontidae.

Some of the important published papers concerning phylogeny in this gomphothere group are: Belyaeva and Gabunia (1960); Maglio (1969*a*); Tobien (1973*a, b*); Tassy (1983, 1987); Tobien *et al.* (1986, 1988); Gaziry (1987). Width and the inner structure of the lower incisors were emphasized by these authors. Tobien (1972) incorporated *Torynobelodon* and *Serbelodon* into *Platybelodon* and *Amebelodon*, respectively. Tassy (1983, 1986) contributed to the phylogeny of shovel-tusked gomphotheres,

Table 13.1 Biostratigraphic subdivisions of the Neogene and early Pleistocene Land Mammal Ages in China (modified after Chiu *et al.* 1979; Qui *et al.* 1987*a*; Tobien *et al.* 1988; Qiu and Qiu 1990; Harrison *et al.* 1991).

Age (Ma)	MN Zone	European Land Mammal Age	Chinese Land Mammal Age	Faunal Units	Theriofaunal Units	Localities with * Chinese gomphotheres	Qiu's 1990's units Yushe Basin
1							
			Nihewanian			Nihewan Shagou	
2	MN17				*'Equus*		V (5)
		Villafranchian		VI	*Fauna'*		
3	MN16		Youhean			Youhe	
					Equus	Zhaotong	
4	MN15				last	Yushe II	IV (4)
		Ruscinian	Jinglean	V	*Hipparion*	(Gaozhuang)	III (3)
5	MN14				Fauna		II (2)
					↑	Ertemte	
6						Lantian	
	MN13					Lufeng (Shihuibe)	I (1)
		Turolian	Baodean			Qing Yang	
7						Yushe I (Baode)	
	MN12				*'Hipparion*	Fugu	
					Fauna'		
8	MN11			IV			
9	MN10						
						Bahe	
		Vallesian	Bahean				
10	MN9				↓		
11	MN8				↑	Zhongning	
12	MN7					Tung Gur	
						Junggar Basin	
13						Guanghe/Qinan/	
		Astaracian	Tunggurian	III		Huangzhong	
	MN6					Lengshuigou	
15						Tongxin	
						Xiaolongtan	
16					*'Anchitherium*	Hsi Shui (Buiuk)	
					Fauna'	Shanwang	
						Xiacaowan	
17	MN5			II			
18	MN4						
		Orleanian	Shanwangian				
19							
	MN3			I			
20					↓		

* For simplicity, localities are juxtaposed with approximately equivalent zones or units, and are listed only once although some embrace more than one MN zone.

focusing on the remains from Siwalik and East Africa. Since then, new discoveries have been made in other localities/countries. Several examples are: *Serridentinus floridanus* (Leidy 1886) (*sensu* Osborn 1936) was allocated to *Amebelodon floridanus* by Turner (1975); *Platybelodon tongxinensis* by Ye *et al.* (1989, 1990) which is referred to as *P. danovi* in this chapter; *A.* (*Konobelodon*) *britti* by Lambert (1990); *Amebelodon tobieni* and *Serbelodon zhongningensis* by Guan (1986, 1988); and *Platybelodon* sp. ('*P. huangzhongensis*') (Guan 1986). A new genus from the Old World, *Archaeobelodon*, was recently established by Tassy (1984). Most of the comparative material used by these authors included dental characters, some cranial and little postcranial.

Following Osborn's (1929*b*, 1936) and Osborn and Granger's (1932) description of *Platybelodon grangeri* from Inner Mongolia, few papers dealt with shovel-tusked gomphotheres from China until the 1960s. Zhai (1959, 1961) reported on a piece of lower tusk of *Platybelodon* sp. collected from Qinan of Gansu Province in north-western China, and Tobien (1973*a*) discussed the *P. grangeri* from this province and referred to it as *Amebelodon*. It was the first report of *Amebelodon* in the Old World. Chen (1978) described some

Table 13.2 Correlation of proboscidean taxa in southern and northern China during a span of about 18 million years

Stages	Southern China	Northern China
Early middle Pleistocene	*Sinomastodon intermedius* *Stegodon zhaotungensis* *Stegotetrabelodon* sp.	
Youhean (Villafranchian)	? cf. *Stegotetrabelodon* sp. *Stegolophodon* cf. ?*latidens*	*Anancus sinensis*
Jinglean (Ruscinian)	*Sinomastodon intermedius* *Mammut borsoni*	*Sinomastodon intermedius* *Mammut borsoni* *Anancus sinensis*
Baodean (Turolian)	*Stegodon* sp. *Stegolophodon* sp. *Gomphotherium* sp. *Zygolophodon* sp. ?'*Sinomastodon huananensis*'	*Choerolophodon* sp. *Tetralophodon* cf. ' *T. hasnotensis*' *Stegotetrabelodon exoletus* *Anancus sinensis* *Sinomastodon intermedius* *Gomphotherium shensiense* *Mammut borsoni* *Stegotetrabelodon exoletus*
Bahean (Vallesian)	'*Zygolophodon gobiensis*' *Choerolophodon* (*Synconolophus*) sp.	*Gomphotherium* sp.
Tunggurian (Astaracian)	'*Zygolophodon chinjiensis*' *Tetralophodon xiaolongtanensis*	*Platybelodon grangeri* *Amebelodon tobieni* *Serbelodon zhongningensis* *Stegotetrabelodon exoletus* *Choerolophodon* sp. '*Zygolophodon gobiensis*' ?'*Zygolophodon nemongoensis*' *Gomphotherium wimani*
Shanwangian (Orleanian)	*Tetralophodon xiaolongtanensis* '*Zygolophodon chinjiensis*' *Stegolophodon hueiheensis*	*Gomphotherium connexum* *Gomphotherium shansiense* *Platybelodon danovi* *Platybelodon* ?'*huangzhongensis*'

molars of *Platybelodon* from Zhongning, Ningxia, and also referred some of the teeth from Tongxin, Ningxia of north-western China to *Gomphotherium* ('*G. tongxinensis*'). This name was changed (to *Platybelodon tongxinensis*), then was reinstated (as '*G. tongxinensis*') when shovel incisor materials were discovered at the site adjacent to the locality where *G. tongxinensis* was found (Ye and Jia 1986; Ye *et al.* 1989, 1990). Guan *et al.* (1981) Guan (1986, 1988, 1991) and Guan and Zhang (1993) described specimens of *Platybelodon, Amebelodon*, and *Serbelodon* from Tongxin, Zhongning of Ningxia Hui Autonomous Region, Huangzhong of Qinghai Province, Guanghe and Qinan, Gansu Province (Fig. 13.1). Guan (1988) established a new species of *Amebelodon* (*A. tobieni*). Tobien *et al.* (1986, 1988) discussed a piece of lower incisor (*Amebelodon*) from Qinan and, simultaneously, the *Platybelodon grangeri* from Inner Mongolia. These authors also reported an incomplete lower jaw and a piece of shovel-like incisor from Xinjiang Uygur Autonomous Region of northern China and referred them to *Gomphotherium* (Chěn 1988; Tobien *et al.* 1986, 1988). Most recently, Qi (1992) reported on *Zygolophodon* from Hilong Jiang (or Heilung Jiang) Province, which is the easternmost locality of Tertiary proboscidean or mammalian distribution in China.

DISTRIBUTION AND STRATIGRAPHIC DATA

Based on available information, it appears that there was a concentration of shovel-tuskers in north-western China (Fig. 13.1), centralized in Gansu, Ningxia, Shaanxi, and Qinghai, as well as Tung Gur of Inner Mongolia. At present, there is no record of shovel-tuskers from eastern and southern China.

A new lithostratigraphic subdivision of the Neogene sequence was elaborated by Li *et al.* (1984) and Qiu (1990), who improved the earlier subdivision applied by Teilhard de Chardin and Trassaert (1937, p. 3). We follow the improved subdivisions and also that of Chiu *et al.* (1979; Table 13.1). Although the stratigraphic record of gomphotheres in China is often poor, it is possible to get a general idea from the locality on a contour distribution map of shovel-tusked gomphotheres in China (Fig. 13.1). Data for this figure were retrieved from Zhou and Zhang (1974), Chow and Zhang (1978), Qiu and Jia (1980), Li *et al.* (1984), Tobien *et al.* (1986, 1988), Guan (1988, 1991), and Guan and Song (1991).

The elephantoid taxa arranged in Table 13.2 cover a time span of about 20 million years, from the Shanwangian (=base of Orleanian in middle and western Europe (Steininger *et al.* 1985,

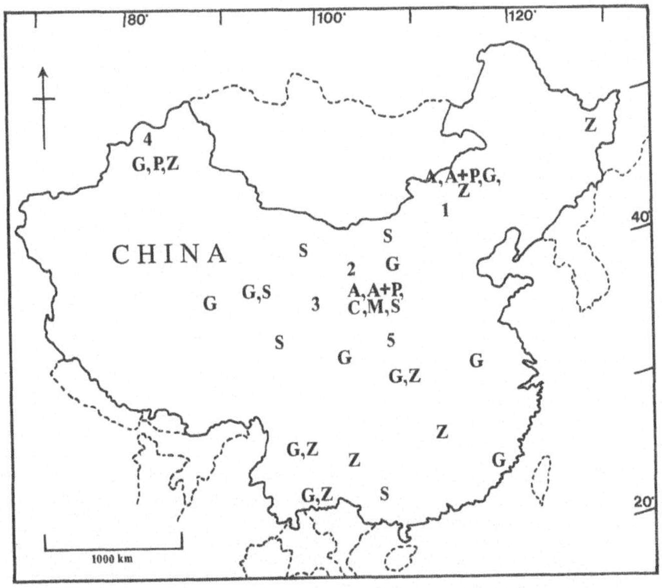

Fig. 13.1 Map showing main shovel-tusked elephantoid localities of Neogene age in People's Republic of China. Key to numbered localities: 1, Tung Gur (Inner Mongolia, 44°N, 114°E), age: middle middle Miocene, *c.* MN7–8; 2, Tongxin (Ningxia Hui Autonomous Region, 37°N, 104°E), age: early to middle Miocene, *c.* MN5–6; 3, Guanghe (Gansu Province, 35.6°N, 102°E), age: early middle Miocene, *c.* MN6–7; 4, Junggar Basin (Xinjiang Province, 46.7°N, 87.8°E), age: early to late of middle Miocene, *c.* MN5–8; and 5, Lengshuigou fauna (Shaanxi Province, 34°N, 109°E), age: early middle Miocene, *c.* MN6. Key to letters: A, *Anancus*; A+P, *Amebelodon* and/or *Platybelodon*; C, *Choerolophodon*; G, *Gomphotherium*; M, *Mammut*; S, *Sinomastodon*; Z, *Zygolophodon*.

fig. 26.1)) until the early Pleistocene (*Gigantopithecus* fauna from a cave in Liucheng County, Guangxi Province), with *Sinomastodon intermedius* (Tobien *et al.* 1986).

FAUNA WITH SHOVEL-TUSKED ELEPHANTOIDS FROM CHINA

Tung Gur (Inner Mongolia; 44°N, 114°E), age: middle middle Miocene, c. MN7–8

As many as 27 species of extinct animals, including the well-known *Platybelodon grangeri*, were recovered at this locality in the 1920s. In 1959 a Sino-Soviet palaeontological expedition attempted unsuccessfully to work on this site. After a long hiatus of collecting from this locality, an excavation team from IVPP worked in this area and a new fauna was advanced by Qiu and his colleagues (Qiu *et al.* 1987*b*). The most important finding of the Third Central Asiatic Expedition and the Sino-Soviet Palaeontological Expedition was that of a 'graveyard' of about 20 platybelodons and rhinoceroses at locality 86021, called Aorigele. The 59 mammalian species found in Tung Gur belong to 8 orders and 28 families, i.e. 32 species more than were previously known. It is possible that the Tung Gur fauna should be correlated with MN7 in the European Neogene mammalian biochronology scale (=late middle Miocene).

Tongxin (Ningxia Hui Autonomous Region; 37°N, 104°E), age: early to middle Miocene, c. MN5–6

The Miocene sediments are exposed in an area of about 350 km^2, situated about 15 km north-east of Tongxin in the north-eastern part of the North-west Plateau of China (Fig. 13.1). Bones excavated from this site by local residents were indentified as 'dragon bones' and were sold in drugstores as a traditional medicine for over one hundred years (Guan and Rice 1990). In most of the 15 sites found in this area, two fossiliferous horizons of Miocene deposits have been recognized (Guan 1986, 1988). They consist of fluviolacustrine sediments in which *Platybelodon*, *Amebelodon*, and *Serbelodon* were found (Fig. 13.1, Table 13.1).

The vertebrate fauna from Tongxin is somewhat older than the extensive fauna of Tung Gur, Inner Mongolia, correlating well with other faunas of broadly comparable age, such as Lengshuigou, Shaanxi and Chetuoguo, Qinghai, from north-western China (Li *et al.* 1984; Guan 1988; Qiu 1990). The fauna from Tongxin also shows a similarity with that of Kuban (Byelometchetskaya), in the former Soviet Union, which is equal to MN6 (Qiu and Guan 1986; Mein 1990). On this basis, the Tongxin fauna can be considered early Astaracian (early middle Miocene) in age. None the less, according to Guan and van der Made (1993), Tongxin fauna may be dated as MN5–6.

Guanghe (Gansu Province; 35.6°N, 102°E), age: early middle Miocene, c. MN6–7

As in Tongxin, 'dragon bones' were used as traditional medicine for many years. Scientific excavations were carried out by a team from the BNHM and the Gansu Provincial Museum. The team discovered two shovel-tusked gomphotheres, *P. danovi* and *Serbelodon zhongningensis* (Fig. 13.1, Plates 13.1– 13.4, Table 13.1).

The fauna from Guanghe can be correlated with that from Tongxin as more than five species identical to those from Guanghe were discovered at the Dingjiaergou site near Tongxin. At present, 19 species of Miocene mammals are known from the Guanghe area (Guan 1988). These include *Platybelodon*, *Serbelodon*, *Caementodon*, *Anchitherium*, and *Kubanochoerus*, taxa which were also found in Tongxin. They were considered as typical middle Miocene elements (Li *et al.* 1984; Mein 1990). There are other elements from Guanghe which were considered from a younger horizon (late middle Miocene, MN7–8) such as *Amphicyon*, *Oioceros*, and *Serbelodon*.

Junggar Basin (Xinjiang Uygur Autonomous Region, China; 46.7°N, 87.8°E), age: early to late middle of middle Miocene, c. MN5–8

Chen (1988) described some material of gomphotheres associated with *Amblycastor tunggurensis*, *Amphicyon*, *Anchitherium cf. aurelianense*, *Brachypotherium* sp., and *Stephanocemas* which are typical middle Miocene elements. Five species of elephantoids were described in her paper: *Gomphotherium* cf. *shensiensis*, *Gomphotherium* sp., *Platybelodon sp.*, *Zygolophodon* (?) *junggarensis* (new species), and *Zygolophodon* sp. This fauna shares some of

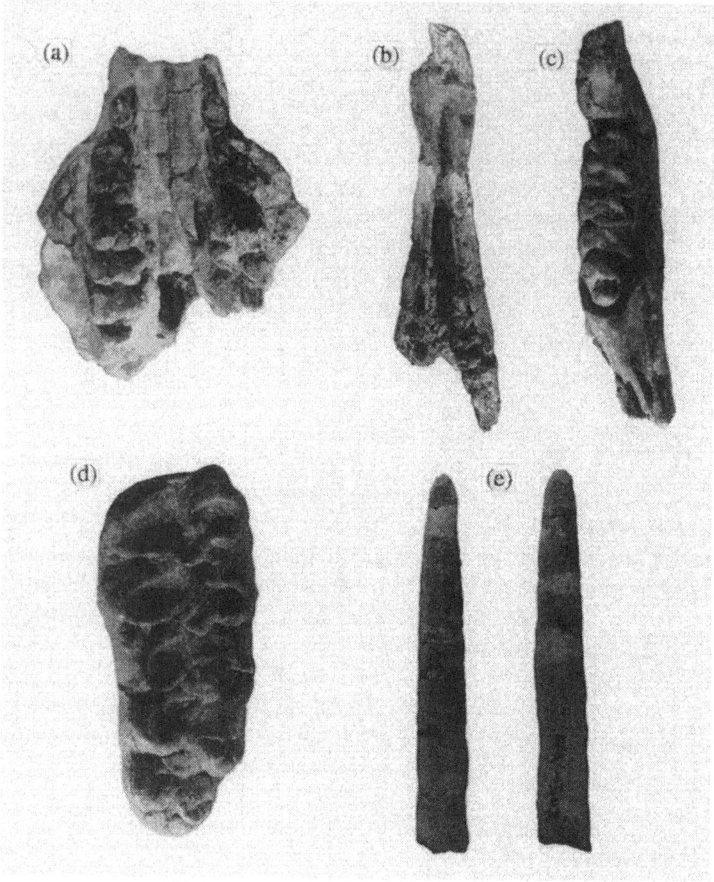

Plate 13.1 *Platybelodon danovi* (a–d): photographs of skull fragments and teeth. (a) Juvenile maxillae and teeth in occlusal view (shown are dp^2, dp^3, and dp^4; length of dp^3 = 4.5 cm, width = 3.1 cm); (b) juvenile mandibular fragment with right incisor (shown are dp_3, dp_4, and part of M_1; length of dp_3 = 4.4 cm, width = 2.5 cm); (c) adult left mandibular ramus in occlusal view (shown are part of M_2, and M_3; length of M_3 = 25.5 cm, width = 7.8 cm; negative is reversed); (d) right M^3 (length = 15.5 cm, width = 6.5 cm; negative is reversed). Fronts of animals are toward top of page. *Amebelodon tobieni* (e): a pair of juvenile upper incisors with enamel bands (length of tusk on left = 20.4 cm, width = 3.2 cm). Specimens depicted in Figs (a), (b), (c), and (e) are from Tongxin (Ningxia Hui Autonomous Region), age: early to middle Miocene *c.* MN5–6, and that shown in Fig. (d) is from Guanghe (Gansu Province), age: early middle Miocene *c.* MN6–7.

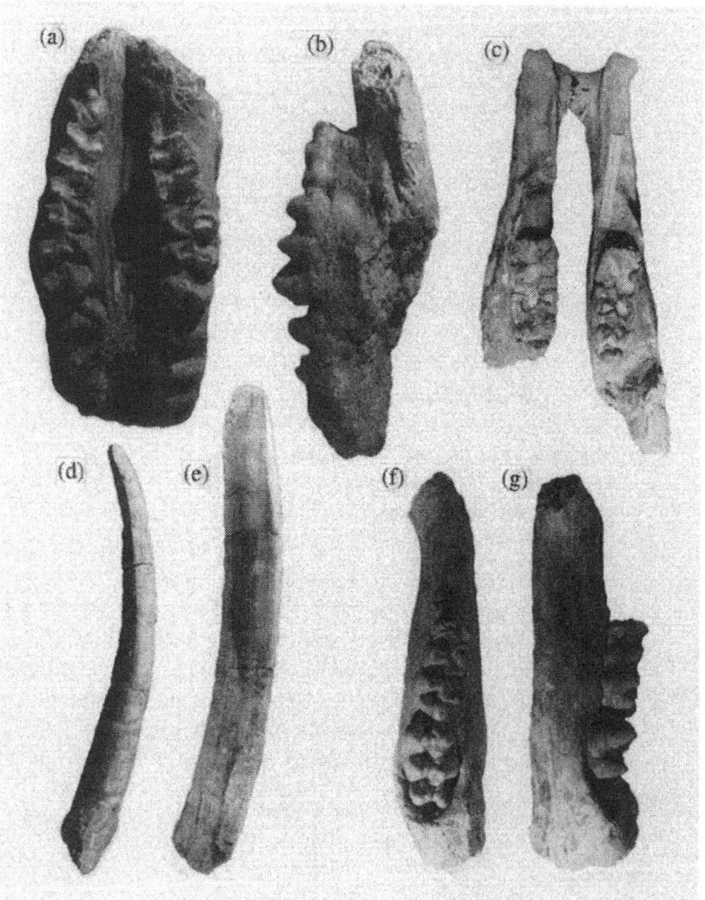

Plate 13.2 *Amebelodon tobieni*: photographs of skull fragments and tusks. (a) Maxillae and teeth in occlusal view (shown are P^4, M^1, M^2, and M^3; length of M^1 = 7.8 cm, width = 5.6 cm; negative is reversed; i.e. P^4 is on left); (b) as in (a), in lateral view, (c) mandibular rami and symphysis in occlusal view of an adult with M_3 in wear (length of M_3 = 16.5 cm, width = 7.1 cm), (d) right I_2 in medial view (length in straight line = 75 cm; negative is reversed); (e) as in (d), dorsal view (negative is reversed); (f) left mandibular ramus in occlusal view (shown are M_2 and M_3; length of M_2 = 12.2 cm, width = 6.5 cm; negative is reversed); (g) same as in (f), in lingual view. Fronts of animals are toward top of page. Specimens depicted in Figs (a–g) are from Tongxin (Ningxia Hui Autonomous Region), age: early to middle Miocene *c.* MN5–6.

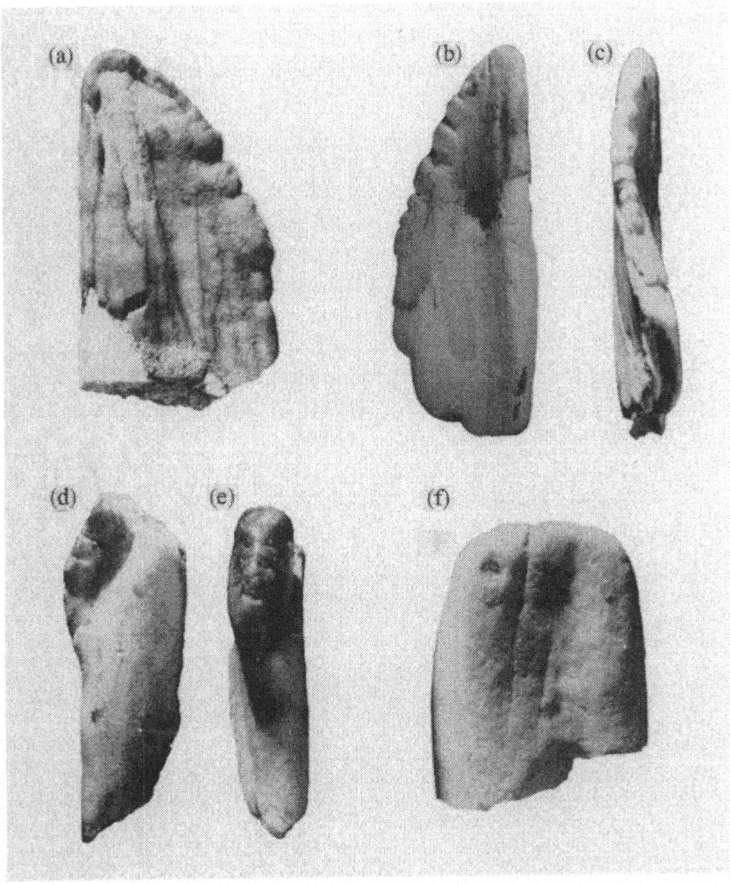

Plate 13.3 Photographs of fragments of juvenile lower incisors. (a) *Platybelodon danovi*; (b–f), *Amebelodon tobieni*. All except (f) are from juvenile individuals, and all except (c) and (e) are in dorsal views; (c) are (e) are in lateral view. (a) Right side (length = 64 mm, width = 36 mm, thickness = 9 mm); (b, c) left side (length = 71 mm, width = 31 mm, thickness = 8 mm); (d, e) left side (length = 60 mm, width = 23 mm, thickness = 11 mm); (f) right side (length = 52 mm, width = 40 mm, thickness = 12 mm). Specimen shown in Fig. (a) is from Zhongning (Ningxia Hui Autonomous Region), age: early to middle Miocene *c.* MN7–8, and those depicted in Figs (b–f) are from Tongxin (Ningxia Hui Autonomous Region), age: early to middle Miocene *c.* MN5–6.

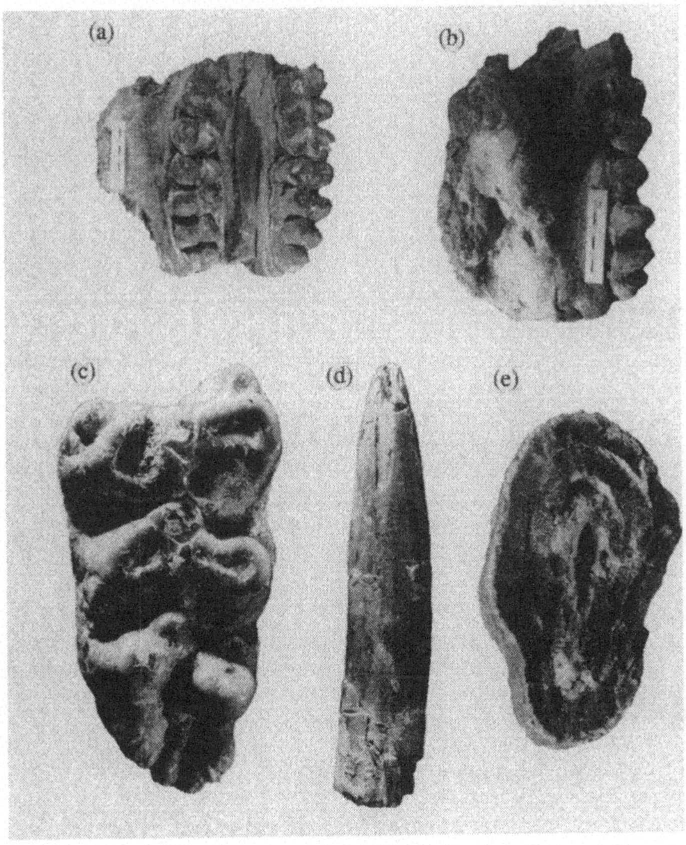

Plate 13.4 *Serbelodon zhongningensis*: photographs of teeth (mostly). (a) Maxillary fragments and teeth in occlusal view (shown are M^2 and M^3; length of M^3 = 12.3 cm, width = 6.0 cm; negative is reversed); (b) as in (a), in lateral view; (c) left M^3 in occlusal view (length = 10.2 cm, width = 5.4 cm); (d) left I_2 in dorso-medial view (length = 41.5 cm, maximum width = 8.3 cm); (e) transverse cross-section of left I_2, in proximal view (medial view toward top of page, ventral toward right; medio-lateral measurement = 7.1 cm, dorso-ventral = 4.5 cm). Specimens depicted in Figs (a) and (b) are from Tongxin (Ningxia Hui Autonomous Region), age: early to middle Miocene *c.* MN5–6, and those in Figs (c–e) are from Zhongning (Ningxia Hui Autonomous Region), age: early to middle Miocene *c.* MN7–8.

the elements with the genera *Gomphotherium, Platybelodon, Anchitherium, Amphicyon, Amblycastor,* and *Stephanocemas* from Guanghe.

Lengshuigou fauna (Shaanxi Province; 34°N, 109°E), age: early middle Miocene, c. MN6

Chow (1978) summed up the information from this locality, indicating that it includes typical middle Miocene fossils such as *Listriodon, Gomphotherium, Paleotragus,* and *Oioceros.* Varying opinions of the age of the Lengshuigou fauna have been discussed by different authors (Young and Liu 1948, p. 289, probably Pontian; Li 1976, p. 67, Pliocene; Chang and Zhai 1978, p. 142, probably middle Miocene). Chiu *et al.* (1979), using the old stratigraphic subdivision, placed the Lengshuigou formation in the middle Miocene, the same age as Tung Gur.

To summarize this section, the local fauna with shovel-tusked elephantoids from China were concentrated in north-western China, west of 34°N and north of 113°E (Fig. 13.1). According to Sung (1958), these areas were a part of the temperate forest distribution (see also Harrison *et al.* 1991). The second horizon with shovel-tusked elephantoids from Tongxin is considered as the oldest (early to middle Miocene) 'shovel-tusker' fossiliferous bed in China (Lengshuigou included); the Guanghe, Qinan, Junggar, and Huangzhong are to be correlated to early middle Miocene and the Tung Gur from Inner Mongolia into middle to late middle Miocene. Representatives of skeletal remains are depicted in Figs 13.2–13.4 and in Plates 13.1–13.4.

TAXONOMY OF SHOVEL-TUSKED GOMPHOTHERE TAXA

The following are brief taxonomic notes on shovel-tusked gomphothere taxa from China (other elephantoid genera and species are given in Table 13.2). Much work awaits description. The intent is to list known taxa on the species level, and only those which were published; no new ones are given here. In addition, only generic and species names are listed, because the taxonomy on the familial level is not yet certain (details in Shoshani, this volume, Chapter 16).

Platybelodon *Borissiak, 1928*

Type species

Platybelodon danovi Borissiak, 1928

Comments

Guan (1991, p. 10, footnote) recognized two subspecies: *P. d. danovi,* and *P. d. cheni.*

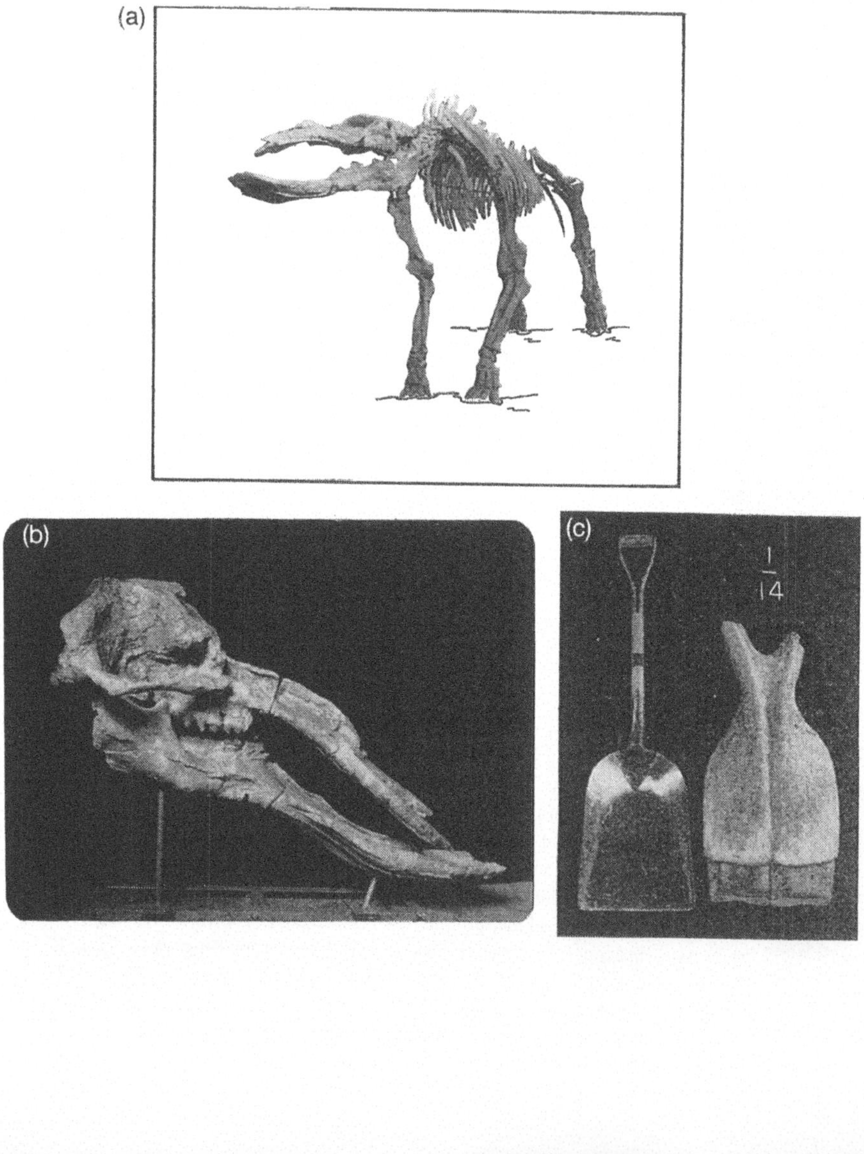

Fig. 13.2 Top: a skeleton of *Platybelodon danovi* (BPV–100) from Tongxin, now at the BNHM (mounted and photographed by J. Guan; approximate shoulder height: 2.5 m, total length: 4.7 m). Bottom left: skull of the same individual, BPV–1000, in lateral view (photograph by J. Guan; length: 1.92 cm, width: 55.5 cm). Bottom right: ventral view of distal end of *P. grangeri*, photographed with a shovel to stress the resemblance (after Osborn and Granger 1932, p. 8, fig. 6; reproduced from Osborn 1936, p. 468, fig. 438, courtesy of the AMNH). Specimens depicted in Figs (a) and (b) are from Tongxin (Ningxia Hui Autonomous Region), age: early to middle Miocene *c.* MN5–6, and the one shown in Fig. (c) is from Tung Gur (Inner Mongolia), age: middle middle Miocene, *c.* MN7–8.

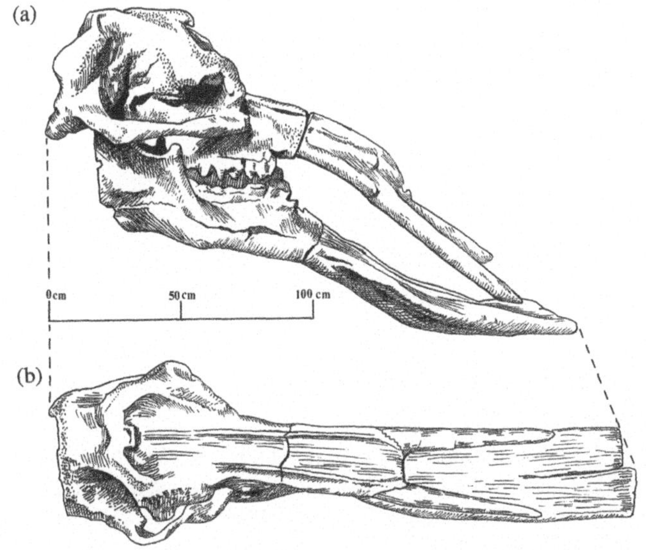

Fig. 13.3 (a) Drawing of *Platybelodon danovi* skull (BPV–1000) of adult individual in lateral view. Parts of M2s and M3s are shown. (b) Same individual as in (a) in dorsal view.

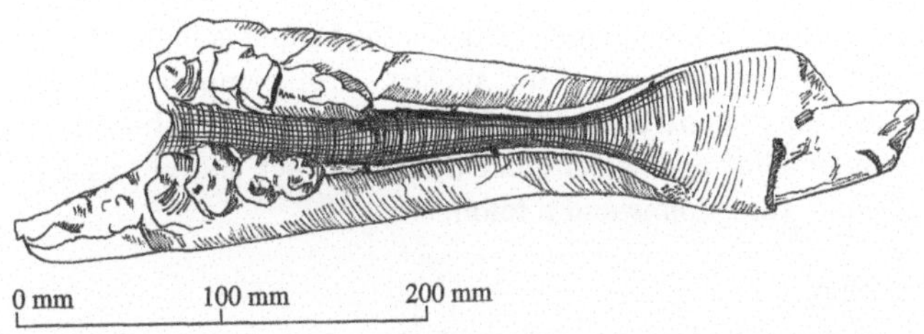

Fig. 13.4 Drawing of *Platybelodon danovi* mandibular fragment with right lower incisor, dp_3, dp_4, and part of M_1 (total length: 44 cm — same individual as shown in Plate 13.1(b)).

Other recognized species

P. grangeri Osborn, 1929*b*; *P. loomisi* (Barbour, 1929); '*P. huangzhongensis*' (Guan 1986). [Editors' note: *P. loomisi* is considered *P. (Torynobelodon) loomisi*; see Apendix C.]

Original diagnosis for Platybelodon

See Borissiak (1928, p. 120).

Emended diagnosis for P. danovi

Cranium elongated and compressed laterally with an extremely long premaxilla in adult individuals, and with a narrow, long mandibular symphysis. Plate-like lower tusk with visible rod–cone structure in cross-section; ladle-shaped symphysis of mandible; index (H/W) of the lower incisor is 23–25; cylindrical superior incisive tusks reduced in size and rather straight, out-turned, slightly downturned, shorter than lower incisor, and without enamel band. Molars are bunodont-like, intermediate molars with three lophs and talons or with the tendency for four lophs. High-crowned molars with abundant cement, especially in the transverse valleys.

Referred and new material

A complete skull with a very well-preserved mandible (BPV-1000) from Dingjiaergou, Tongxin (Figs 13.3–13.4). Paratypes and referred material are in the BPV; they include about 40 crania (many fragmented) and postcranial specimens.

Age

Early to middle Miocene, MN5–6 (Chiu *et al.* 1979; Tobien *et al.* 1986; Guan 1988; Guan and van der Made 1993).

Localities

Type material from Maerzuizigou site of Dingjiaergou, Tongxin; others from 15 sites of Tongxin, Ningxia Hui Autonomous Region, and Maijiaxiang site of Guanghe, Gansu Province.

Species: Platybelodon grangeri Osborn, 1929b

Original diagnosis

See Osborn (1936, pp. 463–70).

Emended diagnosis

Skull is narrow with a long premaxilla in adults, but not as narrow as in *P. danovi*. Shovel mandibular symphysis with plate-like lower tusks; index (H/W), 16–20; superior incisive tusks much reduced in size and without enamel band. Lower incisors with rod–cone structure distinct and strong. High-crowned molars with abundant cement. M2s tend to be tetralophodont, whereas M1s preserve the original trilophodont type.

Holotype

Mandible of a senile individual, posterior portion broken, molars missing (AMNH 26200). Paratypes and referred materials are at the AMNH and BPV.

Age

Late middle or middle middle Miocene, i.e. Chinjian or late Aragonian (Chiu *et al.* 1979, p. 266, table 1; Li *et al.* 1984, p. 174, table 5; Tobien *et al.* 1986, p. 149).

Localities

Dabuhaer site (1930) and Huergulajin site (1986) of Tung Gur (Inner Mongolia); Heijiagou, Zhongning (Ningxia); Baima, Zhongwei (Ningxia); Yushe (Shanxi) (Chang 1964).

Comparison

Differences between *P. grangeri* and *P. danovi* were summed up by Tobien (1973*a*, p. 254), Ye and Jia (1986), and Guan (1991).

Amebelodon *Barbour, 1927*

Type species

Amebelodon fricki Barbour, 1927

Original diagnosis

See Barbour (1927, pp. 131–4).

Other recognized species

A. barnumbrowni Barbour, 1931; *A.* (*Konobelodon*) *britti* Lambert, 1990; *A. floridanus* (Leidy 1886), *A. sinclairi* Barbour, 1930; and *A. tobieni* Guan, 1988. [Editors' note: *A. barnumbrowni* is considered *P.* (*Torynobelodon*) *barnumbrowni*, and *A. sinclairi* as a possible junior synonym of *A. floridanus*; see Appendix C.]

Species: *Amebelodon tobieni* Guan, 1988

Etymology

Named after the late Heinz Tobien who first recognized the existence of *Amebelodon* from Qinan and for his outstanding contribution to the research of gomphotheres from China.

Diagnosis

The cranium is compressed laterally and elongated anteriorly. The mandibular symphysis extremely long (Plate 13.2). The upper tusk is elongated, cylindrical, out-turned, and downturned, with a wide enamel band; it stretches forward over the tip of the lower tusk and mandibular symphysis. Inferior incisive tusks with laminar structure, trough-shovel-shaped mandible; index (H/W), 36–41. Grinding teeth brachyodont to moderately high and bunodont with pretrite trefoils. Trilophodont M2 shows the tendency to be tetralophodont. Cement is absent or in small amounts. The dental formula is: dp4 3/3; P4 2/2; M1 3/3+tid; M2 3+t/3+tid, 4; M3 5, 5+t/4+tid, 5+tid (Guan 1988).

Holotype

An incomplete upper jaw with a mandible in which the horizontal ramus and lower incisors are separated but complete (BPV 533, Plate 13.2*d–e*). Paratypes and referred material are in Qinan Museum, BPV, and OAT.

Age

Early middle Miocene (MN6–7).

Localities

Anjiawan site near Haodi, Qinan, Gansu Province; Wudaoling site of Tongxin, Ningxia Hui Autonomous Region (Li *et al.* 1984; Guan 1988).

Comparison

This species is different from the known *Platybelodon* by its sub-trough mandibular symphysis and the lamellar inner structure of lower tusk. In addition, the comparatively low crown of molars and the bigger upper tusk with enamel band of *A. tobieni* are different from those of *Platybelodon* in the New and Old World. This species also differs from *Amebelodon* in the New world; *A. fricki* and *A. floridanus* have an extremely long symphysis and massive molars which easily separate them from specimens from China.

Serbelodon *Frick, 1933*

Type species

Serbelodon barbourensis Frick, 1933

Original diagnosis

See Frick (1933, p. 594).

Other recognized species

S. burnhami Osborn, 1933; and *S. zhongningensis* (Guan 1986, 1988). [Editors' note: *S. burnhami* is considered *Amebelodon* sp.; see Appendix C.]

Species: *Serbelodon zhongningensis* Guan, 1988 (Plate 13.4)

Diagnosis

A wide and comparatively short skull with short upper and lower incisors. The upper jaw stands high and wide with a short premaxilla as compared with *Platybelodon* and *Amebelodon*. The lower tusk has a slender, elongate, somewhat narrower contour, with the normal lamellar structure in the transverse section. The lower tusks are dorso-ventrally compressed, forming a concave longitudinal shallow groove. The shovel-tusked index (H/W) is high 60–64. The shovel-like mandibular symphysis and lower tusk form a blunt shovel. The round or subround upper incisors with an enamel band stretch forward out-downturned, elongated over the tip level of the lower tusk. The angle between the palate and the basisphenoid is higher than in other shovel-tuskers. The molars are comparatively low crowned, cement is visible. The third molar is sub-zygodont, being smaller and wider than those of other shovel-tuskers. It has only three lophs and a talon.

Holotype

Part of a right lower incisor (BPV 590, Plate 13.4*d*). Paratypes and other materials are at the BPV.

Age

Middle Miocene (perhaps MN7–MN8).

Localities

Longkengkou site of Baima, Zhongning (type) and Wudaoling, Tongxin, Ningxia. Maijiaxiang, Guanghe of Gansu (paratype).

Comparison

Osborn (1936) emphasized the obvious differences between *Serbelodon* and *Amebelodon*, and placed *Serbelodon* in the family Serridentidae, subfamily Serridentinae based on the following characters:

(1) sub-shovel-tusked;

(2) the serrated structure and the trefoils are similar to *Serridentinus*;

(3) wide upper jaw with a short lip; and

(4) shovel-tusked with lamellar structure.

Tobien (1972, 1973*a*) referred some specimens (e.g. *S. burnhami*) to *Amebelodon* because of the similarities in thickness of the lower tusks. *Serbelodon zhongningensis* is the first report of this genus in the Old World.

THE SKELETON OF SHOVEL-TUSKED ELEPHANTOIDS

There are six known mounted or disarticulated skeletons of shovel-tusked gomphotheres in these locations: four partial skeletons of *P. danovi* from Tongxin (housed in BNHM, Fig. 13.2), a skeleton of *P. grangeri* from Tung Gur (recently mounted in IVPP), and a skeleton of *Archaeobelodon filholi* ('*Mastodon angustidens*') from Sansan (housed in the Paris Museum, 'old' catalogue number BVI-5). These skeletons await detailed updated descriptions. Parts of the postcranial material of *A. filholi* were described and figured by Blainville (1845) under the name '*Mastodon angustidens*'.

CHARACTERS USED TO EVALUATE ELEPHANTOIDS FROM CHINA AND PHYLOGENETIC ANALYSIS

Cranial characters of eight shovel-tusked gomphotheres were studied; they were: *A. (Konobelodon) britti*, *A. floridanus*, *A. fricki*, *A. sinclairi*, *A. tobieni*, *P. danovi*, *P. grangeri*, *P. loomisi* (Editors' note: see Appendix C for synonymies). The inner structure of the inferior incisors can be used to distinguish between *Amebelodon* and *Platybelodon* (see Lambert 1990 for a different opinion). They are noted as having unique, commonly derived characters not seen in *Phiomia*, *Amebelodon*, and *Serbelodon*. Analysis in this chapter is confined to skull and dental characteristics, some of which are summarized in Table 13.3.

Characters of shovel-tusked elephantoids have been discussed by Frick (1933), Osborn (1936), Belyaeva and Gabunia (1960), Tobien (1973*a*), and Tassy (1986). The terminology of molars is after Tobien (1973*b*) and Tassy (1983; this volume, Chapter 3). State (0) below is considered to be the primitive condition for the elephantoid/ gomphothere taxa studied (*Phiomia* was usually used as an outgroup), all other states are derived:

DENTAL CHARACTERS

- Number of loph(id)s on second molar, upper and/or lower: (0)=three, (1)=four.
- Number of loph(id)s on third lower molar: (0)=three, (1)=four, (2)=five, (3)=six.
- Pretrite trefoils on the third molars: (0)=not well delineated, (1)=well delineated.
- Cement on molars: (0)=no cement, (1)=cement visible, (2)=much cement.
- Crown height of the third molar: (0)=less than 20 mm, (1)=20 mm or more.
- Index of height/width (H/W) lower tusks: (0)=less than 31, (1)=31 or more. This index was used to allow inclusion of young individuals. Result shows a stable index for each species.
- Inner structure of lower tusk: (0)=no definitive structure, (1)=laminae structure, (2)=rod–cone structure. A deciduous lower incisor of *Platybelodon* from Tongxin and specimens of juveniles show mostly lamellar structure with very small rod–cone structure (adults are characterized by a rod–cone structure, but always a lamellar structure close to the surface). *Amebelodon* juvenile and adult specimens show lamellar structure in the cross-section.
- Enamel band in the upper tusk: (0)=present, (1)=absent.
- Upper incisor is longer than lower incisor: (0)=yes, (1)=no.
- Transverse shape of the upper incisors: (0)=flattened-shape, e.g. *Phiomia*, (1)=triangle or 'D'-shape, (2)=round. When the enamel band is present, its section is mostly pyramidal or ovoid; in the absence of the enamel band, the upper tusk often shows sub-round or round in section.
- Grinding surface at the end of lower tusks: (0)=blunt grinding, (1)=grinding surface on upper and lower, (2)=grinding surface on upper, (3)=lower grinding surface, producing a 'chisel'.

Table 13.3 A summary of characters in selected shovel-tusked gomphotheres

Character	*Platybelodon grangeri*	*Platybelodon danovi*	'*Amebelodon*[a] *barnumbrowni*'	*Amebelodon tobieni*
I_2 inner structure	rod-cones	rod-cones	lamellae	lamellae
Mandibular symphysis	shovel	ladle	blunt	trough
Index of H/W of lower incisor	16–20	27–34	16	36–41
Expanding rate of symphysis	0.57	0.50	0.25	0.38
Grinding surface at tip of I_2	chisel	chisel	chisel	superior polish
Cement on molar	extreme	rich	less	less
Section of upper incisors	round	sub-round	roundish	roundish
Enamel band on upper incisor	absent	absent	present	present

[a] Editors note: In this volume the name *Platybelodon (Torynobelodon) barnumbrowni* is used.

NON-DENTAL CHARACTERS

- The shape of the mandibular symphysis and lower tusks: (0)=trough shape, (1)=ladle shape, (2)=shovel shape (3)= blunt-shovel shape (Table 13.3).
- Growth rate of the symphysis (i.e. lenght of horizontal ramus divided by length of mandibular symphysis, multiplied by 100): (0)=less than 25, (1)=between 25 and 45, (2)=between 45 and 60.
- Index of symphysial length divided by length of mandible: (0)=less than 0.5, (1)=between 0.5 and 0.6, (2)=larger than 0.6.
- Crista orbitalis ventralis and crista orbito temporalis: (0)=slanting gently, crossing at the low position, (1)=slanting steeply, crossing at the high position (definition of character states after Saegusa 1987).
- Temporal line: (0)=type A, (1)=type B, (2)=type C. Primitive proboscideans, e.g. *Phiomia*, exhibit type A of temporal line (i.e. the sagittal crest is present), whereas advanced taxa exhibit type C (sagittal crest absent; definition of character state after Saegusa 1987).
- Individual size: (0)=small, e.g. *Phiomia*; (1)=larger, e.g. most amebelodonts; (2)=still larger, e.g. *A. fricki*.

OTHER CHARACTERS EXAMINED

- Development of upper tusks. Results show that *Amebelodon*, with narrow lower incisors, had big upper tusks (longer than lower tusks), but *Platybelodon*, with small upper incisors, had stronger lower shovel-tusks (superior incisor is shorter than inferior incisor).
- Length of premaxilla. In general, specialized species (e.g. *Platybelodon grangeri, P. danovi*) had long, narrow premaxillae. This character is more informative when the age of the individual is taken into consideration.
- The angle between the basicranial and the palate (Fig. 13.5, after Tassy 1986).
- Three mandibular curvatures were identified: K1 = lateral, viewed dorsally; K2 = ventral at mid mandible, viewed laterally; K3 = curvature of symphysis at distal end, viewed laterally (Guan 1986).
- Lower jaw's width divided by length at increments of 10, 20, 30, ... 100 (giving the full length of lower jaw = 100).
- Index of minimum (narrowest) and maximum (widest) measurements of symphysis (including the lower incisor).
- Plotting the combined length and width of dp2, dp3, dp4 against the length and width of M1, M2, and M3 for each species.

SELECTED SYNAPOMORPHIES

The following characters were observed in shovel-tusked proboscideans: (1) the mandible and lower tusk manifest the ladle, shovel, and blunt-shovel shape; (2) lower tusks with a rod–cone inner structure, especially in adult *Platybelodon*; (3) the index of H/W of the lower tusks is lower than 50; (4) the expansion rate of the symphysis is higher than 40; (5) curvature of the mandible is large; (6) the lower incisor tip has a lower grinding surface.

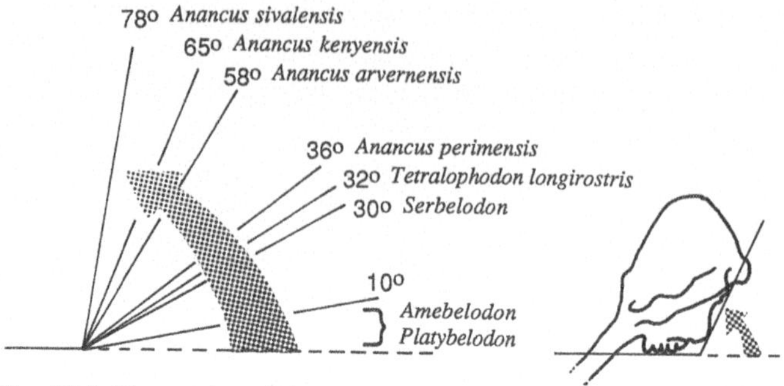

Fig. 13.5 Measured angle between molar alveoli and occipital condyle for selected gomphothere genera (after Tassy 1986, p. 91), except for *Serbelodon*, *Amebelodon* and *Platybelodon*.

A lower angle (below 10 degrees) between the alveoli of cheek teeth and the occipital condyles (Fig. 13.5) is found in *Platybelodon* and *Amebelodon*. A larger angle is displayed in *Serbelodon*. This is an important distinction, therefore a suggestion is made to designate *Serbelodon* as a valid genus.

Amebelodons and platybelodons described in this chapter comprise a monophyletic group sharing the following characters: (1) the mandible and lower tusk are shovel, ladle, and blunt-shovel type; (2) the mandible is largely expanded and elongated; (3) the upper tusk is reduced especially in platybelodons; and (4) the premaxilla is extremely elongated.

P. grangeri and *P.* (*Torynobelodon*) *loomisi* are considered to be related to *P. danovi* based on these characteristics: (1) a rod–cone inner structure in the lower tusk; (2) lower grinding surface on the tip of the tusk; (3) expanded rate of mandible shows large (>0.4); and (4) the shape of the mandible is of the expanded ladle and shovel type.

P. (*T.*) *loomisi* appears to be the most primitive *Platybelodon* among these four taxa because of its low H/W index of the lower tusk. *P. grangeri* is the most advanced species in this genus, possessing the following characters: (1) large lateral curvature of the mandible; (2) shovel-type tusk; (3) abundant cement on molars; and (4) a high expansion rate of the symphysis.

Members of *Amebelodon* share the following characters: (1) a thick symphysis; (2) a small expansion rate of the symphysis; (3) a lower tusk with the lamellar inner structure; (4) the H/W index of the lower tusk is between 35 and 50; (5) a sub-triangular or triangular upper tusk; and (6) a larger upper tusk with an enamel band.

A. tobieni and *A. sinclairi* [Editors' note: *A. sinclairi* is considered an invalid name in this volume.] are considered to have the following morphology: (1) a ladle-shaped mandible and lower tusk; and (2) an H/W index of the lower tusk of about 40. *Amebelodon fricki* presents the following characters: (1) the symphysis is longer than its mandibular alveoli; and (2) a trough-shape mandible and lower tusk. *A.* (*T.*) *barnumbrowni* presents these characters: (1) the H/W index of the lower tusk is less than 20; (2) a blunt shovel mandible and lower tusk; and (3) the symphysis is not curved on its ventral side.

CONCLUDING NOTES ON ORIGIN, EVOLUTION, AND DISTRIBUTION OF SHOVEL-TUSKED ELEPHANTOIDS

Osborn (1936, p. 236), Tobien (1973*a*, p. 261), Tobien *et al.* (1988, p. 206), and Guan (1991, p. 20) noted that the origin of the shovel-tusked gomphotheres may be traced to *Phiomia* from the Oligocene of the Fayûm. These comments were made because of similarities in the lower tusks, probably due to symplesiomorphy (Tassy 1982).

The record of *Platybelodon* lineage in Africa is documented by an incisor plate fragment from the Miocene of the Loperot area, north-western Kenya (Maglio 1969*a*). The age of Loperot is considered as late early Miocene or middle Miocene. *Platybelodon* material from China was dated to about the same date as that from Africa; it is thus premature to envision an African origin for this genus.

Platybelodon grangeri was conjectured not to have survived into the late Miocene (Tobien *et al.* 1986, p. 158) but a piece of the lower incisor and a molar fragment from Fugu (Lao Gaochuan) ought to belong to a shovel-tusked elephantoid of Baode (Baho) age. This material, with the *P. grangeri* characters (slender dentition and rich cement, low rate of H/W of the lower incisor and with a plentiful rod–cone structure in transverse section), is the latest material of the shovel-tuskers in China.

Amebelodon tobieni showed the very typical shape of the shovel-tusker. The contour of the mandible of *A. tobieni* is similar to that of *Phiomia* '*osborni*' from the Oligocene. Most of the *Amebelodon* material in North America demonstrates a specialty of the mandible, such as *A. fricki* and *A. floridanus* with an extremely long trough-like symphysis. As noted earlier, the other 'shovel-tuskers' (e.g. *Serbelodon* and *Torynobelodon*) have different characters in the mandibular symphysis. The 'shovel-tuskers' from Africa and Europe (e.g. *Archaeobelodon* and *Protanancus*), with lamellar structure, show some distinct characters of the mandible and the incisors. With the latest data in hand, the age of the earliest *Archaeobelodon* from Europe is estimated at MN4, whereas the materials from Africa are estimated to be of late early Miocene age (*c.* 17 Ma). It appears, then, that the African and the European specimens are of an earlier geological age than *A. tobieni* from China (middle Miocene, MN6–7). It is possible that *A. tobieni* was the earliest member of the genus *Amebelodon*, which had derived from the previous shovel-tuskers. Assuming this scenario is correct, then the missing 'key-link' could be found in early Miocene sediments.

ACKNOWLEDGEMENTS

With financial aid from the Research Foundation and the City University of New York, the Weatherhead Foundation from American Museum of Natural History, and the Windway Foundation, the author was able to conduct his research on fossil proboscideans in the United States. J. Shoshani made valuable comments, edited and corrected the 'Chinglish' in earlier versions of this chapter; without his tremendous effort, this manuscript would have collected museum dust. H. Semken, P. Tassy, and E. Marsac also helped with the editing. I thank P. Andrews, J. Carter, E. Delson, D. Hunt, R. Johnson, C. T. Madden, H. Semken, R. Tedford, M. Voohries, and R. William for their valuable help. Some of the figures were drawn by Zhoa Yiemu. The photographs were taken by Zhang Hongjie. Rufen Song, Shusheng Hu, Li Mei, and Jann Grimes assisted in other aspects of this chapter.

FOURTEEN

Gomphotheres and mammutids from the Iberian Peninsula

ANA V. MAZO

INTRODUCTION

Remains of gomphotheres and mammutids are abundant in the Iberian Peninsula and are found within a temporal interval from middle Miocene (*c.* 18 Ma) to Villafranchian (*c.* 3 Ma; Table 14.1). Earlier studies of these proboscideans reflect, in Spain and Portugal, multiple genera, species, and varieties. In this study, the taxa considered valid include four gomphotheres and two mammutids. This chapter analyses the temporal distribution of these taxa in the Iberian Peninsula, based on the changes reflected by the incisors and cheek teeth of these mammals through time, taking as reference the biozonation of Neogene mammals proposed by Mein (1979, 1990; Table 14.1).

GOMPHOTHERES: BUNODONT TRILOPHODONS

Gomphotheres are characterized by their cheek teeth, which consist of round cusps. In Spain, bunodont gomphotheres with intermediate trilophodont teeth (dp4, M1, and M2), are known from the early Astaracian (= early Aragonian biozone MN4 proposed by Mein (1979, 1990) to the end of the late Astaracian (biozone MN7). These specimens present different morphotypes, corresponding to temporal intervals that can be defined precisely. In Portugal, evidence of the first gomphotheres has been found at the base of biozone MN4 (Quinta do Narigao, Lisboa), division IVb of Cotter in Dollfus *et al.* (1903), although the register of the late Astaracian interval is incomplete (Fig. 14.1, Table 14.1).

Papiol (Barcelona) and Buñol (Valencia) are some of the oldest Spanish deposits containing gomphotheres. From Papiol originated a mandible with an M_2, an M_3, and two isolated M^3s, all from the same individual; the mandible, however, is not reliable because it has been reconstructed. The M_3s have $4\frac{1}{2}$ lophids and the M^3s have four lophs and a talon with crescentoids. The morphology is very simple. The four crests are transverse, those at the back tend to have an angular inclination, the main tubercles are wide at the base with binary division, but not a clear antero-posterior compression; the interlophs exhibit projections among the pretrite elements, and the complementary part is absent (Plate 14.1). In the front posttrite elements of the M3, there are rudimentary crenellations that descend toward the bottom of the valleys, and these crenellations are homologous to what in German terminology are called zygodont crests (Schlesinger 1917, 1922; Tobien 1975, this volume, Chapter 9). This characteristic added to the amplitude of the valleys, confers upon these specimens a certain zygodontoid aspect. This point has already been mentioned by Bergounioux and Crouzel (1958).

Among the Buñol remains, there are bunodont molars with some zygodont features, molars of an intermediate morphology, and others tending more toward zygodontia. Nevertheless, these specimens should be included in *Gomphotherium angustidens* and not in what Tassy (1985) called a robust morphotype of *Zygolophodon turicensis*, because the main tubercles do not have antero-posterior compression and their walls are convex when viewed laterally. These molar characters do not concur with the Mammutidae's diagnosis. There were also two lower incisor fragments found that can be labelled as *Gomphotherium* because of their pyriform transverse cross-section.

In Spain, there are few deposits in the MN4 zone. The molars recovered from this temporal interval are generally simple bunodonts, or those classically defined as *G. angustidens* of intermediate morphology. This is the case of the Córcoles deposits, Guadalajara (Plate 14.1b) (see also Maldonado *et al.* 1983).

Molars very similar to those of Papiol have been found in Portugal, for example, at Quinta das Pedreiras (the Va division of Cotter, in Dollfus *et al.* 1903) and at Courelas do Covao (the Vb division) (see Bergounioux *et al.* 1953, plate XIX, fig. 157 and plate XXIV, fig. 200). Note that in some of the Lisbon deposits that pertain to the Vb division of Cotter, approximately corresponding to the limit between the MN4 biozone and MN5 biozone (Antunes *et al.* 1987), there appear for the first time intermediate trilophodont teeth with talons that, because of their development, may be interpreted as a fourth initial loph (Antunes and Mazo 1978). Since in various deposits, lower incisors have appeared with dorso-ventral flattening whose lateral wall is not sub-horizontal as is usual in most specimens but becomes vertical as it approaches the alveoli, it is Tassy's (1985) opinion that these dental elements pertain to amebelodons, not gomphotheres.

Tobien (1972, 1973*a*) suggested that taxa related to Eurasian platybelodons existed in western Europe. He referred certain mandibles and lower incisors from Sansan and La Grive (France) to *Platybelodon danovi filholi* (Frick). This hypothesis was accepted by Tassy (1984, 1985) who attributed these and

Table 14.1 Biostratigraphical positions of Iberian localities mentioned in this chapter

European Land Mammal Ages (after Mein 1979, 1990)		MN Zones Mein	Localities in the Iberian peninsula[a]	*Gomphotherium angustidens*	Indeterminate gomphotheres	*Tetralophodon longirostris*	*Anancus arvernensis*	*Zygolophodon turicensis*	*Mammut borsoni*
Villafranchian	Late	17	Mata (8)						●
	Early	16	Villarroya (5)				●		
			Las Higueruelas (17)				●		
Ruscinian	Early	15							
	Early	14							
Turolian	Late	13							
	Middle	12	Concud (14)			●	●	●	
	Early	11	Crevillente (18)				●		
Vallesian	Late	10							
	Early	9	Can Llobateres (10)			●			
			Polinyá (9)			●			
			Fuentidueña (7)			●			
Astaracian	Late	7 & 8	Monte Abadesa (2)		●				
			Benavente (4)					●	
			Saldaña (1)			●			
			Cerro del Cristo (3)	●					
	Middle	6	Paracuellos (13)	●					
		5	Yuncos (15)		●				
			Tarazona (6)	●	●				
		4b	Quinta da Farinheira (21)	●	●				
			Quinta Grande (21)	●	●				
			Courelas do Covao (21)	●					
			Quinta das Pedreiras (21)	●					
	Early	4a	Papiol (11)	●					
			Córcoles (12)	●					
			Buñol (16)	●					
			Quinta do Narigao (21)	●					

[a] Numbers in ( ) refer to sites in Fig. 14.1. Editors' note: compare this table to Appendix D for differences between European Land Mammal Ages and MN zones.

other fossils from Spain and Portugal to *Archaeobelodon filholi* (Frick), which would include *Platybelodon danovi filholi* as a synonym.

This issue is controversial, because neither the isolated molars of the platybelodons, nor those of the *Platybelodon danovi* (Borissiak) differ from those of many gomphotheriids. As for the inci-

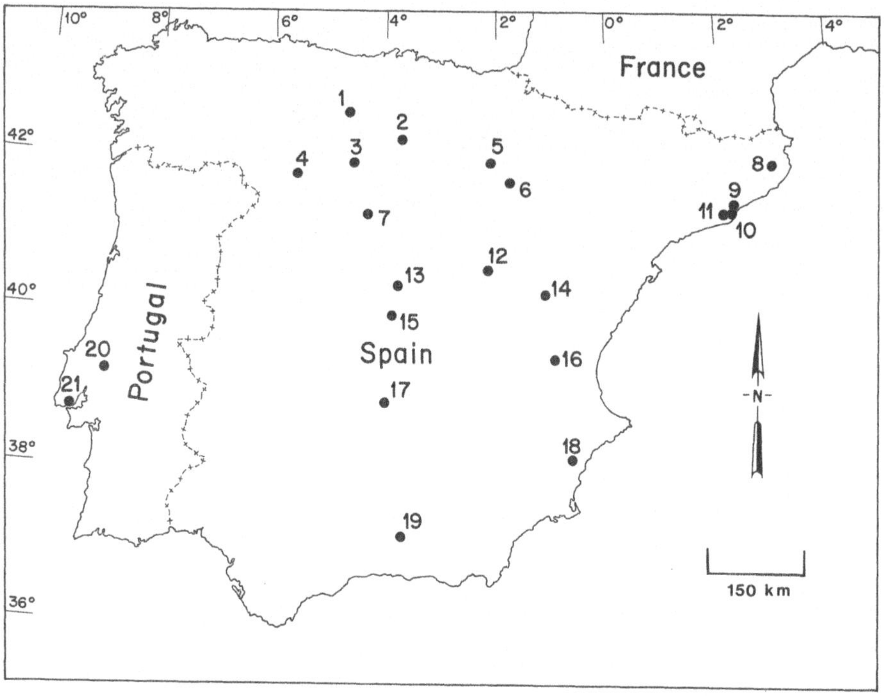

Fig. 14.1 Map of the Iberian peninsula, showing sites where gomphothere and mammutid specimens were found. Key: 1, Saldaña (Palencia); 2, Monte de la Abadesa (Burgos); 3, Cerro del Cristo del Otero (Palencia); 4, Benavente (Zamora); 5, Villarroya (Logroño); 6, Tarazona de Aragón (Zaragoza); 7, Los Valles de Fuentidueña (Segovia); 8, Mata (Gerona); 9, Polinyá (Barcelona); 10, Can Llobateres (Barcelona); 11, Papiol (Barcelona); 12, Córcoles (Guadalajara); 13, Paracuellos (Madrid); 14, Concud (Teruel); 15, Yuncos (Toledo); 16, Buñol (Valencia); 17, Las Higueruelas (Ciudad Real); 18, Crevillente (Alicante); 19, Alfacar (Granada); 20, Santarem (Portugal); 21, Quinta da Noiva, Quinta da Farinheira, Quinta Grande Charneca, Quinta das Pedreiras, and Courelas do Covao (Lisboa, Portugal).

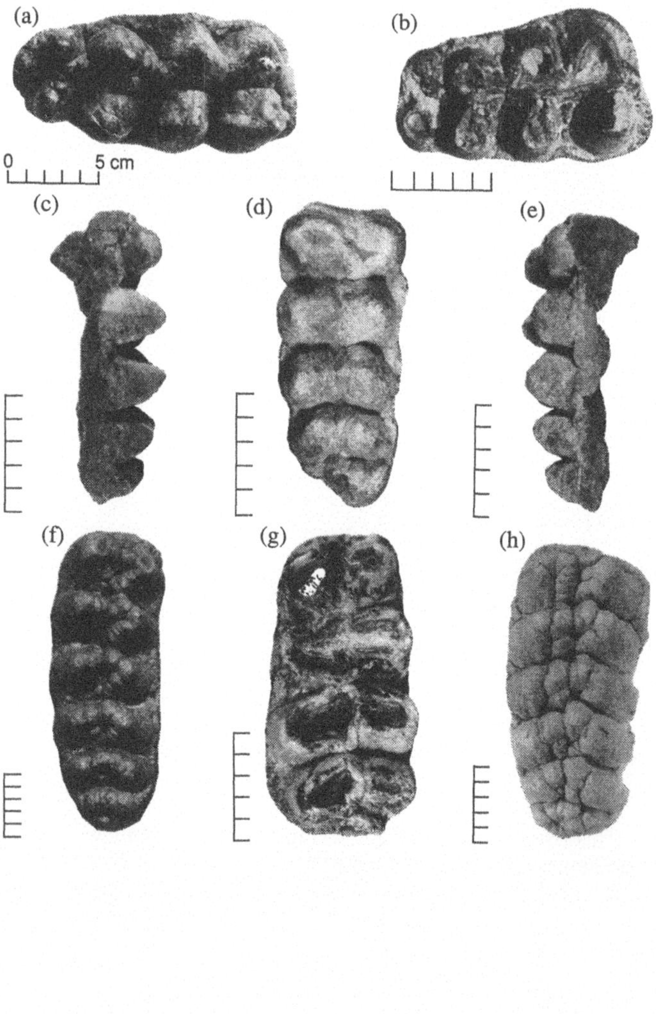

Plate 14.1 Gomphothere specimens from Spain. (In (a) and (b) the anterior is to the right, in (c)–(h) the anterior is towards the top of the page; scales are in centimetres.) (a) *Gomphotherium angustidens*, left M^3 from the early Astaracian of Papiol (Barcelona) in an occlusal view; (b) *Gomphotherium angustidens*, left M^3 from the early Astaracian of Córcoles (Guadalajara) in an occlusal view; (c) *Gomphotherium angustidens*, right M^3 from the early Astaracian of Buñol (Valencia) in a labial view; (d) the same specimen as in (c), but in an occlusal view; (e) the same specimen as in (c) and (d), but in a lingual view; (f) *Gomphotherium angustidens 'grandincisivoid'*, right M_3 from the late Astaracian from Monte de la Abadesa (Burgos) in an occlusal view; (g) *Tetralophodon longirostris*, right M_2 from the late Astaracian of Saldaña (Palencia) in an occlusal view; (h) *Tetralophodon longirostris*, left M^3 from the early Vallesian of Los Valles de Fuentidueña (Segovia) in an occlusal view.

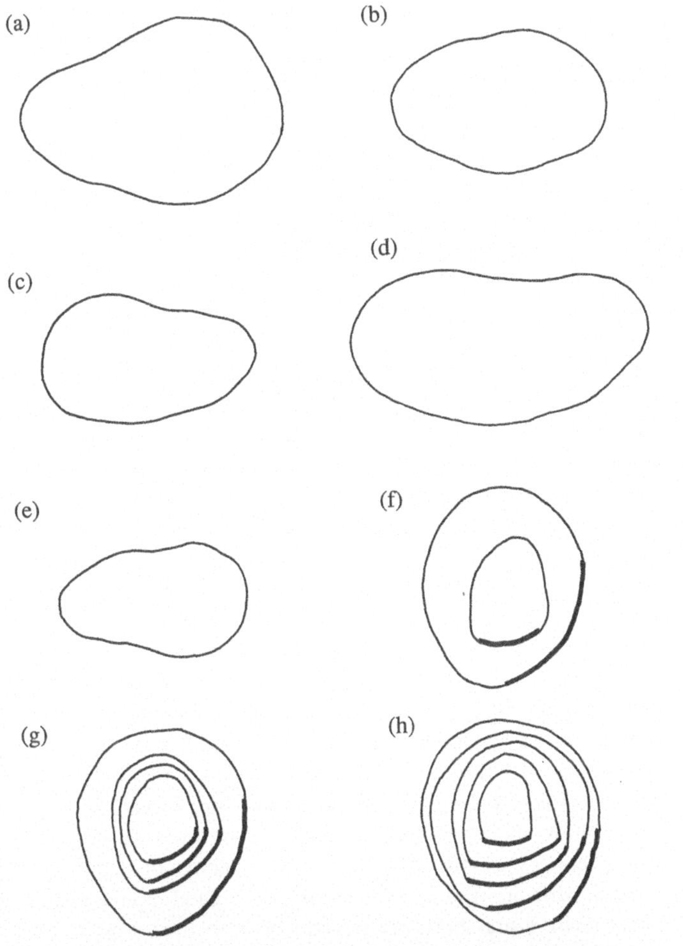

Fig. 14.2 Cross-sections of upper and lower incisors of gomphotheriids from western Europe (in a posterior view. (a) Lower incisor of *Gomphotherium* from the middle Astaracian of Quinta Grande, Charneca (Portugal); (b) Lower incisor of *Gomphotherium* from the middle Astaracian of Quinta da Farinheira, Chelas (Portugal); (c) Lower incisor of *Gomphotherium* from the middle Astaracian of Yuncos, Toledo (Spain); (d) Lower incisor T5–18 from the middle Astaracian of Tarazona de Aragón, Zaragoza (Spain); (e) Lower incisor T–51 from the middle Astaracian of Tarazona de Aragón, Zaragoza (Spain); (f) Upper incisor of *Gomphotherium angustidens* from the middle Astaracian of Yuncos, Toledo (Spain); (g) Upper incisor SEP–410 of *Gomphotherium angustidens* from the late Astaracian of En Péjouan (France) (redrawn from Tassy 1985); (h) Upper incisor SEP–214 of *Gomphotherium angustidens* from the late Astaracian of En Péjouan (France) (redrawn from Tassy 1985).

sors, when in the mandibles, it is evident that their implantation is different from those of the typical *G. angustidens*. The problem arises with the interpretation of the fragments. Because its cross-section is similar in both of the above, it is difficult to determine the correct anatomical position of the larger diameter.

One helpful character is the size, since many of these problematic incisors are larger than usual in *G. angustidens*. The dimensions of one specimen collected in the Quinta Grande, Charneca are, for example, 77 × 52 mm (Fig. 14.2a), and those of the others originating from the Quinta da Farinheira are 65 × 44 mm (Fig. 14.2b). Inferior incisors of this type have also been found in Spain. Lower incisors associated with the complex bunodont molars have been discovered in Tarazona de Aragón (Zaragoza), pertaining to the MN4 zone. These teeth were attributed by Astibia and Mazo (1988) to *G. angustidens*, but some of these lower incisors could be interpreted as belonging to amebelodons. The T5-18 specimen (Fig. 14.2d), whose diameter is 78 × 46 mm, has an extensive dorsal wear facet and is highest in the lateral half. The distal tip fragment T-51 (52.5 × 29.6 mm; Fig. 14.2e) has an obvious bevelled wear facet, and the T2-148 has a pyriform cross-section and is also highest at the lateral half. In none of these incisors is the maximum height situated near the medial border, as occurs in the supposed platybelodons from France (Tobien 1973*a*, p. 255).

The only fragment of upper incisor from Tarazona that conserves a band of enamel shows that it is latero-ventral, with a cross-section much more similar to those indicated by Tassy (1985, fig. 130) as being *Gomphotherium* than to those described by him as being *Archaeobelodon*.

Another interesting fragment of a lower incisor is from a Yuncos (Toledo) individual (Fig. 14.2c and Plate 14.2) possibly from the MN5 zone, which Mazo and Alberdi (1974) referred to *Gomphotherium lusitanicum* (Bergounioux *et al.* 1953) because of the complexity of its molars. Tassy (1985) refers this fragment to *Archaeobelodon filholi*. The incisor is clearly wide and dorso-ventrally flattened, but the upper right tusk of this specimen, anatomically positioned, exhibits at the exit of the alveolus a lateral band of enamel which becomes ventral toward the apex. This is contrary to the diagnosis established by Tassy (1985, p. 570) for *Archaeobelodon*. The position of the enamel in the Yuncos example compares almost exactly to that of several of the incisors of *G. angustidens* from En Péjouan (France), which was identified by Tassy (1985, fig. 130) (see Figs 14.2g and 14.2h). The Yuncos specimen is from a gomphotheriid.

During the middle Astaracian (specifically, in the MN4b and MN5 biozones), in the Iberian Peninsula, 'serridentinoid' teeth appear to be associated with the broad and flattened lower incisors whose implantation is different from that of the normal *G. angustidens*. Nevertheless, several of the characters that Tassy (1984, 1985) considered as diagnostic of *Archaeobelodon filholi* do not seem significant and the possible relationship of these

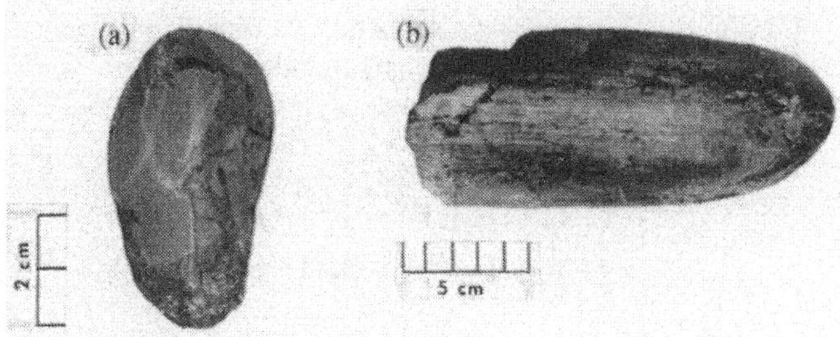

Plate 14.2 Gomphothere specimens. (a) *Gomphotherium*, lower incisor from the Orleanian of Yuncos (Toledo), in transverse view; (b) lower incisor from the Orleanian of Madrid, in dorsal view.

examples from western Europe with the Eurasian platybelodons is doubtful. Possibly there are two types of gomphotheriids already differentiated in these levels; but, at this time, it cannot be determined whether the differences are generic or specific.

Beyond the MN6 biozone, in Spain, no lower incisors with the before-mentioned characteristics have been found. Some of the specimens from France (e.g. Sansan 1887 or Sansan 349) attributed by Tobien (1973*a*) to platybelodons seem to be only variants of *Gomphotherium*, even though it is true that the cross-section of the Sansan 74 incisor (see Tobien 1973*a*, fig. 8) suggests a greater proximity or parallel to the group of the platybelodons *sensu stricto*.

The latest intermediate trilophodont teeth appear in Spain approximately in the MN7/MN8 zone, such as in the deposit of Cerro del Cristo del Otero, Palencia (Hernandez Pacheco and Dantin 1915). Of a similar period must be Monte de la Abadesa in Burgos, where an upper incisor, palate, and mandible of a young animal attributed to '*Trilophodon*' *olisiponensis* (Zbyszewski) were found. Their M2s are very large and have three lophids and a talonid that could almost be considered a fourth lophid. Despite the fact that these features resemble *Tetralophodon*, this animal is a *Gomphotherium*, since the upper incisor has a band of enamel. From the same region are two apex fragments of lower incisor and the M_2 and M_3 of the same individual, but of greater size, which were designated by Bergounioux and Crouzel (1958) '*Geisotodon hispanicus*'. The M_2s have, as in the previous case, a fourth rudimentary crest, and the M_3s are serridentinoids (Plate 14.1f). The incisors are very large (83.5 × 62 mm) and have a sub-oval cross-section with dorsal convexity and a hardly defined upper furrow. Tobien's (1978*a*) hypothesis is that this is a *G. angustidens* 'grandincisivoid', related to the '*Mastodon* (*Bunolophodon*) *grandincisivum*' (Schlesinger) from Germany, Austria, Hungary, and Iran. While the upper incisors of the central European *grandincisivum* all have a more or less pyriform, flattened, dorso-ventral cross-section, those from Burgos are comparatively high.

Lower incisors found in the late Astaracian (MN8) or Vallesian (MN9) of Santarem (Portugal) are comparable to *grandincisivus* of western Europe (see Antunes and Mazo 1983, plate 1, figs 2d and 2g). Both the large intermediate teeth with an initial fourth crest and the robust pyriform incisors are relatively abundant in Spain in the late Astaracian and in the early Vallesian. The morphology and the age of these specimens corroborate what Mottl (1969), after studying the *G. angustidens* of the late Astaracian of Estiria, detected in them: two mutational tendencies, one longirostrine and the other grandincisivoid.

GOMPHOTHERES: BUNODONT TETRALOPHODONS

In Spain, the oldest known intermediate teeth with four crests have been found in the late Astaracian (MN8). An example is the M_2 of the *Tetralophodon* of Saldaña, Palencia (Plate 14.1g) denomination, which should also include the palate that Bergounioux and Crouzel (1955, 1958) referred to *Stegolophodon saldañensis*. These teeth, whose talons are reduced to small crescentoids, are clearly much simpler than the *T. longirostris* type from the Vallesian (MN9) of Eppelsheim (Germany). *Tetralophodon longirostris* is also known from France in the MN8 biozone, specifically in the Montrejeau deposit (Crouzel and Debeaux 1957).

Many authors (Schlesinger 1917; Lehmann 1950; Mottl 1969; Tobien 1978*a*, among others) accept that *T. longirostris* is a descendant of *Gomphotherium angustidens*, a hypothesis now disputed. The fundamental tendencies of the evolutionary process would have been a reduction in the length of the mandibular symphysis, a change in the cross-section of the lower incisors, the loss of the enamel band on the upper tusks, and the acquisition by the intermediate teeth of a fourth crest for the growth of the talon. Nevertheless, Tassy (1985, p. 722) did not accept a direct relationship between *G. angustidens* and *T. longirostris* because, in his opinion, the first possesses more autapomorphies than the second.

T. longirostris has not, as of yet, been discovered together with *G. angustidens* in Spain, Portugal, or France. This association, however, certainly exists in the Vallesian deposit of Eppelsheim, Germany (Lehmann 1950; Tobien 1980*a*, 1986*b*).

In the Iberian Peninsula, there are a considerable number of *T. longirostris* remains, with most of their teeth isolated. The most complete individual is that at the early Vallesian at Polinyá (Barcelona), described by Alberdi (1971), considered by Tobien (1978*a*) to be *G. angustidens*, since when compared to the *T. longirostris* type from Eppelsheim, it has a long symphysis which curves only slightly. The wear prevents us from knowing the characteristics of the M2, but other significant features of *T. longirostris* are: the upper incisors lack a band of enamel, and lower incisors are short and straight with an oval cross-section. Lower tusks with this morphology have not been found in the Iberian Peninsula associated with intermediate trilophodont molars. Another point concerns the dimensions of the postcranial remains, which in gomphotheres of this type usually exceed those of *G. angustidens*. The mandibular symphysis curvature could be due to sexual or individual variation. The great variability demonstrated by the mandibles of *T. longirostris*, gathered in the Vallesian from the Rheinhessen (Germany) region, is one more reason to include the Polinyá mandible in *T. longirostris*. Most of the *T. longirostris* specimens from Spain correspond to the Vallesian (zones MN9 and MN10), but are present in several Turolian sites.

Tobien (1980*a*, 1986*b*) recognizes in the Vallesian of Rheinhessen two types of tetralophodons: *Tetralophodon longirostris* and *Stegotetrabelodon gigantorostris* (Klähn, 1922), in which are included those specimens named by Schlesinger (1917) as '*Mastodon* (*Bunolophodon*) *grandincisivum*'. Ignoring the distinction and the possible relationship between *Stegotetrabelodon* and the African *Stegotetrabelodon* — which seem problematic — the specimens from Spain are insufficient to establish taxonomic differentiations and their relationships. It is evident, however, that the molars found do not correspond to a single morphology; some are simple structures with a small number of cuspids (Plate 14.3d), while others of a very complex morphology are characterized by multiple conules (Plate 14.1h). This is unlikely to be attributed to age differences among the sites or origin, because both types may be present in the same deposit, as occurs in Los Valles de Fuentidueña (Segovia), or in Can Llobateres (Barcelona). It is logical, however, to assume that *T. longirostris* had a variability more or less comparable to that known to occur in *G. angustidens*.

Anancus arvernensis is the latest surviving tetralophodont gomphothere from western Europe, with a short mandibular symphysis, with an early loss of the incisors, and large upper incisors often more than 3 m in length. It has been found in Spain from the early Turolian (Crevillente 2) to the early Villafranchian (Las Higueruelas and Villarroya). Although less numerous in Spain than *G. angustidens*, *Anancus arvernensis* is more common than *T. longirostris*.

Two specimens originating from Concud (Teruel) and Alfacar (Granada) were identified by Bergounioux and Crouzel (1958) as *Choerolophodon pentelici* because of their choerodontia and ptychodontia. In reality, they are two deciduous teeth from *Anancus arvernensis*. *Ch. pentelici* have not been encountered in the Iberian peninsula.

MAMMUTIDS

Mammutids (or 'zygodont mastodons') have wedge-shaped crest molars with antero-posterior compression. They have been found in Africa, Eurasia, and North America, and are less common than the bunodont forms. Members of Mammutidae were less diversified than bunodont gomphotheres maintaining trilophodont teeth throughout their entire evolutionary history.

The distinction between zygodont forms in the early and middle Astaracian and in Europe is difficult because the molars do not often present clear characteristics, but rather a mixture of zygodont and bunodont features. It is not a coincidence that specimens of this age were those that prompted Schlesinger (1917) to define them as '*Mastodon angustidens subtapiroidea*'. Lehmann (1950) and later authors interpreted them as an intermediate form between *G. angustidens* and *Z. turicensis*. The identification of the zygodonts in these levels carries an implied subjectivity that does not facilitate taxonomy or systematization.

In the Iberian Peninsula, perhaps the most interesting specimen in this aspect is the hemimandible from Paracuellos V (Madrid) from the late Astaracian, which has P_4, M_1, and M_2 in anatomical position (Mazo 1985). While the P_4 and the M_1 have zygodont morphology, the M_2 is clearly bunodont (Plate 14.3a–c). According to Mazo (1985), it is *G. angustidens*, that is, a bunodont. It is mentioned here since this is an appropriate place in the text.

The M_1 (62 × 42 mm) is not even a specimen of intermediate morphology, but it presents many of the diagnostic characteristics of *Zygolophodon turicensis*: principal tubercles with antero-postero compression, lophids with an oblique disposition, a slightly defined furrow, and valleys with only two small projections between the pretrite elements. The morphology of the M_2 (95 × 51 mm) is very different from that of the earlier described M_1 since, in this specimen M_2, the cuspids are rounded and of considerable complexity (Plate 14.3b,c).

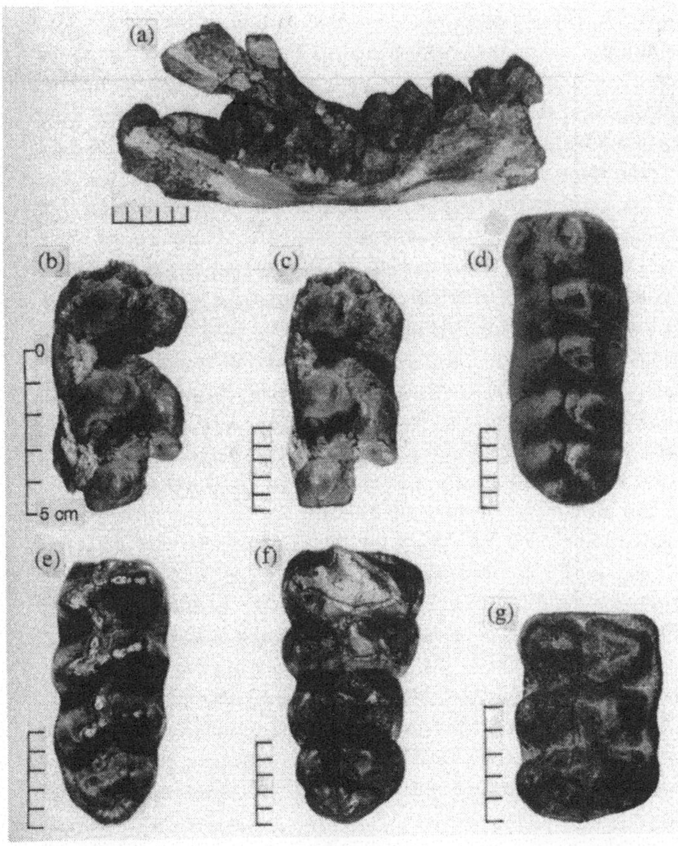

Plate 14.3 Gomphothere and mammutid specimens from Spain (in (a)) the anterior is to the right, in Figs (b)–(g) the anterior is towards the top of the page; scales are in centimetres). (a) *Gomphotherium angustidens*, left hemimandible of the late Astaracian of Paracuellos V (Madrid) in a semi-occlusal view; (b) *Gomphotherium angustidens*, right M_2 of the individual shown in (a), in a semi-occlusal view; (c) the same specimen as in (b) in an occlusal view; (d) *Tetralophodon longirostris*, right M_3 of the early Vallesian of Can Llobateres (Barcelona) in an occlusal view; (e) *Zygolophodon turicensis*, left M_3 of the late Astaracian of Benavente (Zamora) in an occlusal view; (f) *Zygolophodon turicensis*, right M^3 of the middle Turolian of Concud (Teruel) in occlusal view; (g) *Zygolophodon turicensis*, right M^1 of the middle Turolian of Las Pedrizas (Teruel) in an occlusal view.

If these teeth had not been found in the same mandible, it is more than probable that they would not only have been attributed to two different genera, but to two different families. This specimen supports the opinion that many supposedly zygodont dental elements from the early Astaracian of central and western Europe are only examples of the plasticity of these proboscideans. Such variability can be generated by many factors, one of which could be diet, as Saunders (1977) suggested.

Tassy (1985) did not agree with this interpretation, but suggested that *Zygolophodon* was present in Europe from the MN4 biozone with two morphotypes: the 'slender' morphotype and the 'robust' one. The intermediate specimens would correspond to this latter type. In Spain, the robust molars are more abundant in the Orleanian (*c.* 18 Ma) than in higher levels. This personal observation is indirectly corroborated by previously identified specimens, since most of the teeth attributed to *Zygolophodon* originated from the MN4 and MN5 biozones.

In Tassy's opinion (1985, p. 508), *Zygolophodon turicensis* existed in the Miocene of France and Spain, but not in the Miocene fauna of Portugal. The Portuguese Orleanian is extraordinarily rich in material, which I have studied, including molars of intermediate morphology similar to those found in Spain. Therefore, the taxonomic interpretation must be the same for the whole Iberian Peninsula, especially since no other absences of other faunal elements are detected in Portugal.

The first specimen from Spain with clear zygodont morphology is the M3 from Benavente (Zamora) (Plate 14.3e), determined by Bergounioux and Crouzel (1958) to be *Z. pyrenaicus*. The exact age is unknown, but it is highly probable that it is late Astaracian MN7 or MN8. The situation is comparable to that in other European countries since the Astaracian specimens from Malartic or Montrejeau (France) present a notable zygodontia. It should be remembered that the species *turicensis* was defined precisely from molars originating from the late Vindobonian of Elgg (Switzerland).

The Spanish record of zygodonts has been sparse; they are found again in the middle Turolian of Concud (Teruel), MN12, where various dental elements originate (Plate 14.3f). The increase in size of these specimens with respect to the M3 of the Benavente locality is notable; also a slight reduction in size exists in the conules among the pretrite elements. Establishing whether *Z. turicensis* existed in Europe since the early or late Astaracian is very difficult; if *Zygolophodon* and *Gomphotherium* were really differentiated already in the MN4 biozone, a specimen such as that of Paracuellos V would be difficult to explain, even when resorting to evolutionary convergences or parallelisms.

The discovery of one mammutid in eastern Africa from approximately 22 Ma (Tassy and Pickford 1983), with a degree of zygodontia comparable to that exhibited by the specimens from the Vallesian and Turolian of Europe, is one of the most important findings of recent years. The remains, labelled by Tassy and Pickford (1983) as *Eozygodon morotoensis*, show that zygodont differentiation already existed in the early Miocene. These facts, apparently contrary to those established in Europe, prompted Tassy and Pickford (1983) to believe that perhaps *Eozygodon morotoensis* is a sister-group of the other mammutids.

Mammut borsoni, the latest of the mammutids of Europe, is represented in Spain by only one molar fragment, found in the Pliocene of Mata, Gerona (Juliá and Villalta 1974). It is not known with certainty whether *M. borsoni* descended directly from *Z. turicensis* or whether it had its origin in another form of zygodont. According to Tobien (1976), the '*Mammut praetypicum*' from the Odessa region described by Kubiak (1972), whose age Gabunia (1979) suggests could correspond to a final Turolian or a Ruscinian, could be an intermediate taxon in the evolutionary process between *Z. turicensis* and *M. borsoni*. This is a reasonable hypothesis which at the moment, however, cannot be confirmed.

CONCLUSIONS

1. In the Iberian Peninsula, I recognize a bunodont group (gomphotheres) and a zygodont group (mammutids). The gomphotheres include *Gomphotherium angustidens*, a second gomphotheriid whose generic or specific category cannot be precisely determined, *Tetralophodon longirostris*, and *Anancus arvernensis*. The mammutids include *Zygolophodon turicensis* and *Mammut borsoni*.
2. In the Iberian Peninsula *G. angustidens* is a typical Astaracian taxon which is found from the MN4 to the MN7 biozone. *T. longirostris* is recorded from the late Astaracian, MN8 biozone, to the middle Turolian, MN12. *Anancus arvernensis* is found from the early Turolian, MN11, to the Villafranchian, MN16. *Z. turicensis* is known from the late Astaracian, MN7/MN8, to the middle Turolian, MN12 biozone. There is only one reference of *M. borsoni* in the Pliocene (Villafranchian).
3. In Spain and Portugal, *G. angustidens* and *T. longirostris* are not known to be associated in the same deposit. This association, however, is known in central Europe.
4. There are no early amebelodons in the Iberian Peninsula. The incisors and molars from the MN4 and MN5 biozones that have been interpreted as early amebelodons belong to a second group of gomphotheriids whose category cannot be determined precisely.
5. The first specimen with a clear zygodontid morphology appears in the late Astaracian. Molars from the middle and early Astaracian in the Iberian Peninsula that have been interpreted as zygodonts seem, in reality, to pertain to gomphotheriids of an intermediate morphology.

ACKNOWLEDGEMENTS

E. C. Marsac and S. Wolak retyped this chapter. The editors and W. D. Lambert made suggestions for improvements. Photographs are by the Departamento de Fotografía del Museo Nacional de Ciencias Naturales, Madrid; J. S. Grimes helped with figure arrangements.

FIFTEEN

The biogeography of the gomphotheriid proboscideans of North America

W. DAVID LAMBERT

INTRODUCTION

The gomphothere proboscideans *sensu lato*, consisting of the Gomphotheriidae including Amebelodontinae were arguably the most successful of all the proboscideans in terms of their longevity and geographical distribution. They lasted at least from the earliest Miocene to the end of the Pleistocene and reached all continents except the island continents Antarctica and Australia (Tobien 1973*a*). North America played a significant role in the history of gomphothere biogeography and diversity (Lambert and Shoshani, in press). From the middle Miocene to the Pleistocene, North America received numerous immigrant taxa from the Old World via Beringia (and possibly the converse as well). It was also a centre for major autochthonous evolution and was the staging ground for the dispersal of gomphotheres into South America.

This chapter briefly summarizes the complex biogeographical history of the gomphotheres of North America and examines its palaeoecological significance. Systematics are discussed only as necessary for understanding gomphothere biogeography. The species level systematics of the gomphotheres is in desperate need of revision.

FAMILY GOMPHOTHERIIDAE HAY, 1922

The gomphotheriids comprise the generally conservative gomphotheres *sensu lato* as typified by *Gomphotherium*, although there is considerable diversity within the group. This family must be considered to be at best paraphyletic, as the constituent taxa have been lumped together on the basis of shared plesiomorphies.

GOMPHOTHERIUM BURMEISTER, 1837

Synonyms

Trilophodon Falconer, 1857; *Serridentinus* Osborn, 1923; *Trobelodon* Frick, 1933; *Ocalientinus* Frick, 1933; *Tatabelodon* Frick, 1933; *Megabelodon* Barbour, 1914*b*. (*Eubelodon* Barbour, 1914*a* and *Gnathabelodon* Barbour and Sternberg, 1935 are considered valid genera, see Appendix C which also includes five other junior synonyms of *Gomphotherium* — editors.)

Gomphotherium is the most morphologically conservative of all of the gomphotheriid proboscideans, with few diagnostic characters to distinguish it clearly from more advanced gomphotheriids. It embodies these features: trilophodont intermediate cheek teeth (i.e. dp4–M2); M3 with three to four and a fraction loph(id)s; trefoils on the posttrite hemilophs (e.g. secondary trefoils) absent or poorly developed; upper tusks with well-developed enamel bands on their lateral surfaces; elongated mandibular tusks that are circular, oval, or pyriform in cross-section; and elongated mandibular symphyses. This character complex in proboscideans is commonly referred to as the longirostrine condition (Tobien 1973*a*).

There are a number of relatively conservative gomphothere genera in the literature, including *Serridentinus*, *Tatabelodon*, *Trobelodon*, and *Ocalientinus* which are considered synonyms (see above). Tobien (1973*a*) studied these taxa and suggested that they are morphological variants of *Gomphotherium*, possibly all belonging to a single species. Among the more aberrant taxa included by Tobien (1973*a*) in *Gomphotherium* were *Gnathabelodon*, *Eubelodon*, and *Megabelodon*; all these forms lack mandibular tusks and have strangely shaped mandibular symphyses (Barbour 1914*a*, 1914*b*; Barbour and Sternberg 1935).

Gomphotherium was the first of the gomphotheriids to reach North America, entering from Asia during the Barstovian (middle Miocene) along with mammutid proboscideans (i.e. true 'mastodons'). Tedford *et al.* (1987) used the stratigraphic record of *Gomphotherium* in the Great Plains to define the beginning of the early late Barstovian. For example, *Gomphotherium* is unknown from the early Barstovian Lower Snake Creek deposits of Nebraska (the Olcott Formation) but abundant in the overlying early late Barstovian Valentine Formation (Voorhies 1990). Downs (1952), none the less, described *Gomphotherium* material from the Mascall Local Fauna of Oregon, which Tedford *et al.* (1987) date as early Barstovian. *Gomphotherium* does not appear in eastern North America until the latest Barstovian, although this may reflect the rarity or absence of sediments of early late Barstovian age locally, rather than biogeographical reality. This genus is known from the undescribed Kingsford Mine Local Fauna in the Bone Valley region of Central Florida and the Calvert Cliffs Local Fauna in the Calvert Cliffs Formation of Maryland (Gazin and Collins 1950), both of which are dated as latest Barstovian.

After its appearance in the early to middle Barstovian, *Gomphotherium* became widespread and abundant throughout North America. It is known from far too many sites to mention even a significant fraction of them, but some of the more important include: Devil's Gulch, Nebraska, late Barstovian (Barbour 1914*a*; Tedford *et al.* 1987); Love Bone Bed, Florida, late Clarendonian (late Miocene) (erroneously referred to *Amebelodon* by Webb *et al.* 1981); and the series of sites near Santa Fe, New Mexico referred to collectively as the Santa Fe Badlands or Marls, late Clarendonian and latest Hemphillian, earliest Pliocene (Frick 1933; Mebrate 1987) (Fig. 15.1).

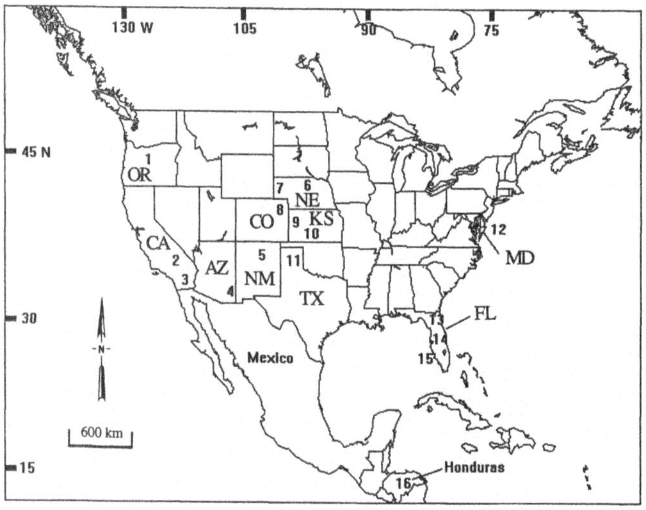

Fig. 15.1 Map showing selected important gomphotheriid including amebelodontine sites in North America. Key: 1, Mascall, Oregon (early Barstovian); 2, Ricardo, California (early Clarendonian); 3, Mount Eden, California (latest Hemphillian); 4, Santa Fe Marls, New Mexico (Clarendonian-latest Hemphillian); 5, Curtis Ranch, Arizona (early Irvingtonian); 6, Devil's Gulch (late Barstovian) and Sand Draw (Blancan), Nebraska; 7, Cambridge (= *Amebelodon*) Quarry, Nebraska (early Hemphillian); 8, Wray, Colorado (early Hemphillian); 9, Rhinoceros Hill, Kansas (late Hemphillian); 10, Deer Park (middle Blancan) and Rexroad (early Blancan), Kansas; 11, Coffee Ranch (late Hemphillian), Cita Canyon (late Blancan), and Clarendon (Clarendonian), Texas; 12, Calvert Cliffs, Maryland (late Barstovian); 13, Love Bone Bed (late Clarendonian), Mixon's Bone Bed (early Hemphillian), and Moss Acres Racetrack (early Hemphillian), Florida; 14, Bone Valley region, Florida (late Barstovian-latest Hemphillian); 15, Punta Gorda, Florida (middle Irvingtonian); and 16, Gracias Formation, Honduras (late Clarendonian or early Hemphillian). (Note that although no specific sites in it are listed, Mexico has a number of sites from which gomphotheriids are known, mostly Pliocene or Pleistocene in age.)

Gomphotherium survived into the latest Hemphillian, being known from the Chamita Formation of the Santa Fe Badlands and the upper Bone Valley Formation (Lambert, in preparation; Mebrate 1987; Tedford *et al.* 1987). It is not known to have survived into the Blancan, early to latest Pliocene (Kurtén and Anderson 1980).

RHYNCHOTHERIUM FALCONER, 1868

Synonyms

Blickotherium Frick, 1933; *Aybelodon* Frick, 1933

Rhynchotherium broadly resembles *Gomphotherium*, differing in the following respects. The enamel bands of its upper tusks spiral to varying degrees, whereas its mandibular tusks commonly, but not always, have enamel bands (a unique feature among the gomphotheres). Its mandibular symphysis is distinct from *Gomphotherium* in being significantly deflected downward (Miller 1980, 1990; Webb and Perrigo 1984).

Rhynchotherium, unlike *Gomphotherium*, is reliably known only from North America and probably represents an autochthonous radiation from *Gomphotherium* stock (Tobien 1973*a*). In the United States and Mexico, *Rhynchotherium* is unknown earlier than the late Hemphillian, and its appearance is used to help define the late Hemphillian (Tedford *et al.* 1987). However, Webb and Perrigo (1984) described a substantial population of *Rhynchotherium* individuals from the Gracias Formation of Honduras that dates, based on the presence of certain key taxa (e.g. the borophagine dog *Osteoborus cyonoides* and the gelocid artiodactyl *Pseudoceras skinneri*) from the late Clarendonian to the early Hemphillian (late Miocene), substantially older than any *Rhynchotherium* finds to the north. Webb and Perrigo (1984) suggested on this basis that *Rhynchotherium* evolved from *Gomphotherium* in Central America during the late Clarendonian or early Hemphillian, and then later dispersed northward. Central America aside, *Rhynchotherium* is generally considered to be diagnostic of faunas from late Hemphillian to Blancan in age. It is represented at the late Blancan Santa Fe River IB Local Fauna in Florida, but evidently did not survive into the Pleistocene (Webb 1974*b*; Kurtén and Anderson 1980).

Specimens referred to *Rhynchotherium* range from California east to Florida and from Kansas south to Central America, with several specimens known from Mexico, including the genotypic specimen (Frick 1933; Osborn 1936; Olsen 1957; Kurtén and Anderson 1980; Miller 1980, 1990). Thus, it can be considered to be a relatively southern genus, possibly due to a restriction of woodland savanna to this region (see below). Important faunas containing *Rhynchotherium* include: Mount Eden, California, latest Hemphillian (Frick 1933; May and Repenning 1982); Coffee Ranch, Texas, late Hemphillian (latest Miocene) (Dahlquest 1983); Palmetto Local Fauna of the Bone Valley region, Florida, latest Hemphillian (Olsen 1957); and Deer Park, Kansas, middle Blancan (Kurtén and Anderson 1980) (see Fig. 15.1).

CUVIERONIUS OSBORN, 1923

Synonym

Cordillerion Osborn, 1926*a*

Cuvieronius is distinct from both *Gomphotherium* and *Rhynchotherium* in having an abbreviated (i.e. brevirostrine) mandibular symphysis and lacking mandibular tusks. Its upper tusks resemble those of *Rhynchotherium* in having spiralled enamel bands. Its molars vary greatly in trefoil morphology, ranging from extreme ptychodonty (i.e. plications within the enamel lophs) and well-developed secondary trefoiling to a condition so simple as to be indistinguishable from that of *Gomphotherium*. The M3 has four and a fraction to five and a fraction loph(id)s (Simpson and Paula Couto 1957; Kurtén and Anderson 1980; Miller 1990).

Cuvieronius is endemic to the New World. Its evolutionary origins are uncertain, but it has commonly been proposed that *Cuvieronius* is closely related to, or directly descended from, *Rhynchotherium* (e.g. Tobien 1973*a*; Webb and Perrigo 1984; Miller 1990), based in part on the shared spiralled enamel bands on their upper tusks. *Cuvieronius* first appeared in the Blancan, and in North America persisted until near the end of the Rancholabrean (late Pleistocene; Kurtén and Anderson 1980) and perhaps into the Wisconsinan in areas now submerged (Dudley, this volume, Chapter 29). Unlike any of the previously mentioned gomphotheriids, *Cuvieronius* was involved in the Great American Interchange, entering South America during the Uquian (roughly the late Blancan/early Irvingtonian of North America, or the Plio-Pleistocene) (Dudley, this volume, Chapter 29; Webb 1978; Simpson 1980).

Cuvieronius is essentially neotropical in its distribution; within North America it is known from Mexico, Arizona, New Mexico, Texas, and Florida. It is relatively common in Florida, with remains found at sites scattered throughout the state. One of the best samples of *Cuvieronius* known is from the Punta Gorda Local Fauna in Florida, which is Irvingtonian in age (Webb 1974*b*; Kurtén and Anderson 1980).

STEGOMASTODON POHLIG, 1912

Stegomastodon is distinctive in both its tusks and cheek dentition. Its upper tusks differ from those of typical gomphotheres *sensu lato* in lacking enamel bands. Like *Cuvieronius*, it is brevirostrine and lacks mandibular tusks. Unlike most gomphotheres, its cranium is distinctly high domed, reminiscent of elephantids. Its M2 may be trilophodont or tetralophodont, depending on the species (Tobien (1973*a*) calls the M2s of advanced *Stegomastodon* 'quasi-tetralophodont'), while its M3 may have from five and a fraction to seven and a fraction loph(id)s. Its molars are characterized by secondary trefoiling, ptychodonty, choerodonty (i.e. the presence of numerous tubercles within the lophs), and an abundance of cementum (Osborn 1936; Savage 1955; Simpson and Paula Couto 1957; Tobien 1973*a*).

Stegomastodon, like *Cuvieronius*, is known from both North and South America. Its origins are disputed. Savage (1955), Simpson and Paula Couto (1957), and Tobien (1973*a*) considered *Stegomastodon* to be part of a North American evolutionary complex including *Cuvieronius* and *Rhynchotherium* on the basis of an overall phenetic similarity. Nevertheless, Hoffstetter (1952) and Mebrate (1987) allied *Stegomastodon* with Old World progressive gomphotheriids such as *Anancus* and *Synconolophus*/*Choerolophodon*. Whatever its origins, *Stegomastodon* was a rather short-lived genus in North America, ranging from the early Blancan to the middle Irvingtonian. In the early Irvingtonian (Uquian), however, it entered South America with *Cuvieronius*, where it survived into the Lujanian (the equivalent of the Rancholabrean in North America, or latest Pleistocene) (Kurtén and Anderson 1980; Simpson 1980).

Unlike *Cuvieronius* and *Rhynchotherium*, *Stegomastodon* was not restricted to southern North America but rather ranged into the Great Plains as far north as Nebraska and Colorado. In addition, the distribution of *Stegomastodon* differed from that of both *Cuvieronius* and *Rhynchotherium* in being distinctly western, with *Stegomastodon* unknown east of Texas. This restriction of *Stegomastodon* to the western half of the continent may reflect a dependence on open grasslands; *Stegomastodon* is commonly believed to have been more specialized for grazing than most gomphotheres (Kurtén and Anderson 1980), though whether it was specialized as the elephantid mammoths is open to debate. Within western North America, *Stegomastodon* has been found in Mexico, Texas, Kansas, Nebraska, Colorado, and Arizona (Osborn 1936; Kurtén and Anderson 1980; Madden 1986). Important *Stegomastodon* sites include: Cita Canyon, Texas, late Blancan (Savage 1955); Rexroad, Kansas, early Blancan (Woodburne 1961); Sand Draw (referred to by Osborn as the *Stegomastodon* Quarry), Nebraska, Blancan (Osborn 1936, p. 726; Kurtén and Anderson 1980); and Curtis Ranch, Arizona, early Irvingtonian (Gidley 1926).

TETRALOPHODON FALCONER, 1857

Tetralophodon (and taxa that evolved from it) embodies the following features: (1) tetralophodonty in all of the intermediate cheek teeth; (2) more loph(id)s in the M3 than the $4–4\frac{1}{2}$ typical of *Gomphotherium*; (3) reduction in the mandibular symphysis; and (4) a tendency towards trefoiling on the posttrite hemiloph(id)s (Tobien 1978*b*) (it should be noted that conditions (2)–(4) are highly prone to parallelism within gomphotheres *sensu lato*). This diagnosis is difficult to apply to referred North American specimens because the loph numbers of their dp4 and M1 are unknown, with the result that proper assessment of this genus in North America is extremely difficult (see below).

Although *Tetralophodon* is known primarily from the Old World, a small number of North American specimens have been referred to this genus. Which, if any, of these specimens can be properly referred to *Tetralophodon* is debatable. A mandible and isolated M3s were referred to *Tetralophodon* (*Morrillia*) by Osborn (1924*a*, 1936), but these specimens were referred to *Stegomastodon* or possibly a *Stegomastodon* cousin by Tobien (1978*b*). An isolated third molar was referred by Osborn (1936) to *Tetralophodon*, but this referral in the absence of such characters as an intermediate cheek tooth loph number was considered highly suspect by Tobien (1978*b*). The taxonomy of the remaining specimens, based on the presence of tetralophodont M2s, has been questioned as well. Tobien (1978*b*) argued that these additional North American *Tetralophodon* individuals may represent one or two parallelisms from native North American *Gomphotherium* stock rather than Old World *Tetralophodon* immigrants, based on ignorance of the number of lophs in the dp4 and M1, a key diagnostic feature for *Tetralophodon*. Thus, the geographical origin and taxonomic assignment of North American *Tetralophodon* is uncertain. The two specimens referred to *Tetralophodon* on the basis of tetralophodont M2s are Clarendonian in age, one from the Sappa Creek Local Fauna in Rollins County, Kansas (Osborn 1936, p. 369) and the other from the Clarendon Local Fauna, Texas (Frick 1933) (Fig. 15.1).

SUBFAMILY AMEBELODONTINAE BARBOUR, 1927

The amebelodontines are loosely referred to as the 'shovel-tusked gomphotheres' or 'shovel-tuskers' and are considered by Tassy and Shoshani (1988) to be the sister-group of the gomphotheriids, elephantids, and stegodontids. Tassy (1986) lists dental and cranial synapomorphies that unite the group into a monophyletic lineage. The best-known members of the group possess distinctly broad and flattened mandibular tusks, hence the term 'shovel-tusker'.

AMEBELODON BARBOUR, 1927

Synonyms

'*Trilophodon*' in part (Cook 1922) (since Cook's taxa were transferred to *Amebelodon*).

Amebelodon is, by far, the most common amebelodontine in North America. Its cranial features are virtually unknown, and hence its diagnosis is based strictly on mandibular features: a mandibular symphysis more elongated than that of *Gomphotherium*; flattened (though less so than *Platybelodon*) and elongated mandibular tusks; and mandibular tusks with dorsal (and often ventral) concavities in cross-sectional profile. Features that traditionally have been used to diagnose the genus, trilophodont intermediate cheek teeth and the absence of internal dentinal rods within the mandibular tusk, have been shown by Lambert (1990) to vary within the genus, and are the basis for his division of the genus into the subgenera *A.* (*Amebelodon*, M3 with 4–5 ridges) and *A.* (*Konobelodon*, M3 with 5–6 ridges).

The biogeographical origin of *Amebelodon* is uncertain. If, however, the amebelodontine gomphotheres are indeed monophyletic, as argued by Tassy (1986), then *Amebelodon*'s roots must stem ultimately from the Old World, since the earliest amebelodontine gomphothere known is *Platybelodon* from the early Miocene of Africa (Maglio 1969*a*). *Amebelodon* is known from the Old World, but it is either of uncertain age within the Miocene (Tobien 1986*a*) or later than its last appearance in North America (Gazíry 1987*b*). This statement is based on a rejection of the synonymy between *Amebelodon* and *Platybelodon* first proposed by Osborn (1929*b*) and followed by Gazíry (1987*b*). The date of the first appearance of *Amebelodon* in North America is controversial, in part because of disagreement over the diagnosis of the genus. Tobien (1972) united *Serbelodon*, another North American amebelodontine, with *Amebelodon* and in doing so effectively extended *Amebelodon*'s chronological range back to the early Clarendonian (see below). However, traditional *Amebelodon*, which I recognize here, does not appear in North America until the early Hemphillian (contrary to Tedford *et al.* (1987), who claim that the first appearance of *Amebelodon* is diagnostic for the late Clarendonian based on a synonymy with *Serbelodon*). *Amebelodon* is a relatively short-lived genus, ranging from the early to the late Hemphillian, with the late Hemphillian date based on a single site (Bennett 1977; Lambert 1990).

Geographically *Amebelodon* is known from the northern and southern Great Plains states, including Nebraska, Kansas, Colorado, and northern Texas, and Florida on the Gulf Coastal Plain (Cook 1922; Barbour 1927; Gregory 1945; Lambert 1990). Important *Amebelodon* sites include: Mixon's Bone Bed, Florida, early Hemphillian (Leidy 1886); Cambridge (equals *Amebelodon*) Quarry, Nebraska, early Hemphillian (Barbour 1927); Wray, Colorado, early Hemphillian (Cook 1922); Moss Acres Racetrack site, Florida, early Hemphillian (Lambert 1990); Bone Valley region, Florida, probable Hemphillian (Lambert 1992); and Rhinoceros Hill, Kansas, late Hemphillian (Barbour and Hibbard 1941; Bennett 1977) (Fig. 15.1).

SERBELODON FRICK, 1933

Synonym

Amebelodon was suggested a senior synonym of *Serbelodon* by Tobien (1972). *Serbelodon* is diagnosed mainly on the basis of mandibular features. In its mandibular tusk cross-section, it resembles *Amebelodon*, having a dorsal concavity and a proportionately similar degree of flattening, and was synonymized with *Amebelodon* by Tobien (1972) on this basis. These characters are variable and plastic in gomphotheres *sensu lato*, however. For example, two *Gomphotherium* mandibular tusks from the large Love Bone Bed sample have been found with cross-sections identical to those of *Amebelodon*, although in all other respects they resemble the other *Gomphotherium* specimens from the site. Thus, I believe that these shared features alone are insufficient bases for uniting these taxa. Beyond their mandibular tusk profiles, *Serbelodon* and *Amebelodon* bear little striking resemblance to each other. The mandibular tusks and symphyses of *Serbelodon* are much shorter than those of *Amebelodon* and are even shorter than those of typical *Gomphotherium* specimens. The cheek teeth are conservative, resembling those of *Gomphotherium*, and the intermediate cheek teeth are trilophodont (Frick 1933).

Serbelodon is a poorly known genus described from North America (*S. barbourensis*; Frick 1933) and from China (*S. zhongningensis*; Guan, this volume, Chapter 13). In North

America it has been found at two localities spanning the Clarendonian (Fig. 15.1): Ricardo, California, early Clarendonian (Osborn 1933); and Christmas Quarry, Nebraska, late Clarendonian (Frick 1933). Webb *et al.* (1981) and Tedford *et al.* (1987) included *Serbelodon* (referred to as *Amebelodon*) in the faunal list of the late Clarendonian Love Bone Bed, Florida; however, this inclusion was based on misidentification of *Gomphotherium* specimens. Note, however, that Shoshani (this volume, Chapter 16) questions the validity of *Serbelodon* and concludes that 'based on the available evidence, *Serbelodon* is a valid taxon, either as genus or perhaps as a subgenus'.

TORYNOBELODON BARBOUR, 1929

Torynobelodon as properly defined is known with certainty on the basis of a single specimen (a distal tusk tip) from Sand Canyon, Nebraska, a locality of uncertain age. Barbour (1929) dated this locality as middle Pliocene in age, which in modern terms roughly corresponds to the Hemphillian. This specimen resembles *Platybelodon* in being highly flattened (more so than *Amebelodon*) and having internal dentinal rods with a distinctive morphology, whereas it resembles *Amebelodon* in being distinctly narrow and having a shallow dorsal concavity (Barbour 1929; Lambert 1990). Fragmentary mandibular tusk material that has been tentatively referred to *Torynobelodon* on the basis of its apparent narrowness has been found at the Withlacoochee River 4A site of Florida, which has been dated as early Hemphillian (Tedford *et al.* 1987; Lambert 1990). *Torynobelodon* appears to be closely related to *Platybelodon*, and almost certainly represents an Old World immigrant. (*Torynobelodon* is listed as a junior synonym of *Platybelodon* in Appendix C — editors.)

PLATYBELODON BORISSIAK, 1928

Synonyms

Torynobelodon Barbour, 1929; *Amebelodon* of Osborn (1929); *Selenolophodon* Chang and Zhai, 1978 (see also Appendix C).

Though an amebelodontine, *Platybelodon* is morphologically distinct from *Amebelodon* and *Torynobelodon* proper. Its mandibular symphysis is considerably shorter, and its mandibular tusks are shorter, broader, and far more flattened than those of *Amebelodon*. It differs from *Torynobelodon* in having much broader mandibular tusks, and differences in mandibular tusk wear patterns indicate that *Platybelodon* had different feeding habits than *Torynobelodon* (Lambert 1992). Unlike any of the other amebelodontines, all of the intermediate cheek teeth of *Platybelodon* are tetralophodont. Like some of the more advanced *Amebelodon* species, the molars show strong secondary trefoiling. The upper tusks are quite reduced relative to those of other gomphotheres *sensu lato* and show extreme sexual dimorphism, the upper tusks of the females being vestigial (Osborn 1936, pp. 463–73; Lambert 1992). *Platybelodon* is typically an Old World genus, and like the gomphothere *Tetralophodon* is known in North America on the basis of extremely limited material, a single mandible from late Clarendonian sediments along the Snake River in Nebraska (Barbour 1931; Skinner and Johnson 1984). The specimen referred to *Platybelodon* by Shotwell (1963) does not represent *Platybelodon* based on the complete absence of dentinal rods in its mandibular tusks, but rather appears to be a new amebelodontine genus, possibly with *Amebelodon* affinities. Specimens from the early Hemphillian Withlacoochee River 4A site in Florida may also represent *Platybelodon*, though a referral to *Torynobelodon* may be somewhat more plausible based on the apparent narrowness of the tusk fragments (see above).

THE PALAEOECOLOGICAL SIGNIFICANCE OF GOMPHOTHERE BIODIVERSITY AND ABUNDANCE IN NORTH AMERICA

Gomphotheres *sensu lato* existed in North America from the middle Miocene to the end of the Pleistocene (late Barstovian to the end of the Rancholabrean), a span of about 15 million years. They were, however, a major part of the continental fauna for only 8 of these 15 million years, having been both a common component of local faunas and widespread geographically from the late Barstovian through the early Hemphillian. Gomphotheres diversity reached its peak during this span as well, with six genera known from the late Clarendonian to the early Hemphillian: *Gomphotherium*, *Rhynchotherium* (in Central America), *Amebelodon*, *Serbelodon*, *Platybelodon*, and *Torynobelodon*. The North American gomphotheres were greatly reduced in numbers and generic diversity by the late Hemphillian, with only *Rhynchotherium*, *Gomphotherium*, and *Amebelodon* present. Of these, only *Rhynchotherium* was reasonably abundant, with *Amebelodon* known only from a single late Hemphillian locality. Gomphotheres became considerably less common in the Blancan than in the Hemphillian, though three genera were represented: *Rhynchotherium*, *Cuvieronius*, and *Stegomastodon*. Gomphotheres survived into the Pleistocene but remained uncommon; *Stegomastodon* became extinct near the middle Irvingtonian, leaving *Cuvieronius* as a neotropical holdover that became extinct near or at the end of the Rancholabrean (Kurtén and Anderson 1980).

The success of the gomphotheres from the late Barstovian to the early Hemphillian and their sudden decline in abundance and diversity by the late Hemphillian broadly mirrors the diversity pattern observed in the overall ungulate community of North America. From the beginning of the Barstovian to the end of the Clarendonian the ungulate community of North America experienced an enormous diversification, mostly from autochthonous evolution, culminating at 40 genera during the Clarendonian. During the late Clarendonian/early Hemphillian, however, this diversity began to decrease, dropping to 30 genera by the beginning of the late Hemphillian and to 18 by the latest Hemphillian (Webb 1983). This diversification and subsequent decline of the North American ungulate fauna during the Miocene has been explained on the basis of the continent-wide replacement of forests by more open forested savannas that began in the Hemingfordian, and subsequent transformation of this savanna into open grassland that began in the late Clarendonian/early Hemphillian (Webb 1977, 1983). Since woodland savannas can support a greater diversity of herbivores than either closed forest or open grassland, this situation had the effect of enlarging and then shrinking available herbivore niche space (Webb 1977, 1983).

The diversity of gomphotheres in North America was the result of both immigration and autochthonous evolution, rather than autochthonous evolution alone as for most of the North American ungulates during the Miocene. Nevertheless, the success of both the immigrant and endemic gomphothere taxa in North America through the early Hemphillian and their decline with the rest of the North American ungulate fauna by the late Hemphillian suggests that whatever their origins, these animals in general strongly preferred savanna habitat to forest or grassland, and thus probably were opportunistic browsers. Lambert (1989, 1992) provided evidence for such browsing in both *Gomphotherium* and *Amebelodon*.

Of the three gomphotheres that survived into the late Hemphillian, *Amebelodon*, *Rhynchotherium*, and *Gomphotherium*, at least *Amebelodon* and *Gomphotherium* appear to have been dependent on savanna habitats. The habitat preferences of *Rhynchotherium* are uncertain, though it is known from the grassland-dominated Coffee Ranch community (Dahlquest 1983). Thus, the rarity of *Gomphotherium* and *Amebelodon* in the late Hemphillian is consistent with a widespread loss of woodland savanna from North America. By the Blancan, woodland savanna had disappeared from most of North America, though it survived in a broad band south of latitude 30°N (Webb 1977). It is noteworthy that during most of the Pliocene and Pleistocene *Rhynchotherium* and *Cuvieronius* were most common in this region, and thus these taxa may have had a preference for woodland savanna as well.

ACKNOWLEDGEMENTS

Gary Morgan, Ron Edwards, Jeheskel Shoshani, Pascal Tassy, Ana V. Mazo, and Daniel C. Fisher critically read this paper, and offered numerous helpful comments.

SIXTEEN

Para- or monophyly of the gomphotheres and their position within Proboscidea

JEHESKEL SHOSHANI

INTRODUCTION

The systematics of Proboscidea in general, and of gomphotheres in particular, has undergone considerable modifications during the past 50 years. Osborn's (1936) discussion of gomphothere taxa and review of earlier literature may be considered a starting point for modern studies. Simpson (1945) reviewed and amended Osborn's work, and Tobien's (1972, 1973*a*, 1973*b*, 1975, 1976, 1978*b*, 1978*c*) and Tobien *et al.*'s (1986, 1988) studies provided good background information on gomphotheres. Employing cladistic methods on proboscidean taxa, Tassy (1982, 1988*a*, 1990*a*, this volume, Chapter 6) concluded that gomphotheres were a paraphyletic group. Similar results were obtained by Tassy and Shoshani (1988). None of these authors conducted a detailed cladistic analysis of all the gomphothere taxa in one study; many questions on affinities among the gomphotheres remain unresolved.

Viewed in a geological time frame, the groups of gomphotheres and stegodontids may be considered part of the 'second radiation' within the Proboscidea (Shoshani and Tassy, this volume, Chapter 34). It appears that *Stegolophodon* may be a paraphyletic taxon whose roots may be sought among gomphotheres (Tassy 1983, 1988*a*; discussions in Kalb *et al.*, this volume, Chapter 12A, and below). Studying relationships among gomphothere lines seems to be a never-ending process, for the more we study them the more intricate the problems seem to be. The present cladistic study was undertaken to help resolve some of the questions raised by various authors about this enormously complicated Neogene core of extinct proboscideans.

GOMPHOTHERES: THE 'WASTEBASKET' OF PROBOSCIDEA

Neogene proboscidean genera and species that do not fit easily into Stegodontidae, Elephantidae, or any other contemporaneous taxon are usually placed in a group called 'gomphotheres' (or sometimes 'mastodonts'). These mammals were widespread throughout most of the geographical range of Proboscidea, i.e. on all the continents except Australia and Antarctica. Most of these gomphothere taxa have been assigned to Gomphotheriidae or Amebelodontidae — both families are used here *sensu lato*, as the names have been applied inconsistently in the literature. Osborn (1936) preferred the name 'Bunomastodontidae' for some genera from both families, but Simpson (1945) and many other authors have employed the term Gomphotheriidae (e.g. Tobien *et al.* 1986, 1988). During the past 10–30 years, many proboscidean genera of uncertain taxonomic position (but which show resemblance to the general plan of gomphotheres) have been classified in the Gomphotheriidae, and thus this family has become the 'wastebasket' of the Proboscidea. This situation has much in common with the 'wastebasket' function 'Insectivora' has served within Eutheria (implied by Simpson 1945), or 'Condylarthra' for ungulate-related taxa (implied by Prothero *et al.* 1988).

The subdivision of gomphotheres into trilophodont and tetralophodont groups dates to Falconer (1857). This division (which is based on whether dp4, M1, and M2 have three or four lophs or lophids) has been popular, but the usage of the terms has become confusing because some authors have used these names in a vernacular sense, while others use them in a strictly taxonomic sense (e.g. Tobien 1973*a*, 1978*b*; Tassy 1988*a*, 1990*a*, this volume, Chapter 6).

Abbreviations. AMNH, American Museum of Natural History (New York, USA); BMNH, British Museum (Natural History) (London, England — new name 'Natural History Museum'); BUGM, Bristol University, Geology Museum (Bristol, England); LPVPH, Laboratoire de Paléontologie des Vertébrés et de Paléontologie Humaine, Université P. & M. Curie (Paris, France); MNHN, Museum National d'Histoire Naturelle (Paris, France); PAUP, Phylogenetic analysis using parsimony (Swofford 1991).

METHODS EMPLOYED

QUESTIONS

The current study addresses several significant issues, including two major questions:

1. Are the gomphotheres a monophyletic or a paraphyletic group?
2. Do members of the family Amebelodontidae belong in a monophyletic Gomphotheriidae?

Validity of various gomphothere genera had to be considered in the early stages of this study because synonymies necessarily affect the choice of taxa for analysis. Appendix A provides synonyms of proboscidean taxa included in this study. It may be noted that for some entries, e.g. *Serbelodon* and *Serridentinus*, the taxonomic status was decided only after the cladistic analysis was conducted (details below).

CHARACTERS EXAMINED

Gomphothere specimens examined and re-examined were housed primarily at the AMNH, BMNH, LPVPH, MNHN, and BUGM. This study reports the results of analysis of these specimens together with data taken from selected literature reports.

A total of 123 osteological and dental characters were examined. Many of the characters were employed by Tassy (1990*a*, this volume, Chapter 6); others were used by Shoshani (1986*b*, and personal observations 1991–92) or other authors (Appendix 16.1). Of these 123 characters, 96 were binary (only '0' and '1' codes) and 27 were multistate characters (codes 0, 1, 2, 3, 4, and 5 used). Missing data were coded as '?'. None of the characters employed in this paper was taken from Kalb *et al.*, Guan, or Saegusa (this volume, Chapters 12A, 13, and 17, respectively), although several characters are similar or identical since we used similar sources.

TAXA STUDIED

Terminal taxa are genera, either monophyletic or paraphyletic. Species-level analysis of each genus, especially for elephantoids, is out of the scope of this study. A list of the genera and the outgroup employed is given in Appendix 16.2 in the context of the data matrix upon which this paper is founded. The 19 gomphothere taxa, which represent a period of approximately 22 million years of evolution, were a part of a set of 37 proboscidean genera that span from *Anthracobune* to *Mammuthus*. Two hypothetical taxa were used as outgroups: Outgroup A (primarily 'condylarths', Artiodactyla, Perissodactyla) and Outgroup B (primarily Hyracoidea, Sirenia, and Desmostylia). For simplicity — since the focus is on Proboscidea, and specifically on 'gomphotheres' — 'Outgroup A' and 'Outgroup B', were coded with a '0' (the primitive condition) in all characters, except for a few which were coded with a '1' (Appendix 16.2).

CLADISTIC ANALYSIS

Computer analysis was conducted using PAUP version 3.0s (Swofford 1991), and MacClade (versions 2.97.96 and 3.0, Maddison and Maddison 1987, 1992, respectively).

In both PAUP and MacClade analyses, I used the standard specifications except for the following (in PAUP): optimization was conducted using DELTRAN (delayed transformation), MAXTREE was set to 100 and to 'Automatically increase by 100' intervals. All characters were considered of equal weight. Two sets of tests were run:

(1) all character types were considered unordered (for each multistate character, all transformations cost only one step); and

(2) all characters were ordered except for character numbers 15, 20, 55, 89, 94, and 121.

The 'ordered' option (determining the transformation sequence for each multistate character) for 21 of the 27 multistate characters was carefully chosen by Pascal Tassy and me after detailed discussion on each of these characters.

The heuristic search option of PAUP was employed to identify equally parsimonious shortest trees (exact methods (e.g. 'Branch and Bound') failed due to the size of the data matrix and constraints on computer memory). Strict and 50 per cent majority-rule consensus trees were constructed from the equally parsimonious trees following each analysis. One of each of the equally parsimonious trees (ordered and unordered) was then inputted into the MacClade program to test phylogenetic hypotheses as explained below (all options used were the Default ones, except for 'Multiple speciation events', under the Tree menu).

RESULTS AND DISCUSSION

RESULTS WITH 21 (OF 27 MULTISTATE) CHARACTERS 'ORDERED'

On the whole, results obtained with the 'ordered' option (for 21 of the 27 multistate characters (Figs 16.1a, 16.2a, 16.3a)) were similar, and in some cases identical, to those obtained with the 'unordered' option. PAUP identified 90 equally parsimonious trees from these data (each with 239 steps, consistency index (CI) = 0.68). A 50 per cent majority-rule consensus tree (length = 240 steps, CI = 0.68) and a strict consensus tree (length = 251 steps, CI = 0.65) were constructed from these 90 trees. Naturally, the length of the most parsimonious tree with the ordered characters was higher than the score for the tree with the unordered characters (by 12 steps, test no. 0 in Appendix 16.3). In all the ordered trees the Mammutidae and gomphotheres appeared paraphyletic (Figs 16.1–16.3).

RESULTS WITH ALL CHARACTERS 'UNORDERED'

Six hundred (600) equally parsimonious trees were obtained from an analysis in which all of the characters were unordered (Figs 16.1b, 16.2b, 16.3b). Each of these trees had a length of 227, CI = 0.72. A 50 per cent majority-rule consensus tree (length = 230 steps, CI = 0.71) and a strict consensus tree

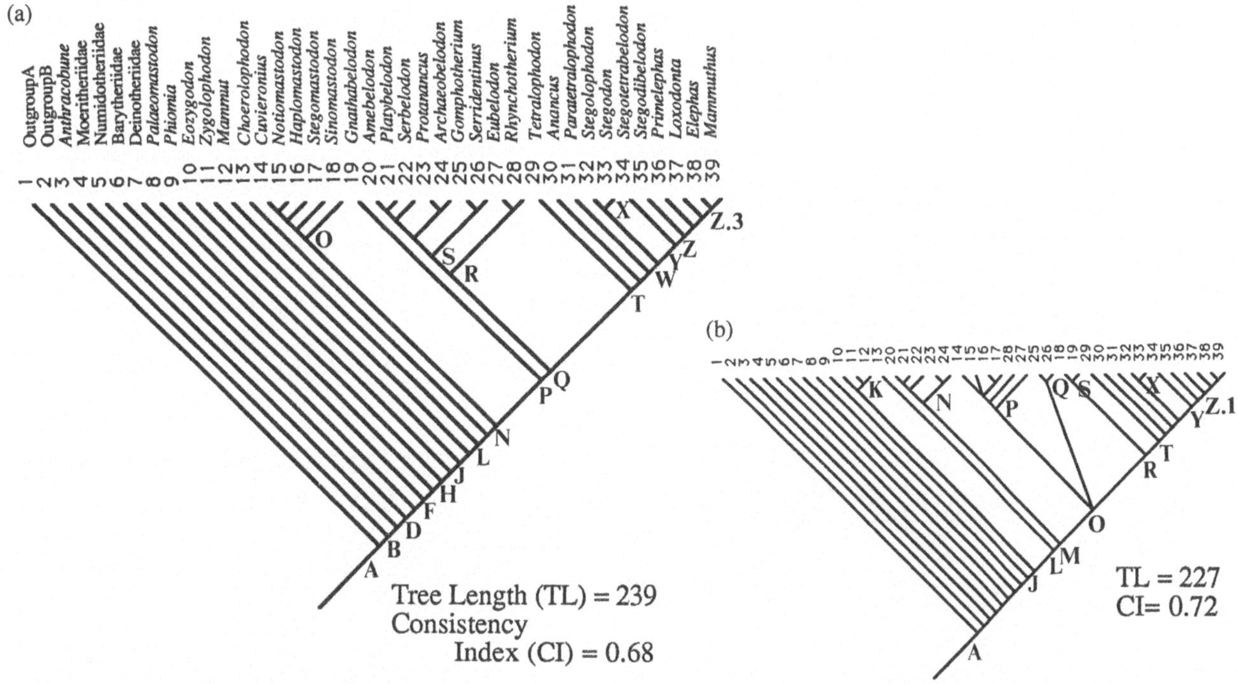

Fig. 16.1 (a) This is one of 90 equally most parsimonious cladograms obtained with PAUP using the ordered option (21 of the 27 multistate characters were ordered, the rest were unordered — see Appendix 16.2) — tree length = 239, consistency index (CI) = 0.68. This cladogram was used to test phylogenetic hypotheses explained in Appendix 16.3, and depicted in Fig. 16.5. Numbers 1–39 at the top are associated with taxa which are also labelled in Figs 16.2–16.5. Letters "A–Z" are designations for nodes in this figure only; they designate different nodes from those of Figs 16.2–16.5. In some figures only the numbers for taxa are given at the top; numbers in the range of 1–7 were omitted because their phylogenetic positions do not change. Since this chapter is focused on gomphothere genera, their placement on the cladograms changes throughout Figs 16.1–16.5, but the numbers remain the same. (b) This is one of 600 equally most parsimonious cladograms obtained with PAUP using unordered characters — tree length = 227, CI = 0.72. This cladogram was used to test phylogenetic hypothesis explained in Appendix 16.3. (Numbers stand for the same taxa as in Figs 16.1(a)–16.5.)

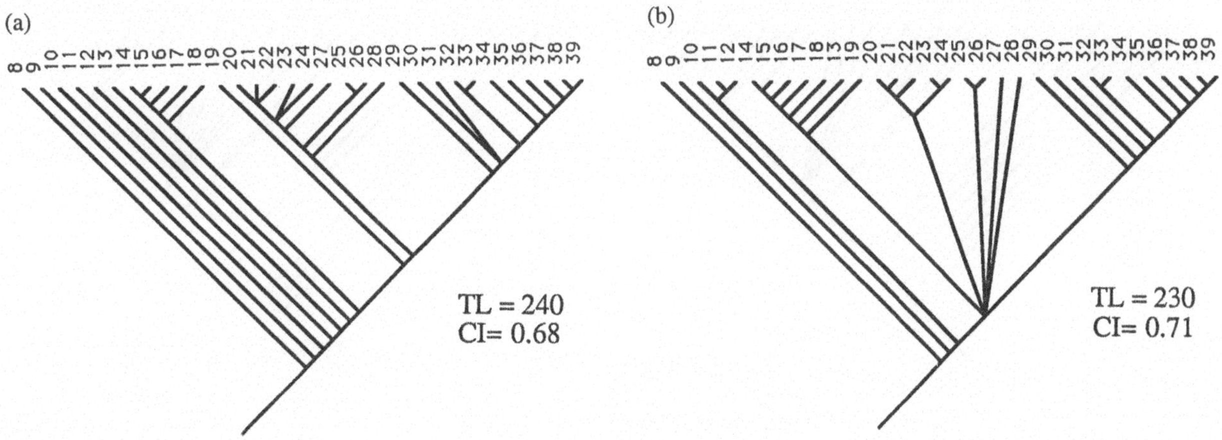

Fig. 16.2 These are 50 per cent majority-rule consensus trees obtained from the equally most parsimonious trees in the ordered (Fig. 16.2a) and the unordered (Fig. 16.2b) options — tree length for (a) = 240, CI = 0.68; tree length for (b) = 230, CI = 0.71. (Numbers stand for the same taxa as in Fig. 16.1a.)

(length = 257 steps, CI = 0.63) were constructed from these 600 trees. In about half of these trees, the Mammutidae appeared paraphyletic, and in all the cladograms the gomphotheres turned out paraphyletic (Figs 16.1–16.3). Then, using MacClade and testing various phylogenetic hypotheses to force some of the gomphothere taxa together (by switching some branches of gomphothere taxa; see test nos 13 and 14 in Appendix 16.3), a tree of 228 steps was obtained. In this figure, the gomphotheres, in the broadest sense, were still paraphyletic. None the less, 10 of the 19 gomphotheres taxa were grouped in one clade, and with two additional steps, 15 of the 19 gomphotheres were grouped in one clade (Test no. 15 in Appendix 16.3). This score (230) is the same as that for the 50 per cent majority-rule consensus tree.

There were three major differences between the ordered (Figs 16.1a, 16.2a) and the unordered cladograms (Figs 16.1b, 16.2b). The first involves the placement of gomphothere taxa found in South America (*Cuvieronius, Haplomastodon, Notiomastodon,* and *Stegomastodon*), either distant from or closer to the clade comprising tetralophodont gomphotheres, Stegodontidae and Elephantidae. Note that in the ordered strict topology, *Sinomastodon* is the sister-group to the South American gomphotheres (Fig. 16.3a), whereas in the unordered cladogram it is not (Fig. 16.3b). The third difference involved resolution among South American gomphotheres. In the unordered tree, this clade is polytomous, whereas in the ordered tree the topology was resolved as follows: (((*Notiomastodon, Cuvieronius*), *Haplomastodon*), *Stegomastodon*).

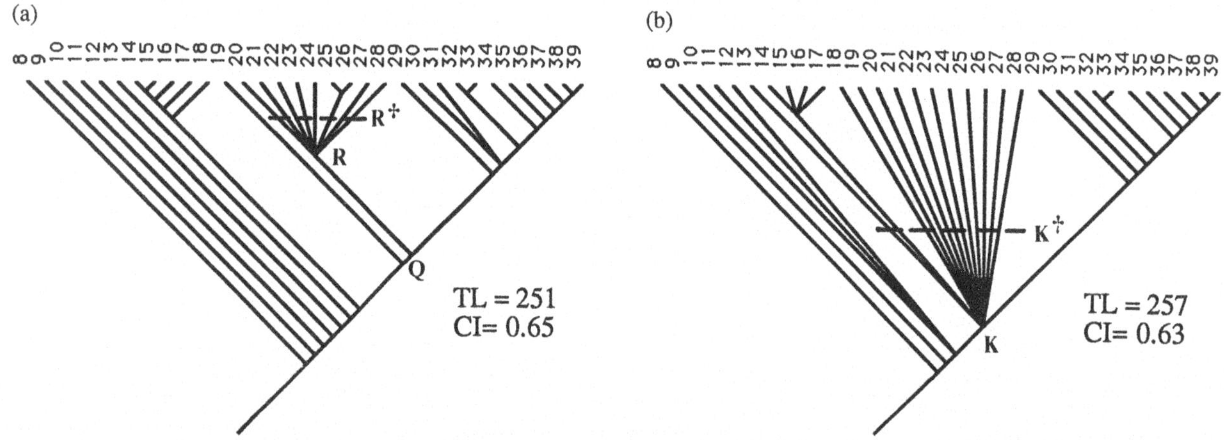

Fig. 16.3 These are strict consensus trees obtained from the equally most parsimonious trees in the ordered and the unordered options — tree length for (a) = 251, CI = 0.65; tree length for (b) = 257, CI = 0.63. Clades marked with 'R†' in (a) were part of clades embraced in Node R in Fig. 16.1a. Note that Node K in (b) is polytomous with 14 clades radiating from one point — extremely uninformative. Thirteen of these 14 clades (marked with 'K†') were parts of clades similar to N, P, Q, and S in Fig. 16.1b, which is one of the 600 equally parsimonious trees obtained with the unordered option in PAUP. (Numbers stand for the same taxa as in Fig. 16.1a.)

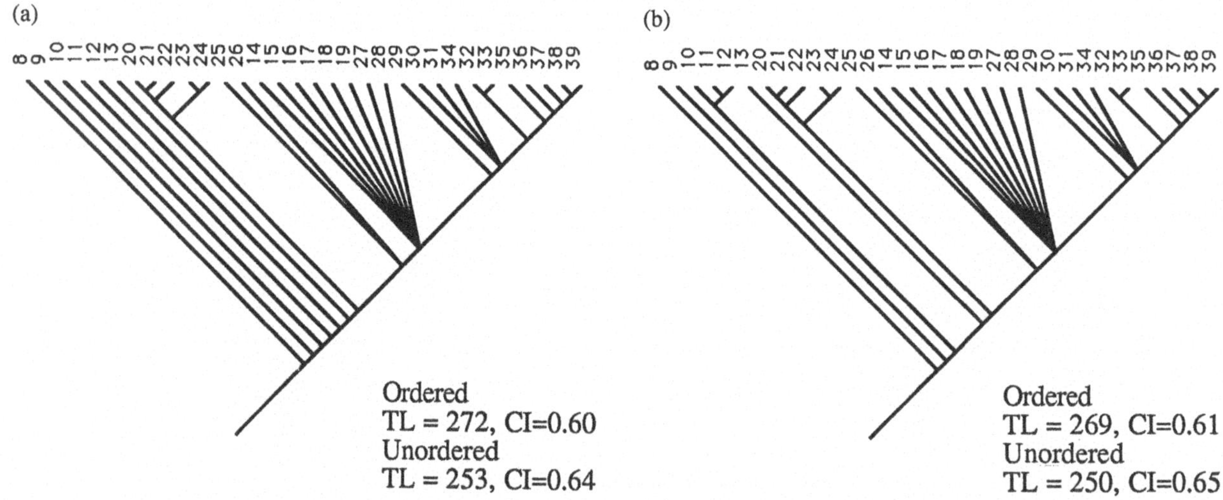

Fig. 16.4 Tassy's phylogenetic reconstruction of the Proboscidea is tested against the updated data matrix employed in this chapter. (a) Family Mammutidae paraphyletic: using ordered characters, tree length = 272, CI = 0.60; or using unordered characters, tree length = 253, CI = 0.64. (b) Mammutidae monophyletic: using ordered characters, tree length = 269, CI = 0.61; or using unordered characters, tree length = 250, CI = 0.65.

EQUALLY PARSIMONIOUS TREES

As expected, there were many (600) equally parsimonious trees with the multistate characters unordered, whereas there were only 90 equally parsimonious trees when these characters were ordered. The multiple equally parsimonious trees resulted not only from lack of data (e.g. when an incomplete cranium was found), but also from changes due to evolutionary transformation in a taxon. For example, when discussing the presence or absence of enamel on the lower tusks (I_2), character no. 6, a taxon such as *Elephas* which does not possess I_2 receives a '?' (because this character 'does not apply' — Swofford 1991, p. 19) as does a taxon which may have possessed lower tusks (e.g. *Stegodibelodon*) but which are lacking in a particular specimen examined.

Rowe (1988) coded missing data due to non-preservation or lack of information as '?'; missing data due to evolutionary change was coded 'N' (non-applicable). Regardless of what coding methods one uses, analysis of taxa examined for characters with missing information may produce indecisive or false results. To remedy such a problem, it is best to use as many characters as possible (especially those without much missing data) and include as many taxa as possible in each analysis. Such a tactic helps to mask unforeseeable false or ambiguous results. Alternatively, it is also helpful to conduct separate tests with multistate characters 'ordered' and 'unordered' and scrutinize the results (cf. Appendix 16.3; Ridley 1986; Simmons 1993).

It is beyond the scope of this chapter to give all the equally parsimonious trees; a few examples will suffice. The polytomous clades are probably the best examples because the positions of some of those lineages stemming from a bifurcation point can be altered within this node without causing change in the score of the tree. For example, *Eubelodon* and *Gnathabelodon*, which have many '?'s in the data matrix, could easily be joined on other branches within their respective nodes without changing the length of the trees (test 26 in Appendix 16.3).

DISCUSSION

Data matrices that contain extinct species are seldom complete, unless the numbers of taxa and characters are limited and the characters examined are on well-preserved specimens. The matrix analysed here is no exception; results presented for certain taxa must be considered tentative. In a case in which the multistate characters were ordered, members of the family Mammutidae, *sensu* Tassy (1990*a*) (*Eozygodon*, *Zygolophodon*, and *Mammut*) turned out to comprise a paraphyletic assemblage, possibly because of the lack of data for *Eozygodon* and *Zygolophodon*. Forcing the three taxa into one clade in the most parsimonious trees obtained with the 'ordered' characters required five additional evolutionary changes — from 239 to 244 (only the ordered option is considered because Tassy's analysis (1990*a*) was conducted on ordered characters; see test no. 8 in Appendix 16.3). In Fig. 16.5, clade K, these taxa are shown united, based on the results obtained with the 50 per cent majority-rule with the unordered characters.

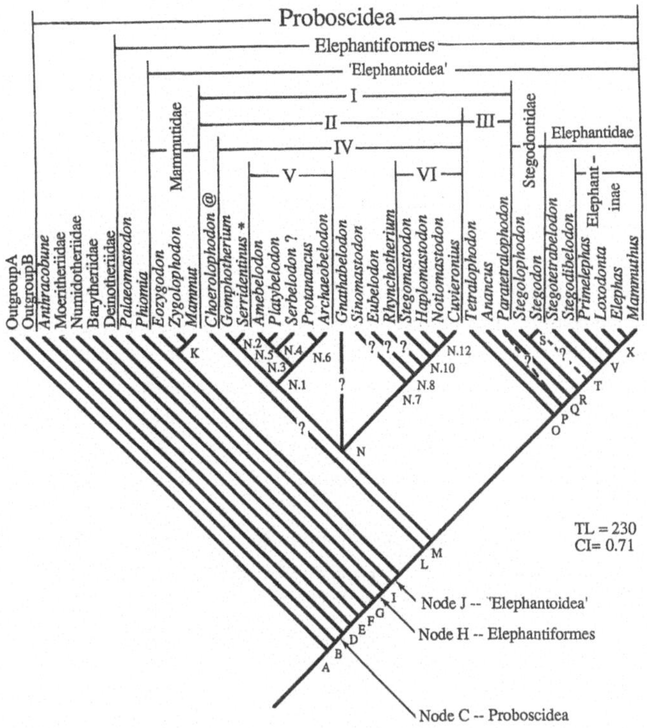

Fig. 16.5 A cladogram of the Proboscidea based on the score of the unordered 50 per cent majority-rule consensus tree, tree length = 230, CI = 0.71 (cf. Fig. 16.2b and test nos 35 and 35.1 in Appendix 16.3). Until additional analyses are conducted, these results *must* be considered as a tentative working hypothesis. A classification was constructed (Table 16.1) based on this cladogram. Note that there is some evidence to include *Choerolophodon* (designated with @) within clade N; see text for details. *Serridentinus* (designated with *) is considered a junior synonym of *Gomphotherium*, and *Serbelodon* (designated with ?) is tentatively considered a valid genus — details in Appendix 16.3. Dashed lines and '?' imply uncertain phylogenetic positions. Key for Roman numerals: I = 'gomphothere' taxa, II = trilophodont 'gomphotheres', III = tetralophodont 'gomphotheres', IV = a group classified in the family Gomphotheriidae, V = subfamily Amebelodontinae, VI = subfamily Cuvieroniinae.

Expectedly, the 'wastebasket hypothesis' — in which all 19 gomphothere taxa were placed in a single clade — required too many changes to be considered likely under parsimony criteria (16–17 steps, test no. 19 in Appendix 16.3) and thus this hypothesis was rejected. Another hypothesis (that of Savage 1955; Simpson and Paula-Couto 1957; Miller 1990; test no. 20 in Appendix 16.3) placed *Rhynchotherium* as a sister-group of the South American clade (*Stegomastodon*, *Haplomastodon*, *Notiomastodon*, and *Cuvieronius*). This hypothesis required 0–3 steps more; it was accepted as part of the most parsimonious and a 'near-most parsimonious' solution.

It would definitely make classification simpler if we joined *Choerolophodon* to other trilophodont gomphotheres, but such possible arrangements required between two and six steps (test nos 9–12 and 16 in Appendix 16.3) depending on whether the test was with ordered or unordered characters. I did not adopt this tree topology for two reasons. First, throughout this study, a conservative approach has been taken and gomphothere-related hypotheses were accepted only when they required not more than three additional changes (at least in one of the options 'unordered' or 'ordered') from a most parsimonious cladogram and did not violate accepted hypotheses. In the case of *Choerolophodon*, although it was classified within the Gomphotheriidae (subfamily Choerolophodontinae, e.g. by Tobien *et al.* 1986, p. 140), Tassy (1985, 1988*a*, 1990*a*, and this volume, Chapter 6) showed *Choerolophodon* belonging in a separate lineage as a sister-group of gomphothere and amebelodontid assemblages, stegodontids, and elephantids. The second reason for not accepting *Choerolophodon* as a member of clade N in Fig. 16.5 (even though it was a member of the clade of gomphotheres and elephantids in Figs 16.2b and 16.3b) was related to character number 80 (the same as character number 43 of Tassy 1990*a*). *Choerolophodon* and all other clades of more basal taxa in Fig. 16.5 (Mammutidae to *Anthracobune* and the outgroups) have the plesiomorphic condition of this character (nasal fossa which is narrow and does not extend laterally beyond the width of the rostrum), whereas all members of clade M in Fig. 16.5 have wide nasal fossae which extend laterally beyond the width of the rostrum. This, I believe, is a functionally complex character associated with vocalization (in conjunction with the architecture of the hyoid apparatus; Shoshani 1991) and should be studied in greater detail (see, however, Chapter 6, this volume, footnote 2 on p. 45; cf. Tassy 1994*a*).

The synonymy of gomphotheres taxa was a tedious but necessary process. Appendix A includes a summary of these synonymies. Decisions on the 'fate' of *Serbelodon* and *Serridentinus* were delayed until after the computer analyses were conducted, because their taxonomic statuses have been heavily discussed recently and often with different conclusions (Simpson 1945; Tobien 1972, 1973*a*; Guan, this volume, Chapter 13; Lambert, this volume, Chapter 15). In Appendix 16.4, I provided explanations for synonymizing *Serridentinus* with *Gomphotherium*, and for the undecided taxonomic position of *Serbelodon*. The grouping of *Tetralophodon*, *Anancus*, and *Paratetralophodon* with Stegodontidae and Elephantidae is not surprising since this study employed many of the characters of Tassy (this volume, Chapter 6) who obtained similar results. Kalb *et al.*'s cladogram (this volume, Chapter 12A) also shows close association among the last three taxa.

The decision on whether or not to accept one cladogram over another should not be based strictly on which of them is more parsimonious, but the final decision should consider the degree of differences between the length of the trees, and whether or not the data matrix is complete or close to complete. In the case of the gomphothere under consideration, the data matrix is certainly not complete, and it takes only between one and three steps to convert most parsimonious cladograms shown in Figs 16.1a and 16.1b to the 50 per cent majority-rule trees shown in Fig. 16.2a and 16.2b, depending on whether the characters are 'ordered' or 'unordered'. I maintain that these differences are within the range of possible solutions because of the nature of the data.

Computer programs are limited as to what they can or cannot generate, depending on the assumptions incorporated in these programs. As McKenna (1987, p. 57) pointed out, 'it is difficult to train a computer' not to assume that fruit bats are descended from whales, or that pangolins are not the closest relatives of carnivores. In my study, the most parsimonious computer-generated solutions separate the gomphotheres into at least eight branches. The taxonomy of gomphotheres has been in desperate need of revision for many years, and thus we may not know which of them or other proboscideans are the 'fruit bats', 'whales', 'pangolins', or 'carnivores'. If we would accept only the most parsimonious results, as some staunch cladists do, then what about a group like the Mammutidae, a well-established monophyletic family (based on zygodont characters; Tobien, this volume, Chapter 9, Tassy 1990*a*; this volume, Chapter 6 — note that Tassy used ordered characters), which, based on the data matrix studied, turned out to be paraphyletic in all the ordered trees and in about 50 per cent of all the unordered ones? To what extent do we allow technology to dictate biological unities? Because the data are homoplasic in many lineages, and because neither the most parsimonious cladograms (examples shown in Figs 16.1a and 16.1b) nor the strict trees (Figs 16.3a and 16.3b) convey the information in the near-most parsimonious trees (based on hypothesis testing given in Appendix 16.3), I adopted a synchretic approach and combined some ideas from both sets of cladograms and from the 50 per cent majority-rule consensus tree (Fig. 16.2a) as shown in Fig. 16.5. The classification, Table 16.1, is based on this latter cladogram. Skeletons of proboscideans representing evolutionary stages are shown in Plates 16.1–16.2.

Before closing this discussion, I want to explain briefly my reasoning for not adopting the hypothesis of Kalb *et al.* (this volume, Chapter 12A) and what my position is on Tassy's hypothesis (1990*a*; this volume, Chapter 6). Kalb *et al.*'s hypothesis for the breaking up of the traditionally held Stegodontidae (*Stegolophodon* and *Stegodon*) is *contra* to that of Saegusa's (this volume, Chapter 17) traditional argument for a unified Stegodontidae. It is possible that the five synapomorphies given by Saegusa for this family are an interrelated mosaic of complex and functionally integrated characters. Kalb *et al.*'s study is limited; it considered only dental and cranial features, and included only four gomphothere taxa. A larger number of gomphothere genera would, plausibly, have influenced the polarities of the characters, and consequently the relationships within the clade composing *Stegolophodon* and *Stegodon*. Results presented in this chapter are based on approximately twice the number of taxa and close to four times the number of characters of Kalb *et al.*, and thus I believe that the results obtained with a larger data set will have a better chance of withstanding criticism. Results obtained in this study support the traditional hypothesis that *Stegolophodon* and *Stegodon* are a part of a monophyletic lineage. Note, however, that this finding is not as strong as other results, and thus the hypothesis proposed by Kalb *et al.* remains a viable alternative (see test nos 30–32 in Appendix 16.3 for details).

Table 16.1 A suggested classification of the Proboscidea with emphasis on the positions of gomphothere taxa based on results presented in Fig. 16.5, and Appendix 16.3 [a]

- Class Mammalia Linnaeus, 1758
  - Subclass Theria Parker and Haswell, 1897 (rank after Gregory 1910)
    - Supercohort Eutheria Gill, 1872 (rank after McKenna 1975)
      - Cohort Epitheria McKenna, 1975
        - Superorder Ungulata Linnaeus, 1766 (rank after Novacek 1986)
          - Grandorder Paenungulata Simpson, 1945 (rank after Novacek 1986) [b]
            - Mirorder Tethytheria McKenna, 1975
              - Order Proboscidea Illiger, 1811
                - Plesion *Anthracobune* Pilgrim, 1940 † [c]
                - Family Moeritheriidae Andrews, 1906 †
                  - *Moeritherium* Andrews, 1901d †
                - Family Numidotheriidae new family † [d]
                  - *Numidotherium* Mahboubi, Ameur, Crochet, and Jaeger, 1986 †
                - Family Barytheriidae Andrews, 1906 †
                  - *Barytherium* Andrews, 1901b †
                - Family Deinotheriidae Bonaparte, 1845 †
                  - *Prodeinotherium* Ehik, 1930 † [e]
                  - *Deinotherium* Kaup, 1829 †
                - Suborder Elephantiformes Tassy, 1988 *a*
                  - Plesion *Palaeomastodon* Andrews, 1901*a* †
                  - Plesion *Phiomia* Andrews and Beadnell, 1902 †
                  - Plesion *Hemimastodon* Pilgrim, 1912 † [f]
                  - Unnamed taxon (node J — 'superfamily Elephantoidea Gray, 1821' of Tassy 1988*a*) [g]
                    - Family Mammutidae Hay, 1992 † [h]
                      - *Eozygodon* Tassy and Pickford 1983 †
                      - *Zygolophodon* Vacek, 1877 †
                      - *Mammut* Blumenbach, 1799 †
                    - Unnamed taxon (node L) [i]
                      - Plesion *Choerolophodon* Schlesinger, 1917 (*sedis mutabilis*) †
                      - Unnamed taxon [node M]
                        - Family Gomphotheriidae Hay, 1922 [node N] [j]
                          - Unnamed taxon (node N.1)
                            - Plesion *Gomphotherium* Burmeister, 1837 †
                              - (includes *Serridentinus* Osborn, 1923) (node N.2) †
                            - Subfamily Amebelodontinae Barbour, 1927 (*incertae sedis=i.s.*) (node N.3) †
                              - *Amebelodon* Barbour, 1927 †
                              - *Platybelodon* Borissiak, 1928 †
                              - *Serbelodon* Frick, 1933 † [k]
                              - *Protanancus* Arambourg, 1945 †
                              - *Archaeobelodon* Tassy, 1984 †
                          - Plesion *Gnathabelodon* Barbour and Sternberg, 1935 (*i.s.*) †
                          - Unnamed taxon (node N.7)
                            - Plesion *Sinomastodon* Tobien, Chen and Li, 1986 (*i.s.*) †
                            - Plesion *Eubelodon* Barbour, 1914*a* (*i.s.*) †
                            - Plesion *Rhynchotherium* Falconer, 1868 (*i.s.*) †
                            - Subfamily Cuvieroniinae Cabrera, 1929 (*i.s.*) (node N.10) †
                              - *Stegomastodon* Pohlig, 1912 †
                              - *Haplomastodon* Hoffstetter, 1950 †
                              - *Notiomastodon* Cabrera, 1929 †
                              - *Cuvieronius* Osborn, 1923 †
                        - Unnamed taxon (Node O)
                          - Plesion *Tetralophodon* Falconer, 1857 †
                          - Plesion *Anancus* Aymard, 1855 †
                          - Plesion *Paratetralophodon* Tassy, 1983 †
                          - Unnamed taxon (Node R)
                            - Family Stegodontidae Osborn, 1918 (Node S) †
                              - *Stegolophodon* Schlesinger, 1917 (*i.s.*) †
                              - *Stegodon* Falconer, 1857 †
                            - Family Elephantidae Gray,1821 (Node T)
                              - Plesion *Stegotetrabelodon* Petrocchi, 1941 (*i.s.*) †

Table 16.1 *(contd)*

Plesion *Stegodibelodon* Coppens, 1972 (*i.s.*) †
Subfamily Elephantinae Gray, 1821 (Node V)
Plesion *Primelephas* Maglio, 1970*b* †
Genus *Loxodonta* Anonymous, 1827 [l]
Genus *Elephas* Linnaeus, 1758
Plesion *Mammuthus* Brookes, 1828 † [m]

† Extinct taxon.

[a] Arrangements of genera, families, or higher categories in this classification (from top to bottom), corroborate the general sequence of the branching pattern from left to right in Fig. 16.5. This particular cladogram is a combination of topologies obtained with the unordered options in PAUP: that which was obtained from the '50 per cent majority-rule' consensus tree (Fig. 16.2b) and a combination of hypothesis testing given in Appendix 16.3. The score of this tree (Fig. 16.2b) is the same as obtained with the '50 per cent majority-rule' which is three steps higher than any of the 600 equally parsimonious trees, but 27 steps lower than the strict consensus tree. Each taxon is followed by a name or names of the author(s) who coined the name and the year it was published. McKenna, Bell, *et al.* (1994) employ an alternative classification of the Proboscidea; see also Kinman (1994) who coined the name 'Probosciformes' for 'elephants and allies'.

[b] Phylogenetic relationships and classification of Proboscidea above the ordinal level were discussed in Fischer (this volume, Chapter 5), Fischer and Tassy (1993), and by Shoshani (1986*a*, 1992*b*, 1993*c*).

[c] The taxonomic status of anthracobunid genera and species was reviewed briefly by Shoshani *et al.* (this volume, Chapter 8) and, in detail, by Fischer and Tassy (1993). At one extreme, there were five genera classified in Anthracobunidae, Proboscidea (Wells and Gingerich 1983 — *Anthracobune, Ishatherium, Jozaria, Lammidhania*, and *Pilgrimella*). West (1984), on the other hand, considered only two or three genera valid — *Anthracobune* (with *Pilgrimella* and *Ishatherium* as synonyms), *Jozaria*, and *? *Lammidhania* — and placed them in Moeritheriidae, Proboscidea; * he noted '*Lammidhania* as closely related to and doubtfully separate from *Anthracobune*' (West 1984, p. 188). Kumar (1991) concluded that Anthracobunidae includes four genera (*Anthracobune Jozaria, Lammidhania*, and *Pilgrimella*), and that this family should be placed in the Proboscidea 'but still with some reservations' (Kumar 1991, p. 236). Results of cladistic analysis by Fischer and Tassy (1993) indicate that *Lammidhania* is not an anthracobunid, and *Jozaria* is close to Proboscidea. It appears that the taxonomic status of anthracobunid taxa is continuously changing and sometimes in opposite directions; it is becoming confusing. Moreover, *Anthracobune* is the only genus for which there are dental and non-dental elements with diagnostic characters. For these reasons, I employed only *Anthracobune* in the data matrix (Appendix 16.2). The phylogenetic positions of other anthracobunid genera cannot be ascertained at the present time; until additional evidence becomes available, I place *Anthracobune* in Proboscidea (without familial designation because this genus, and maybe others, may be placed in Moeritheriidae), and all other taxa which were assigned to Anthracobunidae I place in Tethytheria or Proboscidea *incertae sedis*.

[d] This new familial assignment is provisional and it is based on data by N. Court in Chapter 8 (this volume). Independent tests (e.g. test no. 1, Appendix 16.3) shows that it costs four evolutionary steps to join *Numidotherium* and *Barytherium*.

[e] Ehik (1930, p. 14) spelled it '*Prodinotherium*', but Harris (1973, p. 293) justifiably emended it to '*Prodeinotherium*'. According to the International Code of Zoological Nomenclature, Article 50d, the authorship of 'Ehik, 1930' is valid.

[f] *Hemimastodon* was not included in the cladistic analysis of this study because there are only two teeth (left M^3, IM catalogue no. A446 and left M^3, BMNH catalogue no. 10496) of this genus available, and there has been some doubt whether *Hemimastodon* should have been included in the Proboscidea (e.g. Pickford 1987*b*; Shoshani and McKenna 1991). Since the latter paper was presented, Shoshani is now convinced that *H. crepusculi* may belong within Proboscidea. In this classification, based on the studies of Tassy (1982, 1985, 1988*a*, *b*, 1990*a*, this volume, Chapter 6), *Hemimastodon* is placed within the Elephantiformes, between *Phiomia* and 'Elephantoidea' (of Tassy 1988*b*).

[g] This unnamed node 'J' may be of an infraordinal rank which could embrace superfamilies Mammutoidea, Gomphotherioidea, and Elephantoidea (exact assignments for the last two are uncertain, see also note [i] below). Tassy (1988*a*, p. 46) named members of this clade (not in the same arrangement) 'Elephantoidea', a superfamily, in accordance with the International Code of Zoological Nomenclature Recommendation 29a (since the classification is based on extant species; for the same reason we use 'subfamily Elephantinae' and 'family Elephantidae'). The name Elephantoidea was used by Osborn (1921*a*, p. 2) as a suborder, and in 1942 (Osborn 1942, p. 1582) as a superfamily.

[h] According to the International Code of Zoological Nomenclature, Article 36a, the author and date of the first taxon in a family group become the author and date for all other taxa in that family group that are based upon the same type genus. Thus, even though Cabrera (1929, p. 74) established the name Mammutidae, the authorship for the family should be Hay, 1922, since Hay (1922, p. 101) was the author of Mammutinae.

[i] This unnamed node 'L' of undetermined rank includes members of Gomphotherioidea and Elephantoidea whose exact assignments are uncertain. Coppens *et al.* (1978, p. 339) employed the term 'Gomphotherioidea' as a suborder of Proboscidea to embrace Gomphotheriidae and Elephantidae; Stegodontidae, according to them, was included in the suborder Mammutoidea. In this study, Stegodontidae is a sister-group to Elephantidae (node T in Fig. 16.5).

[j] This node N includes gomphothere taxa which were classified in the family Gomphotheriidae Hay, 1922 *sensu lato*. Discussion on the content of the heterogeneous family Gomphotheriidae (node N in Fig. 16.5) is given in the text, based on tests shown in Appendix 16.3.

[k] Within the subfamily Amebelodontinae, the listing of genera are not necessarily in phylogenetic sequence. Special effort was made to test whether or not *Serbelodon* Frick, 1933, is a junior synonym of *Amebelodon* Barbour, 1927 — based on the available evidence, *Serbelodon* is a valid taxon, either as genus or perhaps as a subgenus.

[l] After Domning and Shoshani (1982). See also Note *L2* in Appendix C.1 where N. Vigors was considered as the author of *Loxodonta*.

[m] Garutt *et al.* (1990) have requested the conservation of *Mammuthus* Brookes, 1828, pp. 73, 74 and it has been accepted (International Commission on Zoological Nomenclature 1991).

As expected, the most parsimonious cladograms of Proboscidea (Figs 16.1a and 16.1b) contain certain branching patterns that are similar to those obtained in Tassy's studies (1990*a*, this volume, Chapter 6) because, of the 123 characters, 97 are identical or similar to those used by Tassy. Four tests were conducted to test Tassy's phylogeny of the Proboscidea against the updated data matrix. Two tests were with multistate characters ordered and two with them unordered. In each case, the Mammutidae taxa were considered as either paraphyletic or monophyletic. Results of these tests are summarized in test no. 35.4 (Appendix 16.3) and Figs 16.4a and 16.4b. Although the scores of Tassy's trees were higher than those in Fig. 16.5, the topology of Tassy's cladogram and those presented in Fig. 16.3b (based on unordered characters) are similar. For these reasons, and until additional evidence becomes available, the hypothesis of Tassy (1990*a*; this volume, Chapter 6) should be considered equally likely as the one presented in Fig. 16.5.

CONCLUSIONS

Studying relationships among gomphotheres must not be conducted in isolation; it should be performed within the framework of as many proboscidean taxa as possible, and with one or two outgroups for the Proboscidea. The final cladogram presented here has the same tree length as the 50 per cent majority-rule consensus tree obtained from PAUP with unordered characters, and it is based on iterative hypotheses testing (Appendix 16.3). The suggested classification for the Proboscidea (Table 16.1) is based on Fig. 16.5. Figure 16.1b has a branching arrangement for gomphothere phylogeny considerably different from that in Fig. 16.5, yet this difference increases the length of the tree only by three steps — relatively small differences for such a fundamentally different result! For this reason, I include the results depicted in Fig. 16.5 and consider them to be within the range of possible solutions. It seems clear that in cladistic analysis, one should not restrict oneself to examining only the shortest trees, because, in certain cases, well-established hypotheses are not supported by a particular data set. For example, if we have accepted the strict parsimony results either with ordered (Fig. 16.3a) or unordered (Fig. 16.3b) characters, then the family Mammutidae (*Eozygodon*, *Zygolophodon*, and *Mammut*) would be paraphyletic, not monophyletic, *contra* Tassy (1990*a*; this volume, Chapter 6). Similarly, consensus trees are not necessarily the best-fitting hypothesis; a most parsimonious tree, a 50 per cent majority-rule consensus tree, or a combination of them, may be a better solution (cf. Figs 16.1–16.3 to Fig. 16.5 for the placement of members of the subfamily Amebelodontinae). Note that Fig. 16.5, in which Mammutidae is monophyletic, costs 27 steps less than the strict consensus tree, Fig. 16.3b, where Mammutidae is paraphyletic.

The current analysis considered 123 morphological characters and 39 taxa (two outgroups and 37 proboscideans, of which 19 were gomphothere genera). The conclusions of the analyses conducted are as follows:

1. The traditional hypothesis that *Phiomia* was a direct ancestor to gomphothere taxa is strongly rejected (it requires nine steps in 'unordered' options and 21 steps in 'ordered'; test no. 7 in Appendix 16.3).
2. The family Mammutidae is a monophyletic taxon even though *Eozygodon*, *Zygolophodon*, and *Mammut* are not always united (cf. Fig. 6.1 (Tassy, this volume) and Figs 16.1–16.3 to Fig. 16.5 (this chapter)) — the differences are probably due to an incomplete data matrix; *Mammut* is a sister-group to *Zygolophodon*, and *Eozygodon* is a sister-group to both.
3. Mammutidae is not a sister-group to Stegodontidae (see test no. 33 in Appendix 16.3 and conclusion nos 12 and 13 below). Mammutidae is the plesiomorphic sister-group of node L, i.e. the clade comprising all gomphotheres, stegodontids and elephantids.
4. The gomphotheres, *sensu lato* (including all 19 taxa listed in Appendix 16.2 and labelled as 'gomphothere taxa' in Fig. 16.5) are definitely not a monophyletic assemblage — there are five separate clades.
5. Fifteen of the 19 so-called gomphotheres may be grouped in one clade (a feeble group, labelled N in Fig. 16.5). Clade N is feeble because it is supported only by one synapomorphy — character no. 34, state 4, which is homoplasic (see Appendix 16.4 under node N).
6. Members of clades N.3 in Fig. 16.5 (*Amebelodon*, *Platybelodon*, *Serbelodon*, *Protanancus*, *Archaeobelodon*) are considered a subset of node N; they were classified by other authors in the family Amebelodontidae. Here, they are classified in the subfamily Amebelodontinae (Table 16.1).
7. The paraphyletic (Figs 16.1–16.3) and the monophyletic (Fig. 16.5) clades of the 15 gomphotheres differ in the placements of the gomphotheres found in South America (*Cuvieronius*, *Haplomastodon*, *Notiomastodon*, and *Stegomastodon*).
8. Taxa grouped in clade N.10 in Fig. 16.5 (*Cuvieronius*, *Notiomastodon*, *Haplomastodon*, and *Stegomastodon*) are classified in the subfamily Cuvieroniinae (Table 16.1).
9. Based on parsimony analysis, *Rhynchotherium* or *Sinomastodon* is a sister-group of Cuvieroniinae (cf. Figs 16.1 and 16.2–16.3). In Fig. 16.5, however, *Rhynchotherium* and *Eubelodon* are closer to Cuvieroniinae than *Sinomastodon*.
10. *Serridentinus* and *Serbelodon* were included in the analysis to test the hypothesis that they are synonyms of *Gomphotherium* and *Amebelodon*, respectively — I conclude that *Serridentinus* is, and therefore I suggest that there should be only 18 valid genera of the so-called gomphotheres. In Fig. 16.5, they are clustered into these five clades: (a) *Choerolophodon*; (b) one clade, node N with 14 genera; (c) the branch of *Tetralophodon*; (d) the branch of *Anancus*; and (e) the branch of *Paratetralophodon*. It is noted that branch (b) includes *Serridentinus* as synonym, that *Serbelodon* is tentatively considered a valid genus, and branches (c–e) appear to be more closely related to the clade of Stegodontidae and Elephantidae than to other gomphothere taxa.
11. The phylogenetic position of *Choerolophodon* is tentative; reanalysis of characters may confirm its current position or help to place it elsewhere on the cladogram, probably within clade N in Fig. 16.5 (cf. Tassy 1994*a*).

Plate 16.1 A sample of skeletons of proboscideans. (a) *Moeritherium*, front view (CGM, late Eocene, early Oligocene, el Fayûm, Egypt; prepared by E. L. Simons and others — note that the mani and pedes are casts and do not belong with this skeleton (shoulder height, about 1 m; head and body length, about 3m)); (b) *Moeritherium*, side view (CGM, same specimen as in (a)); (c,d) two views (side and 3/4) of a mounted skeleton of '*Deinotherium bavaricum*' (now *Prodeinotherium bavaricum*) of the late Miocene of Franzensbad, Bohemia, in the Royal Museum of Natural History Vienna (after Osborn 1936, pp. 99–100) (the length of the femur is 105.5 cm; other details about this specimen are in Osborn 1936, pp. 99–102); (e) *Archaeobelodon filholi* (MNHN BVI–5, late Middle Miocene, Sansan, France; mounted by C. L. Laurillard 1851–52 (shoulder height, about 2 m; head, excluding tusks and body length, about 3.5 m) (f) from left to right: *Elephas maximus* (UNSM 1252, Holocene, Asia; female, note small slender tusks; shoulder height about 2.5 m; head and body length, about 3 m; mounted by H. Reider and E. H. Colbert, directed by E. H. Barbour, 1928); *Elephas* (*Palaeoloxodon*) *falconeri* next to a kneeling man (UNSM 2540, cast, late Pleistocene, Sicily and Malta (shoulder height, maximum 1 m; head, excluding tusks and body length, about 1.5 m) mounted by C. Messenger and K. Kreycik, directed by C. B. Schultz 1970); type specimen of *Elephas maibeni* Barbour 1925*b*, later '*Archidiskodon maibeni*', or '*A. imperator maibeni*', now *Mammuthus imperator or M. columbi* (UNSM 1294, type skeleton no. 5-9-22, late Pleistocene, North America (shoulder height, about 3.75 m; head, excluding tusks and body length, about 4 m mounted by H. Reider and F. Bell, directed by E. H. Barbour, 1933); '*Mammuthus jeffersoni*' (M.R. Voorhies, personal communication 1993) (UNSM 1318, late Pleistocene, North America (shoulder height, about 3m; head, excluding tusks and body length, about 3.5 m) mounted by H. Reider and F. Bell, directed by E. H. Barbour, 1941; see also Barbour 1925*b*, 1926, and Osborn 1942, pp. 1020–3). In the above notes, 'shoulder height' refers to 'skeletal shoulder height', according to Osborn (1942, p. 1022) the 'height in the flesh' is approximately the 'skeletal shoulder height' plus 6.33 per cent of that height. (Photographs by J. Shoshani.)

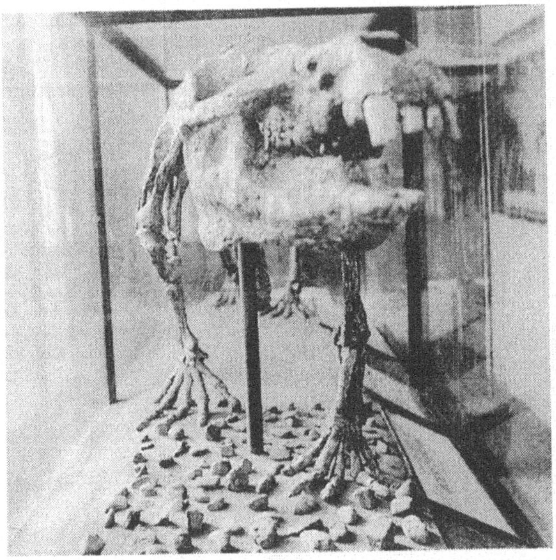

(a)

(b)

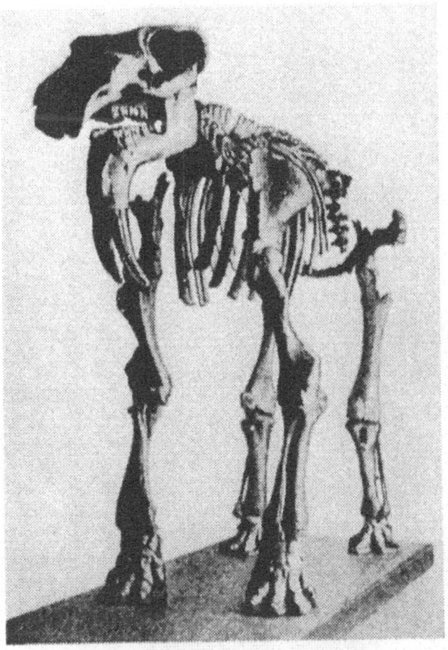

(c)

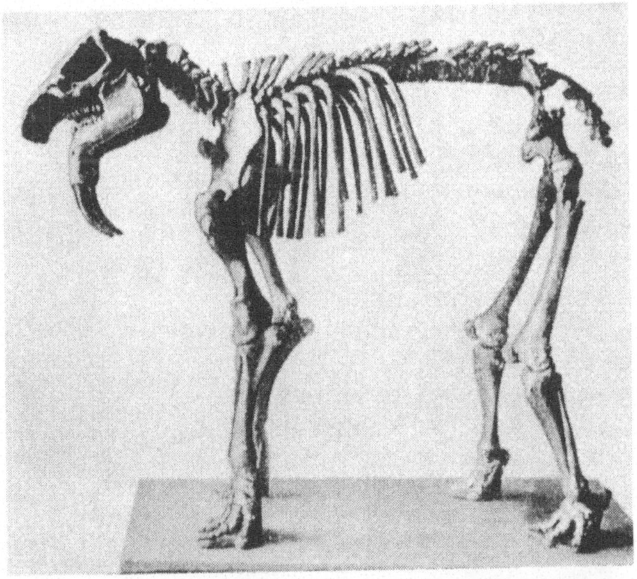

(d)

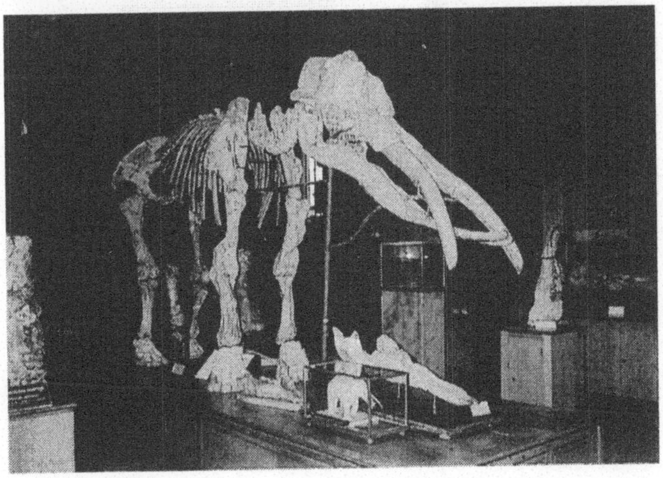

(e)

(f)

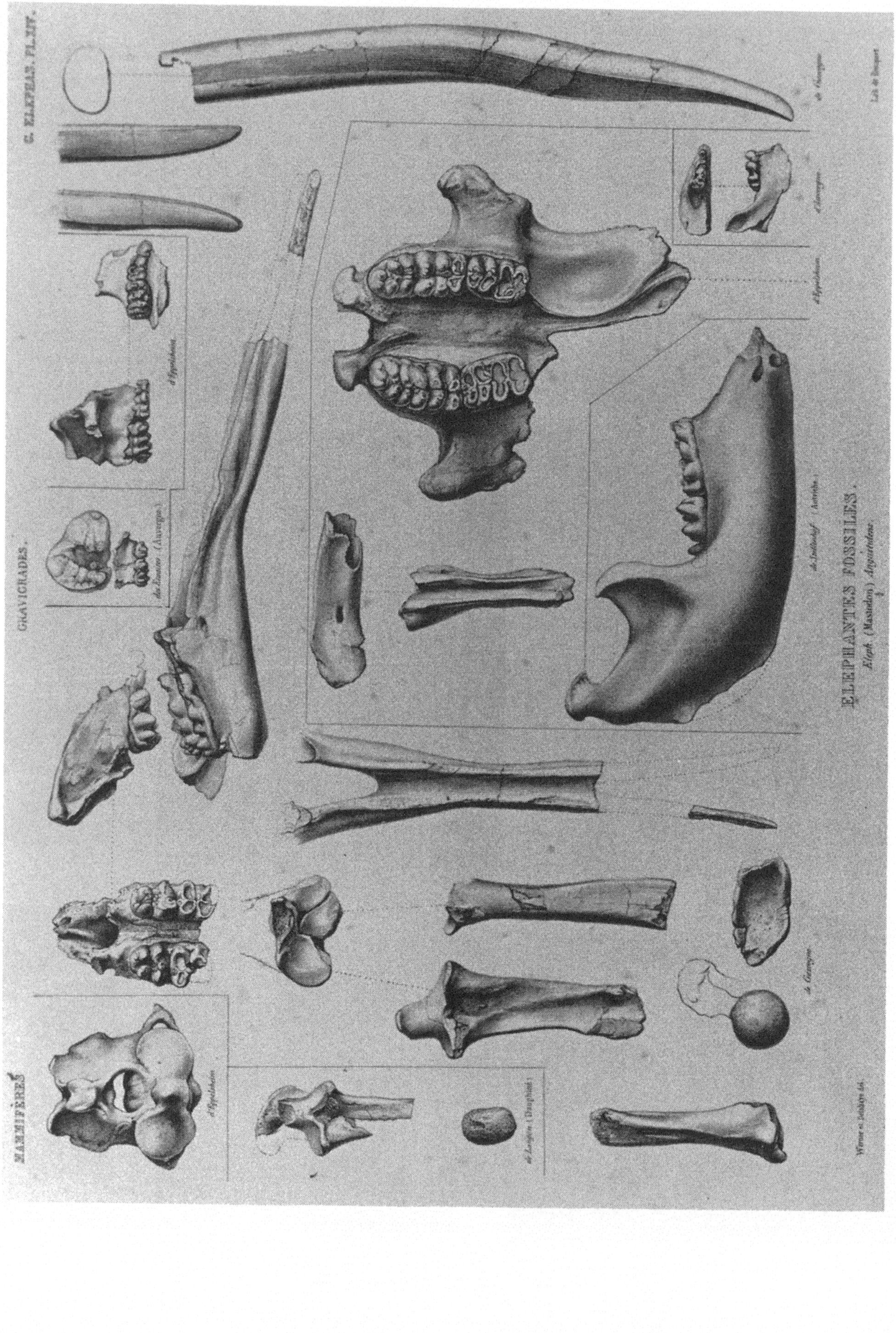

Plate 16.2 Another extraordinary plate from Blainville (1845): Plate XIV, *'Elephas* (*Mastodon*) *Angustidens'*, now called *Gomphotherium angustidens* (material on this plate belongs not only to *G. angustidens*, but also to *Anancus arvernensis* and *Tetralophodon longirostris* (Tassy, personal communication 1993).

12. The close association of *Tetralophodon*, *Anancus*, and *Paratetralophodon* with Stegodontidae and Elephantidae corroborates the hypothesis that Stegodontidae is more closely related to gomphotheres and elephantids than to mammutids.

13. *Stegodon* and *Stegolophodon* are grouped in the family Stegodontidae but this is a loose association because with either one or two evolutionary steps (with the unordered and the ordered options), they can be separated and *Stegodon* will join the Elephantidae clade. The hypothesis proposed by Kalb *et al.* (this volume, Chapter 12A) remains a viable alternate to the topologies shown in Fig. 16.5, a hypothesis which requires between one and three steps (with the unordered) or between zero and two steps (with the ordered); test nos 30–32 in Appendix 16.3.

14. Tassy's phylogenetic reconstruction of the Proboscidea can be employed as an equally valid alternative to that which is presented in Fig. 16.5.

15. A suggested classification, which is based on the results presented in Appendix 16.3 and Fig. 16.5, is given in Table 16.1; note that *Anthracobune* is included in the Proboscidea. Until additional analyses (preferably to include functional components) are conducted, these results must be considered as a tentative working hypothesis.

ACKNOWLEDGEMENTS

Heartfelt thanks to these individuals who helped me in various aspects of this study: J. Alexander, V. Barriel, S. K. Bell, J. Davis, D. Domning, M. Donoghue, D. J. Froehlich, W. T. Golden, E. Goldfinger, M. Goodman, J. E. Kalb, A. Kluge, M. C. McKenna, W. Maddison, E. C. Marsac, J. Meng, M. Norell, M. J. Novacek, J. L. Pierce, H. Saegusa, N. B. Simmons, D. Swofford, P. Tassy, R. Tedford, M. R. Voorhies, and X. Wang. Among these, M. C. McKenna is singled out for his mentorship, sponsorship, and scientific advisorship. Tassy's characters were invaluable and discussions on their polarities were pivotal; his constructive criticism is much appreciated. This chapter was also reviewed by S. K. Bell, J. E. Kalb, M. C. McKenna, and N. B. Simmons; Simmons' editing is particularly noted.

Appendix 16.1: Proboscidean characters examined to test hypotheses presented in Appendix 16.3 and Figs 16.1–16.5

This list of characters is adapted from, or slightly modified after many sources, especially Tassy (1979, 1981, 1982, 1985, 1986, 1988*a*, 1990*a*), Tassy and Shoshani (1988), Shoshani (1986*b*, personal observations, 1991–92; list prepared, by J. Shoshani, in its present form on 14 May 1991), and those references given inside [ ] at the end of each character. Within the [ ], references are listed chronologically and alphabetically for the same year; note that Andrews (1906) has been selected as a starting point in this character listing, although some characters, e.g. number 74, could be traced to Cuvier (1817) or even earlier. It is very difficult, and in some cases, impossible, to trace the exact source of reference for the original observation(s). The key below includes only selected citations which themselves include references of the original observations. For example, Tassy (1988*a*, 1990*a*) include many characters which appeared in Tassy (1981, 1982) and which, in turn, refer to earlier studies. Similarly, Tassy and Shoshani (1988) includes a list of original references for some characters. For simplicity, for many of Tassy's characters only the most recent reference (Tassy 1990*a*) is given.

Keys to references

A, Andrews (1906, pp. 99–177 and plates); C+, Coppens *et al.* (1978); D, Domning *et al.* (1986, pp. 36–8); F+T, Fischer and Tassy (1993); G, Gingerich *et al.*, (1990, pp. 74–7); Gr34, Gregory (1934, pp. 4–12); H, Harris (1978, pp. 315–32); Mg, Maglio (1973); Mh, Mahboubi *et al.* (1986, p. 42); Mi, Miller (1990, pp. 57–67); Mn, Manning (personal communication (numbered characters), 1977); Os36, Osborn (1936); Os42, Osborn (1942); Sa, Savage (1969, p. 170); S, Shoshani (1986*b*, pp. 224–35, 298, 314, 399–410); S+, Shoshani *et al.* (1989*a*, p. 160); Si+PC, Simpson and Paula-Couto (1957); T79, Tassy (1979, p. 86); T81, Tassy (1981, p. 117); T82, Tassy (1982, pp. 227–34); T88, Tassy (1988*a*, p. 55); T90, Tassy (1990*a*, pp. 198–201); T+S, Tassy and Shoshani (1988, pp. 292–3); To64, Tobien (1964, p. 64); To73, Tobien (1973*a*); To75, Tobien (1975); To76, Tobien (1976); To+86, Tobien *et al.* (1986); To+88, Tobien *et al.* (1988); W, Walker (1980); ws, 'white sheet' dated 21 April 1991, data by J. Shoshani.

Numbers following the colon (':') in the listing below are usually character numbers of that citation; a page number is given when the citation above contains more than one page. Sometimes a reference to a plate or a figure was provided: pl., plate; F9, Fig. 9; 167F, a figure on page 167.

Addendum

A thorough comparison among characters must be made to determine whether any can be eliminated since their information overlaps. For example: compare nos 37, 38, 45, and 55; also nos 50–61. In addition, the presence of a sustentacular notch in the calcanea of *Mammut americanum* (Shoshani 1986*b*, p. 405) may also be present in other mammutid taxa and help to strengthen the monophyly of this clade (see conclusion no. 2 p. 157). In Appendix 16.2, character 10 for *Mammut* should have been coded 3 instead of 4 since *Pliomastodon* is considered here to be a junior synonym of *Mammut* (see Appendix A).

Note

When comparisons were made, e.g. size, presence or absence, they included at least these outgroups: Outgroup A (primarily 'condylarths', Artiodactyla, Perissodactyla) and Outgroup B (primarily Hyracoidea, Sirenia, and Desmostylia). Except for a few instances, all characters of the outgroups were coded with a '0' (the primitive condition), the exceptions were coded with a '1' (see Appendix 16.2). For this reason, I did not use some of the ordinal characters employed by Tassy (1990a; this volume, Chapter 6). In every case, most of what is now believed to be a plesiomorphic taxon was used in the comparison. Whenever possible, characters were observed on adult individuals, and at least three specimens were examined for each taxon. Characters below are arranged from anterior to posterior of body, beginning with incisors.

Feature	Primitive (only state '0')	Derived (all states except '0')
1) I^1: [A, p. 99; D:47; T88:9; T90 :78]	(0)=present	(1)=absent
2) I_1: [D:39; Mh, p. 30; T90:70, 82]	(0)=present	(1)=absent
3) dI^2/I^2(upper incisors): [T90:71]	(0)=short	(1)=elongate
4) I^2: [A, pp. 99, 100F; D:24; Mh:2; T90:78]	(0)=average size	(1)=enlarged
5) I_2: size [A, p. 99, 100F; D:20; T90:69]	(0)=as large as other incisors	(1)=enlarged
6) I_2: enamel [T90:73]	(0)=surrounds tooth	(1)=partly surrounds tooth (2)=as band of enamel (3)=absent
7) UPPER TUSKS: viewed laterally [T90:76; 10 ws]	(0)=curve ventrally	(1)=straight (2)=curve dorsally (3)=curve dorsally, much
8) UPPER TUSKS: [Si+PC, p. 183; Mi:57; 12a ws]	(0)=not helicoidal	(1)=helicoidal
9) UPPER TUSKS: viewed anteriorly [Si+PC, p. 183; Mi:58–59; 12b ws]	(0)=close to each other	(1)=diverged
10) UPPER TUSKS: enamel [T+S:38; T90:72; 11 ws]	(0)=surrounds tooth	(1)=partly surrounds tooth (2)=a longitudinal band of enamel (3)=little or usually absent (4)=enamel absent
11) LOWER TUSKS: viewed laterally [T90:77; 18 ws]	(0)=curve dorsally	(1)=straight (2)=curve ventrally
12) LOWER TUSKS: viewed anteriorly [17a ws]	(0)=wear at tip, anteriorly	(1)=wear at tip, dorsally and ventrally (2)=wear at tip, ventrally only

Appendix 16.1. *(contd)*

Feature	Primitive (only state '0')	Derived (all states except '0')
13) LOWER TUSKS: protruding portion, viewed dorsally [19 ws]	(0)=small (less than 10 cm long)	(1)=longer than approximately 10 cm
14) LOWER TUSKS: protruding portion, viewed dorsally [19 ws]	(0)=small (less than 10 cm wide)	(1)=wider than approximately 10 cm
15) LOWER TUSKS: [To73, pp. 208F, 251F, 253F, 263F; T+S:36; T90:74; 14 ws]	(0)=cross-section (c.s.) flat	(1)=c.s. pyriform (2)=round or oval
16) LOWER TUSKS: medial edge: [To73, pp. 251F, 253F, 263F; T90:75]	(0)=round	(1)=straight
17) UPPER AND LOWER TUSKS: [Mn:28b, 18b]	(0)=without cement	(1)=with cement
18) I^3: [A, p. 99; Sa, p. 170; D:36; T90:78, 84]	(0)=present	(1)=absent
19) I_3: [A:100F; D:33; T90:82]	(0)=present	(1)=absent
20) C^1: [A:100F; Sa, p. 170; D:23, 38; T88:10; T90:83]	(0)=average size	(1)=reduced (2)=absent
21) C_1: [A:100F; D:22; T88:2; T90:81]	(0)=present	(1)=absent
22) dp^2: parastyle [T90:120; F+T:21]	(0)=not reduced	(1)=reduced
23) dp^3: posterior loph (oblique, i.e. metacone displaced posterior to hypocone) [T90:122]	(0)=not oblique	(1)=oblique
24) dp^3: oblique contact between pre- and posttrite central conule [T90:124]	(0)=absent	(1)=present
25) dp_3: oblique contact between pre- and posttrite central conule [T90:125]	(0)=absent	(1)=present
26) P^1(and dp^1), and P_1 (and dp_1): [D:21; T88:1; T90:85]	(0)=present	(1)=absent
27) P^2 and P_2 through P^4 and P_4: [T+S:21; T90:86]	(0)=P2/2, P3/3, P4/4 present	(1)=P2/2 absent (2)=P2/2, P3/3 absent (3)=P2/2, P3/3, P4/4 absent
28) P^3: [T90:89]	(0)=without hypocone	(1)=with hypocone
29) P^4: [T88:4; T90:87]	(0)=with two cusps (protocone & paracone and post paracrista)	(1)=with outline of metacone (2)=distinct metacone (3)=with hypocone (4)=with two lophs and slight opening at interlophs (5)=molariform (with distinct opening at interlophs)
30) P_4: bilophodont [T88:8; T90:88]	(0)=absent	(1)=present
31) P3/3–P4/4: posterior supplementary loph(id)s [T90:90]	(0)=absent	(1)=present
32) M^1: [T90:101]	(0)=bilophodont	(1)=third loph outlined (2)=third loph complete (3)=fourth loph complete
33) M^2: [T+S:11; T90:103]	(0)=bilophodont	(1)=trilophodont outlined (2)=trilophodont distinct (3)=tetralophodont
34) M^3: postentoconule [T+S:6; T90:99] [postentoconule is the third pretrite main cusp of upper molars, and it is the counterpart of postentoconulid(=entoconid II) of lower molars]	(0)=reduced or absent	(1)=enlarged (2)=M^3 with two large cusps posterolabially (3)=M^3 is trilophodont (i.e. two valleys formed) (4)=M^3 is tetralophodont (5)=M^3 is pentalophodont or more
35) M^3: [Mn:14j; T90:115]	(0)=large, wide [robustness index I=100/L) is between 70 and 100]	(1)=$I<70$ (2)=$I<60$
36) M^1 and M^2 with distocrista: [T90:97]	(0)=absent	(1)=present
37) M^{1-3}: central conules [T90:134]	(0)=not enlarged	(1)=enlarged
38) M^{1-3}: central conules on the second interloph [T90:112]	(0)=absent	(1)=present
39) M_1: [T90:102]	(0)=bilophodont	(1)=trilophodont (2)=tetralophodont
40) M_2: [T90:104]	(0)=bilophodont	(1)=trilophodont (2)=tetralophodont or more

Appendix 16.1. (*contd*)

Feature	Primitive (only state '0')	Derived (all states except '0')
41) M_3 with two definite cuspids at rear: a labially displaced hypoconulid and a large entoconid II [D:16; T90;95]	(0)=absent	(1)=present
42) M_3: [T+S:11; T90:105]	(0)=bilophodont	(1)=trilophodont (2)=4th lophid with central conule in 3rd interlophid
43) M_3: number of plates [Mg, pp. 20–2, 66; T90:109]	(0)=1–7	(1)=8 or more
44) M_1 and M_2: [T90:96]	(0)=hypoconulid and post-entoconulid absent	(1)=hypoconulid and post-entoconulid on M_1 only (2)=same on M_1 and M_2
45) M_1 and M_2: [Mg, p. 89; C+:349; T90:100]	(0)=without central conules in the second interlophid	(1)=with central conules (small or large)
46) M_{1-3}: anterolingual cingulum on posttrite side [T90:117]	(0)=not reduced	(1)=reduced
47) M_{1-3}: [T+S:7; T90:111]	(0)=with small protoconulid	(1)=with enlarged protoconulid
48) M_{1-3}: [e.g. Os36, p. 613; Os43, pp. 1548–9; T90:114]	(0)=2nd and 3rd lophids perpendicular to longitudinal axis of tooth	(1)=2nd & 3rd lophids oriented postero-labially in relation to longitudinal axis
49) M1: elephantine plates (Mg, pp. 22–3, 67]	(0)=absent	(1)=present
50) M2: number of elephantine laminae [Mg, p. 23]	(0)=5–6	(1)=6–8
51) LOWER MOLARS: crista obliqua in the first valley [T88:3; T90:92; F+T:0]	(0)=situated diagonally from hypoconid to tooth centre	(1)=situated entirely on labial side
52) MOLARS: buno- or bi-lophodont [A, p. 99; T79:1; Mh:16; T90:93]	(0)=absent	(1)=present
53) MOLARS: with true lophodonty (protocone-paracone crest and hypocone-metacone crest) [T90:135]	(0)=absent	(1)=present, i.e. uninterrupted lophs and there is no median sulcus
54) MOLARS: pretrite cusps [T+S:5; T90:94]	(0)=not enlarged transversely	(1)=enlarged transversely
55) MOLARS: trefoils [Mn:28d; T90:113; 3 ws]	(0)=none	(1)=on pretrite plus quasi posttrite (2)=on pretrite and posttrite
56) UPPER MOLARS: cingulum on medial side [Mn:15a; 9 ws]	(0)=present	(1)=absent
57) MOLARS: [T90: 108]	(0)=brachyodont, non-accentuated	(1)=brachyodont, accentuated
58) MOLARS: anterior pretrite central conule and adaxial conelet (mesoconelet) [T90:134]	(0)=not fused	(1)=fused
59) MOLARS: abaxial and adaxial cones (and their conelets) [To73, pp. 217F, 219F, 245F; Mg, pp. 128F, 130F, 141F]	(0)=are of unequal size	(1)=are of about equal size
60) MOLARS' PATTERN [T90:106]	(0)=with cusps or lophs	(1)=with laminae (2)=laminae with height/width=1, 1+ or more
61) MOLARS: with thinner and more numerous laminae [Mg, pp. 31–3, 50–1, 67; T+S:53]	(0)=absent	(1)=present
62) PREMOLARS and MOLARS: Note: 'zygodont' refers to the outline of ridges in anterior view; they are 'yoke-shaped' and sharp when worn [To64; To75, p. 196–207; T90:128]	(0)=non zygodont	(1)=zygodont
63) PREMOLARS and MOLARS: crowns of [T+S:30; T90:116]	(0)=cement absent	(1)=little cement present (2)=abundant cement present
64) PREMOLARS and MOLARS: hypsodonty index [Mg, pp. 105–11; 2 ws; and from other workers]	(0)=brachyodonty	(1)=hypsodonty
65) PREMOLARS and MOLARS: central conule in a form of a crest and does not block the valley ['acr' and 'pcr' of To75, p. 196F]	(0)=absent	(1)=present

Appendix 16.1. *(contd)*

Feature	Primitive (only state '0')	Derived (all states except '0')
66) PREMOLARS and MOLARS: zygodont crest: on posttrite [To75, p. 196–207; T90:129]	(0)=absent	(1)=present
67) CHEEK TEETH: horizontal displacement and vertical succession associated with remodelling of adjacent bones [T+S:20; T90:80]	(0)=absent	(1)=present
68) CRANIUM: external auditory meatus [T79:8; Mh:3; S:5; T88:19; T90:4]	(0)=only the original 'tube'	(1)=closed ventrally by post-tympanic and postglenoid parts of squamosal
69) CRANIUM: glenoid cavity [Mh:3, 9; T90:60]	(0)=at about the level of tooth row, or slightly dorsal to it	(1)=elevated cavity and external auditory meatus
70) CRANIUM: mastoid process [Mh:7; T90:8]	(0)=present	(1)=much reduced or absent
71) CRANIUM: pneumatization (development of diploë) [Mh:9; S:2; T88:22; T90:20]	(0)=absent	(1)=present
72) CRANIUM: supraoccipital bone: deep nuchal fossa with a median crest for the insertion of the *ligamentum nuchae* [Mh:15; T81:19; T90:19]	(0)=absent	(1)=present
73) CRANIUM: [T+S:13; T90:37]	(0)=swelling absent (sagittal crest present)	(1)=swelling present with loss of sagittal crest (2)=wide cerebral area
74) CRANIUM: large dorsal parietal bulges and fronto-parietal concavity [T+S:46; T90:58]	(0)=absent	(1)=present
75) PREMAXILLA: facial anterior concavity pronounced widened laterally and expanded dorsally, see fig. 11.6 in T+S, p. 294 [T+S:8; T90:35]	(0)=absent	(1)=present
76) PREMAXILLA: with posterodorsal process in the midline, at the floor of the nasal fossa [T+S:8; T90:32]	(0)=absent	(1)=present, but does not protrude vertically (2)=protrudes vertically
77) MAXILLA: ridges prominent and close to midline in ventral view [S:9; T+S:48; T90:56]	(0)=absent	(1)=present
78) NARIS: [D:34; Mh:15]	(0)=not retracted	(1)=retracted to above orbit
79) NAREAL OPENING: (see fig. 11.6 in T+S, p. 294) [T+S:28; T90:43]	(0)=narrow (lateral limits are vertical)	(1)=wide (lateral limits are diagonal) (2)=deep depressions laterally
80) NAREAL OPENING: lateral limits of [T90:43, state 2]	(0)=do not extend laterally	(1)=extend laterally up to or beyond width of rostrum
81) PALATINE: ascending process [T79:6; D:27; T90:2]	(0)=present in the orbito-temporal fossa	(1)=absent from orbito-temporal fossa (not visible in lateral view)
82) PALATINE: spina nasalis posterior above choanae [T82, p. 232; T90:31; T+S:3]	(0)=absent	(1)=present
83) ORBIT: anterior border of [A:100F, 131F; T79:2; D:1; T+S:2; T90:30; F+T:56]	(0)=situated above or posterior to M^1	(1)=forward of M^1
84) ORBIT: bones contributing to it ventrally [D:25; S, pp. 227, 314; T90:0; F+T:31]	(0)=mostly jugal	(1)=zygomatic process of maxilla, a little by jugal
85) ORBIT: rim with angular corners [S:7; T+S:47; T90:57]	(0)=absent	(1)=present
86) OPTIC FORAMEN and SPHENO-ORBITAL FISSURE [T90:7]	(0)=in their normal position	(1)=retracted towards ventral of orbito-temporal fossa
87) CONDYLAR FORAMEN: [A, p. 101; T79:9; D:28; Mh:6; S, 229, 298; T90: 6]	(0)=separate	(1)=confluent with foramen lacerum medium
88) CANALIS TEMPORALIS (= squamosal-sinus canals): [T81:11; T+S:18; T90:39]	(0)=present	(1)=absent
89) INFRAORBITAL FORAMEN: duplication, on maxilla [T+S:9; T90:33]	(0)=absent	(1)=always present (2)=sometimes present
90) CRANIAL FORAMINA (on lateral–ventral–posterior end of the orbito-temporal fossa): [T81:10″; T90:34]	(0)=not compressed	(1)=compressed in a restricted posterior notch
91) ETHMOID FORAMEN [D:48; S:86?; T88:23; ?T90:21]	(0)=not retracted	(1)=retracted
92) FORAMEN OVALE and FORAMEN LACERUM MEDIUM: [T81:9; T+S:19; T90:41]	(0)=not confluent	(1)=confluent

Appendix 16.1. *(contd)*

Feature	Primitive (only state '0')	Derived (all states except '0')
93) MANDIBULAR FORAMEN: process at the anterior border [S:10; T+S:49; T90:53]	(0)=absent	(1)=present
94) MANDIBULAR SYMPHYSIS [Gr34, pp. 4–12; T+S:40; T90:52; 1c ws]	(0)=reduced symphysis	(1)=long (2)=short, spout-like
95) DENTARY: length of diastema and position of anterior premolar relative to symphysis [T90:50]	(0)=diastema short; first premolar at symphysis	(1)=diastema short or long; first premolar anterior or at symphysis (2)=diastema long, 1st premolar posterior to symphysis
96) MANDIBULAR ANGULAR PROCESS: [Mn:15b; T90:49]	(0)=distinct	(1)=reduced
97) PARAOCCIPITAL PROCESS: [T81:23; T90:36]	(0)=present	(1)=absent
98) BULLA (ectotympanic area): [T82, p. 240; T+S:17; T90:40]	(0)=not enlarged	(1)=slightly enlarged posterolaterally
99) STYLOHYOID: posterior ramus [S:11; T+S:35; T90:47]	(0)=absent	(1)=present
100) STYLOHYOID: reduced shelf at the base of the inferior ramus [S:12; S+T:42]	(0)=absent	(1)=present
101) STYLOHYOID: more gracile [S:13; T+S:50]	(0)=absent	(1)=present
102) THORACIC VERTEBRAE: no. of [S:20; T+S:51]	(0)=20–21	(1)=18–19, 20
103) SCAPULA: coracoid process [A, pp. 117, 174; D:41; S:23; S+; T, this volume no. 137]	(0)=average size	(1)=enlarged and 'hooked'
104) SCAPULA: infraspinous fossa: [A, pl. XI, F5, pl. XVII, F9; S, p.227, T88:11; T90:11]	(0)=small	(1)=larger than supraspinous fossa
105) HUMERUS: lateral supracondylar crest [A, p. 117, pl. XI, F6; Mh, p. 34F; T88:13; T90:13; W:106]	(0)=at level of condyles	(1)=in line with or dorsal to olecranon fossa
106) HUMERUS: deep bicipital groove and higher trochanter [S:28; T+S:31; T90:44]	(0)=absent	(1)=present
107) RADIUS: [Mh:12]	(0)=can be rotated	(1)=positioned or in fixed pronation
108) ULNA: [Mh:11]	(0)=as large as radius	(1)=larger than radius
109) ULNA: proximal epiphysis wide and thick olecranon [A:pl. XI, F7; T88:14, 15; T90:14]	(0)=absent	(1)=present
110) ULNA: olecranon process tilted medially and extends proximally beyond the trochlear notch [S:30; T+S:43; T90:55]	(0)=absent	(1)=present
111) CARPUS and TARSUS: low and wide, i.e. less 'cursorial' [T+S:32; T90:45]	(0)=absent	(1)=present
112) MAGNUM: reduced lateral bulging of the articular facet of scaphoid [T+S:52]	(0)=absent	(1)=present
113) PELVIS: well-delineated distal obturator notch [S:42; T+S:44; T90:54]	(0)=absent	(1)=present
114) FEMUR: flattening antero-posteriorly [A, p. 120; Mh:13; T90:24]	(0)=absent	(1)=present
115) FEMUR (F) and TIBIA (T): relative sizes [Mh:14; T88:16; T90:15]	(0)=F is smaller or about the same size as T	(1)=F is longer than T
116) TIBIA: crest [T90:25]	(0)=of average size	(1)=enlarged with cranial depression
117) ASTRAGALUS: alternate v. serial (i.e. head articulates distally with navicular only) [A, p. 149; S:51; S+; T82, pp. 227–228; T90:9]	(0)=alternate	(1)=serial
118) ASTRAGALUS: head [Mh, p.39F; S:52; S, p. 228; T90:27]	(0)=globular, long neck	(1)=not globular, semi-circular, neck short
119) ASTRAGALUS: head [G, pp. 75–6]	(0)=*crista capitalis*, present	(1)=*crista capitalis*, absent
120) ASTRAGALUS: fibular facet viewed laterally [T88:25; T90:22]	(0)=long/large and flared	(1)=short/small, not flared
121) ASTRAGALUS: protruding tuberculum mediale [A, 167F; S:53; S+; T82, p. 228; T90:16; G, p. 75]	(0)=absent	(1)=present, large (2)=present, reduced
122) ASTRAGALUS: lateral (ectal) v. sustentacular (sust.) facet [A:167F; Mh, pp. 39, 42; T82:228F; T88:17; T90:10]	(0)=ectal about the size of sustentacular facet and elongate	(1)=ectal is larger than sust. circular, oval, triangular, or teardrop-shaped
123) ASTRAGALUS: foramen [G, p. 74; T, this volume no. 136]	(0)=present	(1)=absent

Appendix 16.2: Proboscidean taxa (37) and two outgroups and their data matrix employed in this study; characters are defined in Appendix 16.1

Character --> / Taxon	1	2	3	4	5	6	7	8	9	10	11	12	13	14	15	16	17	18	19	20	21	22	23	24	25	26	27	28	29	30	31
OutgroupA	0	0	0	0	0	0	0	0	0	0	0	0	0	0	0	0	0	0	0	0	0	0	0	0	0	0	0	0	0	0	0
OutgroupB	0	0	0	0	0	0	0	0	0	0	0	0	0	0	0	0	0	0	0	0	0	0	0	0	0	0	0	0	0	0	0
Anthracobune	0	0	0	?	?	0	?	?	?	?	0	0	0	0	?	?	?	0	0	0	0	?	?	?	?	0	0	0	?	0	0
Moeritheriidae	0	0	0	1	1	1	0	0	0	1	0	0	0	0	?	?	?	0	1	1	1	0	0	0	?	1	0	0	2	0	0
Numidotheriidae	0	0	0	1	?	0	0	0	0	1	0	0	0	0	?	?	1	0	?	1	1	0	0	0	0	1	0	0	1	0	0
Barytheriidae	0	1	0	1	1	1	0	0	0	1	1	0	0	0	?	?	?	1	1	2	1	?	?	?	?	1	0	0	1	0	0
Deinotheriidae	1	1	0	?	1	3	?	?	?	?	2	?	1	0	?	?	?	1	1	2	1	0	0	0	0	1	?	1	3	1	0
Palaeomastodon	1	1	0	1	1	3	?	?	?	?	0	?	?	?	?	?	?	1	1	2	1	?	?	?	?	1	0	1	4	1	0
Phiomia	1	1	1	1	1	3	0	0	1	2	0	1	0	0	0	1	?	1	1	2	1	0	0	0	0	1	0	1	5	1	0
Eozygodon	?	?	1	1	1	3	0	0	?	2	0	1	1	0	0	0	?	?	?	?	1	1	?	?	?	1	?	?	5	?	?
Zygolophodon	?	?	1	1	1	3	0	0	1	2	1	1	0	0	2	0	?	?	?	2	1	1	1	?	?	1	1	1	5	1	?
Mammut	1	1	1	1	1	3	2	0	1	4	1	0	0	0	2	0	?	1	1	2	1	1	1	?	?	1	3	?	?	?	?
Choerolophodon	1	1	1	1	?	?	2	0	1	4	?	?	?	?	?	?	?	1	1	2	1	1	1	0	0	1	3	?	?	?	0
Gomphotherium	1	1	1	1	1	3	0	0	1	2	0	1	1	0	1	0	1	1	1	2	1	1	1	0	0	1	1	1	5	1	0
Serridentinus	1	1	1	1	1	3	0	0	1	2	0	1	1	0	2	0	1	1	1	2	1	1	1	0	0	1	1	1	5	1	0
Amebelodon	1	1	1	1	1	3	0	0	1	2	0	1	1	1	0	1	1	1	1	2	1	1	1	0	0	1	1	1	5	1	0
Platybelodon	1	1	1	1	1	3	0	0	1	2	0	2	1	1	0	1	1	1	1	2	1	1	1	?	?	1	1	1	5	1	0
Serbelodon	1	1	1	1	1	3	0	0	1	2	0	1	1	1	0	1	1	1	1	2	1	1	1	0	0	1	1	1	5	1	0
Protanancus	1	1	1	1	1	3	0	0	1	2	0	1	1	0	0	1	1	1	1	2	1	?	1	?	?	1	1	?	5	?	?
Archaeobelodon	1	1	1	1	1	3	0	0	1	2	0	1	1	0	0	1	1	1	1	2	1	1	1	?	?	1	0	1	5	1	?
Gnathabelodon	1	1	1	1	?	?	2	0	1	4	?	?	?	?	?	?	?	1	1	2	1	1	1	?	?	1	1	?	5	?	0
Sinomastodon	1	1	1	1	?	?	2	0	1	4	?	?	?	?	?	?	?	1	1	2	1	?	?	?	?	1	2	?	5	1	?
Eubelodon	1	1	1	1	?	?	0	0	1	2	?	?	?	?	?	?	?	1	1	2	1	?	1	?	?	1	?	?	5	?	0
Rhynchotherium	1	1	1	1	1	2	0	1	1	2	2	0	1	0	2	0	?	1	1	2	1	1	1	?	?	1	2	?	5	?	?
Stegomastodon	1	1	1	1	?	?	1	1	1	4	?	?	?	?	?	?	1	1	1	2	1	1	1	?	?	1	3	?	?	?	?
Haplomastodon	1	1	1	1	?	?	2	1	1	3	?	?	?	?	?	?	1	1	1	2	1	1	1	?	?	1	3	?	?	?	?
Notiomastodon	1	1	1	1	?	?	?	1	1	2	?	?	?	?	?	?	?	1	1	2	1	?	1	?	?	1	3	?	?	?	?
Cuvieronius	1	1	1	1	?	?	2	1	1	2	?	?	?	?	?	?	?	1	1	2	1	1	1	?	?	1	3	?	?	?	?
Tetralophodon	1	1	1	1	1	3	0	?	1	4	1	1	1	0	2	0	?	1	1	2	1	1	1	1	1	1	1	1	5	1	1
Anancus	1	1	1	1	?	?	1	0	1	4	?	?	?	?	?	?	?	1	1	2	1	1	1	1	1	1	2	?	5	?	1
Paratetralophodon	1	?	1	1	?	?	0	0	1	4	?	?	?	?	?	?	?	1	1	2	?	?	?	?	?	1	?	?	?	1	?
Stegolophodon	1	?	1	1	1	3	0	0	1	2	0	1	1	0	1	0	?	1	1	2	1	1	1	?	?	1	1	?	5	?	1
Stegodon	1	1	1	1	?	?	2	0	1	3	?	?	?	?	?	?	?	1	1	2	1	1	1	?	?	1	3	?	?	?	?
Stegotetrabelodon	1	1	1	1	1	3	1	0	1	4	1	1	1	0	2	0	?	1	1	2	1	?	?	?	1	1	2	?	5	?	1
Stegodibelodon	1	1	?	?	?	?	?	?	?	?	?	?	?	?	?	?	?	1	?	?	1	?	?	?	?	1	?	?	?	?	?
Primelephas	1	1	?	1	1	3	?	0	?	?	1	?	?	0	2	0	?	1	1	?	1	1	?	?	1	1	2	?	5	?	1
Loxodonta	1	1	1	1	?	?	2	0	1	4	?	?	?	?	?	?	?	1	1	2	1	1	1	?	?	1	3	?	?	?	?
Elephas	1	1	1	1	?	?	2	0	1	4	?	?	?	?	?	?	?	1	1	2	1	1	1	?	?	1	2	?	5	?	1
Mammuthus	1	1	1	1	?	?	3	0	1	4	?	?	?	?	?	?	?	1	1	2	1	1	1	?	?	1	2	?	5	?	1

Character --> / Taxon	32	33	34	35	36	37	38	39	40	41	42	43	44	45	46	47	48	49	50	51	52	53	54	55	56	57	58	59	60	61	62
OutgroupA	0	0	0	0	0	0	0	0	0	0	0	0	0	0	0	0	0	?	?	0	0	0	0	0	0	0	0	0	0	0	0
OutgroupB	0	0	0	0	0	0	0	0	0	0	0	0	0	0	0	0	0	?	?	0	0	0	0	0	0	0	0	0	0	0	0
Anthracobune	0	0	0	0	0	0	0	0	0	1	0	0	0	0	0	0	0	?	?	0	0	0	0	0	0	0	0	0	0	?	0
Moeritheriidae	0	0	0	0	1	0	0	0	0	0	0	0	2	0	0	0	0	?	?	1	1	0	0	0	0	0	0	0	0	?	0
Numidotheriidae	0	0	0	0	1	0	0	0	0	0	0	0	0	0	0	0	0	?	?	1	1	1	0	0	0	0	0	0	0	0	0
Barytheriidae	0	0	0	0	1	0	0	0	0	0	0	0	0	0	0	0	0	?	?	1	1	1	0	0	0	0	0	0	0	?	0
Deinotheriidae	2	0	0	0	1	0	0	1	0	0	0	0	1	0	0	0	0	?	?	1	1	1	0	0	1	0	0	0	0	?	0
Palaeomastodon	1	1	1	0	1	0	0	1	1	1	0	0	2	1	0	1	1	?	?	1	1	0	1	0	0	0	0	0	0	?	0
Phiomia	2	2	2	0	1	0	1	1	1	1	1	0	2	1	0	1	1	?	?	1	1	0	1	0	0	0	0	0	0	?	0
Eozygodon	2	2	3	1	1	0	1	1	1	1	2	0	2	1	1	1	1	?	?	1	1	0	1	0	0	0	0	0	0	?	1
Zygolophodon	2	2	3	1	1	0	1	1	1	1	2	0	2	1	1	1	1	?	?	1	1	0	1	0	0	0	0	0	0	?	1
Mammut	2	2	3	1	1	0	1	1	1	1	2	0	2	1	1	1	1	?	?	1	1	0	1	0	0	0	0	0	0	?	1
Choerolophodon	2	2	3	2	1	1	1	1	1	1	2	0	2	1	1	1	1	?	?	1	1	0	1	0	0	0	0	0	0	?	0
Gomphotherium	2	2	3	2	1	0	1	1	1	1	2	0	2	1	1	1	1	?	?	1	1	0	1	0	0	0	0	0	0	?	0
Serridentinus	2	2	3	2	1	0	1	1	1	1	2	0	2	1	1	1	1	?	?	1	1	0	1	0	0	0	0	0	0	?	0
Amebelodon	2	2	4	2	1	1	1	1	1	1	2	0	2	1	1	1	1	?	?	1	1	0	1	2	0	0	0	0	0	?	0
Platybelodon	2	2	4	2	1	1	1	1	1	1	2	0	2	1	1	1	1	?	?	1	1	0	1	2	?	0	0	0	0	?	0
Serbelodon	2	2	4	2	1	1	1	1	1	1	2	0	2	1	1	1	1	?	?	1	1	0	1	0	0	0	0	0	0	?	0
Protanancus	2	2	4	2	1	1	1	1	1	1	2	0	2	1	1	1	1	?	?	1	1	0	1	1	?	0	0	0	0	?	0
Archaeobelodon	2	?	4	2	1	1	1	1	1	1	2	0	2	1	1	1	1	?	?	1	1	0	1	1	0	0	0	0	0	?	0
Gnathabelodon	2	2	4	2	1	1	1	1	1	1	2	0	2	1	1	1	1	?	?	1	1	0	1	0	1	0	0	0	0	?	0
Sinomastodon	2	2	4	2	1	1	1	1	1	1	2	0	2	1	1	1	1	?	?	1	1	0	1	0	0	0	0	0	0	0	0
Eubelodon	2	2	4	2	1	1	1	1	1	1	2	0	2	1	1	1	1	?	?	1	1	0	1	2	?	0	0	0	0	?	0
Rhynchotherium	2	2	4	2	1	1	1	1	1	1	2	0	2	1	1	1	1	?	?	1	1	0	1	1	0	0	0	0	0	?	0
Stegomastodon	2	2	5	2	1	1	1	1	1	1	2	0	2	1	1	1	1	?	?	1	1	0	1	2	?	0	0	0	0	?	0
Haplomastodon	2	2	5	2	1	1	1	1	1	1	2	0	2	1	1	1	1	?	?	1	1	0	1	2	?	0	0	0	0	?	0
Notiomastodon	2	2	?	2	1	1	1	1	1	1	2	0	2	1	1	1	1	?	?	1	1	0	1	2	?	0	0	0	0	?	0
Cuvieronius	2	2	5	2	1	1	1	1	1	1	2	0	2	1	1	1	1	?	?	1	1	0	1	2	0	0	0	0	0	?	0
Tetralophodon	3	3	5	2	1	1	1	2	2	1	2	0	2	1	1	1	1	?	?	1	1	0	1	1	0	0	0	0	0	?	0
Anancus	3	3	5	2	1	1	1	2	2	1	2	0	2	1	1	1	1	?	?	1	1	0	1	1	1	0	0	0	0	?	0
Paratetralophodon	3	3	5	2	1	1	1	?	?	1	2	?	?	1	1	?	?	?	?	1	1	0	1	1	1	0	0	0	0	?	0
Stegolophodon	3	3	5	2	1	1	1	2	2	1	2	0	2	1	1	1	1	?	?	1	1	0	1	0	1	1	1	1	1	?	0
Stegodon	3	3	5	2	1	0	1	2	2	1	2	0	2	1	1	1	1	?	?	1	1	0	1	0	1	1	1	1	1	?	0
Stegotetrabelodon	3	3	5	2	1	1	1	2	2	1	2	0	2	1	1	1	1	0	0	1	1	0	1	0	1	0	0	1	1	?	0
Stegodibelodon	3	3	5	2	1	1	?	2	2	1	2	1	?	?	1	1	0	?	?	1	1	0	1	0	1	0	0	?	1	?	0
Primelephas	3	3	5	2	1	0	1	2	2	1	2	1	2	1	1	1	0	0	0	1	1	0	1	0	?	0	0	1	1	?	0
Loxodonta	3	3	5	2	1	0	1	2	2	1	2	1	2	1	1	1	0	1	1	1	1	0	1	0	1	0	0	1	2	?	0
Elephas	3	3	5	2	1	0	1	2	2	1	2	1	2	1	1	1	0	1	1	1	1	0	1	0	1	0	0	1	2	1	0
Mammuthus	3	3	5	2	1	0	1	2	2	1	2	1	2	0	1	1	0	1	1	1	1	0	1	0	1	0	0	1	2	1	0

Character --> / Taxon	63	64	65	66	67	68	69	70	71	72	73	74	75	76	77	78	79	80	81	82	83	84	85	86	87	88	89	90	91	92	93
OutgroupA	0	0	0	0	0	0	0	0	0	0	0	0	0	0	0	0	0	0	0	0	0	0	0	0	0	0	0	0	0	0	0
OutgroupB	0	0	0	0	0	0	0	0	0	0	0	0	0	0	0	0	0	0	0	0	1	1	0	0	1	0	0	0	0	0	0
Anthracobune	0	0	0	0	0	?	?	?	?	0	?	0	?	?	0	0	?	?	?	?	?	?	0	?	?	?	?	?	?	?	0
Moeritheriidae	0	0	0	0	0	1	1	1	0	0	0	?	0	0	?	0	0	0	1	0	1	1	?	1	1	0	0	0	1	0	?
Numidotheriidae	0	0	0	0	0	1	1	1	1	0	0	0	0	0	?	0	0	0	1	0	1	1	0	1	1	0	0	0	1	0	0
Barytheriidae	0	0	0	0	0	1	1	1	1	?	?	?	?	?	?	0	?	0	?	?	1	?	?	?	?	?	?	?	1	?	?
Deinotheriidae	0	0	0	0	0	1	1	1	1	1	1	0	0	0	0	1	0	0	1	0	1	1	0	1	1	0	0	1	1	0	0
Palaeomastodon	0	0	0	0	0	?	1	1	1	1	?	?	?	?	?	1	?	0	?	1	0	1	?	?	?	?	?	?	1	?	?
Phiomia	0	0	0	0	0	1	1	1	1	1	0	0	1	1	0	1	0	0	1	1	0	1	0	1	1	0	2	1	1	0	0
Eozygodon	0	0	1	1	1	?	?	?	?	?	1	?	?	?	?	?	0	?	?	1	?	?	?	?	?	?	?	?	?	?	?
Zygolophodon	1	0	1	1	1	1	1	?	?	?	1	?	?	?	?	?	0	?	1	1	?	1	?	1	?	?	?	?	?	?	?
Mammut	1	0	1	1	1	1	1	1	1	1	1	0	1	2	0	1	0	0	1	1	0	1	0	1	1	1	2	1	1	1	0
Choerolophodon	1	0	0	0	1	1	1	1	1	1	1	0	1	2	0	1	1	0	1	0	0	1	0	1	?	1	2	1	1	1	0
Gomphotherium	1	0	0	0	1	1	1	1	1	1	2	0	1	2	0	1	2	1	1	0	0	1	0	1	1	1	2	1	1	1	0
Serridentinus	1	0	0	0	1	1	1	1	1	1	2	0	1	2	0	1	2	1	1	0	0	1	0	1	1	1	2	1	1	1	0
Amebelodon	1	0	0	0	1	1	1	1	1	1	1	0	1	2	0	1	2	1	1	0	0	1	0	1	1	1	2	1	1	1	0
Platybelodon	1	0	0	0	1	1	1	1	1	1	1	0	1	2	0	1	2	1	1	0	0	1	0	1	1	1	2	1	1	1	0
Serbelodon	1	0	0	0	1	1	1	1	1	1	1	0	1	2	0	1	2	1	1	0	0	1	0	1	1	1	2	1	1	1	0
Protanancus	1	0	0	0	1	1	1	1	?	1	?	0	1	2	0	1	2	1	1	0	0	?	0	1	?	1	2	1	1	1	0
Archaeobelodon	1	0	0	0	1	1	1	1	1	1	1	0	1	2	0	1	2	1	1	0	0	1	0	1	1	1	2	1	1	1	0
Gnathabelodon	?	0	0	0	1	1	1	1	1	1	2	0	1	2	0	1	2	1	1	0	0	1	0	1	?	1	?	1	1	1	0
Sinomastodon	0	0	0	0	1	1	1	1	1	1	2	0	1	2	?	1	2	?	?	0	0	?	0	?	?	?	?	1	?	?	?
Eubelodon	1	0	0	0	1	1	1	1	1	1	2	0	1	2	0	1	2	1	1	0	0	1	0	1	1	1	?	1	1	1	0
Rhynchotherium	1	0	0	0	1	1	1	1	1	1	2	0	1	2	0	1	2	1	1	0	0	1	0	1	1	1	?	1	1	1	0
Stegomastodon	2	0	0	0	1	1	1	1	1	1	2	?	1	2	?	1	2	1	1	0	0	1	?	1	1	1	?	1	1	1	?
Haplomastodon	1	0	0	0	1	1	1	1	1	1	2	?	1	2	?	1	2	?	1	0	0	1	?	1	1	1	?	1	1	1	?
Notiomastodon	1	0	0	0	1	?	1	?	?	?	2	?	1	?	?	?	?	?	?	0	0	1	?	?	?	?	?	1	?	?	?
Cuvieronius	1	0	0	0	1	1	1	1	1	1	2	0	1	2	0	1	2	1	1	0	0	1	0	1	1	1	?	1	1	1	0
Tetralophodon	1	0	0	0	1	1	1	1	1	1	2	?	1	2	?	1	2	1	1	0	0	1	?	1	?	1	2	1	1	1	?
Anancus	1	0	0	0	1	1	1	1	1	1	2	?	1	2	?	1	2	1	1	0	1	1	?	1	?	1	2	1	1	1	?
Paratetralophodon	2	0	0	0	1	1	1	1	1	1	2	?	1	2	?	1	2	1	1	0	1	1	?	1	?	1	1	1	1	1	?
Stegolophodon	1	0	0	0	1	1	1	1	1	1	2	0	1	2	0	1	2	1	1	0	1	1	0	1	1	1	2	1	1	1	0
Stegodon	2	0	0	0	1	1	1	1	1	1	2	0	1	2	0	1	2	1	1	0	1	1	?	1	1	1	2	1	1	1	?
Stegotetrabelodon	2	1	0	0	1	?	1	1	1	1	?	?	?	?	?	?	?	?	1	0	1	1	?	1	?	1	?	1	?	1	?
Stegodibelodon	2	1	0	0	1	?	?	?	?	?	?	?	?	?	?	?	?	?	?	?	?	?	?	?	?	?	?	?	?	?	?
Primelephas	2	1	0	0	1	?	?	?	?	?	?	?	?	?	?	?	?	?	?	?	?	?	?	?	?	?	?	?	?	?	?
Loxodonta	2	1	0	0	1	1	1	1	1	1	2	0	1	2	0	1	2	1	1	0	1	1	0	1	1	1	1	1	1	1	0
Elephas	2	1	0	0	1	1	1	1	1	1	2	1	1	2	1	1	2	1	1	0	1	1	1	1	1	1	1	1	1	1	1
Mammuthus	2	1	0	0	1	1	1	1	1	1	2	1	1	2	1	1	2	1	1	0	1	1	1	1	1	1	?	1	1	1	1

Character --> Taxon	94	95	96	97	98	99	100	101	102	103	104	105	106	107	108	109	110	111	112	113	114	115	116	117	118	119	120	121	122	123
OutgroupA	0	0	0	0	0	0	0	0	0	0	0	0	0	0	0	0	0	0	0	0	0	0	0	0	0	0	0	0	0	0
OutgroupB	0	0	0	0	0	0	0	0	0	0	1	0	0	0	0	0	0	0	0	0	0	0	0	1	0	0	0	0	0	0
Anthracobune	0	0	?	?	?	?	?	?	0	1	1	?	?	?	?	?	?	?	?	?	?	?	?	1	0	0	0	1	0	0
Moeritheriidae	0	0	0	0	0	?	?	?	?	1	1	0	0	1	?	0	?	0	?	?	1	0	?	1	0	?	0	1	1	1
Numidotheriidae	0	0	0	1	0	?	?	?	?	1	1	1	0	1	1	1	1	0	?	?	1	1	1	1	1	0	1	1	1	1
Barytheriidae	0	0	?	?	?	?	?	?	?	1	1	1	?	1	1	1	?	?	?	?	1	1	1	1	1	1	1	?	1	1
Deinotheriidae	0	0	0	0	0	0	?	0	?	1	1	1	0	1	1	1	?	0	?	?	1	1	1	1	1	1	1	1	1	1
Palaeomastodon	1	2	0	?	?	?	?	?	?	1	1	1	?	1	1	1	?	?	?	?	1	1	1	1	1	1	1	1	1	1
Phiomia	1	1	0	1	0	?	?	?	?	1	1	1	0	1	1	1	?	0	?	?	1	1	1	1	1	1	1	1	1	1
Eozygodon	?	1	?	1	?	?	?	?	?	?	?	?	0	?	?	?	?	0	?	?	?	?	?	1	1	1	1	2	?	1
Zygolophodon	1	1	0	1	?	?	?	?	?	?	?	?	0	?	?	?	?	0	?	?	?	?	?	1	?	?	?	?	?	?
Mammut	2	1	0	1	1	0	?	0	?	1	1	1	0	1	1	1	?	0	?	?	1	1	1	1	1	1	1	2	1	1
Choerolophodon	1	1	1	1	1	?	?	?	?	1	?	1	1	1	1	1	?	1	?	?	1	1	?	1	1	1	1	2	1	1
Gomphotherium	1	1	1	1	1	1	?	0	?	1	1	1	1	1	1	1	?	1	?	?	1	1	1	1	1	1	1	1	1	1
Serridentinus	1	1	1	1	1	1	?	0	?	1	1	1	1	1	1	1	?	1	?	?	1	1	1	1	1	1	1	1	1	1
Amebelodon	1	1	1	1	1	1	?	0	?	1	1	1	1	1	1	1	?	1	?	?	1	1	1	1	1	1	1	2	1	1
Platybelodon	1	1	1	1	1	?	?	?	?	1	1	1	1	1	1	1	?	1	?	?	1	1	1	1	1	1	1	2	1	1
Serbelodon	1	1	1	1	1	1	?	0	?	1	1	1	1	1	1	1	?	1	?	?	1	1	1	1	1	1	1	2	1	1
Protanancus	?	1	?	1	1	?	?	?	?	?	?	?	?	?	?	?	?	?	?	?	1	1	?	?	?	?	?	?	?	?
Archaeobelodon	1	1	1	1	1	?	?	?	?	1	1	1	?	1	1	1	?	?	?	?	1	1	1	1	1	1	1	2	?	1
Gnathabelodon	1	1	1	1	1	?	?	?	?	1	?	1	1	1	1	1	?	1	?	?	1	1	1	1	1	1	1	?	1	1
Sinomastodon	2	1	1	1	?	?	?	?	?	1	1	1	1	1	1	1	1	1	?	?	1	1	1	1	1	1	1	2	1	1
Eubelodon	1	1	1	1	1	?	?	?	?	1	1	1	1	1	1	1	?	1	?	?	1	1	1	1	1	1	1	?	1	1
Rhynchotherium	1	1	1	1	1	?	?	?	?	1	1	1	?	?	?	?	?	?	?	?	?	1	1	?	1	1	1	?	?	1
Stegomastodon	2	1	1	1	1	?	?	?	?	1	?	1	?	?	?	?	?	?	?	?	?	?	1	1	1	1	1	2	1	1
Haplomastodon	2	1	1	1	1	?	?	?	?	1	?	1	?	1	1	?	?	?	?	?	1	1	1	1	1	1	1	2	1	1
Notiomastodon	2	1	1	1	?	?	?	?	?	?	?	?	?	?	?	?	?	?	?	?	?	?	?	?	?	?	?	?	?	?
Cuvieronius	2	1	1	1	1	1	?	0	?	1	?	1	?	1	1	1	?	?	?	?	?	?	1	?	?	1	?	?	?	1
Tetralophodon	1	1	1	1	1	?	?	?	?	1	1	1	1	1	1	1	?	1	?	?	1	1	1	1	1	1	?	?	?	1
Anancus	2	1	1	1	1	?	?	?	?	1	1	1	1	1	1	1	?	1	?	?	1	1	1	1	1	1	1	2	1	1
Paratetralophodon	?	1	?	1	1	?	?	?	?	?	?	?	?	?	?	?	?	?	?	?	?	?	?	?	?	?	?	?	?	?
Stegolophodon	?	?	1	1	1	?	?	?	?	?	?	?	?	?	?	?	?	?	?	?	?	?	?	?	?	?	?	?	?	?
Stegodon	2	1	1	1	1	1	?	0	?	1	1	1	1	1	1	1	?	1	?	1	1	1	1	1	1	1	1	2	1	1
Stegotetrabelodon	1	1	1	1	1	?	?	?	?	?	1	1	1	1	1	1	?	?	?	?	?	?	?	?	?	?	?	?	?	?
Stegodibelodon	2	1	1	1	?	?	?	?	?	?	?	?	?	?	?	?	?	?	?	?	?	?	?	?	?	?	?	?	?	?
Primelephas	?	?	?	1	?	?	?	?	?	?	1	?	?	?	?	?	?	?	?	?	?	?	?	?	?	?	?	?	?	?
Loxodonta	2	1	1	1	1	1	1	0	?	1	1	1	1	1	1	1	1	1	?	1	1	1	1	1	1	1	1	2	1	1
Elephas	2	1	1	1	1	1	1	1	1	1	1	1	1	1	1	1	1	1	1	1	1	1	1	1	1	1	1	2	1	1
Mammuthus	2	1	1	1	1	1	1	1	1	1	1	1	1	1	1	1	1	1	1	1	1	1	1	1	1	1	1	2	1	1

Appendix 16.3: Hypotheses testing among proboscidean taxa[a]

Test no.	Hypotheses tested	Ordered [b]	Unordered [b]
0.	One of the most parsimonious trees; Figs 16.1a and 16.1b [1]	239[b]	227[b]
1.	*Numidotherium* + *Barytherium* [2.1]	243 (4)	231 (4)
2.	*Numidotherium* + *Barytherium* + *Deinotherium* [2.1]	252 (13)	238 (11)
3.	*Deinotherium* + *Barytherium* [2.2]	249 (10)	233 (6)
4.	*Deinotherium* + *Barytherium* + *Numidotherium* [2.1]	254 (15)	239 (12)
5.	All three (*Deinotherium* + *Barytherium* + *Numidotherium*) collapsed [2.1]	255 (16)	239 (12)
6.	*Phiomia* + *Palaeomastodon* [3]	245 (6)	229 (2)
7.	*Phiomia* as a sister-group to node M (in Figs 16.1a and 16.1b) [3.1]	260 (21)	236 (9)
8.	Mammutidae one family (*Eozygodon* + *Zygolophodon* + *Mammut*)[4]	244 (5)	227 (0) &
9.	*Choerolophodon* as a sister-group of node O (in Fig. 16.1b) [5]	243 (4)	229 (2)
10.	*Choerolophodon* as a sister-group of node O (in Fig. 16.1a) [5]	243 (4)	229 (2)
11.	*Choerolophodon* as a sister-group of node R (in Fig. 16.1a) or P (in Fig. 16.1b) [5]	244 (5)	229 (2)
12.	*Choerolophodon* as a sister-group of node S (in Fig. 16.1a) or Q (in Fig. 16.1b) [5]	244 (5)	229 (2)
13.	*Gnathabelodon* + clade R (in Fig. 16.1a) or S+Q (in Fig. 16.1b) [5]	240 (1)	228 (1)
14.	Test no. 13, *Sinomastodon* with R (in Fig. 16.1a) or S+Q+P (in Fig. 16.1b) [5]	240 (1)	228 (1)
15.	Test no. 14, R joined to O (in Fig. 16.1a) or S+Q+P+N (in Fig. 16.1b) [5, 6]	240 (1)	230 (3) &
16.	Test no. 15 + *Choerolophodon* [5]	245 (6)	231 (4)
17.	Test no. 16 + *Tetralophodon* [5]	254 (15)	241 (14)
18.	Test no. 17 + *Anancus* [5]	256 (17)	244 (17)
19.	Test no. 18 + *Paratetralophodon* (= 'the wastebasket hypothesis') [5]	256 (17)	243 (16)
19.1	As in test 19, but *Paratetralophodon*, *Anancus*, and *Tetralophodon* grouped	254 (15)	242 (15)
19.2	As in test 0, but *Paratetralophodon*, *Anancus*, and *Tetralophodon* grouped	242 (3)	230 (3)
20.	*Rhynchotherium* as sister-group of O (in Fig. 16.1a) [5]	242 (3)	N/A
21.	*Rhynchotherium* as sister-group of O (in Fig. 16.1b), including *Sinomastodon* [5]	242 (3)	227 (0)
22.	*Amebelodon* + *Serbelodon* [7]	239 (0)	228 (1)
23.	*Amebelodon* + *Serbelodon* + *Platybelodon* collapsed [8]	239 (0)	228 (1)
24.	Test no. 23 + *Archaeobelodon* + *Protanancus* (or subfamily Amebelodontinae) [8]	239 (0)	228 (1)
25.	Test no. 24 all five amebelodontine taxa collapsed [9]	242 (3)	231 (4)
26.	*Eubelodon* joins Amebelodontinae	239 (0)	229 (2)
27.	((Am + S) (Ar + Pr) (P1)) (= Amebelodontinae, alternative arrangement)	240 (1)	229 (2)
28.	Nodes M and N in Figs 16.1a and 16.1b collapsed	243 (4)	229 (2) ? &
29.	*Anancus* + *Paratetralophodon* + *Stegodon* + *Stegolophodon* + *Tetralophodon* collapsed [10]	251 (12)	238 (11)
30.	*Stegodon* and *Stegolophodon* separated [11]	241 (2)	228 (1) ?&
31.	*Stegodon* as a sister-group of nodes Z.1 (in Figs 16.1a and 16.1b) [12]	241 (2)	230 (3)
32.	Test no. 31, and also *Anancus* and *Paratetralophodon* switch places [13]	239 (0)	228 (1)
33.	Stegodontidae as a sister-group of Mammutidae [14]	263 (24)	242 (15)
34.	Subfamily Stegotetrabelodontinae (= *Stegotetrabelodon* + *Stegodibelodon*) [15]	241 (2)	229 (2) ?&
35.	A cladogram combining tests with & (test nos 0 and 15 for unordered option)[16]	NA	230 (3) [c]
35.1	A cladogram combining tests with & (test nos. 0, 8, and 15 for ordered option) [17]	245 (6)	NA
35.2	As in test no. 35, but Mammutidae is paraphyletic [18]	NA	230 (3)
35.3	Test 35, but *Haplomastodon* joins *Cuvieronius*, or *Stegomastodon* (within node N) [19]	241 (2)	230 (3)
35.4	Test 35 rearranged to test Tassy's hypothesis (cf. Figs 16.4a, 16.4b to Fig. 16.5) [20]	269, 272	250, 253

a The sequence of testing is in the direction from outgroups towards *Mammuthus*. The 39 tests presented in this appendix are a sample of over 250 sets conducted 'manually' with the MacClade program.

b 'Unordered' and 'Ordered' refer to the sequence of transformations from one multistate character to another (unordered, it costs only one step in all characters; ordered, it may cost more than one step, depending on the root of the tree). The numbers in both columns are the length of the cladogram (or the minimum number of evolutionary steps required for a given tree topology) obtained when running PAUP, and the numbers in ( ) are the differences between this tree topology and one of the most parsimonious trees, test no. 0. Each hypothesis testing was performed on the original tree; in other words, after one test was completed, the branching arrangement was returned to its original topology before conducting the next test. Other details are given under 'Methods'.

& This tree topology was incorporated in the near-most parsimonious cladograms (Fig. 16.5).

c This is a near-most parsimonious cladogram (incorporating tests designed with '&', not '?&') depicted in Fig. 16.5; the score for this tree is the same as that of the 50 per cent majority-rule tree from unordered characters. A classification, Table 16.1, was constructed based on Fig. 16.5 (details in the text).

1 The consensus tree obtained with the 'unordered' option was extremely uninformative — one node (K) was polytomous with 14 clades radiating from one point (Fig. 16.3b), whereas in the ordered option (Fig. 16.3a), this 14-split polytomy was rearranged into four clades, one of which (node R) is also uninformative. For this reason, I adopted the branching arrangement shown in Figs 16.1a and 16.1b as starting points for testing phylogenetic hypotheses, hoping to resolve the polytomous nodes.

2.1 A hypothesis which could have been interpreted from the discussion of Mahboubi *et al.* (1984, 1986). 'Collapsed' means that there is a polytomous bifurcation at the node where the three taxa are joined. In some cases, a polytomous fork for a given tree topology costs less than the most parsimonious solution for the tree (Maddison 1989).

2.2 A hypothesis suggested or implied by Harris (1978).

[3] Hypothesis suggested or implied by Andrews (1906), Lehmann (1950), and also by Coppens *et al.* (1978); see also Matsumoto (1924*b*).
[3.1] Towards understanding or testing the hypothesis that *Phiomia* was on the line of ancestry to gomphothere taxa (e.g. Tobien 1976; Coppens *et al.* 1978).
[4] Node K in Fig. 16.5 — hypothesis proposed by Tassy (1990*a*; this volume, Chapter 6). With the unordered option Mammutidae is a monophyletic taxon in most of the 600 equally parsimonious trees, as expressed in the 50 per cent majority-rule consensus tree (Fig. 16.2b) but not in the strict consensus tree (Fig. 16.3b).
[5] Towards grouping all gomphothere taxa — suggested/proposed by many authors, or combination of ideas — 'the wastebasket hypothesis' (e.g. Falconer 1857; Gaudry 1866; Vacek 1877; Cope 1889; Matsumoto 1924*a*, 1925, Osborn 1936, 1942; Tobien 1972, 1973*a*, 1973*b*, 1975, 1976, 1978*b*; Tobien *et al.* (1986, 1988).
[6] The union of these nodes will become a part of node N in Fig. 16.5.
[7] Testing the hypothesis of Tobien (1972) that *Serbelodon* is a synonym of *Amebelodon*. Note that even though there is no change in the score of the tree in the ordered option, there are no synapomorphies assigned for this clade in this particular branching arrangement.
[8] The beginning of forming a clade to include taxa listed in note [9] below.
[9] Amebelodontinae (formerly Amebelodontidae) is composed of *Amebelodon*, *Platybelodon*, *Serbelodon*, *Archaeobelodon*, and *Protanancus* — a hypothesis suggested by Tassy (1986); see also Tobien *et al.* (1986, 1988).
[10] Towards understanding hypotheses tested in notes [11], [12], and [13] below.
[11] *Stegodon* as a sister-group of Elephantidae (nodes Y in Figs 16.1a and 16.1b).
[12] A part of a hypothesis proposed by Kalb *et al.* (this volume, Chapter 12A).
[13] A hypothesis proposed by Kalb *et al.* (this volume, Chapter 12A).
[14] A hypothesis suggested by Maglio (1973) and Tobien (1988).
[15] A hypothesis proposed by Maglio (1973).
[16] Results from this test are incorporated in Fig. 16.5 (see also notes under [c] above).
[17] Note the large difference (six evolutionary steps) as compared to the small difference (three evolutionary steps) between this test and test no. 35.
[18] Note that there is no difference in the score for the final tree when Mammutidae is monophyletic or paraphyletic (test no. 35).
[19] This possible 'shuffling' of genera within Cuvieroniiae was predicted by Simpson and Paula-Couto (1957).
[20] The first number in each set refers to testing when Mammutidae is monophyletic, and the second set refers to testing when Mammutidae is paraphyletic. Although the differences between these tests and those in test no. 35 are large (in the ordered options), until detailed studies are conducted, I consider the hypothesis of Tassy (1990*a*, and this volume, Chapter 6; cf. Figs 16.4a,b) equally valid to the one presented here because much data are lacking in the matrix.

Appendix 16.4: Synapomorphies for selected clades in Fig. 16.5

A complete list of apomorphies (characters on lines to terminal taxa) and synapomorphies (characters on lines shared by sister-groups) and their transformations appear at the end of this appendix. Selection of nodes and discussion in this appendix are focused on gomphotheres and those taxa and/or clades which may have had characters in common with them. Also note that discussion is limited to those characters which produced conflicting results between or among results presented here and published reports. To avoid duplication, I also limited my discussion on characters treated by Tassy (this volume, Chapter 6) and by Tassy and Shoshani (1988). Appendix 16.1 provides a complete list of characters employed in this study, their primitive and derived conditions, and references consulted. Appendix 16.2 contains the data matrix upon which cladistic analyses were conducted. Sometimes the change of character state for a particular node is given, e.g. '(12:3→4)' means that, for this node, the synapomorphy for multistate character 12 is a change from state 3 to state 4; in binary characters the change is either from 0 → 1, or from 1 → 0 in the case of a reversal, *underlined numbers* refer to homoplasic characters. Nodes below are grouped according to those which appear on Fig. 16.5.

Nodes A–B

Characters for these nodes have been discussed by Tassy and Shoshani (1988) and by Fischer and Tassy (1993).

Node C:

Anthracobune and all taxa to the right of it. Characters 103:0 → 1, large and hooked scapular coracoid process; and (121:0 → 1), presence of astragalar *tuberculum mediale*. A homoplasic character (41:0 → 1, third lower molars with two definite cuspids at the posterior end (a labially displaced hypoconulid and a large postentoconulid, or entoconid II) may be assigned to this node and also to node H. A discussion on the taxonomic position of *Anthracobune* is given in Shoshani *et al.* (this volume, Chapter 8); see also note [c] in Table 16.1.

Nodes D–G

Characters for these nodes have been discussed by Tassy (1990*a*; this volume, Chapter 6) and by Tassy and Shoshani (1988).

Node H

Elephantiformes, i.e. *Palaeomastodon* and all taxa to the right of it. Eleven characters (40, 41, 44, 45, 47, 48, 53, 54, 82, 83 and 94) two of which are reversals, see, however, Lehmann and Kraus (1994).

Note I

Phiomia and all taxa to the right of it. Eleven characters (3, 9, 10, 12, 29, 33, 38, 75, 89, 95, and 97).

Node J

'Elephantoidea' of Tassy (1988*a*), i.e. Mammutidae and all taxa to the right of it. Thirteen characters (22, 23, 27, 34, 42, 46, 67, 73, 76, 88, 92, 98, and 121).

Node K

Family Mammutidae (*Eozygodon*, *Zygolophodon*, and *Mammut*). Characters 35, 62, 65, and 66. Tobien's treatment of zygodont molars (this volume, Chapter 9) and details in Tassy (1990*a*; this volume, Chapter 6) provide a better understanding of these synapomorphies.

Node L

Choerolophodon and all taxa to the right of it. Eight characters (10, 35, 37, 63, 82, 96, 106 and 111), one is a reversal.

Node M

All gomphotheres (except *Choerolophodon*), Stegodontidae, and Elephantidae. Characters (13, 15, 73, 79, 80, and 99). Three of these characters have a consistency index (CI) of 1.00, the other three have a CI of less than 1.00 (i.e. they occur elsewhere on the cladogram). Character 80 (the nasal fossa extends laterally up to or beyond the width of the rostrum) appears to be a complex feature.

Node N

Fifteen trilophodont gomphotheres. Character (34.3 → 4), i.e. M^3 changes from trilophodont to tetralophodont (Tassy, personal communication 1993) noted that this is a homoplasic character since it does not apply to *Gomphotherium annectens* Group which means that there is a reversal from node N to node N.2 in Fig. 16.5). At the same time, the intermediate cheek teeth (dp4, M1, M2) have only three lophs, hence the name 'tri-lophodont'. Note that *Gnathabelodon* has 27 out of 123 characters marked with '?' (missing data), and that joining this genus with other gomphothere taxa within node N costs only one evolutionary step (test no. 13 in Appendix 16.3). In Fig. 16.5, *Gnathabelodon* can be either a sister-group to node N.1 or to node N.7.

Node N.1

Gomphotherium and *Serridentinus* (node N.2), and Amebelodontidae (node N.3). Character (<u>10</u>.4 → 2), longitudinal band of enamel on upper tusks. Note that state 4, lack of enamel on upper tusks, is a derived condition, since primitively, the enamel surrounds the tusk; thus 4 → 2 is a reverse condition.

Appendix 16.4 *(contd)*

Node N.2

Gomphotherium and *Serridentinus*. Characters (34:4 → 3), 37:1 → 0, and (121:2 → 1). All three characters are reversals (no. 34 occurred in node N above), based on this particular branching arrangement. There is only one character which distinguishes *Serridentinus* from *Gomphotherium* (number 15; in the former, the cross-section of I_2 is roundish, whereas in the latter it is pyriform). Except for a few of the 600 'unordered' equally parsimonious trees (and consequently the strict consensus tree with unordered characters), in all other unordered equally parsimonious trees and in all parsimonious and consensus trees with 'ordered' option, these two taxa always grouped. These results led me to conclude that *Serridentinus* is a junior synonym of *Gomphotherium*, a hypothesis proposed by Tobien (1972, p. 176).

Node N.3

Amebelodontidae. (After Tassy 1990*a*; this volume, Chapter 6.) Characters (15:2 → 0), 16:0 → 1, and (73:2 → 1). Character 55 (55:0 → 1) may also be assigned to this node. Included in this family are *Amebelodon, Platybelodon,* (?) *Serbelodon, Protanancus,* and *Archaeobelodon.* Character number 73:2 → 1 (loss of sagittal crest), also occurs in *Choerolophodon* and Mammutidae. Number 15:2 → 0 is a reversal to the primitive condition (flat lower tusks) because *Mammut* and *Zygolophodon,* which possessed round or oval lower tusks, occur before clade N on the cladogram. The derived condition for character 16 — straight medial edge of lower tusks — also occurs in *Phiomia*, one of the features which led investigators to suggest that *Phiomia* was on the line of ancestry to gomphothere taxa (e.g. Tobien 1976; Coppens *et al.* 1978). Tassy (1990*a*; this volume, Chapter 6) noted that characters 16 and 55 — the presence of pretrite and posttrite conules on the molars — are synapomorphies for this family, although there are intra- and interfamilial homoplasies (the same problem applies to M^3 with three or four lophs, see notes under node N above). Taxa which exhibit derived states, either '1' or '2', for character 55 are: *Archaeobelodon, Protanancus, Platybelodon, Amebelodon, Eubelodon, Rhynchotherium, Stegomastodon, Haplomastodon, Notiomastodon, Cuvieronius, Tetralophodon, Anancus,* and *Paratetralophodon.*

The synonymy of *Serbelodon* has been discussed by Tobien (1972, pp. 179–80), who concluded that *Serbelodon* is not a valid genus and should be synonymized with *Amebelodon*. I tested this hypothesis by forcing *Serbelodon* and *Amebelodon* in one clade to see which synapomorphies would be assigned to this clade — there were none. Nevertheless, when I forced these two genera in one clade, the score of the tree either did not change (with the unordered option) or was raised by one evolutionary step (with the ordered option; test no. 22 in Appendix 16.3). These results are inconclusive, and, for the present time, I tentatively accept *Serbelodon* as a valid genus.

Based on the above discussion, there is one character which distinguishes *Serbelodon* from *Amebelodon* (number 55 — the former has no posttrite central conules). Future investigators may wish to pursue the question of synonymy between *Serbelodon* and *Amebelodon* and between *Serridentinus* and *Gomphotherium*, and to consult these references: Frick (1933, p. 594–5, figs 14, 16, 27), Osborn (1936, p. 453), Tassy (1983, p. 99–136), and Tobien (1972, pp. 162, 179–80).

Node N.7

Sinomastodon and all taxa to the right of it. Character (27:1 → 2). According to Tobien *et al.* (1986, 1988), *Sinomastodon* is closely related to gomphotheres found in South America (here subfamily Cuvieroniinae), a hypothesis which was tested and corroborated only if a limited number of characters for *Sinomastodon* is included. None the less, once the matrix is denser for *Sinomastodon* (data after Zong *et al.* 1989), results show that *Rhynchotherium* and *Eubelodon* are closer to Cuvieroniinae than *Sinomastodon* (Fig. 16.5).

Node N.8

Eubelodon, Rhynchotherium, and Cuvieroniinae. Characters (10:4 → 2), and (55:0 → 2). Character 10 appeared at node N.1, and 55 can be assigned to node N.3.

Node N.9

Rhynchotherium and Cuvieroniinae. Character 8:0 → 1 (helicoidal upper tusks) appears to be a very good synapomorphy for *Rhynchotherium* and Cuvieroniinae. Note that the term 'spiral' which is often used in the literature to describe the twisting of the upper tusks (e.g. Simpson and Paula-Couto 1957; Miller 1990) is here replaced by 'helicoidal' which is a better term according to R. Hoffstetter (personal communication 1991, via P. Tassy).

Node N.10

Cuvieroniinae (emended family of Cabrera, 1929). Characters (27:2 → 3), (34:4 → 5), and (94:1 → 2). Character 27 (the absence of second, third, and fourth premolars) occurs elsewhere on the cladogram. Character number 94 (mandibular symphysis short and spout-like), even though it is homoplasic within Proboscidea as a whole, is a good synapomorphy for this clade.

Originally Cabrera's subfamily included *Cuvieronius* and *Notiomastodon* (Simpson 1945, p. 133). Simpson and Paula-Couto (1957) added *Stegomastodon* and *Haplomastodon* to this assemblage, a concept which is corroborated in this study, as shown in Fig. 16.5 (see also Casamiquela *et al.*, this volume, Chapter 31). The closest taxon to this clade is *Rhynchotherium* (based on the 50 per cent majority-rule consensus tree). It is also noted that the exact relationship within the South American complex is not entirely clear, because certain changes in the branching arrangements either slightly change the score, or do not alter the score of the final tree (cf. test nos 35 and 35.3; observations also made by Simpson and Paula-Couto 1957).

Appendix 16.4 (*contd*)

Further, based strictly on parsimony analysis, the phylogenetic position of the clade containing the South American gomphothere within the Proboscidea appears to be uncertain. There were two major possibilities of branching arrangements of Cuvieroniinae within Proboscidea in all the most parsimonious solutions and the consensus trees. Cuvieroniinae was either close to all gomphotheres (excluding *Choerolophodon*), Stegodontidae, and Elephantidae (Fig. 16.1a), or it was a sister-group to the clade containing tetralophodont gomphotheres, Stegodontidae and Elephantidae (Fig. 16.1b). As expected, the consensus trees (Figs16.3a and 16.3b) also depict these alternatives.

After the analysis for this chapter was conducted, Haruo Saegusa and the author agreed that it would help to investigate taxa grouped in node J of Fig. 16.5 — specifically those within nodes N.10 and Node O — for the possible presence of three or more loph(id)s in dp3. This character was observed in members of Cuvieroniinae (node N.10) by Boule and Thevenin (1920, pp. 37, 39, and plate IV, figs 4, 10) who called the dp^3 '*avant-dernière molaire de lait*'. This particularity in South American gomphotheres was recalled by Tassy (1990*a*, p. 215).

Node O

Tetralophodont gomphotheres (*Tetralophodon, Anancus, Paratetralophodon*), Stegodontidae, and Elephantidae. Nine characters (24, 25, 31, 32, 33, 34, 39, 40, and 55), of which only two have CIs less than 1.00. Overall, this clade is based on solid characters, some of which are good or very good (numbers ?24, 25, 31, 33, 39, and 40 — all are dental characters and a few involve the number of lophs on M^1s and M^2s). These findings are not new; they were observed by Tassy (1990*a*; this volume, Chapter 6). Kalb *et al.* (this volume, Chapter 12A) also discussed the relationships within this clade and reached similar conclusions, although they noted that Stegodontidae is a paraphyletic group. Results presented here and those of Tassy and Kalb *et al.* corroborate the hypothesis that Stegodontidae is more closely related to gomphotheres and elephantids than to mammutids (*contra* Maglio 1973; Tobien 1988).

Node P

Anancus, Paratetralophodon, Stegodontidae, and Elephantidae. Characters are: <u>56</u>, <u>83</u>, and (<u>94</u>:1 → 2), all with CIs of 0.333. Character 56 (the absence of a cingulum on the medial side of the upper molars) also occurs in Deinotheriidae and *Gnathabelodon*, and character 83 (the anterior border of the orbit situated forward of M^1) also occurs in *Moeritherium, Numidotherium*, Barytheriidae, and Deinotheriidae. These two examples show that when the taxa and characters employed are only a part of the whole picture, it may not be possible to 'discover' homoplasies, and further, it is also possible that the polarities of certain characters may be changed and thus alter the results.

Node Q

Paratetralophodon, Stegodontidae, and Elephantidae. Character (<u>63</u>:1 → 2) (the presence of abundant cement in the crown of premolars and molars), which also appeared as an autapomorphic for *Stegomastodon* (within node N.10). On the other hand, *Stegolophodon*, a taxon within node S, has only little cement present (<u>63</u>:2 → 1). It appears that the phylogenetic position of *Stegolophodon* needs to be investigated further.

Node R

Stegodontidae and Elephantidae. Characters (<u>55</u>:1 → 0), 59, (60:0 → 1), and 113. Character state 0 of number 55 (absence of trefoils in molars) is a reversal of the primitive condition because this trefoil pattern has disappeared on these taxa. To the best of my knowledge, there is no 'objective' method of coding this progression (from presence to absence of trefoils in advanced taxa) without influencing the polarity of the character and thus the results. Characters 59 (adaxial and abaxial cones and their conelets are of about equal size) and 60 (molars with laminae) are good characters (see Appendix 16.1 for references). Character 113 (the presence of a well-delineated distal obturator notch on the pelvis) is not known in *Stegolophodon, Stegotetrabelodon, Stegodibelodon*, or *Primelephas*.

Nodes S

Stegodontidae. Characters 57 and 58. These characters — accentuated brachyodont molars (57), and fusion of anterior pretrite central conule and adaxial conelets (58) — appear to be very good synapomorphies for the Stegodontidae. It is noted that the grouping of *Stegodon* and *Stegolophodon* in Stegodontidae, although strongly defended by Saegusa (this volume, Chapter 17), based on my parsimony analysis, is not strongly supported because with either one or two evolutionary steps (with the unordered and the ordered options), *Stegodon* and *Stegolophodon* can be separated and *Stegodon* will join the Elephantidae clade (test no. 30 in Appendix 16.3). The hypothesis proposed by Kalb *et al.* (this volume, Chapter 12A) remains a viable alternative to the topology shown in Fig. 16.5, a hypothesis which requires two (with the ordered) or three steps (with the unordered) (test nos 31 and 32 in Appendix 16.3).

Node T

Elephantidae. Three characters (11, 27, and 64), of which 64 (hypsodont molars) appears to be the best. Character 11 (lower tusks are straight when viewed laterally) is extremely homoplasic. For example, straight lower tusks were present in *Barytherium, Zygolophodon, Mammut*, and *Tetralophodon*. The content of the Elephantidae was discussed by Maglio (1973), Tassy and Shoshani (1988), Froehlich and Kalb (1995), Tassy (this volume, Chapter 6), Kalb *et al.* (this volume, Chapter 12A), and Todd and Roth (this volume, Chapter 18).

Nodes U–X

Characters for these nodes have been discussed by Tassy (1990*a*; this volume, Chapter 6), Kalb *et al.* (this volume, Chapter 12A), and by Tassy and Shoshani (1988).

Branch (or Node)	Character	Steps	CI	Change
node B	83	1	0.333	0 ==> 1
	84	1	1.000	0 ==> 1
	87	1	1.000	0 ==> 1
	104	1	1.000	0 ==> 1
	117	1	1.000	0 ==> 1
node C	103	1	1.000	0 ==> 1
	121	1	0.667	0 ==> 1
C --> *Anthracobune*	41	1	0.500	0 ==> 1
node C --> node D	4	1	1.000	0 --> 1
	5	1	1.000	0 --> 1
	10	1	0.400	0 --> 1
	20	1	1.000	0 ==> 1
	1	1	1.000	0 ==> 1
	26	1	1.000	0 ==> 1
	36	1	1.000	0 ==> 1
	51	1	1.000	0 ==> 1
	52	1	1.000	0 ==> 1
	68	1	1.000	0 --> 1
	69	1	1.000	0 --> 1
	70	1	1.000	0 --> 1
	1	1	1.000	0 --> 1
	6	1	1.000	0 --> 1
	91	1	1.000	0 --> 1
	107	1	1.000	0 --> 1
	114	1	1.000	0 --> 1
	122	1	1.000	0 ==> 1
	123	1	1.000	0 ==> 1
D --> Moeritheriidae	6	1	0.750	0 --> 1
	19	1	0.500	0 ==> 1
	29	1	1.000	0 --> 2
	44	1	0.667	0 ==> 2
node D --> node E	2	1	1.000	0 ==> 1
	17	1	1.000	0 --> 1
	29	1	1.000	0 --> 1
	53	1	0.500	0 ==> 1
	71	1	1.000	0 ==> 1
	105	1	1.000	0 ==> 1
	108	1	1.000	0 --> 1
	109	1	1.000	0 ==> 1
	110	1	1.000	0 --> 1
	115	1	1.000	0 ==> 1
	116	1	1.000	0 --> 1
	118	1	1.000	0 ==> 1
	120	1	1.000	0 ==> 1
E --> Numidotheriidae	97	1	0.500	0 --> 1
node E --> node F	1	1	1.000	0 ==> 1
	18	1	1.000	0 ==> 1
	20	1	1.000	1 ==> 2
	119	1	1.000	0 ==> 1
F --> Barytheriidae	6	1	0.750	0 --> 1
	11	1	0.333	0 ==> 1
node F --> node G	6	1	0.750	0 --> 3
	19	1	0.500	0 ==> 1
	28	1	1.000	0 ==> 1
	30	1	1.000	0 ==> 1
	32	1	1.000	0 ==> 2
	39	1	1.000	0 ==> 1
	72	1	1.000	0 --> 1
	78	1	1.000	0 ==> 1

Branch (or Node)	Character	Steps	CI	Change
	90	1	1.000	0 --> 1
G --> Deinotheriidae	11	1	0.333	0 ==> 2
	13	1	0.333	0 --> 1
	29	1	1.000	1 --> 3
	44	1	0.667	0 --> 1
	56	1	0.333	0 ==> 1
	73	1	0.500	0 --> 1
node G--> node H	40	1	1.000	0 ==> 1
	41	1	0.500	0 ==> 1
	44	1	0.667	0 --> 2
	45	1	0.500	0 ==> 1
	47	1	1.000	0 ==> 1
	48	1	0.500	0 ==> 1
	53	1	0.500	1 ==> 0
	54	1	1.000	0 ==> 1
	82	1	0.500	0 ==> 1
	83	1	0.333	1 ==> 0
	94	1	0.333	0 ==> 1
H --> *Palaeomastodon*	29	1	1.000	1 --> 4
	32	1	1.000	2 ==> 1
	33	1	1.000	0 --> 1
	34	1	0.714	0 --> 1
	95	1	1.000	0 --> 2
node H --> node I	3	1	1.000	0 ==> 1
	9	1	1.000	0 --> 1
	10	1	0.400	1 --> 2
	12	1	0.500	0 --> 1
	29	1	1.000	1 --> 5
	33	1	1.000	0 --> 2
	38	1	1.000	0 ==> 1
	75	1	1.000	0 --> 1
	89	1	0.667	0 --> 2
	95	1	1.000	0 --> 1
	97	1	0.500	0 --> 1
I --> *Phiomia*	16	1	0.500	0 ==> 1
	34	1	0.714	0 --> 2
	42	1	1.000	0 --> 1
	76	1	1.000	0 --> 1
node I --> node J	22	1	1.000	0 ==> 1
	23	1	1.000	0 ==> 1
	27	1	0.300	0 ==> 1
	34	1	0.714	0 --> 3
	42	1	1.000	0 --> 2
	46	1	1.000	0 ==> 1
	67	1	1.000	0 ==> 1
	73	1	0.500	0 --> 1
	76	1	1.000	0 --> 2
	88	1	1.000	0 ==> 1
	92	1	1.000	0 ==> 1
	98	1	1.000	0 ==> 1
	121	1	0.667	1 ==> 2
node J --> node K	35	1	1.000	0 --> 1
	62	1	1.000	0 ==> 1
	65	1	1.000	0 ==> 1
	66	1	1.000	0 ==> 1
K --> *Eozygodon*	13	1	0.333	0 --> 1
node K --> node K.1	11	1	0.333	0 ==> 1
	15	1	0.400	0 --> 2
	63	1	0.333	0 --> 1

Appendix 16.4

Branch (or Node)	Character	Steps	CI	Change
K.1 --> *Mammut*	7	1	0.273	0==>2
	10	1	0.400	2==>4
	12	1	0.500	1==>0
	27	1	0.300	1-->3
	94	1	0.333	1==>2
node J --> node L	10	1	0.400	2==>4
	35	1	1.000	0-->2
	37	1	0.250	0==>1
	63	1	0.333	0-->1
	82	1	0.500	1==>0
	96	1	1.000	0==>1
	106	1	1.000	0==>1
	111	1	1.000	0==>1
L--> *Choerolophodon*	7	1	0.273	0==>2
	27	1	0.300	1-->3
	79	1	1.000	0-->1
node L--> node M	13	1	0.333	0-->1
	15	1	0.400	0-->2
	73	1	0.500	1==>2
	79	1	1.000	0-->2
	80	1	1.000	0==>1
	99	1	1.000	0-->1
node M --> node N	34	1	0.714	3-->4
node N --> node N.1	10	1	0.400	4==>2
node N.1 --> node N.2	34	1	0.714	4==>3
	37	1	0.250	1==>0
	121	1	0.667	2==>1
N.2 --> *Gomphotherium*	15	1	0.400	2==>1
node N.1 --> node N.3	15	1	0.400	2==>0
	16	1	0.500	0==>1
	73	1	0.500	2==>1
node N.3 --> node N.4	14	1	1.000	0==>1
node N.4 --> node N.5	55	1	0.333	0==>2
N.5 --> *Platybelodon*	12	1	0.500	1==>2
node N.3 --> node N.6	55	1	0.333	0==>1
N.6 --> *Archaeobelodon*	27	1	0.300	1==>0
N --> *Gnathabelodon*	7	1	0.273	0==>2
	56	1	0.333	0==>1
node N --> node N.7	27	1	0.300	1==>2
N.7 --> *Sinomastodon*	7	1	0.273	0==>2
	63	1	0.333	1==>0
	94	1	0.333	1==>2
node N.7 --> node N.8	10	1	0.400	4==>2
	55	1	0.333	0==>2
node N.8 --> node N.9	8	1	1.000	0==>1
N.9 --> *Rhynchotherium*	6	1	0.750	3-->2
	11	1	0.333	0-->2
	12	1	0.500	1-->0
	55	1	0.333	2==>1
node N.9 --> node N.10	27	1	0.300	2==>3
	34	1	0.714	4==>5
	94	1	0.333	1==>2
N.10 --> *Stegomastodon*	7	1	0.273	0-->1
	10	1	0.400	2==>4
	63	1	0.333	1==>2
nodeN.10 --> node N.11	7	1	0.273	0-->2
N.11 --> *Haplomastodon*	10	1	0.400	2==>3
node M --> node O	24	1	1.000	0==>1
	25	1	1.000	0==>1
	31	1	1.000	0==>1
	32	1	1.000	2==>3
	33	1	1.000	2==>3
	34	1	0.714	3-->5
	39	1	1.000	1==>2
	40	1	1.000	1==>2
	55	1	0.333	0==>1
O --> *Tetralophodon*	11	1	0.333	0-->1
node O --> node P	56	1	0.333	0==>1
	83	1	0.333	0==>1
	94	1	0.333	1==>2
P --> *Anancus*	7	1	0.273	0==>1
	27	1	0.300	1-->2
node P --> node Q	63	1	0.333	1==>2
Q --> *Paratetralophodon*	89	1	0.667	2-->1
node Q --> node R	55	1	0.333	1==>0
	59	1	1.000	0==>1
	60	1	1.000	0==>1
	113	1	1.000	0-->1
node R --> node S	57	1	1.000	0==>1
	58	1	1.000	0==>1
S --> *Stegolophodon*	10	1	0.400	4-->2
	15	1	0.400	2-->1
	63	1	0.333	2==>1
S --> *Stegodon*	7	1	0.273	0-->2
	10	1	0.400	4-->3
	27	1	0.300	1-->3
	37	1	0.250	1==>0
node R --> node T	11	1	0.333	0-->1
	27	1	0.300	1-->2
	64	1	1.000	0==>1
T --> *Stegotetrabelodon*	7	1	0.273	0-->1
	94	1	0.333	2==>1
node T --> node U	43	1	1.000	0==>1
	48	1	0.500	1==>0
node U --> node V	37	1	0.250	1==>0
node V --> node W	7	1	0.273	0-->2
	49	1	1.000	0==>1
	50	1	1.000	0==>1
	60	1	1.000	1==>2
	89	1	0.667	2-->1
	100	1	1.000	0-->1
W --> *Loxodonta*	27	1	0.300	2==>3
node W --> node X	61	1	1.000	0-->1
	74	1	1.000	0==>1
	77	1	1.000	0==>1
	85	1	1.000	0==>1
	93	1	1.000	0==>1
	101	1	1.000	0==>1
	102	1	1.000	0-->1
	112	1	1.000	0-->1
X --> *Mammuthus*	7	1	0.273	2==>3
	45	1	0.500	1==>0

Note: 'Double-lined arrows mean that a change is unambiguous (occurs in all possible reconstructions). Single-lined arrows mean that a change is ambiguous (occurs under some reconstructions but not others).' (Swofford 1991, p. 77).

SEVENTEEN

Stegodontidae: evolutionary relationships

HARUO SAEGUSA

INTRODUCTION

The family Stegodontidae is composed of the two genera, *Stegolophodon* and *Stegodon*. This family first appeared in Asia during the early Miocene and became one of the dominant faunas of Asia during the Pliocene and Pleistocene. Because of elephant-like similarities, the systematic position of the stegodontids has developed with the prevailing ideas of elephant evolution (e.g. Osborn 1942; Maglio 1973; Tobien 1976; Tassy 1982; Tassy and Shoshani 1988)

Two hypotheses on the systematic position of the Stegodontidae within Elephantoidea are:

(1) the family is related to the zygodont mastodon group Mammutidae (Maglio 1973; Tobien 1976);

(2) the family is allocated to the group comprising gomphotheres and Elephantidae (Tassy 1982; Tassy and Shoshani 1988).

The second hypothesis is supported by cladistic studies, as well as by recently recovered fossil material. Detailed relationships between elephants, *Stegodon*, and gomphotheres are still not clear because the monophyly of the Stegodontidae is refuted by Kalb *et al.* (this volume, Chapter 12A). In this chapter, however, I maintain the traditional view that this family is monophyletic because it seems to be premature to deny the monophyly of this group (see below).

Studies of *Stegodon* cranial morphology by Saegusa (1987) recognized three monophyletic groups within the genus. In this chapter, I first describe the basic molar morphology of *Stegolophodon* and *Stegodon*. Secondly, I make some comments on the systematics of *Stegodon* and *Stegolophodon*. Thirdly, I review briefly East Asian stegodontids as well as the major species of this group from other areas, together with a description of their geographical diversification. Lastly, I discuss the possible evolutionary history of stegodontids, which allows for their comparison with the evolutionary history of elephants.

Abbreviations

BNHM, Beijing Natural History Museum (Beijing, China); IVPP, Institute of Vertebrate Paleontology and Paleoanthropology (Beijing, China); KU, Kyoto University (Kyoto, Japan); THP, Tianjin Natural History Museum at Paiho (Museum of Huangho and Paiho, Tianjin, China) (=present-day Tianjin Natural History Museum).

MOLAR STRUCTURE OF STEGODONTIDAE

HOMOLOGY OF CUSPS OF MOLARS

As with proboscideans in general (Tobien 1975; Tassy 1982), the mesial ridges of *Stegodon* molars are homologous with cusps of primitive ungulates, and the rest of the ridges are incomplete copies of mesial ridges (Taruno 1985). Figure 17.1 is a composite of the overall cusp homology in a generalized mammal and dental evolution and progression in advanced proboscideans. Figure 17.2 provides terminology for molar structures in occlusal surfaces of *Stegolophodon* and *Stegodon* (cf. Proboscidea Research Group 1977).

RIDGE STRUCTURE IN STEGODONTIDAE

The reduction of the central conule of pretrite is well known in *Stegolophodon* (Tobien 1975) and fusion of the anterior central pretrite conule and corresponding mesoconelet (= adaxial cone) seems to be a common characteristic of the Stegodontidae (Tassy 1983, p. 327). There is wide variation in the degree of reduction of the posterior central pretrite conule among the *Stegolophodon* genera. In a sample of M^3 of *Stegolophodon* cf. *stegodontoides* from the Siwaliks of Pakistan described by Tassy (1983), the structure of the posterior central pretrite conule varies from fully developed and isolated to very small and fused with mesoconelets. In *Stegolophodon* from Japan (*Stegolophodon pseudolatidens, Sl. tsudai*), however, the posterior central pretrite conule of M^3 is far more reduced. The morphological differences between species from Japan and the Siwaliks is greater in the lower molars than in the upper molars. In *Stegolophodon* cf. *stegodontoides* described by Tassy (1983), the posterior ridges of the M_3 show a marked chevron arrangement and bear distinct posterior central pretrite conules. In species from Japan, the posterior ridges are straight in the lingual–buccal direction; the posterior central pretrite conule is found only in the first interlophid region.

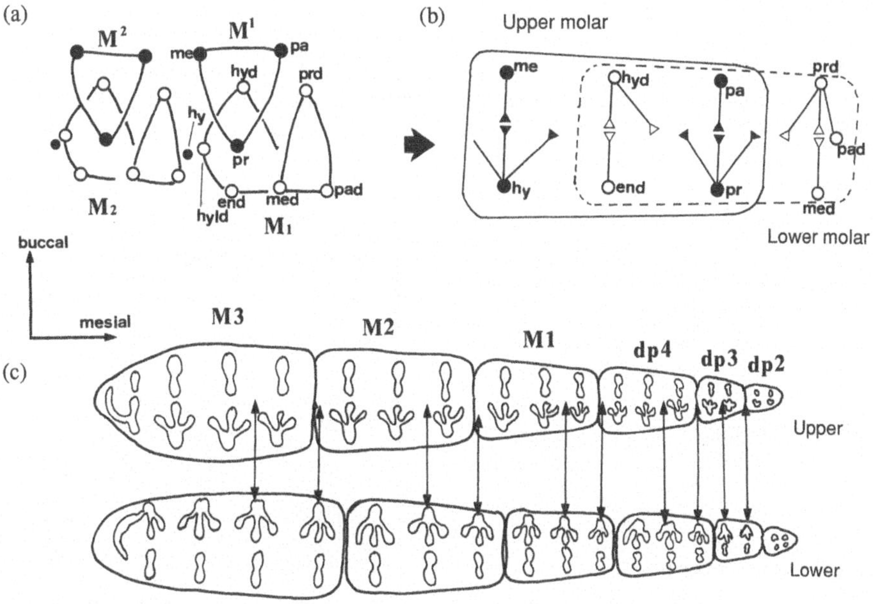

Fig. 17.1 Diagrammatic representation of occlusion of upper and lower molars of (a) generalized mammal and (b) and (c) advanced Proboscidea, Stegodontidae included. Despite its highly modified cusp morphology, the relative position of cusps of upper and lower molars of Elephantiformes in occlusal view (b) had changed only a little from that of the basic primitive mammals (a), and this occlusion pattern is held throughout the cheek teeth row (c). Thus, in derived members of Elephantiformes which exhibit horizontal tooth displacement, the rate of eruption of teeth in both upper and lower molars is regulated so as to maintain the occlusal relationship of homologous cusps in both jaws, the same as in generalized mammals where there is no horizontal tooth displacement. Key: end, entoconid; hy, hypocone; hyd, hypoconid, hyld, hypoconulid; me, metacone; med, metaconid; pa, paracone; pad, paraconid; pr, protocone; prd, protoconid. (Drawing by H. Saegusa, modified after Frick (1926), Osborn (1936, 1942).)

Faint choerodonty is recognized in *Stegolophodon hueiheensis* from the middle Miocene of China. The small tubercles between interlophs of this species might not be central pretrite conules but additional tubercles brought about by the choerodonty condition.

The central pretrite conules also vary in primitive *Stegodon*. In a sample of *S. zdanskyi* from the Yushe Basin, one out of 12 specimens of M^3 has anterior central pretrite conules on the first and second ridges. Also, lophs(id)s are divided into buccal and lingual half-loph(id)s by median sulci on mesial ridges; they are usually further divided into two portions by a lateral fissure. The median portion corresponds to mesoconelets, and the outer portion to the principal pretrite and posttrite cone.

In more derived species, cusp arrangement becomes further modified. The median sulci become indistinct and disappear, except on the first ridge. The degeneration of median sulci may be caused by enlargement of mesoconelets followed by fusion of mesoconelets and the principal cone. The fusion of the principal cone and mesoconelets takes place more frequently in the lingual side and the buccal side of the upper and lower molars, respectively. This

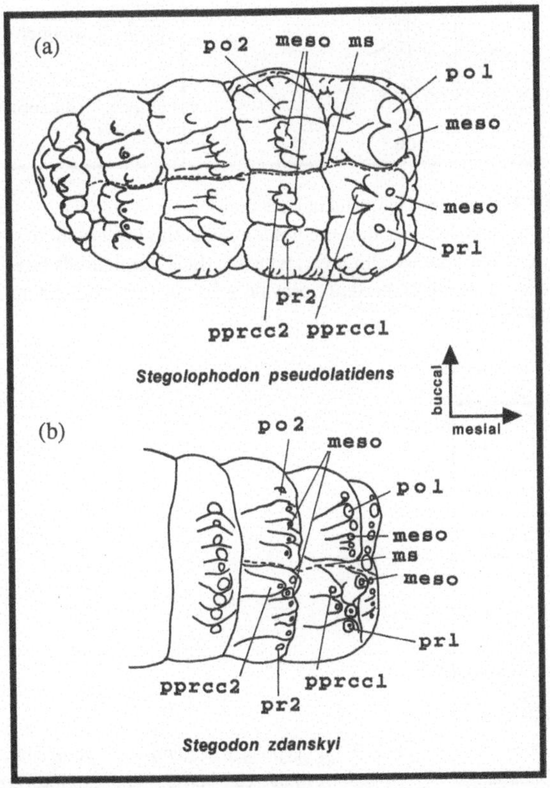

Fig. 17.2 Occlusal surface morphology of right upper M3 of *Stegolophodon* (a) and *Stegodon* (b). Key: meso, mesoconelets; ms, median sulcus; pol, po2, main posttrite cusp of first and second lophs; pprccl, pprcc2, posterior pretrite central conule of first and second lophs, pr1, pr2, main posttrite cusp of first and second lophs.

trend appears to be concordant with *Stegolophodon* in which the pretrite cone is wider than the posttrite cone.

The shape of a valley between ridges has previously been used to distinguish elephantids from *Stegodon*. In elephantids, the valley takes a U- or V-shape from the buccal or lingual view. In *Stegodon*, the valley is Y-shaped (Maglio 1973, p. 16). The distribution of characters between *Stegodon* and elephants is not simple; in *S. zdanskyi* and primitive *Stegodon* from Yunnan, however, the valley is seen clearly as being V-shaped and, in this respect, there is no difference between the primitive *Stegodon* and elephantids. On the other hand, elephantids have not developed the Y-shaped valley which is typically seen in derived *Stegodon* (e.g. *S. orientalis, S. insignis, S. ganesa*). In *S. aurorae*, a U-shaped valley is developed, together with V- and Y-shaped valleys in the same molar; a U-shaped valley is dominant in the distal portion, and a V- or Y-shaped valley is observed on the proximal portion of the molar.

WORN SURFACE OF ENAMEL LOOP

It has long been recognized that *Stegodon* has step-like worn surface reliefs of the enamel layer (Janensch 1911; Shikama 1936; Makiyama 1938). Molar enamel can be divided into inner and outer layers (hereafter, IE and OE, respectively) (Fig. 17.3). Under the scanning electron microscope, it can be seen that the direction of the enamel prism changes abruptly at the boundary between IE and OE (Kamiya and Taruno 1986). The OE is thicker and softer than the IE and, after wear, its worn surface forms a broad shelf below the shearing blade at the boundary between the two enamel layers. Thus, on the worn surface of *Stegodon* molars, a step-like structure is formed.

Most distinct step-like reliefs, which I call 'type 4 *stüfenbildung*', are typically seen in *S. aurorae* and *S. trigonocephalus*. A low 'cliff' separating two 'terraces' is developed at a boundary between two enamel layers in the above species (Fig 17.4). The shearing blade bordering those terraces is distinct and the lower terrace extends outward and almost horizontally from this shearing blade to a junction between enamel and cement. There are no marked gaps of height between the worn surface of the cement layer and the OE at the boundary between two layers. A high-angle cliff emerges at the enamel–cement junction (ECJ) after post-mortem removal of cement.

In *S. orientalis*, the worn surface of the OE slopes steeply from the boundary between the IE and OE to the ECJ. Thus, the surface of the boundary between the IE and OE is exposed widely

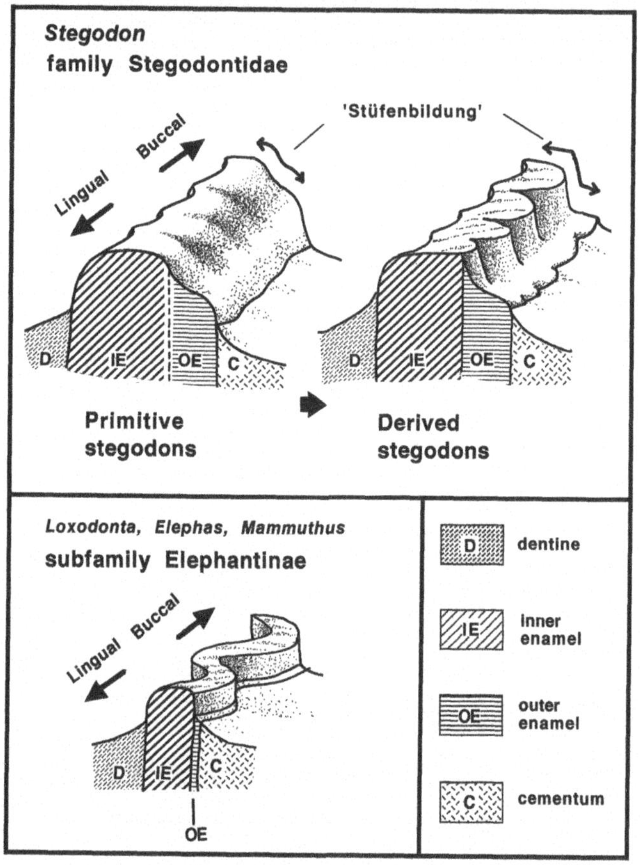

Fig. 17.3 Worn surface relief of stegodont molar. A step-like profile is formed in stegodons because the inner enamel (IE) layer is more resistant than the outer enamel (OE) layer, which is thick. Elephantids have a very thin layer of OE which does not provide space for the development of a step-like structure.

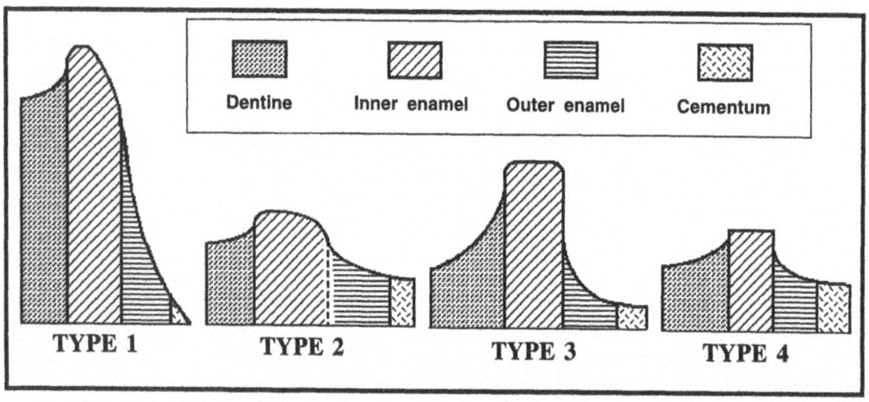

Fig. 17.4 Typology of step-like structure of stegodont molar can be classified into four types.

on the surface facing the valley between ridges. I call this type of step-like structure 'type 3'. Type 3 corresponds to '*Die Stüfenbildung am Schmelts*' of *S. trigonocephalus* described by Janensch (1911; cf. von Koenigswald *et al.* 1993). The most primitive step-like structure, type 2, is observed in primitive *Stegodon* such as *S. zdanskyi*. The shearing blade is obliterated and rounded. Judging from observation with the naked eye, directional changes in enamel prism appears to be rather gradual in type 2. Type 1 is seen on the lightly worn ridges in all species of *Stegodon*. The worn surfaces of the IE and OE form a steep slope extending from the enamel–dentine junction (EDJ) to the ECJ, and the slope is most steep at the boundary between IE and OE.

In elephants, there are also two layers of enamel, an inner and outer layer. A shearing blade of dentine is also developed at the boundary between the two layers, but the situation is different from that of *Stegodon* (Fig. 17.3). The OE is very thin in elephants and consequently its worn surface does not show a step-like structure.

The pattern of enamel folding also eliminates the difference between *Stegodon* and elephantids (Kamiya 1991). The enamel plication can be classified into two types (Fig. 17.3). In type I, the horizontal cross-section of plication at EDJ is composed of a series of parabolic curves. The junction between the parabolic curves is V- or U-shaped. This pattern is copied at the boundary between the IE and OE, but not at the ECJ. Coppens *et al.* (1978, p. 359) called this type of enamel plication a 'scalloped' enamel figure. In type II, the plication of the enamel layer takes the form of parallel foldings or resembles geological ptygmatic folds. It is noted, however, that the evolutionary development of enamel folding of stegodontids seems to differ from that of elephantids. In primitive stegodonts (e.g. *S. zdanskyi*), enamel folding is, though very faint, distributed throughout the whole length and height of the ridge, except for its apex. On the contrary, in early elephants (e.g. *Elephas recki atavus* and *Loxodonta exoptata*), enamel folding is rough and its distribution is centred on the median portion of the plate at the middle height of the plate. In *Stegolophodon*, most specimens are devoid of folding but an intermediate molar of a *Stegolophodon* from Japan shows fine folding distributed the whole length of the enamel loop and type 2 *stüfenbildung*. This may suggest a close relationship between *Stegodon* and *Stegolophodon* from Japan (Kamiya and Taruno 1986).

MASTICATION

Elephantids and *Stegodon* frequently show a shallow, longitudinal trough, running distally from the gap between slightly worn conelets at the distal portion of the lower molar (Fig. 17.5). Formation of such a trough on the worn surface of cement can be explained by the unidirectional jaw movement observed in recent elephants. When movement of food is directed distally in the lower jaw, conelets are expected to block the flow of food at their mesial side. A portion of food flow, however, will escape from the gap between the conelets and make a shallow trough on the cement surface distal to the conelets. The trough has not been observed on the upper molar because the cement surface anterior to the lamellae is too narrow to provide space for a trough. This trough shape configuration is clearly observed on cement surfaces of the lower molar in several specimens of *Stegodon* and *Stegolophodon*. This tells us that *Stegodon* and some *Stegolophodon* had already attained the level of propalineal jaw movement; also, the direction of jaw movement is a little oblique to the axis of the lower molar in both groups.

The aforementioned criterion works well in the material covered with thick cement, but such cement is absent in the majority of samples of *Stegolophodon*. Conversely, oblique facets in the moderately worn ridges of mammutids and certain gomphotheres, which are thought to be brought about by the interlocking of upper and lower molars during mastication (Lehmann 1944; Tobien 1975, 1976) reveal the mode of mastication in *Stegolophodon*. There are two types of oblique facets in the slightly worn ridges of right side teeth of mammutids. In Figure 17.6, one type is inclined mesially (anteriorly) and the other distally; the former facet being labelled 'M' and the latter facet 'D'. Between the upper and lower molars, facet M of the *n*th ridge of the upper molar contacts facet D of the *n*th ridge of the lower molar, and facet D of the *n*th ridge of the upper molar contacts facet M of the (n+1)th ridge of the lower. This indicated *n*th ridge of the upper molar interlocks the valley between the *n*th and (n+1)th ridge of the lower molar (Fig. 17.1). It is impossible to make contact between facet M of the upper molar and facet D of the lower molar simultaneously with that of facet D of the upper molar and facet M of the lower molar. This may indicate that there were at least two phases of contact of upper and lower molar ridges during mastication. Two phases in the mastication of mammutids may correspond to the buccal and lingual phase of mammals in general. Numerous fine striations are found to be running linguało-bucally on facets M and D. The striations are obviously caused by lateral movement of the lower molar.

Thus, the morphology of facet on the lightly worn ridges are evidently brought about by lateral movement of the lower molar. The proboscidean groups which hold fully developed facet M and facet D must retain lateral jaw movement. The Miocene stegodontid, *Stegolophodon*, has lost facets M and D. This indicates that the lateral movement of the lower jaw has already been lost in this group. This fact is important in the interpretation of the controversies concerning the phylogenetic relation of *Stegolophodon* and *Stegodon*. This point will be discussed below.

TAXONOMY OF STEGODONTIDAE

STATUS OF THE GENUS *STEGODON*

At present, the taxonomic unit that can be considered definitely monophyletic is the unit composed of seven species of *Stegodon* (*S. zdanskyi*, *S. aurorae*, *S. trigonocephalus*, *S. ganesa*, *S. bombifrons*, *S. insignis*, *Stegodon* sp. A). Based on the synapomorphic characters of these seven species (listed in the caption of Fig. 17.7), I would like to propose a subgroup of

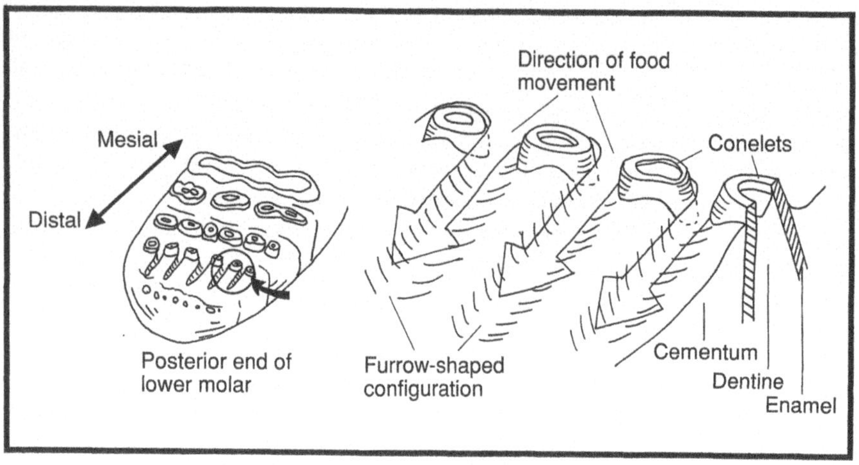

Fig. 17.5 Development of a shallow groove on the cement surface of stegodont and elephant molars. This relief is formed by unidirectional food flow during mastication.

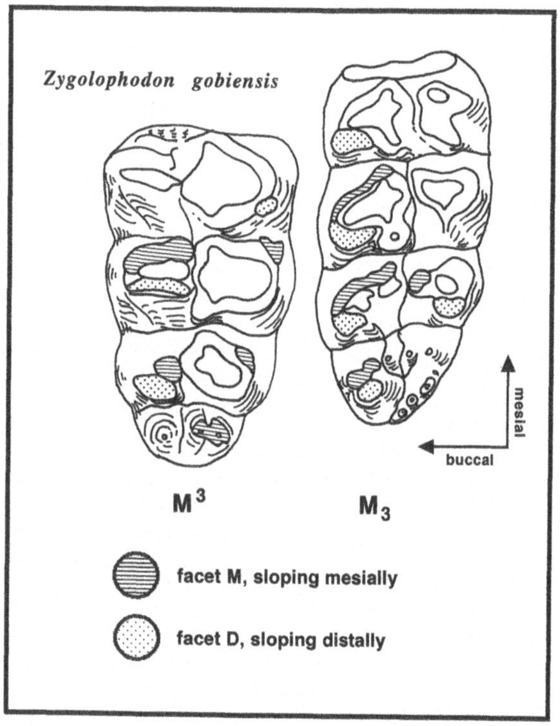

Fig. 17.6 Wear facets of the third molar of *Zygolophodon gobiensis*. Facets labelled M (for mesial or anterior) and D (for distal) can contact with their counterparts on opposing teeth only when the valleys and ridges of opposing molars come into an interlocking position. This indicates that lateral elements of jaw movement might be necessary for the formation of these facets. Such wear facets are a common character of Neogene Elephantoidae. The lower molar is placed half-ridge mesial to the upper molar to show the relative positions of upper and lower molars during interlocking: the *N*th ridges of the upper molar interlocks with the *n*th valley of the lower molar (cf. these drawings to those of Z. *turicensis* shown in Tassy 1985, volume 3, fig. 167).

Stegodon, Stegodon sensu stricto (*s.s.*), for the convenience of discussion, because it is not certain whether all of the *Stegodon* share the synapomorphic characteristics of the above seven species (Figs 17.7 and 17.8).

The cranial morphology of *S. orientalis* Owen, 1887 and *S. pinjorensis* Osborn, 1929 have not been revised or treated cladistically, but they are certainly allocated to *Stegodon s.s.* because their published figures (Osborn 1929*a*, 1942) clearly show a derived character of *Stegodon s.s*, the loss of facial crest from the zygomatic process of the maxilla. Other stegodons having more than eight ridges on the lower and seven ridges on the upper molars might also be allocated to *Stegodon s.s.* (Fig. 17.8).

The species-level taxonomy of primitive *Stegodon* from Yunnan still remains in quite an unsatisfactory state because of the paucity of samples available for comparative study. From the samples currently available, it is certain that primitive *Stegodon* from Yunnan is remarkable from the stegodon systematic point of view. Traditionally, the genus is defined mainly by the characters of the molars, such as roof-shaped loph(id), fine and rather equal-sized conelets, and low and wide molar crown (Osborn 1942). Based on those classic definitions of the genus and presence of type 2 *stüfenbildung* on the worn surfaces of enamel, all of the primitive samples from Yunnan can be allocated to *Stegodon*. Some derived characters of *Stegodon s.s.* are not seen in some specimens of *Stegodon* from Yunnan (see below). Primitive *Stegodon* from Yunnan might be a sister-group of *Stegodon s.s* (Fig. 17.7). For synapomorphic characters of primitive *Stegodon* from Yunnan and *Stegodon s.s.*, see the caption of Fig. 17.7. *Stegodon officinalis* (Hopwood 1935) may also be included in primitive stegodon from Yunnan and can be a sister-group of *Stegodon s.s. Stegodon* sp. B from southern China also cannot be allocated in *Stegodon s.s.* because of the absence of *stüfenbildung* on the heavily worn first ridge and its unique combination with type 3 *stüfenbildung*.

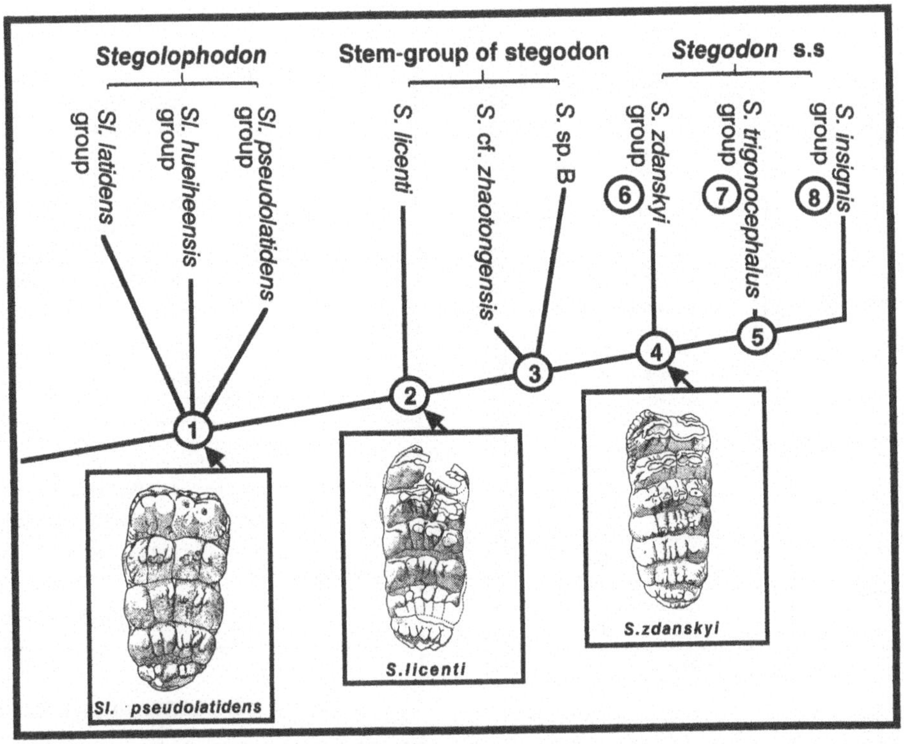

Fig. 17.7 A cladogram of Stegodontidae based on an incomplete set of characters. Molars in the boxes are all upper third, right side. Not to scale. Synapomorphic characteristics of the eight nodes are as follows. Node 1: (1) extreme brachyodonty; (2) mesoconelets are as high as principal cones; (3) loss of facets M and D (propalineal mastication); (4) fusion of anterior pretrite central conule and mesoconelets; (5) reduced posterior pretrite central conule of upper molar. Node 2: (6) ridges are compressed mesio-distally; (7) more than 6 ridges in upper M3; (8) more than 5 ridges in intermediate molars; (9) angular process of mandible rounded; (10) occlusal surface of molar convex; (11) lower molar lost chevron arrangement of ridges on occlusal surface. Node 3: (12) scalloped-shaped enamel figure on worn surface of molar; (13) step-like structures on enamel loop; (14) mammilae (conelets) 6 or more in first ridge of upper M3. Node 4: (15) lower tusks (I2) lost; (16) upper tusk is slightly twisted laterally (direction of the twist is reverse to that of elephantids); (17) long and huge tusk, proximal two-thirds of tusk uniform in diameter; (18) enamel band lost from upper tusks; (19) lower M3, 8 ridges or more; (20) upper M3, 7 ridges or more; (21) facial crest lost from zygomatic process of maxilla; (22) narrow palate; (23) mandibular symphysis short; (24) short cervical vertebrae (shorter than in elephantids); (25) loss of lateral deflection of stylohyoid. Node 5: (26) orbit and infraorbital foramen high; (27) crista orbitotemporalis and crista orbitalis ventralis steep. Node 6: (28) shallow and wide depression present above naris. Node 7: (29) presence of intraorbital depression. Node 8: (30) reduction of upper tusks; (31) basilar tubercle of basioccipital present.

Stegodon licenti can be allocated to the genus *Stegodon* based on the classic definition of the genus. It shows, however, an absence of *stüfenbildung* on the worn surface of the enamel and faint but distinct posterior central pretrite conules on the distal slope of loph(id)s. Such characteristics indicate that their inclusion in the genus *Stegodon* is rather conditional.

In summary, genus *Stegodon* can be divided into four groups; crown taxa of the genus *Stegodon s.s.* and three stem-groups, primitive *Stegodon* from Yunnan, *Stegodon* sp. B, and *Stegodon licenti*.

DIVERSIFICATION IN THE GENUS STEGOLOPHODON

Isolated dentition of *Stegolophodon* cannot provide us with information concerning diversity within the group because of the extremely simplified ridge structure of its molars. To avoid this difficulty, it is necessary to combine the characteristics of upper and lower molars. Only three species of *Stegolophodon, Sl. (Eostegodon) pseudolatidens, Sl.* cf. *stegodontoides*, and *Sl. (Rulengchia) hueiheensis*, are known from the upper and lower jaw of the same individual. Other species, known only from isolated molars, are excluded from the present discussion.

Kamei and Kamiya (1981) discussed, and Taruno (1985) revived the genus name '*Eostegodon*' proposed originally by Yabe (1950) for *Stegolophodon* (*Sl. psuedolatidens, Sl. tsudai, and Sl. miyokoae*) of Japan, arguing that *Stegolophodon* of Japan has the closest relationship with *Stegodon*, also that other *Stegolophodon* should be allocated to the paraphyletic 'gomphotheres'. Separation of *Stegolophodon* of Japan from other *Stegolophodon*, however, is not supported because the former also share synapomorphic characteristics of *Stegodon* and *Stegolophodon*, such as enlargement of the mesoconelet, very wide molar crowns, and propalineal mastication (see above).

Zhou and Zhang (1983, p. 58) proposed as a new genus the name *Rulengchia* for *Stegolophodon hueiheensis* from the middle

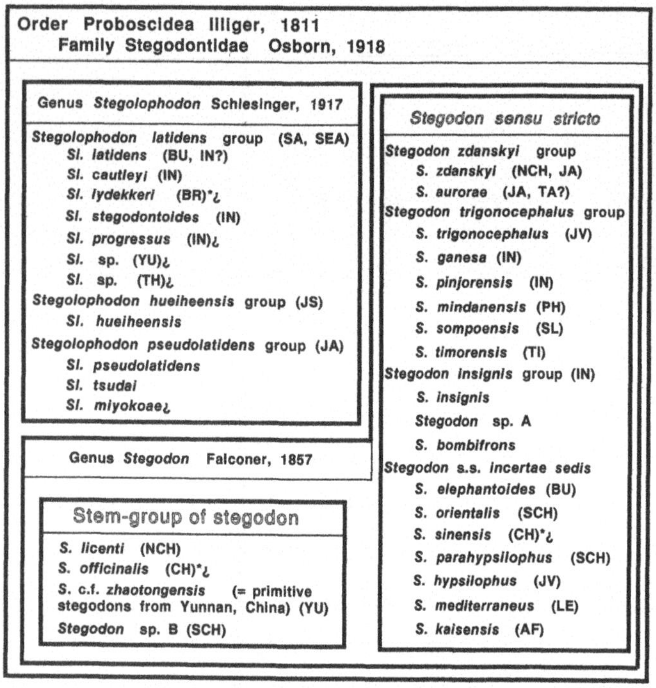

Fig. 17.8 A classification scheme of Stegodontidae proposed in this work. Key to geographical or faunal abbreviation: AF, Africa; BR, Borneo; BU, Burma (now Myanmar); IN, India; JA, Japan; JS, Jiangsu; JV, Java; LE, Levant; NCH, Northern China; PH, Philippine; SA, South Asia; SCH, southern Chinese cave fauna; SEA, South-east Asia; SL, Sulawesi; TA; Taiwan; TH, Thailand; T, Timor; YU, Yunnan. *, exact locality or horizon unknown; ?, presence of the taxon in given area uncertain; ¿, possible synonym of other species.

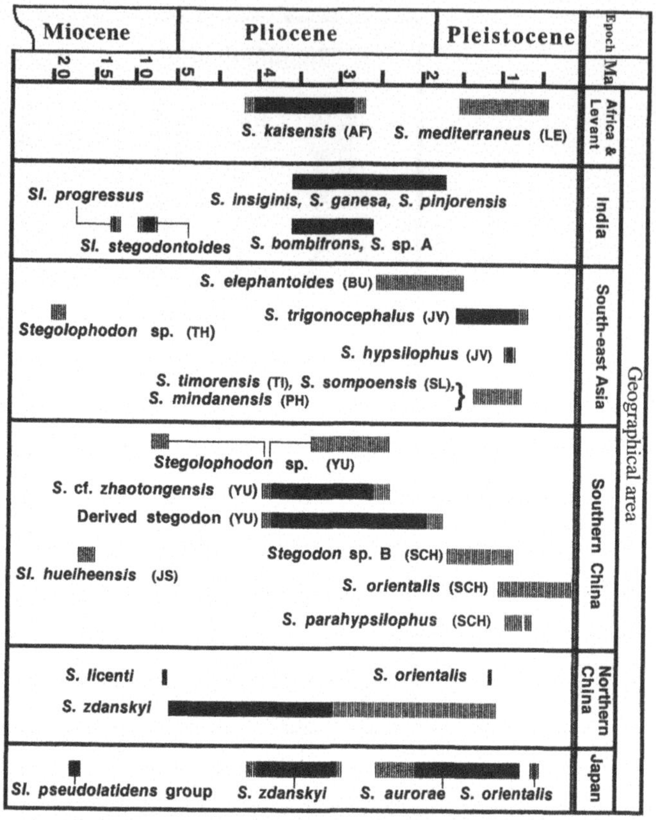

Fig. 17.9 Stratigraphic distribution of selected species of Stegodontidae. Heavy solid bars indicate the range of the species calibrated by magnetostratigraphy or radiometric datings. Broken bars indicate the range of the species calibrated by biostratigraphy only. The following species are not included in this chart because of the lack of recent stratigraphic revision or uncertainty of the locality of the specimen: *Sl. latidens, Sl. lydekkeri, Sl. cautleyi, S. officinalis,* and *S. sinensis.* Key to geographical or faunal abbreviations: AF, Africa; BU, Burma (now Myanmar); JS, Jiangsu; JV, Java; LE, Levant; PH, Philippines; SCH, southern Chinese cave fauna (= Pleistocene fauna of southern China, northern Vietnam, and Laos); SL, Sulawesi; TH, Thailand; TI, Timor; YU, Yunnan. Jiangsu is placed between southern and northern China in this chart because of its intermediate zoogeographical character of southern and northern China. The figure was compiled from Coppens (1967), Cooke and Coryndon (1970), Beden (1976), Huang (1979), Opdyke *et al.* (1979), Azzaroli and Napoleone (1982), Leinders *et al.* (1985), Watanabe and Kadar (1985), Ji (1987), Tobien *et al.* (1988), Zhou (1988), An and Ho (1989), Ratanasthien (1990), and Saegusa and Taruno (1991).

Miocene of China (Figs 17.7–17.9). Later, Tobien *et al.* (1988) stated that *Stegolophodon* should be used for *hueiheensis*. The latter shares with other *Stegolophodon* several derived characters, such as a wide and very low molar crown, reduced central conule, loss of M and D facets (see above), and enlarged mesoconelets. These features suggest that *Sl. hueiheensis* should be allocated to the genus *Stegolophodon. Stegolophodon hueiheensis*, however, has the following derived characters not shared with *Stegolophodon* of Japan or India: brevirostrine with shortened deflected symphysis, narrow gutter, concentric crenellation on the worn surface of enamel, and numerous interloph accessory tubercles. This suggests the separation of *hueiheensis* from other *Stegolophodon* in the genus *Stegolophodon.* I propose the group name '*hueiheensis* group' for *hueiheensis*.

Stegolophodon of India, Myanmar (formerly Burma), and Borneo, e.g. *Sl. latidens*, the type species, *Sl. stegodontoides, Sl. (Tetrazygodon) progressus, Sl. cautleyi, and Sl. lydekkeri*, represent the classic members of the genus. Unfortunately, most of them are known from isolated molars; only *Stegolophodon* cf. *stegodontoides*, described by Tassy (1983), provides an exceptional case. As previously mentioned, this species is more plesiomorphic than those of Japan in the ridge structure of the upper, and especially lower, molars, despite the higher number of ridges in the upper and lower molars. I propose the group name '*latidens* group' for the above classic and plesiomorphic stem-group of *Stegolophodon*.

The above example suggests that at least two monophyletic groups are included in the genus *Stegolophodon*; also, that diversification at certain higher levels had already occurred in the Stegodontidae during the Miocene. Character distribution in this genus is still not well known because of the paucity of samples, but the *Sl. pseudolatidens* group from Japan appears to be the best candidate for the sister-group of *Stegodon*, because this group shows an increase of conelets, fine folding of the enamel loop, and type 2 *stüfenbildung* (Taruno 1985; Kamiya and Taruno 1986).

SYSTEMATIC POSITION OF *STEGOLOPHODON* AND *STEGODON* WITHIN THE SUPERFAMILY ELEPHANTOIDEA

It has long been suggested that the Stegodontidae (*Stegolophodon* plus *Stegodon*) has a close relationship with the Mammutidae (Maglio 1973; Tobien 1976), an idea recently challenged by Tassy (1982, 1983). Based on osteological characters, the latter author suggested that Stegodontidae is closely related to elephantids and gomphotheres rather than Mammutidae.

The main discrepancy between Tobien (1975), Maglio (1973), and Tassy (1983) is whether the characteristics of reduced central conules and strong lophodonty shared by *Stegolophodon* and Mammutidae are homologous. Preliminary analysis shows that reduction of central conules and strong lophodonty in *Stegolophodon* has a different functional background from that of Mammutidae. This may preclude the possibility that observed similarity between *Stegolophodon* and mammutids is homologous.

As discussed above, Mammutidae retained well-developed M and D facets whereas *Stegolophodon* lost these facets (Fig. 17.6). It is known that mammutids had strong lophodonty, even in early members (Tassy 1986). In Mammutidae, therefore, reduced central conules and lophodonty have no relationship with the loss of the lateral jaw movement. On the other hand, the loss of central conules and lophodonty are accompanied with the loss of lateral jaw movement in *Stegolophodon*. Thus, the functional weakening of central conules and of lophodonty in Mammutidae is different from *Stegolophodon*. This supports Tassy's (1982) view that *Stegolophodon* has no close phylogenetic relationship to Mammutidae.

There are several problems remaining concerning the detailed relationship between gomphotheres, *Stegolophodon*, *Stegodon*, and elephantids. Illuminating this problem are recent findings of *Stegodon* stylohyoids, which have the following derived characters; posterior ramus, gracile bone, and a reduced shelf at the base of the inferior ramus. The presence of the posterior ramus of the stylohyoid is a synapomorphy shared with gomphotheres and elephants (Shoshani 1986*b*), and therefore allocation of *Stegodon* in the group with gomphotheres and elephants proposed by Tassy (1982) is supported by this character. Slenderness of the stylohyoid and a reduced shelf at the base of the inferior ramus are seen in elephants (Shoshani 1986*b*) and *Stegodon*. If they are a synapomorphy uniting both groups, the relationship of *Stegodon* to elephants becomes very close. Lack of well-defined lateral deflection of the stylohyoid in specific *Stegodon* specimens may have some relationship with the narrowness of palate in *Stegodon* and the derived character of the pharyngeal region of this genus (Shoshani, personal communication 1992).

Tassy's arguments on the systematic position of the Stegodontidae in the group comprising elephants and gomphotheres may be correct. There remains, however, some problem concerning the detailed relationship among these taxa. The difficulty of determining relationships among *Stegodon*, *Stegolophodon*, elephantids, and gomphotheres is due to the paucity of comparative data available. Re-evaluation of facets on molars may compensate for the paucity of specimens available.

As mentioned in the introduction, monophyly of the Stegodontidae is far from resolved. Tassy and Darlu (1987, p. 487) reported that among 29 equally parsimonious trees (15 taxa, 40 characters), five support the monophyly of Stegodontidae, and they concluded that 'the monophyly of Stegodontidae is not refuted on the sole basis of parsimony'. Kalb *et al.* (this volume, Chapter 12) concluded that *Stegodon* and *Stegolophodon* do not belong in one clade. Their cladistic analysis is based on a character matrix which contains more detailed molar characteristics than that of Tassy and Darlu. Kalb *et al.* (this volume, Chapter 12) may be correct on the ground of cladistic methods only; I question, however, whether it is scientifically proper to conduct cladistic analysis based mostly on a single morphological character complex, that of the molar, without functional or morphogenetic backgrounds. In order to gain confidence in cladistic analysis of elephantids and Stegodontidae, it is preferable to introduce more morphological analyses (cf. Shoshani, this volume, Chapter 16). Until such research is conducted, I maintain the traditional view that Stegodontidae is monophyletic.

SPATIOTEMPORAL DISTRIBUTION PATTERN OF *STEGODON* SPECIES

Below is a brief summary of the specific status and distribution of *Stegodon* species (Fig. 17.8). For convenience, I divide their distribution into the following areas: north-eastern Asia including North China, Japan, and Taiwan; South China; Yunnan; Indo-China and South-East Asian islands; India; and the Middle East, and Africa. A detailed review of stegodon species will be published in the near future.

NORTH CHINA, JAPAN, AND TAIWAN

Very primitive *Stegodon*, *S. licenti* are known only from holotypes and paratypes derived from the Yushe Basin, North China (Teilhard de Chardin and Trassaert 1937). The species status is firm because it can be distinguished easily from other *Stegodon* in primitive molar and jaw structure.

In 1935, Hopwood described the new species, *S. zdanskyi*, based on a partial molar thought to have come from Yushe Basin, Shanxi Province, northern China. Two years later, Teilhard de Chardin and Trassaert (1937) reported additional specimens from this basin and confirmed the type locality of this species. A sample of *S. zdanskyi* from Yushe Basin gives a standard for the comparison of *Stegodon* of North China.

Characteristic figures are as follows: ridge number in the M^3 varies from six to seven, but when the distal talon and ridges are counted together, the sum is constantly eight in M^3 and nine in M_3, enamel layers show type 2 *stüfenbildung*, the folding of EDJ is weak and indistinct.

According to Saegusa (1989), based on material from Yushe Basin, the following three species from North China can be synonymized with *S. zdanskyi*: *S. yushensis*, *S. chiai*, *S. huanghoensis* (the latter species is after Zheng *et al.* 1975, p. 8). Likewise, based on a similar comparison, all of the gigantic stegodons from the Pliocene of Japan (e.g. '*S. elephantoides*' from the Tokai Group, Mie Prefecture, and *S. shinshuensis* from the Sarumaru Formation, Nagano Prefecture) can be allocated to *S. zdanskyi* (Saegusa 1989).

There have been approximately 120 localities from which Stegodontidae specimens were collected; most of them yielded only one specimen. Most of the *S. aurorae* (total *c.* 200) have been from dredged material from Akashi Strait, south-western Japan. In a few localities, footprints of stegodons were discovered (Shimizu *et al.* 1990; Kamiya *et al.* 1991).

The late Pliocene–early Pleistocene mammalian faunas of Japan are characterized by the presence of the small-body-sized subhypsodont *Stegodon*. Matsumoto (1918) described a subhypsodont M^3 from Mt. Tomuro, Kaga Prefecture, central Japan and named it '*Elephas aurorae*'. Later, he established a new genus *Parastegodon* based on this species, implying that it represents a stem species of elephants (Matsumoto 1924*b*, 1929). Five species were proposed under this genus during the 1930s: *Parastegodon kwantoensis, P. sugiyamai, P. akashiensis, and P. infrequensis*. Synonymy among species of *Parastegodon* was warranted from the start (e.g. Shikama 1936; Makiyama 1938; Matsumoto 1938, 1941). They argued that there were two species of subhypsodont *Stegodon* in Japan that could be distinguished from each other mainly by cranial characters.

Taruno (1972) and Saegusa (1987) suggested that cranial features referred to by Makiyama (1938) were the result of heavy distortion. Thus, the concept of subhypsodont *Stegodon* of Japan made by Makiyama (1938) and his followers, Takai (1939) and Matsumoto (1941), is invalid. It is clear that the diagnoses used for the distinction of five species of the subhypsodont *Stegodon* have no basis and that those species are synonymous with *S. aurorae*. (Taruno 1972; Saegusa 1989). Diagnostic features of *S. aurorae* are as follows: shallow depression above the naris, subhypsodonty, type 4 *stüfenbildung*, and a vertical wall at the lingual side of the ridge. Subhypsodont *Stegodon* are known from Taiwan (Shikama *et al.* 1975). Their molars are very similar to those from Japan, but it is not clear whether their similarity is due to descent from common ancestry or convergence.

During the Pliocene, *S. zdanskyi* was distributed in both North China and Japan (Saegusa 1989). The distribution suggests that a land connection was present between the Japanese islands and the mainland of Asia during this age. About 2 Ma, pre-existing *S. zdanskyi* of Japan were replaced by endemic species of *S. aurorae*. Saegusa (1987) suggested a close phylogenetic relationship between *S. zdanskyi* (= *S. huanghoensis*) and *S. aurorae* (= *S. akashiensis*). In *S. aurorae*, body size was reduced to about probably half that of its probable ancestor. It is highly possible that *S, aurorae* was derived from the population of *S. zdanskyi* in Japan by the isolation of the Japanese islands from the mainland.

STEGODON FROM CAVE FAUNA OF SOUTH CHINA

The area between the Qinling Mountains and Vietnam is characterized by Pliocene and Pleistocene mammalian fauna derived from cave and fissure deposits. Three species of *Stegodon* constitute the main elements of this fauna: *S. orientalis*, *S.* sp. B, and *S. subhypsilophus*.

The so-called *S. preorientalis* appeared frequently in the faunal lists of early Pleistocene cave fauna of South China (Kahlke 1961; Pei 1962; Han *et al.* 1975; Huang 1979; Li 1981; see also Zhao *et al.* 1981). Originally, *S. preorientalis* was named on the basis of a mesial fragment of a lower molar (IVPP-V.868) from Babu, Guangxi (Young 1938). Teilhard de Chardin and Leroy (1942) suggested that *S. preorientalis* is a junior synonym of *S. orientalis*. Later, a similar opinion was expressed by Young and Liu (1950). Pei (1987) thought that a primitive *Stegodon* sample from Liucheng Gigantopithecus cave, Guangxi shared a number of diagnostic characters with the holotype of *S. preorientalis* and suggested that the sample validated the specific status of *S. preorientalis*. Concerning the Liucheng sample alone, its specific status is firm but its allocation to *S. preorientalis* has some problems. The holotype of *S. preorientalis* cannot be distinguished from *S. orientalis* and probably belongs to the latter species. To avoid confusion, a new name is suggested for primitive *Stegodon* represented by the Liucheng sample, tentatively named *Stegodon* sp. B.

The type specimen of *S. orientalis* consists of two fragments of a molar obtained at a drug store in Shanghai. Later, this species was confirmed by a large sample from Yanjinggou of Sichuan (Colbert and Hooijer 1953). At about 0.5 Ma *S. orientalis* was found in Japan. *Stegodon sinensis* was established from a fragment of dp^3 (Owen 1870). Later, several permanent molars of *Stegodon* from China and Japan were referred to as this species, but now they are safely allocated to *S. orientalis*. Concerning the type specimen itself, it cannot be assigned to any known species of *Stegodon*. Thus, the specific name *S. sinensis* should be retained only for the holotype.

He (1984) described an upper molar of a dwarf subhypsodont *Stegodon* from Deyang in Sichuan and named it *S. parahypsilophus*. Li and Wen (1986) gave a new specific name, *S. guizhouensis*, for a lower molar of a small hypsodont *Stegodon* from Guanyindon Cave in Guizhou. The latter species is probably a junior synonym of the former species. Distribution of dwarf subhypsodont *Stegodon* on the continent is puzzling because their characteristic features are usually seen in the dwarf *Stegodon* from the South-East Asian islands. It is hard to imagine that they are returning immigrants from the islands, because, by comparison, there are no fossil records of the well-known Mediterranean dwarf returning to the mainland. Another possibility is that the dwarfism was caused by the particular habitat of the dwarf form on the continent. Such a case has been found in the Miocene rhinoceroses from the Gulf Coast of North America (Prothero and Sereno 1982). In summary, *S. orientalis* coexisted during the early Pleistocene with *Stegodon* sp. B and during the middle Pleistocene with *S. parahypsilophus*, respectively, in southern China.

STEGODONS FROM YUNNAN PROVINCE, CHINA

The recent samples discovered of the earliest stages of *Stegodon* are from the Pliocene and early Pleistocene deposits of Yunnan Province, southern China. These species have been named: *S zhaotongensis, S. primitium, S. baoshanensis, S. yuxiensis*, and *S. yuanmouensis*. Zhou and Zhang (1983) removed *S. primitium* from the genus *Stegodon*, suggesting that this species was possibly allocated to the genus *Stegotetrabelodon* Petrocchi. Except for this, all other species are safely allocated to *Stegodon*.

The problem of the taxonomy of these primitive *Stegodon* still remains unsatisfactory. Only two points are clear from the available data: one is the presence of an extremely primitive *Stegodon* in Yunnan; and another is the coexistence of a derived *Stegodon*, together with this primitive *Stegodon*, during the late Pliocene to early Pleistocene.

A juvenile skull fragment of a primitive *Stegodon* (BNHM-BPV277) was recovered from a lignite bed of Zhaotung Basin (Shi *et al.* 1981). The zygomatic process of the maxilla of BNHM-BPV277 has a well-developed facial crest which is degenerated in other *Stegodon* (Saegusa 1987). Yun (1975) reported a set of postcranial elements and teeth from Baoshan Basin and named it *S. baoshanesis*. These specimens have remarkably primitive characters (such as alveolar lower tusks and enamel bands on the upper tusks) and, consequently, most of them are probably conspecific with IVPP-V5316 from Member 1 of the Yuanmou Formation and the holotype of *S. zhaotongensis* (Chow and Zhai 1962) (IVPP-V.2647). A primitive *Stegodon* persists in Member II of the Yuanmou Formation: *S. yuanmouensis* (You *et al.* 1978). Based on an M_3 which has only six ridges plus a posterior talon, this specimen is certainly very primitive because of the low ridge number, the presence of a median cleft on the anterior five ridges, and well-developed central conules on the anterior three ridges. You and Qi (1973) also described a distal fragment of a lower molar of a primitive stegodon from Member II of the Yuanmou Formation.

The confirmation for the coexistence of derived and primitive lineages is possible for two specimens, IVPP-V.5316 and -V.5317, from the basal conglomerate of Member I of the Yuanmou Formation. IVPP-V.5316 is characterized by very plesiomorphic characters, while IVPP-V.5317 has some distinct, although fragmented, derived features (e.g. high ridge frequency, thick cement). Lin *et al.* (1978) allocated this specimen to *S. elephantoides*, but the few specimens limit any distinction between Burmese and Yunnan samples of this species.

Such derived *Stegodon* have also been reported from Member IV of the Yuanmou Formation (Lin *et al.* 1978). They noted its close similarity to IVPP-V.5317 from Member I and allocated this specimen to *S. elephantoides* based on the derived characters of a molar having 10 ridges and thick cement, but this view seems to be invalid because such derived features tend to be afforded by parallelism. Nevertheless, it is certain that a derived *Stegodon* was found from Members I to IV of the Yuanmou Formation.

As a result of the sample from Yuanmou Basin, it is apparent that primitive *Stegodon* is present in Member I and Member II, and derived *Stegodon* is present from Member I to Member IV. A similar situation is expected in other sedimentary basins of the Yunnan region.

S. officinalis Hopwood 1935, whose type is a molar fragment obtained from a drugstore, was synonymized with *S. zdanskyi*, based on its supposed provenance from Yushe and its suggested similarity to later species (Teilhard de Chardin and Trassaert 1937). None the less, a lower molar of a primitive *Stegodon* from Yunnan shows close similarity to *S. officinalis*. This may suggest that the type specimen of *S. officinalis* came from Yunnan and that *S. officinalis* is a valid species.

Tobien *et al.* (1988, pp. 197–8) allocated some molar material of primitive Stegodontidae to *Stegolophodon* cf. *latidens*. One upper molar specimen (IVPP-V4247) shows a step-like configuration on the worn surface of the enamel, whereas a lower molar shows a similar structure to *Sl. stegodontoides*, in the chevron arrangements of lophids and development of a central conule. Thus, they seem to be quite heterogeneous; one seems to be a very primitive *Stegodon* and the other belongs to the *Sl. stegodontoides*-like form. Tobien *et al.* (1988) also reported *Stegolophodon* cf. *latidens* from the Baodean of Yunnan.

INDO-CHINA AND THE SOUTH-EAST ASIAN ISLANDS

Stegodon elephantoides from Myanmar (formerly Burma) is the most classic species, but there has been no revised study of this species. *Stegodon insignis birmanicus* was named based on a mandible fragment with an M_3 from the upper Irrawaddy series (Osborn 1942). The systematic and stratigraphical relations, however, between classic *S. elephantoides* and *S. insignis birmanicus* remain unresolved.

There have been poor finds in this area, except for Java and the South-East Asian island archipelagos. Hooijer (1955) synonymized three species of *Stegodon* (*S. trigonocephalus. S. bondolensis*, and *S. airawana*) from Java into *S. trigonocephalus*. His results have been supported by work on cranial morphology (Saegusa 1987). In Java, however, there is one problematic dwarf species of *Stegodon, S. hypsilophus*. There has long been controversy concerning the generic allocation of this species because of its very elephant-like characters, such as hypsodonty and the very few conelets (Maglio 1973; Hooijer 1974; Sondaar 1984). Close examination of the enamel loop structures, however, indicates a number of distinct *Stegodon* characteristics, such as a thick outer layer of enamel and a scalloped enamel pattern. Thus, although the molar of this species is superficially very similar to that of elephants, *S. hypsilophus* is safely allocated to the genus *Stegodon*. Unfortunately, its sister-group or ancestral morphotype is not clear because of its highly specialized molar morphology.

Island archipelagos of South-East Asia yield a number of dwarf *Stegodon*, such as *S. timorensis* from Timor, *S. mindanensis* from Mindanao Island in the Philippines, *S. trigonocephalus florensis* from Flores, and *S. sompoensis* from Sulawesi. Likewise, there have been controversies on the problem of immigration routes of these dwarf *Stegodon* (see van den Bergh *et al.*, this volume, Chapter 23). Routes of *S. trigonocephalus florensis* and *S. timorensis* do not seem to be a big problem because there has been a recognized morphological transition between the above two species along the islands between Java and Timor, suggesting that immigration took place along this island arc eastwards. Concerning the dwarf *Stegodon* from Sulawesi and the Philippines, there are two competing hypotheses: one is via Taiwan; the other is via the Sunda Shelf. Recent vertebrate fauna of this area suggest that the former possibility is unlikely because there is no close similarity in faunal composition between Taiwan and the Philippines. Recent findings from the small island of Sangihe, located between Sulawesi and the Philippines, support this view (Aziz 1990). A *Stegodon* upper molar found in Sangihe clearly shows type 2 enamel figure (see above) which is found exclusively in the population of *S. trigonocephalus* lineage as a morphotype. Thus, this finding suggests the Sunda immigration route is most likely and that all dwarf *Stegodon* are derived from *S. trigonocephalus*.

THE INDIAN SUBCONTINENT

In recent years, the Siwalik Group has been reinvestigated by a joint team of Pakistani and Western scientists. As a result, there have been new finds of a large number of *Stegodon* from well-defined stratigraphical levels. Two species of *Stegodon* are recognized (Opdyke *et al.* 1979): *S. bombifrons* and *S. insignis*. However based on the cranial morphology, Saegusa (1987) concluded, that three, possibly four, species of *Stegodon* (*S. ganesa*, *S. insignis*, *S. bombifrons*, and *S.* sp. A) exist in the Siwalik Group. Lydekker (1876*a*), Osborn (1942), and Hooijer (1955) interpreted the great difference between the skulls of *S. insignis* and *S. ganesa* as sexual dimorphism. Detailed investigation on the ontogenetic transformation series, however, made by Saegusa (1987) revealed that it is impossible to derive a *S. ganesa* type of skull morphology from that of *S. insignis*. The latter species might be an endemic species of the Indian subcontinent because of its highly specialized cranial character. *Stegodon ganesa* may be an invader from South-East Asia. Concerning *S. bombifrons* and *S.* sp. A. there is another possibility. One cannot conclude that the morphological character suggested by Saegusa (1987) as specific is indeed a polymorphological character. At any rate, it is necessary to reinvestigate a large number of tooth samples obtained by recent intensive stratigraphical studies in order to test the conclusion made by Saegusa (1987).

AFRICA AND THE LEVANT

After the comprehensive revision of African elephants by Maglio (1973), except for *Stegodon kaisensis* (Coppens *et al.* 1978; Sanders 1990, in press) which were retained in Stegodontidae, most stegodontid specimens from Africa were incorporated into the Elephantidae, (e.g. *Stegolophodon* of Singer and Hooijer 1958, from South Africa, and *Stegolophodon* of Hooijer 1963, from the Congo). *Stegodon kaisensis*, known only from molars, is more derived than *S. zdanskyi* from North China and very similar to *S. bombifrons* from the Siwaliks (additional data on African gomphotheres are given by Kalb *et al.*, this volume, Chapter 12A).

The Middle East also has yielded some *Stegodon* fossils (Tchernov and Shoshani 1989; this volume Chapter 21), but these forms are more progressive than those of Africa. Their morphology is reported to be similar to that of *S. trigonocephalus* from Java.

DISCUSSION

DISTRIBUTION AND DIVERSFICATION OF STEGOLOPHODONS

Geologically, the oldest specimen of *Stegolophodon* is known from the early to middle Miocene Mae Long Formation of the Li Basin, Thailand (Ginsburg and Tassy 1985; Ratanasthien 1990) (Fig. 17.9). The first appearance of this genus is striking if we compare it with the first appearance of tetralophodont gomphotheres. During the late–early and middle Miocene, the coastline of East Asia seems to have harboured the invasion of *Stegolophodon* from South-East Asia (Fig. 17.9). Despite the good Neogene record of China, there is no record of *Stegolophodon* except for *Sl. hueiheensis* from the early middle Miocene of Xiacaowan, Jiangsu and *Stegolophodon* cf. *latidens* from Yunnan (Chow 1959; Tobien *et al.* 1988). In Japan, stratigraphic levels bearing *Stegolophodon* are restricted to the N7 and N8 zone of Blow (Taruno 1985), These facts may suggest that a northward immigration of *Stegolophodon* during the late–early and early–middle Miocene has some relationship with the thermal maximum of the Miocene.

The above data suggest that diversification of *Stegolophodon* at the subgeneric level may have already occurred during the Miocene. Unfortunately, paucity of samples precludes determination of the closest relative of *Stegodon*. Morphological evolution of the group had stagnated until the latest Miocene, despite very early acquisition of propalineal mastication. Generally, a change of mastication mode in elephants is associated with change of vegetation type in Africa (Maglio 1973). In this case, there seems to be no relationship between a change of vegetation and transformation of the mastication mode from that of the gomphothere type to that of Stegodontidae. Close inspection of changes of vegetation in South-East Asia during the Miocene and functional morphology of *Stegolophodon* molars is necessary.

CLADOGENESIS AND DISTRIBUTIONAL PATTERN OF STEGODONS

From the brief revision of *Stegodon* presented above, it becomes evident that several species can be synonymized. This leads to a fairly simple distribution pattern of *Stegodon* (Figs 17.9, 17.10).

It becomes obvious that *S. zdanskyi* is quite conservative throughout its history in North China and Japan. Speciation is limited to the case of *S. aurorae* in island environments. On the contrary, in South China, South-East Asia, and India, the coexistence of more primitive and derived species is recognized. This suggests that speciation may have taken place at a higher rate in those areas, resulting in more that two lineages.

In some groups of mammals, the place of origin of the group has been found among those locations in which specific or genetic diversification of the group is high. Wide distribution of a sister-group of *Stegodon s.s.* in Yunnan may indicate that Yunnan is ideal for the dispersion of the group. This consideration may explain the high specific and genetic diversification of the group in Yunnan and its surrounding area.

Conclusively, the following synopsis on Stegodontidae evolution may be possible (Fig. 17.10). In the early Miocene, *Stegolophodon* originated in South-Eastern Asia, and during the thermal maximum of the Miocene, they were distributed in a wider area, including Japan. Though they attained propalineal mastication at the beginning, the evolution of the molar character stagnated during Miocene times. At the beginning of the Pliocene, *Stegodon* diversified rapidly and dispersed to other regions of Asia, including China.

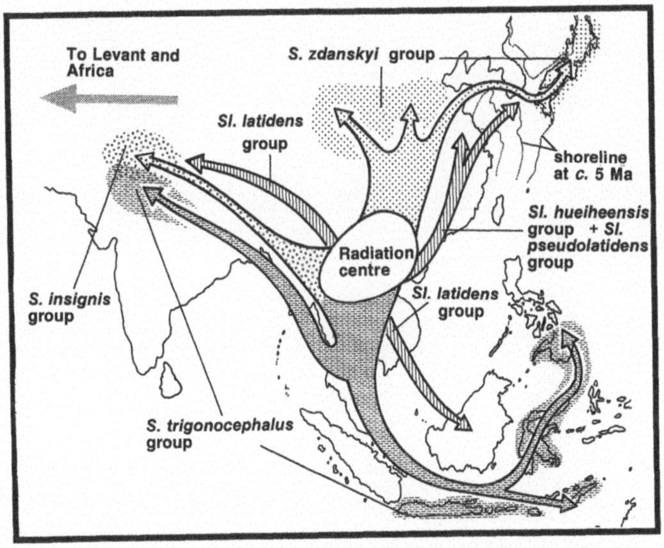

Fig. 17.10 Possible dispersal route of Stegodontidae. Stegodontidae are believed to have derived from gomphothere species in South-East Asia at the beginning of the late Early Miocene. *Stegodon* probably appeared in South China at the beginning of the Pliocene. Three major lineages of *Stegodon* were established and they seem to have dispersed outside of South China.

SUMMARY AND CONCLUSION

Morphological characters of *Stegodon* molars and their chronological and biogeographical distribution are the foundation for the taxonomy and evolution of the Stegodontidae. Molar structure, such as cusp arrangement, composition of ridges, and morphological features of worn surfaces, is reviewed. Functional properties of the wear facet, suggesting an early deviation of propalineal jaw movement in Stegodontidae, indicate a close relationship to the elephants and gomphotheres rather than to Mammutidae.

Based upon diagnostic molar characters of the taxonomic groups of the family, only *Stegodon s.s.*, (which includes classic *Stegodon* species) is definitely monophyletic, while other groups, such as *Stegolophodon* and Yunnan *Stegodon*, are either polyphyletic or paraphyletic groups within the family. Although it is almost impossible to establish a systematic framework in the group, incongruence of the character distribution of derived characters among the *Stegolophodon* species suggest that at least two monophyletic groups were already present within the so-called *Stegolophodon*.

Stegodon zdanskyi of the Pliocene is considered to be a single representative, common in the area of North China and Japan. From the posterity of this species, *S. aurorae* developed in Japan very close to the Pliocene/Pleistocene boundary, while in North China, *S. zdanskyi* was extant to the early Pleistocene. On the other hand, it becomes clear that *Stegodon* in South China developed a high diversity compared with those in the areas of North China and Japan. As to the cave fauna of South China, three lineages of *Stegodon* are represented, and yet the fauna of Yunnan Province is characterized by the coexistence of derived and primitive groups. Spatiotemporal consideration of the biogeographical distribution of *Stegodon* suggests that the place of origin of the genus was in South China.

Based on available evidence, stegodons and stegolophodons are united by five synapomorphies: extreme brachyodonty, mesoconelets are as high as principal cones, loss of facets M and D (propalineal mastication), fusion of the anterior pretrite central conule with mesoconelets, and reduced posterior pretrite central conule of the upper molar. *Stegodon* is monophyletic whereas *Stegolophodon* is a paraphyletic group within the family Stegodontidae. Other synapomorphies are given in Fig. 17.7.

ACKNOWLEDGEMENTS

Special thanks go to the following individuals for their help in various aspects of this study: F. Aziz (Quaternary Laboratory, Bandung, Indonesia), M. M. Chang, C. K. Li, W. P. Huang, Z. C. Jin (Institute of Vertebrate Paleontology and Paleoanthropology, Academia Sinica), J. de Vos (Rijksmuseum van Nattulijke Historie, Leiden, The Netherlands), J. Guan (Beijing Natural History Museum), W. L. Huang (Tianjin Natural History Museum), T. Kamei (Kyoto University), H. Taruno (Osaka Museum of Natural History, Osaka. Japan), Y. Tomida (National Science Museum,

Tokyo). This paper was retyped by S.M. Wolak and edited by D. Froehlich, J. E Kalb, E. C. Marsac, J. L. Pierce, and P. Tassy. I appreciate very much J. Shoshani's kind suggestions for improvement of my text and figures; without his help, stegodons would still be hidden behind the thick bamboo forest of East Asia. This study was supported, in part, by grants from the Japanese Ministry of Education, Science and Culture (No. 01790348).

ADDENDUM

In November 1992, I visited the American Museum of Natural History (AMNH, New York) and made the following observations.

1. *Stegodon pinjorensis* can be safely allocated to the *S. trigonocephalus* group (see Fig. 17.8).
2. Skulls of *S. orientalis* show enigmatic autapomorphies, such as a pair of strong domes on the cranial vault and a vertical shallow depression on the forehead.
3. Skulls of *S. orientalis* in AMNH are reconstructed and partially covered with plaster — examination of the original surfaces may provide characteristics which could affect the cladistic relationship within the genus *Stegodon*, (Saegusa 1993).

PART V

The third radiation: Elephantidae

EIGHTEEN

Origin and radiation of the Elephantidae

NANCY E. TODD AND V. LOUISE ROTH

INTRODUCTION

The emergence of the family Elephantidae in the late Miocene of Africa is thought to be correlated with a major adaptive shift in the masticatory apparatus. This shift was a transition from a combined grinding and shearing molar function with some lateral jaw movement, which is characteristic of gomphotheres, to a mostly horizontal shearing dental function involving fore and aft movement of the jaw, which is present in elephantids. Major structural modifications in skull and mandibular morphology accompanied this transition, which increased the shearing efficiency of the molars, and which was the principal factor in the rapid, early radiation of this family (Maglio 1972, pp. 640, 654).

Although there is consensus on many of the main features of evolution within the Elephantidae, some key issues remain either disputed or unresolved. There is widespread agreement on the constituents of the family from the late Pliocene onward, during which time the Elephantidae comprised the three principal lineages *Elephas*, *Loxodonta*, and *Mammuthus*, but earlier lineages are more difficult to discriminate. *Elephas* and *Mammuthus* are believed to share a more recent common ancestor than either has with *Loxodonta*, but the precise cladistic relationship between these genera and a variety of more primitive forms is in dispute (Beden 1983; Tassy 1985, 1990*a*; Tassy and Shoshani 1988), which makes specifying the point of origin of the Elephantidae difficult. Later diversification within the family (i.e. within these three genera) is also difficult to trace, as authors vary in their tendencies toward taxonomic lumping or splitting. Differing views on the delimitation of species have produced recent estimates of fewer than 30 species in the family (Maglio 1973; Madden 1981*b*), where Osborn (1942) had proposed in excess of 100. Yet despite some taxonomic uncertainty, certain morphological trends are widely acknowledged. Regardless of which phylogenetic hypothesis is accepted, abundant and widespread parallelism is recognized as a major feature of evolutionary diversification within the Elephantidae.

Because of its abundant fossil record, the family Elephantidae has had an important influence on macroevolutionary theory, providing theorists having specialties outside of proboscidean palaeontology with examples of both explosive speciation and rapid phyletic change (Simpson 1953; Stanley 1979; Vrba 1987). The study of the origin and radiation of the Elephantidae lost two important young leaders in the 1970s and early 1980s with the withdrawal of V. J. Maglio from active participation in proboscidean palaeontology, and with the untimely death of M. Beden. These setbacks occurred at a time of rapid conceptual advance in the theory and practice of systematics, and an increasing appreciation of allometry and variability in the interpretation of elephant morphology. In this paper we outline some of the points of agreement as well as the unresolved questions that make the origin and diversification of the Elephantidae a topic ripe for renewed study.

CHARACTERISTICS AND FUNCTIONAL MORPHOLOGY OF THE ELEPHANTIDAE

The following summary account of the evolutionary transformations that accompanied the origin and radiation of the Elephantidae draws heavily on the work of Maglio (1972, 1973). For convenience, our taxonomy follows his, although we (emphatically) acknowledge that some of the groupings have been, or will be, found to be inappropriate cladistically. Phylogenetic hypotheses provide the context for identifying evolutionary trends, yet many aspects of elephantid phylogeny remain unclear. Until more information becomes available, we prefer not to give premature endorsement to any particular cladistic treatment and instead will use traditional groupings, acknowledging their provisional status. Even in the absence of precise information about the point of origin of the family, or branching relationships within the group, when the closest relatives and earliest and most primitive members of the Elephantidae are compared with what are clearly later, more derived forms (according to any of the competing phylogenetic scenarios yet suggested), certain trends are apparent in broadest outline.

Maglio (1972, 1973) described mastication in the gomphotheres in terms of both grinding and shearing function. Molars bear a series of paired cones that are separated by a median cleft (sulcus); the three or four longitudinally successive pairs of cones are separated by V-shaped transverse valleys. Median accessory columns to the buccal cones of lower molars and lingual cones of the uppers created a trefoil (three-lobed) pattern in the teeth with wear. Conical cusps of upper molars fit into the valleys or worn dentine basins of lowers (and vice versa), acting as mortars and pestles with slight circular, lateral, or fore-and-aft motions of the jaws. Forward motion of the mandible would drag the cusps up out of the basins and across the sharp enamel edges of worn cusps, creating a shearing action. As teeth wore down with increasing age, they may have undergone a functional transition, from a predominance of grinding (with conical cusps) to shearing (with worn edges of enamel) (Maglio 1973, p. 89).

The evolution of the Elephantidae was marked by an enhancement of shearing and a reduction in grinding function. Permanent premolars were lost in most taxa, the molars elongated, and the jaw shortened. Paired cusps of the teeth became compressed into plates (lamellae) that incorporated the accessory columns. The lamellae are formed of pockets of dentine encased in enamel, and the valleys between adjacent lamellae are filled with cementum. At their bases the valleys are U-shaped in longitudinal section, an arrangement that permits the molars of elephantids to maintain shearing function through most of the height of the tooth (and which contrasts with the increasing compression of lamellae toward the base of the tooth in gomphotheriids, whose valleys are V-shaped; Maglio 1972). Wear on the occlusal surface of the tooth horizontally sections the series of lamellae into a series of broad transverse enamel loops surrounding dentine, separated and surrounded by cementum. Enamel, dentine, and cementum differ in hardness, and their differential wear leaves in relief a set of transverse, parallel enamel ridges, which act as cutting surfaces well adapted for shearing in a posterior–anterior direction. Shearing is defined as 'cutting by passage of sharp cutting edges on a lower tooth across a corresponding edge on an upper tooth', whereas grinding is a 'process of crushing, between two broad surfaces that are moving over each other' (Maglio 1972, p. 640). Interlocking occlusion between upper and lower dentitions is lost, and the cranium of elephantids shows modifications for a forward shearing stroke of the masticatory muscles (Maglio 1972, p. 638).

Vestigial premolars are present in some primitive species of elephantids, but in general the functional cheek teeth are homologous to the last three deciduous premolars (dp2, dp3, dp4) and the three permanent molars (M1, M2, M3). Morphologically, these teeth are similar to each other, differing mainly in size and in numbers of lamellae (Roth and Shoshani 1988). Elephantid cheek teeth are extremely large — the largest of any vertebrate known — and there is space within the jaw for at most two to be in wear at any time within a single quadrant. Accompanying this enlargement has been a heterochronic change (Roth 1989) — specifically, a delay in the eruption of the most posterior teeth in the toothrow. An unusual and characteristic tooth replacement pattern has evolved: following dp2 and dp3, which are in place one behind the other in the jaws of a neonate, and which erupt vertically, teeth dp4, M1, M2, and M3 form sequentially in the posterior region of the jaws, and subsequently move forward, erupting in the process. Each successive tooth tends to be larger and to have more lamellae than the one preceding it. In the absence of the permanent premolars, the horizontal progression of the teeth is uninterrupted. As the cheek teeth wear on their occlusal surfaces, they continue to erupt and move forward, fragmenting and losing lamellae at the anterior end of the toothrow. A diastema separates the maxillary cheek teeth from the upper incisors (I^2), which are modified into tusks; in the most derived elephantids the mandible lacks functional incisors (Luckett, this volume, Chapter 4).

Elephantids share with other proboscideans specializations of cranial morphology that are associated with prominent development of the tusks, and of the musculature of the nostrils and upper lip that are involved with the trunk. Cranial features that are unique to the Elephantidae are functionally related to their distinctive mastication. In particular, the temporalis muscle is repositioned in such a way that the force exerted by the muscle is essentially constant throughout the power stroke, and the main axis of its fibres runs vertically and is situated directly above the coronoid process of the mandible (Maglio 1972). Thus, the cranial vault in elephantids is 'tall or very tall', and the occipital condyles are situated high on the skull, aligned with the top of the orbits (Beden 1983, p. 40). The tall cranium, with its extensively pneumatized bone, may also serve to absorb shock to the head during sparring with other elephants; to counterbalance the downwardly directed tusks; and to provide a surface for the attachment of nuchal muscles (Kingdon 1979). The glenoid articular surfaces are 'very elongate transversely and strongly convex in sagittal section' (Beden 1983, p. 40), their shape reflecting the path of the mandibular condyles during the masticatory cycle. After rotating about a pivot point near the insertion of the lateral pterygoid muscles (which is displaced to a relatively ventral position on the condylar neck), the mandible then sweeps forward through a downwardly convex arc that follows the curve of the occlusal surfaces; as the jaw returns in the recovery phase, the condyles rotate and elevate into the postglenoid fossa. During elephantid phylogeny, with the reduction and loss of mandibular tusks and a shortening of the mandible, the centre of gravity of the lower jaw shifted posteriorly, its ascending ramus broadened, and the coronoid process became positioned vertically above the mandible's centre of gravity, bringing the jaw into 'nearly perfect balance' (Maglio 1972, p. 653).

PHYLOGENETIC RELATIONSHIPS OF ELEPHANTIDAE AT THE INTERGENERIC LEVEL

For nearly 50 years, Simpson's (1945) practice of recognizing three main genera, *Loxodonta*, *Elephas*, and *Mammuthus* for the Elephantidae proper has been in widespread use (Beden 1983). Maglio (1973) later proposed a new genus, *Primelephas*, for some very primitive elephantid teeth discovered in East Africa.

The traditional techniques of evolutionary systematics — the piecing together of morphoclines from fossils, and their integration with geographical and stratigraphical information — have produced a variety of scenarios for the origin of the Elephantidae.

An African origin for Elephantidae remains undisputed, but the ancestral taxon and/or sister-group are still uncertain.

The Stegodontidae, which (for the sake of discussion) we will consider here to include *Stegodon* and *Stegolophodon* (Maglio 1973), have been moved in and out of elephant ancestry many times in the past century (Saegusa, this volume, Chapter 17; Kalb *et al.*, this volume, Chapter 12A). Believed to be Asiatic in origin (but see Kalb *et al.*, this volume, Chapter 12A), the stegodontids paralleled the elephants in the development of a fore-and-aft shearing action during mastication. Stegodontids have relatively short mandibles with low-crowned molars and thick enamel, and the molars are divided into lamellar plates. With wear, the enamel surfaces of the plates appear in highly folded and scalloped configuration (Saegusa, this volume, Chapter 17). The many similarities of the dentition shared with elephantids prompted many authors to consider one or both genera of stegodontids to be close relatives of, or even ancestral to, the Elephantidae (e.g., Osborn 1942; Simpson 1945; Aguirre 1969*a*; Saegusa, this volume, Chapter 17, Kalb *et al.*, this volume, Chapter 12A), yet morphologically they appear to have specialized too early (and with too many other distinctive features) for many of the derived characters they share with the Elephantidae to be interpreted as homologous.

Because he believed that neither *Stegolophodon* nor any members of the Mammutidae are found in Africa (but see Tassy and Pickford 1983), Arambourg (1942) suggested an Asiatic origin for *Elephas*, with migrations of *Elephas recki* into Africa. *Loxodonta*, according to this scenario, evolved from an African *Stegodon* in a lineage independent of *Elephas*. This hypothesis implies that the Elephantidae is polyphyletic, whereas Maglio (1970*a*) and others have supported a monophyletic origin for the family. Nevertheless, a specimen found in 1959 at Langebaanweg, Cape Province, was attributed to *Stegolophodon*, and, based on this original identification, Aguirre (1969*a*) hypothesized that *Stegolophodon* was ancestral to the Elephantidae (Fig. 18.1). Thus, the Langebaanweg material, as well

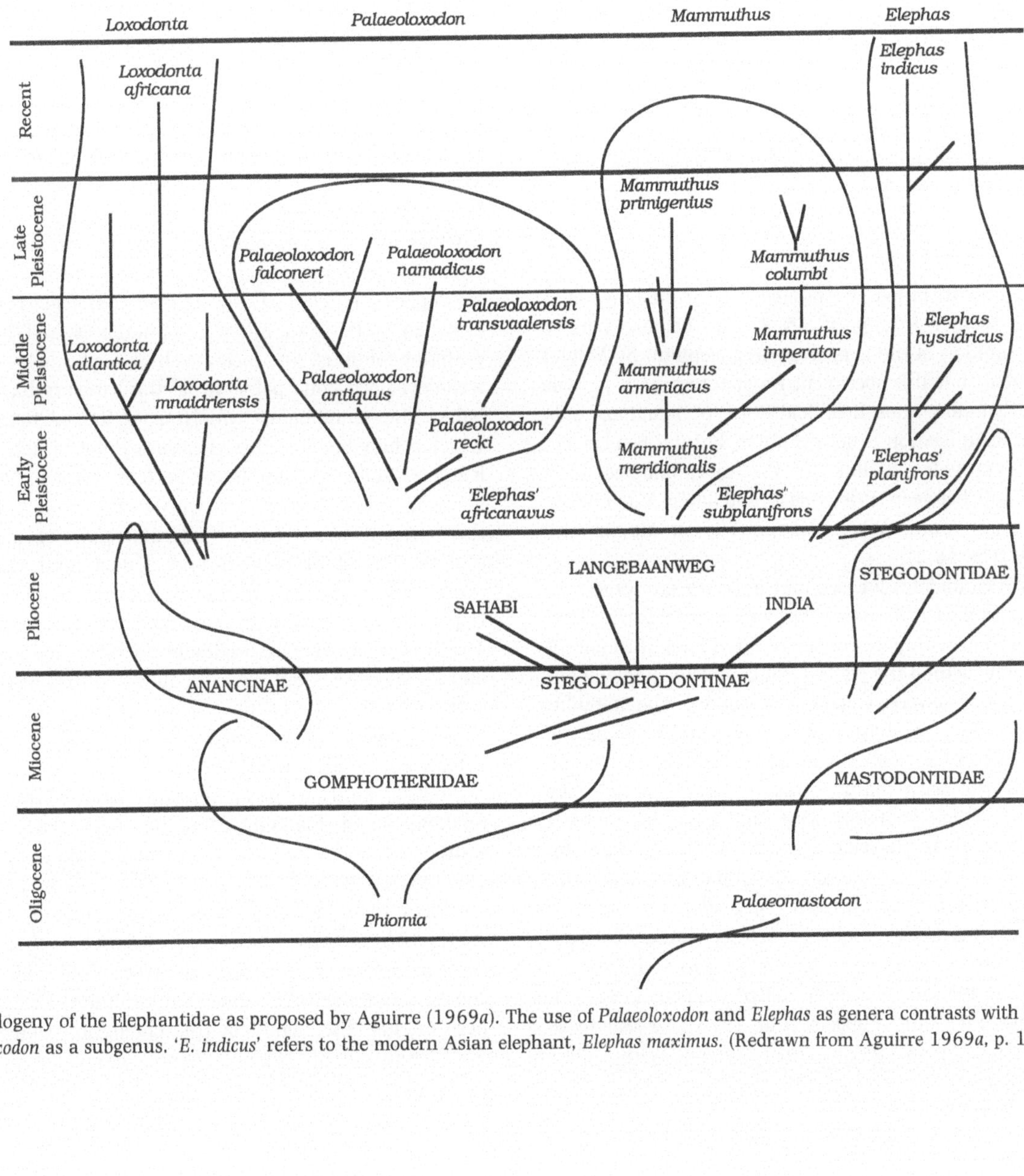

Fig. 18.1 Phylogeny of the Elephantidae as proposed by Aguirre (1969*a*). The use of *Palaeoloxodon* and *Elephas* as genera contrasts with Beden's (1983) use of *Palaeoloxodon* as a subgenus. '*E. indicus*' refers to the modern Asian elephant, *Elephas maximus*. (Redrawn from Aguirre 1969*a*, p. 1374.)

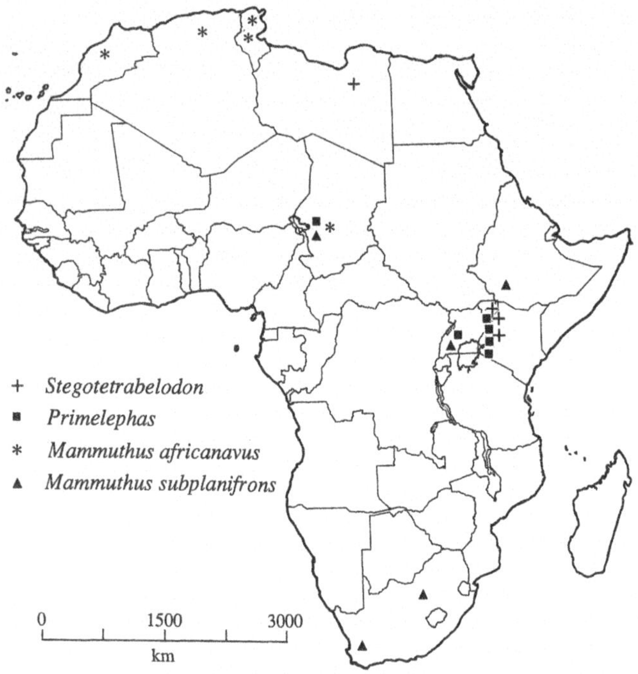

Fig. 18.2 The earliest relatives of the Elephantidae are found in Africa during the Miocene and Pliocene. Many of these fossils are very fragmentary, making absolute identification difficult. Sources consulted: Coppens *et al.* (1978, pp. 346–7), Maglio (1973, pp. 16–18, 20–2, 51, 53).

as other material attributed to *Stegolophodon* from Libya and Central Africa, seemed to support an African origin. A more thorough examination of the Langebaanweg elephant by Maglio (1970*a*) indicated that this specimen, fragmentary as it is, was actually more similar morphologically to *Mammuthus subplanifrons*. This identification is not definitive, but if it is a correct assignment, it provides a link in the morphological and chronological gap between *Primelephas* and later *Mammuthus subplanifrons* from South Africa (Maglio 1970*a*; Maglio and Hendey, 1970) (Fig. 18.2).

Other possible candidates for elephant ancestry are *Stegotetrabelodon* and *Stegodibelodon*. These are large elephantoids with low-crowned molars consisting of relatively few plates. Prominent accessory columns behind the plates form complete enamel loops with wear and the transverse valleys are U-shaped. The mandible has a long protruding symphysis and there is usually one pair of lower incisors present in addition to upper incisors. *Stegotetrabelodon* and *Stegodibelodon* both possess relatively progressive elephant-like characters, combined with what are considered less progressive gomphothere characters, which seems to place them in an intermediate position between the Gomphotheriidae and the Elephantidae (Coppens *et al.* 1978). *Stegodibelodon* (which is monotypic, as *Stegodibelodon schneideri*) is very similar to *Stegotetrabelodon* (which includes two species, *S. syrticus* and *S. orbus*), but it lacks lower incisors, which prompted Coppens (1972) to place it in a different genus (Coppens *et al.* 1978). *Stegodibelodon* also has a shorter mandibular symphysis and is more progressive toward an elephantid condition. Recent work indicates that the two genera are unlikely to constitute a monophyletic group (Tassy 1988*a*; 1990*a*; this volume, Chapters 3, 6, 10, 11; and Kalb *et al.*, this volume, Chapter 12A).

Beden (1983) supported Tobien's (1978*b*) original proposal of a dichotomy within the family. One lineage consisted of *Primelephas*, *Elephas*, and *Mammuthus*, which he further divided into *Elephas* (*Elephas*) and *Elephas* (*Palaeoloxodon*), and a *Loxodonta* lineage which remained in Africa. Beden considered *Stegodibelodon schneideri*, which he referred to as *Loxodonta schneideri*, to be the most ancient taxon in the *Loxodonta* lineage (Fig. 18.3). Beden's use of the generic name *Loxodonta* for this taxon is most likely based on its bearing only two incisors, but it is a confusing taxonomic designation.

Maglio (1973) has suggested that *Primelephas* arose from *Stegotetrabelodon*, which he considered ancestral to all elephantids and derived in turn from a gomphotheriid ancestor (Fig. 18.4). Beden (1983) subsequently argued that *Primelephas* was something of a wastebasket taxon whose constituent lineages would be clarified as information from cranial material was added to the teeth. He proposed that *Elephas* and *Mammuthus* shared a common ancestor in *Primelephas*, but that jointly their common ancestor with *Loxodonta* was to be found in *Stegotetrabelodon*. On the basis of new specimens of *Primelephas* he described of late Miocene age, Tassy (1986) followed Maglio's taxonomy and definition of *Primelephas*.

Other forms of gomphotheres that may be closely related to ancestral elephants include *Tetralophodon* and *Anancus*. According to Coppens *et al.* (1978), African *Tetralophodon* has higher-crowned teeth than other gomphotheres and has prominent posterior columns. Intermediate cheek teeth (dp4 – M2) have four well-developed cone pairs, while third molars have at least 5.5 pairs. The mandibular symphysis is longer than in *Anancus*, which is fully brevirostrine. *Anancus* exhibits a suite of progressive features similar to those found in *Tetralophodon*. Members of this genus have bunodont molars with a subhypsodont tendency, and there are four to five pairs of alternate cones on the intermediate molars. The skull is high and domed, and the mandibular symphysis is greatly shortened, lacking mandibular incisors. Upper incisors are straight and lack enamel bands (Coppens *et al.* 1978). The Anancinae originally included two genera, *Anancus* and *Pentalophodon*, but these are now regarded as conspecific and both are within the genus *Anancus* (Coppens *et al.* 1978). Tassy (1983) proposed *Paratetralophodon hasnotensis* as a possible Asian sister-group of African elephantoids, including *Stegotetrabelodon*.

Cladistic analyses have not (yet) resolved the phylogenetic relationships of this family, but they have clarified some of the sources of difficulty and have made explicit the nature of character conflicts. In a parsimony analysis of 22 proboscidean and three (Sirenia, Desmostylia, anthracobunids) outgroup taxa, using 136 additive (ordered) unweighted characters, Tassy (1990*a*) obtained a consensus tree of branch length 225, consistency index 0.72, and retention index 0.89. We will focus here on just the region of this cladogram that concerns the

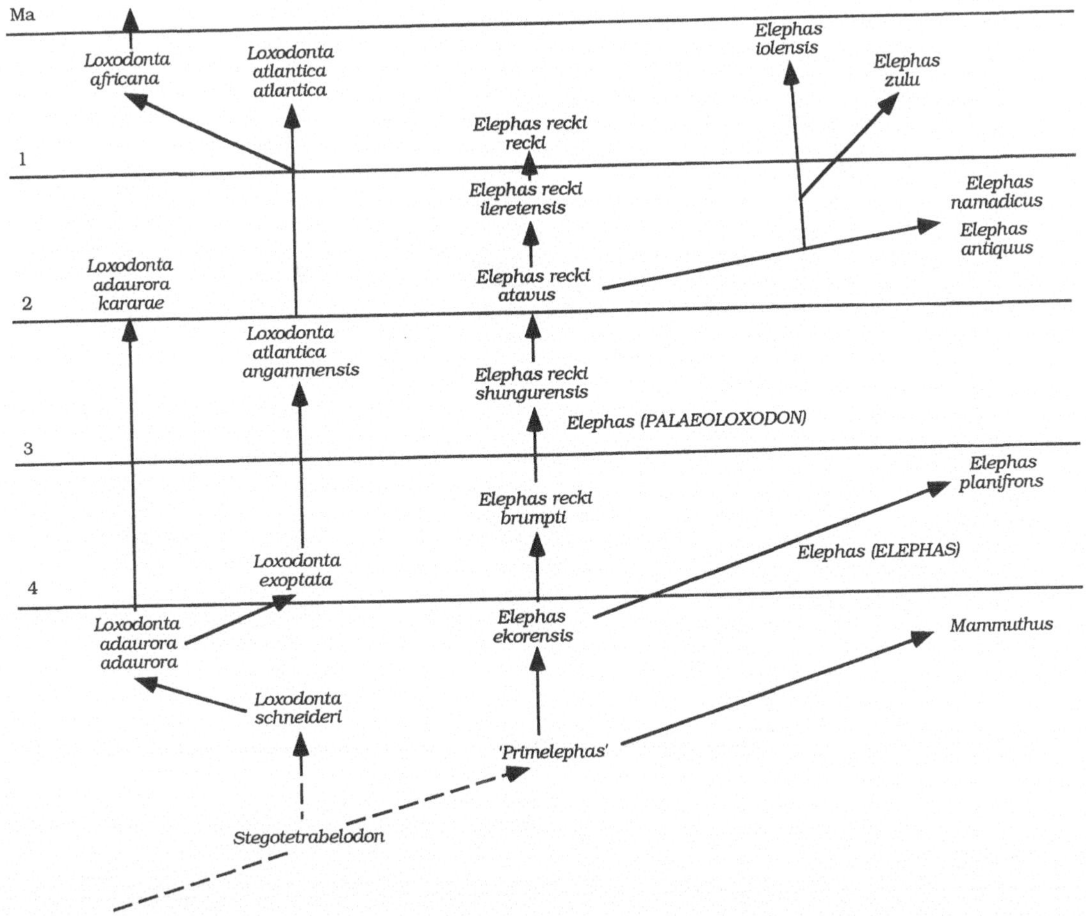

Fig. 18.3 Phylogenetic relationships of African Elephantidae proposed by Beden (1983). In this scheme, *Elephas recki* progresses through five evolutionary stages, contrasting with Maglio's (1973) view of only four stages which were not assigned subspecific names. This lineage presents intriguing problems for determining parameters of variation within fossil species. Dashed lines represent tentative inferences; solid lines imply more probable affinities. (Redrawn from Beden 1983, p. 99.)

Elephantidae and their close relatives (Fig. 18.5). *Loxodonta* and *Elephas* + *Mammuthus* were united as sister taxa (at node 16); the next node (15) joins this clade with *Stegodibelodon schneideri* and *Primelephas gomphotheroides* in a tritomy; these in turn are joined in a pentatomy (at node 13) with *Anancus*, *Stegolophodon* + *Stegodon* (the stegodontids), *Stegotetrabelodon*, and *Paratetralophodon hasnotensis*; with *Tetralophodon* as an outgroup to them all. Most of the ambiguity within the cladogram is concentrated here; the remainder of the Proboscidea is resolved into 11 bifurcations and a single tritomy.

In Tassy's (1990*a*) cladogram, the origin of the Elephantidae would be located somewhere in the area of node 13. This node is supported by nine characters, one of which is a reversal, and two more of which are considered equivocal (of uncertain polarity because of incomplete data). The remaining characters include shortening of the mandibular symphysis, absence of P2 and P3, straightened upper tusks, dental crowns with thick deposits of cementum, and an erect basicranium (which is associated with greater height of the skull and a secondary shift in the orbits anterior to the toothrow). Four equally parsimonious trees are produced at this node. As Tassy (1990*a*) pointed out, the one supported by the most stable set of characters unfortunately has the maximum number of gaps when it is drawn in stratigraphical context, and even the trees that are better in this respect have important lacunae (such as a 6 million year gap between the first appearance of stegodontids and the first appearance of members of their sister-group).

The Elephantinae, whose origin Tassy (1990*a*) identifies with the tritomous node 15, is characterized in this analysis by three synapomorphies. Only one of these — M3 with eight or more lamellae — is exclusive to the Elephantinae; the transformation of dental lophs into lamellae is seen also in *Stegotetrabelodon*, and a reorientation of the lamellae (which is functionally associated with the modification of mastication) produces superficial similarity to the condition of dental lophs in primitive proboscideans (Tassy 1990*a*).

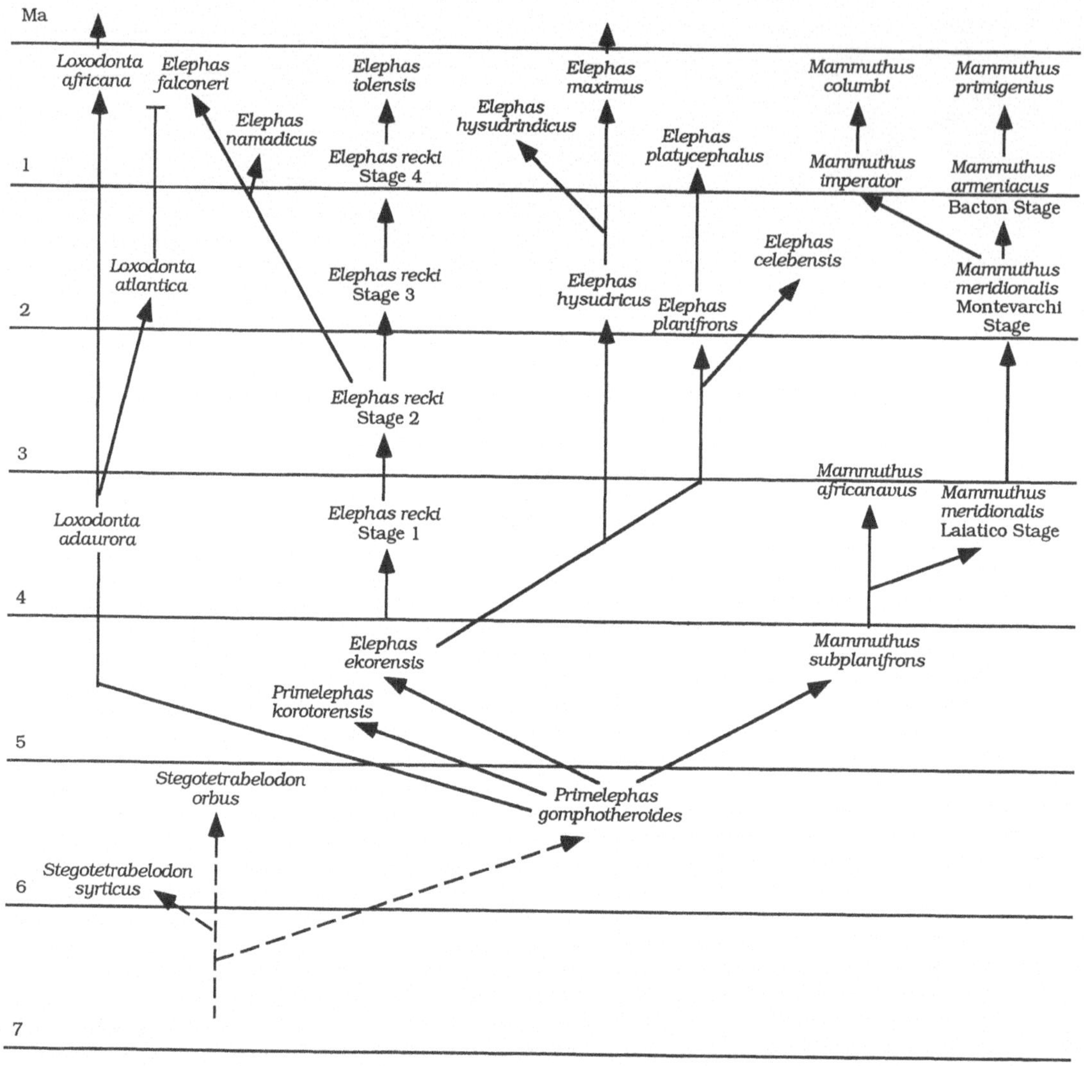

Fig. 18.4 Phylogenetic relationships within the Elephantidae according to Maglio (1973). Solid lines imply probable affinities; dashed lines are more tentative inferences (redrawn from Maglio 1973, p. 77, and Coppens *et al.* 1978, p. 361). For a clearer indication of stratigraphic ranges and inferred splitting times of lineages, see Maglio (1973, p. 79).

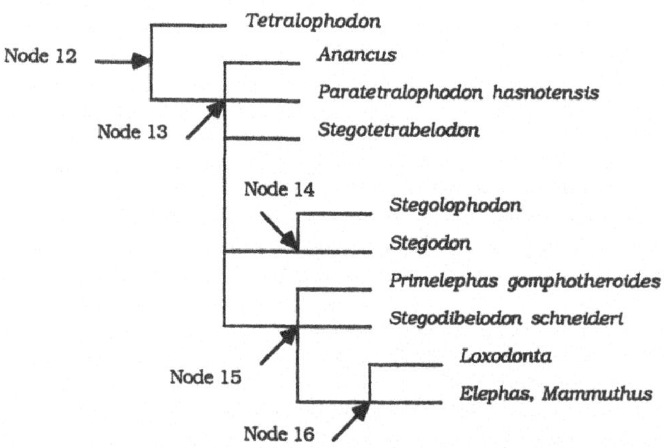

Fig. 18.5 The Elephantidae and their closest relatives according to Tassy's (1990*a*) consensus cladogram of the Proboscidea; see text for discussion. (Redrawn from Tassy 1990*a*, p. 204.)

Clearly, there is room for improvement in our understanding of elephantid relationships. Fortunately, several directions for additional work can readily be identified, A large quantity of osteological (as opposed to dental) material has been overlooked, largely because it is poorly understood (Beden 1983, pp. 41, 42; Shoshani 1986*b*), but with additional collection and further study more fossils may be found to represent new taxa, provide new characters, or bridge stratigraphical gaps. Although because of their intra-taxon variability and rampant parallelism, the Elephantidae present special challenges for phylogenetic reconstruction. Cladistic analysis offers promise, both for a clear identification of contradictory information and for the actual unravelling of the problem itself.

EVOLUTIONARY TRENDS WITHIN THE FAMILY

Within the Elephantidae, *Loxodonta*, *Elephas*, and *Mammuthus* appear to have continued and extended evolutionary trends in the function and morphology of the masticatory apparatus in parallel. In each lineage, the shearing capability of the molars increased with the addition of more lamellae. Even if it is not directly ancestral, as Maglio (1972) suggested, Pliocene *Primelephas gomphotheroides* may be used to illustrate the primitive condition in early elephantids or their forebears. M3 in this species bears seven lamellae, in contrast to 30 for the homologous tooth in Pleistocene *Mammuthus primigenius* (which represents a derived extreme). The shearing index (total length in metres of the leading edges of enamel in the upper and lower jaws that cross each other during a 10 cm stroke) for these same species was, respectively, 4 and 27. As the lamellae increased in number, they also became packed more densely in the tooth, and the thickness of the enamel was reduced (from almost 5 mm in *P. gomphotheroides*, to less than, respectively, 4 mm in *Loxodonta*, 3 mm in *Elephas*, and 2 mm in *Mammuthus*). A broadening of the molars (in *Mammuthus*) and an increase in their height (in all three lineages) enhanced durability and longevity of the teeth. The greater height of the molars is accommodated in heightened maxillae, and so the position of the palate is depressed. With ventral direction of the tusk-bearing premaxillaries, and with the heightening of the cranium, the elephantid skull is flexed in the facial area and foreshortened on its basicranial axis (Maglio 1973).

The most highly derived condition for the *Mammuthus* lineage, exhibited by *Mammuthus primigenius*, was described above. Trends in *Elephas* paralleled those in *Mammuthus*, but reached a lesser extreme: for example, the third molars tripled in plate number, attaining a shearing index of 15 in modern *Elephas maximus*. As the enamel thinned in this lineage, it also became more folded (Maglio 1973).

By Maglio's (1973) account, although *Loxodonta* initiated changes in jaw and skull architecture 2 million years earlier than the *Elephas* and *Mammuthus* lineages, it has retained a relatively long mandibular symphysis, and its molars have remained very conservative. A threefold increase in molar crown height, and an increase in the shearing index to 5, constitute the main changes observed in the dentition of *Loxodonta*.

Throughout its evolutionary history, *Loxodonta* has remained in Africa. *Elephas* apparently dispersed twice out of Africa, in the mid-Pliocene into Asia where it evolved as a lineage terminating in *Elephas maximus*, and in the late Pliocene into Asia and Europe as lineages now extinct (Maglio 1973, pp. 79, 114, 115) (Fig. 18.4). *Mammuthus* migrated out of Africa in the late Pliocene and subsequently spread rapidly throughout Europe, Asia, and North America (Fig. 18.6). Maglio (1973) attributed the more rapid and progressive evolution of the dentitions of *Elephas* and *Mammuthus* to their entry into the fluctuating climates and new environments of the northern continents.

A noteworthy evolutionary trend in the Elephantidae that is known to characterize virtually all large mammals is a reduction of body size in insular forms. On many occasions, populations of elephantids that colonized islands without large carnivores have rapidly declined in body size, in some instances to full stature at a shoulder height of 1.5 m (*Mammuthus exilis*, of Santa Rosa Island, California USA) or even less than 1 m (*Elephas falconeri*, on the island of Sicily). The dwarfism is probably in part a consequence of natural selection for survival and reproduction in conditions of limited resources. Morphologically, the dwarfed forms display a combination of geometric scaling, paedomorphosis, and unique specialization (depending upon the characteristic) (Roth 1992*a*). Dwarfed populations on islands are discussed in detail in later chapters of this volume.

The last glaciation in the northern hemisphere coincided with the disappearance of *Elephas* from Africa and the extinction of *Elephas antiquus* and *Elephas falconeri* in Europe. The extinction of all remaining taxa of the Elephantidae, except for *Loxodonta africana* and *Elephas maximus*, occurred early in the Holocene.

Elephantids are valuable for correlation of Plio-Pleistocene deposits because they underwent rapid evolution over a wide geographical area, but the use of elephant taxa as biochronological indicators is complicated by the lack of an updated taxonomic revision or of general taxonomic consensus. For example, the name *Archidiskodon* is used for early members of the *Mammuthus* lineage of northern Eurasia, but it has also been synonymized with *Palaeoloxodon* for certain specimens in China. *Palaeoloxodon* is used at the generic level in Asia for several species (*E. namadicus*, *E. naumanni*) that seem to be attributed to *Elephas* elsewhere (Ronglin 1983). The *Palaeoloxodon* problem is further complicated by the use of the name as a genus by Aguirre (1969*a*) for African representative of *Elephas* and as a subgenus by Beden (1983) for species within the African *Elephas* lineage. To sort out this complicated taxonomic problem, cooperation between researchers on different continents will be necessary, and the specimens in Africa and Asia that are designated as *Palaeoloxodon* will need to be re-examined to determine the features that characterize *Palaeoloxodon* and distinguish it from *Elephas* and *Mammuthus*.

Disagreement is not limited to generic assignments, but has also arisen over intraspecific variation and evolutionary stages.

Fig. 18.6 World map showing reported distribution of the Elephantidae during the Plio-Pleistocene. *Elephas* and *Mammuthus* experienced a rapid radiation into northern climates, whereas *Loxodonta* remained in more tropical zones. Localities represented are illustrative, and by no means exhaustive. A, *Loxodonta adaurora*; B, *L. atlantica*; C, *Elephas ekorensis*; D, *E. recki*; E, *E. iolensis*; F, *E. namadicus*; G, *E. falconeri*; H, *E. planifrons*; I, *E. celebensis*; J, *E. platycephalus*; K, *E. hysudricus*; L, *E. hysudrindicus*; M, *Mammuthus imperator*; N, *M. meridionalis*; O, *M. armeniacus*; P, *M. primigenius*; Q, *M. columbi*; R, *M. africanavus*; S, *M. subplanifrons*. Sources consulted: Osborn (1942, pp. 912–1361), Maglio (1973, pp. 113, 115, 116), Coppens *et al.* (1978, pp. 351–8), Agenbroad (1984*b*, pp. 93–7). For simplicity, Pleistocene localities of *Loxodonta africana* and *Elephas maximus* were omitted; refer to Maglio (1973) for details, and to Fig. 34.6 in this volume for the distribution of these species.

Elephas recki in East Africa has been divided into four evolutionary stages by Maglio (1973) and into five evolutionary subspecies by Beden (1983). There is no clear consensus about the points at which each stage or subspecies grades into another. The number of junior synonyms for each taxon recognized by Maglio (1973) is astounding, and some of these synonyms have been divided among several taxa because of varying identification of specimens.

SPECIES-LEVEL DIVERSIFICATION

The latest and most comprehensive systematic revision of the Elephantidae is that of Maglio (1973). Research on proboscideans is often constrained by the sheer size of the specimens: it is usually cheaper and easier to send a researcher to various field and museum sites than to assemble a large series of specimens in one place. Maglio's study was unusual in that he had the resources to borrow and assemble a large collection of elephantid specimens from different localities in one place in North America, where he was able to compare them directly (Maglio, personal communication 1987). Such working conditions are unlikely to be duplicated any time soon, and Maglio's monograph remains a landmark not only because it is a beautifully executed study, but also because of its comprehensive nature. The delimitation of species allowed Maglio to calculate taxonomic rates of evolution within the Elephantidae as a counterpart to his analysis of phyletic evolution, which was summarized above.

Maglio (1973) noted that early in the evolution of the family, species diversity was low; new species appeared rapidly through cladogenesis, and rates of extinction were also low. With the emergence of more progressive lineages of *Loxodonta*, *Elephas*, and *Mammuthus*, both speciation and phyletic change contributed to rapid turnover. In the late Pliocene and early Pleistocene, a time when the Elephantidae greatly expanded their geographical range, morphological change was slow, and rates of diversification were modest and accompanied by virtually no extinction. Maglio identified a second phase of evolution in the family in the mid-Pleistocene. Both speciation and phyletic evolution accelerated at this time when elephantids (especially *Elephas* and *Mammuthus*) were subjected to great climatic change and subdivided environments accompanying glaciation and possibly intensified competition between lineages. In the late Pleistocene, rates of extinction rose suddenly and were sustained at eight species per million years. Even now the two surviving species are at risk.

Stanley (1979, p. 240) used Maglio's conclusions about phyletic and taxonomic evolution in elephants in building more general arguments for interpreting the majority of change in evolution as punctuational. Not all of Maglio's primary systematic inferences have remained unchallenged, however. Beden (1983, 1987*a*, *b*) suggested that fossil material from East Africa could accommodate subdivision into a greater number of taxa and stages, to which he provisionally applied subgeneric and subspecific names. Maglio had not directly examined North American *Mammuthus* material, so his views on the systematics of the later members of this genus must be considered tentative.

Madden (1981*b*) undertook the difficult task of systematic revision of the North American mammoths, with somewhat unconventional conclusions (Table 18.1). This revision has not yet stood formal review for publication, and Madden's own views have apparently been subject to some change without explicit discussion (cf. Madden, 1981*a*, *b*, with reference to *Mammuthus imperator*, *M. columbi*, and *M. exilis*). For North American elephantids, Maglio's work remains the most widely cited, with some proviso for uncertainty in the specific status and relationships of *M. imperator* and of *M. jeffersonii* (e.g. Agenbroad 1984*b*; Graham 1986*b*).

The very basis for delimiting species of elephantids may be due for reconsideration. An appreciation of intrapopulational and intraspecific variability is of central importance in the recognition of species differences, and yet systematic accounts and alpha taxonomies of the fossils have incorporated little in the way of information about variability in the two extant species of elephantids. Species-level taxonomy in the Elephantidae is based upon metrical characteristics of the molars, yet by comparison with the teeth of other mammals, elephant teeth develop in an extremely dynamic context. In elephantids, tooth formation continues in the posterior part of the jaw at the same time that other teeth (or even the more anterior portions of the same tooth) endure the forces of mastication, while the toothrow as a whole undergoes eruption and progression. The morphological consequences of such mechanical forces on developing teeth may be deformation, and deformation of varying degrees may not only account for the relatively high frequencies of certain types of dental anomalies in elephants (Roth 1989), but also enhance more subtle forms of intraspecific variation (Roth 1992*b*). Within populations and across the full geographical ranges of species, variability in dental dimensions has been documented to be higher in modern *Elephas maximus* and *Loxodonta africana* than in other mammals. The variability observed in the modern forms thus may raise questions about the validity of some of the nominal species of fossils recognized by Maglio, Madden, and others, providing some guidelines for determining sample sizes necessary for identifying and separating groups (Roth 1992*b*). A re-evaluation of elephant taxonomy that takes account of the heightened phenotypic plasticity in this group will yield a sounder basis for evolutionary inferences.

CONCLUSIONS

The Elephantidae represent the last major radiation of the Proboscidea. What have been interpreted as rapid adaptive shifts resulted in seemingly high diversity, allowing radiations into every part of the world except Antarctica, Australia, and South

Table 18.1 North American mammoths. The species of North American mammoths recognized in the two most recent comprehensive revisions are listed to illustrate some of the complexity of species-level taxonomy within the Elephantidae. Limitations in space prevent our reproducing here the detailed synonymies presented in the original reference sources, which vary in their groupings of individual specimens, but a comparison of Maglio's (1973) list with the common names associated with Madden's (1981*b*) listing gives some indication of how disparate these views are. The tendency most recently has been to decrease the number of recognized species — Osborn (1942), for example, had recognized 16 species of North American mammoths which he grouped into three genera — but simplification has not entirely eliminated confusion. '*Mammuthus jeffersonii*' (Osborn 1922*b*), for example (cf. Madden's *M. jacksoni*, the 'Jeffersonian mammoth'), has been widely regarded (Maglio 1973; Kurtén and Anderson 1980; Agenbroad 1984*b*; Graham 1986*b*) as a progressive, southerly derivative of the *M. imperator–M. columbi* lineage that is difficult to distinguish from *M. primigenius*, a distinct lineage (Figs 18.1, 18.4) with which it has converged.

Maglio (1973)	Madden (1981*b*)
Mammuthus meridionalis (Nesti, 1825)	*M. gromovi* (Alexeeva and Garutt, 1965) Archaic, Eurasian mammoths *M. hayi* (Barbour, 1915) Hay's mammoths
M. imperator (Leidy, 1858)	*M. imperator* (Leidy, 1858) imperial mammoths *M. exilis* (Stock and Furlong, 1928) dwarf mammoths
M. columbi (Falconer, 1857)	*M. columbi* (Falconer, 1857) Columbian mammoths *M. jacksoni* (Mather, 1838) Jeffersonian mammoths
M. primigenius (Blumenbach, 1803)	*M. mammonteus* (Cuvier, 1799) woolly mammoths

America. Major evolutionary trends in morphology were correlated with a functional shift in masticatory function, with specializations in the cranium, mandible, and dentition relating to dietary changes.

Specific details concerning the origin — the identification of ancestors and sister taxa — and radiation of the Elephantidae — how many species have there been and what is their cladistic relationship? — remain unresolved, but not for lack of well-argued scenarios. What is most needed is clear documentation, beginning at the very primary levels of identifying and clearly defining characters (which we may take to include standardizing techniques of measurement), noting unambiguously which characters are observed in which specimens and which specimens are assigned to which taxa, and recognizing and accounting for intrataxonomic variability. Ontogeny in elephants is lengthy, and no view of elephantid morphology is complete without consideration of ontogenetic variation. Osteological (as opposed to dental) material has been all but ignored, but with appropriate study could be put to better use. A picture is now beginning to emerge of the variability manifest in modern species of elephants (Laws 1966; Roth 1984, 1990, 1992*b*, 1993; Roth and Shoshani 1988; Haynes 1991; Lister, this volume, Chapter 19), and this will doubtless assist in delimiting taxa. Studies of variability will require not only review and synthesis, however, but also additional material, to improve sample sizes and to fill gaps in stratigraphy and in taxonomic and geographical sampling. A reassessment of species-level variability is an essential basis for analysis of phylogeny at higher levels, especially where individual specimens have received varying taxonomic assignments.

For phylogenetic reconstruction, cladistic analysis offers the most promising approach for examining the interrelationship of taxa. Although observations about the amounts of morphological plasticity and character conflict raise questions about whether morphological characters have indeed retained historical information to a degree sufficient to permit a full reconstruction of phylogeny, it is only by using such explicit methods that either resolution will be obtained, or the sources of difficulty will be clearly identified.

A call for more material, more data, more analysis, and more synthesis is hardly a novel conclusion, but we believe it is the most reasonable one at the current stage. It is only through further development of these kinds of data that we will begin to understand the morphological transformations of Plio-Pleistocene elephants and to evaluate these animals as models for examining migration, diversification, ecology, and patterns of adaptation in the late Cenozoic.

ACKNOWLEDGEMENTS

We thank Daryl Domning, Jon Kalb, and the editors of this volume for comments on the manuscript. Dan Cole constructed the base maps for the world and Africa, and Gary Aronsen added new sites.

NINETEEN

Evolution and taxonomy of Eurasian mammoths

ADRIAN M. LISTER

INTRODUCTION

Mammoths (genus *Mammuthus*) first appeared in continental Eurasia in the late Pliocene, about 2.5 Ma, and persisted until the end of the Pleistocene, around 10 000 years BP. Samples of mammoth fossils through the Eurasian Plio-Pleistocene show a series of anatomical trends, including changes in body size, heightening of the cranium and molar teeth, shortening of the cranium and mandible, thinning of molar enamel, and increase in the number of lamellae (enamel plates). These trends have long been known from studies of typical samples of the three main nominal chronospecies of the lineage: early Pleistocene *Mammuthus meridionalis* (Nesti), middle Pleistocene *M. trogontherii* (Pohlig), and late Pleistocene *M. primigenius* (Blumenbach) (Soergel 1912; Aguirre 1968, 1969*b*; Maglio 1973; here Fig. 19.1). New material and improved dating techniques now allow many more samples to be placed in chronological order and the evolutionary sequence examined in more detail. Further details of many of the topics summarized in this paper will be given elsewhere.

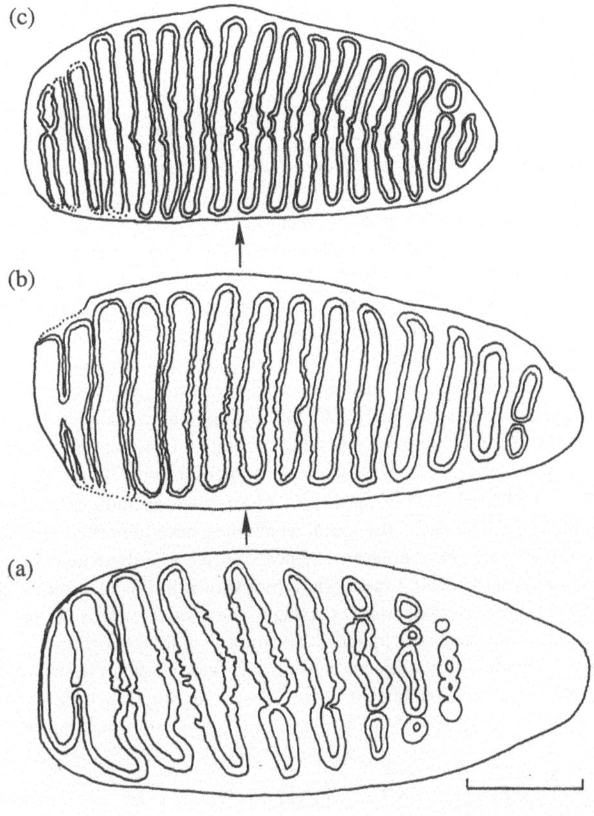

Fig. 19.1 Right third upper molars of *Mammuthus* in occlusal view, illustrating changes in lamellar frequency and enamel thickness. Scale bar 5 cm (a) *M. meridionalis*, early Pleistocene, Siniaya Balka, Russia (Palaeontological Institute, Moscow, no. 1249/234); (b) *M. trogontherii*, early Middle Pleistocene, Mosbach III, Germany (Natural History Museum, Mainz); (c) *M. primigenius* late Pleistocene, Siberia (Palaeontological Institute, Moscow).

METHODOLOGY

The new data presented here are restricted to provenanced samples with some independent evidence of dating. In addition, morphological comparisons are based as far as possible on valid sample sizes. Variation is particularly marked in elephant molars, as a result both of developmental effects, and the state of wear or fragmentation of the fossil (Adam 1988; Roth 1989). By seeking statistically suitable samples, and using a scheme of measurement which takes account of variability due to wear or breakage, observed differences reflect meaningful patterns of variation between biological populations.

The six molariform teeth are named dp2, dp3, dp4, M1, M2, and M3, but in the text are all referred to as 'molars'. Identification of teeth to their position in the row was by size, shape, hypsodonty, and plate (lamellar) count; very few teeth could not be clearly allocated in this way (see also Sher and Garutt 1985*b*; Roth and Shoshani 1988). Techniques of dental measurement are described and illustrated in Fig. 19.2, and are modified after those of Aguirre (1968, 1969*b*) and Maglio (1973). In citing plate count, 'x' indicates anterior and posterior talons. Cranial data have been taken mostly from published sources; measurements are illustrated in Fig. 19.3. All measurements are quoted in millimetres unless otherwise stated.

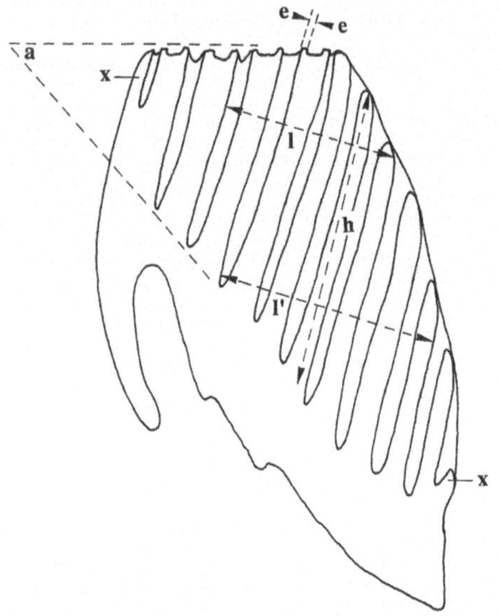

Fig. 19.2 Measurements and features of upper molars, modified after Aguirre (1968, 1969*b*) and Maglio (1973). a, angle of eruption (angle between occlusal surface in early wear, and crown base); e, enamel thickness, measured parallel to the long axis of a plate, and recorded as the mode of a series of measurements on plates across the occlusal surface; h, crown height, measured on unworn plates in the region of maximum height (e.g. between plates 5 and 12 in *M. primigenius*) (Sher and Garutt 1985*a*). The hypsodonty index is the ratio between crown height and width. l and l′, Example points of measurement (five plates shown) for lamellar frequency (LF), taken normal to the lamellae at four points (base and preserved top of crown, labial and lingual sides) and averaged (Maglio 1973). For each measurement, $LF = N'/L' \times 100$, where L′ is the length of N′ plate+cement intervals, excluding the first few and last few plates of the tooth, which may have abnormal LFs. x, Talons (the single most anterior and posterior plates if these do not attach fully to the root). Crown width (not shown in this diagram) is the width of the widest plate, taken parallel to the transverse axis of the plates. This measurement includes cement, and in fossil specimens lacking cement, an estimate for cement thickness, based on the observed cement thickness in equivalent teeth of the same sample, is added.

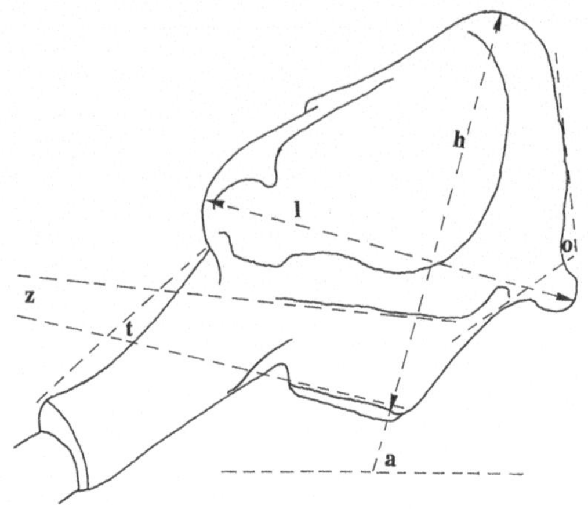

Fig. 19.3 Measurements on crania, after Palombo (1972). a, Angle between dorso-ventral axis of cranium and the horizontal; h, height of cranium, from cranial vertex to alveolar plane of molars; l, length of cranium, from preorbital process to occipital condyles; o, 'occipital angle', the angle between the occipital and basicranial planes; t, angle between tusk alveoli and molar alveoli; z, angle between zygomatic arch and molar alveoli.

DIAGNOSIS AND NOMENCLATURE OF THE MAMMOTH LINEAGE

Diagnostic features of *Mammuthus* include the high, short skull, with dorsally expanded parietals; the lack of a sagittal depression at the cranial vertex; a very obtuse occipital angle; closely spaced and little-divergent tusk alveoli; strongly curved and twisted tusks; and simple, unfolded molar enamel loops (Osborn 1942; Maglio 1973).

The valid generic name for the late Pleistocene Eurasian mammoth is *Mammuthus* Brookes 1828, with *M. primigenius* fixed as its type species (Garutt *et al.* 1990; ICZN 1991). The name *Archidiskodon* Pohlig 1888 (type species *Elephas meridionalis* Nesti 1825) is also taxonomically valid as a genus or subgenus, and is available for use by those who wish to differentiate *meridionalis* and other early species from *Mammuthus* (Azzaroli 1977; Garutt *et al.* 1990). This usage is most justifiable for authors who regard the early species as forming a separate lineage, not directly ancestral to *M. primigenius* (e.g. Osborn 1942; Azzaroli 1977). Others (e.g. Dubrovo 1977) see a continuous lineage yet prefer a taxonomic separation of *meridionalis* as *Archidiskodon* on the basis of its retention of primitive characters, a non-cladistic decision which renders *Archidiskodon* paraphyletic. In this chapter, I will use *Mammuthus*, and the common name 'mammoth', for all Eurasian Pliocene and Pleistocene representatives. Quite apart from the above considerations, the application of a single genus name avoids problems of the generic designation of intermediate forms, e.g. *trogontherii*, which have troubled authors in the past (Garutt and Nikolskaja 1988; see below). In addition, the use of *Archidiskodon* for the early species would create problems for the taxonomy of the North American lineage, derived in parallel from the early Eurasian species, since *Mammuthus*, commonly used for the terminal North American forms (e.g. *M. columbi*) would then be biphyletic (Lister 1993*a*).

THE 'SOUTHERN ELEPHANT': *MAMMUTHUS MERIDIONALIS*

Characteristics of *Mammuthus meridionalis*, compared with later mammoths, were listed by Aguirre (1968, 1969*b*), Maglio (1973), Azzaroli (1966, 1977): broad, low-crowned molars with 11–14 plates in M3; low lamellar frequency and thick enamel; lamellae with occasional median loops (a vestige of the gomphothere central conule) and often deep subdivisions; long, low skull; relatively sharp occipital angle; bulging parietal domes; concave forehead; wide external naris; tusk alveoli projecting forwards; relatively short tusks; long, relatively low mandible; long mandibular rostrum; 'serial' carpal structure (see later). *M. meridionalis* is known mainly from Europe, with rare records from Siberia, the most distant being in the Vily'uj valley, Yakutia (Dubrovo 1964; Dubrovo and Mol 1991). The spread of this species into north-eastern Siberia on at least one occasion is expected in view of its appearance in North America at about 1.5 Ma (Lister 1993*a*).

Among the earliest Eurasian mammoth remains are those from the British Red Crag, now regarded as dating to *c.* 3.0–2.5 Ma (Zalasiewicz *et al.* 1988). A complete third lower molar from this deposit has only x10x plates (Lister and van Essen, unpublished observations 1992). More abundant early material has come from the Khapry faunal complex, especially the Liventsovka quarry, near Rostov, south-western Russia. These late Pliocene deposits date to *c.* 2.5–1.8 Ma on the basis of lithostratigraphy and palaeomagnetism (Dubrovo 1977; Vangengeim and Pevsner 1991). Measurements of molar parameters are plotted in Fig. 19.4, and illustrate a slightly more 'primitive' condition (lower lamellar frequency, higher enamel thickness) than in *M. meridionalis* from its type site of the Upper Valdarno (*c.* 1.6 Ma). Plate number on third molars is generally x12x–x13x.

Alexeeva and Garutt (1965) named the species '*Archidiskodon*' *gromovi* for the Khapry material, on the basis not only of dental morphology, but also cranial differences from typical *M. meridionalis*. According to Garutt and Baigusheva (1981), the best-preserved cranium from Khapry (Fig. 19.5a) has more primitive proportions (length to height ratio 86 per cent, compared to 80 per cent in a later skull of *M. meridionalis* from Nogaisk, Azov Sea region). Azzaroli (1977, p. 153) noted that the Khapry skulls are closely similar to typical *M. meridionalis*, but pointed out that the parietal domes are less prominent, and the occipital/basicranial angle less obtuse, and followed the designation '*A.*' *gromovi*. Dubrovo (1977, 1989; Gabunia and Dubrovo 1990) has sought to refute these differences (and hence the validity of '*A. gromovi*') by showing that the Khapry morphology falls within the range of *M. meridionalis* as a whole; for example, she gives a length/height index of 89 per cent for the lectotype *M. meridionalis* from the Upper Valdarno. In mandibular morphology, the Khapry material shows the long, shallow horizontal ramus and prominent rostrum typical of *M. meridionalis* (Fig. 19.6a).

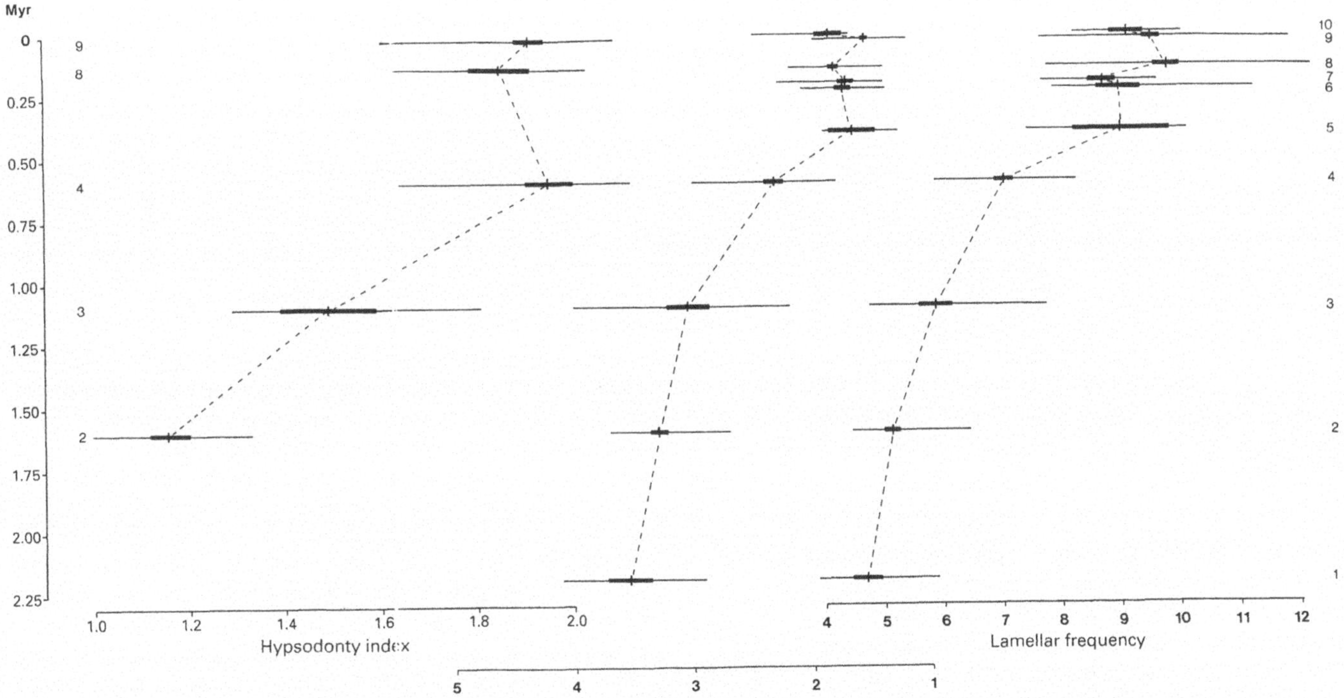

Fig. 19.4 The evolution of hypsodonty, enamel thickness, and lamellar frequency in selected samples of third upper molars of the *Mammuthus* lineage in Europe. Bar symbols indicate the range, mean, and standard error of the mean. Dashed lines are for ease of viewing, and do not necessarily imply gradual evolution between the sampling points. In the following list, sample sizes for hypsodonty index, enamel thickness, and lamellar frequency are given in that order. *M. meridionalis*: 1, Khapry/Liventsovka, Russia (–/7/7); 2, Upper Valdarno, Italy (8/22/19); 3, Siniaya Balka, Russia (5/12/12). *M. trogontherii*: 4, Mosbach III, Germany (10/20/18). *M. primigenius*: 5, Homersfield, England (–3/3/); 6, Tourville, France (–/8/8); 7, Ilford, England (–/13/11); 8, Balderton, England (6/21/23); 9, Předmostí, Czechoslovakia (20/40/42); 10, Kostenki, Russia (–7/6). All data original.

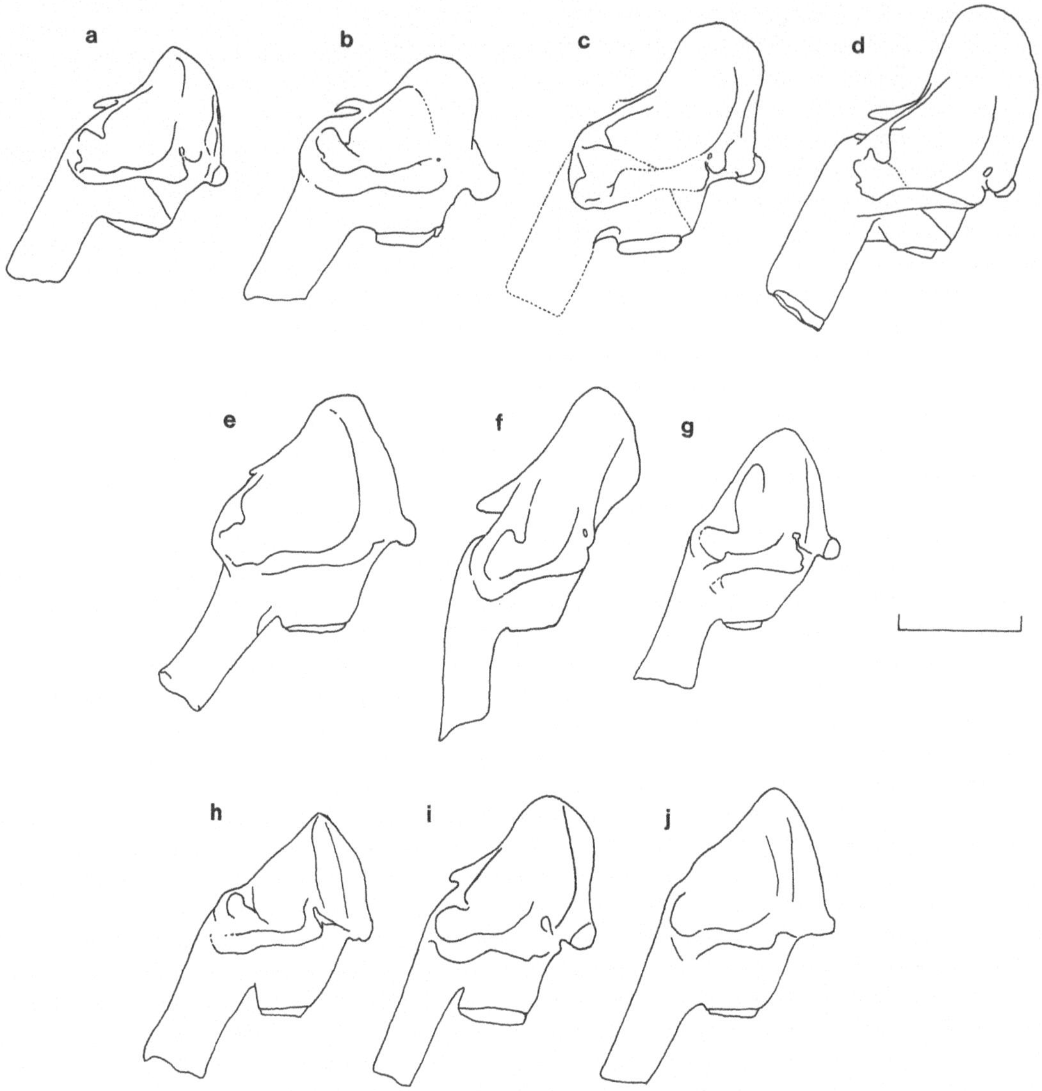

Fig. 19.5 Crania of the *Mammuthus* lineage. *M. meridionalis*: (a) Liventsovka, Russia (after Azzaroli 1977); (b) Chilhac, France (c. 1.9 Ma, after Boeuf 1990); (c) Upper Valdarno, Italy (after Azzaroli 1966); (d) Scoppito, Italy (after Maccagno 1962). *M. trogontherii*: (e) Gelsenkirchen, Germany (after Palombo 1972); (f) Azov, Russia (distorted; after Garutt and Nikolskaja 1988). Early *M. primigenius*: (g) Cherny Yar, Russia (after Dubrovo 1966); (h) Ilford, England (after Adams 1877–1879, 1881); (i) Tarquinia, Italy (after Ambrosetti 1964). Late *M. primigenius*: (j) Debica, Poland (after Kubiak 1980). Scale bar 50 cm.

The most notable character apparently differentiating the Khapry mammoths from later samples is the presence, in one of the skulls, of a permanent fourth upper premolar (Fig. 19.7), generally found only in more primitive proboscideans. Dubrovo (1977, 1989), however, pointed out that as only one skull of appropriate ontogenetic age is known, the presence of a premolar could be an individual atavism; and that in any case, skulls of this ontogenetic age are unknown in typical *M. meridionalis*, so that the presence or absence of premolars there is strictly unknown.

In summary, the molar data presented in this study (Fig. 19.4) indicate that the Khapry sample is statistically slightly more primitive than typical *M. meridionalis*. The questions about skull morphology and permanent premolars must be regarded as open, pending the discovery of further material. Following Aguirre (1968, 1969*b*) and Maglio (1973), the Khapry material can at present be regarded as a somewhat primitive stage of *M. meridionalis*.

The type material of *Mammuthus meridionalis* Nesti (1825) from the Upper Valdarno, Italy was described by Weithofer (1891) and Azzaroli (1977). Azzaroli (1977) and Azzaroli *et al.* (1988) show that while the precise horizon of the older finds is uncertain, almost all came from the 'Montevarchi group', covering the Olivola and Tasso faunal units, and dating to *c.* 1.7–1.4 Ma on the basis of fauna and lithostratigraphy. This 'typical' level of *M. meridionalis* evolution was named the 'Montavarchi Stage' [sic] by Maglio (1973). Ranges of important

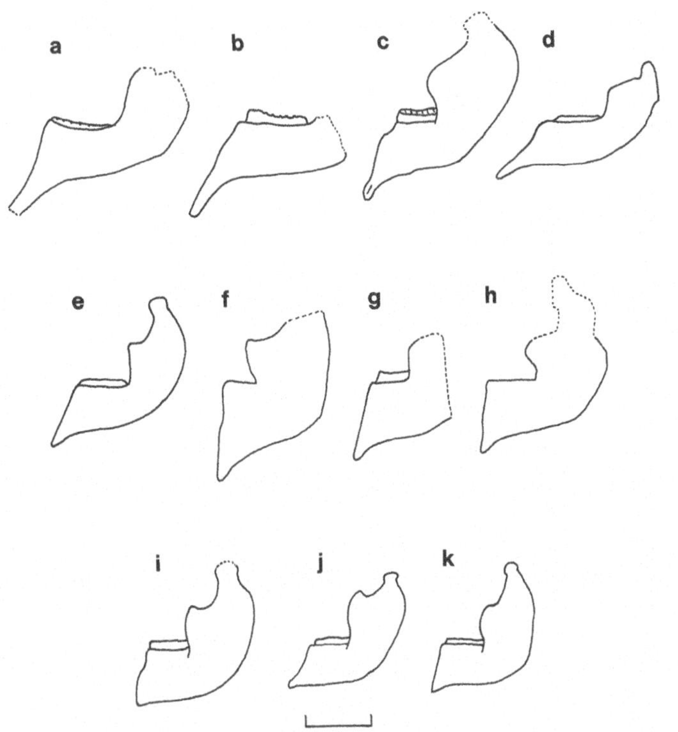

Fig. 19.6 Mandibles of the *Mammuthus* lineage. *M. meridionalis*: (a) Liventsovka, Russia (after Garutt and Baigusheva 1981); (b) Senèze, France (after Depéret *et al.* 1923); (c) Upper Valdarno, Italy (after Azzaroli 1966); (d) Port Katon, Russia (Tamanian; after Dubrovo 1964). *M. trogontherii*: (e) Gelsenkirchen, Germany (after Siegfried 1956); (f) Azov, Russia (after Garutt and Nikolskaja 1988); (g) Mosbach III, Germany (Mainz Museum, original); (h) Tiraspol, Moldavia (after Dubrovo 1971). Early *M. primigenius*: (i) Cherny Yar, Russia (after Dubrovo 1966). Late *M. primigenius*: (j) and (k) Berelekh, Russia (showing longer and shorter rostra, respectively; after Vereshchagin 1977*b*). Scale bar 25 cm.

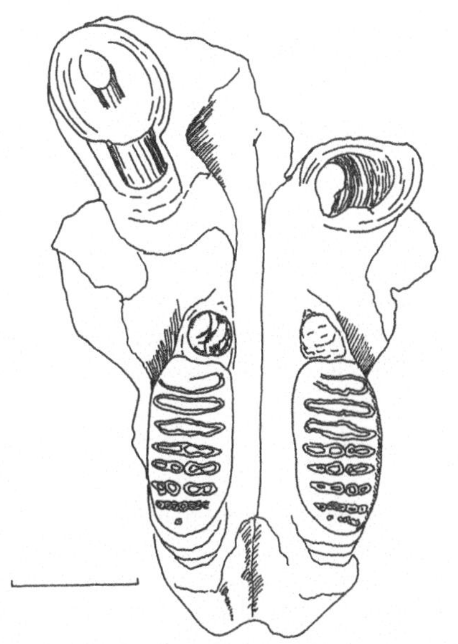

Fig. 19.7 Cranium of *Mammuthus meridionalis* from Liventsovka, Russia, in ventral view (after Gromov 1977). Note that dp4 has been lost by wear, but in its place a permanent premolar P4 can be seen anterior to the right M1. The left P4 alveolus is empty. Scale bar 10 cm.

dental measurements are shown in Fig. 19.4, and the type cranium and a mandible are illustrated in Figs 19.5c and 19.6c. The plate number is typically x12x–x13x.

An important later sample of *M. meridionalis* came from the Siniaya Balka locality in the Taman peninsula, southern Russia, the type locality of the Taman faunal complex, dated by palaeomagnetism and microfauna to *c.* 1.0 Ma (Vangengeim *et al.* 1991). These remains have been described in detail by Dubrovo (1964, 1977), who named them '*Archidiskodon*' *meridionalis tamanensis*. Lamellar frequency and enamel thickness (Fig. 19.4) show clear advancement over the type sample from the Upper Valdarno; plate number on third molars ranges from x14x to x18x (Lister and Joysey 1992). Particularly significant is the increase in crown height, as this implies deeper (more 'advanced') crania and mandibles. Adult mandibles continue to show a strong rostrum (Fig. 19.6d).

Fossils classified as 'advanced' *M. meridionalis* have come from a number of other sites in the range *c.* 1.2–0.6 Ma. Coppens and Beden (1982) described material from St. Prest, France, as more advanced than the type *meridionalis*, and named it *M. meridionalis depereti*. A skeleton from Scoppito (L'Aquila) (Maccagno 1962) and some remains from Farneta, Italy, were referred by Azzaroli (1977) to '*A.*' *meridionalis vestinus*, described as being of large size, and short, high cranium. The 'Farneta faunal unit' is believed to be of similar age to Taman, *c.* 1.0 Ma (Azzaroli *et al.* 1988). Depéret *et al.* (1923) named '*Elephas*' *meridionalis cromerensis* on the basis of material from deposits at Kessingland in the Cromer Forest-bed Formation (CF-bF), England, now regarded as close in age to the type Cromerian, *c.* 0.7–0.6 Ma (Lister 1993*d*). Of similar age (Stuart 1981) are the remains from Voigtstedt, Germany, described by Dietrich (1965) as '*A.*' *meridionalis voigtstedtensis*. Maglio defined late *meridionalis* as the 'Bacton Stage', named after a skull from the CF-bF at Bacton, although there is nothing particularly 'advanced' in this specimen relative to typical remains of the species (Lister and van Essen, unpublished observations 1992). The relationship of these samples to succeeding ones, referred to the chronospecies *Mammuthus trogontherii*, is discussed in the following section.

In the postcranial skeleton, a feature which has been given some importance in distinguishing *M. meridionalis* from later species is the alignment of the carpal bones (Fig. 19.8). In skeletons of *M. meridionalis* from Nogaisk, Azov region (Garutt 1954*b*; here Fig. 19.8a) and Stavropol (Dubrovo and Mol 1991), the carpus has a serial structure, i.e. the lunar, the central bone of the proximal row, fits only over the underlying magnum, and does not extend over the unciform or trapezoid bones. In later species (e.g. *M. primigenius*), the lunar is relatively larger and overlaps these bones, especially the trapezoid (Fig. 19.8d); this is the 'aserial' structure (Garutt 1954*b*).

In other specimens of *M. meridionalis*, however, the carpus does not appear to show a fully serial structure. Examples can be

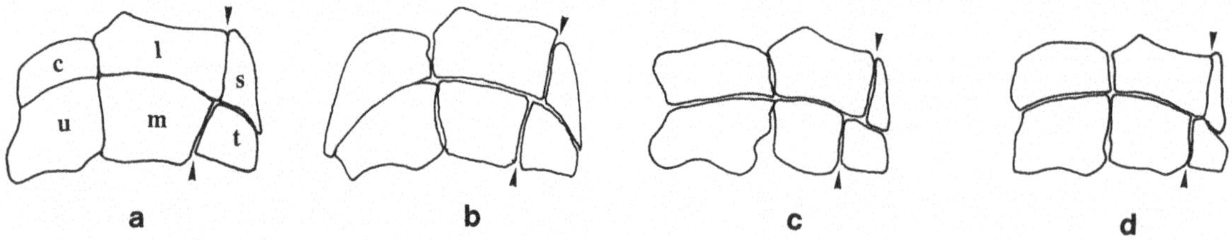

Fig. 19.8 Right carpal bones of the *Mammuthus* lineage. (a) *M. meridionalis*, Nogaisk, Russia (after Garutt 1954*b*; reversed); (b) *M. meridionalis*, Upper Valdarno, Italy (after Weithofer 1891; reversed); (c) *M. meridionalis* or *M. trogontherii*, Edersleben, Germany (Sangerhausen Museum, original); (d) *M. primigenius* (after Olsen 1972). From left to right, the bones of the top row are: c, cuneiform; l, lunar; and s, scaphoid; and of the bottom row: u, unciform; m, magnum, and t, trapezoid. Arrows highlight the degree of overlap between the lunar and trapezoid: (a) serial; (b)–(d) 'aserial'.

seen in material from the Valdarno (Fig. 19.8b; Weithofer 1891), and from Scoppito (Maccagno 1962, pp. vii, 77, 89–91). Rather than a simple transition from serial to aserial at the *meridionalis/trogontherii* boundary, a range of morphologies appears to have occurred in *meridionalis*, with exclusively aserial carpals in later species.

Mammuthus meridionalis was a large elephantid. Published shoulder heights of mounted adult skeletons range from 335 to 410 cm (Table 19.1).

THE 'STEPPE MAMMOTH': *MAMMUTHUS TROGONTHERII*

The nomenclature of *Mammuthus trogontherii* has been the subject of debate. The species '*Elephas*' *trogontherii* was founded by Pohlig (1885) for fossils from Süssenborn; see Garutt (1971), Adam (1988) and Kahlke (1990) for a discussion of the type material. Earlier, however, Falconer (1857) had described the species '*Elephas*' *armeniacus* on the basis of a molar tooth from the

Table 19.1 Shoulder heights (in cm) of adult *Mammuthus* skeletons. Genders taken mainly from Lister (this volume, Chapter 25)

Locality	Gender	Shoulder height (cm)	References
M. meridionalis			
Nogaisk, Russia	Male	410	Garutt (1954*b*)
Georgievsk, Georgia	Male	400	Garutt (1964)
Durfort, France	Male	383	Osborn (1942)
Farneta, Italy	Male	370	Azzaroli (1977)
Borro al Querco, Italy	Male	335	Azzaroli (1977)
M. meridionalis / M. trogontherii			
Edersleben, Germany	Female	345	Garutt and Nikolskaja (1988)
M. trogontherii			
Azov, Russia	Male	450	Garutt and Nikolskaja (1988)
Odessa, Russia	?	>370	Garutt (1954*b*)
M. primigenius (early)			
Steinheim, Germany	Male	370	Dietrich (1912)
Pfannerhall, Germany	Male?	300	Toepfer (1957)
M. primigenius (late)			
Siegsdorf, Germany	Male	~365	Mol *et al.* (1988)
Ahlen, Germany	Male	324	Siegfried (1959); personal observation 1992
Lena R., Siberia	Male?	320	Garutt (1964)
Borna, Germany	Male	320	Felix (1912)
Condover, England	Male	~308*	Coope and Lister (1987); personal observation 1992
Praz Rodet, Switzerland	Male	280	Weidmann (1969); personal observation 1992
Aa, France	Female	280	Pontier (1913); Astre (1929)
Beresovka, Siberia	Male	265*	Zalensky (1903)
Taimyr, Siberia	Male	265	Garutt (1964)
Atrikanova, Siberia	Male	260*	Coppens (1958)

Specimens marked with an asterisk have some unfused long bone epiphyses and so may not quite have completed growth region of Erzurum, Turkey, which in morphology is similar to *M. trogontherii*, and referred some material from the British Cromer Forest-bed Formation to his taxon. On the basis of priority, Aguirre (1968, 1969*b*) and Maglio (1973) regarded *armeniacus* as a senior synonym and applied it to all the European material. Although *armeniacus* cannot be regarded as a *nomen oblitum* (it was used also by Osborn (1942)), *trogontherii* is much more widely dispersed in the literature, and for this reason, Beden and Guérin (1975) and Kotsakis *et al.* (1978) have recommended its retention. There are further points in favour of this policy. *Mammuthus trogontherii* was based on material from Süssenborn, a very extensively studied site in the middle of the species' main geographical range (Kahlke 1969) and yielding the largest known sample of these mammoths, which has been the subject of detailed statistical characterization (Guenther 1969; Lister and Joysey 1992). *Mammuthus armeniacus*, by contrast, was based on material from deposits of uncertain age, outside the main area of discovery of middle Pleistocene mammoths. Further, the fossils are extremely sparse, making the species hard to characterize; indeed, Adam (1988) claims that we have insufficient morphological grounds even for synonymizing it with *M. trogontherii*, and suggests instead that the Erzurum material might represent a westerly population of Asian elephant, *Elephas maximus*. For these reasons, the material described here (all of which is from the European area) is named *Mammuthus trogontherii*.

There has also been some debate concerning the generic attribution of *trogontherii*. Garutt and others (Gromov and Garutt 1975; Baigusheva and Garutt 1987; Garutt and Nikolskaja 1988) have sought to place it in *Archidiskodon* on the basis of supposedly shared characters with *meridionalis*, such as a pronounced mandibular rostrum in some specimens (see above), and the form of lamellar subdivision (short central cusp and two long lateral ones) seen in some molars. Most authors, even those who retain the genus *Archidiskodon* for *meridionalis*, prefer to place *trogontherii* in *Mammuthus*, regarding its morphology as closer to *primigenius* than to *meridionalis* (e.g. Dubrovo 1971; Azzaroli 1977). This would appear to be valid: the major transitions of the mammoth lineage in hypsodonty, skull structure, and carpal arrangement appear to have been completed or at least initialized between the *meridionalis* and *trogontherii* stages, rather than between *trogontherii* and *primigenius*. The problem is avoided altogether, however, by applying *Mammuthus* to the entire series, as explained earlier.

The type sample of *M. trogontherii* from Süssenborn, Germany comprises a very large number of molars and some cranial and postcranial material (Guenther 1969; Lister and Joysey 1992). Another important sample has come from the upper gravels of Mosbach, Germany (Mosbach III). On the basis of lithostratigraphy and microfauna, a date within a cool phase of the Cromerian Complex (*c.* 0.7–0.5 Ma) is likely for both Süssenborn (Steinmüller 1972; Maul 1990) and Mosbach (Brüning 1978; Igel 1984; von Koenigswald and Tobien 1987), with Mosbach possibly slightly later on the basis of microfauna. In hypsodonty, lamellar frequency and enamel thickness, the two samples are statistically indistinguishable (author's unpublished data), but are clearly advanced over that of the late *meridionalis* from Siniaya Balka (Fig. 19.4). Lamellar number in *M. trogontherii* ranges from x17x to x21x in M3 (Lister and Joysey 1992). Aguirre (1968, 1969*b*, fig. 6) indicates angles of molar eruption of 34°–58° in *M. trogontherii*, compared to 25°–37° in *M. meridionalis*, corresponding to the greater height of the crown and implying concomitant cranial modification. A further important sample of *M. trogontherii*, from deposits of similar age at Tiraspol, Moldavia, was named '*Elephas wüsti*' by Pavlova (1910), but shown to be dentally identical to Süssenborn by Dubrovo (1971).

The only partly satisfactory adult crania of *M. trogontherii* are those from Gelsenkirchen, Germany (Siegfried 1956; Fig.19.5e) and Azov, Russia (Baigusheva and Garutt 1987; Fig. 19.5f), although the Azov skull appears somewhat distorted, and that from Gelsenkirchen partially reconstructed. Caution must be attached to comparisons based on such a small sample of uncertain dating. None the less, in some key features, the skulls do seem to be intermediate in morphology between *meridionalis* and *primigenius* (Siegfried 1956).

Mandibular structure in *M. trogontherii* is variable, but is in general more 'advanced' than in *M. meridionalis*. Gromov and Garutt (1975) indicated a well-developed rostrum in some specimens from Süssenborn. However, Dubrovo (1971) drew attention to Süssenborn mandibles with short rostra and illustrated specimens from Tiraspol also showing this condition (Fig. 19.6h). Mandibles from Mosbach are short and deep, with a moderately developed rostrum (personal observations 1992; Fig. 19.6g).

Dubrovo (1977) indicated aserial carpal structure in *M. trogontherii*, based on a skeleton from Edersleben, Germany (Fig. 19.8c; Garutt and Nikolskaja 1988), but the attribution of this specimen to *M. trogontherii* or *M. meridionalis* is uncertain (its M^3 has × 14 × plates), and no other associated forefeet of *M. trogontherii* appear to exist. In body size, *M. trogontherii* was as large as, or slightly larger, on average than *M. meridionalis*; the skeletal heights of two adult males are >370 and 450 cm (Table 19.1).

The status of *Mammuthus trogontherii* as an intermediate link between *M. meridionalis* and *M. primigenius*, first proposed by Pohlig (1885), has been accepted by most authors. Azzaroli (1966, 1977), while agreeing with the derivation of *M. primigenius* from *M. trogontherii*, did not accept the origin of *M. trogontherii* from *M. meridionalis*. He based this chiefly on the '*A. meridionalis vestinus*' remains from Scoppito and Farneta, Italy, regarded as later than the type material from the Upper Valdarno. The expanded parietals and short, deep cranial vault of these specimens (Fig. 19.5d), and the large size of the Scoppito skeleton, were regarded as specializations not in the direction of *M. trogontherii*, and therefore precluding ancestry to the latter. This implies an origin of *M. trogontherii* by allopatric speciation, the two lineages coexisting for a while before the extinction of *M. meridionalis*. The occurrence of advanced *M. trogontherii*-like elephants in north-eastern Siberia, apparently contemporaneous with European *M. meridionalis* (Sher 1986), is consistent with such a model (Sher and Lister, unpublished observations 1995). In addition, there is recent evidence from Britain and Germany suggesting temporal 'overlap' between the latest *meridionalis* and earliest *trogontherii* morphologies in Europe, some specimens typical of the latter species appearing in lower strata than those of the former (Lister unpublished observations 1995). On the other hand, such a reversal, if it can be substantiated, might represent populational variation on a level which at the present day would be perceived as intraspecific. In this case, late *M. meridionalis* might have been differentiated into local populations or subspecies, of which *vestinus* was one, while one or more others, perhaps less specialized, could have been ancestral to *trogontherii*. Whether the variation was inter- or intraspecific, the alternation of morphotypes in the central to western European realm could have been due to population movements, perhaps in accordance with climate (*meridionalis* in more temperate conditions, *trogontherii* in cooler).

THE 'WOOLLY MAMMOTH': *MAMMUTHUS PRIMIGENIUS*

The species *Mammuthus primigenius* is best known from deposits of the Last Cold Stage, *c.* 115 000–10 000 years BP, and was widely distributed across northern Eurasia and northern North America. Typical remains of this species are characterized by the following diagnostic features in comparison with earlier species: thinner enamel and higher lamellar frequency, with x20x–x27x plates in M3; high angle of molar eruption; skull very compressed antero-posteriorly; cranial vertex very high, pointed, and anteriorly placed; very open occipital angle; forehead convex laterally; tusk alveoli close to vertical; short, high mandible with rostrum short or absent; 'aserial' carpals; short tail; relatively small body size (Aguirre 1968, 1969*b*; Maglio 1973).

Following an unsatisfactory situation with regard to the type material of the species, Garutt *et al.* (1990) have designated a male skeleton from the Taimyr peninsula, northern Siberia, as the neotype. These authors also confirm the validity of the name *primigenius* Blumenbach 1799.

Cranial changes in the transition from *M. trogontherii* to *M. primigenius* have been described mainly with reference to a single *M. trogontherii* specimen, partly reconstructed, the Gelsenkirchen skull (Fig. 19.5e). These comparisons should therefore be interpreted with some caution. The most significant difference is in the height/length ratio, given by Siegfried (1956) as 1.09 in *M. trogontherii*, compared to a minimum of 1.2 in a range of adult *M. primigenius* crania (e.g. Figs 19.5i and j). Mandibles are described as showing a relatively shorter horizontal ramus, of more circular section, in *M. primigenius* than *M. trogontherii* (Osborn 1942; Siegfried 1956), although biometric data have not been presented. The rostrum is of variable length in *M. primigenius* (e.g. Vereshchagin 1977*b*; here Figs 19.6j and k) as it is in *M. trogontherii* (see above), but the longest rostra in the latter species have no parallels among *primigenius*.

Dental differences between *Mammuthus trogontherii* and typical *M. primigenius* are illustrated in Fig. 19.4 by comparison of samples 4 (Mosbach *M. trogontherii*) and 9–10 (Last Cold Stage *M. primigenius*). Lamellar frequency has increased and enamel thickness decreased, but there is no further change in hypsodonty index. According to Aguirre (1968, 1969*b*), the angle of molar eruption has increased from 34°–58° in a sample of *M. trogontherii* to 39°–66° in *M. primigenius*.

Typical remains of *Mammuthus trogontherii* are dated *c.* 0.7–0.5 Ma (e.g. Süssenborn, Tiraspol, Mosbach). In the period *c.* 0.4–0.1 Ma, a series of samples illustrates the period of transition to late *M. primigenius*. The picture is a complex one and will be treated in more detail elsewhere. It is clear that there is no simple, unidirectional morphological sequence in mammoth fossils during this period. Dental remains almost as advanced as typical *M. primigenius* appear in Europe as early as *c.* 0.3 Ma or possibly even 0.4 Ma (Fig. 19.4), exemplified at Tourville, France, dated by lithostratigraphy and thermoluminescence (Carpentier and Lautridou 1986) and Homersfield, England, dated by lithostratigraphy (Coxon 1982; Stuart 1982). By *c.* 0.3–0.2 Ma there is evidence of advanced crania too, such as Cherny Yar, Russia (Fig. 19.5a), dated by lithostratigraphy and fauna (Dubrovo 1966; personal communication 1988). Typical *M. primigenius*-type teeth occur again in the period 0.19–0.13 Ma, exemplified at Balderton, England (Fig.19.4), dated by electron spin resonance and lithostratigraphy (Lister and Brandon 1991). In the Last Cold Stage, all samples appear advanced: examples plotted in Fig. 19.4 are from Předmosti, Czechoslovakia, radiocarbon dated to *c.* 26 000 years BP (Musil 1968) and Kostenki I, Russia, radiocarbon dated to *c.* 14 000 years BP (Klein 1969).

On the other hand, more primitive remains, forming a morphological link between *M. trogontherii* and *M. primigenius*, are evidenced at sites such as Steinheim, Germany (cf. early Saalian, *c.* 0.4 or 0.3 Ma) and Ilford, England (Figs 19.4 and 19.5h), dated by fauna, lithostratigraphy, and amino-acid epimerization to *c.* 0.2 Ma (Bowen *et al.* 1989; Lister and Joysey 1992). The molars in two skulls from Italy dated to *c.* 0.12 Ma (Tarquinia and via Flaminia) are also rather primitive, although their cranial morphology appears advanced (Fig. 19.5i; Kotsakis *et al.* 1978).

The *trogontherii–primigenius* transition could be accounted for by cladogenetic speciation, or by intra-specific dynamics (see also Kotsakis *et al.* 1978). The speciation model would be supported by the finding of two species together in a single level. Some evidence of a bimodal distribution in plate number is found in a British sample from Marsworth, *c.* 0.2 Ma (author's unpublished observations 1995). On the other hand, the single-species, population-variation model cannot be ruled out on existing data, which reveal considerable variability among mammoth samples in the period after 0.5 Ma, rather than a simple two-way split (Lister 1993*b*).

Many authors have attempted to crystallize this period of transition in taxonomic terms. Jourdan (1861) created the taxon '*Elephas*' *intermedius* for some remains from loess deposits around Lyon, France, and the name was used by Aguirre (1969*b*, p. 351). Kotsakis *et al.* (1978), however, point out that Jourdan, who intended his taxon as a general intermediate between

M. meridionalis and *M. primigenius*, gave no clear diagnosis for it. Aguirre (1968, 1969*b*, fig. 44) resurrected Pohlig's (1888) name '*Elephas*' *primigenius leith-adamsi* for *trogontherii–primigenius* intermediates, but this taxon does not seem appropriate as it was founded by Pohlig (1888, p. 232) explicitly for a supposedly dwarf form of *primigenius*. Dietrich (1912) founded *M. primigenius fraasi* for the Saalian skeleton from Steinheim, Germany.

The name which has gained widest currency is Dubrovo's (1966) *M. trogontherii chosaricus*, given specific status as *Mammuthus chosaricus* by Garutt (1972) and Kotsakis *et al.* (1978). This taxon was founded on a cranium and mandible from supposedly Holsteinian to early Saalian deposits (*c.* 0.4–0.2 Ma) of the Khasar faunal assemblage at Cherny Yar, southern Russia. Further remains which have been assigned to *M. chosaricus* include the skulls from Tarquinia and Flaminia, Italy (Kotsakis *et al.* 1978). Gromov and Garutt (1975) regard '*Mammuthus chosaricus*' as characteristic of the first half of the Saalian (*c.* 0.3–0.2 Ma), while an 'early form of *M. primigenius*' was present in the later Saalian (*c.* 0.19–0.13 Ma), and a 'late form of *M. primigenius*' in the Weichselian (*c.* 0.115–0.01 Ma). Kotsakis *et al.*, on the other hand, use '*M. chosaricus*' to cover all supposed intermediate forms of the Saalian and Eemian (*c.* 0.3–0.115 Ma). Even if some of the material referred to *M. chosaricus* or similar taxa does indeed represent populations intermediate between typical *M. trogontherii* and *M. primigenius*, the complex, fluctuating pattern of change (Fig. 19.4; Lister 1993*c*) suggests that, for the time being at least, all such nomenclatorial devices cloud rather than clarify the issue.

Numerous skeletons of *M. primigenius* are known; shoulder heights of a selection are given in Table 19.1. Although this species was generally smaller than its predecessors, the very small size of some of the Siberian skeletons (e.g. the Beresovka and Taimyr males, each 265 cm) is not typical of the species as a whole. Some of the earlier representatives (e.g. the Steinheim male, 370 cm) are quite large; of European Last Cold Stage mammoths a typical figure is 300–320 cm. The forefoot of *M. primigenius* always shows a strongly aserial carpal structure (Fig. 19.8d).

THE LATEST MAMMOTHS: ADVANCEMENT OR DEGENERATION?

According to current evidence, *Mammuthus primigenius* became extinct in Europe at around 12 000 years BP, in North America at around 11 000 years BP, and survived in parts of continental northern Siberia until *c.* 10 000 years BP (Martin and Klein 1984; Lister 1991; Stuart 1991). Recent evidence indicates late survival of an isolated population on Wrangel Island in the Arctic Ocean as late as 4000 years BP (Vartanyan *et al.* 1993; Lister 1993*c*; Lister and Bahn 1994). It has been suggested, on different occasions, that on the one hand the latest mammoths show the peak of evolutionary advancement, and on the other that they had begun to 'degenerate' toward a more primitive condition.

The idea of extreme advancement in the latest mammoths derives firstly from a model of continuous, unidirectional evolution through the lineage, progressing even into the Last Cold Stage, as implied by, for example, Adam (1960) and Guenther (1981). The data presented here do not support such a model. Some authors (e.g. Aguirre 1969*b*, p. 353) have seen more direct evidence in the molar morphology of very late samples from Siberia and Alaska, which have occasionally been distinguished as *M. primigenius sibiricus* (e.g. Trevisan 1949). However, it seems likely that in some, if not all, cases, this is an artificial effect of small tooth size. An example is the Berelekh site, radiocarbon dated to *c.* 11 000 years BP (Vereshchagin 1977*b*). The Berelekh sample shows very high lamellar frequencies, but this can be entirely accounted for by the small size of the teeth, squeezing the lamellae close together (Lister and Joysey 1992).

The opposing idea is of 'degeneration' as a prelude to extinction. Musil (1968) suggested that after 30 000 years BP, *M. primigenius* populations show augmented variability, an increase in the number of 'diminutive' individuals, reduced tusks and phalanges, malproportioned crania, and a reversion to more primitive molars. Little firm data have been presented in support of these ideas, however. Musil (1968, pp. 146–7) mentions some 'trogontherioid' molars from Siberia, but admits that they are poorly dated. Kubiak (1980) described a mammmoth skull from Bzianka, Poland, radiocarbon dated to *c.* 14 000 years BP, whose small size and somewhat malformed left tusk led him to suspect degeneration. However, the specimen is not so small or malformed as to preclude a small, slightly pathological female from a normal population. Bosinski (1979) has drawn attention to the prehistoric engravings of mammoths from the Gönnersdorf site, Germany (*c.* 12 500 years BP), in which the tusks are underemphasized or absent. His suggestion that this reflects tusk reduction in late European mammoths needs to be examined with caution, as prehistoric mammoth representations of greater antiquity also often lack tusks (Bahn and Vertut 1988, p. 128).

These suggested indications of 'degeneration' are by themselves insufficient to establish this phenomenon. Moreover, the dental morphology of late samples, whether from Europe (e.g. Kostenki and other Russian Plain sites: Vereshchagin and Kuzmina 1977; here Fig. 19.4) or from Siberia (e.g. Berelekh) do not suggest a significant reversal to 'trogontherioid' morphology. In the two latest European mammoth skeletons, the males from Condover, England (Coope and Lister 1987) and Praz Rodet, Switzerland (Weidmann 1969), radiocarbon dated *c.* 12 500– 12 000 years BP, the tusks appear well-developed, with basal tusk diameters of *c.* 20 cm and 15 cm, respectively (personal observation 1992).

The only possibly valid morphological feature which might be seen as a precursor of extinction is the small size of some late populations. In Siberia, the Berelekh mammoths, based on a large sample of remains, are of relatively small size (Zherekhova 1977; Sher and Garutt 1985*a*), and those from Wrangel, the latest of all, almost qualify as 'dwarfs' (Vartanyan *et al.* 1993). These reduced dimensions might have been caused by unfavourable environmental conditions which, in each case, soon after were to lead to local extinction.

DISCUSSION

CHANGES OF HABITAT AND ADAPTATION

In general, there was a shift in habitat through the Eurasian mammoth sequence, from temperate, wooded conditions to a cold, open environment. The earliest Eurasian mammoths appear to have lived in a 'wooded steppe' habitat, to judge from flora and fauna associated with the Khapry/Liventsovka remains. Palynological studies of the deposit indicate 22 per cent arboreal pollen, especially birch, pine, and oak, and xerophytic grasses and herbs (Garutt and Baigusheva 1981). There is also evidence that even the latest representatives of *M. meridionalis* lived, in at least part of their range, in temperate, partially wooded habitats. At Voigtstedt, for example, palynological and faunal studies indicate interglacial conditions (Erd 1965; Stuart 1981).

Faunal and vegetational studies of sites where *M. trogontherii* is most abundant (e.g. Mosbach and Süssenborn, Germany), suggest a habitat of cool, dry steppe with scattered trees (Kahlke 1969). Its successor *M. primigenius* is generally associated with cold, largely treeless conditions, stomach contents indicating a diet dominated by grasses with subordinate mosses, ferns, and herbs (Ukraintseva 1986), although occasionally the species has been found in temperate, at least partly wooded, contexts (e.g. Aveley, England: Stuart 1982).

The suite of morphological changes in the mammoth lineage forms an integrated whole, with adaptive significance which has been discussed, particularly for the cranium and dentition, by Maglio (1973). The increased number of lamellae in the molars provided greater shearing power and tooth life, enabling the animals to exploit progressively coarser diets in the move to largely treeless habitats. As lamellar packing increased, the enamel thickness of each lamella was reduced, for spatial reasons and to maintain shearing efficiency (Maglio 1973). From early *M. meridionalis* to typical *M. primigenius*, lamellar number approximately doubled and enamel thickness halved (Fig. 19.4). Even more significant was the roughly twofold increase in hypsodonty, providing much greater tooth life, and having concomitant effects on the skull and mandible. As the teeth increased in crown height, the palatine and maxillary region of the skull, and the horizontal ramus of the mandible, deepened to accommodate them. In keeping with the deepened palate, the vertical ramus of the mandible also heightened (Fig. 19.9). To accommodate their greater hypsodonty, unerupted molars were held higher in the skull and mandible, contributing to the antero-posterior shortening of these elements and increasing the angle at which the teeth came into wear (Aguirre 1968, 1969*b*).

The general shortening of the skull was preceded by the forward contraction of the occipital and parietal regions, and was accompanied by anterior displacement of the cranial vertex and temporal fossa (Fig. 19.9). The temporal muscles could thus pull more directly above the molars, and the coronoid process expanded forwards to provide their attachment, adding to the relative shortening of the horizontal ramus. The orientation of the tusk sockets shifted downwards, and the occipital surface and cranial

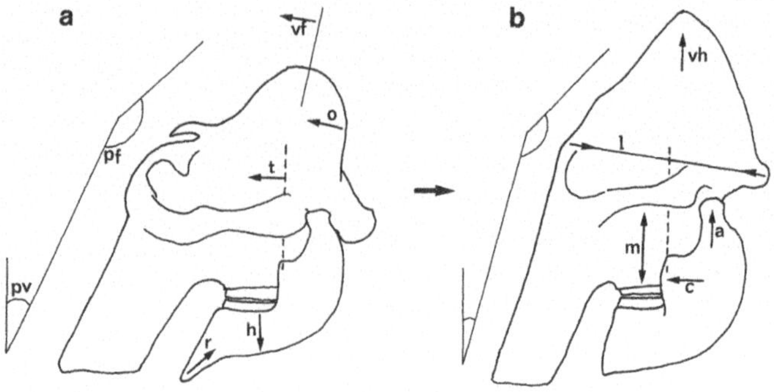

Fig. 19.9 Summary of cranial and mandibular changes in the mammoth lineage (after Maglio 1973). (a) Typical *M. meridionalis*; (b) typical *M. primigenius*. All changes occurred in the transition between the two species, and have been divided between the two diagrams for ease of illustration. Key: a, articular region elevated; c, coronoid process lengthened; h, horizontal ramus deepened; l, length of cranium shortened; m, maxilla deepened; o, occipital region shortened; pf, pv, premaxilla (tusk alveolus) steepened with respect to the forehead and the vertical; r, rostrum shortened; t, temporal muscle shifted anteriorly; vf, vh, vertex shifted forward and heightened.

vertex heightened, moving the centre of gravity backwards and giving the nuchal muscles and tendons greater mechanical efficiency, perhaps as a counterbalance to an increase in size of the tusks (Maglio 1973).

The furry coat of the late Pleistocene Eurasian mammoth was clearly an adaptation to cold climate, as were, in all likelihood, the small ears and tail to protect against heat loss or frostbite (Kubiak 1982). Changes in body size through the lineage are erratic, but *M. meridionalis* and *M. trogontherii* were clearly larger in general than *M. primigenius*, the peak size being reached in *trogontherii* (Table 19.1). The adaptive reasons for these size changes are obscure.

The stimulus for evolution in the mammoth lineage was presumably, at least in part, the cooling of climate in the Pleistocene and the great expansion of open habitats. As suggested by Maglio (1973), however, an additional cause, at least in the second half of the sequence, may have been the entry of *Elephas* (*Palaeoloxodon*) into Eurasia in the early middle Pleistocene, outcompeting *Mammuthus* in warmer, forested areas.

EVOLUTIONARY RATE

The rate and pattern by which the evolutionary changes occurred can be examined in most detail for the dentition, where sampling is most complete. The concept of continuous gradualistic evolution in the mammoth lineage was most explicitly expressed by Adam (1961, table 2), who listed a hypothetical series of transitional forms from early *M. meridionalis* to late *M. primigenius*. In Fig. 19.4, the five samples between *c.* 2.5 and 0.4 Ma appear to show approximately constant-rate, directional evolution in dental characters. Particularly interesting is the intermediacy in hypsodonty, the most anatomically significant feature of the

Siniaya Balka sample (sample 3 in Fig. 19.4; see also Dubrovo 1977). For this period, then, at least in the dentition, the data are consistent with gradual evolution. There are still gaps of up to 0.5 Ma between sampling points, which may hide more complex patterns of change, and there are hints of this in minor reversals revealed where additional material increases sampling density. This requires further investigation, especially for the *meridionalis–trogontherii* transition, where a speciation event is possible.

In the last *c.* 0.4 million years of mammoth evolution, the interpretation of a simple pattern of continuing directional change is not vindicated (Fig. 19.4; Lister 1993*b*). Dentally advanced mammoths first appear as early as 0.4–0.3 Ma, not as late as the 0.12 or 0.03 Ma date suggested by earlier authors. None the less, between 0.4 and 0.2 Ma (possibly up to 0.12 Ma in southern Europe) they alternate in the record with more 'primitive' samples. Finally, from *c.* 0.15 Ma onwards, at least in northern and central Europe, all samples are fully advanced, and there is no significant further evolution between then and extinction, although as always, some variation in mean values is picked up where resolution is fine. The period of change after 0.4 Ma could technically qualify as 'stasis', in that there is fluctuation, but no net change, between samples at the beginning and end of this sequence. However, this disguises the possibility of continuing dynamic interplay of populations leading to 'modernization' of the species as a whole (Lister 1993*c*).

The evolutionary progression in Eurasian mammoths, particularly its supposed uniformly gradualistic mode, has been extensively utilized as a biostratigraphic tool. The apparently gradual sequence between *c.* 2.5 and 0.5 Ma (Fig. 19.4) suggests that the morphology of European mammoth molars can be used with caution as one element of a stratigraphic scheme, to a resolution of about 0.5 Ma (Lister 1992), although even here, indications of possible reversals may complicate the situation. Within the upper part of the sequence (*c.* 0.5–0.01 Ma), the complex pattern of change suggests that the dating of mammoth remains by their morphology is inadvisable.

ACKNOWLEDGEMENTS

I am grateful to K. A. Joysey for support and encouragement, D. Mol and H. van Essen for useful discussion, and the curators of numerous museums for access to collections. The research was funded by the Natural Environment Research Council (grant no. GR/6294A).

TWENTY

Plio-Pleistocene proboscideans from the southern bight of the North Sea and the Eastern Scheldt, The Netherlands

HANS VAN ESSEN AND DICK MOL

INTRODUCTION

The Netherlands and the adjacent part of the North Sea are well known for their abundant remains of late Pleistocene mammals. On the other hand, fossils of Pliocene and early Pleistocene age are comparatively rare in this area. This chapter deals with dredged remains of proboscideans from the Plio-Pleistocene in the North Sea area.

Dredging operations in the floodplains of the rivers Rhine, Waal, IJssel and Meuse (Fig. 20.1) still yield many mammalian fossils. In most cases they are typical of the late Pleistocene (Weichselian, 100 000–12 000 years BP) and characterized by the woolly mammoth, *Mammuthus primigenius* and the woolly rhinoceros, *Coelodonta antiquitatis*. Other elements of this assemblage are the bovids, *Bison priscus* and *Bos primigenius*; giant deer, *Megaloceros giganteus*; red deer, *Cervus elaphus*; reindeer, *Rangifer tarandus*, and carnivores such as hyaena, *Crocuta spelaea*, and lion, *Panthera spelaea*. Age determinations of such collections are necessarily based on the results of geological survey. As yet there are no ^{14}C data on these fossils.

A second category is formed by the mostly well-documented early Pleistocene mammalian remains from the Tegelen Clay (*c.* 1.6 Ma). The species list of this interglacial fauna includes *M. meridionalis*, which is represented by only 11 molar fragments and even fewer postcranial parts. The molars were repeatedly studied, e.g. by Bernsen (1930), Schreuder (1945), and Guenther

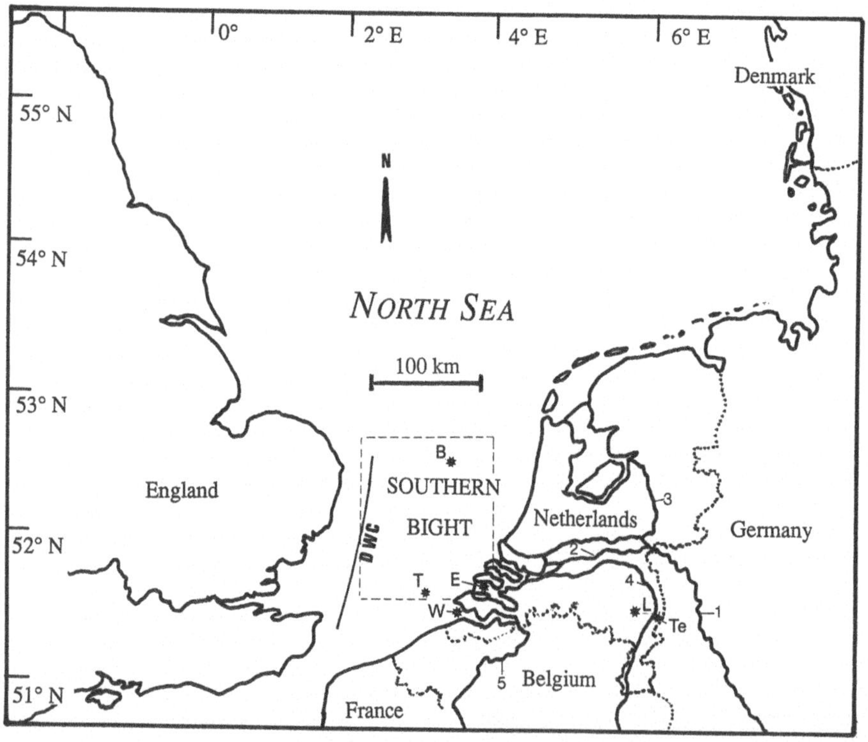

Fig. 20.1 The southern bight of the North Sea and the approximate delineation of the study area (dashed line). Key: B, Brown Bank; E, Eastern Scheldt; DWC, Deep Water Channel; L, Liessel; Te, Tegelen; T, Thornton Bank; W, Western Scheldt; 1, Rhine; 2, Waal; 3, IJssel; 4, Meuse; 5, Scheldt.

(1986). Guenther (1986) concludes that the Tegelen molars are more primitive than those of the Valdarno (Tuscany, Italy). *Anancus arvernensis* should not be included in the Tegelen species list (see below).

A third category of fossil terrestrial mammals comprises those remains that are brought in by fishing boats from the bottom of the Eastern Scheldt (province of Zeeland, The Netherlands) and the southern bight of the North Sea between England and The Netherlands. Detailed accounts of the terrestrial fossils from the Scheldt estuary were published by, among others, de Man (1875), Schreuder (1944, 1945), and Hooijer (1953*a*, 1984*a*, *b*). Critical remarks on their interpretations were made by Drees (1986). Excluding younger faunal elements, one notes that the terrestrial fossils from the Eastern Scheldt have an early Pleistocene age (Tiglian, according to Drees 1986). Among these are *A. arvernensis* and *M. meridionalis*. It is on this early assemblage and a similar section from the North Sea material that the present text concentrates. The Western Scheldt, which has mainly yielded marine mammalian fossils and a few proboscidean specimens, is not considered here. The occurrence of *Mammut borsoni* in the Scheldt estuary, as reported by von Koenigswald (1950), must be considered doubtful.

Over the past 40 years the RGM (see Abbreviations below) at Leiden has been employing a fishing boat once every year especially for the purpose of bringing up fossils from the Eastern Scheldt. Smaller collections from this area are kept in museums at Souburg and Zierikzee.

Much of the material employed in this study was trawled from the North Sea bottom, more specifically the area south-west of the Brown Bank and the borders of the Deep Water Channel (50.00°N/02.00°–03.00°E; Fig. 20.1). The early Pleistocene fossils are mainly characterized by a high degree of mineralization. The late Pleistocene specimens, including *M. primigenius*, are not mineralized, but are mostly well preserved despite their salt-water environment. They must, however, be desalinated because salt would otherwise erode their outer layers.

The less numerous group within the North Sea material is constituted by early to middle Pleistocene faunal elements, such as *M. meridionalis*, the deer genus *Eucladoceros*, the large horse *Equus bressanus*, and *Hippopotamus antiquus*. They are attributed to various stages ranging from the Tiglian to the Cromerian Complex, a span of nearly 1.5 million years (Fig. 20.2).

One of the earliest proboscideans within the North Sea assemblage is *A. arvernensis*, represented by one molar fragment and possibly two incisor fragments from the Thornton Bank 12 nautical miles off the Walcheren coast (province of Zeeland; Fig. 20.1). Apart from *M. meridionalis* and *M. primigenius* previously mentioned, the transitional middle Pleistocene form *M. trogontherii* is also represented, albeit by very few molars. In contrast, *M. primigenius* molars are counted by the thousands.

A fourth elephantid reported from the North Sea is the straight-tusked elephant, *Elephas antiquus*. This species, too, is represented by only a few molars. Eastern Scheldt molars previously ascribed to *Archidiskodon planifrons* are now considered archaic forms of *M. meridionalis*.

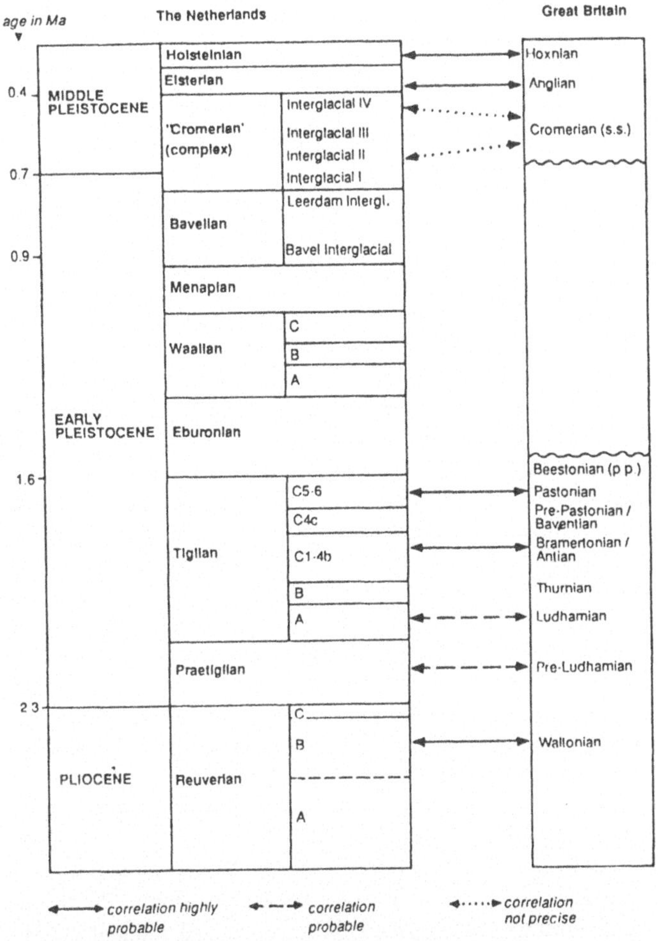

Fig. 20.2 Late Pliocene, early Pleistocene and middle Pleistocene stages in The Netherlands and Great Britain and their suggested correlation (after Gibbard *et al.* 1991).

Abbreviations

a, Attributed number; specimens originally without number in collections; BOK, Van der Bok collection (Ouddorp, The Netherlands); C, complete; ESS, Van Essen collection (Dieren, The Netherlands); ET, enamel thickness; F, fragment; GMZ, Municipal Museum of Zierikzee (The Netherlands); H, height of crown; HI, hypsodonty index; IGF, Geological and Palaeontological Museum of Florence University (Florence, Italy); IVA, Institute for Earth Sciences, Utrecht State University (Utrecht, The Netherlands); JAG, Jager collection (Goes, The Netherlands); L, length of crown; LF, lamellar frequency; MOL, Mol collection (Hoofddorp, The Netherlands); MUL, Mulder collection (Nieuwleusen, The Netherlands); N, number of specimens; NMR, Natuurmuseum Rotterdam (Rotterdam, The Netherlands); R, complete remains; RGM, Rijksmuseum van Geologie en Mineralogie (now 'National Natural History Museum', Leiden, The Netherlands); P, number of lamellae; VTS, Van Tuyll van Serooskerken collection (Oostkapelle, The Netherlands); W, width of crown; ZGW, collection of the Koninklijk Zeeuwsch Genootschap der Wetenschappen (Souburg, The Netherlands).

GEOLOGICAL, PALAEOGEOGRAPHICAL, AND PALAEOCLIMATOLOGICAL ASPECTS OF THE NORTH SEA AND THE EASTERN SCHELDT

The North Sea is a shallow sea on the continental shelf between the European mainland and Great Britain. Its depth ranges approximately from 27 to 45 m. The southern bight of the North Sea and the central part of the adjacent Dutch territory form the southern edge of the North Sea Basin. It has been an area of subsidence since Palaeozoic times, with an axis in and to the north-east of the study area (Heybroek 1974). During the early Pleistocene subsidence was much intensified; deposits representing the local base of the Pleistocene outcrop at the British east coast (top of the Red Crag Formation, Balson and Cameron 1985), but dip towards the basinal axis and appear at approximately 450 m below mean sea-level near the Dutch coast at 53.00°N–04.00°E (top of the Brielle Ground Formation; Cameron *et al.* 1984). In a north-west–south-easterly direction, i.e. at right angles to the gradient, differences are much smaller, to the effect that the base of the Pleistocene in the Eastern Scheldt area is less than 100 m below mean sea-level (van Rummelen 1970).

Within the Dutch sequence, the Plio-Pleistocene boundary is defined as the transition between the Reuverian and Praetiglian stages on the basis of pollen analysis, and dated at about 2.3 Ma. This date for the onset of the Praetiglian cold stage was corroborated by oxygen isotope and palaeomagnetic analysis of a Central Pacific deep-sea core (Shackleton and Opdyke 1977) and by the record of ice-rafted glacial debris in the North Atlantic Ocean (Shackleton *et al.* 1984). On the palaeomagnetic time scale this boundary virtually coincides with the transition from the younger normal Gauss to the reversed Matuyama chron at 2.47 Ma.

The Tertiary North Sea covered most of the present-day Netherlands and part of Belgium. The deposition of marine sediments in the study area (Westkapelle Ground Formation in the North Sea, Oosterhout and Maassluis Formations in the western part of The Netherlands) went on well into the Pleistocene (Tiglian C5). The effects of intensified early Pleistocene basinal subsidence — that lasted until the late Cromerian — were balanced by a much increased rate of fluvial sedimentation caused by uplift of the hinterland (Zagwijn and Doppert 1978; Kasse 1988). Deltaic fans caused a gradual westward shift of the coastline, even beyond its present configuration (IJmuiden Ground Formation and its lateral equivalent the Smith's Knoll Formation, Winterton Shoal Formation, and Yarmouth Roads Formation; Cameron *et al.* 1984). In the Dutch province of Zeeland, Tiglian fluvial sediments filled a wide erosional valley (Tegelen Formation).

During the Plio-Pleistocene cold stages, starting with the Praetiglian, eustatic sea-level falling must have occurred, moving the zone of marine sedimentation away from the study area to the edge of the continental shelf; cool temperature influences are not found in marine fossil assemblages in the southern North Sea Basin (Long *et al.* 1988). The magnitude of early Pleistocene glaciations is estimated to have been at least two-thirds that of the late Pleistocene glacial maxima (Shackleton and Opdyke 1977). According to Zagwijn (1979), coastline reconstructions indicate that the study area was partly dry land and partly submerged during the Pleistocene warm stages. In the early and middle Pleistocene, the southern North Sea Basin was not connected with the Channel through the Straits of Dover, with a possible exception for the Holsteinian warm stage.

In several places, e.g. the northern edge of the Brown Bank, sediments of the Yarmouth Roads Formation down to the Tiglian IJmuiden Ground Formation (Laban, personal communication 1989) were upturned by the thrust of the Elsterian ice sheet (Cameron *et al.* 1984).

The possibility that terrestrial faunal remains were washed into environments where marine sedimentation still took place cannot be ruled out. The English Red Crag Formation offers a good example of such a situation. Certain species, e.g. *M. meridionalis* and *E. antiquus*, are linked with the climatic and environmental conditions prevailing in interglacials. During these warmer stages the central part of the North Sea study area, in all probability, was invariably submerged. At least some of the extreme quantitative differences between finds of *M. primigenius* and of the other species mentioned can be explained from the fact that *M. primigenius* was the only species adapted to periglacial conditions and was flourishing at a time when the North Sea was a wide land bridge between Europe and the British Isles.

MATERIAL STUDIED

MAMMUT BORSONI

Material: Fragment of a lower M3 in the collection of the Natuurmuseum Rotterdam (NMR 698, formerly from Rotterdam Zoo). Measurements in mm: L, 100+; W, 92; H, 62.5; ET, 3.2–4.5.

This specimen was first reported by von Koenigswald (1950), who regarded it as a fragment of an upper M3. Its provenance is unknown, but from its colour von Koenigswald inferred that it originated from the Scheldt estuary and had a Pliocene or early Pleistocene age. In many places throughout the European continent *A. arvernensis* is found associated with *Mammut borsoni*, e.g. in the Triversa faunal unit in Italy (Azzaroli 1983). Since this unit is correlated with the middle of the Gauss magnetic chron (Lindsay *et al.* 1980) it has a Pliocene age. In later Italian faunal units *Mammut borsoni* is absent, whereas *A. arvernensis* lived on into the Pleistocene, accompanied by the genus *Mammuthus* already in the Montopoli fauna of Plio-Pleistocene age. Other European localities where *Anancus* and *Mammut* appeared together are Sulzfeld and Kaltensundheim in eastern Germany (Kahlke *et al.* 1984) and Wolfersheim/Wetterau, Hessen, Germany (Tobien 1977). Nearest to the study area is the late Pliocene marine Red Crag Nodule Bed (East Anglia, Great Britain), where *Mammut borsoni* is among a mixture of very diverse marine and terrestrial elements, including *M. meridionalis* and *A. arvernensis* (Stuart 1982).

The recent discovery of two lower M3 of *Mammut borsoni* at Liessel, The Netherlands (Fig. 20.1) possibly establishes the same association since *A. arvernensis* was reported from this locality in 1987 (Mol and van Essen 1990). The only alternative explanation for the presence of *Mammut borsoni* in the Scheldt estuary could be that the specimen had been washed into the local Tegelen Formation from eroded older deposits.

ANANCUS ARVERNENSIS

The gomphothere *A. arvernensis* had a wide Eurasian geographical range. It is described from various late Turolian localities in Spain (Aguirre *et al.* 1973; Aguirre 1974) and from several other Pliocene localities in central and western Europe (Tobien 1978*b*). It became extinct during the early Pleistocene. As previously observed by Hooijer (1953*a*), the molars from the study area (Fig. 20.3) cannot safely be connected with any particular evolutionary stage of the species on morphological grounds; very few specimens are complete, e.g. the worn lower M3 in a jaw fragment (ZGW a001) found in 1972 (Dumon Tak 1973). When compared with specimens cited by Tobien (1975, p. 218) they seem smaller. Schreuder (1944) reached the same conclusion after comparison with data from older literature. Features normally well developed in *A. arvernensis*, such as the alternate arrangement of the half-loph(id)s (anancoidy) and their oblique positioning (chevroning), can also be observed in the molars from the study area. Traces of cement covering part of the crown (cementodonty) are preserved in a number of specimens.

A. arvernensis is often cited in faunal lists of the Tegelen type locality, which would call for comparison. The one much worn molar from that area is, however, problematical. It was found in the Van Cleef clay pit near Maalbeek, 3 km south of Tegelen. Pollen analysis of the clay in the pit and on the molar was said to indicate deposition during the Eburonian cold stage that followed the warm-temperate Tiglian (Zagwijn 1960, 1963). The occurrence of *A. arvernensis* under deteriorated climatic conditions and at such a late date seems unlikely. The same is true of the tapir *Tapirus arvernensis*. Both species should therefore be excluded from the Tegelen faunal list. Postcranial elements attributed to *A. arvernensis* are considerably smaller than those of most other Pleistocene and recent proboscideans (pygmy elephants of the Mediterranean islands are excepted from this comparison). Manus and pes are particularly suited to demonstrate this point as their parts are very often found complete because of their comparatively small size and compact shape. In *A. arvernensis* these parts are characteristically short and compressed. Eastern Scheldt *A. arvernensis* bones show a high degree of mineralization. Their colour varies from brown to jet black.

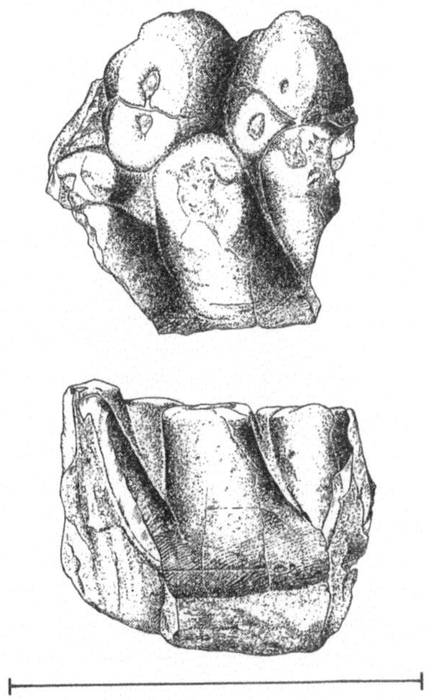

Fig. 20.3 *Anancus arvernensis* (MOL 1752). Fragment of lower left M2 (anterior is to the right). Southern bight of the North Sea, Thornton Bank. Occlusal and buccal view, scale bar 10 cm. (Drawing: H. van Essen.)

The molar from the Thornton Bank (catalogue no. Mol 1752; Fig. 20.3) is still the only specimen from the North Sea. The short geographical distance between this locality and the Eastern Scheldt suggests that there might be a geological correlation. Apart from *M. meridionalis* (discussed below), the Eastern Scheldt bones referred to *A. arvernensis* are accompanied by other terrestrial fossils of broadly early to middle Pleistocene age (Table 20.1).

ELEPHAS ANTIQUUS

E. antiquus of the middle and late Pleistocene is as yet unambiguously represented by only three molar specimens from the North Sea (ESS 138, lower M3 F; ESS 354, upper M3 F; ESS 400, worn upper dp4). Because of the scarcity of this species in the study area and its late age, it is included here for reference of material from the North Sea (Mol 1989).

MAMMUTHUS MERIDIONALIS

Dentition measurements

The morphological features examined are plate number (P), length (L), width (W), height (H), laminar frequency (LF), enamel thickness (ET), and hypsodonty index (HI) as defined by Maglio (1973). All measurements are in mm. As an exception to the rule, W is measured without cover cement, because in the majority of cases there is no cover cement adhering to the widest lamella. This is due either to the fact that cement deposition did not proceed to the lamella measured or to later damage. As a

Table 20.1 Early to middle Pleistocene mammalian taxa from the southern bight of the North Sea and the Eastern Scheldt

Taxon	NS	ES	Collection	References
Carnivora				
Canis sp. (cf. *Canis etruscus*; Forsyth Major, 1877)	X	–	MOL	
Pachycrocuta perrieri (Croizet and Jobert, 1828)	–	X	RGM	Kortenbout van der Sluijs (1985)
Homotherium sp.	–	X	RGM	Hooijer (1962*b*)
Perissodactyla				
Equus sp.	–	X	ZGW	
Equus bressanus Viret, 1954	X	–	MOL	Hooijer (1984*b*)
Equus cf. *stenonis* Cocchi, 1867	X	–	MOL	
Dicerorhinus etruscus (Falconer, 1859)	X	–	VTS	Drees (1986)
Dicerorhinus etruscus or *Dicerorhinus jeanvireti* Guerin, 1973	–	X	GMZ	Hooijer (1981)
Artiodactyla				
Sus sp. (cf. *Sus strozzii* Meneghini; Forsyth Major, 1881)	X	–	MOL	
Hippopotamus antiquus Desmarest, 1822	X	–	MOL	
			RGM	van Kolfschoten and Vervoort-Kerkhoff (1985)
Eucladoceros sp.	X	–	MOL	
Eucladoceros falconeri (Dawkins, 1868)	–	X	ZGW	Kunst (1937)
			RGM	Kunst (1937)
Cervus perrieri Croizet and Jobert, 1828	X	–	MOL	Hooijer (1984*b*)
Cervalces cf. *latifrons* (Johnson, 1874)	X	–	MOL	
Bison sp.	X	–	RGM	Drees (1986)
Bos primigenius Bojanus, 1827	X	–	RGM	
Proboscidea				
Mammut borsoni (Hays, 1834)	–	?	NMR	Von Koenigswald (1950); Mol and van Essen (1990)
Anancus arvernensis (Croizet and Jobert, 1828)		X	RGM	Hooijer (1953*a*)
				Schreuder (1944)
		X	ZGW	Dumon Tak (1973)
	X		MOL	Mol (1989)
				Mol and van Essen (1990)
Mammuthus meridionalis (Nesti, 1825)	X		BOK	
	X		ESS	
		X	GMZ	Hooijer (1984*a*)
	X		AG	
	X		MUL	Hooijer (1984*a*)
	X		MOL	Hooijer (1984*a*)
		X	RGM	Hooijer (1953*a*)
				Schreuder (1944)
	X		VTS	Hooijer (1984*a*)
		X	ZGW	Hooijer (1984*a*)
Mammuthus trogontherii (Pohlig, 1881)	X		BOK	
			ESS	

Abbreviations: NS, North Sea; ES, Eastern Scheldt; see also p. 215 of the text.

result, the HI may be influenced since in *M. meridionalis* the cement cover can be very thick; the increase in HI introduced by measuring bare lamellae may amount to 15 per cent of the figure that includes cement.

A tooth is indicated as complete (C) if the total P can be determined even though as much as 50 per cent of the tooth volume may have been abraded. Teeth showing advanced wear leaving less than the original P are called complete remains (R). The figures for ET represent both extremes of the range measured in each tooth. Averages are sometimes added in brackets.

The primary aim of this section is to ascribe the North Sea and Eastern Scheldt material to morphologically defined evolutionary stages and, if possible, to relate the latter to any of the stages of

Fig. 20.2. Maglio's 1973 data are used as a frame of reference. Maglio, as other authors did before him, divided the morphological range of *M. meridionalis* into three groups, which are labelled 'Laiatico stage', 'Montevarchi stage', and 'Bacton stage' ('primitive', 'typical', and 'advanced', respectively). The progressive character of each of these stages with respect to the preceding ones is mainly expressed by an increase in P, HI, and LF, as well as by a decrease in ET. Lister (this volume, Chapter 19) provides further evidence of the 'gradualistic' nature of this sequence. As the study area has not yielded many complete specimens so far, the values of P are not used in a scatter diagram. The HI, as a ratio not influenced by size, is plotted against LF and ET instead (Figs 20.4 and 20.5). LF and ET are, in part, influenced by individual size and, in part, by the evolutionary stage of the phyletic line. Individual size may, to some extent, be interpreted as sex-related. In order to establish the nature of these relations on a single evolutionary level, a HI/LF diagram is drawn up on the late Pleistocene *M. primigenius* population of the North Sea as well.

M. meridionalis is commonly, though not universally, accepted as the ancestor of *M. trogontherii* and thence *M. primigenius*. The distinction between *M. meridionalis* and *M. trogontherii* on the basis of tooth morphology is hard to establish. A small number of teeth, however, are referred to *M. trogontherii* on account of P, HI, ET, and LF. Only one of these (ESS 196, lower M3) is not heavily mineralized but preserved as *M. primigenius* teeth are. Advanced *M. meridionalis* from an inland site of Bavelian age (RGM 85541) is added for comparison.

From the diagrams (Figs 20.4 and 20.5), the following observations can be made. The HI in the North Sea *M. primigenius* population increases and then decreases with individual age. In upper molars the highest average value is that of M2, in lower ones that of M1. Beginning with dp4, the upper molars of *M. primigenius* have a higher average HI than the corresponding lower ones; only the lower half of the M1–M3 clusters is within the limits of the graph. The configuration of the clusters for *M. meridionalis* is almost analogous; only upper M1 have a somewhat lower average HI than lower M1. The *M. meridionalis* clusters are much more compact than those for *M. primigenius*, even more so because they may represent more than one evolutionary stage. The ET of *M. meridionalis* upper and lower M1 and M2 is confined to the same range. The range of upper M3 is much narrower than that of lower M3. The *M. meridionalis* M3 clusters comprise rather few specimens, but seem to indicate a tendency of Eastern Scheldt molars to occupy the lower halves of the clusters on account of their HI. A number of Eastern Scheldt and North Sea specimens are indistinguishable on the basis of important morphological features and are therefore probably of the same age.

Evolutionary stages

The earliest or 'Laiatico' stage of *M. meridionalis* as outlined by Maglio (1973) has many morphological features reminiscent of its North African ancestor, *M. africanavus*, i.e. P, 11–12; HI, *c.* 110; LF, 3.5–4.0; ET, 3.5–4.0. No total P is preserved in the Eastern Scheldt specimens. On the basis of the other parameters, only RGM 20033 may fit the description, with HI, 106; LF, 3.9; ET, 2.8–5.0 (3.9). This specimen was described as '*Archidiskodon planifrons*' by Schreuder (1944), which shows that its primitive appearance attracted early attention. Posterior median enamel loops are well developed. Some other lower M3 have a very low HI, but do not correspond in every detail or cannot be measured objectively. For example, RGM 170064 is very low (HI, 100–105), has an LF of 4.5 but an ET of only 2.5–2.8 (2.65). This type of molar was repeatedly found in deposits of pre-Pastonian to Pastonian age on the coast of East Anglia (Great Britain). The maximum ET of other low-crowned lower M3 is 3.7.

From the North Sea a large fragment of a lower M3 (ESS 163) is similar to RGM 20033, i.e. HI, 105; LF, 4.3; ET; 2.5–4.0 (3.25). The LF would have been lower if the posterior portion of the tooth had been preserved. Median enamel loops are weakly developed and the enamel is weakly folded. It seems likely, therefore, that *M. meridionalis* teeth with primitive characters as described above are to be referred to the Tiglian stage.

The material of Chilhac (Haute-Loire, France) is useful for comparison. Its age, *c.* 1.9 Ma (Bout 1970), is somewhat greater than that of the Tegelen type site (1.6 Ma). Data on seven M3 from four crania are: W (including cement), 120; ET, 3.0–5.0; P, 12–13 (measurements in mm). The total range of the LF in upper and lower M3 is 4.15–4.3 (Boeuf 1983). These measurements are reminiscent of the primitive specimens from the study area, although nothing can be stated regarding the P of the Eastern Scheldt material. Boeuf concludes that the evolutionary stage of the Chilhac molars is slightly lower than that of the typical form from the Valdarno Superiore (Tuscany, Italy). HI values are not available because all specimens are still in their jaws. All in all, the Eastern Scheldt M3 as represented by RGM 145333 appear only slightly differentiated with regard to the primitive forms: their LF is somewhat higher, their ET a little lower (Table 20.2). In addition, the HI of lower molars such as RGM 20033 allows the assumption of somewhat higher values in the corresponding upper ones.

As mentioned above, data on elephant teeth can be influenced by individual variation, including size/sex differences; the crania described by Boeuf (1983) are exactly contemporaneous, but the ET in skull B is 3.0–3.8 as against 4.0–5.0 in skull C. The skulls A, B, and C have M3 with 13 lamellae, whereas in skull D, they have 12. Finally, data in the literature may be biased by the method of measurement. Because such allowances must be made, it seems safe to assume a late Tiglian age for the Eastern Scheldt molars on account of both morphological similarities and geological setting, which would mean that they are slightly predated by the Chilhac skulls.

The North Sea material

Drees (1986) referred the earliest terrestrial fossils from the North Sea to the Deep Water Channel and arrived at the Waalian warm stage as the lower limit of their occurrence. Because of the morphological similitude of the Eastern Scheldt material and a

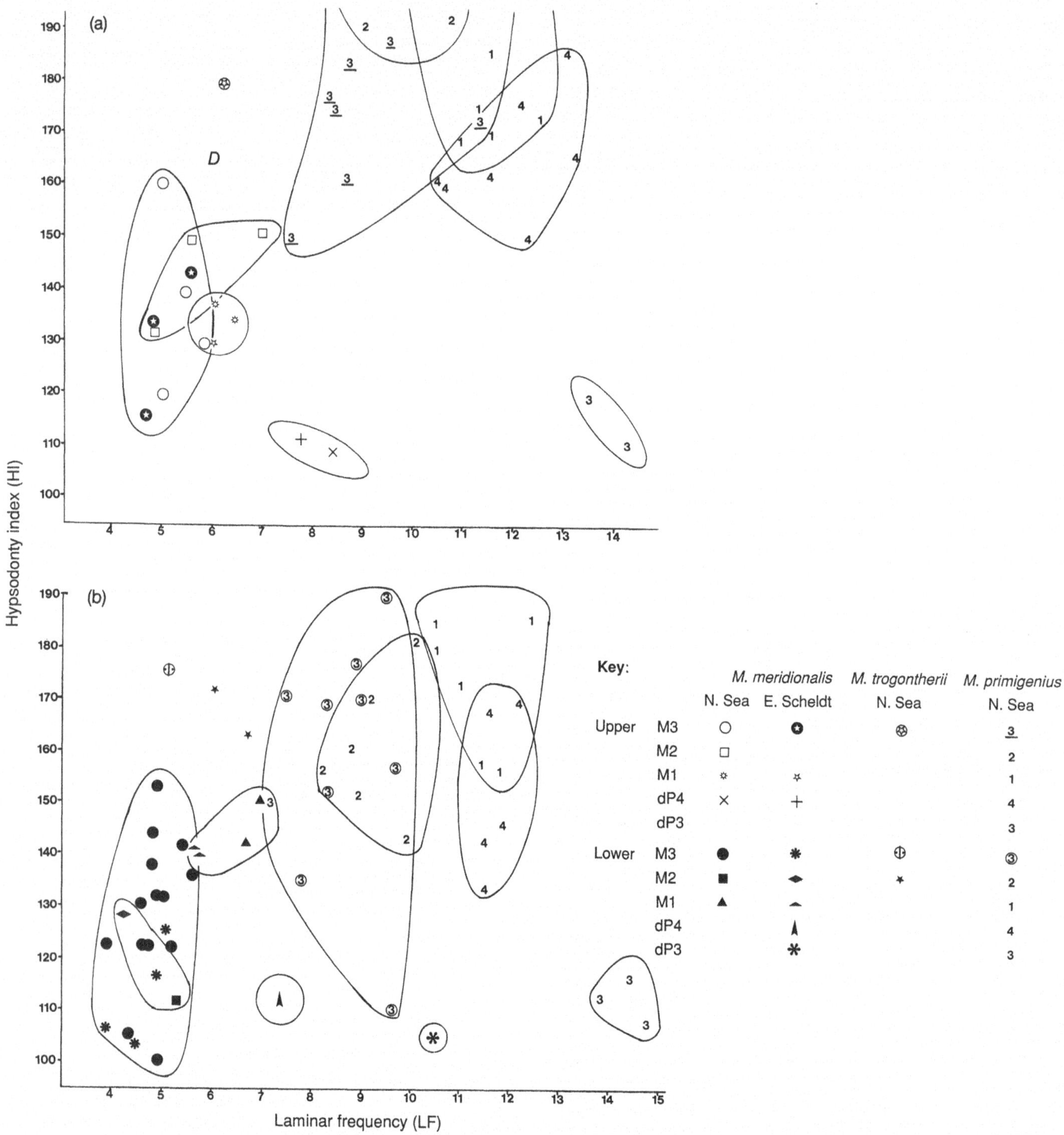

Fig. 20.4 HI/LF scatter diagrams of upper (a) and lower (b) teeth of *Mammuthus meridionalis*, *M. trogontherii*, and *M. primigenius* from the study area. D, RGM 85541, fragment of upper M2/M3 of *M. meridionalis* from Dorst, The Netherlands (Leerdam interglacial, Bavelian stage).

number of North Sea specimens, however, the latter can also be referred to an earlier stage, the late Tiglian. This is underlined by the plate number of some complete upper M3s, which is the same as that of most of the Chilhac upper M3s (P = 13). Table 20.2 and Figs 20.6 and 20.7 may illustrate this further. As the evolution of *M. meridionalis* is considered gradualistic (Lister, personal communication 1992), more advanced forms may be expected to have occurred during the Waalian stage. At present,

Table 20.2 *Mammuthus meridionalis*. Dentition elements from the southern bight of the North Sea and the Eastern Scheldt. RGM 85541 (Dorst, The Netherlands; Bavelian stage) is cited for comparison

Collection	Number	EL	P	W	H	LF	HI	ET
Eastern Scheldt								
RGM	145333	u.M3	–	111.5	106.5⁺	4.6	119	3.0–3.2
RGM	118836	l.M3	–	100.0	116.0	4.8	116	3.0–3.7
North Sea								
ESS	324	u.M3	13	117.5	140.0*	4.9	119*	2.5–3.9
ESS	345	u.M3	13?	91.0	117.0	5.8	129	3.0–3.5
ESS	163	l.M3	–	105.5	111.0	4.3#	105	2.5–4.0
MOL	1999	l.M3	–	89.0	97.0*	5.6	136	3.0–4.3
MOL	1700	l.M3	–	103.2	126.0	4.7	122	3.0
ESS	334	l.M3	–	109.0	110.0*	5.0	100*	2.0–3.2
ESS	165	l.M3	–	102.0	135.0	5.0	132	3.5–4.0
MUL	a007	l.M3	–	102.0	135.0	5.0	132	2.8–3.5
ESS	358	l.M3	–	97.5	125.0*	4.6	130*	2.7–3.3
MOL	1021	l.M3	–	80.0⁺	104.0*	5.4	142*	2.5–4.0
BOK	a027	u.M2	–	101.0	135.0*	4.8	134*	3.0–3.5
BOK	a032	u.M2	10?	104.5	–	4.6	–	2.8–3.0
ESS	346	u.M2	–	108.5	–	4.7#	–	2.8–3.2
ESS	344	u.M2	9	93.0	–	4.5	135*	1.7–2.7
Advanced:								
MUL	a005	u.M3	–	93.0	149.0	4.8	160	2.3–3.5
ESS	119	u.M2/M3	–	88.5	131.5	5.5	149	2.0–2.5
BOK	a034	l.M3	–	84.0	140.0	4.9	153	3.0
ESS	348	u.M2	10	79.0	119.0	6.9	151	2.0–2.8
Dorst, The Netherlands								
RGM	85541	u.M2/M3	–	87.0	142.0	5.7	163	2.0–2.5

⁺ Figure lowered because of wear; # figure too high because of damage; * approximation; u., upper; l., lower; EL, element; W, width without cover cement. Other explanations are given under Abbreviations.

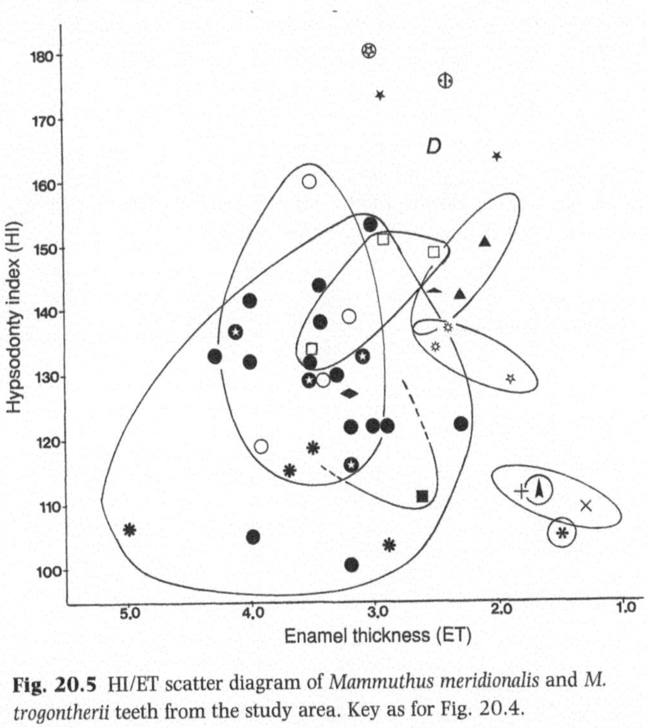

Fig. 20.5 HI/ET scatter diagram of *Mammuthus meridionalis* and *M. trogontherii* teeth from the study area. Key as for Fig. 20.4.

however, there is no means of determining whether or not Waalian *M. meridionalis* is represented among the fossils from the North Sea. An upper M2 shows a marked reduction in ET and antero-posterior flattening of the lamellae, whereas these are widely spaced so that the LF is still low: P, 9; W, 93.0; H,–; LF, 4.5; ET, 1.7 - 2.7 (ESS 344); 50 per cent of the tooth is abraded, so that the HI is uncertain (it may have been about 135). Reasons for retaining such specimens in the Tiglian group (as females?) could be their relatively low P (the total range of upper M2s is 8–11 according to Maglio, 1973) and the fact that a very similar specimen is in the Eastern Scheldt assemblage, i.e. W, 93.0; HI, 138; LF, 5.0; ET, 1.5–2.3 (RGM 104725, upper M2/M3 F). Furthermore, teeth such as these are encountered in British coastal deposits of comparable age.

Because the species *M. meridionalis* and *M. trogontherii* grade into each other, it may be difficult to refer molars representing the transitional phase to either of them. Summary measurements published by Maglio (1973) and graphs in Lister (this volume, Chapter 19) show some overlap in their ranges. The total P on M3 in *M. meridionalis* varies with different authors: Maglio (1973) gives 11–14 in upper and 10–14 in lower M3. Guenther (1986) mentions 16. The data on the North Sea specimens referred to *M. trogontherii* are listed in Table 20.3.

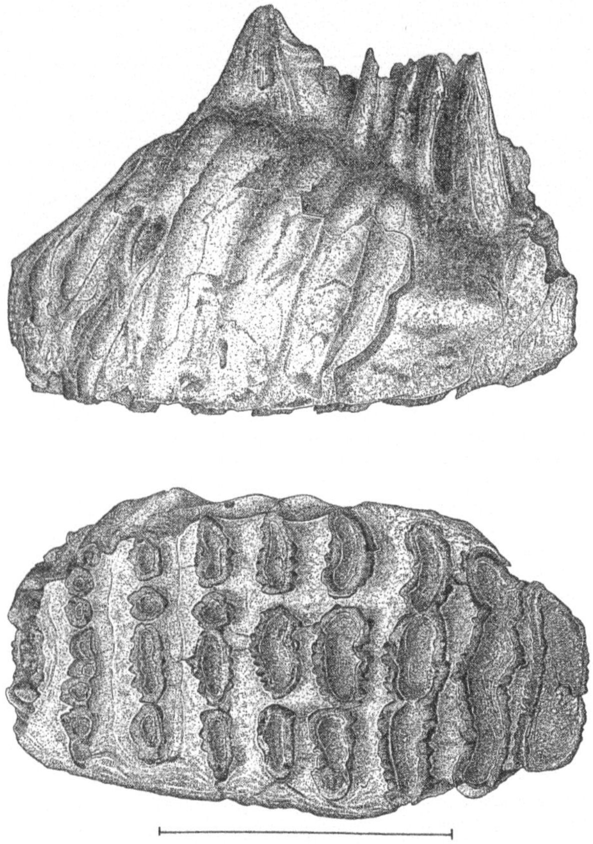

Fig. 20.6 *Mammuthus meridionalis* (typical form, BOK a027). Upper right M2 (anterior is to the right). Southern bight of the North Sea. Buccal and occlusal view, scale bar 10 cm. (Drawing: H. van Essen.)

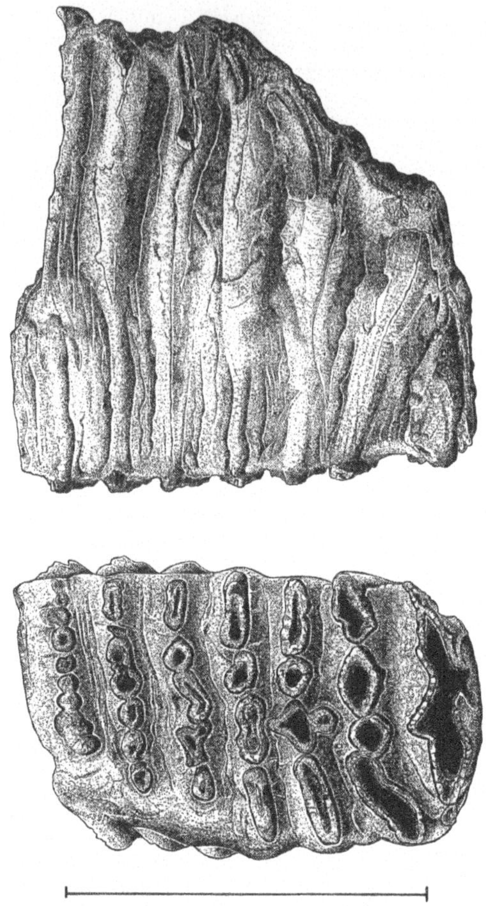

Fig. 20.7 *Mammuthus meridionalis* (advanced form, ESS 119). Fragment of upper right M2 or M3 (anterior is to the right). Southern bight of the North Sea. Buccal and occlusal view, scale bar 10 cm. (Drawing: H. van Essen.)

The postcranial remains from the North Sea are referred to large individuals of *M. meridionalis* or *M. trogontherii*. They can easily be distinguished from the short and robust bones of *A. arvernensis*. The skeletal elements from the Eastern Scheldt are mainly in the collection of the RGM in Leiden. Almost all specimens are fragments of large bones, such as scapula, pelvis, humerus, ulna, femur, and tibia. The parts of manus and pes are complete in some cases. The Mol collection (Hoofddorp) is comprised of 29 postcranial elements of *M. meridionalis* collected from the borders of the Deep Water Channel by the Trawler GO 27. The majority of these specimens belongs to manus and pes. They indicate medium-sized and very large individuals. *M. meridionalis* was one of the largest elephant species of the European Pleistocene. The Nogaisk skeleton in the Leningrad Zoological Museum stands 410 cm at the shoulder. A number of specimens are juvenile; they lack the epiphyses that had not yet fused with the bones.

Table 20.3 *Mammuthus trogontherii*. Dentition elements from the southern bight of the North Sea

Collection	Number	EL	P	W	H	LF	HI	ET
ESS	121	u.M3	14+	104.8	187.0*	6.2	178*	2.0–3.0
BOK	a028	u.M3	12+	82.0	–	8.3	–	1.9–2.4*
ESS	196	l.M3	12+	90.0	160.0*	5.1	175*	2.4
ESS	241	l.M2	–	82.0	132.5	6.7	163	2.0
ESS	243	l.M2	–	102.0	151.7	6.0	172	2.5–2.8
MUL	a004	l.M2/M3	–	77.5	125.0*	5.0	160*	2.2

+, Figure lowered because of wear; *, approximation; u., upper; l., lower; EL, element; W, without cover cement. Other explanations are given under Abbreviations.

MAMMUTHUS TROGONTHERII[1]

Because of the paucity of material available for comparison, it is very doubtful whether skeletal elements of *M. trogontherii* can be distinguished from those of *M. meridionalis*. Since its presence in the southern bight of the North Sea was demonstrated by some teeth (Fig. 20.8), the possibility exists that some postcranial elements referred to *M. meridionalis* were misinterpreted.

M. trogontherii, the steppe mammoth, is usually considered to have been not only the largest of the mammoths but the largest of all proboscideans as well. A skeleton of a male in the Azov Museum appears to be one of the tallest ever recorded; it reaches 450 cm at the shoulder (Garutt and Nikolskaja 1988).

MAMMUTHUS PRIMIGENIUS

Because this study focuses on the Plio-Pleistocene species, *M. primigenius* is not treated in full detail but merely cited for comparative purposes.

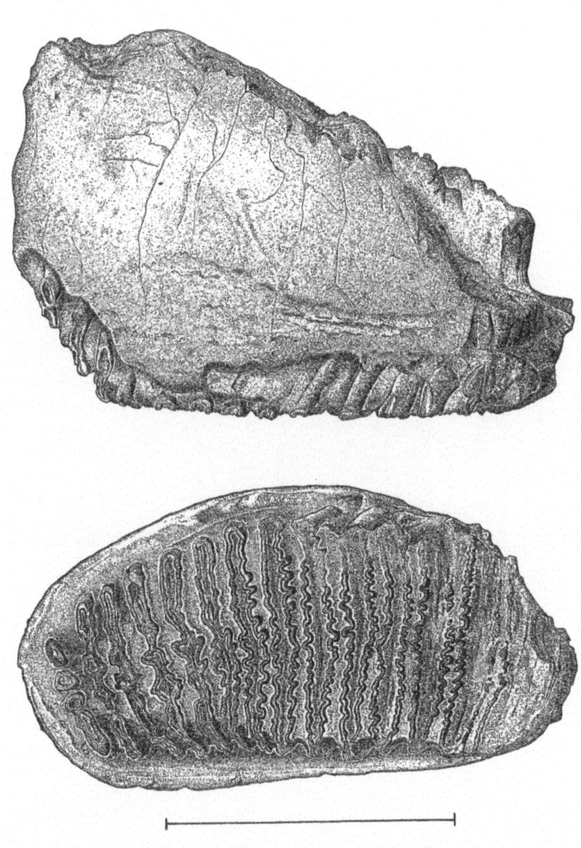

Fig. 20.8 *Mammuthus trogontherii* (BOK a028). Upper right M3 (anterior is to the right). Southern bight of the North Sea. Buccal and occlusal view, scale bar 10 cm. (Drawing: H. van Essen.)

SUMMARY AND CONCLUSIONS

As additional finds are to be expected and they may shed new light on the problems, the following are preliminary conclusions, especially with regard to proboscideans found in the study area.

THE EASTERN SCHELDT ASSEMBLAGE

The Eastern Scheldt assemblage that includes *A. arvernensis* and *M. meridionalis* can be referred to the top of the Tegelen Formation of late Tiglian age. This is in accordance with Drees (1986). The Tegelen Formation locally represents the first continental deposits. The 'Praetiglian beds' to which earlier authors attributed the material (e.g. Maglio 1973) are marine and not likely to contain reworked terrestrial fossils. There are no obvious traces of reworking on these specimens and the evolutionary stage of the *M. meridionalis* teeth is that of the typical form as represented by many specimens of late middle to early late Villafranchian age (collection of the Geological and Palaeontological Museum in Florence, Italy). Postcranial elements of *A. arvernensis* and *M. meridionalis* can be separated on morphological grounds.

THE NORTH SEA ASSEMBLAGE

The geological and palaeogeographical history of the southern bight of the North Sea does not preclude the presence of terrestrial mammals in the margins of the area now submerged; this holds true for interglacials ranging from the late Tiglian to the Cromerian (Zagwijn 1979). The presence of *M. meridionalis* in the area is probably linked to these (warm-) temperate stages. During glacial periods the southern bight of the North Sea was accessible to any terrestrial species with a tolerance for cool to cold climatic conditions because of eustatic sea-level fall. The current list of fossil proboscideans from the North Sea is: *Anancus arvernensis, Mammuthus meridionalis, Mammuthus trogontherii, Mammuthus primigenius,* and *Elephas antiquus*. The occurrence of *M. meridionalis* in the Thornton Bank area, where *A. arvernensis* was found, is not yet established. A late Tiglian age for part of the North Sea assemblage is claimed because of morphological similarity of a number of *M. meridionalis* molars to those of the Eastern Scheldt.

Evolutionary differences between *M. meridionalis* molars from the Tiglian and Waalian stages cannot be confirmed. Similarly, the morphological boundary between Waalian and Bavelian specimens remains undetermined. The separation of teeth of advanced *M. meridionalis* and *M. trogontherii* is in part provisional. As yet, postcranial elements of *M. meridionalis* and *M. trogontherii* cannot be separated on morphological grounds.

[1] Editors' note: *Mammuthus trogontherii* was synonymized with *M. armeniacus* by Maglio (1973). In this volume, these species are considered separate until additional evidence is gathered, see Lister (this volume, Chapter 19) and Appendix C.

CONCLUSIONS

Because there were no *in situ* finds, the study of fossil proboscideans from the North Sea and the Scheldt estuary required a detailed morphological approach and an understanding of the geology of the area. Taxa identified from the study area include *Anancus arvernensis*, *Mammuthus meridionalis*, *M. trogontherii*, and *M. primigenius*. The occurrence of *Mammut borsoni* in the Scheldt estuary is doubtful. Data presented in this paper appear to corroborate the results gathered from geological formations on adjacent terrestrial localities of Plio-Pleistocene to early middle Pleistocene age.

ACKNOWLEDGEMENTS

We thank A. Azzaroli (IGF), C. van der Bok (Ouddorp), P. and A. van Es (Stellendam), C. F. Franssen (Bennekom), M. Freudenthal (RGM), the late C. de Giuli (IGF), C. Jager (Goes), N. C. and A. Kerkhoff (Schiedam), M. van Kolfschoten (IVA), P. de Koningh (Hall), C. Laban (Rijks Geologische Dienst, Haarlem), Th. W. Lammers (Asten), the library personnel of Dieren Public Library, A. M. Lister (Dept of Zoology, Cambridge University), M. Mazzini (IGF), J. M. Moraal (Vlissingen), J. Mulder (Nieuwleusen), G. Nijboer (Bedum), P. Y. Sondaar (IVA), A. van Splunder (Stellendam), C. Strang (RGM), J. Veurink and colleagues (Akzo Research Laboratories, Arnhem), J. de Vos (RGM), and an anonymous reviewer for their help in various aspects of this study.

TWENTY-ONE

Proboscidean remains in southern Levant

EITAN TCHERNOV AND JEHESKEL SHOSHANI

INTRODUCTION

Palaeontological, archaeological, and historical records suggest that proboscideans may have roamed in the Levant for almost 20 million years, from the early Miocene until a few thousand years ago (Hooijer 1962*a*; Savage and Tchernov 1968; Clutton-Brock 1981; Tchernov *et al.* 1987). King Solomon was known to have had a craving to possess ivory, believed to have been brought from the south (possibly northern Africa): 'once every three years came the navy of Tarshish, bringing gold, and silver, ivory, and apes, and peacocks' (I Kings 10: 22). The possibility exists that other ivory pieces which date as late as 800 BC may have been from local elephants which lived in the Levant (Weiss 1985).

BOUNDARIES WITHIN GLOBAL CONTEXT

The southern Levant, on the eastern Mediterranean, is delineated according to Goldberg (1986), Por (1975), and Tchernov and Yom-Tov (1988), as the area between the Taurus–Zagros mountain chain to the north and the Arabian and the Sinai peninsulas to the south and west (Fig. 21.1h). In this chapter we focus on the area of the Jordan Rift Valley (= Dead Sea transform) which cuts through this region, and provide a brief summary of the finds belonging to Proboscidea. Remains of proboscideans in adjacent areas, north and west of the Levant, are discussed by Caloi *et al.* (this volume, Chapter 22) and by Tassy (this volume, Chapter 10).

Proboscidean remains in the southern Levant must be viewed in the overall biogeographical and geomorphological context. Recent neontological and palaeontological studies indicate that the Levantine biotas constitute 'one of the most complicated ecosystems in the world, comprising plants and animals virtually from most of the Old World and even North America. Ethiopian and Oriental taxa coexist with Euro-Siberian, European, North American, central Asiatic and Mediterranean species, mixed with Saharan and Arabian desert floral and faunal elements, giving it a unique biotic feature; a living monument to a fascinating animal history and ecology' (Tchernov and Yom-Tov 1988, p. 1).

This extremely diversified composition is better understood in the light of its strategic geographical location and the geotectonic events that occurred in this region during the Neogene and Quaternary. Situated on the cross-roads of three continents — Africa, Asia, and Europe — the southern Levant has played a crucial role as a corridor and a dynamic filter for fauna and flora for millions of years, similar to the land-bridge between North and South America. Crustal earth movements and mountain formation resulted in climatic changes which added to the already diversified habitats. Palaeontological and archaeological evidence provide only a clue to the myriad species which might have inhabited the Levant.

TAXONOMIC DIVERSITY AND ITS ORIGINS

'Many' fossil remains of animal species have been discovered in southern Levant, among them these proboscidean taxa: *Prodeinotherium* sp., amebelodontid and/or gomphotheriid, *Mammuthus meridionalis*, *M. trogontherii*, *Elephas namadicus*, *E. planifrons*, *E. maximus*, and *Stegodon mediterraneus* (cf. *S. trigonocephalus*), representing at least five families and distinct lineages (Table 21.1). Some of these taxa may have immigrated to the Levant from the south, others may have arrived from Eurasia. That gomphothere species almost certainly inhabited the area known today as Lebanon, Syria, Jordan, and Israel is evidenced by the finding of *Choerolophodon chioticus*, *C. pentelici*, and *Anancus* sp. in Turkey during the middle Miocene (Astaracian) through Pliocene (late Villafranchian) (Savage and Russell 1983; Tassy 1994*b*). The most important proboscidean-producing sites in the southern Levant are: Latamne (Syria), the Rotem and Yeroham basins, the Golan Heights, Bethlehem, 'Ubeidiya, Evron, and Gesher Benot Ya'akov (the latter six sites are in Israel or in Israeli-held territories) (Hooijer 1958; Haas 1966; Horowitz 1979; Tchernov 1984; Tchernov and Shoshani 1989; cf. Figs 21.2, 21.3). The most recent historical records of elephants in the Levant are dated from about 3000 years ago (Clutton-Brock 1981); most likely they belonged to the extant *Elephas maximus*, which lived in areas known today as parts of Syria and Iraq.

TAXONOMIC PROBLEMS ENCOUNTERED

Maglio (1973) hypothesized that the late Pliocene African species *Elephas ekorensis* gave rise to all 10 elephantine forms, two of which were discovered in the eastern Levant: *E. namadicus* and

Table 21.1 Proboscidean taxa discovered in the southern Levant

Taxon	Nature of material	Locality(ies)	Land Mammal Age(s)
Deinotheriidae			
Prodeinotherium sp.	Teeth fragments	Yeroham and Rotem (Israel)	Early Miocene
Family undetermined			
amebelodontid or gomphotheriid	Teeth fragments	Yeroham and Rotem (Israel)	Early Miocene
gomphotheres	Fragmentary teeth, tusks and bones	Golan Heights (southern region)	Neogene
Stegodontidae			
Stegodon sp.	Teeth fragments	Gesher Benot Ya'akov (Israel)	Middle Pleistocene
	Teeth fragments	'Ubeidiya (Israel)	Early middle Pleistocene Early Biharian[a]
	Teeth fragments	Evron (northern Israel)	Middle Pleistocene[c]
Stegodon cf. *trigonocephalus*	? teeth fragments	Latamne (Orontes (=Asi), Syria)	Middle Pleistocene [a,c,d]
Stegodon mediterraneus	Fragmentary teeth,	Gesher Benot Ya'akov (Israel)	Middle Pleistocene[a,c,d]
	teeth fragments	'Ubeidiya (Israel)	Early middle Pleistocene Early Biharian [a,b,c]
Elephantidae			
Mammuthus meridionalis (cf. *M. m. tamanensis*; Fig. 21.4	Teeth fragments	'Ubeidiya (Israel)	Early middle Pleistocene Early Biharian[a,b,c,d]
Mammuthus trogontherii	? Teeth fragments	Latamne (Orontes (Asi), Syria)	Middle Pleistocene[a,c,d]
Mammuthus primigenius	??	??	Reported on p.92 of Tchernov (1981)
Elephas sp.	Teeth fragments	Evron (northern Israel)	Middle Pleistocene
Elephas planifrons	Teeth and fragments	Bethlehem	Pliocene–early Pleistocene[a]
Elephas namadicus	Skull and teeth	Gesher Benot Ya'akov (Israel)	Middle Pleistocene[e]
Elephas sp. or ? *Elephas maximus*	Cranial fragment	Beth-Shean (found with other elements at Roman ruins)	? Recent/late Pleistocene[f]

[a] After Bar-Yosef and Tchernov (1972, table 4).
[b] After Beden (1986).
[c] After Tchernov (1984, p. 536).
[d] After Tchernov (1981, pp. 86, 88, 92).
[e] Identification of teeth was made by Lister (personal communication 1992); see, however, Tchenov *et al.* 1994, p. 332.
[f] It is possible that this piece belongs to a different (older) excavation layer and it became mixed with more recent material.

E. planifrons. Specimens of the latter species were found in Bethlehem, and according to Maglio (1973, pp. 45–6) 'the Bethlehem specimens represent the earliest stage of *Elephas* recorded outside Africa and fill the morphological and geographical gap between *E. ekorensis* of Africa and *E. planifrons* of Asia'. Maglio's interpretation dates *E. planifrons* as a late Pliocene taxon, *contra* Tchernov (1981, 1984) who considers the Bethlehem assemblage to date to early Pleistocene or preglacial deposits.

Other problems encountered include specific identification of specimens which are important to our understanding of Levantine zoogeography. For example, Latamne material of *Mammuthus trogontherii* (originally identified as *E. trogontherii*; Hooijer 1961, p. 120) was synonymized with *M. armeniacus* by Maglio (1973, pp. 42, 57). Similarly, the *M. trogontherii* from the Benot Ya'akov Formation of Hooijer (1959) and Tchernov (1984) has been re-identified as *E. namadicus* by Maglio (1973, p. 42). The presence of *E. namadicus* in the Middle East is, according to Maglio (1973, pp. 114–17), an important component in establishing migratory routes from Africa to Asia and Europe during the early Pleistocene.

FAUNAL CORRELATION AND DISPERSALS

Correlation of southern Levantine fossil mammalian communities shows affinities with the early Biharian of Europe, the Tamanian Faunal Complex of Moldavia (formerly part of the USSR), Irvingtonian of North America, early Ensenadan of South America, Ain-Hanech of North Africa, and Olduvai Lower and Upper Bed II of East Africa, dating from the early to the middle Pleistocene epoch, about 1.7 to 1.4 Ma (Savage and Russell 1983; Tchernov 1984, 1988; Fig. 21.4). Based on available evidence, we suggest that the proboscidean remains from the Levant represent two main 'waves' of immigrants from Africa. The first occurred at the beginning of the Miocene (*c.* 20 Ma) and the second during the late Pliocene and early Pleistocene, some 2.5 to 1.5 Ma.

The second 'wave' included *Stegodon* sp. (possibly *S. mediterraneus* from the Benot Ya'akov Formation; Hooijer 1960) and *S.* cf. *trigonocephalus* (from Latamne, Orontes (= Asi), Syria; Hooijer 1961). The survival of a member of the family Stegodontidae in the Levant during the middle or late Pleistocene is significant for it represents an entirely distinct lineage of

Proboscidea with an earlier phylogenetic history (Tassy and Shoshani 1988) than the family Elephantidae (to which many other Levantine proboscidean taxa belong).

THE RISE AND FALL OF THE TETHYS

The extant biotic composition of the Levant bears little resemblance to the biotic distribution that was known here even during the last glacial stages. This composition is the outcome of very complex chains of geological events and cannot be understood without careful understanding of this past history. The section of the Great Rift Valley between the Red Sea and southern Turkey was a crucial link between at least two biogeographical realms, the Palaearctic and the Ethiopian, and thus was vital to their floral and faunal evolution. Geological, palaeontological, and archaeological traces of this dynamic evolution can be found within and around the Levant.

A summary of the tectonic events which shaped and reshaped the Levantine landscape over the past 20 million years is given in Tchernov (1988; see also Thomas 1985). Here we recap these events. Figure 21.1 comprises a series of palaeogeographical maps, showing the extent of the epicontinental sea complex which connected the Tethys Sea with the Indian Ocean. A brief description of each of the major palaeogeological events follows (cf. Figs 21.5 and 21.6). The Mediterranean Neogene, or MN, biostratigraphical subdivisions are after Mein (1976, 1990).

EARLY BURDIGALIAN, OR EARLY ORLEANIAN MAMMAL AGE (MN 2–3A; 21–19 MA)

At the end of the Oligocene and beginning of the Miocene, Eurasia and Afro-Arabia were separated by the epicontinental sea, the Tethys, which joined the areas of the present-day Mediterranean Sea and the Persian Gulf (Fig. 21.1a). This situation is believed to have existed for 3–5 million years, and acted as a barrier between the faunistic and floristic elements of what is today Europe and Asia to the north, and Africa and Arabia to the south.

LATE BURDIGALIAN, OR MIDDLE ORLEANIAN MAMMAL AGE (MN 3A–3B; 18–17 MA)

Topographic changes in the Levant during this period (Fig. 21.1b) were a part of wide-scale global events during the Miocene epoch (24–5 Ma), e.g. Alpine-Himalayan mountain-building and the beginning of the Great Rift Valley (Fig. 21.1l). At about the early Miocene (late Burdigalian–Langhian), the Tethyian barrier mentioned above was 'broken' at the north-east edge of the Afro-Arabian plate along the present Anatolian–Iranian suture. The gateway formed allowed interchange of terrestrial biota between the two realms.

Proboscidean remains from the Levant and vicinity during this period included essentially African elements, such as *Prodeinotherium* sp., *Deinotherium* sp., *Gomphotherium* sp., amebelodon, *G. angustidens*, *G. cooperi*, *G. pygmaeum*, *G. spenceri*, and *Platybelodon* sp. These were not isolated immigrants; they were part of an assemblage of herbivore (Hyracoidea, Artiodactyla, Perissodactyla, Rodentia, Lagomorpha), carnivore (Carnivora, Cereodonta), and omnivore (Insectivora, Primates) taxa, details of which are given in Tchernov (1988, table 2).

Except for *Gomphotherium cooperi* from Ad Dabtiyah, Saudi Arabia (Gentry 1987) which may be assigned to the '*Gomphotherium annectens* group', there is no evidence that other gomphothere material from the Middle East belongs to this group. The '*Gomphotherium annectens* group' was also found in the Far East (e.g. Japan: Matsumoto, 1925, 1926; Tassy, this volume, Chapter 10). One of us (JS), believes that data on gomphothere geochronology and palaeo-distribution (Wang and Wang 1991; Tassy, this volume, Chapter 10) may be interpreted as indicating that the presence of advanced gomphotheres in Europe may have been the result of one of two events. Advanced gomphotheres, e.g. *Gomphotherium angustidens*, may have evolved from the more generalized '*Gomphotherium annectens* group'. Alternatively, advanced gomphotheres may have evolved in Europe (also from the '*Gomphotherium annectens* group') as an autochthonous event. In either case, the Levant seems to have played an important transitory corridor for these early Miocene migrants.

LANGHIAN, OR LATEST ORLEANIAN MAMMAL AGE (MN 4–5; 16–15 MA)

The palaeogeographical evidence shows that during this time span, the Tethys and Paratethys seas reconnected, thus cutting the terrestrial passageway between Eurasia and Afro-Arabia (Fig. 21.1c). Further, the Arabian peninsula was increasingly isolated due to the developing Red Sea Rift. This temporary reisolation of the Eurasian and Afro-Arabian biotic units may have initiated autochthonous radiation of certain taxa.

EARLY SERRAVALLIAN, OR ASTARACIAN MAMMAL AGE (MN 6: 14 MA)

The Levantine and adjacent areas' landscapes did not change much during this short interval. There was a small land-bridge, or marshland corridor across the Anatolian border (Fig. 21.1d). This corridor permitted the re-establishment of biotic interchange, especially in the southerly direction. In addition, the Tethys had shrunk and begun to take the shape of the Mediterranean Sea, which is its remnant. It may be noted that intermittent land-bridges and isthmuses did not begin and end as definite events, but rather they may have been episodal events.

EARLY TORTONIAN, OR LATE VALLESIAN MAMMAL AGE (MN 10–11; 11–9 MA)

Palaearctic and tropical savanna regions were fully joined through a wide corridor, whereas the Arabian peninsula was

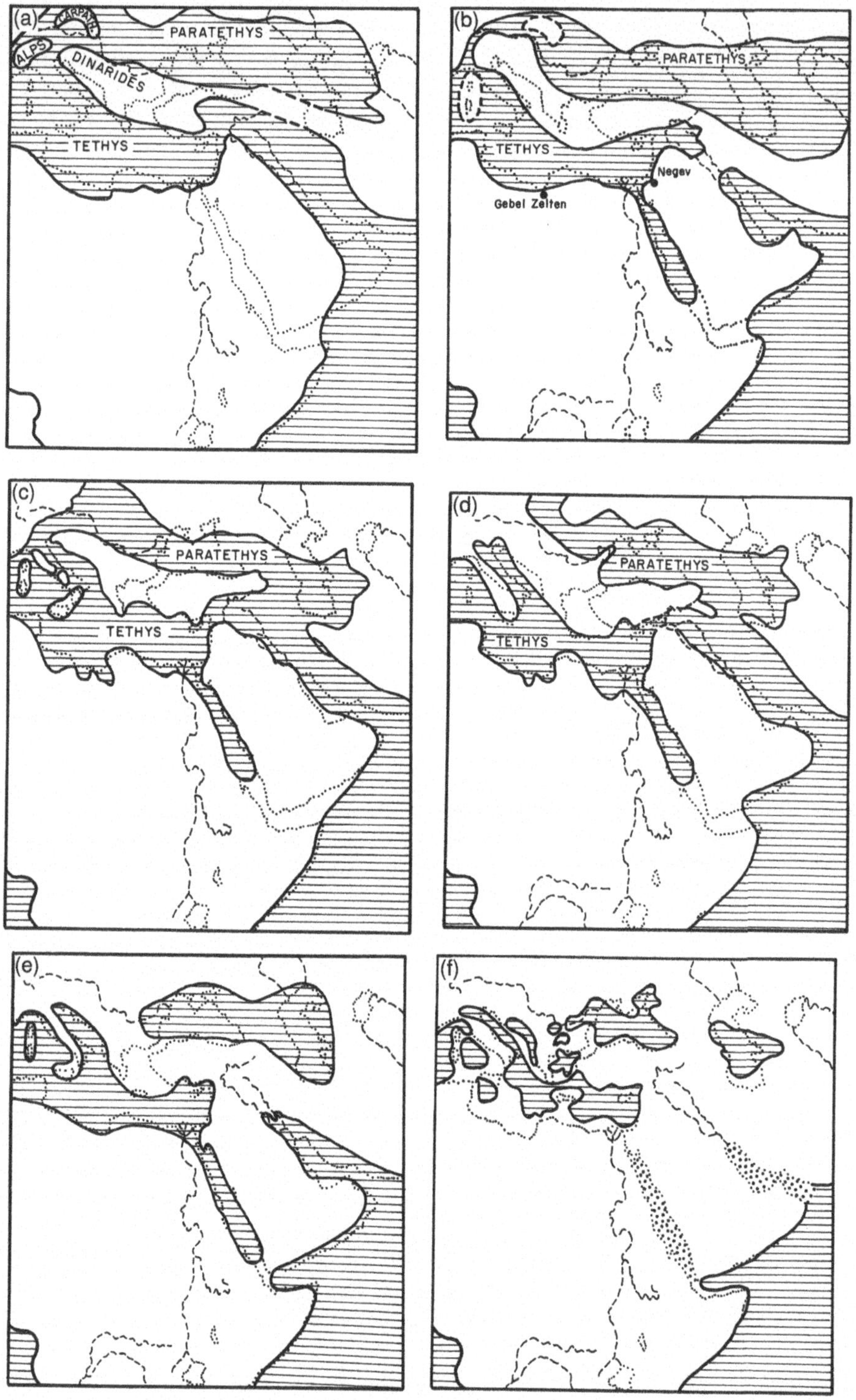
(a)
ALPS
PARATETHYS
DINARIDES
TETHYS
(b)
PARATETHYS
TETHYS
Negev
Gebel Zelten
(c)
PARATETHYS
TETHYS
(d)
PARATETHYS
TETHYS
(e)
(f)

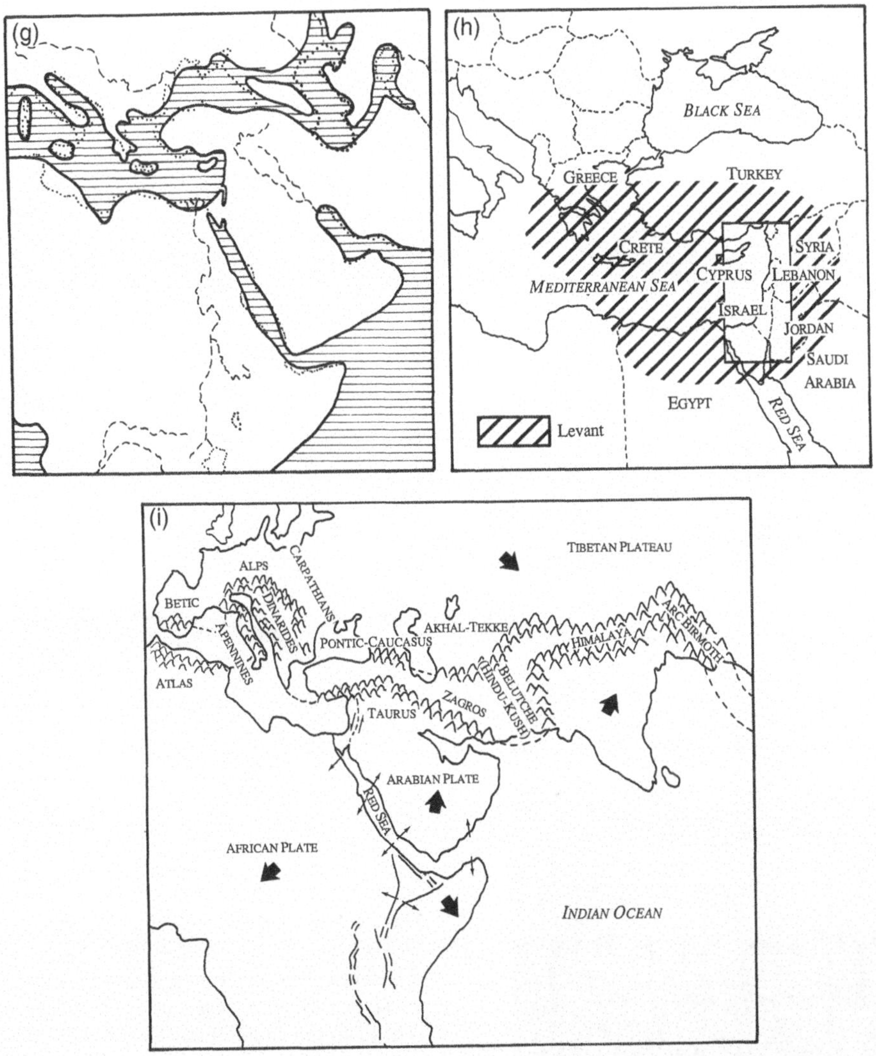

Fig. 21.1 A series of simplified palaeogeographical maps covering a span of about 20 million years, showing the extent of the epicontinental sea complex (also known as the Indo-Mediterranean Sea Way) which connected the Tethys Sea with the Indian Ocean (slightly modified after Tchernov 1988, 1992). (a) Early Burdigalian, or early Orleanian Mammal Age (MN 2–3a; 21–19 Ma); (b) late Burdigalian, or middle Orleanian Mammal Age (MN 3a–3b; 18–17 Ma); (c) Langhian, or latest Orleanian Mammal Age (MN 4–5; 16–15 Ma); (d) early Serravallian, or Astaracian Mammal Age (MN 6; 14 Ma); (e) early Tortonian, or late Vallesian Mammal Age (MN 10–11; 11–9 Ma); (f) Messinian, or late Turolian Mammal Age (MN 13; 6–5.5 Ma); (g) Zanclean–Piacenzian, or Ruscinian–Villafranchian Mammal Ages (MN 14–17; 5–2 Ma); (h) a map showing the general area of the Levant (the area in the rectangle is shown in detail in Fig. 21.2). Much of the Levantine landscape during the Pleistocene–Holocene (2–0 Ma) may not have been different from the present. (i) A generalized map showing the direction of crustal plate movements and the Eurasian and Himalayan orogenic fold belt. These mountain belts formed a barrier between the Irano-Turkish plateau and the Levant and limited the biotic exchanged during the late Miocene.

gradually isolated from the African complex, leaving a narrow isthmus for limited interchange (Fig. 21.1e). In fact the areas close to this isthmus are nowadays considered a part of the Ethiopian Region faunal division (Smith 1974, p. 532).

Adding to this freedom of biotic interchange was the complete isolation of the Paratethys Sea from the Mediterranean Basin during the Vallesian or Turolian. The land connections not only allowed exchange of fauna and flora between the Levant and adjacent land complexes, but also opened new niches for the North American (Nearctic) immigrant equid taxa, e.g. *Hipparion*.

MESSINIAN, OR LATE TUROLIAN MAMMAL AGE (MN 13: 6–5.5 MA)

Overall, this time span of less than 1 million years was dominated by considerable desiccation of the Mediterranean and Paratethyan basins (Fig. 21.1f), a process which terminated a general trend of cooling and drying. Much of the topography of the region, which includes deep canyons, wadis, and mountain erosions, was sculptured during this period.

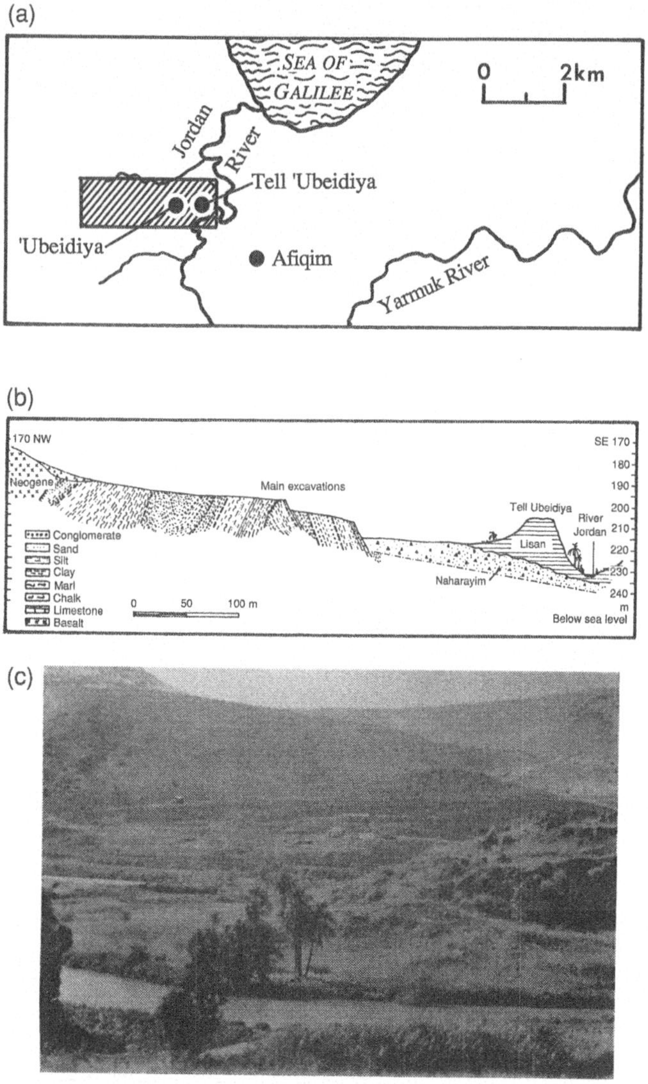

Fig. 21.2 (a) Central Jordan Valley with the site of 'Ubeidiya within the shaded area. (b) Geological cross-section of the area boxed above. (c) General view of the 'Ubeidiya site depicting the western escarpment of the Jordan Valley (background), excavation trenches around the field laboratory, and Tell 'Ubeidiya to the right (middle), and the Jordan River (foreground). (All illustrations after Bar-Yosef and Tchernov (1972), photograph by A. Nive.)

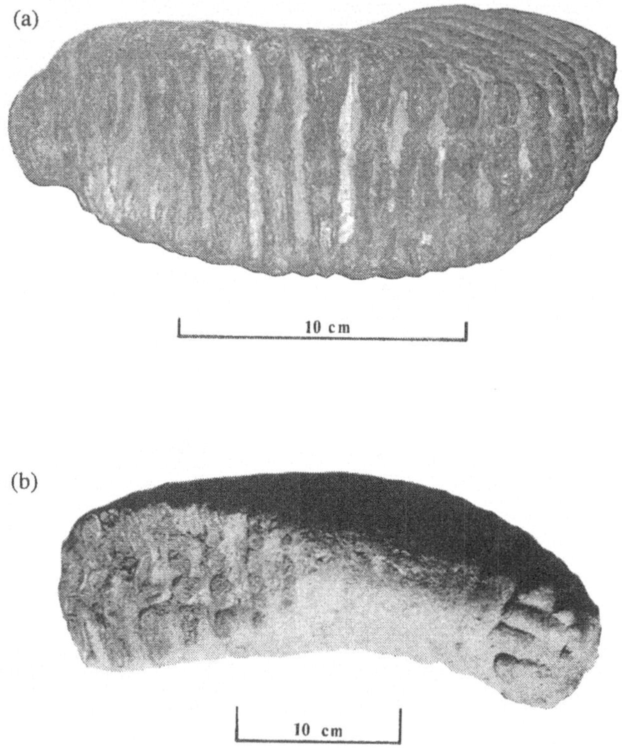

Fig. 21.3 (a) Occlusal view of a complete right upper third molar of *Elephas namadicus* from Gesher Benot Ya'akov, a middle Pleistocene site. (b) Occlusal view of a complete lower left third molar of *Mammuthus meridionalis* cf. *M. m. tamanensis* from 'Ubeidiya, an early Pleistocene site. (The tooth shown in (b) was identified by M. Beden in a letter to J. Shoshani, 23 June 1983 (cf. Maglio 1973; Mazo *et al.* 1990); photographs by J. Shoshani.)

On the global scale, the 'Messinian Crisis' (or 'Salinity Crisis'), as it is called, was manifested in land connections between Africa and Europe via the Gibraltar Strait to the west and between Africa and Asia through the entire stretch of the areas known today as the Gulf of Aden and the Suez Canal. Such extensive land 'highways' permitted corresponding faunal and floral interchanges, and maximal biotic similarities were attained during this period.

ZANCLEAN–PIACENZIAN, OR RUSCINIAN–VILLAFRANCHIAN MAMMAL AGES (MN 14–17; 5–2 MA)

The Mediterranean 'Salinity Crisis' mentioned above began to show signs of recovery at the beginning of the Pliocene. This short period of peripheral desiccation, due to an abrupt regression of the Mediterranean, was followed by a re-establishment of

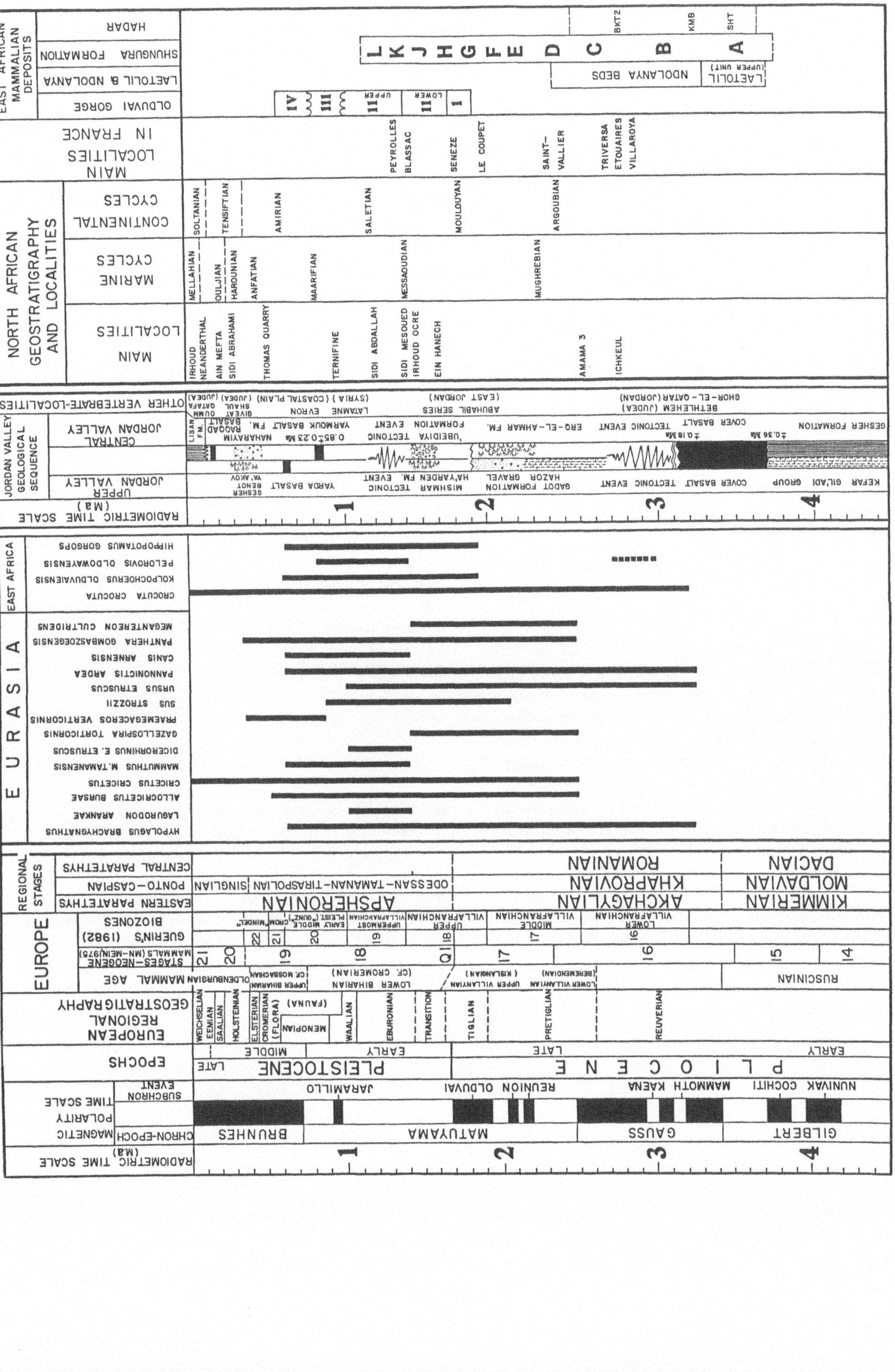

Fig. 21.4 Geostratigraphical and chronological framework of main events of the Plio-Pleistocene in the eastern Levant and temporal correlations among eastern Levant, Europe, and East Africa (after Tchernov 1988, pp. 186–7, fig. 10).

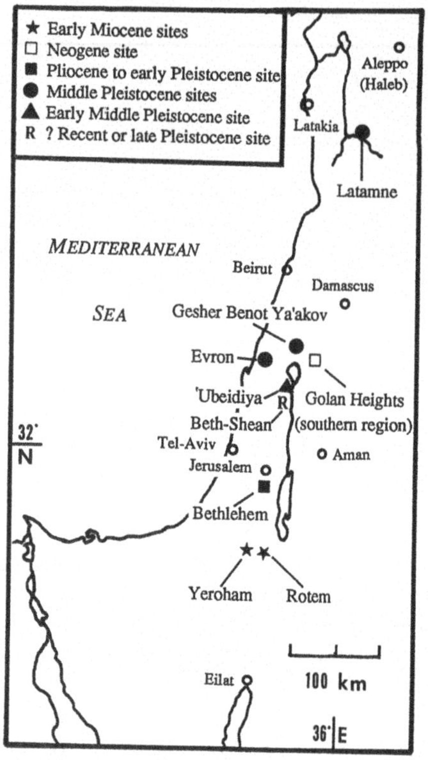

Fig. 21.5 Neogene and Pleistocene proboscidean sites in the Levant (cf. Table 21.1). Other Levantine sites which produced proboscidean materials are depicted in Figs 22.2 and 34.5.

open Atlanto-Mediterranean marine conditions (Fig. 21.1g). It is believed that this major (climatic) change was distinguished by a rapid decrease of about 18 °C at about 5.0 Ma. Such a plummeting in environmental temperatures was considered a direct consequence of a glacial impact and a world-wide marine transgression. The dawn of the Pliocene also brought a sudden flooding of the Mediterranean Basin, causing major shifts in the peri-Mediterranean climate and biotas. Diversified Mediterranean floral taxa were largely represented in the Pliocene of southern Europe by Pinaceae (*Picea*, *Pinus*), but also by Cypressaceae, Betulaceae (*Alnus*), Fagaceae (*Quercus*), and Ulmaceae. Another consequence of the early Pliocene marine transgression was the insular isolation of the biotas which were left on islands within the Mediterranean Basin (details in Tchernov 1988; see also Reese 1989; Theodorou 1990).

PLEISTOCENE–HOLOCENE AGES

The topography of the Levant by about 2 Ma may have been similar to the present-day landscape (Por 1975). Proboscidean taxa discovered during this time span include: *Mammuthus meridionalis*, *M. trogontherii*, (?)*M. primigenius*, *Elephas namadicus*, *E. planifrons*, *E. maximus*, and *Stegodon mediterraneus* (Table 21.1). Unlike the hypothesis of 'post-Pleistocene desiccation' (Guilday 1967; see also Stuart 1991) as the main cause for world-wide extinction of terrestrial megafauna, the large animals in the Middle East exhibited 'more stability during the critical period. Their disappearance from the area was caused mainly through the agency of man during historical times; some species disappeared only during the nineteenth and twentieth centuries' (Tchernov 1984, p. 548).

CONCLUSIONS

Based on the available evidence, it is suggested that during most of its entire geological Cenozoic history, the Levant has been a pivotal link between major biogeographical regions with extremely varied habitats, ranging from montane forests with alpine tundra complex to grasslands, savannas, and deserts.

The proboscidean specimens discovered in the Middle East — though a small sample and not completely sorted out — represent several diverse groups, typical of what one would expect from a 'random catch' in a land-bridge filter such as the southern Levant, which has intermittently connected three continents and two biogeographic realms, Palaearctic and Ethiopian.

ACKNOWLEDGEMENTS

We thank Dr W. Junk Publishers (Dordrecht, The Netherlands) for permitting us to reprint Figs 21.1a–i from Tchernov (1988). J. S. Grimes helped with the figures. A. M. Lister, E. C. Marsac, J. L. Pierce, D. S. Reese, and P. Tassy made helpful suggestions to improve the manuscript. We are grateful to N. Goren Inbar and A. M. Lister for providing us with unpublished information on the proboscidean remains from the Gesher Benot Ya'akov Formation, Israel.

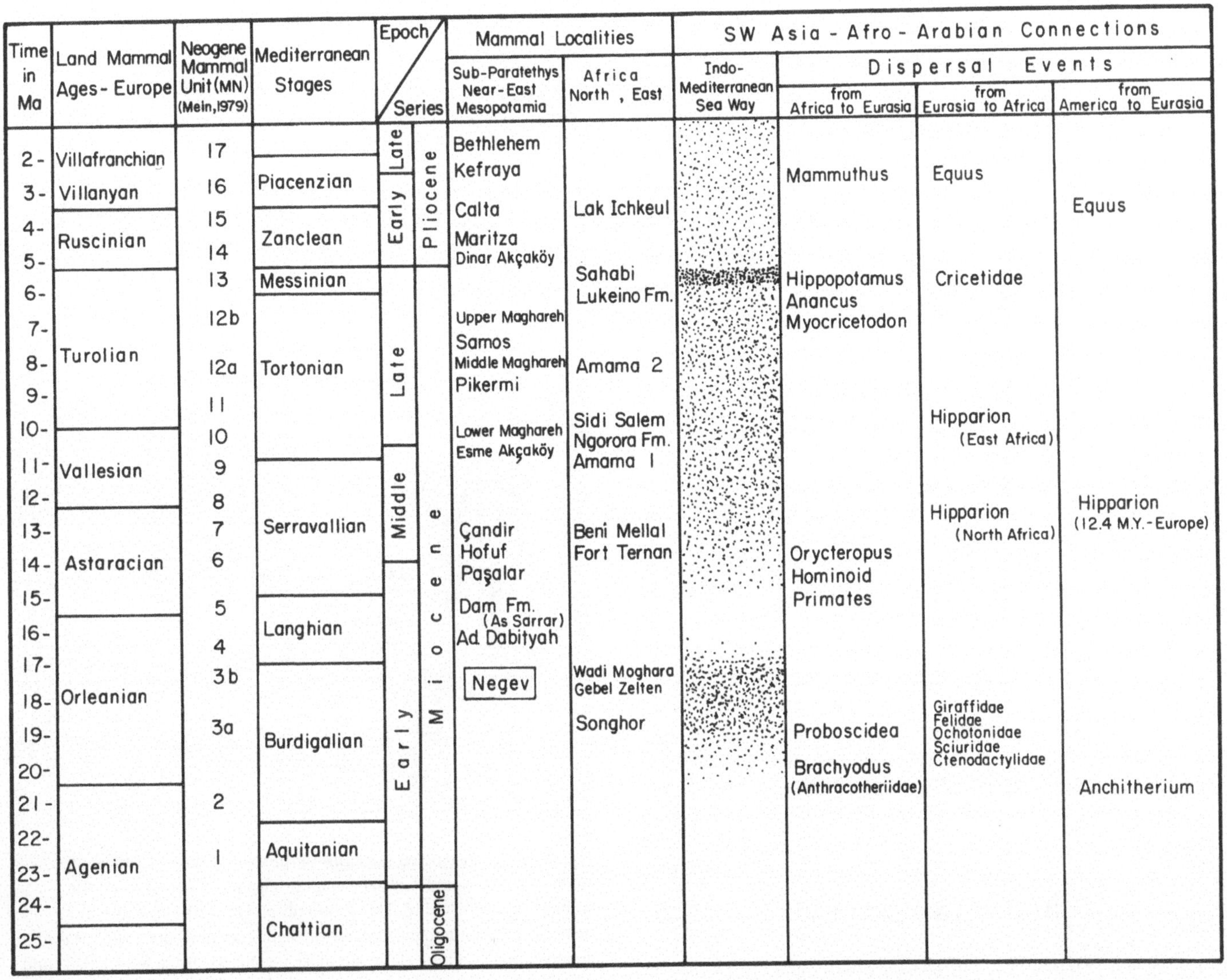

Fig. 21.6 Biostratigraphy and possible correlation of mammalian dispersals among Afro-Arabia, Eurasia, and North America (after Tchernov 1988, p. 163, table 1; the intensity of dots in the stippled column indicates the relative absence of the Indo-Mediterranean Seaway, e.g. it was entirely absent during the 'Messinian Crisis', cf. Fig. 21.1f).

TWENTY-TWO

The Pleistocene dwarf elephants of Mediterranean islands

LUCIA CALOI, TASSOS KOTSAKIS, MARIA R. PALOMBO, AND CARMELO PETRONIO

INTRODUCTION

The discovery of 'dwarf' elephants (Fig. 22.1) on the Mediterranean islands (Fig. 22.2) dates back to the first half of the nineteenth century. There were two schools of thought regarding the classification of these forms. Generally, a single colonization of each island or group of islands was followed by a process of dwarfism. Now, it is believed that, on some islands, there was more than one colonization and the dwarfing processes were repeated. This hypothesis has been demonstrated on the island of Crete, where micro-mammals permit a biostratigraphical control (Mayhew 1977), and also on Sicily and Malta.

Two hypotheses are also suggested for possible migration and colonization of elephants: land bridges and swimming capability (Azzaroli 1982; Dermitzakis and Sondaar 1979). Both models are possible because of the palaeogeography of the Mediterranean Basin.

Fig. 22.1 Reconstructed skeletons of the extinct (late Pleistocene) Mediterranean islands' dwarf elephant 'Elephas falconeri' from the Spinagalo Cave in Sicily. This group (displayed at the Instituto di Geologia e Paieontologia, Universita degli Studi di Roma, Italy), is said to include a male (on the right, shoulder height 90 cm), a female (left) and two young males. (Photo credit: L. Caloi, T. Kotsakis, M. R. Palombo.)

SICILY AND MALTA

In the Sicilian–Maltese area, three forms with different degrees of endemism are known (*Elephas antiquus leonardii*[1] Aguirre, *E. mnaidriensis*, and *E. falconeri*). These are interpreted as the result of progressive body reduction starting from the ancestor for *E. antiquus* Falconer and Cautley. The specific identity of *E. melitensis*, originally believed to be intermediate in size between *E. mnaidriensis* and *E. falconeri*, is now considered as a synonym of *E. falconeri* (Ambrosetti 1968). The largest taxon found in Sicily shows only slight body reduction. The subspecies *E. a. leonardii* has been erected from remains found in post-Sicilian marine deposits at Via Liberta, Palermo (Aguirre 1968), and recorded also in conglomerate levels of the 'Messina Formation' (Bonfiglio and Berdar 1986; Bonfiglio and Kotsakis 1987). The relatively short and narrow hypsodont molars of *E. a. leonardii*, with thin enamel and small folds, could correspond to an evolutionary stage comparable to that of some Italian specimens of the middle–late Pleistocene (Palombo 1986).

The medium-sized species, *E. mnaidriensis*, has been erected based on the largest elephant bones found on Malta. The molars seem to be characterized by two morphotypes: pachyganal (large, slightly folded molars with thick enamel; lower laminar frequencies; laminae slightly to markedly subtriangular) and endoganial (thinner, more folded enamel, higher laminar frequencies; reduced loxodontoid expansion). The two morphotypes have been variously interpreted as belonging to distinct forms (Osborn

[1] The authors prefer to retain the name *Elephas antiquus* for the specimens described here (instead of *E. namadicus*), first because *E. antiquus* was traditionally employed for the material from the Mediterranean islands and, secondly, because skulls of *E. antiquus* have not been thoroughly studied to justify synonymity with *E. namadicus* (Maglio (1973, p. 40) considered *E. antiquus* Falconer and Cautley, 1845, pl. 13A, figs 4, 5, as a junior synonym of *E. namadicus* Falconer and Cautley, 1845, pl. 13, figs 1–3).

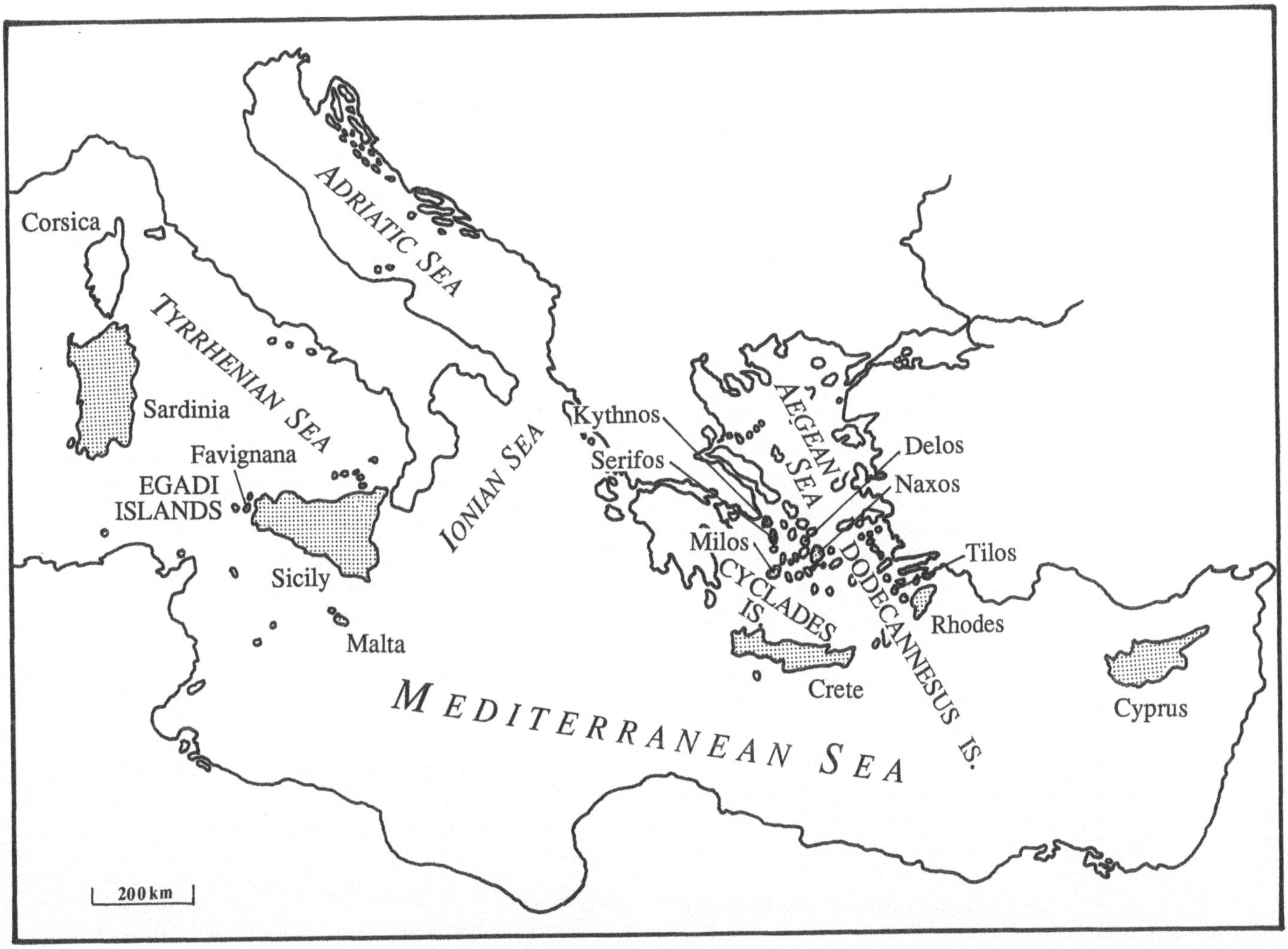

Fig. 22.2 Map of the Mediterranean Basin showing islands (shaded) where proboscidean remains have been discovered since the nineteenth century.

1942; Aguirre 1968) or considered as an expression of a marked sexual dimorphism (Ambrosetti 1968).

Indeed, the molars ascribed to *E. mnaidriensis* show a broad range of variation; both of the morphotypes show varieties with intermediate features, which are difficult to ascribe to either sex and a broader variation than seems to be present in the continental populations. This variability, recognized in the Puntali Cave (Carini, Palermo) (Pohlig 1893; Burgio *et al.* 1983), the richest in remains known up to now, makes difficult a comparison among the molars of the various localities of Malta and Sicily, thus hampering the possibility of inferences on their evolutionary degree. The cranial features of the Puntali elephant allow us to recognize two morphotypes (Ambrosetti 1968). Both morphotypes show peculiar traits, partly due to allometric variations (and, in any case, independent from size reduction), which distinguish them from *E. antiquus*. None the less, the basic structure of the skull shows certainly, in the case of both morphotypes, stronger affinities with *E. antiquus* than with any other elephantine form. There is general agreement in admitting the specific identity between the middle-sized elephants from Sicily and *E. mnaidriensis* from Malta.

E. melitensis was the first species described for smaller-sized elephants from Malta. Only later, in the material originally ascribed by Falconer (1868) to his new species, Busk (1867) separated the smaller-sized specimens, indicating for them a dental morphology similar to '*E. indicus*' (= *Elephas maximus* Linnaeus). The taxonomical status of the remains originally ascribed to *E. mnaidriensis* and *E. falconeri* has been interpreted in many ways (e.g. Vaufrey 1929; Osborn 1942; Aguirre 1968; Ambrosetti 1968). *E. melitensis* appears inadequately defined and the mere dimensional criterion does not allow separation of the teeth remains ascribed by Busk (1867) respectively to *E. melitensis* and *E. falconeri*. Ambrosetti (1968) considered the two species as synonyms and suggests retaining *E. falconeri*; he referred to this species as the remains of the rich Spinagallo (Siracusa) population (cf. Petronio 1970). Dimensions of most molars from Luparello (Palermo), ascribed by Vaufrey (1929) to '*E. a. falconeri*' and '*E. a. melitensis*', seem to fall into the range of the Spinagallo population, while the limb bones ascribed to *E. a. melitensis*' are larger in size. Based on the available evidence, the existence of elephants of intermediate size between the Spinagallo and Puntali populations in the Sicily–Malta district is possible. The taxonomical rank of these forms could be established only after a re-examination of all remains ascribed to *E. melitensis* and the definition of the morphological and dimensional variation range of the various Sicilian–Maltese small-sized elephantine taxa.

In Sicily and in Malta, very small forms occur, among the smallest forms found on the Mediterranean islands. These are generally referred to as *E. falconeri* and considered for decades as the last step of a progressive size reduction. The progressive stages started with *E. antiquus* and continued through smaller forms (*E. mnaidriensis* to *E. melitensis*, according to some authors; *E. a. leonardii* to *E. melitensis* to others; or finally exclusively *E. mnaidriensis*). These changes led from the graviportal continental species to relatively slender-limbed forms, reaching, at most, 90 cm withers height. In Malta, the smallest elephant is prevalently represented by teeth remains collected mostly from the Zebbug, Mnaidra, and Benghisa deposits (Vaufrey 1929), which, according to the iconographical data, have a wide morphological variation range. Morphotypes with not very thick, somewhat folded, enamel prevail.

In Sicily, the two deposits which yielded the richest populations are Luparello (Vaufrey 1929; Imbesi 1956) and Spinagallo (Ambrosetti 1968). The molars of the two localities do not seem to differ morphologically. In these deposits both morphotypes are recognizable, attributed to males and females (Ambrosetti 1968), and it is difficult to hypothesize, on a statistical basis, whether the two populations display the same evolutionary level and degree of differentiation.

The derivation of *E. falconeri* from *E. antiquus* has been generally accepted (Ambrosetti 1968). *E. falconeri* was considered as descendant from a taxon derived, in turn, from *E. antiquus*. Because of merely osteological evidence, the derivation of smaller-sized elephants from *E. antiquus* still remains questionable. Although dental morphology does not exclude a connection with a 'palaeoloxodontine' line, the extreme modification of the skull and the morphostructural variation of limb bones make it difficult to recognize the ancestral form. The Spinagallo skull (Ambrosetti 1968) shows neotenic features such as little or no pneumatization, great development of the brain case, with the disproportionality between neurocranium (portion of cranium which forms the braincase) and splanchnocranium (portions of the cranium derived from the primitive skeleton of the gill apparatus), reduction of the incisive zones, and frontal shift to orbitae. In the limb region, the body reduction implies a reduction of load, and it is difficult to establish to what extent, respectively, size reduction, different ambulation type and possibly different phyletic relationships, may have influenced the modifications in respect to *E. antiquus*.

In Sicily, remains attributed to *E. falconeri* have been recorded in oligotypic, extremely endemic faunas: Luparello (Vaufrey 1929; Imbesi 1956), Spinagallo (Ambrosetti 1968), and Poggio Schinaldo (Esu *et al.* 1986). These were generally attributed to the last glacial (Würm) and considered as more recent than remains associated with *E. mnaidriensis*, more diversified, and with a less pronounced degree of endemism (e.g. Kotsakis 1979; Caloi and Palombo 1985; Palombo 1986; Caloi *et al.* 1988, in press; Bonfiglio and Insacco 1992). As already underlined by Bonfiglio and Kotsakis (1987), Caloi *et al.* (1988), Caloi and Palombo (1989), and Burgio and Cani (1989), new stratigraphical data and absolute dates show the need for new biostratigraphical hypotheses.

The existing evidence may be summarized as follows:

1. Racemization dating (Belluomini and Bada 1985) indicates an older age for the Spinagallo elephant than the Puntali.
2. In the Alcamo travertines (Burgio and Cani 1989), elephants the size of *E. falconeri* occur in the layers in which the sand component is higher; from fissure infill deposits in the same travertine, elephant remains the size of *E. mnaidriensis* have been found.
3. In the Messina Formation Gravels, three species of different sizes are recorded (Bonfiglio and Berdar 1979).
4. It had been hypothesized that the Spinagallo Cave had been first filled with '*E. mnaidriensis*' breccias, later eroded by the Tyrrhenian Sea and finally filled with '*E. falconeri*' deposits (Ambrosetti 1968). The remains of the small elephant, however, came from a cavity below the main cave, separated from the latter by a pavement which, although containing fragments of the vault, does not contain clasts of the ossiferous '*E. mnaidriensis*' breccia, which occurs as residual sheet in the main cave.
5. Remains of '*E. melitensis*' size occur in the '*E. falconeri*' deposit at Spinagallo (Caloi, unpublished data), at Favignana, in a faunal assemblage which includes also *Megaceroides carburangelensis* (De Gregorio) (Capasso Barbato *et al.* 1989) and may occur in the basal layers at Luparello (Vaufrey 1929).
6. At Poggio Schinaldo (Esu *et al.* 1986), the endemic fauna including the small elephant seems to be older than the one containing *Megaceroides carburangelensis*.

From the above brief summary on Sicily and Malta, we outline these two hypotheses:

1. During the middle Pleistocene, *E. antiquus* colonized the Sicilian–Maltese area, originating the endemic subspecies *E. a. leonardii*. This latter taxon gave rise to a species of *E. melitensis* size (e.g. in the lowermost levels at Luparello, also represented by sparse remains at Spinagallo), which in turn gave rise to the small-sized elephant, *E. falconeri*. A second arrival of *E. antiquus*, during the latest middle Pleistocene or in the earliest late Pleistocene, may have been the origin of elephants the size of *E. mnaidriensis*.
2. In this alternative hypothesis, two arrivals are also postulated, but *E. a. leonardii* is suggested as the ancestor of

elephants of the size of *E. mnaidriensis*, while the ancestor of the middle and smaller forms remains to be determined (possibly *E. antiquus*, or *Mammuthus (Archidiskodon) meridionalis*).

Other hypotheses which postulate even more arrivals than proposed or adaptive radiation do not appear to be satisfactory. New discoveries in eastern Sicily (Bonfiglio and Burgio 1990), however, indicate the presence of three contemporaneous species of elephants (*E. a. leonardii*, '*E. mnaidriensis*', and '*E. falconeri*')[2].

The lack of a precise knowledge of stratigraphical relationships among the various Maltese deposits, as well as the impossibility of making certain correlations between Sicilian and Maltese faunas, do not allow a clear systematic and stratigraphic framework of the Sicilian–Maltese elephants. These problems are compounded by the scarcity of data concerning the morphological and dimensional range of each elephant population, especially in Malta. The scenario of colonization of the two islands is still uncertain, but it is evident that the hypothesis of a progressive reduction of body size from *E. antiquus* to *E. falconeri* is unacceptable. In addition, the specific designations cannot be accepted as such without first establishing the succession of different elephantine forms and the possible identity of *E. mnaidriensis* from Malta with medium-sized forms from Sicily.

SARDINIA

The only endemic elephant belonging to the mammoth line has been found in Sardinia: *Mammuthus lamarmorae* (Major). To it may be attributed some tarsal and carpal bones from the Wurmian eolian deposits of Fontana Morimentu (Gonnesa), described first (together with some long bones, which have been lost) by Acconci (1881). Using these bones, Major (1883) described his new species, without illustration (Comaschi Caria 1965). Recently, two molars have been found, one in post-Tyrrhenian breccias at Tramariglio (Alghero) (Malatesta 1954), the other from the pre-Würmian eolian formation of S. Giovanni in Sinis (Oristano) (Ambrosetti 1972; Caloi *et al.* 1980). The molar's size seems smaller than in *E. mnaidriensis*. The morphology of these molars excludes designation to the 'palaeoloxodontine' line, but it does not allow us to distinguish as a possible ancestor between an advanced *Mammuthus (Archidiskodon) meridionalis* form and a less advanced *Mammuthus armeniacus* one (Ambrosetti 1972; Palombo 1986). It is likely that the ancestor of the elephant in Sardinia arrived between the early Pleistocene and the early middle Pleistocene, although other possibilities cannot be excluded (Palombo 1986; cf. Caloi *et al.* 1988).

A direct re-examination of the carpal and tarsal bones could shed light upon the problem of the origins and the possible adaptations of this insular form. From the illustrations of Comaschi Caria (1965, plates 1, 2, 3), it is possible to make some inferences. These carpal and tarsal bones are not very long. In the carpus, the magnum is very wide in frontal view, more or less as much as the unciform. In the tarsus, in frontal view, the astragalus is very wide, almost completely impending over the fourth metatarsal. This corresponds to the considerable extension of the articular portion of the calcaneum, in which the tuber is greatly extended transversely thus providing a large attachment surface for *gastrocnemius* and *soleus* muscles. The lower external profile of the calcaneum is very expanded. The ectocuneiform seems to articulate only with the third metatarsal and not with the fourth. Finally, these carpals and tarsals seem to be arranged in a weakly arcuated way. Several of these morphologies seem to occur also in the dwarf elephant from Spinagallo, although with peculiar and often more accentuated features. In conclusion, in the Sardinian elephant the limb posture appears to have been quite different from that of the continental forms, being adapted to both forest and more open environments, thus allowing a better support for the body on the ground, and a better thrust in movements.

CRETE

The occurrence of endemic elephants on this island was first mentioned by Bate (1905), who encountered an extremely small elephantine type at Maleka Cape, and ascribed it (Bate 1907) to *Elephas creticus*. The same author, in 1905, also recorded remains of *E. antiquus* from Kharoumbes, similar to the continental species.

The elephant bones found at Grida Avlaki described as *Elephas priscus* by Simonelli (1908) have an intermediate size between the two species. Vaufrey (1929) did not accept this medium-sized species and referred to it as *E. antiquus*, and accordingly, considers *E. creticus* synonymous with '*E. a. melitensis*' of the Sicilian–Maltese archipelago. Kuss (1965, 1966) described a new species of middle-sized elephant, *Loxodonta creutzburgi* Kuss, on the basis of some remains found at Kalo Chorafi, and ascribed to this species, also, the fossils described by Simonelli (1908). Note, however, that the genus *Loxodonta*, according to Maglio (1973), had never left the African continent. As will be explained in the following two paragraphs, '*L. creutzburgi*' was renamed *E. creutzburgi*, or *P. creutzburgi*.

Kuss (1965, 1966) maintained that three other species also occur in Crete: *E. antiquus*, *Loxodonta cretica*, and *Hesperoloxodon antiquus falconeri*. Sondaar and Boekschoten (1967) do not accept the species *creutzburgi* and reduced the Cretan elephantine species to: *E. creticus* and *E. antiquus*. Kuss (1970), re-examining the systematics of Cretan elephants, accepts three species: *E. creticus* (believed by the author to be derived from '*E. meridionalis*'), *E. antiquus*, and *Loxodonta creutzburgi*; the latter species was then moved, in 1971, to the genus *Elephas* (Kuss 1973). Kotsakis *et al.* (1980), in agreement with the phylogenetic hypothesis proposed

[2] In a paper published by Bonfiglio and Insacco (1992), the presence of a middle-sized elephant morphologically different from *E. mnaidriensis*, associated with *E. falconeri*, in a few localities of south-western Sicily has been confirmed.

by Maglio (1973), believed that three species of *Palaeoloxodon* occur in Crete: *P. antiquus* (similar to the continental species), *P. creutzburgi* (about the size of *E. mnaidriensis* from Sicily and Malta), and *P. creticus* (the same size as *E. falconeri*). The first species is the one found by Bate (1905) at Kharoumbes, the second at Kalo Chorafi (Kuss 1965), also mentioned from the Simonelli Cave and other localities near Rethymnon (Kotsakis 1980), and the third, the smallest, from Capo Maleka (Bate 1907). De Vos (1984), Dermitzakis and de Vos (1987), Palombo and Petronio (1989), and Kotsakis (1990) share substantially the same opinion, although they ascribe the endemic elephantine species to the genus *Elephas* and subgenus *Palaeoloxodon*.

In light of the available evidence, it seems possible to identify two or three elephantine species in Crete: the oldest, testifying to a first migration phase, is the smallest (*E. creticus*), found in association with *Kritimys kiridus* (Bate), and thus attributed to the end of the early Pleistocene or the early middle Pleistocene (Dermitzakis and de Vos 1987; Capasso Barbato 1992). Concerning phylogeny and the geological record, it should be mentioned that the reduced lamellar formula seems to suggest a likely 'palaeoloxodontine' origin of this small elephant. A possible proof of a second migration phase, in the middle Pleistocene (Dermitzakis and de Vos 1987; Capasso Barbato 1992), was given by the discovery of *E. antiquus*, similar to the continental species. Theodorou (1986), however, excludes the presence of an elephant of *antiquus* size on the island. Finally, derived from this middle Pleistocene form, in the late Pleistocene, *E. creutzburgi* occurs in the Simonelli Cave and in many other caves near Rethymnon (Dermitzakis and de Vos 1987). This species is associated with *Mus minotaurus* Bate, allowing its attribution to a younger zone (Capasso Barbato 1992).

CYCLADES ISLANDS

Remains of 'dwarf' elephants are reported from Delos, Naxos, Kythnos, and Milos islands (Cyclades, central Aegean Sea, Greece). The elephant of Naxos (Mitzopoulos 1961), associated with a big murid of the genus *Apodemus* (cf. Sondaar 1971), is similar in size to *E. melitensis* and clearly belongs to an endemic species derived from *E. antiquus*. The age of the remains may be ascribed to the middle Pleistocene (Anastasakis and Dermitzakis 1990). The single molar of an elephant discovered on the Delos island (Cayeux 1908; Vaufrey 1929) may belong to a larger endemic species or to a small *E. antiquus*. Delos and neighbouring islands remained connected to Eubea through most of the Pleistocene (Anastasakis and Dermitzakis 1990). The elephantine fossils from Kythnos, Serifos, and Milos are not known to have been described (Caloi *et al.* 1988; Kotsakis 1990).

DODECANESE ISLANDS

A dwarf elephant as big as *E. mnaidriensis* was found in a cave near Ladiko on the Rhodos island (Fig. 22.2 — one of the group of islands in the Aegean Sea off the south-western coast of Turkey; Marinos and Symeonidis 1973). No teeth or skull fragments have been discovered, but the morphology of the limb bones indicates a relationship with *E. antiquus*. These remains belong to a local species of late Pleistocene age (Symeonidis *et al.* 1974).

In the cave of Charkadio on the Tilos island, numerous remains of a dwarf elephant believed to have derived from *E. antiquus* were found (Symeonidis 1972). Referred to two different forms, they are now considered as belonging to a single species with marked dimorphism. The size of this endemic species is slightly larger than *E. falconeri* and the age of the deposits where it was found spans from the very late Pleistocene to the early Holocene (Theodorou 1983, 1988).

CYPRUS

A small elephant from Cyprus, as large as *E. falconeri*, with a primitive molar morphology, has been described as *E. cypriotes* Bate. Its remains have been collected in a few sites on the island (Bate 1903, 1904; Boekschoten and Sondaar 1972). They show a 'palaeoloxodontine' morphology but their relationships with *E. antiquus* or another form are unclear. The age of this taxon is middle and/or late Pleistocene. Two teeth from Achna (Boekschoten and Sondaar 1972) belong to a larger elephant of 'palaeoloxodontine' type, but their relationships with *E. cypriotes* are unknown.

SUMMARY

Fossil remains of endemic dwarf elephants have been collected in Quaternary deposits of many Mediterranean islands. The phylogenetic relationships of some insular forms are not very clear and a systematic revision is necessary. In this chapter, the present knowledge is summarized and the unresolved problems are indicated.

SICILY AND MALTA

Species of dwarf elephants of the Sicily/Malta area have been identified from material collected on Malta, but, unfortunately, the stratigraphic setting of Maltese caves is unclear. After the classic model of dwarfism, larger elephants are believed to have been older than smaller ones. Thus, three evolutionary stages have been proposed, corresponding to three distinct species: *Elephas* (*Palaeoloxodon*) *antiquus*, *E.* (*P.*) *mnaidriensis*, and *E.* (*P.*) *falconeri*. Recent field and laboratory research, however, indicates that an alternative interpretation to this model may be possible: smaller elephants appear to be older than larger ones. At least two 'invasions' of Sicily/Malta by elephants led to dwarf species and two hypotheses can be postulated:

1. During the middle Pleistocene, *E. antiquus* colonized the Sicilian–Maltese area, originating the endemic subspecies

E. a. leonardii. It is believed that a species of *E. melitensis* size was derived from the latter (e.g. in the lowermost levels at Luparello, also represented by remains at Spinagallo). Similarly, *E. melitensis* gave rise to the smallest elephant. A second arrival of *E. antiquus* during the late middle Pleistocene or in the very early late Pleistocene gave rise to elephants of the size of *E. mnaidriensis*.

2. Two arrivals are again postulated, but *E. a. leonardii* could be the ancestor of elephants of *E. mnaidriensis* size, while the ancestor of the medium- and smaller-sized forms remains to be determined (either *E. antiquus* or *Mammuthus* (*Archidiskodon*) *meridionalis*).

The lack of precise stratigraphical relationships among the various Maltese deposits and the paucity of data on morphology do not allow a clear systematic and stratigraphic framework of the Sicilian–Maltese elephants. In addition, specific designations cannot be accepted as such without first establishing the probable succession of the different elephantine forms and the number of different species.

SARDINIA

Rare remains of small elephants have been collected in Sardinia. They were designated as belonging to a dwarf form, *Mammuthus lamarmorae*. An advanced subspecies of *Mammuthus meridionalis* or, more probably, a primitive *Mammuthus armeniacus* could have been the ancestor of the Sardinian endemic species.

CRETE

As in Sicily, we believe that in Crete, two elephantine 'invasions' took place. An early middle Pleistocene dwarf form, *Elephas creticus*, of palaeoloxodontine affinity, might have been the result of the first colonization. *Elephas antiquus* invaded Crete in the late middle Pleistocene and was the origin of a new endemic form, *Elephas* (*Palaeoloxodon*) *creutzburgi*.

CYCLADES

Remains of fossil elephants of small size were found on the islands of Delos, Serifos, Kythnos, Naxos, and Milos. The published fossils belong to endemic forms derived from *Elephas antiquus*. The age of the remains may be ascribed to the middle Pleistocene.

DODECANESE

In the cave of Charkadio (Tilos Island), numerous remains of a dwarf elephant with marked dimorphism and derived from *Elephas antiquus* have been found. A new specific name is necessary for this endemic form. A small elephant of palaeoloxodontine affinity was reported from Ladiko Cave (Rhodos Island).

CYPRUS

A dwarf elephant of palaeoloxodontine affinity, *Elephas cypriotes*, is present in the middle and late Pleistocene deposits at Cyprus. A few remains of a larger form are also reported from Achna.

ACKNOWLEDGEMENTS

We thank S. Wolak and E. C. Marsac for typing and proofreading, and V. L. Roth and J. Shoshani for editing this manuscript.

TWENTY-THREE

The proboscideans of the South-East Asian islands

GERT D. VAN DEN BERGH, PAUL Y. SONDAAR, JOHN DE VOS, AND FACHROEL AZIZ

INTRODUCTION

Present-day elephants are excellent swimmers (Johnson 1978). The presence of Pleistocene fossil proboscidean remains on several islands surrounded by deep seas suggests that, in the past, proboscideans had a relatively high potential to cross sea barriers. It is not uncommon to find proboscideans in a fossil assemblage of impoverished endemic island faunas, sometimes as the only large mammal. In the Lesser Sunda Arc, Indonesia (Fig. 23.1), only Stegodontidae have been reported as large mammals on the islands Flores, Sumba, and Timor.

The most striking change that large mammals underwent when they colonized an island was a diminution in size. Large size lost its significance as there were no large predators on these islands. On the other hand, small size could be an advantage for the following reasons: (1) less food would be needed, (2) greater mobility, and (3) reduction of the territorial area (Sondaar 1977).

The fossil record of proboscideans of the South-East Asian islands indicates, however, that a reduction in size was not always the case. Remains of a dwarf *Stegodon* were found in the Kedung Brubus Fauna from the middle Pleistocene of Java. This fauna

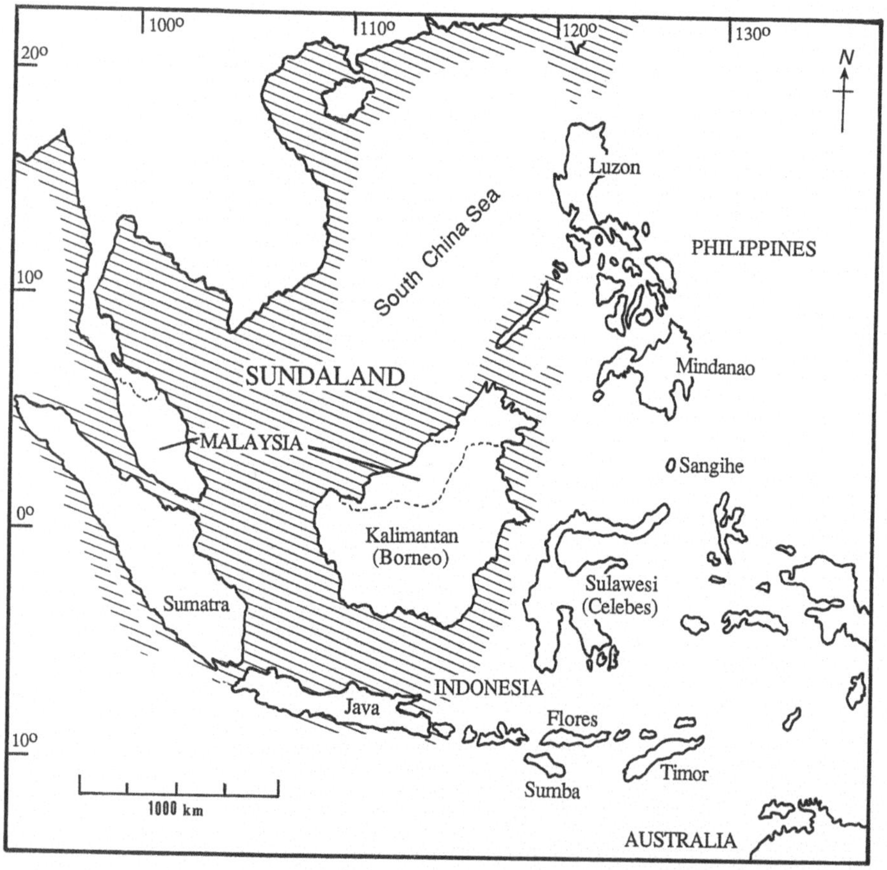

Fig. 23.1 A map of South-East Asia, showing the Greater Sunda islands (Borneo, Java, Sulawesi, and Sumatra), the Lesser Sunda islands (smaller islands east of Java, including Flores, Sumba, Timor, and Sangihe), and the Philippines — islands where the remains of extinct proboscideans have been found. Hatched area indicates the Sunda Continental Shelf.

was clearly of a mainland origin. On the other hand, remains of Pleistocene *Stegodon* of mainland proportions were found on the island Flores in a bed in which stone artefacts were also recovered, indicating the contemporaneous presence of humans (Maringer and Verhoeven 1970; Allen 1991). The dwarf *Stegodon* in a mainland fauna (Kedung Brubus) and a *Stegodon* of mainland proportions in an island fauna seem in contradiction with the evolutionary effect of insularity on large mammals. It appears that the changing palaeogeography of South-East Asia during the Pleistocene had an impact on the proboscidean migration and evolution pattern. The arrival of Pleistocene hominids (large predators) on islands might have been another aspect which needs to be taken into consideration.

Van den Bergh *et al.* (1992) noted that the taxonomy of the proboscideans of Indonesia is controversial either because of incorrect stratigraphic data (de Vos *et al.* 1982; Sondaar 1984; Leinders *et al.* 1985; Allen 1991) or because of different evaluations of morphological characters of fragmentary molar material. The purpose of this chapter is to better understand the evolutionary history of the proboscideans of the Plio-Pleistocene of South-East Asian islands. We will focus on stratigraphy, palaeozoogeography, and evolutionary aspects.

Abbreviations

GRDC, Geological Research and Development Centre (Bandung, Indonesia); NMNH, National Museum of Natural History (Leiden, The Netherlands; formerly 'Rijksmuseum van Natuurlijke Historie').

THE COLONIZING GENERA AND HOW TO RECOGNIZE THEM

Three Plio-Pleistocene proboscidean genera are recognized on the islands of South-East Asia. They are: *Tetralophodon* (from Java and possibly also from Borneo (see footnote, p. 245), *Stegodon* (from Java, Sumba, Flores, Timor, Sulawesi, Sangihe, Luzon, and Mindanao), and *Elephas* (from Java, Luzon, and Sulawesi). It appears that different islands had various endemic species.

Isolated molar fragments are difficult to identify, especially when their morphology is confusing. For example, Osborn (1942, p. 807) states that the bottom of the valleys between adjacent lamellae are closed or V-shaped in what he considered as the superfamily Stegodontoidea, whereas they are open or U-shaped in the Elephantoidea. Later, it appeared that the more primitive elephantids such as *E. planifrons* may have had V-shaped valleys as well, and Hooijer (1964, 1974) and Maglio (1973) consider the presence of Y-shaped valleys typical for Stegodontidae. Yet a single molar may exhibit Y-, V-, or U-shaped valleys, depending on where the tooth is sectioned (details on these features and the morphology of enamel are discussed by Saegusa, this volume, Chapter 17). The picture is further obscured by the large variability in size and number of molar lamellae that can be observed in molars of the same rank within a single species (Roth and Shoshani 1988). Therefore, only complete molars should be taken into account, of which the rank can be ascertained (especially the tapering M3s), eventually in combination with such criteria as the presence in the same locality of large *casu quo* small-sized postcranial material with fully fused epiphyses.

SULAWESI (CELEBES)

INTRODUCTION

In 1949 the first dwarf proboscidean remains from southwestern Sulawesi (near the village of Sompoh, west of the Walanae River in the Sengkang district) were described by Hooijer as *Archidiskodon celebensis*. Other proboscideans found on this island include *S. sompoensis* and *S.* cf. *trigonocephalus* (Hooijer 1949, 1953*b–d*, 1954*a*, 1955, 1972*a*). In 1990 and 1991 some large-sized *Stegodon* molars and postcranial remains were recovered at several places in the Tanrung River, most of them *in situ* in the Tanrung Conglomerate, a Pleistocene deposit of Sulawesi.

STRATIGRAPHY

Sartono (1979*a*) stated that the vertebrate fossils originate from the upper part of the Berru Formation (*sensu* Sartono) and is considered here as a subunit of the Walanae Formation. Both '*E.*' *celebensis* and *S. sompoensis* occur in the upper 150 m of the Berru Member. The Tanrung Conglomerate, which horizontally overlies slightly folded marine deposits in the Walanae Formation, yielded several large-sized *Stegodon* remains but no dwarf proboscideans. This large *Stegodon* is probably younger in age than the fossils from the Berru Member, the latter of which is also folded and has a late Pliocene age (Table 23.1).

Table 23.1 Stratigraphic scheme showing the occurrence of the proboscidean taxa from southwestern Sulawesi

Age	Lithological unit	Taxa
Late Pleistocene	Colluvium and alluvial terraces	No proboscideans
Middle–Late? Pleistocene	Tanrung Conglomerate	Large-sized proboscidean (*Stegodon*)
Late Pliocene (c. 3 Ma)	Upper part of Berru Member of Walanae Formation	'*Elephas*' *celebensis*, *Stegodon sompoensis*

TAXONOMY

'Elephas' celebensis *Hooijer, 1949*

Aziz (1990) described proboscidean material collected in Sulawesi in 1986 and 1987, including a fairly complete adult skull of *'E.' celebensis* found *in situ* in a fluviatile sandstone of the upper part of the Berru Member exposed north of the Lakibong River. Based on the morphology of this skull he proposed to place this species in a separate genus, which is given no generic name so far. The additional diagnosis of *'E.' celebensis* (van den Bergh *et al.* 1992) reads as follows (modified after Hooijer 1954*a*, *b*): skull very short and high, showing antero-posterior convexity of fronto-parietal region; well-developed temporal crests, marking a considerable contraction of the temporal fossae; temporal fossae shifted forward, approximately above the anterior part of the occlusal surface; absence of dorsal parietal swellings; plane of the tusk sockets parallel to but far below the fronto-parietal plane; upward and lateral extension of tusks and alveolar sockets, separated by a deep fossa incisiva; presence of antero-orbital foramen dorsally of the infra-orbital fossae; mandibles possessing a high horizontal ramus with an elongated symphysis, sometimes provided with functional mandibular tusks (males?); general size of the molars approaches half the size of *E. planiforns* (Falconer and Cautley 1846); low number of molar lamellae (10 to 11 in the M3) and moderate hypsodonty (H/W index = 104–145 in the M3); lamellae possess well-marked median pillars that produce a pronounced sinus with wear. Close to the base of the lamellae the wear figures are sub-lozenge to sub-triangular shaped; lamellae separated by broad valleys, U- or V-shaped at their base; relatively thick enamel (2.0–3.5 mm in the M3), very weakly folded, rather undulating, consisting of a thicker inner layer and a thinner outer layer; cement relatively abundant; and, finally, presence of premolars (P3 and P4).

'E.' celebensis shows plesiomorphic as well as apomorphic characters. The plesiomorphic characters are: (1) the presence of premolars, P3 and P4; (2) the occasional occurrence of mandibular tusks; (3) the persistence in the adult stage of antero-orbital foramina placed dorsally of the infra-orbital foramina; (4) the typical well-pronounced median pillars of the molar lamellae; and (5) the only slightly folded rather undulating enamel. Apomorphic characters are a short and high skull, relatively high-crowned molars, and a relatively high number of molar lamellae.

The presence of ante-orbital foramina and mandibular tusks are not primitively shared with *E. planifrons* and suggest a more primitive ancestor for *'E.' celebensis*, possibly one of the earlier members of the Elephantidae, such as *Primelephas* or *Stegotetrabelodon*. *Stegotetrabelodon* is known from Africa and from the Pliocene of Europe and Asia (Maglio 1973; Tobien 1978*b*; Tobien *et al.* 1988). The latter authors, however, place *Stegotetrabelodon* amongst the bunodont tetrabelodont longirostrine gomphotheres, although they agree that this genus is near the source of the elephants, an opinion which is shared by Beden (1979). Alternatively, the primitive aspects of *'E.' celebensis* might have resulted from paedomorphosis, an explanation advocated by Maglio (1973), who considers *'E.' celebensis* as a direct descendant of *E. planifrons*. The advanced characters of *E. celebensis*, such as the short and high skull and the relatively high hypsodonty, could be explained as a parallel evolutionary trend with the Elephantidae. The absence of parietal swelling in *'E.' celebensis* could be explained as due to the reduction of the pneumatic bone tissue in the parietal–occipital region.

Stegodon sompoensis *Hooijer, 1964*

The rank of the fragmentary and worn type specimen figured by Hooijer (1964, plate 3, fig. 1) is hard to assess. It might be a dp3, dp4, or M1 (van den Bergh *et al.* 1992) and is therefore not very convincing for the indication of a dwarf species. Later, Hooijer (1972*a*) refers an entire isolated lower molar from Calio to *S. sompoensis*, which can definitely be determined as an M3, as indicated by tapering in the posterior direction. It carries 10 lamellae and a small posterior cingular lamella. The size of this M3 (L = 140 mm, W = 45 mm) is half as large as the homologue in *S. trigonocephalus* from Java and therefore belong to a dwarf species. Another specimen that can be ascribed to *S. sompoensis* is a mandible with M3, originating from a sandstone bed near Bulucepo and which is figured by Aziz (1990, plate V, figs. 1–2) as *Stegodon* sp. A cf. *trigonocephalus* (see also van den Bergh *et al.* 1994).

An almost complete cranium (Fig. 23.2a) referable to *S. sompoensis* (GRDC no. 3050) was found east of the village of Paroto in 1989. The *in situ* cranium originated from a pebbly sandstone bed of the upper part of the Berru Member. A full description of this cranium will be given in another paper. Briefly, its characters are: both M3s broken at the roots, and the cranium is of similar size to that of *'E.' celebensis*, but differs in morphology (Fig. 23.2b). Posterior to the broken molars there are no succeeding molars and thus we concluded that the molars observed in the maxillae are M3s. The original amount of lamellae can be deduced from the alveolar outline. There were probably eight lamellae and a posterior talon in both M3s.

The cranium from Paroto is relatively low and elongated and is further characterized by: (1) a relatively long and convex fronto-parietal region; (2) a marked contraction between the temporal fossae; and (3) the position of the tusk sockets in the plane of the frontals. The cranium does not show the triangular outline that is typical for crania of *S. trigonocephalus* Martin, 1887, due to the more pronounced contraction between the temporal fossae. Both the lower and the upper M3s of *S. sompoensis* have a lower amount of lamellae than *S. trigonocephalus*, the latter possessing 11–13 lamellae in the lower and 10–12 lamellae in the upper M3s. The plate formula of the M3s of *S. sompoensis* corresponds better with that of *S. bombifrons* Falconer and Cautley, 1846 as given by Osborn (1942): nine lamellae for the upper and lower M3. In morphology the Paroto cranium resembles more the type and referred crania of *S. bombifrons* figured in Osborn (1942, figs 742, 744) than the adult crania from Java referred to as *S. trigonocephalus* by Hooijer (1955), though it lacks the extreme contraction between the temporal fossae that seems characteristic for *S. bombifrons*. Considering the relatively old age of the upper part of the Walanae Formation, and the lower amount of lamellae in the M3s as compared to *S. trigonocephalus*, *S. sompoensis* cannot

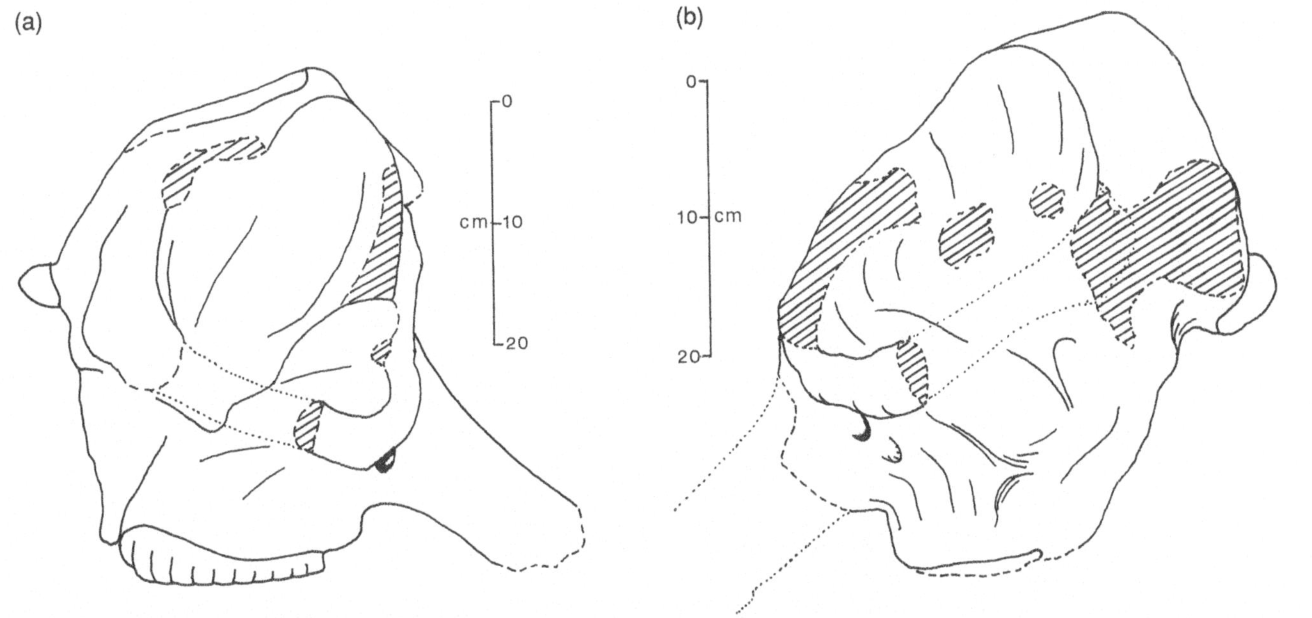

Fig. 23.2 (a) Cranium of *Stegodon sompoensis* Hooijer, 1964 (GRDC no. 3050) found east of the village of Paroto, Sulawesi in 1989. (b) The cranium of adult *Stegodon sompoensis* is similar in size to that of '*E.*' *celebensis*, but differs in morphology.

have been derived from *S. trigonocephalus*, which has its oldest occurrence on Java at around 1.2 Ma (Leinders *et al.* 1985). Theoretically it remains possible that in the process of dwarfing, the number of lamellae in the molars of *S. sompoensis* was reduced.

JAVA

INTRODUCTION

The fossil mammals from Java are unique; they differ in many respects from their contemporaries from continental Asia. *Equus* never reached Java while, on the other hand, many endemics evolved. It is possible that Java, situated at the edge of a faunal region, has been separated several times by sea from the Asian mainland. Java could possibly have functioned as the filter and sweepstake routes for faunal interchange (Simpson 1965).

Two essentially different faunal associations can be recognized on Java. One is unbalanced, lacking large predatory mammals and very poor in species. This type of fauna is well known from the Pleistocene Mediterranean islands. These island faunas are generally characterized by endemic species of hippopotamids, cervids, and elephantids. This is not surprising as members of these families are all known to be excellent swimmers. A fossil fauna consisting of elephants, deer, and hippos clearly points to island circumstances (Sondaar 1977). Such a fauna (Fig. 23.3) is found in the lower part of the Bumiaju section in central Java (van der Maarel 1932; Sondaar 1984) and the lower part of the Sangiran section (Leinders *et al.* 1985). A second type of fauna is clearly continental, though, in the case of Java, somewhat impoverished compared to mainland Asian faunas and containing some endemic species, probably caused by the eccentric zoogeographical position of Java. Most faunas of this type from Java indicate an open woodland environment, dryer than the present-day fauna.

Based on data of the Indonesian–Japanese Joint Research and from older but well-documented collections, the following faunal succession can be established, from old to young.

THE SATIR FAUNA

Age

Combining the faunal similarity between the Satir Fauna from the type locality and the lower part of the Sangiran Formation with the absolute dating results by means of fissiontrack dating in the Sangiran area (Suzuki *et al.* 1985), Leinders *et al.* (1985) were able to assign an age of 1.5 Ma to this fauna.

Interpretation

The *Tetralophodon* and *Hexaprotodon* are both archaic and point to a relatively old age. The unbalanced character of the fauna and the presence of a giant tortoise point to island conditions. First and last occurrence of *Tetralophodon*.

THE CI SAAT FAUNA

Age

Comprising the fauna of Ci Saat Fauna from the type locality and the upper part of the Sangiran Formation to the absolute dating results by means of fissiontrack dating in the Sangiran area (Suzuki *et al.* 1985). Leinders *et al.* (1985) were able to assign an age of 1.2 Ma to this fauna.

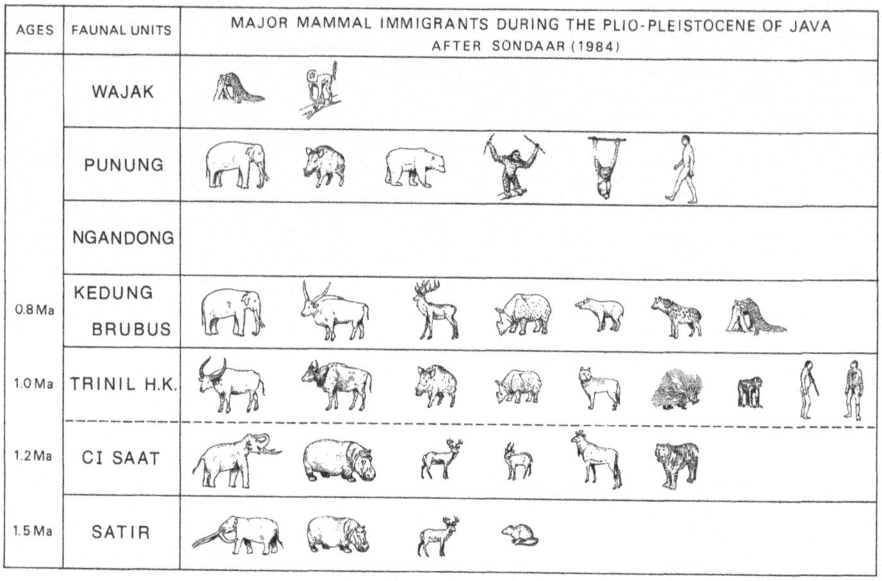

Fig. 23.3 Major mammal immigrants during the Plio-Pleistocene of Java (after Sondaar 1984, table 2). Taxa for each faunal unit (from left to right) are: Satir, *Tetralophodon bumiajuensis* (see footnote 1), *Hexaprotodon simplex*, Cervidae, Muridae; Ci Saat, *Stegodon*, *Hexaprotodon sivalensis*, *Axis*, *Muntiacus*, Bovidae, *Panthera*; Trinil H. K., *S. trigonocephalus* (not shown), *Bos* (= *Bibos*), *Bubalus*, *Sus brachygnathus*, *Rhinoceros sondaicus*, *Mececyon* (one of three carnivore genera), *Hystrix* (= *Acanthion*), *Macaca*, *Meganthropus*, *Homo*; Kedung Brubus, *Elephas hysudrindicus* (one of three proboscidean genera), *Epileptobos*, *Cervus* (= *Rusa*), *Rhinoceros kendengindicus*, *Tapirus*, *Hyaena*, *Manis palaeojavanicus*; Ngandong, fossil material needs to be re-evaluated, possible last occurrence of *S. trigonocephalus* and *E. hysudrindicus*; Punung, *Elephas maximus*, *Sus vittatus*, *Ursus*, *Pongo*, *Hylobates*, *Homo sapiens*; Wajak, *Manis javanica*, *Presbytis*.

Interpretation

The Ci Saat Fauna is poor in species, which might point to isolated conditions, though the presence of a large felid would indicate a land connection with the Asian mainland. First occurrence of *S. trigonocephalus*.

THE TRINIL H. K. FAUNA

Age

1.0 Ma, obtained by combining the faunal similarity between the type locality at Trinil and the Grenzbank at Sangiran with the absolute fissiontrack datings of Suzuki *et al.* (1985).

Interpretation

The Trinil H. K. Fauna is relatively poor in species though much more balanced than the preceding Ci Saat Fauna, and includes three carnivore genera. An endemic genus is represented by *Duboisia*. There is only one species of proboscidean: *S. trigonocephalus*.

KEDUNG BRUBUS FAUNA

Age

By combining the similarity between the type fauna at Kedung Brubus and the fauna from the upper part of the Bapang Formation at Sangiran with the absolute fissiontrack datings of Suzuki *et al.* (1985), an age of 0.8 Ma was inferred by Leinders *et al.* (1985).

Interpretation.

The Kedung Brubus Fauna bears a more mainland character then the Trinil H. K. Fauna and a maximal faunal interchange with the South-East Asian mainland seemed to have occurred. New arrivals of *Hyaena*, *Epileptobos*, *Tapirus*, *Elephas*, and *Rhinoceros kendengindicus* support this hypothesis. First occurrence of *Elephas hysudrindicus*. The pygmy *Stegodon hypsilophus* does not occur in the Kedung Brubus Fauna proper, but is likely to pertain to at least part of this faunal zone because of its presence in the area north of Djetis and Perning. It may have evolved on a separate palaeoisland that subsequently became connected with the emerging Sunda Shelf. There are three species of proboscidean present: *S. trigonocephalus*, *S. hypsilophus*, and *E. hysudrindicus*.

THE NGANDONG FAUNA

Age

Younger than the Kedung Brubus Fauna.

Interpretation

The fossil material needs to be restudied to better understand this fauna. It seems to resemble the Kedung Brubus Fauna in some aspects, pointing to an open woodland environment. Last occurrence of *S. trigonocephalus* and *E. hysudrindicus*.

THE PUNUNG FAUNA

Age

Late Pleistocene, presumably the last interglacial.

Interpretation

The fauna has a recent character and possibly all faunal elements are new immigrants, such as *Pongo*, *Elephas maximus*, *Capricornis*, *Sus vittatus*, *Sus barbatus*, and *Ursus*, whereas more archaic forms, such as *Stegodon*, *E. hysudrindicus*, and *Sus macrognathus*, have disappeared. The large quantity of *Pongo* fossils and the presence of other primates indicates a humid tropical rainforest environment (de Vos 1983). First occurrence of *E. maximus*.

THE WAJAK FAUNA

Age

Holocene.

Interpretation

Although *Elephas maximus* is not present in Wajak, it is still present in the younger locality of Sampung. The Wajak Fauna is impoverished compared to Punung and several rainforest elements have disappeared.

TAXONOMY

Tetralophodon bumiajuensis[1] *van der Maarel, 1932*

Synonymies were given by Sondaar (1984).

Holotype

A cranium fragment with both worn M3s figured by van der Maarel (1932, plate VII, figs 1–2; plate IX, figs 1–3; plate X, figs 1–3).

Hypodigm

The present location of the type is not known. A mandible with M3, a fragmentary maxilla with both M3s and two molar fragments are kept at the GRDC, Bandung.

Age and location

The specimens described by van der Maarel (1932) are from excavations 1–4 and 8 in the Bumiaju area on Central Java. The fauna of the type locality has been named the Satir Fauna (Sondaar 1984), which can be correlated with the oldest mammal-bearing beds at Sangiran (from the so-called 'black clays'), having an age of 1.7–1.5 Ma (Leinders *et al.* 1985; Suzuki *et al.* 1985).

Diagnosis

Molars smaller than those of *T. punjabiensis* (Osborn 1936, p. 365).

Stegodon trigonocephalus *Martin, 1887*

For list of synonyms and description of this large-sized, advanced *Stegodon*, see Hooijer (1955). The first occurrence of *S. trigonocephalus* is in the Ci Saat Fauna with an age of approximately 1.2 Ma. It still occurs in the Kedung Brubus Fauna with an age of 0.8 Ma. The last occurrence is in the Ngandong Fauna, which is insufficiently studied, but younger than the Kedung Brubus Fauna (Sondaar 1984; Leinders *et al.* 1985; de Vos, in press).

Stegodon hypsilophus *Hooijer, 1954b*

Synonymies were given by Sondaar (1984).

Holotype

A right and left M3 of the same individual kept at NMNH.

Hypodigm

Besides the type there are four fragmentary molars and a distal humerus stored at NMNH.

Age and location

Type specimens were found North of Djetis and Perning. Other specimens referred to this species are from Karang Djati and Tritik. A distal portion of a humerus of an adult is from Kedung Brubus. These localities can probably be correlated with the Kedung Brubus faunal zone, which has an age of 0.8 Ma.

Diagnosis

A proboscidean with *Stegodon* affinity with relatively hypsodont molars (H/W index = 73–92 in the M3) and small size, about two-thirds the size of *S. trigonocephalus*; relatively thick (4.4–6.0 mm in the M3) double-layered enamel, of which the inner layer is less wrinkled than in *S. trigonocephalus*; more or less weak median expansions of the lamellae at the intermediate height, especially at the anterior lamellae where they can be developed as median pillars.

Elephas hysudrindicus *Dubois, 1908*

This is the first true *Elephas* that reached Java. Its first occurrence was in the Kedung Brubus Fauna; its last occurrence was in the Ngandong Fauna. In the younger Punung Fauna *E. hysudrindicus* was replaced by *E. maximus*.

[1] Editors' note: there is no conclusive evidence that safely allocates the species '*bumiajuensis*' to the genus *Tetralophodon*, a statement corroborated by H. Saegusa, personal communication 1993.

Supposed presence of Elephas planifrons, 'E.' celebensis, *or* S. ? hypsilophus *on Java*

The paucity of material from Ci Pangglosoran cannot be ascribed with certainty to *E. planifrons*, *'E.' celebensis*, or *S. ? hypsilophus*, and does not justify the creation of a new genus and species. It might represent a species that descended from the same mainland stock from which *'E.' celebensis* evolved, but which developed separately on Java.

FLORES

INTRODUCTION

Verhoeven discovered at Ola Bula, central Ngadu on the island of Flores, fossils of stegodons, which were described by Hooijer (1957, 1972*b*). Near Menge Ruda in the same area, Verhoeven discovered a *Stegodon*-bearing level with a lithic industry. Maringer and Verhoeven (1970) suggested that humans and *Stegodon* lived contemporaneously on the island. According to Hooijer (1972*b*) the age of the *Stegodon* is middle Pleistocene.

STRATIGRAPHY

Stegodon trigonocephalus florensis is described from different localities, viz. Ola Bula, Boa Leza, Mata Menge, and Menge Ruda. At Menge Ruda, the site with artefacts, the giant rat *Hooijeromys nusatenggara* was also found. The assemblage of the locality Bhisu Sau is clearly different. Only a dwarf *Stegodon* was found, together with the giant tortoise. The material of the dwarf *Stegodon* is too scanty to allow a specific designation, but the molars and the atlas suggest a size for this species which is at least one-half the size of *S. trigonocephalus* and belonged most probably to the same species as the molars mentioned by Hartono (1961). At the site Bhisu Sau, not a single molar fragment was found of the large *Stegodon*. The giant tortoise and the dwarf *Stegodon* belong to the same faunal assemblage and are older in age than the large *S. trigonocephalus*, as the site Bhisu Sau is 50 m lower in this section than Ola Bula. In Ola Bula there are no stone artefacts found *in situ*, but the same *Stegodon* might point to a similar age with the site Menge Ruda, where a lithic industry was found.

In conclusion, it is possible to distinguish two different endemic island faunas: the older with a dwarf *Stegodon*, plausibly a new species, and a *Geochelone*, a giant tortoise; the younger with a subspecies of *S. trigonocephalus*, a giant rat *Hooijeromys* and a lithic industry, from which it can be concluded that humans were also a component of this fauna. Near Ola Kile (Fig. 23.4) we find the two fauna assemblages in the same section in the Ola Bula Formation.

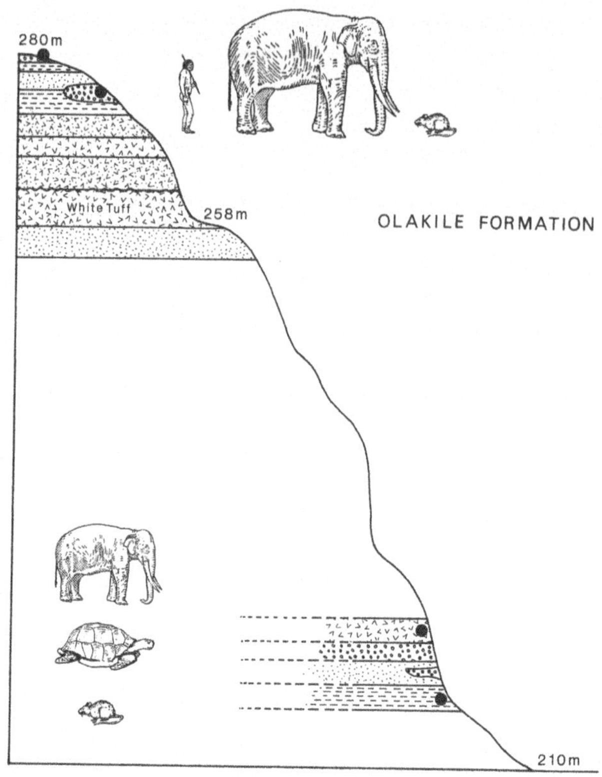

Fig. 23.4 Schematic representation of the faunistic elements at Ola Kile Formation in Flores island (black dot • indicates a locality with vertebrate fossils). Animals at the bottom are of an older age: giant rat, *Hooijeromys nusatenggara*; giant tortoise, *Geochelone* sp.; and dwarf Flores stegodon, *Stegodon* sp. Those at the top are of a younger age: *H. nusatenggara*, *S. trigonocephalus* (of mainland proportions), and *Homo sapiens*.

TAXONOMY

Stegodon trigonocephalus florensis *Hooijer, 1957*

Holotype

A mandible with both M3s.

Hypodigm

The type, two isolated upper and two lower M3s, a mandibular ramus with M1 and dp4, an M1, and a number of fragmental molars.

Age and localities

The type is from Ola Bula; other molar remains are from Boa Leza, Mata Menge, and Menge Ruda. They are from the sandstone member of the Ola Bula Formation, which presumably is of middle Pleistocene age (Hooijer 1972*b*).

Diagnosis

The subspecies is slightly smaller than *S. t. trigonocephalus* from Java and relatively higher crowned than the latter (H/W index varies between 59 and 87 for the upper and between 68 and 113 for the lower M3). Number of lamellae are the same as in *S. t. trigonocephalus* or one lamellae more.

Discussion

S. t. florensis seems a valid taxon.

Stegodon species C, *dwarf form*

Hypodigm

The only material so far described are the two lower dp3s described by Hooijer (1964, fig. 1). Their present depository is not known. An atlas of an adult is kept in the collection of the Institute of Earth Sciences, Utrecht.

Age and localities

The dp3s are from Ola Bula and were partly covered by a tuff crust.

PROBOSCIDEANS FROM SOUTH-EAST ASIAN ISLANDS WITHOUT STRATIGRAPHIC CONTROL

Kalimantan (formerly Borneo)

Lydekker (1885) described a tooth of a pygmy proboscidean ('*Mastodon latidens*') from Kalimantan.

Sumba

Sartono (1979*b*) reported on a discovery of a pygmy *Stegodon* from Sumba.

Timor

INTRODUCTION

The posterior part of an M3 of a dwarf *Stegodon* was first described and figured by Sartono (1969) who named it *S. timorensis*. Hooijer (1969) described a large-sized *Stegodon* from Timor (represented by a slightly worn dp3 from Sadilaun; Hooijer, 1969) as *S. t. florensis*.

TAXONOMY

Stegodon timorensis *Sartono, 1969*

Holotype

The molar portion described by Sartono (1969, fig. 1).

Hypodigm

The type, besides the material described by Hooijer in 1969 (amongst it is a mandible with a complete left M3 and an isolated right M3, and a mandibular ramus with an M2; casts kept at the NMNH) and 1972*b* (fragmentary molars, tusk fragments, and some postcranial elements), and a maxilla and mandibular ramus.

Age and localities

The type is from Weaiwe; other remains are from several localities in the same region, about 3–10 km north-east of Atambua. The remains were collected from gravels of unknown age.

Diagnosis

A dwarf *Stegodon* with molars about half the size of *S. trigonocephalus* from Java; the molars with one lamella fewer than the lowest number of lamellae in *S. trigonocephalus*; lamellae relatively hypsodont (H/W index 77–98 in M3 versus 51–68 in *S. trigonocephalus*).

Discussion

S. timorensis seems to be a valid taxon.

Stegodon trigonocephalus *subspecies D*

The material from Timor referred to this species by Hooijer (1969, 1972*b*) are dp3 and M1, besides some molar fragments.

Sangihe

The small island of Sangihe lies between the northern tip of Sulawesi and the island Mindanao of the Philippines. In 1989 Aziz and Shibazaki collected some stegodon material from this island, including three molar fragments, a tusk, a dextral astragalus, and a phalanx, all kept in the GRDC, Bandung. These specimens originated from the Pintareng Formation in the south-east of Sangihe Island and are thought to be of Pleistocene age (Samodra 1989). The material was described briefly and figured by Aziz (1990, pp. 63–5; plate VI, figs 1–3) and attributed to *S.* sp. B cf. *trigonocephalus*.

Philippines

LUZON

Von Koenigswald (1956) described a large form, *S.* cf. *trigonocephalus* and a new species, *S. luzonensis*, which is a dwarf form. Furthermore, a left molar of an *Elephas*, described by von Koenigswald as a new species, *E. beyeri*, was also considered a dwarf form. There are no stratigraphical data.

MINDANAO

The taxonomic position of *S. mindanensis* Naumann, 1890 is not completely clear.

TRENDS IN EVOLUTION AND PALAEOECOLOGY OF THE PROBOSCIDEANS OF PLIO-PLEISTOCENE SOUTH-EAST ASIA

Trends in evolution and palaeoecology of the proboscideans of Plio-Pleistocene South-East Asia must be viewed in the light of the complex palaeozoogeographical history of the South-East Asian Archipelago (Indonesia and Philippines). Some of the islands may always have been isolated, others sometimes connected with the South-East Asian mainland. Proboscideans, which are excellent swimmers, were usually the first colonizers of islands and evolved locally on the different archipelagos (see also Caloi *et al.* and Roth, this volume, Chapters 22 and 24, respectively). It is known that present-day elephants visit islands in the Indian Ocean off the coasts of India and Sri Lanka in search of food.

Palaeogeography and palaeoecology of proboscideans in the South-East Asian islands can be separated into three categories:

(1) islands which were isolated from the Asiatic mainland, and are of Pliocene age, e.g. Flores, Java, Philippines, Sulawesi, Sumba, and Timor;

(2) islands which were isolated, but occupied by humans, and are of middle or late Pleistocene age, e.g. Flores and probably also Timor, Philippines, and Sulawesi; and

(3) islands which were separated and reconnected several times with the mainland of Asia, and are of latest Pleistocene to Holocene age, e.g. Java.

CATEGORY 1

If a proboscidean colonized an island, a general evolutionary trend was dwarfing. Concomitantly, we observe some allometrical changes in skull morphology and an increase in hypsodonty. The colonizer of Pliocene Java was *Tetralophodon* whose material is insufficiently studied and its ancestor is not known. *Tetralophodon bumiajuensis* (see footnote, p. 245) was not extremely large and it became extinct when Java was connected to South-East Asia during the Pleistocene. The arrival of large predators such as tigers and humans, as well as species which were ecological concurrents, may have contributed to the extinction of the island fauna of Pliocene Java, including dwarf proboscideans.

CATEGORY 2

The impact of the arrival of humans on Flores can be studied because sites on this island are known in one stratigraphic section: the older without the presence of humans, and the younger with artefacts and an insular fauna. Islands with palaeolithic settlements were Flores, Timor, Sulawesi, and the Philippines. On these islands giant rats were and are present. Even now, they are eaten by the people on these islands, and they were probably also an important source of food for palaeolithic humans. The Pleistocene faunas of Flores consisted of only a few taxa, indicating island conditions at that time. These fauna were characterized by the proboscidean *Stegodon*, giant rats, and giant tortoises (cf. Alcover and McMinn (1994) for the presence of 'giant varanids which could have preyed on offspring of dwarf stegodonts, know from fossils on Timor and Flores' — eds). From the Ola Kile section we learned that the stratigraphically older species was a dwarf form, whereas the younger was of mainland proportions, which contradicts the evolutionary pattern of elephants on islands in which we can observe a dwarfing through time (Sondaar 1977). The *Stegodon* with mainland proportions is the most common fossil of Pleistocene Flores. From this we may conclude that this large *Stegodon* was not an incidental visitor, but a mammal which characterized the fauna.

At the locality of Menge Ruda, Verhoeven succeeded in collecting some fossils of a giant rat, which were described by Musser (1981) as *Hooijeromys nusatenggara*. The scanty finds of giant rat remains from the Pleistocene may be due more to taphonomical reasons than that they were not present. In the stratigraphically older site, Bhisu Sau, with the dwarf *Stegodon*, no giant rats were found, but they should have been present. The giant tortoise *Geochelone* was found only at the site Bhisu Sau, and it is believed to have been an easy prey for humans. The same can be said of a dwarf elephant; this mammal had a low reproduction rate and may have been an easy prey for the palaeolithic hunter (Sondaar 1989). The first requirement for palaeolithic settlement on an island is a continuously sufficient stock of food for a hunter– gatherer population. If we accept the presence of giant rats in the Bhisu Sau Fauna, then animals with a high reproduction rate and a constant food supply would supply the basic requirements for permanent human colonization during the Pleistocene on Flores. Arrival of humans on this island would have accelerated the extinction of the dwarf *Stegodon* and the giant tortoise *Geochelone*. Recent discoveries on the island of Cyprus (Simmons 1988) show associations of cultural material with dwarf hippopotamus, approximately 10 000 years BP, and humans may have caused the extinction of these island hippopotami.

Stegodon trigonocephalus is known from *Homo erectus* faunas from Java and this suggests that there was an equilibrium between the hunting activities of *Homo erectus* and the establishment of a successful population of this proboscidean. The *Stegodon*, found in the same stratigraphical level as humans, differs from the population of the same species of Pleistocene Java (Hooijer 1967, 1972*b*), indicating that they were not accidental visitors of Flores, but settled on the islands.

CATEGORY 3

Java is a perfect example of an island which was sometimes connected with the mainland of Asia. As observed earlier in this chapter, after the first connection of Java with the mainland, *Tetralophodon* became extinct and was replaced by *Stegodon*. In the Kedung Brubus Fauna we find the largest diversity in mammallian species, including these proboscideans: *S. trigonocephalus*, *S. hypsilophus*, and *E. hysudrindicus*. The latter species is believed to have been a new arrival, whereas *S. hypsilophus* may have evolved on a separate palaeoisland that became connected with the emerging Sunda Shelf.

ACKNOWLEDGEMENTS

The financial support of the Netherlands Foundation for the Advancement of Tropical Research is acknowledged. We also thank R. Sukamto of the Geological Research and Development Centre and the Lembaga Ilmu Pengetahuan Indonesia for their support. Constructive comments were made by V. L. Roth, H. Saegusa, J. Shoshani, and P. Tassy.

TWENTY-FOUR

Pleistocene dwarf elephants of the California Islands

V. LOUISE ROTH

INTRODUCTION

Mammoths of the Santa Barbara, California, Channel Islands are the only proboscideans documented from islands in the New World. Like proboscidean fossils from islands elsewhere, they tend to be dwarfed by comparison with their mainland ancestors, typically attaining as adults 120–180 cm in shoulder height, with a few extreme individuals broadening the range to 105–243 cm (Roth 1982), when their closest relatives on the mainland could reach 400 cm (Orr 1968). (See also Oldale *et al.* 1987, and Whitmore *et al.* 1967, on relatively small mammoths from the North Atlantic Continental Shelf.)

By far the most extensive collections of mammoth fossils from these islands have been obtained from Santa Rosa Island (Orr 1968; Roth 1982), although adjacent San Miguel Island has also been a notable source of material, and small numbers of specimens are known from other islands (Fig. 24.1). A single tooth has been found on San Nicolas Island (Vedder and Norris 1963), and a total of seven specimens — six of them isolated teeth, three of which were substantial enough to be identified and described — have been reported from Quaternary sediments on the western coast of Santa Cruz Island (Cushing *et al.* 1984). More recently, a large tusk was discovered on Santa Cruz Island and has been placed on display at the Santa Barbara Museum of Natural History (SBMNH). None of the specimens from San Nicolas or Santa Cruz is obviously dwarfed. Cushing *et al.* (1984) argued against the possibility that the fossils on Santa Cruz were brought from the mainland and left on the island by humans, and suggested that mammoth fossils are scarce on Santa Cruz because its Quaternary deposits are greatly restricted in area.

The mammoth fossils from the California Islands have sparked the interest of specialists from a wide range of disciplines because they bear upon questions in anthropology, geography, and evolutionary biology.

DISCOVERY OF THE CHANNEL ISLAND MAMMOTHS

During a geodetic survey of Santa Rosa Island, California in the early 1850s, W. G. Blunt found a fossil tooth of an elephant which he subsequently presented to the California Academy of Sciences. A brief note in the *Proceedings of the California Academy of Sciences* in 1873 marks the discovery of mammoth remains on the island (Stearns 1873).

Orr (1968) reviewed the history of work on Santa Rosa Island. Although the late nineteenth-century literature records sporadic references to the finding of mammoth material on the island (including suggestions that it was a hoax), the mammoths of Santa Rosa Island apparently received little attention until D. Irma Cooke discovered a tusk during an SBMNH botanical expedition to the island in the late 1920s. According to Orr (1968), SBMNH archaeologist David Banks Rogers called these fossils to the attention of Chester Stock of the California Institute of Technology (CIT) in 1927. Stock and Furlong (1928) credit Spencer Atkinson and J. A. Barbieri of Pasadena with securing the evidence presented to CIT.

Stock and Furlong undertook excavation on the north-western coast of Santa Rosa during 1927 and 1928. The products of these excavations formed a substantial collection at CIT, a collection subsequently acquired by the Los Angeles County Museum of Natural History (LACM). Unfortunately Stock's field notes covering these expeditions were destroyed after his death, but a file of photographs survives in the Paleontology Department archives at LACM, and supplies some documentation of the expeditions. In 1928 Stock and Furlong described a new species, *Elephas exilis*, from newly unearthed material.

Beginning in 1948 and extending over a period of 20 years, Phil Orr of the Archaeology and Geology departments of SBMNH established camps on the north-western end of Santa Rosa and continued excavating fossil mammoths. Orr's collection at SBMNH approximately tripled the volume of material gathered. Orr (1968) estimated that his collection represented over 200 individuals, but he considered differences in locality in his estimation (Orr, personal communication 1978), and in LACM and SBMNH combined, there are fewer than 50 of even the most numerous single element (teeth sorted by jaw and side).

Since Orr's (1968) semi-popular account surveyed the pre-history of Santa Rosa Island, dwarfed mammoths have been discussed in the primary literature of several scientific specialties. This work is reviewed briefly below.

STRATIGRAPHY AND DATING

On Santa Rosa, mammoth fossils occur primarily in the north-western portion of the island. The only other terrestrial mammalian fossils reported in association with these bones are of the extinct *Peromyscus nesodytes* (Wilson 1936), a deer mouse larger than any modern *Peromyscus* from the United States (Wilson 1936; White 1966; Walker 1980), though the island fox (*Urocyon littoralis*) may also have been contemporaneous (Orr 1968; Collins 1991). The sea cliffs from which most of the collections of fossil mammoths have been obtained consist of Miocene shales capped by unconsolidated Pleistocene alluvium consisting of clays, sands, and gravels, which in places attains a thickness greater than 25 m (Stock and Furlong 1928; Orr 1960*a*). Santa Rosa has a badland topography, deeply eroded into small canyons along stream courses, and the fossils are well exposed, weathering out of the cliffs. Both above these cliffs and below them (submerged in the channel) extend a series of wave-cut platforms which presumably reflect eustatic changes in sea-level. It is mainly the lowest of the subaerial terraces, the sea cliffs, which have been the source of mammoth fossils. Orr (1960*a*) termed the cover on this terrace the Santa Rosa Island Formation, and recognized a division into three members: Garañon, Fox, and Tecolote, in order from oldest to most recent. Uranium–radium dating of shells placed a 'tentative' (Orr 1968) age of 200 000 years or more on the marine facies underlying the Garañon member. Whereas Madden (1981*a*) suggested an earlier middle Pleistocene or earlier Kansan origin for *Mammuthus exilis*, Orr believed:

(1) that the mammoths reached the island subsequent to this, during the Illinoian glaciation;

(2) that the Fox and Garañon members represent the Sangamon Interglacial; and

(3) that the Tecolote member was deposited during the Wisconsinan (Orr 1967).

Portions of this stratigraphic sequence were radiocarbon dated during the 1950s and 1960s. The youngest date available that is associated with the mammoth material is 11 800 ± 800 (UCLA-106; Fergusson and Libby 1962). Collagen extraction techniques yielded a date of 8000 ± 250 (UCLA-705; Berger *et al.* 1965) on material for which charcoal dated at 12 500 ± 250 (L-290-T; Broecker and Kulp 1957), but Orr distrusted the younger estimate (Orr 1964, written communication to Berger, SBMNH archives) and it is not generally cited. Except for the discounted collagen date, these and other dates that Orr commissioned are based on material associated with mammoth remains, but one date, 29 700 ± 3000 (L-290-R, the average of four measurements; Broecker and Kulp 1957) was obtained directly from apparently charred mammoth bone in the Tecolote member (Orr 1960*b*). More recently, McKinney (1980) obtained uranium–thorium ages for two full-sized molars from Santa Rosa Island: 60 000 ± 8000 and 25 000 ± 5000, respectively.

L. Marcus and S. Miller (unpublished data; personal communication 1979) obtained $^{13}C/^{12}C$ ratios from collagen extracted from a rib fragment in the same quarry as the L-290-R dated material. Like much of the mammalian material analysed from Rancho La Brea on the mainland (Marcus and Berger 1984), this Santa Rosa Island mammoth material showed the sorts of ^{13}C values expected for animals that consume C_3, as opposed to C_4, plants (Ambrose and DeNiro 1989; Lee-Thorp *et al.* 1989). C_4 plants tend to be well adapted to high water stress and arid habitats. In reviewing palaeoecological data for the past 128 000 years, including oxygen isotope ratios and pollen frequencies, Johnson (1983) concluded that the climate of the California borderland fluctuated but remained Mediterranean and somewhat cooler and moister than today during the late Quaternary, until approximately 8000 years ago, when a rise in the percentage of oak (as opposed to conifer) pollen appearing in cores signals the onset of drier conditions.

The accuracy and resolution of radiometric dating has improved greatly over the past three decades, and it will be important to obtain additional dates, especially if they can be based directly on mammoth material. Cushing *et al.* (1986), citing evidence that mammoth bones have been affected by chemical processes occurring in groundwater, have called for the re-evaluation of all existing radiocarbon dates for the Northern Channel Islands (see also Wenner *et al.* 1991). The possibility that some of the fossil deposits have been reworked, and some of the material transported and redeposited, makes a chronology based on dating, rather than on position in an assumed sequence and layering, especially desirable.

GEOGRAPHY

As Pleistocene sea-levels fluctuated with glacial advance and retreat, the California Channel Islands were alternately exposed, at times as a single island, and flooded. Highest sea-levels occurred in the Aftonian or First Interglacial Period when the ocean completely covered all of the islands but Santa Rosa. Wave-cut platforms on the island record at least two other interglacials (Orr 1960*a*, 1967). From oxygen isotope data, sea-level curves, and studies of marine terraces, Johnson (1983) has deduced to a first approximation the numbers of islands and their areas exposed along the southern California borderland during the past 128 000 years.

It has been widely believed that mammoths first reached the islands from the mainland by way of a peninsular extension of the Santa Monica Mountains exposed during the periods of lowered sea-level (Orr 1968). Indeed, the presence of mammoth bones has been adduced as primary evidence for the existence of such a bridge (Stock and Furlong 1928; Chaney and Mason 1934; Stock 1935). Nevertheless, the swimming powers of elephants are certainly adequate to have permitted colonization and occasional recolonization of the islands whether a land bridge existed or not (Sondaar 1977; Johnson 1980). Even extraordinarily rare events, such as elephants crossing the channel during higher sea-stands, become likely occurrences given time spans of tens and hundreds of thousands of years. Vedder and Howell

(1980) concluded that there is sufficient geological and geophysical evidence to demonstrate that land bridges were not primary features during the last half of the Pleistocene epoch, and that the link between the Northern Channel Islands and the mainland was ephemeral if it existed at all. Johnson and his co-authors (Johnson 1972, 1978; Junger and Johnson, 1980; Wenner and Johnson, 1980) accumulated further biological, geographical, and biogeographical evidence to challenge the contention that a land bridge did exist. Bathymetric readings indicate that the channel between the islands and mainland is too deep (236 m at shallowest) to provide dry-land connection during the lowest (~ 125 m) depression of sea-level, if one assumes that the deepest immersed wave-cut platform preserves a record of the lowest level the sea attained. Dotted lines in Fig. 24.1 show the configuration of shorelines Vedder and Howell (1980) inferred from bathymetric profiles for lowest sea-stands during the late Pleistocene. Still, some have argued that the history of the tectonically active Santa Barbara Basin floor is not sufficiently well known to rule out the existence of a land bridge (Azzaroli 1981).

Johnson (1972), in examining the role of periodic fires and drought upon the landscape evolution of San Miguel Island, implicated mammoths in the edaphic changes in the history of this island. Highly concentrated populations of elephants are known to overbrowse and destroy habitat, interrupt succession, and convert woodland to grassland. Laws (1970) proposed that elephants might have been a factor in the post-Pleistocene elimination of forested areas in Somalia and Turkana today, although opinions vary on how reversible habitat destruction by elephants is (Myers 1973; cf. Croze *et al.* 1981; Laws 1981).

ARCHAEOLOGY

What appears to be charred mammoth bone has been the source of controversy in anthropological circles. Peculiar reddened 'fire' areas have been found close to mammoth remains on the islands, and it was Orr and Berger's belief (Berger and Orr 1966; Orr and Berger 1966) that these circular depressions represent red-baked barbecue pits in which early native American inhabitants of Santa Rosa roasted their mammoth quarry. These authors attributed some of the disarticulation and breakage among the mammoth fossils to human butchers; the extinction of mammoths may have been the result of hunting by humans.

Berger (1980) described one pit on Santa Rosa Island in which what he interpreted as charred mammoth remains and some coarse stone tools were found together. Charcoal samples showed no measurable radiocarbon activity and were therefore dated as older than 40 000 years BP (the upper limit for the technique). He suggested (Berger 1980) that this site may be one of the oldest mammoth kills in America. Indeed, the earliest dates known to be associated directly with human remains in the Americas are on the order of 10–12 000 years (Berger 1980), although there have been recurrent, controversial suggestions in anthropological literature that the first human occupation of the

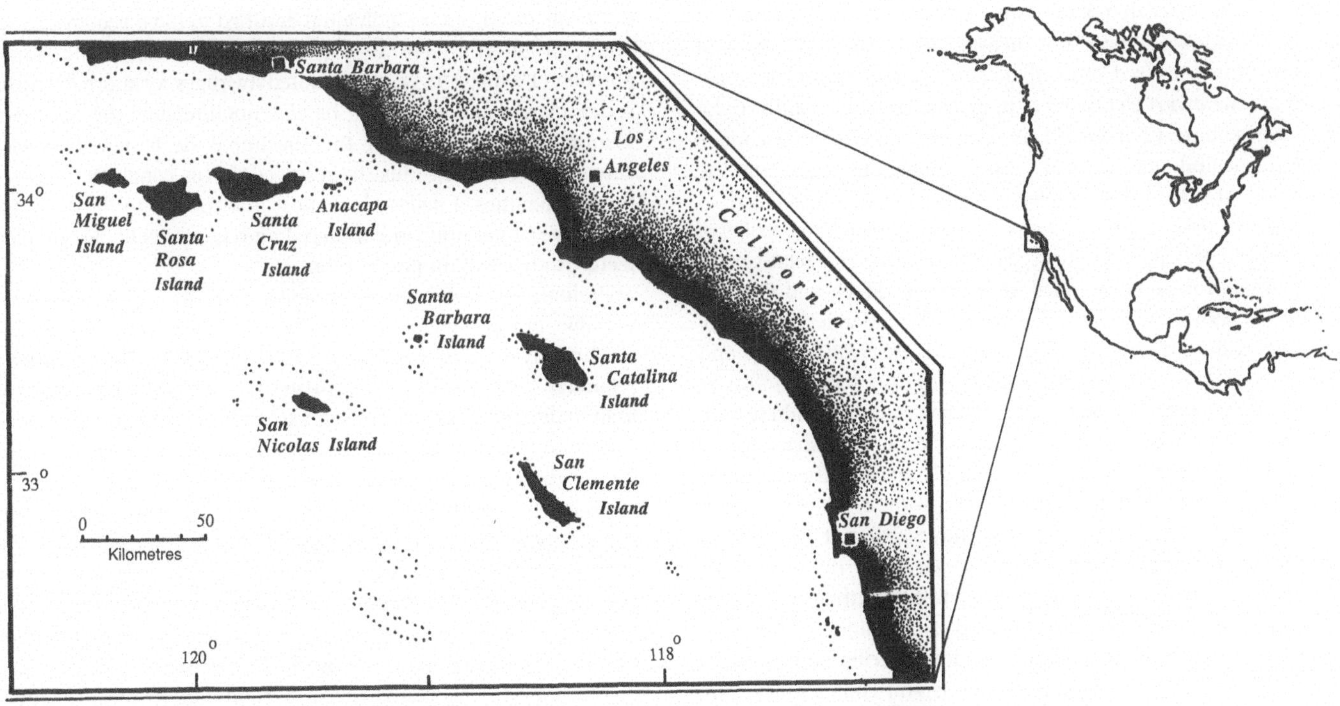

Fig. 24.1 The California Channel Islands. Dotted line shows–120 m depth contour and the shoreline at lowest sea stands in the late Pleistocene (Vedder and Howell 1980). Fossils of dwarfed mammoths are most abundant on Santa Rosa and San Miguel islands.

New World was much earlier (Berger and Orr 1966; MacNeish 1982; Morell 1990). Establishing an early arrival for man on Santa Rosa Island would raise new problems, however. As indicated above, the youngest dates on island mammoths indicate survival until at least 12 000 years BP, and it seems unlikely that predatory man and dwarf mammoth could have co-existed on the islands for tens of thousands of years (Berger and Orr 1966).

Some critics have suggested that the 'barbecue pits' Orr and Berger described are in fact charred depressions left by overturned tree stumps from forests demolished by natural fires (Johnson 1972; P. Walker, personal communication 1979; Wendorf 1982). More recently, in a thorough analysis of the 'fire' areas, Cushing *et al.* (1986) concluded that the observed discolouration of sediment and bone could not have occurred at high temperatures, and was instead due to manganese and iron minerals precipitated (possibly through microbial action) in groundwater. Thus there remain serious doubts not only that the mammoth bones were charred in fires caused by humans, but also whether some were charred in fires at all.

SYSTEMATICS

For nomenclatural clarity and convenience, the dwarfed mammoths from the California Channel Islands are referred to here collectively as *Mammuthus exilis*; large individuals indistinguishable from those on the mainland are considered *M. columbi* (Madden 1981*b*). It is nevertheless important to recognize the likelihood that the island residents fluctuated in size over time — Orr's (1968) stratigraphy suggested that some of the smallest specimens occurred temporally intermediate between samples of larger individuals — and that invasion from the mainland and introgression occurred more than once (a hypothesis also proposed for the dwarf elephants of the Mediterranean islands; Caloi *et al.* this volume, Chapter 22). To distinguish the large specimens from the island taxonomically from individuals found on the mainland, Madden (1981*b*) referred to them as *M. columbi orri*.

By different authors, *M. exilis* has been considered to be of subspecific, specific, or multispecific status; its closest relatives on the mainland have been called *M. imperator* (Stock and Furlong 1928; Orr 1968; Maglio 1973; Hooijer 1976; Madden 1981*b*, p. 80; each defines the species differently), *M. jeffersonii* (Kurtén and Anderson 1980), *M. columbi* (Madden 1977*a*, 1981*a*; Johnson 1981; Guenther 1987), or *Elephas primigenius* (Stock and Furlong 1928). Much of the confusion arises because the systematics of North American mammoths is itself highly problematic (for overviews, see Kurtén and Anderson 1980; Roth 1982; Agenbroad 1984*b*; Graham 1986*b*). Not only have different authorities recognized different numbers of species, but divisions and synonymies between taxa vary from author to author even when the same specific epithets are accepted. An additional source of problems is that our appreciation for intraspecific variability in even the modern species of elephants is still rudimentary, but with enhanced understanding, the number of fossil species of elephants recognized as valid is likely to decline (Roth 1989, 1992*b*).

SIZE, SHAPE, BIOLOGICAL FUNCTION, AND EVOLUTION

To judge from dimensions of limb bones with fully fused epiphyses, adult mammoths of Santa Rosa Island vary in size even more than (1) their closest relative on the mainland; (2) either of the modern species of elephants; or (3) *Elephas falconeri*, the smallest insular dwarf elephant from Sicily (Roth 1993; Caloi *et al.* this volume, Chapter 22). Most fossils from Santa Rosa are clearly dwarfed, but a few individual specimens are similar in size to those on the mainland. Unfortunately, inadequate information about sampling techniques for existing collections makes it difficult to interpret any distributions of body sizes one might construct: the variation could either represent high intrapopulational variability, or temporal heterogeneity.

Postcranial dimensions of the California Island mammoths have been tabulated by Madden (1981*a*, *b*) and Hooijer (1976); Hooijer (1976) and Guenther (1987) have published data and summary statistics on small ($n < 10$) samples of teeth. These accounts are useful indicators of some of the variation found on the islands, but the specimens represented were not the most informative of those available, and, for example, I differ in my identification (maxillary, rather than mandibular) of at least one specimen tabulated by Guenther. It is difficult to identify with certainty the third, fourth and fifth (i.e. dp4, M1, and M2) of the sequence of six cheek teeth in elephants, even for the modern species (e.g. Roth and Shoshani 1988). The task is all the more challenging when individuals in a sample vary greatly in size, for the serial categories of teeth are distinguished in part on the basis of lengths and widths. A small tooth can be interpreted as an *early* category from an individual destined to have become a *large* adult, or alternatively as a *later* stage of a *small* individual. Madden (1980*a*, 1981*a*, *b*) tabulated averages for more extensive series of specimens, but as he did not indicate the specimen identification numbers for the specimens he assigned to each category, judgement must be reserved on his conclusions. Given these difficulties, I believe it is most informative to illustrate the full range of variation encountered among all available complete teeth (Roth 1982, in preparation).

Preliminary observations suggest that *M. exilis* had more variable, but also generally more flexed limb posture than its full-sized relatives (Roth 1982, in preparation). Flexed limb posture is relatively common in medium-sized (as opposed to graviportal) mammals, and is associated with greater agility and locomotion that may involve leaping and more rapid acceleration and deceleration (Gregory 1929; Biewener 1989).

Skeletal proportions can give clues to the evolutionary processes involved in the evolution of dwarfed mammoths. The shapes of single limb bones suggest that achondroplasia — a genetic anomaly, often a dominant autosomal mutation, responsible for certain dwarfing syndromes in a wide variety of mammalian species — was not involved (Roth 1993). Bony anatomy and intermembral proportions reveal a combination of ontogenetic and geometric scaling in dwarfed mammoths that suggests a role for paedomorphic heterochrony, and progenesis in particular (Roth 1982, 1984, 1992*a*). A scenario involving stunting combined with progressive selection for small size is most likely the cause of size reduction in the island mammoths (Roth 1992*a*), while sporadic secondary invasion of forms from the mainland would account for the discoveries of larger forms (Johnson 1978, 1981).

Regression analysis of body mass on skeletal dimensions makes it possible to estimate body masses for fossils; from this one concludes that adult *M. exilis* ranged in weight between 200 kg and 2 metric tons, by contrast with *M. columbi* at 5 to 15 metric tons (Roth 1990). Body mass correlates in turn with a large number of other biological variables (e.g. lifespan, age at reproductive maturity, and maximum possible population densities), permitting further inferences about basic biology, natural history, and selective forces (Roth 1992*a*).

SUMMARY

Dwarf mammoths from the California Channel Islands, like other insular dwarf elephants, represent extreme examples of the island rule of mammalian body size (Van Valen 1973; Roth 1992*a*), and are prime subjects for the study of morphological variability, allometry, heterochrony, and body size evolution. Apart from their relevance to research on such intrinsically biological issues, these animals have provided information important to the reconstruction of late Pleistocene environments, the geography of the southern California borderland, and the peopling of North America. While our understanding of all of these topics continues to advance with comparative study of osteological and dental specimens, of key importance will be improvements in our ability to place these specimens accurately in a temporal context.

ACKNOWLEDGEMENTS

I thank L. Marcus and S. Miller for sharing unpublished data, and S. Miller, J. Cushing, and D. Mol for bringing useful references to my attention.

TWENTY-FIVE

Sexual dimorphism in the mammoth pelvis: an aid to gender determination

ADRIAN M. LISTER

INTRODUCTION

Determining the gender of fossil proboscidean skeletons can make an important contribution to understanding their taphonomy. This is particularly the case among members of the Elephantidae, where fossil species can be assumed to have had a broadly similar social structure to that of the living elephants (Haynes 1991). For example, mammoth assemblages such as Lehner Ranch, Arizona, thought to result from the slaughter of matriarchal groups, are expected to be dominated by females, as in living elephant species (Saunders 1980). At others, such as Hot Springs, South Dakota, the predominance of young adult males is believed to result from a gradual accumulation of individuals as these roving, adventurous animals became caught in a natural trap (Agenbroad and Mead 1987*a*).

Sexing proboscidean skeletons is also of value in evolutionary and anatomical studies, where sexually dimorphic effects need to be taken into account when assessing variation in body size and morphology.

There are four main categories of potential information about the gender of proboscidean skeletons: (1) preserved genitalia, (2) skeletal size and robusticity, (3) skull and tusk morphology, and (4) pelvic morphology. The first three categories are discussed by Averianov (this volume, Chapter 26). Genitalia are preserved in only very few frozen carcasses. Body size is useful for separating males and females within large samples, but must be used with caution on individual specimens because of potential temporal and geographical variation in mean size between populations. Skull and tusk form is of great value for separating male and female adult animals, but skeletons are often found with these elements lacking. In these instances pelvic morphology is of particular importance, and is, in any case, a valuable line of additional evidence even when skull and tusks are preserved. Pelvic form has been applied successfully to the sexing of other mammalian taxa (Tyler 1987).

MATERIALS AND METHODS

This study is based on pelvic girdles of mammoths (*Mammuthus* spp.), mostly from Eurasia. The lineage of the woolly mammoth evolved through three chronospecies, *M. meridionalis* (Nesti), *M. trogontherii* (Pohlig), and *M. primigenius* (Blumenbach) (Maglio 1973). The present sample comprised five pelves of *M. meridionalis*, one of *M. trogontherii*, and 15 of *M. primigenius*. This reflects the availability of material in the collections. The anatomy of the pelvic girdle is illustrated in Fig. 25.1. Using flexible tape and fixed calipers, a series of measurements was taken on each specimen (Fig. 25.2). Where appropriate, measurements were taken on both the left and the right side of the pelvis. If the two halves differed, their mean was used in tabulation and graph plotting. The resulting data are presented in Table 25.1. Each pelvis was also photographed in antero-ventral aspect (e.g. Fig. 25.3), using precisely the same viewpoint for each specimen to avoid the considerable apparent differences of form resulting from photographs taken at different angles.

The essential difference between the male and female pelvis is that in the male, the pelvic aperture (birth canal) is relatively smaller, and the ilium relatively broader, than in the female. This is illustrated in Fig. 25.3 for typical male and female pelves of *M. primigenius*. A similar difference has been illustrated for the Asian elephant (*Elephas maximus*) by Deraniyagala (1955, plate 27) (Fig. 25.4), and for the African elephant *Loxodonta africana* by Haynes (1990) (Fig. 25.5). A qualitative indication of gender difference between proboscidean pelves can be obtained by making outline drawings from photographs, adjusting specimens to the same overall width or aperture width, and superimposing the drawings to visualize differences in proportion (Fig. 25.3c). For reliable gender determination, however, it is essential to quantify the difference.

The size of the pelvic aperture is satisfactorily quantified as its greatest width (Fig. 25.2, measure 3). This measurement is easily taken when both left and right halves of the pelvis are present, and are joined ventrally or via the sacrum dorsally. If only one half of the girdle is preserved, an estimate of aperture width can be obtained by projecting an imaginary line dorsally from, and parallel to, the pubic symphysis, measuring at right angles from this line to the widest part of the aperture, and doubling the measurement. Alternatively, the size of the pelvic aperture can be satisfactorily quantified by a different measure, the diagonal height of the aperture from the pubic symphysis to the lowest point of the sacral scar (measurement 2 in Fig. 25.2).

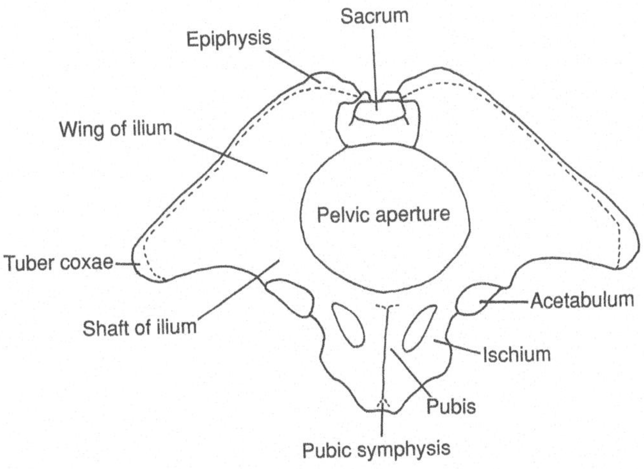

Fig. 25.1 Anatomical features of the mammoth pelvis.

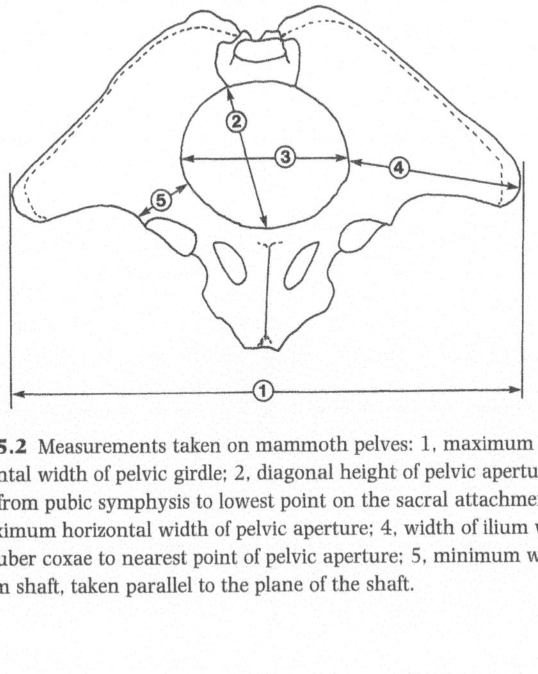

Fig. 25.2 Measurements taken on mammoth pelves: 1, maximum horizontal width of pelvic girdle; 2, diagonal height of pelvic aperture, taken from pubic symphysis to lowest point on the sacral attachment; 3, maximum horizontal width of pelvic aperture; 4, width of ilium wing, from tuber coxae to nearest point of pelvic aperture; 5, minimum width of ilium shaft, taken parallel to the plane of the shaft.

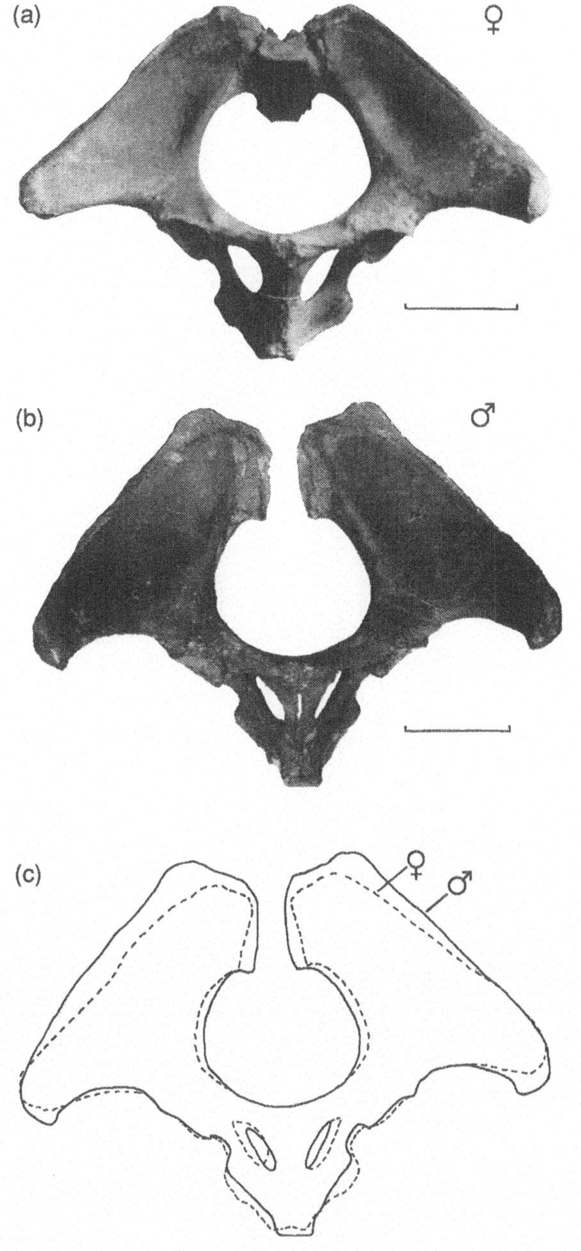

Fig. 25.3 Mammoth pelves photographed in antero-ventral view and adjusted to the same maximum width. (a) Pelvis of the Oyosh skeleton (female) in the Siberian Academy of Sciences, Novosibirsk; (b) pelvis of the Taimyr skeleton (male) in the Zoological Institute, St Petersburg; (c) superimposed outlines of the two pelves, aligned at the acetabula. (Photographs courtesy of V. E. Garutt; scale bars 25 cm.)

Quantifying the width of the ilium is more problematical. The width of the wing of the ilium is clearly dimorphic (Fig. 25.3), but it proved impossible to define an unambiguous and repeatable point of measurement in this region. The maximum width of the ilium (measurement 4 in Fig. 25.2) is a clearly definable measure, but is strongly influenced by the form and size of the lateral part of the wing and the tuber coxae, which show considerable variation not clearly linked to gender. This measurement also varies markedly depending on whether the lateral epiphysis of the ilium is fused to the wing, or has been lost in the fossil. Finally, maximum width of ilium is an unsatisfactory choice of measurement because the lateral part and tuber coxae are very prone to damage in fossil specimens. These considerations also apply to the total width of the pelvis (measurement 1, Fig. 25.2).

A measurement which has proved to be both clearly definable, and adequately to reflect sexual dimorphism in ilium width, is the minimum width of the shaft of the ilium, the constriction between the acetabulum and the wing (measurement 5, Fig.

Table 25.1 Data on mammoth pelves examined in this study (measurements are in centimetres)

Species and site	Museum	Gender/hd[b]	Measure[a] 1	2	3	4	5	Ratio 5 to 1	3 to 1	3 to 2	3 to 4	3 to 5	2 to 5	Reference
Mammuthus meridionalis														
Valdarno, Italy 14838	Florence	male/ c(p)t	160	40	44		27.0	0.169	0.275	1.10		1.63	1.48	Azzaroli (1977)
Valdarno, Italy 10791	Florence	male/ c(p)t	188	47	54		28.4	0.151	0.287	1.15		1.90	1.65	Azzaroli (1977)
Valdarno, Italy 1057	Florence	(male)/ (p)	178	47	49		25.1	0.141	0.275	1.04		1.95	1.87	no reference
Valdarno, Italy 1050	Florence	female/ c(p)t	130	45	47		17.0	0.131	0.362	1.04		2.76	2.65	Azzaroli (1977)
Valdarno, Italy 13730	Florence	female/ c(p)t	138	45	44		15.6	0.113	0.319	0.98		2.82	2.88	Azzaroli (1977)
Mammuthus trogontherii														
Edersleben	Sangerhausen	female/(p)t	154	63	57	63	19.2	0.125	0.370	0.90	0.904	2.97	3.28	Garutt and Nikolskaja (1988)
Mammuthus primigenius														
Praz Rodet, Switzerland	Lausanne	male/ c(p)t			42		20.0					2.10		Weidmann (1969)
Steinheim, Germany	Stuttgart	male/ bc(p)t		48	51		25.0			1.06		2.04	1.92	Dietrich (1912)
Siegsdorf, Germany	Siegsdorf	male b(p)t	160	44	50		26.0	0.163	0.313	1.14		1.92	1.69	Mol *et al.* (1988)
Ahlen, Germany	Münster	male/ b t	137	40	50	62			0.365	1.25	0.806			Siegfried (1959)
Condover, England	Shropshire	male/ b(p)t	143	43	46	[49]	20.2	0.141	0.322	1.07	0.939	2.28	2.13	Coope and Lister (1987)
Gewande, Holland	Hoofddorp	(male)/(p)	(119)	42	42	(46)	19.0	(0.160)	(0.353)	1.00	(0.913)	2.21	2.21	Mol (1984)
Beresovka, Russia	St. Petersburg	male/ cg(p)t	134	33	41	52	18.3	0.137	0.306	1.24	0.788	2.24	1.80	Zalensky (1903)
Taimyr, Russia	St. Petersburg	male/ c(p)t	125	41	43	47	19.4	0.155	0.344	1.05	0.915	2.22	2.11	Garutt *et al.* (1990)
Liakhov Island, Russia	Paris	male/ c(p)t	(112)	33	43	(47)	20.8	(0.186)	(0.384)	1.30	(0.915)	2.07	1.59	Coppens (1958)
Gydan, Russia	Moscow	male/ bc(p)t	130		46	46	19.3	0.148	0.354		1.000	2.38		Garutt (1964)
Lena R. (Adams'), Russia[c]	St. Petersburg	male/ b	137	44	48	55	18.2	0.133	0.350	1.09	0.872	2.64	2.42	Garutt (1964)
Aa, France	Boulogne	female/(p)t	(134)		54	(40)	19.4	(0.145)	0.403		(1.350)	2.78		Pontier (1913)
Oyosh, Russia	Novosibirsk	female/ bc(p)	117	39	44	[45.2]	[15.4]	0.132	0.377	1.13	0.973	2.86	2.53	Averianov, this volume, Chapter 26
Rochester, Indiana, USA	New York	female / c(p)t		46	50		19.0			1.09		2.63	2.42	Osborn (1942)
Kerkdriel, Holland	Empel	(female)/(p)	128	43	45	42	16.0	0.125	0.352	1.05	1.07	2.81	2.69	Mol, pers. comm., 1988

[a] Measures 1 through 5 are defined in Fig. 25.2. Measurements in '( )' are minimum measures because epiphyses were absent, measurements in '[ ]' were taken from photographs.

[b] 'hd' refers to how the gender was determined. Key: b, body size; c, cranial morphology; g, genitalia; p, pelvic morphology; t, tusk morphology. Body size is invoked only for specimens which are particularly large (male) or particularly small (female). Pelvic morphology is based on results presented here, and it is given in parentheses as it is not an independent determinant of gender in this study. The genders of isolated pelves are given in parentheses for the same reason. Blank cells imply that no data are available.

[c] The gender of the Lena (Adams') mammoth is discussed in the text.

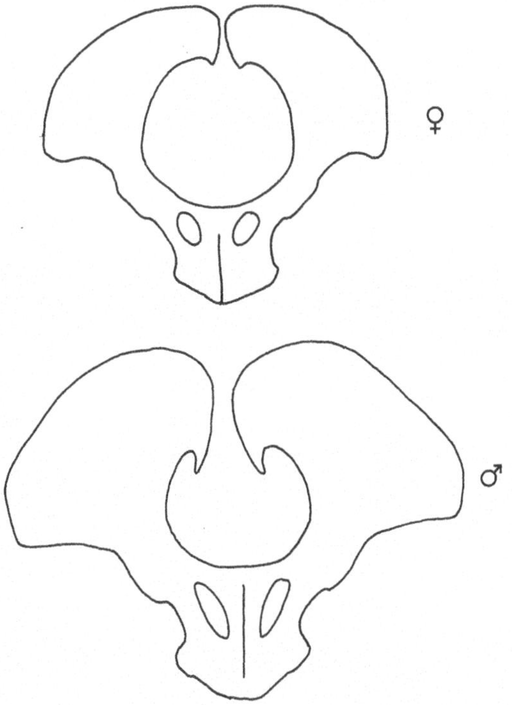

Fig. 25.4 Male and female pelvic girdles of Asian elephant, *Elephas maximus*, in posterior view. Traced from Deraniyagala (1955, plate 27). Scale not given.

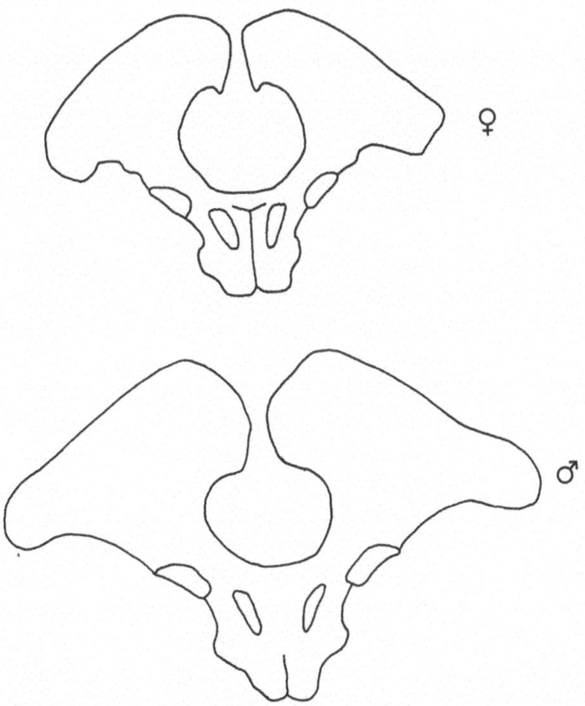

Fig. 25.5 Male and female pelvic girdles of African elephant, *Loxodonta africana*, in anterior view. From Haynes (1990, fig. 1). Scale not given.

25.2). This measurement also has the important advantage that it is very rarely subject to abrasion, and can be measured even on pelves where large parts of the ilium are broken away.

RESULTS

The data are presented in Table 25.1. Each of the measurements was first plotted on a univariate axis to test for sexual dimorphism, using specimens whose gender was known independently from other body parts. It was found that none of the pelvic measures, taken alone, provides clear discrimination between males and females: there was overlap in the ranges of the two sexes. This applies to aperture width and shaft width as well as to the other measures, and is due at least in part to individual size variation independent of gender (e.g. a large male may have as wide a pelvic aperture as a small female).

Ratios between measures therefore provide clearer indicators of gender differences. Several such ratios have been tested on the independently sexed pelves (Table 25.1). Most of the ratios show considerable overlap in the ranges of males and females, and are therefore not suitable for discrimination. However, the ratio between aperture width and shaft width (measurement 3 to measurement 5) shows almost complete separation between the sexes, at least as far as sample sizes allow (Fig. 25.6). Samples are very small for *M. meridionalis* and *M. trogontherii*, but the specimens available appear to follow the same pattern as the larger *M. primigenius* sample, suggesting that the distinction may be valid for all *Mammuthus* species. A similar pattern has been found for North American *M. columbi* by Lister and Agenbroad (1994). The samples presented here suggest that a ratio of less than about 2.6–2.7 in aperture width/shaft width indicates a male, while a ratio of greater than this value indicates a female. If the Adams' mammoth is indeed a male (see discussion below), male and female ratios approach each other closely in the region of 2.6–2.7, so caution is required in sexing pelves with ratios in this region.

An alternative means of quantifying aperture to ilium ratio, useful if only half of the pelvis is preserved, is the ratio between measurements 2 and 5 (Fig. 25.2). In the sample available, this also gives barely overlapping ranges between independently sexed males and females, the dividing line falling at values of approximately 2.4–2.5 (Fig. 25.7).

Coope and Lister (1987) (Fig. 25.8) suggested that qualitative morphological variation among mammoth pelves could be used to determine gender even where aperture width cannot be measured, for example in isolated ilia. In two juvenile ilia from the same population of *M. primigenius*, of identical size and ontogenetic age, they found a clear difference in length of the ilium shaft, and a complementary difference in shape of the inner edge of the aperture (Fig. 25.8). The longer and shorter shafts were regarded as pertaining to male and female animals, respectively. The specimens of adult *Mammuthus* pelves scored in the present study have been examined for this feature. This has demonstrated a tendency for males and females to differ in the

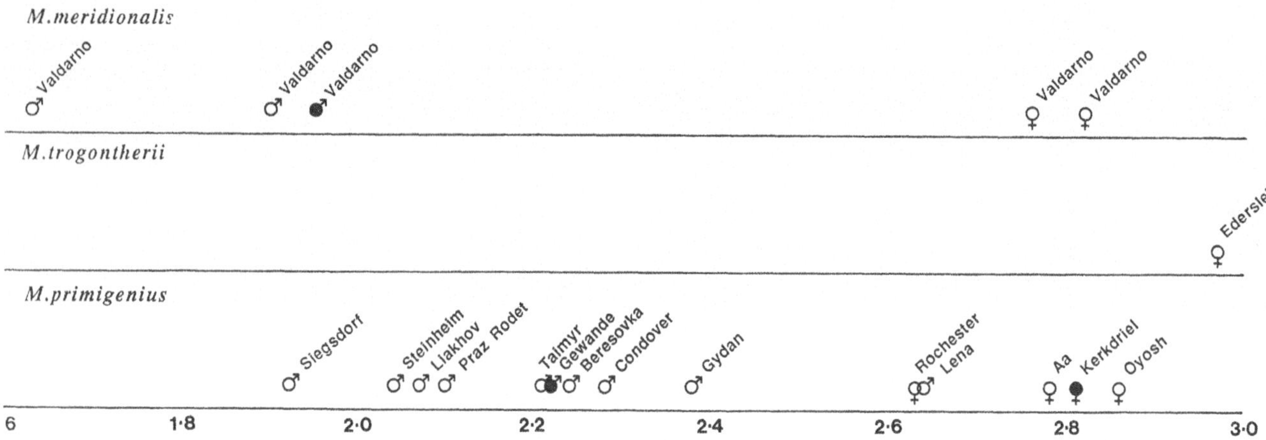

Fig. 25.6 Ratio between pelvic aperture width and ilium shaft width (measurements 3:5). Filled symbols: pelves for which there is no independent evidence of gender.

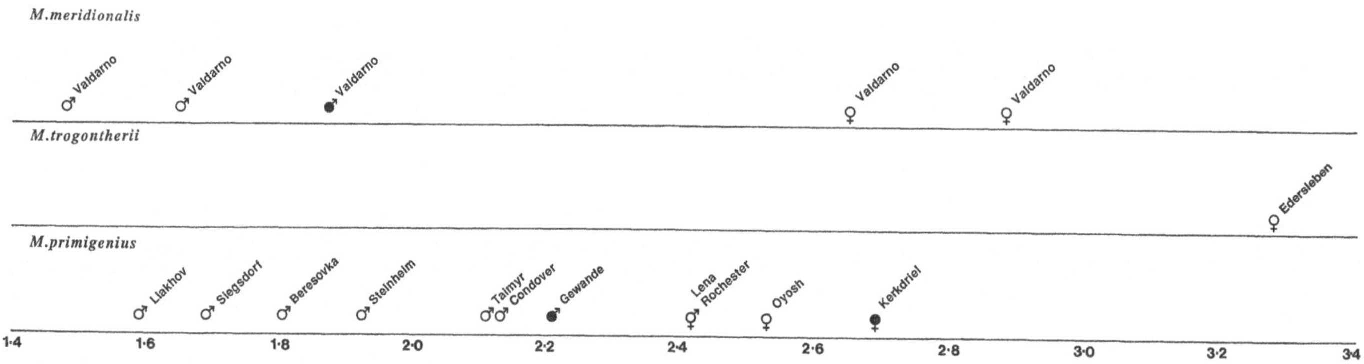

Fig. 25.7 Ratio between pelvic aperture height and ilium shaft width (measurements 2:5). Filled symbols: pelves from which there is no independent evidence of gender.

manner described, but the ranges of variation of the two sexes overlap to the extent that the feature is not an infallible indicator of gender. None the less, in cases such as the specimens from Condover, England described by Coope and Lister (1987), it is likely that the distinction is due to gender difference. The morphological difference is very clearly marked, and all other variables (size, ontogenetic age, geological age, and locality) are constant.

CONCLUSIONS

The most practicable and reliable indicator of gender in *Mammuthus* pelves is the ratio between aperture diameter and ilium shaft width. This ratio gives higher values in females than in males, with barely overlapping ranges for the two sexes in the samples studied here.

Most of the *M. primigenius* pelves examined here are from male animals. This may be linked to their mostly being from isolated skeletons. Many of the sites may have been natural traps such as crevasses or sink holes, which solitary male animals, roving and adventurous, would be more likely to enter than the more cautious, group-living females. For one of the sites at least, a natural trap situation (a kettle hole) is clearly indicated by the geology (Condover: Coope and Lister 1987).

Female specimens of *M. primigenius* are those from Rochester, Indiana, USA (already recognized as female on the basis of the skull and tusks: Osborn 1942, p. 1157); the Aa River at Arques, France (Pontier, 1913; confirmed by the slender form of the tusks); and the isolated Kerkdriel pelvis, from the Kreftenheye Formation of the River Meuse, The Netherlands (Mol, personal communication 1989). Another pelvis, from the young skeleton found at Yuribei, western Siberia, was unfortunately not available for measurement in this study. However, a photograph of the pelvis (Dubrovo 1982, plate 7) strongly suggests a female — it shows a very wide pelvic aperture and narrow shaft of the ilium.

The famous skeleton discovered at the Lena River in 1799 and collected by Adams in 1806 is enigmatic, as its pelvic dimensions better fit a female than a male (Figs 25.6 and 25.7). This

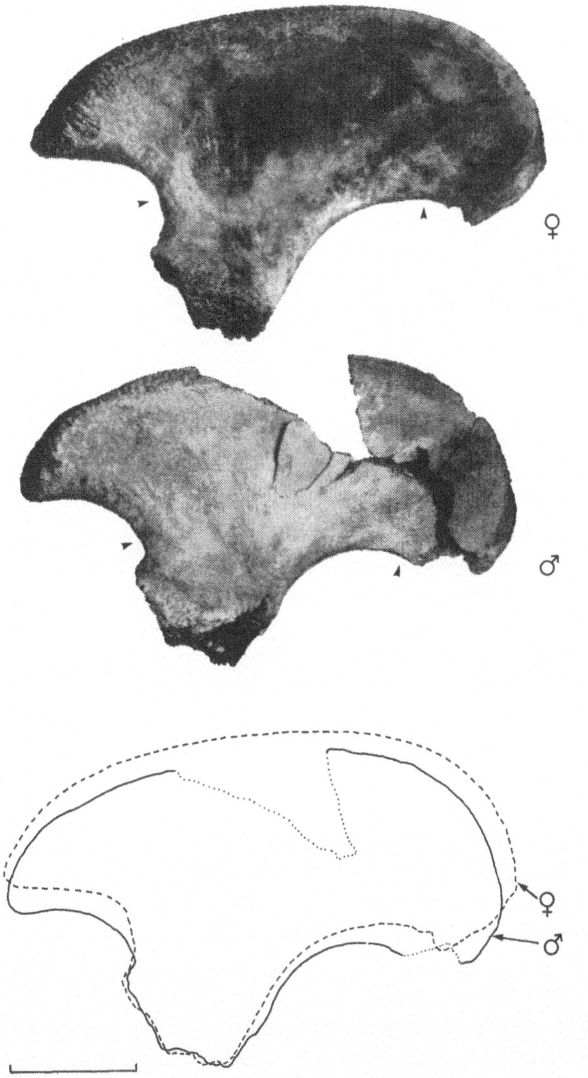

Fig. 25.8 Right ilia of two juvenile mammoths from Condover, England, in postero-lateral view (from Coope and Lister 1987), and superimposed outlines aligned at the acetabular suture. Arrows indicate regions of significant difference, the upper arrow in the region of the pelvic aperture, the lower on the ventro-lateral edge of the ilium shaft. Scale bar 10 cm.

specimen has always been regarded as of male gender, but by the time Adams reached the specimen in 1806, most of the flesh had decomposed (Garutt 1964), and Tilesius (1815) described it as a male without explicitly mentioning genitals. The large tusks on the skeleton in St. Petersburg are very likely those of a male, but we cannot be entirely certain they belong to the skeleton, as the tusks belonging to the carcass were removed in 1804, sold in Yakutsk, and the existing pair (presumed the same ones) repurchased in 1806 (Garutt 1964). The alveolar region of the skull is too damaged for gender determination, so that the only unambiguous evidence for male gender is the rather large size of the skeleton, especially in comparison with other Siberian mammoths (measurements in Garutt 1964; Averianov, this volume, Chapter 26).

The *M. trogontherii* skeleton from Edersleben, Germany (Garutt and Nikolskaja 1988), one of only two or three known of this species, is indicated as a female, its pelvic morphology corroborated by the slender tusks of the animal.

Of the four skeletons of *M. meridionalis* from the Upper Valdarno, Italy, preserved at the University of Florence, two are males and two females. Specimen 14838, regarded by Azzaroli (1977) as a female, is here redetermined as male.

ACKNOWLEDGEMENTS

I am grateful to the curators of the various museums for allowing me to study fossil material in their care. Thanks are also due to D. Mol and I. Foronova for their help in measuring the Edersleben and Oyosh specimens, respectively. V. E. Garutt kindly provided the photographs for Fig. 25.2. This research was funded by the Natural Environment Research Council (grant no. GR3/6294).

TWENTY-SIX

Sexual dimorphism in the mammoth skull, teeth, and long bones

ALEXANDER O. AVERIANOV

INTRODUCTION

Sex determination in mammals is important for understanding their ecology and population biology. In large mammals, sexual dimorphism is expressed by the fact that males are bigger than females and have strongly developed secondary sex features, such as horns, or enlarged canines or incisors. In recent African elephants, *Loxodonta africana* (Blumenbach 1797), the males are taller than females, (males may reach 4.0 m in height, as distinguished from females that range from 2.0 to 3.0 m). Tusks of males are also thicker and longer that those of females (Sikes 1971). Direct sex determination in extinct animals is possible only in exceptionally rare cases, as in frozen carcasses of Pleistocene mammals with preserved genitalia. The sexes of three individual woolly mammoths, *Mammuthus primigenius* (Blumenbach 1799), all males, were determined in this manner. They are: the Adams' mammoth (1799; Fig. 26.1a, b), the Berezovka mammoth (1901; Fig. 26.2a, b), and the Magadanian mammoth calf (1977; Fig. 26.3). The sex of the Adams' mammoth (ZIN N 71911; Fig. 26.1a, b), discovered in 1797 by a Tungusian hunter, Osip Schumachov, and a merchant, Roman Boltunov, (Tilesius 1815), was easily determined because the carcass contained preserved identifiable tissues and organs. The penis of the Berezovka mammoth (ZIN N 77340; Fig. 26.2a, b) can be observed on the mounted specimen at the Zoological Museum in St. Petersburg. On the other hand, the penis of 'Dima' (ZIN N 70188; Fig. 26.3) was separated from the carcass (fig. 18 of Vereshchagin 1981).

More usual discoveries include separate bones of mammoths and it is very difficult to determine the sex of individuals from such material. The absence of reliable published data about sexual dimorphism and ages of individual mammoths has limited our knowledge about the palaeoecology, population dynamics, peculiarities, and systematic position of some forms of this species.

Bones and teeth of adult female woolly mammoths were often mistaken for the remains of special forms of dwarf mammoths, which were marked by different scientific names: *Elephas pygmaeus* Fisher de Waldheim, 1829; *Elephas stenotoechus* Brandt, 1831; *Elephas primigenius leith-adamsi* Pohlig, 1888; *Elephas primigenius minor* Gromov, 1939. Only by studying sexual dimorphism can the problem of the existence of dwarf mammoth as a special systematic form of *M. primigenius* be solved.

This study was undertaken to determine diagnostic biometric and morphological features of different parts of skeletons of woolly mammoths for sex determination. Data on relatively complete skeletons of *M. primigenius* from some Soviet museums are discussed below. The sex of these individuals can be determined with confidence by existing genitalia or by a complex of other features (Tables 26.1–26.3 provide some of these data). I hope that this comparative information will be useful for future workers in determining the sex of individuals from separate bones.

Abbreviations

KGM, Geological Museum of Kazan University (Kazan, Russia); NKM, Novosibirsk Museum of Regional Ethnography (Novosibirsk, Russia); PIN, Paleontological Institute, Russian Academy of Sciences, (Moscow, Russia); ZIN, Zoological Institute, Academy of Sciences USSR (St. Petersburg, Russia; new name 'Zoological Museum, Russian Academy of Sciences', abbreviated as ZM).

RESULTS AND DISCUSSION

Below is a summary of diagnostic features of different parts of a skeleton for sex determination.

TUSKS

Male and female mammoths are distinguished clearly by the sizes and shapes of their tusks. Vereshchagin and Tikhonov (1986) noted that females had relatively more slender, straight, and gracile tusks, with the diameter near the ends of the alveoli less than 90 mm in adult animals, and their tusks turned medially only slightly. Tusks of males were more massive, with a diameter at the end of the alveoli up to 200 mm. Old males usually had

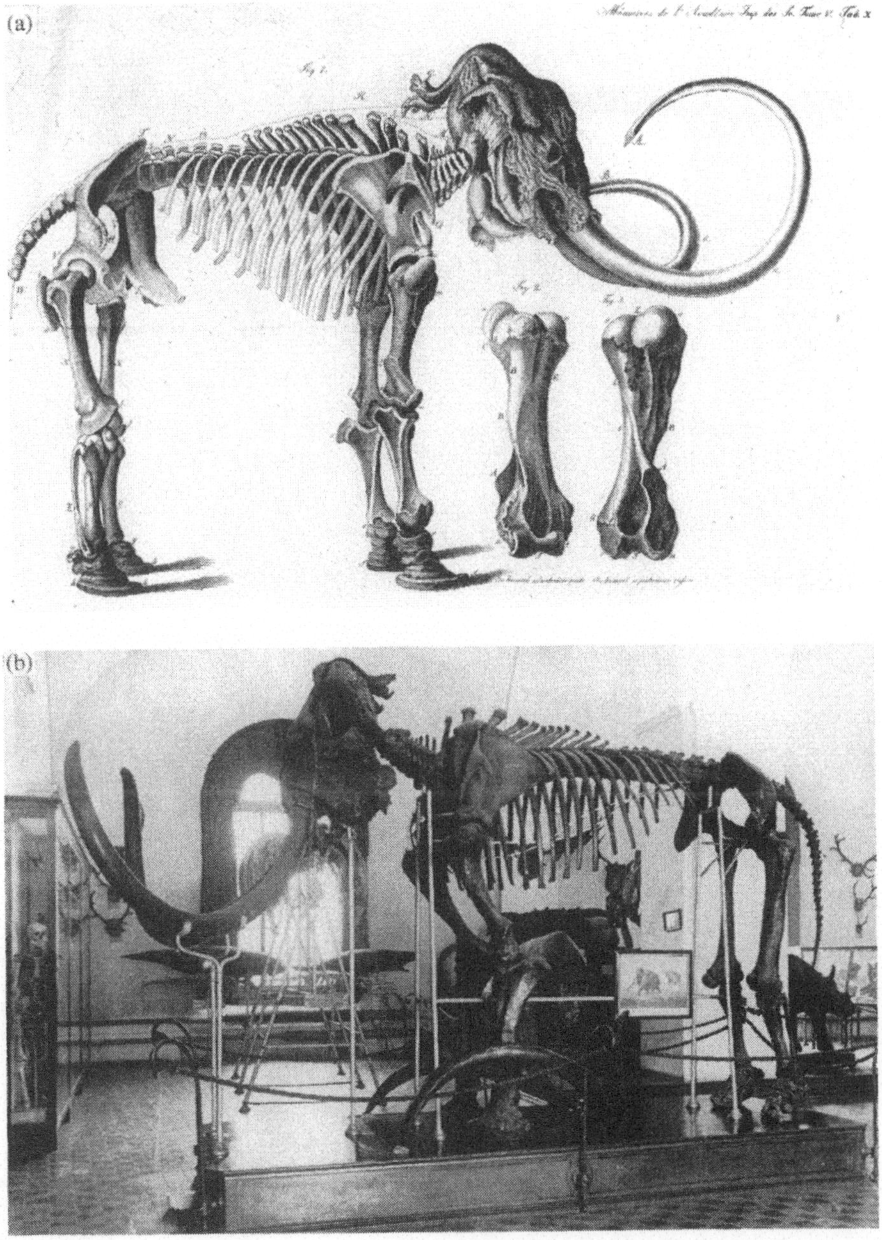

Fig. 26.1 Skeleton of the Lena (Adamsi) mammoth, ZIN N 71911, male, 50 or more years old (note: soft tissues on skull; left and right tusks are transposed). (a) Etching from Tilesius (1815, plate 10), which contains confirmation of the sex of this individual (see also Osborn 1942, pp. 1147–9 for additional information on this 'famous' skeleton). (b) Skeleton now in the Zoological Museum, Russia Academy of Sciences, St. Petersburg, Russia. (Photograph by V. E. Garutt.)

tusks turned medially very strongly and even crossed at their ends. The value of tusks for the determination of sex depends also on the fact that the ages of individuals may be established by growth rings, but this is open to criticism because: (1) not all rings are annual, and (2) distal wear removes rings (Vereshchagin and Tikhonov 1986).

The extent of tusk development in elephants may be associated with the concentration of sex hormones. In Asian elephants (*E. maximus* Linnaeus, 1758), tusks are weaker than in African elephants (*L. africana*). Females of *E. maximus* have short, straight tusks, if any. Those of males are longer and more curved, and most are shorter than in males of *L. africana*. Some

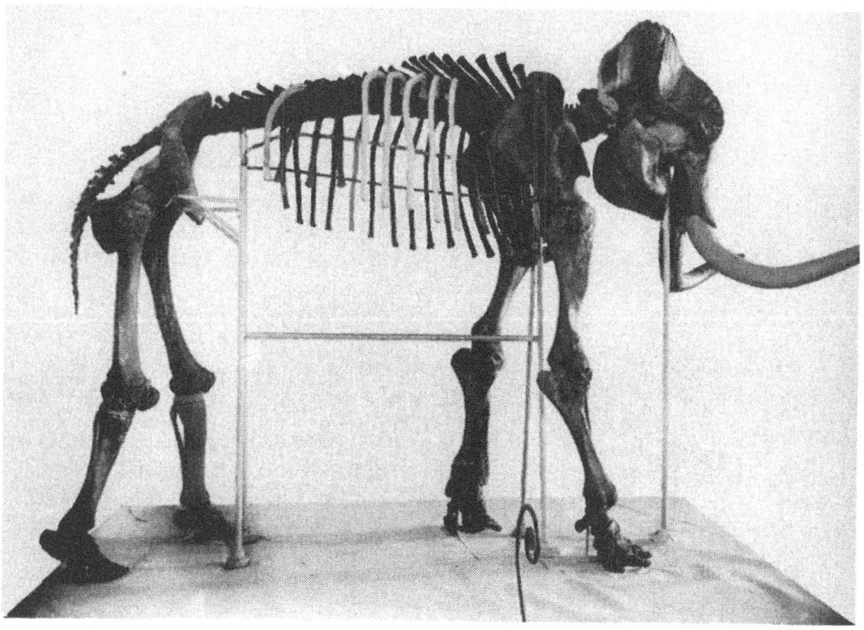

Fig. 26.2 (a) Mounted Berezovka mammoth, ZIN N 77340, male, about 40 years old. Penis can be observed between hind and forelegs (photograph by V. E. Garutt). (b) Skeleton of the Berezovka mammoth, ZIN N 5315, male, about 40 years old (after Zalensky 1903, plate 24). Printed from the original negative conserved in the Zoological Institute, USSR Academy of Sciences.

males of *E. maximus* lack tusks (and are known as 'mukna'). Some females of *L. africana* have larger tusks than those of Asian males.

Male elephants of both genera were observed to have more massive muscle masses in their heads and necks than females (Sikes 1971; Shoshani, personal communication 1991). It is therefore hypothesized that mammoth males required powerful muscles for long, massive tusks. They had relatively more pronounced muscle ridges on the bones of the shoulder girdle and on the cervical vertebrae (cf. Averianov 1994).

CRANIA

The crania of male mammoths were relatively more massive than those of females, with wider and more robust alveolar processes of premaxillary bones. Sometimes the lateral ends of the bony nasal

Table 26.1 Sex, dental age, and form of mandible symphysial process in some mammoth individuals

Individual mammoths	Sex	Teeth generation	Age (years)	Symphysial process
1	Male[a]	dp2 and dp3	1[b]	Long
2	Female	dp4 and M1	10–12[c]	Very short
3	Female	M2 and M3	35–40	Short
4	Female	M2 and M3	35	Long[d]
5	Male[a]	M2 and M3	40	Long
6	Male	M3	45–50	Long
7	Female	M3	50–60	Almost absent
8	Male	M3	45–50	Short
9	Male[a]	M3	50+	Long

Information on individual mammoths: 1, Magadanian mammoth calf 'Dima', ZIN N 70188 (Fig. 26.3), Chokotka; 2, Uribean mammoth, PIN N 3941, Gydanian peninsula; 3, Oyesh mammoth, NKM without number; 4, Kamskoe Ustye mammoth ZIN N 30873, Tataria; 5, Berezovka mammoth, ZIN N 5315, skeleton (Fig. 26.2b) and N 77340, stuffed animal (Fig. 26.2a), Yakutia; 6, Kozlovo mammoth KGM N 116/261, Perm oblast; 7, Sanga-Yuriakh mammoth, ZIN N 31738 (Fig. 26.5), Yakutia; 8, Taimyr mammoth, ZIN N 2710 (Fig. 26.4a,b), Taimyr peninsula; 9, Lena (Adamsi) mammoth, ZIN N 71911 (Fig. 26.1a,b), Yakutia.

[a] Confirmed by genitalia.

[b] After Vereshchagin (1981).

[c] After Dubrovo (1982).

[d] In this individual, possibly belongs to an earlier form of mammoth, it may be an atavistic feature.

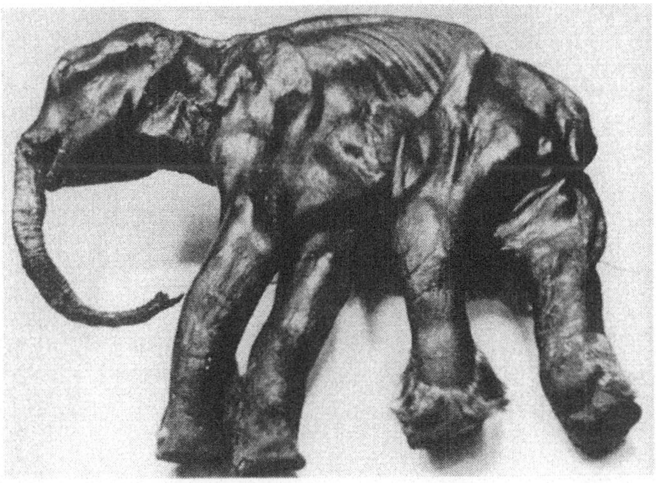

Fig. 26.3 Magadanian mammoth baby 'Dima', ZIN N 70188, male, approximately 1 year old (photograph by V. E. Garutt).

cavity are sharp and somewhat inclined downward in males, and rounded in females (Table 26.2; Figs 26.4b, 26.5a, 26.6).

TEETH

Using an assemblage of isolated teeth, and teeth in crania and mandibles, from the Berelekh 'cemetery' of mammoths (East Siberia, Yakutia), Zherekhova (1977) ascribed to males one-half of teeth from each tooth generation (there were more long teeth; figs 2 and 3 of Zherekhova 1977). I do not concur with her statement and believe that her method of investigating sexual dimorphism is doubtful. In her diagrams, only two separate groups of points can be recognized as M3 (fig. 3 of Zherekhova 1977). But she did not separate upper and lower teeth, so even for last teeth generations, her conclusions are doubtful.

There are some records of small teeth without a trace of pressure from the following tooth. Such teeth are often recognized as M3 of females or dwarfish individuals of mammoths. But the small number of lamellae on these teeth (14–19) show their M2 nature rather than M3. Absence of a pressure trace on these teeth may be explained by records of mammoth mandibles with bone layers up to 3 cm thick between teeth (Alexeeva 1980). Based on this information, I believe that isolated mammoth teeth are not useful for sex determination.

MANDIBULAE

Kornietz (1962) proposed the length of symphysial process for sex determination. Two forms of mandible were recognized by her, based on mammoth material from the palaeolithic site of Mesin (Ukraine). These types are: (1) a low mandible, narrow at its anterior end and having a long and narrow symphysial process; and (2) a wide and high mandible with a short, narrow process. Mandibles of the first type were ascribed by her to males; mandibles of the second type to females (Kornietz 1962). These two forms were also found in mammoth material from the Berelekh mammoth 'cemetery' (Vereshchagin 1977*a*), from the palaeolithic site Kostenki, Voronezhskaya oblast (Urbanas 1980) (Fig. 26.7), and from the late Pleistocene mammals locality, Myssy on the Kama River, Tataria (author's data).

ATLAS AND EPISTROPHAEUS

An adult male mammoth had a relatively larger atlas and epistrophaeus (the area of cervical muscle attachment) than a

Table 26.2 Measurements (in mm) of crania of mammoth individuals numbered 3–8 in Table 26.1 (see Fig. 26.6 for locations of measurements 1–15)

Measurements	Individuals 3	4	5	6	7	8
1	541	—	580	550	510	589
2	700	716	700	753	667	786
3	935	1077	1170	1050	931	1172
4	853	922	830	930	787	978
5	371	—	470	410	313	522
6	550	—	720	670	530	655
7	518	—	680	695	524	661
8	513	—	660	654	520	676
9	250	—	320	302	259	270
10	420	438	470	410	405	432
11	429	—	610	—	445	490
12	290	—	420	454	347	422
13	256	326	330	322	236	307
14	325	452	330	450	285	445
15	75	126	150	156	89	221

Table 26.3 Measurements (in mm) of some postcranial elements in mammoth individuals numbered 1–9 as in Table 26.1

Element	Mammoth individuals 1	2[b]	3	4	5[c]	6	7	8	9
A	—	275	—	379	—	—	267	374	388
B	250[a]	745	770	—	840	810	726	838	995
	–/–	–/–	+/+	–/–	+/+	+/+	+/+	+/+	+/+
C	260[a]	—	575	—	642	660	548	674	761
	–/–	+/–	+/–	+/–	+/–	+/–	+/+	+/+	+/+
D	210[a]	—	592	—	683	—	593	711	824
	–/–	+/–	+/–	–/–	+/–	+/–	+/+	+/+	+/+
E	—	—	945	—	1097	1050	916	1070	1214
	–/–	–/–	+/+	–/–	–/–	+/+	+/+	+/+	+/+
F	—	491	536	—	626	582	—	522	629
	–/–	–/–	+/+	–/–	–/–	+/+	?/?	+/+	+/+
G	—	460	535	476	585	—	—	556	657
	–/–	+/–	+/+	–/–	+/–	?/?	?/?	+/+	+/+

Key to elements illustrated below:
A, atlas, maximal width;
B, humerus, oblique length (from caput humeri to distal epiphysis lateral edge);
C, ulna, length (from semilunare incisura edge to distal end);
D, radius, total length;
E, femur, length (from caput to distal epiphysis lateral edge);
F, tibia, total length (on medial side);
G, fibula, total length.

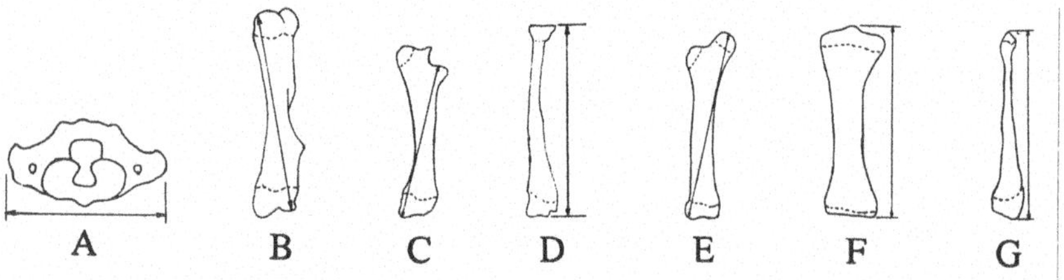

[a] Diaphyses only, after Vereshchagin (1981).
[b] After Dubrovo (1982).
[c] After Zalensky (1903).
The plus and minus signs refer to skeletal maturation in mammoth individuals numbered 1–9 as in Table 26.1. The plus sign indicates that the epiphysis is fused. The minus sign indicates that the epiphysis is not fused, or is separated by a suture. The first sign concerns the upper epiphysis, the second sign, the lower one.

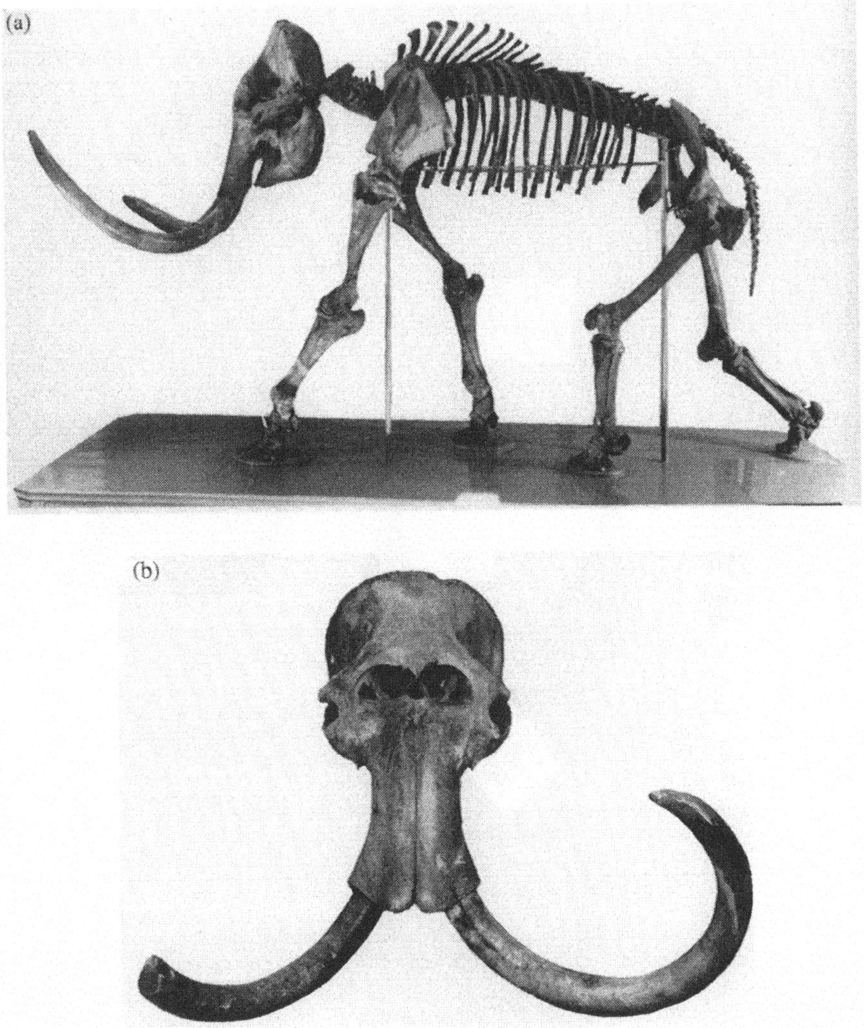

Fig. 26.4 The Taimyr mammoth, ZIN N 2710, male, 45–50 years old. (a) Skeleton; (b) cranium (anterior view). (Photographs by V. E. Garutt.)

female of the same age, a robust dorsal tubercle on the arcus dorsalis (atlas), and an extensive sculpture on the epistrophaeus process (Table 26.3).

LONG BONES OF THE FORELIMBS AND HIND LIMBS

Sex determination from the long bones of the extremities is based on the fact that fusion of the epiphyses (maturation) takes place at approximately the same time in males and females, with achievement of the maximal bone size. According to Baryshnikov *et al.* (1977), relatively small bones with fused epiphyses belong to the skeletons of females, and similar bones without epiphyses, or large bones with fused epiphyses belong to the skeletons of males (my data are given in Table 26.3). A small tibia with fused epiphyses from the palaeolithic site of Eliseevichi, Bryanskaya oblast, attributed to the dwarfish mammoth form *E. primigenius minor* Gromov (Scheglova 1958), possibly belonged to a female of typical mammoth form. The extent of muscle ridge development is also used in sex determination. On the bones of males, muscle ridges are more pronounced than in females of the same age.

PELVES

Female mammoths had broader pelves with larger apertures than males (Lister 1989; personal observations). Lister (this volume, Chapter 25) provides detailed observations on sexing mammoths based on their pelves.

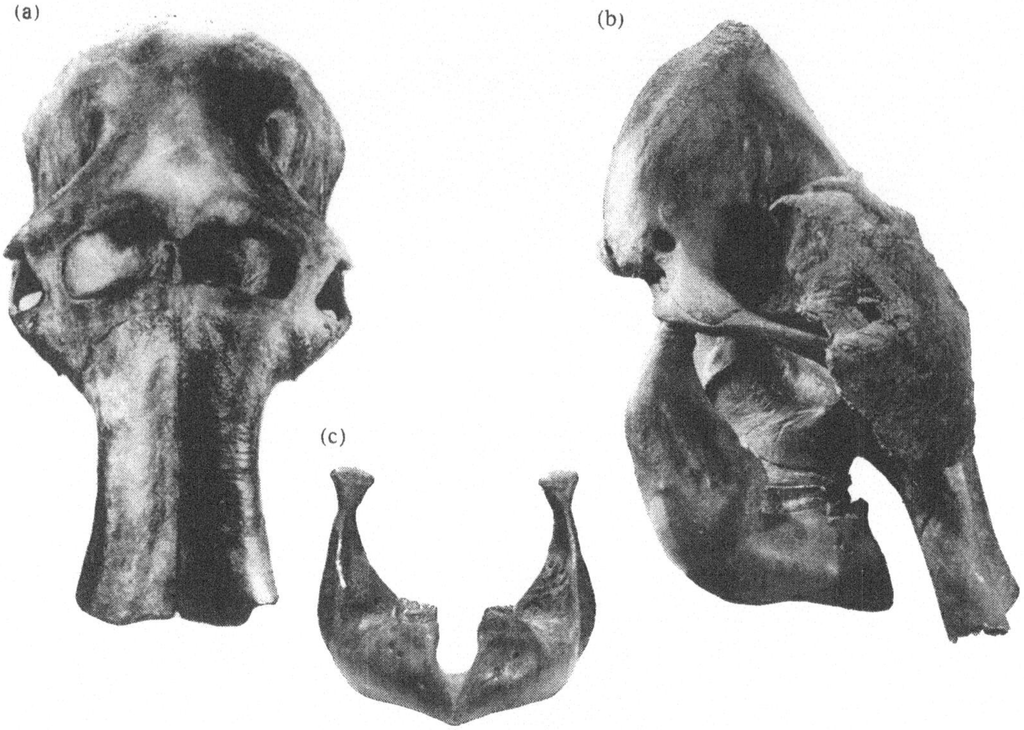

Fig. 26.5 Cranium, (a) anterior view and (b) lateral view, and (c) mandible (anterior view) of the Sanga-Yuriakh mammoth, ZIN N 31738, female 50–60 years old (photograph by V. E. Garutt).

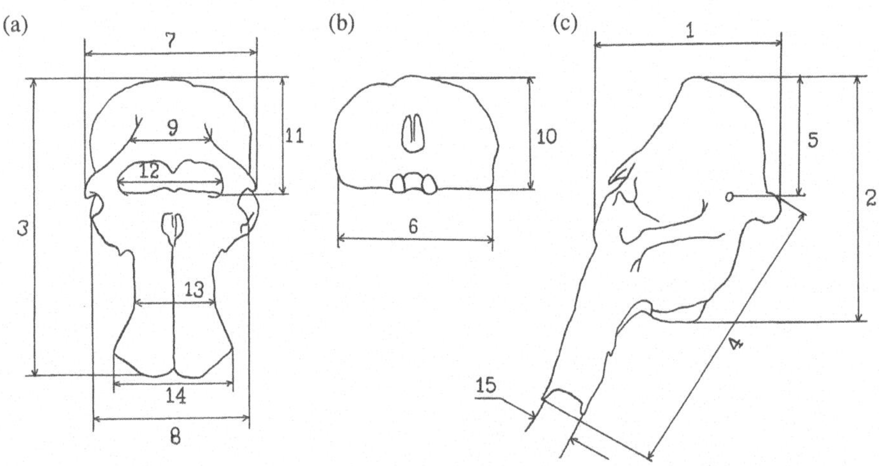

Fig. 26.6 Crania of mammoths, showing locations of measurements 1–15 taken on *M. primigenius* specimens numbered 3–8 in Table 26.2; (a) anterior view; (b) posterior view; (c) lateral view.

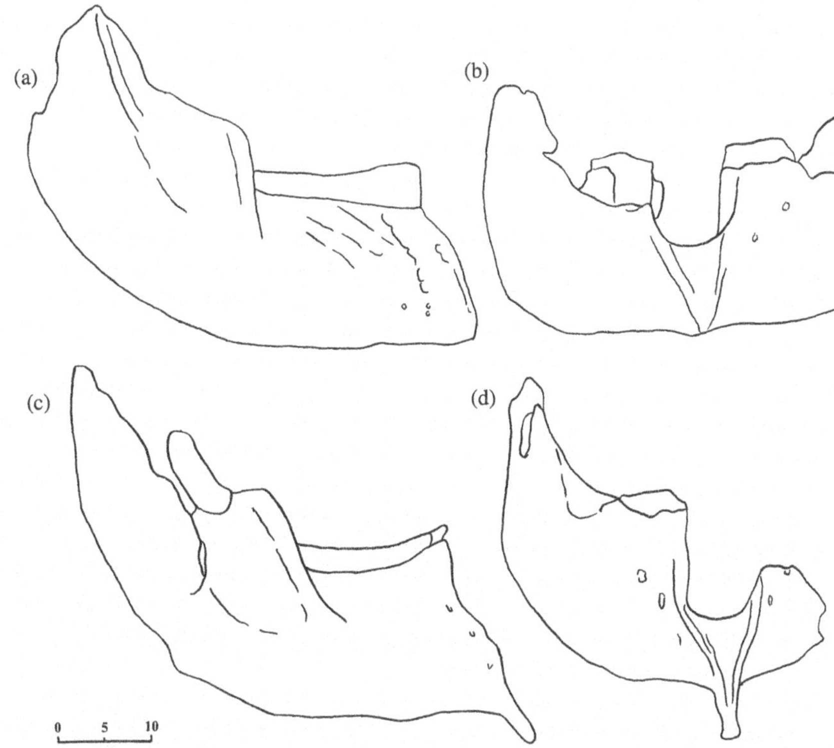

Fig. 26.7 Two types of mammoth mandibles from the palaeolithic Kostenki site (in Voronezh region, Russia) (a), (b) ZIN N31269, with a short symphysial process, possibly a female; (c), (d) ZIN N31274, with a long and narrow process, possibly a male; (a) and (c) are lateral views, (b) and (d) are anterior views. (Drawn after the photographs in fig. 3 of Urbanas 1980.)

SUMMARY AND CONCLUSIONS

Sex determination is more reliable on the basis of a complex of features; some are presented in Tables 26.1–26.3 and Figs 26.1–26.7. Therefore, the sex of an individual represented by a whole skeleton may be determined with a very high degree of accuracy. None the less, isolated skeletal elements such as crania, tusks, atlas vertebrae (all are discussed in this paper), and the pelvic girdle (discussed by Lister, this volume, Chapter 25) have proved to be useful in determining the gender of mammoths.

ACKNOWLEDGEMENTS

I am very grateful to V. E. Garutt for assistance, permission to use some of his data, and for providing photographs (Figs 26.1, 26.2a, 26.3–26.5). J. Shoshani had the original manuscript retyped (by Susan Wolak) and patiently corrected the English. This paper was reviewed by the editors, A. M. Lister, and J. L. Pierce.

PART VI

Palaeoecology, taphonomy, extinction, and conservation

TWENTY-SEVEN

North American Mammutidae

JEFFREY J. SAUNDERS

INTRODUCTION

The North American diversification of elephantoids included representatives of the families Mammutidae, Amebelodontidae, Gomphotheriidae, and Elephantidae (Simpson 1945; Tassy 1985). North American records of these families resulted from one immigration of gomphotheriids (*Gomphotherium*), two of amebelodontids (*Amebelodon*, *Platybelodon*), and one of mammutids (*Miomastodon*) in the Miocene and two of elephantids (*Mammuthus meridionalis*, *Mammuthus primigenius*) in the Pleistocene.

This chapter, focusing on North American mammutids, is a synthesis based in part on recent and ongoing work by the author and collaborators on late Pleistocene *Mammut americanum* (Saunders 1977; King and Saunders 1984; Saunders and Tassy 1989). (Hereafter, *Mammut americanum* is abbreviated as *M. americanum*.) Present in North Africa and Europe in the early Miocene (Rusingan and Orleanian mammal ages, respectively; Savage and Russell 1983; Tassy 1990*b*), gomphotheriids and mammutids appear in North America as immigrants utilizing the Bering Strait in the middle Miocene, *c.* 15.5 Ma, during the early (Green Hills Fauna phase) Barstovian Mammal Age (e.g. *Gomphotherium calvertense*, *Miomastodon merriami*) (Tedford *et al.* 1987). After a consideration of distinguishing features, this chapter presents a brief overview of the preservation mode of North American Tertiary mammutids and contemporary elephantoids and discusses trends in the history and evolving role of later mammutids that are represented by abundant, well-preserved material.

The Mammutidae include *Eozygodon* (early and middle Miocene, Africa), *Zygolophodon* (early Miocene–late Pliocene, Eurasia), *Miomastodon*[1] (middle Miocene, North America), *Pliomastodon* (?middle Miocene–late Miocene, North America), and *Mammut* (late Pliocene–late Pleistocene, North America). Mammutids possess a broad and low cranium and a progressively brevirostrine mandible with a laterally compressed rostrum containing persistent rod-like lower tusks. Typical upcurved and outcurved (and terminally incurved in latest forms) upper tusks are browsing adapted structures suitable for the manipulation of standing trees. In North American mammutids, an enamel band present on the upper tusks of *Miomastodon* is vestigial or absent in *Pliomastodon* and absent in *Mammut*. Cheek teeth are brachyodont in *Miomastodon*, somewhat elevated in *Pliomastodon*, and subhypsodont in *Mammut*; in all forms the transverse valleys are open. Intermediate teeth (dp4–M2) are progressively trilophodont. The M^3 is progressively tetralophodont and the M_3 progressively exceeding true tetralophodonty. In *Mammut americanum* the M_3 may approach the pentalophodont condition (Fig. 27.1). With wear, mammutid teeth present an occlusal surface composed of both enamel and dentine, which when used together contribute to hypsodonty (Fig. 27.1) (Pfretzschner 1990). Enamel and dentine combining as occlusal tissues is a generalized elephantoid trait. This supports the view of Lepper *et al.* (1991), from dietary evidence, that *Mammut americanum* (and earlier mammutids displaying prominent enamel figures, e.g. a mammutid in the Myers Farm fauna, Webster County, south-central Nebraska, of late Barstovian age) may also have been a facultative grazer. Nevertheless, relative to gomphotheriids, trefoil spurs exposed by wear on transverse crests are reduced and transverse crests are simple.

Gomphotheriids possess an abbreviated cranium and elongated rostrum, facial region, and lower jaw. The dentition includes downcurved (occasionally horizontal) and outcurved upper tusks with a broad enamel band and persistent, progressively spatulate lower tusks lacking enamel. Cheek teeth are primarily bunodont. Conules occur in transverse valleys and occlusal surfaces are complex. Intermediate teeth dp4–M2 are trilophodont (e.g.

[1] It is the view of Tassy (personal communication 1992), from the perspective of Eurasian workers, that *Miomastodon* Osborn 1922 is probably a junior synomym of *Zygolophodon* Vacek 1877. Other current workers also view *Pliomastodon* Osborn 1926 to be a junior synonym of *Mammut* Blumenbach 1799. These concerns represent a return to Simpson's view (1945) that *Zygolophodon* and all North American mammutids represent subgenera of *Mammut*. Because recent references of importance to this paper (Savage and Russell 1983; Woodburne 1987 and contributions therein) present stratigraphic information for the forms authored by Osborn as well as for those authored by Blumenbach and Vacek, this chapter also presents them as differentiated at the generic level. (Editors' note: Appendix A provides a list of synonyms of proboscidean taxa.)

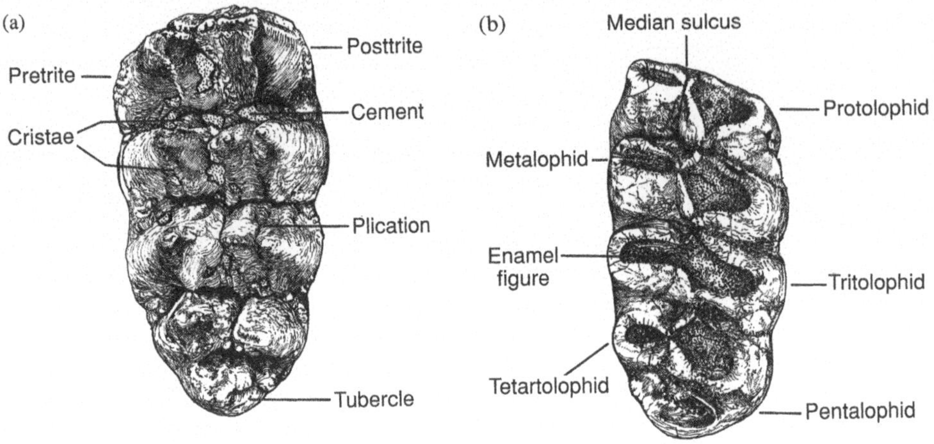

Fig. 27.1 (a) Left M^3 (207BS71) and (b) right M_3 (10UM23HI-210) of *Mammut americanum* (crown views), Boney Spring (a) and Trolinger Spring (b), Missouri. dp4–M2 are trilophodont and M3 is tetralophodont. The fusion of cuspules, which are small projections on the enamel crown, produces cingula on all cheek teeth and tubercles on M3s. A fifth transverse crest may be moderately developed on M_3 (b), or may, especially on M^3 (a), be only incipient, represented by a cone or cones not forming a distinct crest. Transverse crests are formed by a pretrite (internal on upper molars and external on lower molars) and posttrite cusp. Each principal cusp is composed of two conelets: in the pretrite these are the entoconelet (in the upper teeth) or ectoconelet (in the lower teeth) and mesoconelet; in the posttrite these are the ectoconelet (in the upper teeth) or entoconelet (in the lower teeth) and mesoconelet. A median sulcus extends the length of the crown and separates the principal cusps of each transverse crest. Transverse valleys between the crests are generally unrestricted, although distinct valley morphologies have evolved. Pretrites and posttrites are buttressed by cristae arising from the entoconelet in the upper teeth and from the ectoconelet in the lower teeth (cristae are vestiges of conules in antecedent gomphotheriids). At the bottom of each transverse valley the crista descending from the pretrite in front approaches that descending from the pretrite behind, producing a longitudinal ridge running from front to rear over the cusps. This ridge is variably interrupted at the bottom of the transverse valleys and where this interruption is slight, the valleys are obstructed. As teeth wear, the buttressed pretrite cusps are primarily affected and deleted. Wear is a composite of abrasion with mastication and attrition as chewing surfaces are rejuvenated. Wear truncates the pretrite cristae to form enamel figures enclosing simple trefoils. Wear of the posttrite cusps results in enamel figures usually enclosing simple oblongs but plication of the enamel in rugged mastodont teeth may produce small buttresses that form trefoil patterns in late wear stages.

Gomphotherium) with single trefoils, or tetralophodont with double trefoils and subhypsodonty to hypsodonty. M3 may contain 5+ loph(id)s (6+ in *Stegomastodon*).

BARSTOVIAN MAMMAL AGE

Middle Miocene elephantoid samples are of small size. Widely dispersed remains are primarily fragmentary, although whole isolated specimens and a few concentrations of material occur.

As noted by Frick (1933), *Miomastodon* was wide ranging but known examples are rare. Isolated material has been reported from Colorado, California, Montana, and Nevada. The type of *Miomastodon merriami* (Osborn 1921*b*) (early Barstovian) is representative of a better preservational mode. This consists of portions of one individual, including upper tusks with broad enamel bands on the outer surfaces, fragmentary upper molars, two well-preserved lower molars, and bone fragments, from the Virgin Valley Formation, Humboldt County, north-western Nevada. Referred material, including representative dental specimens of three individuals, comes from Pawnee Creek, north-eastern Colorado.

CLARENDONIAN MAMMAL AGE

Late Miocene elephantoid records continue to be isolated teeth and jaws but several partial gomphotheriid skeletons are known. Clarendonian gomphotheriids include *Gomphotherium* and *Eubelodon*. *Miomastodon* continues on the basis of unsatisfactory material and *Pliomastodon* (Shotwell and Russell 1963) appears. Characters differentiating *Pliomastodon* from Barstovian *Miomastodon* include vestigial (or absent) enamel on the upper tusks and transverse crests wearing to expose expanded trefoils on pretrite cusps, in addition to elevated enamel crowns. With the appearance of amebelodontids in the late Clarendonian, there is well-preserved material representative of this elephantoid, including skeletons, and associated faunas.

Inferences of palaeoecological roles depend on the resolution of taphonomic pathways, and a broad similarity of occurrence pattern of amebelodontids and later mammutids is suggested. Modes of occurrence of amebelodontids (see, for example, Andrews 1932, pp. 429–36), often associated with aquatic forms (e.g. *Amblycastor*, Stirton 1934), reflect unique taphonomic pathways and presumably ecology, including, at times, a feeding strategy incorporating aquatic shovelling (Lambert 1992). Amebelodontids possess a shortened cranium and rostrum and elongate lower jaws that are moderately (*Amebelodon*) or greatly (*Platybelodon*) expanded distally. Upper tusks bear enamel and are reduced. Lower tusks, lacking enamel, are elongated, narrow, and closely appressed (*Amebelodon*) or abbreviated, broadened, and transversely truncated or 'scoop-shaped' (*Platybelodon*). The cheek teeth are brachyodont and bunodont, with single trefoils and central conules. Intermediate teeth dp4–M2 are trilophodont and tetralophodont; M_3s exceed true tetralophodonty.

Amebelodontids include *Amebelodon barbourensis* and *Platybelodon barnumbrowni*. *A. barbourensis* (Frick, 1933) (late Clarendonian) is based on a mandible from Christmas Quarry, near Ainsworth, north-central Nebraska. Referred material includes one whole and one partial mandible with teeth and four partial skulls and teeth of individuals of different age classes.

Platybelodon barnumbrowni (Barbour, 1931) is based on a mandible with tusks and molars recovered from gravels of late Clarendonian age along Snake River, Cherry County, north-central Nebraska. Referred material from the type site includes three (?four) skulls and articulated mandibles with complete and well-preserved dentitions, as well as postcranial material comprising three nearly complete skeletons.

HEMPHILLIAN MAMMAL AGE

Later Miocene samples often are of abundant fossil material recovered throughout a nearly complete sequence of exposures from the central Great Plains to Texas, New Mexico, Arizona, and California. Hemphillian records include the continuation of *Gomphotherium* (last appearance is in the late Hemphillian), the appearance of *Cuvieronius* (*C. edensis*, late Hemphillian), earliest certain appearance of *Rhynchotherium* (late Hemphillian), partial skeletons of *Pliomastodon*, entire amebelodontid skeletons, and associated faunal assemblages.

Early Hemphillian material from Snake Creek B Horizon (upper), Sioux County, Nebraska, includes carry-over *G. nebrascensis* and the specific type specimens of *Rhynchotherium anguirivale* (a worn M_3, Osborn, 1926*a*) and *Pliomastodon matthewi* (a right M^3, Osborn, 1921*b*). Rhynchotheres, known primarily from widely scattered, fragmented and/or isolated teeth are, from their first North American appearance (by the late Hemphillian but perhaps as early as late Barstovian, Tedford *et al.* 1987) to their last (late Blancan Mammal Age), among the rarest of all gomphotheriids.

Cuvieronius edensis (Frick, 1921) is based on a fragmentary rostrum and well-preserved upper tusks from the Eden Formation, San Bernadino County, southern California. A spiralling enamel band on the upper tusks, the low and elongate (mammutid-like) cranium, and cheek teeth with single trefoils distinguish the genus (Simpson and Paula-Couto 1957).

Mammutids include *Pliomastodon matthewi* (Osborn, 1921*b*) and *P. vexillarius* (Matthew, 1930). As noted, *P. matthewi* (early Hemphillian) is based on a right M^3 from Snake Creek Formation, level B (upper), Sioux County, Nebraska; paratypic material includes fragmentary molars from two other individuals. Associations include *Procamelus* and *Hipparion*. *P. vexillarius* (late Hemphillian) is based on a partial skeleton, including skull portions and horizontal rami with teeth, from near Coalinga, Fresno County, southern California. Associations include *Plesippus proversus* and *Castor californicus*. Hemphillian amebelodontids include *Amebelodon fricki* (Barbour, 1927) and *A. hicksi* (Cook, 1922). *A. hicksi* (early Hemphillian), from Wray, Yuma County, north-eastern Colorado, is based on a perfectly preserved mandible with M_3s and lower tusks. Recovered skull portions and upper tusk fragments are probably associated with the type individual. The Wray quarry is a rich deposit of well-preserved faunal material. The fossils are concentrated in an area of approximately 6×23 m in a fluvial deposit about 1 m thick. A high diversity of taxa is represented, including other species of elephantoids, the three-toed horses, the amphibious rhinoceros *Teleoceras*, a large species of *Aphelops*, large carnivores, tayassuids, cervids, and camelids. *A fricki* (middle Hemphillian = Hh2, Tedford *et al.* 1987) is also based on a perfectly preserved mandible with tusks and pentalophodont M_3s recovered near Freedom, Frontier County, south-central Nebraska.

Pliocene samples are of well-preserved material associated with diverse local faunal assemblages (Lundelius *et al.* 1987). Of interest in the Blancan is the appearance (early Blancan, *Stegomastodon rexroadensis* Woodburne, 1961) and role (Blancan to early Irvingtonian) of *Stegomastodon*, the continuation of *Cuvieronius* (last appearance in the Rancholabrean Mammal Age), and the origination of *Mammut* (*M. americanum*, earliest Blancan). *Stegomastodon* is characterized by an elongate palate, high but abbreviated skull and jaw (elephantid-like), and upturned tusks without enamel. The molar teeth possess ptychodont crowns with infolded enamel, subhypsodonty, and double trefoils; M3s have progressively 4+ to 7+ loph(id)s. The complex occlusal configuration and subhypsodonty of the cheek teeth and the abbreviated mandibular rostrum are grazing adaptations.

IRVINGTONIAN MAMMAL AGE

Early Pleistocene (*c.* 1.9–0.5 Ma) records include partial and complete skeletons of *Stegomastodon*, *Mammut*, and *Mammuthus*, and many diverse associated faunas. *Mammuthus*, appearing in North America during the early Irvingtonian (Webb *et al.* 1989) seems to have replaced *Stegomastodon* (last occurrence in the early Irvingtonian, e.g. Lindsay *et al.* 1984) but mammutids were unaffected.

RANCHOLABREAN MAMMAL AGE

Late Pleistocene records (*c.* 500 000–10 000 years BP) include *Cuvieronius*, *Mammut*, and *Mammuthus*. Sample sizes, associated with special habitus and taphonomic histories, including cultural death assemblages for each form (e.g. Cotter 1937; Graham *et al.* 1981; Dillehay 1989) pointing to very late Pleistocene occurrences, are often large for the latter two taxa, with high degrees of specimen completeness and excellent preservation.

THE POSITION OF *MAMMUT* IN THE *MIOMASTODON–PLIOMASTODON–MAMMUT* LINEAGE

As noted by Simpson (1945, p. 246) the mammutid lineage may have arisen from a long-jawed, bunolophodont gomphotheriid similar to *Palaeomastodon* (e.g. *P. beadnelli*) from the earliest of the 'Fayûm faunas (early Oligocene)' in Egypt (Savage and Russell 1983, p. 161). Diversification of each lineage occurred in Africa during the Oligocene and the forms were clearly differentiated by their early Miocene appearance in Europe (*Gomphotherium*, *Zygolophodon*).

The earliest North American mammutid, *Miomastodon merriami*, is a migrant stemming from *Zygolophodon turicensis* in Eurasia or from contemporary allied forms (*Z. atavus*, *Z. metachinjiensis*, and *Z. gobiensis*; Tassy 1985, 1990*b*) in that region. In association with other mammals, notably with gomphotheriids, these mammutids may have utilized the Bering Strait during the middle Miocene.

Early mammutid differentiation is exhibited by the lower jaw. Stemming from a Eurasian basal stock in the early Miocene possessing a long mandibular symphysis (*Zygolophodon turicensis* in Europe and close contemporaries in Asia), divergence within the mammutid lineage included a decrease in length of the symphysis and size as well as persistence of the lower tusks. This divergence occurred in parallel during the middle and late Miocene and Pliocene in Eurasia (*Zygolophodon borsoni*) and in North America (*Miomastodon merriami*, *Mammut americanum*) (Saunders and Tassy 1989).

Characters differentiating *Miomastodon*, *Pliomastodon*, and *Mammut* are not considered significant by some authorities (Simpson (1945), viewed the entire evolving lineage to be distinguished by subgenera of *Mammut*) but are given continued emphasis by others. These characters include such primitive traits as the number of crests on M^3 and M_3 of *Miomastodon merriami*, compared with *Mammut*, as well as the form and orientation of the upper tusks of the former, which also bear enamel. Other characters distinguishing *Miomastodon–Pliomastodon–Mammut* have been observed.

As previously noted by Saunders and Tassy (1989), if *Miomastodon merriami* is considered valid, *Mammut americanum* can be viewed to have evolved in place, and the evolution of North American mammutids then parallels that of mammutids of the Old World. This hypothesis, of a single migration of mammutids to North America is most commonly accepted (e.g. Saunders and Tassy 1989). An alternative hypothesis, proposed by Schlesinger (1922, *fide* Kubiak 1972), has been revived by Kubiak (1972). In their view, *Mammut praetypicum*, contemporaneous with *Zygolophodon borsoni* in the late Miocene of Eurasia and sharing with it certain evolved characters of *M. americanum* (morphology of the tusks, brevirostry), but not characters of the cheek teeth of *Z. borsoni*, which are distinct, provides the direct ancestor of *Mammut americanum*. In this competing hypothesis, *Mammut praetypicum*, migrating to North America, replaced *Miomastodon merriami*. The North American Hemphillian (later Miocene) species *Pliomastodon matthewi* can then be viewed to be the descendant of *Mammut praetypicum* and the precursor of *M. americanum*. However, *M. praetypicum* has been considered by some authors to be only a morphological variant of *Zygolophodon borsoni* (Rakovec 1968, *fide* Kubiak 1972), a view supported by at least partial contemporaneity of all known occurrences of the two species and by their frequent association. Furthermore, other authors (Tobien 1976) recognize the differentiated status of *M. praetypicum* but do not view it to be ancestral to *M. americanum*. In this state of understanding, therefore, the second migration of mammutids to North America in the late Miocene remains to be demonstrated.

MAMMUT AMERICANUM

Kerr (1792, *fide* Osborn 1936) described what is now *Mammut americanum* from material recovered 'on the banks of the Ohio [River] ... in a marshy place called Big-bone-swamp'. Named by others before him 'the American Elephant' and given by him the name '*Elephas americanus*', Kerr (1792, *fide* Osborn 1936) noted that the unique cheek teeth were 'covered uniformly with enamel, and furnished with a double row of high conic processes'. Emendations to this characterization (Osborn 1936) included a broad and low cranium, simple upper tusks that were upturned upon leaving the skull, out-turned in the middle, and in-turned at the tips, a brevirostrine mandible with variably persistent lower tusks, and sharp-crowned teeth consisting of two (dp2, dp3), three (dp4–M2), or four + (M3) transverse crests composed of two major cusps. Transverse valleys between the crests were open and, compared to those of gomphotheriids, unobstructed. A prominent median sulcus separates and accentuates the major cusps. The body was low, long, and stocky, of a type called 'mastodontine' by Osborn (1936) (Fig. 27.2).

Fig. 27.2 *Mammut americanum* in artist's reconstruction by R. G. Larson. The head is elongate, low, and broad and the body is long, low, and stocky. Simple upturned tusks, often with wear at the tips and occuring as facets along the curvatures, are browsing adapted.

The Ringold Formation at the White Bluffs, south-central Washington, of earliest Blancan age (Repenning 1987; Tedford *et al.* 1987) contains the earliest record (*c.* 3.75 Ma) of *Mammut americanum*, known from parts of two tusks, right and left M^{1-2} and left M^3, right and left M_{2-3}, and mandible portions in the United States National Museum (USNM19101) that are stated to be indistinguishable from late Pleistocene samples of this taxon (Gustafson 1978). The Glenns Ferry Formation, west of Hagerman, south-western Idaho, contains a slightly later but also early Blancan record of the species (*c.* 3.5 Ma). Other Blancan records are from Santa Fe River IB (late Blancan), northern Florida (Webb 1974*a*), and perhaps from Keefe Canyon, Rexroad Formation (early Blancan) of Kansas (Hibbard and Riggs 1949, where *Mammut adamsi* = *M. americanum*).

During the Irvingtonian, *Mammut americanum* continued in Washington, Idaho, and Florida but expanded northward and eastward to Nebraska, Maryland, and Pennsylvania. In the Rancholabrean the species was widespread, ranging in the west from Fairbanks, Alaska, to Pueblo, central Mexico, and from east to west across the United States, including occurrences on the Atlantic Continental Shelf (Whitmore *et al.* 1967). In the east, the species ranged from Florida northward to New Brunswick and Nova Scotia (Harington *et al.* 1993) The taxon was most numerous, increasingly so with time, in regions bordering the Great Lakes in Illinois, Indiana, Michigan, Ohio, New York, and Ontario (King and Saunders 1984; McAndrews and Jackson 1988; Saunders and Tassy 1989; Shoshani 1989). *Mammut americanum* was associated with man during the very latest Pleistocene (Graham *et al.* 1981). The latest date for the species given in a compilation of last occurrences is 10 395 ± 100 years BP (from Pleasant Lake, south-eastern Michigan; Meltzer and Mead 1983).

ASSOCIATED PLANT REMAINS

Floral indicators of environment associated with *Mammut americanum* are abundant in late Wisconsinan 'heartland' sites in Michigan, Ohio, and Indiana (Dreimanis 1968). Pollen from the skull of the Smith mastodon, central Michigan, dated at 10 700 ± 400 years BP, was 42 per cent spruce (*Picea*) and 35 per cent pine (*Pinus*); other taxa have values of less than 5 per cent (Oltz and Kapp 1963). Pollen associated with the Pitt mastodon in the same region was 27 per cent spruce and 59 per cent pine. Greater pine abundance suggested a later age for this undated specimen (Oltz and Kapp 1963). Pollen and plant macrofossils associated with the Pontiac mastodon, south-eastern Michigan, dated at 11 900 ± 350 years BP, indicate spruce dominance, with low amounts of larch (*Larix*), pine, and some deciduous species (Stoutamire and Benninghoff 1964). Floral remains associated with the Shelton mastodon (Shoshani *et al.* 1989*b*) in the same region suggest the presence of a coniferous forest dominated by spruce, with birch (*Betula*), sedge (Cyperaceae), grass (Gramineae), larch, and cattail (*Typha*), but not pine, dated to 12 320 ± 110 to 11 740 ± 175 years BP. An associated fauna includes beaver (*Castor canadensis*), canid (*Canis* sp.), moose (*Alces alces*), and stag-moose (*Cervalces scotti*), as well as fish, amphibians, birds, a vole, and muskrat.

Pollen from the Orleton Farms mastodon (Sears and Clisby 1952), central Ohio, included a spruce–pine–oak (*Quercus*)–grass assemblage like that associated with the Smith mastodon in Michigan. Two other late-glacial mastodons from Ohio, dated 13 180 ± 520 years BP and 10 654 ± 188 years BP, are also associated with spruce pollen dominance and lesser amounts of pine, fir (*Abies*), sedge, grass, and deciduous taxa (Ogden and Hay 1965, 1967).

At Christensen Bog, central Indiana (Whitehead *et al.* 1982), pollen indicating a spruce-dominated boreal forest containing deciduous species is associated with giant beaver (*Castoroides ohioensis*), mastodon, deer (*Odocoileus virginianus*), and caribou (*Rangifer tarandus*) (Graham *et al.* 1983). Radiocarbon dates indicate this record to be 14 000–12 000 years BP. The Wells mastodon, north-central Indiana, dated 12 000 ± 450 years BP (Gooding and Ogden 1965), occurs with pollen dominated by 35–60 per cent spruce, with fir, birch, ash (*Fraxinus*), and other deciduous taxa represented.

These records in the Great Lakes region show *Mammut americanum* to be associated with spruce dominance, with some fir, larch, and pine pollen. Small percentages, but less than 15 per cent, of deciduous tree pollen are indicated, and grass occurs.

Fine-grained dietary information is provided by samples from *M. americanum* intestines/gut contents and teeth. Other dietary information has been forthcoming from isotopic analyses of dental and skeletal tissues. Intestinal contents of the Burning Tree mastodon, central Ohio (Lepper *et al.* 1991), included twigs of non-coniferous species exhibiting sheared ends and seeds of clover (*Trifolium*), sedges (*Carex* spp.), and naiads (*Najas* spp.) with fewer seeds of pond weed (*Potamogeton*), water-lily (*Nymphaea*), and other aquatic plants. Radiocarbon ages determined from these intestinal contents are 11 660 ± 120 years BP and 11 450 ± 70 years BP. Pollen extracted from molar cementum and caries included sedge and grass, supporting their inclusion in the diet reconstructed from intestinal contents. The joint occurrence of twigs and herbaceous vegetation shown by these analyses suggests that the Burning Tree mastodon was, at least seasonally, a selective feeder, having both browsing and grazing habits.

Preliminary study of the isotopic ecology of *Mammut americanum* and *Mammuthus jeffersonii*, on the basis of extracted collagen reported by Koch (1991), suggests as a point of dietary distinction that *M. americanum* ingested higher amounts of nitrogen-fixing alder (*Alnus*). Consumption of this plant links *M. americanum* to the spruce–sedge–alder parkland inferred for southern Michigan, northern Indiana, northern Ohio, and western New York by 11 000 years BP (Koch 1991). Alder is a pioneer on recently deglaciated terrain (E. C. Grimm, personal communication 1992), supporting the suggestion by Saunders and Tassy (1989) that *Mammut americanum* occupied expansive wet areas in the Great Lakes region and Atlantic Canada from 14 000–10 000 years BP during the terminal Wisconsinan (Fulton *et al.* 1984; Karrow and Calkin 1985), where it may have had semiaquatic habits (*vide* Hallin 1983).

ASSOCIATED FAUNAL REMAINS

At Saltville, south-western Virginia, in the Atlantic Coastal region, a late Pleistocene fauna of browsing ground sloth (*Megalonyx*), mastodon, horse, ?stag-moose, caribou, bison, and woodland musk ox (*Symbos*) occurs with spruce and pine pollen, indicating a boreal parkland at 13 450 ± 420 years BP (Ray *et al.* 1967; Guilday *et al.* 1975).

Sites such as Saltville, Shelton, Christensen Bog, and others indicate that in late Pleistocene deposits in the Great Lakes region and elsewhere, *Mammut americanum* is often associated with known or presumed browsers, suggesting the occupancy of woodlands or parklands as well as aquatic environments: beavers (*Castor canadensis, Castoroides ohioensis*), deer (*Cervus elaphus, Odocoileus, Alces alces, Cervalces scotti*), and musk oxen (*Bootherium bombifrons* and *Symbos cavifrons*). In the western marginal areas of their distribution (e.g. California, Arizona, Texas, central Mexico, and south-western Missouri) as well as in Florida (Webb and Newsom 1991), reported mastodons are associated with grazers, suggesting the occupancy of savannas, woodlands, and parklands: grazing ground sloths (*Glossotherium harlani*), Jefferson's mammoth (*Mammuthus jeffersonii*), horses (*Equus* spp.), bison (*Bison antiquus*), and antilocaprids (*Capromeryx arizonensis*).

OZARK HIGHLAND RECORDS

Records of *Mammut americanum*, increasing knowledge of trends in their ecology and evolution, occur in the Ozark Highland, a karstic region in the central United States. In this region during the Pleistocene, artesian springs, initiated during wet stadial intervals, formed ponds on developing terrace surfaces. During subsequent drier interglacials and interstadials, these ponds succeeded, with diminishing rates of spring discharge, to peat bogs in which the remains of contemporary plants and animals accumulated. Five mastodon-containing biotic assemblages excavated by the author and colleagues from three artesian springs along an 11 km section of the Pomme de Terre River in the western Ozark Highland, Missouri provide a punctuated record of geological and biological history during the past 130 000 years (King 1973; Saunders 1977, 1988; King and Saunders 1984; Haynes 1985) in the western marginal region of the mastodon's late Pleistocene distribution. These records permit our examination of environmental changes through time *at one place* that are tied to the local evolution and ecology of *Mammut americanum* away from its later centre of distribution.

The earliest record of *M. americanum* in the Ozark Highland sequence is from the lowest of three superimposed peat lenses in Jones Spring (unit c_1). Associated fauna includes the large box turtle (*Terrapene carolina putnami*), alligator (*Alligator mississipiensis*), grazing ground sloth, giant beaver, sabertooth (*Smilodon fatalis*), Jefferson's mammoth, horses (*Equus complicatus* and *E. calobatus* or *E. hemionus*), tapir (*Tapirus veroensis*), camel (*Camelops* sp.), and giant bison (*Bison latifrons*) (Saunders 1988). Warmth-loving alligators and *B. latifrons* occurring elsewhere in North America in the last interglacial suggest that the unit c_1 biotic assemblage, with a radiocarbon age of >60 000 years BP, is of Sangamonian age. Macrofossils of maple (*Acer*), plum (*Prunus*), oak, and hickory (*Carya*) indicate a temperate deciduous vegetation.

Unit c_1 in Jones Spring is succeeded by unit c_2, containing box turtle, mastodon, Jefferson's mammoth, horses, smaller bison (*Bison antiquus*), and woodland musk oxen (*Symbos cavifrons* or *Bootherium bombifrons*). Radiocarbon ages for associated wood fragments of 51 000 ± 1500 years BP and 48 900 ± 900 years BP indicate depositon during the early to middle Wisconsinan. A pollen record of 20–30 per cent pine, 10 per cent oak, sedge, and grass indicates cool conditions. The herbs suggest parkland areas.

Unit e succeeds unit c_2. Unit e contained mastodons (18 individuals, 46 per cent of total mastodon sample in Jones Spring), Jefferson's mammoth (one individual, 10 per cent of total mammoth sample in Jones Spring), horses (*E. complicatus*), smaller bison, and woodland musk oxen (58 per cent of the total woodland musk oxen sample in Jones Spring). A radiocarbon age of >35 320 years BP for entombing peat is also applied to the fauna. Floral indicators were not recovered in unit e but the fauna suggests coniferous parkland and cooler or cold conditions at this time.

Fig. 27.3 The bone bed in Boney Spring, Benton County, Missouri. The sediment body (unit E, alluvial clay) containing the faunal accumulation is lenticular in section and *c.* circular in plan, with a diameter of *c.* 10 m and a maximum thickness of *c.* 0.8 m; the sediment body and the fossil concentration are coincident. Clastic particles in unit E are either chert, wood, or moss fragments, and bones. The faunal material in the bone bed occurred like 'logs in a jam'. Bones rested directly on each other and the uniform clay matrix was often restricted to thin interstitial spaces among the bones. The bone-free inner and western areas are not due to the absence of bones but to unsystematic excavations conducted prior to 1971 (from Saunders 1977). This and other mass occurrences of *M. americanum* (Riggs 1936) recall *Platybelodon grangeri* records in Tung Gur deposits (Astaracian = middle Miocene) in central Mongolia described by Andrews (1932) and suggests ecological similarities. Tung Gur records are primarily from quarried bog deposits. One of these contained 16 jaws, six skulls, and much postcranial material. All but at most three of the individuals were adult and faunal associates were unrepresented. This contrasted with the occurrence in another Tung Gur quarry of as many as eight juveniles and one adult *Platybelodon* in association with the remains of rhinocerids, bovids, carnivores, rodents, and horses. In contrasting the recovery of *P. grangeri* and *Gomphotherium gobiensis* (= *Zygolophodon gobiensis* of the Z. *'turicensis* group', Tassy 1990*b*) from Tung Gur deposits, Andrews (1932) noted that the type jaw of the latter is the only other proboscidean material found in the region (up to 1931), suggesting that either the gomphotheriid was rare or, more likely, that its habitat was different. *Platybelodon* is frequently associated with freshwater bivalves most commonly occurring in the shallow water (R. E. Warren, personal communication 1992) and its broad, flattened lower tusks are inferred to have been adapted for exploiting aquatic vegetation in shallow zones (Andrews 1932, plate CIV). The small, rounded tusks of *G. gobiensis*, coupled with its relative rarity, suggest other adaptations and preferred habitat. The occurrence of 31 American mastodons in Boney Spring is not necessarily a function of habitat but of the availability of water, a limiting resource for extant elephantoids. However, some data suggest that *M. americanum*, like *Platybelodon*, occupied and exploited wetlands.

This cooler or cold late phase of the middle Wisconsinan interstadial is also recorded by a peat lens in Trolinger Spring, 150 m east of Jones Spring. Radiocarbon ages for the non-diverse assemblage are between 38 280 ± 3430 years BP and 29 340 ± 900 years BP. Mastodons (15 individuals), stilt-legged deer (*Sangamonia fugitiva*), and woodland musk oxen (*Symbos cavifrons*) are represented. Associated pollen and plant macrofossils indicate a pine parkland.

The latest Pleistocene accumulation encountered in our Ozark Highland sequence occurred at Boney Spring, 5 km north-west of Jones Spring (Fig. 27.3). The late glacial fauna included 31 mastodons, in association with extinct (*Glossotherium harlani*, *Castoroides ohioensis*, *Equus* sp., and *Tapirus veroensis*) and modern forms. Two radiocarbon dates on organic material from the pulp cavities of dispersed mastodon tusks at the top of the deposit date the dissociation episode to 13 625 ± 500 years BP. Well-preserved botanical remains in these tusk fillings indicate a spruce woodland with deciduous species.

MORPHOLOGICAL RESPONSE TO ENVIRONMENT

The record of punctuated accumulational episodes revealed in the Ozark Highland excavations provides the basis for a model of morphological response to changing environment in *Mammut*

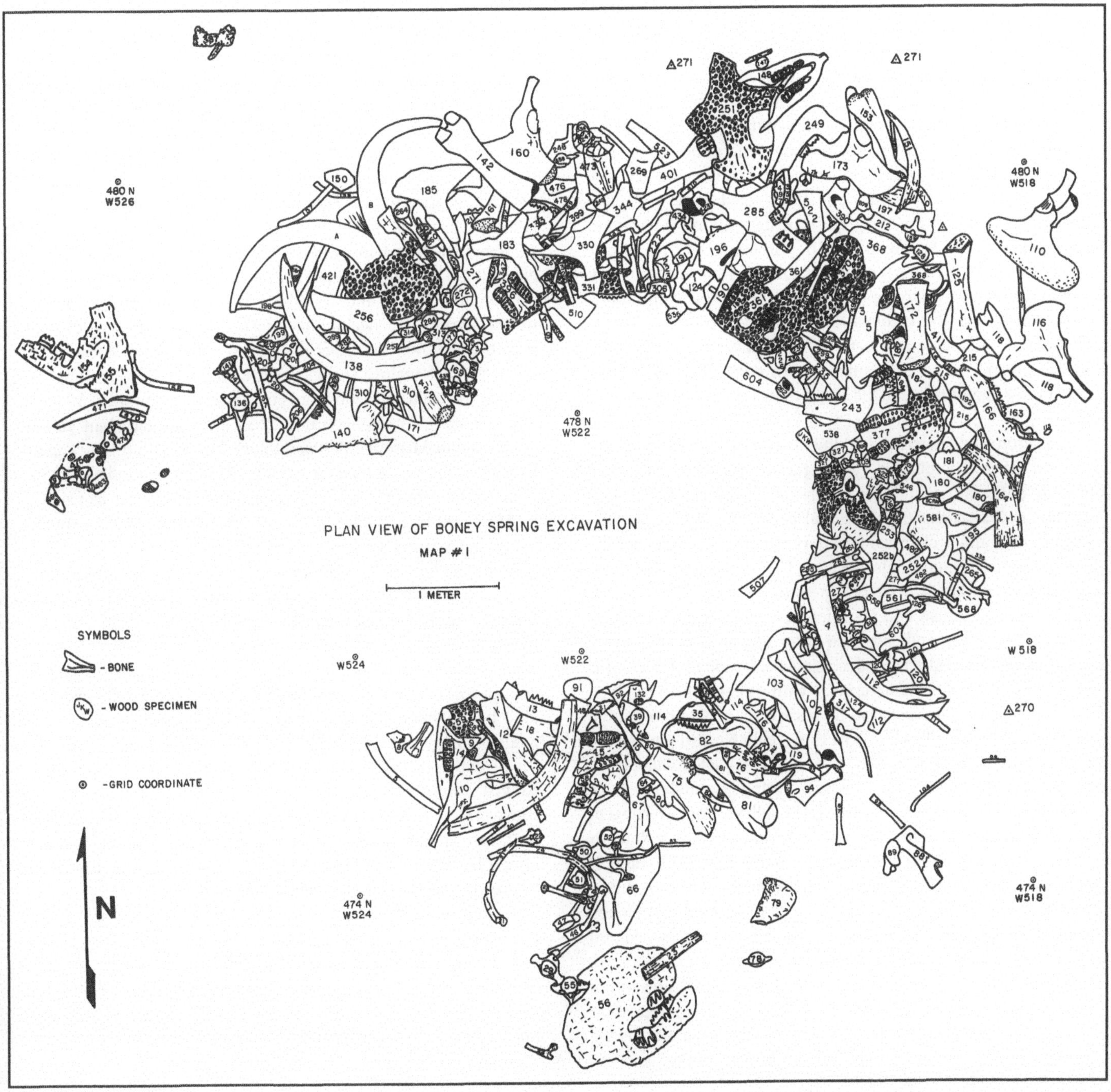

americanum that is presented in this section. The model is corroborated by records from other geological intervals and regions, including records of other elephantoid taxa. The model is, moreover, falsifiable.

Of all elephantoids, the mammutids are the most conservative (Tassy 1985). From their earliest (*Eozygodon morotoensis*, early Miocene, *c.* 22 Ma, East Africa, Tassy 1990*b*) to their most recent occurrence (*M. americanum*, Pleasant Lake, North America), mammutid molar morphology is little changed, compared to considerable dental transformation in gomphotheriid and elephantid lineages.

Leidy osberved (1869, p. 242) two morphological extremes or varieties among large series of *Mammut americanum* teeth: (1) a 'smooth' variety with unobstructed transverse valleys, smooth enamel, and weak or only moderately developed cingula, contrasted with (2) a 'rugged' variety with transverse valleys that are obstructed in their bottoms by the joining of cristae descending from the anterior and posterior faces of the primary cusps; in addition, the enamel crown is plicate and choerodont and cingula are well developed. From direct observations, teeth of the earliest *Mammut americanum* are of the rugged variety. A fragmentary right dp^4 from early Blancan deposits at Hagerman, Idaho (University of Michigan 49946), although unerupted, was sturdily formed. The anterior and posterior margins exhibit very strong cingula and a strong cingulum is also present lingually. Pretrite cristae on the tritoloph are well developed and strong. Compared to homologous teeth in later mastodons from the Ozark Highland the tooth is small (width across the tritoloph = 50.3 mm), suggesting that the Hagerman mastodons were also small.

Pollen from the Hagerman sediments reported by Wright and Leopold (1982) is dominated by pine (50–80 per cent) with small amounts of fir, spruce, juniper (*Juniperus*), hemlock (*Tsuga*), and Douglas fir (*Pseudotsuga*). Occasional peaks of grass, sagebrush (*Artemisia*), and chenopods–amaranths (Chenopodiaceae–Amaranthaceae undiff.) also occur. These indicate that pine woodland or parkland occurred in south-western Idaho in the middle Pliocene. Pollen and plant fossils from Miocene sediments in the same region indicate that drier steppe occurred there during earlier intervals. This succession from xeric to mesic floras suggests that *Mammut americanum* with rugged teeth evolved from antecedent *Pliomastodon* in response to pine-dominated woodland or parkland habitats in the Pacific North-west, where the earliest records of the taxon occur. *Mammut americanum* teeth in Jones Spring unit e and Trolinger Spring assemblages have rugose enamel and relatively obstructed valleys, and are rugged. They contrast with teeth from Boney Spring possessing smooth enamel and unobstructed valleys. *M. americanum* in Trolinger Spring deposits and, from inference those in Jones Spring unit e as well, occupied pine parkland *c.* 39 000 years BP. At Boney Spring 13 500 years BP, mastodons inhabited spruce woodland (primary full-glacial) with deciduous species (emergent Holocene).

Rugged teeth from middle Wisconsinan interstadial sites in the Ozark Highland exhibit greater wear relative to their progression through the jaws than do smooth teeth in jaws from late glacial stadial deposits. In a current study (J. J. Saunders, J. R. Purdue, B. K. Dawson, unpublished data) this is associated with the accelerated wear of teeth in the mastication of pine or plants associated with pine.

The absolute size of individual *M. americanum* also varies with the inferred environment of Ozark Highland fossil assemblages (King and Saunders 1984). Individuals from late Wisconsinan stadial deposits were, on average, larger than individuals from middle Wisconsinan interstadial deposits, on the unambiguous basis of recovered postcranials and on the inferred basis of larger teeth. Length and width measurements of $M^2/_2$ of *M. americanum* reported by King and Saunders (1984, figs 3–4) from Jones Spring (primarily unit e) and Trolinger Spring are, on average, smaller than those from Boney Spring.

Length and width measurements of M2 of *M. americanum* from late Wisconsinan stadial localities in eastern Missouri (Kimmswick, Jefferson County) and western Tennessee (Nonconnah Creek, Selby County) support this distinctiveness. The undated Kimmswick record of a Clovis culture– *M. americanum* association (Graham *et al.* 1981) presumably dates to the interval 11 200–10 750 years BP, on the basis of cultural association. At Nonconnah Creek, pollen and plant macrofossils associated with *M. americanum* are dated at 17 195 ± 505 years BP (Delcourt *et al.* 1980). The pollen, 75 per cent spruce, 4 per cent pine, <2 per cent fir and larch, and 7–16 per cent oak, is interpreted as a mixed spruce and deciduous forest. Macrofossils include temperate deciduous taxa: beech (*Fagus*), tulip-tree (*Liriodendron*), hickory, black walnut (*Juglans nigra*), and oak.

Based on tooth measurements obtained by the author, *Mammut americanum* from Nonconnah Creek was larger than middle Wisconsinan Missouri *M. americanum*, but is within the range of the Boney Spring sample reported by King and Saunders (1984). A mastodon from the Clovis level at Kimmswick was as small as middle Wisconsinan mastodons in the Ozark Highland sequence (King and Saunders 1984).

From the foregoing comparisons, it appears that rugged teeth in pine-dominated environments, a trend since early Blancan, indicate that *Mammut americanum* fared less well in these habitats relative to more optimum conditions (smooth teeth, large size, e.g. Boney Spring) provided by late Pleistocene spruce-dominated habitats. Rugged teeth in mastodons occupying high-pine habitats compensated partially for the accelerated wear of enamel in these environments. Because rugged teeth represent the primitive condition (cristae, which contribute prominently to rugged tooth morphology, are vestiges of conules in the transverse valleys of antecedent primitive elephantoids), it follows that spruce-dominated coniferous forests or woodlands of the later Pleistocene were distinctive, compared to the coniferous forests or woodlands of the middle Pliocene.

Midcontinent *Mammut americanum* were of small size during the middle Wisconsinan interstadial. *M. americanum* increased in size during the Wisconsinan full-glacial (a dramatic shift to increased size presumably occurred between *c.* 39 000 and 18 000 years BP, based on samples from Trolinger Spring/Jones Spring unit e and Nonconnah Creek), and seem, on the basis of inadequate samples only, to have begun decreasing in size by the terminal portion of the late-glacial (Kimmswick), just prior to their extinction. Heintz and Garutt (1965) interpret a trend in forelimb stature of Siberian mammoths (*Mammuthus primigenius*) during the same dated interval, 47 500–11 200 years BP, as a morphological response to favourable (large size) and less favourable (small size) environments. Although these Siberian data have been criticized on the bases of small sample size and developmentally early radiocarbon analyses, the Ozark Highland record of *Mammut americanum* supports Heintz and Garutt's conclusions for Siberian *Mammuthus primigenius*.

CONCLUSIONS

Mammut americanum was the last surviving member of a lineage first appearing *c.* 22 Ma in Africa. *M. americanum* appears to have evolved in place in North America, from clearly conservative differentiation initiated in the middle Miocene by immigration across the Bering Strait of *Miomastodon*, similar to, and stemming from, a mammutid inhabiting Asia *c.* 15 Ma. Divergence within North American mammutids paralleled that of the Eurasian mammutids (e.g. *Zygolophodon turicensis*, *Mammut (Z.) borsoni*) in such morphological characters as progressive brevirostry, reduction and loss of the mandibular tusks, and in browsing-adapted features of the upper tusks.

North American mammutid records predating the Blancan are rare and palaeobiologically unsatisfactory. By the Blancan a taphonomic pattern is clearly evident: independent of time period, elephantoid specimens and the diversity of associated local faunal assemblages increase as inferred aquatic habits become established (amebelodontids, ?Pleistocene mammutids). This representation by well-preserved material is a primary feature enabling the resolution of trends in the history and evolving role of, for example, later mammutids.

Mammutids did not occur beyond the late Pliocene in Eurasia while in North America divergence within *M. americanum* continued until the late Pleistocene with microevolutionary patterns of change.

Mammut americanum appears primarily to have been a browser on trees, shrubs, and/or herbs associated with Pleistocene forests, woodlands, parklands, savannas, and wetlands, although facultative grazing is also indicated from isotopic analyses. Pollen and plant macrofossils associated with *M. americanum* in spring-deposited sequences in the western Ozark Highland amplify presumed habitat preference suggested by previous studies in the mastodon heartland of the Great Lakes region. These western records clearly associate the species with pine parkland and mixed deciduous and pine woodland or parkland during Wisconsinan interstadials and with deciduous forest, woodland, or savanna during the Sangamonian interglacial. During the terminal late Wisconsinan, *M. americanum* in the Ozark Highland was associated with spruce forest or woodland with deciduous species.

Morphological response to environmental differences resulted in local evolutionary changes in later mammutids. Middle Wisconsinan *M. americanum* associated with pine parkland in the Ozark Highland sequence possessed rugged teeth and small stature (Jones Spring unit e, Trolinger Spring); *M. americanum* associated with late Wisconsinan full-glacial spruce woodlands or forests in the Ozark Highland and Central Mississippi Valley had smooth teeth and were large (Boney spring, Nonconnah Creek). By the terminal late-glacial, *Mammut americanum* in the Central Mississippi Valley may have been tending toward a smaller size (Kimmswick). Small size, and trends towards smaller size, in elephantoids are associated with less favourable habitat in this study and others. This morphological response hypothesis is falsified if (1) smooth teeth/small mastodon- or rugged teeth/large mastodon-dominated assemblages are recovered, or (2) if smooth teeth-dominated assemblages and rugged teeth-dominated assemblages are recovered in association with floral indicators of dominantly pine and dominantly spruce vegetations, respectively.

Considering the broader implications of mastodon habitats, morphologies, and adaptive strategies revealed in Ozark Highland excavations, environmental insularity explains the extinction of *M. americanum*. Biotic records from spring sites in this region include discrete temporal slices of mastodon populations that illuminate the time- and space-transgressive events culminating in the extinction of *Mammut americanum*.

ACKNOWLEDGEMENTS

This chapter was reviewed by D. C. Fisher, W. D. Lambert, and the editors.

TWENTY-EIGHT

Distribution and palaeoecology of central and western North American *Mammuthus*

LARRY D. AGENBROAD AND JIM I. MEAD

NORTH AMERICAN MAMMOTH: AN OVERVIEW

North American mammoths not only endured varied climatic conditions during the Pleistocene, they radiated biologically and geographically. The early Irvingtonian species *Mammuthus meridionalis* (Nesti, 1825) was followed by *M. imperator* (Leidy, 1858) in the late Irvingtonian, and succeeded by *M. columbi* (Falconer, 1857), *M. primigenius* Blumenbach, 1803, and *M. exilis* (Stock and Furlong, 1928) in the Rancholabrean (following Maglio 1973; Agenbroad and Barton 1991). These latter species were contemporary but adapted to different habitats, with their distribution covering most of the North American continent. Highly mobile and migratory, adapted to grasslands (which were expanding) or steppe-tundra (which was diminishing), it seems highly unlikely that habitat mosaic change alone could cause extinction of the genus at the end of the Pleistocene.

Two localities in South Dakota (Fig. 28.1) and multiple localities in the Colorado Plateau portion of Utah (Fig. 28.2) have recently provided interpretive data. These data sets enhance our understanding of the mammoth palaeoecology during mid-Wisconsinan and late Wisconsinan time in central and western North America.

The 26 000 years BP Hot Springs, South Dakota site provides a large sample of mammoths clustered within a tight elephant age-group (Agenbroad and Mead 1987*a*, 1994; Agenbroad 1990) in the setting of a unique geological/hydrological natural trap. In addition to the mammoths, a large sample of microfauna, invertebrates, and pollen provide sufficient data to reconstruct a palaeoenvironmental model at this locality, approximately 26 000 years ago.

The Colorado Plateau deposits provide multiple, dated localities containing mammoth dung from +28 000 years BP to within 700 years of the extinction of this genus at ±11 000 years BP. The fossil dung deposits allow dating of discrete samples and reconstruction of the local environment at that time.

MAMMOTHS IN SOUTH DAKOTA

The Mammoth Site of Hot Springs, South Dakota (Fig. 28.1) was discovered in 1974 and partially excavated in the 1974–79; 1983; and 1986–91 field seasons. To date, approximately 30 per cent of the deposit has been excavated (Agenbroad 1984*a*; Agenbroad *et al.* 1990). The deposit consists of a sink-hole trap containing an extinct artesian spring and pond. Surficial geology, forming the walls of the sink-hole, was Permian–Triassic Spearfish Shale. A greater than 60° slope of the wallrock and submoistening of the 2 m interval above the pond surface was apparently sufficient to entrap mammoths, which expired by starvation, or drowning. Deposition at the site was active over a period of 500 to 1000 years of accumulation and in-filling, until the trap was no longer functional.

At the close of the 1991 field season, 42 individual *M. columbi*, two individual *M. primigenius*, one short-faced bear (*Arctodus simus*) and a variety of smaller fauna had been recovered. We have estimated that more than 100 individual mammoths will be discovered *in situ*.

Analysis of the age distribution of the mammoth population (44 by tusk count) at the close of the 1991 field season provided individual ages for up to 52 animals (by dentition) at the time of death (Fig. 28.3). This sample indicates that the sinkhole trap was selective for subadult and young adult mammoths (Agenbroad 1990; Agenbroad and Mead 1987*a*; Haynes 1987, 1991). This high incidence of entrapment and mortality for that age-group is most probably due to behavioural habits and activities (Moss 1988).

The deposits at the Mammoth Site represent the first known Pleistocene deposits within the Black Hills uplift. Prior reconstructions for the region, from distant (*c*. 260 to 650 km) data sources, have concluded that a spruce–pine forest was present in this area, which lay just south of the Wisconsinan ice sheet. Analyses of the invertebrates, microvertebrates, and pollen from the sinkhole fill allow us to reconstruct the local palaeoenvironment at 26 000 years BP for the Southern Black Hills physiographical province of South Dakota, providing an alternative

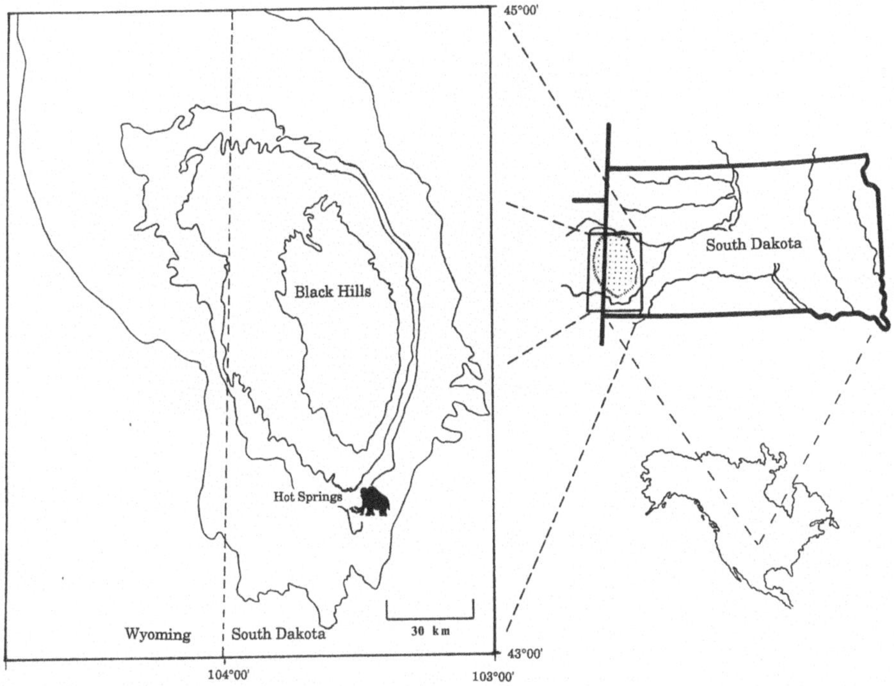

◀ **Fig. 28.1** Location map for the Mammoth Site of Hot Springs, South Dakota, USA.

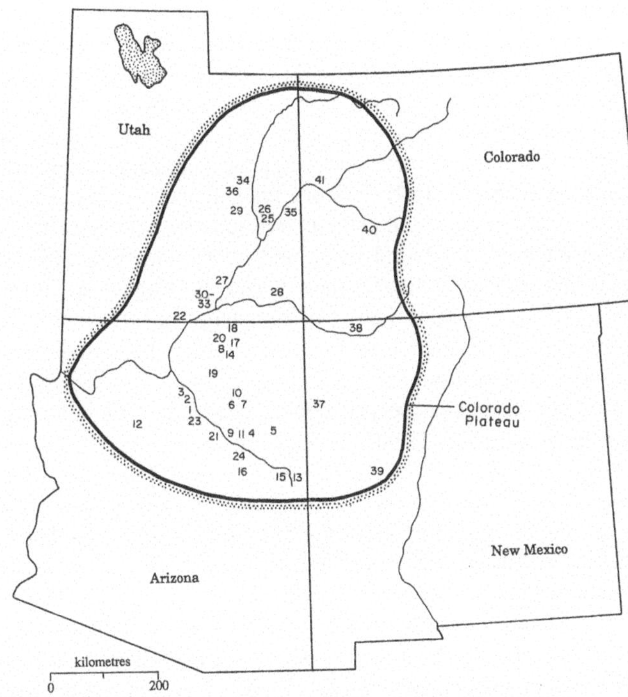

Fig. 28.2 Location map for mammoth localities on the Colorado Plateau.

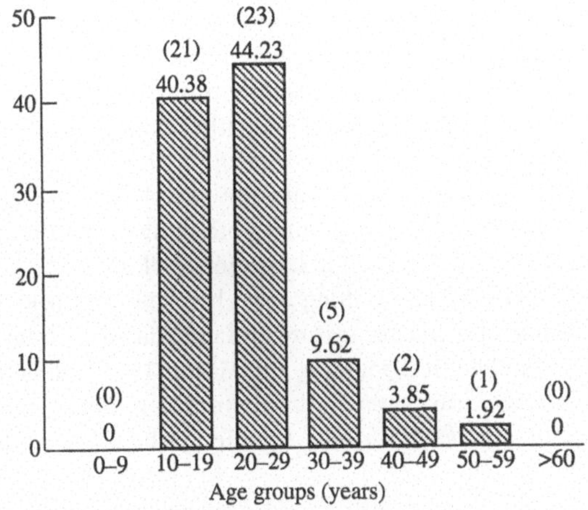

Fig. 28.3 Age–structure analysis of *M. columbi* from Hot Springs, South Dakota (after Laws 1966 and Jachmann 1985).

conclusion, i.e. that the area adjacent to Hot Springs was a steppe–grassland.

Grasses, sedge, and aquatic plants comprise ±70 per cent of the pollen types recovered. Of the grasses, 75 per cent are terrestrial in origin. Other non-arboreal pollen grains include long- and short-spine composites, sagebrush, Cheno-Ams, mustard, parsley, rose, crowfoot, saxifrage, and rumex. Arboreal pollen includes both deciduous trees and evergreens. Ferns, bryophytes, and free-sporing vascular plants were also observed (Mead *et al.* 1990*b*).

Molluscs represent both freshwater (predominate) and terrestrial varieties. Combined temperature tolerance data for plants, molluscs, and fish from the deposit indicate a water temperature of less than 40 °C, which may reflect the western, low energy (cooler) portion of the deposit. Laury (1980) estimated a temperature of at least 35 °C from ostracods and chemical data.

In addition to the Hot Springs Mammoth Site, the recently reported Lange–Ferguson Site documents a mammoth kill at 10 670 years BP (Hannus 1985, 1990; Martin 1987) in the nearby Badlands of the White River. Combining these sites and several others for Wyoming (Walker 1987), South Dakota, and Iowa (Semken and Falk 1987), it appears that a steppe–tundra or steppe–grassland environment changed to a grassland or montane environment about 11 000 to 12 000 years ago, then to grasslands at 10 000 to 11 000 years ago in most localities across this transect. Modern conditions were established as late as 8500 years ago to the east in Iowa.

MAMMOTHS ON THE COLORADO PLATEAU

Beginning in 1983, numerous deposits containing mammoth remains were recovered from the central region of the Colorado Plateau (Agenbroad and Mead 1987*b*, 1989), an area from which little evidence of mammoths had previously been reported (Fig. 28.2). Investigation of Bechan Cave in the rugged canyon country of the central Colorado Plateau, uncovered a large deposit of mammoth dung (Agenbroad *et al.* 1984; Davis *et al.* 1984, 1985; Mead *et al.* 1984, 1986), containing the first pristine and unquestionable mammoth dung boluses from the New World.

Attention of the exploring party was drawn to dry dung fragments exposed in the walls of shallow holes dug by pot hunters in the sandy floor of the cave. Inspection revealed that the fragments were rich in graminoid culms, rather than in browse fragments typical of *Nothrotheriops shastensis* (the Shasta ground sloth) dung. In shape and texture the sample was unlike the numerous samples of Shasta ground sloth dung in the University of Arizona palaeoenvironmental laboratory collection.

Finely chewed ruminant dung could be eliminated from comparison. While of a coarser texture and dominated by fragments 0.3–2.0 mm in length (Spaulding and Martin 1979), equid droppings lack large grass culms. The unknown sample resembled droppings of *Loxodonta africana* (Blumenbach, 1797) (African elephant).

Bechan Cave is 52.8 m deep, 31.4 m wide, up to 9.1 m in height and well lit during the daytime through its cavernous mouth. The dung blanket proved extensive; to determine its thickness a small pit was dug close to the north wall of the cave. The contents of the pit included two large boluses of unbroken dung. The larger, designated M-1 (Fig. 28.4), measures 230 × 170 × 85 mm; the smaller, M-2, is subspherical and 225 mm in diameter. Both closely resemble boluses of African elephant dung and they lack the short-clipped twigs typical of dung balls of the Shasta ground sloth.

These remarkable specimens may not be the first record of fossil proboscidean dung found in Utah, but they are the first complete boluses. A similar dung blanket was found in Cowboy Cave located just north-west of Canyonlands National Park, Utah (Jennings 1980). Faecal remains from a variety of herbivores, living and extinct, (predominantly *Bison*) including mammoth dung fragments (as indicated in a size-fraction analysis of plant fragments in the dung bed) were reported by Hansen (1980) from Cowboy Cave.

Fig. 28.4 Mammoth bolus (M-1) from Bechan Cave, Utah, USA.

Subsequent excavations (1984) and screening of test pits at Bechan Cave have yielded fragments and an additional 15 intact boluses of the quality of M-1 and M-2. We also recovered faecal pellets comparable to *Oreamnos harringtoni* (Harrington's mountain goat), in addition to some items 26 × 19 × 18 mm in size which resemble the unknown herbivore droppings ('elk-camel') reported in Hansen (1980, p. 182). It is our preliminary conclusion that these pellets represent the extinct *Euceratherium collinum* (shrub ox). In addition, we found twigs and other plant macrofossils apparently transported by *Neotoma cinerea* (bush-tailed woodrat) and a strand of coarse hair 150 mm long and 1.15 mm in diameter, too long and coarse to be attributed to living members of the Utah megafauna.

The cave floor was mapped (Fig. 28.5a), a test pit was dug adjacent to a pot-hunter pit (Pit M), contents of the test pit were screened with window screen (1 mm mesh), and samples were removed for radiocarbon dating and palaeobotanical study. At 35 cm depth, Test Pit 1 yielded a ground sloth dung bolus packed with twigs, and containing an acorn of *Quercus gambelii* (Gambel oak).

The thickness of overburden and thickness of the dung blanket were measured in a series of 49 auger holes positioned on a 5 m grid, yielding an isopach map (Fig. 28.5b). A second test pit was dug to sterile sand near the back of the cave and its contents screened. Two trenches and a 2 × 2 m pit were excavated through the dung deposit.

Based on auger probes, the dung blanket of Bechan Cave is estimated to contain 300 m^3 of organic materials. Overburden near the cave walls constitutes 10–20 cm of loose sand, thickening to 90 cm or more toward the centre of the cave and exceed-

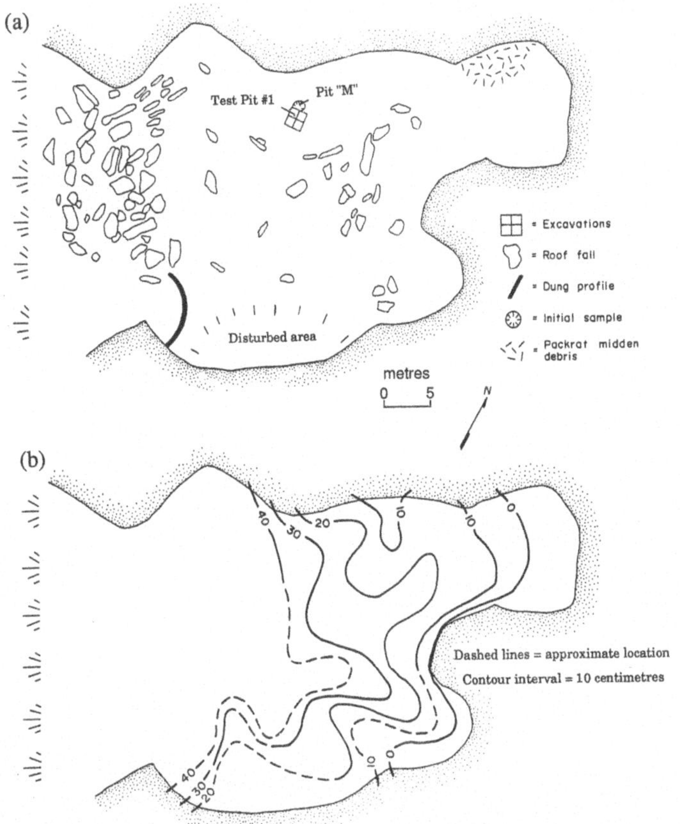

Fig. 28.5 (a) Plan view of the floor of Bechan Cave, Utah. (b) Isopach (equal thickness contour) map of the buried Pleistocene dung bed in Bechan Cave, Utah. Contour interval is 10 cm.

ing that near the mouth where dislodged sandstone blocks prevented augering. In 2.1 m^3 of screened test pit material (less than 1 per cent of the deposit) and in auger probes, only two diagnostic bones of large mammals were found. Large bone fragments within the dung blanket near the mouth of the cave were not diagnostic for taxonomic purposes. Numerous small mammal, bird, reptile, and amphibian remains were collected.

In the absence of quantifiable diagnostic large bones, we sought to identify the ancient occupants by two means: comparison of dung with faecal samples of living and extinct megaherbivores, and by identification of hair. Several dozen individual hairs were removed during screening of the dung blanket for macrofossils, and occasional hairs or hair masses were found during testing of the dung blanket. Preliminary study of the collection by Charles Bolen, a former graduate student in anthropology at Northern Arizona University, has revealed the presence of hair comparable in structure and thickness to mammoth, hair of two types of ground sloths (cf. *Mylodon harlani* and *Nothrotheriops*), several artiodactyls, *Equus* sp. (horse), and various small mammals.

The radiocarbon samples from the dung deposit cluster from 11 700 to 12 900 years in age; the dung blanket appears to have been deposited, perhaps intermittently, over as little as 1000 years (Mead *et al.* 1986). Various authors (Long and Martin 1974; Thompson *et al.* 1980; Mead *et al.* 1981; Martin *et al.* 1985) have stressed the ninth millennium BC (11 000 years BP) as the critical time for extinction of Shasta ground sloths and other animals in this region. On the basis of the radiocarbon dates in Table 28.1, it would appear that Bechan Cave (Table 28.2) was not occupied by Pleistocene herbivores after 11 700 years BP, which is slightly older than the time of extinction of Shasta ground sloths in the south-west (Thompson *et al.* 1980; Martin *et al.* 1985).

Abandonment of Bechan Cave by megaherbivores can be explained in at least two ways; either temporary access to the cave ended with erosion of a dune deposit left as a stabilized remnant 100 m south-west of the cave, or the vegetation adjacent to the cave may have been attractive to the megafauna only for a brief interval.

To gain the information needed to reconstruct the palaeoenvironment, mammoth dung samples from Bechan Cave were soaked in water for 24 h, then screened for their contents. Identifiable macrofossil remains were compared with the University of Arizona herbarium and palaeoenvironmental laboratory collections. Pollen was extracted from the dung layer, the large boluses, and at 2 cm intervals around the walls of the 2 m × 2 m test pit, as well as from the sand above and below the dung layer (Davis *et al.* 1984).

The dung samples are composed primarily of crushed culms and leaves of graminoids (Gramineae and Cyperaceae). Apart from cactus spines, nutlets of sedges are the most abundant type of fossil in the dung layer. The concentration of graminoids is greatest in the uppermost samples, which is true for the concentrations of most other macrofossil types (Davis *et al.* 1984).

With a few notable exceptions, the macrofossil composition in mammoth dung samples from different depths is very similar. The lack of substantial differences among samples may result from the short period of accumulation of the dung layer, which is suggested by the narrow range of radiocarbon dates. Lack of a clear-cut stratigraphic sequence within the millennium embraced by the radiocarbon dates could reflect trampling and disturbance of the dung blanket by those mammals that deposited it.

Dietary information gained from the dung column macrofossils provides evidence of the local vegetation. The fossil assemblage includes plants characteristic of several types of recent habitats. As is true today, the riparian community was an important vegetation type in the narrow canyon. None the less, the local species were different from those outside the cave today. Three species of sedges were identified, based on their perigynia and nutlets; the present southern limits of two of these, *Carex lasiocarpa* and *C.* cf. *lenticularis*, are currently in Idaho and Montana (Davis *et al.* 1985). The ancient community contained *Betula occidentalis* (water birch), which at the latitude of the cave occurs today only at higher elevations.

Several of the shrub species found in the mammoth dung were probably more abundant in the local riparian community than in

Table 28.1 Radiocarbon (TAMS and conventional) dates (years BP) for Bechan Cave mammoth dung

TAMS	
AA-1109	12 470 ± 140
AA-1110	12 390 ± 120
AA-1111	13 040 ± 280
AA-1112	12 430 ± 300
AA-1113	12 090 ± 210
AA-1114	12 320 ± 160
AA-1116	12 610 ± 140
AA-1117	12 620 ± 130
AA-1119	12 430 ± 150
AA-1120	12 880 ± 140
AA-1121	12 570 ± 130
AA-1122	12 390 ± 140
AA-1123	11 870 ± 140
AA-1124	12 570 ± 100
Conventional	
A-3212	11 670 ± 300
A-3297	12 400 ± 250
A-3213	12 900 ± 160
A-3296	11 850 ± 160
A-3298	12 620 ± 220
GX-9371	13 505 ± 580

TAMS, Tandem accelerator–mass spectrometer; AA, University of Arizona accelerator date; A, University of Arizona conventional date; GX, isotopes laboratory.

Table 28.2 Faunal list: Bechan Cave (from hair analysis by C. W. Bolen, September 1983)

Species	Common name
Large mammals	
Mammuthus columbi[a]	Columbian mammoth
Mylodon harlani (Glossotherium)[a]	Mylodont ground sloth
Nothrotheriops shastensis[a]	Shasta ground sloth
Equus sp.[a]	Horse
Ovis sp. (*canadensis*)	Big horn sheep
Odocoileus hemionus	Mule deer
Ursus americanus	Black bear
Canis latrans	Coyote
Small mammals	
Neotoma cinerea	Pack rat
Myotis leibii	Short footed bat
Peromyscus maniculatus	Deer mouse
Sciurus aberti[b]	Aberts squirrel
Sorex sp.	Shrew
Perognathus fasciatus[b]	Olive backed mouse

[a] Denotes extinct species.
[b] Denotes extant species no longer found in the cave region.

the uplands: *Sambucus* (elderberry), *Symphoricarpos* (snowberry), *Rosa* (wild rose), *Rubus* (raspberry), and *Ribes* (currant). Given the small amounts of *Picea* (spruce) pollen in the dung samples and the tendency for the species to reach its lowest elevational limits along stream banks, we suspect that spruce, too, may have favoured the riparian community. We have collected *Picea pungens* (blue spruce) in stream bottoms in the nearby Henry Mountains at about 1890 m, where it associates with *Betula*. *Juniperus communis* (dwarf juniper), present in the uppermost fossil sample, is also presently restricted to higher elevations in the region.

Dune vegetation represents another type of habitat reflected in the fossil assemblage. Several species characteristic of arenaceous habitats today were recovered in the ancient dung layer (Davis *et al.* 1985). Fossil members of this community include *Opuntia polyacantha* (prickly pear), (a different species occurs in the modern vegetation), *Sclerocactus* sp. (fishhook cactus), *Atriplex canescens* (saltbush), and *Oryzopsis hymenoides* (Indian rice grass).

The actual upland vegetation growing on Navajo Sandstone during the time of dung accumulation is difficult to reconstruct from the macrofossil remains. Today the regional vegetation at the elevation of the cave is *Coleogyne ramosissima* (black brush) steppe. The relatively high percentages (30.9 per cent) of *Artemisia* (sagebrush) pollen in the dung layer may indicate that sagebrush steppe was the dominant regional vegetation type at the time the dung accumulated. *Artemisia* cf. *tridentata* macrofossils are present in the dung layer; the species is not now found near Bechan Cave.

Although the dung layer may have experienced some mixing by trampling, some trends in the macrofossil distributions deserve mention because they may indicate vegetation change during the period of dung accumulation. The abundance of aquatic plants at the top of the dung layer is greater than in lower samples. The threefold increase in sedge achenes in the 0–10 cm sample is far greater than the increase in concentration of any other type, and this sample contains two other aquatic taxa: *Scirpus* sp. (sedge) and *Potamogeton diversifolium* (pondweed) that do not occur in lower samples. Other taxa, including *Symphoricarpos* (wolfberry), *Quercus*, and *Picea pungens*, show the opposite trend and are more abundant in the three lower samples. *Betula occidentalis* is most abundant in the two middle samples (Davis *et al.* 1984).

Preliminary analysis of pollen from the dung unit only partially confirms this trend. The percentage of *Picea* pollen is greater in the sample from the bottom of the dung, as is the abundance of *Betula* pollen. *Quercus* pollen is more abundant in the top sample, and *Symphoricarpos* pollen was not recovered (Davis *et al.* 1984).

Although macrofossils and pollen from the dung layer provide limited evidence for vegetation change, preliminary analysis of pollen in the cave fill above and below the dung layer suggests strong temporal differences when compared to the dung bed. Differences in *Picea* pollen percentages above and below the dung layer are similar to the trend of the macrofossil concentrations and dung pollen percentages. Immediately below the dung layer, *Picea* pollen reaches 8.9 per cent (by number) and above the layer, the percentages drop to 0.2 per cent (Davis *et al.* 1984).

Several other taxa show changes, with *Ostrya* (hop hornbeam), *Acer negundo* (box elder), Gramineae, and Cyperaceae all more abundant below the dung layer than above. *Juniperus*, *Quercus*, and *Ambrosia* show the opposite trend and are more abundant above the dung layer than below. Although the pollen data are preliminary, they are sufficient to indicate that the dung layer accumulated during an interval of major vegetation change when spruce and birch diminished and oak, juniper, and ambrosia became more abundant.

CHRONOLOGY

A chronology for dated mammoth occurrences in North America, post-15 000 years BP was presented in Agenbroad (1984*b*). Agenbroad and Mead (1989) published a chronology for mammoth on the Colorado Plateau. In the 1984 paper, an average of the nine most recent dates on a variety of dated material, provided a date of 11 424 years BP as a terminal date for the species. Using man–mammoth associations, the average date is 11 200 ± 100 years BP. Our Colorado Plateau research has provided dates produced from mammoth dung, preserved in dry caves and alcoves, which provides a more precise dating of the mammoth occupancy than does non-dietary, associated organic material or radiometric dates from skeletal remains (due to possible chemical contamination, or collagen deletion). Using radiocarbon dates from three of the youngest dung boluses, plus an accelerator mass spectrometer (AMS) date from tusk fragments at an alluvial site, a weighted average date for the most recent occurrence of mammoth on the plateau is 11 270 ± 65 years BP. Using just the three youngest dung boluses, the average is 11 820 ± 80 years BP. Using the data from Mead and Meltzer (1984), it appears that a continental average for the extinction of mammoth is 10 550 years BP.

Dated mammoth dung deposits from the Colorado Plateau provide us with dietary information, and therefore a palaeoenvironmental model for the area less than 700–1000 years prior to extinction. The quality of data from these dung deposits is unrivalled. We are currently experimenting with the use of a nuclear magnetic resonance (NMR) technique for positive identification of the producer of the fossil dung. Initial results suggest that we may be able to differentiate to genus level for several varieties of extinct fauna (Mead *et al.* 1990*a*). Such refinement, added to absolute dating and histological analyses promises detailed dietary information from these late Pleistocene deposits.

In 1984 an isochron map of post-15 000 years BP mammoth die-off was presented (Agenbroad 1984*b*). A plot of man–mammoth associations, temporally and geographically, (Agenbroad 1988) provides a match of data that is too great to be attributed to chance. Humans had a decided role in North American mammoth extinction, though not necessarily the only role.

DISCUSSION AND SUMMARY

The Mammoth Site of South Dakota provides a large sample of a local population of late Pleistocene mammoths. It was a natural trap which was selective for juvenile and young adult mammoths, as interpreted from the faunal remains, and from the age structure of modern African elephant populations.

Palaeoenvironmental interpretations for this portion of the northern High Plains have previously been extrapolated from distant (250–650 km) data sources. The Mammoth Site provides the first palaeoecological data set within the Black Hills of South Dakota. Microfauna recovered from the deposits indicate it was not a tundra environment. Molluscan and palynological evidence indicate a warm spring-fed pond within the sinkhole trap. Our model extrapolated from on-site data, indicates a cold steppe–grassland with local warm-water oases, in the Southern Black Hills region. The sample of mammoths present, plus associated Pleistocene fauna, indicate that the local habitat supported large mammoth populations.

Grasses (Poaceae) were the most abundant pollen type recovered from the sediments of the Mammoth Site. The second most abundant pollen type was the sedges (Cyperaceae). Additional aquatic pollen types included pondweed (*Potamogeton*), water smartweed (*Polygonum*), parrotweed (*Myriophyllum*), and sweet flag (*Acorus*). Other non-arboreal taxa included sagebrush, Cheno-Ams, mustard, parsley, rose, and saxifrage.

Arboreal pollen types include *Pinus* (pine), *Picea* (spruce), *Larix* (larch), *Abies* (fir), *Juniperus*, *Betula*, *Celtis* (hackberry), *Fraxinus* (ash), *Juglans* (walnut), *Populus* (poplar), *Quercus*, *Ulmus* (elm), and *Salix* (willow).

The suggested local environment at the Mammoth Site, 26 000 years ago, was a cold steppe–grassland with scattered riparian tree cover. This environment was locally modified by warm water springs, ponds and streams, fed by artesian systems which are still extant in the Southern Black Hills.

Within a few hundred years of the last records of Shasta ground sloths and Columbian mammoths in the region (around 11 000 years BP), several species of large extinct herbivores utilized Bechan Cave in southern Utah. They left a distinct stratigraphic unit within the cave fill, a dung blanket containing abundant plant material and accumulated animal droppings up to 40 cm in thickness and estimated at 300 m^3 in volume. Few diagnostic bones of large mammals have been found to date; however, based on the preliminary examination of dry boluses of herbivore dung and collections of hair, we are convinced that the cave was visited by mammoths. At least one, and probably two, kinds of ground sloth and various species of ungulates ranging in size from that of mountain goats to bison were also present. Subsequent use of Bechan Cave by large mammals was largely restricted to humans; charcoal from hearths (younger than 8000 years BP) commonly occurs above the dung blanket (Agenbroad *et al.* 1989). The dung blanket at Bechan Cave is larger than a similar deposit of comparable age found beneath abundant archaeological material at Cowboy Cave in eastern Utah. Neither site has yielded evidence of human occupation during the time of dung blanket deposition.

As determined by pollen and plant macrofossil analysis, the environment outside Bechan Cave approximately 12 000 years ago consisted of a riparian gallery of water birch, blue spruce, elderberry, snowberry, currant, and sedges. Other species present in the sandy valley or along the cliff-rimmed canyon beyond the cave include (in order of abundance of fossil remains), *Artemisia* cf. *tridentata*, *Juniperus communis*, *Sclerocactus* sp., *Opuntia* sp., *Atriplex* sp., *Quercus* sp., *Oryzopsis hymenoides* and other grasses, and *Ostrya* sp. Many of these species do not occur in the locality today. Extra-local species can be found presently at higher latitudes or higher elevations (e.g. 2440 m in canyons of the Henry Mountains or Aquarius Plateau). The regional vegetation around 12 000 years ago may have been primarily sagebrush-steppe rather than *Coleogyne*-steppe as it is now.

Bechan Cave and possibly Cowboy Cave, plus at least three other recently discovered localities, provide direct evidence of mammoth diet (from dung) from the temperate portions of North America. A recent discovery of mammoth hair and dung with skeletal remains in the Alaskan permafrost (D. Guthrie, personal communication) provides another source of evidence. The discovery of *Mammuthus* dung boluses and an extensive mat of trampled mammoth dung allows the first comparison of the diet of temperate North American mammoths with the dietary information gained from pollen and gastrointestinal remains from specimens from the permafrost of the Soviet Union (Table 28.3). Ukraintseva (1981, 1985) has summarized the data of earlier Soviet researchers, as well as her own works.

The Soviet localities are at a more northerly latitude, (+30°), so it is surprising that there is such correspondence of food plants eaten by the mammoths of North American steppe–grassland and sagebrush–steppe habitats and those eaten by the mammoths of the Soviet Union. A comparison at the family level of plant taxonomy indicates that some 42 per cent of the plants in the diet belong to the same plant families.

Continued exploration in protected alcoves of remote canyons in the Colorado Plateau uncovered additional dung deposits for mammoths and contemporary megafauna, including extinct mountain goat, shrub ox, bison, camel, horse, and sloth (Mead and Agenbroad 1986, 1989; Agenbroad and Mead 1987*b*). Analyses of mammoth dung and selected other species provide a dietary and palaeoenvironmental reconstruction for the past 30 000 years on the Plateau. Thermoluminescence dates (TL) indicate ages of over 115 000 years sampled in the same deposit, but below the dung layers. Initial pollen samples from this earlier horizon indicate a much different pollen spectra than present from either Wisconsinan, or Holocene environments, suggesting an absence of conifers.

A composite model for the 18 000–11 000 years BP interval for the central Colorado Plateau, based on the analysis of fossil dung from mammoth and associated fauna, indicates an environment rich in grasses with riparian forests and bodies of water in many of the canyons. Geological and palaeohydrological evidence indicate an aggradational history from the Sangamon interglacial to the Pleistocene–Holocene boundary, with a repetitive cycle of aggradation and degradation on ±3000 year intervals, or less, during the Holocene. In all cases, the Pleistocene alluvium is at higher elevation above base level (up to 58 m) than Holocene deposits (18 m), indicative of a different climatic–depositional environment.

It appears that for an interval of 115 000–50 000 years BP, aggradation has taken place due to damming of narrow canyons in the Navajo Sandstone by mass wasting or aeolian deposits. High energy flows contributed large volumes of sediments to the impoundments, gradually filling the canyons with nearly 67 m of alluvium. The water table also rose, creating relatively wide lush pasture in what are currently much deeper, narrow, canyons. It appears that similar events at differing temporal intervals took place in adjacent canyons, creating strip-like patches of high quality and high yield pasturage for late Pleistocene herbivores.

Grass, sedge, and common reeds predominate in mammoth dung from the cave and alcove deposits, although woody species such as sagebrush, saltbush, rose, cacti, birch, and spruce are present in the dung (Davis *et al.* 1984; Agenbroad and Mead 1989; Mead and Agenbroad 1989). The sedge and reeds indicate standing water or marshy ground in some areas of the canyon floor; the trees indicate riparian and, perhaps, marginal (near the canyon wall) forests. Sagebrush and grass parklands are envisioned for much of the canyon floor. Dung analyses of smaller herbivores (mountain goat and shrub ox) indicate oak scrub was also locally abundant.

A pre-Wisconsinan pollen sample indicates a variety of deciduous trees with no conifers present. This initial sample may be indicative of a different set of conditions for the Sangamonian interglacial than presently exist at these localities, where juniper and piñon currently exist.

In summary, the northern High Plains and the Colorado Plateau portion of the south-western United States supported relatively substantial mammoth populations during the late Pleistocene (<26 000 years BP to ±11 000 years BP). In both regions, the evidence from pollen and dung are indicative of grassland habitats with adequate water sources. These palaeoenvironmental elements have changed markedly in the ensuing 11 000 years. In both regions there has been warming and change in plant communities since the extinction of the mammoths. Particularly on the Colorado Plateau, the late Pleistocene mammoth habitat has been replaced by more xeric plant communities. In South Dakota a cold-tolerant grassland remains but a Ponderosa pine invasion has created a more patchy grassland mosaic within the southern Black Hills.

There has been a documented change in the composition of plant communities in the late Wisconsinan and early Holocene. There is no compelling evidence that this happened at such a rapid rate that it caused nutritional stress leading to mammoth extinction. On the contrary, mammoths became extinct at the time when grasslands should have been expanding. Mammoths

Table 28.3. Botanical remains from mammoth dung and gastrointestinal tracts from the former Soviet Union and the United States

Locality & Latitude	**Berezovka, USSR (70°N)**		**Shandrin, USSR (71°N)**		**Yuribei, USSR (71°N)**		**Cowboy Cave, USA (38°N)**		**Bechan Cave, USA (37°N)**	
Species	*M. primigenius*		*M. primigenius*		*M. primigenius*		*M. columbi*		*M. columbi*	
C-14 Age	44,000 B P		41,470 B P		11,450 B P		13,040–11,020 B P		13,505–11,670 B P	
Reference	Ukraintseva, 1981		Ukraintseva, 1981		Stanischeva, 1982		Jennings, 1980		Agenbroad et al., 1984	
Taxon	Macro	Pollen	Macro	Pollen	Macro	Pollen	Macro	Pollen	Macro	Pollen
Ulmaceae										●
Fagaceae							●	●	●	●
Cornaceae							●	●	●	
Polygonaceae		●								
Chenopodiaceae		●					●	●	●	●
Caryophyllaceae		●		●		●	●	●	●	●
Ranuculaceae	●	●				●				
Brassicaceae		●				●				●
Papaveraceae						●				
Cleomaceae							●			●
Saxifragaceae				●		●				
Rosaceae		●		●	●	●	●		●	●
Leguminosae	●	●								●
Primulaceae						●				
Malvaceae										●
Cactaceae							●		●	
Geraniaceae										●
Umbelliferae		●								●
Ericaceae			●	●						
Gentanaceae		●								
Menthaceae				●						●
Polemoniaceae										●
Scrophulanaceae				●						
Valerianaceae				●						
Asteraceae (Compositae)										●
Artemisia				●		●	●	●	●	●
Cadifoliaceae									●	
Rhamnaceae										●
BRYOPHYTA										
Musci	●	●	●	●	●	●				
Lycophyta		●				●				
Sphenophyta				●			●			
Pterophyta						●				
Coniferophyta										
Pinaceae										
Larix	●		●	●	●	●				
Abies										●
Picea				●			●	●	●	●
Pinus		●		●			●	●		●
Cupressaceae										
Juniperus							●	●	●	●
GNETOPHYTA										
ANTHOPHYTA										
Monocotyledonae										
Gramineae	●	●	●	●	●	●		●	●	●
Juncaceae						●				
Cyperaceae	●	●	●	●	●	●	●		●	●
Liliaceae							●			
Typhaceae										●
Potamogetonaceae						●				●
Dicotyledonae										
Salicaceae		●	●	●	●	●				●
Betulaceae										
Alnus	●	●	●	●						
Betula	●	●	●	●	●	●		●	●	●
Ostrya								●	●	●

and other Pleistocene megaherbivores are capable of lengthy migrations in search of new pasture. They are not doomed by local plant mosaic change as may be the case for microherbivores (e.g. rodents). The most visible (profound) vegetation changes occurred in middle and late Holocene *after* the extinction of most Pleistocene herbivores.

Continued research on dung deposits of extinct megaherbivores may provide new clues to the rate of plant community change, or the possibility of nutrient change during the transition. Physical, botanical, and chemical analyses of these unique Pleistocene resources, preserved in hyperarid regions such as the Colorado Plateau, have the potential of providing testable hypotheses for other regions of the continent where such preservation has not been possible.

ACKNOWLEDGEMENTS

The South Dakota research has been funded by The Center for Field Research, The National Geographic Society, The High

Plains Research Center, The Nebraska Legislature, and Northern Arizona University. Research on the Colorado Plateau has been supported by the National Geographic Society, The National Science Foundation, The National Park Service, and Northern Arizona University. J. P. Dudley and J. Shoshani made constructive comments.

TWENTY-NINE

Mammoths, gomphotheres, and the Great American Faunal Interchange

JOSEPH P. DUDLEY

INTRODUCTION

The mammoths (*Mammuthus*: Elephantidae) were the most highly specialized of all proboscidean taxa in their dentitions and morphological/physiological adaptations to survival in cold climates. The Elephantidae was one of the most rapidly evolving families of mammals during the Plio-Pleistocene (Simpson 1944). The Elephantidae radiated explosively during the late Cenozoic but were decimated by the extinction of several genera during the late Pleistocene (Simpson 1944; Aguirre 1969*a*; Coppens *et al.* 1978).

Mammuthus primigenius possessed the most highly grazing-adapted dentition of any proboscidean, while the ancestral early Pleistocene *M. meridionalis* was comparable in its degree of dental specialization to modern *Elephas maximus* (Osborn 1942; Maglio 1970*a*; Hooijer 1980; Lister, this volume, Chapter 19). Stomach contents and fossil dung of latest Pleistocene mammoths (e.g. *M. primigenius, M. columbi*) indicate that graminoids comprised a high proportion of the diet, as much as 90 per cent or more seasonally (Farrand 1961; Ukraintseva 1985; Mead *et al.* 1986). The inferred habitat of early mammoths in the Northern Hemisphere appears to have ranged from mesic parkland to semi-arid savanna and scrub habitats.

By the classical definition, the southern mammoth *Mammuthus* (= *Archidiskodon* as used by Osborn 1932*a*, 1934*c*, 1942) *meridionalis* entered North America in the earliest Pleistocene; current records date this event to about 1.9–2.0 Ma (Kurtén and Anderson 1980; Agenbroad 1984*a*, *b*; Churcher 1986), which corresponds to the generally accepted date for the Plio-Pleistocene boundary in North America established through other criteria. Subsequent colonization of the continental interior appears to have been rapid, with mammoths known from sites well into the southern latitudes (e.g. central New Mexico and peninsular Florida) dating back to at least 1.5 Ma (Falconer 1863; Logan *et al.* 1984; Dudley 1987).

While recent and fossil elephantids have been reported to cross extensive water barriers (Williams 1950; Johnson 1978), mammoth dispersal into North America most probably occurred across the Beringian land bridge during a eustatic low sea-level interval. The conditions necessary for the dispersal of the early southern mammoths (*M. meridionalis*) into North America would have been provided by an early interglacial or exceptionally warm interstadial phase at the Plio-Pleistocene boundary. These conditions include:

(1) Beringian land bridge — an interglacial-phase eustatic sea-level depression opening a terrestrial habitat corridor across the continental shelf region of the Beringian Straits;

(2) temperate climatic regime — ambient temperatures warm enough to enable a temperate-adapted proboscidean grazer/browser to maintain thermoregulatory capability and avoid disabling frostbite on the vulnerable extremities, particularly the trunk and ears;

(3) forage availability — a savanna, steppe, scrub, or parkland landscape; grazing resources and the availability of browse and forbs to provide forage of sufficient quality to maintain protein intake at or above maintenance levels; and

(4) water — availability of surface water for drinking.

TIMING AND PATTERNS OF MAMMOTH DISPERSAL INTO NORTH AMERICA

I hypothesize that North American mammoth populations south of the Beringian region were the results of two discrete dispersal events. The early or southern mammoth *M. meridionalis* apparently first entered North America at the time of the Pliocene/Pleistocene transition, between about 1.9 and 2.0 Ma. The timing of dispersal for *M. primigenius* into eastern Beringia (i.e. arctic North America) is uncertain; however, woolly mammoths are believed to have penetrated into North America south of Beringia during or just prior to the start of the last major glaciation of the Pleistocene (North America: Wisconsinan; Eurasia: Würm).

Three species of mammoths are believed to have resulted from an independent adaptive radiation among *M. meridionalis* populations within temperate steppe and savanna habitats in central and southern North America. The endemic North American mammoths recognized by most authors, with some disagreement on nomenclature and chronological range, are: *M. imperator*, *M. columbi* (includes *M. c. jeffersonii*), and *M. exilis*.

At least two alternate hypotheses can be proposed from the available evidence. It can be argued that the diversity of North American mammoths is a product of repeated dispersals of increasingly progressive Eurasian mammoth taxa (*M. meridionalis*; *M. trogontherii* cf. *imperator*; *M. armeniacus* cf. *columbi*; *M. primigenius*) during the course of the Pleistocene. Under this hypothesis, no *in situ* radiation would be required to produce the observed diversity among North American mammoths (with the exception of the dwarf Channel Island mammoths). Within this scenario, the serial invasions of more progressive Eurasian mammoth lineages could have caused the replacement of more conservative American mammoth populations through genetic swamping and/or competitive displacement.

A second alternative could include multiple dispersals in conjunction with an endemic radiation. An adaptive radiation of mammoths might be recognized within a conservative *M. meridionalis* → *M. imperator* → *M. i. scotti* (Osborn, 1942) lineage (Dudley 1987). The more classically progressive southern mammoths (*M. columbi*, *M. c. jeffersonii*) could be attributed to a dispersal of *M. armeniacus* along with *Bison* into North America at the beginning of the Rancholabrean Land Mammal Age (Harington 1984; Guthrie 1990*a*). Unequivocal evidence of mammoths other than *M. primigenius* from the central and western Beringian region is presently lacking (A. Sher, personal communication 1991); without such evidence, the hypothesis of additional dispersals of temperate-adapted mammoths into North America cannot be supported.

The reproductively isolated populations of early North American and African/Eurasian mammoths underwent parallel and convergent evolution in dental characters throughout the Pleistocene (Osborn 1942; Maglio 1973). In general, the dominant trends were a thinning of the enamel layer in molar lamellae and a corresponding increase in the hypsodonty (molar height/ depth: lengthening of the constituent tooth lamellae) of the permanent molars (Maglio 1973; Fortelius 1985). The taxonomy of *Mammuthus* in North America remains somewhat uncertain due to variations within the genus on this continent.

THE BERINGIAN CONNECTION

Eurasian mammoths, both temperate-adapted *M. meridionalis* and arctic-adapted *M. primigenius*, entered North America from Siberia across the (then) Beringian Coastal Plain and the northern rim of the Eurasian/North American supercontinent. The Beringian land bridge has emerged repeatedly during eustatic sea-level depressions of the late Cenozoic (Bloom 1983). The Yukon and Mackenzie River Valleys of central Alaska and north-western Canada would have provided mammoths with the simplest and most direct route into the middle latitudes of North America.

Early mammoths probably migrated southward from Beringia through an ice-free temperate steppe or parkland habitat corridor. An ice-free corridor is believed to have been present intermittently within the upper Yukon and Mackenzie Valley regions of Canada during the Pleistocene (Guthrie 1990*a*). The founder population of early mammoths would have been isolated and driven southward into the middle latitudes by climatic deterioration during the onset of the following stadial.

The most extensive known glaciation of the Brooks Range and the North Slope of Alaska occurred during early Pleistocene time (Walker 1973). It is hypothesized that the onset of the early Pleistocene glaciation effectively closed the Beringian corridor to *M. meridionalis* following the dispersal of these early temperate-adapted mammoths into North America. It must be noted, however, that eustatic sea-level rises associated with prolonged interglacial conditions would have proved an equally effective isolating mechanism for these early North American mammoths.

Records from Siberia indicate a progressive decline in dominance of forest tree-species within Beringian landscapes beginning in the late Pliocene and continuing through the Pleistocene. Palynological data from Kutuyakh in northern Siberia indicate that Beringian floras of the latter Pliocene were dominated by a *Picea*/*Larix*/*Pinus* taiga with few graminoids (A. Sher, personal communication 1991). This forested landscape would not have provided adequate forage for mammoths. The Pliocene forest taiga of western Beringia was replaced at the time of the Plio-Pleistocene transition by parkland mosaic of grasslands and *Larix*-dominated riparian gallery woodlands (A. Sher, personal communication 1991).

The available evidence indicates that *M. meridionalis* dispersed into North America through a sub-mesic *Larix* parkland landscape quite different from the xeric Beringian steppe of the late Pleistocene. Hence, the term 'mammoth steppe' of Guthrie (1990*a*) may be somewhat confusing if considered (as Guthrie most surely does not) within the context of early Pleistocene mammoth dispersal. The early Pleistocene *Larix* parkland is believed to have been replaced by steppe and/or xeric steppe–tundra communities sometime during the middle to late Pleistocene.

The gradual desiccation of Arctic and sub-Arctic latitudes during the Plio-Pleistocene follows the pattern of drying climates described for temperate and subtropical latitudes of North America during the late Cenozoic (Webb 1977, 1989). By the late Pleistocene (*c.* 500 000 years BP) a xeric steppe biome, the 'mammoth steppe' of Guthrie (1990*a*), completely spanned the higher latitudes of the Holarctic region. This vast mammoth steppe biome extended in a more-or-less continuous belt from the Atlantic coastal plain of North America westward and northward across Beringia, Asia, and Europe and into the British Isles (Vereshchagin and Baryshnikov 1982; Sutcliffe 1985; Guthrie 1990*a*).

Eastward movements of wandering mammoth herds across the Beringian Coastal Plain and up the valleys of the Yukon and Mackenzie Rivers would have brought mammoths past the barrier of the northern Canadian Rockies to the plateau fringing the north-western rim of the American Great Plains region. The Yukon and Mackenzie River valleys probably served as the principal dispersal corridors southward from Beringia into the Great Plains (Fig. 29.1).

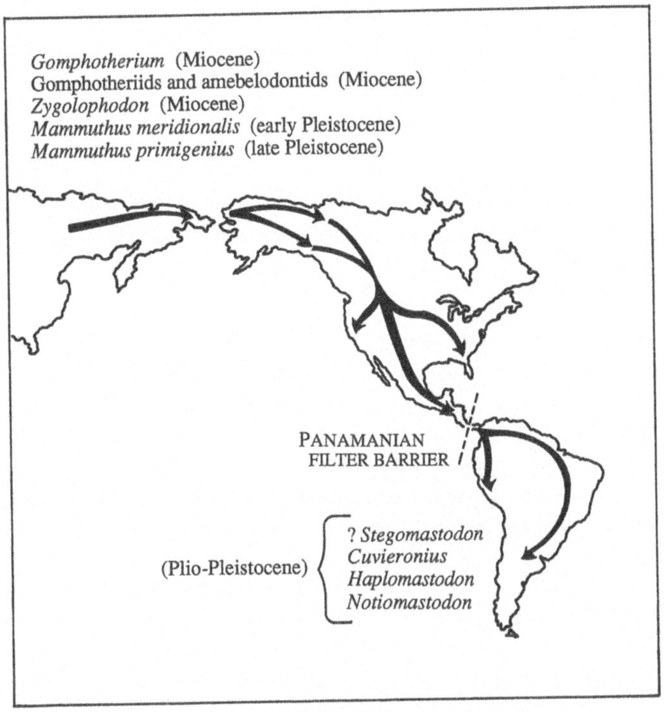

Fig. 29.1 New World proboscidean dispersals (modified after Osborn 1936, 1942; Webb 1978, 1991; Guthrie 1990*a*; see notes on p. 295).

Mammoths could have overcome the mountain barrier via an inland route (the 'high road') through central Alaska or via a coastal route ('low road') following the Arctic coastal plain. The high road would have taken mammoths up the Yukon valley by following the river eastward and southward (upstream) to its headwaters. The low road would have followed the Mackenzie Valley southward (upstream) from its lower reaches into central Canada. Mammoths may also have migrated between the Mackenzie and Yukon River valleys through the Crow River basin of Canada, which has proved rich in fossil mammoths (Churcher 1986).

ASSUMPTIONS REGARDING THE LONG-DISTANCE MIGRATION CAPABILITIES OF EARLY MAMMOTHS

The potential for long-distance migrations in mammoths has been the subject of some speculation in the literature (Churcher 1980). Studies of African savanna elephants (*Loxodonta africana africana*) in arid southern African habitats have shown that elephantids can travel considerable distances within a relatively short time. A radiotelemetry study in Botswana has recorded a straight-line one-way migration of more than 230 km for an adult female *L. a. africana* during a single season (G. Calef, personal communication 1988). Viljoen and Bothma (1990) found that adult *L. a. africana* in the deserts of western Namibia travelled an annual mean distance of 27.5 km/24 h period (range 9–68 km) in the course of their daily and seasonal movements. More recent observations indicate that these desert elephants can exceed 70 km/day and movements of up to 195 km/day were recorded (Viljoen 1992, p. 132).

Although it is thus conceivable that mammoths could have covered linear distances of 2000 km or more within a single year (Churcher 1980), it seems unlikely that such megadistance migrations were actually achieved by any individual mammoth during the course of its lifetime. Migratory ungulates whose young achieve virtually adult levels of endurance and mobility within weeks or months of their birth (e.g. *Saiga, Rangifer, Bison, Connochaetes*) can successfully undertake arduous seasonal or annual migrations without undue hindrance from the juvenile cohort. The seasonal movements of matriarchal elephant herds, on the other hand, are considerably constrained by the limited physical capabilities of the highly altricial youngest calves (0–5 years) and the higher nutritional requirements of the lactating cows and juveniles.

Mammoths are assumed to have been similar in behaviour to their living relatives (Olivier 1982). It seems reasonable, therefore, to assume that mammoths and other palaeo-elephantids, like the living species, travelled no further than was required to meet their needs for forage, minerals, water, and social/sexual interaction. In this light, it seems unlikely that North American mammoth populations would have undertaken seasonal migrations or dispersals of 1000 km more under any but the most extraordinary of circumstances.

THE WAY SOUTH

Southward dispersal of nomadic or semi-nomadic mammoth herds from eastern Beringia into the Great Plains region of North America could have been facilitated by the response of mammoths to other ecological factors. These include:

(1) upstream movements within the major river drainages to track temporal shifts in peak primary productivity across altitudinal and/or topographic gradients;

(2) upwind and upvalley migrations to avoid plagues of biting and parasitic flies in lowland regions (Curatolo 1975; Dau 1986); and

(3) climatic deterioration caused by the early Pleistocene shift back toward full-glacial conditions and expansion of the Alaskan ice-shield, isolating *M. meridionalis* populations of the eastern Beringian region and perhaps causing them to move southward in conjunction with a latitudinal shift in suitable climate and habitat.

THE GATEWAY TO THE SOUTHERN GREAT PLAINS

Once east of the Canadian Rocky Mountains and into the uppermost (= southernmost) reaches of the Mackenzie drainage, mammoths would have reached the threshold of the Great Plains region (Guthrie 1990*a*). Beyond the northern Rocky Mountains and the great northern glaciers lay the vast Great Plains region of North America, an unbroken expanse of flatlands stretching from central Canada southward into Mexico. Early mammoths almost certainly followed the same routes in colonizing the North American continent as those used by middle Miocene mammutids and gomphotheriids and the late Pleistocene woolly mammoths (*M. primigenius*) (Fig. 29.1), albeit under radically different climatic and vegetation regimes from those of the Miocene and later Pleistocene (Madden and Storer 1985; Guthrie 1990*a*).

The appearance of mammoths in fossil faunas marks the beginning of the Irvingtonian Land Mammal Age and sets the classical lower limit for the Pleistocene of North America (Stehli and Webb 1985). The Rancholabrean Land Mammal Age of the late Pleistocene (500 000–10 000 years BP) is likewise defined by the presence of *Bison*. The increasing aridity of climates during glacial periods of the late Cenozoic greatly increased the extent of arid steppe and scrub (but not desert) biomes within North America, with a corresponding decrease in the extent of savanna, woodland, and forest habitats. The steppe biomes of the continental interior of North America in Pleistocene time were both far more extensive and of a somewhat different character than those of Recent time (Webb 1977; Guthrie 1984).

In the continental interior, woodlands and forests contracted westward into the Rocky Mountains or persisted as gallery woodlands and alluvial lowland forests within major drainages of the eastern Great Plains and Mississippi Valley region. A vast Pleistocene steppe and scrub biome spread across North America during the Irvingtonian, ultimately reaching southward deep into peninsular North America (= Central America) and eastward across the Gulf Coastal Plain into the Florida peninsula (Webb 1977). A suite of steppe-adapted species, including jackrabbits (*Lepus*), burrowing owls (*Speotyto*), and pocket gophers (*Geomys*), dispersed into habitats within the Florida peninsula during the Irvingtonian (Webb 1974*a*).

By the end of the Pleistocene, mammoths had occupied habitats virtually throughout unglaciated North America (Osborn 1942; Agenbroad 1984*b*). Mammoths are known to have ranged from Beringia southward to the Nicaraguan Basin and the tip of peninsular Florida, and from now-submerged regions of the Atlantic continental shelf westward to the Northern Channel Islands of California and the San Juan Islands of Puget Sound (Osborn 1942; Whitmore *et al.* 1967; Johnson 1978; Agenbroad 1984b; Webb and Perrigo 1984; Mead *et al.* 1986; Guthrie 1990a).

Mammuthus appears to have been extraordinarily successful in North America considering its relatively brief tenure on the continent. Mammoths ranged throughout most of unglaciated North America. Specimens of *Mammuthus imperator* and *M. columbi* from North America at estimated shoulder heights of 4 m or more (Barbour 1913; Osborn 1942; S. D. Webb, personal communication 1985) rank among the largest proboscideans ever recorded. The genus included the continent's largest contemporary terrestrial herbivore taxa, and during the latest Pleistocene, mammoths ranged from the high arctic of Beringia into Central America. It appears that the only obstacle to mammoth dispersal within the New World was the forested tropical lowland region of the Nicaraguan Basin and the Panamanian isthmus.

CASE STUDIES OF EARLY MAMMOTHS FROM THE SOUTHERN LATITUDES

FLORIDA

A recent study of *Mammuthus* molars in collections of the British Museum (Natural History) by the author has confirmed the view of Webb (1974*a*) and the findings of a previous study (Dudley 1987, 1988) referring early Pleistocene mammoths to the taxon *M. meridionalis*. The Leisey Shell Pit site in central Florida (Hillsborough County, near Tampa), dated at 1.5 Ma, produced a wealth of superbly preserved mammoth dental material (Dudley 1987, 1988). Unfortunately, much of the best mammoth dental and cranial material from this site is known to be in private collections, and has been inaccessible for scientific research. The Leisey *Mammuthus* specimens in the collections of the Florida Museum of Natural History are none the less extensive (minimum number of individuals = 35); remarkably, the dentitions present span virtually the entire possible range of dental ontogeny from neonate to extreme senility (Dudley 1987).

While the Leisey mammoths may, in fact, represent an evolutionary stage intermediate between Falconer's (1857) type specimen of *M. armeniacus* and those specimens which he referred to *M. meridionalis*, it appears most reasonable to assign the Leisey specimens to *M. meridionalis* on the basis of the dental metrics for this taxon presented in Maglio (1973). The dental characters of Falconer's type specimens of *M. armeniacus* (Falconer, 1857) are definitely much more derived than those of early Pleistocene mammoths (1.5 Ma) from the Leisey Shell Pit site in central Florida. Had the type specimen of *M. armeniacus* (Falconer) been recovered from a site in the Western Hemisphere, this specimen could be referred without hesitation to the taxon *M. columbi*.

NEW MEXICO

An unusual early Pleistocene mammoth specimen was recovered from the Tijeras Arroyo site, Santa Fe Formation, near Albuquerque, New Mexico (Logan 1984). The peculiar metrical characters of these molars result from an extremely advanced stage of tooth wear (A. Sher, personal communication 1991). Dental metrics cannot, therefore, provide a reliable means for assigning this specimen to any known taxon. Were it not for an underlying volcanic ash layer dated at <1.5 Ma (Logan *et al.* 1984; Sobus and Logan 1984), these molars might be assumed on the basis of metrical characters to represent an unknown and extremely archaic mammoth of Blancan (Pliocene) age (Dudley 1987).

SYMPATRY AND PARAPATRY OF MAMMOTHS AND OTHER NEW WORLD PROBOSCIDEANS

The hypothesis of Kurtén and Anderson (1980) that competition from the early mammoths caused the displacement of the gomphothere *Stegomastodon* from its south-western North American habitats in the early Pleistocene seems overly simplistic. Cooling climates and the changing character of the region from savanna to steppe/prairie during the Plio-Pleistocene transition would probably have displaced *Stegomastodon* from the northern portions of its range, even in the absence of competition by invading early mammoths.

Stegomastodon apparently shared habitats in northern Arizona with *Mammut americanum* without any indication of competitive displacement, as did sympatric populations of *Cuvieronius* and *Mammut* within the Florida peninsula (Dudley 1987). The apparent ecological niche segregation (parapatry) of mammoths and gomphotheres within habitats of peninsular Florida would seem to support a hypothesis of ecological rather than competitive displacement of *Stegomastodon* from habitats in south-western North America during the early Pleistocene.

Riparian gallery forests are inferred as the primary habitat for the ubiquitous *Mammut* within the Florida peninsula. *Mammut* was also quite common in boreal forest and parkland landscapes of periglacial regions and is known from Pleistocene deposits as far north as central Alaska (Dreimanis 1968; King and Saunders 1984; Shoshani 1990*b*).

Mammoth (*M. meridionalis, M. c. columbi*), American mastodon (*Mammut americanum*), and the gomphothere *Cuvieronius gratum* (Hay, 1917) (= *Cordillerion gratum* of Osborn 1936, p. 559) are known to have co-existed on the Florida peninsula during the Pleistocene. Mammoth and mastodon remains are commonly recovered in coastal and inland river and sinkhole deposits (Simpson 1929). *Mammut americanum* was widely distributed within peninsular Florida and its remains are frequently found in association with those of both *Mammuthus* and *Cuvieronius*. Mammoth and gomphothere are only rarely recovered within the same deposits, and where the two do co-occur (Leisey Shell Pit (Irvingtonian), Ormond Beach (Rancholabrean)), their relative representation within local faunas are typically highly skewed (Lambert, this volume, Chapter 15).

Taphonomic data suggest that *Mammuthus* and *Cuvieronius* were parapatric, with mammoths inhabiting more xeric upland savanna and scrub habitats while gomphotheres seem to have favoured subtropical mesic coastal lowland habitats (Webb 1974*a*; Dudley 1987). This degree of niche segregation is seen as equivalent to that postulated for parapatric populations of *M. c. jeffersonii* and *M. primigenius* in northern latitudes.

The gomphothere *Cuvieronius* apparently occupied inland areas of peninsular Florida (i.e. areas above Recent sea-level) only during the warmer and wetter interglacial periods of the Pleistocene, where its remains are found in association with mammals of predominantly hydric and mesic tropical lowland affinities (Dudley 1987). *Cuvieronius* in Florida are believed to have favoured mesic coastal savanna/scrub habitats. Hence, gomphotheres would have been largely restricted to areas now lying 100–150 m below present sea-level during the eustatic sea-level depressions associated with glacial episodes.

NORTH AMERICAN TERTIARY PALAEOENDEMIC PROBOSCIDEANS

The morphologically conservative mammutids and gomphotheriids had an exceedingly long and colourful history in North America, with early records for these families dating back to the middle Miocene of Saskatchewan and Florida (Madden and Storer 1985; Lambert, this volume, Chapter 15). Later forms included the beak-tusked *Rhynchotherium* and the massive shovel-tusker *Amebelodon* (*Konobelodon*) *britti* (Lambert 1990). Several genera of progressive gomphotheres successfully colonized South America following the establishment of the Panamanian land bridge in latest Pliocene time (Simpson and Paula Couto 1957; Simpson 1980).

During the Miocene and Pliocene gomphotheres ranged northward well into the middle latitudes of North America. Pliocene gomphotheres (*Cuvieronius* or a related taxon) inhabited the Atlantic Coastal Plain, ranging at least as far north as the southern shore of the Chesapeake Bay (Lambert, this volume, Chapter 15). *Mammut* and at least one genus of gomphotheriids survived into the latest Pleistocene and possibly early Holocene in New World habitats (Meltzer and Mead 1983; Gruhn and Bryan 1984).

GOMPHOTHERIID DIVERSITY IN THE NEOTROPICS

At least three genera of gomphotheres (*Cuvieronius, Stegomastodon*, and *Haplomastodon*) inhabited South America and the southern latitudes of North America during the Pleistocene. Simpson and Paula Couto (1957) proposed that all known gomphotheres from South America derived from an independent radiation in peninsular North America (i.e. Central America/Mexico). The genus *Cuvieronius* is believed to have evolved in southern North America from a progressive *Rhynchotherium* stock during the middle Pliocene (Webb and Perrigo 1984).

Remains of *Haplomastodon* have been found in possible association with human artefacts (Gruhn and Bryan 1984), and the genus is therefore believed to have survived in South America until latest Pleistocene time. Human predation may ultimately have been responsible for the extinction of New World gomphotheres (Martin 1967, 1984; Fisher, this volume, Chapter 30; Casamiquela *et al.*, this volume, Chapter 31). Certainly it seems reasonable that, unlike North America, substantial areas with suitable climate and habitat for gomphotheres could have persisted as refugia within the known South American range of gomphotheres during the Pleistocene/Holocene transition.

The evolutionary diversity of New World proboscideans during the Pleistocene was truly astounding. The New World proboscidean fauna included mammutids and gomphotheriids from several Tertiary endemic lineages as well as an assortment of relatively late-arriving mammoth taxa. At least five genera (*Mammuthus, Mammut, Cuvieronius, Stegomastodon,* and *Haplomastodon*), including seven recognized species, vanished from the Americas in the course of the Pleistocene megafaunal extinction. The extinction of Pleistocene megafaunal elements during the Pleistocene/Holocene transition may have had profound consequences on the structure and composition of New World floras (Janzen and Martin 1982).

MAMMOTHS AND THE GREAT AMERICAN INTERCHANGE

There are apparently no known *verifiable* records of mammoths from any site within the South American continent (Simpson 1980). The provenance of the mammoth molar fragments reportedly recovered from Cayenne, French Guiana in the early nineteenth century (Osborn 1942) is considered uncertain and suspect. Pending the independent confirmation of the presence of *Mammuthus* in South America, it is reasonable to assume that *Mammuthus* did not colonize the southern continent (Simpson, 1980).

The establishment of the Panamanian land bridge about 3 Ma opened a habitat corridor for dispersal of terrestrial plants and animals into and out of South America, and precipitated an event known as the Great American Interchange (Webb 1976, 1978; Simpson 1980; Marshall *et al.* 1982; Stehli and Webb 1985). A number of widely distributed mammalian genera, including *Mammuthus* and *Mammut*, which might be expected to have crossed the Panamanian land bridge, did not reach South America. This phenomenon is considered highly significant in the light of the multitude of species from both continents which made the crossing.

The Panamanian land bridge functioned as an ecologically selective dispersal corridor or 'filter barrier' (Webb 1978; Simpson 1980). Eurytopic, savanna, and scrub mammal taxa dispersed into South America, whereas exclusively steppe or desert mammal taxa did not. Biogeographical data indicate an alternation among three major types of Plio-Pleistocene habitat corridors on the Panamanian land bridge: mesic tropical forest, mesic savanna, and xeric scrub/savanna (Webb 1978, 1991).

The dominant vegetation of the Panamanian corridor would have changed according to variation in precipitation regimes associated with climatic changes due to crustal uplifting and glacial–interglacial cycles. Xeric habitats would have dominated upland landscapes during full-glacial periods, while coastal areas on the continental shelf exposed by eustatic sea-level depression (Bloom 1983) probably maintained substantial areas of mesic savanna and mangrove forest habitats. During late Pliocene pluvial episodes, and perhaps throughout most of the Pleistocene, tropical lowland rain forests would have reclaimed the landscape of the Panamanian isthmus.

Mammals representing some 29 genera of North American/Eurasian lineages dispersed southward, while 17 genera of South America palaeoendemic mammals dispersed into peninsular North America (Stehli and Webb 1985). Canids, camelids, cervids, felids, gomphotheres, horses, rodents, tapirs, and ursids occupied new habitats in South America. Bizarre South American autochthons including glyptodonts, rhinoceros-like toxodonts, giant armadillos, and the 6000 kg ground sloth *Eremotherium* dispersed northward, accompanied in the trek by more prosaic forms such as capybaras, opossums, primates, and porcupines (Marshall *et al.* 1982; Webb and Perrigo 1984; Webb and Wilkins 1984; Webb 1991). Many of what are today characterized as typical South American mammals are in fact descendants of Plio-Pleistocene immigrant taxa of Nearctic and Eurasian origins (Eisenberg 1989).

Major faunal overlaps were established between North and South America by early Pleistocene time. Pleistocene faunas of the Florida peninsula and the Argentine pampas are remarkably similar in terms of constituent taxa and community structure (Dudley 1987). Shared taxa were largely savanna-adapted forms and included species of hydric, mesic, and xeric affinities (Webb 1978). More than half of the mammalian genera known from early Pleistocene local faunas in Argentina are recorded in contemporaneous fossil sites from peninsular Florida.

The Pleistocene Pan-American Fauna included several megaherbivores (giant sloths, toxodonts, and gomphotheres), numerous large ungulates (tapirs, camels, and horses), large carnivores (cave lions, jaguars, dire wolves), as well as capybaras, glyptodonts, and armadillos (Simpson 1929, 1980; Churcher 1959; Webb 1974a, 1978, 1991; Gillette and Ray 1981; Lundelius *et al.* 1983).

Temperature regimes in the isthmian region probably were not significantly altered even during peak full-glacial cycles; the climate of the Panamanian isthmus, whether wet or dry, would have remained tropical. High ambient temperatures possibly could have proved a major limiting factor for temperate-adapted mammals having limited physiological capacities for thermoregulation in tropical climates. The failure of *Mammuthus* and *Mammut* to break through the filter barrier of the Panamanian isthmus may have been due as much to constraints on thermoregulation capabilities as to a lack of suitable habitat.

The most likely explanation for the absence of mammoths in South America appears to be closure of the Panamanian savanna corridor by an expansion of the tropical rain forest biome at the time of the Plio-Pleistocene transition. This factor may have been coupled with limitations in the thermoregulatory capability of the temperate-adapted mammoths at elevated ambient temperatures, as both living species of elephants must utilize a combination of behavioural and physiological mechanisms to maintain thermoregulation and avoid hyperthermia in tropical climates (Eisenberg 1981). Competitive exclusion of mammoths from South American habitats by resident gomphothere populations seems unlikely in view of the apparent differences in niche space among these taxa, as discussed above.

ACKNOWLEDGEMENTS

This research was supported by assistantships from the Florida Museum of Natural History, the Department of Wildlife and Range Sciences, University of Florida, and the Department of Biology and Wildlife, University of Alaska-Fairbanks, and grants from the Theodore Roosevelt Memorial Fund (American Museum of Natural History), the Otto Geist Fund (University of Alaska Museum), and Sigma Xi. John F. Eisenberg, S. David Webb, and R. Dale Guthrie provided critical insight, direction, and encouragement. Thanks are also due to the staff of the many institutions (too numerous to list) who provided the author with access to mammoth materials in their collections. D. C. Fisher, A. M. Lister, and the editors made constructive comments.

APPENDIX 29.1
A NEW PROBOSCIDEAN FROM AMAZONIAN PERU
(Carl D. Frailey, Kenneth E. Campbell, Jr., and Lidia Romero-Pittman)

Recently, a proboscidean was found along the Madre de Dios River in southeastern Peru. This discovery came too late for adequate discussion in this volume, but full description and discussion are forthcoming. The fossil was found in a part of the Tertiary Solimões Formation that underlies, and is separated by a major unconformity from a stratum dated faunally as upper Miocene (Huayquerian) (Frailey 1986). Dental features of the lower teeth identify this specimen as an advanced gomphothere with a typical trefoil pattern on the labial (pretrite) side, and accessory central conules (crescentoids) on the lingual (posttrite) valleys, especially prominent for the first lophid and weakly developed for the other lophids. Based on the thickness and medial inflation of the ramus, the lower jaw was intermediate between brevirostrine (i.e. *Rhynchotherium*) and longirostrine (i.e *Gomphotherium*) conditions.

This new gomphothere is temporally and geographically situated to be derived from the rhynchorostrine complex of Savage (1955) and to be an early representative of the Cuvieroniinae. The relative complexity of the teeth (compared to *Gomphotherium*) can be viewed as plesiomorphic for this group. The dental characters of later genera of Cuvieroniinae (*Cuvieronius, Haplomastodon, Notiomastodon* and *Stegomastodon*) could be derived from this pattern.

With this discovery, proboscideans may be the earliest known North American immigrants found in South America. It is reasonable to postulate that much, if not all, of the evolution of the South American gomphotheres occurred in South America. Instead of having been a one-way, north-to-south range extension, one, two or perhaps even three of the South American gomphotheres may have re-invaded North America during the Pliocene.

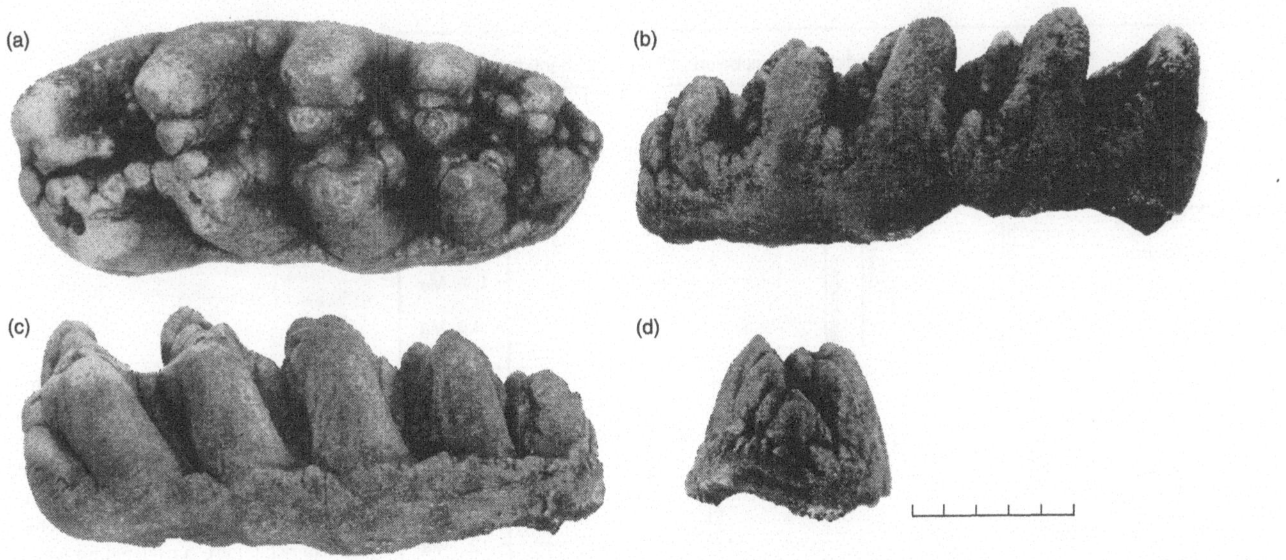

Fig. A29.1 Left M_3 of a new, yet unnamed, proboscidean taxon from sediments interpreted (based on the paleofauna recovered from overlying beds; Campbell and Frailey (1995)) as the upper Miocene of Peru, South America. This specimen was part of a partial articulated skeleton. Views are: (a), occlusal; (b), lingual; (c), buccal; and (d), distal. Scale bar in cm. Photographs are of a cast in the Vertebrate Paleontology Collections, Natural History Museum of Los Angeles County, LACM 140398.

THIRTY

Extinction of proboscideans in North America

DANIEL C. FISHER

INTRODUCTION

The history of proboscideans in North America begins with the immigration of mammutids and gomphotheriids from Eurasia in the middle Miocene (*c.* 15.5 Ma). It is marked subsequently by phyletic change, additional immigrations (of amebelodontids and elephantids), origins of new lineages, and extinctions of existing ones, but this pattern came full circle about 10 000 years BP, with the final disappearance of proboscideans from this continent. In some respects it is artificial to treat this history apart from that of South America, whose proboscidean fauna was derived from North America, or apart from that of Africa and Eurasia, where proboscidean faunas experienced somewhat parallel changes, but extinction is an ecological process, taking place within spatiotemporally bounded systems. The North American segment of the proboscidean record may thus be treated as an informative subset of their broader history.

This chapter summarizes knowledge of the extinction events experienced by North American proboscideans and reviews some of the causes hypothesized for these events. It also discusses some new directions in palaeobiological research which promise to shed additional light on the cause of proboscidean extinction. These include stable isotope studies of climate and proboscidean diet, dietary and taphonomic inferences based on analysis of preserved intestinal contents, and investigations of dentinal growth lines that may allow reconstruction of growth rates and reproductive life histories. Finally, a method is proposed by which assessments of life history parameters may be used to test climatic v. overkill hypotheses for late Pleistocene proboscidean extinctions.

TIMING OF NORTH AMERICAN PROBOSCIDEAN EXTINCTIONS

The stratigraphic ranges for proboscidean genera shown in Fig. 30.1 are based on data presented by Kurtén and Anderson (1980), Webb (1984), Lundelius *et al.* (1987), and Tedford *et al.* (1987), modified following Saunders (this volume, Chapter 27) and Lambert (this volume, Chapter 15). These data indicate two major episodes of extinction: the late Hemphillian (*c.* 4.5 Ma;

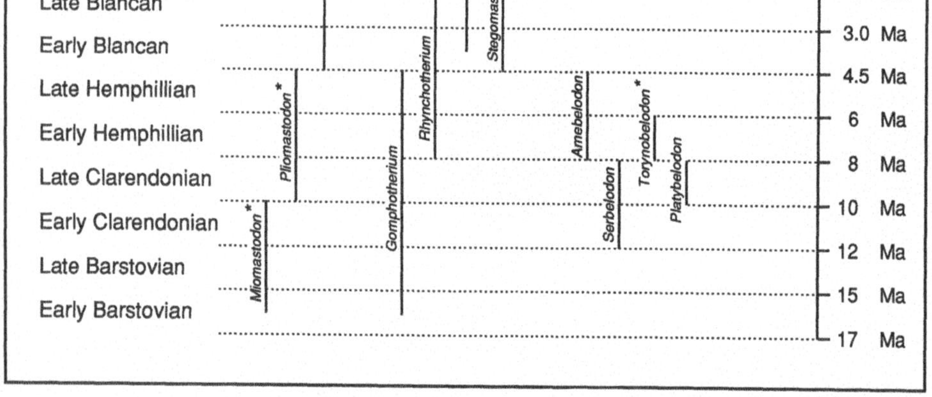

Fig. 30.1 Stratigraphic ranges of North American proboscidean genera. Data from Kurtén and Anderson (1980), Webb (1984), Lundelius *et al.* (1987), and Tedford *et al.* (1987), modified following Saunders (this volume, Chapter 27) and Lambert (this volume, Chapter 15). Genera marked by an asterisk (*) are considered junior synonyms of other genera in Appendix C. The higher taxa marked with an asterisk are handled somewhat differently by Shoshani (this volume, Chapter 16).

with some earlier extinctions that may be related), and the Rancholabrean (*c.* 10 000 years BP). The late Hemphillian extinction brought the disappearance of *Pliomastodon* (considered a junior synonym of *Mammut* by Shoshani, this volume, Chapter 16), *Gomphotherium*, and *Amebelodon*, following upon earlier loss of *Serbelodon*, *Torynobelodon*, and *Platybelodon*. *Rhynchotherium* and *Stegomastodon* dropped out at different times, both prior to the next main episode. Finally, near the end of the Rancholabrean, or terminal Pleistocene, *Mammut*, *Cuvieronius*, and *Mammuthus* (newly arrived from Eurasia *c.* 1.9 Ma) became extinct. Some of these losses represent regional extirpation, rather than final extinction, but this does not mute their testimony to regional ecological change. Moreover, expanding treatment to a global scale does not necessarily alter the broad outlines of the pattern. For example, *Stegomastodon* survived longer in South America (Kurtén and Anderson 1980), but only to join other taxa in the terminal Pleistocene extinction. Also, *Mammuthus* was widespread in Eurasia, but populations there appear to have survived no longer than in North America.

An important limitation of this record of extinctions, recognized fully by those who have compiled it, is the low temporal resolution available for all but the late Pleistocene events. Many radiocarbon-dated specimens of *Mammut americanum* and *Mammuthus* sp. represent the last few thousand years of their range, allowing us to establish time of extinction relatively precisely (Meltzer and Mead 1983). In contrast, for earlier extinctions we have fewer specimens and sites, and dating methods are less precise. Late Hemphillian proboscidean extinctions thus do not necessarily represent the degree of synchroneity demonstrated for *Mammut* and *Mammuthus* in the late Pleistocene (Webb 1984).

Another limitation is that a comprehensive extinction history of North American proboscideans is not yet available below the generic level. Especially for gomphotheriids and amebelodontids, additional work is needed to determine how many species-level taxa are involved (Kurtén and Anderson 1980). Associated with this are unresolved problems in phylogenetic history affecting our ability to identify termination of species-level lineages. The record of taxonomic extinction may overrepresent actual extinction by including cases of pseudoextinction (termination of a named taxon, reflecting morphological change, without termination of the name-bearing species-level lineage). For instance, *Pliomastodon* and *Mammut* may represent ancestral and descendant portions of a single lineage (Simpson 1945, p. 246). More often, however, working at the generic level underestimates occurrence of lineage termination. For example, there were clearly at least three species-level lineages of *Mammuthus* in North America by the end of the Pleistocene (*M. primigenius*, *M. columbi*, and *M. exilis*; *M. jeffersonii* may represent a fourth; cf. Dudley, this volume, Chapter 29), all of which became extinct at nearly the same time.

CAUSES OF LATE HEMPHILLIAN EXTINCTIONS

Reservations regarding temporal resolution notwithstanding, the earliest episode of North American proboscidean extinction involved multiple lineages, and these losses coincided with extinctions of many other elements of the mammalian fauna (Webb 1983, 1984). In the number of mammalian genera affected, the late Hemphillian extinction was the largest in North America in the past 10 million years (Webb 1984). As Webb points out, characterizing the magnitude of an extinction episode in terms of the total number of taxa affected, without dividing by the length of the time interval over which extinctions are summed, is appropriate only if the extinctions occurred simultaneously, which we do not know. The late Hemphillian extinctions, however, would stand out even if prorated in this way, suggesting that they represent a time when the mammalian fauna was indeed stressed on a continental scale. These extinctions occurred before human activity can have had any role, but they were associated with a late Cenozoic trend toward cooler, drier climates in temperate North America (Wolfe 1978) and coincided with a transition from widespread savanna habitats to a dominance of steppe habitats (Webb 1983). This transition would have severely stressed both browsers and grazers and probably brought about the collapse of what is recognized as the 'Clarendonian chronofauna', which may be construed as including the proboscideans going extinct around this time (Webb 1983). Climate change and resultant vegetational change seem to be the only well-documented phenomena of this time interval that would have affected continental biotas on this broad a geographical and taxonomic scale.

CAUSES OF LATE PLEISTOCENE EXTINCTIONS

Coincidence of multiple proboscidean extinctions and extinctions in the rest of the mammalian fauna also characterizes the terminal Pleistocene episode. This perspective, with late Hemphillian extinctions as a backdrop, gives the impression that late Pleistocene extinctions could have been just the next 'surge' in a much longer-term pattern of transformation of the North American fauna in response to climatic and environmental change (Webb 1984). There are, of course, differences between these two extinction episodes. As implied by the epithet 'megafaunal extinction', the late Pleistocene extinctions claimed a greater proportion of taxa of large body size (Gingerich 1984; Webb 1984). They also took place at a time when human involvement was a potential contributing factor. Yet such similarities and differences, especially when considered at the large temporal and taxonomic scale to which most comparisons are constrained, cannot be translated easily into similarities or differences in underlying cause.

The challenge of explaining late Pleistocene extinctions has attracted sustained interest from a broad cross-section of the scientific community (Martin and Wright 1967; Martin and Klein 1984; Agenbroad *et al.* 1990). This extinction episode was a global event, despite the focus here on its North American manifestation. There may, of course, be no single solution to such a large problem. Each affected species and geographical region could have experienced a subtly or a profoundly different mix of ecological stresses and biotic responses, each of which, moreover, may have changed through time. We therefore face not a single 'search for a cause' (Martin and Wright 1967), but a set of parallel searches that may or may not yield similar answers. As pointed out by Guthrie (1990*b*), extinction is only part of the total suite of biotic changes associated with this time interval; many other taxa showed rapid evolutionary change or drastic range shifts. Yet the irreversibility of extinction makes it in many respects the most impressive of the changes that occurred.

The two main classes of explanatory hypotheses appeal, respectively, to climate change and hunting by humans. Hypotheses depending on climate change are based on recognition that the late Pleistocene was a time of increasing climatic seasonality and instability associated with deglaciation (Webb 1984; COHMAP Members 1988; Webb and Bartlein 1988). From this starting point, however, climatic hypotheses diverge to consider a range of proximate mechanisms. For instance, Pleistocene vertebrate communities often contained combinations of taxa that are not sympatric today (Hibbard 1960), and the same generalization applies to floral communities (Mehringer *et al.* 1970). These observations, coupled with consideration of physiological tolerances of certain taxa, are in fact part of the empirical support for the realization that climate change at the end of the Pleistocene involved not just shifts in mean annual temperature or precipitation, but a more complex change in the distribution of precipitation and thermal energy within and between years. The physical conditions responsible for the co-occurrence of taxa in given communities were essentially 'reshuffled' (Graham 1986*a*) rather than being displaced smoothly along latitudinal gradients (Lundelius 1967; Guthrie 1968). The 'individualistic' nature of species' responses to environmental change resulted in the breakup of associations of taxa that had co-occurred at least throughout most of the preceding glacial cycle, and this is postulated to have produced a 'coevolutionary disequilibrium' (Graham and Lundelius 1984). If proboscideans were dependent on the availability of particular combinations of plants, either to satisfy nutritional needs, or to avoid excessive accumulation of the toxins produced by some plants as defenses from herbivores (Guthrie 1984), and if plant communities did not remain intact, shifting their ranges in a coherent manner, there might have been no simple resource tracking option open to proboscideans. Without adequate food resources, they might show reduced rates of growth and recruitment, leading eventually to extinction (Graham and Lundelius 1984). Megaherbivore tolerance of low-quality forage might moderate this effect (Owen-Smith 1988), but the outcome of such opposed factors is difficult to predict.

Reconstructions of the pattern and scale of environmental variation also suggest that there was a change from a relatively complex mosaic of habitats, during full glacial times, to a more homogeneous, latitudinally zoned pattern during the late glacial (Guthrie 1984, 1990*b*). Such a change would be more likely to affect large mammals, whose large home ranges might earlier have included several habitat types, than it would small mammals, and it might have left some large mammals without a sufficiently diverse resource base. Guthrie (1984, 1990*b*) has also proposed that climatic changes led to a shorter growing season throughout large portions of North America, and to a shift in the composition of grassland floras from dominance by species using the C_3 photosynthetic pathway, which tend to be more nutritious for mammalian grazers, to greater abundance of species using the C_4 pathway, which tend to be less nutritious. Guthrie suggests that this depressed both the quantity and the quality of food available for proboscideans (and other large herbivores), leading again to reduced rates of growth and reproduction.

While some habitats may have increased in area, others may well have decreased. In a study of the extinction of the American mastodon, *M. americanum*, King and Saunders (1984) point out that the type of environment in which these animals appear to have been most common, an open, spruce parkland with abundant ponds and marshy areas, became progressively restricted toward the end of the Pleistocene. Even if some such areas remained, their decreased size and increased isolation ('environmental insularity') would have resulted in smaller proboscidean populations, more subject to stochastic effects and less able to repopulate areas in which local extirpation had occurred.

A more direct effect of late Pleistocene climate change is that a shorter growing season and increased seasonality could have made it more difficult for large mammals, with their long gestation times, to synchronize reproductive cycles with seasonal peaks in availability of resources (Kiltie 1984). The most dramatic effect might be on species that, before the change, had barely managed to complete one reproductive cycle per year; with a shorter growing season, they might be limited to one cycle every two years. If extinct proboscideans had reproductive cycles similar to those of their extant relatives — a gestation period of 18 to 22 months (Laursen and Bekoff 1978; Shoshani and Eisenberg 1982), followed by a period of lactational anoestrus of 2 years or more — a decrease in the length of the growing season would probably not have halved their reproductive potential, but it would have tended to depress long-term reproductive output. In addition to the increased number of years it might have taken for calves to mature and for females to come into oestrus again, any increased severity or unpredictability of winter storms, which some climate models associate with the late Pleistocene, could have put calves and/or pregnant females at additional risk (Kiltie 1984).

Another direct effect is that increased seasonality in precipitation could have led to higher incidence of drought stress. In parts of south-western North America, late Pleistocene climates induced major changes in alluvial systems that imply reduced availability of surface water (C. V. Haynes 1991). Late Pleistocene proboscideans almost certainly needed regular access to surface water, and either seasonal droughts or longer term and less predictable reductions in the availability of water could have severely restricted the area habitable by proboscideans (G. Haynes 1991).

Turning to hypotheses involving human hunting, the best known is Martin's model of 'prehistoric *blitzkrieg*' (Martin 1967; Mosimann and Martin 1975), in which extinction is produced by an intense but locally brief period of hunting, associated with a rapidly expanding 'front' of human population density. These earliest Americans, Paleoindians, apparently came from northwestern Asia, through Beringia. The timing of this immigration is controversial, but Martin and others interpret Paleoindians as having arrived south of the ice sheet *c.* 12 000 years BP, shortly before the first widely accepted archaeological evidence of human presence in North America. Martin explains extinction as due to a big-game focus of Paleoindian subsistence practices, the megafauna's lack of experience with such an effective predator, and the low rates of population recovery to which large-bodied prey are limited. Like many of the climate-based hypotheses, this was formulated as a general explanation for late Pleistocene megafaunal extinction, and yet proboscideans, specifically mammoths, occupy a central place in this hypothesis in that Clovis–mammoth associations (listed in Saunders 1990) provide some of the most widely accepted evidence of human predation on megafauna. Although some of these occurrences may represent human-scavenged natural deaths (Sellards 1952), sites such as Naco (an unbutchered mammoth with several associated Clovis projectile points, interpreted as an 'escapee' who later died; Haynes 1966), and studies of the physical properties of Clovis fluted points (Frison 1990) combine to make a strong case that Paleoindians actively hunted healthy mammoths, uncompromised by accident or injury. Thus, in at least some instances, humans represented an agent of mortality independent of 'natural causes'.

Association of Paleoindians with *M. americanum* is more controversial. A fluted point was apparently found with the Boaz mastodon, but its significance was only recognized later, and documentation of this association had to be largely reconstructed (Palmer and Stoltman 1976). Multiple claims of human association are found in older literature, but these either involve lithic artefacts that are now viewed as Holocene in age (side- or corner-notched projectile points of apparent Archaic affinity; Sanford 1935 and others, discussed by Williams 1957), or rest on taphonomic evidence that has been considered inconclusive (e.g. Wittry 1965). A partly healed injury from a bone projectile tip embedded in a mastodon rib at the Manis site (Gustafson *et al.* 1979) indicates hunting of the live animal (which must have been healthy enough to escape), and discovery of Clovis projectile points, other artefacts, and lithic debitage at the Kimmswick site (Graham *et al.* 1981; Graham and Kay 1988) suggests mastodon carcass processing, and probably hunting, by Paleoindians. In addition, I have argued elsewhere that there is a rich record of human utilization of and predation on mastodons in the Great Lakes region (Fisher 1984, 1987; Shipman *et al.* 1984; Fisher *et al.* 1991).

Although Martin's *blitzkrieg* is the most dramatic 'overkill' hypothesis, there is a wide range of overkill models with less extreme values of human population density, kill rates, and prey recruitment rates (Whittington and Dyke 1984). These lead to more protracted versions of overkill, in which Paleoindian/megafaunal interaction was sustained, both locally and regionally, over longer intervals of time.

Several multifactorial models have also been considered (Marshall 1984). In these, typically, some combination of climate stress and human activity puts megafaunal populations at risk (e.g. Guilday 1984; Haynes 1984). In a recent example of this approach, G. Haynes (1991) has suggested that mammoths, and possibly other North American proboscideans, may have been stressed by drought and constrained to congregate around the few remaining sources of water, where they were then easy prey for Paleoindian hunters. There have also been suggestions of factors unrelated to climate or hunting that may have played a part in the extinction process. For instance, Guilday (1984) and Guthrie (1984) mentioned competition with cervids and/or bovids as a potential source of stress on late Pleistocene proboscideans. Shoshani (1991) has recently proposed, based on the morphology of the hyoid apparatus, that mammutids were not capable of the infrasonic communication documented in extant elephantids (Payne *et al.* 1986). This deficit might have represented a liability for mammutids, but the fact that *Mammuthus* had a hyoid system (and presumably infrasonic vocalization) comparable to that of extant elephantids did not preserve it from a similar fate. Finally, whatever the cause of proboscidean extinction, their loss may have profoundly changed the balance of forces maintaining the physical structure of Pleistocene ecosystems (Owen-Smith 1987). In Africa, the browsing and tree-felling activities of extant elephants open up areas that would otherwise be taken over by forest, thus maintaining habitat heterogeneity and indirectly supporting a high diversity of other herbivores. Proboscidean extinction could have been a cause, rather than an effect, of some late Pleistocene environmental changes and thus could have contributed to extinction of other elements of the megafauna. Although this does not explain proboscidean extinction, it illustrates the potential complexity of organism–environment interactions that need to be considered.

PROBLEMS WITH HYPOTHESIZED CAUSES OF EXTINCTION

I do not presume, in this brief review, to offer a comprehensive evaluation of proposed hypotheses for late Pleistocene proboscidean extinction in North America. It is, none the less, important to comment on a few of the more salient 'sticking points' of particular causal hypotheses. Perhaps the main question raised by climatic hypotheses is why comparable extinctions did not occur with earlier episodes of deglaciation. The obvious answer is that the rate and/or magnitude of climatic change may have been less extreme for deglaciations not associated with extinction (Graham 1986*a*). There is some independent support for this explanation in midcontinent palynological records (King and Saunders 1986) and in the rate and magnitude of fluctuations in the marine oxygen isotopic record of benthic foraminifera (Shackleton and Opdyke 1973; Ruddiman and McIntyre 1982). Nevertheless, more detailed comparisons would be helpful in deciding whether the climatic uniqueness of the late Pleistocene is commensurate with its impact on the mammalian fauna.

There are also problems associated with particular proposed linkages between climatic change and stress experienced by proboscidean populations. A general increase in seasonality is well supported, but by how much, and what were the effects on local climate patterns? Milankovitch models (i.e. calculations of the Earth's orbital parameters at times in the past) and the marine oxygen isotopic record yield broad indications of past climates (Imbrie *et al.* 1984), and palaeobotany and physiological ecology help to provide constraints of a more local nature (Graham and Grimm 1990), but factors such as the incidence and severity of drought or the length of the growing season are still several inferential steps removed. Likewise, disruption of reproductive schedules or development of nutritional stress have not been directly measured in the fossil record.

One of the problems of overkill hypotheses is why there are not more sites showing evidence of human association with extinct megafauna. Uncontested Clovis proboscidean sites number fewer than two dozen, and comparable sites for most of the rest of the megafauna are rare or non-existent. Martin's *blitzkrieg* model explains this as a result of the short duration of human/megafaunal interaction in any one area; he suggests that the physical evidence of overkill should be imperceptible relative to the presumably larger contribution to the fossil record made during the rest of the late Pleistocene. Yet some archaeologists argue for a substantial pre-Clovis occupation of North and South America (e.g. Bryan *et al.* 1978; Adovasio *et al.* 1980; Dillehay and Collins 1988). Although controversial, and not incompatible with more protracted overkill models, this is acknowledged by Martin as a potential falsifier of *blitzkrieg* (Martin 1967).

Even if human association with proboscideans or other megafauna is demonstrated, another hurdle concerns whether utilized carcasses were procured by hunting, or only by scavenging. And even if hunting were demonstrated or assumed, how should we measure the actual impact of that activity on populations of an extinct species? What inferential springboard will ever take us from a finite number of dead animals to an understanding of the cause of extinction of a whole species or group of related species?

Grayson (1984) has argued that part of the difficulty in determining the cause of late Pleistocene extinctions is the tendency of many workers to operate from a stance of 'advocacy' for one view or another. The problem is that many facts can be accommodated in a rational (if sometimes *ad hoc*) manner by both sides. Rarely are tests devised in which divergent predictions are associated with competing hypotheses. Whether because of this, or because of the inherent complexity of the problem, recent work on megafaunal extinction (e.g. as sampled in Agenbroad *et al.* 1990) has shown little tendency to gravitate toward one end or the other of the interpretive spectrum.

RECENT DIRECTIONS OF INVESTIGATION

SPATIOTEMPORAL PATTERNS OF DISTRIBUTION OF ORGANISMS

Documentation of the temporal and biogeographical patterns on which we base our understanding of late Pleistocene biotic change has been improving for decades, but usually in the sporadic fashion dictated by 'accidents' of discovery. In contrast, R. W. Graham, assisted by a team of regional co-ordinators, has initiated a much more comprehensive and consistently organized compilation of such data. This project, known as FAUNMAP, covers much of North America and focuses on a wide range of taxa, both extant and extinct (Graham *et al.* 1991). Its product will be a large-scale database that will link numerous attributes involving individual taxa, specimens, sites, and stratigraphic units, allowing all to be placed within a geographical and temporal framework. Although still in an early stage, it will clearly yield a greatly improved picture of the nature and timing of biotic change against which to compare the record of climatic change and human activity.

Palynologists and palaeobotanists have explored similar approaches, enhancing the usefulness of the floral record. This is important for the primary documentation of response patterns in another segment of biotic communities, and for the insight it offers into the changing resource base for herbivores. Unfortunately, the plant taxa about which most can be said are often not those that would have been most important nutritionally for megafaunal herbivores. There has also been strong interaction between palaeobotanists and climate modellers, using patterns of floral distribution to suggest regional values for climate parameters, and using changes in those patterns as 'ground truth' against which to test models of climate dynamics (Webb and Bartlein 1988). The plant record can serve in this capacity because of the density of sites at which fossil pollen has been sampled, the long and taxonomically rich profiles that are typically obtained, and the ecological informativeness of quantitative data on abundance. Improved models of climate dynamics integrate and flesh out our knowledge of the late Pleistocene (Kutzbach and Guetter 1986) and also provide better tools for evaluating current global change (COHMAP Project Members 1988).

The timing and geographical pattern of local extirpation is a potential source of insight into the cause of extinction (Alford 1974). This approach may not yield clearly divergent predictions associated with climate and overkill models, but comparisons between the spatiotemporal pattern of extinction and particular aspects of climate change, human activity, or proboscidean biology might help isolate specific causal relationships. Meltzer and Mead (1983), Agenbroad (1984*a*), and Grayson (1989) have documented aspects of the pattern of extinction in time and space, but additional investigation of these patterns is needed. A problem with much of the radiocarbon dating done in the past is that bone dates (both apatite and whole collagen), though ideal in providing direct assessment of the age of faunal material, have been subject to a variety of sources of contamination, often resulting in dates that are younger than the probable age of the material. In contrast, wood or charcoal can be preprocessed in ways that greatly reduce the threat of contamination, but they leave uncertainties concerning the association of dated material with faunal remains. Recent advances in purification of bone collagen (Stafford *et al.* 1991) promise significant improvement in removal of contaminants and increased dating accuracy. Previous radiocarbon dating has demonstrated extinction patterns with a broad brush — unfortunately too broad to address critical issues of temporal order and simultaneity. These new advances therefore provide the first real opportunity to evaluate whether taxa in a given region became extinct at the same time or at different times, whether extinctions showed the same patterns in different regions, and when extinctions occurred relative to other relevant events.

TAPHONOMIC STUDIES

Several proboscidean sites have recently been reported that display features of post-mortem bone modification interpreted as evidence of carcass processing by humans (e.g. Hansen *et al.* 1978; Fisher 1984, 1987; Steele and Carlson 1989; Fisher *et al.* 1991). These have been considered controversial by some (Meltzer 1988; Haynes 1991), perhaps in part because most lack unambiguous associations with lithic artefacts. Yet there are other features whose origin seems to call for human involvement: matching marks (of a type that only human activity seems capable of producing) on conarticular surfaces of disarticulated pairs of bones, cutmarks (presumably made during meat removal) identified on the basis of scanning electron microscopy and anatomical context, burned bone heated to at least 440 °C (too high a temperature to be explained by natural fires), distinctive patterns of gouging and breakage at some points of muscle attachment (traces of meat removal different from those left by non-human predators or scavengers), and use wear and secondary flaking on some bone fragments interpreted as tools (Fisher 1984, 1987; Shipman *et al.* 1984). Many of these sites occur in low-energy depositional settings, often in ponds or bogs, with minimal subaerial exposure of bone and limited opportunities for trampling damage. Most workers evaluating such sites are aware of the complexity of interpreting many of these features (Behrensmeyer *et al.* 1986) but rely on multiple lines of evidence and both intersite and intrasite comparisons to constrain taphonomic inferences. Unfortunately, space does not permit detailed reflection on the debate surrounding interpretation of these sites, but *if* they are accepted as evidence of human utilization of proboscideans, the record of such activity may be much richer than suspected previously.

A new perspective on at least some of these occurrences was suggested by excavations at the Heisler site (Calhoun County, Michigan), where bones of a young male *M. americanum* showing evidence of butchery were located in sediments of a late Pleistocene open-water pond, at 1–2 m palaeodepth (Fisher 1987). Associated with some of the bones were 'clastic anchors' and vertical posts (Fisher 1989). Clastic anchors appear to have been sections of intestine (from the mastodon) loaded with sand and cobbles derived from the margin of the pond in which the bones were preserved. The intestinal origin of the anchors has been demonstrated by palynological analysis of associated plant material interpreted as residual intestinal contents (Kapp and Snyder, personal communication 1988) and by recovery of living, enteric bacteria (Tiedje, personal communication 1991) comparable to those from mastodon intestinal contents at the Burning Tree site (Lepper *et al.* 1991). The function of clastic anchors was apparently to hold parts of the carcass on the pond bottom, where they were intentionally placed as a strategy for meat storage. The locations of parts of this cache were marked by the vertical posts. Subaerial meat caching has been suggested at other Paleoindian sites (Frison and Todd 1986; Gramly 1988), and subaqueous meat caching is known ethnographically (Taylor 1969) and may have played an important role in Paleoindian subsistence. Other aspects of the evidence supporting this interpretation will be presented elsewhere, but a corollary is that butchery took place at some other location — not where the meat was stored. This is probably part of the reason why lithic artefacts are rarely encountered in these contexts. Each site will have to be interpreted on its merits, but it may be that many such carcasses preserved in wetland settings represent unrecovered (or recovered, utilized, and abandoned) meat caches.

If butchered proboscideans were procured by scavenging carcasses of animals that had died of natural causes (i.e. unrelated to human activity), or by killing animals that were already close to death, the interaction implied would have been incidental to the extinction process. It is therefore relevant to ask whether butchered animals were hunted, i.e. whether human activity represented an agent of mortality independent of natural causes. To evaluate hunting v. scavenging procurement hypotheses, Fisher (1987) compared the sex ratio, age profile, and distribution of season of death for mastodons that appeared to have been butchered, to the same types of data (discussed further below) for a separate group of mastodons that appeared not to have been butchered. In each comparison there were differences, suggesting that butchered and non-butchered animals represented different sampling processes. Although alternatives could not be ruled out, the observed patterns seemed more likely to be outcomes of hunting than of biased scavenging, and to this degree, hunting was supported over scavenging.

DIRECT EVIDENCE OF DIET

The diets of proboscideans are clearly relevant to some climate-dominated extinction models, and there have recently been substantial advances in the quality and quantity of dietary data available. Several late Pleistocene dry cave sites in western USA have dung deposits attributed to *Mammuthus*, and analysis of this material suggests a combination of browsing and grazing habits, featuring plant species that are, for the most part, still abundant in that general region (Davis *et al.* 1984; Agenbroad and Mead 1987*b*). 'Stomach contents' of *M. americanum* have been reported several times in the past (e.g. Dreimanis 1968) but have rarely been analysed in detail. The discovery of intestinal contents at the recently excavated Burning Tree site, in Ohio (Lepper *et al.* 1991), and at the Heisler site, in Michigan (Fisher 1989), has therefore provided renewed opportunities to study the diet of the American mastodon. Quantitative analyses are underway, but it is already clear that a mixture of browsing and grazing is represented. Deciduous trees and wetland vegetation predominate in mid-autumn food remains (Lepper *et al.* 1991), but conifer material may become more important in late autumn (Ford, personal communication 1992). The generality of these results is at present unclear, but several other reports of possible intestinal contents or faecal material (Garland and Cogswell 1985; Laub 1990; Webb and Newsom 1991; Saunders, personal communication 1991; Laub *et al.* 1994) suggest that new information may soon be available. Dietary data is unlikely to reveal the cause of extinction directly, but it will help to evaluate proposed linkages between diet and ecological stress.

STABLE ISOTOPIC EVIDENCE OF DIET AND CLIMATE

Preserved food remains may provide the most highly resolved evidence of diet, but another source of information is the stable isotopic composition of collagen in bone or in tusk dentine (Koch 1991). Isotopic values can be altered by diagenesis, but for collagen there are independent monitors of isotopic integrity, such as amino acid profiles and C/N ratios (Tuross *et al.* 1988). When well-preserved collagen is available, as is frequently the case for late Pleistocene material, the ratio of $^{13}C/^{12}C$ reflects the carbon isotopic composition of the animal's food. For a herbivore, this distinguishes mainly between food plants that use the C_3 photosynthetic pathway (trees, shrubs, and most grasses and herbs in cool/wet climates) and those that use the C_4 photosynthetic pathway (sedges, grasses, and some herbs from warm/dry climates). This is not exactly equivalent to the distinction between browsing and grazing, respectively, but it is an informative categorization and one that figures in Guthrie's (1984) proposal of extinction mechanisms. The possibility of a shift from C_3 to C_4 plants in proboscidean diets has not been investigated extensively, but Koch's (1991) analyses indicate 100 per cent C_3 diets for late Pleistocene *Mammuthus 'jeffersonii'* and *Mammut americanum* from the Great Lakes region. Krueger (1991), working on molars dredged from the Atlantic coastal plain, reported similar results for mastodons but suggested 50 per cent C_4 diets based on analyses of two mammoth specimens.

Another source of insight into diet is the ratio of $^{15}N/^{14}N$ in collagen. Nitrogen isotope values distinguish between plants that do, and do not, fix nitrogen, but in this case there is a complicating factor. Animals may respond to periods of nutritional stress by metabolizing their own tissues, which would result in a fractionation that shifts nitrogen isotope values positively, an effect similar to that of lower dietary abundance of nitrogen fixing plants (Sealy *et al.* 1987). Koch (1991) found that late Pleistocene Great Lakes region *Mammuthus 'jeffersonii'* had nitrogen values that were higher than those of *Mammut americanum* from the same time period and geographical area. It was not clear, however, whether this reflected differences in the composition of their diets, or a higher level of nutritional stress for the mammoths. Krueger's (1991) mastodons and mammoths were indistinguishable in nitrogen isotope composition and intermediate to those analysed by Koch.

The collagen required for isotope analyses is most easily obtained from bones, but this averages carbon and nitrogen ingested over an unknown span of time, with the potential for modification by later remodelling of bone. Tooth dentine (also enamel) is an accretionary system in which there is no remodelling, but temporal control of sampling depends on the development of incremental lamination. Relative to most other tooth tissues, proboscidean tusk dentine provides extraordinary temporal resolution in that a large amount of material is laid down per unit time, and incremental lamination provides temporal markers on an annual and subannual scale (see below). Temporal resolution is limited by the amount of sample needed for an analysis, but with improvements in technology this continues to decline. This offers an opportunity for seasonally stratified sampling capable of tracking changes in diet within and between years. Moreover, it allows isotopic composition to be related to rate of tusk growth (by measuring the thickness of periodic growth increments in the zone from which each isotope sample is derived). Rate of tusk growth is probably related to nutritional status (Fisher 1991), making it possible in principle to tease apart some of the dietary composition and nutritional stress components of the nitrogen isotope signal. Ambiguities may remain for some short-term fluctuations, in some phytogeographical settings. None the less, suggestions such as G. Haynes' (1991), that late Pleistocene Clovis-associated mammoths in western USA were in starving condition, 'tethered' to water holes by drought, should be testable with isotopic data.

Data on oxygen isotopes have also contributed recently to understanding proboscidean palaeobiology and palaeoenvironment (Koch *et al.* 1989). These data were derived from the carbonate fraction of hydroxyapatite in tusk dentine, although analytical advances may soon allow results to be obtained from the more diagenetically resistant phosphate fraction. Sampling was conducted in the temporally stratified manner described above and demonstrated regular patterns of change in $^{18}O/^{16}O$, corroborating both the annual nature of the largest-scale growth increments in mammoth and mastodon dentine and the seasonal identifications that had originally been based on inferred apposition rate of tusk dentine (Fisher 1987). Because mammalian homeothermy rules out environmental temperature as a component of this variation (dentine is secreted at core body temperature), the changes in oxygen isotope values measured in tusks mainly reflect changes in the isotopic composition of ingested water. The composition of the water is itself largely a function of the temperature at which it condensed (Dansgaard 1964). If ingested water were dominated by recent precipitation, as it would be if proboscideans acquired most of their water from surface runoff and plant water (perhaps a better approximation for some small mammals), the oxygen isotope variation in tusks would record local variation in temperature of condensation, providing a quantitative record of seasonality. In all likelihood, hydrological factors (e.g. mixing of waters from different sources, with different residence times in near-surface environments) will complicate this simple picture, but even if the signal is damped, its amplitude still provides a lower limit on seasonality. Moreover, there may be parts of the year that preserve a relatively unadulterated signal, such as if proboscideans ate snow or ice during winter months (Koch *et al.* 1989). Comparative studies (e.g. looking at the same species in different regions, or at taxa of different body size and presumed habits in the same region) would be needed to sort out confounding factors, but there may be an opportunity to document the history of changing seasonality in some detail. This is especially important for mid-continent regions because other sources of climatic information are either far removed geographically (e.g. the marine oxygen isotopic record) or less direct (e.g. palaeobotanical evidence).

STUDIES OF GROWTH RATE AND LIFE HISTORY

The goal of many studies of late Pleistocene extinction has been to evaluate extinction hypotheses by investigating the dynamics of putative causal factors and the temporal relationship between their action and the extinction event. It may be more fruitful, however, to focus on the other end of the cause–effect vector, i.e. the actual patterns of response of the biota. This is already being done in studies of range shift and evolutionary change in morphology and body size (King and Saunders 1984; Graham 1986*a*; Guthrie 1990*b*), but there has been little investigation of biotic response on the level of population dynamics and life history. The main requirement for such studies, detailed information on growth, development, and reproductive biology, has simply not been available. Recent work on growth increments in proboscidean tusk and molar dentine (Fisher 1987, 1988, 1990; Koch *et al.* 1989), however, shows promise of providing the information needed to implement this approach. Procedures have been proposed for recovering data on season of death, individual age, sex, male age at sexual maturity, and age-specific data on other aspects of male and female growth and development. Female age at sexual maturity and calving records (age at first calving, calving intervals, and number of offspring) may also be available. While such data are routinely collected for extant organisms (e.g. Clutton-Brock *et al.* 1982), this level of detail would generally not be considered accessible for fossil vertebrates. The principal 'recording device' that provides this unexpected yield is the permanent tusk, which begins to deposit dentine near birth and continues to grow throughout life. The 'record' involves the number and spacing of periodically formed, incremental laminae in tusk dentine, which directly measure tusk growth rate, but which indirectly reflect nutritional status and significant reproductive events.

I argue below that the principal factors thought to have caused late Pleistocene extinction should have had contrasting effects on proboscidean life history parameters. At the risk of oversimplifying, if climatic and/or vegetational change was the main cause of extinction, proboscideans approaching extinction might show declining body size, in this case associated with decreased growth rates, increased age at sexual maturity (possibly to a value of about 20 years), and increased calving intervals (to a value of about 8–10 years). On the other hand, if hunting by humans was the prime cause, body size might again decline, but in this case it would be associated with stable or even increased growth rates, a stable or decreased age at sexual maturity (to about 10 years), and stable or decreased calving intervals (to about 4 years). These contrasting predictions may provide a means of testing climatic and overkill hypotheses. I do not presume at this time to be able to document actual changes in life history parameters in enough detail to consider the problem solved, but articulation of a design for a test is a first step. Before taking this up, it may be useful to review the mode of growth of proboscidean tusks and the ways in which life history may be reflected in tusk growth.

TUSK GROWTH AND INCREMENTAL LAMINATION

Elephantid and mammutid tusks have a structure and pattern of growth that appears to be typical of most proboscideans (the principal exception being the dentinal rods of some amebelodontids; Osborn 1936). Most of the volume of the tusk is dentine that is deposited proximally, along the surface of a conical pulp cavity, and the tusk grows throughout life by apposition of dentine increments and correlated eruption (Fig. 30.2). Enamel is present initially at the tip of the tusk but is worn away early in life. Cementum is deposited within the alveolus, on the outer surface of the elongating dentinal portion of the tusk, but when a given part of the tusk has erupted, the layer of cementum, and subsequently the underlying dentine, are subject to abrasion. Material may be lost from the tusk tip by abrasion or fracture, but this is usually at a rate greatly exceeded by eruption, so that the exposed portion of the tusk continues to elongate.

Some of the first advances in understanding the growth of proboscidean tusks and the biological data they record were made by Osborn (1910). On the outer surface of the tusks of the

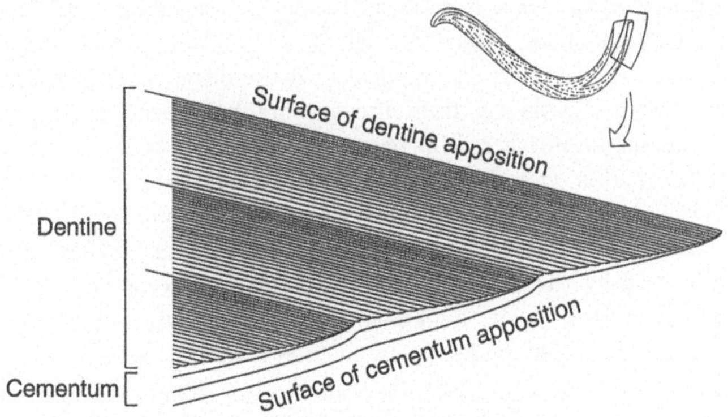

Fig. 30.2 Schematic diagram of elephantid and mammutid tusks. Index drawing shows the lateral half of a longitudinally sectioned, right upper tusk. The conical pulp cavity is evident at the tusk base, and dentine increments are exposed in section as imbricate, v-shaped traces. At this scale, the cementum layer on the outside of the tusk is too thin to show clearly. Below is an enlarged view of dentine increments (first- and second-order) and cementum increments (first-order only) near the proximal end of the tusk.

Warren mastodon, Osborn noted regular topographic features that clearly reflected growth increments. His description somewhat confounds topographic and temporal issues by using the term 'annular', meaning ringlike, in a context where he obviously meant 'annual', or yearly (reprinted in Osborn 1936, p. 181–2):

> In repairing the tusks, the outer sheathing of the dentine was found in large part absent. The inner sheathing exposed a series of concentric constrictions and expansions which were observed to be approximately symmetrical on the two sides... Secondly, it was noted that the intervals between these constrictions are broader in the middle and fore part of the tusks (corresponding with the youthful stage of growth of the tusk) and become narrower toward the base of the tusk (corresponding with the mature or adult stage of growth). Eighteen of these rings are preserved on one side and thirteen on the other. They are faintly indicated also in the waving surface of the dentine of the tusk... On the hypothesis that these are actual annular increments of growth, the right tusk...consisted of about twenty-eight segments, which allowing for the milk dentition and for the part worn off at the tip would assign to the Warren Mastodon an age of perhaps thirty years.

By 'outer sheathing' I suspect Osborn meant the layer of cementum that covers the outer surface of the proximal portion of the tusk, although in fact, as shown in his accompanying drawing, this only partly obscures the regular, topographic features that Osborn recognized. It is less clear what he means by 'inner sheathing', but I suspect that this refers to the outermost layers of dentine, exposed by spalling of cementum and themselves somewhat subject to exfoliation. '[T]he waving surface of the dentine' probably refers to that part of the tusk surface that is formed by dentine, where the cementum has worn away. Since only eighteen topographical features are preserved, it is not clear how he came to suggest the presence of 'twenty-eight segments'. It is not unusual for these incremental features to be expressed partly as colour banding in the dentine (and/or the cementum), but Osborn does not note this for the Warren mastodon (it may be suggested in tusks of *Elephas* that he illustrates; 1936, p. 183), and in my experience it is difficult to trace these colour bands on the external surface near the tusk tip. Osborn may simply have estimated the number of 'segments' in the distal portion of the tusk by extrapolating from the spacing he observed for the 'segments' he numbered (in reverse chronological order) 17 and 18, but this would imply no independent support for his characterization of intervals in the 'fore part of the tusks'.

Osborn intended to do additional work on dental histology, as witnessed by the thin-section illustrations and comments published posthumously as an appendix to Volume II of his 'Proboscidea' (Osborn 1942). In addition to details of enamel and cementum structure, these illustrations show fine-scale lamination within the dentine, but there is no suggestion of the temporal significance of these laminae or their relation to the surface topography thought to represent annual features. Dentinal lamination in proboscideans was also studied briefly by Schour and Hoffman (1939), who noted, in a mastodon molar, the same type of fine laminae illustrated by Osborn. They proposed that these were daily increments, but the scale of these features corresponds to what are recognized below as fortnightly laminae.

As described elsewhere (Fisher 1984, 1987, 1988, 1990), proboscidean dentine is characterized by a hierarchical system of growth increments representing first-order (annual), second-order (generally fortnightly), and third-order (daily) physiological-depositional cycles. First-order features are what Osborn noted on the Warren mastodon. Their topographical expression is a result of seasonal changes in the relative rates of dentine apposition, measured normal to the incremental surface, and dentine transgression (i.e. extension of the proximal margin of the tusk, or cheek tooth root, here analogized to stratigraphic transgression). Typically, the surface of dentine apposition dips shallowly relative to the external surface of the tusk (or cheek tooth root), and so the average linear rate of dentine apposition must be lower than the average linear rate of dentine transgression. During winter, both rates decline, but the rate of transgression declines disproportionately, producing a constriction, or annulation. During summer, both increase, but the rate of transgression increases disproportionately, making up for its winter lag. Progressive deposition of cementum (thin near the growing margin of the tusk, but with greater accumulated thickness toward the alveolar border) subdues this topography, but traces of its relief are exposed again on the abraded surface of the erupted portion of the tusk, which often cuts down to the dentine-cementum interface and into the dentine. On well-preserved tusks (or cheek tooth roots) first-order features may be more or less conspicuous depending on the seasonality of the climate to which an individual has been exposed, individual physiological variation, and the amount of cementum formed. Frequently, however, desiccation-induced spalling of cementum (cementum shrinks less than dentine) removes this latter variable by exposing the topography of the dentine–cementum interface directly to view.

First-order features may also be characterized by colour variation, although appearance varies with depositional setting and banding is often subtle in the best-preserved specimens. Although some colour variation is clearly primary (and seen in recent material), I suspect most represents diagenetic enhancement of zonal structural variation. Winter is usually represented by darker material, and spring–autumn by lighter material. A comparable zonation may develop within cementum, but because of the mirror-image directions of apposition of dentine (inward, onlapping proximally) and cementum (outward, onlapping proximally), the three-dimensional configuration of colour zones, and their relation to the tusk surface, is complex. Dentine increments resemble a series of imbricated 'cones', and cementum increments resemble a series of imbricated 'sleeves'. Figure 30.3 illustrates this with highly idealized models, in which thicknesses of first-order features are constant (rather than varying with ontogeny and year-to-year 'noise') and cementum thickness is exaggerated. In Figs 30.3a and b, longitudinally sectioned tusk-models show the interior organization and exterior appearance of colour bands in cementum and dentine, respectively. These models ignore abrasion, but this allows them to show more clearly how the three-dimensional structure of the tusk develops. The proximal, 'unbanded' portion of the outer surface of the cementum

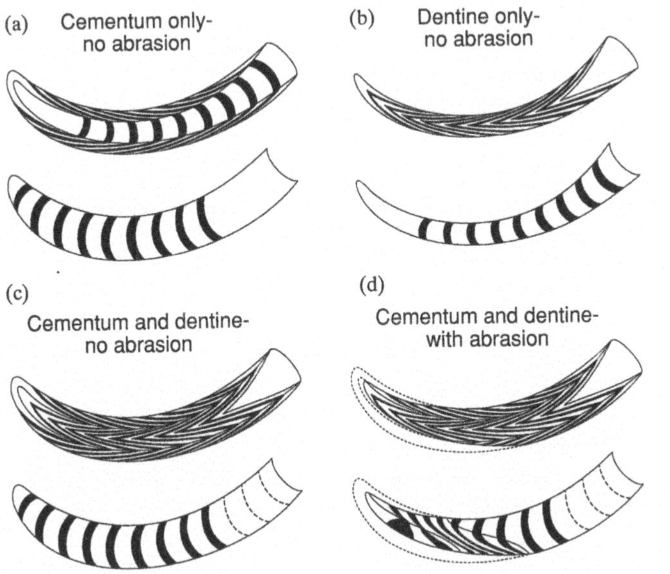

Fig. 30.3 Schematic diagram of elephantid and mammutid tusks, showing first-order colour bands in dentine and cementum. Each set of diagrams shows a tusk sectioned along its longitudinal axis, with halves displaced to show the external aspect below and the axial cross-section above. Tusks are simplified in having curvature confined to a plane (to avoid foreshortening) and first-order features of constant thickness. (a) Tusk cementum only (dentine not shown), in exaggerated thickness (to clarify the pattern of cementum bands) and as it would look if no abrasion had removed material from the outer surface of the tusk. (b) Tusk dentine only (cementum not shown), again as it would look if no abrasion occurred. (c) Dentine and cementum combined as in real tusks, as they would look if no abrasion occurred; the topography of first-order features in dentine is indicated by dashed lines where it can be perceived beneath the cementum of the alveolar portion of the tusk. (d) Cementum and dentine abraded, and the outcrop pattern of bands shifted accordingly (outline of material removed shown by dotted line). This is a fairly typical pattern, although in many cases, cementum is removed completely from the distal quarter or third of the tusk. Nine first-order, dark (winter) bands are shown. Because cementum forms as a series of proximally imbricating, geometrically similar 'sleeves', increments of cementum prior to the last-formed ones outcrop no nearer the proximal margin of the tusk than the alveolar border (near the gumline); because dentine forms as a series of proximally imbricating, geometrically similar cones, some of the very early dentine outcrops well back from the tusk tip.

represents the finite length over which cementum is more or less simultaneously deposited within the alveolus (i.e. the length of the 'sleeve'); the distal, 'unbanded' portion at the dentine–cementum interface represents the finite length of the initial cone of dentine. Figure 30.3c combines these models and shows how first-order topographic features of the underlying dentine, in the alveolar portion of the tusk, may 'show through' on the unbanded, appositional surface of the cementum. Figure 30.3d adds abrasion, showing how the generally low, but locally variable, angular relation between colour bands and the tusk surface on which they are exposed can lead to complication of their outcrop pattern. Apparent complexity may be even greater in real tusks because their cementum is thinner and because abrasion is often more irregular than shown here. This can make bands difficult to follow and count on the tusk surface, especially if cementum and dentinal bands are not consistently distinguished.

Returning to dentine, second- and third-order increments are demarcated in thin section by distinctive types of 'growth lines' (zones of vacuities formed under altered conditions or rate of dentine deposition), second-order increments are expressed topographically at the dentine–cementum interface, and first-order increments are marked by density differences, spacing of higher-order increments, and patterns of isotopic variation (Koch *et al.* 1989). Fourier analysis of second- and third-order increment spacing has confirmed their spatially periodic nature (Koch 1989) and the fundamental numerical relation that first-order increments usually contain about 26 second-order increments and that second-order increments usually contain about 14 third-order increments (Fisher 1987). This corroborates Osborn's suggestion that annulations on the tusk of the Warren mastodon represent years, and suggests that second- and third-order features are temporally as well as spatially periodic, representing fortnights and days, respectively.

That annual features might appear in this context is not surprising, given the frequency of annually repetitive, seasonal influences on mammalian physiology and the appearance of similar incremental features in other taxa (Klevezal and Kleinenberg 1969). Likewise, daily features may simply reflect the influence of circadian rhythms on mammalian biology and are matched by daily features (von Ebner lines) in other taxa (e.g. Miani and Miani 1972; Rosenberg and Simmons 1980). Second-order laminae are more problematic, partly because they can be expressed variably, even within individuals. Their periodicity is usually fortnightly, but in a few instances it is twice this length (twice as many third-order features included) — apparently equivalent to lunar monthly (though whether it corresponds to any lunar phenomenon is an open question). By checking their number and thickness relative to first-order lamination, it is usually clear whether fortnightly or lunar monthly increments are at hand, but it is not clear what cyclical aspect of proboscidean biology they record or why proboscideans would shift from one to the other mode. Based on reports of a two-week oestrous cycle in extant elephants (Short 1966; Eisenberg *et al.* 1971), behavioural observations on elephant mating and feeding activity, and a small sample of cases in which I could specify the age–sex distribution of fortnightly and lunar monthly second-order lamination, I originally proposed that fortnightly laminae were a facultative reflection of mating activity and its perturbation of normal feeding patterns, and that lunar monthly laminae represented a background dependence of feeding activity on the level of nocturnal (lunar) illumination (Fisher 1987, 1988). I now consider this hypothesis refuted by three classes of data:

(1) additional cases where I could specify the age–sex distribution of the two types of second-order lamination did not replicate the patterns observed originally;

(2) a zoo-reared *Elephas maximus* which had never been mated showed normal fortnightly patterns (Koch 1989); and

(3) more recent data on elephant oestrous cycles indicate that they are closer to four months in length (Schmidt 1992).

The biological basis for second-order lamination is not yet clear, but a strong role for temporal (as opposed to purely structural) control is supported by observation that the number of second-order increments per first-order increment remains about constant despite variation in thickness of first-order increments. It is also interesting (Koch 1989) that patterns of this general periodicity have been observed in other groups of mammals, e.g. dugongs (monthly, Marsh 1980), dolphins (monthly, Hohn 1980), and sea lions (fortnightly, Laws 1962). In addition to these marine examples, there are weekly increments in human tooth enamel (Bromage and Dean 1985). Further work on all these patterns is necessary, but in the meantime, second-order incremental features in proboscideans may be used as temporal markers, even though their cause is not understood.

Methods for collecting data on proboscidean dentinal lamination include measurements and counts made on the external surface of the tusk, on the dentine–cementum interface where cementum has spalled off, on an axially positioned longitudinal section of a tusk, or on transverse sections from a series of positions along the length of a tusk (Fisher 1987, 1988, 1990). The most challenging of these approaches is longitudinal sectioning of complete tusks, but it offers the most comprehensive documentation of tusk growth and has the added benefit, if done when the tusk is freshly excavated and not yet dried out, of preventing much of the radial and concentric fracturing that can reduce some tusks to rubble in a matter of months or years. Although work on first-order increments can sometimes be done without thin sections, analysis of second- and third-order increments almost always requires them. Thin sections are examined with a petrographic microscope, and increment counts and data on spacing have been collected from photographs and camera lucida drawing, or more recently by using the microdensitometric capabilities of an image analysis system.

General identification of seasons is based on colour band analogy with extant mammals, variation in increment spacing and the implied rate of dentine apposition, and oxygen isotopic signature. Quantitative assessment of temporal relations on a subannual scale is based on a convention for locating annual boundaries. The winter–spring boundary is the time of year with the most consistent association of properties and is taken as the end of the last thin second-order increment within a group of thin second-order increments associated with a dark (winter) colour band. The number of included fortnightly increments may vary from year to year either because of real variation in the timing of onset of spring or because of errors in identifying increments, but variation between the expected and observed number of fortnights is usually small ($c. \pm 2$).

LIFE-HISTORY DATA

As should be clear from the above discussion, proboscidean dental materials, especially tusks, provide an unusually complete record of growth, along with an internal 'metronome', the production of daily and fortnightly growth lines, ticking off equal intervals of time. If the proximal end of a tusk is available, with even a partly intact pulp cavity, season of death may be determined by counting the number of second-order increments, in thin section, since the last winter–spring boundary (Fig. 30.4). With a complete tusk, age at death can be estimated by counting the number of first-order increments. As Osborn did, this can be attempted from the tusk exterior, but it is usually only practical after sectioning. In extant elephants, the permanent tusks begin to form near birth ('allowing for the milk dentition' may therefore be unnecessary; Osborn 1910), erupt at 2–3 years, and continue growing throughout life (Sikes 1971). As noted above,

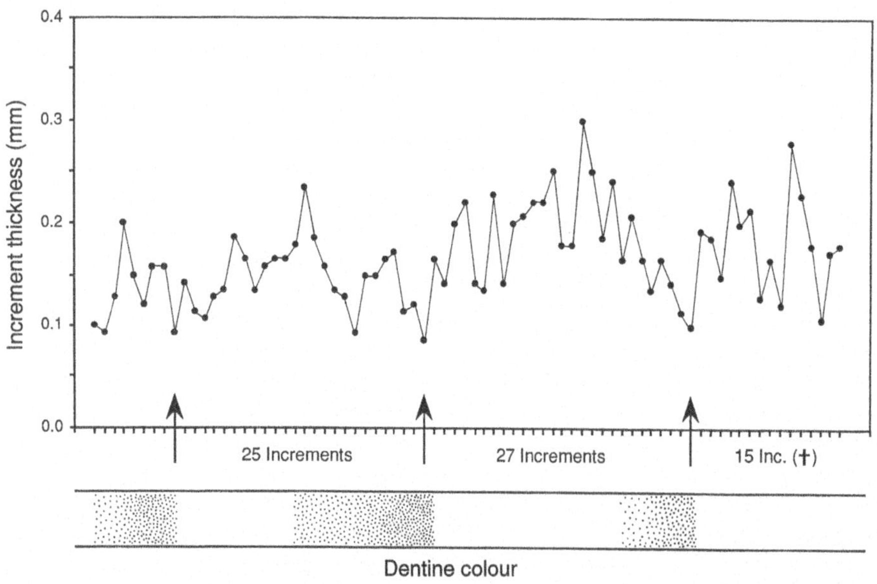

Fig. 30.4 Profile of second-order (fortnightly) increment thickness for the last several years of life of the Grandville mastodon (♂). The profile terminates at death; arrows mark winter–spring boundaries, and the season of death (autumn) is evaluated relative to the last winter–spring boundary. The bar below the graph shows dentine colour in reflected light (after Fisher 1990).

material may be lost by abrasion or fracture of the tusk tip, but this does not necessarily imply a significant loss of time from the record of life history. The length of the first year's cone of dentine may be 20–30 cm, and since it is symmetrical about the tusk axis, while abrasion is usually asymmetrical (G. Haynes 1991), the tusk tip can recede substantially without losing all of the first year's dentine. Larger-scale breakage sometimes occurs, but is generally evident from the tusk profile. If both tusks, particularly of a young individual, are acutely pointed and of the same length, the chances of having lost a large amount of time from either are small. Taking the first year in the permanent tusk as the first year of life may turn out to err slightly, but it is a convenient provisional reference point and can be tested by comparison with other reference points in life history (see below), by recognition of a neonatal line, and by nitrogen isotope analyses documenting a lactational signature (Koch, personal communication 1990).

The record in tusk dentine can also be correlated with that of cheek tooth dentine. Single cheek teeth have a more restricted interval of dentine deposition, but because of the protracted schedule of tooth formation and replacement typical of proboscideans (Laws 1966; Roth and Shoshani 1988), teeth at one or more positions in the sequence (within each quadrant) are likely to be depositing dentine in all but the oldest individuals. This provides multiple checks and/or multiple opportunities to compile a record of life history, even when dealing with incomplete preservation.

Sexual dimorphism in most proboscideans is strong and is reflected mainly in tusk size and body size (Osborn 1936). Although it needs further analysis, the dimorphism parallels that of extant elephants (Elder 1970). Figure 30.5 shows the general pattern for *M. americanum*, with males having tusks that are longer and more robust for their age and stage of tooth eruption and wear. This is also reflected in thicker annual dentine increments. Bone lengths and schedules of epiphysis fusion also display sexual dimorphism (Roth 1984; G. Haynes 1991).

Data on reproductive biology appear to be recorded in patterns of tusk growth rate. Because dentine is laid down relatively continuously (at least over time intervals longer than a day), the thickness of periodic (usually second-order) increments is an indication of average rate of dentine apposition — one measure of tusk growth rate. The pattern of seasonal variation in tusk growth rate, measured in this way, suggests that tusk growth is highly correlated with nutritional status and, for much of the lifespan, with overall somatic growth rate (Fisher 1990): low during winter and higher during spring–autumn. For instance, in the last full year in Fig. 30.4, from a mature male *M. americanum*, dentine is laid down in the summer at a rate as high as 0.3 mm/fortnight and in late winter at a rate as low as 0.1 mm/fortnight, for a total yearly accretion of 5.007 mm at an average rate of 0.185 mm/fortnight. By counting annual increments along a complete tusk, the record of growth rates provided by any portion of the tusk can be placed within the context of the whole life, and the lifelong pattern of variation in growth rate is itself an informative record of life history.

Another approach to studying tusk growth rate is to measure increase in length. Increments of tusk dentine crop out obliquely on the dentine–cementum interface and add length as well as 'stratigraphic' thickness to the tusk. Another measure of tusk growth rate is thus the amount by which tusk length is extended with the addition of each increment. If annual increments are clear enough, this technique can be used on the exterior and/or longitudinally sawed surfaces of tusks, but thin sections provide confirmation and greater precision in the distal, early portions of tusks, where expression of annual increments is often faint. Where I have compared rates of dentine apposition to rates of length increase for the same year (Fisher 1990), they seem to be giving the same signal, and this is not surprising given the basic geometry of tusks.

Profiles of tusk growth rate throughout life might be expected to give information on normal year-to-year variability in climate, available food, or health status. Comparing profiles for same-sex individuals, however, shows striking, age-related regularities that I propose reflect life-history parameters involving reproductive biology. Figures 30.6a–c show profiles of annual increments in tusk length for three male mastodons. The overall pattern is one

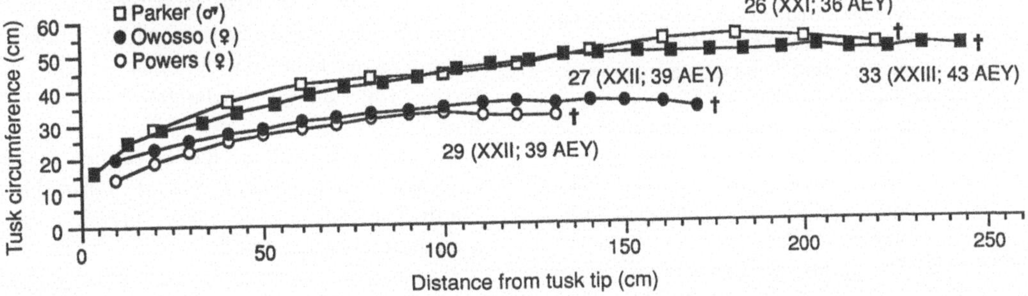

Fig. 30.5 Sexual dimorphism of American mastodon tusks demonstrated by graphs of tusk circumference v. distance from tusk tip, at 10 or 20 cm intervals, on the Grandville (♂), Parker (♂), Owosso (♀), and Powers (♀) mastodons. Tusk length and basal circumference at death are indicated by †. Toward the end of each profile is indicated the age at death, and in parentheses, the Laws (1966) age group (based on cheek tooth eruption and wear) and corresponding age estimate in 'African-equivalent years' (AEY; Saunders 1980). Females have shorter and more slender tusks for a given stage of cheek tooth development.

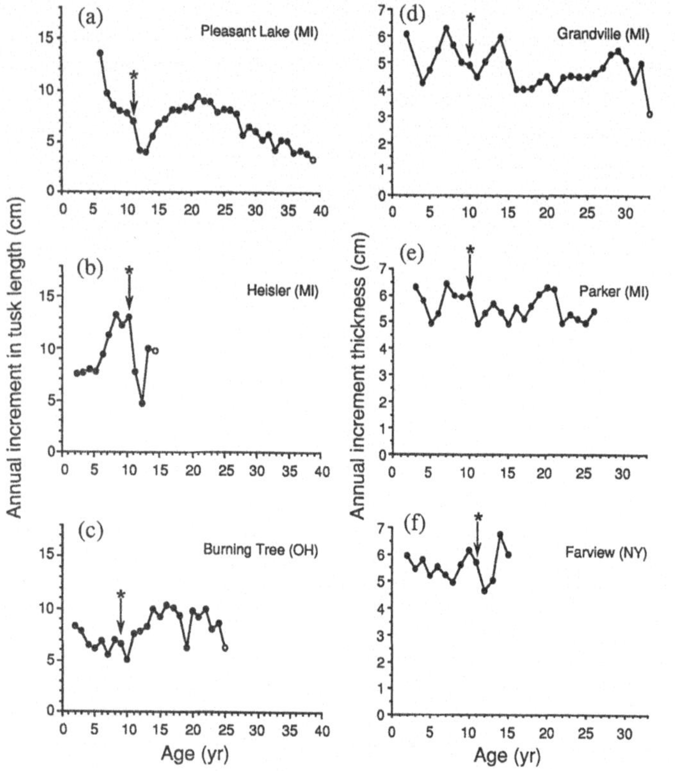

Fig. 30.6 Tusk growth profiles for male American mastodons. An asterisk marks the inferred age at sexual maturity, immediately preceding a conspicuous drop in tusk growth rate interpreted as a reflection of nutritional stress in novice males recently excluded from their natal family unit. (a–c) graphs of annual increment in tusk length, the distance between the (x-1)th and xth winter–spring boundaries, measured along the outside curvature of the tusk. Measurements between well demarcated winter–spring boundaries are shown as solid dots; open circles report data for an incomplete year, generally an autumn death; no data are shown where boundaries are obscure (positions of all winter–spring boundaries could be determined by analysis of thin sections). Boundaries were not clear in early years of the Pleasant Lake mastodon, but alignment of its profile relative to the others was estimated by registration of its tusk circumference v. length profile to those of several other males. (d–f) Graphs of annual increment thickness measured on transverse sections cut from longitudinally sawed tusks. Annual increment thicknesses were measured in the plane of section but were converted trigonometrically to 'true' thicknesses. Measurements near the tusk axis and the dentine–cementum interface are problematic, because annual increments thin slightly near these loci. Close spacing of samples covering years 5–15 represents these years at full thickness and prevents this from influencing identification of age at sexual maturity, but some of the local minima elsewhere on these profiles are artefacts of inadequate sample coverage. Tip breakage has removed an estimated 1 year from Parker; profiles for Grandville and Parker extend to death, but only about the distal half of one Farview tusk was recovered.

of variable juvenile and young-adult rates, declining in older individuals. Most detailed aspects of the pattern differ from one individual to another, but within about one year of age 10, all males experience a marked drop in tusk growth rate, which then recovers over a period of several years. This same pattern is shown in profiles of annual increment thickness for three additional males (Figs 30.6d–f; these graphs include an additional component of 'noise' in their adult portions, explained in the caption). I have suggested (Fisher 1990, 1991) that the drop in growth rate is a response to expulsion of males from a matriarchal family group at the time they become sexually mature and that the following years of low growth rate reflect the nutritional stress of novice males before they are acclimated to life independent of the matriarchal group in which they were nurtured as calves. This implies a social structure comparable to that of extant elephants, as has been proposed elsewhere (G. Haynes 1990, 1991). The implied age of sexual maturity is near the lower end of the range observed for extant elephants (Eltringham 1982; Moss 1988) but is thoroughly plausible. Behavioural observations of African elephants support the concept of stress and loss of condition in young males after expulsion, leading in fact to an increase in mortality rate (Moss 1988, p. 268) corresponding to what is seen here as the interval of depressed tusk growth rate. Although no one has analysed elephant tusks in the manner of Fig. 30.6, Hanks (1972, p. 265) showed graphs of tusk circumference v. age that have an inflection located at about the same position and that I suspect represent the same phenomenon. In short, by noting the age at which this pronounced drop occurs, it may be possible to determine age of sexual maturity for male mammoths and mammutids.

Additional insight into tusk growth rate in males may be gained by looking at a finer temporal scale. Figure 30.7 shows fortnightly increment thicknesses for the Grandville mastodon from years 8–14. Late in the summer of year 10, after an excellent growing season, comes a dramatic drop in tusk growth rate to a level even lower than many late winter intervals. A brief recovery flickers, but then consistently low values set in for winter. The following spring is uncharacteristically sluggish, and the summer brings only one brief peak. The succeeding spring is likewise slow, though its summer shows some improvement. Overall, it is not until the spring of year 13 that the pattern returns to normal. In this male, the 'bad' years immediately following maturation do not show unusually low winter rates; instead, annual growth rates are low because of failure to attain all of the high short-term rates typical of most spring–autumn intervals. This may be because all late winters are stressful, and being a novice male does not make this time of year appreciably worse, while lack of matriarchal guidance or protective status may make access to resources during the growing season less predictable. Although it might be easy to overinterpret any single sequence such as this, it represents a plausible scenario for male maturation, and its generality may be easily tested by examination of other individuals.

Vereshchagin and Tikhonov (1987) have proposed a different approach to assessing the timing of male sexual maturity. They too have measured features that they treat as annual increments in tusk length, but they propose that sexual maturation can be recognized as an interval showing especially large growth increments. While males approaching prime reproductive age may

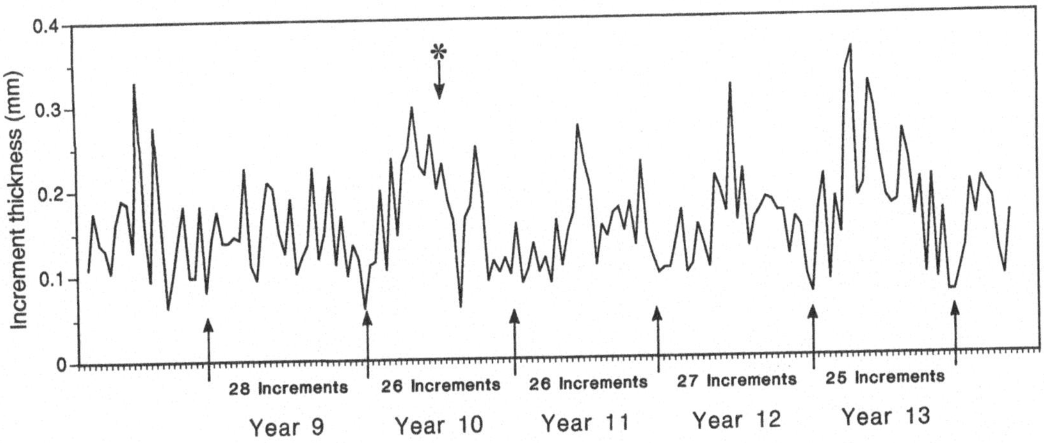

Fig. 30.7 Profile of second-order (fortnightly) increment thickness on the Grandville mastodon (♂), spanning an interval from before the inferred age of sexual maturity (asterisk), through a period of depressed growth rates, to after resumption of a normal growth pattern. Arrows mark winter–spring boundaries.

show large growth increments, this phase of life history does not occur in extant elephants until well after sexual maturity. Furthermore, a male's highest reproductive status and success is associated with the onset of musth, and I suspect that this will be marked by repeated (if not exactly periodic) *decreases* in tusk growth rate (because of the nutritional stress associated with musth; Moss 1988), superimposed on a seasonal growth cycle. A further uncertainty concerns the ages that Vereshchagin and Tikhonov (1987) attribute to their animals and thence to inferred life-history events. G. Haynes (1991) has noted that the maximum ages Vereshchagin and Tikhonov record for Siberian mammoths are unexpectedly high, and he interprets this as due to counting something other than annual features. While this is possible, an alternative interpretation is that they counted only first-order features but somehow combined dentinal and cementum features. Not having examined their material, I do not wish to defend or attack any particular interpretation, but as an example of one type of possible error, in a tusk from an old individual, with progressively shorter annual increments in its alveolar portion, counting the equivalent of dashed *and* solid first-order features in Fig. 30.3d could add 15 or more years to apparent age.

In keeping with observed sexual dimorphism, female tusk growth rates are mostly lower than those of males, but they show the same broad pattern of variable juvenile and young-adult rates followed by a general decline in older individuals (Fig. 30.8). Females do not show the same type of drop in tusk growth rate around age 10 as do males, but since female elephants remain within their matriarchal units (Moss 1988), this is to be expected. At about the same age, however, females do initiate a pattern different from that seen in males — a rather regular fall and rise of tusk growth rate that may continue at fairly constant intervals for the rest of their life. These patterns are subtle, and I recognize the danger of reading too much into every jig and jag of a time-series, but I propose that these profiles are a record of calving for each individual. Before about age 10 no consistent pattern is evident, but at that age (essentially the same as male maturity, as is usually true for elephants) the undulatory pattern begins. Although any time-series will have its ups and downs, in these tusk profiles many of the undulations seem to have a period of about 3–4 years, which is within (near the low end of) the range of observed calving intervals for extant elephants. Furthermore, these patterns seem to be sustained in a regular manner within individuals (e.g. Powers and Alma in Fig. 30.8). Variation in climate and/or food availability, such as we may in fact be seeing in the adult lives of males (Figs 30.4 and 30.6), would be expected to be more irregular. Figure 30.9, comparing the inferred calving histories for females from Fig. 30.8 to equivalent summaries abstracted from the males of Figs 30.6a–c, shows that this expectation is borne out. If the rate fluctuations in female tusk growth do reflect calving, conception probably occurred during the year just preceding a marked drop in tusk growth rate; the drop would reflect the depressed serum calcium levels associated with pregnancy (Luciano *et al.* 1978) and increased investment of calcium and phosphate in fetal growth and ossification, to the detriment of tusk growth rate. Lactation (which in elephants may continue into the following pregnancy) would also have an impact on tusk growth rate, but probably not as much as fetal growth. If American mastodons, like elephants, had a gestation period of 18–22 months (Laursen and Bekoff 1978; Shoshani and Eisenberg 1982), we might expect about 1.5–2 years of low tusk growth rate followed by some recovery during a period of lactational anoestrus. In elephants this lasts another 2 years or more, giving the 4 year, or greater, calving interval.

As in males, the significance of these patterns may be investigated further by examining them on a fortnightly scale, though as of yet, I have done only spot checks to gain a sense of general similarities and differences. In contrast to the pattern noted above in a novice male, the low annual growth rates identified here as pregnancy show somewhat subdued, though more or less typical levels of intra-annual variation, but the whole pattern is shifted to lower values. This is consistent with a more continuously acting influence on growth rates such as pregnancy would represent.

Returning to profiles of annual increments, the Owosso mastodon (Fig. 30.9) shows an interesting departure from the pattern of the Powers and Alma mastodons in that after what

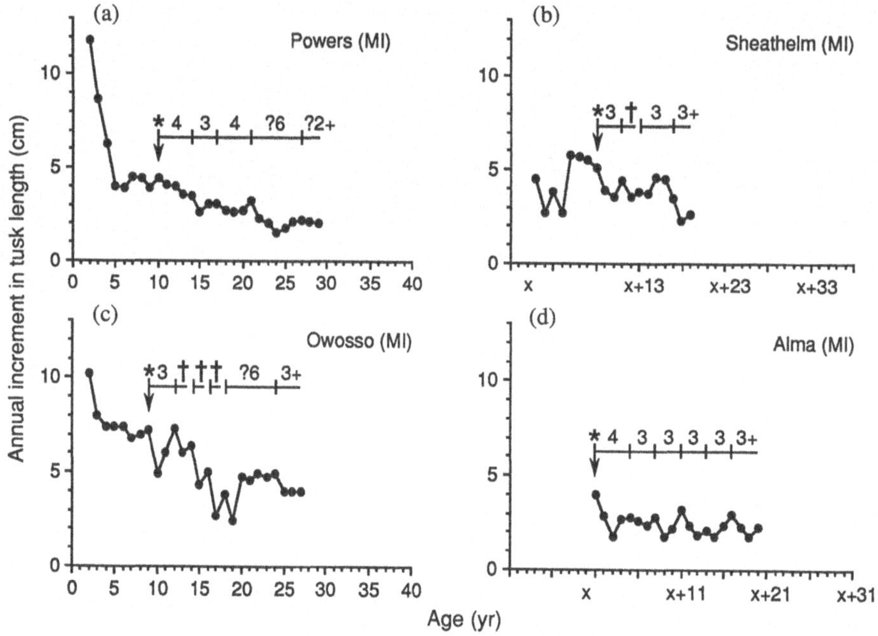

Fig. 30.8 Tusk growth profiles for female American mastodons. (a–d) Graphs of annual increment in tusk length, the distance between the (x-1)th and xth winter–spring boundaries, measured along the outside curvature of the tusk. Measurements between well-demarcated winter–spring boundaries are shown as solid dots; no data are shown where boundaries are obscure (positions of all winter–spring boundaries could be determined by analysis of thin sections). Boundaries were not clear in the early years of Sheathelm and Alma, but the alignment of their profiles relative to others was estimated by registration of their tusk circumference v. length profiles to those of other females. The flatter circumference v. length profile of females (relative to males) makes this estimate less precise than for males (acknowledged by use of a relative scale on the abscissa), but age is probably still precise to ± 2 yr. An asterisk marks the inferred age at sexual maturity, the beginning of a series of decreases and increases in tusk growth rate which may reflect calving. The line to the right of the asterisk represents inferred calving history. Conception (bars on the history line) is marked by the beginning of a 1 or 2 year decline in tusk growth rate, generally followed by a recovery until the next conception. Calving intervals are indicated by numbers along the history line, except where † marks the inferred death of a calf soon after birth. All profiles extend to death (in the late winter or early spring).

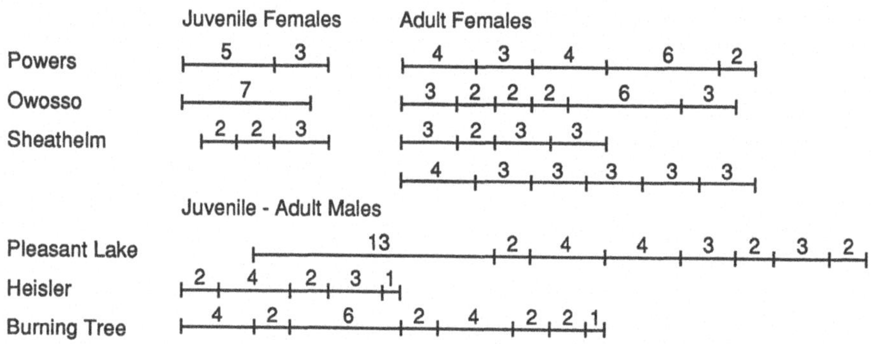

Fig. 30.9 Inferred calving histories of adult female American mastodons, compared to growth histories of juvenile females and juvenile–adult males for which comparable data are available. Criteria for reconstruction of adult female life history are explained in the text and in the caption to Fig. 30.8. For this comparison, only interval length (not the growth profile within intervals) is considered. Interval lengths for juvenile females and juvenile–adult males are taken strictly as the number of years between local maxima on growth profiles. Mean interval length does not differ significantly between adult females and juvenile females + males, but the variance of the former (1.123) is significantly smaller than that of the latter (2.402; Bartlett test for homogeneity of group variances; $P = 0.001$). Data on males in Fig. 30.6d–f are excluded because their profiles include artefacts of sample spacing (see caption).

seems like one normal (if somewhat short — 3 year) calving cycle, this female experienced a 'cascade' of 2 year 'truncated cycles' (i.e. with a growth profile like the first 2 years of a normal 3–4 year cycle) that could represent the death of a series of calves, soon after birth, each followed immediately by another conception. This rapid 'recycling' has been observed in African elephants as a response to the death of a calf (Moss 1988), but such a sequence of deaths as shown by Owosso implies much greater calf mortality than is reported for extant elephants (Moss 1988; Owen-Smith 1988). Even the interval interpreted as a 6 year (successful) calving interval may instead represent a 2 year interval — death of another calf — followed by 4 years without conception (chemical signatures of pregnancy, discussed below, might decide this). Similarly, the Sheathelm mastodon shows what could be the death of a calf soon after her second pregnancy. Could some of this calf mortality be due to human hunting? Because of the season of calf hunting (spring?) or the small body size of calves, they may not have been cached in ponds in the fashion of adults, and thus would not generally be preserved. Whether this level of calf mortality is unusual can only be judged relative to patterns in earlier populations or by comparison with extant elephants. These interpretations of female reproductive history are admittedly speculative, but they are susceptible to further evaluation and, if confirmed, hold great promise for future application.

Finally, there are additional developmental patterns whose timing can be used to define stages in life history: the schedule of formation, eruption, and wear of cheek teeth; growth profiles of individual bones; and schedules of epiphysis fusion. None of these have the potential for yielding detailed inferences of the type discussed above, but they do bear on the important issues of adult body size and the timing of maturation. Although data on these aspects of life history are not presented at this time, they are useful as auxiliary references in reconstructing life history, and should be integrated into the test of extinction hypotheses proposed here.

Before a rigorous test of late Pleistocene extinction hypotheses can be carried out, it will be essential to develop some independent assessment of methods of reconstructing life history from the tusk record. One such approach is to examine, in thin section, the distribution of higher-order increment thicknesses underlying variation in first-order (annual) increment thickness, as done here in Fig. 30.7. Comparisons should be made among individuals within each sex, between the two sexes, between juvenile and adult stages of life history (Fig. 30.9), and even between mastodons and mammoths. If I am interpreting life history correctly, the broad patterns should persist within categories homogeneous for taxon, sex, or stage, even if their boundaries shift one way or another. Ontogenetic reordering or sex-switching of patterns would refute the interpretation. Statistical tests for comparison of time series will be important for evaluating quantitative aspects of these patterns.

Another obvious test is to try the analyses on extant elephants of known life history, either captive individuals or ones observed in the wild. This is, of course, not a simple proposition. One potential problem is that the artificial diet of captive elephants may subdue the contrasts in nutritional status thought to mediate some of the informative differences in growth rate. Using wild animals would circumvent this problem, but because of the endangered status of elephants and the sensitivity of all facets of the disposition of tusks, this route entails other complications. Nevertheless, these are both matters of high priority. Offering new data on life history and population dynamics could even assist in conserving extant elephant populations.

There may also be chemical methods of evaluating reproductive status. For example, measurement of sodium within hydroxyapatite may be useful, because sodium levels typically increase during pregnancy (Luciano *et al.* 1978). This could be done on both fossil and extant material, with males and young females as controls. Diagenetic modification of sodium values is possible, but comparisons within and between individuals, with surrounding sediment, and with fluorine concentrations (which show negatively correlated diagenetic variation; Parker and Toots 1980) should clarify whether any observed patterns are primary.

Finally, although profiles of bone growth and schedules of epiphysis fusion do not allow precise reconstruction of life history, they are causally related to maturation. Sexual maturity in mammals is associated with an increased production of androgens in both males and females, inducing an accelerated growth phase during adolescence and eventual fusion of epiphyses (both processes occur on an unusually protracted schedule in proboscideans; Roth 1984). By censusing bone sizes and status of epiphyses in individuals of a range of ages, these patterns of development can be reconstructed. Because they are related, through endocrine changes, to maturation, a change in age of maturation should cause a parallel shift in patterns of osteological development. Thus, differences in age of sexual maturity deduced from tusks can be checked by analysing postcrania.

LIFE-HISTORY PARAMETERS AND EXTINCTION HYPOTHESES

The study of life history is a major area of research in ecology and evolutionary biology (Boyce 1988; Southwood *et al.* 1991). Because of the many variables involved, and the potential for complex interactions among them, it is not a trivial matter to predict life-history responses to changes in ecological context — indeed, simple *r*- and *K*-selection arguments have been criticized on this account. Life, death, and reproduction represent complex economies to say the least. None the less, in cases where values of some of the variables are known or can be determined, problems become more tractable. Broad features of the ecological context in which mammoths and mastodons existed can be inferred both from knowledge of late Pleistocene environments and the behaviour and ecology of extant elephants (G. Haynes 1991). Both mammoths and mastodons were large-bodied herbivores with a fairly generalist feeding strategy (Davis *et al.* 1984; Lepper *et al.* 1991) — probably hindgut fermenters relying on a high intake and high rate of passage to subsist on food of mixed, often low, quality (Guthrie 1984). Their size, sexual dimorphism, and secondary sexual characters (such as tusks) suggest a structured social environment (probably involving matriarchal family units as in extant elephants) and polygynous mating system in which variance in reproductive success is much greater for males than for females (Clutton-Brock *et al.* 1982). Many social interactions would have been related to rank in a dominance hierarchy; male body size was probably a strong determinant, or correlate, of reproductive success and access to resources, and female body size would have influenced both access to resources and the status and reproductive success of offspring of both sexes. A high premium would have been placed on early maximization of growth rate, because its effects are cumulative. Although early reproduction might offer a potential reward of increased effective reproductive output (through the 'compound interest' of offspring bearing offspring), it carries the penalty of reduced growth rate and smaller adult body size, which in this social environment tends to preclude realization of any advantage. The long maturation times of elephants have the effect of increasing reproductive success when the 'effort' is finally made. Mortality from non-human predation would usually have been limited to juveniles, but humans could have hunted any size-age class.

Let us now suppose that the dominant stress on late Pleistocene proboscidean populations, i.e. the cause of their extinction, was climatic change and associated deterioration of their effective environment, whether by shortened growing season (Kiltie 1984), progressive drying (G. Haynes 1985, 1991), reorganization of floral communities (Guthrie 1984; King and Saunders 1984), or some other effect (including human-induced habitat destruction). This would have tended to increase the importance of density-dependent mechanisms of population regulation (Fowler 1981; even though some mortality would still be density-independent) and would also have had proximate effects such as decreased somatic growth rates. These conditions are associated with a syndrome of life-history modifications including increased age at sexual maturity and increased interval between offspring (Stearns 1977). Precisely these changes have been observed in similarly stressed populations of African elephants (Hanks and McIntosh 1973; Smith and Buss 1973). These effects may be manifested on an ontogenetic, ecological, or evolutionary time scale, and they result from combinations of individual ecophenotypic response, intrapopulational interactions (which need not involve heritable transformation), and natural selection. Elephants in stressed environments may delay maturation more than 10 years, becoming mature at age 22 or older (Laws *et al.* 1975), and may have calving intervals on the order of 9 years. Even if a female loses a calf before weaning, she is usually in too poor condition to come immediately into oestrus and conceive again (Moss 1988). Mortality caused by environmental deterioration may be significant, affecting individuals roughly in reverse order of dominance (G. Haynes 1985, 1991; Moss 1988), but suppression or delay of reproduction is, in a sense, the most profound threat to the population.

In contrast, if there were a relatively uncompromised environment, but increased mortality due to predation by humans, the stress on proboscideans would be more density-independent. In this setting the proboscidean population would not be resource-limited, and growth rates could remain constant, or even increase, if the increased mortality led to reduced intraspecific competition. These conditions are commonly associated with a syndrome of changes, including decreased age at sexual maturity and decreased interval between offspring. If hunting were focused on mature males (as the Great Lakes region record suggests for mastodons; Fisher 1987), it would tend to reduce size-based variance in male reproductive success (even a male large for his age could be killed, while still young, by hunters) and reduce male–male competition for mates or resources. This would decrease the importance of attaining large size relative to that of maturing early in order to beat the odds of surviving to reproduce. If females were hunted, they would experience the same effect directly, but even if they were not, the changes in male survivorship would favour females who reproduced earlier, even if at smaller body size. The same dynamic would drive calving intervals down, even at the cost of smaller calves, and if a female lost an unweaned calf, whether to human or non-human predation (or any other cause), she would tend to cycle again immediately (Moss 1988).

Perhaps more realistic are scenarios involving both environmental deterioration and increased human predation, but although this might rule out certain responses (e.g. increased somatic growth rate), the same tendencies would operate. Age at maturity and calving interval must either decrease, increase, or stay the same, and the latter is only likely if opposing forces balance exactly — an informative outcome in its own right. In general, one of the principal ecological stresses is likely to dominate, and different patterns of biotic response are expected, depending on whether environmental deterioration or human hunting was more important. One trait for which there is a non-divergent prediction is adult body size, for it may decline when either environmental deterioration or human hunting dominate. In the former case the decline would be produced by lowered growth rates (King and Saunders 1984), and in the latter, it would be a consequence of accelerated reproductive schedules (i.e. progenesis; Edwards 1967; McDonald 1984). This is especially relevant because there is already evidence that the adult size of American mastodons declined between about 13 500 years BP and their time of extinction (King and Saunders 1984), and there is a similar size decrease for mammoths (Guthrie 1984). The pattern for mastodons was interpreted by King and Saunders as support for their 'environmental insularity' model of extinction, and the same perception has been central to several previous explanations of dwarfing of island lineages (including mammoths; Sondaar 1977) and even some continental faunas (Kurtén 1968). Size reduction alone, however, does not discriminate between overkill and environmental change hypotheses. The critical, and previously unresolvable, issue is the mechanism by which size reduction is produced. If this can be specified, as proposed here, one or the other of the competing hypotheses will be favoured. Fortunately, predicting response in this system is more tractable than in some others (e.g. shelled molluscs; Kitchell 1990) where size may be as much a refuge from predation as early reproduction is a hedge against it.

One requirement of this test of extinction mechanisms is that they operate long enough for biotic response to emerge and be represented in the fossil record. Size reduction alone suggests that this has happened, and the physiological and social nature of many of the effects considered above make it likely that change could occur over time scales that are short in evolutionary terms. Calving interval and age of sexual maturity for African elephants have shown a significant response to environmental change within decades (Hanks 1979; Eltringham 1982; Moss 1988). Even 'rapid' climatic change in the late Pleistocene should be recorded, but what about Martin's *blitzkrieg*? Although he has argued (Mosimann and Martin 1975) that its visibility in the record might be minimal, in the Great Lakes region, butchered and cached mastodons appear to span more than a millennium (10 395 ± 100 years BP (Beta-1388) for the Pleasant Lake mastodon to 11 770 ± 110 years BP (NSRL-282; AA-6979) for the Heisler mastodon). While I would not want to rule out *blitzkrieg* prematurely, this suggests that at least in the Great Lakes region there would have been time for life histories to respond to hunting pressure.

The ideal approach to testing extinction hypotheses would be to track species' responses through time, in representative regions. This would allow values of life history parameters to be compared strictly within species, maintaining optimal taxonomic control. Reaching even further back in time, it may be possible to do parallel studies on earlier proboscidean extinctions. In addition to studying sequences of change within species, it is relevant to compare life-history attributes for fossil proboscideans near their time of extinction, to published data for extant elephants. This takes advantage of the extant analogues that are in fact available, and by characterizing the endpoint of the extinction process, gives some indication of what the more complete picture is likely to reveal. Studies of ecology and reproductive biology have been conducted on both African (*Loxodonta africana*) and Asian (*Elephas maximus*) elephants, in a variety of geographical regions and environments (e.g. Laws 1969; Eisenberg *et al.* 1971; Hanks and McIntosh 1973; Hess *et al.* 1983; McNeilly *et al.* 1983; Owen-Smith 1988; Sukumar 1989), and in some cases encompassing periods of significant environmental change (e.g. Buss and Smith 1966; Laws *et al.* 1975; Moss 1988). Since size differences among mastodons, mammoths, and elephants are modest relative to those within other orders of mammals, I anticipate few confounding allometric effects. Less easy to evaluate is the issue of how appropriate interspecific comparisons are in this case. This problem may be minimal for mammoths, since they are cladistically bracketed by Asian elephants and African elephants (Maglio 1973; Tassy 1982), which resemble one another closely in life-history patterns, but mammutids are less closely related to any elephantid than any of the latter are to each other. None the less, available life-history data for sirenians (e.g. Marsh 1980) show interesting similarities to elephantids, tending to support the broad comparability of mammutids and elephantids. Based on these considerations, I predict that if mammoths and mastodons were seriously stressed by climatic and/or vegetational change, they would have become sexually mature at about 17–22 years, would have had calving intervals greater than 7 years, would rarely have become pregnant immediately after losing a calf, and would have had growth rates comparable to or lower than those of extant elephants. On the other hand, if mammoths and mastodons were primarily stressed by hunting, they would have become sexually mature in 9–12 years, would have had calving intervals of 3–5 years, would frequently have conceived immediately after losing a calf, and would have grown at comparable or higher rates than extant elephants.

The preliminary characterizations of life history reported above are by no means silent with respect to these predictions. Late Pleistocene male and female mastodons seem to have matured at ages similar to the minimum reported for elephants, and calving intervals seem as short as any known for elephants (Figs 30.6–30.8). There may be evidence of conception soon after loss of an unweaned calf, and absolute ages are mostly younger (indicating faster development) than for African elephants of comparable dental stage (Fig. 30.10; Table 30.1). Granting the provisional nature of these analyses, these are not hallmarks of a population

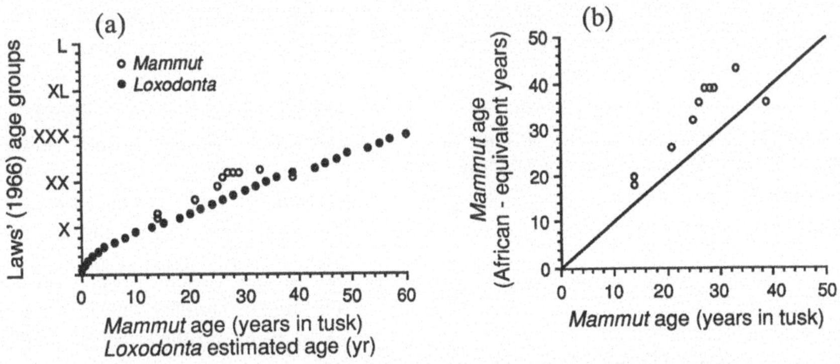

Fig. 30.10 Absolute and relative age estimates (data from Table 30.1). (a) Relationship between age and stage of dental development (Laws 1966) in American mastodons and African elephants. For *Mammut*, age is determined by counting annual increments in tusk dentine (complete years only for spring deaths; complete years + 1 for autumn deaths). For *Loxodonta*, age estimates are those given by Laws (1966). (b) Relationship between two sources of estimates of *Mammut* age: stage of cheek tooth eruption and wear (and analogy to *Loxodonta*, as in Saunders 1977) and number of annual increments in tusk dentine. If these were equivalent, points would lie along the solid line, but with the exception of the Pleasant Lake mastodon (below line), tusk ages for mastodons are younger than analogy with *Loxodonta* would suggest.

Table 30.1 Specimens and associated data used in this study

Site	Repository[a]	Sex[a]	Age[b]	Laws Group[c]	AEY[c]	Craig Scale[d]
Heisler	UM61888	M	14	XII	18	~14–15
Shelton	CIS	F	(14)	XIII	20	~15.5
Sheathelm	MSUVP1355	F	(21)	XVI	26	~19–21
Burning Tree	Private	M	25	XIX	32	~29
Parker	AC	M	(26)	XXI	36	~33–35
Farview	Private	M	–[e]	XXI	36	~33–35
Owosso	UM23498	F	27	XXII	39	~36–38
Warren	AMNH9951	M	(28)	XXII	39	~36–38
Powers	WMU	F	29	XXII	39	~36–38
Alma	AC	F	(29)	–[f]	–[f]	–[f]
Grandville	GRPM	M	33	XXIII	43	~39–42
Pleasant Lake	UM57705	M	39	XXI	36	~33–35

[a] Abbreviations: AC, Alma College; AMNH, American Museum of Natural History; CIS, Cranbrook Institute of Science; F, female; GRPM, Grand Rapids Public Museum; M, male; MSU, Michigan State University; UM, University of Michigan; WMU, Western Michigan University.

[b] Age estimated from number of complete years in tusk, for winter–spring deaths, or number of complete years + 1, for autumn deaths. Values in parentheses indicate some uncertainty due to less clearly demarcated increments in the juvenile portion of the tusk.

[c] Relative age groups and associated absolute ages (AEY, 'African-equivalent years'; Saunders 1980) as assigned to *L. africana* by Laws (1966).

[d] Craig Scale of absolute equivalents for Laws (1966) relative age groups (G. Haynes 1991).

[e] Only the distal half of one Farview tusk is available; work in progress may allow this to be correlated to first-order increments in the cheek teeth for an absolute age estimate.

[f] Associated cheek teeth not yet analysed.

suffering environmental deterioration (Saunders (1980) drew a similar conclusion from the age profile of mammoths from the Dent site). None the less, I do not presume that these patterns should be considered any more than suggestive of the kind of information that may be accessible. Any test that seriously grapples with the problem of late Pleistocene extinctions will have to be based on more thorough study of individual specimens and larger sample sizes.

The research outlined here allows specimens to be treated as samples of a population from a given region and time period, independent of whether human activity was involved in their death or post-mortem history. The aspects of reproductive biology being assessed were determined years before the specific circumstances leading to death and preservation and reflect interactions in a much larger ecological theater than the small stage on which an individual's last act was played. The approach thus avoids some of the difficulties that beset investigations of individual sites as case studies in human/megafaunal interaction. Thus, despite the shortcomings of data now in hand, an approach to solving the problem of late Pleistocene proboscidean extinctions — one that could even be generalized to some extent to other victims of megafaunal extinction — may be taking shape.

SUMMARY AND CONCLUSIONS

1. Most North American proboscidean extinctions cluster temporally in two episodes, the first in and shortly before the late Hemphillian, about 4.5 Ma and the second at the close of the late Pleistocene, about 10 000 years BP. This latter episode was part of the late Pleistocene megafaunal extinction and occurred during a time of widespread biotic and environmental change.

2. Late Hemphillian extinctions probably occurred because of climatic and vegetational change associated with the spread of steppe conditions throughout much of temperate North America. Late Pleistocene extinctions may also have been caused ultimately by climate change, but hunting by humans may be an alternative, or additional, causal factor.

3. Climate-based hypotheses for late Pleistocene extinctions involve a wide variety of mechanisms that might have mediated the effects of climate change and acted as more proximal causes of extinction. These include direct effects of climate change such as increased drought and unpredictability of storms, a shortened effective growing season for plants, floral changes that reduce the nutritive value of food available for herbivores, changes in the scale of heterogeneities in floral communities, disruption of coevolved plant–animal communities (coevolutionary disequilibrium), contraction and isolation of habitats (environmental insularity), and shortened seasonal 'windows' for cycling of mammalian reproductive systems.

4. The most dramatic overkill hypothesis for late Pleistocene extinctions is *blitzkrieg*, in which an expanding wave of humans extirpate game locally within very short intervals, continuing the process as they spread into adjacent areas. Alternative models of overkill, however, involve more protracted interaction between humans and megafauna, in which human hunting takes a smaller, but still significant, toll on prey populations.

5. Climatic hypotheses for late Pleistocene extinction face the challenge of explaining why this deglaciation event had consequences for the fauna that were so much more drastic than earlier events. They also must hold up under scrutiny of individual causal linkages between climate change and ecological stress. Overkill hypotheses have to deal with what has long seemed a rather tenuous record of human interaction with extinct megafauna.

6. Climatic hypotheses may be integrated into a much longer-term perspective on the evolution of the North American land mammal fauna, and there is no dispute that climatic change occurred near the time of the late Pleistocene extinction. In contrast, although there is also no question but that humans were present in North America at the time of the extinction, the extent to which they represented an agent of mortality independent of 'natural' causes is a matter of debate.

7. New directions of research on this problem are providing improved resolution of biogeography, improved accuracy of absolute dating, taphonomic analyses of occurrences of extinct megafauna, dietary information for extinct megafauna, and stable isotopic data on diet and local climate. Most of these new sources of information do not address the problem of late Pleistocene extinctions 'head-on', but rather relate to the efficacy of one or another of the proposed linkages between large-scale causal factors and the eventual outcome.

8. Tusks and cheek teeth of elephantids and mammutids exhibit dentinal lamination that records details of individual life history. Successive orders of lamination represent annual, fortnightly, and daily cycles in accreted dentine. Analyses of these patterns yield data on seasonal mortality, age, sex, growth rates, age at sexual maturity, and possibly calving.

9. Life-history traits such as age and size at sexual maturity, and calving interval, are sensitive indicators of the nature and degree of ecological stress in that they show predictable patterns of response to altered nutritional status and/or predation pressure. Patterns of dentinal lamination may be used to quantify these life-history traits for North American proboscideans approaching their time of extinction. The values observed may also be compared to those documented for extant elephants from a range of environments. The outcome of these comparisons will comprise a test of the two major, competing causal hypotheses for late Pleistocene extinction: overkill and climatic/vegetational change.

ACKNOWLEDGEMENTS

I thank the editors of this volume, Hezy Shoshani and Pascal Tassy, for inviting this contribution. They, along with Paul Koch, David Lambert, and Jeffrey Saunders, improved the text greatly through their comments. I have also benefitted from discussions with Philip Gingerich, Russell Graham, the late Ronald Kapp, Paul Koch, Jeffrey Saunders, Holly Smith, John Speth, Dennis Stanford, and Henry Wright. Specimens for conducting much of the work described here were generously loaned by Sherm Byers, Richard Flanders, Elizabeth Garland, Alan Holman, Paul Hooge, the late Ronald Kapp, Bradley Lepper, George McIntosh, Richard Redding, and Hezy Shoshani. Early stages of this research were supported by NSF grants BNS-8521097 and BSR-8605310. Scott Beld, Paul Koch, Jim Dawson, and Jeanne DeLanois provided steadfast technical assistance, Bonnie Miljour drafted the figures, and volunteers from the Michigan Archaeological Society made all the difference in being able to conduct controlled excavations at several mastodon sites. Finally, landowners Walter and Ellen Egler, Lester and Mary Heisler, and Jim and Cindy Heisler, donated specimens without which this research would never have been conceived.

THIRTY-ONE

South American proboscideans: general introduction and reflections on Pleistocene extinctions

RODOLFO M. CASAMIQUELA, JEHESKEL SHOSHANI, AND THOMAS D. DILLEHAY

GENERAL BACKGROUND (J. Shoshani)

Based on available palaeontological, archaeological, and stratigraphic evidence, the proboscideans of South America (*Cuvieronius, Haplomastodon, Notiomastodon*, and *Stegomastodon*) were either descendants of gomphothere stock originating in North America or had evolved autochthonously, a scenario summarized in Fig. 29.1 of Dudley (this volume, Chapter 29). Immigrants to South America probably arrived via the Panamanian filter during the Pliocene or Plio-Pleistocene boundary. Those, in turn, may have been intermittently isolated and evolved on an 'island continent'. Little systematic research on South American proboscideans has been carried out. Osborn's (1936, 1942) information on South American gomphotheres was augmented by Hoffstetter's (1950, 1952, 1955), and additional observations were made by Simpson and Paula-Couto (1957), especially on the material from Brazil and other countries in South America. None of the more recent publications (e.g. Patterson and Pascual 1972; Webb 1978, 1991; Simpson 1980; Marshall and Cifelli 1990; Marshall *et al.* 1984) have expanded on these works but have only examined it from different perspectives.

For the purpose of summarizing the major observations made by these and other authors, we have included a classification of gomphothere remains found in South America (Table 31.1). This table is provided mainly as a reference for taxa mentioned in the text below, and for comparison with the classification of gomphotheres given by Shoshani (this volume, Chapter 16).

Relationships among South American gomphotheres have been discussed in the excellent, almost 'prophetic', summary of Simpson and Paula-Couto (1957, p. 185). Excerpts from their summary are as follows:

1. and 2. a large collection of proboscidean remains from Brazil seems to represent a single population of a species identified as *Haplomastodon waringi*;

3. the material from Ecuador appears to belong to the same species;

4. recognized South American species are *Cuvieronius hyodon, Haplomastodon waringi*, and *Stegomastodon platensis*. 'All these names have numerous synonyms'. *Stegomastodon superbus* and *Notiomastodon ornatus* are also recognized but of doubtful status;

5. '*Haplomastodon* is believed to be about as closely related to *Cuvieronius* as to *Stegomastodon*';

6. 'All the South American mastodonts seem to represent end terms of divergence within a single broad stock. *Cuvieronius, Stegomastodon*, and probably *Haplomastodon* were already differentiated before they spread from North America to South America. The North American *Rhynchotherium* may belong to the same complex. All may ultimately have arisen from early Old World Anancinae, and all are tentatively referred to that subfamily'.

Table 31.1 A simplified 'outline of classification' of gomphotheres found in South America (modified after Simpson and Paula-Couto 1957, p. 133)

Family Gomphotheriidae Hay, 1922, p. 101
Stegomastodon Pohlig, 1912, p. 193
Stegomastodon platensis (Ameghino, 1888, p. 7)
Stegomastodon superbus (Ameghino, 1888, p. 7)
Notiomastodon Cabrera, 1929, p. 90
Notiomastodon ornatus Cabrera, 1929, p. 91
Haplomastodon Hoffstetter, 1950, p. 24
Haplomastodon waringi (Holland, 1920, p. 229)
Cuvieronius Osborn, 1923, p. 1
Cuvieronius hyodon (Fischer, 1814, p. 341)

According to Simpson and Paula Couto, each of these taxa had a few or numerous synonyms. (cf. Table 16.1, Appendix B, and Appendix C).

Tobien *et al.* (1986) described a new genus, *Sinomastodon*, from China. They believed it was related to the South American complex; in fact they classified it in the family Gomphotheriidae, subfamily Notiomastodontinae. Shoshani's results (this volume, Chapter 16) seem to concur with Tobien *et al.*'s (1986) assessments, but this finding may change once additional materials of sinomastodons are discovered.

Webb (1978) provided a good summary on the South American faunal assemblages, the interamerican interchange, and some aspects of extinction. Fisher (this volume, Chapter 30) also discusses proboscidean extinction in South America but in a very broad sense. Discussed below are some of the observations and questions on Pleistocene extinctions of gomphotheres at four palaeontological and archaeological sites in South America (Fig. 31.1). Apart from the observations on these sites, T. D. Dillehay (personal communication 1992) noted that there are at least two other archaeological localities of terminal Pleistocene age with gomphothere remains, one in Colombia and one in Brazil.

Fig. 31.1 Map of South America with Pleistocene sites mentioned in the text. Shaded parts represent generalized areas where gomphothere remains were discovered. Pleistocene localities by Casamiquela and Dillehay; gomphotheres after Osborn (1936, pp. 536, 567, 734), Hoffstetter (1950, p. 43) Simpson and Paula-Couto (1957, pp. 159, 175), Kurtén and Anderson (1980, p. 346–8), Anderson (1984, p. 85), Gruhn and Bryan (1984, pp. 132–3), Marshall *et al.* (1984, p. 20) Rancy (1991, p. 113), and Webb (1991, p. 275). According to Webb (1991, and personal communication 1993), *Stegomastodon* was not represented in the South American fauna during the Pleistocene. (Artwork by J. S. Grimes.)
Editors' note: a gomphothere interpreted from the Huayquerian (late Miocene) of Peru may be the oldest in South America (see also Appendix 29.1 on p. 295), and will possibly alter our understanding of the Great Faunal Interchange.

OBSERVATIONS ON THE ROLE OF HUMANS IN PLEISTOCENE EXTINCTION

(R. M. Casamiquela)

Information concerning South American proboscidean extinction is sparse and not always reliable. A critical revision of such evidence is indispensable, but at the present time, not feasible. The following comments focus specifically on the possible role of humans in the extinction of gomphotheres, as evidenced at a few late Pleistocene archaeological sites (Taima-Taima in Venezuela, and Quereo, Tagua-Tagua, and Monte Verde in Chile) that I know of as a vertebrate palaeontologist. I should note that my analysis of the material from Monte Verde was limited to laboratory taxonomic inspection of a portion of the bone assemblage. I was not able to participate in the excavation and in the *in situ* taphonomic analysis of bones. Nor did I study the artefactual remains from Monte Verde.

TAIMA-TAIMA

In this well-known site, dated *c.* 13 000 years BP, the skeleton of a semi-articulated young individual (*Haplomastodon*, or perhaps *Stegomastodon*), was associated with a fragment of a lithic point lodged in the pubic cavity, and with clear cutmarks on several bones (Fig. 31.2), most distinguishable on the humerus. The artefacts associated with the pubic cavity belonged to the El Jobo complex (Gruhn and Bryan 1984; Fig. 31.3). In addition to *Haplomastodon* (or *Stegomastodon*), *Equus*, edentate (*Glyptodon* and *Glossotherium*), and carnivore (ursid? and felid?) remains were recovered from the same stratigraphic unit (Bryan *et al.* 1978; Casamiquela 1979; Gruhn and Bryan 1984, p. 133).

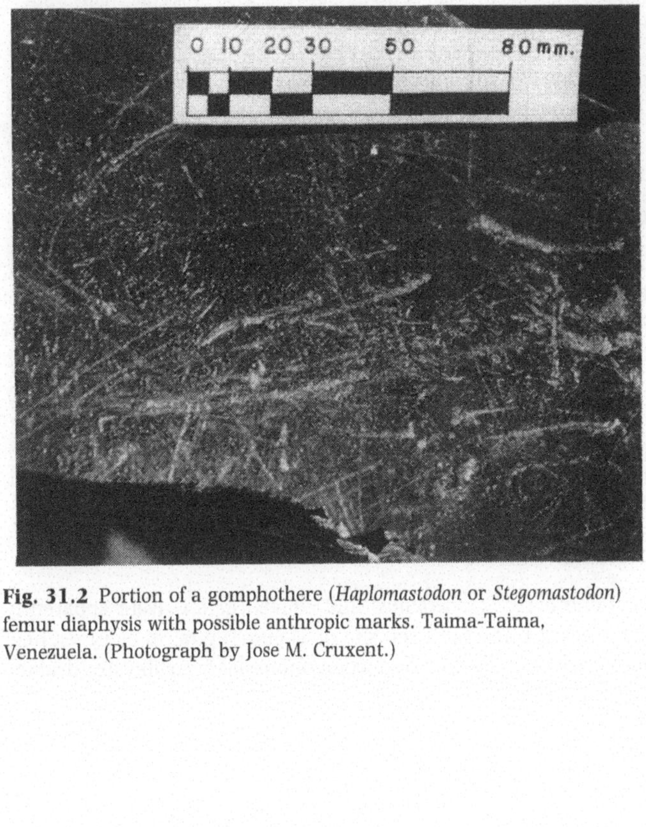

Fig. 31.2 Portion of a gomphothere (*Haplomastodon* or *Stegomastodon*) femur diaphysis with possible anthropic marks. Taima-Taima, Venezuela. (Photograph by Jose M. Cruxent.)

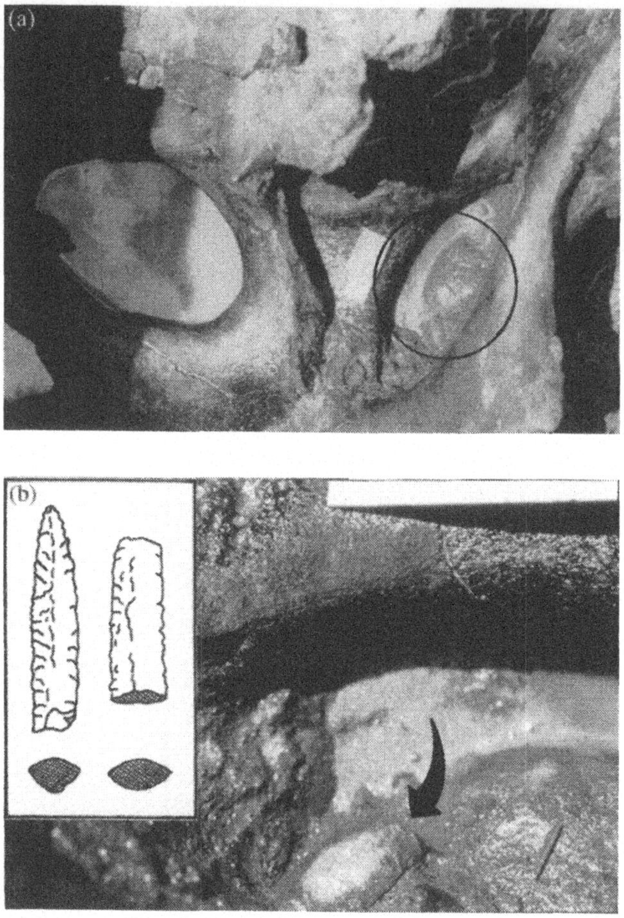

Fig. 31.3 An El Jobo projectile point within the pelvis of a young gomphothere, possibly *Haplomastodon waringi*, at Taima-Taima, Venezuela (projectile point is within a circle in (a), and (identified by the arrow in (b); photographs by Alan L. Bryan). Inset: El Jobo projectile points from the type site in the Rio Pedregal valley (Venezuela); one is almost complete and the other is a midsection — 'The midsection is similar to the specimen found within the body cavity of the juvenile mastodont at Taima-Taima' (from Gruhn and Bryan 1984, fig. 5.3, p. 132).

QUEREO

The archaeological evidence at this site (Núñez *et al.* 1983; Fig. 31.4) consists of possible cutmarks on humerii (mostly of gomphotheres and horses) and other bones of a skeleton stored at the Santiago National Museum of Natural History. The presence of horse bones, possibly the result of human agency (dated *c.* 11 000 years BP), indirectly suggests a linkage between proboscideans and humans at this site. More recent excavations yielded few proboscidean bones, none of which had cutmarks.

TAGUA-TAGUA

Gomphotheres (*Cuvieronius*) and equids (*Equus*) were apparently the prey of hunters in the site of Tagua-Tagua, in central Chile, located in the basin of an eponymous lake, artificially drained during the nineteenth century (Casamiquela *et al.* 1967; Montané 1967, 1968*a*, *b*; Mostny 1968; Casamiquela 1972, 1976; Varela Barbagelata 1976).

Tagua-Tagua was re-excavated in 1990–91 by a multidisciplinary staff headed by Lautaro Núñez. On this occasion, bones of several individuals of different ages, and at least one tusk (Fig. 31.5) were found. Clear human intervention is demonstrable by the presence of cutmarks on bone and by the direct association of classic 'fishtail' projectile points. If we assume that the sample of bones found to date at the ancient lake are only a fraction of the total population, it may be possible that the actual number of gomphotheres scavenged and/or killed by the Paleoindians was larger.

MONTE VERDE

The presence of nearly 400 bone fragments of *Cuvieronius* at Monte Verde (dated *c.* 12 800–12 400 years BP) represents from five to seven individual animals. Based on a preliminary inspection of some of these bones, it is not possible to determine whether humans killed these animals or utilized the carcasses (Casamiquela 1981, 1983, 1985).

CONCLUDING REMARKS

In regard to proboscidean extinction, I could concur with Martin's hypothesis of faunal 'overkill' (Martin 1984). This hypothesis agrees with claims of a Pleistocene antiquity for the peopling of the Americas, as was suggested earlier by Osborn (1936, p. 574). Palaeontological and archaeological evidence, as reviewed here, seems to demonstrate that faunal overkill (of gomphotheres and horses, at least) was a cultural event.

I compare the situation with the consequences of a long, hard winter in Patagonia (an example familiar to me): guanaco and 'ostrich' populations, the prey of historic Patagonian hunters, were numerically decreased, but the same happened to the puma and the fox, and the faunal equilibrium remained unchanged. This was presumably the case at the end of the last glaciation: faunal equilibrium remained intact, but game populations largely decreased until the arrival of a new, unexpected predator — the specialized human hunter!

MONTE VERDE (T. D. Dillehay)

Monte Verde, located on a small tributary of the Maullin River in a cool, temperate rainforest, Chile, is an archaeological site where proboscidean and other organic remains have been preserved by an overlying mud layer (Dillehay 1984, 1989; Pino 1989). The site is defined by one well-documented cultural episode, radiocarbon dated between approximately 12 800 and 12 400 years BP, and another possible one at 33 000 years BP. Bone remains of *Cuvieronius* (Casamiquela 1981; Casamiquela and Dillehay 1989) are associated only with the younger episode, which also is characterized by stone and wood artefacts, various features, and abundant numbers of economically useful plants. To date, nearly

Fig. 31.4 An overview of the excavation site at Quereo, Chile, with bones *in situ*. (Photograph by Lautaro Núñez.)

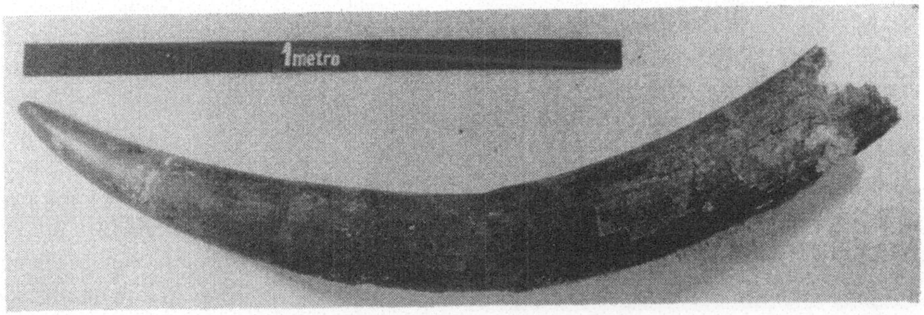

Fig. 31.5 A gomphothere (*Cuvieronius*) tusk from Tagua-Tagua, Chile (collection and photography, Museo Nacional de Historia Natural, Chile). Casamiquela (1972; personal communication to J. Shoshani 1992) noted that cuvieronine proboscideans from northern Chile, southern Bolivia, and southern Peru had slender twisted tusks with a conspicuous enamel band, whereas individual from central and southern Chile had thick, slightly curved tusks, without the enamel band.

400 bones or bone fragments have been recovered; a large majority of these are from ribs. The site has also yielded 38 small pieces of animal hide and muscle tissue, some still preserved on bones of *Cuvieronius*. Pieces of hide also were recovered from hearth areas, living floors, and wooden structural remains. Some pieces were still attached to wooden poles, possibly suggesting the presence of hide-draped huts (Fig. 31.6). Pathological (Cibull 1988; Wolf 1989) and other (Tuross and Dillehay 1991) analyses of these pieces suggest that they are also of a proboscidean.

Controversy over the exact nature of Monte Verde stemmed from the pre-12 000 years BP (pre-Clovis) dates and whether or not gomphothere bone fragments had been worked by humans (Dillehay *et al.* 1983; Dillehay 1989; Lynch 1990; Marshall 1990; Haynes 1991; Wolkomir 1991; Meltzer 1993). Thorough systematic taphonomic, archaeological, geological, and pathological studies of remains at Monte Verde have provided sufficient interdisciplinary information to demonstrate that this site presents *in situ* deposition and preservation of a cultural-related collection of animal bone and soft tissue. This evidence included burned bones recovered from hearth and activity areas, modified tusk and bone fragments (Dillehay 1992; Shipman 1992), burin-like implements made of tooth enamel, and conjoining pieces of bone scattered among distinctive activity areas across the site.

SUMMARY AND CONCLUSIONS

(R. M. Casamiquela, J. Shoshani, and T. D. Dillehay)

Proboscidean fossil remains from South America span from the Pliocene (Chapadmalalan or Uquian South American Land Mammal Age) to the latter part of the late Pleistocene (Lujanian South American Land Mammal Age). Palaeontological data suggest that *Haplomastodon* and *Notiomastodon* occurred only in South America, whereas *Stegomastodon* and *Cuvieronius*

Fig. 31.6 Section of a wooden pole from architectural structure at Monte Verde showing attached proboscidean skin fragment, probably from *Cuvieronius*. (a) A surficially burned (see arrows) timber, radiocarbon dated around 12 800 years BP. The dashed lines isolate a fragment of attached skin identified as that of a proboscidean. (b) A microscopic amplification (300 X) of a thin section cut across the timber and attached skin. The structure of the decomposed skin tissue is shown on the upper part of the photo; the wood tissue is on the bottom. This skin fragment was compared at the cellular and amino acid levels to pieces of skin attached to bones of *Cuvieronius* found at this site; tests were conducted by M. Cibull and N. Tuross, respectively. (Photographs by T. D. Dillehay.)

were found in North and South America (Anderson 1984; Lambert, this volume, Chapter 15; Dudley, this volume, Chapter 29). Further, based on vegetational distribution maps of South America during the glacial and interglacial phases, and the geographical distribution of South American gomphotheres, it appears that *Cuvieronius* inhabited an arid landscape, whereas *Haplomastodon* and *Notiomastodon* seem to have predominated in areas identified as savanna and rainforest (Webb 1991, figs 3 and 4). An exception appears to be at Monte Verde in south-central Chile where *Cuvieronius* adapted to a cool rainforest.

Cladistic analysis (Shoshani, this volume, Chapter 16) results in the grouping of *Cuvieronius* and *Notiomastodon* which, in turn, are joined successively by *Haplomastodon* and *Stegomastodon*. Alternatively, with zero or one evolutionary step, the first three genera can form a polytomous clade. These results evoke the question of vicariance biogeography and how it relates to phylogenetic relationships, especially when one Asian taxon, *Sinomastodon*, is thought to be closely related to the South American taxa (Tobien *et al.* 1986; cf. Fig. 31.1 in this chapter to Figs 13.1 and 16.5). A detailed re-examination of gomphotheres found in South America, as well as related taxa found in North America, is urgently needed, in order to shed light on the questions raised here.

In reflecting on the possible role of humans in the extinction of proboscideans in South America, the sample size of archaeological sites is so limited that we cannot make any reliable statements. It is the opinion of Shoshani and Dillehay that until more conclusive palaeontological and archaeological evidence is available from different palaeo-landscapes and cultural settings across the continent, we can only remark that there was contact between humans and proboscideans at a few sites in different environmental settings. Moreover, until modern multidisciplinary taphonomic studies have been carried out on the bone assemblages from Taima-Taima, Quereo, and Tagua-Tagua, we must be very cautious in drawing any conclusions about the nature of this contact at these sites. The present evidence from Taima-Taima and Tagua-Tagua is sufficiently convincing to demonstrate human exploitation of proboscideans, and perhaps hunting. Quereo, on the other hand, provides little, if any, demonstrable evidence of human activity, and even less for consideration of human–proboscidean contact. We must also be cautious in our interpretation of Tagua-Tagua, which was an animal water-hole and possible kill-site in late Pleistocene times. Numerous bones of gomphothere and other extinct animals have been recovered from the shoreline of this old lagoon. Some bones exhibit fractures and cutmarks which seem to be the results of human action. None the less, it is probable that some bones are associated with natural deaths at this arid water-hole, and that many fractures and scars on bones can be attributed to animal trampling and other localized natural processes. Similar patterns have been observed by several investigators (e.g. Haynes 1986*a*, 1988) at arid water-hole sites in Africa.

The extinction of gomphotheres in South America before the arrival of humans may be attributed to a combination of factors, including climate and competition for available resources (Fisher, this volume, Chapter 30). After the arrival of humans, the rate of extinction may have been expedited. Evidence for such a hypothesis in some places, appears to be more suggestive than in other areas.

ACKNOWLEDGEMENTS

Retyping of this article was done by S. Wolak; comments by E. C. Marsac, J. L. Pierce, and P. Tassy.

THIRTY-TWO

Loxodonta africana: range and distribution, past and present

IAIN DOUGLAS-HAMILTON AND FRANCES MICHELMORE

HISTORY

Despite its great size and ecological requirements, the African elephant has been a supremely successful survivor amongst Africa's megafauna. The species, until recently, occupied almost every habitat south of the Sahara, with the exception of extreme desert and high alpine montane. This is in contrast to other parts of the world where certain proboscideans disappeared with the ascent of man.

The recorded history of the African elephant is a tiny fraction of its real history. Systematic quantitative estimates on a continental scale were not made until the 1970s (Douglas-Hamilton 1979), although earlier records of elephant distribution in the interior of the continent were made by the first European explorers. On the basis of probable range and density, a conservative estimate of elephant numbers before the nineteenth century would be of the order of 10 million. At that time, human densities were very low and there would have been no obvious factor to prevent elephants from occupying double their present range at an average density of one elephant per km^2. If there were yet earlier elephant population fluctuations, the evidence has been lost for ever.

By the middle of the 1800s, however, the ivory trade had become a major factor affecting their status. Hundreds of tons of ivory were exported, which one major ivory trader estimated to be equivalent to 50 000 elephants every year (Kunz 1916). The overall picture became one of rapid decline in the face of the burgeoning demand for ivory. At the turn of the century the conventional wisdom was that the elephant would be exterminated or reduced to very low numbers in most of its range (Selous 1890; Bryden 1903; Conniff 1987). Early colonial game laws in many countries put an end to the more blatant forms of commercial hunting for ivory, restricting the offtake to carefully controlled quotas for sport hunting, which restricted the human predation largely to mature bulls.

In many areas of southern, east, and central Africa, elephant populations staged a recovery in the first six decades of the twentieth century (Percival 1924; Blunt 1933; Pitman 1934; Spinage 1973, 1991; Laws *et al.* 1975; Cumming and Jackson 1984; Hall-Martin 1991; Lindeque, 1992). In many other parts of their range, however, elephants were excluded by the increasing human population. The effect of this competition was to compress elephants into game reserves, national parks, and other protected areas (Laws 1970; Sikes 1971; Burrill and Douglas-Hamilton 1987).

THE AFRICAN ELEPHANT DATABASE

Studies of continent-wide elephant status began in 1976 and have resulted in the accumulation of a database which is unusually comprehensive for any species of large mammal. The database is conceived as an inclusive repository and depends on input from a wide range of sources, in particular the technical network of the African Elephant Specialist Group (AESG) of the International Union for Conservation of Nature and Natural Resources (IUCN). The data are of variable quality and reliability, and are graded by AESG on a scale of 1 to 3. Census methods include aerial surveys, dropping counts, registration of known individuals, and extrapolations. The AESG, originally formed in 1975, has met almost every year (Douglas-Hamilton 1979, 1989; Cumming and Jackson 1984; Cumming *et al.* 1990; Douglas-Hamilton 1989).

The African elephant database is housed at the United Nations Environment Programme (UNEP) Headquarters in Nairobi, in a computerized form which stores elephant estimates geographically (Michelmore *et al.* 1994). These estimates are available as a resource to management authorities and other users. They are periodically updated with current estimates, the older ones being archived for future reference. Reviews of data have been made in 1979, 1981, 1987, and 1989. The latest estimates were reviewed by the AESG in 1991 and later revised by range states at the African Elephant Range States and Donors meeting held in Nairobi at UNEP in 1992. The most recent estimates are given in Table 32.1, and indicate a continental elephant population of the order of 550 000–650 000 animals (Douglas-Hamilton 1992; Douglas-Hamilton and Douglas-Hamilton 1992; Douglas-Hamilton *et al.* 1992).

This latest revision indicates an elephant range of 5 188 000 km^2, much of it rarely visited by elephants and shrinking rapidly

Table 32.1 African elephant numbers: estimated by country and region

	ITRG 1989	Elephant Conservation Plans, 1991 (a)	AESG 1991	Regional sub-total, 1991
CENTRAL AFRICA				
Cameroon	22,000	24,000	23,000	
Central African Republic	23,000		19,000	
Chad	2,100	2,300	2,300	
Congo	42,000	49,000	48,000	
Equatorial Guinea	500	500-600	600	
Gabon	74,000	87,000	85,000	
Zaire	112,000	90,000	91,000	
Subtotal	277,000			268,000
EASTERN AFRICA				
Ethiopia	8,000	2,450	2,000-2,500	
Kenya	16,000	20,000	26,000	
Rwanda	50	80-100	90	
Somalia	2,000	1,000-2,000	no return	
Sudan	22,000	26,000-45,000	no return	
Tanzania	61,000	57,000	52,000	
Uganda	1,600	1,200-2,000	1,900	
Subtotal	110,000			102,000-122,000
SOUTHERN AFRICA				
Angola	18,000	10,000-50,000	no return	
Botswana	68,000		49,000-64,000	
Malawi	2,800	2,200	2,500	
Mozambique	17,000	13,000	no return	
Namibia	5,700	4,500-6,500	3,500-6,000	
South Africa	7,800	9,000	9,000	
Zambia	32,000		20,000	
Zimbabwe	52,000		50,000-68,000	
Subtotal	204,000			155,000-233,000
WESTERN AFRICA				
Benin	2,100	1,100-2,400	no return	
Burkina Faso	4,500	2,000	1,700-2,500	
Ghana	2,800	640-2,900	no return	
Guinea	560	300	800-1,200	
Guinea Bissau	40	10-40	20-50	
Ivory Coast	3,600	2,000	1,300-2,400	
Liberia	1,300	1,800	1,800	
Mali	840	600	360-630	
Mauritania	100	0	no return	
Niger	440	250-500	700	
Nigeria	1,300	1,200-1,500	no return	
Senegal	140	20-40	30-60	
Sierra Leone	380	250-500	no return	
Togo	380	200	200	
Subtotal	19,000			10,000-17,000
Grand total				549,000-652,000

Notes: Numbers above 5000 are rounded to the nearest thousand, numbers between 100 and 5000 are rounded to the nearest hundred, below 100 are rounded to the nearest ten. The 'ITRG 1989' estimates are based on Douglas-Hamilton (1989). Regional sub-totals for 1991 and the continental total are derived from combinations of AESG and Elephant Conservation Plan data.
(a) African Elephant Ranges States and Donors meeting, UNEP Nairobi, 1992.

in the face of human development. Elephant range is divided between the regions as follows: 49 per cent, central Africa; 21 per cent, eastern Africa; 25 per cent southern Africa and 4 per cent western Africa. It is likely, however, that the continental range is overestimated as estimates in many of the countries have not been revised for many years.

It can be observed that elephant range in West Africa is very fragmented, with the East African range also beginning to break up (Figs 32.1, 32.2). Large, unbroken elephant ranges are confined to the central and southern African regions. The largest and most intact population is found in northern Botswana and ranges across into western Zimbabwe, into Namibia across the Caprivi Strip and into southern Angola and south-western Zambia. Other large populations are found in the Zambezi Valley in Zimbabwe, the Luangwa Valley in Zambia, the Kruger National Park in South Africa, the Selous Game Reserve, Ruaha Rungwa complex and the Tarangire National Park in Tanzania, and the Tsavo National Park in Kenya. The main forest populations are found in the rain forests of central Africa in Gabon, Zaire, Central African Republic, Congo, and Cameroon.

TRENDS

Analyses of regional and continental trends have been made by Douglas-Hamilton (1984, 1987, 1989), Milner-Gulland and Mace (1991), and Caughley *et al.* (1990). Based on specific trends in key elephant populations and computer modelling, the currently accepted view is that elephants were severely depleted by the ivory trade (Cobb 1989) and lost at least half their population in the decade 1978–89. There were believed to have been some 1.3 million elephants in Africa in 1979 (Douglas-Hamilton 1979). This figure is thought to have declined to 760 000 in 1987 (Cumming *et al.* 1989) and approximately 609 000 by 1989 (Douglas-Hamilton 1989, 1992). The drop in elephant numbers is matched in the ivory offtakes calculated for the same period by TRAFFIC (Luxmore *et al.* 1989) when ivory equivalent to some 700 000 elephants was exported from Africa.

Three lines of evidence convincingly demonstrated a continental elephant population crash:

(1) population trend data from most of the African continent showed widespread falls in elephant numbers, accompanied

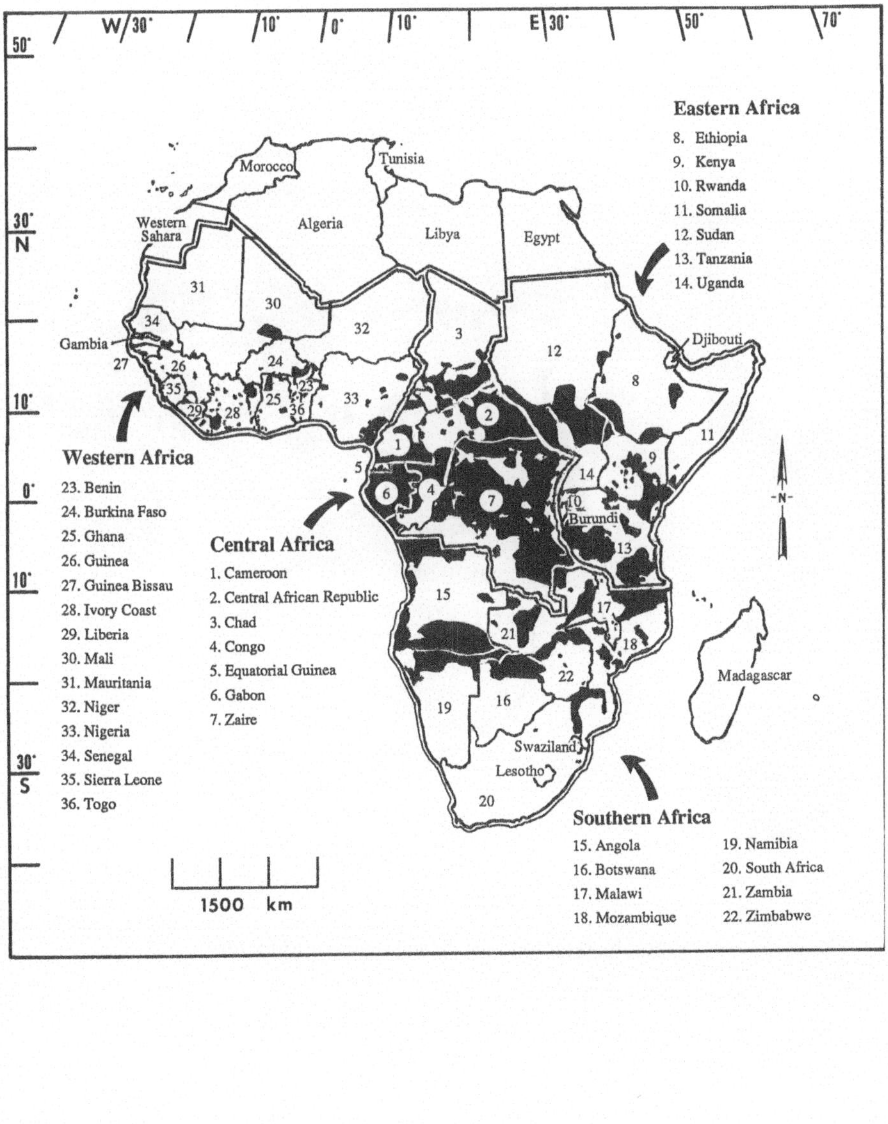

Fig. 32.1 Distribution map of the African elephant, *Loxodonta africana* (darkened areas). Regions in Africa and the countries within these divisions appears as they are listed in Table 32.1.

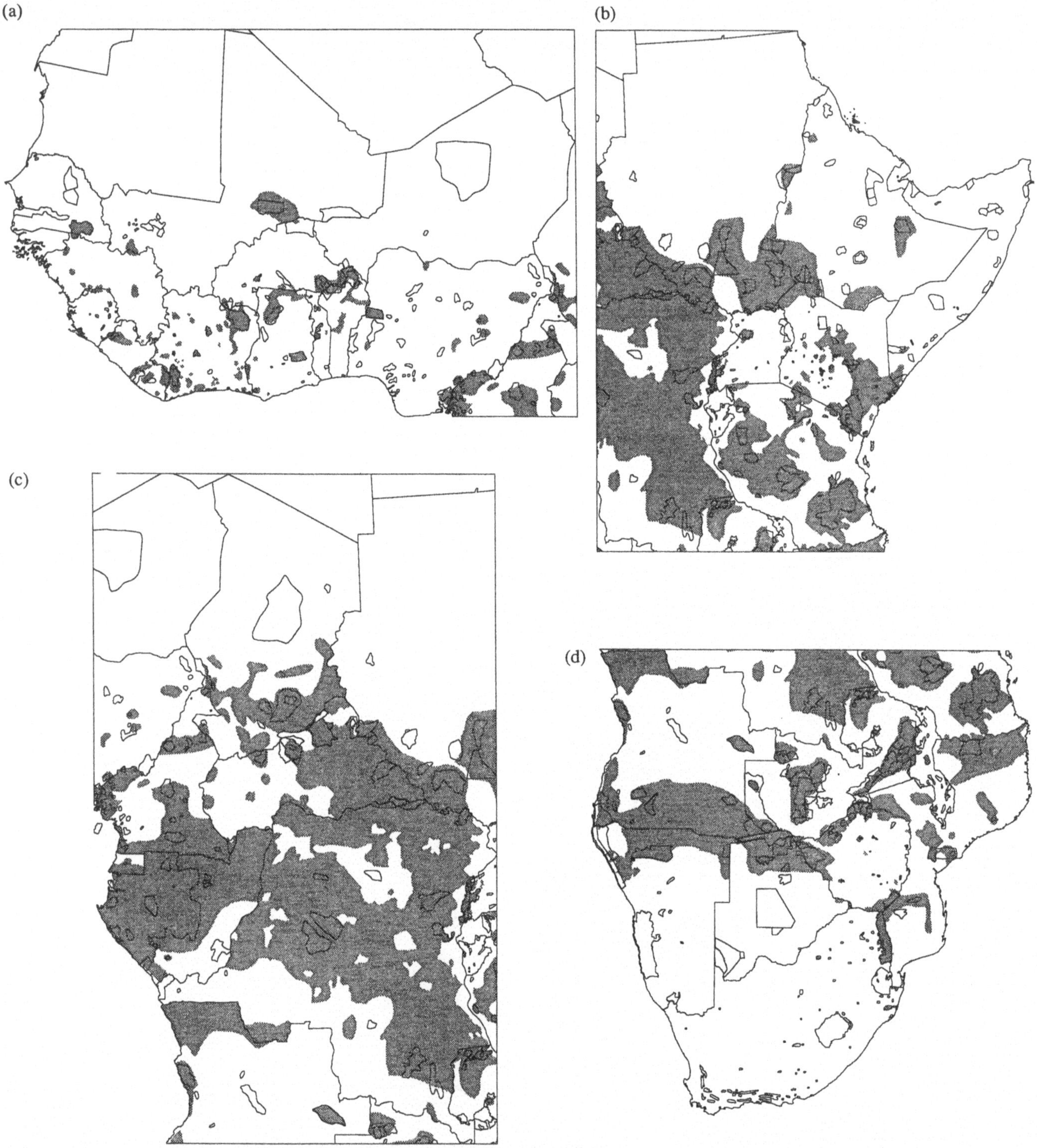

Fig. 32.2 Protected areas and elephant range in western Africa (a), eastern Africa (b), central Africa (c), and southern Africa (d).

by high carcass ratios — ratio of dead to dead plus live animals (Douglas-Hamilton 1989; Douglas-Hamilton and Burrill 1991);

(2) ivory export curves increased until the mid-eighties and then rapidly collapsed in what appeared to be a classic case of 'overfishing', and the calculated quantities of tusks exported were of the same order as the overall population decline (Luxmore *et al.* 1989); and

(3) data from samples of ivory seized from poachers in Africa and from ivory imports abroad, indicated a major drop in mean tusk weight. This reflected destruction of the males and older females (Luxmore *et al.* 1989). In East Africa, this was backed up by age structure records for key populations which, with the exception of the well-protected Amboseli population in Kenya, all showed a distortion in the sex ratio and a lack of older age classes (Poole 1989).

These data definitively contradict the alternative view that the ivory trade was not responsible for widespread declines (Parker and Amin 1983).

At a population level, quantitative information on elephant trends is limited, but existing data have been used to estimate regional trends. In summary, these showed massive declines of the order of 80 per cent in the northern savannah zones, which stretch across Chad, the Central African Republic and Zaire, to Sudan and Somalia. In the central African forests, Barnes (1989) calculated that forest elephants have been reduced by some 43 per cent, but the time span over which this has taken place is not certain.

In Kenya, Uganda, and Tanzania, the total drop from 1977 to 1989 was at least 185 000, thought to be about 74 per cent. The largest single population drop recorded was in the Selous Game Reserve, which lost 75 000 elephants from 1976 to 1989, or about 75 per cent. As elephants diminished, the rate of decline accelerated, consistent with Caughley *et al.*'s (1990) model of the effects of the ivory trade in East Africa. In southern Africa, rates of change measured in the Luangwa Valley in Zambia were similar to the East African scenario, and qualitative accounts of poaching in the region suggested similar or more severe rates of decline in Mozambique and Angola.

Exceptions to the general decline were to be found in the southern African region in South Africa, Namibia, Botswana, and Zimbabwe, where populations were thought to be stable or expanding. It appeared that in Botswana and Zimbabwe the shared population may have increased by the order of some 50 000 animals in the decade 1979–89.

Experience from 1975 to 1989 showed that when stocks of elephants were exhausted in one locality, the killing moved to another. With the ivory trade moratorium introduced in 1989, however, the offtake did not continue at the same level and poaching did not spread. Rates of poaching declined dramatically in East Africa, although the evidence was more equivocal elsewhere on the continent (Jachmann and Dublin 1991).

The downward trend in most of Africa is closely associated with the adequacy of protection. Only 1.5 per cent of the elephants' total range is adequately protected. Where the wildlife services have been inadequately funded, elephants are killed inside as well as outside the protected areas. The decline of elephants in most of Africa is associated with fragmentation of their range. None the less, experience gained in the Addo Elephant National Park in South Africa gives hope that recovery can be rapid if protection is restored. The Addo population increased from about 16 animals in 1954 to a population of over 120 by the late 1980s. The gain was attributed to better protection and improved relations with neighbours due to an elephant-proof fence which prevented crop-raiding by elephants (Hall-Martin 1979). The future challenge for conserving the species across the continent is to weave elephant conservation into a framework of human development where conflicts between the two demanding species are reconciled.

SUMMARY AND CONCLUSIONS

Elephant numbers and trends have often been disputed, and the figures given here are merely a summary of the best available data at the time of writing. They may yet be revised with the benefit of hindsight and new methods of analysis, but the overall picture is clear. There was a massive and unprecedented decline of elephants in the last twenty-five years.

These recent disturbances were caused by two main factors, the killing of elephants for ivory and human population increase. Predation by human beings has been exacerbated by an ivory price rise in the early 1970s, and by civil wars and the immense proliferation of automatic weapons in the hands of ordinary people (Douglas-Hamilton 1983). If the ivory harvest of the past decade had continued at the same rate, the vast majority of elephants would have disappeared even in supposedly protected areas.

The downward trend of elephant populations in most of Africa is in places associated with the fragmentation of their range caused by competition with man. The break-up of continuous range in west Africa is now occurring increasingly elsewhere (Douglas-Hamilton 1992). Only 1.5 per cent of the elephants' total range is adequately protected.

The recent decision adopted at the 1989 CITES (Convention on International Trade in Endangered Species of Wild Fauna and Flora) meeting to place the African elephant on Appendix I prohibits all trade in ivory among countries which are signatories to CITES. Although the devastating elephant poaching of the 1970s and 1980s has been checked, it has not totally disappeared. Ivory is still moving in small quantities in Africa (Milliken, personal communication 1993). The ivory consumer market may be dead in the West, but it is alive in the East.

The ban on ivory is fragile and the ceasefire gained for elephants would be reversed if the ivory trade ever regained its previous levels. With many areas suffering from wars or civil disturbance, controls are non-existent. Calls for limited trading of legal ivory to be resumed are unwarranted, as they send out a signal to illegal traders and poachers that the trade is open again.

Under current conditions illegal ivory would inevitably be mixed with legal stock.

The ivory trade ban, however, will not work on its own to conserve elephants, as there is gross underfunding of the national parks and reserves, exacerbated by Africa's chronic debt stricken economies. A consequence of the International Monetary Fund's austerity rules can often be the axing of funds to wildlife departments with consequent reduction in wildlife law enforcement, as happened in Zimbabwe with a consequent increase in elephant poaching in 1993.

Conserving healthy elephants means conserving healthy ecosystems and the basic life-support processes of nature. The question of whether or not enough space can be preserved in the face of exponential human population increase remains a moot point. Africa's population now (in 1994) stands at about 548 million people and optimistically 550 to 650 thousand elephants. If demographers are right there may be 2.9 billion people by 2050, under which circumstances there will be little space left for elephants.

ACKNOWLEDGEMENTS

The authors thank the African Elephant Specialist Group members for the field data on which this overview is based. The United Nations Environment Programme headquarters have housed and supported the database since it was set up in 1986. This work was also funded by the Commission of the European Communities, the Elsa Wild Animal appeal, and Elefriends. J. Shoshani, M. P. Kahl, and R. I. Orenstein reviewed this paper.

THIRTY-THREE

Elephas maximus: status and distribution

RAMAN SUKUMAR AND CHARLES SANTIAPILLAI

INTRODUCTION

The present-day distribution of the Asian elephant (*Elephas maximus*) (Fig. 33.1) covers only a tiny fraction of its former extensive range. Six thousand years ago, its range extended from the Tigris–Euphrates Basin in Western Asia, eastward up to the Yangtze-Kiang, and perhaps even beyond this, in northern China. This distribution covered present-day Iraq and nearby countries, southern Iran, Pakistan, the entire Indian subcontinent south of the Himalayas, continental South-East Asia, a substantial part of China, and islands such as Sri Lanka, Sumatra, Borneo, and possibly Java.

Since then, it has disappeared from western Asia, a major part of the Indian subcontinent, substantial areas of South-East Asia, and almost entirely from China. The domestication of the elephant is believed to have been first achieved by the people of the Indus Valley civilization about 4000 years ago, and its regular capture in large numbers on the Asian continent may have played a major role in the decline of wild elephants. The loss of habitat due to expansion of human settlements and agriculture has been the single most important factor causing a decline in elephant population. As settlements spread along river valleys and plains, the elephant was forced into the forested hills which were relatively inaccessible to people. During the twentieth century, even these last strongholds have been settled in many regions.

Elephas maximus is believed to comprise three subspecies (Shoshani and Eisenberg 1982): *E. m. maximus* (habitat, the island of Sri Lanka), *E. m. indicus* (habitat, mainland Asia), and

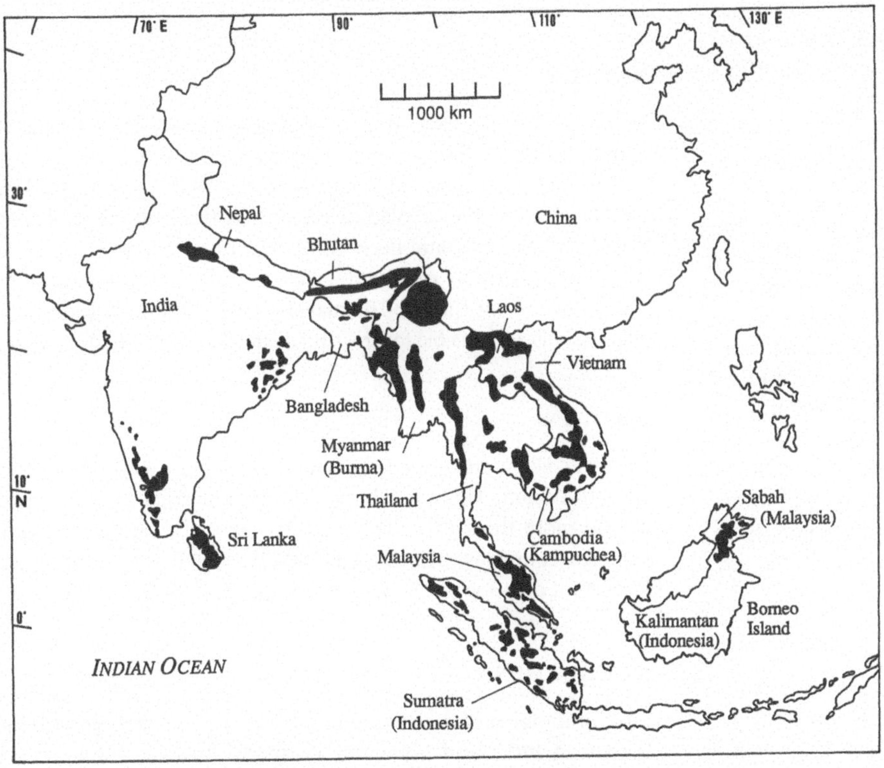

Fig. 33.1 The present-day distribution of the Asian elephant (*Elephas maximus*).

E. m. sumatranus (habitat, the island of Sumatra). Currently, the Asian elephant distribution covers an estimated 500 000 km^2, holding a total wild population of from 37 000 to 55 000 individuals. The population estimates are largely educated guesses. Objective estimates are available only from parts of India and Sri Lanka, and from one or two places in Sumatra and Thailand.

The account, by country, given below is based largely on Olivier (1978), Daniel (1980), Sukumar (1989), and Santiapillai and Jackson (1990), who, in turn, have drawn upon a variety of published and unpublished material, based on the work of the Asian Elephant Specialist Group of the International Union for Conservation of Nature and Natural Resources (IUCN), World Wide Fund for Nature (WWF), and wildlife administrators and researchers in the respective countries. The order in which the countries are discussed follows geographical proximity.

ACCOUNTS OF ELEPHANTS IN ASIA

INDIA

India holds the largest number of Asian elephants in the wild. There are between 20 000 and 25 000 elephants distributed over four widely separated regions — north-western, north-eastern, central, and southern India.

North-western India

The moist forests along the foothills of Himalayas in the state of Uttar Pradesh hold an estimated 750 elephants. Among the important elephant areas are Corbett and Rajaji National Parks and the Lansdowne Forest Division (Singh 1978). The habitat has been fragmented by an irrigation canal and recent human settlements including an army cantonment (Johnsingh *et al.* 1990).

North-eastern India

With 9000–12 000 elephants, this region is believed to hold the largest number in the country (Lahiri-Choudhury 1986). One large population of 3800–5800 elephants ranges along the Himalayan foothills from northern West Bengal, eastward through Assam and Arunachal Pradesh. The vegetation includes moist deciduous and evergreen forest. Key conservation areas include the Manas Tiger Reserve and Namdapha National Park. South of the Brahmaputra River, another large population of nearly 2000 elephants inhabits the alluvial floodplains of Kaziranga National Park and Nagaland hills. A third population of 2500–3500 elephants in the Garo hills and Khasi hills of Meghalaya is under severe threat from habitat loss due to extensive shifting cultivation. Small scattered populations have been observed in the Jainti hills, Tripura, Manipur, and Mizoram.

Central India

The elephants are largely confined to the states of Bihar and Orissa (Shahi and Chowdhury 1986), although, in recent years, they have also been moving into Madhya Pradesh and southern West Bengal, where they have been coming into serious conflict with people. In Bihar about 65 elephants are found in the Palamau Tiger Reserve, 200 in the Singhbum tract, and 70 in the Dalbhum tract. Orissa has an estimated 1300 elephants in a number of isolated populations, of which only 375 found in Simplipal Tiger Reserve and 300 in Satkosia Gorge Sanctuary seem to have some measure of security. The elephants of central India face severe threats to their habitat from degradation, shifting cultivation, mining, and replacement of natural forest by monoculture plantations.

Southern India

The elephant is distributed in the hill forests of the Western Ghats and adjoining portions of the Eastern Ghats in the states of Karnataka, Kerala, and Tamilnadu (Nair and Gadgil 1978; Nair *et al.* 1980; Sukumar 1986, 1989). About 10 000 elephants are found distributed as a series of at least 10 distinct populations. The largest population, conservatively estimated at 6000, inhabits the Nilgiri hills and Eastern Ghats. Sizeable numbers are also found in the Anamalai hills (800–1000) and the Periyar plateau (1000+). Elephant densities in some locations are among the highest in Asia and comparable to those for the African bush elephant. Nagarhole National Park, Bandipur Tiger Reserve, and Mudumalai Sanctuary in the Nilgiris, and Periyar Tiger Reserve have densities in the range of 1–3 elephants/km^2. The diverse vegetation types — evergreen forest, deciduous forest, and thorn scrub — available to elephants within a relatively small area are no doubt an important factor in supporting these high elephant densities. The main threats to the habitat come from developmental projects such as dams, tea and coffee plantations, agriculture, railway lines, and roads.

NEPAL

Nepal has few elephants that are resident. The present estimate is about 50–100 elephants in the country, of which most are migratory from the adjacent Indian state of Uttar Pradesh (Mishra 1980). Shukla Phanta Wildlife Reserve in the extreme west and Royal Chitwan National Park in the south are two areas where it may be possible to maintain viable populations. Expansion of agriculture along the sub-Himalayan tract in recent decades has been mainly responsible for the contraction of the elephant's range.

BHUTAN

Bhutan's elephants are partly or entirely shared with the Indian state of Assam. There has been no survey carried out but there are no more than 150 elephants that cross the border between the two countries. The Manas Wildlife Sanctuary is the most important area for elephants (Santiapillai and Jackson 1990).

BANGLADESH

Most of Bangladesh's elephants are found in the Chittagong Hills bordering India and Myanmar (formerly Burma). Estimates of elephant numbers vary from 200 (Reza Khan 1980) to about 350 (Gittins and Akonda 1982). Some of these elephants undoubtedly move across international borders. Two reserves, one in the Chittagong tract and another in Cox's Bazaar, are being planned by the government for the protection of elephants.

SRI LANKA

As in other Asian countries, the elephant population on the island of Sri Lanka has declined drastically during the past century. The most recent estimate by A. B. Fernando (personal communication) is between 2500 and 3200. Elephants are found in the drier habitats to the north and eastern half of the country. The vegetation includes dry deciduous woodland, scrub, grassland, and marshes called *villus* in the floodpans of rivers. The highest densities of elephants probably occur in the south-east of the country in places such as the Ruhuna National Park and its environs.

The decline in elephant numbers can be traced largely to agricultural development, which has led to the pocketed herd phenomenon in many areas. The on-going Accelerated Mahaweli Development Scheme, involving the construction of a number of dams and the opening up of new areas for agriculture, is expected to further constrict the habitat for elephants. To cope with the anticipated fragmentation of habitat, the country has established a network of protected areas, linked wherever possible by corridors. Well-known reserves, where elephants can be seen, are Wilpattu in the north-west, Somawathiya, Wasgamuwa, and Maduru Oya in the Mahaweli Basin to the east, and Gal Oya, Uda Walawe, Lahugala, and Ruhuna in the south-east.

MYANMAR (FORMERLY BURMA)

Elephants are widespread in Myanmar as a large proportion (about 50 per cent) of the country is still under forest. Although elephants are seen over a wide variety of habitats, they are most plentiful in teak and bamboo forests that are extensive in many regions. Precise information on distribution is not available, but elephants are known to occur in five regions (Blower 1985). The Northern Hill ranges probably support the largest populations, including one of about 1600 in Myitkynia–Bhamo and another of about 1800 in the Irrawady and Chindwin Valleys. Sizeable numbers are also seen in the Western Hill ranges, particularly the Arakan Yoma and Chin hills, the Pegu Yoma in the central region, the eastern Shan states, and the Tenasserim Yoma in the south-west, bordering on Thailand.

Although information on population sizes is subjective, it is certain that Myanmar has the largest number of elephants after India. Considering the relatively extensive area still available to elephants, a conservative estimate of 5000–8000 elephants can be made for the wild population. One estimate by the government, however, puts the figure at 6520 elephants.

Myanmar still continues to capture elephants and this seems to be a cause for depletion of the wild population. In recent years the poaching of elephants on a large scale and the smuggling of their products, including ivory and skin, into Thailand for sale seem to be common.

CHINA

The elephant which once extended further north of the Yangtze-Kiang has retreated entirely into the southern part of the province of Yunnan along the border with Myanmar and Laos (Gao 1981). The habitat here consists of broad-leaved mixed forest, bamboos, and grasslands. Only about 300–500 elephants seem to survive, largely in the Xishuangbanna Reserve (Santiapillai *et al.* 1991).

THAILAND

Thailand has lost a substantial amount of forest cover during recent decades, from 80 per cent of the country's area during the 1930s to only 30 per cent at present. The elephant populations have also declined and suffered in the process. Since Lekagul and McNeely's (1977) estimate of 2600–4450 elephants made in 1977, it seems that they have been further reduced (Dobias 1987).

The largest numbers of elephants occur in the north and west in the Tenasserim hill range along Thailand's border with Myanmar. The estimate for this region is 1300–2100 elephants. The hills have a diversity of vegetation types, including evergreen forest, but dry deciduous forests, bamboo forests, and grasslands predominate. Two protected area complexes, one in the north comprising Om Koi, Maetuen, and Mai Ping Reserves, and another in the west including the Huai Kha Khaeng, Thung Yai, Sri Nakarin, and Erawan Falls Reserves, seem to offer the best hope for the elephants.

In the Petchabun mountains to the north, a population of 200–500 elephants is estimated to range over a mosaic of temperate and tropical forests. The important protected area complex for elephants includes Phu Kradung, Nam Nao, and Phu Khieo.

With 325–425 elephants, the Dangrak mountain range along the border with Cambodia (Kampuchea) to the south-east is another important elephant belt. At the western edge of the range, the complex of Khao Yai, Thapan Lan, and Pang Sida National Parks is one of the best protected. Khao Yai is probably the only place in Thailand where some objective information on elephant ecology and numbers is available. A density of 0.15 elephants km^2 or 225 elephants for the park has been estimated by Dobias (1985).

A significant population of 900–1500 elephants also occurs in peninsular Thailand to the south. These are believed to be scattered as numerous isolated populations. The Khlong Naka and Khlong Saeng Sanctuaries and Khao Sok National Park seem to be the most important elephant reserves.

Dobias (1987) estimates that only 1300–1700 of Thailand's elephants are found in the protected area network extending over some 25 000 km^2. The future of elephants seems largely tied to this network.

LAOS

About 40 per cent of Laos is still under forest, and thus it could be expected that elephants are still widespread in the country. There have been, however, no surveys of elephant distribution in Laos. It is believed that elephants are more widely distributed in the south, which has better forest cover (Sayer 1983). Significant populations also occur in central Laos, particularly in the provinces of Sayabouri (west of the Mekong River), Vientiane (Phou Khao and Khouly Massif), Balikhamsay, and Khammouane (Nakai plateau). Some of these are shared with Vietnam. In the north, where shifting cultivation is extensive, elephants are seen only around Hong Sa, along the border with China.

A recent estimate by Venevongphet (1988) puts the wild elephant population in Laos at 2000–3000. This should only be taken as a rough approximation as it is not based on any objective census (Phanthavong and Santiapillai 1992).

CAMBODIA (ALSO KNOWN AS KAMPUCHEA)

As with some of the other countries, very little is known about the status of elephants in Cambodia. Elephants are found in three regions. The forests to the east, along its border with Vietnam, seem to be the most important habitat for them. Potentially, the Dangrak mountains to the west, along the Thailand border, are also a good elephant habitat. Similarly, one would expect elephants to be present in the rainforests of the Cardamon and Chuor Phum Damrei (Elephant) mountains to the south. Only an estimate of 1000–2000 elephants is available for this country.

VIETNAM

Elephants are found mainly along the western boundaries with Laos and Cambodia, and a series of isolated populations exists in interior southern Vietnam (Tuoc and Santiapillai 1991). As in the two neighbouring countries, both elephants and people face a risk from unexploded bombs and mines in the forests. Estimates range from 1500–2000 elephants (Vu Khoi 1988) to 500+ elephants (S. Dawson, unpublished).

PENINSULAR MALAYSIA

The clearing of peninsular Malaysia's equatorial rainforests for the commercial rubber and oil palm plantations has severely fragmented the elephant's habitat. Fewer than 1000 elephants may survive in the peninsula according to one estimate (Mohd Khan 1985). Many of these seem to exist as isolated herds which have no hope for the future. A well-organized programme of relocating such small elephant herds to more viable areas is in operation. Taman Negara, with its estimated 150–200 elephants in the central region, and Endau Rompin to the south offer the best prospects for conservation.

SUMATRA (INDONESIA)

Though widespread, the elephant is believed to be distributed as 44 distinct, small populations on the island of Sumatra (Blouch and Haryanto 1984; Blouch and Simbolon 1985; Santiapillai and Suprahman 1985). The habitat is almost entirely dense rainforest, often in hilly, inaccessible terrain. As in Malaysia, agricultural development in the form of rubber and oil palm, and the migration of people from Java, have been largely responsible for the reduction in forest cover.

Only 15 populations in Sumatra are believed to consist of over 100 elephants. The total population on the island is estimated at 2800–4800 elephants with a maximum of 2500 in the 28 protected areas. The most important of these reserves for elephant conservation are Gunung Leuser, Way Kambas, Kerinci-Seblat, and Barisan Selatan.

BORNEO (MALAYSIA AND INDONESIA)

Fossil evidence indicates that elephants once occurred in Borneo (van den Bergh *et al.*, this volume, Chapter 23), but it is not clear if the present-day elephants are native or are descended from captive elephants brought to the island in 1750 and later set free. Today, elephants are confined to the north-east, largely in Sabah (Malaysia) and a small area in Kalimantan (Indonesia). Between 500 and 2000 elephants may occur in the dense rainforests in hilly country (Andau and Payne 1985).

A NOTE ON CONSERVATION AND MANAGEMENT

The best way to manage smaller isolated populations in the short term would be to ensure a genetically effective population size of at least 50 elephants in order to keep loss of heterozygosity below 1 per cent per generation (Franklin 1981). In many cases, this may be achieved by artificial translocation of individuals between populations (Sukumar 1989). Very small herds may have to be captured and retained in captivity or released in more viable habitats.

Poaching of males for ivory has been a problem in southern India and parts of South-East Asia, including Myanmar and Thailand. Control of the illegal trade of tusks in Asia has to be linked to the international efforts to curb the ivory trade under the umbrella of CITES (Convention in International Trade in Endangered Species of Wild Fauna and Flora).

Elephants also come into conflict with people by damaging crops and houses, and killing people. This creates antagonistic feelings towards elephant conservation among people. It is thus important to promote schemes such as digging trenches or erecting electrified barriers along the forest–cultivation boundary or compensating farmers for losses in the elephant ranges (Sukumar 1989).

Table 33.1 Population estimates for wild Asian elephants (arranged as they are discussed in the text)

Country/region	Population		Minimum habitat area (km^2)
	Minimum	Maximum	
India	20 000	25 000	82 000
Nepal	50	100	2500
Bhutan	100	150	2000
Bangladesh	200	350	2000
Sri Lanka	2500	3200	8000
Myanmar (formerly Burma)	5000	8000	100 000
China	300	500	2500
Thailand	1500	3000	35 000
Laos	2000	3000	20 000
Cambodia	1000	2000	15 000
Vietnam	500	1000	15 000
Peninsular Malaysia	1000	1500	20 000
Sumatra (Indonesia)	2800	4800	100 000
Borneo (Malaysia and Indonesia)	500	2000	35 000
Total	37 450	54 600	439 000

SUMMARY

Estimates of the wild Asian elephant population range between 37 000 and 57 000 individuals, and their habitat covers an area of about half a million km^2 (Table 33.1). The elephants are found, however, in perhaps as many as a hundred or more distinct populations, with little or no possibility of natural genetic exchange. For many Asian elephant populations, viability is thus severely constrained by small population numbers and insufficient habitat for expansion. Only about 10 populations seem to consist of over 1000 elephants each. These include Nilgiri–Eastern Ghats in southern India; Arunachal–Assam, Kaziranga and Garo–Khasi hills in north-eastern India; south-eastern Sri Lanka; Myitkynia–Bhamo and Irrawady–Chindwin in Myanmar (formerly Burma); Tenasserim mountains along the Myanmar–Thailand border; and possibly Anamalai–Periyar in southern India, Laos–Cambodia–Vietnam borders, and Sabah in Borneo. In addition, some of the smaller populations, including Taman Negara in peninsular Malaysia and those in Sumatra, are important as they may represent genetically distinct subpopulations that have evolved in the species-rich rainforests. Initiatives for long-term conservation should pay special attention to these populations.

ACKNOWLEDGEMENTS

This paper was typed by Susan Wolak and edited by J. Shoshani and E. C. Marsac.

PART VII

Summary and conclusions

THIRTY-FOUR

Summary, conclusions, and a glimpse into the future

JEHESKEL SHOSHANI AND PASCAL TASSY

'One of the most spectacular stories in mammalian evolution is that of the order Proboscidea — the mastodons, elephants, and related types.'

Alfred S. Romer 1966, p. 248

INTRODUCTION

Skeletal remains of extinct proboscideans from all over the world have been the *sine qua non* for our understanding of this mammalian order. This prerequisite was noted by Simpson (1945, p. 244) when he wrote 'Classification within the order [Proboscidea] is obviously a task for paleontology rather than neontology'. These fossil remains, together with features of the living species, provide us with the foundation upon which researchers have been constructing the phylogeny and classification of the Proboscidea.

Summaries of results presented in this volume and data compiled from the literature are presented in Figs 34.1–34.8, Tables 34.1 and 34.2, and Appendices A–D, most of which deal with taxonomy, phylogeny, and distribution. A special attempt has been made to provide the reader with a classification and 'a family tree' of the Proboscidea (Appendix B, Fig. 34.3), for they can be used as working hypotheses for further research. The views expressed in these figures and tables do not represent a consensus among students of Proboscidea; some of the disagreements are discussed later in this chapter.

CURRENT TAXONOMY

NUMBER OF FAMILIES, GENERA AND SPECIES

Comparison of proboscidean taxa recognized by Osborn in 1936 and those we recognize today, 1995 (Table 34.1, Fig. 34.1), reveals that five of Osborn's eight families are no longer recognized; of the 8+ (?9) families listed for 1995, 1+ (?8) are new. In addition, 16 of Osborn's 44 genera are not used today; of the 38+ genera listed, at least 10 are new or were not recognized as proboscideans in 1936 (*Anthracobune, Numidotherium, Eozygodon, Protanancus, Archaeobelodon, Haplomastodon, Paratetralophodon, Stegotetrabelodon, Stegodibelodon,* and *Primelephas*). The estimate on the species and subspecies level is based on counting those which have been published and/or based on reviewers' comments, and it should be used as a guideline only — the 162 species listed for 1995 indicate a reduction of more than 50 per cent since Osborn's count 59 years ago. This number still needs to be re-evaluated. Overall, the number of genera in the history of the Proboscidea has been reduced, whereas the number of species has been condensed by about one-half across all proboscidean families.

Not only were gomphotheres the most diversified proboscideans, but they were also the largest group in terms of absolute number of taxa during the Cenozoic era; this trend is simplified in Fig. 34.2. Note that there are 10 proboscidean species which are presently known from the Eocene. An almost tenfold increase in the number of species occurred during the Miocene, an increase which seemed to have been maintained during the Pliocene. Sharp declines appeared to have taken place during the Pleistocene and Holocene, at which time only two species remained of this once-successful mammalian order.

PHYLOGENY AND DISTRIBUTION

Observations made in Table 34.1 and Figs 34.1 and 34.2 are incorporated in a single diagram (Fig. 34.3). The simplified phylogenetic relationships are based on the cladograms shown in Figs 6.1 and 16.5. Note that *Anthracobune* is included in this phylogram because, based on cladistic analysis, it is a proboscidean. All the earliest proboscideans, which were discovered in Africa, come from rocks of the Eocene through Oligocene ages. In Fig. 34.3 gomphotheres and amebelodontids are grouped together to make the illustration simpler. It must be kept in mind that, in the most parsimonious solution of Tassy (cf. Fig. 6.1), they form a paraphyletic assemblage, a stem-group for stegodontids and elephants (not their sister-group). Yet, certain members of this same paraphyletic assemblage are united in the most parsimonious solutions in Shoshani's study (cf. Figs 16.1a, b).

The proportion of the number of taxa in the group of gomphotheres and amebelodontids compared to other groups is such that it was about equal to the number of all other proboscidean taxa combined. This large group was also extremely diversified and its

Table 34.1 Comparison of proboscidean taxa recognized by Osborn in 1936 and those recognized today, 1995

Number of:	Super-families	Families	Subfamilies	Genera	Species and subspecies
Osborn (1936)	5	8	21	44	352
Current status (1995)	–	8+	–	38+	162[d]
New taxa since 1936	–[a]	5[b]	–[a]	10[c]	? ~ 50[e]

[a] Super- and subfamilies: these are excluded from the present analyses.

[b] Families: five of Osborn's eight families are not used today; of the eight families + ?nine possible families listed for 1995, at least one family plus possibly 8 families are new — a rate of about one family discovered or named every 7 years (cf. Tables 1.1, 34.2, and Appendix C.1 for details).

[c] Genera: 16 of Osborn's 44 genera are not used today; of the 38+ genera listed for 1995, at least 10 are new (or were not recognized as proboscideans in 1936) — a rate of about one genus discovered or named every 6 years.

[d] Estimate based on a range of 136–188 (see Appendix C). A new species, *Numidotherium savagei* Court 1995, p. 651, has been described since this chapter was written.

[e] Species and subspecies: the 162 mostly species listed for 1995 are approximately 50 per cent of the total recognized in 1936; this number needs to be re-evaluated.

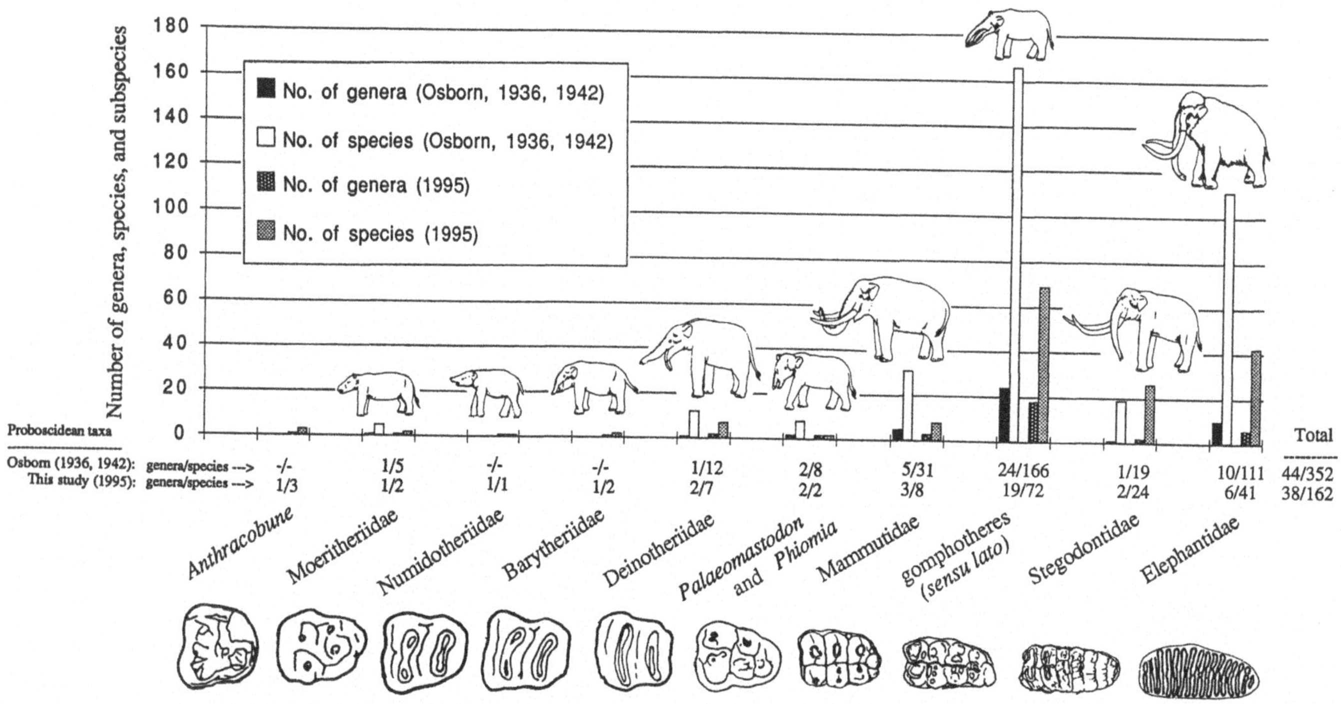

Fig. 34.1 A histogram showing the relative numbers of genera, species, and subspecies of proboscidean taxa as recognized by Osborn (1936, 1942) and compared to the numbers recognized by present-day (1995) workers. Numbers on the *x*-axis represent genera/species given by Osborn (top) and those recognized today (bottom); data for the three families of Osborn's superfamily 'Mastodontoidea' (Bunomastodontidae, Humboltidae, and Serridentidae) are grouped together and compared to 'gomphotheres' (*sensu lato*). Animal sketches are similar to those in the Frontispiece. Teeth at the bottom of the figure are upper third molars, not to scale and not necessarily the most primitive of each taxon, but they represent evolutionary changes in tooth morphology, size and number of ridges. Anterior is to the left. Approximate (or minimum) lengths × widths (in cm) in occlusal view/possible maximum number of ridges or plates per tooth of the taxa depicted are as follows: *Anthracobune* ('*Pilgrimella pilgrimi*'), 1.7 × 2.0/2; *Moeritherium* sp. 3.1 × 2.5/2; *Numidotherium koholense*, 3.7 × 3.7/2; *Barytherium grave*, 8.2 × 7.3/2; *Deinotherium giganteum*, 7.0 × 7.0/2; *Phiomia serridens*, 6.1 × 4.5/3; *Eozygodon morotoensis*, 10.4 × 7.0/3+; *Gomphotherium angustidens*, 16.0 × 8.0/4+; *Stegodon zdanskyi*, 26.7 × 13.8/7+; *Mammuthus primigenius*, 28.5 × 11.3/27+. Molars of the first six taxa are from Fig. 8.2, this volume; *E. morotoensis* is from Tassy and Pickford (1983, p. 61); *G. angustidens* is from Tassy (1985, Tome 3. Fig. 98); *S. zdanskyi* is from Fig. 17.7, this volume:; *M. primigenius* is from Fig. 19.1 this volume.

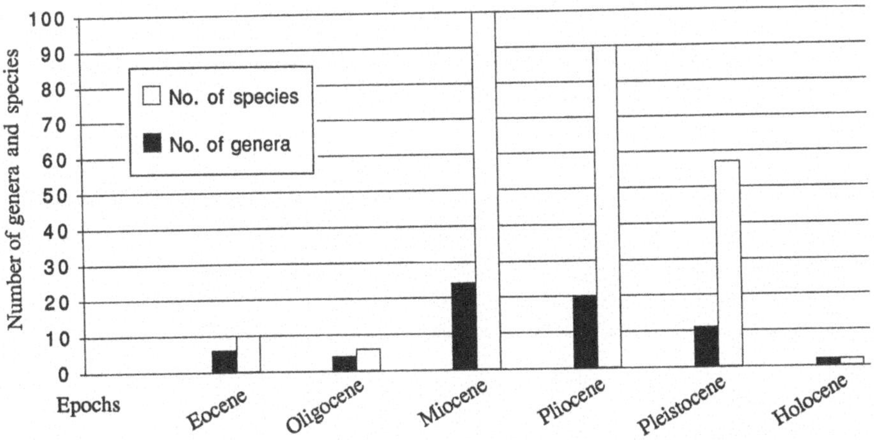

Fig. 34.2 A histogram depicting approximate numbers of proboscidean genera and species from the Eocene through the Holocene (epochs are not to scale). Note that adding the numbers at the top of each bar does not give a total of 38 genera or 162 species (the current estimate as of 1995), because some taxa lived during more than one epoch.

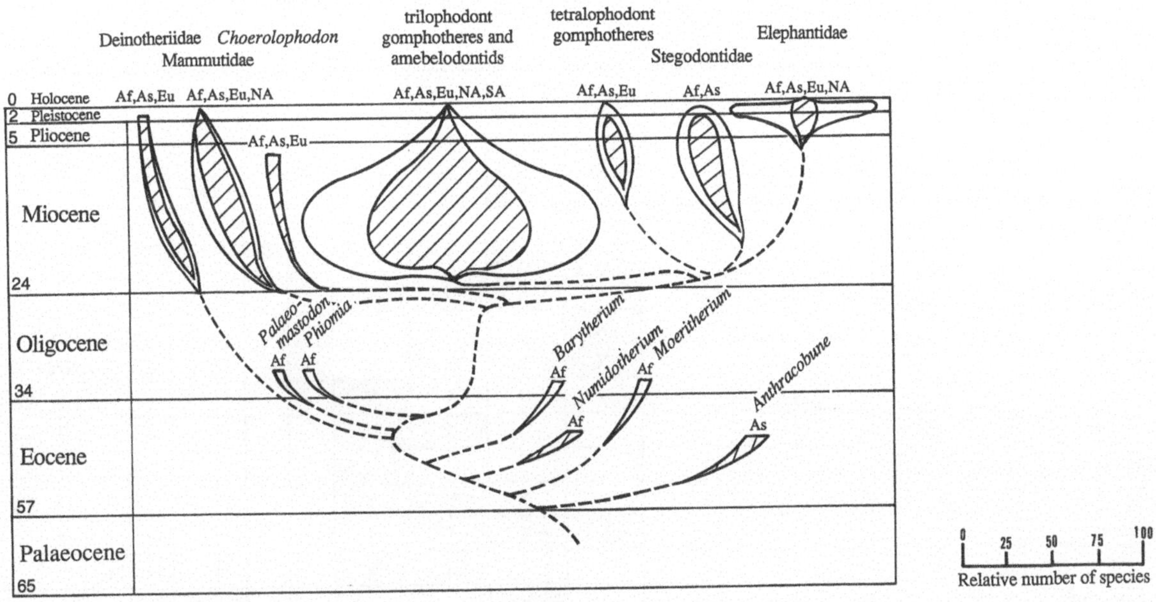

Fig. 34.3 A simplified phylogram of proboscidean taxa drawn in approximate proportion (scale at bottom of chart represents number of species and subspecies). Blank areas refer to numbers as recognized by Osborn (1934*b*, 1936); hatched areas are those recognized in the present study (1995); see also Figs 34.1, 34.2 and Table 34.1). Note that *Hemimastodon* is lacking, trilophodont gomphotheres and amebelodontids are grouped for simplicity, and tetralophodont gomphotheres are assumed to be closer to stegodontids and elephantids, but do not form a monophyletic unit. Solid lines are based on the known fossil record; dashed lines are hypothetical connections. Af, Africa; As, Asia; Eu, Europe; NA, North America; SA, South America; Ma, million years ago. (Artwork by J. S. Grimes and G. H. Marchant.) Addendum: Court (1995) placed *Numidotherium* earlier than *Moeritherium*. Note, however, that *Barytherium* was not included in Court's analysis.

remains were found on all continents except Australia and Antarctica (Table 34.2, Figs 34.4–34.6; Osborn 1936, 1942; Mahoney and Ride 1975; Vickers-Rich *et al.* 1991; Stucky and McKenna 1993).

One might view the Proboscidea as a complex assembly of lineages; yet, in a general sense, it may have undergone three major radiation events — an image upon which the organization of this volume is based (cf. Shoshani 1992*e*). Some workers have suggested that at least the third of these radiations occurred in a relatively short time and exemplifies punctuated evolution (Stanley 1979, 1987). These radiation events were at the early or middle Eocene (*c.* 53–37 Ma), at the end of the Oligocene/ beginning of the Miocene (*c.* 27–23 Ma), and at about the boundary between the Miocene and the Pliocene (*c.* 7 Ma). The first major radiation is the least defined; it is included here more for convenience of discussion than for its phylogenetic content. It

Table 34.2 Known geological distribution of proboscidean genera (cf. Appendices A–D, Tables 13.2, 14.1, and Figs 16.5, 30.1, 34.2–34.4).

Taxon	eE	mE	lE	O	eM	mM	lM	Pl	Ps	eH	lH	Continent (2)				
	<------ Geological Record (1) ------>															
Anthracobune	X	X											As			
Moeritherium		?	X	X								Af				
Numidotherium	?	X										Af				
Barytherium			X	X								Af				
Prodeinotherium					X	X	X					Af	As	Eu		
Deinotherium						X	X	X	X			Af	As	Eu		
Palaeomastodon			X	X								Af				
Phiomia			X	X								Af				
Hemimastodon					X								As			
Eozygodon					X	X						Af				
Zygolophodon					X	X	X	?				Af	As	Eu	NA	
Mammut						X	X	X	X	X		Af	As	Eu	NA	
Choerolophodon					X	X	X	?				Af	As	Eu		
Gomphotherium					X	X	X	?				Af	As	Eu	NA	
Amebelodon						X	X	X				Af	As		NA	
Platybelodon					X	X	X	?				Af	As	Eu	NA	
Serbelodon						X	X						As	?	NA	
Protanancus						X	X					Af	As			
Archaeobelodon					X	X						Af		Eu		
Gnathabelodon						X	X								NA	
Sinomastodon							X	X	X				As			
Eubelodon						X									NA	
Rhynchotherium						X	X	X							NA	
Stegomastodon								?	X						NA	SA (3)
Haplomastodon								X	X						?	SA
Notiomastodon									X							SA
Cuvieronius							?	X	X						NA	SA
Tetralophodon						X	X	X	?			Af	As	Eu	?	
Anancus							X	X	X			Af	As	Eu		
Paratetralophodon							X						As			
Stegolophodon					X	X	X	X	?				As			
Stegodon							X	X	X	X		Af	As			
Stegotetrabelodon							X	X				Af	?	?		
Stegodibelodon							X	X				Af				
Primelephas							X	X				Af				
Loxodonta							X	X	X	X	X	Af				
Elephas								X	X	X	X	Af	As	Eu		
Mammuthus								X	X	X		Af	As	Eu	NA	

1. Key to geological records: eE, early Eocene; mE, middle Eocene; 1E, late Eocene; O, Oligocene; eM, early Miocene; mM, middle Miocene; 1M, late Miocene; Pl, Pliocene; Ps, Pleistocene; eH, early Holocene; 1H, late Holocene.
2. Key to continents on which remains of these genera were found: Af, Africa; As, Asia; Eu, Europe; NA, North America (including Central America for *Zygolophodon, Mammut, Rhynchotherium, Stegomastodon, Cuvieronius*, and *Mammuthus*); SA, South America (including Central America for *Haplomastodon*).
3. According to Webb (1991, and personal communication, 1993), *Stegomastodon* was not found in South America; see Chapter 31, however.

Editors' note: The earliest proboscideans known to date are from the early Eocene of Asia (*Anthracobune*), or from the (?) early or middle Eocene of North Africa (*Numidotherium*). Recently, two partial maxillae were recovered from the Palaeocene of Morocco, and are being described as a new taxon (Gheerbrant *et al.*, in preparation 1995). If this taxon is confirmed to be a *bona fide* proboscidean, then we will have to reevaluate the contents of the Proboscidea and its synapomorphies. Although the material assigned to this taxon is poor, it is probably the most important discovery in proboscidean research since that of *Moeritherium*.

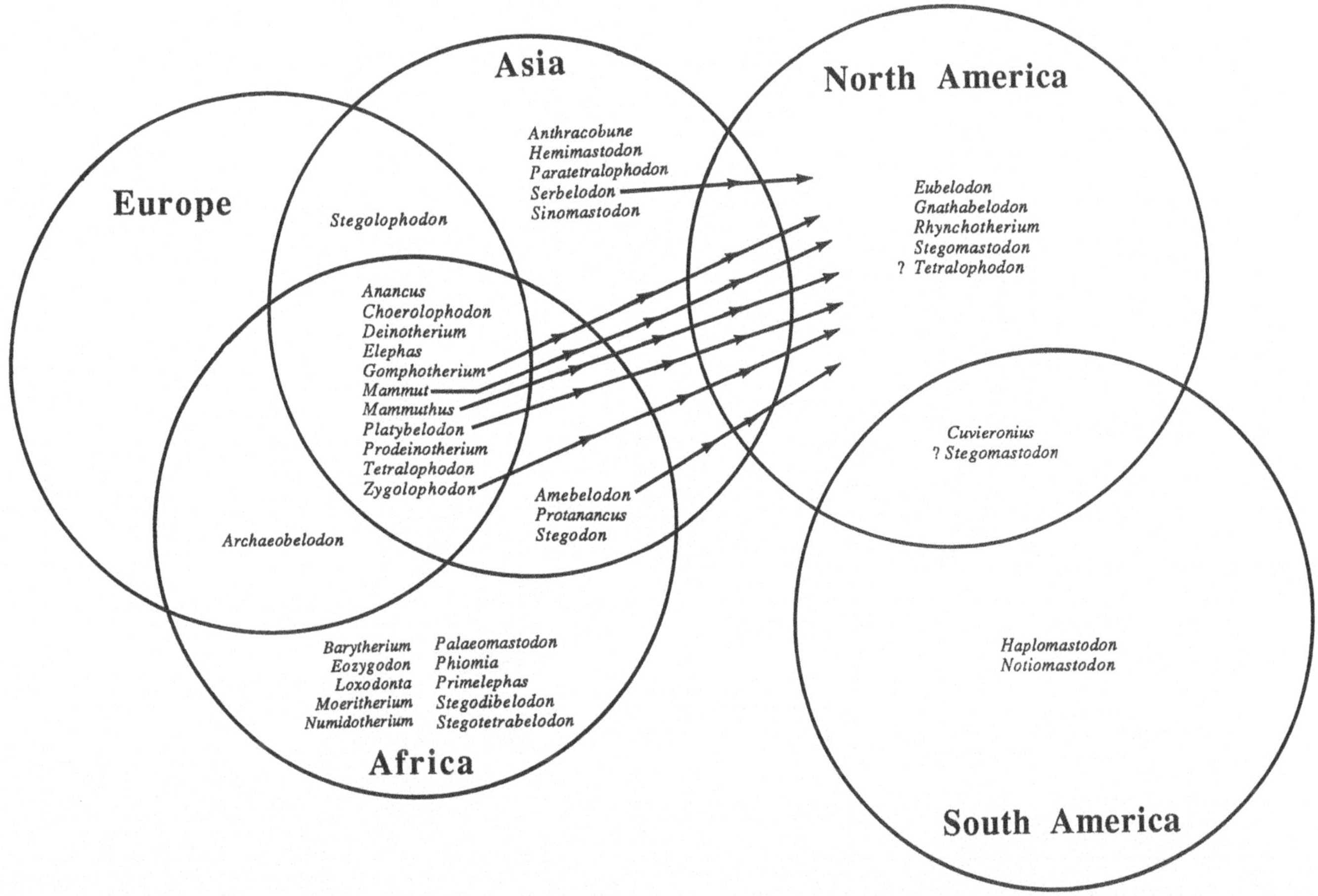

Fig. 34.4 A simplified Venn (bubble) diagram, depicting palaeozoogeographical relationships among proboscidean taxa (listed in alphabetical order) shown in Fig. 34.3. A taxon within a bubble means that it was found there. A taxon shown in a shared area means it was found on two or more continents. An arrow implies that this proboscidean migrated to that continent (cf. to Table 34.2).

includes particularly the earliest proboscideans — anthracobunids, moeritheres, numidotheres, barytheres, deinotheres, palaeomastodons, and mammutids. We have only a very rough idea of the importance of the first radiation. It should be noticed that recent findings from the early Oligocene of the Sultanate of Oman produced four isolated proboscidean teeth and several fragments which do not match those of any described species (Thomas *et al.* 1989*a,b*). We can surmise that the diversity of Palaeogene proboscideans was higher than currently known. The second radiation constitutes the gomphotheres (*sensu lato*) and the stegodontids. The third radiation, which is the most delineated group, is the Elephantidae.

BONES OF CONTENTION

WHAT CHARACTERS MAKE AN ANIMAL A PROBOSCIDEAN?

Earlier in this volume, we listed and briefly discussed the characters which define the boundaries of Proboscidea (Tassy, this volume, Chapter 6; Shoshani *et al.* this volume, Chapter 8; Shoshani, this volume, Chapter 16). For the purpose of merging our ideas, we select shared-derived characters for the Proboscidea, with or without the Asian genus *Anthracobune*.

The synapomorphies for *Anthracobune* and other proboscideans are:

(1) large and hooked scapular coracoid process;

(2) astragalus with a *tuberculum mediale*;

(3) lower third molars with two definite cuspids at the posterior end (this character is homoplasic because the dominant morphology in *Moeritherium* is a single cusp).

Selected synapomorphies for taxa from *Moeritherium* to *Mammuthus* are:

(1) near closure of the external auditory meatus by the enfolding of the postglenoid process and the squamosal;

(2) confluence of the condylar foramen with the foramen lacerum posterius;

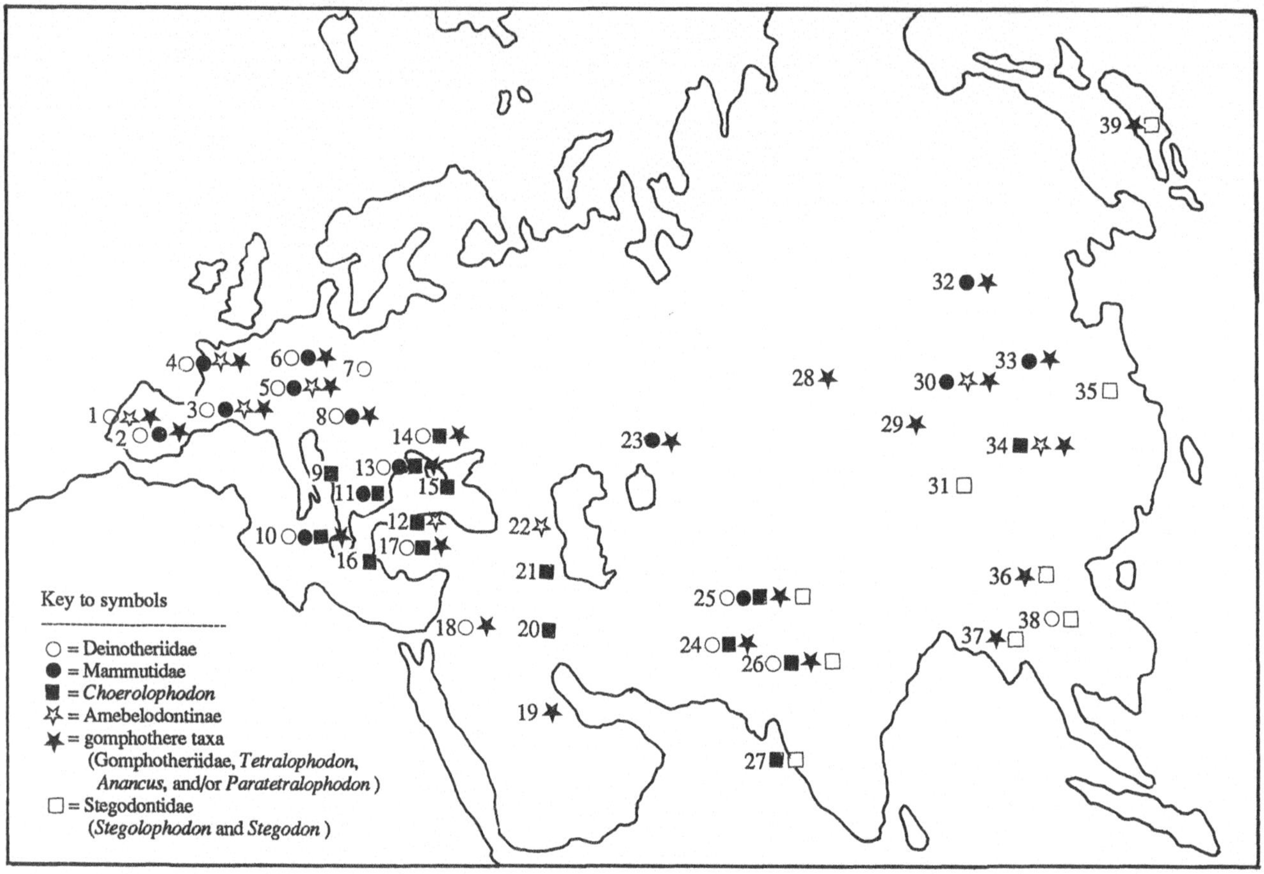

Fig. 34.5 Main Eurasiatic areas/localities which have yielded Neogene proboscideans (mostly after Thenius 1969; Savage and Russell 1983; Tassy 1985; Tobien 1986*a*, 1988; Tobien *et al* 1988). Other maps in this volume with gomphothere localities/distributions are Figs 10.3, 13.1, 14.1, 15.1, 17.10, 21.5, 23.1, 29.1, and 34.6. African gomphothere (*sensu lato*) localities are given in Coppens *et al.* (1978); Kalb *et al.* (1982); Savage and Russell (1983), Tassy (1986), and Kalb *et al.* (this volume, Chapter 12A).
Key to localities: 1. Lisbon area (Portugal); 2. Madrid area, Segovia, Ciudad Real, Catalayud-Teruel (Spain); 3. Aquitaine Basin (France); 4. Loire Basin (France); 5. Oberessüsswassermolasse (Germany); 6. Dinotheriensande (Germany); 7. Franzensbad (Austria); 8. Steiermark area (Austria); 9. Veles (formerly Yugoslavia); 10. Pikermi (Greece); 11. Macedonia (Greece); 12. Northern Turkey; 13. Bulgaria; 14. Odessa area (southern Ukraine); 15. Crimea (southern Ukraine); 16. Aegean Islands (Chios and Samos in Greece); 17. Anatolia (Esme-Akcakoy, Yassiören, Kayadibi, and Pasalar in Turkey); 18. Negev area (Israel); 19. Ad Dabtiyah, As-Sarrar (Saudi Arabia); 20. Gebel Hamrin (Iraq); 21. Maragha (Iran); 22. Belometchetskaya, Caucasus (Russia); 23. Djilancik Beds, Turgai (Kazakhstan); 24. Bugti Hills, Baluchistan (Pakistan); 25. Potwar Plateau (Pakistan); 26. Siwalik Hills (India); 27. Perim Island (India); 28. Xinjiang Uygur Autonomous Region (China); 29. Qinghai Province (China); 30. Ningxia Hui Autonomous Region (China); 31. Gansu Province (China); 32. Tung Gur (China); 33. Yushe Basin (China); 34. Shaanxi Province (China); 35. Jiangsu (China); 36. Yunnan Province (China); 37. Yenangyaung, Irrawaddy (Burma); 38. Northern Thailand; 39. Northern and central Japan.

(3) enlarged second incisors (perhaps also in *Anthracobune*);

(4) loss of first premolar;

(5) radius positioned or fixed in pronation position;

(6) antero-posterior flattening of femur;

(7) loss of astragalar foramen;

(8) loss of crista capitalis on astragalar head.

Characters 1, 2, 5, and 6 are not known in *Anthracobune*.

These two lists emphasize the problem of the extent of the order Proboscidea. Any fossil taxa (such as *Anthracobune*) which appear to be closer to extant elephants than to their sister-group (order Sirenia) can be classified among the proboscideans. This is a pure cladistic conception of classification. In the present case, the traits that define the order Proboscidea are few: scapular and astragalar characters, and a homoplasic feature on the lower M_3. More characters are added if one restricts the Proboscidea to a traditional concept (from *Moeritherium* up to *Mammuthus*). This is partly why the classification of anthracobunid taxa within Tethytheria remains unresolved. Tassy (this volume, Chapter 6) and Fischer and Tassy (1993), although recognizing close cladistic relationships, exclude *Anthracobune* from the Proboscidea *sensu stricto*, whereas Shoshani *et al.* (1989*a*) classified it in the

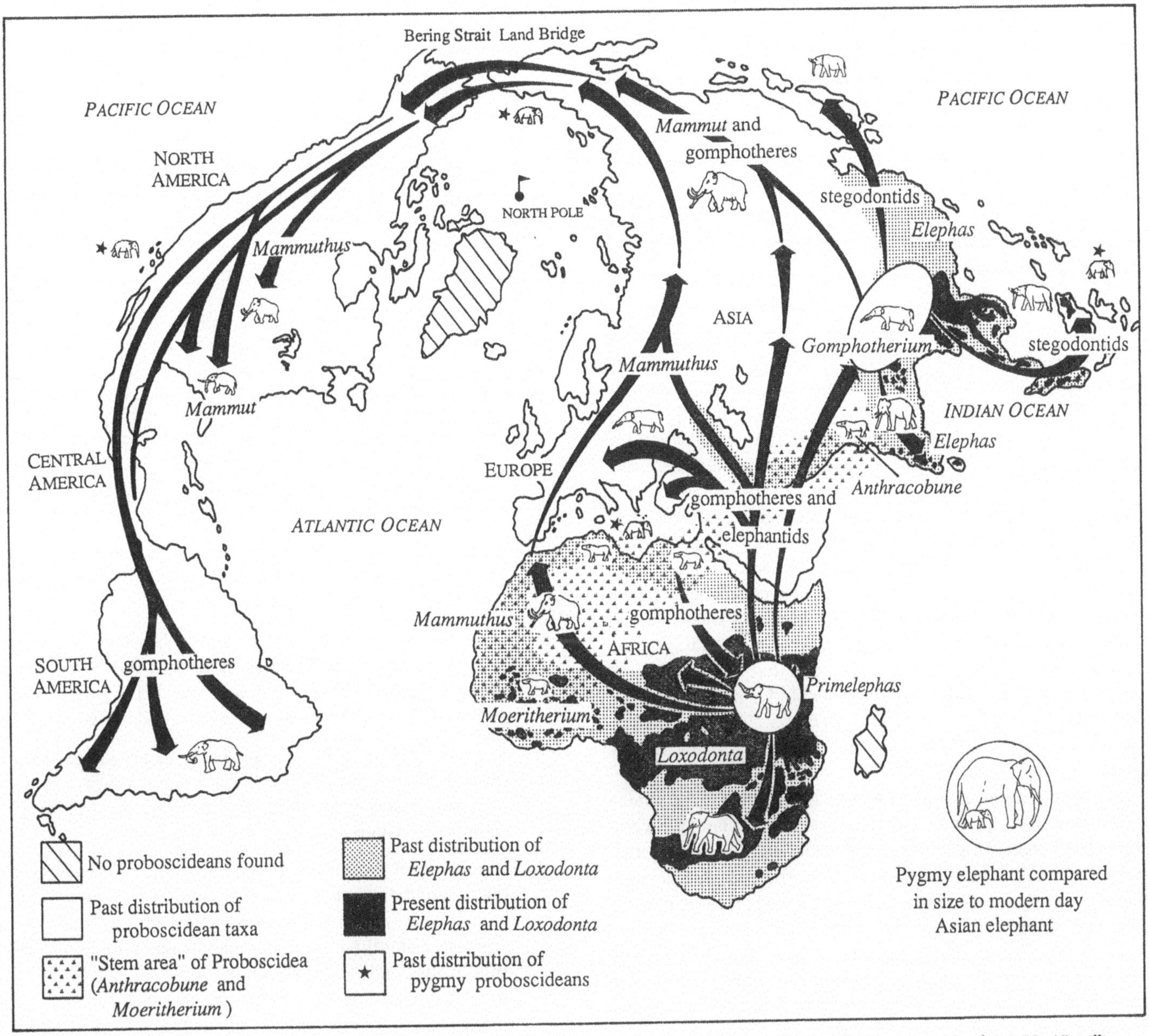

Fig. 34.6 A reconstruction of proboscidean world distribution pattern. Ancestors of modern elephants (during the Eocene epoch, about 50–45 millions years ago) are believed to have inhabited the shores of the Tethys Sea. From there some have spread to Asia, Europe, and the New World. Stippled and darkened areas in Africa and Asia show past and present distribution of *Loxodonta* and *Elephas*, respectively. The dwarf mammoths of Wrangel Island (north of the Bering Strait; Vartanyan *et al.* 1993) have been assigned to a new subspecies — *Mammuthus primigenius vrangeliensis* (Garutt *et al.* 1993; Averianov *et al.* 1995). The 'pygmy elephant' depicted in the lower right corner compared to the Asian elephant is *Elephas (Palaeoloxodon) falconeri*. (Based on the data from Osborn 1936, 1942, and other sources; artwork by J. S. Grimes and G. H. Marchant.)

Proboscidea, a position held by Wells and Gingerich (1983), West (1984), Gingerich *et al.* (1990), and Kumar (1991). Other related genera are tentatively placed in the Tethytheria or Proboscidea *incertae sedis* (discussion in Chapters 6, 8 and 16), whereas Fischer and Tassy (1993) consider cladistically, *Jozaria* closer to Proboscidea *sensu stricto* than are *Anthracobune* and *Pilgrimella*. The controversy over the characters defining Anthracobunidae (whose content is a 'bone of contention') is somehow reminiscent of the controversy over the inclusion or exclusion of moeritheres, barytheres, and deinotheres in the Proboscidea (Harris 1978; Tassy 1990*a*). Now that these latter problems appear to have been resolved, the Anthracobunidae dilemma has surfaced. Based on studies of the authors mentioned above, *Anthracobune* is no longer a proboscidean-like mammal, but a *bona-fide* proboscidean. It is noted that the sister-group relationship between *Anthracobune* and the rest of the proboscideans has not only phylogenetic and systematic implications, but it also focuses on the stem area of this order (Fig. 34.6). For decades, we stated that Eocene proboscideans were confined to northern Africa; now their geographical distribution includes south-western Asia. In either case, the known habitats of tethytheres (including Proboscidea, Sirenia, and Desmostylia) were near-shore, or estuarine, shallow-water environments, probably on the coasts of the Tethys Sea.

TAXA WHOSE PHYLOGENETIC POSITIONS ARE BEING QUESTIONED

When there is inadequate evidence, authors are reluctant to make firm decisions on the classification of particular taxa. For example, Harris (1978), following Maglio (1973), excluded Barytheriidae from the Proboscidea based on paucity of material (a maxillary fragment, a mandible, a humerus, a scapula, and a radius); it has been placed in and out of that order since the beginning of the century. The saga of *Hemimastodon crepusculi*, not figured in Fig. 34.1, is also complicated. In brief, there are only two M^3s allocated to this species (Pilgrim 1908, 1912). Since its discovery in early Miocene sediments of Bugti Hills, Pakistan, the species has been reclassified at least 17 times. Most recently, it has been transferred to Kubanochoerinae (Suidae, Artiodactyla) by Pickford (1987*b*), and back to Proboscidea by Tassy (1982, 1983, 1988*b*, 1990*a*, this volume, Chapter 6). Shoshani and McKenna (1991) concluded that 'until additional material of *H. crepusculi* becomes available for comparison, we assign this holotype to Ungulata *incertae sedis*'. Nevertheless, since this paper was presented, Shoshani believes that it may belong within Proboscidea.

A relatively recent argument arose whether the Stegodontidae is monophyletic or paraphyletic. Traditionally, this family encompasses the genera *Stegolophodon* and *Stegodon*, a view supported by Saegusa (this volume, Chapter 17). On the other hand, using cladistic analysis of 34 characters and 18 taxa, Kalb *et al.* (this volume, Chapter 12A) have observed that *Stegolophodon* is more closely related to certain gomphothere taxa than to *Stegodon*. Results from cladistic analysis of the whole order Proboscidea (Tassy, this volume, Chapter 6; Shoshani, this volume, Chapter 16) show that *Stegodon* and *Stegolophodon* are grouped. But this clade of the Stegodontidae is weak; it costs only 1–3 evolutionary steps to separate *Stegodon* from *Stegolophodon* in Shoshani's analysis (cf. Shoshani 1994; Lister 1994). A new species of *Stegolophodon* from the early Miocene of Thailand shows that the molars of early stegolophodons do not include all the characters displayed by other stegolophodons and stegodons (Tassy *et al.* 1992), and thus, the paraphyly of Stegodontidae cannot be excluded. Additional research is required with emphasis on resolving the taxonomy of tetralophodont gomphotheres, for they may hold the key to the problem of Stegodontidae. Attention should also be given to the Irrawaddy formation of Myanmar (formerly Burma, the type area of the type species of the genus *Stegolophodon* = *S. latidens* (Clift, 1828)). It appears that this question is not going to be resolved in the near future.

EVOLUTIONARY TRENDS OF PROBOSCIDEANS

We cannot resist the pleasure of summarizing the main trends which characterize the remarkable evolution of the Proboscidea. It has long been known that evolutionary changes which resulted in the diversification of proboscideans were marked by certain dominant trends. Perhaps the most significant trend among proboscidean species was the progressive increase in size on the phylogenetic scale, as well as continuous growth on the ontogenetic level. Concomitant with the increase in size came the enlargement of the tusks and the elongation of the trunk, with considerable differences between deinotheres and elephantiformes. The summary given below is modified after Colbert (1955, 1969; the sequence in which the trends are listed below does not necessarily imply importance):

1. Increase in size. Except for the earliest forms, most later proboscideans became giants, reaching over 4 m in height at the shoulders; dwarfism is observed in certain lineages, however (Chapters 22–24).
2. Lengthening of limb bones and the development of short, broad feet. This has been a common evolutionary trend among very large mammals.
3. Growth of the skull to extraordinarily large size (Chapter 11). This growth was particularly noticed in the cranium, where air cells (diploë) developed; the diploë pattern is a feature which provides strength, without added weight.
4. Shortening of the neck. The skull and its associated structures (tusks and trunk) became large and heavy and the neck was reduced in length. This trait appeared in Oligocene elephantiformes.
5. Elongation of the lower jaw and secondary shortening of jaws. Lengthening of the jaw was an early primary trait among proboscideans. Secondary shortening of the lower jaw (especially the area of the mandibular symphysis) and shift in the centre of gravity of the head posteriorly, was a trend associated with considerable parallel evolution in many of the later (advanced) proboscideans (cf. illustrations in Spinage 1972, p. 154; Maglio 1973, p. 105; Kingdon 1979, p. 9; Fig. 34.7).
6. Development of a proboscis. This observation is based on the elevated position of the external naris (the single cranial opening where the trunk begins in living elephants), the connection between frontal and premaxilla bones, and the shape of the premaxilla. It is believed that the elongation of the upper lip and the nostril have evolved along with elongation and later shortening of the lower jaw. Subsequently, the proboscis is further elongated to form a very mobile trunk. It is noted, however, that the development of an elephant-like proboscis may have evolved independently in different lineages (elephantids, mammutids, several times among gomphotheres).

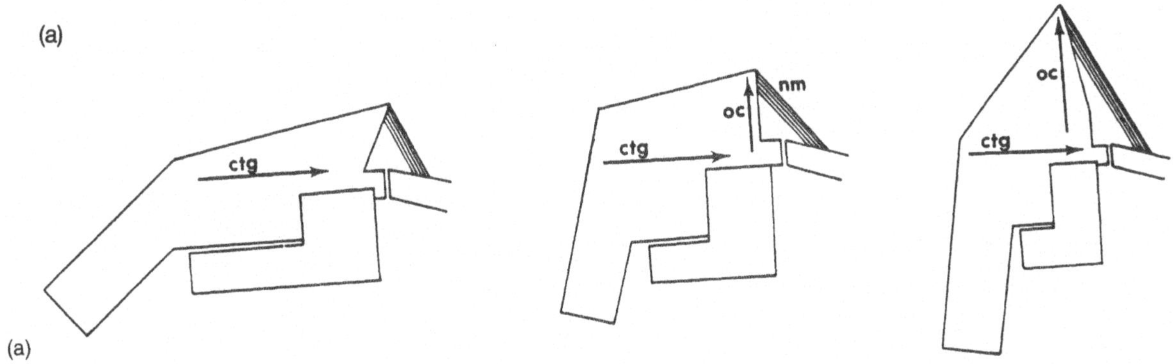

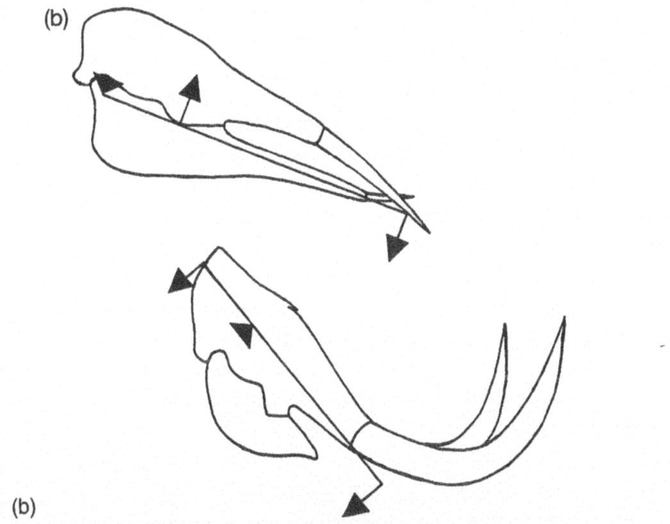

Fig. 34.7 (a) Maglio's (1973, p. 105) generalized, diagrammatic representation of changes in proboscidean skulls. Note that as the center of gravity (ctg) of the skull is shifted posteriorly, the insertion of the nuchal muscles (nm) on the occipital region (oc) is elevated above the level of the occipital condyles. (b) Principles of leverage exhibited in proboscidean evolution: Top: the third order of lever of *Gomphotherium angustidens*. Bottom: the first order of lever of *Mammuthus primigenius*. The latter also depicts how shortening of the weight arm is effected by the upwardly curving tusks (after Spinage 1972, p. 154).

7. Forward/horizontal displacement of cheek teeth (premolars and molars). This feature is present in all known Neogene taxa (character no. 67 in Appendix 16.1; Node J—'Elephantoidea' of Tassy 1988*a*—in Fig. 16.5). It is believed to manifest itself earlier in proboscidean evolution in lineages whose early members possessed components of vertical replacement (of deciduous premolars by premolars) and horizontal displacement (further loss of premolars and drift of molars), which were present simultaneously. Premolars were lost independently in many lineages, such as in more advanced proboscideans, and tooth displacement was in a horizontal fashion, reminiscent of a very slow 'conveyor belt' system.

8. Reduction in number of teeth from the full eutherian dentition. This trend started with *Moeritherium*. Throughout the history of the Proboscidea, there is a decrease in the numbers of premolars, canines, and incisors.

9. Hypertrophy (excess growth) of the second incisors to form tusks. Some of these were straight, curved downwards, or curved upwards and spiralled; they functioned in food-gathering, defense, offense, and display. Enamel covering of tusks decreased to a longitudinal band and then disappeared. Tusks greatly increased in length and diameter; tusks of proboscideans are the largest known teeth of animals, living or extinct.

10. Enlargement and specialization of the cheek teeth. By increasing the number of central conules, conelets, and the numbers of cross-lophs, or lamellae (from the simple two lophs in the earliest members to 30 lophs in the most advanced taxa); this trend was accompanied by molarizing the deciduous premolars and thinning of enamel, and began at the early stages of proboscidean evolution. Parallel evolution in increasing the number of lamellae is found among elephants (within the three genera of Elephantinae), and in *Stegodon*.

PARALLEL V. CONVERGENT EVOLUTION

Perhaps the best example of parallel evolution among proboscideans is that between stegodons and advanced elephantids. For instance, in the more specialized stegodons (e.g. *Stegodon aurorae*, family Stegodontidae) which lived into the Pleistocene epoch, we note that their skulls were deep and the upper tusks were extremely long and curved upward. The mandibles were short and bore no tusks. The cheek teeth grew so large that the simultaneous arrangement of all the teeth in longitudinal rows in the maxilla and mandible was no longer possible (as was observed in *Moeritherium* through *Phiomia*), and horizontal displacement took place. Similar, if not identical processes, have been noted among members of the family Elephantidae, e.g. *Elephas maximus* and *Mammuthus primigenius*.

Examples of convergence among proboscideans may be found in deinotheres and certain gomphotheres. Deinotheres (family Deinotheriidae) and rhynchotheres (family Gomphotheriidae, in part, suborder Elephantiformes) are believed not to have had a recent common ancestor (see Figs 16.5 and 34.3); both had M^1

with three lophs, and both had relatively long lower tusks which were directed downward (more so in Deinotheriidae). These features were not present in the earliest members of Elephantiformes (*Palaeomastodon* and *Phiomia*), and long, downcurved lower tusks were confined to deinotheres and to some extent in rhynchotheres.

BODY SIZE AND TOOTH COMPLEXITY

As the size of the animal increased from *Moeritherium* to *Gomphotherium*, and to *Mammuthus*, its growth in size and height did not always correspond proportionately to the lengthening of the neck. Adaptations to possible new conditions involved the lengthening of the skull and/or the development of a proboscis. The lengthening of the skull was particularly noticed in the edentulous symphysial part of the mandible. This elongation was even more pronounced when the lower jaw housed the spoon-shaped tusks protruding forward (e.g. *Platybelodon* and *Amebelodon*), a feature which probably led to the evolution of the extraordinary shovel-tuskers such as the *Platybelodon grangeri* (Chapter 13).

With the increase in body size came the increase in tooth size, an increase in complexity of the chewing surfaces, and the gradual development of horizontal tooth displacement. This process is known in early Miocene proboscideans and it may have begun at an earlier geological time.

Differences in the skull architecture from the most primitive to the most advanced proboscideans seem to have been associated with dental changes and interrelated jaw mechanics — apparently, to convert from the simpler grinding movements (seen in earliest proboscideans) to the highly efficient shearing action accomplished by fore and aft jaw movements (seen in elephantids). Other changes in dental and jaw apparati included: increased number of cusps, development of ridges (or lamellae), increased number of lamellae and thus lengthening of enamel surface, thinning of enamel, and increase in hypsodonty index or having very high-crowned cheek teeth (details in Chapters 2, 3, 8, 9, 11, 17, and 18).

RECENTLY EXTINCT PROBOSCIDEANS AND A GLIMPSE INTO THE FUTURE

MAMMUT V. *MAMMUTHUS*

Focusing on Pleistocene proboscideans in the New World (Fig. 34.8), the overlap in distribution of Elephantidae and Mammutidae (specifically *Mammuthus* and *Mammut*) is emphasized. Agenbroad (1984*b*) provided distribution maps for mammoths in the New World. It is evident that mammoths were more common in the western than in the eastern portion of North America. When we superimpose the distribution map of *Mammut americanum* (after Shoshani 1990*b*) onto that of *Mammuthus* (after Agenbroad 1984*b*), as in Fig. 34.8, centre, we note that they overlap mostly in the areas along the Mississippi River drainage basin, and that overlap appears to be less as we move away from the river to the east or west, except for the peninsula of Florida and adjacent areas, and certain areas on the west coast (Fig. 34.8). The minimum estimated number of *Mammuthus* individual remains in the New World (based on Agenbroad 1984*b*, fig. 3.2, p. 93; personal communication 1992; see also Harington and Ashworth 1986) is between 1500 and 2000. The estimated number of *M. americanum* individual remains found in the New World was 1473 (Shoshani 1990*b*), plus a few specimens which were found recently (Abraczinskas 1993; D.C. Fisher, personal communication 1994), giving a minimum estimated total of 1500.

A subject which was not discussed previously in this volume, but which is directly related to the above observations, is the ecological relationships between the American mastodon (*M. americanum*) and mammoths (*Mammuthus* sp.). Using specific numbers of individual specimens of *M. americanum* and *Mammuthus* discovered in the United States and data from other sources, it is possible to develop an ecological model which can be used to predict one ecological parameter if two others are known (the parameters are: relative numbers of paired taxa, their anatomical structures, and their sympatric habitats; Shoshani 1988, 1989, 1990*b*, 1992*c* (the last provides detailed explanations of this hypothesis)).

An intriguing observation relating to the spatial and geological distribution of *M. americanum* is made by Saunders (this volume, Chapter 27). According to his hypothesis, pine-dominated parkland of the middle Wisconsinan included specimens of the American mastodon with rugged teeth, whereas spruce-dominated woodlands of the late Wisconsinan contained specimens of the same species, but with smooth teeth which were slightly larger than those from the pine-dominated areas. Overlaying distribution maps of *Pinus* and *Picea* in North America (e.g. King and Saunders 1984; Saunders and Tassy 1989; Grimm and Jacobson 1992) on the distribution map of *M. americanum* (Fig. 34.8, left) appears to support Saunders' hypothesis.

Discussion on *Mammut* and *Mammuthus* in the New World would be incomplete without brief notes on their origins. Two hypotheses were proposed for the origin of the American mastodon, *Mammut americanum*. In Eurasia, during the early to middle Miocene, *Zygolophodon* gave rise to *Mammut borsoni* and *Mammut americanum*; the latter migrated during the Pliocene to North America via the Bering Bridge. Another hypothesis, more widely accepted by some investigators, states that parallel mammutid evolution occurred in the Old and the New Worlds. In Eurasia, *Zygolophodon* gave rise to *Mammut borsoni* (perhaps via an intermediate form, '*Mammut praetypicum*'), and concomitantly, during the middle Miocene, *Zygolophodon* (e.g. '*Z. merriami*', now *Z. proavus*) migrated to North America where it gave rise to the lineage of *Mammut americanum* (details in chapters 6, 9, and 27 by Tassy, Tobien, and Saunders, respectively; see also Appendix C under note Z2).

Mammuthus species in North America are believed to have derived from two independent migrations. In one scenario, *Mammuthus meridionalis* reached North America from Eurasia at about 2.0 Ma and it or *M. hayi* gave rise to the endemic *M. columbi* (including *M. c. inperator* or *M. imperator* and *M. c. jeffersonii* or

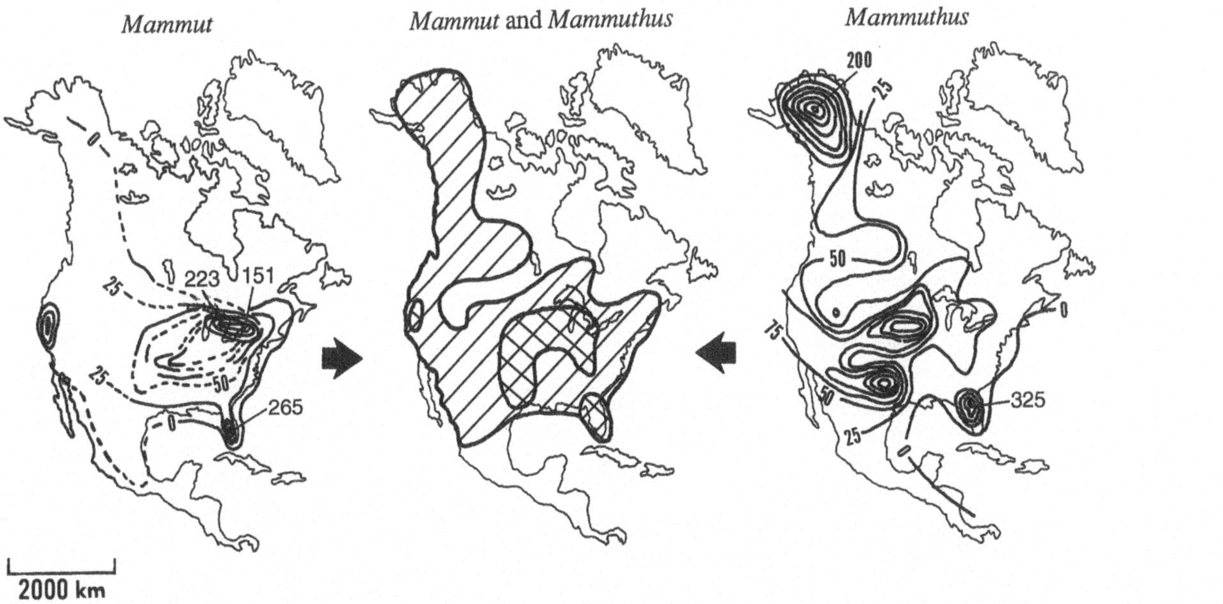

Fig. 34.8 Distribution of *Mammuthus* (right) and *Mammut* (left), and the primary overlapping range (centre). Right, slightly modified after Agenbroad (1984*b*); left, modified after Shoshani (1990*b*). The southernmost distribution of *Mammut* is Costa Rica (Lucas and Alvarado 1991*a,b*). Contour intervals for maps on the left and right equal 25 individuals per area unit. A rough minimum total estimate of remains of *Mammuthus* and *Mammut* individuals in the New World is between 1500 and 2000 each (Agenbroad 1984*b*, personal communication 1992; Shoshani 1990*b*, personal observation 1992). This minimum estimate is based on reported specimens in publications and in some, but not all, museums and private collections.

M. jeffersonii — see Appendix C.1 notes Mth1–5). *M. exilis* was apparently derived from *M. columbi*. *M. primigenius* is believed to have migrated from Eurasia during the late Pleistocene, and was possibly derived from *M. armeniacus* which, in turn, was derived from *M. meridionalis* in Eurasia. Alternatively, mammoth species found in North America (with the exception of *M. exilis*) might have originated in the Old World and arrived independently and some gave rise to autochthonous radiation (details in Chapter 29, this volume, by Dudley; see also Chapters 18 and 24, this volume, by Todd and Roth, and by Roth, respectively).

ELEPHANT: THE SUPER-KEYSTONE SPECIES

Field observations during the past few decades have enriched our knowledge and brought us to realize that elephants play a pivotal role in their respective Asian and African ecosystems. Western (1989, 1991) described elephants as helping to maintain biodiversity, and as species which are capable of shaping the environment in which they live and, by doing so, other animals benefit — that is, they are keystone-species. Seed dispersal, the transformation of savannas into grasslands (by clearing large areas of young trees and underbrush), and the creation of 'oases' in time of drought (by digging waterholes which other animals also use) are examples of the crucial role elephants play in creating habitats (even though temporary) for dozens, or perhaps hundreds, of species. Shoshani (1992*d*, p. 229; 1993*a*) amplified on this idea and called elephants 'super-keystone-species', because, in addition to the growing list of elephants' attributes and their role in their environment, one has to consider the absolute total biomass of all the individual elephants in a given area. Larger species require larger quantities of food which, in turn, require larger home ranges than smaller species. It follows, then, that preserving an area large enough for sustaining an elephant — the largest living terrestrial mammal — can automatically support other species as long as they are not directly competing for the same food source.

DEDUCING THE PALAEOECOLOGICAL IMPACT OF GOMPHOTHERES

It has been suggested that we apply ecological relationships observed in living elephants to the extinct gomphotheres in the New World. Referring to the work of Janzen and Martin (1982), Western (1991) noted that the loss of Pleistocene megafauna 'has resulted in the impoverishment of so much of Central America's forests. Trees like the guanacaste have big seeds that were once dispersed by the gomphothere, an extinct relative of the elephant, and other large mammals. Today they have no seed dispersers. Those important large plants have therefore been reduced to a few areas. The forests have become simplified' (Western 1991, pp. 91–2).

The principle of habitat size and the type and number of species it can support (noted in the previous section) can be easily applied to extinct proboscidean taxa. As a result of such analysis, the crucial ecological role they played in the past is beginning to emerge, based on remains of animals and plants found in association with proboscidean species in palaeontological and archaeological excavations.

HUMANS AND PROBOSCIDEAN SPECIES

It appears that early humans and elephants and possibly other proboscideans shared some ecological similarities, including adaptation to a wide range of habitats and the ability to modify their habitats (Laws 1978). Recent elephants and humans also share an ability to use and make tools (Gordon 1966).

It is apparent that early humans scavenged on dead elephantids or perhaps even hunted them. Prehistory, as a science, emerged in Europe and has helped us better understand the early idea that man hunted mammoths, an idea based partly on human artefacts (carved ivory) and paintings. Evidence for such activities in the New World was summarized by Fisher, and by Casamiquela *et al.* (this volume, Chapters 30 and 31, respectively; see also Fox *et al.* 1992). In Africa, study of certain localities where elephantid and deinothere remains were found also reveals signs of human activities and/or human remains, e.g. in Olduvai Gorge and Laetoli, both in Tanzania (Harris 1978; Howell 1978; Klein 1984; Leakey 1987). We are just beginning to understand the complex factors which may have contributed to extinction (Donovan 1989), and how the '*blitzkrieg* hypothesis' (Martin 1984) fits into these schemes (details in Chapters 15, 29, 30, and 31).

NUMBERS OF PROBOSCIDEANS AND STEPS TO CONSERVE THEM

Regardless of how many species and subspecies of proboscideans lived in the past — the number ranges from 352 (Osborn 1936, 1942) to about 162 (a 1995 estimate which needs re-examination) — it is still a large number of species that have become extinct and it does not lessen or improve the endangered status of some populations of the two remaining genera and species. Based on the available information, there are close to 500 000 wild African elephants, *Loxodonta africana* (Douglas-Hamilton and Michelmore, this volume, Chapter 32), and only about one-tenth of this number of wild Asian elephants, *Elephas maximus* (Sukumar and Santiapillai, this volume, Chapter 33; cf. Sukumar 1994). These estimates demonstrate the dramatic decrease in the wild elephant population in the past several decades. The explanations given for such a trend include fragmented habitats, due to agriculture/ human development, and also due to poaching. To curb this decline of elephants, the greatest effort has been directed at fighting poachers and at supporting research projects which aim to preserve the remaining populations. There has been little to no discussion among conservationists as to how to solve the real problem — the current human population explosion. Human developments imply fewer areas for wildlife, elephants included. In some parts of Africa, elephants have exceeded the carrying capacities of their habitats, and wildlife management authorities have to choose between laissez-faire and culling programmes (cf. Hanks 1979; Redmond 1987; Buss 1990; Chadwick 1992; Orenstein 1991).

Reducing poachers' activities and providing technical and financial support to protect the remaining elephants in Asia or in Africa, are only temporary solutions, analogous to 'an aspirin solution' to lessen the pain of a patient who suffers from a headache due to a disease, a topic discussed in detail by Shoshani (1990*a*). A long-term solution to these accumulating problems is to help reduce the birth rate of the ever-growing human population.

PALAEOICHNOLOGY

During the past few years, there have been some very exciting discoveries of proboscidean footprints preserved as fossils. One of these finds was made even as the last few sentences of this chapter were being written. Trails attributed to *Deinotherium bozasi* and *Loxodonta exoptata* were found in Tanzania (Leakey 1987, p. 467) in tuff dating to the Pliocene, *c.* 5–4 Ma. In Japan, trails of *Stegodon aurorae* were discovered in mudstone dating to the late Pliocene-earliest Pleistocene, *c.* 2 Ma (Kamiya *et al.* 1991; Fig. 34.9), and also trails of *Elephas* (*Palaeoloxodon*) *naumanni* were discovered in volcanic ash dating to *c.* 43 000 years BP (Taruno *et al.* 1992). In addition, H. Matsumoto discovered numerous elephantid footprints preserved in siltstone dating to the early Pleistocene in central Japan (Matsumoto *et al.* 1968). In the United States, Laury (1980) and Agenbroad (1984*a*, pp. 116–17) reported on *Mammuthus columbi* tracks, dated to approximately 26 000 years BP, discovered in sand and silt sediments at the Mammoth Site of Hot Springs, South Dakota. In mid-1992, footprints of the American mastodon (*Mammut americanum*) were uncovered, for the first time in Michigan and possibly in North America, in sandy sediments of an ancient pond bed dating to *c.* 11 000 years BP (Fisher 1994). Fossilized *Mammuthus* scats (coprolites) were reported from North America (Agenbroad 1984*a*, *b*) and from China (Liu and Li 1984).

FUTURE INVESTIGATIONS

The following are suggestions for conducting research in areas on proboscideans and related taxa. Only major issues are addressed here (details are given in the summaries of each chapter). It is hoped that future investigators will concentrate on these topics.

EMBRYOLOGY

Surprisingly, relatively few embryological studies have been conducted on elephants. A carefully documented series of fetuses of known ages (easily obtained today because of the culling operations in Africa) would be useful in providing data for detailed investigations. Recent data on the morphogenesis of the inner ear (Fischer 1990; this volume, Chapter 5) and on the homology and development of tusks (Luckett, this volume, Chapter 4) are examples of this under-researched important scientific field.

Fig. 34.9 Two million-year-old fossil footprints of an extinct *Stegodon aurorae* (family, Stegodontidae) preserved in mudstone, Kobiwako Group, central Japan. Front is toward the top of the page. Impressions of 'toes' (nail-like structures) can be observed, five on the forefoot and at least four on the hind foot. The length of the stride is between 200 and 220 cm and the width of the pace (distance between centres of footprints on right and left side of body, measured perpendicular to length of stride) is about 40 cm; diameter of the forefoot is approximately 30 cm and dimensions of the hind foot are about 30 cm long and 25 cm wide (photograph by H. Kamiya, M. Tamura, T. Kamei, D. Shimizu, S. Ishigaki, N. Kuga, C. Matsuoka, Y. Okamura, and the Research Group for the Fossil Footprints of Yasu-gawa 1995; see also Okamura *et al.* 1995). Other known fossil proboscidean footprints are those of *Deinotherium bozasi*, *Mammuthus columbi*, *Elephas* (*Palaeoloxodon*) *naumanni*, *Loxodonta exoptata*, and *Mammut americanum* (see text for details).

TAXONOMY

The question of which living mammalian order is most closely related to Tethytheria (Proboscidea and Sirenia) is intriguing and must be pursued (Fischer, this volume, Chapter 5). This problem is directly connected to the identification of mammalian interrelationships at the ordinal level, an old challenge and still an enormous task (Novacek 1992). Similarly, a better knowledge of proboscidean characters will help resolve the question as to whether or not remote taxa (such as anthracobunids other than *Anthracobune*) belong in the Proboscidea. It must be kept in mind, however, that the 'proboscidean astragalus' could be a tethythere character since no hind foot bones of Eocene sirenians are presently known. Further, the mammalian order Desmostylia which has been considered a tethytherian (e.g. Domning *et al.* 1986; Ray *et al.* 1994), has an extremely autapomorphic carpus (Novacek and Wyss 1987), a morphological condition which needs evaluation on the intra- and interordinal levels. Within Proboscidea, a special effort should be made to elucidate the evolutionary relationships among gomphotheres (*sensu lato*), and their relationships to other proboscideans. The identification of early gomphotheres and early amebelodontids would necessitate future discoveries of well-preserved fossils from the early Miocene. Emphasis should be given to the question of the monophyly of Stegodontidae, and how it is related to certain gomphothere taxa. Refinement in molecular techniques may help resolve some of the taxonomic questions, even of extinct species (cf. Lowenstein and Shoshani, this volume, Chapter 7; Hagelberg *et al.* 1994; Hoss *et al.* 1994; Yang *et al.* 1996)

Still, synonymy on the genus and species level is truly needed in all proboscidean lineages, especially among gomphotheres, *sensu lato*, and elephantids (cf. Todd and Roth, this volume, Chapter 18; Lister, this volume, Chapter 19). To ensure that variations within a taxon have been accounted for, as many specimens as possible must be compared before reaching a final conclusion because, as usual, once a name has been published and is later found to be invalid, it is difficult to untangle taxonomic and synonymy-related problems. A significant observation made by Mazo (this volume, Chapter 14) illustrates the importance of accounting for variations, not only among specimens, but on teeth of the same specimens: 'If these teeth [of the hemimandible from Paracuellos V] had not been found in the same mandible, it is more than probable that they would not only have been attributed to two different genera, but to two different families.' In addition, when describing a new specimen, it is imperative to include illustrations and/or photographs in as many views as possible. Most authors in this volume stress the need for additional data and need for revisions. In the words of Lambert (this volume, Chapter 15): 'The species level systematics of the gomphotheres is in desperate need of revision.' Todd and Roth (this volume, Chapter 18) write: 'A call for more material, more data, more analysis, and more synthesis is hardly a novel conclusion...'. An enormous number of undescribed gomphotheres from North America await study — many are stored in museums, still in their original crates, from decades ago.

STRATIGRAPHY, GEOCHRONOLOGY, AND TAPHONOMY

Detailed interdisciplinary approaches, including stratigraphy and geochronology, are essential when collecting data from palaeontological and archaeological sites, especially when taphonomical investigations are conducted (Fisher 1984, this volume, Chapter 30; Laub *et al.* 1988; Shoshani *et al.* 1989*b*, Graham, personal communication 1992; Agenbroad and Mead, this volume, Chapter 28). There is need for an accurate count of specific animals found in a given habitat, because specific numbers and ratios provide a better understanding of the ecological densities and biodiversities. Furthermore, detailed examination of accumulation of bones of Recent elephant populations provides a better understanding of taphonomic processes in palaeontological and archaeological excavations (Haynes 1986*b*, 1987, 1991).

MUSEUM RESEARCH

A few museums house hundreds, if not thousands, of proboscidean specimens collected over many decades; they were stored and, in some cases, never touched again. There is much data to be retrieved from these collections. It will be a great service to science if this proboscidean material is unwrapped and studied in detail. As we tried to point out in editing this volume, Osborn's monograph of 1936, 1942 did not exhaust the topic. Fifty-three years later, we feel that future palaeontological investigations promise a considerable increase of knowledge on proboscidean evolution and, more generally, about evolutionary patterns and processes.

CONCLUDING REMARKS

When we initially embarked on this proboscidean project in 1988, we noted that 46 years had passed since the last comprehensive book on this subject was written (the reference was to Osborn's two-volume set of *Proboscidea*, published in 1936 and 1942), and that these volumes are no longer in print. Our aim has been to enhance knowledge of the evolutionary biology of the Proboscidea and its relationships to other mammals, incorporating as many fields of interest as possible in a multidisciplinary approach. We hope that, with the help of the contributors, we have achieved the major goal which we set.

We also have learned that increasing our knowledge of a certain group does not always mean that we are able to achieve clearer resolution on phylogenetic questions. For example, the complexity among proboscidean taxa, and especially among gomphothere lineages, is becoming more and more perplexing as we study them in greater depth.

The pioneers in the field of elephantology and proboscideanology during the eighteenth century (e.g. Blair and Blumenbach) and the nineteenth century (e.g. Blainville, Cuvier, Falconer, Gaudry, Gray, Kaup, and Vacek) provided the 'building blocks' for workers in the first half of the twentieth century (e.g. Andrews, Barbour, Depéret, Matsumoto, Osborn, and Schlesinger), and those, in turn, broadened the foundations for workers in the latter part of the twentieth century (e.g. Aguirre, Beden, Garutt, Maglio, Tobien, and Vereshchagin). In addition to the section by Colbert at the beginning of this volume, students of Proboscidea may wish to glean further pearls of wisdom about Henry Fairfield Osborn from the following publications: Lull (1936), Watson (1946), D'Erasmo (1948), Gould (1977), Simpson (1984), Rainger (1991), Colbert (1992) and Shoshani (in press). Selected technical and popular books which discuss proboscidean evolution and palaeoecology include Osborn (1936, 1942), Simpson (1945, 1984), Deraniyagala (1955), Carrington (1958), Colbert (1969), Sikes (1971), Savage and Long (1986), Pfeffer (1989), Delort (1990), Eltringham and Ward (1991), Haynes (1991), Shoshani (1992*d*), and Tassy (1993).

It is hoped that the results presented in this volume will serve as catalysts for future studies by subsequent generations of scientists who will build on previously accumulated knowledge to reconstruct the past history of the Proboscidea.

ACKNOWLEDGEMENTS

Versions of this chapter were reviewed by N. Court, D. C. Fisher, R. W. Graham, J. E. Kalb, J. M. Lowenstein, V. J. Maglio, E. C. Marsac, J. L. Pierce, S. L. Shoshani, and R. M. West.

APPENDIX A

Synonyms of proboscidean taxa, genera and subgenera, J. Shoshani and P. Tassy with reference to McKenna, Bell, *et al.* (1994).

Taxon (=junior synonym) [author(s), year of publication, page]	**Synonym with (=senior synonym)**
Aleamastodon Hoffstetter, 1952, p. 208	*Haplomastodon* Hoffstetter, 1950, p. 24
Antelephas Sarwar, 1979, p. 133	*Stegolophodon* Schlesinger, 1917, p. 228[a]
Archidiskodon Pohlig, 1888, p. 138	*Mammuthus* Brookes, 1828, pp. 73–4 [b]
Aybelodon Frick, 1933, p. 532	*Rhynchotherium* Falconer, 1868, pp. 74–5
Blickotherium Frick, 1933, p. 529	*Rhynchotherium* Falconer, 1868, pp. 74–5
Bradytherium Andrews, 1901*d*, p. 4	*Barytherium* Andrews, 1901*b*, p. 577
Bunolophodon Vacek, 1877, p. 45	*Gomphotherium* Burmeister, 1837, p. 795
Cheirolites von Meyer and Göppert, 1848, p. 286	*Mammuthus* Brookes, 1828, pp. 73–4
Cordillerion Osborn, 1926*a*, p. 15	*Cuvieronius* Osborn, 1923, p. 1
Cryptomastodon von Koenigswald, 1933, p. 112	*Stegodon* Falconer, 1857, p. 318[c]
Dibelodon Cope, 1884*b*, p. 2	*Rhynchotherium* Falconer, 1868, pp. 74–5
Dibunodon Schlesinger, 1917, p. 124	*Anancus* Aymard 1855, p. 507
Dicyclotherium Geoffroy Saint-Hilaire, 1837, p. 119	*Mammuthus* Brookes, 1828, pp. 73–4
Dinotherium Kaup, 1831, p. 18	*Deinotherium* Kaup, 1829, p. 401 [d]
Elephantus Geoffroy and Cuvier, 1795, p. 189	*Elephas* Linnaeus, 1758, p. 33
Emmenodon Cope, 1889, p. 194	*Stegodon* Falconer, 1857, p. 318
Eoelephas Garutt, 1986, p. 23 [cf. Garutt 1992, p. 37]	*Elephas* Linnaeus, 1758, p. 33
Eostegodon Yabe, 1950, p. 65	*Stegolophodon* Schlesinger, 1917, p. 228 [a,e]
Euelephas Falconer, 1857, p. 318	*Elephas* Linnaeus, 1758, p. 33
Gamphotherium Gloger, 1841, p. 119	*Gomphotherium* Burmeister, 1837, p. 795
Geisotodon Bergounioux and Crouzel, 1955, p. 1488	*Tetralophodon* Falconer, 1857, p. 316 [a]
Genomastodon Barbour, 1917, p. 512	*Gomphotherium* Burmeister, 1837, p. 795
Harpagmotherium Fischer de Waldheim, 1808, pp. 19–20	*Mammut* Blumenbach, 1799, pp. 697–8 [f]
Hemilophodon Kretzoi, 1942, p. 139	*Gomphotherium* Burmeister, 1837, p. 795 [a]
Hesperoloxodon Osborn, 1931, p. 21	*Elephas* Linnaeus, 1758, p. 33
Hypselephas Osborn, 1936, p. 12	*Elephas* Linnaeus, 1758, p. 33 [g]
Ishatherium Sahni and Kumar, 1980, p. 133	*Anthracobune* Pilgrim, 1940, p. 129 [h]
Jozaria Wells and Gingerich, 1983, p. 125	*Anthracobune* Pilgrim, 1940, p. 129 [i]
Konobelodon Lambert, 1990, p. 1033	*Amebelodon* Barbour, 1927, p. 131 [j]
Leith-Adamsia Matsumoto, 1928, p. 214	*Elephas* Linnaeus, 1758, p. 33 [k]
Leviathan Koch, 1841, p. 17	*Mammut* Blumenbach, 1799, pp. 697–698 [l]
Lydekkeria Osborn, 1924*a*, p. 2	*Tetralophodon* Falconer, 1857, p. 316
Mammonteum Blainville, 1864, p. 237	*Mammuthus* Brookes, 1828, pp. 73–4, [b,m]
Mastodon Rafinesque, 1814, p. 182	*Mammut* Blumenbach 1799, pp. 697–8 [n]
Mastodontum Blainville, 1817, p. 276	*Mammut* Blumenbach, 1799, pp. 697–8 [o]
Mastolophodon Chakravarti, 1957, p. 84	*Zygolophodon* Vacek, 1877, p. 45
Mastotherium Fischer de Waldheim, 1814, p. 337	*Mammut* Blumenbach, 1799, pp. 697–8
Megabelodon Barbour, 1914b, p. 217	*Gomphotherium* Burmeister, 1837, p. 795
Metarchidiskodon Osborn, 1934*c*, p. 12	*Mammuthus* Brookes, 1828, pp. 73, 74
Miomastodon Osborn, 1922*a*, p. 4	*Zygolophodon* Vacek, 1877, p. 45 [a,p]
Missourium Koch, ?1840, pp. 1–2	*Mammut* Blumenbach, 1799, pp. 697–8 [q]
Morrillia Osborn, 1924*a*, p. 1	*incertae sedis*; see note Mo1 in Appendix C.1

Taxon (=junior synonym) [author(s), year of publication, page]	Synonym with (=senior synonym)
Ocalientinus Frick, 1933, p. 579	*Gomphotherium* Burmeister, 1837, p. 795
Omoloxodon Deraniyagala, 1955, p. 25	*Elephas* Linnaeus, 1758, p. 33
Palaeoloxodon Matsumoto, 1924*a*, p. 257	*Elephas* Linnaeus, 1758, p. 33 [r]
Paraloxodonta Garutt, 1986, p. 19 [cf. Garutt 1992. p. 37]	*Loxodonta* Anonymous, 1827, p. 140
Parastegodon Matsumoto, 1929, p. 13	*Stegodon* Falconer, 1857, p. 318 [s]
Parelephas Osborn, 1924*b*, p. 4	*Mammuthus* Brookes, 1828, pp. 73, 74
Pentalophodon Falconer, 1857, p. 314	*Anancus* Aymard 1855, p. 507 [t]
Phanagoroloxodon Garutt, 1957, p. 333	*Elephas* Linnaeus, 1758, p. 33 [a]
Pilgrimella Dehm and Oettingen-Spielberg, 1958, p. 33	*Anthracobune* Pilgrim, 1940, p. 129 [u]
Pilgrimia Osborn, 1924*b*, p. 2	*Elephas* Linnaeus, 1758, p. 33
Platelephas Osborn, 1936, p. 12	*Elephas* Linnaeus, 1758, p. 33 [v]
Platystegodon Deraniyagala, 1954 (cf. Deraniyagala 1955, p. 24)	*Stegodon* Falconer, 1857, p. 318
Pliomastodon Osborn, 1926*a*, p. 1	*Mammut* Blumenbach 1799, pp. 697–8 [a]
Prodinotherium Ehik, 1930, p. 14	*Prodeinotherium* Ehik, 1930, p. 14 [w]
Prostegodon Matsumoto, 1923 [no page number]	*Stegolophodon* Schlesinger, 1917, p. 228 [x]
Protelephas Garutt, 1954*a*, p. 189	*Elephas* Linnaeus, 1758, p. 33 [a]
Protoloxodonta Garutt, 1986, p. 19 [cf. Garutt 1992, p. 37]	*Loxodonta* Anonymous, 1827, p. 140
Protostegolophodon Gaziry, 1994*a*, p. 130	'gomphotheres' *incertae sedis*
Rhabdobunus Hay, 1914, p. 373	*Stegomastodon* Pohlig, 1912, p. 193
Rulengchia Zhou and Zhang, 1983, p. 57	*Stegolophodon* Schlesinger, 1917, p. 228 [a,e]
Selenolophodon Chang and Zhai, 1978, p. 136	*Platybelodon* Borissiak, 1928, p. 120 [a]
Serridentinus Osborn, 1923, p. 2	*Gomphotherium* Burmeister, 1837, p. 795 [a,y]
Sivalikia Osborn, 1924*b*, p. 2	*Elephas* Linnaeus, 1758, p. 33
Stegoloxodon Kretzoi, 1950, p. 405	*Elephas* Linnaeus, 1758, p. 33
Sulcicephalus Deraniyagala, 1955, p. 216	*Stegodon* Falconer, 1857, p. 318
Synconolophus Osborn, 1929*a*, p. 9	*Choerolophodon* Schlesinger, 1917, p. 181
Tatabelodon Frick, 1933, p. 581	*Gomphotherium* Burmeister, 1837, p. 795
Teleobunomastodon Revilliod, 1931, p. 21	*Cuvieronius* Osborn, 1923, p. 1
Tetrabelodon Cope, 1884*b*, p. 4	*Gomphotherium* Burmeister, 1837, p. 795
Tetracaulodon Godman, 1830, p. 484	*Mammut* Blumenbach, 1799, pp. 697–8 [z]
Tetrazygodon Tobien, 1978*a*, p. 244	*Stegolophodon* Schlesinger, 1917, p. 228
Torynobelodon Barbour, 1929, p. 147	*Platybelodon* Borissiak, 1928, p. 120 [y]
Trilophodon Falconer, 1857, p. 316	*Gomphotherium* Burmeister, 1837, p. 795
Trobelodon Frick, 1933, p. 580	*Gomphotherium* Burmeister, 1837, p. 795
Turicius Osborn, 1926*a*, p. 3	*Zygolophodon* Vacek, 1877, p. 45
Wanoloxodon Liu and Zhen, 1981, p. 220	*Elephas* Linnaeus, 1758, p. 33 [a]

General comments: this synonymy table is an attempt to solve some of the taxonomic problems encountered in the literature. It is beyond the scope of this volume to provide or give all the first published synonymies for each of the genera. Those given are based on several sources, including the classification of mammals by McKenna, Bell, *et al.* ms., 1994.

[a] Based on examination of specimens and/or illustration and description in the literature; study conducted by J. Shoshani in 1992 (notes on synonymies are on file in M. C. McKenna's office, Department of Vertebrate Paleontology, American Museum of Natural History, New York, USA).

[b] Garutt *et al.* (1990) have requested the conservation of *Mammuthus* Brookes, 1828 (type species *Elephas primigenius* Blumenbach, 1799), and this request has been accepted by the International Commission on Zoological Nomenclature (1991); previously '*Mammuthus* Burnett, 1830' was used (e.g. Simpson 1945, p. 134). Lister (this volume, Chapter 19) noted: 'The name *Archidiskodon* Pohlig 1888 (type species *Elephas meridionalis* Nesti 1825) is also taxonomically valid as a genus or subgenus, and is available for use by those who wish to differentiate *meridionalis* and other early species from *Mammuthus* (Garutt *et al.* 1990).' Stated otherwise, if *Archidiskodon* and *Mammuthus* are considered to belong in the same lineage (Dubrovo 1964, 1977; Dubrovo and Mol 1991), then only *Mammuthus* is a valid genus for this monophyletic unit.

[c] '*Cryptomastodon martini* Von Koenigswald must be sunk as it is based upon two already named taxa, *Stegodon trigonocephalus* Martin and *Geochelone atlas* (Falconer and Cautley). In the end, then, it is a true martini, a cocktail: two-thirds mammal and one-third reptile' (Hooijer 1984*c*, p. 230).

[d] Page number for *Dinotherium* in Kaup, 1831, may be either 18 or 78; see also Kaup 1833, and Osborn 1936, pp. 84–6, 782, and 1942, p. 1387.

[e] Saegusa (this volume, Chapter 17) considers *Eostegodon* and *Rulengchia* subgenera of *Stegolophodon*. Nearly all other authors (too numerous to mention) also considered this and perhaps other related genera as synonyms of *Stegolophodon*.

[f] See notes in Osborn, 1942, p. 1372.

[g] In Osborn 1936, p. 12, *Hypselephas* is listed only, not described; description appears in 1942, p. 1340.

[h] After West (1984, p. 186).

[i] After Domning *et al.* (1986, p. 43).

[j] *Konobelodon* was designated as a subgenus of *Amebelodon* by Lambert (1990). Also note that Tobien (1972) considered *Serbelodon* a junior synonym of *Amebelodon*. In this study, there is insufficient evidence to support this hypothesis, but Shoshani (this volume, Chapter 16; see also notes in Table 16.1) leaves this option open.

[k] The separate article in which *Leith–Adamsia* was described lacks publication date and page number. Osborne (1942, p. 1381) gives 1927 as the year of publication. Publication date and page corrected to 1928 and 214, respectively (S.K. Bell, personal communication 1996).

[l] See notes in Osborn, 1942, p. 1374.

[m] Osborn (1924b) used '*Mammonteus*' after Camper (1788); it was reconstructed from the vernacular '*Mammonteum*' used by Camper, not following the Linnaean system, thus '*Mammonteum*' of Camper (1788) is invalid (Simpson, 1945, pp. 134–259). Instead, *Mammonteum* Blainville, 1864, p. 237 should be used (McKenna, Bell, *et al.* 1994).

[n] In 1806, G. Cuvier coined the name 'Mastodonte' (French spelling, not italicized) for the American mastodon, a specimen which was found in the type locality of Big Bone Lick, Kentucky, USA (where fossil bones were discovered in 1739; Osborn 1936, p. 136). Cuvier also allocated other species under the umbrella of 'Mastodonte', for example, 'Mastodonte à dents étroites', 'Mastodonte des cordillères', 'Mastodonte tapiroïde', and 'petit Mastodonte'. In 1814, Rafinesque coined the scientific name '*Mastodon*' for the same animal from Big Bone Lick, and allocated all of Cuvier's 'Mastodonte' species under this new genus, *Mastodon*. This new name was endorsed by Cuvier (1817, pp. 232–3). It must be emphasized that the list of junior synonyms which may be included under 'Mastodonte' cannot be ascertained because Cuvier probably included in it such taxa as *Gomphotherium*, *Anancus*, and others (e.g. *Mastodon* Rafinesque, 1814, p. 182 = *Cuvieronius* Osborn 1923, p. 1 – other details on Cuvier's 'mastodonts' are given in Osborn 1936, pp. 122–5). Under the 'Principle of Priority' rule of Zoological Nomenclature, neither 'Mastodonte' nor '*Mastodon*' are valid; *Mammut* of Blumenbach, 1799, pp. 697–8 takes precedence.

[o] See notes in Osborn 1942, p. 1373.

[p] After Tobien (1964, p. 64); see also Tobien (1975, 1976) and Madden (1980*b*).

[q] See notes in Osborn, 1942, p. 1374.

[r] *Palaeoloxodon* has been employed as a subgenus of *Elephas*, written as '*Elephas* (*Palaeoloxodon*) *antiquus*' (see, for example, Caloi *et al.*, this volume, Chapter 22, and note *d* in Appendix C.2).

[s] Osborn (1942, p. 1380) gives 1924 as the year of publication; according to McKenna, Bell, *et al.* (1994) it should be 1929, p. 13, because *Parastegodon* of Matsumoto, 1924, pp. 256–7, and Matsumoto, 1926, p. 1, is *nomen nudum*.

[t] See also Zong (1979).

[u] After West (1983, p. 362). In addition, note that Kumar (1991) considered four valid anthracobunid genera: *Anthracobune*, *Jozaria*, *Lammidhania*, and *Pilgrimella* (see also note *c* in Table 16.1; Domning *et al.* (1986), and Fischer and Tassy (1993)).

[v] In Osborn 1936, p. 12, *Platelephas* is listed only, not described; description appears in 1942, p. 1358.

[w] Ehik (1930, p. 14) spelled it '*Prodinotherium*'. Harris (1973, p. 293) justifiably emended it to '*Prodeinotherium*'. According to the International Code of Zoological Nomenclature, Article 50d, 'a justified emendation is attributed to the author of the name in its original incorrect spelling'.

[x] After Osborn (1923, p. 2). Osborn (1942, p. 840) noted: 'Thus *Prostegodon* Matsumoto-Osborn, 1923, 1924, 1926, becomes a synonym of *Stegolophodon* Schlesinger, 1917'. This idea was endorsed by Tassy, 1985, p. 759.

[y] Lambert (this volume, Chapter 15) considers *Serridentinus* and *Torynobelodon* valid names. Shoshani (this volume, Chapter 16), however, regards *Serridentinus* as a junior synonym of *Gomphotherium*, a hypothesis proposed by Tobien (1972). Tobien (1973, pp. 254–5) considered *Torynobelodon* a subgenus of *Platybelodon*.

[z] See notes in Osborn 1942, p. 1373; cf. Kaup 1832*b*.

Note: In addition to the synonymies above, the following names have been listed as junior synonyms in Osborn 1942, pp. 1372, 1375–8. They are not included in the above synonymies as they may be invalid names because they have no type species, they were preoccupied, or any other taxonomic related problems (S. K. Bell, personal communication 1993).

Ohio Blumenbach, 1797, Heft 2, no. 19.	*Mammut* Blumenbach, 1799, pp. 697–8
Loxodon Falconer and Cautley, 1847, plate XLII, XLIV	*Loxodonta* Anonymous, 1827, p. 140
Synodontherium Costa, 1850, pp. 271–5	*Mammuthus* Brookes, 1828, pp. 73–4
Leptodon Gunn, 1867, p. 422	*Elephas* Linnaeus, 1758, p. 33
Antoletherium Falconer, 1868, p. 416	*Deinotherium* Kaup, 1829, p. 401
Archidiskodonten Pohlig, 1885, p. 1027	*Mammuthus* Brookes, 1828, pp. 73–4 [b]
Loxodonten Pohlig, 1885, p. 1027	*Loxodonta* Anonymous, 1827, p. 140

APPENDIX B

A suggested classification of the Proboscidea based on combined results by P. Tassy (cf. Fig. 6.1) and J. Shoshani (cf. Fig. 16.5 and Appendix 16.3; notes [a–m] are given in Table 16.1; see McKenna, Bell, *et al.* (1994) for alternative classification).

- Order Proboscidea Illiger, 1811
  - Plesion *Anthracobune* Pilgrim, 1940† [c]
  - Family Moeritheriidae Andrews, 1906†
    - *Moeritherium* Andrews, 1901*d*†
  - Family Numidotheriidae new family† [d]
    - *Numidotherium* Mahboubi, Ameur, Crochet, and Jaeger, 1986†
  - Family Barytheriidae Andrews, 1906†
    - *Barytherium* Andrews, 1901b†
  - Family Deinotheriidae Bonaparte, 1845†
    - *Prodeinotherium* Ehik, 1930† [e]
    - *Deinotherium* Kaup, 1829†
  - Suborder Elephantiformes Tassy, 1988*a*
    - Plesion *Palaeomastodon* Andrews, 1901*a*†
    - Plesion *Phiomia* Andrews and Beadnell, 1902†
    - Plesion *Hemimastodon* Pilgrim, 1912† [f]
    - Unnamed taxon (node J — 'superfamily Elephantoidea Gray, 1821' of Tassy, 1988*a*)[g]
      - Family Mammutidae Hay, 1922† [h]
        - *Eozygodon* Tassy and Pickford, 1983†
        - *Zygolophodon* Vacek, 1877†
        - *Mammut* Blumenbach, 1799†
      - Unnamed taxon (node L)[i]
        - Plesion *Choerolophodon* Schlesinger, 1917 (*sedis mutabilis*)†
        - Unnamed taxon (node M)
          - Family Gomphotheriidae Hay, 1922 (node N)[j]
            - Unnamed taxon (node N.1)
              - Plesion *Gomphotherium* Burmeister, 1837†
                - (includes *Serridentinus* Osborn, 1923) (node N.2)†
              - Subfamily Amebelodontinae Barbour, 1927 (*incertae sedis=i.s.*) (node N.3)†
                - *Amebelodon* Barbour, 1927†
                - *Platybelodon* Borissiak, 1928†
                - *Serbelodon* Frick, 1933† [k]
                - *Protanancus* Arambourg, 1945†
                - *Archaeobelodon* Tassy, 1984†
            - Plesion *Gnathabelodon* Barbour and Sternberg, 1935 (*i.s.*)†
            - Unnamed taxon (node N.7)
              - Plesion *Sinomastodon* Tobien, Chen, and Li, 1986 (*i.s.*)†
              - Plesion *Eubelodon* Barbour, 1914*a* (*i.s.*)†
              - Plesion *Rhynchotherium* Falconer, 1868 (*i.s.*)†
              - Subfamily Cuvieroniinae Cabrera, 1929 (*i.s.*) (node N.10)†
                - *Stegomastodon* Pohlig, 1912†
                - *Haplomastodon* Hoffstetter, 1950†
                - *Notiomastodon* Cabrera, 1929†
                - *Cuvieronius* Osborn, 1923†
          - Unnamed taxon (node O)
            - Plesion *Tetralophodon* Falconer, 1857†
            - Plesion *Anancus* Aymard, 1855†
            - Plesion *Paratetralophodon* Tassy, 1983†
            - Unnamed taxon (node R)

Family Stegodontidae Osborn, 1918 (node S)†
Stegolophodon Schlesinger, 1917 (*i.s.*)†
Stegodon Falconer, 1857†
Family Elephantidae Gray, 1821 (node T)
Plesion *Stegotetrabelodon* Petrocchi, 1941 (i.s.)†
Plesion *Stegodibelodon* Coppens, 1972 (i.s.)†
Subfamily Elephantinae Gray, 1821 (node V)
Plesion *Primelephas* Maglio, 1970*b*†
Genus *Loxodonta* Anonymous, 1827 [l]
Genus *Elephas* Linnaeus, 1758
Plesion *Mammuthus* Brookes, 1828† [m]

---

†, Extinct taxon.

APPENDIX C.1

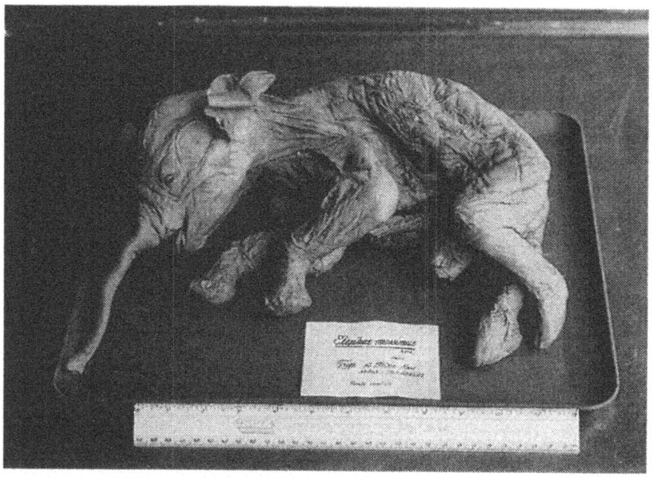

Fig. C.1 Naturhistoriska Riksmuseet (Stockholm, Sweden) specimen number S-104 05 — the original *Loxodonta africana* fetus used by Carolus Linnaeus as the Type Specimen for the species *Elephas maximus* Linnaeus, 1758 (photographed by J. Shoshani, 13 August 1982). This same specimen was described 250 years ago by Seba, in 1734. A footnote in Osborn (1942, p. 1323) reads: 'Seba (1734, p. 175, Pl. CXI) gives a very full description of the Foetus Elephantis Africani ineditus, which Lönnberg refers to as the type. In this early stage of zoology no one dreamed of selecting any particular specimen and designating it as the *type*' (see also note E1 in this appendix).

Appendix C.1 A summary of proboscidean taxa (footnotes, key to references, and abbreviations are at the end of this appendix)

A	B	C	D	E	F	G	H	I
Family	Genus	Type species	Other species	Total genera	Total species	Total of genera/species	Reference for column D	Museum(s)
Family uncertain [A1]	*Anthracobune*	*A. pinfoldi*	*A. aijiensis*				Kumar 1991 [for A.a.]	W(An)
	Pilgrim 1940, p. 129	Pilgrim 1940, p. 129	*?A. pilgrimi*	1	2-3	1 / 2-3	details in {8} [?A.p.]	W(An)
Family Moeritheriidae	*Moeritherium*	*M. lyonsi*	*M. trigodon*				An(04a) [M.t.]; C&B(78):334	W(Mo)
Andrews 1906, p. 99	Andrews 1901d, p. 4	Andrews 1901d, p. 4		1	2	1 / 2	{8}	
Family Numidotheriidae [N1]	*Numidotherium*	*N. koholense*	*N. savagei* [N1]				{8} Court 1995: 651	LPULM, BUGM
	Mahboubi *et al.* 1986, p. 25	Mahboubi *et al.* 1986, p. 25		1	1	1 / 1		
Family Barytheriidae	*Barytherium*	*B. grave*	*B. species novum*				{8}; Ha(78):329	W(Ba)
Andrews 1906, p. 172	Andrews 1901b, p. 577	Andrews 1901d, p. 4		1	2	1 / 2	Sho+(89a) [B.s.n.]	
Family Deinotheriidae [D1]	*Prodeinotherium*	*P. bavaricum*	*P. hobleyi*				{8}; Ha(78):320	W(D-idae), W(De)
Bonaparte 1845, p. 4	Ehik 1930, p. 14	(von Meyer 1831*a*, pp. 296-7)	*P. pentapotamiae*				{8}; Lydekker 1876*b*: 72	W(D-idae), W(De)
			P. ?orlovii	1	3-4		Sahni and Tripathi 1957 [P. ?o.]	W(D-idae), W(De)
	Deinotherium	*D. giganteum*	*D. bozasi*				{8}; Arambourg 1934: 86	W(De), W(D-idae)
	Kaup 1829, p. 401	Kaup 1829, p. 401	*D. indicum*	1	3	2 / 6-7	{8}; Falconer 1845: 361	W(De), W(D-idae)
Family Palaeomastodontidae	*Palaeomastodon*	*P. beadnelli*	none				{8}; C+(78):341	W(P&P)
Andrews 1906, p. 130	Andrews 1901a, p. 319	Andrews 1901a, p. 318, 319		1	1	1 / 1		
Family uncertain	*Phiomia*	*P. serridens*	none				{8}; To(78c):179	W(P&P)
	Andrews and Beadnell 1902, pp. 1-9	Andrews and Beadnell 1902, pp. 1-9		1	1	1 / 1		
Family uncertain [He1]	*Hemimastodon*	*H. crepusculi*	none				Pilgrim 1908, 1912	IM, NHM
	Pilgrim 1912, p. 17	(Pilgrim 1908, p. 157)		1	1	1 / 1		
Family Mammutidae	*Eozygodon*	*E. morotoensis*	none				{6}; {8}; {16}	W(Eo), W(M-idae)
Hay 1922, p. 101	Tassy and Pickford 1983, p. 58	(Pickford and Tassy 1980, p. 242)		1	1			
	Zygolophodon	*Z. turicensis* [Z1]	*Z. gromovae*				Dubrovo 1970 [Z.g.]; {9}	W(Zy), W(M-idae)
	Vacek 1877, p. 45	(Schinz 1824, p. 278)	*Z. proavus* [Z2]	1	3		Cope 1873; M&S(85) [Z.p.]	W(Zy), W(M-idae)
	Mammut	*M. americanum*	*M. borsoni* [Z3]				To+(88):212 [M.b.]	NA(M&M), W(Ma)
	Blumenbach 1799, pp. 697-8	(Kerr 1792, p. 116)	*M. matthewi*				Osborn 1921b [M.m.]	NA(M&M), W(Ma)
		incertae sedis -->	*M. praetypicum*				Schlesinger 1919 [M.p.]	W(M-idae)
		incertae sedis -->	*M. shansiensis*	1	3-5	3 / 7-9	Chow and Chang 1961 [M.s.]	W(M-idae)
Family uncertain	*Choerolophodon*	*C. pentelici*	*C. chioticus*				Tobien 1980*b* [C.c.]; T(86):77	W(Ch), W(gom)
	Schlesinger 1917, p. 181	(Gaudry and Lartet 1856, p. 273)	*C. corrugatus*				Pilgrim 1913 [C.c.]; T(83):236	W(Ch), W(gom)
			C. ngorora				Maglio 1974b [C.n.]; T(86):77	W(Ch), W(gom)
			C. palaeindicus [C1]				Lydekker, 1884*a* [C.p.]; T(86):77	W(Ch), W(gom)
			? C. spenceri	1	5-6	1 / 5-6	Fourtau 1918 [?C.s.]; T(90*b*):248	W(Ch), W(gom)
Family Gomphotheriidae	*Gomphotherium* [G1]	*G. angustidens*	*G. annectens* [G2]				Matsumoto 1925; T(85):678	W(Go), W(gom)
Hay 1922, p. 101	Burmeister 1837, p. 795	(G. Cuvier 1817, p. 233)	*G. browni*				Osborn 1926*a* [G.b.]; T(83):262	W(Go), W(gom)
			G. connexum				Hopwood 1935; To+(88):212	W(Go), W(gom)
			G. ?inopinatum				B&B(28); Os(36):278	W(Go), W(gom)
			? G. intermedium				Sarwar 1977 [?G.i.]	W(Go), W(gom)
			G. osborni				Barbour 1916; Os(42):1405	W(Go), W(gom)

Appendix C.1 (*continued*)

A	B	C	D	E	F	G	H	I
			G. productum [G3]				Cope 1875; Os(36):404, 770	W(Go), W(gom)
			G. republicanum				To(72):191	W(Go), W(gom)
			G. shensiense				C&Z(78): 139; To+(88):212	W(Go), W(gom)
			G. steinheimense				Klähn 1922; T(85):711	W(Go), W(gom)
			G. tongxinense				{13}	W(Go), W(gom)
			G. wimani				To+(88):212, {13}	W(Go), W(gom)
		incertae sedis -->	*G. angustidens austrogermanicum*				Wegner 1913 [G.a.a.]	W(Go), W(gom)
		incertae sedis -->	*G. angustidens libycum*				Fourtau 1918 [G.a.l.]	W(Go), W(gom)
		incertae sedis -->	*G. angustidens subtapiroideum*				To(72):191	W(Go), W(gom)
		incertae sedis -->	*G. mongoliense*				Osborn 1924c [G.m.]	W(Go), W(gom)
		incertae sedis -->	*G. sendaicum* [G4]	1	11-18		Matsumoto 1925 [G.s.]	W(Go), W(gom)
Family Gomphotheriidae	*Amebelodon* [Am1]	*A. fricki*	*A. (Konobelodon) britti*				Lm(90):1033	W(Am), W(gom)
Subfamily Amebelodontinae	Barbour 1927, p. 131	Barbour 1927, p. 131	*A. floridanus*				{15}; Turner 1975	W(Am), W(gom)
Barbour 1927, p. 131			*A. tobieni*	1	4		{13}	W(Am), W(gom)
	Platybelodon	*P. danovi* [P1]	*P. cyrenaicus* [P2]				(Gaziry 1987*b*) [P.c.]	W(Pl), W(gom)
	Borissiak 1928, p. 120	(Borissiak 1928, p. 120)	*P. grangeri*				Os(29b); To+(88):212; {13}	W(Pl), W(gom)
			P. ?spectabilis [P3]				Chang and Zhai 1978, p. 136	W(Pl), W(gom)
			P. (Torynobelodon) loomisi				{13}; Barbour 1929 [P.(T.)l.]	W(Pl), W(gom)
			? P. (T.) barnumbrowni	1	4-6		Barbour 1931 [?P.(T.)b.]	W(Pl), W(gom)
	Serbelodon	*S. barbourensis*	*S. zhongningensis*				{13}	W(Se), W(gom)
	Frick 1933, p. 594	Frick 1933, p. 594		1	2			
	Protanancus	*P. macinnesi*	*P. chinjiensis*				Pilgrim 1913 [P.c.]	W(Pro), W(gom)
	Arambourg 1945, p. 493	Arambourg 1945, p. 494		1	2		see also T(86):28, 55, 56	
	Archaeobelodon	*A. filholi*	*A.* aff. *filholi*				T(86):28 [*A.* aff. *filholi*]	W(Ar), W(gom)
	Tassy 1984, p. 462	(Frick 1933, pp. 509, 535)		1	2			
Family Gomphotheriidae	*Gnathabelodon* [Gn1]	*G. thorpei*	*?G. buckneri*				Sellards 1940 [?G.b.]	NA(Gn), W(gom)
no subfamilial assignment	Barbour and Sternberg 1935, p. 395	Barbour and Sternberg 1935, p. 395		1	1-2			
	Sinomastodon	*S. intermedius*	*S. hanjiangensis* [Sin1]				To+(88):212 [S.i.]	China(Si)
	Tobien et al. 1986, p. 160	(Teilhard and Trassaert 1937)		1	1-3		Zong et al. 1989 [S.h.]	
	Eubelodon	*E. morrilli* [Eu1]	none				Barbour 1914a	NA(Eu), W(gom)
	Barbour 1914a, p. 186	Barbour 1914a, p. 186 or 194		1	1			
	Rhynchotherium [R1]	*R. tlascalae*	*R. blicki*				see W&T(68); W&P(84); Mil(90)	W(Rh), W(gom)
	Falconer 1868, pp. 74-5	Osborn 1918, pp. 134-5	*R. ?simpsoni*				Olsen 1957; To(73a):242 [R.?s.]	W(Rh), W(gom)
			R. falconeri	1	3-4		see Mil(90):58	W(Rh), W(gom)
Family Gomphotheriidae	*Cuvieronius*	*C. hyodon*	*C. tropicus* [Cuv2]				(Cope 1884*b*), following W&D(ip)	W(C-inae), W(gom)
Subfamily Cuvieroniinae [Cuv1]	Osborn 1923, p. 1	(Fischer de Waldheim 1814, p. 341)	*C. ?oligobunis*	1	2-3		(Cope 1893); C-C&M(87) [C.?o.]	
Cabrera 1929, p. 76	*Haplomastodon*	*H. waringi*	*H. ?guayasensis* [H1]				Hoffstetter 1952	W(C-inae), W(gom)
	Hoffstetter 1950, p. 24	(Holland 1920, p. 229)		1	1-2			
	Notiomastodon	*N. ornatus*	none				{31}	W(C-inae), W(gom)
	Cabrera 1929, p. 90	Cabrera 1929, p. 91		1	1			
	Stegomastodon	*S. mirificus*	*S. platensis*				{31}	W(St), W(C-inae)
	Pohlig 1912, p. 193	(Leidy 1858, p. 10) Stm1 -->	*S. primitivus or S. rexroadensis*				Sav(55):67-69; L&O(86) [S.r.]	W(St), W(C-inae)
			S. superbus	1	4-5	14 / 40-54	{31}	W(St), W(C-inae)
Family uncertain	*Tetralophodon* [T1]	*T. longirostris*	*T. atticus*				Wagner 1857 [T.a.]; T(85):742	W(Te), W(gom)
	Falconer 1857, p. 316	(Kaup 1832*b*, p. 420)	*T. ?bumiajuensis*				{23} [T.?b.]	W(Te), W(gom)
			T. xiaolongtanensis				To+(88):104, 212	W(Te), W(gom)
		incertae sedis -->	*T. longirostris gigantorostris*	1	3-5	1 / 3-5	Klähn 1922; Os(36):282	W(Te), W(gom)

Appendix C.1 (*continued*)

A	B	C	D	E	F	G	H	I
Family uncertain	*Anancus*	*A. arvernensis* [*An1*]	*A. cuneatus/paisuyensis*				T(86):103	W(Ana), W(gom)
	Aymard 1855, p. 507	(Croizet and Jobert 1828, p. 138)	*A. kenyensis*				MacInnes 1942; {12A} [A.k.]	W(Ana), W(gom)
			A. osiris				Arambourg 1945; T(86):103	W(Ana), W(gom)
			A. perimensis				F&C(47); T(86):103 [A.p.]	W(Ana), W(gom)
			A. petrocchii				Coppens 1965; {12A,.12B}	W(Ana), W(gom)
			A. sinensis				To+(88):212	W(Ana), W(gom)
			A. sivalensis				Cautley 1836; T(86):103	W(Ana), W(gom)
			A. sp. (Langebaanweg-type)				{12A}	W(Ana), W(gom)
			A. sp. (Sagantole-type)	1	10	1 / 10	{12A}	W(Ana), W(gom)
Family uncertain	*Paratetralophodon*	*P. hasnotensis*	none					W(Pa), W(gom)
	Tassy 1983, p. 273	(Osborn 1929a, pp. 2-3)		1	1	1 / 1		
"gomphotheres" (*Choerolophodon*, Gomphotheriidae, *Tetralophodon*,		*incertae sedis* -->	*Gomphotherium florescens*				Osborn 1929b	W(gom)
Anancus, Paratetralophodon) --	*incertae sedis*	*incertae sedis* -->	*Mastodon pandionis*				Falconer 1857; T(83):107	NHM, W(gom)
		incertae sedis -->	*Mastodon punjabiensis* --				Lydekker 1886b; T(83):107, 271	IM, W(gom)
			(*? Amebelodontinae incertae sedis*)					
		incertae sedis -->	*Mastodon (Trilophodon) pyrenaicus*				Lartet in Falconer 1857	MHNT; MNHN; MO
		incertae sedis -->	*Morrillia barbouri* [*Mo1*]				McKenna, Bell, *et al.* 1994	W(gom)
		incertae sedis -->	*Tetralophodon wumingensis* [*T1*]			(0-1) / 0-6	Zhao 1980	W(gom)
Family Stegodontidae	*Stegolophodon* [*Slp1*]	*Sl. latidens*	*Sl. cautleyi* (*i.s.*)				Lydekker 1886a [Sl.c.]; {17}	W(S-idae)
Osborn 1918, p. 135	Schlesinger 1917, p. 228	(Clift 1828, p. 371)	*Sl. hueiheensis*				To+(88):212; {13}; {17}	W(S-idae)
		[from Osborn 1942, pp. 825, 842-6]	*Sl. lydekkeri* (*i.s.*)				{17}	W(S-idae)
			Sl. miyokoae (*i.s.*)				{17}	W(S-idae)
			Sl. nasaiensis				Tassy *et al.* 1992 [Sl.n.]	W(S-idae)
			Sl. praelatidens				vK(59) [Sl.p.]	W(S-idae)
			Sl. progressus (*i.s.*)				Osborn 1929a [Sl.p.]	W(S-idae)
			Sl. pseudolatidens				Yabe 1950 [Sl.p.]; {17}	W(S-idae)
			Sl. stegodontoides (*i.s.*)				Pilgrim 1913 [Sl.s.]; {17}	W(S-idae)
			Sl. ?tatrotensis [*Slp2*]				Sarwar 1979 [Sl.?t.]	W(S-idae)
			Sl. tsudai (*i.s.*)	1	5-12		{17}	W(S-idae)
	Stegodon [*Sg1*]	*S. elephantoides* (=*E. cliftii* *)	*S. aurorae*				{17}	Japan(Sa), W(S-idae)
	Falconer 1857, p. 318	* fide Falconer and Cautley 1846, p. 47	*S. bombifrons* (*i.s.*)				{17}; {23}	W(S-idae)
		(= *Mastodon elephantoides*	*S. ganesa* (*i.s.*) [*Sg2*]				{17}	W(S-idae)
		Clift 1828, p. 372, in part)	*S. huanghoensis* [*Sg3*]				{17}	China(Sh), W(S-idae)
		[from Osborn 1942, pp. 825, 861-3]	*S. hypsilophus* (*i.s.*)				{23}	W(S-idae)
			S. insignis				{17}	W(S-idae)
			S. kaisensis (*i.s.*)				{17}	W(S-idae)
			S. licenti [*Sg4*]				{17}	W(S-idae)
			S. ?luzonensis				{23}	W(S-idae)
			S. mediterraneus (*i.s.*)				{21}	W(S-idae)
			S. mindanensis (*i.s.*)				{17}; {23}	W(S-idae)
			S. officinalis (*i.s.*) [*Sg4*]				{17}	W(S-idae)
			S. orientalis (*i.s.*)				{17}	W(S-idae)
			S. parahypsilophus (*i.s.*)				{17}	W(S-idae)
			S. pinjorensis (*i.s.*)				{17}	W(S-idae)
			S. sinensis (*i.s.*)				{17}	W(S-idae)
			S. sompoensis (*i.s.*)				{23}	W(S-idae)
			S. timorensis (*i.s.*) [*Sg5*]				{17}; {23}	W(S-idae)
			S. trigonocephalus				{17}	W(S-idae)

Appendix C.1 (*continued*)

A	B	C	D	E	F	G	H	I
			S. zdanskyi *[Sg6]*				{17}	W(S-idae)
			S. zhaotongensis	1	8-22	2 / 13-34	{17}	W(S-idae)
Family Elephantidae	*Stegotetrabelodon* *[St1]*	*S. syrticus* *[St2]*	*S. exoletus*				To+(88):212; {13} [S.e.]	W(Stb), W(E-idae)
Gray 1821, p. 305	Petrocchi 1941, p. 110	Petrocchi 1941, p. 110	*S. orbus*				Maglio 1970b [S.o.]	W(Stb), W(E-idae)
Subfamily Stegotetrabelodontinae		*incertae sedis* -- >	*S. maluvalensis*	1	3-4		Sarwar 1977 [S.m.]	W(Stb), W(E-idae)
Aguirre 1969*a*, p. 1370	*Stegodibelodon*	*S. schneideri*	none				{12A}; {18}	?MNHN, W(E-idae)
	Coppens 1972, p. 2964	Coppens 1972, p. 2964		1	1			
Subfamily Elephantinae	*Primelephas*	*P. gomphotheroides*	*P. korotorensis*				M(73):22 [P.k.]	W(Pri), W(E-idae)
Gray 1821, p. 305	Maglio 1970b, p. 10	Maglio 1970b, p. 10		1	2			
L1 -->	*Loxodonta*	*L. africana*	*L. "cf. Loxodonta"*				{12A}; T(86); K&M(93)	NMK, W(La)
L2 -->	[Anonymous 1827, p. 140]	(Blumenbach 1797, p. 125)	(Lukeino-A type)					
			L. adaurora				{12A}; {18}	W(Lad), W(E-idae)
			L. atlantica				{12A}; {18}	W(Lat), W(E-idae)
			L. exoptata	1	5		{12A}; {18}	NMK, NME
E1 -->	*Elephas*	*E. maximus* E2 -->	*E. (Palaeoloxodon) antiquus*				{22}; A&C(28); Os(31); vK(86)	W(Em), W(Ea)
	Linnaeus 1758, p. 33	Linnaeus 1758, p. 33	*E. (P.) creutzburgi*				{22}	W(E-idae)
			E. (P.) falconeri *[E3]*				{22}	W(E-idae)
			E. (P.) mnaidriensis				{22}	W(E-idae)
			E. (P.) naumanni				Makiyama 1924; Iz+(75)	Japan(En), W(E-idae)
			E. ?beyeri				{23}	W(E-idae)
			E. celebensis				{18}; {23}	W(E-idae)
			E. creticus or E. (P.) creticus				{22}	W(E-idae)
			E. cypriotes or E. (P.) cypriotes				{22}	W(E-idae)
			E. ekorensis				{18}	W(E-idae)
			E. hysudricus				{18}	W(E-idae)
			E. hysudrindicus				{18}	W(E-idae)
			E. iolensis				{18}	W(E-idae)
			E. melitensis				{22}	W(E-idae)
			E. namadicus *[E2]*				{18}	W(E-idae)
			E. planifrons *[E4]*				{18}	W(E-idae)
			E. platycephalus				{18}	W(E-idae)
			E. recki	1	18-19		{18}	W(E-idae)
Mth1 -->	*Mammuthus*	*M. primigenius*	*M. africanavus*				{18}	EA(Mp); NA(M&M)
	Brookes 1828, pp. 73-4	(Blumenbach 1799, p. 697)	*M. armeniacus* *[Mth2]*				{18}	NHM, W(E-idae)
			M. columbi *[Mth3, 4]*				{18}	NA(M&M), W(E-idae)
			M. exilis				{18}; {24}	SBMNH, LACM
			M. hayi *[Mth5]*				Barbour 1915, cf. W&D(ip)	W(E-idae)
			M. ?imperator *[Mth4, 5]*				{18}	W(E-idae)
			M. ?jeffersonii *[Mth3]*				{18}; {27}; Os(24b); Os(36)	W(E-idae)
			M. lamarmorae *[Mth6]*				{22}	W(E-idae)
			M. meridionalis				{18}; Mazo *et al.* 1990 [M.m.]	EA(Mm), W(E-idae)
			M. subplanifrons				{18}	W(Ms), W(E-idae)
			M. trogontherii *[Mth2]*	1	10-12	6 / 39-43	{18}; {19}	EA(Mt), W(E-idae)
				Subgrand	total-->	38 / 134-186 **	<--- Estimated total, not final	
Proboscidea *incertae sedis*	*incertae sedis*	*incertae sedis*	not applicable			2		
(see notes below)				Grand	total-->	38 / 136-188 **	<--- Estimated total, not final	

** Taxa listed in this appendix and those species given in the synonymy section below are not intended to be exhaustive. The estimated total of 162 species and subspecies of proboscideans is an average of a range of 136–188, and does not include many of the subspecies listed in the second part of this appendix (Appendix C.2, synonymy), nor does it include the 21 taxa listed below under '*Nomina dubia*' (12), '*Nomina nuda*' (1), '*Nomina oblita*' (4), and '*Nomina vana*' (4). This total also does not include the species listed under a 'group' (see notes C1, G2, and Z1 below), a possible addition of up to 11 species. There are only two living proboscidean species, with a possibility of five subspecies (see comments under notes E1 and L1 below).

General notes

- Overall, the arrangement of families and genera in this appendix follows that of Table 16.1 and Appendix B. Each entry consists of a 'Family' or 'Family uncertain' (column A), genus or genera (column B), type species (column C), and other species (column D). These are followed by total number of genera (column E), total of species (column F), and accumulated total for genera and species (column G). The total in column G is summed for each entry (Family, or Family uncertain), and this total includes taxa labeled with '*incertae sedis*' and '?', which are incorporated in the upper range of counts. The grand total of genera and species is given at the end of this table. Authorship(s), year of publication, and page number(s) are given in columns A–C, and references for 'Other species' (column D) are given in column H (the latter references are mostly for species which are abbreviated in square brackets), followed by an abbreviated list of museums (column I) where specimens of those taxa may be found (details below).
- To save space, page numbers in the Reference column (H) are displayed following a colon (:), unlike the format elsewhere in this volume where page numbers are displayed following 'p.' and 'pp.'. Also, due to space limitations, and, in some cases, uncertainty of exact authorship, when a reference is provided in column 'H', authorship is not in (). For example, the citation for *M. matthewi* should have been '(Osborn 1921*b*)' because the original description was for *Pliomastodon matthewi* — see Osborn 1936, pp. 137, 157, for details and Appendix A for generic synonymies.
- Citations for authorship which are given in this appendix, but may not appear in the References for this volume, may be found in Maglio (1973), Osborn (1936, 1942), Simpson (1945), and Simpson and Paulo Couto (1957).
- Key to museums where specimens may be found is given at the end of this appendix (note: only a general guideline as to representatives of a taxon — a genus and above, sometimes a species — is provided). This key begins with 'Grouped institutions' followed by 'individual institutions'.

Key to references in column H

{*x*} = Chapter *x* in this volume;
A&C(28), Andrews and Cooper (1928);
An(04a), Andrews (1904*a*);
B&B(28), Borissiak and Belyaeva (1928);
C&B(78), Coppens and Beden (1978);
C-C&M(87), Carranza-Castañeda and Miller (1987);
C&Z(78), Chang and Zhai (1978);
C+(78); Coppens *et al.* (1978);
F&C(47), Falconer and Cautley (1847);
Ha(78), Harris (1978);
Iz+(75), Inuzuka *et al.* (1975);
K&M(93), Kalb and Melbrate (1993);
Lm(90), Lambert (1990);
L&O(86), Lucas and Oakes (1986);
M(73), Maglio (1973);
M&S(85), Madden and Storer (1985);
Mil(90), Miller (1990);
Os(24a), Os(24b), Os(29b), Os(31), Os(36), Os(42), Osborn (1924*a*, 1924*b*, 1929*b*, 1931, 1936, 1942);
Sav(55), Savage (1955);
Sho+(89a), Shoshani *et al.* (1989*a*);
T(86), Tassy (1986);
To(72), To(78c), Tobien (1972, 1978*c*);
To+(86), To+(88), Tobien *et al.* (1986, 1988);
vK(59), vK(86), von Koenigswald (1959, 1986);
W&D(ip) = Webb and Dudley (in press);
W&P(84), Webb and Perrigo (1984);
W&T(68), Webb and Tessman (1968).

Footnotes

A1 *Anthracobune* is the only anthracobunid genus considered a *bona fide* proboscidean (the taxonomic statuses of *Ishatherium*, *Jozaria*, *Lammidhania*, and *Pilgrimella* are discussed in Chapters 8 and 16; see note *c* in Table 16.1).

Am1 Until additional studies are conducted, *Amebelodon hicksi* Cook, 1922, *A. paladentatus* Cook, 1922, and *A. sinclairi* Barbour, 1930, are all considered junior synonyms of *A. floridanus* (Leidy, 1886). These synonymies are in accordance with Turner's (1975) abstract; none the less, W.D. Lambert (personal communication 1994) believes that *A. paladentatus* may be closely related to *A. fricki*. In addition, according to Lambert (personal communication 1994) '*Amebelodon robertsoni*' (of Mebrate 1987), is *Amebelodon* (*Konobelodon*) *britti* Lambert 1990, and '*Amebelodon* cf. *barbourensis*' of Webb *et al.* (1981) is probably *Gomphotherium* sp.

An1 Synonymy of *Anancus arvernensis* was given by Tobien (1978*b*, p. 168); see also von Meyer (1831*b*), and Osborn (1936, pp. 630–40). Until detailed comparisons are conducted, the species '*Anancus alexeevae*', Bajguscheva (or Baigusheva), 1971, p.13, and '*Anancus kazachstanicus*' Aubekerova, 1974, p. 65 (Aubekerova (1974), later renamed as '*Anancus kazachstanensis*' (Tleuberdina, 1982, p. 17 — note: Aubekerova and Tleuberdina are the same person)) are considered junior synonyms of either *A. arvernensis* or *A. sinensis* (details on these latter species are given in Tobien 1978*b*, and Tobien *et al.* 1988).

C1 *Choerolophodon* '*palaeindicus* group' includes these species: *C. palaeindicus* (Lydekker, 1884*a*), *C. kisumuensis* (MacInnes, 1942; originally '*Platybelodon kisumuensis*'), and perhaps *C. zaltaniensis* Gaziry, 1987*a* (p. 70). Gaziry 1987*b*, noted that *C. zaltaniensis* is different from *C. palaeindicus* and *C. kisumuensis*. Tobien *et al.* (1986, p. 140) listed *Gomphotherium changzhiensis* Zhai, 1963 as a junior synonym of *Choerolophodon* sp. Similarly, Tobien *et al.* (1988, p. 212) listed *G. changzhiensis* and 60 other taxa as invalid. Maglio's (1974*b*, p. 700) *G. ngorora* is now *C. ngorora.*

Cuv1 Tobien (1973*a*, p. 243) considered all four genera — *Cuvieronius, Haplomastodon, Notiomastodon*, and *Stegomastodon* — valid, and noted (p. 248) that they should be united in a subfamily (not Anancinae, as suggested by Simpson and Paula Couto (1957, pp. 133, 185). Simpson and Paulo Couto (1957, pp. 165–6) synonymized *Cuvieronius humboldtii* ('*Mastodonte humboldien*') (Fischer de Waldheim, 1814 or Cuvier, 1824), and '*Mastodon Andium*' Cuvier, 1824, with *Cuvieronius hyodon* (Fischer de Waldheim, 1814). In addition, Simpson and Paulo Couto 1957, pp. 171–2, synonymized *Haplomastodon chimborazi* (Proaño, 1922) with *H. waringi* (Holland, 1920). Until additional information is available, we accept these synonymies.

Cuv2 Webb and Dudley (1995) employed *Cuvieronius tropicus* (Cope's (1884*b*) *Dibelodon tropicus*) as the oldest and valid name for cuvieronine taxa in North America; they also listed *Cordillerion oligobunis* Cope, 1893 as a junior synonym. In this volume we consider *Cuvieronius gratum* and all other species names of cuvieronine taxa in North America as junior synonyms of *Cuvieronius tropicus* (Note: Cope 1893, p. 59, used *Mastodon oligobunis*, whereas Osborn 1936, p. 554, employed *Cordillerion oligobunis*). We do, however, leave the possibility open that *Cuvieronius oligobunis* may be a valid species.

D1 Deinothere taxa listed by Mein (1990, pp. 83–6) as valid are considered in this volume as synonyms of the following species: *Deinotherium cuvieri* — junior synonym of *Prodeinotherium bavaricum*; *Deinotherium laevius* (in part) — junior synonym of *Prodeinotherium bavaricum*; and *Prodeinotherium laevius* (in part) — junior synonym of *Deinotherium giganteum*. Tobien (1990, p. 465) gave '*Deinotherium* aff. *levius* (Jourdan, 1861)' as a valid name. According to J. M. Harris (1970), the spelling of '*levius*' has seniority over '*laevius*' (as in *Deinotherium laevius* and *Prodeinotherium laevius* from Mein (1990). Many workers, among them Mayet (1908) and Graf (1957), considered *Deinotherium levius* a late Miocene species that was intermediate between *Prodeinotherium bavaricum* and *Deinotherium giganteum*. Laskarev (1944), on the other hand, thought that *D. levius* was an immigrant to Europe and closely related to *P. hobleyi*. John M. Harris (1970; personal communication 1993) agrees with Bergounioux and Crouzel (1962) that *D. levius* is a junior synonym of *D. giganteum*.

E1 *Elephas maximus* is the only living species of the genus *Elephas*, with three possible subspecies recognized: *E. m. maximus* Linnaeus, 1758 (confined to the island of Sri Lanka); *E. m. indicus* Cuvier, 1798 (found throughout IndoChina mainland, the island of Borneo, and the Andaman Islands); and *E. m. sumatranus* Temminck, 1847 (confined to the island of Sumatra) (Shoshani and Eisenberg 1982; see also notes under L1). From detailed accounts in Osborn (1936, pp. 1308–11 and 1322–3), and personal observations by J. Shoshani of the original material studied by Carolus Linnaeus at the Naturhistoriska Riksmuseet (Stockholm, Sweden), it appears that the taxon *Elephas maximus* of Linnaeus (1758) is based on one, or possibly two, type specimens: a fetus of an African elephant (*Loxodonta africanus*, studied by Seba 1734 and Linnaeus 1758; Fig. C1), and possibly that of an Asian elephant (*E. maximus*) described by Ray (1693). Osborn (1936, pp. 1309, 1323) preferred the name *E. indicus* (of Linnaeus 1754) rather than *E. maximus* (of Linnaeus 1758). He also noted: 'We must remember that Linnaeus believed that the number of created species was limited; in his mind the Indian and African elephants were of the same species. This is indicated in the following letter from Professor Einar Lönnberg (July 31, 1929). . .' (Osborn 1936, pp. 1309–10).

E2 *Elephas namadicus* vs. *E. antiquus* — Maglio (1973, p. 40) considered *Elephas antiquus* Falconer and Cautley, 1845, pl. 13, figs 1–3 1847, as a junior synonym of *Elephas namadicus* Falconer and Cautley, 1845, pl. 13A, figs 4, 5 1846. None the less, Caloi *et al.* (this volume, Chapter 22) employ the name *E. antiquus* (cf. Pilgrim, 1905). Todd and Roth (this volume, Chapter 18) seem to prefer *E. namadicus*. Lister (personal communication 1992) noted that one should not rely only on the date of authorship but also on the source of type specimens (the type for *E. antiquus* was from Europe, and for *E. namadicus* from India) and the number of detailed comparative studies which were conducted — in this case none. Until additional evidence becomes available, we retain both species as valid.
Note: The subgeneric designation of *Palaeoloxodon* for *Elephas* (*Palaeoloxodon*) *antiquus* and possibly for *E.* (*P.*) *namadicus* is used here to acknowledge the possibility that *Palaeoloxodon* may be elevated to a generic level (see note *d* in Appendix C.2). In addition, the validity of two species from China — '*Wanoloxodon chiai*' of Liu and Zhen, 1981 (*Wanoloxodon* is considered a junior synonym of *Elephas*, see Appendix A), and '*Elephas* (*Palaeoloxodon*) *huaiheensis*' (J. Guan, personal communication 1992) — remains uncertain.

E3 Alternatively, *Elephas* (*Palaeoloxodon*) *falconeri* may be designated as ?*Mammuthus falconeri* (see note *l* in Appendix C.2).

E4 According to Maglio (1973, pp. 42–4), *Elephas praeplanifrons* (or '*Archidiskodon praeplanifrons*') is a junior synonym of *Elephas planifrons*.

Eu1 Tobien (1973*a*, pp. 233–7) questioned the validity of *Eubelodon morrilli* and concluded that it '...is perhaps a transitional stage to the North American b-3-brevirosterine mastodonts' (see also note G3 below).

G1 Some of the *Gomphotherium* species and subspecies names may be invalid. They are listed here for further investigation. Not included in this appendix is a gomphothere taxon from China mentioned in Guan (1988, pp. 3, 15, 20) spelled *Gomphotherium yinnanensis* or *G. yinanensis*; this taxon was not described, only named, and its generic or species validity requires further investigation. Chang (1980) described *Gomphotherium yongrensis* as a new Pleistocene species from Yongren, Yunnan, China — based on the illustrations included (right M2 and right M3), these molars may belong to already existing species of sinomastodontid (*Sinomastodon*). Proboscidean materials from China and adjacent areas, although recently studied, are still in need of complete investigation; a review of the literature (especially obscure publications and those in languages of the region) for possible synonymies is paramount, e.g. Deraniyagala (1965), Zhou and Zhang (1974), Chow (1978), Chow *et al.* (1979), Tobien *et al.* (1986, 1988), Li (1988), Xie (1989), Xue and Zhang (1991), Ma and Tang (1992), and Guan (this volume, Chapter 13). Recently, Sarwar and Akhtar (1992) described a new genus and species, *Kunatia giganteus*, based on what they believe to be a right upper second molar of 'an oldest Siwalik record of the family Hippopotamidae' (order Artiodactyla). In a letter (dated April 5, 1994) to J. Shoshani, Malcolm C. McKenna suspected that the tooth Sarwar and Akhtar (1992) described 'might be the milk tooth of a proboscidean'. Tassy (letter dated June 7, 1994) to J. Shoshani, also believed 'that *Kunatia giganteus* is a proboscidean, more probably an upper P4 of a gomphothere (a conjecture: ?*Gomphotherium browni*)'.

G2 *Gomphotherium* '*annectens* group' includes these species: *G. annectens* Matsumoto, 1925, *G. cooperi* (Osborn, 1932*b*) and *G. sylvaticum* Tassy, 1985. The date of publication for the authorship of *G. annectens* is 1924 (Osborn, 1936, pp. 457, 788), but should be 1925 because Matsumoto's article was issued in 1925, although dated December 20, 1924 (therefore, citation is '*Gomphotherium annectens* Matsumoto, 1925' — see also Appendix C.3).

G3 *Gomphotherium productum*, according to Tobien (1973*a*, pp. 223, 231–2, 236–7) includes these species and/or subspecies: *G. calvertensis*, *G. cimarronis*, *G. cingulatum*, *Mastodon obscurus*, *Megabelodon minor*, *Ocalientinus ojocaliensis*, *Serridentinus progressus*, *Tetrabelodon serridens cimarronis*, *Trilophodon* (*Genomastodon*) *obsorni*, *T.* (*Genomastodon*) *willistoni*, *T.* (*Megabelodon*) *lulli*, *T.* (*Tatabelodon*) *gregorii*, *T.* (*Tatabelodon*) *riograndensis*, *T. abeli*, *T. cruziensis*, *T. fricki*, *T. giganteus*, *T. joraki*, *T. phippsi*, *T. pojoaquensis*, *T. simplicidens*, *T. simpsoni*, and *Trobelodon taoensis*. Tobien (1973*a*, pp. 232–7) also included species of *Eubelodon* and *Gnathabelodon* under possible synonymies with *Gomphotherium productum* — in this study *Eubelodon* and *Gnathabelodon* are considered valid genera (see also this volume, Chapters 15 and 16 by Lambert and Shoshani, respectively). Note that some of the species listed by Tobien (1973a) as junior synonyms of *Gomphotherium productum* are now considered synonyms of other species. For example, '*Trilophodon* (*Genomastodon*) *osborni*' is considered a junior synonym of *Gomphotherium osborni* (original name is *Tetrabelodon osborni* of Barbour 1916).

G4 Original binomen after Matsumoto (1925) is *Trilophodon sendaicus*, here it is listed under *Gomphotherium incertae sedis* (as *G. sendaicum*, cf. notes in Taxa Index) until detailed examination is conducted. P. Tassy believes that '*T. sendaicus*' is not a zygolophodon (as suggested in the volume by Kamei, 1991, p. 15, given as '*Zygolophodon sendaicus*'; a photograph is found in the front of that volume) but a gomphothere, and it is closely related to *Sinomastodon intermedius* (see Tobien *et al.* 1986, p. 160) based on these characters: central conules not inflated, wide valleys, and small posttrite trefoils (at least on the second lophid).

Gn1 Tobien (1973*a*, pp. 233–7) questioned the validity of *Gnathabelodon* and concluded that it is 'nearer to *Amebelodon*' (see also note G3 above).

H1 In the Royal Ontario Museum (Toronto, Canada), all specimens of *Haplomastodon* from Talara, Peru are labelled *H. guayasensis* Hoffstetter, 1952, p. 208. Simpson and Paula-Couto (1957, p. 172), however, synonymized *H. guayasensis* with *H. waringi*.

He1 Details on the phylogenetic position of *Hemimastodon* are given in Chapters 8 and 16.

L1 *Loxodonta africana* is the only living species of the genus *Loxodonta*, with two possible subspecies recognized: *L. a. africana* (Blumenbach, 1797), distributed in sub-Saharan Africa, and *L. a. cyclotis* (Matschie, 1900), found in forests of central and west Africa, (Laursen and Bekoff, 1978; see also notes under E1). Based on morphometrics, Groves *et al.* (1993) concluded that there are two living species of *Loxodonta*: *L. africana* and *L. cyclotis*. Shoshani (1993*b*), however, left open the question of how many living African species exist. Georgiadis *et al.* (1994) concluded that *Loxodonta a. africana* in eastern and southern Africa maintain sufficient mitochondrial DNA (mtDNA) similarities indicating, that although populations are subdivided, they are not strictly allopatrically isolated. Georgiadis (personal communication 1994) noted, based on mtDNA results, that *L. a. africana* and *L. a. cyclotis* are very different from one another.

L2 Original French spelling 'Loxodonte' by F. Cuvier (1825) is not available because the name was not latinized [Article 11(b) of International Commission on Zoological Nomenclature, 1985, p. 19]. Domning and Shoshani, 1982, employed *Loxodonta* Anonymous, 1827, but since then, the editors 'discovered' that Kretzoi (1950, pp. 406, 408) and Frade (1955, p. 774) credited N. A. Vigors as the author of *Loxodonta* because, as editor of the journal, he or one of the journal's staff reviewed the chapter by F. Cuvier (1825) and latinized the name *Loxodonta*, although Vigors may not have signed the review. The name '[N. Vigors]' was also credited for *Loxodonta* by Heider (1932). Many thanks to Walter Bock and Daryl P. Domning for their help in researching this taxonomic problem.

Mo1 McKenna and Shoshani (in McKenna, Bell, *et al.* 1994) employed *Morrillia* Osborn, 1924*a* as a generic, not a subgeneric category (type species is *Tetralophodon barbouri* Osborn, 1921*a*). This change was made at a late date prior to publication and thus the total number of proboscidean genera remains at 38.

Mth1 Note that the authorship for *Mammuthus* is Brookes, 1828, not Burnett, 1830 (details in Appendix A, note *b*). Species of *Mammuthus* were divided among three genera in Osborn (1936, 1942) where he employed '*Archidiskodon*' for the more primitive species, e.g. '*A. subplanifrons*', '*A. meridionalis*', and '*A. imperator*'; '*Paralephas*' for the intermediate species, e.g. '*P. columbi*' and '*P. armeniacus*'; and '*Mammonteus*' for the more progressive species '*M. primigenius*'.

Mth2 *Mammuthus armeniacus* vs. *M. trogontherii*: Maglio (1973, p. 57) considered *Mammuthus trogontherii* Pohlig, 1885, p. 1027, a junior synonym of *M. armeniacus* (Falconer, 1857, p. 319). Lister (this volume, Chapter 19) provides an explanation as to why *M. trogontherii* is a valid taxon (based on the uncertain synonymy with *armeniacus*, due to very limited material of *M. armeniacus* — making diagnosis difficult — compared to large quantities of *M. trogontherii* and the type of *M. armeniacus* coming from outside the main range of *trogontherii* (the type locality of *M. armeniacus* is Erzurum region, Turkey, and that of *M. trogontherii* is Süssenborn, Germany); in addition, *M. armeniacus* and *M. trogontherii* have been used continuously in the literature as separate names). For these reasons, in this volume, we retain both species. Lister further strengthens retaining the use of *Mammuthus* for all European species, thus avoiding the use of *Archidiskodon*. (See also note *b* in Appendix C.2 where Lister suggested this nomenclature for Eurasian mammoths: *Mammuthus meridionalis*, *Mammuthus armeniacus* (only for specimens from Erzurum, Turkey), *Mammuthus trogontherii*, and *Mammathus primigenius*.)

Mth3 *Mammuthus columbi* vs. *M. jeffersonii* — Maglio (1973, p. 63) considered *M. jeffersonii* Osborn, 1922*b*, p. 11, a junior synonym of *M. columbi* (Falconer, 1857, p. 319; see also Graham, 1986*b*). Saunders (this volume, Chapter 27) uses *M. jeffersonii*, whereas Dudley (this volume, Chapter 29) suggested using *M. c. jeffersonii* a subspecies name adopted here.

Mth4 *Mammuthus imperator* (Leidy 1858) vs. *M. columbi* (Falconer, 1857, p. 319) — Maglio (1973, p. 63) considered both species valid. Stock (1956), Harris and Jefferson (1985), and Shaw and Quinn (1986), all considered the mammoth species whose remains were found at the Rancho La Brea Tar Pits as the imperial mammoth (*M. imperator*), but according to Marcus and Berger (1984, p. 164) and Stock (1992), the mammoth species is the Columbian mammoth (*M. columbi*). Agenbroad and Mead (this volume, Chapter 28) consider *M. imperator* and *M. columbi* as two separate species. V. Louise Roth (personal communication 1992) suggested using a combined term *Mammuthus columbi/imperator* which reflects that the two species were very close (see also note Mth5 where *M. c. imperator* is considered).

Mth5 Webb and Dudley (1995) employed *Mammuthus hayi* (originally '*Elephas hayi*' Barbour, 1915) as the oldest *Mammuthus* in North America (not *M. meridionalis*); these authors also listed three mammoth species — *M. scotti* (originally '*Elephas scotti*' Barbour, 1925*a*), *M. haroldcooki* (originally '*Elephas haroldcooki*' Hay, 1928), and *M. sonoriensis* (originally '*Archidiskodon sonoriensis*' Osborn, 1929*a*) — as junior synonyms of *M. Hayi*. If one accepts this taxonomy with emendation (see notes above), North American mammoth species would comprise two major lineages: (1) *M. primigenius*, and (2) *M. hayi* and *M. columbi* with three subspecies (*M. c. imperator, M. c. jeffersonii, M. c. exilis*). Each of these subspecies has been classified as a species, and pending further investigation, *M. c. exilis* should perhaps be reinstated to a species category (discussions in Todd and Roth, Lister, Agenbroad and Mead, and Dudley, this volume, Chapters 18, 19, 28, 29, respectively; Graham (1986*b*), Agenbroad (1994), and Webb and Dudley (1995)). Should the above taxonomy and nomenclature be adapted, they will affect names of species (e.g. the use of *M. hayi* instead of *M. meridionalis* for North American specimens) now employed in Chapters 18, 27, 28 and 29.

Mth6 Caloi *et al.* (this volume, Chapter 22) appear to consider *Mammuthus lamarmorae* a valid taxon. Maglio (1973, p. 42) referred to it as *Elephas lamarmorae* and noted that the validity and relationships of '...dwarf elephants from Mediterranean islands are less certain at this time'.

N1 The familial assignment of *Numidotherium* in Numidotheriidae is by Court (see Chapter 8). *N. savagei* was added in proof and thus not included in the total count.

P1 Guan (1991, p. 10) recognized two subspecies of *Platybelodon danovi* (*P. d. danovi* and *P. d. cheni*); sufficient evidence permitting, *P. d. cheni* may be elevated to a species category. '*P. huangzhongensis*' of Guan (1986) is considered here as *Platybelodon* sp. until further examination. Belyaeva and Gabunia (1960) described *Platybelodon jamandzhalgensis* that lived at the same time and same place as *P. danovi* (the former from Byelometcheskaya-Yaman Dzhalga ravine, and the latter from the Chokrak beds of Byelometcheskaya, Cisaucasia). Tobien (1973*a*, p. 253) believed that it is difficult to accept two different species based on similar characters, living at the same time in the same place. Therefore, we consider *P. jamandzhalgensis* a junior synonym of *P. danovi*.

P2 *Amebelodon cyrenaicus* Gaziry, 1987*b* named for specimens from the Sahabi of Libya is, according to Tassy (1990*b*, fig. 4, p. 244), *Platybelodon* sp.; in this volume we emend the original name to *Platybelodon cyrenaicus*. This emendation was done for the following reasons. Gaziry (1987*b*) considered *Amebelodon* and *Platybelodon* on a subgeneric level and possibly a synonym of each other, thus when he allocated *cyrenaicus* to *Amebelodon*, in essence it was also allocated to *Platybelodon*. Also, the lower tusks of *A. cyrenaicus* possess dentinal rods, a character of *Platybelodon* only (Guan this volume, Chapter 13). Note, however, that Lambert (1990) treated *A. cyrenaicus* as a valid species.

P3 '*spectabilis*' was the type species of *Selenolophodon* Chang and Zhai, 1978 (see Appendix A for generic synonymies).

R1 According to Tobien (1973*a*, p. 242), species of *Rhynchotherium* may be allocated to three groups: (1) '*Aybelodon hondurensis*', '*Blickotherium blicki*', and *R. tlascalae*; (2) *R. francisi, R. anguirivale, R. rectidens, R. shepardi,* and *R. brevidens*; and (3) *R. browni, R. falconeri, R. edense,* and *R. simpsoni*. Among those from Group 1, *A. hondurensis* is considered a *Gomphotherium* sp. or *Gomphotherium* aff. *productum* (Tobien 1973*a*, p. 238), whereas '*B. blicki*' and *R. tlascalae* are considered here valid species of *Rhynchotherium*. All members of Group 2 are, according to Tobien (1973*a*, p. 242), classified in *Rhynchotherium* sp. or in other bunodont or zygodont groups. Among those from Group 3, *R. browni* and *R. edense* are, as suggested by Miller (1980, 1990, p. 63), junior synonyms of *R. falconeri*, and *R. simpsoni* is treated here as *R. ?simpsoni* (details in Tobien 1973*a*; see also notes by Savage 1955, pp. 52–3). *R. edensis* (formerly *Dibelodon edensis*) was transferred to the genus *Cuvieronius* (May and Repenning, 1982) but Webb and Perrigo (1984, pp. 242–3) disagreed. *Rhynchotherium euhypodon* is possibly *Rhynchotherium* sp. or *Gomphotherium* sp. (Tobien 1973*a*, pp. 238–9). Tobien (1973*a*) and Miller (1990) considered *Rhynchotherium* to be restricted to the New World. *Rhynchotherium chinjiense* from India (see Osborn 1942, p. 1414) is considered here *nomen vanum*. Tobien *et al.* (1988, p. 212) noted that *R. huananensis* (of Chow and Zhang, 1978) from China is not valid, an opinion supported by Miller (1990, p. 62); see also note *Sin1* below.

Sg1 All *Stegodon* species marked with "(i.s.)" are either members of a group or may be invalid (see Fig. 17.8), therefore the estimated number of species for this genus varies from 8 to 22.

Sg2 *Stegodon ganesa* may be a synonym of *S. insignis* (see Matsumoto 1915, and Colbert 1943 for other synonyms of stegodontids).

Sg3 *Stegodon huanghoensis* may be a junior synonym of *Stegodon zdanskyi* (H. Saegusa, personal communication 1993). A synonymy of *S. huanghoensis* with *S. zdanskyi* will probably not be accepted by Chinese scholars for scientific reasons (*S. huanghoensis* is, according to J. Guan, personal communication 1993, a valid taxon based on a number of characters indicating that it is more primitive than *S. zdanskyi*) as well as for historical and political reasons. *S. huanghoensis* translates into 'The Yellow River Stegodon' and students in primary schools throughout China read about this stegodon; textbooks stress the significance of 'huanghoensis' in the literature. The skeleton of *S. huanghoensis* — perhaps the most complete skeleton in China — about 4 meters tall, was described in detail by Zheng *et al.* (1975). As of 1993, the original skull with tusks is housed at the Beijing Natural History Museum (BNHM), and the rest of the skeleton is in boxes at the Institute of Vertebrate Paleontology and paleoanthropology (IVPP), also in Beijing. Casts of this skeleton are found in Gansu Provincial Museum (Lanzhou, Gansu Province, China), Tianjin Natural History Museum (Pian-Ho, near Tianjin, Hebei Province, China), and BNHM. Two additional skeletons were discovered recently near Xian (Shaanxi Province; J. Guan, personal communication 1993).

Sg4 Zong (1992, p. 294) considered *Stegodon licenti* and *S. officinalis* to belong in the genus *Stegolophodon*. Saegusa (personal communication 1994) however, disagrees; he suggests that these species should remain in *Stegodon*.

Sg5 *Stegodon timorensis* may be a synonym of *S. sompoensis*.

Sg6 *Stegodon zdanskyi*, according to Zong (personal communication 1993), includes *S. shinshuensis*

Sin1 Zong *et al.* (1989, p. 71) included these taxa in the genus *Sinomastodon*: *Sinomastodon* (*Gomphotherium*) *wimani*, *S.* (*G.*) *yangziensis*, *S.* (*G.*) *guangxiensis*, *S.* (*G.*) *wufengensis*, *S.* (*G.*) *serridenstoides*, *S.* (*G.*) *liuchengensis*, *S.* (*Rhynchotherium*) *huananensis*, *S. yanyaunensis* [*sic*] — should be *yanyuannensis*, and *S. hanjiangensis*. With the exception of the last two species which appeared in print after 1986, Tobien *et al.* (1986, pp. 160–1) considered the species above as junior synonyms of *S. intermedius*. In this volume, we consider only two species of *Sinomastodon* valid: *S. intermedius* and *S. hanjiangensis*. *S. yanyuannensis* is either a junior synonym of *S. hanjiangensis* or *Sinomastodon* sp. (because only two molars were used as the basis for *S. yanyuannensis* by Zong 1987, and Zong *et al.* 1989, p. 84 noted that *S. hanjiangensis* 'seems very close to *S. yanyaunensis*'). With regard to *S. (G.) wimani* (Tobien et al. 1986, p. 160, listed as '*Trilophodon* cf. *wimani*'), Saegusa (personal communication 1994), noted that the name *Sinomastodon wimani* in Zong *et al.* (1989) cannot be accepted because the status of the holotype of *G. wimani* had not been discussed by these authors.

Slp1 All *Stegolophodon* species marked with '(i.s.)' are either members of a group or may be invalid (see Fig. 17.8), therefore the estimated number of species for this genus varies from 5 to 12. The higher estimate may increase once a comparison is made with those species recognized by Zong (1992). According to Tassy (1977*b*, p. 1391) and Mazo (this volume, Chapter 14) '*Stegolophodon saldanensis*' is a *Tetralophodon longirostris*.

Slp2 '*tatrotensis*' was the type species of *Antelephas* Sarwar, 1979 (see Appendix A for generic synonymies).

St1 According to Zong (personal communication 1993) *Elephas yuheensis* (or '*E. youheensis*') may be included in *Stegotetrabelodon*; in this volume we designate *E. youheensis* as 'Elephas sp.' or 'indeterminate'.

St2 Coppens *et al.* (1978, p. 350) listed *Stegotetrabelodon lybicus* and *S. sahabianus* as junior synonyms of *S. syrticus*, whereas Boaz *et al.* (1979) included *S. lybicus* as a valid taxon.

Stm1 W.D. Lambert (personal communication 1994) believes that *Stegomastodon primitivus* and probably also *S. rexroadensis* are valid taxa.

T1 Fossil remains of *Tetralophodon* are known from Eurasia (Falconer 1857; Osborn 1936; Tobien *et al.* 1988), Africa (Osborn 1936; Kalb *et al.* this volume, Chapter 12B; Madden 1977*c*), and possibly North America (details in Table 34.2 and Fig. 34.4). *T. wumingensis* (of Zhao, 1980) is placed in *incertae sedis* because the holotype, the sole specimen (mandible) has only one M3, but there are no immediate teeth (M1, M2) associated with it. Thus, there is no evidence that this species belonged to tetralophodont grade. It is possible that this taxon is misidentified; '? cf. *Sinomastodon* sp.' has been suggested.

Z1 *Zygolophodon* '*turicensis* group' includes these species: *Z. turicensis* (Schinz, 1824); *Z. metachinjiensis* (Osborn, 1929*a*); *Serridentinus chinjiensis* (Osborn, 1929*a*); *Z. gobiensis* (Osborn and Granger, 1932); *Z. atavus* (Borissiak, 1936); ?*Z. nemonguensis* Chow and Chang, 1961; and ?*Z. tongxinensis* Chen, 1978. P.A. Tleuberdina (personal communication 1993 to J. Shoshani) believes that '*Z. atavus*' is a gomphothere, not a zygodontid, whereas P. Tassy disagrees. Also note that '*Z. tongxiensis*' has been renamed as '*G. tongxinensis*' (other specimens from the same locality — Gujiazhuang, near Dingjiaergou — have been named as *Platybelodon tongxinensis*; Guan, personal communication, 1994; cf. Wang *et al.* 1994; see also notes under '*Mastodon tapiroides*' below). *Zygolophodon lufengensis* Zhang, 1982 from Lufeng, near Kunming, southern China (also listed by Qi 1993, p. 7, among the proboscidean taxa found in Lufeng) is considered a possible junior synonym of *Z. gobiensis*, discussed in Tobien *et al.* 1988.

Z2 *Zygolophodon proavus* is a senior synonym of *Z. merriami* (Madden 1977*b*, p. 29; Madden and Storer 1985, pp. 1345–6).

Z3 Mein (1990, pp. 88–9) listed *Zygolophodon borsoni* as a valid taxon; in this volume it is considered a junior synonym of *Mammut borsoni* (Hays, 1834).

Proboscidea *incertae sedis* (total of 2 species)

(1) '*Mastodon*' *grandincisivus* Schlesinger, 1917 — according to Osborn (1936, pp. 360–2), this taxon belongs in the genus *Tetralophodon*. Tobien (1978*b*, pp. 177–94), however, placed '*grandincisivus*' in *Stegotetrabelodon*; whereas, this species, according to Tassy (1986, p. 55, fig. 18), is an amebelodontid. Tassy and Darlu (1987, pp. 491–2) discussed the implications of inclusion and exclusion of '*Mastodon*' *grandincisivus* in their analyses. Cladistic analysis by Kalb *et al.* (this volume, Chapter 12A) places '*M.*' *grandincisivus* as a sister-group to the clade containing: *Tetralophodon*, *Paratetralophodon*, *Anancus*, *Stegolophodon*, *Stegotetrabelodon*, *Stegodibelodon*, *Stegodon*, and Elephantinae. See also Mazo (this volume, Chapter 14), who uses the names '*Mastodon* (*Bunolophodon*) *grandincisivus*' and *Gomphotherium angustidens* '*grandincisivoid*'; these taxa we also considered *incertae sedis*.

(2) *Tetralophodon falconeri* (Lydekker, 1877) — Tobien (1978*b*, p. 168), listed '*Anancus falconeri*' Osborn as one of the '...diverse species, which have been created around *Anancus arvernensis*...'; a molar of *A. falconeri* is depicted in plate 16, fig. 10 of Tobien (1978*b*). *Tetralophodon falconeri* was first named by Lydekker (1877) as '*Mastodon falconeri*' not *Anancus falconeri*. The holotype of *Mastodon falconeri* is a tetralophodont non-anancoid M2 in a mandible from the Siwaliks, whereas the holotype of *Anancus falconeri* Osborn, 1926*a*, is a lower M3 from the Crag of England, later synonymized with *Anancus arvernensis* as can be seen in Tobien (1978*b*, noted above).

Nomina dubia (total of 12 species)

(1) '*Archidiskodon vanalpheni*' Dart, 1929; (2) '*Archidiskodon milletti*' Dart, 1929; (3) '*Archidiskodon loxodontoides*' Dart, 1929; and (4) '*Loxodonta griqua*' Haughton, 1922 — these four species, according to Maglio (1973, pp. 63–4) are 'Elephantinae, *gen. indet.*', *nomina dubia*, all probably date to late Pliocene, and '...all are similar enough to be included in a single species and may be conspecific with *Loxodonta adaurora* or with *Mammuthus africanavus*'.

(5) *Gomphotherium chabbariensis* — Sarwar, 1977.

(6) *Gomphotherium pasalarense* Gaziry, 1976 — The original definition by Gaziry (1976, p. 24) is that *pasalarense* (from

Pasalar, Turkey) is a subspecies of *angustidens* (i.e. *G. a. pasalarense*). Tassy (1985, p. 701) considered *G. a. pasalarense* a junior synonym of *G. angustidens*, but Tobien (1990, p. 467) designated the subspecies *G. a. pasalarense* as a species *G. pasalarense*, and noted, among other observations, that the lower incisors of '*G. pasalarense*' do not exhibit '...the piriform section, typical for the earlier and later *Gomph. angustidens*, as for example the many specimens excavated by Tassy...'. These conflicting characters require additional studies to verify whether or not the variations in tusk morphology of '*G. pasalarense*' are consistent in large sampling.

(7) *Mastodon angustidens pygmaeus* Depéret, 1897 — originally labelled (p. 520), *Mastodon angustidens* Cuvier, *mut. pygmaeus* (*mut.* = mutation). It could be an abnormal tooth.

(8) *Mastodon tapiroides* Desmarest, 1822 — *Mastodon tapiroides* and *Mastodon turicensis* were considered synonyms by Vacek (1877). *M. tapiroides* was translated by Desmarest (1822) from 'le mastodonte tapiroïde' defined by Cuvier of what he formerly called 'le petit mastodonte'. Cuvier (1824, p. 527) wrote '*Mastodon tapiroides*' two years after Desmarest 1822, p. 386; see Tassy 1985, pp. 477–8.

Mastodon tapiroides is a *nomen dubium*; see comments by Gaudry (1862) and Mayet (1908). Tassy (1977*a*, 1985) agrees that Matthew's decision (1918) to choose *M. tapiroides* as type species of *Zygolophodon* was unfortunate. Hence, following Vacek' synonymy, *M. turicensis* should be considered as the type species (the holotype of which is non-ambiguous), even if it does not conform to the rules of the Code (Tassy 1985, p. 479, followed by Tobien this volume, Chapter 9). If we do not follow this gentlemen's agreement, *Zygolophodon* disappears from the nomenclature and must be replaced by *Turicius* Osborn, 1926*a*. Simpson's (1945, pp. 247–8) footnote corroborates these suggestions.

Note that *Mastodon tapiroides* Desmarest, 1822 is antedated by *Mastodon microdon* Rafinesque, 1814, as well as *Mastotherium microdon* Fischer de Waldheim, 1814. These species names have never been used; consequently they are *nomina oblita* (see below).

(9) *Tetrabelodon macrognathus* Pilgrim, 1913.

Nomina dubia (for 3 taxa of Australian origin)

(10) *Dinotherium Australe* Owen, 1843, p. 332. Listed in Mahoney and Ride (1975, p. 86).

(11) *Mastodon australis* Owen, 1844, p. 4. Listed in Mahoney and Ride (1975, p. 153).

(12) *Notelephas* Owen, 1882, p. 777. Mahoney and Ride (1975, p. 153), gave the following information: 'Species described from material said to be Australian, but seeming on zoogeographical grounds to be from elsewhere'. Yet, below that, under generic names, they listed '*Notelephas* Owen, 1882' and provided details on page 154.

Colin Groves (personal communication March 23, 1991, via David Ride) noted: '(1) The types of *Dinotherium australe* and *Mastodon australis* are lost, (2) The holotype of *Notelephas australis* is still in the BM(NH)'. Groves also noted that 'He (D. Ride) reiterates that it seems very unlikely that any of these really came from Australia; even the reported locality of one of them (*D. australe*) is so confused that it cannot be correct'. This information corroborates Vickers-Rich *et al.* (1991) who did not mention proboscidean taxa in Australia.

According to Branagan (1992), Leichhardt, in a letter to Owen, took issue with Owen for placing the 'gigantic Pachyderm' among the *Dinotherium*, suggesting that the specimen Leichhardt sent to Owen 'will show you that it is an animal with different incisors and really more allied to the dentition of the kangaroo than to any other animal' (quotes from Leichhardt (1844, cited in Branagan 1992, p. 98)).

Nomina nuda (total of 1 species)

(1) *Mastelephas* Barbour, 1917 (no definition and genotypic species given; designated as *nomen nudum* by Osborn, 1942, p. 1380).

Nomina oblita (total of 4 species)

(1) *Mastodon microdon* Rafinesque, 1814

(2) *Mastodon senodon* Rafinesque, 1814

(3) *Mastotherium microdon* Fischer, 1814 (Note: Fischer = Fischer de Waldheim)

(4) *Mastotherium leptodon* Fischer, 1814 (Note: Fischer = Fischer de Waldheim)

Nomina vana (total of 4 species)

(1) *Rhynchotherium chinjiense* Osborn, 1929*a*; cf. Miller (1990), p. 63

(2) *Serridentinus prochinjiensis* Osborn, 1929*a*

(3) *Stegolophodon nathotensis* Osborn, 1929*a*

(4) *Trilophodon hasnotensis* Osborn, 1936

Museum abbreviations (Column I)

Included in the list below are the locations where skeletons and type specimens of proboscideans are housed (mounted skeletons (some are composites) are marked with '(S)' and some of the 'bones' may be casts). Excluded from this list are private collections and small museums, especially if they are given in chapters of this book (see, for example, Chapters 13, 20, 30). This list of museums and collections has been compiled from those listed in the chapters of this volume and from other sources, such as Osborn (1936, pp. 745–60; 1942, pp. 1609–21) and Maglio (1973, p. 8; specimens studied by Maglio (1973) are housed in these museums: MCZ, NMK, PU, and YPM (abbreviations are defind under 'Individual institutions' below)). Note that the listing in this appendix does not include all of Osborn's list of collections, and that a summary of observations on skeletal material mentioned in Osborn's *Proboscidea* is given in Osborn (1942, pp. 1600–5). To save space, some entries are listed under 'Grouped institutions', for example, 'Japan(Sa)' includes museums which house mounted skeletons of *Stegodon aurorae* (most listings are on the generic level, sometimes down to species). Within these groupings, a museum name is written in full if it is not abbreviated under 'individual institutions'. This summary of proboscidean genera and species and their locations, we hope, will stimulate research to update the taxonomy of Proboscidea including a current listing of institutions. The following people helped us in updating names of museums and/or made valuable comments on the taxonomy in this appendix: Alexander O. Averianov, Susan K. Bell, Walter Bock, Daryl P. Domning, Daniel C. Fisher, Jian Guan, John M. Harris, Adrian M. Lister, Malcolm C. Mckenna, V. Louise Roth, Haruo Saegusa, Jeffrey J. Saunders, Keiichi Takahashi, Heinz Tobien, and Robert M. West. Singled out is Susan K. Bell for her meticulous editing of this appendix.

Note added in proof: recent publications by Mol and Van Essen (1992) and Lister and Bahn (1994) provide lists of museums and their addresses. Some of these museums are included in the list below; others are not included because the volume had gone to press. In addition, Keiichi Takahashi (of Lake Biwa Museum Project Office, Otsu, Japan) sent us a detailed list of Japanese museums and other institutions where proboscidean skeletal material is held; some of these places are included below, others are not.

Grouped institutions

Abbreviations are defined under 'Individual institutions' below.

China(Sh), includes museums which house mounted skeletons of *Stegodon huanghoensis*: BNHM (cast of a cast), GPM (cast), IVPP; THP (cast): Xunyi County Museum (Xunyi, Shaanxi Province, China).

China(Si), includes museums in China which house mounted skeletons, bones, or teeth of *Sinomastodon*: BNHM; Geological Museum of Shaanxi Province (Xian, China — skeleton of *S. hanjiangensis*); IVPP; THP.

China(E-idae), includes museums in China which house mounted skeletons, skulls or isolated bones and teeth of members of the family Elephantidae: APM; BNHM; CCG; GPM; IVPP; THP.

EA(Mm), includes Eurasian museums which house mounted skeletons of *Mammuthus meridionalis*: Castle of L'Aquila (L'Aquila, Italy); GMZ; IGF; MNHM; ZM. These museums hold other material: NHB; NHM.

EA(Mp), includes Eurasian museums which house mounted skeletons of *Mammuthus primigenius*: Geological Museum (Lausanne, Switzerland); Geological-Palaeontological Institute (Münster, Germany); KGM; Mátra Museum (Gyöngyös, Hungary); Moravian Government Museum of Brno (Brünn, Czechoslovakia, now Czech Republic); Musee Royal d'Histoire Naturelle de Belgique (Brussels, Belgium); NKM; PIN; Regional Prehistory Museum (Halle, Germany); State Ethnographical Museum or Grassi Museum (Leipzig, Germany; also known as Museum für Volkerkunde); ZM (includes soft tissues and stomach contents). Lectotype, left M_3 (according to Maglio 1973, p. 60) is in the Blumenbach collection of the Zoological Institute of the Museum of the University of Göttingen (Germany). See also listing under 'NA(M&M)'.

EA(Mt), includes Eurasian museums which house mounted skeletons or indicated material of *Mammuthus trogontherii*: Azov Museum (Azov, Russia); NHB (skeletal material); NHM (skeletal material); NMM (includes important Mosbach collection, no skeleton); Spengler Museum (Sangerhausen, Germany); WM (includes type material from Germany, no skeleton).

EA(Mth.af.), includes Eurasian museums which house bones and teeth of *Mammuthus africanavus*: MNHN.

Japan(En), includes museums which house mounted skeletons of *Elephas* (*Palaeoloxodon*) *naumanni*: Historical Museum of Hokkaido (Sapporo, Japan); National Science Museum (Tokyo, Japan); Nojiri Museum (Kamiminochi-gun, Nagano Prefecture, Japan); TMTMNH.

Japan(Sa), includes museums which house mounted skeletons of *Stegodon aurorae*: ACM (also includes casts of footprints; see Chapter 34); MNHA; Osaka Museum of Natural History (Osaka, Japan); Saitama Museum of Natural History (Chichibu-gun, Saitama Prefecture, Japan).

NA(Eu), includes North American museums which house skeleton, bones, or teeth of *Eubelodon*: AMNH; UNSM.

NA(Gn), includes North American museums which house bones, or teeth of *Gnathabelodon*: AMNH; Fort Hays State College Museum (Kansas State College, Hays, Kansas, USA — *G. thorpei*); TMM (*G. buckneri*); UNSM.

NA(M&M), includes North American museums which house mounted skeletons (S), bones (B), or teeth (T) of *Mammut americanum* and *Mammuthus* sp. When not specified, the collection may include material from either or both *M. americanum* or *Mammuthus* sp.: AMNH (S, B, T — (S) of *M. americanum*, known as the 'Warren mastodon'; Warren 1852); ANSP (?S, B, T); Blackwater Draw Museum (Portales, New Mexico, USA — ?S, B, T of *Mammuthus columbi*); Cincinnati Museum of Natural History (Cincinnati, Ohio, USA — S, B, T); CIT (?S, B, T); Cleveland Museum of Natural History (Cleveland, Ohio, USA — S, B, T); CMNH (S, B, T); Cranbrook Institute of Science (Bloomfield Hills, Michigan, USA — B, T); DMNH (S, B, T); FMNH (S, B, T); GMUW (?S, B, T); Idaho Museum of Natural History (Pocatello, Idaho, USA — ?S, B, T); ISM (S, B, T); LACM (S, B, T); MCZ (S, B, T — (S) of *M. americanum*, known as the 'Harvard mastodon'; Warren 1852); Milwaukee Public Museum (Milwaukee, Wisconsin, USA — S, B, T); MNHUK (?S, B, T); MSHS (S, B, T); MSUM (B, T); NMNH (S, B, T); New York State Museum (Albany, New York, USA — ?S, B, T); Oakland Community College, Highland Lakes Campus (Union Lake, Michigan, USA — (S) of *M. americanum*, known as 'Elmer'; Dorr *et al.* 1982); Ohio State University Geological Museum (Columbus, Ohio, USA — (S) of *M. americanum*); SBMNH (S, B, T); SDNHM(? S, B,T); TMM(?S,B, T); UCB (S, B, T,); UF (B, T); UMMP (S, B, T); University of Alaska Museum (Fairbanks, Alaska, USA — ?S, B, T); UNSM (S, B, T); YPM (S, B, T — (S) of *M. americanum*, known as the 'Otisville mastodon').

W(Am), includes world-wide museums which house bones or teeth of *Amebelodon*: AMNH; BNHM; IVPP; LPVPH; MCZ; MNHN; MNHUK; NHM; UF; UNSM; YPM.

W(An), includes world-wide museums which house bones or teeth of *Anthracobune*: Howard University-Geological Survey of Pakistan Project (Washington, DC, USA); IM; Lucknow University Vertebrate Paleontology Collection (Lucknow, India); NHM; UMMP; WIHG.

W(Ana), includes world-wide museums which house bones or teeth of *Anancus*: AMNH; BNHM; IGF; IGPB; IM; IVPP; LPVPH; MNHN; NHM; NME; NMK; THP; VM; YPM.

W(Ar), includes world-wide museums which house bones or teeth of *Archaeobelodon*: LPVPH; MNHN; ?NHM; NMK.

W(Ba), includes world-wide museums which house bones and teeth of *Barytherium*: BUGM; CGM; HL; LPVPH; MNHN; NHM; SM; SM(St); YPM.

W(C-inae), includes world-wide museums which house skeletons, bones, or teeth of members of the subfamily Cuvieroninae (*Cuvieronius*, *Haplomastodon*, *Notiomastodon*): AMNH, FMNH; LPVPH: MACN; MINGM; MNHN; MNLP; MUCQ; NHM; ROM; UCB; UF; YPM; see also under W(St).

W(Ch), includes world-wide museums which house bones or teeth of *Choerolophodon*: AMNH; BNHM; IVPP; KUJC; LPVPH; MNHN; NHM; NMK; THP; YPM.

W(D-idae), includes world-wide museums which house mounted skeletons or isolated bones and teeth of members of the family Deinotheriidae: same as in 'W(De)', and IGPB.

W(De), includes world-wide museums which house mounted skeletons or isolated bones and teeth of *Deinotherium*: AMNH; Bohemian Museum (Prague, Czechoslovakia); Bucharest University Museum (Bucharest, Romania); BUGM; HL; Institut für Paläontologie und Historisches Geologie, Munich, Germany; LPVPH; MNCN; MNHN; NHM; NMK; SM; SM(St); UCB; VM; YPM.

W(E-idae), includes world-wide major museums which house mounted skeletons, skulls or isolated bones and teeth of members of the family Elephantidae (i.e. any species or subspecies classified in these genera: *Stegotetrabelodon*, *Stegodibelodon*, *Primelephas*, *Loxodonta*, *Elephas*, and *Mammuthus*): museums listed under China(E-idae); EA(Mm); EA(Mp); EA(Mt); EA(Mth.af.); Japan(En); NA(M&M); W(Ea); W(Em); W(La); W(Lad); W(Lat); W(Ms); W(Pri); W(Stb), and also MO; NMT; SMNH; TMP; UKPD; and YPM.

W(Ea), includes world-wide major museums which house mounted skeletons or isolated bones and teeth of *Elephas* (*Palaeoloxodon*) *antiquus*: AMNH; BNHM; DZHU; IGF; IVPP, MNHN; NHM.

W(Em), includes world-wide major museums which house mounted skeletons, bones, or teeth of *Elephas maximus maximus*, *E. m. indicus*, and *E. m. sumatranus*: AMNH; CNM; IM; MCZ; MNHN; MSUM; NHM; NMK; UNSM; WSUMNH (the last is not a 'major' museum but it includes a mounted skeleton and a preserved proboscis of 'Iki' (Shoshani *et al.* 1982; Riha *et al.* 1991). (Note: museums which hold material from other species of *Elephas* are given in column I, or may be given in chapter(s) listed in column H.)

W(Eo), includes world-wide museums which house bones or teeth of *Eozygodon*: AMNH; BNHM; LPVPH; MNHN; NHM; NMK.

W(Go), includes world-wide museums which house skeletons, bones, or teeth of *Gomphotherium*: many of the institutions given under 'W(gom)'; mounted skeletons may be found in these institutions: AMNH; DMNH; UNSM.

W(gom), includes world-wide museums which house skeletons, bones, or teeth of 'gomphotheres' (that is, all taxa belonging in *Choerolophodon*, Gomphotheriidae, *Tetralophodon*, *Anancus*, *Paratetralophodon* (separate listing may be given for certain entries): AMNH; ANSP; BNHM; BUGM; CGM; CMNH; CMNHN; DMNH; DZHU; FMNH; FSA; GMUW; GRDC; HL; IGF; IGP; IGPB; IM; ISEA; ISM; IVPP; KUJC; LACM; LPULM; LPVPH; MACN; MCN; MCZ; MGL; MINGM; MN; MNCN; MNH(K); MNHN; MNHT; MNHUK; MNLP; MP; MUCQ; NHB; NHM; NME; NMK; NMNH; NMNH(L); NMR; NMT; PMU; ROM; SBMNH; SDNHM; SM; SM(St); THP; UCB; UF; UMMP; UNSM; VM; WIHG; YPM; ZM.

W(La), includes world-wide major museums which house mounted skeletons, bones, or teeth of *Loxodonta a. africana* and/or *L. a. cyclotis*: AMNH; MCZ; MNHN; MRAC; MSUM; NHM; NMK; UNSM; YPM. (Note: museums which hold material from other species of *Loxodonta* are given in column I, or may be given in chapter(s) listed in column H.)

W(Lad), includes world-wide museums which house bones or teeth of *Loxodonta adaurora*: IPUB; MCZ; NHM; NMK.

W(Lat), includes world-wide museums which house bones or teeth of *Loxodonta atlantica*: MNHN; PU.

W(M-idae), includes world-wide museums which house bones or teeth of members of family Mammutidae, excluding *Mammut americanum*, which is treated under 'NA(M&M)' and 'W(Ma)': AMNH; BNHM; IVPP; LPVPH; MCZ; MNHN; NHM; NME; NMK; PMU; THP; YPM.

W(Ma), includes world-wide museums (except in North America; see under 'NA(M&M)') which house mounted skeletons [S], bones, or teeth of *Mammut*: HL (S of Peale's mastodon, *M. americanum*); MNHA (S of *M. americanum*); NHM (S of Koch's mastodon).

W(Mo), includes world-wide museums which house mounted skeletons [S], bones [B], or teeth [T] of *Moeritherium*: AMNH [B, T]; CGM [S, B, T]; LPVPH [B, T]; MNHN [B, T]; NHM [B, T]; SM[B, T]; UMMP [B, T]; UNSM [S, B, T]; YPM [S, B, T]. Note: a partial cranium, now housed at the Centre for Advanced Studies in Geology at Punjab University in Chandigarh, India, was considered by Bajpai *et al.* (1989) to have 'distinct moerithere-like affinities'. Examination of the cranium by M.C. McKenna, 1994, revealed that it appears to belong to the base of the Proboscidea and might be an anthracobune-like mammal.

W(Ms), includes Eurasian museums which house skeletal material of *Mammuthus subplanifrons*: BCUL; MMK; NHB; NMB; SAM; YPM.

W(P&P), includes world-wide museums which house bones or teeth of *Palaeomastodon* and *Phiomia*: AMNH; CGM; LPVPH; MNHN; NHB; NHM; SM; YPM.

W(Pa), includes world-wide museums which house bones or teeth of *Paratetralophodon*: AMNH; GSP; LPVPH; MNHN.

W(P1), includes world-wide museums which house mounted skeletons [S], bones, or teeth of *Platybelodon*: AMNH; BNHM [S]; IVPP [S]; LPVPH; MNHN; NHM; THP; UNSM; ZM.

W(Pri), includes world-wide museums which house bones or teeth of *Primelephas*: BCUL; MCZ; MNHN; NHM; NMK; YPM.

W(Pro), includes world-wide museums which house bones or teeth of *Protanancus*: AMNH; LPVPH; MNHN; NHM; NME; NMK.

W(Prob-b), includes world-wide museums which house casts of brains of proboscideans: AMNH; LACM; MCZ; NHM; ?NMNH.

W(Prob-f), includes world-wide museums which house footprints or casts of footprints of proboscideans: ACM; KUJC; MSHS; NMK; NMSM; UMMP.

W(Rh), includes world-wide museums which house bones or teeth of *Rhynchotherium*: AMNH; Brigham Young University (Provo, Utah, USA); Geneva Natural History Museum (Geneva, Switzerland); Instituto de Geologia, Universidad Nacional Autonoma de Mexico (Mexico City, Mexico); Las Vegas Museum of Natural History (Las Vegas, Nevada, USA); UCB; YPM.

W(S-idae), includes world-wide museums which house skeletal material of Stegodontidae (*Stegolophodon*, *Stegodon*): AMNH; BNHM; CCG; DGNU; GRDC; GSP; IGP; IM; IVPP; KUJC; LPVPH; MMMM; MNHN; MYP; NHM; NMNH(L); NSM; PCZ; THP; TMTMNH; YPM; see also under China(Sh) and Japan(Sa).

W(Se), includes world-wide museums which house bones or teeth of *Serbelodon*: AMNH; BNHM; IVPP; ?NMK; UF; UNSM.

W(St), includes world-wide museums which house bones or teeth of *Stegomastodon*: AMNH; ?MNHN; MNHUK; NHM; Panhandle Plains Historical Museum, West Texas State College (Canyon, Texas, USA); UCB; UF; UNSM; YPM; see also under W(C-inae).

W(Stb), includes world-wide museums which house bones or teeth of *Stegotetrabelodon*: BCUL; IGR; MCZ; MNHT; NMK; PMU; YPM.

W(Te), includes world-wide museums which house bones or teeth of *Tetralophodon*: AMNH; GRDC; IVPP; LPVPH; MNHN; NHM.

W(Zy), includes world-wide museums which house bones or teeth of *Zygolophodon* (including *Miomastodon*; see Appendix A): AMNH; BNHM; IVPP; LPVPH; MNHN; NHM.

Individual institutions

See also 'Grouped institutions' above.

ACM, Akashi Cultural Museum (Akashi City, Honshu, Japan).

AMNH, American Museum of Natural History (Frick Collection, Department of Vertebrate Paleontology and Department of Mammalogy, New York, New York, USA).

ANSP, Academy of Natural Sciences (Philadelphia, Pennsylvania, USA).

APM, Anhuei Provincial Museum (Hofei, China).

BCUL, Bedford College, University of London (London, England).

BM(NH), see under NHM.

BNHM, Beijing Natural History Museum (Beijing, China).

BUGM, Bristol University, Geology Museum (Bristol, England).

CCG, Chendu College of Geology (Chendu, Sichuan Province, China).

CGM, Cairo Geological Museum (Cairo, Egypt).

CIT, California Institute of Technology, Vertebrate Paleontology (Pasadena, California, USA).

CMNH, Carnegie Museum of Natural History (Pittsburgh, Pennsylvania, USA).

CMNHN, Chile Museum of Natural History (Museo Nacional de Historia Natural, Santiago, Chile).

CNM, Ceylon National Museum (Colombo, Sri Lanka).

DGNU, Department of Geology, Northwest University (Xian, Shaanxi Province, China).

DMNH, Denver Museum of Natural History (Denver, Colorado, USA).

DZHU, Department of Zoology, Hebrew University (Jerusalem, Israel).

FMNH, Field Museum of Natural History (Chicago, Illinois, USA).

FSA, Faculté des Sciences (Algiers, Algeria).

GMUW, Geological Museum, University of Wyoming (Laramie, Wyoming, USA).

GMZ, Municipal Museum of Zierikzee (Zierikzee, The Netherlands).

GPM, Gansu Provincial Museum (Lanzhou, Gansu Province, China).

GRDC, Geological Research and Development Centre (Bandung, Indonesia).

GSP, Geological Survey of Pakistan (Islamabad, Pakistan).

HL, Hessisches Landesmuseum (Darmstadt, Germany) (formerly 'Museum of Darmstadt', includes Peale's skeleton — Simpson and Tobien 1954; Ostrom 1983–84).

IGF, Geological and Paleontological Museum, University of Florence (Florence, Italy).

IGP, Institute of Geology and Paleontology (Tohoku University, Sendai, Japan).

IGPB, Istituto di Geologia e Paleontologia, University of Bologna (Bologna, Italy) (also known as Bologna Museum).

IGR, Istituto di Geologia e Paleontologia (Universita Degli Studi di Roma, Italy).

IM, Indian Museum (Calcutta, India).

IPUB, Institute of Paleontology, Humboldt University (Berlin, Germany).

ISEA, Institute of Systematics and Evolution of Animals (Kraków, Poland).

ISM, Illinois State Museum (Springfield, Illinois, USA).

IVPP, Institute of Vertebrate Paleontology and Paleoanthropology (Beijing, China).

KGM, Geological Museum of Kazan University (Kazan, Russia).

KNM, see under NMK.

KUJC, Kyoto University Japanese Cenozoic (Kyoto, Japan — includes casts of footprints of *Stegodon aurorae*; see Chapter 34).

LACM, Los Angeles County Museum of Natural History (Los Angeles, California, USA). (Note: LACM includes the George C. Page Museum of La Brea Discoveries, which houses material from the Rancho La Brea Tar Pits at which mammoths are represented only by the species *Mammuthus columbi* (Stock 1992; J. M. Harris, personal communication.))

LPULM, Laboratoire de Paléontologie, Université des Sciences et des Technique du Languedoc (Montpelier, France).

LPVPH, Laboratoire de Paléontologie des Vertébrés et de Paléontologie Humaine (Université P. et M. Curie, Paris, France).

MACN, Museo Argentino de Ciencias Naturales (Buenos Aires, Argentina).

MCN, Museo de Ciencias Naturales (Caracas, Venezuela).

MCZ, Museum of Comparative Zoology, Harvard University (Cambridge, Massachusetts, USA).

MEO, see under NSM.

MGL, Musée Guimet (Lyon, France).

MHNT, Muséum d'Histoire Naturelle de Toulouse (Toulouse, France).

MINGM, Museo Instituto Nacional Geologico Mexico (Mexico City, Mexico).

MMK, McGregor Memorial Museum (Kimberley, South Africa).

MMMM, Museum of Mae Moh Mine (E.G.A.T., Lampang, Thailand).

MN, Museu Nacional (Rio de Janeiro, Brazil).

MNCN, Museo Nacional de Ciencias Naturales (Madrid, Spain).

MNH(K), Museum of Natural History, Tribhuven University (Kathmandu, Nepal).

MNHA, Museum of Nature and Human Activities (Sanda City, Hyogo, Japan).

MNHN, Muséum National d'Histoire Naturelle (Paléontologie et Anatomie Comparée, Paris, France).

MNHT, Museum of Natural History (Tripoli, Libya).

MNHUK, Museum of Natural History, University of Kansas (Lawrence, Kansas, USA).

MNLP, Museo Nacional de La Plata (La Plata, Argentina).

MO, Musée d'Orléans (Orléans, France).

MP, Museum of Paleontology (Pisa, Italy).

MRAC, Musée Royale de l'Afrique Centrale (Tervuren, Belgium).

MSHS, Mammoth Site of Hot Springs (South Dakota, USA — includes footprints of *Mammuthus columbi*; see Chapter 34).

MSUM, Michigan State University Museum (East Lansing, Michigan, USA).

MUCQ, Museos de la Universidad Central del Ecuador (Quito, Ecuador).

MYP, Museum of Yunnan Province (Kunming, Yunnan Province, China).

NHB, Naturhistorisches Museum (Basel, Switzerland).

NHM, Natural History Museum (Zoology and Palaeontology; London, England) (formerly 'BM(NH)', British Museum (Natural History)).

NKM, Novosibirsk Museum of Regional Ethnography (Novosibirsk, Russia).

NMB, National Museum (Bloemfontein, South Africa).

NME, National Museum of Ethiopia (Addis Ababa, Ethiopia).

NMK, National Museum of Kenya (Nairobi, Kenya) (sometimes abbreviated as KNM (Kenya National Museum)).

NMM, Naturhistorisches Museum (Mainz, Germany).

NMNH, National Museum of Natural History, Smithsonian Institution (Washington DC, USA) (sometimes abbreviated as USNM (United States National Museum)).

NMNH(L), National Museum of Natural History (Leiden, The Netherlands) (formerly 'RMNL', Rijksmuseum van Natuurlijke Historie; this museum incorporates the 'Rijksmuseum van Geologie en Mineralogie').

NMR, Natuurmuseum Rotterdam (Rotterdam, The Netherlands).

NMSM, Nagaoka Municipal Science Museum (Nagaoka City, Niigata Prefecture, Japan — includes footprints of proboscideans, some of which were discovered by H. Matsumoto; see Chapter 34).

NMT, National Museum of Tanzania (Dar es Salaam, Tanzania)

NRS, Naturhistoriska Riksmuseet (Stockholm, Sweden); formerly Museum Adolfi Frederici Regis.

NSM, National Science Museum (Tokyo, Japan); formerly Museum of Education, Ochanomizu (Tokyo).

PCZ, Paleontological Collection in the Zoology Department, University of the Punjab (Lahore, Pakistan).

PIN, Paleontological Institute, Russian Academy of Sciences (Moscow, Russia).

PMU, Paleontological Museum (Upsala, Sweden).

PU, Department of Geological and Geophysical Sciences, Princeton University (Princeton, New Jersey, USA).

RMNL, see under NMNH(L).

ROM, Royal Ontario Museum (Toronto, Ontario, Canada).

SAM, South African Museum (Cape Town, South Africa).

SBMNH, Santa Barbara Museum of Natural History (Santa Barbara, California, USA).

SDNHM, San Diego Natural History Museum (San Diego, California, USA).

SM, Senckenberg Museum (Frankfurt, Germany) (also known as 'Frankfurt Museum' (Osborn 1936, p. 180)).

SM(St), Stuttgart Museum (Stuttgart, Germany).

SMNH, Swedish Museum of Natural History (Stockholm, Sweden).

THP, Tianjin Natural History Museum at Paiho (Tianjin, China) (Museum of Huangho and Paiho, China).

TMM, Texas Memorial Museum (Austin, Texas, USA).

TMP, Transvaal Museum (Pretoria, South Africa).

TMTMNH, Tokyo Metropolitan Takao Museum of Natural History (Hachioji, Japan; partial skeleton of *Stegodon zdanskyi*).

UCB, Paleontological Museum, University of California (Berkeley, California, USA).

UF, University of Florida (Gainesville, Florida, USA). (Note: the official name is 'Florida Museum of Natural History', but the abbreviation for the collection is 'UF'.)

UKPD, University of Kinshasa, Paleontology Department (Kinshasa, Zaire)

UMMP, University of Michigan, Museum of Paleontology (Ann Arbor, Michigan, USA).

UNSM, University of Nebraska State Museum (Lincoln, Nebraska, USA).

USNM, see under NMNH.

VM, Vienna Museum (Vienna, Austria).

WIHG, Wadia Institute of Himalayan Geology (Dehra Dun, Uttar Pradesh, India).

WM, Weimar Museum, Institute for Quaternary Paleontology (Weimar, Germany).

WSUMNH, Wayne State University, Museum of Natural History (Detroit, Michigan, USA — includes a preserved proboscis of *Elephas maximus*; see Chapter 2).

YPM, Yale Peabody Museum, Yale University (New Haven, Connecticut, USA).

ZM, Zoological Museum, Russian Academy of Sciences (St. Petersburg, Russia) (formerly 'ZIM', Zoological Institute Museum).

ADDENDUM TO APPENDIX C.1

The following is a list of proboscidean taxa, presented in the left column, as junior synonyms or unknown synonymies of names in the right column. In square brackets, after the name in the left column, is the source — the most recent if there is more than one — and page number where the name (the junior synonym) was encountered. When space permits, a second source may be included. For this addendum to Appendix C.1, we do not provide authors of junior synonyms. Junior synonyms of living species which are given in Laursen and Bekoff (1978) and Shoshani and Eisenberg (1982) are not repeated here. Synonymies are carried on the generic and species levels; rarely, subspecies designations are given. In a few cases (e.g. *Gomphotherium dialensis*), the author of the name is also the source — Sarwar 1977. In square brackets, after the name in the right column, is the source consulted for the senior synonym or notes. The vast majority of the sources in square brackets postdate Osborn's monograph (1936, 1942), *Proboscidea*. For this reason, we include this list in this appendix (Appendix C.1) and not in Appendix C.2. Entries to this list were 'discovered' in a number of publications, many of which are in Chinese, after the major portion of this appendix (the table and notes A.1 through Z.3) was completed, and thus, this list is separate from them. Ignoring these names may cause confusion among future researchers because of inconsistencies in the literature concerning their spellings and synonymies. The lists of taxa included in this and the accompanying appendix (Appendix C.2) are not exhaustive; they provide a basis for future investigators who wish to pursue this important matter.

Table of proboscidean junior synonyms and their senior synonyms and notes, in addition to those given in Appendix C.2. Note: 'sources' in square brackets in this table may not be the authors of these taxa but the references where they were encountered (see explanations in the preceding paragraph).

Species/taxon (= junior synonym) [source(s) and page number(s)]	**Senior synonym employed in this volume and notes [or source(s) consulted]**
Amebelodon grangeri [Tobien *et al.* 1986, p. 148]	*Platybelodon grangeri* [Tobien *et al.* 1986, p. 148]
Anancus osborni [Sarwar 1977, p. 118]	?*Paratetralophodon hasnotensis* [Tassy 1983, p. 272]
Anancus paisuyensis [Chow *et al.* 1979, p. 445]	*Anancus cuneatus/paisuyensis* [this volume]
Archidiskodon aktjubiensis [Jilkibayev 1975, p. 47] Note: Other Jilkibayev Elephantidae taxa whose synonymies are uncertain have been referred to elsewhere in this volume.	*Mammuthus* sp. or *M. trogontherii* [this volume]
Archidiskodon meridionalis [Chow *et al.* 1979, p. 454]	*Mammuthus meridionalis* [this volume]
Archidiskodon meridionalis kazachstanensis [Jilkibayev 1975, p. 33]	*Mammuthus meridionalis* [this volume]
Archidiskodon meridionalis meridionalis [Dubrovo 1994, p. 445]	*Mammuthus meridionalis* [this volume, Chapter 19; see note Mth2 above] With few exceptions, subspecies of valid extinct taxa were avoided in this volume.
Archidiskodon meridionalis tamanensis [Dubrovo 1994, p. 445]	*Mammuthus meridionalis* [this volume, Chapter 19]
Archidiskodon meridionalis taribanensis [Dubrovo 1994, p. 445]	*Mammuthus meridionalis* [this volume, Appendix C.1]
Archidiskodon planifrons [Chow *et al.* 1979, p. 453]	*Elephas planifrons* [this volume]
Archidiskodon rumanus: *A.* ex gr. *rumanus* [Dubrovo 1994, pp. 434–5, 445]	*Mammuthus meridionalis* [this volume, Appendix C.1, Appendix C.2]
Archidiskodon weifangensis [Jin 1983, p. 256]	indeterminate [this volume]
Archidiskodon youheensis [Xue 1981, plt. 1]	?*Elephas* sp. or indeterminate [note St1 above]
Bunolophodon grandidens [Gaziry 1994*a*, p. 117]	*Gomphotherium steinheimense/G. angustidens austrogermanicum*
Bunolophodon subtapiroides [Lehmann and Krause, 1994, p. 164]	indeterminate or *Gomphotherium* sp.
Elephas indicus ceylanicus [Deraniyagala 1955, p. 41]	*Elephas maximus maximus* [this volume]
Elephas irshadi [Sarwar 1977, p. 156]	*Elephas* sp. or indeterminate [this volume]
Elephas jacksoni [Maglio 1973, p. 63]	*Mammuthus primigenius* [Maglio 1973, p. 63]; *M. columbi* or *M. c. jeffersonii* [this volume] see also Table 18.1 of Todd and Roth, this volume.
Elephas kiangnanensis [Huang 1979, p. 340]	?*Elephas* sp.
Elephas kustanajensis [Zhilkibaev 1982, p. 84] (also spelled Zhylkibaev or Jilkibayev)	*Mammuthus trogontherii* [A. O. Averianov, personal communication 1995 (AOA95)]
Elephas lyrodon [Dubrovo 1994, p. 436]	*Mammuthus meridionalis* [this volume, Appendix C.1]
Elephas maximus asurus [Deraniyagala 1958, p. 98]	*Elephas maximus indicus* [this volume]
Elephas maximus bengalensis [Deraniyagala 1958, p. 98]	*Elephas maximus indicus* [this volume]
Elephas maximus birmanicus [Deraniyagala 1958, p. 98]	*Elephas maximus indicus* [this volume]
Elephas maximus borneensis [Deraniyagala 1958, p. 98]	*Elephas maximus indicus* [this volume]

Elephas maximus dakhunensis [Deraniyagala 1958, p. 98]	*Elephas maximus indicus* [this volume]
Elephas maximus hirsutus [Deraniyagala 1958, p. 98]	*Elephas maximus indicus* [this volume]
Elephas maximus rubridens [Deraniyagala 1958, p. 98]	*Elephas maximus indicus* [this volume]
Elephas maximus sinhaleyus [Deraniyagala 1958, p. 98]	*Elephas maximus maximus* [this volume]
Elephas maximus sondaicus [Deraniyagala 1958, p. 98]	*Elephas maximus indicus* [this volume]
Elephas maximus vilaliya [Deraniyagala 1958, p. 98]	*Elephas maximus maximus* [this volume]
Elephas maximus zeylanicus [Deraniyagala 1955, p. 41]	*Elephas maximus maximus* [this volume]
Elephas planifrons groznensis [Scherstjukov 1954, p. 228]	*Mammuthus meridionalis* ['*Archidiskodon*' of AOA95]
Elephas primigenius jatzkovi [Golowko 1958, p. 319]	*Mammuthus primigenius* [AOA95]
Elephas youheensis [Xue 1981, p. 35]	*Elephas* sp. or indeterminate [note St1 above]
Elephas yuheensis [Zong, G., personal communication 1993]	*Elephas* sp. or indeterminate [note St1 above]
Eoelephas ekorensis [Garutt 1986, p. 23]	*Elephas ekorensis* [cf. Maglio 1973, p. 33]
Gomphotherium chinjiensis [Chen 1988, p. 275]	*Zygolophodon 'turicensis* group' [see note Z1 above]
Gomphotherium dialensis [Sarwar 1977, p. 66]	*Protanancus chinjiensis* [Tassy 1983, p. 116]
Gomphotherium elegans [Chow *et al.* 1979, p. 442]	*Choerolophodon* sp. [Tobien *et al.* 1986, p. 140]
Gomphotherium guangsiensis [Chow and Chang 1961, p. 252]	*Sinomastodon intermedius* [Tobien *et al.* 1986, p. 160]
Gomphotherium guangxiensis [Chow *et al.* 1979, p. 443; Zhou and Zhang 1983, p. 56]	*Sinomastodon intermedius* [Tobien *et al.* 1986, pp. 160–1; note Sin1 above]
Gomphotherium hannibali [Welcomme 1994, p. 137]	*Gomphotherium 'annectens* group' [see note G2 above]
Gomphotherium hopwoodi [Chow *et al.* 1979, p. 441]	*Platybelodon grangeri* [Tobien *et al.* 1986, p. 148]
Gomphotherium hypsodontum [Tobien *et al.* 1986, p. 148]	*Platybelodon grangeri* [Tobien *et al.* 1986, p. 148]
Gomphotherium luociensis [on display at the BNHM 1994] (BNHM, Beijing Natural History Museum)	indeterminate or *Stegotetrabelodon* sp. [Guan, personal communication 1994]
Gomphotherium lusitanicus [Savage and Russell 1983, p. 228]	*Gomphotherium angustidens* [Tassy 1977*b*, p. 1390]
Gomphotherium obscurum [Gazin and Collins 1950, p. 7]	*Gomphotherium productum* [Tobien 1973*a*, p. 232]
Gomphotherium olisiponensis [Savage and Russell 1983, p. 228]	*Gomphotherium angustidens* (Tassy 1977*b*, p. 1390]
Gomphotherium pygmaeum [Tchernov and Shoshani, this volume Chapter 21]	*Gomphotherium* sp. or *incertae sedis* [this volume]
Gomphotherium qinanensis [Chow and Chang 1961, p. 245]	*Stegotetrabelodon exoletus* [Tobien *et al.* 1988, p. 140; see, however, footnote *u* in Appendix C.2]
Gomphotherium quinanensis [Chow and Chang 1961, p. 254; Chow *et al.* 1979, p. 442]	*Stegotetrabelodon exoletus* [Tobien *et al.* 1988, p. 140; see, however, footnote *u* in Appendix C.2]
Gomphotherium serridenstoides [Zhou and Zhang 1974, p. 67]	*Sinomastodon intermedius* [Tobien *et al.* 1986, pp. 160–1; note Sin1 above]
Gomphotherium serridentoides [Chow *et al.* 1979, p. 443; Huang 1979, p. 330]	*Sinomastodon intermedius* [Tobien *et al.* 1986, pp. 160–1; note Sin1 above]
Gomphotherium shensiensis [Chen 1988, p. 267]	*Gomphotherium shensiense* [Tobien *et al.* 1988, p. 212]
Gomphotherium spectabilis [Zhai 1963, p. 60]	*Platybelodon grangeri* [Tobien *et al.* 1986, p. 148]
Gomphotherium subtapiroides [Savage and Russell 1983, p. 217]	*Gomphotherium* sp. or '*Gomphotherium subtapiroides*' [this volume]
Gomphotherium tologojensis [Savage and Russell 1983, p. 217]	*Gomphotherium* sp. or *Platybelodon* sp. [this volume] [the original binomen is *Serridentinus tologojensis*]
Gomphotherium watzeensis [Zhou and Zhang 1974, p. 67; Chow *et al.* 1979, p. 442]	*Tetralophodon* (*Paratetralophodon*) cf. *hasnotensis* [Tobien *et al.* 1988, p. 108]
Gomphotherium wufengensis [Zhou and Zhang 1974, p. 67; Chow *et al.* 1979, p. 443]	*Sinomastodon intermedius* [Tobien *et al.* 1986, pp. 160–1; note Sin1 above]
Gomphotherium xiaolongtanensis [Yun 1975, p. 232; Chow *et al.* 1979, p. 442]	*Tetralophodon xiaolongtanensis* [Tobien *et al.* 1988, p. 104]
Gomphotherium yangziensis [Zhou and Zhang 1974, p. 67; Chow *et al.* 1979, p. 443]	*Sinomastodon intermedius* [Tobien *et al.* 1986, pp. 160–1; note Sin1 above]
Gomphotherium (*Mastodon*) *pygmaeus* [Chen 1988, p. 267]	*Mastodon angustidens* [*nomen dubium*, this volume]
Gomphotherium (*Serridentinus*) *obscurum* [Gazin and Collins 1950, p. 7]	*Gomphotherium productum* [Tobien 1973*a*, p. 232]
Gomphotherium (*Trilophodon*) *elegans* [Zhai 1963, pp. 59–60]	*Choerolophodon* sp. [Tobien *et al.* 1986, p. 140]

Gomphotherium (Trilophodon) quinanensis [Chow and Chang 1961, p. 250]	*Stegotetrabelodon exoletus* [Tobien *et al.* 1988, p. 140; see, however, footnote *u* in Appendix C.2]
Leith-Adamsia siwalikiensis [Shoshani and Eisenberg 1982, p. 1]	*Elephas planifrons* [this volume]
Mammuthus aldanensis [Russanov 1968, p. 50]	*Mammuthus primigenius* [AOA95]
Mammuthus jacksoni [Madden 1981*b*]	*Mammuthus primigenius* [Maglio 1973, p. 63]; *M. columbi* or *M. c. jeffersonii* [this volume] see also Table 18.1 of Todd and Roth, this volume.
Mammuthus primigenius liupanshanensis [Zhou and Zhang 1974, p. 68; Chow *et al.* 1979, p. 455; Liu and Li 1984, p. 517]	*Mammuthus primigenius* [this volume]
Mammuthus sungari [Chow and Hsu 1959, p. 145; Zhou and Zhang 1974, p. 68; Liu and Li 1984, p. 517]	*?Mammuthus primigenius* or indeterminate [Saegusa, personal communication 1994]
Mammuthus suolensis [Russanov 1968, p. 81]	*Mammuthus primigenius* [AOA95]
Mammuthus trogontherii chosaricus [Dubrovo 1994, pp. 438, 445]	*Mammuthus trogontherii* [this volume, Chapter 19]
Mammuthus trogontherii trogontherii [Dubrovo 1994, pp. 438, 445]	*Mammuthus trogontherii* [this volume, Chapter 19]
Mammuthus (Archaeomammuthus) hayi [Agenbroad 1994, p. 160]	*Mammuthus hayi* [this volume, see note Mth5 above]
Mammuthus (Parelephas) columbi [Agenbroad 1994, p. 160] [Note: *Parelephas* is misspelled on p. 160]	*Mammuthus columbi* [this volume]
Mammuthus (Parelephas) exilis [Agenbroad 1994, p. 160]	*Mammuthus exilis* or *M. columbi exilis* [this volume]
Mammuthus (Parelephas) imperator [Agenbroad 1994, p. 160]	*Mammuthus ?imperator* or *M. c. imperator* [this volume]
Mammuthus (Parelephas) jacksoni [Agenbroad 1994, p. 160]	*Mammuthus primigenius* [Maglio 1973, p. 63]
Mammuthus (Parelephas) sungari [Chow *et al.* 1979, pp. 454–5]	*?Mammuthus primigenius* or indeterminate [Saegusa, personal communication 1994]
Mammuthus (Parelephas) trogontherii [Chow *et al.* 1979, p. 454]	*Mammuthus trogontherii* [this volume]
Mastodon angustidens forma subtapiroides [Chen 1988, p. 271; cf. Gaziry 1994*b*, p.107]	*Gomphotherium angustidens* [this volume; see also *Mastodon tapiroides* under *nomina dubia*]
Mastodon angustidens tapiroidea [Chen 1988, p. 276]	*Mastodon tapiroides* [*nomen dubium*, see above]
Mastodon angustidens tapiroides [Chen 1988, p. 271]	*Mastodon tapiroides* [*nomen dubium*, see above]
Mastodon angustidens turicensis [Chen 1988, p. 271]	indeterminate [this volume; see also *Mastodon tapiroides* under *Nomina dubia*]
Mastodon angustidens typica [Chen 1988, p. 276]	*Gomphotherium angustidens* [this volume; see also *Mastodon tapiroides* under *Nomina dubia*]
Mastodon borsoni [Chow and Chang 1961, p. 253]	*Mammut borsoni* [Tobien *et al.* 1988, p. 157]
Mastodon chapmani [Gazin and Collins 1950, p. 6]	indeterminate [Gazin and Collins 1950, p. 7, stated that *M. chapmani* Hays. '. .must be ignored, as there is a strong probability that this specimen is of foreign origin']
Mastodon giganteus [e.g. Gazin and Collins 1950, p. 2]	*Mammut americanum* [this volume]
Mastodon intermedius [Tobien *et al.* 1986, p. 160]	*Sinomastodon intermedius* [Tobien *et al.* 1986, p. 160]
Mastodon mirificus [Gazin and Collins 1950, p. 6]	*Stegomastodon mirificus* [this volume]
Mastodon obliquelophus [Mukha 1980, p. 20]	*Mammut borsoni* [this volume]
Mastodon perimensis var. *sinensis* [Koken 1885, p. 34; Osborn 1942, p. 1398]	*Sinomastodon intermedius* [Tobien *et al.* 1986, p. 160]
Mastodon pyrenaicus [Lartet 1859, p. 513; Osborn 1942, p. 1394]	'gomphotheres' *incertae sedis* [this volume]
Mastodon sublatidens [Zhou and Zhang 1983, p. 53]	*Tetralophodon longirostris* [this volume]
Miomastodon tongxinensis [Tobien *et al.* 1988, p. 147]	*Gomphotherium tongxinense* [see note Z1 above and Guan this volume, Chapter 13]
Palaeoloxodon antiquus [Dubrovo 1994, pp. 442, 445] [Note: *Palaeoloxodon* is misspelled on p. 445]	*Elephas (Palaeoloxodon) antiquus* [this volume, Appendix C.1]
Palaeoloxodon dzhergalanicus [Jilkibayev 1975, p. 103]	*Elephas (Palaeoloxodon)* sp. [this volume]
Palaeoloxodon germanicus [Dubrovo 1994, pp. 442, 445]	*Elephas* sp. or *E. (Palaeoloxodon) ?antiquus* [this volume, ApC.1, ApC.2]
Palaeoloxodon mengchengensis [on display at the BNHM 1994]	*Elephas (Palaeoloxodon)* sp. [Guan, personal communication 1994]
Palaeoloxodon meridionaloides [Dubrovo 1994, pp. 442, 445]	*Elephas* sp. or *E. (Palaeoloxodon) ?antiquus* [this volume, ApC.1]
Palaeoloxodon naumanni [Chow *et al.* 1979, p. 456]	*Elephas (Palaeoloxodon) naumanni* [this volume]

Palaeoloxodon namadicus; *P.* ex gr. *namadicus* [Dubrovo 1994, pp. 442, 445]	*Elephas namadicus* or *E.* (*P.*) *namadicus* [this volume, Appendix C. 2]
Palaeoloxodon pingliangensis [Zhang *et al.* 1983, p. 65]	*Elephas* sp. or indeterminate [see under *Elephas youheensis*]
Palaeoloxodon planifrons [Zhang *et al.* 1983, p. 64]	*Elephas* sp. or *E. planifrons* [this volume]
Palaeoloxodon turkmenicus [Dubrovo 1994, pp. 441–2, 445]	*Elephas* sp. or *E.* (*Palaeoloxodon*) *?antiquus* [this volume, Appendix C.1]
Palaeoloxodon youheensis [Zhang *et al.* 1983, pp. 64, 66–8]	*Elephas* sp. or indeterminate [note St1 above]
Paraloxodonta atlantica [Garutt 1986, p. 19]	*Loxodonta atlantica* [cf. Maglio 1973, p. 25]
Pentalophodon cuneatus [Chang 1964, p. 38]	*Anancus cuneatus/paisuyensis* [this volume]
Pentalophodon longdongensis or *P. loungdongensis* [Zong 1979, pp. 82–3]	*Anancus sinensis* [Tobien *et al.* 1988, p. 115]
Pentalophodon pingliangensis [Zong 1979, p. 81]	*Anancus sinensis* [Tobien *et al.* 1988, p. 115]
Pentalophodon qinyangensis [Zong 1979, p. 83]	*Anancus sinensis* [Tobien *et al.* 1988, p. 115]
Pentalophodon qingyangensis [Huang 1975, p. 235]	*Anancus sinensis* [Tobien *et al.* 1988, p. 115]
Pentalophodon simplex [Chow *et al.* 1979, p. 446; Zong 1979, p. 83]	*Anancus sinensis* [Tobien *et al.* 1988, p. 115]
Pentalophodon sinensis [Chang 1964, p. 38]	*Anancus sinensis* [Tobien *et al.* 1988, p. 114]
Pentalophodon yusheensis [Tobien *et al.* 1988, p. 115]	*Anancus sinensis* [Tobien *et al.* 1988, p. 115]
Pentalophodon yushensis [Zong 1979, p. 83]	*Anancus sinensis* [Tobien *et al.* 1988, p. 115]
Pentalophodon yüsheensis [Chow *et al.* 1979, p. 446]	*Anancus sinensis* [Tobien *et al.* 1988, p. 115]
Pentalophodon yüshensis [Zhou and Zhang 1974, p. 67]	*Anancus sinensis* [Tobien *et al.* 1988, p. 115]
Platybelodon danovi filholi [Ye and Jia 1986, p. 147]	*Archaeobelodon filholi* [this volume]
Primelephas gomphotherioides [Zhou and Zhang 1983, p. 56]	*Primelephas gomphotheroides* [Maglio 1970*b*, p. 10]
Protelephas planifrons [Dubrovo 1994, p. 426]	*Mammuthus meridionalis* [this volume, Appendix C.1]
Protelephas planifrons groznensis [Dubrovo 1994, p. 426]	*Mammuthus meridionalis* [this volume, Appendix C.1]
Protoloxodonta adaurora [Garutt 1986, p. 19]	*Loxodonta adaurora* [cf. Maglio 1973, p. 23]
Protostegolophodon pyrenaicus [Gaziry 1994*a*, p. 129]	'gomphotheres' *incertae sedis* [this volume]
Rhynchotherium huananensis [Yun 1975, p. 232; Chow *et al.* 1979, p. 446]	*Sinomastodon intermedius* [Tobien *et al.* 1986, pp. 160–1; note Sin1 above]
Rhynchotherium praecursor [Savage and Russell 1983, p. 362]	zygodont group [this volume]
Rulengchia hueiheensis [Zong 1992, p. 294]	*Stegolophodon hueiheensis* [Tobien *et al.* 1988, p. 212]
Selenolophodon spectabilis [Ye and Jia 1986, p. 150]	*Platybelodon grangeri* [Tobien *et al.* 1986, p. 148]
Serridentinus chinjiensis [Osborn 1942, p. 1413]	*Zygolophodon 'turicensis* group' [see note Z1 above]
Serridentinus floridanus [Gazin and Collins 1950, p. 10]	*Amebelodon floridanus* [this volume]
Serridentinus obliquidens [Gazin and Collins 1950, p. 7]	*Gomphotherium productum* [see note G3 above]
Serridentinus obscurus [Gazin and Collins 1950, p. 7]	*Gomphotherium productum* [Tobien 1973*a*, p. 232]
Sinomastodon yanyuannensis (or *S. yanyaunnensis*) [Zong 1987, pp. 137, 145]	*Sinomastodon hanjiangensis* or *Sinomastodon* sp. [this volume; note Sin1 above]
Sinomastodon (*Rhynchotherium*) *huananensis* [Yun 1975, p. 232]	*Sinomastodon intermedius* [Tobien *et al.* 1986, pp. 160–1; note Sin1 above]
Stegodon chiai [Chow *et al.* 1979, p. 452]	*Stegodon zdanskyi* [this volume]
Stegodon dhokawanensis [Sarwar 1977, p. 47]	*Stegodon* sp. or indeterminate [this volume]
Stegodon preorientalis [Chow *et al.* 1979, p. 452]	*?Stegodon orientalis* [this volume]
Stegodon primitium [Chow *et al.* 1979, p. 452; Zhou and Zhang 1983, p. 54]	*Stegolophodon* cf. *latidens* [Tobien *et al.* 1988, p. 197] or indeterminate [Saegusa, personal communication 1994]
Stegodon sardhokensis [Sarwar 1977, p. 50]	*Stegodon* sp. or indeterminate [this volume]
Stegodon yushensis [Saegusa, this volume, Chapter 17]	*Stegodon zdanskyi* [this volume]
Stegodon yuxiensis [Saegusa, this volume, Chapter 17]	*Stegodon zdanskyi* or indeterminate [this volume]
Stegodon? yüshiensis [Savage and Russell 1983, p. 284]	*Stegodon zdanskyi* [this volume]
Stegodon (*Sulcicephalus*) *szechuani* [Zhou and Zhang 1974, p. 67]	*?Stegodon* sp. [this volume]
Stegolophodon banguoensis [Chow *et al.* 1979, p. 450]	*Stegolophodon* cf. *latidens* [Tobien *et al.* 1988, p. 197]
Stegolophodon caementifer [Gaziry 1994*b*, p. 101]	*nomen nudum* [Gaziry 1994*a*, p. 129]
Stegolophodon cristatus [Zong 1992, p. 290]	*Stegolophodon* sp. [this volume]
Stegolophodon daretensis [Zong 1992, p. 290]	indeterminate [Tassy 1983, p. 107]
'*Stegolophodon sahabianus*' [Kalb *et al.*, this volume, Chapter 12B]	*Stegotetrabelodon syrticus* [Maglio 1970*b*, p. 6]

Stegoloxodon indonesicus [Shoshani and Eisenberg 1982, p. 1]	*Elephas celebensis* [this volume]
Stegotetrabelodon gigantorostris [Savage and Russell 1983, p. 263]	*Tetralophodon longirostris gigantorostris* [this volume]
Stegotetrabelodon lehmanni [Lehmann and Krause, 1994, p. 166]	indeterminate [original after Gaziry in press, not seen]
Stegotetrabelodon primitium [Zhou and Zhang 1983, p. 55]	indeterminate [see under *Stegodon primitium* and Tobien *et al.* 1988, p. 198]
Synconolophus fukienensis [Zhou and Zhang 1974, p. 67; Chow *et al.* 1979, p. 446]	gomphothere *incertae sedis* [Tobien *et al.* 1986, p. 146 noted that 'For the time being it (*S. fukienensis*) might be classified as a gomphotheriid inc. sedis,']
Synconolophus obsorni [Sarwar 1977, p. 83]	*Choerolophodon corrugatus* [Tassy 1983, p. 236]
Synconolophus serridentinoides [Savage and Russell 1983, p. 263]	*Choerolophodon pentelici* [this volume]
Tetralophodon buzdarensis [Sarwar 1977, p. 105]	cf. *Anancus perimensis* [Tassy 1983, p. 273]
Tetralophodon curvirostris [Bergounioux and Crouzel 1960, pl. 6]	*Tetralophodon longirostris* [Tobien 1978*b*, p. 162]
Tetralophodon exoletus [Hopwood 1935, pp. 35–40; Chow *et al.* 1979, p. 442; on display at the BNHM 1994]	*Stegotetrabelodon exoletus* [Tobien *et al.* 1988, p. 140; see also Tobien *et al.* 1988, p. 104, and footnote *u* in Appendix C.2]
Tetralophodon liuchengensis [Zhou and Zhang 1974, p. 67; Chow *et al.* 1979, p. 444; Li 1981, p. 76]	*Sinomastodon intermedius* [Tobien *et al.* 1986, pp. 160–1; note Sin1 above]
Tetralophodon sinensis [Chow *et al.* 1979, p. 444; Savage and Russell 1983, p. 284]	*Sinomastodon intermedius* [Tobien *et al.* 1986, p. 161]
Trilophodon elegans [Zhai 1963, pp. 59–60]	*Choerolophodon* sp. [Tobien *et al.* 1986, p. 140]
Trilophodon hopwoodi [Tobien *et al.* 1986, p. 148]	*Platybelodon grangeri* [Tobien *et al.* 1986, p. 148]
Trilophodon serridenstoides [Pei 1962, p. 213; Huang 1979, p. 331]	*Sinomastodon intermedius* [Tobien *et al.* 1986, p. 161]
Trilophodon wufengensis [Pei 1965, p. 213]	*Sinomastodon intermedius* [Tobien *et al.* 1986, p. 160]
Turicius pyrenaicus [Chow and Chang 1961, p. 254]	'gomphotheres' *incertae sedis* [this volume]
Zygolophodon borsoni [Chow *et al.* 1979, p. 448]	*Mammut borsoni* [Tobien *et al.* 1988, p. 156]
Zygolophodon chinjiensis [Yun 1975, pp. 232–3]	*Z. gobiensis* [Tobien *et al.* 1988, p. 146] under *Z.* '*turicensis* group' [this volume]
Zygolophodon intermedius [Chow *et al.* 1979, p. 448]	*Sinomastodon intermedius* [Tobien *et al.* 1986, p. 160]
Zygolophodon jiningensis [Chow *et al.* 1979, p. 448]	*Z. gobiensis* [Tobien *et al.* 1988, p. 146] under *Z.* '*turicensis* group' [this volume]
Zygolophodon morotoensis [Pickford and Tassy 1980, p. 242]	*Eozygodon morotoensis* [this volume]
Zygolophodon nemonguensis [Zhou and Zhang 1974, p. 67; Chow *et al.* 1979, p. 448]	*Zygolophodon gobiensis* [Tobien *et al.* 1988, p. 146] under *Zygolophodon* '*turicensis* group' [this volume]
Zygolophodon pyrenaicus [Savage and Russell 1983, p. 228]	'gomphotheres' *incertae sedis* [this volume]
Zygolophodon shansiensis [Zhou and Zhang 1974, p. 67; Chow *et al.* 1979, p. 448]	*Mammut shansiensis* [this volume] or *M. borsoni* [Tobien *et al.* 1988, p. 157]
Zygolophodon yunnanensis [on display at the BNHM 1994]	indeterminate or *Stegotetrabelodon* sp. [Guan, personal communication 1994]
Zygolophodon (Turicius) jiningensis [Zhou and Zhang 1974, p. 67]	*Z. gobiensis* [Tobien *et al.* 1988, p. 146] under *Z.* '*turicensis* group' [this volume]

APPENDIX C.2

Synonymies of species and subspecies listed chronologically by Osborn [1942, in right columns of pages 1382–420, or on pages 1523–43; pages in Osborn are given in parentheses ()]. Synonymies are for names and not necessarily for particular specimens; they are not intended to be final, but to stimulate further research. See below for notes and explanation of abbreviations.

Species and/or subspecies given by Osborn (1942) and page numbers	**Name or synonym employed in this volume, sources consulted, and notes**
Amebelodon fricki (1412) Δ	*Amebelodon fricki* {15; ApC}
Amebelodon sinclairi (1415) Δ	*Amebelodon fricki* {ApC}
Amebelodon (Trilophodon) hicksi (1407) Δ	*Amebelodon fricki* {ApC}
Amebelodon (Trilophodon) paladentatus (1407) Δ	*Amebelodon fricki* {ApC}
Anancus arvernensis (1386, 93) Δ	*Anancus arvernensis* {ApC}
Anancus arvernensis brevirostris (1392) Δ	*Anancus arvernensis* {To(78b):168}
Anancus arvernensis dissimilis (1394; NI 1534)	*Anancus arvernensis* {To(78b):168}
Anancus arvernensis progressor (1403) Δ	*Anancus arvernensis* {To(78b):168}
Anancus falconeri (1411) Δ	*Anancus arvernensis* {To(78b):168}
Anancus gigantarvernensis (1407) Δ	*Anancus arvernensis* {To(78b):168}
Anancus intermedius (1387) Δ	*Anancus arvernensis* {To(78b):168}
Anancus minutoarvernensis (1407) Δ	*Anancus arvernensis* {To(78b):168}
Anancus perimensis (1392) Δ	*Anancus perimensis* {ApC}[a]
Anancus properimensis (1419) Δ	*Protanancus chinjiensis* {T(83):116}
Anancus sinensis (1418) Δ	*Anancus sinensis* {ApC}
Archidiskodon broomi (1412) Δ	*Elephas iolensis* {M(73):37, 39}
Archidiskodon exilis (1412) Δ	*Mammuthus exilis* {18; 24; ApA; ApC}
Archidiskodon haroldcooki (1412) Δ	*Mammuthus meridionalis* {M(73):63}
Archidiskodon hayi (1405) Δ	*Mammuthus meridionalis* {M(73):63}
Archidiskodon imperator (1394, 1407) Δ	*Mammuthus columbi* or *M. c. imperator* {ApA; ApC}
Archidiskodon imperator falconeri (1407) Δ	*Mammuthus ?imperator* {ApA; ApC}
Archidiskodon imperator maibeni (1410) Δ	*Mammuthus columbi* or *M. c. imperator* {16; ApC}
Archidiskodon imperator scotti (1410) Δ	*Mammuthus columbi* or *M. c. imperator* {ApA; ApC}
Archidiskodon imperator silvestris (1407) Δ	*Mammuthus columbi* or *M. c. imperator* {ApA; ApC}
Archidiskodon loxodontoides (1414) Δ	*nomen dubium* {ApC}
Archidiskodon maibeni (1410)	*Mammuthus columbi* or *M. c. imperator* {16; ApC}
Archidiskodon meridionalis (1386, 94, 99) Δ	*Mammuthus meridionalis* {M(73):53, 63}
Archidiskodon meridionalis cromerensis (1407) Δ	*Mammuthus meridionalis* {19; cf. M(73):53}
Archidiskodon meridionalis nebrascensis (1416) Δ	*Mammuthus meridionalis* {M(73):63}
Archidiskodon milletti (1414) Δ	*nomen dubium* {ApC}
Archidiskodon paramammonteus (ND; 1420)	*Mammuthus* sp.[b]
Archidiskodon planifrons (1391, 1412) Δ	*Elephas planifrons* {L(93); cf. M(73):44}
Archidiskodon planifrons rumanus (1408) Δ	*Mammuthus meridionalis* {M(73):55}
Archidiskodon proplanifrons (1418) Δ	*Mammuthus subplanifrons* {M(73):51}

Archidiskodon sonoriensis (1414) Δ	*Mammuthus columbi* or *M. c. imperator* {M(73):63}
Archidiskodon subplanifrons (1412) Δ	*Mammuthus subplanifrons* {M(73):51}
Archidiskodon vanalpheni (1414) Δ	*nomen dubium* {ApC}
Archidiskodon yorki (1415) Δ	*Elephas iolensis* {M(73):37}
Aybelodon hondurensis (1416) Δ	*Gomphotherium* sp. or *G.* aff. *productum* {ApC.1; To(73a):238}
Blickotherium blicki (1416) Δ	*Rhynchotherium blicki* {ApA; ApC}
Blickotherium euhypodon (1398) Δ	*Rhynchotherium* sp. or *Gomphotherium* sp. {ApA; ApC}
Bunolophodon yokotii (ND; 1420)	?*Gomphotherium* sp. {ApA}
C. [Caenobasileus] tremontigerus (1377, 1396)	?*Mammut americanum* {ApC}
Cordillerion andium (1384-86) Δ	*Cuvieronius hyodon* {S&P(57):133; note Cuvl above}
Cordillerion andium kraglievichii (1412; NI 1534)	?*Cuvieronius hyodon* {S&P(57):166}
Cordillerion bensonensis (1411) Δ	*Cuvieronius tropicus* or *C.* ?*oligobunis* {ApA; ApC}
Cordillerion bolivianus (1400) Δ	*Cuvieronius hyodon* {S&P(57):166}
Cordillerion (?)*bolivianus* (1415)	*Cuvieronius* ?*hyodon* {ApC}
Cordillerion defloccatus (1411) Δ	?*Cuvieronius tropicus* {ApA; ApC}
Cordillerion edensis (1419) Δ	*Rhynchotherium falconeri* {ApC, note R1}
Cordillerion gratum (1405) Δ	*Cuvieronius tropicus* {ApA; ApC}
Cordillerion oligobunis (1400) Δ	*Cuvieronius tropicus* {ApA; ApC}
Cordillerion oligobunis antiquissimus (1407) Δ	?*Cuvieronius tropicus* {ApA; ApC}
Cordillerion(?) *oligobunis felicis* (1407) Δ	?*Cuvieronius tropicus* {ApA; ApC}
Cordillerion(?) *oligobunis intermedius* (1407) Δ	?*Cuvieronius tropicus* {ApA; ApC}
Cordillerion oligobunis progressus (1407) Δ	?*Cuvieronius tropicus* {ApA; ApC}
Cordillerion orarius (1411) Δ	*Cuvieronius tropicus* {ApA; ApC}
Cordillerion tarijensis (1401) Δ	*Cuvieronius hyodon* {S&P(57):166}
Cordillerion tropicus (1398) Δ	*Stegomastodon mirificus* {Sav(55):54}
Cryptomastodon martini (1417)	*Stegodon trigonocephalus* {ApA, note c}
Cuvieronius ayorae (1413) Δ	*Haplomastodon waringi* {S&P(57):133, 171–2}
Cuvieronius bonaerensis (=*superbus*) (1399) Δ	*Stegomastodon platensis* {S&P(57):176; ApC}
Cuvieronius brasiliensis (1390) Δ	*Haplomastodon waringi* {S&P(57):133, 171–2}
Cuvieronius chilensis (1400) Δ	*Cuvieronius hyodon* {S&P(57):166}
Cuvieronius humboldtii (1384-86, 93) Δ	*Cuvieronius hyodon* {note Cuvl above}
Cuvieronius (?)*maderianus* (1400) Δ	*Stegomastodon platensis* {S&P(57):176; ApC}
Cuvieronius pirayuiensis (1405) Δ	*Stegomastodon superbus* {S&P(57):176; ApC}
Cuvieronius platensis (1399) Δ	*Stegomastodon platensis* {S&P(57):176; ApC}
Cuvieronius postremus (1415) Δ	*Haplomastodon waringi* {S&P(57):133, 171–2}
Cuvieronius rectus (1399) Δ	*Stegomastodon platensis* {S&P(57):176; ApC}
Cuvieronius superbus (1399) Δ	*Stegomastodon superbus* {S&P(57):176; ApC}
D. [Dinotherium] bozasi (ND; 1417)	*Deinotherium bozasi* {Ha(93); cf. 8; ApC}
Deinotherium bavaricum (1387) Δ	*Prodeinotherium bavaricum* {Ha(78):320}
Deinotherium cuvieri (1387) Δ	*Prodeinotherium bavaricum* {ApC, note D1}
Deinotherium giganteum (1386-87, 91) Δ	*Deinotherium giganteum* {8}
Deinotherium gigantissimum (1400) Δ	*Deinotherium giganteum* {Ha(78):324}
Deinotherium hobleyi (1403) Δ	*Prodeinotherium hobleyi* {Ha(93); cf. 8; ApC}
Deinotherium hopwoodi (1419) Δ	*Deinotherium bozasi* {Ha(93); Ha(78):323}
Deinotherium hungaricum (1415) Δ	*Deinotherium giganteum* {Ha(93); cf. 8; G(57)}[c]
Deinotherium indicum (1391, 95-96) Δ	*Deinotherium indicum* {8}
Deinotherium indicum gajense (1402-03) Δ	*Deinotherium indicum* {Ha(93); cf. 8; S&T(57):20}
Deinotherium intermedium (1391) Δ	*Deinotherium giganteum* {W(83); G(57)}[c]
Deinotherium levius (1395; NI 1529)	*Deinotherium* cf. *giganteum* {ApC, note D1}
Deinotherium medium (1388; NI 1529)	?*Deinotherium giganteum* {W(83); G(57)}[c]
Deinotherium podolicum (1389) Δ	*Deinotherium giganteum* {W(83); G(57)}[c]
Deinotherium proavus (1389; NI 1529)	*Deinotherium giganteum* {W(83); G(57)}[c]
Deinotherium secundarium (1389; NI 1529)	*Deinotherium giganteum* {W(83); G(57)}[c]
Deinotherium sindiense (1397; NI 1529)	*Deinotherium indicum* {S&T(57):6, 8}
Deinotherium uralense (1389) Δ	*Deinotherium giganteum* {W(83); G(57)}[c]

Dibelodon edensis (1406)	*Rhynchotherium falconeri* {ApA; ApC, note R1}
Dinohyus petersoni (1403)	a suid
Dinotherium australe (1390)	*nomen dubium* {ApC; ?marsupial}
Dinotherium königii (1390)	*Deinotherium giganteum* {W(83); G(57)}[c]
Dinotherium minutum (1390)	?*Deinotherium giganteum* {W(83); G(57)}[c]
Dinotherium pentapotamicum (1395)	*Prodeinotherium pentapotamiae* {8; S&T(57):20}
Dinotherium styriacum (ND; 1404)	*Prodeinotherium bavaricum* or *Deinotherium giganteum* {8}
Diprotodon australis (1397)	*nomen dubium* {ApC; ?marsupial}
E. [Elephas] antiquus mut. *ruthenensis* (ND; 1420)	indeterminate or ?*Elephas (Palaeoloxodon) antiquus*[d]
E. [Elephas] antiquus var. *minor* (1398)	?*Elephas (Palaeoloxodon) antiquus* {cf. M(73):40, 115}[d]
E. [Elephas] (Eueleph.) antiquus (1394)	*Elephas (Palaeoloxodon) antiquus* {ApC}[d]
E. [Elephas] primigenius americanus (1399)	*Mammuthus primigenius* {M(73):60}
E. [Elephas] primigenius germanicus (1391)	*Mammuthus primigenius* {ApC}[e]
E. [Elephas] primigenius matsumotoi (ND; 1411)	*Mammuthus* sp.[b]
E. [Elephas] primigenius meridionalis (1391)	*Mammuthus primigenius* {ApC}[e]
Elephas affinis (1388)	?possibly, not a fossil {cf. Os(42):1388}
Elephas affinis (1393)	?*Mammuthus primigenius* {cf. Os(42):1393}
Elephas africanus priscus (1391)	*Elephas (Palaeoloxodon)* sp.[f]
Elephas antiquus var. *insularis* = *nomen nudum* (1403)	indeterminate {cf. Os(42):1182, 1187, 1257, 1269}
Elephas [Archidiskodon] meridionalis (1382)	*Mammuthus meridionalis*
Elephas homotaphrus (1387)	?*Mammuthus primigenius* {cf. Os(42):1123}
Elephas homotaphrus (1387)	*nomen nudum* {cf. Os(42):1136}
Elephas hysudricus primitivus (ND; 1404)	indeterminate {cf. Os(42):1319, 1403}
Elephas indianapolis (1396)	?*Mammuthus primigenius* {cf. Os(42):1070}[g]
Elephas indicus (1382–84) Δ	*Elephas maximus* {S&E(82):1}
Elephas indicus bengalensis (1391, 1405) Δ	*Elephas maximus indicus* {S&E(82):1}[e]
Elephas indicus ceylanicus (1391, 1402) Δ	*Elephas maximus maximus* {D(55):43}[e]
Elephas indicus heterodactylus (ND; 1389)	*Elephas maximus indicus* {S&E(82):1}
Elephas indicus hirsutus (1404) Δ	*Elephas maximus indicus* {S&E(82):1}
Elephas indicus isodactylus (ND; 1389)	*Elephas maximus indicus* {S&E(82):1}
Elephas indicus sumatranus (1393, 1401) Δ	*Elephas maximus sumatranus* {D(55):121; S&E(82):1}
Elephas indicus var. Dauntela (1392) Δ	*Elephas maximus indicus* {S&E(82):1}
Elephas indicus var. Mukna (1393) Δ	*Elephas maximus indicus* {S&E(82):1}
Elephas meridionalis antiquitatis (1400)	?*Elephas (Palaeoloxodon) antiquus* {Os(42):1404, 1414}[d]
Elephas minimus (1384, 92)	indeterminate
Elephas minutus (1395)	indeterminate
Elephas namadicus namadi (1409)	?*Elephas (Palaeoloxodon) naumanni* {K(91):14–15}[d]
Elephas (Palaeoloxodon) namadicus setoensis (ND; 1413)	?*Elephas (Palaeoloxodon) naumanni* {K(91):14–15}[d]
Elephas primigenius Blum var. *trogontherii* (1404)	*Mammuthus* ?*trogontherii* {cf. Os(42):1404, 1407}
Elephas primigenius Blumenbach var. *n. pachyganalis* (ND; 1412)	?*Mammuthus primigenius* {cf. Os(42):1412}
Elephas primigenius columbi (1399)	?*Mammuthus columbi* {cf. Os(42):1004, 14, 43, 1399}
Elephas (primigenius) leith-adamsi Pohlig var. *minor* (1399)	*Mammuthus primigenius* {19; cf. Os(42):1150, 1399}
Elephas (primigenius) trogontherii (1399)	*Mammuthus trogontherii* {19; ApC}
Elephas primordialis (1387)	*Mammuthus primigenius* {Os(42):1136}
Elephas primus (ND; 1403)	indeterminate
Elephas socotrus (ND; 1384)	*Elephas maximus (for E. gigas)* {cf. Os(36):794; D(55):41}
Eubelodon morrilli (1404) Δ	*Eubelodon morrilli* {ApC}
Gnathabelodon thorpei (1418) Δ	*Gnathabelodon thorpei* {ApC}
Gomphotherium conodon (1403)	a suid (*Dinohyus petersoni*)
Gomphotherium serpentirivale (1415)	?'*nomen nudum*' {Os(42):251, 1415}
Harpagmotherium canadense (1408)	*Mammut americanum* {ApC}
Harpagonotherium canadense (1408)	*Mammut americanum* {ApC; cf. Os(36): 165}
Hemimastodon crepusculi (1403)	*Hemimastodon crepusculi* {ApC}
Hesperoloxodon antiquus (1392, 94, 97, 1400, 16) Δ	*Elephas (Palaeoloxodon) antiquus* {ApA; M(73):31}[d]
Hesperoloxodon antiquus ausonius (1396) Δ	*Elephas (Palaeoloxodon) antiquus* {ApC}[d, h]

Hesperoloxodon antiquus germanicus (1386, 1408) Δ	*Elephas (Palaeoloxodon) antiquus* {L(93); cf. D(55):26; M(73):31}[d]
Hesperoloxodon antiquus italicus (1416) Δ	*Elephas (Palaeoloxodon) antiquus* {L(93); cf. D(55):26; M(73):31}[d]
Hesperoloxodon antiquus nanus (1397) Δ	*Elephas (Palaeoloxodon) antiquus* {L(93); cf. D(55):26; M(73):31}[d]
Hesperoloxodon antiquus platyrhynchus (1400) Δ	*Elephas namadicus* {M(73):40}
Hypselephas hysudricus (1391, 1414) Δ	*Elephas hysudricus* {M(73):32; 48; ApC}
Loxodonta africana (1383, 85) Δ	*Loxodonta africana africana* {L&B(78):1; ApC}
Loxodonta africana albertensis (1402) Δ	*Loxodonta africana cyclotis* {L&B(78):1}
Loxodonta africana angolensis (1412) Δ	*Loxodonta africana africana* {D(55):29; L&B(78):1}
Loxodonta africana capensis (1383) Δ	*Loxodonta africana africana* {D(55):29; L&B(78):1}
Loxodonta africana cavendishi (1402) Δ	*Loxodonta africana africana* {D(55):31; L&B(78):1}
Loxodonta africana cottoni (1402) Δ	*Loxodonta africana cyclotis* {D(55):31; L&B(78):1}
Loxodonta africana fransseni (1404) Δ	*Loxodonta africana cyclotis* {D(55):31; L&B(78):1}
Loxodonta africana knochenhaueri (1401) Δ	*Loxodonta africana africana* {D(55):29; L&B(78):1}
Loxodonta africana mocambica (1409) Δ	*Loxodonta africana africana* {D(55):29; L&B(78):1}
Loxodonta africana orleansi (1402) Δ	*Loxodonta africana africana* {D(55):31; L&B(78):1}
Loxodonta africana oxyotis (1401-02) Δ	*Loxodonta africana africana* {D(55):31; L&B(78):1}
Loxodonta africana peeli (1402) Δ	*Loxodonta africana africana* {D(55):29; L&B(78):1}
Loxodonta africana pumilio (1402) Δ	*Loxodonta africana cyclotis* {L&B(78):1}
Loxodonta africana selousi (1402) Δ	*Loxodonta africana africana* {D(55):29; L&B(78):1}
Loxodonta africana toxotis (1402) Δ	*Loxodonta africana africana* {D(55):29; L&B(78):1}
Loxodonta africana var. *obliqua* (1415) Δ	*Loxodonta atlantica* {D(55):27; cf. M(73):25–7}
Loxodonta africana zukowskyi (1409)Δ	*Loxodonta africana africana* {D(55):29; L&B(78):1}
Loxodonta capensis cyclotis (1401) Δ	*Loxodonta africana cyclotis* {D(55):231; L&B(78):1}
Loxodonta cornaliae (1396) Δ	*Elephas (Palaeoloxodon)* sp.[f]
Loxodonta prima (1415) Δ	*Loxodonta africana africana* {M(73):29; L&B(78):1}
Loxodonta subantiqua (1416) Δ	*Elephas iolensis* {M(73):37}
Loxodonta zulu (1402) Δ	*Loxodonta atlantica* {D(55):27; M(73):25–7}
M. (Mastodon) parvus (1388)	*Gomphotherium angustidens* {ApA; ApC}
Mammonteus primigenius (1383-89, 91, 97) Δ	*Mammuthus primigenius* {ApC}[e]
Mammonteus primigenius alaskensis (1417) Δ	*Mammuthus primigenius* {ApA; ApC}
Mammonteus primigenius americanus (1390, 1410) Δ	*Mammuthus primigenius* {ApA; ApC}
Mammonteus primigenius astensis (1407) Δ	*Mammuthus ?trogontherii* {cf. Os(42):1407; ApA; ApC}
Mammonteus primigenius compressus (1409) Δ	*Mammuthus primigenius* {ApA; ApC}
Mammonteus primigenius fraasi (1403) Δ	*Mammuthus primigenius* {ApA; ApC}
Mammonteus primigenius hydruntinus (1399) Δ	*Mammuthus primigenius* {ApA; ApC}
Mammonteus(?) primigenius leith-adamsi Pohlig var. *minor* (1399) Δ	*Mammuthus primigenius* {ApA; ApC}
Mastodon acutidens (1419) Δ	*Mammut americanum* {ApA; ApC}
Mastodon americanus (1382-91, 95) Δ	*Mammut americanum* {ApA; ApC}
Mastodon americanus alaskensis (1417) Δ	*Mammut americanum* {ApA; ApC}
Mastodon americanus plicatus (1410) Δ	*Mammut americanum* {ApA; ApC}
Mastodon americanus rugosidens (1399) Δ	*Mammut americanum* {ApA; ApC}
Mastodon americanus rupertianus (1393) Δ	*Mammut americanum* {ApA; ApC}
Mastodon arvernensis Croiz. et Job. var. *conservativus* (1403)	*Anancus arvenensis* {ApC}
Mastodon atavus (ND; 1418)	*Zygolophodon 'turicensis* group' {ApC}[i]
Mastodon aureliense (ND; 1401)	indeterminate
Mastodon australis (1390)	*nomen dubium* {ApC; ?marsupial}
Mastodon grangeri (1417) Δ	indeterminate {?not *Platybelodon grangeri*}
Mastodon intermedius (ND; 1420)	*Sinomastodon intermedius* {To+(86):160}
Mastodon moodiei (1416) Δ	*Mammut ?americanum* {ApC}
Mastodon oregonensis (1410) Δ	*Mammut americanum* {ApC}
Mastodon pavlowi (1419) Δ	*Mammut borsoni* {ApC}
Mastodon progenius (1404) Δ	*Mammut americanum* {ApC}[l]
Mastodon raki (1417) Δ	*Mammut americanum* {L(87):32; ApC}
Mastodonte de Alangasi– not italicized (1412)	*Haplomastodon waringi* {S&P(57):133, 171–2}
Megabelodon cruziensis (1417) Δ	*Gomphotherium productum* {To(73a)}[k]

Megabelodon joraki (1417) Δ	*Gomphotherium productum* {To(73a)}[k]
Megabelodon lulli (1404) Δ	*Gomphotherium productum* {To(73a)}[k]
Megabelodon phippsi (1412) Δ	*Gomphotherium productum* {To(73a)}[k]
Metachidiskodon griqua (1406) Δ	*nomen dubium* {ApC}
Miomastodon depereti (1419) Δ	*Gomphotherium angustidens* {T(85):701; ApA; ApC}
Miomastodon merriami (1406) Δ	*Zygolophodon proavus* {ApC, note Z2}
Miomastodon matthewi (1406)	*Mammut matthewi* {ApA; ApC}
Miomastodon tapiroides americanus (1406) Δ	*?Zygolophodon turicensis* {ApA; ApC}
Moeritherium ancestrale (1407) Δ	*Moeritherium lyonsi* {8; ApC}
Moeritherium andrewsi (1403) Δ	*Moeritherium trigodon* {8; ApC}
Moeritherium gracile (1401) Δ	*Moeritherium lyonsi* {8; ApC}
Moeritherium lyonsi (1401) Δ	*Moeritherium lyonsi* {8; ApC}
Moeritherium trigodon (1401) Δ	*Moeritherium trigodon* {8; ApC}
Moeritherium trigonodon (1402)	*Moeritherium trigodon* {8; ApC}
Morrillia barbouri (1406, 08) Δ	*incertae sedis* {ApA}
Notiomastodon argentinus (1399) Δ	*Cuvieronius hyodon* {S&P(57):166}
Notiomastodon ornatus (1414) Δ	*Notiomastodon ornatus* {ApC}
Ocalientinus emmonsi (1415) Δ	amebelodontid or brevirostrine {To(72):191}
Ocalientinus ojocaliensis (1417) Δ	*Gomphotherium productum* {To(73a)}[k]
Ocalientinus (Serridentinus) bifoliatus (1413) Δ	*incertae sedis* {To(72):191}
Ocalientinus (Serridentinus) florescens (1413) Δ	*Gomphotherium* sp. {To(72):191} or *G. florescens* {ApC.1T, *i.s.*}
Ocalientinus (Serridentinus) floridanus (1398) Δ	*Amebelodon floridanus* {Tu(75); ApC.1, note *Am1*}
Ocalientinus (Serridentinus) floridanus leidii (1411) Δ	*Amebelodon floridanus* {Tu(75); ApC.1, note *Am1*}
Ocalientinus (Serridentinus) obliquidens (1410) Δ	*incertae sedis* {To(72):191}
Ocalientinus (Serridentinus) republicanus (1410) Δ	*Gomphotherium* sp. {To(72):191}
Palaeoloxodon(?) andrewsi (1415) Δ	*Mammuthus subplanifrons* {M(73):51; cf. Os(42):993; 1278}
Palaeoloxodon aomoriensis (ND; 1419)	*Elephas (Palaeoloxodon) naumanni* {K(91):14-5}[d]
Palaeoloxodon archidiskodontoides (1416) Δ	*Elephas iolensis* {M(73):37}
Palaeoloxodon (?Archidiskodon) tokunagai mut. *junior* (1414) Δ	*?Elephas namadicus* or *E. (P.) naumanni* {K(91):12; M(73):40}
Palaeoloxodon atlanticus (1397-98) Δ	*Loxodonta atlantica* {M(73):25}
?Palaeoloxodon buski (1412) Δ	*Elephas (Palaeoloxodon) naumanni* {K(91):14-5}[d]
Palaeoloxodon creticus (1402) Δ	*Elephas (Palaeoloxodon) creticus* {22}[d]
Palaeoloxodon cypriotes (1401) Δ	*Elephas (Palaeoloxodon) cypriotes* {22}[d]
Palaeoloxodon darti (ND; 1420)	*Elephas iolensis* {M(73):37}[d]
Palaeoloxodon falconeri (1395) Δ	*?Mammuthus falconeri* {L(93); cf. 22; ApC}[l]
Palaeoloxodon hanekomi (1415) Δ	*Elephas iolensis* {M(73):37}
Palaeoloxodon hysudrindicus (1403) Δ	*Elephas hysudrindicus* {M(73):49}
Palaeoloxodon iolensis (1400) Δ	*Loxodonta atlantica* {cf. M(73):25; Os(42):1274}
Palaeoloxodon kuhni (1415) Δ	*Elephas iolensis* {M(73):37}
Palaeoloxodon lamarmorae (1398) Δ	*?Mammuthus lamarmorae* {L(93); cf. 22; ApC}
Palaeoloxodon melitensis (1395) Δ	*Elephas melitensis* {L(93); cf. 22; ApC}[f, d]
Palaeoloxodon mnaidriensis (1395) Δ	*Elephas (Palaeoloxodon) mnaidriensis* {22, ApC}[f, d]
Palaeoloxodon namadicus (1392) Δ	*Elephas namadicus* {M(73):40}
Palaeoloxodon namadicus namadi (1542) Δ	*?Elephas (Palaeoloxodon) naumanni* {this volume}[d]
Palaeoloxodon namadicus naumanni (1408) Δ	*?Elephas namadicus* or *E. (P.) naumanni* {K(91): 14–15; M(73):40}
Palaeoloxodon namadicus yabei (1414) Δ	*Elephas (Palaeoloxodon) naumanni* {K(91):14–15}[d]
Palaeoloxodon priscus var. *bosei* (ND; 1418)	*Elephas (Palaeoloxodon)* sp. {cf. Os(42):1118, 1624}[f, d]
Palaeoloxodon protomammonteus (1409, 11) Δ	*Elephas namadicus* {M(73):40} *or 'Mammuthus armeniacus'* {K(91):14–15}
Palaeoloxodon protomammonteus proximus (1411) Δ	*Mammuthus* sp.[b]
Palaeoloxodon recki (1405) Δ	*Elephas recki* {M(73):34–5; cf. Os(42):1275–6}
Palaeoloxodon sheppardi (1412) Δ	*Elephas iolensis* {M(73):37}
Palaeoloxodon tokunagai (1409) Δ	*Elephas namadicus* {M(73):40} or *E. (P.) naumanni* {K(91):14–15}
Palaeoloxodon transvaalensis (1412) Δ	*Elephas iolensis* {M(73):37}
Palaeoloxodon wilmani (1415) Δ	*Elephas iolensis* {M(73):37}

Palaeoloxodon yokohamanus (ND; 1417) Δ	*Elephas namadicus* or *E. (P.) naumanni* {K(91):152; M(73):40}
Palaeoloxodon yorki (1415) Δ	*Elephas iolensis* {M(73):37}
Palaeomastodon barroisi (1402)	*Palaeomastodon beadnelli* {C+(78):341}[m]
Palaeomastodon beadnelli (1401) Δ	*Palaeomastodon beadnelli* {8; ApC}
Palaeomastodon intermedius (1406) Δ	*Palaeomastodon beadnelli* {C+(78):341}
Palaeomastodon parvus (1401) Δ	*Palaeomastodon beadnelli* {C+(78):341}
Parastegodon akashiensis (ND; 1420)	*Stegodon aurorae* {17; cf. K(91):14–15}
Parastegodon infrequens (ND; 1420)	*Stegodon aurorae* {17; cf. K(91):14–15}
Parastegodon? kwantoensis (ND; 1418)	*Stegodon aurorae* {17; cf. K(91):14–15}
Parastegodon sugiyamai (ND; 1539)	*Stegodon aurorae* {this volume}
Parelephas armeniacus (1394) Δ	*Mammuthus armeniacus* {19; cf. M(73):57}[n]
Parelephas columbi (1394) Δ	*Mammuthus columbi* {M(73):57}
Parelephas columbi cayennensis (1414) Δ	indeterminate {cf. Os(42):1083; 1105}[o]
Parelephas columbi felicis (1407) Δ	?*Mammuthus columbi*
Parelephas(?) eellsi (1411) Δ	*Mammuthus columbi* {M(73):63}
Parelephas floridanus (1414) Δ	*Mammuthus columbi* {M(73):63; W&D(95)}
Parelephas intermedius (1395) Δ	*Mammuthus trogontherii* {L(93); cf. M(73):57}
Parelephas jacksoni (1389) Δ	*Mammuthus primigenius* {M(73):63}
Parelephas jeffersonii (1407) Δ	*Mammuthus columbi* or *M. c. jeffersonii* {ApA; ApC}
Parelephas(?) mississippiensis(?) (1396)	?*Mammuthus primigenius* {cf. Os(42):1070}[g]
Parelephas progressus (1409) Δ	*Mammuthus columbi* {M(73):63}
Parelephas protomammonteus (Matsumoto) *matsumotoi* (ND; 1416)	*Mammuthus* sp.[b]
Parelephas proximus uehataensis (ND; 1420)	*Mammuthus* sp.[b]
Parelephas roosevelti (1541) Δ	*Mammuthus columbi* {M(73):63}
Parelephas trogontherii (1398) Δ	*Mammuthus trogontherii* {19; ApC}[p]
Parelephas(?) trogontherii nestii (1400) Δ	*Mammuthus trogontherii* {L(93); cf. M(73):57}
Parelephas trogontherioides (1404) Δ	?*Elephas (Palaeoloxodon) antiquus* {Os(42):1400, 04}[d]
Parelephas washingtonii (1408) Δ	*Mammuthus columbi* {M(73):63}
Parelephas wüsti (1403); or *P. wüstii* {M(73):57} Δ	*Mammuthus trogontherii* {19; cf. M(73):57}
Pentalophodon cuneatus (ND; 1420)	*Anancus sinensis* or *A. cuneatus/paisuyensis* {To+(88):114; T(86):103; ApC}
Pentalophodon falconeri (1419) Δ	*Anancus sivalensis* {ApC}
Pentalophodon sivalensis (1389) Δ	*Anancus sivalensis* {T(86):103; ApC}
Phiomia minor (1401) Δ	*Phiomia serridens* {C+(78):341}[m]
Phiomia osborni (1406) Δ	*Phiomia serridens* {C+(78):341}
Phiomia pygmaeus ['*Mastodon angustidens*'] (1400) Δ	*nomen dubium* {ApC}
Phiomia serridens (1401) Δ	*Phiomia serridens* {8; ApC}
Phiomia wintoni (1402) Δ	*Phiomia serridens* {C+(78):341}[m]
Phiomia wintoni (cf. *serridens*) (1533) Δ	*Phiomia serridens* {this volume}[m]
Platybelodon danovi (1413) Δ	*Platybelodon danovi* {13; ApC}
Platybelodon grangeri (1413) Δ	*Platybelodon grangeri* {13; ApC}
Platelephas platycephalus (1414) Δ	*Elephas platycephalus* {M(73):47}
Pliomastodon americanus praetypica (1406) Δ	?*Mammut borsoni* or ?'*Mammut praetypicum*' {9; 34; ApC}[q]
Pliomastodon matthewi (1406) Δ	*Mammut matthewi* {ApA; ApC}
Pliomastodon nevadanus (ND; 1418)	*Mammut matthewi* {ApA; Md(77b):29; ApC}
Pilomastodon sellardsi (1415) Δ	*Mammut matthewi* {ApA; Md(77b):29; ApC}
Pliomastodon vexillarius (1415) Δ	*Mammut matthewi* {ApA; Md(77b):29; ApC}
Rhynchotherium anguirivale (1411) Δ	*Rhynchotherium* sp., bunodont or zygodont group {ApC; To(73a):242}
Rhynchotherium brevidens (1399) Δ	*Rhynchotherium* sp., bunodont or zygodont group {ApC; To(73a):242}
Rhynchotherium browni (1419) Δ	*Rhynchotherium falconeri* or *R.* sp. {ApC}
Rhynchotherium chinjiense (1414) Δ	*nomen vanum* {ApC}
Rhynchotherium falconeri (1408) Δ	*Rhynchotherium falconeri* {ApC}
?*Rhynchotherium francisi* (1410) Δ	*Rhynchotherium* sp., bunodont or zygodont group {ApC;

	To(73a):242}
Rhynchotherium rectidens (1408) Δ	*Rhynchotherium* sp., bunodont or zygodont group {ApC; To(73a):242}
Rhynchotherium shepardi (1396) Δ	*Rhynchotherium* sp., bunodont or zygodont group {ApC; To(73a):242}
Rhynchotherium shepardi edense (1406, 11, 13) Δ	*Rhynchotherium falconeri* or *R.* sp. {ApA; ApC}
Rhynchotherium(?) spenceri (1405) Δ	?*Choerolophodon spenceri* {ApA; ApC}
Rhynchotherium tlascalae (1405) Δ	*Rhynchotherium tlascalae* {ApA}
Serbelodon barbourensis (1417) Δ	*Serbelodon barbourensis* {6; 13; 15; ApC}
Serbelodon burnhami (1417) Δ	*Amebelodon* sp. {To(72):179–80}
Serbelodon(?) praecursor (1400) Δ	zygodont group {To(72):191}
Serridentinus anguirivalis (1410) Δ	zygodont group {To(72):191}
Serridentinus annectens (1409) Δ	*Gomphotherium annectens* {K(91):14-5; ApC.1, note G2}
Serridentinus barstonis (1417) Δ	*incertae sedis* {To(72):191}
Serridentinus brewsterensis (1410) Δ	amebelodontid or brevirostrine {To(72):191}
Serridentinus browni (1410) Δ	*Gomphotherium browni* {To(72):174; 191; T83:262}
Serridentinus chinjiensis (1413) Δ	*Zygolophodon 'turicensis* group' {ApC.1, note Z1; T(83):113}
Serridentinus filholi (1417) Δ	*Archaeobelodon filholi* {T(84):466; ApC}
Serridentinus gobiensis (1416) Δ	*Zygolophodon gobiensis* {To(72):176; ApC.1, note Z1}
Serridentinus guatemalensis (1411) Δ	?brevirostrine group {To(72):191}
Serridentinus hasnotensis (1413) Δ	*Paratetralophodon hasnotensis* {(T83):273–6; ApC}
Serridentinus lydekkeri (1401) Δ	zygodont group {To(72):191}
Serridentinus metachinjiensis (1413) Δ	*Zygolophodon 'turicensis* group' {ApC.1, note Z1; T(83):113}
Serridentinus mongoliensis (1408) Δ	*Gomphotherium* sp. {To(72):191} or *G. mongoliense* {ApC.1T, *i.s.*}
Serridentinus (Ocalientinus?) nebrascensis (1410) Δ	amebelodontid or brevirostrine {To(72):191}
Serridentinus proavus (1396) Δ	*Zygolophodon proavus* {M&S(85):1346}
Serridentinus prochinjiensis (1410) Δ	*nomen vanum* {ApC}
Serridentinus productus (1396) Δ	*Gomphotherium productum* {To(72):191}
Serridentinus progressus (1408) Δ	*Gomphotherium productum* {To(73a)}[k]
Serridentinus serridens (1398) Δ	zygodont group {To(72):191}
Serridentinus serridens cimarronis (1400) Δ	?*Gomphotherium productum* {To(73a)}[k]
Serridentinus subtapiroideus (1405; cf. p. 395) Δ	*Gomphotherium* sp., '*G. subtapiroides*', or *G. angustidens* {cf. To(72):191; Kowalski and Kubiak 1993, p. 278}
Serridentinus wimani (1418) Δ	*Gomphotherium wimani* {To(72):175}
Stegodon airâwana (1399, 1403) Δ	*Stegodon trigonocephalus* {17; ApC; H(55)}
Stegodon (Archidiskodon?) mindanensis (1399) Δ	*Stegodon mindanensis* {17; 23}
Stegodon aurorae (1405) Δ	*Stegodon aurorae* {17; ApC}
Stegodon bombifrons (1392) Δ	*Stegodon bombifrons* {17; ApC}
Stegodon bondolensis (1416; ND 1539)	*Stegodon trigonocephalus* {17; ApC; H(55)}
Stegodon elephantoides (1386) Δ	*Stegodon elephantoides* {17; ApC}
Stegodon elephantoides (=*cliftii*) (1392) Δ	*Stegodon elephantoides* {17; ApC}
Stegodon ganesa (1539) Δ	*Stegodon ganesa* {ApC}
Stegodon ganesa javanicus (1539) Δ	*Stegodon trigonocephalus* {this volume}
Stegodon insignis (1539) Δ	*Stegodon insignis* {ApC}
Stegodon insignis birmanicus (1414) Δ	?*Stegodon insignis* {D(55):24; H(55); 17; ApC}
Stegodon insignis-ganesa (1391–2) Δ	?*Stegodon insignis* {D(55):24; H(55); 17; ApC}[r]
Stegodon javanoganesa (1403)	*Stegodon trigonocephalus* {17; ApC; H(55)}
Stegodon licenti (ND; 1420)	*Stegodon licenti* {17; ApC}[s]
Stegodon officinalis (1418; ND 1539)	*Stegodon officinalis* {17; ApC}[s]
Stegodon orientalis (1395) Δ	*Stegodon orientalis* {17; ApC}
Stegodon orientalis grangeri (1414) Δ	*Stegodon orientalis* {17; ApC; C&H(53)}
Stegodon orientalis shôdoënsis (1409) Δ	*Stegodon orientalis* {ApC; K(91):14–15}
Stegodon pinjorensis (1414) Δ	*Stegodon pinjorensis* {17; ApC}
Stegodon shôdoënsis akashiensis (ND; 1420)	*Stegodon aurorae* {K(91):14–15}
Stegodon sinensis (1395) Δ	*Stegodon sinensis* {17; ApC}

[*Stegodon(?) sugiyamail*] (1418) Δ	*Stegodon aurorae* {K(91):14–15}
Stegodon trigonocephalus (1398) Δ	*Stegodon trigonocephalus* {17; ApC; H(55)}
Stegodon trigonocephalus praecursor (1417; ND 1539)	*Stegodon trigonocephalus* {17; ApC; H(55)}
Stegodon yüshensis (ND; 1418)	*Stegodon zdanskyi* {17; ApC}
Stegodon zdanskyi (1418; ND 1539)	*Stegodon zdanskyi* {17; ApC}
Stegolophodon cautleyi (1398) Δ	*Stegolophodon cautleyi* {17; ApC}
Stegolophodon cautleyi progressus (1414) Δ	?*Stegolophodon cautleyi* {17; ApC}
Stegolophodon latidens (1386) Δ	*Stegolophodon latidens* {17; ApC}
Stegolophodon lydekkeri (1419) Δ	*Stegolophodon lydekkeri* {17; ApC}
Stegolophodon nathotensis (1414) Δ	*nomen vanum* {ApC}
Stegolophodon stegodontoides (1404) Δ	*Stegolophodon stegodontoides* {17; ApC}
Stegolophodon sublatidens (1405) Δ	cf. *Tetralophodon longirostris* {ApC; see also Os(42):846}
Stegomastodon aftoniae (1409) Δ	?*Stegomastodon mirificus* {cf. Os(36):683 to Sav(55):72}
Stegomastodon arizonae (1410) Δ	*Stegomastodon mirificus* {Sav(55):66}
Stegomastodon chapmani (1388) Δ	indeterminate
Stegomastodon mirificus (1394, 1403) Δ	*Stegomastodon mirificus* {Sav(55):54}
Stegomastodon priestleyi (1415) Δ	?*Stegomastodon mirificus* {Os(36):684}
Stegomastodon primitivus (1419) Δ	*Stegomastodon mirificus* {Sav(55):66}
Stegomastodon successor (1400) Δ	*Stegomastodon mirificus* {Sav(55):54}
Stegomastodon texanus (1409) Δ	*Stegomastodon mirificus* {Sav(55):54}
Synconolophus corrugatus (1404) Δ	*Choerolophodon corrugatus* {T(83):236; ApC}
Synconolophus dhokpathanensis (1414) Δ	*Choerolophodon corrugatus* {T(83):236; ApC}
Synconolophus hasnoti (1404) Δ	*Choerolophodon corrugatus* {T(83):236; ApC}
Synconolophus propathanensis (1414) Δ	*Choerolophodon corrugatus* {T(83):236; ApC}
Synconolophus ptychodus (1414) Δ	*Protanancus chinjiensis* {T(83):116}
Tetrabelodon crepusculi (1402)	*Hemimastodon crepusculi* {cf. Pilgrim 1908, 1910, 1912:17}
Tetralophodon bumiajuensis (1416) Δ	?*Tetralophodon bumiajuensis* {23; ApC}[t]
Tetralophodon campester (1397) Δ	an ambelodontid {6}
Tetralophodon elegans (1405) Δ	?an ambelodontid {cf. Os(36):373}
Tetralophodon exoletus (1418) Δ	*Stegotetrabelodon exoletus* {To+(88):140}[u]
Tetralophodon fricki (1419) Δ	?an ambelodontid, ?*Amebelodon fricki* {cf. Os(36):375}
Tetralophodon gigantorostris (1406) Δ	aff. *Tetralophodon longirostris gigantorostris* {ApC}
Tetralophodon grandincisivus (1405) Δ	*incertae sedis* {ApC; see notes there}
Tetralophodon longirostris (1387–89) Δ	*Tetralphodon longirostris* {6; 9; 12A; 14}
Tetralophodon (Lydekkeria) falconeri (1396) Δ	*incertae sedis* {ApC; see notes under *incertae sedis*}
Tetralophodon (Lydekkeria) sinensis (1398) Δ	?'*Mastodon perimensis* var. *sinensis*' {Koken 1885} or ?*Sinomastodon intermedius* {To+(86):160}
Tetralophodon punjabiensis part. (1398) Δ	*Paratetralophodon hasnotensis* {T(83):273-276; ApC}[v]
Tetralophodon sp. (1400)	indeterminate
Torynobelodon barnumbrowni (1416) Δ	?*Platybelodon (Torynobelodon) barnumbrowni* {?13; ApC}
Torynobelodon loomisi (1413) Δ	*Platybelodon (Torynobelodon) loomisi* {6; 13; ApC}
Trilophodon abeli (1410) Δ	*Gomphotherium productum* {To(73a)}[k]
Trilophodon angustidens (1384-85, 93) Δ	*Gomphotherium angustidens*
Trilophodon angustidens cuvieri (1393) Δ	*nomen nudum* {ApA; ApC}
Trilophodon angustidens gaillardi (1413) Δ	*Gomphotherium angustidens* {ApA; ApC}
Trilophodon angustidens gaujaci (1393) Δ	*nomen nudum* {ApA; ApC}
Trilophodon angustidens libycus (1405) Δ	*Gomphotherium* sp. {ApA; ApC}
Trilophodon angustidens minutus (1384-86, 89) Δ	*Gomphotherium angustidens*
Trilophodon angustidens var. *austrogermanicus* (1403) Δ	?*Gomphotherium angustidens* {cf. Os(36):259–60}
Trilophodon chinjiensis (1404) Δ	*Protanancus chinjiensis* {T(83):116; ApC}
Trilophodon (Choerolophodon) pentelicus (1393) Δ	*Choerolophodon pentelici* {ApC; cf. Os(36):263; 266}
Trilophodon connexus (1418) Δ	*Gomphotherium connexum* {To+(86):127}
Trilophodon cooperi (1416) Δ	*Gomphotherium* '*annectens* group' {ApA; ApC}
Trilophodon dinotherioides (1403) Δ	?*Gomphotherium productum* {cf. Os(36):291, 293}
Trilophodon engelswiesensis (1406) Δ	?*Gomphotherium angustidens* {cf. Os(36):281}

Trilophodon esselbornensis (1407) Δ	*Tetralophodon longirostris* {cf. Os(36):281}
Trilophodon fricki (1412) Δ	*Gomphotherium productum* {To(73a)}[k]
Trilophodon (Genomastodon) osborni (1405) Δ	*Gomphotherium osborni* {ApC.1T}
Trilophodon (Genomastodon) willistoni (1404) Δ	*Gomphotherium productum* {To(73a)}[k]
Trilophodon giganteus (1406) Δ	*Gomphotherium productum* {To(73a)}[k]
Trilophodon hasnotensis (1419) Δ	*nomen vanum* {ApC}
Trilophodon inopinatus (1412) Δ	*Gomphotherium ?inopinatum* {ApC; cf. Os(36):278–9}
Trilophodon ligoniferus (1405) Δ	*?Gomphotherium productum* {cf. Os(36):298}
Trilophodon macrognathus (1404) Δ	*nomen dubium* {ApC}
Trilophodon obscurus (1395) Δ	*Gomphotherium productum* {To(73a)}[k]
Trilophodon palaeindicus (1398) Δ	*Choerolophodon palaeindicus* {T(83):259; ApC; cf. Os(36):268}[w]
Trilophodon pandionis part. (1394–95) Δ	*Choerolophodon corrugatus* {T(83):236; ApC}[x]
Trilophodon pojoaquensis (1411) Δ	*Gomphotherium productum* {To(73a)}[k]
Trilophodon pontileviensis (1405) Δ	*?Gomphotherium angustidens* {cf. Os(36):283}
Trilophodon sendaicus (1409) Δ	?*'Zygolophodon sendaicus'* {K(91):14–15}
Trilophodon simplicidens (1408) Δ	*Gomphotherium productum* {To(73a)}[k]
Trilophodon spectabilis (1418) Δ	*Platybelodon grangeri* {To+(86):148}
Trilophodon steinheimensis (1406) Δ	*Gomphotherium steinheimense* {T(85):702; cf. Os(36):282}
Trilophodon (Tatabelodon) gregorii (1417) Δ	*Gomphotherium productum* {To(73a)}[k]
Trilophodon (Tatabelodon) riograndensis (1417) Δ	*Gomphotherium productum* {To(73a)}[k]
Trilophodon (?Tetralophodon) brazosius (1408) Δ	?an amebelodontid {cf. Os(36):374}
Trobelodon taoensis (1417) Δ	*Gomphotherium productum* {To(73a)}[k]
Turicius atticus (1394) Δ	*Tetralophodon atticus* {6; ApA; ApC}
Turicius tapiroides (1384–86) Δ	*nomen dubium* {ApA; ApC}
Turicius tapiroides-minus (1389; NI 1530)	*nomen dubium* {ApA; ApC}
Turicius turicensis (1386) Δ	*Zygolophodon turicensis* {ApA; ApC}
Turicius turicensis simorrensis (1410) Δ	*Zygolophodon turicensis* {ApA; ApC}
Turicius virgatidens (1395) Δ	*?Zygolophodon turicensis* {ApA; ApC}
Turicius wahlheimensis (1407) Δ	*?Zygolophodon turicensis* {ApA; ApC; cf. Os(36):282}
Zygolophodon borsoni (1388) Δ	*Mammut borsoni* {ApC}
Zygolophodon borsoni affinis (1397; NI 1530)	*Mammut borsoni* {ApA; ApC}
Zygolophodon borsoni buffonis (1393; NI 1530)	*Mammut borsoni* {ApA; ApC}
Zygolophodon borsoni vellavus (1393; NI 1530)	*Mammut borsoni* {ApC}
Zygolophodon borsoni vialetii (1393; NI 1530)	*Mammut borsoni* {ApC}
Zygolophodon borsoni zaddachi (1397) Δ	*Mammut borsoni* {ApC}
Zygolophodon pyrenaicus (1394) Δ	*incertae sedis* {ApA; ApC}
Zygolophodon pyrenaicus aurelianensis (1410) Δ	*nomen dubium* {ApA; ApC}
Zygolophodon sp. (1401)	indeterminate

Left-hand column

- Δ designates taxa considered valid by Osborn (1936, 1942). According to Osborn (1942, p. 1527), a total of 352 valid species and subspecies of proboscideans were distributed among five superfamilies: five belonged to Moeritherioidea, 12 to Deinotherioidea, 205 to Mastodontoidea, 19 to Stegodontoidea, and 111 to Elephantoidea (our count totalled 206 instead of 205, and 18 instead of 19; other totals were the same). Contents of the five superfamilies are given on p. 1526 of Osborn 1942 (eight families and 21 subfamilies), and a list of the 44 genera considered valid by Osborn is given on p. 1527. A complete list of 552 species and subspecies which were described since the year 1754 and their synonymies according to Osborn (1936, 1942) appear on pp. 1382–420. Our total of 352 species is based on this list (pp. 1382–420), and on that given on pp. 1523–43. There are possible inconsistencies with regard to 'not determined' taxa and to total numbers when comparing the data on pp. 1382–420 and pp. 1523–43. Of the 352 species and subspecies, 'at least 22 are existing species, subspecies, or geographical varieties' (Osborn 1942, p. 1527); our count is a total of 29 species of elephants (22 loxodontine and 7 elephantine), some of which are extinct. On p. 1541 there are only six recognized *Loxodonta* taxa, and on p. 1543 there are five recognized *Elephas* taxa. In this volume only five living subspecies are recognized today; see notes E1 and L1 in Appendix C.1. Tables 1.1 and 34.1, and Fig. 34.1 provide additional information and comparison between Osborn's valid species and those recognized in this volume — 352 versus *c.* 162. Names which appear in this appendix are not included in the Taxa Index (unless they were mentioned elsewhere in the volume) because they are easily accessible in this table.

- Entries in the left column of this table are as in Osborn, except when the first letter of a species was an upper case letter, we changed it to lower case to be consistent with other entries (e.g, '*Dinotherium australe*' was used instead of '*Dinotherium Australe*'). Note: the names of taxa in Osborn (1942, pp. 1382–420) were taken from the right columns on the pages unless there was no entry, in which case we used the name from the left side.
- Whenever possible, we checked the original sources, or at least checked in Osborn for distribution maps, geological ages, and other possible synonymies.
- **ND**, 'Not determined by present author' (Osborn 1942), or 'indeterminate' (Osborn 1942).
- **NI**, Not included as a valid taxon on that page.

Right-hand column

- References consulted and the relevant chapter numbers in this volume are given in curly brackets; see below for explanation of the abbreviations of references used. Numbers following colons indicate the relevant page numbers of the preceding reference.
- ApA refers to Appendix A, this volume.
- ApC refers to Appendix C, this volume, and mainly refers to the tabulated portion of Appendix C.1 which precedes this section.
- ApC.1T and ApC.1T, *i.s.* (*incertae sedis*) refer to Appendix C.1 Table.

Abbreviations of references

C+(78) = Coppens *et al.* 1978
C&H(53) = Colbert and Hooijer 1953
D(55) = Deraniyagala 1955
G(57) = Graf 1957
H(55) = Hooijer 1955
Ha(78) = Harris 1978; Ha(93) = facsimile dated December 13, 1993, from J. Harris to J. Shoshani
K(91) = Kamei 1991
L(87) = Lucas 1987
L(93) = letter dated 19th December 1993 from A.M. Lister to J. Shoshani
L&B(78) = Laursen and Bekoff 1978
M(73) = Maglio 1973
Md(77b) = Madden 1977*b*
M&S(85) = Madden and Storer 1985
Os(36), Os(42) = Osborn 1936, 1942
S&E(82) = Shoshani and Eisenberg 1982
S&P(57) = Simpson and Paula Couto 1957
S&T(57) = Sahni and Tripathi 1957
Sav(55) = Savage 1955
T(83), T(84), T(85), T(86) = Tassy 1983, 1984, 1985, 1986
To(72), To(73a), To(78b) = Tobien 1972, 1973*a*, 1978*b*
To+(86), To+(88) = Tobien *et al.* 1986, 1988
Tu(75) = Turner 1975
W&D(95) = Webb and Dudley, 1995
W(83) = Weinsheimer 1883

Other notes

[a] *Anancus perimensis* was originally described by Falconer and Cautley (1846) and Lydekker (1884*b*) as *Mastodon perimensis*.

[b] According to Kamei (1991, pp. 14–15), this taxon should be named '*Mammuthus armeniacus proximus*'. Lister (this volume, Chapter 19; and letter of 19 December 1993 to J. Shoshani) noted, however, that European middle Pleistocene *Mammuthus* should be named *Mammuthus trogontherii* (those from Erzurum, Turkey, should be called *Mammuthus armeniacus*).

[c] As noted by Sahni and Tripathi 1957, pp. 1–2, and/or Harris 1978, pp. 320–3. Harris' summary (this volume, Chapter 8) of nomenclature of Deinotheriidae taxa is: *Prodeinotherium hobleyi* (in Africa), *P. bavaricum* (in Europe), *P. pentapotamiae* and *P. ?orlovii* (in Asia); *Deinotherium bozasi* (in Africa), *D. giganteum* (in Europe), and *D. indicum* (in Asia).

[d] The genus *Palaeoloxodon* Matsumoto, 1924*a* has been synonymized with *Elephas* Linnaeus 1758 (notes in Appendix A, and in Caloi *et al.* this volume, Chapter 22). According to K. Takahashi (personal communication 1993, to J. Shoshani; see also Takahashi and Sawamura 1985, Takahashi *et al.* 1991, Inuzuka 1977*a*, *b*, and Garutt 1981), *Palaeoloxodon* is distinct from *Elephas*, not only in dental

features, but also in postcranial characters. Until a detailed report is published justifying the validity of *Palaeoloxodon*, we will retain *Palaeoloxodon* as a subgenus of *Elephas* because it is very difficult to distinguish molars with 'palaeoloxodontine' characters of specimens from Africa which are *Loxodonta, Elephas*, or *Mammuthus* (e.g. *Palaeoloxodon iolensis* — a junior synonym of *Loxodonta atlantica*; *Palaeoloxodon darti* — a junior synonym of *Elephas iolensis*; and *Palaeoloxodon*(?) *andrewsi* — a junior synonym of *Mammuthus subplanifrons*).

[e] A footnote in Osborn (1942, p. 1391) reads: 'These perhaps may be regarded as geographic designations rather than as subspecies.'

[f] According to Maglio (1973, 1974*a*), *Loxodonta* is confined to the continent of Africa. Caloi *et al.* (this volume, Chapter 22) noted that there were three forms of elephantids on the islands of Sicily and Malta: *Elephas antiquus leonardii, E. mnaidriensis*, and *E. falconeri*. '*Loxodonta cornaliae*' of Osborn 1942, p. 1396, or '*Elephas Cornaliae*' of Aradas, 1870, is from Sicily (see Osborn 1942, pp. 1204–5). *Elephas africanus priscus*, a European form, which was also considered *Loxodonta* (Osborn 1942, p. 1391), may be '*Elephas priscus*' (Caloi *et al.* this volume, Chapter 22) or *Elephas (Palaeoloxodon) priscus* (see also *Palaeoloxodon priscus* var *bosei*). Colin Groves (personal communication 1993, to J. Shoshani) believes that the tooth of '*Loxodonta cornaliae*' depicted in Osborn (1942, p. 1205) belongs to a loxodontine, not elephantine form, possibly a tame Carthaginian elephant (cf. Deraniyagala 1955, pp. 27–9, 116–17, where he discussed the employment of '*Loxodonta africana pharaohensis*' and '*Elephas maximus asurus*' in the battle by the Carthaginians against the Romans).

[g] Foster (1873; meeting of August 1872) referred to a tooth as '*Elephas Indianapolis*', and in September 1872, he called the specimen '*Elephas Mississippiensis*' (Foster 1872). Osborn (1942, p. 1396) renamed it as '*Parelephas*(?) *mississippiensis(?)*' (see also Osborn 1936, p. 775; 1942, p. 1070).

[h] Details on *Elephas ausonius* are given by Osborn 1942, pp. 1232–3, as '*Hesperoloxodon antiquus ausonius*'; see also Caterini (1923). Synonymy with *Elephas (Palaeoloxodon) antiquus* is after Lister (personal communication 1994).

[i] Osborn 1942, p. 1479, noted: 'This species (*Trilophodon atavus*) is very close to '*T. angustidens*' (assuming that *Mastodon atavus* and *Trilophodon atavus* are the same taxon). In addition, P.A. Tleuberdina (personal communication 1993, to J. Shoshani) insisted that *M. atavus* of Borissiak (1936) is a gomphothere, not a zygodontid. P. Tassy (facsimile to J. Shoshani dated December 15, 1993) stated, however, 'I doubt very much that *M. atavus* is a gomphothere: lower tusks and molars are not those of gomphotheres as far as can be assumed from the published illustrations and descriptions (which both are quite good).' Tobien (1973*a*, p. 213) first noted that *M. atavus* was 'a primitive zygodont'.

[j] Hay 1914, p. 368, spelled it *Mammut progenium*, yet Calvin (1911, p. 213) used *Mastodon americanus* for the same specimen. Hay (1923, 1924, 1927) used *Mammut americanum* for the American mastodon. Thereafter many authors (e.g. Simpson, 1945; Skeels 1962) employed the name *Mammut americanum* (cf. Simpson 1945, pp. 246–7, for discussion on synonymy of *M. americanum*).

[k] After Tobien (1973*a*, pp. 223, 231–2, 236–7).

[l] There are two alternatives for the most recent ancestor of *Elephas falconeri*. The dental characters are more like those of *Elephas (Palaeoloxodon)* yet the skull and tusks resemble those of *Mammuthus*. In this appendix, we follow Lister (personal communication 1994) who believes that the ancestor was probably *Mammuthus*.

[m] Coppens *et al.* (1978, pp. 340–1) considered *Phiomia* a subgenus of *Palaeomastodon*; in this volume they are considered as two separate genera. Note, however, that some of the species names (in the senior synonymies, in the right column) may belong to either of the genera, e.g. *Phiomia minor* may belong with *Phiomia serridens* or with *Palaeomastodon beadnelli* (details in Coppens *et al.* 1978, pp. 340–1; cf. Depéret 1907*b*).

[n] Type locality of '*Parelephas armeniacus*' (from Osborn, 1936, p. 1394) is: 'Near Khanoos, Province of Erzurum, Armenia'. Therefore, following Lister (this volume, Chapter 19), *Mammuthus armeniacus* is the correct name; see also note *b* above.

[o] Osborn's (1942, p. 1414) determination for the presence of *Mammuthus* in South America is based on 'Three and a half ridge-plates of a third right superior molar, r.M^3, collected by Captain Perret in Cayenne (French Guiana), South America, and now preserved in the Muséum d'Historie Naturelle, Marseille, France, as No. 8449 (cast Amer. Mus 21933)'. According to S. David Webb (personal. communication 1993, to J. Shoshani), there are no authenticated records of *Mammuthus* in South America (*Parelephas* = *Mammuthus*; see Appendix A).

[p] See, however, '*Parelephas trogontherii*' given in Kamei (1991, pp. 14–15) as junior synonym for '*Mammuthus armeniacus proximus*' (cf. note *b* in this list).

[q] Description was given by Schlesinger (1919) and also by Kubiak (1989) who seemed to consider *Mammut praetypicum* a valid species.

[r] H. Saegusa (personal communication 1993 to J. Shoshani) noted that *Stegodon insignis-ganesa* includes two species: *S. insignis* and *S. ganesa* (see also comments in Osborn 1942, p. 874).

[s] Zong (1992, pp. 290, 294) recognized these 12 *Stegolophodon* species and subspecies: *Sl. latidens, Sl. cautleyi, Sl. stegodontoides, Sl. stegodontoides progress* (it should have been *'progressus'* — editors), *Sl. pseudolatidens, Sl. hueiheensis, Sl. miyokoae, Sl. cristatus, Sl. daratensis, Sl. bangouensis* [*sic*] — should be *banguoensis, Sl. licenti, Sl. officinalis*. According to Zong (1992, pp. 290, 294), *Stegodon licenti* is *Stegolophodon licenti*, and also *Stegodon zhaotongensis, S. baoshanensis*, and *S. officinalis* are junior synonyms of *Stegolophodon officinalis*. Most of the species recognized by Zong (1992) are included in the tabulated portion of Appendix C.1; a comparison and a

thorough review of the literature (e.g. those cited by Saegusa, this volume, Chapter 17; Osborn 1936, 1942; Buckland 1826) for possible synonymies are needed.

[t] van den Bergh *et al.* (this volume, Chapter 23) acknowledge the presence of *Teralophodon bumiajuensis* van der Maarel, 1932 from the island of Java. There is no conclusive evidence, however, that the species '*bumiajuensis*' belongs in *Tetralophodon*. H. Saegusa (personal communication 1993) noted that 'tetralophodons have never been in Java'.

[u] P. Tassy (personal communication to J. Shoshani, 1993) does not believe that *Tetralophodon exoletus* is a *Stegotetrabelodon*.

[v] The specimen cited in Osborn (1942, p. 1398) is not the holotype (lectotype) for '*Mastodon punjabiensis*'.

[w] Originally described in Lydekker (1884*a*, p. 19) as '*Mastodon (Trilophodon) angustidens*, Cuv. var. *palaeindicus*', now junior synonym of *Choerolophodon palaeindicus*.

[x] The holotype of *Trilophodon pandionis* is indeterminate.

• The following individuals helped us to update Appendix C.2: Eden Akavia, Alexander O. Averianov, Debra B. Colchamiro, Carole S. DeFord, Jian Guan, John M. Harris, R. Anton Hough, Weilong Huang, W. David Lambert, Greg Liggett, Yuqing Li, Adrian M. Lister, Qingwu Lu, Zhengfu Ma, Haruo Saegusa, William J. Sanders, Jingyun Sun, Keiichi Takahashi, Roscoe A. Thompson, Xiaoming Wang, S. David Webb, Michael S. Woodford, Hong Yang, Jianju Zhang, Xing Zhang, and Guanfu Zong.

APPENDIX C.3
Errata, corrigenda, and explanatia for Osborn's monograph *Proboscidea* (1936, 1942)

There are two errata in Osborn's monograph *Proboscidea* (1936, 1942), one on page 802 (with 12 entries), and the other following page 1675 (with 10 entries). Below are additional errata or possible errata items (but not philosophical notes or updates of information) found in Osborn's monograph as of the end of 1995 and their corrigenda or explanatia.

Page 51 fig. 23: the genus *Ancodon*, listed twice for the 'Lower Oligocene', was erroneously assigned, because it is a Eurasian taxon; *Bothriogenys* should have been used instead (P. Holroyd, personal communication 1995, see also p. 58 in this volume).

Page 396 caption to fig. 354: what is labelled as 'four lower grinding teeth' of '*Serridentinus mongoliensis*' should be 'four upper grinding teeth'. In addition, the 'r m_2' is a left tooth.

Page 767 right column: the year of publication for Burnett, G. T. ('1829.1') — should read '(1830)'.

Page 779 right column, and page 1625, right column: there is no reference for Hay (1928) as noted in Osborn (1942, pp. 1029, 1412 for the authorship of '*Elephas haroldcooki*') because credit to Hay as the author of this taxon is given in Hay and Cook (1928, p. 33).

Page 788 right column, and page 1409 (under *Hemimastodon annectens*): the year of publication for Matsumoto, H. ('1924.4') should be (1925) because Matsumoto's article was published in 1925, although dated December 20, 1924.

Page 788 right column, and page 1381 (under *Leith-Adamsia*): the year of publication for Matsumoto, H. ('1927.2') should be (1928) because Matsumoto's article was published in March 1928 (see note *k*, p. 351 in this volume).

Page 796 right column, under Rafinesque, C. S. ('1814.1'): in the title, 'della' — should read 'delle'.

Page 916 caption to fig. 796: it appears as though Osborn's (1942) 'first', 'second', and 'third' cheek teeth in mammals are not homologous with the loci for extant elephants as percieved today (see caption to Fig. 2.3, pp. 11–12 in this volume).

Page 1128 captions to fig. 995: two of the three captions were switched — the caption '*Mammonteus primigenius* Reconstruction' under the top figure should be under the bottom figure and '*Loxodonta africana*' should be under the top figure (caption of the middle figure is correct).

Page 1380 the year of publication for *Parastegodon* should be 1929, not 1934 (see note *s*, p. 351 in this volume.

Page 1527 5th and 6th lines from bottom: 'Stegodontoidea (19)' and 'Mastodontoidea (205)' — should read 'Stegodontoidea (18)' and 'Mastodontoidea (206)' — see also p. 382 in this volume.

Page 1604 8th line from top: 'autopsy' — should read 'necropsy'.

Page 1654 right column: *Mastotherium leptodon* — this taxon does not appear on page 1372 as indicated in the index entry, but it appears on p. 1373, and other pages mentioned.

APPENDIX D

Appendix D Geological time-scale and correlation chart among Land Mammal Ages[a]

Age (Ma)[b]	Era[c]	Period[d]	Epoch[e]	Standard Marine Ages[f]	Land Mammal Ages: North America[g]	South America[h]	MN[i]	Europe[j]	Africa[k]	Age (Ma)[b]
0.004	CENOZOIC	QUATERNARY	HOLOCENE							0.004
0.01			PLEISTOCENE			Lujanian (0.01)			Naivashan (0.01)	0.01
0.5				Emilian (0.6)	Rancholabrean (0.3)	Ensenadan (0.5)		Oldenburgian (0.3)		0.5
1.0				Sicilian (0.8)					Natronian (1.0)	1.0
1.5		QUATERNARY	PLEISTOCENE (1.6)					Biharian (?1.8)		1.5
2.0		NEOGENE	PLIOCENE	Calabrian (1.8)	Irvingtonian (1.8)		17			2.0
2.5						Uquian (2.5)			Turkanan (2.6)	2.5
3.0						Chapadmalalan (2.8)	16			3.0
3.5				Piacenzian (3.4)				Villafranchian (3.4)		3.5
4.0							15		Afarian (4.0)	4.0
4.5					Blancan (4.5)					4.5
5.0			PLIOCENE (5.3)	Zanclian (5.2)			14	Ruscinian (5.3)		5.0
5.5			late			Montehermosan (5.8)				5.5
6.0			MIOCENE							6.0
6.5							13			6.5
7.0				Messinian (7.2)					Kerian (7.0)	7.0
7.5										7.5
8.0					Hemphillian (8.2)		12			8.0
8.5						Huayquerian (8.5)	11	Turolian (8.5)		8.5
9.0										9.0
9.5		NEOGENE					10			9.5
10							9	Vallesian (10.5)	Sugutan (10)	10
11				Tortonian (11)			8			11
12					Clarendonian (11.7)	Chasicoan (12)	7			12
13			medial						Tugenian (13)	13
14			MIOCENE	Serravallian (14.8)						14
15						Friasian (15)	6		Tinderetian (15)	15
16				Langhian (16.4)	Barstovian (16.5)			Astaracian (16.4)		16
17							5			17
18						Santacrucian (18)	4			18
19						Colhuehuapian (19)				19
20					Hemingfordian (20)	hiatus hiatus	3	Orleanian (20)		20

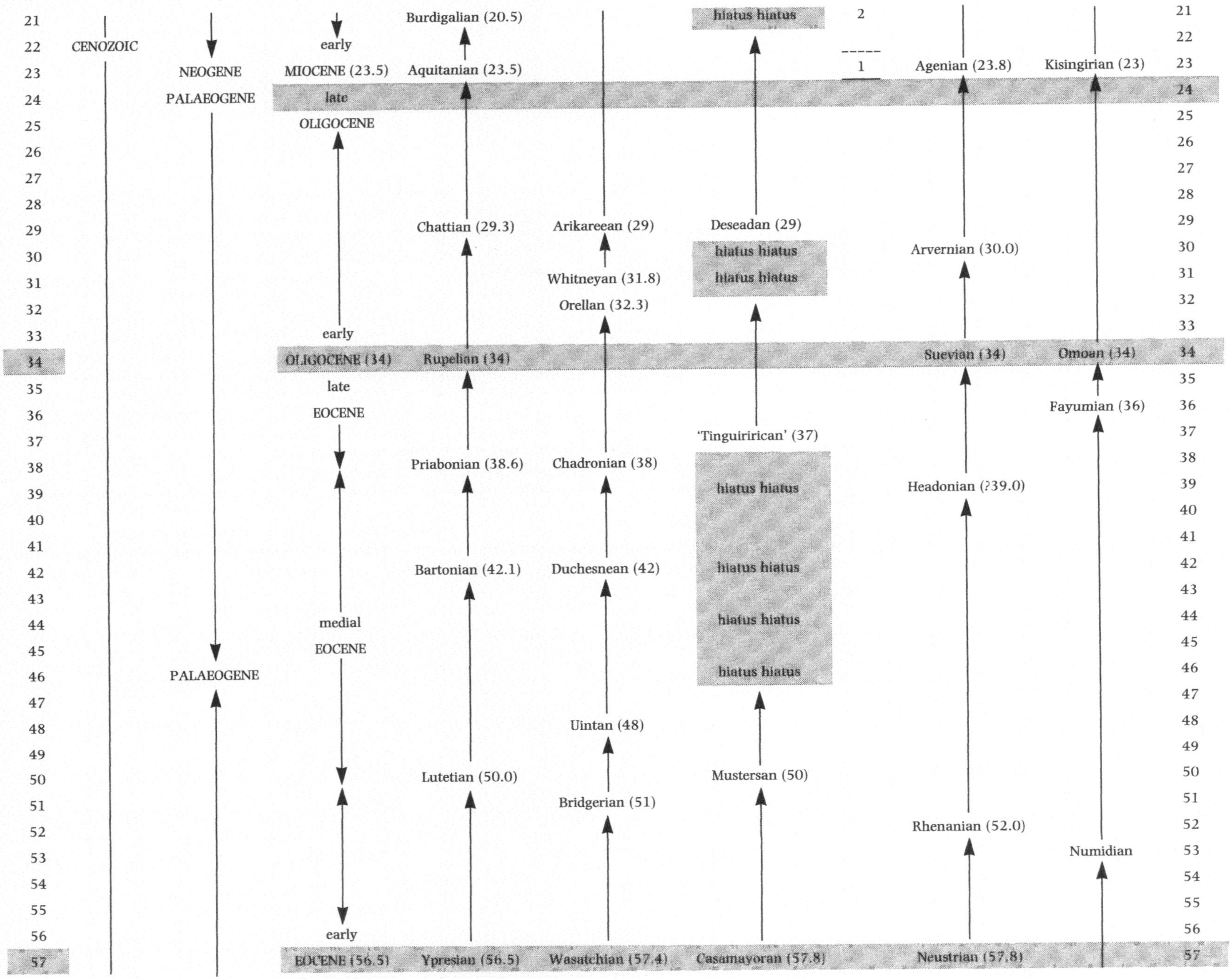
21
22
23
24
25
26
27
28
29
30
31
32
33
34
35
36
37
38
39
40
41
42
43
44
45
46
47
48
49
50
51
52
53
54
55
56
57
CENOZOIC
NEOGENE
PALAEOGENE
PALAEOGENE
early
MIOCENE (23.5)
late
OLIGOCENE
early
OLIGOCENE (34)
late
EOCENE
medial
EOCENE
early
EOCENE (56.5)
Burdigalian (20.5)
Aquitanian (23.5)
Chattian (29.3)
Rupelian (34)
Priabonian (38.6)
Bartonian (42.1)
Lutetian (50.0)
Ypresian (56.5)
Arikareean (29)
Whitneyan (31.8)
Orellan (32.3)
Chadronian (38)
Duchesnean (42)
Uintan (48)
Bridgerian (51)
Wasatchian (57.4)
hiatus hiatus
Deseadan (29)
hiatus hiatus
hiatus hiatus
'Tinguiririican' (37)
hiatus hiatus
hiatus hiatus
hiatus hiatus
hiatus hiatus
Mustersan (50)
Casamayoran (57.8)
2
1
Agenian (23.8)
Arvernian (30.0)
Suevian (34)
Headonian (?39.0)
Rhenanian (52.0)
Neustrian (57.8)
Kisingirian (23)
Omoan (34)
Fayumian (36)
Numidian

Appendix D *(cont.)*

Age (Ma)[b]	Era[c]	Period[d]	Epoch[e]	Standard Marine Ages[f]	Land Mammal Ages: North America[g]	South America[h]	MN[i]	Europe[j]	Africa[k]	Age (Ma)[b]
58			late							58
59			PALAEOCENE		Clarkforkian (58.8)					59
60				Thanetian (60.5)						60
61						Riochican (61)				61
62								Cernaysian (62.3)		62
63					Tiffanian (63)	Itaboraian (63.6)				63
64										64
65			early		Torrejonian (65)					65
65.5	CENOZOIC	PALAEOGENE	PALAEOCENE (65.5)	Danian (65.5)	Puercan (65.5)	Tiupampan (65.5)		Dano-Montian (65+)	Numidian (65)	65.5

[a] Special thanks to Susan K. Bell and Malcolm C. McKenna for helping us to update this geological time-scale (cf. McKenna, Bell, *et al.* 1994). John Van Couvering kindly provided an unpublished version of African Land Mammal Ages, and he and Eric Delson critically reviewed this appendex. None of these people are responsible, however, for any inaccuracies. Note that this chart is a composite based on many references, cited below. Numbers in parentheses refer to the beginning of an epoch or a Mammal Age in millions of years.

[b–e] Ages in million years before present (Ma, Mega annum) are after Harland *et al.* (1990), and Benton (1993).

[d] Some authors (e.g. Harland *et al.* 1990; Benton 1993) employ an additional division to accommodate the Quarternary (Pleistocene only) and Tertiary (Paleogene and Neogene). We follow Cowie *et al.* (1986, 1989) who eliminate the term Tertiary and include the Pleistocene and Holocene in the Quaternary.

[e] After Berggren *et al.* (1985*a,b*) and Berggren and Prothero (1992). Subdivisions of the Pleistocene epoch, including archaeological time-scales, can be found in Sherratt (1980) and Clutton-Brock (1981).

[f] Also known as Stages; after Berggren *et al.* (1985*a,b*), Berggren and Prothero (1992), Harland *et al.* (1990), Steininger *et al.* (1990), Tchernov (1992), and Woodburne and Swisher (1995).

[g] After Berggren *et al.* (1985*a,b*), Berggren and Prothero (1992), Prothero and Swisher (1992), Tedford *et al.* (1987), and Woodburne (1987); see, however, Woodburne and Swisher (1995) for differences in ages.

[h] After Marshall and Cifelli (1990), Marshall *et al.* (1984), Wyss *et al.* (1993), and de Muizon *et al.* (1984). Note: the South American Divisaderan Land Mammal Age was omitted 'for the sake of simplicity' following Wyss *et al.* (1993, p. 435).

[i] After Mein (1975, 1976, 1979, 1990) and input from R. L. Bernor (personal communication 1994), cf. Fahlbusch (1991). Some publications (e.g. Tattersall *et al.* 1988, p. xxxiv; Tchernov 1992, p. 157) employed divisions of 'a' and 'b' for MN2, MN3, MN4, MN12, and also used MN18 zone or higher.

[j] After Berggren *et al.* (1985*b*), Fahlbusch (1976), Savage and Russell (1983), Steininger *et al.* (1990), and Tchernov (1992).

[k] After John Van Couvering and Judith Harris (unpublished data 1994); see also Van Couvering and Harris (1991), Tattersall *et al.* (1988), Gagnon (1992), Gingerich (1992), and Rasmussen *et al.* (1992). The Eocene–Oligocene boundary date of 34 Ma is after Kappelman *et al.* (1992).

References

Where the authors' names have a prefix such as 'de' or 'van' these have been included in the letter-by-letter alphabetization, with the sole exception of 'de Blainville', which is listed under 'B'.

Numbers and abbreviations in { } indicate the sections, chapters, or appendices in which references are cited.

Citations used only for authorship may not appear in this bibliography (e.g. on p. 316, or some references in Appendix C, column H).

Key to abbreviations: plt, plate; tbl, table; fig, figure; AST, abbreviations, symbols and terms; Col, Colbert's contribution: ApA, ApB, ApC, ApD, Appendices A, B, C, and D.

Note that in 1983 the spelling of many Chinese names was changed to accommodate Pinyin spelling, for example 'Chow' and 'Chang' have been changed to 'Zhou' and 'Zhang', respectively. Some articles have appeared which use a mixture of spellings, for example Chow and Zhang (1978).

Abraczinskas, L. M. (1993). Pleistocene proboscidean sites in Michigan: new records and an update on published sites. *Michigan Academician*, **25** (4), 443–90. {34}

Acconci, L. (1881). Sopra alcune ossa fossili di elefante rinvenute nel Quaternario della zona di Marimenta in Sardegna. *Atti della Societa Toscana di Scienze Naturali, Processi verbali, Pisa*, **2**, 266–7. {22}

Adam, K. D. (1960). Das Mammut aus dem Grabental bei Münsingen (Kanton Bern). Ein überfordertes Leitfossil. *Eclogae Geologicae Helvetiae*, **53**, 481–91. {19}

Adam, K. D. (1961). Die Bedeutung der pleistozänen Säugetier-Faunen Mitteleuropas für die Geschichte des Eiszeitalters. *Stuttgarter Beiträge zur Naturkunde*, **78**, 1–34. {19}

Adam, K. D. (1988). Über Pleistozäne Elefanten-Funde im Umland von Erzurum in Ostanatolien. Ein Beiträg zur Namengebung von *Elephas armeniacus* und *Elephas trogontherii*. *Stuttgarter Beiträge zur Naturkunde B (Geologie und Paläontologie)*, **146**, 1–89. {19}

Adams, A. L. (1877). *Monograph on the British fossil elephants*, Vol. 1. The Palaeontographical Society of London. {19}

Adams, A. L. (1879). *Monograph on the British fossil elephants*, Vol. 2. The Palaeontographical Society of London. {19}

Adams, A. L. (1881). *Monograph on the British fossil elephants*, Vol. 3. The Palaeontographical Society of London. {12A, 19}

Adovasio, J. M., *et al.* (1980). Meadowcroft rockshelter — retrospect 1977: part 2. *North American Archaeologist*, **1**, 99–136. {30}

Agenbroad, L. D. (1984*a*). Hot Springs, South Dakota: entrapment and taphonomy of Columbian mammoth. In *Quaternary extinctions: a prehistoric revolution*, (ed. P. S. Martin and R. G. Klein), pp. 113–27. University of Arizona Press, Tucson. {28, 29, 30, 34}

Agenbroad, L. D. (1984*b*). New World mammoth distributions. In *Quaternary extinctions: a prehistoric revolution*, (ed. P. S. Martin and R. G. Klein), pp. 90–108. University of Arizona Press, Tucson. {18, 24, 28, 29, 34}

Agenbroad, L. D. (1988). The human factor in the Pleistocene megafauna extinction equation. In *Americans before Columbus: ice-age origins*, (ed. R. Carlisle). University of Pittsburgh Ethnology Monographs, No. 8, pp. 61–72. {28}

Agenbroad, L. D. (1990). The mammoth population of the Hot Springs Site and associated megafauna. In *Megafauna and man: discovery of America's heartland*, (ed. L. D. Agenbroad, J. I. Mead, and L. W. Nelson). The Mammoth Site of Hot Springs, South Dakota, Inc., Hot Springs, Scientific Papers, Vol. 1, pp. 32–9. {28}

Agenbroad, L. D. (1994). Taxonomy of North American *Mammuthus* and biometrics of the Hot Springs mammoths. In *The Hot Spring Mammoth Site: a decade of field and laboratory research in paleontology, geology, and paleoecology*, (ed. L. D. Agenbroad and J. I. Mead), pp. 158–207. The Mammoth Site of South Dakota, Hot Springs. {ApC}

Agenbroad, L. D. and Barton, B. R. (1991). *North American mammoths: an annotated bibliography 1940–1990*. The Mammoth Site of Hot Springs, South Dakota, Inc., Hot Springs, Scientific Papers, Vol. 2. {28}

Agenbroad, L. D. and Mead, J. I. (1987*a*). Age structure analysis of *Mammuthus columbi*, Hot Springs Mammoth Site, South Dakota. *Current Research in the Pleistocene*, **4**, 101–2. {25, 28}

Agenbroad, L. D. and Mead, J. I. (1987*b*). Late Pleistocene alluvium and megafauna dung deposits of the central Colorado Plateau. In *Geologic diversity of Arizona and its margins: excursions to choice areas*, (ed. G. Davis and E. Vanden Dolder). Arizona Bureau of Geology and Mineral Technology Special Paper, No. 5, pp. 68–84. {28, 30}

Agenbroad, L. D. and Mead, J. I. (1989). Quaternary geochronology and distribution of *Mammuthus* on the Colorado Plateau. *Geology*, **17**, 861–4. {28}

Agenbroad, L. D. and Mead, J. I. (ed.) (1994). *The Hot Springs Mammoth Site: a decade of field and laboratory research in paleontology, geology, and paleoecology*. The Mammoth Site of South Dakota, Hot Springs. {28}

Agenbroad, L. D., Mead, J. I., and Martin, P. S. (1984). Chronology and diet of *Mammuthus* from Eurasia and temperate North America. *Abstracts 27th International Geological Congress*, p. 337. Moscow. {28}

Agenbroad, L. D., Mead, J. I., Mead, E. M., and Elder, D. (1989). Archaeology, alluvium, and cave stratigraphy: the record from Bechan Cave, Utah. *Kiva*, **54**, 335–51. {28}

Agenbroad, L. D., Mead, J. I., and Nelson, L. W. (ed.) (1990). *Megafauna and man: discovery of America's heartland*. The Mammoth Site of Hot Springs, South Dakota, Inc., Hot Springs, Scientific Papers, Vol. 1. {28, 30}

Aguirre, E. E. (1968). Revisión sistemática de los Elephantidae por su morfología y morfometría dentaria. *Estudios Geológicos*, **24**, 109–67 [published as a monograph with Aguirre 1969*b*]. {19, 22}

Aguirre, E. E. (1969*a*). Evolutionary history of the elephant. *Science*, **164**, 1366–76. {12A, 18, 29, ApC}

Aguirre, E. E. (1969*b*). Revisión sistemática de los Elephantidae por su morfología y morfometría dentaria. *Estudios Geológicos*, **25**, 123–77, 317–67. {19}

Aguirre, E. E. (1974). Correlación continental-marino en el Neógeno mediterraneo. Datos actuales y problemas. *Estudios Geológicos*, **30**, 655–60. {20}

Aguirre, E. E., Robles, F., Thaler, L., Lopez, N., Alberdi, M. T., and Fuentes, C. (1973). Venta del Moro, nueva fauna finimiocena de moluscos y vertebradas. *Estudios Geológicos*, **29**, 569–78. {20}

Alberdi, M. T. (1971). Primer ejemplar completo de un *Tetralophodon longirostris* (Kaup) encontrado en España. *Estudios Geológicos*, **14** (2), 181–6. {14}

Alcover, J. A. and McMinn, M. (1994). Predators of vertebrates on islands. *Bioscience*, **44** (1), 12–8. {23}

Alexeeva, E. W. (1980). *Pleistocene mammals of the southeast of western Siberia (Carnivora, Proboscidea, ungulates)*. Nauka, Moscow [in Russian]. {26}

Alexeeva, L. I. and Garutt, V. E. (1965). New data on the evolution of the elephant genus *Archidiskodon*. *Byulleten Komissii po Izucheniyu Chetvertichnogo Perioda*, **30**, 161–6. [Author's name is sometimes spelt 'W. E. Garutt'] {19}

Alford, J. J. (1974). The geography of mastodon extinction. *Professional Geographer*, **26**, 425–9. {30}

Allen, H. (1991). Stegodonts and the dating of stone tool assemblages in island Southeast Asia. *Asian Perspectives*, **30** (2), 243–65. {23}

Ambrose, S. H. and DeNiro, M. J. (1989). Climate and habitat reconstruction using stable carbon and nitrogen isotope ratios of collagen in prehistoric herbivore teeth from Kenya. *Quaternary Research*, **31**, 407–22. {24}

Ambrosetti, P. (1964). L'*Elephas primigenius* di Tarquinia. *Estratto da Geologica Romana*, **3**, 367–82. {19}

Ambrosetti, P. (1968). The Pleistocene dwarf elephants of Spinagallo. *Estratto da Geologica Romana*, **7**, 277–398. {22}

Ambrosetti, P. (1972). L'elefante fossile della Sardegna. *Bollettino della Societa Geologica Italiana, Roma*, **91**, 127–31. {22}

An, Z. S. and Ho, C. K. (1989). New magnetostratigraphic dates of Lantian *Homo erectus*. *Quaternary Research*, **32**, 213–21. {17}

Anastasakis, G. C. and Dermitzakis, M. (1990). Post middle Miocene paleogeographic evolution of the central Aegean Sea and detailed Quaternary reconstruction of the region; its possible influence on the distribution of the Quaternary mammals of the Cyclades Islands. *Neues Jahrbuch für Geologie und Paläontologie Monatshefte*, **1990**, 1–16. {22}

Andau, P. and Payne, J. (1985). *Elephants in Sabah*. WWF Monthly Report, December 1985, pp. 301–5. World Wildlife Fund, Gland, Switzerland. {33}

Anderson, E. (1984). Who's who in the Pleistocene: a mammalian bestiary. In *Quaternary extinctions: a prehistoric revolution*, (ed. P. S. Martin and R. G. Klein), pp. 40–89. The University of Arizona Press, Tucson. {31}

Andrews, C. W. (1901*a*). Fossil vertebrates from Egypt. *The Zoologist*, **(4) V**, 318–19. {8, 16, ApB, ApC}

Andrews, C. W. (1901*b*). A new name for an ungulate. *Nature*, **64**, 577. {8, 16, ApA, ApB, ApC}

Andrews, C. W. (1901*c*). Preliminary note on some recently discovered extinct vertebrates from Egypt. (Part I). *The Geological Magazine*, Decade IV, New Series, **VIII**, 400–9. {Col, 8}

Andrews, C. W. (1901*d*). Über das Vorkommen von Proboscidiern in untertertiären Ablagerungen Aegyptens. *Tageblatt des V Internationalen Zoologischen-Kongresses*, Berlin, **6**, 4–5. {8, 16, ApA, ApB, ApC}

Andrews, C. W. (1902). Preliminary note on some recently discovered extinct vertebrates from Egypt. (Part III). *The Geological Magazine*, Decade IV, New Series, **IX**, 291–5. {8}

Andrews, C. W. (1903). On the evolution of the Proboscidea. *Philosophical Transactions of the Royal Society of London*, Series B, **196**, 99–118. {1, 8}

Andrews, C. W. (1904*a*). Further notes on the mammals of the Eocene of Egypt. Part I. *The Geological Magazine*, Decade V, New Series, **I**, 109–15. {8, ApC}

Andrews, C. W. (1904*b*). Note on Barypoda, a new order of ungulate mammals. *The Geological Magazine*, Decade V, New Series, **I**, 481–82. {8}

Andrews, C. W. (1906). *A descriptive catalogue of the Tertiary Vertebrata of the Fayûm, Egypt*. British Museum (Natural History), London. {1, 3, 4, 6, 8, 16, ApB, ApC}

Andrews, C. W. (1911). On a new species of *Dinotherium* (*Dinotherium hobleyi*) from British East Africa. *Proceedings of the Zoological Society of London*, **1911**, 943–5 + 1 plt. {8}

Andrews, C. W. and Beadnell, H. J. L. (1902). *A preliminary note on some new mammals from the upper Eocene of Egypt*, pp. 1–9. Survey Department, Public Works, Ministry, Cairo. {8, 12A, 16, ApB, ApC}

Andrews, C. W. and Cooper, C. F. (1928). *On a specimen of* Elephas antiquus *from Upnor, with further notes on the teeth and skeleton*. Clowes, London [printed by order of the British Museum (Natural History)]. {ApC}

Andrews, R. C. (compiler) (1932). *The new conquest of central Asia: a narrative of the explorations of the central Asiatic expeditions in Mongolia and China, 1921–1930*. American Museum of Natural History, New York. {27}

Anonymous (1827). A review of Frédéric Cuvier and Étienne Geoffroy Saint-Hilaire *Histoire Naturelle des Mammiféres*. *The Zoological Journal (London)*, **III**, 140–3. {12A, 16, ApA, ApB, ApC}

Anthony, R. (1933). Recherches sur les incisives supérieures des Elephantidae actuels et fossiles (Éléphants et Mastodontes): morphologie, structure, évolution ontogénique, interprétation. *Archives du Muséum National d'Histoire Naturelle*, 6th Series, **10**, 65–124 + 2 plts. {4}

Anthony, R. and Friant, M. (1941). Introduction à la connaissance de la dentition des Proboscidiens. *Mémoires de la Société Géologique et Minéralogique de Bretagne*, **6**, 1–104. {3, 4}

Antunes, M. T. (1990). The proboscideans data, age and paleogeography: evidence from the Miocene of Lisbon. In *European Neogene mammal chronology*, (ed. E. H. Lindsay, V. Fahlbusch, and P. Mein), pp. 253–62, Plenum, New York [copyright date, 1989]. {10}

Antunes, M. T. and Mazo, A. V. (1978). Les plus anciens mastodontes tétralophodontes (Langhien inférieur Vb), évolution et remarques sur la tétralophodontie. *Ciencias da Terra (Universidade Nova de Lisboa), Portugal*, **4**, 9–30. {14}

Antunes, M. T. and Mazo, A. V. (1983). Quelques mastodontes miocènes du Portugal. *Ciencias da Terra (Universidade Nova de Lisboa), Portugal*, **7**, 115–28. {14}

Antunes, M. T., *et al.* (1987). Ensayo de correlación entre el Neógeno de las Areas de Madrid y Lisboa (Cuencas Alta y Baja del Rio Tajo). *Comunicacoes dos Servicos Geológicos de Portugal, Lisboa*, **73** (132), 85–102. {14}

Aradas, A. (1870). Sopra un molare elefantino fossile riferibile a specie Distinta dalle Conosciute. *Atti della Accademia Gioenia Scienze Naturali Catania*, (3) **4**, 231–5. {ApC}

Arambourg, C. (1934). Le Dinotherium des gisements de l'Omo. *Comptes Rendus Sommaire de la Société Géologique de France*, **1934**, 86–87. {ApC}

Arambourg, C. (1942). L'*Elephas recki* Dietrich; sa position systématique et ses affinités. *Bulletin de la Société Géologique de France*, 5th Series, **12**, 73–89. {18}

Arambourg, C. (1945). *Anancus osiris*, un Mastodonte nouveau du Pliocène inférieur d'Égypte. *Bulletin de la Société Géologique de France*, 5th Series, **15**, 479–95 + pl. IX. {ApB, ApC}

Arambourg, C. (1970). Les vertébrés du Pléistocène de l'Afrique du Nord. *Archives du Muséum National d'Histoire Naturelle Paris*, **7**, 1–126. {12A}

Arambourg, C. and Magnier, P. (1961). Gisements de vertébrés dans le bassin tertiaire de Syrte (Libye). *Comptes Rendus Hebdomadaires des Séances de l'Académie des Sciences*, **252**, 1181–3. {8}

Arslanov, K. A., Lyadov, V. V., Tertychnaya, T. A., and Filonov, B. A. (1982). On the absolute age of the Khatanga mammoth. In *The mammoth fauna of the Asian part of the USSR*, (ed. N. K. Vereshchagin and L. A. Kusmina). Proceedings of the Zoological Institute, Vol. 3, pp. 17–18 [in Russian]. {7}

Astibia, H. and Mazo, A. V. (1988). Los Mastodontes (Proboscidea, Mammalia) del Mioceno medio de Tarazona de Aragon (Zaragoza, España). *Estudios Geológicos*, **44**, 329–38. {14}

Astre, G. (1929). Le mammouth de Baulou. *Bulletin de la Société d'Histoire Naturelle de Toulouse*, **58**, 609–33. {19}

Aubekerova, P. A. (1974). A new species of mastodont (*Anancus kazachstanicus aubekerova*) from the location of Esekartkan. *Teriologia*, Tom II (Academy of Sciences of USSR, Siberian Branch), 65–77 [in Russian with English summary]. {ApC}

Augusta, J. and Burian, Z. (1962). *A book of mammoths*. Hamlyn, London. {2}

Averianov, A. O. (1994). Kutomanov mammoth. *Trudy Zoologicheskogo Instituta, AN SSSR*, **256**, 111–35 [in Russian with English summary, name of journal often translated as 'USSR Academy of Sciences. Proceedings of the Zoological Institute' or 'USSR Academy of Sciences']. {26}

Averianov, A. O., Vartanyan, S. L., and Garutt, V. E. (1995). Small woolly mammoth (*Mammuthus primigenius vrangeliensis* Garutt, Averianov et Vartanyan, 1993) from the Wrangel Island (North-Eastern Siberia). *Trudy Zoologicheskogo Instituta, AN SSSR*, **263**, 184–99 [in Russian with English summary]. {34}

Aymard, A. (1855). *Anancus* Aymard, *Anancus macroplus* Aymard. In Notice géologique sur le cratère de Coupet et sur son gisement de gemmes et d'ossements fossiles, (ed. M. J. Dorlhac), p. 505. *Annales de la Societé d'Agriculture, Sciences, Arts et Commerce du Puy (Procès-Verbaux, Janvier)*, **19**, for 1854, 497–517. {16, ApA, ApB, ApC}

Aziz, F. (1990). Pleistocene mammal faunas of Sulawesi and their bearings to paleozoogeography. Unpublished D. Phil. thesis. Kyoto University, Japan. {17, 23}

Azzaroli, A. (1966). La valeur des caractères crâniens dans la classification des éléphants. *Eclogae Geologicae Helvetiae*, **59**, 541–64. {19}

Azzaroli, A. (1977). Evolutionary patterns of Villafranchian elephants in central Italy. *Atti della Accademia Nazionale dei Lincei, Memorie, Classe di Scienze fisiche, matematiche e naturali*. Ser. 8 (Fisica, chimica, geologia, paleontologia e mineralogia), **14**, Section IIa, 149–68 + 2 plts. {19, 25}

Azzaroli, A. (1981). About pigmy mammoths of the Northern Channel Islands and other island faunas. *Quaternary Research*, **16**, 423–5. {24}

Azzaroli, A. (1982). Insularity and its effects on terrestrial vertebrates: evolutionary and biogeographic aspects. In *Palaeontology, essential of historical geology*, (ed. E. M. Gallitelli), pp. 193–213. STEM Mucchi, Modena, Italy. {22}

Azzaroli, A. (1983). Quaternary mammals and the 'end Villafranchian' dispersal event — a turning point in the history of Eurasia. *Palaeogeography, Palaeoclimatology, Palaeoecology*, **44**, 117–39. {20}

Azzaroli, A. and Napoleone, G. (1982). Magnetostratigraphic investigation of the Upper Sivaliks near Pinjor, India. *Rivista Italiana di Paleontologia e stratigrafia*, **87** (4), 739–62. {12B, 17}

Azzaroli, A., de Giuli, C., Ficcarelli, G., and Torre, D. (1988). Late Pliocene to early mid-Pleistocene mammals in Eurasia: faunal succession and dispersal events. *Palaeogeography, Palaeoclimatology, Palaeoecology*, **66**, 77–100. {19}

Bahn, P. and Vertut, J. (1988). *Images of the ice age*. Windward, Leicester. {19}

Baigusheva (or Bajguscheva), V. S. (1971). Fossil teriofauna of the Liventzovka sand-pit. In *Material on the fauna of the Anthropogene of USSR*, (ed. N. K. Vereshchagin), pp. 5–29 + 6 plts. Academy of Sciences of the USSR. Proceedings of the Zoological Institute, Vol. 49, Leningrad [in Russian with English summary]. {ApC}

Baigusheva, V. S. and Garutt, V. E. (1987). A skeleton of the steppe elephant *Archidiskodon trogontherii* (Pohlig, 1885) from the north-east Azov Sea region. *Proceedings of the Zoological Institute (Leningrad)*, **168**, 21–36. Academy of Sciences of the USSR [in Russian]. {19}

Bajpai, S., Srivastata, S., and Jolly, A. (1989). Sirenian–moerithere dichotomy: some evidence from the middle Eocene of Kachchh, western India. *Current Science*, **58** (6), 304–6. {8, ApC}

Bakalov, P. and Nikolov, I. (1962). Les fossiles de Bulgarie: mammifères Tertiaires. *Académie des Sciences de Bulgarie (Sofia)*, **X**, 1–162 + 82 plts [in Russian]. {9, 12A}

Balson, P. S. and Cameron, T. D. J. (1985). Quaternary mapping offshore East Anglia. *Modern Geology*, **9**, 221–39. {20}

Barbour, E. H. (1913). A tooth of the Columbian mammoth. *Nebraska Geological Survey*, **4** (5), 58–65. {29}

Barbour, E. H. (1914*a*). Mammalian fossils from Devil's Gulch. *Nebraska Geological Survey*, **4** (11), 177–90 [Osborn 1936, p. 763 gave the date as 1914 although on p. 190 of Barbour, 'December, 1913' is given as the date]. {15, ApB, ApC}

Barbour, E. H. (1914*b*). A new longirostral mastodon from Cherry County, Nebraska. *Nebraska Geological Survey*, **4**, 213–22. {15, ApA}

Barbour, E. H. (1915). A new Nebraska mammoth, *Elephas hayi*. *American Journal of Science*, **(4)40**, No. 236, 129–34. {ApC}

Barbour, E. H. (1916). A new longirostral mastodon from Nebraska, *Tetrabelodon osborni*, sp. nov. *American Journal of Science*, (4) **41**, 522–9. {APC}

Barbour, E. H. (1917). The Boyd County mastodon, *Tetrabelodon osborni*. *Nebraska Geological Survey*, **4** (30), 499–512. {ApA, ApC}

Barbour, E. H. (1925*a*). *Elephas scotti*. A new primitive mammoth from Nebraska. *Bulletin of the Nebraska State Museum*, **2(1)**, 21–4. {ApC}

Barbour, E. H. (1925*b*). Skeletal parts of the Columbian mammoth *Elephas maibeni*, sp. nov. *Bulletin of the Nebraska State Museum*, **10** (1), 95–118. {16}

Barbour, E. H. (1926). *Archidiskodon maibeni*. *Bulletin of the Nebraska State Museum*, **11** (1), 119–22. {16}

Barbour, E. H. (1927). Preliminary notice of a new proboscidean *Amebelodon fricki*, gen. et sp. nov. *Bulletin of the Nebraska State Museum*, **13** (1), 131–4. {13, 15, 16, 27, ApA, ApB, ApC}

Barbour, E. H. (1929). *Torynobelodon loomisi* gen. et. sp. nov. *Bulletin of the Nebraska State Museum*, **16** (1), 147–53. {13, 15, ApA, ApC}

Barbour, E. H. (1930). *Amebelodon sinclairi* sp. nov. *Bulletin of the Nebraska State Museum*, **17** (1), 155–8. {13, ApC}

Barbour, E. H. (1931). A new amebelodont, *Torynobelodon barnumbrowni* sp. nov. a preliminary report. *Bulletin of the Nebraska State Museum*, **22** (I), 191–8. {13, 15, 27, ApC}

Barbour, E. H. and Hibbard, C. (1941). A shovel-tusked mastodon, *Amebelodon fricki*, from Kansas. *Bulletin of the Nebraska State Museum*, **2**, 35–46. {15}

Barbour, E. H. and Sternberg, G. (1935). *Gnathabelodon thorpei*, gen. et. sp. nov. A new mud-grubbing mastodon. *Bulletin of the Nebraska State Museum*, **42** (I), 395–404. {15, ApB, ApC}

Barnes, R. F. W. (1989). The status of elephants in the forests of central Africa: results of a reconnaissance survey. In *The ivory trade and the future of the African elephant*, (co-ordinator S. Cobb), Vol. 2, section 1.2. The Ivory Trade Review Group, Oxford. {32}

Barnhart, M. I., *et al.* (1980). Tissue vestiges of an ancient Magadan mammoth calf. *Scanning Electron Microscopy*, **1980**, 163–70. {7}

Barry, J. C., Lindsay, E. H., and Jacobs, L. L. (1982). A biostratigraphic zonation of the Middle and Upper Siwaliks of the Potwar Plateau of northern Pakistan, *Palaeogeography, Palaeoclimatology, Palaeoecology*, **37**, 95–130. {12B}

Barry, J. C., Johnson, N. M., Raza, S. M., and Jacobs, L. L. (1985). Neogene mammalian faunal change in southern Asia: correlations with climate, tectonic, and eustatic events. *Geology*, **13**, 637–40. {10, 12B}

Bar-Yosef, O. and Tchernov, E., (1972). On the palaeo-ecological history of the site of 'Ubeidiya. *The Israel Academy of Sciences and Humanities, Jerusalem*, 1–35 + 5 plts. {21}

Baryshnikov, G. F., Kuzmina, I. E., and Hrabryi, V. M. (1977). Results of measurements of tubular bones of mammoths from Berelekh "burial ground". *In Mammoth fauna of the Russian plaine [sic] and eastern Siberia*, (ed. N. K. Vereshchagin). *Proceedings of the Zoological Institute (Leningrad)* **72**, 58–67. Academy of Sciences of the USSR [in Russian]. {26}

Bate, D. M. A. (1903). Preliminary note on the discovery of a pigmy elephant in the Pleistocene of Cyprus. *Proceedings of the Royal Society of London*, **71** (475), 498–500. {22}

Bate, D. M. A. (1904). Further note on the remains of *Elephas cypriotes* from a cave-deposit in Cyprus. *Philosophical Transactions of the Royal Society of London*, Series B, **197**, 347–60. {22}

Bate, D. M. A. (1905). Four and a half months in Crete in search of Pleistocene mammalian remains. *The Geological Magazine*, Decade V, New Series II, 193–202 + 2 plts. {22}

Bate, D. M. A. (1907). On elephant remains from Crete, with description of *Elephas creticus* sp. n. *Proceedings of the Zoological Society of London*, **1907**, 238–50. {22}

Beden, M. (1976). Proboscideans from Omo Group Formations. In *Earliest man and environments in the Lake Rudolf Basin*, (ed. Y. Coppens, F. C. Howell, G. L. Isaac, and R. E. F. Leakey), pp. 193–208. University of Chicago Press. {12A, 12B, 17}

Beden, M. (1979). Les eléphants (*Loxodonta* et *Elephas*) d'Afrique orientale: systématique, phylogénie, intérêt biochronologique. Unpublished Thèse Doctorat ès Sciences. Faculté des Sciences de l'Université de Poitiers. Vols I and II, 1–567. {11, 12A, 23}

Beden, M. (1983). Family Elephantidae. In *Koobi Fora research project*. Vol. 2. *The fossil ungulates: Proboscidea, Perissodactyla, and Suidae*, (ed. J. M. Harris), pp. 40–129. Clarendon, Oxford. {12A, 18}

Beden, M. (1985). Les proboscidiens des grands gisements à Hominidés Plio-Pleistocènes d'Afrique orientale. In *L'environement des Hominidés au Plio-Pléistocène*, (ed. M. Beden, *et al.*), pp. 21–44. Collogue international, Juin 1981, Fondation Singer-Polignac. Masson, Paris. {12A, 12B}

Beden, M. (1986). Le mammouth d'Oubeidyeh (Israel). In *Les mammifères du Pléistocène Inférieure de la vallée du Jourdain à Oubeidiyeh*, (ed. E. Tchernov), pp. 213–34. Mémoires et Travaux du Centre de Recherche Français de Jérusalem, 5. Association Paléorient, Paris. {21}

Beden, M. (1987*a*). Fossil Elephantidae from Laetoli. In *Laetoli: a Pliocene site in northern Tanzania*, (ed. M. D. Leakey and J. M. Harris), pp. 259–300. Clarendon, Oxford. {12A, 12B, 18}

Beden, M. (1987*b*). *Les faunes Plio-Pléistocène de la basse vallée de l'Omo (Éthiopie)*. Vol. 2: *Les Eléphantidés (Mammalia — Proboscidea)*, (directed by Y. Coppens and F. C. Howell), pp. 1–162. Cahiers de Paléontologie —Travaux de Paléontologie est-africaine. Centre National de la Recherche Scientifique (CNRS), Paris. {12A, 18}

Beden, M. and Guérin, C. (1975). Les proboscidiens (Mammalia) du gisement Pléistocène moyen des Abîmes de la Fage à Noailles (Corrèze). *Nouvelles Archives du Muséum d'Histoire Naturelle de Lyon*, **13**, 69–87. {19}

Behrensmeyer, A. K., Gordon, K. D., and Yanagi, G. T. (1986). Trampling as a cause of bone surface damage and pseudo-cutmarks. *Nature*, **319**, 768–71. {30}

Belluomini, G. and Bada, J. L. (1985). Isoleucine epimerization ages of the dwarf elephants of Sicily. *Geology*, **13**, 451–2. {22}

Belyaeva, E. I. (1962). Catalogue of Tertiary fossil sites of land mammals in the U.S.S.R., 1–114. *Trudy paleontologicheskogo Institute*, **15** (3), 3 and 36–116, Moscow–Leningrad, 1948 [translated from Akademya Nauk SSSR by the American Geological Institute, Washington, DC]. {9}

Belyaeva, E. I. and Gabunia, L. K. (1960). New finds concerning Platybelodontinae from the Caucasus. *Trudy Instituta Paleobiologii Akademii Nauk Gruzinskoi, SSR* **5**, 63–105 + 2 plts [in Russian]. {13, ApC}

Benedict, F. G. (1936). *The physiology of the elephant*. Carnegie Institute of Washington Publication No. 474, Washington, DC. {2}

Bennett, D. K. (1977). Paleontology, paleoecology, and biostratigraphy of the Rhinoceros Hill fauna (Hemphillian, latest Miocene) and sedimentology of the enclosing Ogallala Formation, Ash Hollow member, Wallace County, Kansas. Unpublished M.S. thesis. University of Kansas. {15}

Benton, M. J. (ed.) (1993). *The fossil record 2*. Chapman and Hall, London. {ApD}

Berger, R. G. (1980). Early man on Santa Rosa Island. In *The California islands: proceedings of a multidisciplinary symposium*, (ed. D. M. Power), pp. 73–8. Santa Barbara Museum of Natural History. {24}

Berger, R. G. and Orr, P. C. (1966). The fire areas on Santa Rosa Island, California II. *Proceedings of the National Academy of Sciences, USA*, **56**, 1678–82. {24}

Berger, R. G., Fergusson, G. J., and Libby, W. F. (1965). UCLA Radiocarbon dates IV, *Radiocarbon*, **7**, 336–71. {24}

Berggren, W. A. and Prothero, D. R. (1992). Eocene–Oligocene climatic and biotic evolution: an overview. In *Eocene–Oligocene climatic and biotic evolution*, (ed. D. R. Prothero and W. A. Berggren), pp. 1–28. Princeton University Press, Princeton, New Jersey. {ApD}

Berggren, W. A., Kent, D. V., and Flynn, J. J. (1985*a*). Jurassic to Paleogene: Part 2. Paleogene geochronology and chronostratigraphy. In *The chronology of the geological record*, (ed. N. J. Snelling). Geological Society Memoir No. 10, pp. 141–95, Blackwell Scientific Publications, Oxford. {ApD}

Berggren, W. A., Kent, D. V., and Van Couvering, J. A. (1985*b*). The Neogene: Part 2. Neogene geochronology and chronostratigraphy. In *The chronology of the geological record*, (ed. N. J. Snelling). Geological Society Memoir No. 10, pp. 211–60. Blackwell Scientific Publications, Oxford. {ApD}

Bergounioux, F. M. and Crouzel, F. (1955). Quelques nouvelles formes de mastodontes du Miocène de la péninsule ibérique. *Comptes Rendus Hebdomadaires des Séances de l'Académie des Sciences, Paris*, **241**, 1488–90. {14, ApA}

Bergounioux, F. M. and Crouzel, F. (1956). Présence de *Tetralophodon longirostris* dans le Vindobonien inférieur de Tunisie. *Bulletin Société Géologique de France*, **6**, 547–58. {12B}

Bergounioux, F. M. and Crouzel, F. (1958). Les mastodontes d'Espagne. *Estudios Geológicos*, **14** (40), 224–343. {14}

Bergounioux, F. M. and Crouzel, F. (1960). *Tetralophodon curvirostris* n. sp. (Mammalia, Proboscidea) aus dem Unterpliozän (Pontien) von Esselborn (Rheinhessen). *Jahresberichte und Mitteilungen Oberrheinischer Geologischer Verein*, **42**, 109–21. {12A}

Bergounioux, F. M. and Crouzel, F. (1962). Les deinotheridés d'Europe. *Annales de Paléontologie (Vert.-Invert.)*, **48**, 13–56. {ApC}

Bergounioux, F. M., Zbyszewski, G., and Crouzel, F. (1953). Les mastodontes miocènes du Portugal. *Memorias dos Servicos Geológicos de Portugal, Lisbon*, **1**, 1–140. {14}

Bernsen, J. J. A. (1930). Eine Revision der fossilen Säugetierfauna aus den Tonen von Tegelen. I. *Natuurhistorisch Maandblad*, **19** (12), 147–53. {20}

Biewener, A. A. (1989). Scaling body support in mammals: limb posture and muscle mechanics. *Science*, **245**, 45–8. {24}

Biske, S. F. and Baranova, Y. P. (1986). Main paleogeographic features of Beringia in the pre-Quaternary Cenozoic. In *Beringia in the Cenozoic era*, (ed. V. L. Kontrimavichus, Russian translation Series 28), pp. 154–63. Balkama, Rotterdam. {9}

Blainville, H. M. D., de. (1816). Prodrome d'une nouvelle distribution systématique du règne animal. *Bulletin des Sciences, Société Philomathique de Paris*, Series 3, **3**, 105–24. {5}

Blainville, H. M. D., de. (1817). *Dents*. Nouveau Dictionnaire d'Histoire Naturelle (nouvelle édition), ix, pp. + 252–352, synoptic table, Paris [from Osborn 1942, p. 1623]. {ApA}

Blainville, H. M. D., de. (1836). [Classification adopted in 1834, published by P. Gervais, article 'Mammalogie'] in *Dictionnaire Pittoresque d'Histoire Naturelle et des Phénomènes de la Nature*, (ed. F. E. Guérin), pp. 614–40. Cosson, Paris. {1}

Blainville, H. M. D., de. (1845). *Ostéographie, fascicule 16, Des éléphants*. Arthus Bertrand, Paris. [Editors' note: dates from 1839 to 1864 (Osborn 1936, p. 765)] {13, ApA}

Blair, P. (1710). Osteographia elephantina: or, a full and exact description of all the bones of an elephant, which dy'd near Dundee, April the 27th, 1706, with their several dimensions. *Philosophical Transactions*, Part I, **27** (326), 51–116, Part II, **27** (327), 117–68. {2}

Bloom, A. L. (1983). Sea level and coastal morphology of the United States through the late Wisconsinan glacial maximum. In *Late Quaternary environments of the United States*. Vol. 1, *The late Pleistocene*, (ed. S. C. Porter), pp. 215–29. University of Minnesota Press, Minneapolis. {29}

Blouch, R. A. and Haryanto. (1984). *Elephants in southern Sumatra*. Report No. 3, World Wildlife Fund Project 3133, Bogor, Indonesia [no first initial for Haryanto]. {33}

Blouch, R. A. and Simbolon, K. (1985). *Elephants in northern Sumatra*. Report No. 9, World Wildlife Fund Project 3133, Bogor, Indonesia. {33}

Blower, J. (1985). *Elephants in Burma*. WWF Monthly Report, December 1985, pp. 279–85. World Wildlife Fund, Gland, Switzerland. {33}

Blumenbach, J. F. (1779). *Handbuch der Naturgeshichte* (1st edn). Dieterich, Göttingen. {7, 34}

Blumenbach, J. F. (1797). *Handbuch der Naturgeschichte* (5th edn). Dieterich, Göttingen. {ApA, ApC}

Blumenbach, J. F. (1799). *Handbuch der Naturgeschichte*, (6th edn). Dieterich, Göttingen. {ApA, ApB, ApC}

Blunt, D. E. (1933). *Elephant*. Holland Press, London. {32}

Boas, J. E. V. and Paulli, S. (1908). *The elephant's head: studies in the comparative anatomy of the organs of the head of the Indian elephant and other mammals. Part I: the facial muscles and the proboscis*. Fischer, Jena. {2}

Boas, J. E. V. and Paulli, S. (1925). *The elephant's head: studies in the comparative anatomy of the organs of the head of the Indian elephant and other mammals. Part II: the facial muscles and the proboscis*. Fischer, Jena. {2}

Boaz, N. T. (1987). Introduction. In *Neogene paleontology and geology of Sahabi*, (ed. N. T. Boaz, A. El-Arnauti, A. W. Gaziry, J. de Heinzelin, and D. D. Boaz), pp. XI–XV. Alan R. Liss, New York. {12B}

Boaz, N. T., Gaziry, A. W., and El-Arnauti, A. (1979). New fossil finds from the Libyan upper Neogene site of Sahabi. *Nature*, **280**, 137–40. {ApC}

Boekschoten, G. J. and Sondaar, P. Y. (1972). On the fossil Mammalia of Cyprus. I–II. *Proceedings of the Koninklijke Nederlandse Akademie van Wetenschappen, Amsterdam*, B, **75**, 306–38. {22}

Boeuf, O. (1983). Le site Villafranchien de Chilhac (Haute-Loire) France. Etude paléontologique et biochronologique. Thèse de doctorat d'Etat, Université de Paris VII. {20}

Boeuf, O. (1990). Originalité et importance de la faune Plio-Pléistocène de Chilhac (Haute-Loire, France). *Quartärpaläontologie*, **8**, 13–28. {19}

Bolk, L. (1919). *Odontologische Studien III. Zur Ontogenie des Elefantengebisses*. Fischer, Jena. {4}

Bonaparte, C. L. J. L. (1845). *Catalogo metodico dei mammiferi Europei*. Coi tipi di L. di Giacomo Pirola, Milan. {ApB, ApC}

Bonfiglio, L. and Berdar, A. (1979). Gli elefanti delle ghiaie pleistoceniche di Messina. *Quaternaria, Roma*, **21**, 139–77. {22}

Bonfiglio, L. and Berdar, A. (1986). Gli elefanti del Pleistocene superiore di Archi (R. C.): nuove evidenze di insularita della Calabria meridionale durante il ciclo Tirreniano. *Bollettino della Societa Paleontologicia Italiana, Modena*, **25**, 9–34. {22}

Bonfiglio, L. and Burgio, E. (1990). New stratigraphic and palaeoecological data on mammal-bearing Pleistocene deposits of Sicily. *Abstracts of IX Congress of Regional Committee of Mediterranean Neogene Stratigraphy, Barcelona*. {22}

Bonfiglio, L. and Insacco, G. (1992). Palaeoenvironmental, paleontologic and stratigraphic significance of vertebrate remains in Pleistocene limnic and alluvial deposits from southeastern Sicily. *Palaeogeography, Palaeoclimatology, Palaeoecology*, **95** (3/4), 195–208. {22}

Bonfiglio, L. and Kotsakis, T. (1987). Les associations de vertébrés du Pléistocene de Sicile: les peuplements successif. *Documents et Travaux IGAL, Paris*, **11**, 263–66. {22}

Borissiak, A. A. (1928). On a new mastodon from the Chokrak Beds (middle Miocene) of the Kuban region, *Platybelodon danovi*, n. gen., n. sp. *Annual of the Paleontological Society Russie* **7**, 105–20. {13, 15, ApA, ApB, ApC}

Borissiak, A. A. (1936). *Mastodon atavus* n. sp., der primitivste Vertreter der Gruppe *M. angustidens*. *Travaux de Institut de Paléozoologie. Academie des Sciences del' URSS*. **5**, 171–234. {9, ApC}

Borissiak, A. A. and Belyaeva, E. (1928). *Trilophodon (Serridentinus?) inopinatus* N. Sp. from the Jilanc̄ik Beds from the Turgai Region. Bulletin de l'Academic des Sciences, Physico-Mathematiques, pp. 241–252 + 2 plts. {ApC}

Bosinski, G. (1979). The mammoth engravings of the Magdalenian site of Gönnersdorf (Rhineland, Germany). *Colloques de la Société Suisse des Sciences Humaines*, **3**, 295–322. {19}

Boule, M. and Thevenin, A. (1920). *Mammifères fossiles de Tarija*. Imprimerie Nationale, Paris. {12A}

Bout, P. (1970). Absolute ages of some volcanic formations in the Auvergne and Velay areas and chronology of the European Pleistocene. *Palaeogeography, Palaeoclimatology, Palaeoecology*, **8**, 95–106. {20}

Bowen, D. Q., Hughes, S., Sykes, G. A., and Miller, G. H. (1989). Land–sea correlations in the Pleistocene based on isoleucine epimerization in non-marine molluscs. *Nature*, **340**, 49–51. {19}

Bown, T. M., *et al.* (1982). The Fayûm primate forest revisited. *Journal of Human Evolution*, **11**, 603–32. {8}

Boyce, M. S. (ed.) (1988). *Evolution of life histories of mammals: theory and pattern*. Yale University Press, New Haven. {30}

Branagan, D. (1992). Richard Owen in antipodean context [a review]. *Journal and Proceedings, Royal Society of New South Wales*, **125**, 95–102. {ApC}

Broecker, W. S. and Kulp, J. L. (1957). Lamont natural radiocarbon measurements IV. *Science*, **126**, 1324–34. {24}

Bromage, T. G. and Dean, M. C. (1985). Re-evaluation of the age at death of immature fossil hominids. *Nature*, **317**, 525–7. {30}

Brookes, J. (1828). *A catalogue of the anatomical and zoological museum of Joshua Brookes, Esq.*, Part 1. R. Taylor, London. {16, ApA, ApB, ApC}

Brown, F. H., McDougal, I., Davies, T., and Maier, R. (1985). An integrated Plio-Pleistocene chronology for the Turkana Basin. In *Ancestors: the hard evidence*, (ed. E. Delson), pp. 82–90. Liss, New York. {12B}

Brüning, H. (1978). Zur Untergliederung der Mosbacher Terrassenabfolge und zum klimatischen Stellenwert der Mosbacher Tierwelt im Rahmen des Cromer-Komplexes. *Mainzer Naturwissenschaftliches Archiv*, **16**, 143–90. {19}

Bryan, A. L., Casamiquela, R. M., Cruxent, J. M., Gruhn, R., and Ochsenius, C. (1978). An El Jobo mastodon kill at Taima-taima, Venezuela. *Science*, **200**, 1275–7. {30, 31}

Bryden, H. A. (1903). The decline and fall of the South African elephant. *Fortnightly Review (n.s.)*, **79**, 100–8. {32}

Buckland, W. (1826). Geological account of a series of animal and vegetable remains and of rocks, collected by J. Crawford, Esq. on a voyage up the Irawadi to Ava, in 1826 and 1827. *Transactions of the Geological Society of London*, **2** (1), page numbers not present + 9 plts. {ApC}

Buckland, W. (1835). Über den Bau und die mechaniche Kraft des Unterkiefers des *Dinotherium*. *Neues Jahrbuch für Mineralogie, Abhandlungen*, **1835**, 516–18. {8}

Burgio, E. and Cani, M. (1989). Sul ritrovamento di elefanti fossili ad Alcamo (TP). *Il Naturalista Siciliano, Palermo*, **4**, (12) (1988), 87–97. {22}

Burgio, E., Oliva, N., and Scalone, E. (1983). La collezione vertebratologica della Grotta dei Puntali presso Carini (Palermo). *Il Naturalista Siciliano, Palermo*, **4** (7), 67–79. {22}

Burmeister, H. (1837). *Handbuch der Naturgeschichte. Zum Gebrauch bei Vorlesungen entworfen*. Zweite Abteilung, Zoologie, T. C. F. Enslin, Berlin. {15, ApA, ApB, ApC}

Burnett, G. T. (1830). Illustrations of the Quadrupeda, or Quadrupeds, being the arrangement of the true four-footed beasts indicated in outline. *Quarterly of the Journal of Science (London)*, **December** 1829, 336–53. {ApA, ApC}

Burrill, A. and Douglas-Hamilton, I. (1987). *African elephant database project: final report*—phase one, GRID Case Study Series No. 2. UNEP, Nairobi. {32}

Busk, G. (1867). Description of the remains of three extinct species of elephant, collected by Capt. Spratt, C. B., R. N., in the ossiferous cavern of Zebbug, in the island of Malta... partly from the notes of the late H. Falconer, M. D., F. R. S. *Transactions of the Zoological Society of London*, **VI**, (Pt. V), 227–306, plts VLIV–LIII. {22}

Buss, I. O. (1990). *Elephant life: fifteen years of high population density*. Iowa State University Press, Ames. {34}

Buss, I. O. and Smith, N. S. (1966). Observations on reproduction and breeding behavior of the African elephant. *Journal of Wildlife Management*, **30**, 375–88. {30}

Cabrera, Á. (1929). Una revisión de los mastodontes Argentinos. *Revista del Museo de la Plata*, **32**, 61–144 + 2 plts [cover page gives year 1930]. {16, ApB, ApC}

Caloi, L. and Palombo, M. R. (1985). Osservazioni sugli ippopotami nani delle isole del Mediterraneo. *Geologica Romana*, **22** (1983), 45–83. {22}

Caloi, L. and Palombo, M. R. (1989). Oligotipia nelle mammalofaune insulari. In *Atti 3º Simposio di Ecologia e Palecologia delle Comunita Bentoniche*, Catania-Taormina, **12–16 ottobre 1985**, 259–83. {22}

Caloi, L., Kotsakis, T., Palombo, M. R., and Petronio, C. (1980). Il giacimento a vertebrati del Pleistocene superiore di San Giovanni in Sinis (Sardegna occidentale). *Rendiconti dell' Accademia Nazionale dei Lincei, Roma*, **8**, (69), 1851–97. {22}

Caloi, L., Kotsakis, T., and Palombo, M. R. (1988). La fauna a vertebrati terrestri del Pleistocene della isole del Mediterraneo. *Geologica Romana*, **25** (1986), 235–56. {22}

Caloi, L., Capasso Barbato, L., Esu, D., Kotsakis, T., Palombo, M. R., and Petronio, C. (in press). Observations on Pleistocene continental vertebrate and mollusc faunas of Sicily. *Geologica Romana*. {22}

Calvin, S. (1911). Aftonian mammalian fauna II. *Bulletin of the Geological Society of America*, **22**, 207–16, plts 18–24. {ApC}

Cameron, T. D. J., Laban, C., and Schuttenhelm, R. T. E. (1984). *Flemish Bight, sheet 52 °N–02 °E Quaternary geology 1 : 250, 000*. British Geological Survey and Rijks Geologische Dienst, Southampton. {20}

Campbell, K. E. and Frailey, C. D. (1995). A new date for the beginning of the Great American Faunal Interchange. *Journal of Vertebrate Paleontology* (Abstracts of Papers), **15** (Supplement to no. 3), 21A. {29}

Camper, P. (1788). Complementa varia Academiae Imperialis Scientiarum Petropolitanae Communicanda, ad Clar. ac Celeb. Pallas. Nova Acta Academiae Imperialis Scientiarum Petropolitanae, II, for the year 1784, pp. 250–264 [from Osborn 1936, p. 768]. {ApA}

Capasso Barbato, L. (1992). Observations on the biostratigraphy of Cretan Pleistocene vertebrates Il Quaternario, **5** (1), 67–76. {22}

Capasso Barbato, L, Minieri, M. R., and Petronio, C. (1989). Resti di mammiferi endemici nelle grotte del Faraglione di Favignana (Egadi, Trapani). *Il Naturalista Siciliano, Palermo*, **4**, (12) (1988), 99–105. {22}

Carpentier, G. and Lautridou, J.-P. (1986). The low terrace of the Seine: the alluvium, periglacial deposits, interglacial fluviomarine deposits, slope deposits and palaeosols, fauna. In *The Loess and other Pleistocene periglacial deposits of northwest Europe, including their relationships with marine formations and features*, (ed. J.-P. Lautridou, D. H. Keen, and J. L. Monnier), pp. 64–8. Centre National de la Recherche Scientifique (CNRS), Caen. {19}

Carranza-Castañeda, O. and Miller, W. E. (1987). Rediscovered type specimens and other important published Pleistocene mammalian fossils from central Mexico. *Journal of Vertebrate Paleontology*, **7**, 335–41. {ApC}

Carrington, R. (1958). *Elephants: a short account of their natural history, evolution, and influence on mankind*. Penguin Books, London. {34}

Carroll, R. L. (1988). *Vertebrate paleontology and evolution*. Freeman, San Francisco. {5}

Casamiquela, R. M. (1972). Catalogación crítica de algunos vertebrados fósiles chilenos. II Los mastodontes. *Ameghiniana, revista de la Asociación Paleontológica Argentina*, **IX**, (3), 193–208. {31}

Casamiquela, R. M. (1976). *Los vertebrados fósiles de Tagua-Tagua, Chile*. Primer Congreso Geológico Chileno, Santiago, pp. 88–102. {31}

Casamiquela, R. M. (1979). An interpretation of the fossil vertebrates of the Taima-Taima Site. In *South American Quaternary documentation program. Taima-Taima. A late Pleistocene Paleo-Indian kill site in northernmost South America–final reports of 1976 excavations*, (ed. C. Ochsenius and R. Gruhn). Printed in the Federal Republic of Germany. {31}

Casamiquela, R. M. (1981). Informe de los restos oseos del sitio de Monte Verde: 1979. Report on file at the University of Kentucky, Lexington. {31}

Casamiquela, R. M. (1983). Informe de los restos oseos del sitio de Monte Verde: 1983. Report on file at the University of Kentucky, Lexington. {31}

Casamiquela, R. M. (1985). Informe de los restos oseos del sitio de Monte Verde: 1985. Report on file at the University of Kentucky, Lexington. {31}

Casamiquela, R. M. and Dillehay, T. D. (1989). Vertebrate and invertebrate faunal analysis. In *Monte Verde: a late Pleistocene settlement in Chile*, Vol. I, *Palaeoenvironment and site context*, (ed. T. D. Dillehay), pp. 205–10. Smithsonian Institution Press, Washington, DC. {31}

Casamiquela, R., Montané, J., and Santana, R. (1967) Convivencia del hombre con el mastodonte en Chile central. Noticias sobre las investigaciones en la Laguna de Taguatagua. *Noticiero Mensual. Museo Nacional de Historia Natural, Santiago*, **132**, 1–6. {31}

Caterini, F. (1923). Catologo dei Proboscidiani pliocenici e quaternari conservati nel Museo di Geologia dell'Università di Pisa. *Atti della Società Toscana di Scienze Naturali*, **35**, 3–18. [Osborn 1936, p. 768, gave pp. 195–210] {ApC}

Caughley, G., Dublin, H., and Parker, I. (1990). Projected decline of the African elephant. *Biological Conservation*, **54**, 157–64. {32}

Cautley, P. T. (1836). Note on the teeth of the mastodon à dents etroites of Siwálik Hills. *Journal of the Asiatic Society of Bengal*, **V(May)**, 294–6, Plt xi. {ApC}

Cayeux, L. (1908). Découverte de l'*Elephas antiquus* à l'île de Délos (Cyclades). *Comptes Rendus Hebdomadaires des Séances de l'Académie des Sciences, Paris*, **147** (2), 1089–90. {22}

Chadwick, D. H. (1992). *The fate of the elephant*. Sierra Club Books, San Francisco.{34}

Chai, J. (1959), see under Zhai, R. (1959)

Chakravarti, D. K. (1957). A geological, palaeontological and phylogenetic study of the Elephantoidea of India, Pakistan and Burma: Part I. Gomphotheriidae. *Journal of the Paleontological Society of India*, **2**, 83–94. {ApA}

Chaney, R. W. and Mason, H. L. (1934). A Pleistocene flora from Santa Cruz Island, California. In *Studies of the Pleistocene palaeobotany of California*. Contributions to Palaeontology, Vol. 13, Carnegie Institution of Washington Publication, No. **415**, pp. 1–24, Washington, DC. {24}

Chang, H. (1964). New materials of mastodonts from the Yüshe Basin, Shansi. *Vertebrata PalAsiatica*, **8** (1), 33–41 [in Chinese with English summary]. {13, ApC}

Chang, H. and Zhai, R. (1978). Miocene mastodonts of Lantian and Lintung, Shensi. *Professional Papers of Stratigraphy and Palaeontology*, **7**, 136–42 + 4 plts (Geological Publishing House, Peking, China) [in Chinese with English summary]. {13, ApA, ApC}

Chang, X. Y. (1980). Fossil mammals from the early Pleistocene of Yongren, Yunnan. *Vertebrata PalAsiatica*, **18** (1), 45–51 + 2 plts. [In Chinese with English abstract] {ApC}

Chen, G. (1978). Some mastodonts from Miocene of Ningxia. *Vertebrata PalAsiatica*, **16**, 103–10. {13, ApC}

Chen, G. (1988). Mastodont remains from the Miocene of Junggar Basin in Xinjiang. *Vertebrata PalAsiatica*, **26** (4), 265–77 + 4 plts [in Chinese with English text]. {13, ApC}

Chiu, C., Li, C., and Chiu, C. (1979). The Neogene of China: a preliminary review of the mammalian localities and faunas. *Annales Géologiques des Pays Helléniques*, tome hors série, **I**, 263–72 + 1 tbl [VIII International Congress of the Mediterranean Neogene]. {13}

Chow, M. (1959). New species of fossil Proboscidea from south China. *Acta Palaeontologica Sinica*, **7** (4), 251–8 + 4 plts [in Chinese with English summary]. {12A, 17}

Chow, M. (1978). Tertiary mammalian faunas of the Lantian District, Shensi. *Professional Papers of Stratigraphy and Paleontology*, **7**, 98–108. (Geological Publishing House, Peking, China) [in Chinese]. {13, ApC}

Chow, M. and Chang, Y. (1961). New mastodonts from north China. *Vertebrata PalAsiatica*, **3**, 245–61 [in the 1970s, spelling of Chow and Chang was changed to Zhou and Zhang]. {9, ApC}

Chow, M. and Hsu, Y. (1959). A case of abnormal molar of the elephant. *Vertebrata PalAsiatica*, **1** (3), 145–6 + 1 plt [in Chinese]. {ApC}

Chow, M. and Zhai, R. (1962). Early Pleistocene mammals of Chaotung, Yunnan, with notes on some Chinese stegodonts. *Vertebrata PalAsiatica*, **6** (2), 138–49 [in Chinese with English abstract]. {17}

Chow, M. and Zhang, Y. (1978). Notes on some mastodonts from Yunnan. *Professional Papers of Stratigraphy and Palaeontology*, **7**, 68–74 + 5 plts (Geological Publishing House, Peking, China) [in Chinese]. {13, ApC}

Chow, M. *et al.* (1979). *Manual of fossil vertebrate animals from China (revised and enlarged edition)*. Science Press, Peking. [In Chinese, scientific names in English; Proboscidea by Zhang and Yu-Ping, pp. 435–57] {ApC}

Churcher, C. S. (1959). Fossil *Canis* from the tar pits of La Brea, Peru. *Science*, **130**, 564–5. {29}

Churcher, C. S. (1980). Did the North American mammoths migrate? *Canadian Journal of Anthropology*, **1**, 103–6. {29}

Churcher, C. S. (1986). A mammoth measure of time: molar compression in *Mammuthus* from the Old Crow Basin, Yukon Territory, Canada. *Current Research in the Pleistocene*, **3**, 61–4. {29}

Cibull, M. (1988). Report on skin fragments from Monte Verde. Manuscript on file at the University of Kentucky, Lexington. {31}

Cifelli, R. L. (1983). Eutherian tarsals from the late Paleocene of Brazil. *American Museum Novitates*, **2761**, 1–31. {8}

Clark, J. D., Asfaw, B., Assefa, G., Harris, J. W. K., Kurashina, H., Walter, R. C. *et al.* (1984). Palaeoanthropological discoveries in the Middle Awash Valley, Ethiopia. *Nature*, **307**, 423–8. {12B}

Clift, W. (1828). On the fossil remains of two new species of Mastodon, and of other vertebrated animals, found on the left bank of the Irawadi. *Transactions of the Geological Society of London*, (2), II, Part III, 369–75, + plts XXXVI–XLIV. {34, ApC}

Clutton-Brock, J. (1981). *Domesticated animals from early times*. Heinemann and British Museum (Natural History), London. {21, ApD}

Clutton-Brock, T. H., Guinness, F. E., and Albon, S. D. (1982). *Red deer: behavior and ecology of two sexes*. University of Chicago Press. {30}

Cobb, S. (co-ordinator) (1989). *The ivory trade and the future of the African elephant*. Ivory Trade Review Group, Oxford (2 volumes). {32}

COHMAP Members (1988). Climatic changes of the last 18,000 years: observations and model simulations. *Science*, **241**, 1043–52. {30}

Colbert, E. H. (1943). Pleistocene vertebrates collected in Burma by The American Southeast Asiatic Expedition. *Transactions of the American Philosophical Society*, **32** (3), 395–429 + 32 plts. {ApC}

Colbert, E. H. (1955). *Evolution of the vertebrates: a history of the backboned animals through time*, (1st edn). Wiley, New York. {34}

Colbert, E. H. (1969). *Evolution of the vertebrates: a history of the backboned animals through time*, (2nd edn), Wiley, New York. {34}

Colbert, E. H. (1992). My experiences with the late Henry Fairfield Osborn. In *Elephants: majestic creatures of the wild* (consult. ed. J. Shoshani), pp. 34–5. Rodale Press, Emmaus, Pennsylvania. {34}

Colbert, E. H. and Hooijer, D. A. (1953). Pleistocene mammals from the limestone fissures of Szechuan, China. *Bulletin of the American Museum of Natural History*, **102**, 1–134. {17, ApC}

Collins, P. W. (1991). Interaction between island foxes (*Urocyon littoralis*) and Indians on islands off the coast of southern California: I. Morphologic and archaeological evidence of human assisted dispersal. *Journal of Ethnobiology*, **11**, 51–81. {24}

Colyer, F. and Miles, A. E. W. (1957). Injury to and rate of growth of an elephant tusk. *Journal of Mammalogy*, **38**, 243–7. {2}

Comaschi Caria, I. (1965). L'elefante nano del Quaternario di Gonnesa (Sardegna sub-occidentale). *Rendiconti del Seminario dell' Facolta di Scienze dell' Universita di Cagliari, Bologna*, **35**. {22}

Conniff, R. (1987). When the music in our parlors brought death to darkest Africa. *Audubon Magazine*, **July 1987**, 76–93. {32}

Cook, H. J. (1922). Two new bunomastodonts from Colorado. *Proceedings of the Colorado Museum of Natural History*, **4**, (1), 1–15. {15, 27, ApC}

Cooke, H. B. S. and Coryndon, S. (1970). Fossil mammals from the Kaiso Formation and other related deposits in Uganda. In *Fossil vertebrates of Africa*, Vol. 2, (ed. L. S. B. Leakey and R. G. J. Savage), pp. 107–224. Academic Press, Edinburgh. {12A, 12B, 17}

Coombs, M. C. and Coombs, W. (1979). *Pilgrimella*, a primitive Asiatic perissodactyl. *Zoological Journal of the Linnaean Society*, **65**, 185–95. {8}

Coope, G. R. and Lister, A. M. (1987). Late-glacial mammoth skeletons from Condover, Shropshire, England. *Nature*, **330**, 472–4. {19, 25}

Cooper, C. F. (1922). Miocene Proboscidea from Baluchistan. *Proceedings of the Zoological Society of London*, **42**, 602–26. {10}

Cope, E. D. (1873). *Synopsis of new Vertebrata from the Tertiary of Colorado obtained during the summer of 1873*. Government Printing Office, Washington, October 16, pp. 1–19. {ApC}

Cope, E. D. (1875). On a new mastodon and rodent. *Proceedings of the Academy of Natural Sciences of Philadelphia*, **26** (1874), 221–3. [Dated 1874, issued February 2, 1875; Osborn 1936, p. 770] {ApC}

Cope, E. D. (1882). The classification of the ungulate Mammalia. *Proceedings of the American Philosophical Society*, **20**, 438–61. {5}

Cope, E. D. (1884*a*). The Condylarthra. *American Naturalist*, **18**, 790–805. {6, ApC}

Cope, E. D. (1884*b*). The extinct Mammalia of the valley of Mexico. *Proceedings of the American Philosophical Society*, **22**, 1–21. {ApA, ApC}

Cope, E. D. (1889). The Proboscidia. *American Naturalist*, **23** (268), 191–211 [Note: Spelling is 'Proboscidia', not 'Proboscidea']. {1, 12A, ApA}

Cope, E. D. (1893). A preliminary report on the vertebrate paleontology of the Liano Estacado. *4th Annual Report of the Geological Survey of Texas, for the year 1892*, pp. 1–87 + 23 plts. {ApC}

Coppens, Y. (1958). Le mammouth de l'Atrikanova (Sibérie). *Bulletin du Muséum National d'Histoire Naturelle*, Paris, 2nd Series, **30**, 402–6. {19, 25}

Coppens, Y. (1965). Les proboscidiens du Tchad — leur contribution à la chronologie du Quaternaire Africain. *Actes du Ve Congrès Panafricain de Préhistoire et de l'Étude du Quaternaire* (Santa Cruz de Tenerife), **I**, (5), 331–87. {12A, 12B, ApC}

Coppens, Y. (1967). Les faunes de vertébrés quaternaires du Tchad. In *Background to evolution in Africa*, (ed. W. W. Bishop and J. D. Clark), pp. 89–97. University of Chicago Press. {12A, 12B, 17}

Coppens, Y. (1972). Un nouveau proboscidien de Pliocène du Tchad, *Stegodibelodon schneideri* nov. gen. nov. sp., et le phylum des Stegotetrabelodontinae. *Comptes Rendus de l'Académie des Sciences, Paris*, Series D, **274**, 2962–5. {6, 12A, 12B, 18, ApB, ApC}

Coppens, Y. and Beden, M. (1978). Moeritherioidea. In *Evolution of African mammals*, (ed. V. J. Maglio and H. B. S. Cooke), pp. 333–5. Harvard University Press, Cambridge, Mass. {8, ApC}

Coppens, Y. and Beden, M. (1982). *Mammuthus meridionalis depereti* n. subsp. (Mammalia, Elephantidae), nouveau mammouth du Pléistocène inférieur d'Europe occidentale. *Comptes Rendus de l'Académie des Sciences*, Series II, **294**, 291–4. {19}

Coppens, Y. and Koeniguer, J.-C. (1976). Signification climatique des paléoflores ligneuses du Tertiaire et du Quaternaire du Tchad. *Bulletin Société Géologique de France*, **18**, (7), 1009–15. {12B}

Coppens, Y., Maglio, V. J., Madden, C. T., and Beden, M. (1978). Proboscidea. In *Evolution of African mammals*, (ed. V. J. Maglio and H. B. S. Cooke), pp. 336–67. Harvard University Press, Cambridge, Mass. {18, 12A, 12B, 16, 17, 18, 29, 34, ApC}

Corse, J. (1799). Observations on the different species of Asiatic elephants, and their mode of denition. *Philosophical Transactions of the Royal Society of London*, **89**, 205–36 + 7 plts. {4}

Costa, O. G. (1850). Paleontologia del regno di Napoli. *Atti della Accademia Pontaniana*, V, pt. I, pp. 233–433, Tav. i–xv. Napoli [from Osborn 1936, p. 771]. {ApA}

Cotter, J. L. (1937). The occurrence of flints and extinct animals in pluvial deposits near Clovis, New Mexico. Part IV, Report on excavation at the gravel pit, 1936. *Proceedings of the Academy of Natural Sciences of Philadelphia*, **89**, 1–16. {27}

Court, N. (1990). Periotic anatomy of *Arsinoitherium* (Mammalia, Embrithopoda) and its phylogenetic implications. *Journal of Vertebrate Paleontology*, **10**, 170–82. {5}

Court, N. (1992). Cochlea anatomy of *Numidotherium koholense*: auditory acuity in the oldest known proboscidean. *Lethaia*, **25**, 211–15. {8}

Court, N. (1993*a*). A dental peculiarity in *Numidotherium koholense*: evidence of feeding behaviour in a primitive proboscidean. *Zeitschrift für Säugetierkunde*, **58**, 194–6. {8}

Court, N. (1993*b*). Morphology and functional anatomy of the postcranial skeleton in *Arsinoitherium* (Mammalia, Embrithopoda). *Paleontographica*, Abteilung A, Band **226**, 125–169 + 9 plts. {8}

Court, N. (1994*a*). Limb posture and gait in *Numidotherium koholense*, a primitive proboscidean from the Eocene of Algeria. *Zoological Journal of the Linnaean Society* **111**, 297–338. {8}

Court, N. (1994*b*). The periotic of *Moeritherium* (Mammalia, Proboscidea): homology or homoplasy in the ear region of Tethytheria McKenna, 1975? *Zoological Journal of the Linnean Society*, 112, 13–28 {8}.

Court, N. (1995). A new species of *Numidotherium* (Mammalia: Proboscidea) from the Eocene of Libya and the early phylogeny of the Proboscidea. *Journal of Vertebrate Paleontology*, **15(3)**, 650–71. {34}

Court, N. and Jaeger, J.-J. (1991). Anatomy of the periotic bone in the Eocene proboscidean *Numidotherium koholense*: an example of parallel evolution in the inner ear of tethytheres. *Comptes Rendus de l'Académie des Sciences, Paris*, Series II, **312**, 559–65. {5, 8}

Court, N. and Mahboubi, M. (1993). Reassessment of lower Eocene *Seggeurius amourensis*: aspects of primitive dental morphology in the mammalian order Hyracoidea. *Journal of Paleontology*, **67** (5), 889–93. {8}

Cowie, J. W., Ziegler, W., Boucot, A. J., Bassett, M. G., and Remane, J. (1986). Guidelines and statutes of the International Commission on Stratigraphy (ICS). *Courier Forschungsinstitut Senckenberg*, **83**, 1–14. {ApD}

Cowie, J. W., Ziegler, W., and Remane, J. (1989). Stratigraphic commission accelerates progress, 1984 to 1989. *Episodes*, **12** (2), 79–83 + one-page supplement. {ApD}

Coxon, P. (1982). The terraces of the River Waveney. In *Field guide to the Gipping and Waveney valleys, Suffolk*, (ed. P. Allen), pp. 80–94. Quaternary Research Association, Cambridge. {19}

Cring, F. D. (1986). Enamel prism patterns in proboscidean molar teeth. *Elephant*, **2**, (2), 72–9. {2}

Croizet, J. B. and Jobert, A. C. G. (1828). *Recherches sur les ossemens fossiles du Département du Puy-De-Dôme*. Principaux Libraires, Paris. {ApC}

Crouzel, F. and Debeaux, M. (1957). Découverte de *Tetralophodon longirostris* près de Montrejeau (Haute Garonne). *Bulletin de la Société d'Histoire Naturelle de Toulouse*, **92**, 202–4. {14}

Croze, H., Hillman, A. K. K., and Lang, E. M. (1981). Elephants and their habitats: how do they tolerate each other. In *Dynamics of large mammal populations*, (ed. C. W. Fowler and T. D. Smith), pp. 297–316. Wiley, New York. {24}

Cumming, D. H. M. and Jackson, P. (ed.) (1984). *The status and conservation of Africa's elephants and rhinos.* Proceedings of the Joint Meeting of IUCN/SSC African Elephant and African Rhino Specialist Groups at Hwange Safari Lodge, Zimbabwe, 30 July–7 August 1981. IUCN Publications, Gland, Switzerland. {32}

Cumming, D. H. M., Du Toit, R. F., and Stuart, S. N. (compilers) (1990). *African elephants and rhinos: status survey and conservation action plan.* IUCN/SSC African Elephant and Rhino Specialist Group, Gland, Switzerland. {32}

Curatolo, J. A. (1975). Factors influencing local movements and behavior of barren ground caribou (*Rangifer tarandus granti*). Unpublished M.Sc. thesis. University of Alaska, Fairbanks. {29}

Cushing, J., Daily, M., Noble, E., Roth, V. L., and Wenner, A. (1984). Fossil mammoths from Santa Cruz Island, California. *Quaternary Research*, **21**, 376–84. {24}

Cushing, J., Wenner, A. M., Noble, E., and Daily, M. (1986). A groundwater hypothesis for the origin of "fire areas" on the Northern Channel Islands, California. *Quaternary Research*, **26**, 207–17. {24}

Cuvier, F. (1825). Eléphant d'Afrique. In *Histoire naturelle des Mammifères*, avec des figures originales, coloriées, dessinées d'après des animaux vivants; publiée sous l'autorité de l'administration du Museum d'Histoire Naturelle (ed. E. Geoffroy Saint-Hilaire and F. Cuvier, for the years of 1824–1847, 7 volumes), pp. 1–2 with one plate, Volume 3. A. Belin, Paris. {ApC}

Cuvier, G. (1798). *Tableau élémentaire de l'histoire naturelle des animaux.* Baudouin, Paris. {ApC}

Cuvier, G. (1806). Sur différentes dents du genre des Mastodontes, mais d'espèces moindres que celle de l'Ohio, trouvées en plusieurs lieux des deux continents. *Annales du Muséum d'Histoire Naturelle*, **8**, 401–24. {1, 3, ApA}

Cuvier, G. (1817). *Le règne animal*, Vol. 1. Déterville, Paris. {1, 5, 6, 10, 11, 16, ApA, ApC}

Cuvier, G. (1824). *Recherches sur les ossemens fossiles*, (2nd edn), Vol. 5. Dufour et d'Ocagne édit, Paris. {ApC}

Cuvier, G. (1849). *Anatomie Comparée, recueil des planches de myologie.* Published under the auspices of C. Laurillard and Mercier, Paris. {2}

Czelusniak, J., *et al.* (1990). Perspectives from amino acid and nucleotide sequences on cladistic relationships among higher taxa of Eutheria. In *Current mammalogy*, Vol. 2 (ed. H. H. Genoways), pp. 545–72. Plenum, New York. {5}

Dahlquest, W. W. (1983). Mammals of the Coffee Ranch local fauna, Hemphillian of Texas. *Pearce-Sellards Series*, **38**, 1–31. {15}

Daniel, J. C. (1980). The status of the Asian elephant in India. *Elephant*, **1**, (4), 16–28. {33}

Dansgaard, W. (1964). Stable isotopes in precipitation. *Tellus*, **16**, 436–68. {30}

Dart, R. A. (1929). Mammoths and other fossil elephants of the Vaal and Limpopo watersheds. *South African Journal of Science*, **26**, 698–731. {ApC}

Darwin, C. (1859). *The origin of species by means of natural selection.* J. Murray, London. {1}

Dau, J. (1986). Distribution and behavior of barren ground caribou in relation to weather and insects. Unpublished M.Sc. thesis. University of Alaska, Fairbanks. {29}

Daubenton, L. J. M. (1764). Description de la partie du cabinet qui a rapport à l'histoire naturelle du zèbre et de l'hippopotame. *Histoire Naturelle Général et Particulière avec la Description du Cabinet du Roi*, **XII**, 69–78, nos. MLXXXVII–MCXIII. {11}

Davis, O. K., Agenbroad, L. D., Martin, P. S., and Mead, J. I. (1984). The Pleistocene dung blanket of Bechan Cave, Utah. In *Contributions to Quaternary vertebrate paleontology: a volume in memorial to John E. Guilday*, (ed. H. H. Genoways and M. R. Dawson), pp. 267–82. Special Publication of Carnegie Museum of Natural History No. 8, Pittsburgh. {28, 30}

Davis, O. K., Mead, J. I., Martin, P. S., and Agenbroad, L. D. (1985). Riparian plants were a major component of the diet of mammoths of southern Utah. *Current Research in the Pleistocene*, **2**, 81–2. {28}

de Blainville, H. M. D. (1816). See Blainville.

Dehm, R. and Oettingen-Spielberg, T. (1958). Paläontologische und geologische Untersuchungen im Tertiär von Pakistan. 2. Die mitteleocänen Säugetiere von Ganda Kas bei Basal in Nordwest-Pakistan. *Bayerische Akademie der Wissenschaften Mathematisch-Naturwissenschaftliche Klasse*, Neue Folge, **91**, 1–54 + 3 plts. {8, ApA}

de Jong, W. W. and Goodman, M. (1982). Mammalian phylogeny studied by sequence analysis of the eye lens protein x-crystallin. *Zeitschrift für Säugetierkunde*, **47**, 257–76. {5}

de Jong, W. W., Leunissen, J. A. M. and Wistow, G. J. (1993). Eye lens crystallins and the phylogeny of placental orders: evidence for a macroscelid-paenungulate clade? In *Mammal phylogeny: placentals*, (ed. F. S. Szalay, M. J. Novacek, and M. C. McKenna) pp. 5–12. Springer-Verlag, New York. {5}

Delcourt, P. A., Delcourt, H. R., Brister, R. C., and Lackey, L. E. (1980). Quaternary vegetation history of the Mississippi embayment. *Quaternary Research*, **13**, 111–32. {27}

Delort, R. (1990). *Les éléphant piliers du monde.* Découvertes Gallimard, Histoire Naturelles, Evreux (France). [English translation published in 1992 by Abrams, New York, as *The life and lore of the elephant.*] {34}

de Man, J. D. (1875). Beenderen van den mammouth en van het uitgestorven rund, opgevist in den omtrek van Zeeland. *Archief van het Zeeuwsch Genootschap der Wetenschappen*, **3**, (3), 1–22, Middelburg. {20}

de Muizon, C., Marshall, L. G., and Sigé, B. (1984). The mammal fauna from the El Molino Formation (late Cretaceous, Maestrichtian) at Tiupampa, Southcentral Bolivia [in English with French, Spanish, and English abstracts]. *Bulletin du Muséum National d'Histoire Naturelle, Paris*, Series **4** (6), section C (4), 327–51. {ApD}

Depéret, C. (1897). Découverte du *M. angustidens* dans l'étage Cartennien de Kabylie. *Bulletin de la Société Géologique de France*, Series 3, **25**, 518–21. {ApC}

Depéret, C. (1907*a*). *Les transformations du monde animal.* Flammarion, Paris. {12A}

Depéret, C. (1907*b*). Sur l'âge des couches à *Palaeomastodon* du Fayoum. *Bulletin de la Société Géologique de France*, 4 serie, tome VII, 193–4. {ApC}

Depéret, C., Mayet, L., and Roman, F. (1923). Les éléphants pliocènes. *Annales de l'Université de Lyon, Sciences, Médecine*, New Series I, **42**, 1–224. {19}

Deraniyagala, P. E. P. (1951*a*). *Elephas maximus*, the elephant of Ceylon (Part 1). *Spolia Zeylanica*, **26**, 21–69 + 17 plts. {2}

Deraniyagala, P. E. P. (1951*b*). *Elephas maximus*, the elephant of Ceylon (Part 2). *Spolia Zeylanica*, **26**, 161–77 + 8 plts. {2}

Deraniyagala, P. E. P. (1952). *Elephas maximus*, the elephant of Ceylon (Part 3). *Spolia Zeylanica*, **27**, 109–25. {2}

Deraniyagala, P. E. P. (1954). New subgenera of stegodont elephants. *Proceedings of the 10th meeting of Ceylon Association for Advancement of Science* [from Deraniyagala, 1955, pp. 24, 127]. {ApA}

Deraniyagala, P. E. P. (1955). *Some extinct elephants, their relatives and the two living species.* Ceylon National Museums Publication, Colombo. {1, 2, 8, 11, 25, 34, ApA, ApC}

Deraniyagala, P. E. P. (1958). *The Pleistocene of Ceylon.* Ceylon National Museums (Natural History Series), Colombo. {ApC}

Deraniyagala, P. E. P. (1963). Tusk and tail variation in *Elephas maximus maximus* Linné and in *Loxodonta africana* (Blumenbach). *Spolia Zeylanica*, **30**, (1), 85–6 + 1 plt. {2}

Deraniyagala, P. E. P. (1965). A pygmy stegodont elephant from China. *Spolia Zeylanica*, **30** (2), 3–5. {ApC}

D'Erasmo, G. (1948). I proboscidati fossili nello stato attuale della scienza. *Estratto dal Rendiconto dell' Accademia delle Scienze Fisiche e Matematiche della Società di Scienze, Lettere ed Arti*, Napoli, **15** (4), 1–13. {34}

Dermitzakis, M. D. and de Vos, J. (1987). Faunal succession and the evolution of mammals in Crete during the Pleistocene. *Neues Jahrbuch für Geologie und Paläontologie, Abhandlungen, Stuttgart*, **173**, 377–408. {22}

Dermitzakis, M. D. and Sondaar, P. Y. (1979). The importance of fossil mammals in reconstructing paleogeography with special reference to the Pleistocene Aegean Archipelago. *Annales Géologiques des Pays Helléniques, Athènes*, **29** (1978), 808–40. {22}

Desmarest, A. G. (1822). *Mammalogie ou description des espèces de mammifères, seconde partie, contenant les ordres des rongeurs, des édentés, des pachydermes, des ruminans et des cétacés*, viii + 277–530 + supplément, 531–55. Mme. Veuve Agasse, Paris. {ApC}

de Vos, J. (1983). The Pongo faunas from Java and Sumatra and their significance for biostratigraphical and paleo-ecological interpretations. *Proceedings of the Koninklijke Nederlandse Akademie van Wetenschappen, Amsterdam, B*, **86**, (4), 417–25. {23}

de Vos, J. (1984). The endemic Pleistocene deer of Crete. *Verhandelingen der Koninklijke Nederlandse Akademie van Wetenschappen, Amsterdam*, **31**, 1–100. {22}

de Vos, J. (in press). Pleistocene larger mammal faunas of the Indonesian Archipelago and their paleoecological and paleogeographical implications. *Proceedings of the International Conference "Early man in island environments", 25 September–2 October 1988, Oliena, Sardegna.* {23}

de Vos, J., Sartono, S., Hardja-Sasmita, H. S., and Sondaar, P. Y. (1982). The fauna from Trinil, type locality of *Homo erectus*; a reinterpretation. *Geologie en Mijnbouw*, **61**, 207–11. {23}

Dietrich, W. O. (1912). *Elephas primigenius Fraasi*, eine schwäbische Mammutrasse. *Jahresheft des Vereins für vaterländische Naturkunde in Württemberg*, **68**, 42–106. {19, 25}

Dietrich, W. O. (1965). Fossile Elefantenzähne von Voigtstedt in Thüringen. *Paläontologische Abhandlungen, Abteilung A (Paläozoologie)*, **2**, 521–36. {19}

Dillehay, T. D. (1984). A late Ice-Age settlement in southern Chile. *Scientific American*, **251**, 106–17. {31}

Dillehay, T. D. (ed.) (1989). *Monte Verde. A late Pleistocene site in Chile*, Vol. 1. *Paleoenvironmental and site context*. Smithsonian Institution Press, Washington, DC. {27, 31}

Dillehay, T. D. (1992). Bone assemblage at Monte Verde. Manuscript on file at the University of Kentucky, Lexington. {31}

Dillehay, T. D. and Collins, M. B. (1988). Early cultural evidence from Monte Verde in Chile. *Nature*, **332**, 150–2. {30}

Dillehay, T. D., Pino, M., Davis, E. M., Valastro, S. Jr, Varela, A. G., and Casamiquela, R. (1983). Monte Verde: radiocarbon dates from an Early Man site in south–central Chile. *Journal of Field Archaeology*, **9**, 547–9. {31}

Dobias, R. (1985). *Elephants in Thailand.* WWF Monthly Report, December 1985, pp. 307–12. World Wildlife Fund, Gland, Switzerland. {33}

Dobias, R. J. (1987). Elephants in Thailand: An overview of their status and conservation. *Tigerpaper*, **14**, 19–24. {33}

Dollfus, G., Cotter, M., and Gomes, J. P. (1903). Mollusques tertiaires du Portugal. *Memorias e Comunicacoes Servicos Geológicos de Portugal, Lisbon*, 1–48. {14}

Domning, D. P. and Shoshani, J. (1982). Order Proboscidea. In *Mammal species of the world: a taxonomic and geographic reference*, (ed. J. H. Honacki, K. E. Kinman, and J. W. Koeppl), p. 307. Allen Press, and The Association of Systematics Collections, Lawrence, Kansas. {12A, 16, ApC}

Domning, D. P., Ray, C. E., and McKenna, M. C. (1986). Two new Oligocene desmostylians and a discussion of tethytherian systematics. *Smithsonian Contributions to Paleontology*, **59**, 1–56. {3, 5, 8, 34, ApA}

Donovan, S. K. (ed.) (1989). *Mass extinctions: processes and evidence*. Columbia University Press, New York. {34}

Dorr, V. *et al.* (1982). *A guide to the Groleau-White Lake mastodon (Michigan's second mounted mastodon), Mammut americanum*. Oakland Community College, Highland Lakes Campus, Union Lake, Michigan. {ApC}

Douglas-Hamilton, I. (1979). African elephant survey. *Oryx*, **14**, (1), 24. {32}

Douglas-Hamilton, I. (1983). Elephants hit by arms race. Recent factors affecting elephant populations. *AERSG Newsletter*, **2**, 11–13. {32}

Douglas-Hamilton, I. (1984). Trends in key African elephant populations. *Pachyderm*, **4**, 7–9. {32}

Douglas-Hamilton, I. (1987). African elephants: population trends and their causes. *Oryx*, **21**, (1), 11–24. {32}

Douglas-Hamilton, I. (1989). Overview of the status and trends of the African elephant. In *The ivory trade and the future of the African elephant*, (co-ordinator S. Cobb), Vol. 2, section 1.1. The Ivory Trade Review Group, Oxford. {32}

Douglas-Hamilton, I. (1992). The African elephant population today. In *Elephants: majestic creatures of the wild*, (consult. ed. J. Shoshani), pp. 178–83. Rodale Press, Emmaus, Pennsylvania. {32}

Douglas-Hamilton, I. and Burrill, A. (1991). Using elephant carcass ratios to determine population trends. *African Wildlife: Research and Management*, 98–105. {32}

Douglas-Hamilton, I. and Douglas-Hamilton, O. (1975). *Among the elephants*. Viking Press, New York. {2}

Douglas-Hamilton, I. and Douglas-Hamilton, O. (1992). *Battle for the elephants*, (ed. B. Jackman). Doubleday, London. {32}

Douglas-Hamilton, I., Michelmore, F., and Inamdar, A. (1992). *The African elephant database.* Report to the European Commission African Elephant Survey and Conservation Programme. United Nations Environment Programme (UNEP) Publication, Nairobi. {32}

Downs, T. (1952). A new mastodont from the Miocene of Oregon: with remarks on *Gomphotherium*. *University of California Publications in Geological Sciences*, **29**, (1), 1–20. {15}

Drake, R. and Curtis, G. (1987). K–Ar geochronology of the Laetoli fossil localities. In *Laetoli — a Pliocene site in northern Tanzania*, (ed. M. D. Leakey and J. M. Harris), pp. 48–52. Clarendon, Oxford. {12B}

Drees, M. (1986). Kritische kanttekeningen bij de naam "zwarte botten fauna". *Cranium*, **3**, (2), 103–20. {20}

Dreimanis, A. (1968). Extinction of mastodons in eastern North America: testing a new climatic–environmental hypothesis. *Ohio Journal of Science*, **68**, 257–72. {27, 29, 30}

Driak, F. (1935). Studien der Zahnanlagen an einem fotus von *Elephas indicus*. *Morphologisches Jahrbuch*, **75**, 1–14. {4}

Dubois, E. (1908). Das geologische Alter der Kendeng-oder Trinil-Fauna. *Tijdschrift van de Koninklijke Nederlandse Aardrijkskundig Genootschap*, Series 2, **25**, 1235–70. {23}

Dubrovo, I. A. (1964). Elephants of the genus *Archidiskodon* in the USSR. *Paleontologicheskiy Zhurnal*, **3**, 82–94. {19, ApA}

Dubrovo, I. A. (1966). Systematic position of an elephant from the Khozar faunal assemblage. *Byulleten Komissii po Izucheniyu Chetvertichnogo Perioda*, **32**, 63–74. {19}

Dubrovo, I. A. (1970). New data to Miocene mastodonts of Inner Mongolia. In *Materials for the evolution of continental vertebrates*, (ed. K. K. Flerov), pp. 135–40 [in Russian]. {9, ApC}

Dubrovo, I. A. (1971). Order Proboscidea. In *Pleistocene of Tiraspol*, (ed. K. V. Nikoforova, E. I. Belyaeva, E. A. Vangengeim, N. A. Konstantinova, and K. N. Negadaev-Nikonov), pp. 92–108. Academy of Sciences of the Moldavian SSR, Kishinev. {19}

Dubrovo, I. A. (1974). New data on mastodonts of western Mongolia. Paleontologiya i biostratigrafiya Mongolii, (ed. N. N. Kramarenko *et al.*). *Sovmestnaya Sovjet.-Mongolian Paleontologiya Ekspedition*, **1**, 64–73. {9}

Dubrovo, I. A. (1977). A history of elephants of the *Archidiskodon–Mammuthus* phylogenetic line on the territory of the USSR. *Journal of the Palaeontological Society of India*, **20**, 33–40. {19, ApA}

Dubrovo, I. A. (1982). Morphology of the Yuribei mammoth skeleton. In *The Yuribei mammoth*, (ed. V. E. Sokolov), pp. 53–99. USSR Academy of Sciences, Moscow [in Russian]. {25, 26}

Dubrovo, I. A. (1989). Systematic position of the Khapry elephants. *Paleontologicheskiy Zhurnal*, **1989/1**, 78–87. {19}

Dubrovo, I. A. (1994). Fossil elephants from the Commonwealth of Independent States. In *The Hot Springs Mammoth Site: a decade of field and laboratory research in paleontology, geology, and paleoecology*, (ed. L. D. Agenbroad and J. I. Mead), pp. 426–451. The Mammoth Site of South Dakota, Hot Springs. {ApC}

Dubrovo, I. A. and Mol, D. (1991). Fossiele olifanten van de Sovjet-Unie en Nederland. *Cranium*, **8**, (2), 77–87. {19, ApA}

Dudley, J. P. (1987). Wildlife and landscapes of late Pleistocene Florida: the paleoecology of Rancholabrean mammoth–mastodont faunas in southeastern North America. Unpublished M.Sc. thesis. University of Florida, Gainesville. {29}

Dudley, J. P. (1988). Dental metrics of *Mammuthus* in the non-glacial regions of North America. *Current Research in the Pleistocene*, **5**, 68–9. {29}

Dumon Tak, A. M. (1973). Een mastodonkaak uit de Oosterschelde. *Zeeuws Tijdschrift*, **23**, (4), 21–2. {20}

Eales, N. B. (1925). External characters, skin, and temporal gland of a foetal African elephant, *Proceedings of the Zoological Society of London*, **1925**, 445–56. {2, 4}

Eales, N. B. (1926). The anatomy of the head of a foetal African elephant, *Elephas africanus (Loxodonta africana). Transactions of the Royal Society of Edinburgh*, **54**, (3), 491–551 + 12 plts. {2, 4}

Eales, N. B. (1928). The anatomy of a foetal African elephant, *Elephas africanus* (*Loxodonta africana*). Part II. The body muscles. *Transactions of the Royal Society of Edinburgh*, **55**, (3), 609–42 + 5 plts. {2, 8}

Eales, N. B. (1929). The anatomy of a foetal African elephant, *Elephas africanus (Loxodonta africana)*. Part III. The contents of the thorax and abdomen, and the skeleton. *Transactions of the Royal Society of Edinburgh*, **56**, 203–46 + 6 plts. {2, 8}

Eales, N. B. (1931). The development of the mandible in the elephant. *Proceedings of the Zoological Society of London*, **1931**, 115–25. {2}

Edwards, W. E. (1967). The late Pleistocene extinction and diminution in size of many mammalian species. In *Pleistocene extinctions: the search for a cause*, (ed. P. S. Martin and H. E. Wright Jr), pp. 141–54. Yale University Press, New Haven. {30}

Ehik, J. (1930). *Prodinotherium hungaricum* n.g., n. sp. *Geologica Hungarica, Series Paleontologica*, **6**, 1–24. {16, ApA, ApB, ApC}

Eichwald, E. (1835). De pecorum et pachydermorum reliquiis fossilibus in Lithuania, Volhynia et Podolia repertis. *Nova acta Academiae Caesareae Leopoldino-Carolinae Germanicae Naturae Curiosorum*, **17**, 677–780. {9}

Eisenberg, J. F. (1981). *The mammalian radiations: an analysis of trends in evolution, adaptation, and behavior*. University of Chicago Press. {2, 29}

Eisenberg, J. F. (1989). *Mammals of the neotropics*. University of Chicago Press. {29}

Eisenberg, J. F., McKay, G. M., and Jainudeen, M. R. (1971). Reproductive behavior of the Asiatic elephant (*Elephas maximus maximus* L.) *Behavior*, **38**, 193–225. {30}

Elder, W. H. (1970). Morphometry of elephant tusks. *Zoologica Africana*, **5**, 143–59. {30}

Eltringham, S. K. (1982). *Elephants*. Blandford, Poole, Dorset. {30}

Eltringham, S. K. and Ward, D. (consultant eds.) (1991). *The illustrated encyclopedia of elephants*. Salamander, London. {34}

Erd, K. (1965). Pollenanalytische Untersuchungen im Altpleistozän von Voigtstedt in Thüringen. *Paläontologische Abhandlungen A*, **2**, 261–72. {19}

Espinoza, E. O. and Mann, M.-J. (1993). The history and significance of the Schreger pattern in proboscidean ivory characterization. *Journal of the American Institute for Conservation*, **32** (1993), 241–8. {2}

Espinoza, E. O., Mann, M.-J., LeMay, J. P., and Oakes, K. A. (1990). A method for differentiating modern from ancient proboscidean ivory in worked objects. *Current Research in the Pleistocene*, **7**, 81–3. {2}

Esu, D., Kotsakis, T., and Burgio, E. (1986). I vertebrati ei molluschi continentali pleistocenici di Poggio Schinaldo (Palermo, Sicilia) *Bollettino della Societa Geologica Italiana, Roma*, **105**, 233–41. {22}

Fahlbusch, V. (1976). Report on the International Symposium on Mammalian Stratigraphy of the European Tertiary. *Newsletters on Stratigraphy*, **5**, (2/3), 160–7. {ApD}

Fahlbusch, V. (1991). The meaning of MN-zonation: considerations for a subdivision of the European continental Tertiary using mammals. *Newsletters on Stratigraphy*, **24** (3), 159–73. {ApD}

Falconer, H. (1845). Description of some fossil remains of *Dinotherium*, giraffe, and other Mammalia from the Gulf of Cambay, western coast of India. *Quarterly Journal of the Geological Society of London*, 1, 356–372 + 1 plt. {ApC}

Falconer, H. (1857). On the species of mastodon and elephant occurring in the fossil state in Great Britain. Part I. *Mastodon. Quarterly Journal of the Geological Society of London*, **13**, 307–60. {1, 6, 12A, 16, 19, 29, ApA, ApB, ApC}

Falconer, H. (1863). On the American fossil elephant of the regions bordering the Gulf of Mexico (*Elephas columbi* Falc.), with general observations on the living and extinct species. *Natural History Review*, **3**, 43–114. {29}

Falconer, H. (1868). *Paleontological memoirs and notes of the late Hugh Falconer with a biographical sketch of the author*, (ed. C. Murchison). Hardwicke, London. {15, 22, ApA, ApB, ApC}

Falconer, H. and Cautley, P. T. (1846). *Fauna Antiqua Sivalensis, being the fossil zoology of the Sewalik Hills, in the north of India*. Smith, Elder, London (plts only). [Editors' note: range of years is from 1845 to 1849 (Osborn 1936, p. 774). Part I was published in 1845, Part II in 1846, Parts III–VIII in 1847, Part IX in 1849 (Maglio 1973, p. 122)]. {12A, 15, 23, ApA, ApC}

Farrand, W. R. (1961). Frozen mammoths and modern geology. *Science*, **133**, 729–35. {29}

Farris, J. S. (1988). *Hennig86 version 1.5*. Port Jefferson Station, New York [user's manual printed by the author]. {6}

Felix, J. (1912). Das mammuth von Borna. *Veröffentlichungen des Städtischen Museums für Völkerkunde zu Leipzig*, **4**, 1–52. {19}

Fergusson, G. J. and Libby, W. F. (1962). UCLA Radiocarbon dates I. *Radiocarbon*, **4**, 109–14. {24}

Fernando, S. D. A., Jayasinghe, J. B., and Panabokke, R. G. (1963). A study of the temporal gland of an Asiatic elephant. *Ceylon Veterinary Journal*, **11**, 108–11. {2}

Fischer, M. S. (1986). Die Stellung der Schliefer (Hyracoidea) im phylogenetischen System der Eutheria. Zugleich ein Beiträg zur Anpassungsgeschichte der Procaviidae. *Courier Forschungsinstitut Senckenberg*, **84**, 1–132. {5}

Fischer, M. S. (1989). Hyracoids, the sister-group of perissodactyls. In *The evolution of perissodactyls*, (ed. D. R. Prothero and R. M. Schoch), pp. 37–56. Oxford University Press, New York. {5}

Fischer, M. S. (1990). Un trait unique de l'oreille des éléphants et des siréniens (Mammalia): un paradoxe phylogénétique. *Comptes Rendus de l'Académie des Sciences, Paris*, Series III, **311**, 157–62. [English abstract]. {5, 8, 34}

Fischer, M. S. and Tassy, P. (1993). The interrelation between Proboscidea–Sirenia–Hyracoidea–Mesaxonia: the morphological evidence. In *Mammal phylogeny: placentals*, (ed. F. S. Szalay, M. J. Novacek, and M. C. McKenna), pp. 217–34. Springer-Verlag, New York {5, 6, 8, 16, 34, ApA}

Fischer, M. S. and Trautmann, U. (1987). Fetuses of African elephants (*Loxodonta africana*) in photographs. *Elephant*, **2**, (3), 40–5. {2, 4}

Fischer de Waldheim, G. (1808). Programme d'Invit. Séance. *Publication de la Société Impériale des Naturalistes, Moscou*, pp. 19, 20, September [from Osborn 1936, p. 775]. {ApA}

Fischer de Waldheim, G. (1814). *Zoognosia. Tabulis synopticis illustrata, in usum Praelectionum Academiae Imperialis Medicochirurgicae*. Mosquensis Edita, Nicolai Sergeidis Vsevolozsky, Mosque, 13–1814, III, pp. xxiv + 694. [From Osborn, 1936, p. 775] {ApA, ApC}

Fisher, D. C. (1984). Taphonomic analysis of late Pleistocene mastodon occurrences: evidence of butchery by North American Paleo-Indians. *Paleobiology*, **10**, 338–57. {30, 34}

Fisher, D. C. (1987). Mastodont procurement by Paleoindians of the Great Lakes region: hunting or scavenging? In *The evolution of human hunting*, (ed. M. H. Nitecki and D. V. Nitecki), pp. 309–421. Plenum, New York. {30}

Fisher, D. C. (1988). Season of death of the Hiscock mastodonts. *Bulletin of the Buffalo Society of Natural Sciences*, **33**, 115–25. {30}

Fisher, D. C. (1989). Meat caches and clastic anchors: the cryptic record of Paleoindian subsistence in the Great Lakes region. *Geological Society of America, Abstracts with Programs*, **21**, (6), A234. {30}

Fisher, D. C. (1990). Age, sex, and season of death of the Grandville mastodont. *Michigan Archaeologist*, **36**, 141–60. {30}

Fisher, D. C. (1991). Growth profiles in mastodont tusks as an indicator of age at sexual maturity. *Journal of Vertebrate Paleontology*, **11** (3, Suppl.), 28A–29A. {30}

Fisher, D. C. (1994). Late Pleistocene proboscidean trackways in pond-margin sediments in southeastern Michigan. *Journal of Vertebrate Paleontology, Abstracts of Papers*, **14** (3), 25A. {34}

Fisher, D. C., Lepper, B. T., and Hooge, P. E. (1991). Taphonomic analysis of the Burning Tree mastodont. *Current Research in the Pleistocene*, **8**, 88–92. {30}

Flerov, C. C. (1931). A trunk of mammoth (*Elephas primigenius* Blum.) found in the Kolyma District (Siberia). *Bulletin of the Academy of Sciences USSR*, **1931**, 863–70 + 2 plts. {2}

Flower, W. H. and Lydekker, R. (1891). *An introduction to the study of mammals living and extinct*. Black, London. {5}

Flynn, L. J., Pilbeam, D., Jacobs, L. L., Barry, J. C., Behrensmeyer, A. K., and Kappelman, J. W. (1990). The Siwaliks of Pakistan: time and faunas in a Miocene terrestrial setting. *Journal of Geology*, **98**, 589–604. {12B}

Fortelius, M. (1985). Ungulate cheek teeth: developmental, functional, and evolutionary interrelations. *Acta Zoologica Fennica*, **180**, 1–76. {29}

Foster, J. W. (1872). [A new species of fossil elephant, *Elephas Mississippiensis*]. *Nature*, September 26, 1872, vol. 6, p. 443. {ApC}

Foster, J. W. (1873). [*Elephas Indianapolis*: a new species of fossil elephant]. *Proceedings of the American Association for the Advancement of Science*, Meeting of August, 1872. p. 259. {ApC}

Fourtau, R. (1918). *Contribution a l'étude des vertébrés miocènes de l'Egypte*. Ministry of Finance, Egypt: Survey Department, Government Press, Cairo [2nd edn published 1920; Osborn 1936, p. 775]. {ApC}

Fowler, C. W. (1981). Density dependence as related to life history strategy. *Ecology*, **62**, 602–10. {30}

Fox, J. W., Smith, C. B., and Wilkins, K. T. (ed.) (1992). *Proboscidean and Paleoindian interactions*. Baylor University Press, Waco, Texas. {30}

Frade, F. (1955). Ordre des proboscidiens (Proboscidea Illiger, 1811). In *Traité de Zoologie* (ed. P.-P. Grassé), pp. 715–875. Masson et Cie Éditeurs, Paris, **17** (1), 1–1170. {ApC}

Frailey, C. D. (1986). Late Miocene and Holocene mammals, exclusive of the Notoungulata, of the Rio Acre Region, Western Amazonia. Contributions in Science, Natural History Museum of Los Angeles County, **374**, 1–46. {29}

Franklin, I. R. (1981). Evolutionary change in small populations. In *Conservation biology: an evolutionary–ecological perspective*, (ed. M. E. Soule and B. A. Wilcox), pp. 135–49. Sinauer Associates, Sunderland, Massachusetts. {33}

Frick, C. (1921). Extinct vertebrate faunas of the badlands of Bautista Creek and San Timoteo Canon, southern California. *Bulletin Department of Geology, University of California*, **12**, 277–424. {27}

Frick, C. (1926). Tooth sequence in certain trilophodont tetrabelodont mastodons and *Trilophodon* (*Serridentinus*) *pojoaquensis*, new species. *Bulletin of the American Museum of Natural History*, **56**, 122–78. {17}

Frick, C. (1933). New remains of trilophodont–tetrabelodont mastodons. *Bulletin of the American Museum of Natural History*, **59**, 505–652. {6, 11, 13, 15, 16, 27, 34, ApA, ApB, ApC}

Frison, G. C. (1990). Clovis, Goshen, and Folsom: lifeways and cultural relationships. In *Megafauna and man: discovery of America's heartland*, (ed. L. D. Agenbroad, J. I. Mead, and L. W. Nelson). The Mammoth Site of Hot Springs, South Dakota, Inc., Hot Springs, Scientific Papers, Vol. 1, 100–8. {30}

Frison, G. C. and Todd, L. C. (1986). *The Colby mammoth site: taphonomy and archaeology of a Clovis kill in Northern Wyoming*. University of New Mexico Press, Albuquerque. {30}

Froelich, D. J. and Kalb, J. E. (1995). Internal reconstruction of elephantid molars: applications for functional anatomy and systematics. *Paleobiology*, **21**, 379–92. {16}

Fulton, R. J., Karrow, P. F., LaSalle, P., and Grant D. R. (1984). Summary of Quaternary stratigraphy and history, eastern Canada. In *Quaternary stratigraphy of Canada – a Canadian contribution to IGCP project 24*, (ed. R. J. Fulton), pp. 193–210. Geological Survey of Canada Paper 84–10 Ottawa, Ontario, Canada. {27}

Gabunia, L. K. (1979). Biostratigraphic correlations between the Neogene Land Mammal Faunas of the east and central Paratethys. *Annales Géologiques des Pays Helléniques, Athènes*, **1**, 413–23 ['Gabunia', also spelled 'Gabounia' and 'Gabuniya']. {14}

Gabunia, L. K. (1981). Traits essentiels de l'évolution des faunes de mammifères Néogènes de la région mer Noire-Caspienne. *Bulletin du Muséum National d'Histoire Naturelle, Paris*, **3**, (4), section C (2), 195–204. {9}

Gabunia, L. K. and Dubrovo, I. A. (1990). The classification and stratigraphic significance of the genus *Archidiskodon* (Mammalia, Proboscidea). *Byulleten Moskovskogo Obshchestva Ispytatelei Prirody. Moskva. Otdel Geologicheskii*, **65**, 75–82. {19}

Gagnon, M. (1992). Succession of mammalian communitites [*sic*] in the Fayum of Egypt in relation to the age of the Jebel Qatrani formation. *Journal of Vertebrate Paleontology, Abstracts of Papers*, **12** (3), 29A. {ApD}

Gao, Y. (1981). On the present status, historical distribution and conservation of wild elephant in China. *Acta Theriqlogica Sinica*, **1**, (1), 19–26. {33}

Garland, E. B. and Cogswell, J. W. (1985). The Powers mastodon site. *Michigan Archaeologist*, **31**, 3–39. {30}

Garutt, V. E. (1954*a*). Novye dannye o drevneishikh slonakh, rod *Protelephas* gen nov. [New data on the earliest elephants, genus *Protelephas* gen. nov.]. *Doklady An SSSR (Earth Sciences)* **114** (1), 189–91. [From Maglio 1973, p. 122] {ApA}

Garutt, V. E. (1954*b*). Southern elephant *Archidiskodon meridionalis* (Nesti) from the northern coast of the Azov Sea. *Trudy Komissii po Izucheniyu Chetvertichnogo Perioda*, **10**, 3–76. {19}

Garutt, V. E. (1957). O novom Iskopaemom slone *Phanagoroloxodon mammontoides* gen. et sp. nov. S. Kaukaza [A new extinct elephant *Phanagoroloxodon mammontoides* gen. et sp. nov., from the Caucasus]. *Doklady An SSSR*, **112** (2), 333–5. [From Maglio, 1973, p. 123] {ApA}

Garutt, V. E. (1964). *Das Mammut, M.* primigenius *(Blumenbach)*. Franckh'sche Verlagshandlung, Kosmos-Verlag, Stuttgart. {19, 25}

Garutt, V. E. (1971). On the history of discovery of intermediate links in the phylogenetic sequence of elephants *Archidiskodon–Mammuthus*. In *Chronology of the Glacial Age*, (ed. V. A. Zubakov and V. V. Kochegura), pp. 78–88. Pleistocene Commission, Geographical Society of the USSR, Leningrad. {19}

Garutt, V. E. (1972). Skeleton of Chosarian mammoth (*Mammuthus chosaricus* Dubrovo) from middle Pleistocene deposits of the river Or (Kama Basin). In *Aspects of the stratigraphy and correlation of Pliocene and Pleistocene deposits of the northern and southern parts of the Fore-Urals*, vol. 2, pp. 35–55. Institute of Geology, Bashkiria Branch of the Academy of Sciences of the USSR, Ufa. {19}

Garutt, V. E. (1981). Proboscidea Illiger, 1811. In *Catalogue of mammals of the USSR: Pliocene–Recent*, (ed. I. M. Gromov and G. I. Baranova), pp. 304–18. 'Nauka', Leningrad. {ApC}

Garutt, V. E. (1986). Origin of elephants Elephantidae and their phylogeny. *Trudy Zoologicheskogo Instituta, AN SSSR*, **149**, 15–32 [in Russian with English summary]. {ApA, ApC}

Garutt, V. E. (1992). Phanagorian elephant *Phanagoroloxodon mammontoides* Garutt, 1957, and the evolutionary pathways of the subfamily Elephantinae. *Trudy Zoologicheskogo Instituta, AN SSSR*, **246**, 29–40 [in Russian with English summary]. {ApA}

Garutt, V. E. and Baigusheva, V. S. (1981). *Archidiskodon gromovi* Garutt et Alexeeva–der älteste Elefant der Mammutlinie in Eurasien. *Quartärpaläontologie*, **4**, 7–18. {19}

Garutt, V. E. and Nikolskaja, V. N. (1988). Über das Skelett vom Steppenelefanten aus Edersleben. *Beiträge zur Heimatforschung, Sangerhausen, Spengler Museum*, **9**, 3–13. {19, 20, 25}

Garutt, V. E., Gentry, A., and Lister, A. M. (1990). *Mammuthus* Brookes, 1828 (Mammalia, Proboscidea): proposed conservation, and *Elephas primigenius* Blumenbach, 1799 (currently *Mammuthus primigenius*): proposed designation as the type species of *Mammuthus*, and designation of a neotype. *Bulletin of Zoological Nomenclature*, **47** (1), 38–44. {16, 19, 25, ApA}

Garutt, V. E., Averianov, A. O., and Vartanyan, S. L. (1993). On the systematic position of isolated population of mammoth *Mammuthus primigenius* (Blumenbach, 1799) of Wrangel Island (north-eastern Siberia). *Papers of the Russian Academy of Sciences*, **332(6)**, 799–801 [in Russian with English summary OR abstract]. {34}

Gaudry, A. (1862). *Animaux fossiles et géologie de l'Attique*. Savy, Paris. {ApC}

Gaudry, A. (1866). Considérations générales sur les animaux fossiles de Pikermi. *Librairie de la Société Géologique de France*, Savy, 1–68. {1, 12A}

Gaudry, A. (1893). L'éléphant de Durfort. In *Volume Commémoratif du Centenaire du Muséum d'Histoire Naturelle, Paris*, pp. 327–47. Muséum National d'Histoire Naturelle, Paris. {8}

Gaudry, A. and Lartet, É. (1856). Résultats des recherches paléontologiques entreprises dans l'Attique sous les auspices de l'Académie. *Comptes Rendus de l'Académie des Sciences*, **43**, 271–4. {ApC}

Gazin, C. L. and Collins, R. L. (1950). Remains of land mammals from the Miocene of the Chesapeake Bay region. *Smithsonian Miscellaneous Collections*, **116** (20), 1–21. {15, ApC}

Gaziry, A. W. (1976). Jungtertiäre Mastodonten aus Anatolien (Türkei). *Geologisches Jahrbuch*, **B 22**, 3–143. {9, ApC}

Gaziry, A. W. (1980). Weitere Choerodonten (Mastodonten, Proboscidea, Mammalia) aus dem Jungtertiar Anatoliens. *Mainzer Geowissenschaftliche Mitteilungen*, **9**, 117–20. {9}

Gaziry, A. W. (1982). Proboscidea from the Sahabi Formation. *Garyounis Scientific Bulletin*, **4**, 101–8. {12A}

Gaziry, A. W. (1987*a*). New mammals from the Jabel Zaltan site, Libya. *Senckenbergiana Lethaea*, **68**, 69–89. {ApC}

Gaziry, A. W. (1987*b*). Remains of Proboscidea from the early Pliocene of Sahabi, Libya. In *Neogene Paleontology and Geology of Sahabi*, (ed. N. T. Boaz, A. El-Arnauti, A. W. Gaziry, J. de Heinzelin, and D. D. Boaz), pp. 183–203. Alan R. Liss, New York. {12A, 12B, 13, 15, ApC}

Gaziry, A. W. (1994*a*). *Bunolophodon grandidens* n. sp. aus der Oberen Sußwassermolasse Süddeutschlands; zur Phylogenie der trilophodonten Mastodonten (Mammalia, Proboscidea). *Verhandlungen des Naturwissenschaftlichen Vereins in Hamburg*. (NF) **34**, 113–33 {ApA, ApC}

Gaziry, A. W. (1994*b*). Über Mastodonten (Mammalia, Proboscidea) aus den Dinotheriensanden (M-Europa; Ober-Miozän): Materialien des Hessischen Landesmuseums Darmstadt. *Verhandlungen des Naturwissenschaftlichen Vereins in Hamburg*, (NF) **34**, 95–112. {ApC}

Gentry, A. W. (1987). Mastodons from the Miocene of Saudi Arabia. *Bulletin of the British Museum (Natural History) Geology*, **41**, 395–407. {10, 21}

Geoffroy Saint-Hilaire, E. (1837). Encore cet écrit sur le *Sivatherium*, considéré sous le point de vue de ses révélations contestées relativement à la philosophie zoologique. *Comptes Rendus de l'Académie des Sciences*, **4**, 113–22. [From Osborn 1936, p. 776] {ApA}

Geoffroy Saint-Hilaire, E. and Cuvier, G. L. C. F. D. (1795). Mémoire sur une nouvelle division des mammifères, et sur les principes qui doivent servir de base dans cette sorte de travail, lu à la société d'Histoire naturelle le premier floréal de l'an troisième. *Magasin Encyclopédique ou Journal des Sciences, des Lettres et des Arts*, **2**, 164–90. Millin, Noel and Warens, Paris. {5, ApA}

Geoffroy Saint-Hilaire, E. and Cuvier, F. (eds). (1825). *Histoire naturelle des Mammifères avec des figures originales, coloriees, dessinées d'après des animaux vivans; publiee sous l'autorité de l'administration du Museum d'Histoire Naturelle* (for the years of 1824–1847, 7 volumes). Volumes **3** and **6**. A. Belin, Paris. {12A}

Georgiadis, N., Bischof, L., Templeton, A., Patton, J., Karesh, W., and Western D. (1994). Structure and history of African elephant populations: I. Eastern and southern Africa. *Journal of Heredity*, **85 (2)**, 100–4. {ApC}

Geraads, D. (1989). Vertébrés fossiles du Miocène supérieur du Djebel Krechem el Artsouma (Tunisie centrale). Comparaisons biostratigraphiques. *Géobios*, **6** (22), 777–801. {12B}

Gervais, P. (1836). Mammalogie. In *Dictionnaire Pittoresque d'histoire naturelle et des Phénomènes de la Nature*, (ed. F. E. Guérin), pp. 614–40. Cosson, Paris. {5}

Gheerbrant, E., Sudre, J., and Cappetta, H. (in preparation). Oldest known proboscidean from the Paleocene phosphates of Morocco [submitted to *Nature*]. {34}

Gibbard, P. L., *et al.* (1991). Early and middle Pleistocene correlations in the southern North Sea basin. *Quaternary Science Reviews*, **10**, 23–52. {20}

Gidley, J. W. (1926). Fossil Proboscidea and Edentata of the San Pedro Valley, Arizona, *US Geological Survey Professional Paper*, **140B**, 83–95. {15}

Gill, T. (1872). Arrangement of the families of mammals; with analytical tables. *Smithsonian Miscellaneous Collections*, **11** (Article 1, No. 230), i–vi + 1–98. {5}

Gillette, D. D. and Ray, C. E. (1981). Glyptodonts of North America. *Smithsonian Contributions to Paleobiology*, **40**, 1–255. {29}

Gingerich, P. D. (1977). A small collection of fossil vertebrates from the middle Eocene Kuldana and Kohat Formation of Punjab (Pakistan). *Contributions from the Museum of Paleontology, The University of Michigan*, **24**, (18), 190–203. {8}

Gingerich, P. D. (1984). Pleistocene extinctions in the context of origination–extinction equilibria in Cenozoic mammals. In *Quaternary extinctions: a prehistoric revolution*, (ed. P. S. Martin and R. G. Klein), pp. 211–22. University of Arizona Press, Tucson. {30}

Gingerich, P. D. (1992). *Marine mammals (Cetacea and Sirenia) from the Eocene of Gebel Mokattam and Fayûm, Egypt: stratigraphy, age, and paleoenvironments.* University of Michigan, Papers on Paleontology, No. 30, ix + 1–84. {8, ApD}

Gingerich, P. D. and Russell, D. E. (1981). *Pakictus inachus*, a new archaeocete (Mammalia, Cetacea) from the early–middle Eocene Kuldana Formation of Kohat (Pakistan). *Contributions from the Museum of Paleontology, The University of Michigan*, **25**, (11), 235–46. {8}

Gingerich, P. D., Russell, D. E., and Wells, N. A. (1990). Astragalus of *Anthracobune* (Mammalia, Proboscidea) from the early–middle Eocene of Kashmir. *Contributions from the Museum of Paleontology, The University of Michigan*, **28**, (3), 71–7. {6, 8, 16, 34}

Ginsburg, L. (1990). The faunas and stratigraphical subdivisions of the Orleanian in the Loire basin (France). In *European Neogene mammal chronology*, (ed. E. H. Lindsay, V. Fahlbusch, and P. Mein), pp. 157–76. Plenum, New York [copyright date, 1989]. {9, 10}

Ginsburg, L. and Antunes, M. T. (1966). Considérations sur les mastodontes du Burdigalien de Lisbonne et des Sables de l'Orléanais (France). *Revista de Faculdade de Lisboa*, 2nd Series, C, **14**, 135–50. {10}

Ginsburg, L. and Tassy, P. (1985). The fossil mammals and the age of the lignite beds in the intramontane basins of northern Thailand. *Journal of the Geological Society of Thailand*, **8**, (1–2), 13–27. {12B, 17}

Girard, C. (1852). On the classification of Mammalia. *Proceedings of the American Association for the Advancement of Science* (for the year 1851, 6th meeting, Albany, New York), pp. 319–35, 2 diagrams. {1}

Gittins, S. P. and Akonda, A. W. (1982). What survives in Bangladesh? *Tigerpaper*, **9**, 5–11. {33}

Gloger, C. W. L. (1841). *Gemeinnütziges Hand- und Hilfsbuch Naturgeschichte der Thiere*, A. Schulz, Breslau [from Osborn 1936, p. 777; see note there]. {ApA}

Godman, J. D. (1830). Description of a new genus and new species of extinct mammiferous quadruped. *Transactions of the American Philosophical Society*, N.S., **III**, 478–85, plts xvii, xviii [from Osborn 1936, p. 777]. {ApA}

Goldberg, P. (1986). Late Quaternary environmental history of the southern Levant. *Geoarcheology*, **1**, 225–44. {21}

Golowko, W. P. (1958). Fossil mammoth *Elephas primigenius jatzkovi* subsp. nova from the alluvium of Chichiklii River near village Alekseevka. *Prazi Odes'kogo Derzhavnogo Universiteta, Odessa*, **94(148)**, 315–20 [in Ukranian]. {ApC}

Gooding, A. M. and Ogden, J. G., III. (1965). A radiocarbon dated pollen sequence from the Wells Mastodon site near Rochester, Indiana. *Ohio Journal of Science*, **65**, 1–11. {27}

Gordon, J. A. (1966). Elephants do think. *African Wildlife*, **20**, 75–9. {34}

Gould, S. J. (1977). Eternal metaphors of palaeontology. In *Patterns of evolution: as illustrated by the fossil record*, (ed. A. Hallam), pp. 1–26. Elsevier Scientific Publishing, Amsterdam. {34}

Graf, I. E. (1957). Die Prinzipien der Arbestimmung bei *Deinotherium*. *Palaeontographica*, **180(A)**, 131–85. {ApC}

Graham, R. W. (1986*a*). Plant–animal interactions and Pleistocene extinctions. In *Dynamics of extinction*, (ed. D. K. Elliott), pp. 131–54. Wiley, New York. {30}

Graham, R. W. (1986*b*). Taxonomy of North American mammoths. In *The Colby mammoth site — taxonomy and archaeology of a Clovis kill in northern Wyoming*, (ed. G. C. Frison and L. C. Todd), pp. 165–9. University of New Mexico Press, Albuquerque. {18, 24, ApC}

Graham, R. W. and Grimm, E. C. (1990). Effects of global climate change on the patterns of terrestrial biological communities. *Trends in Ecology and Evolution*, **5**, 289–92. {30}

Graham, R. W. and Kay, M. (1988). Taphonomic comparisons of cultural and noncultural faunal deposits at the Kimmswick and Barnhart sites, Jefferson County, Missouri. *Bulletin of the Buffalo Society of Natural Sciences*, **33**, 227–40. {30}

Graham, R. W. and Lundelius, E. L. Jr. (1984). Coevolutionary disequilibrium and Pleistocene extinctions. In *Quaternary extinctions: a prehistoric revolution*, (ed. P. S. Martin and R. G. Klein), pp. 223–49. University of Arizona Press, Tucson. {30}

Graham, R. W., Haynes, C. V., Johnson, D. L., and Kay, M. (1981). Kimmswick: a Clovis–mastodon association in eastern Missouri. *Science*, **213**, 1115–17. {27, 30}

Graham, R. W., Holman, J.A., and Parmalee, P. W. (1983). Taphonomy and paleoecology of the Christensen Bog mastodon bone bed, Hancock County, Indiana. *Ilinois State Museum Reports of Investigations*, **38**, 29. {27}

Graham, R. W., Lundelius, E. L. Jr, Graham, M. A., and Stearley, R. F. (1991). FAUNMAP: computerized data base on the distribution of late Quaternary mammals. *Journal of Vertebrate Paleontology*, **11**, (3, Suppl.), 32A. {30}

Gramly, R. M. (1988). *The Adkins site: a Palaeo-Indian habitation and associated stone structure*. Persimmon, Buffalo, New York. {30}

Gray, J. E. (1821). On the natural arrangements of vertebrose animals. *London Medical Repository*, **15**, (88), 296–310. {1, 16, ApB, ApC}

Grayson, D. K. (1984). Explaining Pleistocene extinctions: thoughts on the structure of a debate. In *Quaternary extinctions: a prehistoric revolution*, (ed. P. S. Martin and R. G. Klein), pp. 807–23. University of Arizona Press, Tucson. {30}

Grayson, D. K. (1989). The chronology of North American late Pleistocene extinctions. *Journal of Archaeological Science*, **16**, 153–65. {30}

Gregory, J. T. (1945). An *Amebelodon* jaw from the Texas Panhandle. *University of Texas Publication*, **4401**, 477–84. {13, 15}

Gregory, W. K. (1903). Adaptive significance of the shortening of the elephant's skull. *Bulletin of the American Museum of Natural History*, **19**, 387–94 + 1 plt. {2}

Gregory, W. K. (1910). The orders of mammals. *Bulletin of the American Museum of Natural History*, **27**, 1–524. {5}

Gregory, W. K. (1929). Mechanics of locomotion in the evolution of limb structure as bearing on the form and habits of the titanotheres and the related odd-toed ungulates. In *The titanotheres of ancient Wyoming, Dakota, and Nebraska*, pp. 727–56. US Government Printing Office Monograph, No. 55, Washington. {24}

Gregory, W. K. (1934). On the significance of the supra-symphysial depression and groove in the shovel-tusked mastodonts. *Journal of Mammalogy*, **15**, 4–12. {16}

Gregory, W. K. (1935). Henry Fairfield Osborn. *Science*, **82**, 452–4. {Col}

Grimm, E. C. and Jacobson, G. L. (1992). Fossil-pollen evidence for abrupt climate changes during the past 18 000 years in eastern North America. *Climate Dynamics*, **6**, 179–84. {34}

Gromov, V. I. (1977). The elephant skull from Khapry. In *Palaeontological basis for the stratigraphy of the Anthropogene*, (ed. K. V. Nikiforova), 10th INQUA Congress, pp. 83–94. Geological Institute, USSR Academy of Sciences, Moscow. {19}

Gromov, V. I. and Garutt, V. E. (1975). Mandibel-Reste einer Frühform des *Mammuthus primigenius* (Blumenbach) von Weimar-Ehringsdorf. *Quartärpaläontologie*, **1**, 453–64. {19}

Groves, C., Grubb, P., and Bricknell, S. (1993). Taxonomy of African elephants. *Abstracts, Sixth International Theriological Congress, Sydney, Australia*, 4–10 July 1993, p. 120. {ApC}

Gruhn, R. and Bryan, A. L. (1984). The record of Pleistocene megafaunal extinction at Taima-taima, northern Venezuela. In *Quaternary extinctions: a prehistoric revolution*, (ed. P. S. Martin and R. G. Klein), pp. 128–37. University of Arizona Press, Tuson. {29, 31}

Guan, J. (1986). On classification and morphology of Amebelodontidae discovered from Ningxia and Longzhong Basin. Unpublished M.S. thesis. China University of Geosciences. [In Chinese with English abstract] {13, ApC}

Guan, J. (1988). The Miocene strata and mammals from Tongxin, Ningxia and Guanghe, Gansu. *Memoirs of Beijing Natural History Museum*, **42**, 1–21. {13, ApC}

Guan, J. (1991). The character analysis and phylogeny discussion on the shovel tusk mastodonts. *Memoirs of Beijing Natural History Museum*, **50**, 1–21. [In Chinese and English] {13}

Guan, J. and Rice, J. (1990). The dragon bones of Tongxin. *Natural History*, **Sept. 1990**, 60–7. {13}

Guan, J. and Song, R. (1991). A remark on statistical analysis on shovel-tusk mastodonts. In *60th anniversary essays for discovery of the first Peking man's skull*, pp. 144–51. Beijing Science and Technology Press. {13}

Guan, J. and van der Made, J. (1993). Fossil Suidae from Dingjiaergou near Tongxin, China. *Memoirs of Beijing Natural History Museum*, **53**, 151–99 [in English]. {13}

Guan, J. and Zhang, X. (1993). The middle Miocene lithic and mammalian records from Ningxia and Lonzhong basins, northwestern China. *Memoirs of Beijing Natural History Museum*, **53**, 229–43 [in English and Chinese]. {13]

Guan, J., Zhang, B., and Ma, Z. (1981). The fossiliferous Miocene deposits survey in Tongxin District, Ningxia. *Memoirs of Beijing Natural History Museum*, **13**, 1–13. {13}

Guenther, E. W. (1969). Die Elefantenmolaren aus den Kiesen von Süßenborn bei Weimar. *Paläontologische Abhandlungen A*, **3**, 711–34. {19}

Guenther, E. W. (1981). Die Backenzähne der Mammute von Salzgitter-Lebenstedt. *Quartärpaläontologie*, **4**, 27–41. {19}

Guenther, E. W. (1986). Funde von *Archidiskodon meridionalis* und von *Trogontherium cuvieri* aus den interglazialen Tegelen-Schichten. *Quartärpaläontologie*, **6**, 53–65. {20}

Guenther, E. W. (1987). Mammute und ihre Kümmerformen von Eurasien bis Mexiko. *Quartär*, **38**, 13–51. {24}

Guilday, J. E. (1967). Differential extinction during late-Pleistocene and recent times. In *Pleistocene extinctions: the search for a cause*, (ed. P. S. Martin and H. E. Wright), pp. 121–40. Yale University Press, New Haven. {21}

Guilday, J. E. (1984). Pleistocene extinction and environmental change: case study of the Appalachians. In *Quaternary extinctions: a prehistoric revolution*, (ed. P. S. Martin and R. G. Klein), pp. 250–8. University of Arizona Press, Tucson. {30}

Guilday, J. E., Hamilton, H. W., and Parmalee, P. W. (1975). Caribou (*Rangifer tarandus* L.) from the Pleistocene of Tennessee. *Journal Tennessee Academy of Science*, **50**, 109–12. {27}

Gunn, J. (1867). [Teeth of elephants from Norfolk.] Note in *The Geological Magazine*, Decade IV, **XXXIX**, 421–2. {ApA}

Gustafson, C. E., Gilbow, D., and Daugherty, R. D. (1979). The Manis mastodon site: early man on the Olympic Peninsula. *Canadian Journal of Archaeology*, **3**, 157–64. {30}

Gustafson, E. P. (1978). The vertebrate faunas of the Pliocene Ringold Formation, south-central Washington. *Bulletin of the Museum of Natural History, University of Oregon*, **23**, 62. {27}

Guthrie, R. D. (1968). Paleoecology of the large mammal community in interior Alaska during the late Pleistocene. *American Midland Naturalist*, **79**, 346–63. {30}

Guthrie, R. D. (1984). Mosaics, allelochemics and nutrients: an ecological theory of late Pleistocene megafaunal extinctions. In *Quaternary extinctions: a prehistoric revolution*, (ed. P. S. Martin and R. G. Klein), pp. 259–98. University of Arizona Press, Tucson. {29, 30}

Guthrie, R. D. (1990*a*). *Frozen fauna of the mammoth steppe: the story of Blue Babe*. University of Chicago Press. {29}

Guthrie, R. D. (1990*b*). Late Pleistocene faunal revolution — a new perspective on the extinction debate. In *Megafauna and man: discovery of America's heartland*, (ed. L. D. Agenbroad, J. I. Mead, and L. W. Nelson). The Mammoth Site of Hot Springs, South Dakota, Inc., Hot Springs, Scientific Papers, Vol. 1, pp. 42–53. {30}

Haas, G. (1966). *On the vertebrate fauna of the lower Pleistocene of 'Ubeidiya*. The Israel Academy of Sciences and Humanities, Jerusalem. {21}

Hagelberg, E., Thomas, M. G., Cook, C. E. Jr., Sher, A. V., Baryshnikov, G. F., and Lister, A. M. (1994). DNA from ancient mammoth bones. *Nature*, **370**, 333–4. {34}

Haileab, B. and Brown, F. H. (1992). Turkana Basin — Middle Awash Valley correlations and the age of the Sagantole and Hadar formations. *Journal of Human Evolution*, **22**, 453–68. {12B}

Hall-Martin, A. (1979). The Addo Elephant Research project: a brief introduction. *Elephant*, **1**, (3), 12–13. {32}

Hall-Martin, A. (1991). Draft elephant conservation plan for South Africa. Unpublished Typescript Report. {32}

Hallin, K. F. (1983). Hair of the American mastodon indicates an adaptation to a semiaquatic habitat. *American Zoologist*, **23**, 949. {27}

Han, D. F., Xu, C. H., and Yi, G. Y. (1975). Quaternary mammalian fossils from Bijiashan, Liuzhou, Guangxi. *Vertebrata PalAsiatica*, **13**, 250–6 + 2 plts [in Chinese]. {17}

Hanks, J. (1972). Growth of the African elephant (*Loxodonta africana*). *East African Wildlife Journal*, **10**, 251–72. {11, 30}

Hanks, J. (1979). *The struggle for survival: the elephant problem*. Mayflower Books, New York. {4, 11, 30, 34}

Hanks, J. and McIntosh, J. E. A. (1973). Population dynamics of the African elephant (*Loxodonta africana*). *Journal of Zoology*, **169**, 29–38. {30}

Hannus, L. A. (1985). The Lange/Ferguson site — an event of Clovis mammoth butchery with the associated bone tool technology: the mammoth and its track. Unpublished D. Phil. thesis. University of Utah. {28}

Hannus, L. A. (1990). The Lange–Ferguson site: a case for mammoth bone-butchering tools. In *Megafauna and man: the discovery of America's heartland*, (ed. L. D. Agenbroad, J. I. Mead, and L. W. Nelson). The Mammoth Site of Hot Springs, South Dakota, Inc., Hot Springs, Scientific Papers, Vol. 1, pp. 86–99. {28}

Hansen, M. C., Davids, D., Erwin, F., Haver, G. R., Smith, J. E., and Wright, R. P. (1978). A radiocarbon-dated mammoth site, Marion County, Ohio. *Ohio Journal of Science*, **78**, 103–5. {30}

Hansen, R. M. (1980). Late Pleistocene plant fragments in the dung of herbivores at Cowboy Cave. In *Cowboy Cave*, University of Utah Anthropological Papers No. 104, (ed. J. Jennings), pp. 179–89. Salt Lake City. {28}

Haq, B. U., Hardenbol, J., and Vail, P. R. (1988). Mesozoic and Cenozoic chronostratigraphy and cycles of sea-level change. In *Sea-level changes: an integrated approach*, (ed. C. K. Wilgus, B. S. Hastings, C. G. Kendall, H. W. Posamentier, C. A. Ross, and J. C. Van Wagoner), pp. 71–108. Society of Economic Paleontologists and Mineralogists, Special Publication 42, Tulsa, Oklahoma. {12B}

Harington, C. R. (1984). Mammoths, bison, and time in North America. In *Quaternary dating methods*, (ed. W. C. Mahaney), pp. 299–309. Elsevier Science Publishers, Amsterdam. {29}

Harington, C. R. and Ashworth, A. C. (1986). A mammoth (*Mammuthus primigenius*) tooth from late Wisconsin deposits near Embden, North Dakota, and comments on the distribution of woolly mammoths south of the Wisconsin ice sheets. *Canadian Journal of Earth Science*, **23**, 909–18. {34}

Harington, C. R., Grant, D. R., and Mott, R. J. (1993). The Hillsborough, New Brunswick, mastodon and comments on other Pleistocene mastodon fossils from Nova Scotia. *Canadian Journal of Earth Science*, **30**, 1242–53. {27}

Harland, W. B., Armstrong, R. L., Cox, A. V., Craig, L. E., Smith, A. G., and Smith, D. G. (1990). *A geologic time scale 1989* (poster and book). Cambridge University Press, Cambridge, England [Cox, A. V. is not a co-author on the poster]. {12B, ApD}

Harris, J. M. (1970). *Prodeinotherium* from Gebel Zelten, Libya. Unpublished D. Phil. thesis. University of Bristol. {ApC}

Harris, J. M. (1973). *Prodeinotherium* from Gebel Zelten, Libya. *Bulletin of the British Museum (Natural History) Geology*, **23**, (5), 285–348 + 5 plts. {3, 8, 16, ApA}

Harris, J. M. (1978). Deinotherioidea and Barytherioidea. In *Evolution of African mammals*, (ed. V. J. Maglio and H. B. S. Cooke), pp. 315–32. Harvard University Press, Cambridge, Mass. {6, 8, 34, ApC}

Harris, J. M. (1987). Summary. In *Laetoli — a Pliocene site in northern Tanzania* (ed. M. D. Leakey and J. M. Harris), pp. 524–32. Clarendon, Oxford. {12B}

Harris, J. M. and Jefferson, G. T. (ed.) (1985). Rancho La Brea: treasures of the Tar Pits. *Los Angeles County Museum of Natural History, George C. Page Museum of La Brea Discoveries, Science Series*, **31**, viii + 1–87. {ApC}

Harrison R. (1847). On the anatomy of the elephant. *Proceedings of the Royal Irish Academy*, **3**, (61), 385–98. {2}

Harrison, T., Delson, E., and Guan, J. (1991). A new species of *Pliopithecus* from the middle Miocene of China and its implications for early catarrhine zoogeography. *Journal of Human Evolution*, **21**, 329–61. {13}

Hartono, H. M. S. (1961). Geological investigations at Olabula, Flores. Unpublished report, Djawatan Geologi, Bandung, Indonesia. {23}

Haughton, S. H. (1922). A note on some fossils from the Vaal River gravels. *Transactions of the Geological Society of South Africa*, **24**, 11–16 + 1 plt. {ApC}

Hay, O. P. (1914). The Pleistocene mammals of Iowa (Mammals of the Pleistocene). *Iowa Geological Survey Annual Report for 1912*, **23**, 1–662 + 75 plts. {ApA, ApC}

Hay, O. P. (1922). Further observations on some extinct elephants. *Proceedings of the Biological Society of Washington*, **35**, 97–101. {16, ApB, ApC}

Hay, O. P. (1923). The Pleistocene of North America and its vertebrated animals from the states east of the Mississippi River and from the Canadian provinces east of longitude 95°. *Carnegie Institution of Washington Publication*, **322**, 1–499. {ApC}

Hay, O. P. (1924). The Pleistocene of the middle region of North America and its vertebrated animals. *Carnegie Institution of Washington Publication*, **322A**, 1–385. {ApC}

Hay, O. P. (1927). The Pleistocene of the western region of North America and its vertebrated animals. *Carnegie Institute of Washington Publication*, **322B**, 1–346 + 12 plts. {ApC}

Hay, O. P. [*in* Hay, O. P. and Cook, H. J.] (1928). Preliminary descriptions of fossil mammals recently discovered in Oklahoma, Texas and New Mexico. *Proceedings of the Colorado Museum of Natural History*, **8**(2), Part 1, 33. {ApC}

Haynes, C. V. Jr (1966). Elephant-hunting in North America. *Scientific American*, **214**, 104–12. {30}

Haynes, C. V. Jr (1984). Stratigraphy and late Pleistocene extinction in the United States. In *Quaternary extinctions: a prehistoric revolution*, (ed. P. S. Martin and R. G. Klein), pp. 345–53. University of Arizona Press, Tucson ['Jr' omitted from this paper]. {30}

Haynes, C. V. Jr (1985). Mastodon-bearing springs and late Quaternary geochronology of the lower Pomme de Terre Valley, Missouri. *Geological Society of America, Special Paper*, **204**, 1–35. {27}

Haynes, C. V. Jr (1991). Geoarchaeological and paleohydrological evidence for a Clovis-age drought in North America and its bearing on extinction. *Quaternary Research*, **35**, 438–50. {30}

Haynes, G. (1985). Age profiles in elephant and mammoth bone assemblages. *Quaternary Research*, **24**, 333–45. {30}

Haynes, G. (1986*a*). Spiral fractures and cut marks — mimic in noncultural elephant bone assemblages. *Current Research in the Pleistocene*, **3**, 45–6. {31}

Haynes, G. (1986*b*). Taphonomic studies of elephant mortality in Zimbabwe. *Elephant*, **2**, (2), 69–71. {34}

Haynes, G. (1987). Proboscidean die-offs and die-outs: age profiles in fossil collections. *Journal of Archaeological Science*, **14**, 659–68. {28, 34}

Haynes, G. (1988). Mass deaths and serial predation: comparative taphonomic studies of modern large mammal death sites. *Journal of Archaeological Science*, **15**, 219–35. {31}

Haynes, G. (1990). The mountains that fell down: life and death of heartland mammoths. In *Megafauna and man: discovery of America's heartland*, (ed. L. D. Agenbroad, J. I. Mead, and L. W. Nelson). The Mammoth Site of Hot Spring, South Dakota, Inc., Hot Springs, Scientific Papers, Vol. 1, 22–31. {25, 30}

Haynes, G. (1991). *Mammoths, mastodonts, and elephants: biology, behavior, and the fossil record*. Cambridge University Press. {18, 25, 28, 30, 31, 34}

Hays, I. (1834). Descriptions of the specimens of inferior maxillary bones of mastodons in the cabinet of the American Philosophical Society, with remarks on the genus *Tetracaulodon* (Godman), etc. *Transactions of the American Philosophical Society*, N.S., **4**, 317–39, plts XX–XXIX. {9, 20, ApC}

He, X. L. (1984). *The vertebrate fossils of Sichuan*. Sichuan Scientific and Technological Publishing House, Chengdu [in Chinese]. {17}

Heider, K. (1932). *Nomenclator animalium generum et subgenerum*. Band III. Im Verlage der Preußischen Akademie der Wissenschaften, Berlin [see Schulze, F. E., Kükenthal, W., and Heider, K. (1932)]. {ApC}

Heintz, A. E. and Garutt, V. E. (1965). Determination of absolute age of the fossil remains of mammoth and woolly rhinoceros from the permafrost in Siberia by the help of radiocarbon (C^{14}). *Norsk Geologisk Tidsskrift*, **45**, 73–9. {27}

Hendey, Q. B. (1970*a*). The age of the fossiliferous deposits at Langebaanweg, Cape Province. *Annals of the South African Museum*, **56**, 119–31. {12B}

Hendey, Q. B. (1970*b*). A review of the geology and palaeontology of the Plio-Pleistocene deposits at Langebaanweg, Cape Province. *Annals of the South African Museum*, **56**, 75–117. {12A, 12B}

Hendey, Q. B. (1978). The age of the fossils from Baard's Quarry, Langebaanweg, South Africa. *Annals of the South African Museum*, **75**, 1–24. {12A, 12B}

Hendey, Q. B. (1982). *Langebaanweg — a record of past life*. Rustica, South Africa. {12A}

Hennig, W. (1966). *Phylogenetic systematics*. University of Illinois, Urbana. {12A}

Hernandez Pacheco, F. and Dantin, J. (1915). Geología y Paleontología del Mioceno de Palencia. *Comision de Investigaciones Paleontologicas y Prehistóricas*. **5**, 296. Junta para la Amplicacion de Estudios e Investigaciones Cientificas, Madrid. {14}

Hershkovitz, P. (1971). Basic crown patterns and cusp homologies of mammalian teeth. In *Dental morphology and evolution*, (ed. A. A. Dahlberg), pp. 95–150. University of Chicago Press. {3}

Hess, D. L., Schmidt, A. M., and Schmidt, M. J. (1983). Reproductive cycle of the Asian elephant (*Elephas maximus*) in captivity. *Biology of Reproduction*, **28**, 767–73. {30}

Heybroek, P. (1974). Explanation to tectonic maps of The Netherlands by Nederlandse Aardolie Maatschappij B. V., Assen. *Geologie en Mijnbouw*, **53**, (2), 43–50. {20}

Hibbard, C. W. (1960). An interpretation of Pliocene and Pleistocene climates in North America. *Papers of the Michigan Academy of Science, Arts, and Letters*, President's Address, 62nd Annual Report, pp. 5–30. {30}

Hibbard, C. W. and Riggs, E. S. (1949). Upper Pliocene vertebrates from Keefe Canyon, Meade County, Kansas. *Bulletin of the Geological Society of America*, **60**, 829–60. {27}

Hildebrand, M. (1974). *Analysis of vertebrates structure*. Wiley, New York. {8}

Hill, A., *et al.* (1985). Neogene palaeontology and geochronology of the Baringo Basin, Kenya. *Journal of Human Evolution*, **14**, 759–73. {12B}

Hill, A., Curtis, G., and Drake, R. (1986). Sedimentary stratigraphy of the Tugen Hills, Baringo, Kenya. In *Sedimentation in the African rifts*, (ed. L. E. Frostick, R. W. Renaut, I. Reid, and J. J. Tiercelin), Geological Society (London), Special Publication No. 25, pp. 285–95. {12B}

Hill, W. C. O. (1953). The anatomy of the African elephant. In *The Elephant in east central Africa*, (ed. R. Ward), pp. 15–60. Rowland Ward, London. {2}

Hoffstetter, R. (1950). *Observaciones sobre los mastodontes de Sud America y especialmente del Ecuador*: Haplomastodon, *subgen., nov. de* Stegomastodon. Publicaciones de la Escuela Politecnica Nacional, no. 1, pp. 1–51. Quito, Ecuador. {6, 31, ApA, ApB, ApC}

Hoffstetter, R. (1952). Les mammifères pléistocènes de la République de l'Équateur. *Mémoires de la Société Géologique de France*. New Series **31**, 1–4 no. 66, 1–391 + 8 plts. {15, 31, ApA, ApC}

Hoffstetter, R. (1955). Remarques sur la classification et la phylogénie des mastodontes Sud-Américains. *Bulletin du Muséum National d'Histoire Naturelle*, Series 2, **27**, (6), 484–91. {31}

Hohn, A. A. (1980). Analysis of growth layers in the teeth of *Tursiops truncatus*, using light microscopy, microradiography, and SEM. In *Age determination of toothed whales and sirenians*, Reports of the International Whaling Commission, Special Issue 3, (ed. W. F. Perrin and A. C. Myrick Jr), pp. 155–60. International Whaling Commission, Cambridge. {30}

Holland, W. J. (1920). Fossil mammals collected at Pedra Vermelha, Bahia, Brazil, by Gerald A. Waring. *Annals of the Carnegie Museum*, **13**, 224–32. {31, ApC}

Holm-Nielsen (1965). See Nielsen.

Hooijer, D. A. (1949). Pleistocene vertebrates from Celebes. IV. *Archidiskodon celebensis* nov. spec. *Zoologische Mededelingen*, **30**, 205–26. {23}

Hooijer, D. A. (1953*a*). On dredged specimens of *Anancus*, *Archidiskodon*, and *Equus* from the Schelde estuary, Netherlands. *Leidse Geologische Mededelingen*, **17**, 335–59. {20}

Hooijer, D. A. (1953*b*). Pleistocene vertebrates from Celebes. V. Lower molars of *Archidiskodon celebensis* Hooijer. *Zoologische Mededelingen*, **31**, 311–18. {23}

Hooijer, D. A. (1953*c*). Pleistocene vertebrates from Celebes. VI. *Stegodon* sp. *Zoologische Mededelingen*, **32**, 107–12. {23}

Hooijer, D. A. (1953*d*). Pleistocene vertebrates from Celebes. VII. Milk molars and premolars of *Archidiskodon celebensis* Hooijer. *Zoologische Mededelingen*, **32**, 221–30. {23}

Hooijer, D. A. (1954*a*). Pleistocene vertebrates from Celebes. XI. Molars and a tusked mandible of *Archidiskodon celebensis* Hooijer. *Zoologische Mededelingen*, **33** (15), 103–20 + 3 plts. {12B, 23}

Hooijer, D. A. (1954*b*). A pygmy *Stegodon* from the middle Pleistocene of eastern Java. *Zoologische Mededelingen*, **33**, 91–102. {23}

Hooijer, D. A. (1955). Fossil Proboscidea from the Malay Archipelago and Punjab. *Zoologische Verhandelingen*, **28**, 1–146. {17, 23, ApC}

Hooijer, D. A. (1957). A *Stegodon* from Flores. *Treubia*, **24**, 119–29. {23}

Hooijer, D. A. (1958). An early Pleistocene mammalian fauna from Bethlehem. *Bulletin of the British Museum (Natural History) Geology*, **3** (8), 267–92. {12A, 12B, 21}

Hooijer, D. A. (1959). Fossil mammals from Jisr Banat Yaqub, south of Lake Huleh, Israel. *Bulletin of the Research Council of Israel*, **8G**, 177–99. {21}

Hooijer, D. A. (1960). A *Stegodon* from Israel. *Bulletin of the Research Council of Israel*, **G9**, 104–8. {21}

Hooijer, D. A. (1961). Middle Pleistocene mammals from Latamne, Orontes Valley, Syria. *Annales Archeologiques de Syrie*, **11**, 117–32. {12B, 21}

Hooijer, D. A. (1962*a*). The middle Pleistocene fauna of the Near East. In *Evolution and hominization*, (ed. G. Kurtén), pp. 81–3. Fischer Verlag, Stuttgart. {21}

Hooijer, D. A. (1962*b*). The sabre-toothed cat *Homotherium* found in The Netherlands. *Lutra*, **4**, (1), 24–6. {20}

Hooijer, D. A. (1963). *Miocene Mammalia of Congo*. Musée Royal de l'Afrique Centrale, Annales, Série 8, Sciences Géologiques, No. 46, 1–77 + 10 plts. {12B, 17}

Hooijer, D. A. (1964). Pleistocene vertebrates from Celebes. XII. Notes on pygmy stegodonts. *Zoologische Mededelingen*, **40**, 37–44. {23}

Hooijer, D. A. (1967). Indo-Australian insular elephants. *Genetica*, **38**, 143–62. {23}

Hooijer, D. A. (1969). The *Stegodon* from Timor. *Proceedings of the Koninklijke Nederlandse Akademie van Wetenschappen*, Series B, **72** (3), 201–10. {23}

Hooijer, D. A. (1970). *Miocene Mammalia of Congo, a correction*. Annales du Musée Royal de l'Afrique — Tervuren, Belgique, Serie In-8, Sciences Geologiques, No. 67, pp. 163–7. {12B}

Hooijer, D. A. (1972*a*). Pleistocene vertebrates from Celebes. XIV. Additions to the *Archidiskodon–Celebochoerus* fauna. *Zoologische Mededelingen*, **46**, 1–16. {23}

Hooijer, D. A. (1972*b*). *Stegodon trigonocephalus florensis* Hooijer and *Stegodon timorensis* Sartono from the Pleistocene of Flores and Timor. *Proceedings of the Koninklijke Nederlandse Akademie van Wetenschappen, Amsterdam*, Series B, **75** (1), 12–33. {23}

Hooijer, D. A. (1974). *Elephas celebensis* (Hooijer) from the Pleistocene of Java. *Zoologische Mededelingen*, **48**, 85–93. {17, 23}

Hooijer, D. A. (1976). Observations on the pigmy mammoths of the Channel Islands, California. In *Athlon: essays on paleontology in honor of Loris Shano Russell*, (ed. C. S. Churcher), pp. 220–5. Royal Ontario Museum of Life Sciences Miscellaneous Publication, University of Toronto Press. {24}

Hooijer, D. A. (1980). Remarks upon the dentition and tooth replacement in elephants. *Netherlands Journal of Zoology*, **30** (3), 510–15. {29}

Hooijer, D. A. (1981). The first rhinocerotid of the Pretiglian 'black bones' fauna from The Netherlands. *Netherlands Journal of Zoology*, **31** (2), 472–5. {20}

Hooijer, D. A. (1984*a*). *Mammuthus meridionalis* (Nesti) and *M. armeniacus* (Falconer) from the North Sea. *Proceedings of the Koninklijke Nederlandse Akademie van Wetenschappen*, Series B, **87** (3), 335–59. {20}

Hooijer, D. A. (1984*b*). A Pleistocene ass *Equus asinus*. L. subspecies from the North Sea between Britain and The Netherlands. *Lutra*, **27**, 193–202. {20}

Hooijer, D. A. (1984*c*). The solution of the *Cryptomastodon* problem. *Netherlands Journal of Zoology*, **34** (2), 228–31. {ApA}

Hopwood, A. T. (1935). Fossil Proboscidea from China. *Palaeontologia Sinica*, Series C, **9** (3), 1–108 + 8 plts. {13, 17, ApC}

Horowitz, A. (ed.) (1979). *The Quaternary of Israel*. Academic Press, New York. {21}

Höss, M., Pääbo, S., and Vereshchagin, N. K. (1994). Mammoth DNA sequences. *Nature*, **370**, 333. {34}

Howard, A. L. (1979). 'Motty' — birth of an African/Asian elephant at Chester Zoo. *Elephant*, **1** (3), 36–41. {7}

Howell, F. C. (1978). Hominidae. In *Evolution of African mammals*, (ed. V. J. Maglio and H. B. S. Cooke), pp. 154–248. Harvard University Press, Cambridge, Mass. {34}

Huang, W. (1979). On the age of the cave-faunas of south China. *Vertebrata PalAsiatica*, **17** (4), 327–43 [in Chinese with English summary]. {17, ApC}

Huang, X. (1975). A new *Pentalophodon* from the Pliocene of Qingyang, Gansu. *Vertebrata PalAsiatica*, **13** (4), 234–8 + 1 plt [in Chinese with English abstract]. {ApC}

Huxley, T. H. (1869). *An introduction to the classification of animals*. Churchill, London. {5}

Igel, W. (1984). Paläogeographie durch sediment-Morphoskopie. Grundlagen, Methodik und Anwendung. Unpublished D. Phil. thesis. University of Mainz. {19}

Illiger, C. D. (1811). *Prodromus systematis mammalium et avium additis terminis zoographicis uttriusque classis*. Salfeld, Berlin. {1, 5, ApB}

Imbesi, M. (1956). Sugli elefanti nani della grotta di Luparello (Palermo). *Actes du IV Congres International du Quaternaire, Roma-Pisa 1953, Roma*, pp. 443–7. {22}

Imbrie, J., *et al.* (1984). The orbital theory of Pleistocene climate: support from a revised chronology of the marine $\delta^{18}O$ record. In *Milankovitch and climate: understanding the response to astronomical forcing*, Part 1, (ed. A. Berger, J. Imbrie, J. D. Hays, G. Kukla, and B. Saltzman), pp. 269–305. Reidel, Dordrecht. {30}

International Commission on Zoological Nomenclature. (1985). *International Code of Zoological Nomenclature* (3rd edn). International Trust for Zoological Nomenclature, London. {ApC}

International Commission on Zoological Nomenclature (1991). *Mammuthus* Brookes, 1828 (Mammalia, Proboscidea): conserved, and *Elephas primigenius* Blumenbach, 1799 designated as the type species. *Bulletin of Zoological Nomenclature*, **48** (3), 279–80. {19, ApA}

Inuzuka, N. (1977*a*). On a fossil skull of *Palaeoloxodon naumanni* from Saruyama, Shimosamachi, Chiba Prefecture, central Japan. *Journal of Geology, Japan*, **83** (8), 523–36 + 4 plts. {ApC}

Inuzuka, N. (1977*b*). On the origin of *Palaeoloxodon naumanni*—a comparative osteology of the cranium. *Journal of Geology, Japan*, **83** (10), 639–55. {ApC}

Inuzuka, N., Hasegawa, Y., Nogariya, H., and Kamei, T. (1975). On the stylohyoid bone of Naumann's elephant (*Elephas naumanni* Makiyama) from Lake Nojiri. *Memoirs of the Faculty of Science, Kyoto University, Series of Geology and Mineralogy*, **41**, 49–65 + 3 plts. {ApC}

Irwin, D. M. and Wilson, A. C. (1993). Limitations of molecular methods for establishing the phylogeny of mammals, with specific reference to the position of elephants. In *Mammal phylogeny: placentals*, (ed. F. S. Szalay, M. J. Novacek, and M. C. McKenna), pp. 257–67. Springer-Verlag, New York. {5}

Jachmann, H. (1985). Estimating age in African elephants. *African Journal of Ecology*, **23**, 199–202. {28}

Jachmann, H. and Dublin, H. (1991). The impact of the ivory ban on illegal hunting of elephants in six range states in Africa. Unpublished report to WWF. {32}

Jacobs, L. L., Flynn, L. J. Downs, W. R., and Barry, J. C. (1990). *Quo vadis, Antemus?* The Siwalik muroid record. In *European Neogene mammal chronology*, (ed. E. H. Lindsay, V. Fahlbusch, and P. Mein), pp. 573–86. Plenum, New York [copyright date, 1989]. {10}

Janensch, W. (1911). Die Proboscidean-Schadel der Trinil-Expeditions-Sammulung. In *Die Pithecanthropus-Schichten auf Java*, (ed. L. Selenka and M. Blanckenhorn), pp. 151–95. Engelmann, Leipzig. {17}

Janis, C. M. (1988). New ideas in ungulate phylogeny and evolution. *Trends in Recent Ecology and Evolution*, **3** (11), 291–7. {5}

Janzen, D. H. and Martin, P. S. (1982). Neotropical anachronisms: the fruits the gomphotheres ate. *Science*, **215**, 19–27. {29, 34}

Jennings, J. D. (1980). *Cowboy Cave*. University of Utah Anthropological Papers No. 104, Salt Lake City. {28}

Jentzsch, C. A. (1883). Ueber einige tertiäre Säugethierreste aus Ost- und Westpreussen. *Schriften der physikalisch-ökonomischen Gesellschaft zu Königsberg*, **23** (2), 201–5. {9}

Jerison, H. J. (1973). *Evolution of the brain and intelligence*. Academic Press, New York. {2}

Ji, H. (1987). Geographical distribution and subdivision of Quaternary mammalian fauna of China. *Journal of Stratigraphy*, **11**, 91–102 [in Chinese]. {17}

Jilkibayev, K. J. (1975). *Fossil elephants of Kazakhstan*. Publishing House Nauka of Kazakh SSR, Alma-Ata (Kazakhstan). [In Russian] {ApC}

Jin, C. (1983). A new fossil elephant from Weixan, Shandong Province. *Vertebrata PalAsiatica*, **21** (3), 255–65 [in Chinese with English summary]. {ApC}

Johanson, D. C. and Edey, M. (1981). *Lucy: The beginnings of humankind*. Simon and Schuster, New York. {12A}

Johnsingh, A. J. T., Narendra Prasad, S., and Goyal, S. P. (1990). Conservation status of the Chila-Motichur corridor for elephant movement in Rajaji-Corbett National Parks area, India. *Biological Conservation*, **51**, 125–38. {33}

Johnson, D. L. (1972). Landscape evolution on San Miguel Island, California. Unpublished D. Phil. thesis, University of Kansas, Lawrence. University Microfilms, Ann Arbor. {24}

Johnson, D. L. (1978). The origin of island mammoths and the Quaternary land bridge history of the Northern Channel Islands, California. *Quaternary Research*, **10**, 204–25. {23, 24, 29}

Johnson, D. L. (1980). Episodic vegetation stripping, soil erosion, and landscape modification in prehistoric and recent historic time, San Miguel Island, California. In *The California Islands: Proceedings of a multidisciplinary symposium*, (ed. D. M. Power), pp. 103–21. Santa Barbara Museum of Natural History. {24}

Johnson, D. L. (1981). More comments on the Northern Channel Islands mammoths. *Quaternary Research*, **15**, 105–6. {24}

Johnson, D. L. (1983). The California continental borderland: landbridges, watergaps and biotic dispersals. In *Quaternary coastlines and marine archaeology: towards the prehistory of land bridges and continental shelves*, (ed. P. M. Masters and N. C. Flemming), pp. 481–527. Academic Press, London. {24}

Johnson, N. M., Stix, J., Tauxe, L., Cerveny, P. F., and Tahirkheli, R. A. K. (1985). Paleomagnetic chronology, fluvial processes, and tectonic implications of the Siwalik deposits near Chinji Village, Pakistan. *Journal of Geology*, **93**, 27–40. {10, 12B}

Jourdan, C. (1861). Des terrains sidérolithiques. *Compte rendus de l'Académie des Sciences, Paris*, **53**, 1009–14. {19, ApC}

Juliá, R. and Villalta, J. F. (1974). El Ampurdan. *Coloquio Internacional sobre Biostratigrafía continental del Neógeno superior y Cuaternario inferior. Montpellier-Madrid, Libro-Guia*, pp. 3–9. {14}

Junger, A. and Johnson, D. L. (1980). Was there a Quaternary land bridge to the Northern Channel Islands? In *The California Islands: proceedings of a multidisciplinary symposium*, (ed. D. M. Power), pp. 33–9. Santa Barbara Museum of Natural History. {24}

Kahlke, H.-D. (1961). On the complex of the *Stegodon–Ailuropoda* fauna of southern China and the chronological position of *Gigantopithecus blacki* V. Koenigswald. *Vertebrata PalAsiatica*, **3**, 83–108 [in Chinese with English summary]. {17}

Kahlke, H.-D. (ed.) (1969). *Das Pleistozän von Süßenborn*. Paläontologische Abhandlungen A (Paläozoologie), Vol. 3. Deutsche Gesellschaft für Geologische Wissenschaften, Berlin. {19}

Kahlke, H.-D., Eismann, L., and Wiegank, F. (1984). Die Neogen/Quartar-Grenze auf dem Territorium der Deutschen Demokratischen Republik. *Zeitschrift für angewandte Geologie*, 30, (1), 44–8. {20}

Kahlke, R.-D. (1990). Zur Festlegung des Lectotypus von *Mammuthus trogontherii* (Pohlig, 1885) (Mammalia, Proboscidea). *Quartärpaläontologie*, **8**, 119–24. {19}

Kalb, J. E. (1993). Refined stratigraphy of the hominid-bearing Awash Group, Middle Awash Valley, Afar Depression, Ethiopia. *Newsletters on Stratigraphy*, **29** (1), 21–62. {12B}

Kalb, J. E. (1995). Fossil elephantoids, Awash paleolake basins, and the Afar triple junction, Ethiopia. *Palaeogeography, Palaeoclimatology, Palaeoecology*, **114**, 357–68. {12B}

Kalb, J. E. and Jolly, C. J. (1982). Late Miocene and early Pliocene formations in the Middle Awash Valley, Ethiopia, and their bearing on the zoo geography of Sahabi. *Garyounis Scientific Bulletin*, **4**, 123–32. {12B}

Kalb, J. E. and Mebrate, A. (1993). Fossil elephantoids from the hominid-bearing Awash Group, Middle Awash Valley, Afar Depression, Ethiopia. *Transactions of the American Philosophical Society*, **83** (1), xvi + 1–114. {12A, 12B, ApC}

Kalb, J. E., *et al.* (1982). Vertebrate faunas from the Awash Group, Middle Awash Valley, Afar, Ethiopia. *Journal of Vertebrate Paleontology*, **2**, 237–58. {12A, 12B, 34}

Kamei, T. (ed.) (1991). *Japanese proboscidean fossils*. Tsukiji-Shokan, Tokyo. {ApC}

Kamei, T. and Kamiya, H. (1981). On the fossil teeth of *Stegolophodon pseudolatidens* (Yabe) from the Miocene bed of the Abukuma mountains. *Memoirs of the Faculty of Science, Kyoto University, Series of Geology and Mineralogy*, **47** (2), 165–78. {17}

Kamei, T., Okazaki, Y., and Nonogaki, I. (1977). On some new material of *Gomphotherium annectens* (Matsumoto) from the Mizunami group, central Japan. *Bulletin of the Mizunami Fossil Museum*, **4**, 509–61. {10}

Kamiya, H. (1991). Development of enamel layar [*sic*] observed in some fossil proboscideans. In *Mechanisms and phylogeny of mineralization in biological systems*, (ed. S. Suga and H. Nakahara), pp. 483–7. Springer-Verlag, Tokyo. {2, 17}

Kamiya, H. and Taruno, H. (1986). Tooth structure in *Stegolophodon*, *Eostegodon*, and *Stegodon* (Proboscidea, Mammalia): their phylogenetic relation. In *Teeth revisited: proceedings of the VIIth international symposium on dental morphology, Paris, 1986*, (ed. D. E. Russell, J.-P. Santoro, and D. Sigogneau-Russell). *Mémoires du Muséum National d'Histoire Naturelle, Paris (Series C)*, No. 53, 233–40. {17}

Kamiya, H., *et al.* (1991). Well preserved fossil mammalian footprints from the Plio-Pleistocene Kobiwako group, central Japan. *Abstracts, XIII International Union for Quaternary Research (INQUA), 2–9 August, Beijing, China*. {17, 34}

Kappelman, J., Simons, E. L., and Swisher, C. C., III. (1992). New age determinations for the Eocene-Oligocene boundary sediments in the Fayum depression, northern Egypt. *The Journal of Geology*, **100**, 647–68. {ApD}

Karrow, P. F. and Calkin, P. E. (1985). *Quaternary evolution of the Great Lakes*. Special Paper of the Geological Association of Canada, No. 30, p. 258. {27}

Kasse, K. (1988). Early Pleistocene tidal and fluviatile environments in the southern Netherlands and northern Belgium. Unpublished D. Phil. thesis. Free University Press, Amsterdam. {20}

Kaup, J. J. (1829). *Deinotherium giganteum*. *Isis*, **22 (4)**, 401–4. {1, 8, ApA, ApB, ApC}

Kaup, J. J. (1831). *Fossil*. Säugethiere Rheinhessens (*Fide* de Blainville, *Dinotherium*, p. 18, of his `Ostéographie'). [From Osborn 1936, p. 782] {ApA}

Kaup, J. J. (1832*a*). *Description d'Ossements fossiles de Mammifères inconnus jusqu'à présent, qui se trouvent au Musée grand-ducal de Darmstadt*. J. G. Heyer, Darmstadt. {6, 12A}

Kaup, J. J. (1832*b*). Ueber zwei Fragmente eines Unterkiefers von *Mastodon angustidens* Cuv., nach welchen diese Art in die Gattung *Tetracaulodon* Godmann gehört. *Isis* **25(6)**, 628–31, Tab. xia. {ApA, ApC}

Kaup, J. J. (1833). [*Dinotherium, Rhinoceros, Dorcatherium, Cervus, Sus*]. *Jahrbuch für Mineralogie, Geologie und Paläontologie*, **1833**, 419–20. {ApA}

Kerr, R. (1792). *The animal kingdom or zoological system, of the celebrated Sir Charles Linnaeus...*, London [complete citation, together with C. Linnaeus and J. F. Gmelin, in Osborn 1936, p. 782, publisher not given]. {1, 6, 27, ApC}

Kiltie, R. A. (1984). Seasonality, gestation time, and large mammal extinctions. In *Quaternary extinctions: a prehistoric revolution*, (ed. P. S. Martin and R. G. Klein), pp. 299–314. University of Arizona Press, Tucson. {30}

King, J. E. (1973). Late Pleistocene palynology and biogeography of the western Missouri Ozarks. *Ecological Monographs*, **43**, 539–65. {27}

King, J. E. and Saunders, J. J. (1984). Environmental insularity and the extinction of the American mastodont. In *Quaternary extinctions: a prehistoric revolution*, (ed. P. S. Martin and R. G. Klein), pp. 315–39. University of Arizona Press, Tucson. {3, 6, 27, 29, 30, 34}

King, J. E. and Saunders, J. J. (1986). *Geochelone* in Illinois and the Illinoian–Sangamonian vegetation of the type region. *Quaternary Research*, **25**, 89–99. {30}

Kingdon, J. (1979). Proboscids, Elephantids. In *East African mammals*, Vol. IIIB, pp. 8–75. Academic Press, London. {2, 18, 34}

Kinman, K. E. (1994). *The Kinman system: toward a stable cladisto-eclectic classification of organisms (living and extinct: 48 phyla, 269 classes, 1719 orders)*. Self-published, Hays, Kansas. {16}

Kitchell, J. A. (1990). The reciprocal interaction of organism and effective environment: learning more about 'and.' In *Causes of evolution: a paleontological perspective*, (ed. R. M. Ross and W. D. Allmon), pp. 151–69. University of Chicago Press. {30}

Klähn, H. (1922). *Die badischen mastodonten und ihre suddeutschen Verwandten*. Gebruder Borntraeger, Berlin. {14, ApC}

Klähn, H. (1931). Rheinhessisches Pliozän im Rahmen des mitteleuropäischen Pliozäns. *Geologische und Palaeontologische*, **18**, 279–339. {12A}

Klein, R. G. (1969). *Man and culture in the late Pleistocene*. Chandler, San Francisco. {19}

Klein, R. G. (1984). Mammalian extinctions and Stone Age people in Africa. In *Quaternary extinctions: a prehistoric revolution*, (ed. P. S. Martin and R. G. Klein), pp. 553–73. University of Arizona Press, Tucson. {34}

Kleinschmidt, T., Czelusniak, J., Goodman, M., and Braunitzer, G. (1986). Paenungulata: a comparison of the hemoglobin sequences from elephant, hyrax, and manatee. *Journal of Molecular Biology and Evolution*, **3**, 427–35. {5}

Klevezal, G. A. and Kleinenberg, S. E. (1969). *Age determination of mammals from annual layers in teeth and bones*. Israel Program of Scientific Translations, Jerusalem TT, 69-55033, pp. 1–128. {30}

Koch, A. K. [1840(?)]. *Fossil remains*. Churchill and Stewart, ?St. Louis, 8 pp and 1 plt. [from Osborn 1936, p. 783]. {ApA}

Koch, A. K. (1841) Description of the Missourium, or Missouri Leviathan; together with its supposed habits. Indian traditions concerning the location from whence it was exhumed; also, comparision with the whale, crocodile, and Missourium, with the Leviathan, as described in the 41st chapter of the book of Job. Charles Keemle, St. Louis, 16 pp. and no plate [from Osborn 1936, p. 783]. {ApA}

Koch, P. L. (1989). Paleobiology of late Pleistocene mastodonts and mammoths from southern Michigan and western New York. Unpublished D. Phil. thesis. University of Michigan. {30}

Koch, P. L. (1991). The isotopic ecology of Pleistocene proboscideans. *Journal of Vertebrate Paleontology*, **11** (Suppl. 3), 40A. {27, 30}

Koch, P. L., Fisher, D. C., and Dettman, D. L. (1989). Oxygen isotope variation in the tusks of extinct proboscideans: a measure of season of death and seasonality. *Geology*, **17**, 515–19. {30}

Koken, E. (1885). Ueber fossile Säugetiere aus China. *Geologie und Paläontologie*, Abhandlungen **3**, 31–114, plt 6–11, Berlin. {ApC}

Kornietz, N. L. (1962). On the causes of mammoth extinction on the Ukrainian territory. In *Extinct faunas of the Ukraine and neighbouring territories*, Vol. 1, pp. 93–169, Naukova Dumka, Kiev [in Ukrainian]. {26}

Kortenbout van der Sluijs, G. (1985). Botten uit de Oosterschelde. *Cranium*, **2** (1), 9–10. {20}

Kotsakis, T. (1979). Sulle mammalofaune quaternarie siciliane. *Bollettino del Servizio Geologico d'Italia, Roma*, **99**, 263–76. {22}

Kotsakis, T. (1980). Short account on the elephant remains of the Simonelli cave. *Quaderni dell'Academia Nazionale dei Lincei, Roma*, **249**, 115–21. {22}

Kotsakis, T. (1990). Insular and non insular vertebrate fossil faunas in the eastern Mediterranean islands. *Proceedings of the international symposium on biogeographical aspects of insularity, Rome 18–22 May 1987, Atti da Conregni Lincei, Roma*, pp. 289–334. {22}

Kotsakis, T., Palombo, M. R., and Petronio, C. (1978). *Mammuthus chosaricus* e *Cervus elaphus* del Pleistocene superiore di Via Flaminia (Roma). *Geologica Romana*, **17**, 411–45. {19}

Kotsakis, T., Petronio, C., and Sirna, G. (1980). The Quaternary vertebrates of the Aegean Islands: palaeogeographical implications. *Annales Géologiques des Pays Helléniques, Athènes*, **30** (1979), 31–64. {22}

Kowalski, K. and Kubiak, H. (1993). *Gomphotherium angustidens* (Cuvier, 1806) (Proboscidea, Mammalia) from the Miocene of Bełchatów and the proboscidean datum in Poland. *Acta Zoologica Cracoviensia*, **36** (2), 275–80. {ApC}

Krause, D. W. and Jenkins, F. A. Jr (1983). The postcranial skeleton of North American multituberculates. *Bulletin of the Museum of Comparative Zoology, Harvard University, Cambridge, Massachusetts*, **150** (4), 199–246. {8}

Krause, H. (1978). *The mammoth in the ice and snow?* Self-Publisher, Stuttgart. {7}

Kretzoi, M. (1942). Ausländische säugetierfossilien der ungarischen museen. *Különlenyomat a Földtani Közlöny*, **72** (1–3), 139–48. {ApA}

Kretzoi, M. (1950). Stegoloxodon nov. gen., a possible Asiatic ancestor of true loxodonts (Stegoloxodon nov. gen., a loxodonta elefántok esetleges ázsiai öse). *Különlenyomat a Földtani Közlöny*, **80**, 405–8. [In Hungarian and English] {ApA, Adm to ApC}

Krueger, H. W. (1991). Exchange of carbon with biological apatite. *Journal of Archaeological Science*, **18**, 355–61. {30}

Kubiak, H. (1972). The skull of *Mammut praetypicum* (Proboscidea, Mammalia) from the collection of the Jagiellonian University in Cracow, Poland. *Acta Zoologica Cracoviensia*, **17**, 305–24. {9, 14, 27}

Kubiak, H. (1980). The skulls of *Mammuthus primigenius* (Blumenbach) from Debica and Bzianka near Rzeszów, south Poland. *Folia Quaternaria*, **51**, 31–45. {19}

Kubiak, H. (1982). Morphological characters of the mammoth: an adaptation to the arctic–steppe environment. In *Paleoecology of Beringia*, (ed. D. M. Hopkins and J. V. Matthews, Jr), pp. 281–9. Academic Press, New York. {19}

Kubiak, H. (1989). Trabowce—Proboscidea. In *History and evolution of the terrestrial fauna of Poland* (ed. K. Kowalski), pp. 203–8. Folia Quaternaria, 59–60, Kraków, Poland. {ApC}

Kumar, K. (1991). *Anthracobune aijiensis* nov. sp. (Mammalia: Proboscidea) from the Subathu Formation, Eocene from NW Himalaya, India. *Géobios*, **24** (2), 221–39. {8, 16, 34, ApA, ApC}

Kunst, C. E. (1937). Die niederlandischen Pleistozänen Hirsche. Unpublished D. Phil. thesis. University of Leiden. {20}

Kunz, G. F. (1916). *Ivory and the elephant in art, in archaeology, and in science*. Doubleday, New York. {32}

Kurtén, B. (1968). *Pleistocene mammals of Europe*. Weidenfeld and Nicolson, London. {30}

Kurtén, B. and Anderson, E. (1980). *Pleistocene mammals of North America*. Columbia University Press, New York. {15, 18, 24, 29, 30, 31}

Kuss, S. E. (1965). Eine pleistozäne Säugtierfauna der Insel Kreta. *Berichte der Naturforschende Gesellschaft Freiburg im Brisgau*, **55**, 271–348. {22}

Kuss, S. E. (1966). Beiträge zur Pleistozän-Fauna der Insel Kreta. I. Die von D. Bate 1904 gesammelten Elefanten-und Cerviden-reste. *Berichte der Naturforschende Gesellschaft Freiburg im Brisgau*, **56**, 169–81. {22}

Kuss, S. E. (1970). Abfolge und Alter der Pleistozänen Säugetierfauna der Insel Kreta. *Berichte der Naturforschende Gesellschaft Freiburg im Brisgau*, **60**, 35–83. {22}

Kuss, S. E. (1973). Die pleistozänen Säugetierfaunen der Insel Kreta. *Pepragmena III Diethnous Kritologikou Synedriou, Rethymnon 18–23 Septemvriou 1971*, Vol. **1**, pp. 154–60. {22}

Kutzbach, J. E. and Guetter, P. J. (1986). The influence of changing orbital parameters and surface boundary conditions on climate simulations for the past 18 000 years. *Journal of Atmospheric Science*, **43**, 1726–59. {30}

Lahiri-Choudhury, D. K. (1986). *Elephants in northeast India*. WWF Monthly Report, January 1986, pp. 7–17. World Wildlife Fund, Gland, Switzerland. {33}

Lambert, W. D. (1989). The morphology and ecology of feeding in the shovel-tusked gomphotheres (Proboscidea, Gomphotheriidae, Amebelodontinae). Unpublished M.S. thesis. University of Florida. {15}

Lambert, W. D. (1990). Rediagnosis of the genus *Amebelodon* (Mammalia, Proboscidea, Gomphotheriidae), with a new subgenus and species, *Amebelodon* (*Konobelodon*) *britti*. *Journal of Paleontology*, **64**, 1032–40. {13, 15, 29, 34, ApA, ApC}

Lambert, W. D. (1992). The feeding habits of the shovel-tusked gomphotheres: evidence from tusk wear patterns. *Paleobiology*, **18** (2), 132–47. {15, 27}

Lambert, W. D. and Shoshani, J. (in press). Proboscidea. In *Tertiary mammals of North America* (ed. C. M. Janis, K. M. Scott, and L. L. Jacobs), Vol. 1. Cambridge University Press, New York. {15}

Langbauer, W. R., Payne, K. B., Charif, R. A., Rapaport, L. and Osborn, F. (1991). African elephants respond to distant playbacks of low-frequency conspecific calls. *Journal of Experimental Biology*, **157**, 35–46. {8}

Lark, R. M. (1984). A comparison between techniques for estimating the ages of African elephants (*Loxodonta africana*). *African Journal of Ecology*, **22**, 69–91. {2}

Lartet, E. (1859). Sur la dentition des Proboscidiens fossiles (*Dinotherium*, mastodontes et éléphants) et sur la distribution géographique et stratigraphique de leurs débris en Europe. *Bulletin de la Société Géologique de France*, Series 2, **16**, 469–516. {3, 11, ApC}

Laskarev, V. (1944). Ueber die Dinotherien reste aus Serbien. *Neues Jahrbuch für Mineralogie Monatsch*, Series B, **1944**, (2/3), 67–77. {ApC}

Laub, R. S. (1990). The Hiscock site (western New York): recent developments of Pleistocene and early Holocene interest. *Current Research in the Pleistocene*, **7**, 116–18. {30}

Laub, R. S., Miller, N. G., and Steadman, D. W. (ed.) (1988). Late Pleistocene and early Holocene paleoecology and archeology of the eastern Great Lakes region. *Bulletin of the Buffalo Society of Natural Sciences*, **33**, 1–316. {34}

Laub, R. S., Dufort, C. A., and Christensen, D. J. (1994). Possible mastodon gastrointestinal and fecal contents from the late Pleistocene of the Hiscock Site, western New York state. In *Studies in stratigraphy and paleontology in honor of Donald W. Fisher* (ed. E. Landing), pp. 135–48. New York State Museum Bulletin **481**. {30}

Laursen, L. and Bekoff, M. (1978). *Loxodonta africana*. *Mammalian Species*, **92**, 1–8. {2, 30, ApC}

Laury, R. L. (1980). Paleoenvironment of a late Quaternary mammoth-bearing sinkhole deposit, Hot Springs, South Dakota. *Bulletin of the Geological Society of America*, Part 1, **91** (8), 465–75. {28, 34}

Laws, R. M. (1962). Age determination of pinnipeds with special reference to growth layers in the teeth. *Zeitschrift für Säugetierkunde*, **27**, 129–46. {30}

Laws, R. M. (1966). Age criteria for the African elephant, *Loxodonta a. africana*. *East African Wildlife Journal*, **4**, 1–37. {2, 11, 12A, 18, 28, 30}

Laws, R. M. (1969). The Tsavo research project. *Journal of Reproduction and Fertility, Supplement* **6**, 495–531. {30}

Laws, R. M. (1970). Elephants as agents of habitat and landscape change in East Africa. *Oikos*, **21**, 1–15. {24, 32}

Laws, R. M. (1978). Deja vu ... *Swara*, **1 (3)**, 126–8. {34}

Laws, R. M. (1981). Experiences in the study of large mammals. In *Dynamics of large mammal populations* (ed. C. W. Fowler and T. D. Smith), pp. 19–46. Wiley, New York. {24}

Laws, R. M., Parker, I. S. C., and Johnstone, R. C. B. (1975). *Elephants and their habitats: the ecology of elephants in North Bunyoro. Uganda*. Clarendon, Oxford. {4, 11, 30, 32}

Leakey, M. D. (1987). Animal prints and trails. In *Laetoli: a Pliocene site in northern Tanzania*, (ed. M. D. Leakey and J. M. Harris), pp. 451–89. Clarendon, Oxford. {34}

Lee-Thorp, J. A., Sealy, J. C., and van der Merwe, N. J. (1989). Stable carbon isotope ratio differences between bone collagen and bone apatite, and their relationship to diet. *Journal of Archaeological Science*, **16**, 585–99. {24}

Lehmann, U. (1944). Über Kaubewegung und Abkauung bei bunodontem und tapiroidem Zahnbau (Bemerkungen zum Zahnbau einiger pflanzanfressender Säugetiere). *Neues Jahrbuch für Mineralogie, Geologie und Paläontologie Abhandlungen*, Series B, **88**, 385–400. {9, 17}

Lehmann, U. (1950). Über Mastodontenreste in der Bayerischen Staatssammlung in München. *Palaeontographica*, **99** (A), 121–228 + Plates X–XXII. {12A, 14, 16}

Lehmann, U. and Kraus, O. (1994). Outline of a phylogenetic system of the 'mastodonts' (Mammalia, Proboscidea). *Verhandlungen des Naturwissenschaftlichen Vereins in Hamburg*. (NF) **34**, 135–76. {16, ApC}

Leidy, J. (1858). Notice of remains of extinct vertebrata, from the valley of the Niobrara River, collected during the exploring expedition of 1857, in Nebraska, under the command of Lieut. G. K. Warren, U. S. Top. Eng., by Dr. F. V. Hayden, geologist to the expedition. *Proceedings of the Academy of Natural Sciences of Philadelphia*, **10**, 20–9. {15, 27, 28, ApC}

Leidy, J. (1869). The extinct mammalian fauna of Dakota and Nebraska, including an account of some allied forms from other localities, together with a synopsis of the mammalian remains of North America. *Journal of the Academy of Natural Sciences of Philadelphia*, 2nd Series, **7**, 1–472. {27}

Leidy, J. (1886). *Mastodon* and *Llama* from Florida. *Proceedings of the Academy of Natural Sciences, Philadelphia*, **38**, 11–12. {13, 15, ApC}

Leinders, J. J. M., Aziz, F., Sondaar, P. Y., and de Vos, J. (1985). The age of the hominid-bearing deposits of Java: state of the art. *Geologie en Mijnbouw*, **64**, 167–73. {17, 23}

Lekagul, B. and McNeely, J. A. (1977). Elephants in Thailand: importance, status and conservation. *Tigerpaper*, **4**, 22–5. {33}

Lepper, B. T., *et al.* (1991). Intestinal contents of a late Pleistocene mastodont from midcontinental North America. *Quaternary Research*, **36** (1), 120–5. {27, 30}

Lessertisseur, J. and Saban, R. (1967). Squelette appendiculaire. In *Traité de Zoologie*, (ed. P.-P. Grassé), Vol. 16, pp. 709–1078, 1110–16. Masson et Cie, Paris. {8}

Li, C., Wu, W., and Qiu, Z. (1984). Chinese Neogene: subdivision and correlation. *Vertebrata PalAsiatica*, **22** (3), 163–78 [in Chinese with English abstract]. {13}

Li, D. (1988). A study of Fuyang `*namadicus*' from the late Pleistocene. Unpublished Master thesis, China University of Geosciences, Beijing, China. [In Chinese with English abstract]. {ApC}

Li, Y. (1976). A new species of *Gomphotherium* from Shanxi. *Vertebrata PalAsiatica*, **14**, 67–70 [in Chinese]. {13}

Li, Y. (1981). On the subdivisions and evolution of the Quaternary mammalian faunas of south China. *Vertebrata PalAsiatica*, **19** (1), 67–76 [in Chinese with English summary]. {17, ApC}

Li, Y. and Wen, B. (1986). *Guanyindong, a lower Paleolithic site at Qianxi County, Guizhou Province*. Cultural Relics Publishing House, Beijing. {17}

Lin, Y. P., Pan, Y. R., and Lu, Q. W. (1978). The early Pleistocene fauna of Yuanmou basin. In *Professional Papers of the Paleoanthropology*, (ed. Institute of Vertebrate Paleontology and Paleoanthropology, Academia Sinica), pp. 101–19. Science Press, Beijing [in Chinese]. {17}

Lindeque, M. (1992). Namibia African elephant action plan. Typescript report by Ministry of Wildlife Conservation and Tourism, Windhoek, Namibia. {32}

Lindsay, E. H., Opdyke, N. D., and Johnson, N. M. (1980). Pliocene dispersal of the horse *Equus* and late Cenozoic mammalian dispersal events. *Nature*, **287**, 135–8. {20}

Lindsay, E. H., Opdyke, N. D., and Johnson, N. M. (1984). Blancan–Hemphillian land mammal ages and late Cenozoic mammal dispersal events. *Annual Review Earth Planetary Science*, **12**, 445–88. {27}

Linnaeus, C. (1754). *Animalia rariora, imprimis, et exotica*. Museum Adolfi Frederici Regis, Holmiae (Stockholm). {ApC}

Linnaeus, C. (1758). *Systema naturae per regna tria naturae, secundum classes, ordines, genera, species cum caracteribus, differentiis, synonymis, locis. Editio decima, reformata*, Vol. I. Laurentii Salvii, Holmiae. {5, 12A, ApA, ApB, ApC}

Lister, A. M. (1989). Pelvic sexing and skeletal maturation in the mammoth. *Abstracts of Papers and Posters Fifth International Theriological Congress, Rome, 22–29 August 1989*, Vol. I, p. 152. {26}

Lister, A. M. (1991). Late Glacial mammoths in Britain. In *The Late Glacial in north-west Europe*, (ed. N. Barton, A. J. Roberts, and D. A. Roe), pp. 51–9. Research Report 77, Council for British Archaeology, London. {19}

Lister, A. M. (1992). Mammalian fossils and Quaternary biostratigraphy. *Quaternary Science Reviews*, **11**, 329–44. {19}

Lister, A. M. (1993*a*). Evolution of mammoths and moose: the Holarctic perspective. In *Morphological change in Quaternary mammals of North America*, (ed. R. A. Martin and A. D. Barnosky), pp. 178–204. Cambridge University Press, New York. {19}

Lister, A. M. (1993*b*). 'Gradualistic' evolution: its interpretation in Quaternary large mammal species. *Quaternary International*, **19**, 77–84. {19}

Lister, A. M. (1993*c*). Mammoths in miniature. *Nature*, **362**, 288–89. {19}

Lister, A. M. (1993*d*). The stratigraphical signficance of deer species in the Cromer Forest-bed Formation. *Journal of Quaternary Science*, **8**, 95–108. {19}

Lister, A. M. (1994). A review of Kalb, J. E. and Mebrate, A. (1993). 'Fossil elephantoids from the hominid-bearing Awash Group, Middle Awash Valley, Afar Depression, Ethiopia.' published by Transactions of the American Philosophical Society, 83(1), xv + 1–114, *Journal of Evolutionary Biology*, **7(4)**, 517–8. {34}

Lister, A. M. and Agenbroad, L. D. (1994). Gender determination of the Hot Springs mammoths. In *The Hot Springs Mammoth Site: a decade of field and laboratory research in paleontology, geology, and paleoecology*, (ed. L. D. Agenbroad and J. I. Mead), pp. 208–14. The Mammoth Site of South Dakota Hot Springs. {25}

Lister, A. M. and Bahn, P. (1994). *Mammoths*. MacMillan, New York. {19}

Lister, A. M. and Brandon, A. (1991). A pre-Ipswichian cold stage mammalian fauna from the Balderton sand and gravel, Lincolnshire, England. *Journal of Quaternary Science*, **6**, 139–57. {19}

Lister, A. M. and Joysey, K. A. (1992). Scaling effects in elephant dental evolution – the example of Eurasian *Mammuthus*. In *Structure, function and evolution of teeth*, (ed. P. Smith and E. Tchernov), pp. 185–213. Freund, Tel-Aviv. {19}

Liu, J. and Zhen, S. (1981). Quaternary mammalian fauna from N. Anhui and a principle in relation to it. *Scientia Sinica*, **24** (2), 216–25 + 2 plts. {ApA, ApC}

Liu, T. and Li, X. (1984). Mammoths in China. In *Quaternary extinctions: a prehistoric revolution*, (ed. P. S. Martin and R. G. Klein), pp. 517–27. University of Arizona Press, Tucson. {34, ApC}

Logan, T. P. (1984). Early Irvingtonian (early Pleistocene) mammals from the upper part of the Santa Fe Group, Albuquerque-Belen Basin, central New Mexico. *Geological Society of America, Abstracts with Programs*, **16**, 245. {29}

Logan, T. P., Lucas, S. G., and Sobas, J. C. (1984). Blancan–Irvingtonian boundary in the Ceja Member of the Santa Fe Formation, Tiejeras Arroyo, Albuquerque area, New Mexico. *New Mexico Geology*, **6**, 14. {29}

Long, A. and Martin, P. S. (1974). Death of American ground sloths. *Science*, **186**, 638–40. {28}

Long, D., Laban, C., Streif, H., Cameron, T. D. J., and Schuttenhelm, R. T. E. (1988). The sedimentary record of climatic variation in the southern North Sea. *Philosophical Transactions of the Royal Society of London*, Series B, **318**, 523–37. {20}

Lortet, L. and Chantre, E. (1878). Recherches sur les mastodontes et les faunes mammalogiques qui les accompagnent. *Archives du Muséum d'Histoire Naturelle de Lyon*, **2**, 285–311. {9}

Lowenstein, J. M. (1980). Immunospecificity of fossil collagens. In *Biogeochemistry of amino acids*, (ed. P. E. Hare *et al.*), pp. 41–51, Wiley, New York. {7}

Lowenstein, J. M. (1985). Molecular approaches to the identification of species. *American Scientist*, **73** (6), 541–7. {7}

Lowenstein, J. M., Sarich, V. M., and Richardson, B. J. (1981). Albumin systematics of the extinct mammoth and Tasmanian wolf. *Nature*, **291**, 409–11. {7}

Lucas, S. G. (1987). American mastodont from the Sandia Mountains, New Mexico. *New Mexico Journal of Science*, **27** (1), 29–32. {ApC}

Lucas, S. G. and Alvarado, G. E. (1991*a*). Comentario sobre la clasificacion del mastodonte de Barra Honda (= Rio Nacaome), Guanacaste, Costa Rica. *Revista Geologica de América Central*, **13**, 97–8. {34}

Lucas, S. G. and Alvarado, G. E. (1991*b*). El hallazgo mas austral de un *Mammut americanum*: el caso del mastodonte de San Pedro Sula, Honduras. *Revista Geologica de América Central*, **13**, 85–9. {34}

Lucas, S. G. and Oakes, W. (1986). Pliocene (Blancan) vertebrates from the Palomas Formation, south-central New Mexico. *New Mexico Geological Society Guidebook, 37th Field Conference, Truth or Consequences*, 249–55. {ApC}

Luciano, D. S., Vander, A. J., and Sherman, J. H. (1978). *Human structure and function*. McGraw-Hill, New York. {30}

Luckett, W. P. (1993). An ontogenetic assessment of dental homologies in therian mammals. In *Mammal phylogeny: Mesozoic differentiation, multituberculates, monotremes, early therians, and marsupials*, (ed. F. S. Szalay, M. J. Novacek, and M. C. McKenna), pp. 182–204. Springer-Verlag, New York. {4}

Lull, R. S. (1936). A review of 'Proboscidea' by Henry Fairfield Osborn, edited by Mabel Rice Percy. *Natural History*, **38**, 357–8. {34}

Lundelius, E. L. Jr (1967). Late-Pleistocene and Holocene faunal history of central Texas. In *Pleistocene extinctions: the search for a cause*, (ed. P. S. Martin and H. E. Wright Jr), pp. 287–319. Yale University Press, New Haven. {30}

Lundelius, E. L. Jr, *et al.* (1983). Terrestrial vertebrate faunas. In *Late Quaternary environments of the United States*. Vol. 1. The late Pleistocene, (ed. S. C. Porter), pp. 311–53. University of Minnesota Press, Minneapolis. {29}

Lundelius, E. L. Jr, *et al.* (1987). The North American Quaternary sequence. In *Cenozoic mammals of North America: geochronology and biostratigraphy*, (ed. M. O. Woodburne), pp. 211–35. University of California Press, Berkeley. {27, 30}

Luxmore, R., Caldwell, J., and Hithersay, L. (1989). The volume of raw ivory entering international trade from African producing countries from 1979 to 1988. In *The ivory trade and the future of the African elephant*, (co-ordinator S. Cobb), Vol. 2, section 2.1. The Ivory Trade Review Group, Oxford. {32}

Lyal, C. H. C. (1985). Phylogeny and classification of the Psocodea with particular reference to the lice (Psocodea: Phthiraptera). *Systematic Entomology*, **10**, 45–65. {5}

Lydekker, R. (1876*a*). Description of a cranium of *Stegodon ganesa*, with notes on the sub-genus and allied forms. *Records, Geological Survey of India*, **9**, 42–9. {17}

Lydekker, R. (1876*b*). Molar teeth and other remains of Mammalia. *Memoirs of the Geological Survey of India Palaeontologia Indica*, **(X)**, **I**, **Pt II**, 19–87 + 7 plts, 1 page appendix. {ApC}

Lydekker, R. (1877). Notices of new or rare mammals from the Siwaliks. *Records, Geological Survey of India*, **10** (2), 76–83. {ApC}

Lydekker, R. (1884*a*). Additional Siwalik Perissodactyla and Proboscidea. *Memoirs of the Geological Survey of India Palaeontologia Indica*, (X), III, Part I, iv + pp. 1–34 + 5 plts. {ApC}

Lydekker, R. (1884*b*). Mastodon teeth from Perim Island. *Memoirs of the Geological Survey of India Palaeontologia Indica*, (X), III, Part 5, 149–54 + 2 plts. {ApC}

Lydekker, R. (1885). Description of a tooth of *Mastodon latidens*, Clift, from Borneo. *Proceedings of the Zoological Society of London*, **1885**, 777–9 + 1 plt. {23}

Lydekker, R. (1886*a*). Addenda to synopsis of Siwalik & Narbada Mammalia. *Memoirs of the Geological Survey of India Palaeontologia Indica* (X) III, xi–xxiv. {ApC}

Lydekker, R. (1886*b*). *Catalogue of the fossil Mammalia in the British Museum (Natural History)*. Part IV, containing the Order Ungulata, Suborder Proboscidea. British Museum (Natural History), London. {ApC}

Lydekker, R. (1907). The ears as a race-character in the African elephant. *Proceedings of the Zoological Society of London*, **26**, 380–403. {2}

Lynch, T. F. (1990). Glacial age man in South America? A critical review. *American Antiquity*, **55**, 12–36. {31}

Ma, A. and Tang, H. (1992). On discovery and significance of a Holocene *Ailuropoda—Stegodon* fauna from Jinhua, Zhejiang. *Vertebrata PalAsiatica*, **30** (4), 295–312 + 3 plts. {ApC.1}

McAndrews, J. H. and Jackson, L. J. (1988). Age and environment of late Pleistocene mastodont and mammoth in southern Ontario. *Bulletin of the Buffalo Society of Natural Sciences*, **33**, 161–72. {27}

Maccagno, A. M. (1962). L'*Elephas meridionalis* Nesti di contrada 'Madonna della Strada', Scoppito (L'Aquila). *Atti della Accademia delle Scienze Fisiche e matematiche, Napoli*, 3rd Series, **4**, (1), 1–132. {19}

McDonald, J. N. (1984). The reordered North American selection regime and late Quaternary megafaunal extinctions. In *Quaternary extinctions: a prehistoric revolution*, (ed. P. S. Martin and R. G. Klein), pp. 404–39. University of Arizona Press, Tucson. {30}

MacInnes, D. G. (1942). Miocene and post-Miocene Proboscidea from East Africa. *Transactions of the Zoological Society of London*, **25**, (2), 33–106 + 8 plts. {6, 12A, 12B, ApC}

McKenna, M. C. (1975). Toward a phylogenetic classification of the Mammalia. In *Phylogeny of the primates: a multidisciplinary approach*, (ed. W. P. Luckett and F. S. Szalay), pp. 21–46. Plenum, New York. {1, 5, 8}

McKenna, M. C. (1987). Molecular and morphological analysis of high-level mammalian interrelationships. In *Molecules and morphology in evolution: conflict or compromise?*, (ed. C. Patterson), pp. 55–93. Cambridge University Press. {16}

McKenna, M. C. (1992). The alpha crystallin A chain of the eye lens and mammalian phylogeny. *Annales Zoologici Fennici*, **28**, 349–60. {5}

McKenna, M. C. and Manning, E. (1977). Affinities and paleobiogeographic significance of the Mongolian Paleogene genus *Phenacolophus*. *Géobios*, Mémoire Spécial, **1**, 61–85. {5}

McKenna, M. C., Bell, S. K. *et al.* (1994). *A classification of mammals*. Manuscript on file at the American Museum of Natural History, New York. {16, ApA, ApB, ApC, ApD}

McKinney, C. R. (1980). Uranium series ages of Columbian mammoth enamel on Santa Rosa Island; implications for low sea stands and man's entry into North America. *Geological Society of America, Abstracts with Programs*, **12**, 481. {24}

McNeilly, A. S., Martin, R. D., Hodges, J. K., and Smuts, G. L. (1983). Blood concentrations of gonadotropins, prolactin and gonadal steroids in males and in non-pregnant female African elephants (*Loxodonta africana*). *Journal of Reproduction and Fertility*, **67**, 113–20. {30}

MacNeish, R. S. (1982). A late commentary on an early subject. In *Peopling of the new world*, (ed. J. E. Ericson, R. E. Taylor, and R. Berger), pp. 311–15. Ballena Press Anthropological Papers, No. 23. {24}

Madden, C. T. (1977*a*). Elephants of the Santa Barbara Channel Islands, southern California. *Geological Society of America, Abstracts with Programs*, **9**, 458–9. {24}

Madden, C. T. (1977*b*). *Pliomastodon* (Proboscidea, Palaeomastodontidae) from the late Miocene of Colorado. *The Mountain Geologist*, **14** (1), 29–32. {ApC}

Madden, C. T. (1977*c*). *Tetralophodon* (Proboscidea, Gomphotheriidae) from subsaharan Africa. *Revue de Zoologie Africaine*, **91** (1), 152–60. {ApC}

Madden, C. T. (1980*a*). Earliest isotopically dated mammoths from North America. *Quaternary Research*, **13**, 147–50. {24}

Madden, C. T. (1980*b*). *Zygolophodon* from subsaharan Africa, with observations on the systematics of palaeomastodontid proboscideans. *Journal of Paleontology*, **54**, 57–64. {6, 9, ApA}

Madden, C. T. (1981*a*). Letters to the editor: origin(s) of mammoths from Northern Channel Islands, California. *Quaternary Research*, **15**, 101–4. {18, 24}

Madden, C. T. (1981*b*). Mammoths of North America. Unpublished D. Phil. thesis. University of Colorado, Boulder, University Microfilms, Ann Arbor. {18, 24, ApC}

Madden, C. T. (1982). Primitive *Stegotetrabelodon* from latest Miocene of Subsaharan Africa (Proboscidea, Gomphotheriidae). *Revue de Zoologie Africaine*, **96**, 782–96. {12B}

Madden, C. T. (1986). *Stegomastodon* associated with *Mammuthus* in Arizona during the Quaternary. *Quaternary Research*, **26**, 266–71. {15}

Madden, C. T. and Storer, J. E. (1985). The Proboscidea from the middle Miocene Wood Mountain Formation, Saskatchewan. *Canadian Journal of Earth Science*, **22**, 1345–50. {29, ApC}

Madden, C. T. and Van Couvering, J. A. (1976). The proboscidean datum event: early Miocene migration from Africa. *Geological Society of America, Abstracts with Programs*, **8**, 992–3. {10, 12B}

Madden, C. T., *et al.* (1982). Stegotetrabelodon (*Proboscidea, Gomphotheriidae) from the Miocene of Abu Dhabi*. US Geological Survey Technical Report, TR-02-00, pp. 1–22. Jiddah, Saudi Arabia. {12B}

Maddison, W. P. (1989). Reconstructing character evolution on polytomous cladograms. *Cladistics*, **5**, 365–77. {16}

Maddison, W. P. and Maddison, D. R. (1987). *MacClade: an interactive, graphic program for analyzing phylogenies and studying character evolution*. Version 2.97.96. Computer program distributed by the authors. University of Arizona Press, Tucson. {16}

Maddison, W. P. and Maddison, D. R. (1992). *MacClade: analysis of phylogeny and character evolution*. Version 3.0. Computer program distributed by Sinauer Associates, Sunderland, Massachusetts. {16}

Maglio, V. J. (1969*a*). A shovel-tusked gomphothere from the Miocene of Kenya. *Breviora*, **310**, 1–10. {13, 15}

Maglio, V. J. (1969*b*). The status of the East African elephant '*Archidiskodon exoptatus*' Dietrich 1942. *Breviora*, **336**, 1–25. {12A}

Maglio, V. J. (1970*a*). Early Elephantidae of Africa and a tentative correlation of African Plio-Pleistocene deposits. *Nature*, **225**, 328–32. {12A, 18, 29}

Maglio, V. J. (1970*b*). Four new species of Elephantidae from the Plio-Pleistocene of northwestern Kenya. *Breviora*, **341**, 1–43. {1, 12A, ApB, ApC}

Maglio, V. J. (1972). Evolution of mastication in the Elephantidae. *Evolution*, **26** (4), 638–58. {12A, 18}

Maglio, V. J. (1973). Origin and evolution of the Elephantidae. *Transactions of The American Philosophical Society of Philadelphia, New Series*, **63** (3), 1–149. {1, 6, 12A, 12B, 16, 17, 18, 19, 20, 21, 22, 23, 24, 25, 28, 29, 30, 34, ApC}

Maglio, V. J. (1974*a*). Late Tertiary fossil vertebrate successions in the northern Gregory Rift, East Africa. *Annals of the Geological Survey of Egypt*, **4**, 269–85. {ApC}

Maglio, V.J. (1974*b*). A new proboscidean from the late Miocene of Kenya. *Palaeontology*, **17** (3), 699–705. {ApC}

Maglio, V. J. (1978). Patterns of faunal evolution. In *Evolution of African mammals* (ed. V. J. Maglio and H. B. S. Cooke), pp. 603–19. Harvard University Press, Cambridge, Massachusetts. {8}

Maglio, V. J. and Hendey, Q. B. (1970). New evidence relating to the supposed stegolophodont ancestry of the Elephantidae. *South African Archaeological Bulletin*, **25** (3–4), 85–7 + 2 plts. {12A, 12B, 18}

Maglio, V. J. and Ricca, A. B. (1978). Dental and skeletal morphology of the earliest elephants. *Verhandelingen der Koninklijke Nederlandse Akademie van Wetenschappen, Afd. Natuurkunde*, 1st Series, **29**, 1–51. {12A}

Mahboubi, M., Ameur, R., Crochet, J. Y., and Jaeger, J. J. (1984). Earliest known proboscidean from early Eocene of north-west Africa. *Nature*, **308**, 543–4. {8, 16}

Mahboubi, M., Ameur, R., Crochet, J. Y., and Jaeger, J. J. (1986). El Kohol (Saharan Atlas, Algeria): a new Eocene mammal locality in northwestern Africa. Stratigraphical, phylogenetic and paleobiogeographical data. *Paleontographica*, Series A, **192**, 15–49 + 6 plts. {3, 4, 6, 8, 16, ApB, ApC}

Mahoney, J. A. and Ride, W. D. L. (1975). *Index to the genera and species of fossil Mammalia described from Australia and New Guinea between 1838 and 1968 (including citations of type species and primary type specimens)*. Western Australia Museum Special Publication 6, pp. 1–250. {34, ApC}

Major, F. C. J. (1883). Die Tyrrhenis: Studien über geographische Verbreitung von Tieren und Pflanzen im westlich Mittelmeergebiet. *Kosmos*, **13**, 1–17, 81–106. {22}

Makiyama, J. (1924). Notes on a fossil elephant from Sahamma, Tôtômi. *Memoires of the College of Science, Kyoto Imperial University*, Series B, **1** (2), 255–64 + 5 plts. [Originally published in Japanese] {ApC}

Makiyama, J. (1938). Japonic Proboscidea. *Memoirs of the College of Science, Kyoto Imperial University*, Series B, **14** (1), 1–59. {10, 17}

Malatesta, A. (1954). Primo dente di elefante fossile rinvenuto in Sardegna. *Quaternaria*, **1**, 97–105. {22}

Maldonado, E., Mazo, A. V., and Alferez, F. (1983). Los mastodontes (Proboscidea, Mammalia) del Orleaniense medio de Córcoles (Guadalajara). *Estudios Geológicos*, **39**, 431–49. {14}

Marcus, L. F. and Berger, R. (1984). The significance of radiocarbon dates for Rancho La Brea. In *Quaternary extinctions: a prehistoric revolution*, (ed. P. S. Martin and R. G. Klein), pp. 159–83. University of Arizona Press, Tucson. {24, ApC}

Mariappa, D. (1986). *Anatomy and histology of the Indian elephant*. Indira Publishing House, Oak Park, Michigan. {2, 8}

Maringer, J. and Verhoeven, T. (1970). Die Steinartefacte aus der *Stegodon*-fossil-schicht von Mengeruda auf Flores, Indonesien. *Anthropos*, **65**, 229–47. {23}

Marinos, G. and Symeonidis, N. K. (1973). Erstmalige Funde von Zwergelefanten auf der Insel Rhodos. *Anzeiger der Mathematisch-Naturwissenschaftiche Klasse der Osterreichse Akademie der Wissenschaften, Wein*, **1973**, 130–1. {22}

Marsh, H. (1980). Age determination of the dugong (*Dugong dugon* Müller) in northern Australia and its biological implications. In *Age determination of toothed whales and Sirenians*, Reports of the International Whaling Commission, Special Issue 3 (ed. W. F. Perrin and A. C. Myrick Jr), pp. 181–201. International Whaling Commission. {30}

Marshall, E. (1990). Clovis counterrevolution. *Science*, **249**, 738–41. {31}

Marshall, L. G. (1984). Who killed Cock Robin? An investigation of the extinction controversy. In *Quaternary extinctions: a prehistoric revolution*, (ed. P. S. Martin and P. G. Klein), pp. 785–806. University of Arizona Press, Tucson. {30}

Marshall, L. G. and Cifelli, R. L. (1990). Analysis of changing diversity patterns in Cenozoic land mammal age faunas, South America. *Palaeovertebrata*, **19** (4), 169–210. {31, ApD}

Marshall, L. G., Webb, S. D., Sepkoski, J. J. Jr, and Raup, D. M. (1982). Mammalian evolution and the great American interchange. *Science*, **215**, 1351–7. {29}

Marshall, L. G. *et al.* (1984). Mammals and stratigraphy: geochronology of the continental mammal-bearing Quaternary of South America. *Palaeovertebrata, Mémoire Extra*, **1984**, 1–76. {31, ApD}

Martin, J. E. (1987). Paleoenvironment of the Lange/Ferguson Clovis kill site in the Badlands of South Dakota. In *Late Quaternary mammalian biogeography and environments of the great plains and prairies*, (ed. R. Graham, H. Semken, and M. Graham). Illinois State Museum, Scientific Papers, Vol. 22, pp. 314–32. {28}

Martin, K. (1887). Fossile Saugethierreste von Java und Japan. *Jaarboek van het Mijnwesen voor Nederlansch Oost Indie, wetenschappelijk gedeelte, XVI, Palaeontologie van Nederlandsch-Indie*, **21**, 1–45. {23}

Martin, P. S. (1967). Prehistoric overkill. In *Pleistocene extinctions: the search for a cause*, (ed. P. S. Martin and H. E. Wright Jr), pp. 75–120. University Press, New Haven. {29, 30}

Martin, P. S. (1984). Prehistoric overkill: the global model. *Quaternary extinctions: a prehistoric revolution*, (ed. P. S. Martin and R. G. Klein), pp. 354–403. University of Arizona Press, Tucson. {29, 31, 34}

Martin, P. S. and Klein, R. G. (ed.) (1984). *Quaternary extinctions: a prehistoric revolution*. University of Arizona Press, Tucson. {8, 19, 30}

Martin, P. S. and Wright, H. E. Jr (ed.) (1967). *Pleistocene extinctions: the search for a cause*. Yale University Press, New Haven. {30}

Martin, P. S., Thompson, R. S., and Long, A. (1985). Shasta ground sloth extinction: a test of the blitzkreig model. In *Environments and extinction: man in late glacial North America*, (ed. J. Mead and D. Meltzer), pp. 5–14. University of Maine, Orono, Maine. {28}

Mather, W. W. (1838). Remarks in addition to and explanation of the review of the report of the Geological Survey of Ohio—in a letter to the editor. *American Journal of Science and Arts*, **34** (2), 362–4. {18}

Matschie, P. (1900). Über geographische Albarten des Afrikanischen elephantens. *Sitzungsberichte Gesellschaft naturforschunde Freunde Berlin*, **8**, 189–97. {ApC}

Matsumoto, H. (1915). Part I: On some fossil mammals from Sze-chuan, China; Part II: On some fossil mammals from Ho-nan, China; Part III: On some fossil mammals from Tsukinoki, Ugo. *Science Reports of the Tôhoku Imperial University*, Second Series, **3** (1), 1–49 + 17 plts. {ApC}

Matsumoto, H. (1918). On a new archetypal fossil elephant from Mt. Tomuro, Kaga. *Science Reports of the Tôhoku Imperial University*, Second Series, **3**, 51–60. {17}

Matsumoto, H. (1922). Revision of *Palaeomastodon* and *Moeritherium*. *Palaeomastodon intermedius*, and *Phiomia osborni*, New species. *American Museum Novitates*, **51**, 1–6. {8}

Matsumoto, H. (1923). A contribution to the knowledge of *Moeritherium*. *Bulletin of the American Museum of Natural History*, **48**, 97–140. {4, 8, ApA}

Matsumoto, H. (1924*a*). Preliminary note on fossil elephants in Japan. *The Journal of the Geological Society of Tokyo*, **31**, (371), 255–72 [in Japanese]. {12A, 16, 17, ApA, ApC}

Matsumoto, H. (1924*b*). A revision of *Palaeomastodon* dividing it into two genera, and with descriptions of two new species. *Bulletin of the American Museum of Natural History*, **50**, 1–58. {1, 6, 12A, 16}

Matsumoto, H. (1925). Preliminary notes on two new species of fossil mastodon from Japan. *The Journal of the Geological Society of Tokyo*, **31**, (375), 395–414 [in Japanese; date and page corrected]. {6, 10, 16, 21, ApC}

Matsumoto, H. (1926). On two new mastodonts and an archetypal stegodon of Japan. *Science Reports of Tôhoku Imperial University*, Second Series, **10**, (1), 1–11 + 5 plts. {10, 12A, 21, ApA}

Matsumoto, H. (1928). On *Leith-Adamsia siwalikiensis*, a new generic and specific name of archetypal elephants. *Japanese Journal of Geology and Geography*, **5** (4), Article XII, 213–14 [date and page corrected]. {ApA}

Matsumoto, H. (1929). On *Parastegodon* Matsumoto and its bearing on the descent of earlier elephants. *Science Reports of the Tôhoku Imperial University*, Second Series, **13**, 13–15. {17, ApA}

Matsumoto, H. (1938). Brief notes of some Japanese fossil mammals. *Zoological Magazine (Japan)*, **50**, 481–7 [in Japanese]. {17}

Matsumoto, H. (1941). On Japanese stegodonts and parastegodonts. *Zoological Magazine (Japan)*, **53**, 385–96. [in Japanese with English abstract]. {17}

Matsumoto, H., Mori, H., and Kitame, T. (1968). On the fossil mammalian footstamps found at Iwashimizu, Tsukanoyama, near Nagaoka City, Niigata Prefecture. In *Aoi Ashiato*, (ed. member of Archaeology Laboratory), pp. 16–21. Archaeology Laboratory, Nagaoka Municipal Science Museum, Nagaoka, Japan [in Japanese with English title]. {34}

Matthew, W. D. (1918). Contributions to the Snake Creek fauna. With notes upon the Pleistocene of western Nebraska — American Expedition of 1916. *Bulletin of the American Museum of Natural History*, **38**, 183–229. {ApC}

Matthew, W. D. (1930). A Pliocene mastodon skull from California, *Pliomastodon vexillarius*, N. Sp. *Bulletin of the Department of Geology, University of California*, **19**, 335–48. {27}

Maul, L. (1990). Biharische Kleinsäugerfunde von Untermaßfeld, Voigtstedt und Süßenborn und ihre chronologische Stellung im Rahmen der biharischen Micromammalia-Faunen Europas. Unpublished D. Phil. thesis. Humboldt University, Berlin. {19}

May, S. R. and Repenning, C. A. (1982). New evidence for the age of the Mount Eden fauna, southern California. *Journal of Vertebrate Paleontology*, **2**, 109–13. {15, ApC}

Mayet, L. (1908). Etude des mammifères miocènes des Sables de l'Orléanais et des Faluns de la Touraine. *Annales de l'Université de Lyon*, **21**, 1–336. {ApC}

Mayhew, D. F. (1977). The endemic Pleistocene murids of Crete. *Proceedings of the Koninklijke Nederlandse Akademie van Wetenschappen, Amsterdam*, B, **80**, 182–214. {22}

Mazo, A. V. (1976). El *Gomphotherium augustidens* de la Ceramica Mirasierra, Tetuan de las Vicorias (Madrid). *Estudios Geológicos*, **32**, 331–47. {11}

Mazo, A. V. (1985). Proboscídeos terciarios de Madrid y su provincia. In *Geología y paleontología del Terciario continental de la provincia de Madrid*, pp. 41–6. Publication especial del Consejo Superior de Investigaciones Cientificas, Madrid. {14}

Mazo, A. V. and Alberdi, M. T. (1974). Estudio descriptivo y taxonómico del mastodonte de Yuncos (Toledo). *Estudios Geológicos*, **30**, 47–61. {14}

Mazo, A. V., Perez-Gonzalez, A., and Aguirre, E. (1990). Las faunas pleistocenas de Fuensanta del Jùcar y El Provencio y su significado en la evolución del Cuaternario de la Llanura manchega. *Boletín Geológico y Minero*, **101–3**, 404–18. {21, ApC}

Mead, J. I. and Agenbroad, L. D. (1986). New locations of extinct megafauna and plant community associations, Rancholabrean, southeastern Utah. *Current Research in the Pleistocene*, **3**, 72–4. {28}

Mead, J. I. and Agenbroad, L. D. (1989). Pleistocene dung and the extinct herbivores of the Colorado Plateau, southwestern USA. *Cranium*, **6**, 29–44. {28}

Mead, J. I. and Meltzer, D. J. (1984). North American late Quaternary extinctions and the radiocarbon record. In *Quaternary extinctions: a prehistoric revolution*, (ed. P. S. Martin and R. G. Klein), pp. 440–50. University of Arizona Press, Tucson. {28}

Mead, J. I., Agenbroad, L. D., Davis, O. K., and Martin, P. S. (1986). Dung of *Mammuthus* in the arid southwest, North America. *Quaternary Research*, **25**, 121–7. {28, 29}

Mead, J. I., Agenbroad, L. D., and Theis, J. H. (1990*a*). Diet, chronology and nuclear magnetic resonance analysis of late Pleistocene herbivores, western North America: a review. In *Chronostratigraphy of the Paleolithic in North, Central, East Asia and America*, pp. 213–29. USSR Academy of Science, Novasibirsk, USSR. {28}

Mead, J. I., Hevly, R. H., and Agenbroad, L. D. (1990*b*). Late Pleistocene invertebrates and plant remains, Mammoth Site, Black Hills, South Dakota. In *Megafauna and man: discovery of America's heartland*, (ed. L.

D. Agenbroad, J. I. Mead, and L. W.Nelson). The Mammoth Site of Hot Springs, South Dakota, Inc., Hot Springs, Scientific Papers, Vol. 1, pp. 9–10. {28}

Mebrate, A. (1987). The long-jawed gomphotheres of North America. Unpublished D. Phil. thesis. University of Kansas. {15, ApC}

Mebrate, A. and Kalb, J. E. (1985). Anancinae (Proboscidea: Gomphotheriidae) from the Middle Awash Valley, Afar, Ethiopia. *Journal of Vertebrate Paleontology*, **5** (1), 93–102. {12A}

Mehringer, P. J. Jr, King, J. E., and Lindsay, E. H. (1970). A record of Wisconsin-age vegetation and fauna from the Ozarks of western Missouri. In *Pleistocene and Recent environments of the central Great Plains*, (ed. W. Dort Jr and J. K. Jones Jr), pp. 173–85. University of Kansas Press, Lawrence. {30}

Mein, P. (1975). Résultats du Groupe de Travail des Vertébrés. In *Abstracts of VI Congress of Regional Committee of Mediterranean Neogene Stratigraphy*, (ed. J. Seneš), pp. 78–81. VEDA, Publishing House of the Slovak Academy of Sciences, Bratislava. {ApD}

Mein, P. (1976). Biozonation du Néogène méditerranéen à partir des mammifères. Unnumbered table in *Proceedings of the VIth Congress, Regional Committee of Mediterranean Neogene Stratigraphy*, Vol. 2 (ed. J. Senes). VEDA, Publishing House of the Slovak Academy of Sciences. Bratislava. {21, ApD}

Mein, P. (1979). Rapport d'activité du groupe de travail des Vertébrés, mise à jour de la biostratigraphie du Néogène basée sur les Mammifères. *Annales Géologiques des Pays Helléniques, Athinae*, **3**, 1367–72. {14, ApD}

Mein, P. (1990). Updating of MN zones. In *European Neogene mammal chronology*, (ed. E. H. Lindsay, V. Fahlbusch, and P. Mein), pp. 73–90. Plenum, New York [copyright date, 1989]. {9, 10, 12B, 13, 14, 21, ApC, ApD}

Meltzer, D. J. (1988). Late Pleistocene human adaptations in eastern North America. *Journal of World Prehistory*, **2**, 1–52. {30}

Meltzer, D. J. (1993). Coming to America. *Discover*, **14** (10), 90–7. {31}

Meltzer, D. J. and Mead, J. I. (1983). The timing of late Pleistocene mammalian extinctions in North America. *Quaternary Research*, **19**, 130–5. {27, 29, 30}

Meyer, G. E. (1978). Hyracoidea. In *Evolution of African mammals*, (ed. V. J. Maglio and H. B. S. Cooke), pp. 284–314. Harvard University Press, Cambridge, Mass. {8}

Miall, L. C. and Greenwood, F. (1878). *Anatomy of the Indian elephant. Studies in comparative anatomy — No. II.* Macmillan, London. {2}

Miani, A. and Miani, C. (1972). Circadian advancement rhythm of the calcification front in dog dentin. *Panminerva Medica*, **14**, 127–36. {30}

Michelmore, F., Beardsley, K., Barnes, R. F. W., and Douglas-Hamilton, I. (1994). A model illustrating the changes in forest elephant numbers caused by poaching. *African Journal of Ecology*, **32**, 89–99. {32}

Miles, A. E. W. and White, J. W. (1960). Ivory. *Proceedings of the Royal Society of Medicine*, **53**, 775–80. {2}

Miller, M. F., II and Wyckoff, R. W. G. (1968). Proteins in dinosaur bones. *Proceedings of the National Academy of Sciences, USA*, **60**, 176–8. {7}

Miller, W. E. (1980). The late Pliocene Las Tunas local fauna from southernmost Baja California, Mexico. *Journal of Paleontology*, **54**, 762–805. {15, ApC}

Miller, W. E. (1990). A *Rhynchotherium* skull and mandible from southeastern Arizona. *Brigham Young University Geology Studies*, **36**, 57–67. {15, 16, ApC}

Milne-Edwards, A. (1844). Considération sur quelques principes relatifs à la classification naturelle des animaux. *Annales des Sciences Naturelle*, 3rd Series, **1**, 65–99. {5}

Milner-Gulland, E. J. and Mace, R. (1991). The impact of the ivory trade on the African elephant *Loxodonta africana* population as assessed by data from the trade. *Biological Conservation*. **55** (2), 215–29. {32}

Mishra, H. R. (1980). Status of the Asian elephant in Nepal. In *The status of the Asian elephant in the Indian sub-continent*. IUCN/SSC Report (ed. J. C. Daniel). Bombay Natural History Society. {33}

Mitzopoulos, M. (1961). Über einen pleistozänen Zwergelefanten von der Insel Naxos (Kykladen). *Praktika Akadimias Athinon, Athina*, **36**, 332–40. {22}

Miyamoto, M. M. and Goodman, M. (1986). Biomolecular systematics of eutherian mammals: phylogenetic patterns and classification. *Systematic Zoology*, **35** (2), 230–40. {5}

Mohd Khan, b. M. K. (1985). *Elephants in peninsular Malaysia*. WWF Monthly Report, December 1985, pp. 297–9. World Wildlife Fund, Gland, Switzerland. {33}

Mol, D. (1984). Postcraniale skeletdelen van één individu van de mammoet uit Nederland. *Cranium*, **1**, 47–9. {25}

Mol, D. (1989). Fossil Proboscidea from the North Sea bottom between England and The Netherlands. *Abstracts of Papers and Posters of the Fifth International Theriological Congress, Rome, 22–29 August 1989*, Vol. I, pp. 154–5. {20}

Mol, D. J. and Van Essen, J. A. (1990). *Mammut borsoni* from The Netherlands. *Lutra*, **33**, (2), 183–6. {20}

Mol, D. and van Essen, H. (1992). *De mammoet: sporen uit de IJstijd.* Uitgeverij BZZTôH, 's-Gravenhage, TheNetherlands (copyright date 1991). {ApC}

Mol, D., Verhagen, A., and van Essen, H. (1988). De mammoet en de prehistorische mens van Siegsdorf. *Cranium*, **5**, (1), 57–64. {19, 25}

Montané J. (1967). Investigaciones interdisciplinarias en la ex laguna Taguatagua, Provincia de O'Higgins, Chile. *Anales de la Academia Chilena de Ciencias Naturales. Universidad Católica*, **52**, (30), 327–31. {31}

Montané, J. (1968*a*). Paleo-Indian remains from Laguna de Tagua Tagua, central Chile. *Science*, **161**, 1137–8. {31}

Montané, J. (1968*b*). Primera fecha radiocarbónica de Taguatagua. *Noticiero Mensual, Museo Nacional de Historia Natural*, **139**, 11. {31}

Morell, V. (1990). Research news: confusion in earliest America. *Science*, **248**, 439–41. {24}

Mosimann, J. E. and Martin, P. S. (1975). Simulating overkill by Paleoindians. *American Scientist*, **63**, 304–13. {30}

Moss, C. (1988). *Elephant memories: thirteen years in the life of an elephant family*. Morrow, New York. {2, 28, 30}

Mostny, G. (1968). Association of human industries with Pleistocene fauna in central Chile. *Current Anthropology*, **9**, 214–15. {31}

Mottl, M. (1969). Bedeutende Proboscidier–Neufunde aus dem Altpliozän (Pannonien) Südost-Österreichs. *Österreichische Akademie der Wissenschaften Mathematisch-Naturwissenschaftliche Klasse Denkschriften*, **115**, 1–22. {12A, 14}

Mottl, M. (1970). Die jungtertiären Säugetierfaunen der Steiermark, Südost-Österreich. *Mitteilungen des Museums für Bergbau Geologie und Technik, am Landesmuseum "Joanneum" Graz*, **31**, 79–168. {12A}

Mukha, B. B. (1980). A new species of ridge-toothed mastodon from the Pliocene of south-western USSR. In *Quaternary and Neogene faunas and flora of the Moldavian SSR* (ed. K. N. Negadaev-Nikonov), pp. 19–26, Kishinen, 'Shtiintza' {ApC}

Musil, R. (1968). Die Mammutmolaren von Predmostí (CSSR). *Paläontologische Abhandlungen. Abteilung A (Paläozoologie)*, **3**, 1–192. {19}

Musser, G. G. (1981). The giant rats of Flores and its relatives. East of Borneo and Bali. *Bulletin of the American Museum of Natural History*, **169**, 69–175. {23}

Myers, N. (1973). Tsavo National Park, Kenya, and its elephants: an interim appraisal. *Biological Conservation*, **5**, 123–32. {24}

Nair, P. V. and Gadgil, M. (1978). The status and distribution of elephant populations of Karnataka. *Journal of the Bombay Natural History Society*, **75**, (Suppl.), 1000–16. {33}

Nair, P. V., Sukumar, R., and Gadgil, M. (1980). The elephant in south India: a review. In *The status of the Asian elephant in the Indian subcontinent*, IUCN/SSC Report, (ed. J. C. Daniel). Bombay Natural History Society. {33}

Nakaya, H., Pickford, M., Nakana, Y., and Ishida, H. (1984). The late Miocene large mammal fauna from the Namurungule Formation, Samburu Hills, northern Kenya. *African Study Monographs*, **2**, 87–131. Kyoto University, Japan. {12B}

Nesti, F. (1825). Sulla nuova specie di elefante fossile del Valdarno all'Illustrissimo sig. Dott. Prof. Ottaviano Targioni Tozzetti. (Lettere sopra alcune ossa fossili del Valdarno non per anco descritte.) *Nuovo Giornale de' Letterati*, **11** (24), 195–216 + 1 table. {18, 19, 20, 25, 28, ApA}

Neuville, H. (1921). On the extinction of the mammoth. *Smithsonian Report for 1919*, 327–38 + 3 plts. {2}

Nielsen, H. E. (1965). Die Muskulature der Vordergliedmasse bei *Elephas indicus*. *Anatomischer Anzeiger*, **117**, 171–92. {2}

Novacek, M. J. (1986). The skull of leptictid insectivorans and the higher-level classification of eutherian mammals. *Bulletin of the American Museum of Natural History*, **183**, 1–111. {5}

Novacek, M. J. (1989). Higher mammal phylogeny: the morphological–molecular synthesis. In *The hierarchy of life: molecules and morphology in phylogenetic analysis*, (ed. B. Fernholm, K. Bremer, and H. Jornvall), pp. 421–35. Elsevier Science Publishers, Amsterdam. {5}

Novacek, M. J. (1992). Mammalian phylogeny: shaking the tree. *Nature*, **356**, 121–5. {34}

Novacek, M. J. and Wyss, A. R. (1986). Higher-level relationships of the Recent eutherian orders: morphological evidence. *Cladistics*, **2**, 257–87. {5}

Novacek, M. J. and Wyss, A. R. (1987). Selected features of the desmostylian skeleton and their phylogenetic implications. *American Museum Novitates*, **2870**, 1–8. {34}

Novacek, M. J., Wyss, A. R., and McKenna, M. C. (1988). The major groups of eutherian mammals. In *The phylogeny and classification of the tetrapods*. Vol. 2: Mammals, (ed. M. J. Benton), pp. 31–71. The Systematics Association, Special Volume No. 35B. Clarendon , Oxford. {5, 8}

Núñez, L., Varela, J., and Casamiquela, R. (1983). *Occupación paleoindio en Quereo. Reconstrucción multidisciplinaria en el territorio árido de Chile*. Universidad del Norte, Antofagasta, Chile. {31}

Obermayer, F. (1881). Beitrag zur Kenntnis des Zahnbeines vom Elefanten, Nilpferd, Walross und Narwal. *Neunter Jahres-Bericht des Vereines der Wiener Handels-Akademie*, **1881**, 102–13. {2}

Ogden, J. G. III and Hay, R. J. (1965). Ohio Wesleyan University natural radiocarbon measurements II. *Radiocarbon*, **7**, 166–73. {27}

Ogden, J. G. III and Hay, R. J. (1967). Ohio Wesleyan University natural radiocarbon measurements III. *Radiocarbon*, **9**, 316–32. {27}

Okamura, Y., Tamura, M., Takahashi, K., and Lake Biwa Museum Research Associates. (1995). An outline of fossil footprints from the Kobiwako group, central Japan. *Lake Biwa Museum Project Office* **Report No. 3**, 135–99 [in Japanese with English abstract]. {34}

Oldale, R. N., Whitmore, F. C. Jr, and Grimes, J. R. (1987). Elephant teeth from the western gulf of Maine, and their implications. *National Geographic Research*, **3**, 439–46. {24}

Olivier, R. (1978). Distribution and status of the Asian elephant. *Oryx*, **14**, 379–424. {33}

Olivier, R. C. D. (1982). Ecology and behavior of living elephants: bases for assumptions concerning the extinct woolly mammoths. In *Paleoecology of Beringia*, (ed. D. M. Hopkins, J. V. Matthews Jr, C. E. Schweger, and S. B. Young), pp. 291–305. Academic Press, New York. {29}

Olsen, S. J. (1957). A new beak-jawed mastodont from Florida. *Journal of the Palaeontological Society of India, Lucknow*, **2**, 131–5. {15, ApC}

Olsen, S. J. (1972). The American mastodon and the woolly mammoth. *Osteology for the archaeologist, No. 3: Papers of the Peabody Museum of Archaeology and Ethnology*, **56**, (3), x + 1–45. {8, 19}

Oltz, D. F. Jr and Kapp, R. O. (1963). Plant remains associated with mastodon and mammoth remains in central Michigan. *American Midland Naturalist*, **70**, 339–46. {27}

Opdyke, N. D., Lindsay, E., Johnson, G. D., Johnson, N., Tahirkheli, R. A. K., and Mirza, M. A. (1979). Magnetic polarity stratigraphy and vertebrate paleontology of the Upper Siwalik subgroup of northern Pakistan. *Palaeogeography, Palaeoclimatology, Palaeoecology*, **27**, 1–34. {17}

Orenstein, R. (ed.) (1991). *Elephants: the deciding decade*. Key Porter Books, Toronto. {34}

Orr, P. C. (1960*a*). Late Pleistocene marine terraces on Santa Rosa Island, California. *Bulletin of the Geological Society of America*, **71**, 1113–20. {24}

Orr, P. C. (1960*b*). *Radiocarbon dates from Santa Rosa Island, II*. Santa Barbara Museum of Natural History Bulletin, No. **3**. {24}

Orr, P. C. (1967). Geochronology of Santa Rosa Island, California. In *Proceedings of the symposium on the biology of the California Islands*, (ed. R. N. Philbrick), pp. 317–25. Santa Barbara Botanic Garden. {24}

Orr, P. C. (1968). *Prehistory of Santa Rosa Island*. Santa Barbara Museum of Natural History, Santa Barbara, California. {24}

Orr, P. C. and Berger, R. (1966). The fire areas on Santa Rosa Island, California *Proceedings of the National Academy of Sciences, USA*, **56**, 1409–16. {24}

Osborn, H. F. (1888). The nomenclature of the mammalian molar cusps. *American Naturalist*, **22** (262), 926–8. {3}

Osborn, H. F. (1898). The extinct rhinoceroses. *Memoirs of the American Museum of Natural History*, **1**, 75–164 + plts 12a–20. {Col}

Osborn, H. F. (1900*a*). The angulation of the limbs of Proboscidia, Dinocerata, and other quadrupeds in adaptation to weight. *American Naturalist*, **34** (398) 89–94 [Note: spelling is 'Proboscidia', not 'Proboscidea']. {Col}

Osborn, H. F. (1900*b*). The geological and faunal relations of Europe and America during the Tertiary Period and the theory of the successive invasions of an African fauna. *Science*, New Series, **11**, (276), 561–74. {Col}

Osborn, H. F. (1907*a*). *Evolution of mammalian molar teeth to and from the triangular type* (ed. W. K. Gregory). MacMillan, New York. {3}

Osborn, H. F. (1907*b*). Hunting the ancestral elephant in the Fayûm Desert. Discoveries, of the recent African expedition of the American Museum of Natural History. *The Century Magazine*, **74**, (6), 815–35. {Col}

Osborn, H. F. (1910). A means of estimating the age of the mastodon and other Proboscidea. Separates issued in 1910. *Proceedings of the Seventh International Zoological Congress*, **1912**, 782–4. {30}

Osborn, H. F. (1918). Equidae of the Oligocene, Miocene, and Pliocene of North America, iconographic type revision. *Memoirs of the American Museum of Natural History*, New Series, **2** (1), 1–330. {Col, ApB, ApC}

Osborn, H. F. (1921*a*). The evolution, phylogeny, and classification of the Proboscidea. *American Museum Novitates*, **1**, 1–15. {Col, 3, 16, ApC}

Osborn, H. F. (1921*b*). First appearance of the true mastodon in America. *American Museum Novitates*, **10**, 1–6. {6, 27, ApC}

Osborn, H. F. (1922*a*). *Dibelodon edensis* (Frick) of southern California, *Miomastodon* of the middle Miocene, new genus. *American Museum Novitates*, **49**, 1–4. {6, ApA}

Osborn, H. F. (1922*b*). Species of American Pleistocene mammoths *Elephas jeffersonii*, new species. *American Museum Novitates*, **41**, 1–16. {18, ApC}

Osborn, H. F. (1923). New subfamily, generic, and specific stages in the evolution of the Proboscidea. *American Museum Novitates*, **99**, 1–4. {15, ApA, ApB, ApC}

Osborn, H. F. (1924*a*). Additional generic and specific stages in the evolution of the Proboscidea. *American Museum Novitates*, **154**, 1–5. {15, ApA, ApC}

Osborn, H. F. (1924*b*). *Parelephas* in relation to phyla and genera of the family Elephantidae. *American Museum Novitates*, **152**, 1–7. {ApA, ApC}

Osborn, H. F. (1924*c*). *Serridentinus* and *Baluchitherium*, Loh formation, Mongolia. *American Museum Novitates*, **148**, 1–5. {13, ApC}

Osborn, H. F. (1925). Final conclusions on the evolution, phylogeny and classification of the Proboscidea. *Proceedings of the American Philosophical Society*, **64** (1), 17–35. {Col}

Osborn, H. F. (1926*a*). Additional new genera and species of the mastodontoid Proboscidea. *American Museum Novitates*, **238**, 1–16. {9, 15, 27, ApA, ApC}

Osborn, H. F. (1926*b*). *Mastodons and mammoths of North America*. American Museum of Natural History, New York, Guide Leaflet, **62**, 1–46. {9}

Osborn, H. F. (1929*a*). New Eurasiatic and American proboscideans. *American Museum Novitates*. **393**, 1–23. {6, 9, 12A, 17, ApA, ApC}

Osborn, H. F. (1929*b*). The revival of central Asiatic life. *Natural History*, **29**, 3–16 + 1 illustration. {13, 15, ApC}

Osborn, H. F. (1929*c*). *The titanotheres of ancient Wyoming, Dakota and Nebraska*. US Geological Survey, Monograph 55, Vol. 1. {Col}

Osborn, H. F. (1929*d*). *The titanotheres of ancient Wyoming, Dakota and Nebraska*. US Geological Survey, Monograph 55, Vol. 2. {Col}

Osborn, H. F. (1930). *Fifty-two years of research, observation and publication. 1877–1929*. Scribner's, New York. {Col}

Osborn, H. F. (1931). *Palaeoloxodon antiquus italicus* sp. nov., final stage in the '*Elephas antiquus*' phylum. *American Museum Novitates*, **460**, 1–24. {ApA, ApC}

Osborn, H. F. (1932*a*). The "*Elephas meridionalis*" stage arrives in America. *Proceedings of the Colorado Museum of Natural History*, **11**, (1), 1–3 + 2 plts. {29}

Osborn, H. F. (1932*b*). *Trilophodon cooperi*, sp. nov., of Dera Bugti, Baluchistan. *American Museum Novitates*, **585**, 1–6. {6, 10, ApC}

Osborn, H. F. (1933). *Serbelodon burnhami*, a new shovel-tusker from California. *American Museum Novitates*, **639**, 1–5. {13, 15}

Osborn, H. F. (1934*a*). Aristogenesis, the creative principle in the origin of species. *American Naturalist*, **68** (716), 193–235. {9, Col}

Osborn, H. F. (1934*b*). Evolution and geographic distribution of the Proboscidea: moeritheres, deinotheres and mastodonts. *Journal of Mammalogy*, **15**, 177–84. {1, 34}

Osborn, H. F. (1934*c*). Primitive *Archidiskodon* and *Palaeoloxodon* of South Africa. *American Museum Novitates*, **741**, 1–15. {29, ApA}

Osborn, H. F. (1936). *Proboscidea: a monograph of the discovery, evolution, migration and extinction of the mastodonts and elephants of the world*. Vol. I: *Moeritherioidea, Deinotherioidea, Mastodontoidea*. The American Museum Press, New York. {Col, 1, 2, 3, 4, 8, 9, 12A, 12B, 13, 15, 16, 17, 23, 27, 29, 30, 31, 34, ApA, ApC}

Osborn, H. F. (1938). Eighteen principles of adaptation in alloiometrons and aristogens. *Palaeobiologica*, **VI**, 273–302. {Col}

Osborn, H. F. (1942). *Proboscidea: a monograph of the discovery, evolution, migration and extinction of the mastodonts and elephants of the world*. Vol. II: *Stegodontoidea, Elephantoidea*. The American Museum Press, New York. {Col, 1, 2, 8, 9, 12A, 12B, 13, 16, 17, 18, 19, 22, 23, 25, 26, 29, 30, 31, 34, ApA, ApC}

Osborn, H. F. and Colbert, E. H. (1931). The elephant enamel method of measuring Pleistocene time. Also stages in the succession of fossil man and Stone Age industries. *Proceedings of the American Philosophical Society*, **70** (2), 187–91. {Col}

Osborn, H. F. and Granger, W. (1931). The shovel-tuskers, Amebelodontinae, of central Asia. *American Museum Novitates*, **470**, 1–12. {13}

Osborn, H. F. and Granger, W. (1932). *Platybelodon grangeri*, three growth stages, and a new serridentine from Mongolia. *American Museum Novitates*, **537**, 1–13. {9, 13, ApC}

Osborn, H. F. and Mook, C. C. (1921). *Camarasaurus, Amphicoelias* and other sauropods of Cope. *Memoirs American Museum of Natural History*, New Series, **3** (3), 247–387 + plts 60–85. {Col}

Ostrom, J. H. (1983–84). Mr. Peale's missing mastodon: the story of an Ice Age mastodon and early American painter Charles Willson Peale. *Discovery*, **17**, (1), 3–9. {ApC}

Owen, R. (1843). Additional evidence proving the Australian *Pachyderm* described in a former number of the 'Annals' to be a *Dinotherium*, with remarks on the nature and affinities of that genus. *The Annals and Magazine of Natural History*, **11**, (71), 329–32. {ApC}

Owen, R. (1844). Description of a fossil molar tooth of a *Mastodon* discovered by Count Strzlecki in Australia. *The Annals and Magazine of Natural History*, 1844, 1–4. {ApC}

Owen, R. (1870). On fossil remains of mammals found in China. *Quarterly Journal of the Geological Society of London*, **26**, 417–33. {17}

Owen, R. (1882). Description of portions of a tusk of a proboscidian [*sic*] mammal (*Notelephas australis*, Owen). *Philosophical Transactions of the Royal Society of London*, **173**, 777–81 + 1 plt. {ApC}

Owen-Smith, N. (1987). Pleistocene extinctions: the pivotal role of megaherbivores. *Paleobiology*, **13**, 351–62. {30}

Owen-Smith, N. (1988). *Megaherbivores: the influence of very large body size on ecology*. Cambridge University Press, Cambridge. {30}

Palmer, H. A. and Stoltman, J. B. (1976). The Boaz mastodon: a possible association of man and mastodon in Wisconsin. *Mid-Continental Journal of Archaeology*, **1**, 163–77. {30}

Palombo, M. R. (1972). Un cranio di *Elephas primigenius* della Via Flaminia (Roma). *Geologica Romana*, **11**, 199–228. {19}

Palombo, M. R. (1986). I grandi mammiferi pleistocenici delle isole del Mediterraneo: tempi e vie di migrazione. *Bollettino della Societa Paleontologica Italiana, Modena*, **24**, (1985), 201–24. {22}

Palombo, M. R. and Petronio, C. (1989). Caratteri morfostrutturali del femore di *Elephas* (*Palaeoloxodon*) *creutzburgi* (Kuss, 1965) di grotta Simonelli (Rethymnon, Creta). *Hystrix, Roma*, New Series **1**, 95–105. {22}

Parker, I. S. C. and Amin, M. (1983). *Ivory crisis*. Chatto & Windus, London. {32}

Parker, R. B. and Toots, H. (1980). Trace elements in bones as paleobiological indicators. In *Fossils in the making: vertebrate taphonomy and paleoecology*, (ed. A. K. Behrensmeyer and A. P. Hill), pp. 197–207. University of Chicago Press. {30}

Patterson, B. and Pascual, R. (1972). The fossil mammal fauna of South America. In *Evolution, mammals, and southern continents*, (ed. A. Keast, F. C. Erk, and B. Glass), pp. 247–309. State University of New York Press, Albany. {31}

Pavlova, M. (1910). Les éléphants fossiles de la Russie. *Nouveaux Mémoires de la Société Impériale des Naturalistes de Moscou*, **27**, 1–56. {19}

Payne, K. B., Langbauer, W. R., and Thomas, E. M. (1986). Infrasonic calls of the Asian elephant (*Elephas maximus*). *Behavioral Ecology and Sociobiology*, **18**, 297–301. {8, 30}

Pei, W. (1962). Quaternary mammals from the Liucheng *Gigantopithecus* cave and other caves of Kwangsi. *Vertebrata PalAsiatica*, **6** (3) 211–18. [In Chinese with English abstract] {17, ApC}

Pei, W. (1965). On a new *Trilophodon* tooth from Wufang of Hupeh Province. *Vertebrata PalAsiatica*, **9**, 209–16. {ApC}

Pei, W. C. (1987). Carnivora, Proboscidea and Rodentia from Liucheng *Gigantopithecus* cave and other caves in Guangxi. *Memoir of the Institute of Vertebrate Paleontology and Paleanthropology. Academia Sinica*, **18**, 5–119 [in Chinese with English summary]. {17}

Perry, J. S. (1953). The reproduction of the African elephant, *Loxodonta africana*. *Philosophical Transactions of the Royal Society of London*, Series B, **643**, 93–149. {2}

Percival, A. B. (1924). *Game ranger's notebook*. Nisbet, London. {32}

Peterson, O. A. (1926). The fossils of the Frankstown Cave, Blair County, Pennsylvania. *Annals of Carnegie Museum*, **16**, 249–315. {4}

Petrocchi, C. (1941). Il giacimento fossilifero di Sahabi. *Bollettino della Societa Geologica Italiana*, **60** (1), 107–14 + 2 plts. {12A, ApB, ApC}

Petrocchi, C. (1943). *Il giacimento fossilifero di Sahabi. Parte II — Paleontologia*. Colliezione Scientifica e Documentaria dell'Africa Italian, Ministero dell'Africa Italian (Verbania), Vol. **XII**, pp. 69–167. {12A, 12B}

Petrocchi, C. (1954). Paleontologia di Sahabi — I Proboscidati di Sahabi. *Rendiconti Accademia Nazionale dei XL*, Series IV, **IV** and **V**. Rome. {12A, 12B}

Petronievics, B. (1923). Remarks upon the skulls of *Moeritherium* and *Palaeomastodon*. *The Annals and Magazine of Natural History*, (9), **12**, 55–61, plt II. {8}

Petronio, C. (1970). I roditori Pleistocenici della grotta di Spinagallo (Siracusa). *Estratto da Geologica Romana*, **9**, 149–93. {22}

Peyer, B. (1968). *Comparative odontology*, (trans. and ed. R. Zangerl). University of Chicago Press. {2}

Pfeffer, P. (1989). *Vie et mort d'un géant, l'éléphant d'Afrique*. Flammarion, Paris. {34}

Pfretzschner, H. U. (1990). Hypsodonty and enamel microstructure in large herbivorous mammals. *Journal of Vertebrate Paleontology*, **10**, (Suppl. 3), 38A. {27}

Pfretzschner, H. U. (1992). Enamel microstructure and hypsodonty in large mammals. In *Structure, function and evolution of teeth* (ed. P. Smith and E. Tchernov), pp. 185–213. Freund, Tel Aviv. {8}

Phanthavong, B. and Santiapillai, K. (1992). Conservation of elephants in Laos. *IUCN/SSC Asian Elephant Specialist Group Newsletter*, **8**, 25–33. {33}

Pickford, M. (1981). Preliminary Miocene mammalian biostratigraphy for western Kenya. *Journal of Human Evolution*, **10**, 73–97. {10}

Pickford, M. (1986). Cainozoic paleontological sites of western Kenya. *Münchner Geowissenschaftliche Abhandlungen*, **8**, 1–151. {12B}

Pickford, M. (1987*a*). The geology and paleontology of the Kanam erosion gullies (Kenya). *Mainzer Geowissenschaftliche*, **16**, 209–26. {12B}

Pickford, M. (1987*b*). Révision des Suiformes (Artiodactyla, Mammalia) de Bugti (Pakistan). *Annales de Paléontologie (Vert.-Invert.)*, **73**, (4), 289–350. {6, 16, 34}

Pickford, M. and Tassy, P. (1980). A new species of *Zygolophodon* (Mammalia, Proboscidea) from the Miocene hominoid localities of Meswa Bridge and Moroto (East Africa). *Neues Jahrbuch für Geologie und Paläontologie Monatshefte*, **1980**, (4), 235–51. {6, ApC}

Pickford, M., Nakaya, H., Ishida, H., and Nakano, Y. (1984). The biostratigraphic analyses of the faunas of the Nachola area and Samburu Hills, northern Kenya. *African Study Monographs*, **2**, 67–72. Kyoto University, Japan. {12B}

Pickford, M., Senut, B., Poupeau, G., Brown, F. H., and Haileab, B. (1991). Correlation of tephra layers from the Western Rift Valley (Uganda) to the Turkana Basin (Ethiopia/Kenya) and the Gulf of Aden. *Comptes Rendus des Séances de l' Académie des Sciences Paris*, **313 (Series II)**, 223–9. {12B}

Pilgrim, G. E. (1905). On the occurrence of *Elephas antiquus (Namadicus)* in the Godavari Alluvium, with remarks on the species, its distribution and the age of the associated Indian deposits. *Records, Geological Survey of India*, **32** (3), 199–218 + 5 plts. {ApC}

Pilgrim, G. E. (1908). The Tertiary and post-Tertiary freshwater deposits of Baluchistan and Sind with notices of new vertebrates. *Records, Geological Survey of India*, **37**, 139–66. {6, 34, ApC}

Pilgrim, G. E. (1910). Preliminary note on a revised classification of the Tertiary freshwater deposits of India. *Records, Geological Survey of India*, **40** (3), 189–205. {ApC}

Pilgrim, G. E. (1912). The vertebrate fauna of the Gaj Series in the Bugti Hills and the Punjab. *Memoirs of the Geological Survey of India Palaeontologia Indica*, **4**, (2), 1–83. {8, ApB, ApC}

Pilgrim, G. E. (1913). The correlation of the Siwaliks with mammal horizons of Europe. *Records, Geological Survey of India*, **43** (4), 264–326, plts XXVI–XXVIII. {ApC}

Pilgrim, G. E. (1940). Middle Eocene mammals from north-west India. *Proceedings, Zoological Society of London*, Series B, **110**, 127–52. {6, 8, ApA, ApB, ApC}

Pino, M. (1989). Regional and site geology. In *Monte Verde: a late Pleistocene settlement in Chile*, Vol. 1, (ed. T. D. Dillehay), pp. 89–132. Smithsonian Institution, Washington, DC. {31}

Pitman, Lt. Col. C. R. S. (1934). *A report on a faunal survey of northern Rhodesia with especial reference to same, elephant control and national parks*. Government Printers, Livingstone. {32}

Pohlig, H. (1885). Über ein Hipparionen-fauna von Maragha in Nord-Persien, über fossile Elefantenreste Kaukasiens und Persiens, und über die Resultate einer Monographie der fossilen Elefanten Deutschlands und Italiens. *Zeitschrift der Deutschen Geologischen Gesellschaft*, **37**, 1022–7. {19, ApA, ApC}

Pohlig, H. (1888). Dentition und Kraniologie des *Elephas antiquus* Falc. mit Beiträgen über *Elephas primigenius* Blum. und *Elephas meridionalis* Nesti. *Nova Acta der Kaiserlichen Leopoldinisch-Carolischen Deutschen Akademie der Naturforscher*, **53**, 1–280. {19, ApA}

Pohlig, H. (1893). Eine Elephantenhole Siciliens und der erste Nachweis des Cranialdomes von *Elephas antiquus*. *Abhandlungen der Bayerische Akademie der Wissenschaften, München*, **18**, 75–109. {22}

Pohlig, H. (1912). Sur une vieille mandibule de "*Tetracaulodon ohiotocum*" Blum, avec défense *in situ*. *Bulletin de la Société Belge Géologique*, **26**, 187–93. {15, ApA, ApB, ApC}

Pontier, G. (1913). Étude sur le mammouth de l'Aa. *Bulletin de la Société Préhistorique de France*, **10**, 621–46. {19, 25}

Poole, D. F. G. (1967). Phylogeny of tooth tissues: enameloid and enamel in recent vertebrates, with a note on the history of cementum. In *Structural and chemical organization of teeth*, Vol. I, (ed. A. E. W. Miles), pp. 111–49. Academic Press, New York. {2}

Poole, J. H. (1989). The effects of poaching on the age structures and social and reproductive patterns of selected East African elephant populations. In *The ivory trade and the future of the African elephant*, (co-ordinator S. Cobb), Vol. 2, section 5.1. The Ivory Trade Review Group, Oxford. {32}

Poole, J. H. and Moss, C. J. (1981). Musth in the African elephant (*Loxodonta africana*). *Nature*, **292**, 830–1. {2}

Por, F. C. (1975). An outline to the zoogeography of the Levant. *Zoologica Scripta*, **4**, (1), 5–20. {21}

Powers, D. W. (1980). Geology of Mio-Pliocene sediments of the lower Kerio River Valley, Kenya. Unpublished D. Phil. Thesis. Princeton University. {12B}

Prager, E. M., Wilson, A. C., Lowenstein, J. M., and Sarich, V. M. (1980). Mammoth albumin. *Science*, **209**, 287–9. {7}

Press, F. and Siever, R. (1974). *Earth*. W. H. Freeman and Company, San Francisco. {8}

Proaño, J. F. (1922). La Virgen del Dios Chimborazo, tradiciones Puruhaes. Riobamba, Ecuador, Imp. *El Observador*, 1–23 + loose unnumbered page. [Not seen; cited in Simpson and Paula Couto 1957] {ApC}

Proboscidea Research Group (1977). Terminology on cranium and tooth of Proboscidea. *Fossil Club Bulletin* Suppl. No. (1977), 1–15. {17}

Prothero, D. P. (1993). Ungulate phylogeny: molecular vs. morphological evidence. In *Mammal phylogeny: placentals*, (ed. F. S. Szalay, M. J. Novacek, and M. C. McKenna), pp. 173–81. Springer-Verlag, New York. {5}

Prothero, D. P. and Sereno, P. C. (1982). Allometry and paleoecology of median Miocene dwarf rhinoceroses from the Texas Gulf coastal plain. *Paleobiology*, **8**, 16–30. {17}

Prothero, D. R. and Swisher, C. C. III. (1992). Magnetostratigraphy and geochronology of the terrestrial Eocene–Oligocene transition in North America. In *Eocene–Oligocene climatic and biotic evolution*, (ed. D. R. Prothero and W. A. Berggren), pp. 46–73. Princeton University Press, Princeton, New Jersey. {ApD}

Prothero, D. R., Manning, E. M., and Fischer, M. (1988). The phylogeny of the ungulates. In *The phylogeny and classification of the tetrapods*. Volume 2: *Mammals*, (ed. M. J. Benton), pp. 201–34. The Systematics Association, Special Volume No. 35B. Clarendon, Oxford. {5, 8, 16}

Qi, G. (1993). The environment and ecology of the Lufeng hominoids. *Journal of Human Evolution*, **24** (1), 3–11. {ApC}

Qi, T. (1992). The east most locality of Tertiary mammalian distribution in China. *Vertebrata PalAsiatica*, **30** (4), 325–6. [in Chinese with English summary]. {13}

Qiu, Z. (1990). The Chinese Neogene mammalian biochronology — its correlation with the European Neogene mammalian zonation. In *European Neogene mammal chronology*, (ed. E. H. Lindsay, V. Fahlbusch, and P. Mein), pp. 527–56. Plenum, New York [copyright date, 1989]. {10, 12B, 13}

Qiu, Z. and Guan, J. (1986). A lower molar of *Pliopithecus* from Tongxin, Ningxia Hui Autonomous. *Acta Anthropologica, Sinica*, **5**, 201–7. {13}

Qiu, Z. and Jia, H. (1980). *Field guide to the Yushe Series in Yushe and Wuxing counties*. Institute of Vertebrate Paleontology and Paleoanthropology. {13}

Qiu, Z. and Qiu, Z. (1990). The sequence and age determination of mammalian local faunas in the late Tertiary of China. *Journal of Stratigraphy*, **14** (4), 241–60. {13}

Qiu, Z., Huang, W., and Guo, Z. (1987*a*). The Chinese hipparionine fossils. *Palaeontologia Sinica*, Series C, **175** (25), 1–250 + 47 plts. [In Chinese and English text] {13}

Qiu, Z., Yan, D., Chen, G., and Qiu, Z. (1987*b*). Preliminary report on paleontological expedition in Tung Gur, Inner Mongolia. *Academia Sinica*, **19**, 1487–91. [In Chinese] {13}

Radinsky, L. B. (1969). The early evolution of the Perissodactyla. *Evolution*, **23**, 308–28. {3}

Rafinesque, C. S. (1814). Compendio delle ricerche del Sig. Cuvier sopra i quadrupedi pachidermi fossili. *Specchio Scienze*, **II**, (No. 12, December), 182–3. Palermo [Osborn 1936, p. 796 gave the name Rafinesque-Schmaltz]. {ApA}

Rainey, W. E., Lowenstein, J. M., Sarich, V. M., and Magor, D. M. (1984). Sirenian molecular systematics — including the extinct Steller's sea cow (*Hydrodamalis gigas*). *Naturwissenschaften*, **71**, 586–8. {7}

Rainger, R. (1991). An agenda for antiquity: Henry Fairfield Osborn and vertebrate paleontology at the American Museum of National History 1890–1935. University of Alabama Press, Tuscaloosa. {34}

Rancy, A. (1991). Pleistocene mammals and paleoecology of the western Amazon. Unpublished D. Phil. thesis. University of Florida, Gainsville. {31}

Rasmussen, D. T. and Simons, E. L. (1988). New Oligocene hyracoids from Egypt. *Journal of Vertebrate Paleontology*, **8**, 67–83. {8}

Rasmussen, D. T., Bown, T. M., and Simons, E. L. (1992). The Eocene-Oligocene transition in continental Africa. In *Eocene–Oligocene climatic and biotic evolution*, (ed. D. R. Prothero and W. A. Berggren), pp. 548–66. Princeton University Press, New Jersey. {ApD}

Rasmussen, L. E. L. and Hulgren, B. (1990). Gross and microscopic anatomy of the vomeronasal organ of the Asian elephant. In *Chemical signals vertebrates* 5 (ed. D. W. McDonald, D. Müller-Schwartze, and S. E. Natynczuk), pp. 154–61. Oxford University Press, Oxford. {2}

Ratanasthien, B. (1990). Mae Long Formation of Li Basin, Thailand. In *Pacific Neogene events: their timing, nature and interrelationship*, (ed. R. Tsuchi), pp. 123–8. University of Tokyo. {17}

Ray, J. (1693). *Synopsis methodica animalium quadrupedum et serpentini generis*. Impensis S. Smith & B. Walford Societatis Regiae Typographorum ad Insignia Principis in Coementerio D. Pauli, Imprimatur, Robert Southwell, R. S. P., London. {ApC}

Ray. C. E. Cooper, B. N., and Benninghoff, W. S. (1967). Fossil mammals and pollen in a late Pleistocene deposit at Saltville, Virginia. *Journal of Paleontology*, **41**, 608–22. {27}

Ray, C. E., Domning, D. P., and McKenna, M. C. (1994). A new specimen of *Behemotops proteus* (order Desmostylia) from the marine Oligocene of Washington. In *Contributions in marine mammal paleontology honoring Frank C. Whitmore, Jr.* (ed. A. Berta and T. A. Deméré). *Proceedings of the San Diego Society of Natural History*, **29**, 205–22. {34}

Raza, S. M. Barry, J. C., Meyer, G. E., and Martin, L. (1984). Preliminary report on the geology and vertebrate fauna of the Miocene Manchar Formation, Sind, Pakistan. *Journal of Vertebrate Paleontology*, **4**, 584–99. {12B}

Redmond, I. (1987). Island of elephants. *Swara*, **10(2)**, 14–19. {34}

Reese, D. S. (1989). Tracking the extinct pygmy hippopotamus of Cyprus. *Field Museum of Natural History Bulletin*, **60/2**, 22–9. {21}

Remane, A. (1960). Zähne und Gebiss. In *Primatologia*, (ed. A. Hofer, A. H. Schultz, and D. Starck), Vol. 3, pp. 527–56. Karger, Switzerland. {3}

Repenning, C. A. (1987). Biochronology of the microtine rodents of the United States. In *Cenozoic mammals of North America — geochronology and biostratigraphy*, (ed. M. O. Woodburne), pp. 236–68. University of California, Berkeley. {27}

Research Group for the Fossil Footprints of Yasu-gawa. (1995). Fossil footprints from the Kobiwako group at River Yasu, Kosei Town, Shiga Prefecture, central Japan. *Lake Biwa Museum Project Office*, **Report No. 3**, 1–134 [in Japanese with English abstract]. {34}

Retzius, A. (1837). Ueber die röhringe Bildung der Zähne nebst dazu gehörigen Präparaten. *Oken, Isis*, columns 519–20. {2}

Revilliod, P. (1931). Sur un mastodon de Bolivie. *Mémoires de la Société Paléontologique Suisse*, **51**, 1–21, plts I, II. [From Osborn 1936, p. 796] {ApA}

Reza Khan, M. A. (1980). On the distribution and population status of the Asian elephant in Bangladesh. In *The status of the Asian elephant in the Indian sub-continent*. IUCN/SSC Report, (ed. J. C. Daniel). Bombay Natural History Society. {33}

Ridley, M. (1986). *Evolution and classification: the reformation of cladism.* Longman, London. {16}

Riggs, E. S. (1936). A Pleistocene bog deposit and its fossil fauna. *Transactions of Illinois State Academy of Science*, **29**, 186–9. {27}

Riha, V. F., Peterson, L. J., Rossmoore, H. W., and Shoshani, J. (1991). Detection and inhibition of fungi from the excised trunk of an Asian elephant. *International Biodeterioration*, **28**,113–24. {2, ApC}

Romer, A. S. (1966). *Vertebrate paleontology*, (3rd edn). University of Chicago Press. {34}

Ronglin, S. (1983). Fossils of *Palaeoloxodon naumanni* from Zhucheng, Linyi and Yixian, Shandong. *Vertebrata PalAsiatica*, **21**, 129–33. {18}

Rose, C. (1894). Über den Zahnbau und Zahnwechsel von *Elephas indicus*. *Morphologisches Arbeiten*, **3**, 173–94. {4}

Rosenberg, G. D. and Simmons, D. V. (1980). Rhythmic dentinogenesis in the rabbit incisor: circadian, ultradian, and infradian periods. *Calcified Tissue International*, **32**, 29–44. {30}

Roth, V. L. (1982). Dwarf mammoths from the Santa Barbara, California Channel Islands: size, shape, development, and evolution. Unpublished D. Phil. thesis. Yale University. University Microfilms, Ann Arbor. {24}

Roth, V. L. (1984). How elephants grow: heterochrony and the calibration of developmental stages in some living and fossil species. *Journal of Vertebrate Paleontology*, **4**, 126–45. {18, 24, 30}

Roth, V. L. (1989). Fabricational noise in elephant dentitions. *Paleobiology*, **15**, 165–79 {12A, 18, 19, 24}

Roth, V. L. (1990). Insular dwarf elephants: a case study in body mass estimation and ecological inference. In *Body size in mammalian paleobiology: estimation and biological implications*, (ed. J. Damuth and B. J. MacFadden), pp. 151–79. Cambridge University Press, Cambridge. {18, 24}

Roth, V. L. (1992*a*). Inferences from allometry and fossils: dwarfing of elephants on islands. In *Oxford surveys in evolutionary biology*, Vol. 8, (ed. D. Futuyma and J. Antonovics), pp. 259–88. Oxford University Press, New York. {18, 24}

Roth, V. L. (1992*b*). Quantitative variation in elephant dentitions: implications for the delimitation of fossil species. *Paleobiology*, **18**, 184–202. {18, 24}

Roth, V. L. (1993). Dwarfism and variability in the Santa Rosa Island mammoth: an interspecific comparison of limb bone sizes and shapes in elephants. In *Third California Islands symposium: recent advances in research on the California Islands*, (ed. F. G. Hochberg), pp. 433–42. Santa Barbara Museum of Natural History. {18, 24}

Roth, V. L. and Shoshani, J. (1988). Dental identification and age determination in *Elephas maximus*. *Journal of Zoology, London*, **214**, 567–88. {2, 11, 18, 23, 24, 30}

Rowe, T. (1988). Definition, diagnosis, and origin of Mammalia. *Journal of Vertebrate Paleontology*, **8**, 241–64. {16}

Ruddiman, W. F. and McIntyre, A. (1982). Severity and speed of northern hemisphere glaciation pulses: the limiting case? *Bulletin of the Geological Society of America*, **93**, 1273–9. {30}

Russanov, B. S. (1968). *Biostratigrafia kainozoiskikh otlozhenii Yuzhnoi Yakutii*. Nauka, Moskva [in Russian]. {ApC}

Saegusa, H. (1987). Cranial morphology and phylogeny of the stegodonts. *The Compass*, **64**, (4), 221–43. {2, 12A, 12B, 13, 17}

Saegusa, H. (1989). Molar structure and taxonomy of east Asian stegodonts. Unpublished D. Phil. thesis. Kyoto University. {17}

Saegusa, H. (1993). A note on cranial materials of *Stegodon orientalis* Owen, 1870 (Stegodontidae, Proboscidea) from Yanjingou, Sichuan, China. *Humans and Nature*, **2**, 95–103 [in Japanese with English abstract]. {17}

Saegusa, H. and Taruno, H. (1991). Stegodonts. In *Japanese proboscidean fossils*, (ed. T. Kamei), pp. 68–110. Tsukiji-Shokan, Tokyo [in Japanese]. {17}

Sahni, A. and Kumar, K. (1980). Lower Eocene Sirenia, *Ishatherium subathuensis*, Gen. et sp. nov. from the type area, Subathu Formation, Subathu, Simla, Himalayas, H. P. *Journal of the Palaeontological Society of India*, **23–24**, 132–5. {8, ApA}

Sahni, M. R. and Tripathi, C. (1957). A new classification of the Indian deinotheres and description of *D. orlovii* sp. nov. *Memoirs of the Geological Survey of India Palaeontologia Indica*, **33**, (4), 1–33 + 3 plts. {ApC}

Samodra, H. (1989). Kedudukan stratigrafi temuan fossil *Stegodon* di pulau Sangihe. *Paper presented at the XVIII annual meeting of the A. I. G., Yogyakarta, December, 1989* [in Indonesian]. {23}

Sanders, W. J. (1990). Fossil Proboscidea from the Pliocene Lusso Beds of the Western Rift, Zaire. In *Evolution of environments and Hominidae in the African Western Rift Valley*, (ed. N. T. Boaz). *Virginia Museum of Natural History Memoir*, No. **1**, pp. 171–87. {12B, 17}

Sanders, W. J. (in press). Fossil Proboscidea from the Manonga-Wembere Formation, Manonga Valley, Tanzania. In *Neogene Paleontology of the Manonga Valley, Tanzania* (ed. T. Harrison). Plenum, New York. {17}

Sanford, J. T. (1935). The Richmond mastodon. *Proceedings of the Rochester Academy of Science*, **7**, 137–56. {30}

Santiapillai, C. and Jackson, P. (1990). *The Asian elephant: an action plan for its conservation*. IUCN/SSC Asian Elephant Specialist Group, IUCN, Gland, Switzerland. {33}

Santiapillai, C. and Suprahman, H. (1985). Elephants in Indonesia (Sumatra). *WWF Monthly Report*, December 1985, pp. 287–96. World Wildlife Fund, Gland, Switzerland. {33}

Santiapillai, C., Zhu, X., Dong, Y. H., and Sheng, Q. Z. (1991). *Distribution of Asian elephant in Xishuangbanna, People's Republic of China*. WWF Project 3194 Report. World Wide Fund for Nature, Gland, Switzerland. {33}

Sartono, S. (1969). *Stegodon timorensis*: a pygmy species from Timor (Indonesia). *Proceedings of the Koninklijke Nederlandse Akademie van Weternschappen, Amsterdam*, Series B, **72**, 192–202. {23}

Sartono, S. (1979*a*). The age of the vertebrate fossils and artefacts from Cabenge in south Sulawesi, Indonesia. *Modern Quaternary Research in Southeast Asia*, **5**, 65–81. {23}

Sartono, S. (1979*b*). The discovery of a pygmy *Stegodon* from Sumba, East Indonesia: an announcement. *Modern Quaternary Research in Southeast Asia*, **5**, 57–63. {23}

Sarwar, M. (1974). Studies on the Siwalik Proboscidea. Unpublished D. Phil. thesis. University of Punjab. {9}

Sarwar, M. (1977). Taxonomy and distribution of the Siwalik Proboscidea. *Bulletin of the Department of Zoology*, University of The Punjab, (New Series), Article 10, 1–172. {9, 12A, 12B, ApC}

Sarwar, M. (1979). A new elephant from the Upper Siwalik of Pakistan. *Pakistan Journal of Science*, **31**, (3–6), 133–8. {ApA, ApC}

Sarwar, M. and Akhtar, M. (1992). An oldest Siwalik record of the family Hippopotamidae Gray. *Pakistan Journal of Zoology*, **24** (3), 219–220. {ApC.1}

Saunders, J. J. (1977). Late Pleistocene vertebrates of the western Ozark Highland, Missouri. *Illinois State Museum, Reports of Investigations*, **33**, x + 1–118. {11, 14, 27, 30}

Saunders, J. J. (1979). A close look at ivory. *The Living Museum*, **41**, (4), 56–9. {2}

Saunders, J. J. (1980). A model for man–mammoth relationships in late Pleistocene North America. *Canadian Journal of Anthropology*, **1**, 87–98. {25, 30}

Saunders, J. J. (1988). Fossiliferous spring sites in southwestern Missouri. *Bulletin of the Buffalo Society of Natural Sciences*, **33**, 127–49. {27}

Saunders, J. J. (1990). Immanence, configuration, and the discovery of America's past. In *Megafauna and man: discovery of America's heartland*, (ed. L. D. Agenbroad, J. I. Mead, and L. W. Nelson). *The Mammoth Site of Hot Springs. South Dakota, Inc., Hot Springs, Scientific Papers*, Vol. 1, pp. 136–43. {30}

Saunders, J. J. and Tassy, P. (1989). Le mastodonte américain. *La Recherche*, **20**, 452–61. {27, 34}

Savage, D. E. (1955). A survey of various late Cenozoic vertebrate faunas of the Panhandle of Texas, Part II. Proboscidea. *University of California Publications in Geological Sciences*, **31**, 51–74. {15, 16, 29, ApC}

Savage, D. E. and Russell, D. E. (ed.) (1983). *Mammalian paleofaunas of the world*. Addison-Wesley, Reading, Mass. {21, 27, 34, ApC, ApD}

Savage, R. J. G. (1969). Early Tertiary mammal locality in southern Libya. *Proceedings of the Geological Society of London*, **1657**, 167–71. {8}

Savage, R. J. G. (1990). The African dimension in European early Miocene mammal faunas. In *European Neogene mammal chronology*, (ed. E. H. Lindsay, V. Fahlbusch, and P. Mein), pp. 587–599. Plenum, New York [copyright date, 1989]. {12B}

Savage, R. J. G. and Court, N. (1989). Skeletal anatomy of *Barytherium*. *Abstracts of Papers and Posters Fifth International Theriological Congress, Rome, 22–29 August 1989*, Vol. I, p. 159. {8}

Savage, R. J. G. and Long, M. R. (1986). *Mammal evolution: an illustrated guide*. British Museum (Natural History), London. {8, 34}

Savage, R. J. G. and Tchernov, E. (1968). Miocene mammals of Israel. *Proceedings of the Geological Society of London*, **1648**, 98–101. {21}

Sayer, J. (1983). Nature conservation priorities in Laos. *Tigerpaper*, **10**, 10–14. {33}

Schaub, S. (1948). Das Gebiss der Elephanten. *Verhandlung Naturforschung Gesellschaft*, **59**, 89–112. {4}

Scheglova, W. W. (1958). On the remains of small mammoth (*Elephas primigenius minor* W. Gromov) from late paleolithic site Eliseevichi. *Reports of the Academy of Sciences of the Belorussian SSR*, **2**, (7), 303–5 [in Russian]. {26}

Schenk, E. T., McMasters, J. H., Keen, A. M., and Muller, S. W. (1956). *Procedure in taxonomy: including a reprint in translation of the Règles Internationales de la Nomenclature Zoölogique (International Code of Zoölogical Nomenclature) with titles and notes on the opinions rendered to the present date (1907 to 1956) completely indexed* (3rd edn). Stanford University Press, Stanford. {AST}

Scherstjukov, N. M. (1954). Description of remains of fossil elephants found in Groznenskaya Province. *Trudy Estesvenno-Istoricheskogo Museya imeni G. Zardabi, Vypusk*, **8**, 227–261 [in Russian]. {ApC}

Schinz, H. R. (1824). *Naturgeschichte und Abbildungen der Säugethiere*, Brodtmanns, Lithographischer Kunstanstalt, Zurich, vi + 417, Atlas, 177 plts. {6, 9, ApC}

Schlesinger, G. (1912). Studien über die Stammesgeschichte der Proboscidier. *Jahrbuch der Kaiserlich-Königlichen geologischen Reichsanstalt*, **62**, 87–182. {12A}

Schlesinger, G. (1917). Die Mastodonten des K. K. Naturhistorischen Hof-Museums. *Denkschriften der K. K. Naturhistorischen Hofmuseums 1, Geologisch-Paläontologische Reihe*, **1**, 1–230. {3, 6, 9, 12A, 14, ApA, ApB, ApC}

Schlesinger, G. (1919). Die stratigraphische bedeutung der europäischen Mastodonten. *Mitteilungen der Geologischen Gesellschaft in Wien*, **11**, 129–66, Plts II–VII. {ApC}

Schlesinger, G. (1922). Die Mastodonten der Budapester Sammlungen. *Geologica Hungarica*, Editio Separata, **2**, 1–284. {9, 12A, 14, 27}

Schlosser, M. (1911). Beiträge zur Kenntnis der Oligozänen Landsäugetiere aus dem Fayum: Agypten. *Beiträge zur Paläontologie und Geologie Österreich-Ungarns*, **24**, 51–167. {8}

Schmidt, M. J. (1992). The elephant beneath the mask. In *Elephants: majestic creatures of the wild*, (consult. ed. J. Shoshani), pp. 92–7. Rodale Press, Emmaus, Pennsylvania. {2, 30}

Schour, I. and Hoffman, M. M. (1939). Studies in tooth development: I. The 16 microns calcification rhythm in the enamel and dentin from fish to man. *Journal of Dental Research*, **18**, 91–102. {30}

Schreuder, A. (1944). Upper-Pliocene Proboscidea out of the Scheldt and the Lower-Rhine. *Leidse Geologische Mededelingen*, **14**, (1), 40–58. {20}

Schreuder, A. (1945). The Tegelen fauna, with a description of new remains of its rare components (*Leptobos, Archidiskodon meridionalis, Macaca, Sus strozzii*). *Archives Néerlandaises de Zoologie*, **7**, 153–204. {20}

Schulze, F. E., Kükenthal, W., and Heider, K. (1932). *Nomenclator animalium generum et subgenerum*. Band III. [Heider is the principal author for this series; for complete citation, see Heider (1932)] {ApC}

Scott, W. B. (1939). *Some memories of a paleontologist*. Princeton, Princeton University. {Col}

Sealy, J. C., van der Merwe, N. J., Lee-Thorpe, J. A., and Lanhan, J. L. (1987). Nitrogen isotopic ecology in southern Africa: implications for environmental and dietary tracing. *Geochemica et Cosmochemica Acta*, **51**, 2707–17. {30}

Sears, P. B. and Clisby, K. H. (1952). Pollen spectra associated with the Orleton Farms Mastodon site. *Ohio Journal of Science*, **52**, 9–10. {27}

Seba, A. (1734). *Locupletissimi rerum naturalium thesauri accurata descriptio, et iconibus artificiosissimis expressio, per universam physiees historiam. Opus, cui, in hoc rerum genere, nullum par exstitit. Ex toto terrarum orbe collegit, digessit, descripsit, et de pingendum curavit Albertus Seba*. Apud Janssonio-Waesbergios [etc.], Amstelaedami (Amsterdam). {ApC}

Sellards, E. H. (1940). New Pliocene mastodon. *Bulletin of the Geological Society of America*, **51**, 1659–64. {ApC}

Sellards, E. H. (1952). *Early man in America: a study in prehistory*. University of Texas, Austin. {30}

Selous, F. C. (1890). *A hunter's wanderings in Africa: a narrative of nine years spent amongst the game of the far interior of South Africa, containing accounts of explorations beyond the Zambesi*. Bentley, London. {32}

Semken, H. A. Jr and Falk, C. R. (1987). Late Pleistocene/Holocene mammalian faunas and environmental changes on the northern plains of the United States. In *Late Quaternary mammalian biogeography and environments of the great plains and prairies*, (ed. R. Graham, H. Semken, and M. Graham). Illinois State Museum Scientific Papers, Vol. **22**, 176–313. {28}

Shackleton, N. J. and Opdyke, N. D. (1973). Oxygen isotope and paleomagnetic stratigraphy of equatorial Pacific core V28–238: oxygen isotope temperatures and ice volumes on a 10^5 year and 10^6 year scale. *Quaternary Research*, **3**, 39–55. {30}

Shackleton, N. J. and Opdyke, N. D. (1977). Oxygen isotope and paleomagnetic evidence for early Northern Hemisphere glaciation. *Nature*, **270**, 216–19. {20}

Shackleton, N. J., *et al.* (1984). Oxygen isotope calibration of the onset of ice-rafting and history of glaciation in the North Atlantic region. *Nature*, **307**, 620–3. {20}

Shahi, S. P. and Chowdhury, S. (1986). *Elephants in central India. WWF Monthly Report* January 1986, pp. 19–22. World Wildlife Fund, Gland, Switzerland. {33}

Shaw, C. A. and Quinn, J. P. (1986). Rancho La Brea: a look at coastal southern California's past. *California Geology*, **39**, (6), 123–33. {ApC}

Sher, A. V. (1986). Olyorian land mammal age of northeastern Siberia. *Palaeontographia Italica*, **74**, 97–112. {19}

Sher, A. V. and Garutt, V. E. (1985*a*). New data on the morphology of the roots of elephant molars. *Doklady Akademiya Nauk SSSR*, **285**, 221–5. {19}

Sher, A. V. and Garutt, V. E. (1985*b*). On the method of identification of molar generations in elephants. *Proceedings of the Zoological Institute*, **131**, 93–103. Academy of Sciences of the USSR. {19}

Sherratt, A. (ed.) (1980). *The Cambridge encyclopedia of archaeology*. Crown Publishers/Cambridge University Press, New York. {ApD}

Shi, M., Guan, J., Pan, R., and Tang, D. (1981). Pliocene mammals collected from lignite in Zhaotung, Yunnan. *Memoirs of Beijing Natural History Museum*, **11**, 1–15 + 3 plts [in Chinese with English summary]. {17}

Shikama, T. (1936). Note on *Parastegodon akashiensis takai* from the Akashi district. *Proceedings of the Imperial Academy of Japan, Tokyo*, **12**, 22–4. {17}

Shikama, T., Otsuka, H., and Tomida, Y. (1975). Fossil Proboscidea from Taiwan. *Science Reports of the Yokohama National University*, **22**, 13–62. {17}

Shimizu, Y., Fujita, T., Kamiya, T., and Isokawa, S. (1960). Anatomy of a female Indian elephant with special reference to its visceral organs. *Acta Anatomica Nippon*, **35**, (3), 261–301. {2}

Shimizu, D., Matsuoka, C., Tamura, M., Kamiya, H., Kuga, N., and Ishigaki, S. (1990). *Mammalian footprints from Yasugawa site, late Pliocene, Japan*. Board of Education of Khosei town, Japan [in Japanese]. {17}

Shindo, T. and Mori, M. (1956). Musculature of the Indian elephant. Part III: Musculature of the trunk, neck and head. *Okajimas Folia Anatomica Japonica*, **29**, 17–40. {2}

Shipman, P. (1992). Observations and remarks on cutmarks on selected bones from Monte Verde. Manuscript on file at the University of Kentucky, Lexington. {31}

Shipman, P., Fisher, D. C., and Rose, J. J. (1984). Mastodon butchery: microscopic evidence of carcass processing and bone tool use. *Paleobiology*, **10**, 358–65. {30}

Short, R. V. (1966). Oestrous behaviour, ovulation and the formation of the corpus luteum in the African elephant, *Loxodonta africana*. *East African Wildlife Journal*, **4**, 56–68. {30}

Shoshani, J. (1986*a*). Mammalian phylogeny: comparison of morphological and molecular results. *Molecular Biology and Evolution*, **3**, 222–42. {5, 16}

Shoshani, J. (1986*b*). On the phylogenetic relationships among Paenungulata and within Elephantidae as demonstrated by molecular and osteological evidence. Unpublished D. Phil. thesis. Wayne State University, Detroit, Ann Arbor, University Microfilms. {2, 6, 8, 17, 18}

Shoshani, J. (1988). A new hypothesis on the numbers of mastodons and mammoths in the USA. *Geological Society of America, Abstracts with Programs*, **20**, (5), 388. {34}

Shoshani, J. (1989). A report on the Shelton mastodon site and a discussion of the numbers of mastodons and mammoths in Michigan. *Michigan Academician*, **21**, (2), 115–32. {27, 34}

Shoshani, J. (1990*a*). The "aspirin" approach to elephant conservation. Presented at the *Eleventh Annual International Elephant Workshop, held at Milwaukee County Zoo, Milwaukee, Wisconsin USA (24–27 October, 1990)*. {34}

Shoshani, J. (1990*b*). Distribution of *Mammut americanum* in the New World. *Current Research in the Pleistocene*, **7**, 124–7. {9, 29, 34}

Shoshani, J. (1991). Capability of producing low frequency calls in *Mammuthus vs. Mammut*. *Geological Society of America, Abstracts with Programs*, **23**, (5), A457. {16, 30}

Shoshani, J. (1992*a*). Anatomy and physiology. In *Elephants: majestic creatures of the wild*, (consult. ed. J. Shoshani), pp. 66–81. Rodale Press, Emmaus, Pennsylvania. {2}

Shoshani, J. (1992*b*). Cuvier *vis-à-vis* Huxley on the relationship of Hyracoidea and an update on an old controversy. In *Ongulés/Ungulates 91*, (ed. F. Spitz, G. Janeau, G. Gonzalez, S. Aulagnier, J. Joachim, and C. S. Mann), pp. 103–12. SFEPM–IRGM (Société Française pour l'Etude et la Protection des Mammiferes — Paris Institut de Recherche sur les Grands Mammiferes), Paris–Toulouse. {5, 16}

Shoshani, J. (1992*c*). Ecology and taxonomy: a new approach using *Heterohyrax* vs. *Procavia* as a case study. *Israel Journal of Zoology*, **38**, (3–4), 399–416. {34}

Shoshani, J. (consult. ed.) (1992*d*). *Elephants: majestic creatures of the wild*. Rodale Press, Emmaus, Pennsylvania. {34}

Shoshani, J. (1992*e*). Evolution of the Proboscidea. In *Elephants: majestic creatures of the wild* (consult. ed. J. Shoshani), pp. 18–33. Rodale Press, Emmaus, Pennsylvania. {34}

Shoshani, J. (1993*a*). Elephants: the super keystone species. *Swara*, **16** (2), 25–9. {34}

Shoshani, J. (1993*b*). How many species or subspecies of forest *Loxodonta* exist? *Abstracts, Sixth International Theriological Congress, Sydney, Australia*, 4–10 July 1993, p. 278–9. {ApC}.

Shoshani, J. (1993*c*). Hyracoidea–Tethytheria affinity based on myological data. In *Mammal phylogeny: placentals*, (ed. F. S. Szalay, M. J. Novacek, and M. C. McKenna), pp. 235–56. Springer-Verlag, New York. {5, 16}

Shoshani, J. (1994). A review of Kalb, J. E. and Mebrate, A. (1993). 'Fossil elephantoids from the hominid-bearing Awash Group, Middle Awash Valley, Afar Depression, Ethiopia.' published by Transactions of the American Philosophical Society, 83(1), xv + 1–114, *The Quarterly Review of Biology*, **69** (3), 392–3. {34}

Shoshani, J. (in press). Henry Fairfield Osborn. In *Biographical dictionary of North American environmentalists*, (ed. K. B. Sterling). Greenwood Press, Westport, Connecticut. {34}

Shoshani, J. and Eisenberg, J. F. (1982). *Elephas maximus*. *Mammalian Species*, **182**, 1–8. {2, 30, 33, ApC}

Shoshani, J. and McKenna, M. C. (1991). *Hemimastodon crepusculi*: is it a proboscidean, an artiodactyl, or Ungulata *incertae sedis*? *Journal of Vertebrate Paleontology*, **11**, (Suppl. 3), 55A. {6, 16, 34}

Shoshani, J., Goodman, M., Barnhart, M. I., Prychodko, W., Vereshchagin, N. K., and Mikhelson, V. M. (1981). Blood cells and proteins in the Magadan mammoth calf: immunodiffusion comparisons of *Mammuthus* to extant paenungulates and tissue ultrastructure. In *Magadan baby mammoth*, Mammuthus primigenius (*Blumenbach*), (ed. N. K. Vereshchagin and V. M. Mikhelson), pp. 191–220. Nauka Publishers, Leningrad [in Russian]. {7}

Shoshani, J. *et al.* (76 coauthors). (1982). On the dissection of a female Asian elephant (*Elephas maximus maximus* Linnaeus, 1758) and data from other elephants. *Elephant*, **2**, (1), 3–93. {2, ApC}

Shoshani, J., Lowenstein, J. M., Walz, D. A., and Goodman, M. (1985*a*). Proboscidean origins of mastodon and woolly mammoth demonstrated immunologically. *Paleobiology*, **11**, 429–37. {7}

Shoshani, J., Walz, D. A., Goodman, M., Lowenstein, J. M., and Prychodko, W. (1985*b*). Protein and anatomical evidence of the phylogenetic position of *Mammuthus primigenius* within the Elephantinae. *Acta Zoologica Fennica*, **170**, 237–40. {2, 7}

Shoshani, J., Hillman, J. C., and Walcek, J. M. (1987). "Ahmed", the logo of the Elephant Interest Group: encounters in Marsabit and notes on his model and skeleton. *Elephant*, **2**, (3), 7–32. {2}

Shoshani, J., Savage, R. J. G., and West, R. M. (1989*a*). A new species of *Barytherium* (Mammalia, Proboscidea) from Africa, and a discussion of early proboscidean systematics and paleoecology. *Abstracts of Papers and Posters Fifth International Theriological Congress, Rome, 22–29 August 1989*, Vol. I, p. 160. {6, 8, 16, 34, ApC}

Shoshani, J., *et al.* (1989*b*). The Shelton mastodon site: multidisciplinary study of a late Pleistocene (Twocreekan) locality in southeastern Michigan. *Contributions from the Museum of Paleontology, The University of Michigan*, **27**, (14), 393–436. {7, 27, 34}

Shotwell, J. A. (1963). The Juntura Basin: studies in earth history and paleoecology. *Transactions of the American Philosophical Society*, **53**, 1–77. {15}

Shotwell, J. A. and Russell, D. E. (1963). Mammalian fauna of the upper Juntura Formation, the Black Butte local fauna. In The Juntura Basin: studies in Earth history and paleoecology, (ed. J. A. Shotwell). *Transactions of the American Philosophical Society*, **53**, 42–69. {27}

Siegfried, P. (1956). Der Schädel eines *Parelephas trogontherii* von Gelsenkirchen. *Paläontologische Zeitschrift*, **30**, 171–79. {19, 25}

Siegfried, P. (1959). Das Mammut von Ahlen. *Paläontologische Zeitschrift*, **33**, 172–84. {19}

Sikes, S. K. (1966). The African elephant, *Loxodonta africana*: a field method for the estimation of age. *Journal of Zoology*, London, **150**, 279–95. {2, 12A}

Sikes, S. K. (1971). *The natural history of the African elephant*. Weidenfeld and Nicolson, London. {2, 4, 11, 12A, 26, 30, 32, 34}

Simmelink, J. W. (1988). Histology of enamel. In *Oral development and histology*, (ed. J. K. Avery), pp. 140–51. Decker, Toronto. {2}

Simmons, A. H. (1988). Pygmy hippos, early man and the initial human occupation of Cyprus. *Abstracts of the International Conference "Early man in island environments", Oliena (Sardegna), 25 September–2 October 1988*. {23}

Simmons, N. B. (1993). The importance of methods: archontan phylogeny and cladistic analysis of morphological data. In *Primates and their relatives in phylogenetic perspective. Advances in primatology series*, (ed. R. D. E. MacPhee), pp. 1–61. Plenum, New York. {16}

Simonelli, V. (1908). Mammiferi quaternari dell'isola di Candia. II. *Memorie dell'Accademia delle Scienze dell'Istituto di Bologna, Bologna*, **6**, (5), 397–405. {22}

Simons, E. L. (1964). Yale Peabody Museum. *Society of Vertebrate Paleontology News Bulletin*, **70**, 14–15. {2}

Simons, E. L. (1968). Early Cenozoic mammalian faunas, Fayûm Province, Egypt, Part I, African Oligocene mammals: introduction, history of study, and faunal succession. *Bulletin of the Peabody Museum of Natural History Yale University*, **28**, 1–21, 103–5. {8}

Simpson, G. G. (1929). *The extinct land mammals of Florida*. 20th Annual Report of the Florida State Geological Survey, pp. 229–79. {29}

Simpson, G. G. (1944). *Tempo and mode in evolution*. Columbia University, New York. {29}

Simpson, G. G. (1945). The principles of classification and a classification of mammals. *Bulletin of the American Museum of Natural History*, **85**, 1–350. {1, 5, 8, 12A, 16, 18, 27, 30, 34, ApA, ApC}

Simpson, G. G. (1953). *The major features of evolution*. Columbia University, New York. {18}

Simpson, G. G. (1965). *The geography of evolution*. Chilton, New York. {23}

Simpson, G. G. (1980). *Splendid isolation: the curious history of South American mammals*. Yale University, New Haven. {15, 29, 31}

Simpson, G. G. (1984). *Discoveries of the lost world: an account of some of those who brought back to life South American mammals long buried in the abyss of time*. Yale University, New Haven. {34}

Simpson, G. G. and Paula Couto, C. de (1957). The mastodonts of Brazil. *Bulletin of the American Museum of Natural History*, **112**, 125–89 + 23 plts. {15, 16, 27, 29, 31, ApC}

Simpson, G. G. and Tobien, H. (1954). The rediscovery of Peale's mastodon. *Proceedings of the American Philosophical Society*, **98**, (4), 279–81. {ApC}

Singer, R. and Hooijer, D. A. (1958). A *Stegolophodon* from South Africa. *Nature*, **182**, 101–2. {17}

Singh, V. B. (1978). The elephant in U. P. (India): a resurvey of its status after 10 years. *Journal of the Bombay Natural History Society*, **75**, 71–85. {33}

Skeels, M. A. (1962). The mastodons and mammoths of Michigan. *Papers of the Michigan Academy of Science, Arts, and Letters*, **47**, 101–33. {ApC}

Skinner, M. and Johnson, W. (1984). Tertiary stratigraphy and the Frick collection of fossil vertebrates from north-central Nebraska. *Bulletin of the American Museum of Natural History*, **178**, 215–368. {15}

Smart, C. (1976). The Lothagam 1 fauna: its phylogenetic, ecological, and biogeographic significance. In *Earliest man and environments in the Lake Rudolf Basin*, (ed. Y. Coppens, F. C. Howell, G. L. Isaac, and R. Leakey), pp. 361–9. University of Chicago. {12B}

Smith, F. (1890). On the histology of the skin of the elephant. *Journal of Anatomy and Physiology*, **24**, 493–503. {2}

Smith, N. S. and Buss, I. O. (1973). Reproductive ecology of the female African elephant. *Journal of Wildlife Management*, **37**, 524–34. {30}

Smith, R. L. (1974). *Ecology and field biology*, (2nd edn). Harper & Row, New York. {21}

Sobus, J. C. and Logan, T. P. (1984). Taphonomy of mammoth occurrences — Ceja member of the Santa Fe Formation, Tijeras Arroyo, Bernalillo County, New Mexico. *New Mexico Geology*, **6**, 14–15. {29}

Soergel, W. (1912). *Elephas trogontherii* Pohlig und *Elephas antiquus* Falconer ihre Stammesgeschichte und ihre Bedeutung für die Gliederung des deutschen Diluviums. *Palaeontrographica*, **60**, 1–114. {12A, 19}

Solario, P. J. M. S. (1864). Letter in *Les Mondes*, 29 September 1864. {8}

Sondaar, P. Y. (1971). Palaeozoogeography of the Pleistocene mammals from the Aegean. *Opera Botanica, Lund*, **30**, 65–70. {22}

Sondaar, P. Y. (1977). Insularity and its effect on mammal evolution. In *Major patterns in vertebrate evolution*, (ed. M. K. Hecht, P. C. Goody, and B. M. Hecht), pp. 671–707. Plenum, New York. {23, 24, 30}

Sondaar, P. Y. (1984). Faunal evolution and the mammalian biostratigraphy of Java. *Courier Forschungsinstitut Senckenberg*, **69**, 219–35. {17, 23}

Sondaar, P. Y. (1989). *Did man reach Australia via the giant rat and dingo route*. Publication of the Geological Research and Development Centre, Paleontogy Series, Vol. 5, pp. 76–83. {23}

Sondaar, P. Y. and Boekschoten, G. J. (1967). Quaternary mammals in the south Aegean island archipelagos: with notes on other fossil mammals from the coastal regions of the Mediterranean. I–II. *Proceedings of the Koninklijke Nederlandse Akademie van Wetenschappen, Amsterdam, B*, **70**, (5), 556–76. {22}

Southwood, T. R. E., Harvey, P. H., and Partridge, L. (ed.) (1991). *The evolution of reproductive strategies*. Cambridge University Press. {30}

Spaulding, W. G. and Martin, P. S. (1979). Ground sloth dung of the Guadalupe Mountains. In *Biological investigations in the Guadalupe Mountains National Park, Texas*. National Park Service Proceedings and Transactions Series No. 4, (ed. H. Genoways and R. Baker), pp. 259–69. Washington, DC. {28}

Spinage, C. A. (1972). Evolution of the Proboscidea. *Säugetierkundliche Mitteilungen*, **20**, (1–2), 152–6. {34}

Spinage, C. A. (1973). A review of ivory exploitation and elephant population trends in Africa. *East African Journal of Wildlife*, **11**, 281–9. {32}

Spinage, C. A. (1991). Botswana's problem elephants. *Pachyderm*, **14**, 46. {32}

Stafford, T. W. Jr, Hare, P. E., Curie, L., Jull, A. J. T., and Donahue, D. J. (1991). Accelerator radiocarbon dating at the molecular level. *Journal of Archaeological Science*, **18**, 35–72. {30}

Stanischeva, O. N. (1982). Environmental surroundings of the region of death of the Yurebei Mammoth based on pollen information. In *Yurebei Mammoth*, (ed. V. E. Sokolov), pp. 30–35. USSR Academy of Science, Moscow [in Russian]. {28}

Stanley, S. M. (1979). *Macroevolution: pattern and process*. Freeman, New York. {18, 34}

Stanley, S. M. (1987). *Extinction*. Scientific American Books, New York. {34}

Stearns, R. E. C. (1873). [no title; paraphrase of comments attributed to Stearns in minutes of the Academy's regular meeting.]. *Proceedings of the California Academy of Sciences*, **5**, 152. {24}

Stearns, S. C. (1977). The evolution of life history traits: a critique of the theory and a review of the data. *Annual Review of Ecology and Systematics*, **8**, 145–71. {30}

Steele, D. G. and Carlson, D. L. (1989). Excavation and taphonomy of mammoth remains from the Duewall–Newberry site, Brazos County, Texas. In *Bone modification*, (ed. R. Bonnichsen and M. Sorg), pp. 413–30. Center for the Study of the First Americans, Orono, Maine. {30}

Stehli, F. G. and Webb, S. D. (ed.) (1985). *The great American biotic interchange. Topics in Geobiology*, Vol. 4, Plenum, New York. {29}

Steininger, F., Rabeder, G., and Rögl, F. (1985). Land mammal distribution in the Mediterranean Neogene: a consequence of geokinematic and climatic events. In *Geological evolution of the Mediterranean basin*, (ed. D. J. Stanley and F. C. Wezel), pp. 559–71. Springer-Verlag, New York. {9, 13}

Steininger, F. F., Bernor, R. L., and Fahlbusch, V. (1990). European Neogene marine/continental chronologic correlations. In *European Neogene mammal chronology*, (ed. E. H. Lindsay, V. Fahlbusch, and P. Mein), pp. 15–46. Plenum, New York [copyright date, 1989]. {12B, ApD}

Steinmüller, A. (1972). Die Schichtenfolgen von Süßenborn und Voigtstedt und die Gliederung des Mittelpleistozäns. *Geologie*, **21**, 129–42. {19}

Stirton, R. A. (1934). A new species of *Amblycastor* from the *Platybelodon* beds, Tung Gur Formation, of Mongolia. *American Museum Novitates*, 694, 1–4. {27}

Stock, C. (1935). Exiled elephants of the Channel Islands, California. *Scientific Monthly*, **41**, 205–14. {24}

Stock, C. (1956). *Rancho La Brea: a record of Pleistocene life in California*, (6th edn, 7th printing). Los Angeles County Museum of Natural History, George C. Page Museum of La Brea Discoveries, Science Series No. 20, Paleontology, Vol. 11, pp. 1–81. {ApC}

Stock, C. (1992). *Rancho La Brea: a record of Pleistocene life in California*, (7th edn, revised by J. M. Harris). Natural History Museum of Los Angeles County, Science Series, No. 37, pp. 1–113. {ApC}

Stock, C. and Furlong, E. L. (1928). The Pleistocene elephants of Santa Rosa Island, California. *Science*, **68**, 140–1. {24}

Stöcker, L. (1957). Trigeminusmuskulatur und Kiefergelenk von *Elephas maximus* L. *Gegenbaurs Morphologisches Jahrbuch*, **98**, 35–76. {2}

Storr, G. C. C. (1780). *Prodromus methodi Mammalium ... inauguralem disputatianem propositus*. Wolffer, Tubingae. {5}

Stoutamire, W. P. and Benninghoff, W. S. (1964). Biotic assemblage associated with a mastodon skull from Oakland County, Michigan. *Papers Michigan Academy of Science, Arts, and Letters*, **49**, 47–60. {27}

Stuart, A. J. (1981). A comparison of the middle Pleistocene mammal faunas of Voigtstedt (Thüringia, GDR) and West Runton (Norfolk, England). *Quartärpaläontologie*, **4**, 155–63. {19}

Stuart, A. J. (1982). *Pleistocene vertebrates in the British Isles*. Longman, London. {19, 20}

Stuart, A. J. (1991). Mammalian extinctions in the late Pleistocene of northern Eurasia and North America. *Biological Reviews*, **66**, 453–562. {19, 21}

Stucky, R. K. and McKenna, M. C. (1993). Mammalia. In *The fossil record 2*, (ed. M. J. Benton), pp. 739–71. Chapman and Hall, London. {34}

Sukumar, R. (1986). The elephant populations of India: strategies for conservation. *Proceedings of the Indian Academy of Sciences (Animal Sciences/Plant Sciences), Supplement*, **November 1986**, 59–71. {33}

Sukumar, R. (1989). *The Asian elephant: ecology and management*. Cambridge University Press, Cambridge. {30, 33}

Sukumar, R. (1994). *Elephant days and nights: 10 years with the Indian elephant*. Oxford University Press, Oxford. {34}

Sung, T. (1958). Tertiary spore and pollen complexes from the red beds of Chiuchuan, Kansu and their geological and botanical significance. *Acta Palaeontologica Sinica*, **6** (2), 159–67 + 7 plts. [In Chinese with English summary] {13}

Sutcliffe, A. J. (1985). *On the track of Ice Age mammals*. Dorset Press, Dorchester. {29}

Suzuki, M., Wikarno, Budisantoso, Saefudin, I. and Itihara, M. (1985). Fission track ages of pumice tuff, tuff layers and javites of hominid fossil bearing formations in Sangiran area, central Java. In *Quaternary geology of the hominid fossil bearing formations in Java*, (ed. N. Watanabe and D. Kadar), pp. 309–31. Geological Research and Development Centre, Special Publication No. 4 [no first initial for Wikarno and Budisantoso]. {23}

Suzuki, N., Tikhonov, A., Vereshchagin, N. K., and Hamada, T. (1992). Extracted heart from the frozen baby mammoth in Siberia. *Scientific Papers of the College of Arts and Sciences, The University of Tokyo*, **42(1)**, 63–78. {2}

Swofford, D. L. (1989). *PAUP: phylogenetic analysis using parsimony*. Version 3.0a. Computer program distributed by the Illinois Natural History Survey, Champaign, Illinois. {12A}

Swofford, D. L. (1991). *PAUP: phylogenetic analysis using parsimony*. Version 3.0s. Computer program distributed by the Illinois Natural History Survey, Champaign, Illinois. {16}

Swofford, D. L. and Olsen, G. J. (1990). Phylogeny reconstruction. In *Molecular systematics*, (ed. D. M. Hillis and C. Moritz), pp. 411–90. Sinauer Associates, Sunderland, Mass. {12A}

Symeonidis, N. K. (1972). Die entdeckung von Zwergelefanten in der Hohle "Charkadio" auf der Insel Tilos (Dodekanes, Griechenland). *Annales Géologiques des Pays Helléniques, Athina*, **24**, 445–61. {22}

Symeonidis, N. K., Bachmayer, F., and Zapfe, H. (1974). Entdeckung von Zwergelefanten auf der Insel Rhodos (Ausgrabungen 1973). *Annalen der Naturhistorischen Museum von Wien*, **78**, 193–202. {22}

Szalay, F. S. (1968). The beginnings of primates. *Evolution*, **22**, 19–36. {3}

Szalay, F. S. (1969). Origin and evolution of function of the mesonychid condylarth feeding mechanism. *Evolution*, **23**, 703–20. {3}

Takahashi, K. and Sawamura, H. (1985). Morphological feature of *Palaeoloxodon naumanni* based upon skeletal restoration of the Hamacho specimen. *Journal of Fossil Research*, supplement, **2**, 26–9. [In Japanese] {ApC}

Takahashi, K., Inuzuka, N., Kawamura, Y., and Nasu, T. (1991). Paleontology of *Palaeoloxodon naumanni*. In *Japanese proboscidean fossils*, (ed. T. Kamei), pp. 11–179. Tsukiji-Shokan, Tokyo. [In Japanese] {ApC}

Takai, F. (1939). On some Cenozoic mammals from Japan (Part I). *The Journal of the Geological Society of Japan*, **46**, 481–9. [in Japanese with English abstract]. {17}

Takai, F. (1954). An addition to the mammalian fauna of the Japanese Miocene. *Journal of the Faculty of Sciences, University of Tokyo*, Section II, **9**, 331–5. {10}

Tarling, D. H. (1980). Continental drift and biological evolution. *Carolina Biological Readers*, **113**, 1–32. {8}

Taruno, H. (1972). A study on *Stegodon akashiensis* Takai. Unpublished master thesis. Kyoto University. {17}

Taruno, H. (1985). Genus *Stegodon* and genus *Stegolophodon* — their criteria and phylogenetic relation — (Mammalia: Proboscidea). *Bulletin of the Osaka Museum of Natural History*, **38**, 23–36 + 4 plts [in Japanese with English abstract]. {12A, 12B, 17}

Taruno, H., *et al.* (1992). Footprint fossils of Naumann's elephant discovered from the upper Pleistocene Nojiri-ko Formation. *Earth Science*, **46**, (6), 385–404 + i–v [in Japanese with English abstract]. {34}

Tassy, P. (1974). *Gomphotherium angustidens* (Proboscidea, Mammalia) du Burdigalien d'Artenay (Loiret). Thèse Dr. 3ème Cycle, Université Paris VI. {9}

Tassy, P. (1977*a*). Découverte de *Zygolophodon turicensis* (Schinz) (Proboscidea, Mammalia) au lieu-dit Malartic à Simorre, Gers (Vindobonien moyen); implications paléoécologiques et biostratigraphiques. *Géobios*, **10**, 655–69. {3, 10, ApC}

Tassy, P. (1977*b*). Les Mastodontes miocènes du Bassin aquitain: une mise au point taxonomique. *Comptes Rendus Hebdomadaires de l'Académie des Sciences, Paris*, **284** (Série D), 1389–92. {ApC}

Tassy, P. (1977*c*). Le plus ancien squelette de gomphothère (Proboscidea, Mammalia) dans la formation burdigalienne des sables de l'Orléanais, France. *Mémoires du Muséum National d'Histoire Naturelle*, New Series C, **37**, 1–51 + 3 plts. {10}

Tassy, P. (1979). Relations phylogénétiques du genre *Moeritherium* Andrews, 1901 (Mammalia). *Comptes Rendus de l'Académie des Sciences, Paris*, Series D, **289**, 85–8. {5, 8, 16}

Tassy, P. (1981). Le crâne de *Moeritherium* (Proboscidea, Mammalia) de l'Éocène de Dor el Talha (Libye) et le problème de la classification phylogénétique du genre dans les Tethytheria McKenna, 1975. *Bulletin du Muséum National d'Histoire Naturelle, Paris*, Series 4, **3**, (C), 87–147. {5, 8, 16}

Tassy, P. (1982). Les principales dichotomies dans l'histoire des Proboscidea (Mammalia); une approche phylogénétique. *Géobios*, Special edition, **6**, 225–45. {3, 8, 9, 12A, 13, 16, 17, 30, 34}

Tassy, P. (1983). Les Elephantoidea miocènes du Plateau du Potwar, Groupe de Siwalik, Pakistan. *Annales de Paléontologie (Vert.–Invert.)*, **69**, (2), 99–136; (3), 235–97; and (4), 317–54. {3, 6, 9, 11, 12A, 12B, 13, 16, 17, 18, 34, ApB, ApC}

Tassy, P. (1984). Le mastodonte à dents étroites, le grade trilophodonte et la radiation initiale des Amebelodontidae. In *Actes du symposium paléontologique Georges Cuvier*, (ed. E. Buffetaut, J. M. Mazin, and E. Salmon), pp. 459–73. Impressions le Serpentaire, Montbéliard. {11, 13, 14, 16, ApB, ApC}

Tassy, P. (1985). La place des mastodontes miocènes de l'ancien monde dans la phylogénie des Proboscidea (Mammalia): hypothèses et conjectures. Unpublished Thèse Doctorat ès Sciences, Université Pierre et Marie Curie, Paris, 85–34, Volumes I–III. {14, 6, 8, 9, 10, 11, 12A, 13, 14, 16, 17, 18, 27, 34, ApA, ApC}

Tassy, P. (1986). Nouveaux Elephantoidea (Proboscidea, Mammalia) dans le Miocène du Kenya: essai de réévaluation systématique. *Cahiers de Paléontologie. Éditions du Centre National de la Recherche Scientifique (CNRS), Paris.* {6, 9, 10, 11, 12A, 12B, 13, 15, 16, 17, 18, 34, ApC}

Tassy, P. (1987). A hypothesis on the homology of proboscidean tusks based on paleontological data. *American Museum Novitates*, **2895**, 1–18. {4, 13}

Tassy, P. (1988*a*). The classification of Proboscidea: how many cladistic classifications? *Cladistics*, **4**, 43–57. {8, 10, 12A, 13, 16, 18, 34, ApB}

Tassy, P. (1988*b*). Le statut systématique de l'espèce *Hemimastodon crepusculi* (Pilgrim): l'éternel problème de l'homologie et de la convergence. *Annales de Paléontologie (Vert.-Invert.)*, **74**, (3), 115–27. {6, 16, 34}

Tassy, P. (1990*a*). Phylogénie et classification des Proboscidea (Mammalia): historique et actualité. *Annales de Paléontologie (Vert.-Invert.)*, **76**, (3), 159–224. {1, 6, 8, 9, 12A, 12B, 16, 18, 34}

Tassy, P. (1990*b*). The "proboscidean datum event:" how many proboscideans and how many events? In *European Neogene mammal chronology*, (ed. E. H. Lindsay, V. Fahlbusch, and P. Mein), pp. 237–52. Plenum, New York [copyright date, 1989]. {9, 10, 12B, 27, ApC}

Tassy, P. (1993). *The Message of fossils*. McGraw-Hill, New York. [French edition published 1991 as `Le Message des fossiles', by Hachette, Paris] {34}

Tassy, P. (1994*a*). Gaps, parsimony, and early Miocene elephantoids (Mammalia), with a re-evaluation of *Gomphotherium annectens* (Matsumoto, 1925). *Zoological Journal of the Linnean Society*, **112**, 101–117. {6, 16}

Tassy, P. (1994*b*). Les gisements de mammifères du Miocène supérieur de Kemiklitepe, Turquie: 7. Proboscidea (Mammalia). *Bulletin du Muséum National d'Histoire Naturelle, Paris*, Series 4, **Vol. 16**, section C, 143–57. {21}

Tassy, P. (1994*c*). Origin and differentiation of the Elephantiformes (Mammalia, Proboscidea). *Verhandlungen Naturwissenschaftlichen Verains in Hamburg* (NF, 34, 73–94. {6}

Tassy, P. (1994*d*). Les proboscidiens (Mammalia) fossiles du Rift Occidental, Ouganda. In *Geology and Palaeobiology of the Albertine Rift Valley, Uganda-Zaire, Vol II: Palaeobiology* (ed. B. Senut and M. Pickford), pp. 217–57. CIFEG Occasional Publications, Orléans. {6}

Tassy, P. and Darlu, P. (1986). Analyse cladistique numérique et analyse de parcimonie; l'exemple des Elephantidae. *Géobios*, **19**, (5), 587–600. {12A, 12B}

Tassy, P. and Darlu, P. (1987). Les Elephantidae: nouveau regard sur les analyses de parcimonie. *Géobios*, **20**, (4), 487–94. {12A, 17, ApC}

Tassy, P. and Pickford, M. (1983). Un nouveau mastodonte zygolophodonte (Proboscidea, Mammalia) dans le Miocène inférieur d'Afrique orientale: systématique et paléoenvironnement. *Géobios*, **16**, (1), 53–77. {9, 10, 14, 18, 34, ApB, ApC}

Tassy, P. and Shoshani, J. (1988). The Tethytheria: elephants and their relatives. In *The phylogeny and classification of the tetrapods*. Vol. 2: *Mammals*, (ed. M. J. Benton), pp. 283–315. The Systematics Association, Special Volume No. 35B. Clarendon, Oxford. {1, 3, 5, 6, 8, 12A, 12B, 15, 16, 17, 18, 21}

Tassy, P., Anupandhanat, P., Ginsburg, L., Mein, P., Ratanasthien, B., and Sutteethorn, V. (1992). A new *Stegolophodon* (Proboscidea, Mammalia) from the early Miocene of northern Thailand. *Géobios*, **25**, (4), 511–23. {6, 34, ApC}

Tattersall, I., Delson, E., and Van Couvering, J. (1988). *Encyclopedia of human evolution and prehistory*. Garland Publishing, New York. {ApD}

Taylor, J. G. (1969). William Turner's journeys to the caribou country with the Labrador Eskimos in 1780. *Ethnohistory*, **16**, 141–64. {30}

Tchernov, E. (1981). *The biostratigraphy of the Middle East*. Colloques Internationaux du CNRS No. 598. Prehistoire du Levant (Editions du Centre National de la Recherche Scientifique, Paris), pp. 67–97. Maison de l'Orient, Lyon. {21}

Tchernov, E. (1984). Faunal turnover and extinction rate in the Levant. In *Quaternary extinctions: a prehistoric revolution*, (ed. P. S. Martin and R. G. Klein), pp. 528–52. University of Arizona Press, Tucson. {21}

Tchernov, E. (1988). The Levant — a biogeographical crossroad. In *The zoogeography of Israel: the distribution and abundance at a zoogeographical crossroad*, (ed. Y. Yom-Tov and E. Tchernov), pp. 159–250. Junk Publishers, Dordrecht, The Netherlands. {21}

Tchernov, E. (1992). The Afro-Arabian component in the Levantine mammalian fauna — a short biogeographical review. *Israel Journal of Zoology*, **38**, (3–4), 155–92. {21, ApD}

Tchernov, E. and Shoshani, J. (1989). Neogene and Pleistocene proboscidean remains in the southern Levant. *Abstracts of Papers and Posters Fifth International Theriological Congress, Rome, 22–29 August 1989*, Vol. I, pp. 164–5. {17, 21}

Tchernov, E. and Yom-Tov, Y. (1988). Zoogeography of Israel. In *The zoogeography of Israel: the distribution and abundance at a zoogeographical crossroad*, (ed. Y. Yom-Tov and E. Tchernov), pp. 1–6. Junk Publishers, Dordrecht, The Netherlands. {21}

Tchernov, E., Ginsburg, L., Tassy, P., and Goldsmith, N. F. (1987). Miocene mammals of the Negev (Israel). *Journal of Vertebrate Paleontology*, **7**, 284–310. {10, 21}

Tchernov, E., Kolska Horowitz, L., Ronen, A., and Lister, A. (1994). The faunal remains from Evron Quarry in relation to other lower Paleolithic hominid sites in the southern Levant. *Quaternary Research* **42**, 328–39. {21}

Tedford, R. H., *et al.* (1987). Faunal succession and biochronology of the Arikareean through Hemphillian interval (late Oligocene through earliest Pliocene epochs) in North America. In *Cenozoic mammals of North America: geochronology and biostratigraphy*, (ed. M. O. Woodburne), pp. 153–210. University of California Press, Berkeley. {15, 27, 30, ApD}

Tedford, R., Flynn, L., Zhanxiang, Q., Opdyke, N., and Downs, W. (1991). Yushe Basin, China; paleomagnetically calibrated mammalian biostratigraphic standard for the late Neogene of eastern Asia. *Journal of Vertebrate Paleontology*, **11**, 519–26. {12B}

Teilhard de Chardin, P. and Leroy, P. (1942). *Chinese fossil mammals, a complete bibliography analyzed, tabulated, annotated and indexed*. Publications de l'Institut de Geo-Biologie, Pekin, Vol. 8, pp. 1–142. {17}

Teilhard de Chardin, P. and Trassaert, M. (1937). The proboscideans of southeast Shansi. *Paleontologia Sinica*, Serie C, **13** (1), 1–58 [8 tbls + 13 plts on pp. 59–85. Chinese abstract]. {6, 13, 17, ApC}

Temminck, C. J. (1847). *Coup-d'oeil général sur les possessions néerlandaises dans l'Inde archipélagique*, Vol. 2, Arnz, Leide. {ApC}

Thenius, E. (1969). *Phylogenie der Mammalia. Stammesgeschichte der Säugetiere (einschliesslich der Hominiden)*. Walter de Gruyter and Co., Berlin. {1, 5, 12A, 34}

Theodorou, G. E. (1983). Die fossilen Zwergelefanten der Höhle "Charkadio" auf der Insel Tilos (Dodekanes, Griechenland). Unpublished D. Phil. Thesis, Athens. {22}

Theodorou, G. E. (1986). Pleistocene elephants from Crete (Greece). *Modern Geology, London*, **10**, 235–42. {22}

Theodorou, G. E. (1988). Environmental factors affecting the evolution of island endemics: the Tilos example from Greece. *Modern Geology, London*, **13**, 183–8. {22}

Theodorou, G. E. (1990). The dwarf elephants of Tilos. *The Athenian*, **May 1990**, 17–19. {21)

Thewissen, J. G. M. and Domning, D. P. (1992). The role of phenacodontids in the origin of the modern orders of ungulate mammals. *Journal of Vertebrate Paleontology*, **12(4)**, 494–504. {5}

Thomas, H. (1985). The early and middle Miocene land connection of the Afro-Arabian plate and Asia: a major event for hominoid dispersal? In *Ancestors: the hard evidence*, (ed. E. Delson), pp. 42–50. Liss, New York. {10, 21}

Thomas, H., Sen, S., Khan, M., Battail, B., and Ligabue, G. (1982). The lower Miocene fauna of Al-Sarrar (Eastern Province, Saudi Arabia). *Atlal*, **5**, 109–36. {10}

Thomas, H., Roger, J., Sen, S., Bourdillon de Grissac, C., and Al-Sulaimani, Z. (1989*a*). Découverte de vertébrés fossiles dans l'Oligocène inférieur du Dhofar (Sultanat d'Oman). *Géobios*, **22**, 101–20. {34}

Thomas, H., Tassy, P., and Sen, S. (1989*b*). Paleogene proboscidean remains from the southern Dhofar (Sultanate of Oman). *Abstracts of Papers and Posters Fifth International Theriological Congress, Rome, 22–29 August 1989*, Vol. I, p. 166. {34}

Thompson, R. S., van Devender, T. R., Martin, P. S., Foppe, T., and Long, A. (1980). Shasta ground sloth (*Nothrotheriops shastense* Hoffsetter) at Shelter Cave, New Mexico: environment, diet, and extinction. *Quaternary Research*, **14**, 360–76. {28}

Tiercelin, J.-J., Michaux, J., and Bandet, Y. (1979). Le Miocene supérieur du sud de la Depression de l'Afar, Ethiopie: sédiments, faunes, ages isotopiques. *Bulletin Société Géologique de France*, **21**, 255–8. {12B}

Tilesius, G. (1815) . De sceleto mammonteo sibirico ad maris glacialis littora anno 1807, effosso cui praemissae elephantini generis specierum distinctiones. Sectio Primo ... Conventui exhibuit die 10 Jan. 1810. *Mémoires de l'Académie Impérial des Sciences de St. Petersbourg*, **5**, 406–513. {25, 26}

Tleuberdina, P. A. (1982). *Late Neogene fauna of south-eastern Kazachstan*. Academy of Science of Kazachstan SSR, Institute of Zoology, Alma Ata [in Russian]. {ApC}

Tobien, H. (1964). Early zygodont mastodonts of western North America and their European relatives. *Geological Society of America (Cordillerian Section), Seismological Society of America, and Paleontological Society (Pacific Coast Section), University of Washington, Seattle, Washington. Program, 60th Annual Meeting*, p. 64. {9, 16, 34, ApA}

Tobien, H. (1971). *Moeritherium, Palaeomastodon, Phiomia* aus dem Paläeogen Nordafrikas und die Abstammung der Mastodonten (Proboscidea, Mammalia). *Mitteilungen, aus dem Geologischen Institut der Technischen Universität Hannover*, **10**, 141–63. {6, 8, 9}

Tobien, H. (1972). Status of the genus *Serridentinus* Osborn 1923 (Proboscidea, Mammalia) and related forms. *Mainzer Geowissenschaftliche Mitteilungen*, **1**, 143–91. {6, 9, 12A, 13, 14, 15, 16, 34, ApA, ApC}

Tobien, H. (1973*a*). On the evolution of mastodonts (Proboscidea, Mammalia). Part 1: The bunodont trilophodont groups. *Notizblatt des Hessischen Landesamtes für Bodenforschung zu Wiesbaden*, **101**, 202–76 + plts 23–26. {6, 9, 10, 11, 13, 14, 15, 16, ApA, ApC}

Tobien, H. (1973*b*). The structure of the mastodont molar (Proboscidea, Mammalia). Part 1: The bunodont pattern. *Mainzer Geowissenschaftliche Mitteilungen*, **2**, 115–47. {3, 9, 12A, 13, 16}

Tobien, H. (1975). The structure of the mastodont molar (Proboscidea, Mammalia). Part 2: The zygodont and zygobunodont patterns. *Mainzer Geowissenschaftliche Mitteilungen*, **4**, 195–233. {2, 3, 6, 9, 12A, 14, 16, 17, 20, ApA}

Tobien, H. (1976). Zur paläeontologischen Geschichte der Mastodonten (Proboscidea, Mammalia). *Mainzer Geowissenschaftliche Mitteilungen*, **5**, 143–225. {6, 9, 12A, 14, 16, 17, 27, ApA}

Tobien, H. (1977). Fauna. In *Jungtertiar und Quatar im Horloff-Graben/ Vogelsberg, Geologisches Abhandlungen Hessen*, **75**, 1–80. {20}

Tobien, H. (1978*a*). An advanced zygodont mastodont (Proboscidea, Mammalia) from the Miocene of the Siwaliks of Pakistan. *Mainzer Geowissenschaftliche Mitteilungen*, **7**, 239–48. {12A, ApA}

Tobien, H. (1978*b*). On the evolution of mastodonts (Proboscidea, Mammalia). Part 2: The bunodont tetralophodont groups. *Geologisches Jahrbuch Hessen*, **106**, 159–208 + plts 10–21. {2, 6, 12A, 12B, 14, 15, 16, 18, 20, 23, ApC}

Tobien, H. (1978*c*). The structure of the mastodont molar (Proboscidea, Mammalia). Part 3: The Oligocene mastodont genera *Palaeomastodon*, *Phiomia* and the Eo/Oliogene paenungulate *Moeritherium*. *Mainzer Geowissenschaftliche Mitteilungen*, **6**, 177–208. {8, 12A, 16, ApC}

Tobien, H. (1980*a*). A note on the mastodont taxa (Proboscidea, Mammalia) of the "Dinotheriensande" (Upper Miocene, Rheinhessen, Federal Republic of Germany). *Mainzer Geowissenschaftliche Mitteilungen*, **9**, 187–201. {9, 14}

Tobien, H. (1980*b*). A note on the skull and mandible of a new choerolophodont mastodont (Proboscidea, Mammalia) from the middle Miocene of Chios (Aegean Sea, Greece). In *Aspects of vertebrate history. Essays in honor of Edwin Harris Colbert* (ed. L. L. Jacobs), pp. 299–307. Museum of Northern Arizona Press, Flagstaff. {ApC}

Tobien, H. (1986*a*). Mastodonts (Proboscidea, Mammalia) from the late Neogene and early Pleistocene of the People's Republic of China. *Mainzer Geowissenschaftliche Mitteilungen*, **15**, 119–81. {12A, 15, 34}

Tobien, H. (1986*b*). Die paläontologische Geschichte der proboscidier (Mammalia) in Mainzer Becken (BRD). *Mainzer Naturwissenschaften Archiv*, **24**, 155–261. {12A, 14}

Tobien, H. (1988). Some major features in the evolution of mastodonts (Proboscidea, Mammalia). *Geological Survey of India, Special Publication* No. 11, Vol. 2, pp. 109–16. {1, 9, 16, 34}

Tobien, H. (1990). Proboscidea: a preliminary note. *Journal of Human Evolution*, **19**, 465–9. {9, ApC}

Tobien, H., Chen, G. F., and Li, Y. Q. (1984). The mastodonts (Proboscidea, Mammalia) of China: evolution, palaeobiogeography, palaeoecology. In *The evolution of the east Asian environment II: palaeobotany, palaeozoology, and palaeoanthropology*, (ed. R. O. Whyte), pp. 689–96. Centre of Asian Studies, University of Hong Kong. {9, 13}

Tobien, H., Chen, G. F., and Li, Y. Q. (1986). Mastodonts (Proboscidea, Mammalia) from the late Neogene and early Pleistocene of the People's Republic of China. Part 1. Historical account; the genera *Gomphotherium, Choerolophodon, Synconolophus, Amebelodon, Platybelodon, Sinomastodon*. *Mainzer Geowissenschaftliche Mitteilungen*, **15**, 119–81. {6, 13, 16, 31, ApB, ApC}

Tobien, H., Chen, G. F., and Li, Y. Q. (1988). Mastodonts (Proboscidea, Mammalia) from the late Neogene and early Pleistocene of the People's Republic of China. Part 2. The genera *Tetralophodon, Anancus, Stegotetrabelodon, Zygolophodon, Mammut, Stegolophodon*. Some generalities on the Chinese mastodonts. *Mainzer Geowissenschaftliche Mitteilungen*, **17**, 95–220. {6, 9, 12A, 12B, 13, 16, 17, 23, 34, ApC}

Toepfer, V. (1957). Die Mammutfunde von Pfännerhall im Geiseltal. *Veröffentlichungen des Landesmuseums für Vorgeschichte in Halle*, **16**, 1–58. {19}

Toldt, K. (1913). Über die äuszere Körpergestalt eines Fetus von *Elephas maximus* (= *indicus*) L., nebst vergleichenden Betrachtungen über sein Integument, insbesondere über die Behaarung. *Denkschriften der kaiserlichen Akademie der Wissenschaften Mathematisch-Naturwissenschaftliche Klasse, Wien*, **90**, 259–300. {4}

Trevisan, L. (1949). Lineamenti dell'evoluzione del ceppo di elefanti eurasiatici nel Quaternario. *La Ricerca Scientifica, Supplement*, **19**, 10. {19}

Tröster, G. (1990). Der Koph von *Hybophthirus notophallus* (Neumann) (Phthiraptera: Anoplura). Eine funktionsmorphologische und konsequent-phylogenetische Analyse. *Stuttgarter Beiträge zur Naturkunde* A, **441**, 1–89. {5}

Tuoc, D. and Santiapillai, C. (1991). The status of elephants in Vietnam. *IUCN/SSC Asian Elephant Specialist Group Newsletter*, **7**, 2–8. {33}

Turner, M. A. (1975). *Amebelodon floridanus* (Leidy), a shovel-tusked gomphothere from Mixson's Bone Bed Levy County, Florida. *Proceedings of Nebraska Academy of Sciences Inc.*, abstract [1 page]. {13, ApC}

Tuross, N. (1989). Albumin preservation in the Taima-taima mastodon skeleton. *Applied Geochemistry*, **4**, 255–9. {7}

Tuross, N. and Dillehay, T. (1991). Molecular and atomic analyses at Monte Verde and their archaeological implications. Manuscript on file at the University of Kentucky, Lexington. {31}

Tuross, N., Fogel, M. L., and Hare, P. E. (1988). Variability in the preservation of the isotope composition of collagen from fossil bone. *Geochemica et Cosmochemica Acta*, **52**, 929–35. {30}

Tyler, N. J. C. (1987). Sexual dimorphism in the pelvic bones of Svalbard reindeer, *Rangifer tarandus platyrhynchus*. *Journal of Zoology, London*, **213**, 147–52. {25}

Ukraintseva, V. V. (1981). Vegetation of warm late Pleistocene intervals and the extinction of some large herbivorous mammals. *Polar Geography and Geology* **4**, 189–203. {28}

Ukraintseva, V. V. (1985). Forage of the large herbivorous mammals of the epoch of the mammoth. *Acta Zoologica Fennica*, **170**, 215–20. {28, 29}

Ukraintseva V. V. (1986). On the composition of the forage of the large herbivorous mammals of the mammoth epoch: significance for palaeobiological and palaeogeographical reconstruction. *Quartärpaläontologie* **6**, 231–8. {19}

Urbanas, E. V. (1980). Teeth of mammoth from late palaeolithic stations of Kostenki village in Voronezh region. In *Mammals of eastern Europe in the Anthropogene*, (ed. N. K. Vereshchagin), Proceedings of the Zoological Institute, Vol. 93, pp. 81–90. Academy of Sciences of the USSR, Leningrad [in Russian]. {26}

Vacek, M. (1877). Über österreichische Mastodonten und ihre Beziehungen zu den Mastodon-Arten Europas. *Abhandlungen der Kaiserlich-Königlichen geologischen Reichsanstalt*, **7**, (4), 1–45. {1, 3, 9, 12A, 16, ApA, ApB, ApC}

Van Couvering, J. and Berggren, W. A. (1977). Biostratigraphical basis of the Neogene Time Scale. In *Concepts and methods of biostratigraphy* (ed. E. G. Kauffman and K. J. F. Hazel), pp. 282–306. Dowden, Hutchinson and Ross, Stroudsbourg. {AST}

Van Couvering, J. and Harris, J. A. (1991). Late Eocene age of Fayum mammal faunas. *Journal of Human Evolution*, **21**, 241–60. {ApD}

Van Couvering, J. and Harris, J. A. (1994). African Land Mammal Ages. Manuscript on file at the American Museum of Natural History, New York. {ApD}

Vandebroek, G. (1961). The comparative anatomy of the teeth of lower and non specialized mammals. In *Colloque international sur l'évolution des mammifères inférieurs et non spécialisés*, (ed. G. Vandebroek), Vol. 1, pp. 215–320. Koninklijke Vlaamse Academie voor Wetenschappen Belgie. {3}

Van den Bergh, G. D., Aziz, F., Sondaar, P. Y., and Hussain, S. T. (1992). Taxonomy, stratigraphy and paleozoogeography of Plio-Pleistocene proboscideans from the Indonesian islands. *Bulletin of the Geological Research and Development Centre, Paleontological Series*, **7**, 28–58. {23}

Van den Bergh, G. D., Aziz, F., Sondaar, P. Y., and de Vos, J. (1994). The first *Stegodon* fossils from central Sulawesi and a new advanced *Elephas* species from south Sulawesi. *Geological Research and Development Centre*, Bulletin no. **17**, 22–39. {23}

Van der Maarel, F. H. (1932). Contributions to the knowledge of the fossil mammalian fauna of Java. *Wetenschappelijke mededeelingen, dienst van den mijnbouw in Nederlandsch-Indië*, **15**, 1–208. {23, ApC}

Vangengeim, E. A. and Pevsner, M. A. (1991). The Villafranchian of the USSR: bio- and magnetostratigraphy. In *Palaeogeography and biostratigraphy of the Pliocene and Anthropogene*, pp. 124–45. Geological Institute, USSR Academy of Sciences, Moscow. {19}

Vangengeim, E. A., Vekua, M. L., Zhegallo, V. I., Pevsner, M. A., Taktakishvili, I. G., and Tesakov, A. C. (1991). Position of the Taman faunal complex on the stratigraphic and magnetochronological scales. *Byulletin Komissii po Izucheniyu Chetvertichnogo Perioda*, **60**, 41–52. {19}

van Kolfschoten, T. and Vervoort-Kerkhoff, Y. (1985). Nijlpaarden van Nederlandse bodem en uit de Noordzee. *Cranium*, **2**, (2), 36–43. {20}

van Rummelen, F. F. F. E. (1970). *Toelichting bij de geologische kaart van Nederland 1 : 50.000. Blad Schouwen-Duiveland.* Rijks Geologische Dienst, Haarlem. {20}

Van Valen, L. (1966). Deltatheridia, a new order of mammals. *Bulletin of the American Museum of Natural History*, **132**, 1–126. {3}

Van Valen, L. (1973). Pattern and the balance of nature. *Evolutionary Theory*, **1**, 31–49. {24}

Varela Barbagelata, J. (1976). Estudio estratigráfico-sedimentológico de los depósitos de Laguna de Taguatagua. Provincia de O'Higgins. Tesis de Prueba para optar al Título de Geólogo. Santiago, Chile. {31}

Vartanyan, S. L., Garutt, V. E., and Sher, A. V. (1993). Holocene dwarf mammoths from Wrangel Island in the Siberian Arctic. *Nature*, **362**, 337–40. {19}

Vaufrey, R. (1929). Les éléphants nains des iles méditerranéennes et la question des isthmes Pléistocènes. *Archives de l'Institut de Paléontologie Humaine, Paris*, **6**, 1–216 + 9 plts. {22}

Vedder, J. G. and Howell, D. G. (1980). Topographic evolution of the southern California borderland during late Cenozoic time. In *The California Islands: proceedings of a multidisciplinary symposium*, (ed. D. M. Power), pp. 7–31. Santa Barbara Museum of Natural History. {24}

Vedder, J. G. and Norris, R. M. (1963). *Geology of San Nicholas Island.* California Geological Survey Professional Paper, No. 369, i–vi, pp. 1–65. {24}

Venevongphet (1988). The status of the elephant in Laos. In *Proceedings of the IUCN/SSC Asian Elephant Specialist Group Meeting, Chiangmai, Thailand, January, 1988* [no first initial for Venevongphet]. {33}

Vereshchagin, N. K. (1977*a*). Berelekh "cemetery" of mammoths. In Mammoth fauna of the Russian plaine [*sic*] and eastern Siberia, (ed. N. K. Vereshchagin). *Proceedings of the Zoological Institute (Leningrad)*, **72**, 5–50. Academy of Sciences of the USSR. [in Russian]. {26}

Vereshchagin, N. K. (ed.) (1977*b*). *Mammoth fauna of the Russian plaine [sic] and eastern Siberia. Proceedings of the Zoological Institute (Leningrad)*, **72**. Academy of Sciences of the USSR. [in Russian]. {19}

Vereshchagin, N. K. (1981). Morphometric description of the mammoth baby. In *Magadon baby mammoth*, Mammuthus primigenius *(Blumenbach)*, (ed. N. K. Vereshchagin and V. M. Mikhelson), pp. 52–80, Nauka, Leningrad [in Russian]. {26}

Vereshchagin, N. K. (1982). New Gydansk (Yuribey) mammoth discovery. *New Zoologie*, **3**, 32–8 [in Russian]. {7}

Vereshchagin, N. K. and Baryshnikov, G. F. (1982). Paleoecology of mammoth fauna in the Eurasian arctic. In *Paleoecology of Beringia*, (ed. D. M. Hopkins, J. V. Matthews Jr, C. E. Schweger, and S. B. Young), pp. 267–80. Academic Press, New York. {29}

Vereshchagin, N. K. and Kuzmina, I. E. (1977). Remains of mammals from Paleolithic sites on the Don and Desna Rivers. *Proceedings of the Zoological Institute (Leningrad)*, **72**, 77–110. Academy of Sciences of the USSR. {19}

Vereshchagin, N. K. and Mikhelson, V. M. (ed.) (1981). *Magadan baby mammoth*, Mammuthus primigenius *(Blumenbach)*. Nauka, Leningrad [in Russian]. {7}

Vereshchagin, N. K. and Tikhonov, A. N. (1986). A study of mammoth tusks. In Mammals of the Quaternary fauna of the USSR, (ed. N. K. Vereshchagin and I. E. Kuzmina). *Proceedings of the Zoological Institute (Leningrad)*, **149**, 3–14. Academy of Sciences of the USSR [in Russian]. {26}

Vereshchagin, N. K. and Tikhonov, A. N. (1987). A study of mammoth tusks from permafrost of northeastern Siberia. *Current Research in the Pleistocene*, **4**, 120–2. {30}

Vickers-Rich, P., Monaghan, J. M. Baird, R. F., Rich, T. H., Thompson, E. M., and Williams, C. (1991). *Vertebrate paleontology of Australasia.* Pioneer Design Studio, Melbourne. {34, ApC}

Vigors, N. A. (1827). A review of Étienne Geoffroy Saint-Hilaire and Frédéric Cuvier *Histoire Naturelle des Mammiféres, avec des figures originales, dessinées d'après des animaux vivans. The Zoological Journal* (London), **III**, 140–3. ['Histoire Naturelle des Mammiféres' is a chapter in Geoffroy Saint-Hilaire and F. Cuvier (1825)]. {16, ApC}

Viljoen, P. J. (1992). The desert elephants of Namibia. In *Elephants: majestic creatures of the wild*, (consult. ed. J. Shoshani), pp. 131–3. Rodale Press, Emmaus, Pennsylvania. {29}

Viljoen, P. J. and du P. Bothma, J. (1990). Daily movements of desert dwelling elephants in the northern Namid Desert. *South African Journal of Wildlife Research*, **20**, (2), 69–72. {29}

von Koenigswald, G. H. R. (1933). Beitrag zur Kenntnis der fossilen Wirbeltiere Javas. I. *Wetenschappelijke mededeelingen-dienst van den Mijnbouw in Nederlandsch-oost Indië*, **23**, 1–178. [From Osborn 1936, p. 783. and Hooijer 1984, p. 231]. {ApA}

von Koenigswald, G. H. R. (1950). Voorlopige mededeling omtrent het voorkomen van Mastodon borsoni in Nederland. *Geologie en Mijnbouw*, New Series, **12**, 14–15. {20}

von Koenigswald, G. H. R. (1956). Fossil mammals from the Philippines. Full text with illustrations, of paper No. 22, *Proceedings of the fourth far-eastern prehistory congress* [originally presented as abstract 45]. Special reprint, with revisions in 1955. National Research Council of the Philippines, University of the Philippines, Diliman, Quezon City.{23}

von Koenigswald, G. H. R. (1959). *A mastodon and other fossil mammals from Thailand.* Royal Department of Mines (Bangkok, Thailand). Report of Investigation No. 2, pp. 25–8 + 5 figs. {ApC}

von Koenigswald, W. (1986). A skeleton of the straight-tusked elephant from upper Pleistocene gravels of the Rhine from Crumstadt near Darmstadt (West Germany). *Cranium*, **3**, (1), 8–13 [in German with English summary]. {ApC}

von Koenigswald, W. and Tobien, H. (1987). Bemerkungen zur Altersstellung der pleistozänen Mosbach-Sande bei Wiesbaden. *Geologisches Jahrbuch Hessen*, **115**, 227–37. {19}

von Koenigswald, W., Martin, T., and Pfretzschner, H. U. (1993). Phylogenetic interpretation of enamel structures in mammalian teeth: possibilities and problems. In *Mammal phylogeny: placentals* (ed. F. S. Szalay, M. J. Novacek, and M. C. McKenna), pp. 303–14. Springer-Verlag, New York. {2, 17}

von Meyer, H. (1831*a*). [Note on *Dinotherium giganteum* and *D. bavaricum*] *Neues Jahrbuch für Mineralogie*, 296–7. [From Osborn 1936, p. 790] {ApC}

von Meyer, H. (1831*b*). Ueber *Mastodon arvernensis* bei Eppelsheim. *Beiträge zur Petrefactenkunde*, **15** (2), 113–24 + 1 plt. {ApC}

von Meyer, H. (1867). Studien über das genus *Mastodon*. *Palaeontographica*, **18**, 1–71. {9}

von Meyer, H. and Göppert, H. R. (1848). *Index Palaeontologicus in Bronn's Handbuch einer Geschichte der Natur*, **3** (1), erste hälfte, pp. viii + lxxxiv + 775. [From Osborn 1936, p. 790] {ApA}

Voorhies, M. R. (1990). *Vertebrate paleontology of the proposed Norden Reservoir area: Brown, Cherry, and Keya Paha counties, Nebraska.* Division of Archaeological Research, University of Nebraska, Lincoln Technical Report, No. **82–9**, pp. 1–105. {15}

Vrba, E. S. (1987). Ecology in relation to speciation rates: some cases histories of Miocene–Recent mammal clades. *Evolutionary Ecology*, **1**, 283–300. {18}

Vu Khoi, L. (1988). Some biological characteristics of elephant and elephant domestication in Vietnam. In *Proceedings of the IUCN/SSC Asian Elephant Specialist Group Meeting, Chiangmai, Thailand, January 1988*. {33}

Wagner, A. (1857). Neue Beiträge sur Kenntniss der fossilen Säugthier-Ueberreste von Pikermi. *Abhandlungen der Bayerische Akademie der Wissenschaften*, München, **II Cl. VIII, I**, 111–58 (or pages 1–50), plts III–IX. {ApC}

Walker D. N. (1987). Late Pleistocene/Holocene environmental changes in Wyoming: the mammalian record. In *Late Quaternary mammalian biogeography and environments of the great plains and prairies*, (ed. R. Graham, H. Semken, and M. Graham). Illinois State Museum, Scientific Papers, Vol. 22, pp. 334–92. {28}

Walker, H. J. (1973). Morphology of the North Slope. In *Alaska Arctic Tundra*, (ed. M. E. Britton). Arctic Institute of North America Technical Paper No. 25, pp. 49–92. {29}

Walker, P. L. (1980). Archaeological evidence for the recent extinction of three terrestrial mammals on San Miguel Island. In *The California Islands: proceedings of a multidisciplinary symposium*, (ed. D. M. Power), pp. 703–12. Santa Barbara Museum of Natural History. {24}

Walker, W. F. (1980). *Vertebrate dissection*, (6th edn). Saunders College, Philadelphia. {8}

Wang, Y. (1989). The study of Fuyang fossil fauna from the late Pleistocene. Unpublished Master thesis, China University of Geosciences, Beijing, China. [In Chinese with English abstract]. {ApC}

Wang, B.-Y. and Wang, P.-Y. (1991). Discovery of early Miocene mammalian fauna from Urtu area, Alxa Left Banner, Inner Mongolia. *Chinese Science Bulletin*, **36**, (2), 135–9. {21}

Wang, H., Chen, W., and Yu, J. (1994). A survey of ancient animal fossils at Tongxin Ningxia and about some connected questions. *Journal of Ningxia University (Natural Science Edition)*, **15** (2), 76–83 [in Chinese with English abstract]. {ApC}

Warren, J. C. (1852). *"Mastodon giganteus" of North America*. John Wilson and Sons, Boston. {ApC}

Watanabe, N. and Kadar, D. (ed.) (1985). *Quaternary geology of the hominid bearing formation in Java*. Special Publication, 4, Geological Research and Development Centre, Bandung. {17}

Watson, D. M. S. (1946). The evolution of the Proboscidea. *Biological Review*, **21**, 15–29. {34}

Webb, S. D. (1974*a*). Chronology of Florida Pleistocene mammals. In *Pleistocene mammals of Florida*, (ed. S. D. Webb), pp. 5–31. University of Florida, Gainesville. {27, 29}

Webb, S. D. (ed.) (1974*b*). *Pleistocene Mammals of Florida*. University of Florida Press, Gainesville. {15}

Webb, S. D. (1976). Mammalian faunal dynamics of the great American interchange. *Paleobiology*, **2**, (3), 220–34. {29}

Webb, S. D. (1977). A history of savanna vertebrates in the New World. Part I: North America. *Annual Review of Ecology and Systematics*, **8**, 355–80. {15, 29}

Webb, S. D. (1978). A history of savanna vertebrates in the New World. Part II: South America and the Great Interchange. *Annual Review of Ecology and Systematics*, **9**, 393–426. {15, 31}

Webb, S. D. (1983). The rise and fall of the late Miocene ungulate fauna in North America. In *Coevolution*, (ed. M. H. Nitecki), pp. 267–306. University of Chicago Press. {15, 30}

Webb, S. D. (1984). Ten million years of mammal extinctions in North America. In *Quaternary extinctions: a prehistoric revolution*, (ed. P. S. Martin and R. G. Klein), pp. 189–210. University of Arizona Press, Tucson. {30}

Webb, S. D. (1989). The fourth dimension in North American terrestrial mammal communities. In *Patterns in the structure of mammalian communities*, (ed. D. W. Morris, B. J. Fox, and M. R. Willig). Special Publication Texas Technical University, No. 28, Lubbock. {29}

Webb, S. D. (1991). Ecogeography and the great American interchange. *Paleobiology*, **17**, (3), 266–80. {29, 31, 34}

Webb, S. D. and Dudley, J. P. (1995). Proboscidea from the Leisy Shell Pits, Hillsborough County, Florida. In *Paleontology and geology of the Leisey Shell Pit, early Pleistocene of Florida*, (ed. R. C. Hulbert, Jr., G. S. Morgan, and S. D. Webb). *Bulletin of the Florida Museum of Natural History*, **37** Part II (20), 645–60. {ApC}

Webb, S. D. and Newsom, L. (1991). Diet of *Mammut americanum* in late Pleistocene of Florida. *Journal of Vertebrate Paleontology*, **11**, (Suppl. 3), 60A–61A. {27, 30}

Webb, S. D. and Perrigo, S. C. (1984). Late Cenozoic vertebrates from Honduras and El Salvador. *Journal of Vertebrate Paleontology*, **4**, 237–54. {15, 29, ApC}

Webb, S. D. and Tessman N. (1968). A Pliocene vertebrate fauna from low elevation in Manatee County, Florida. *American Journal of Science*, **266**, 777–811. {ApC}

Webb, S. D. and Wilkins, K. T. (1984). Historical biogeography of Florida Pleistocene mammals. In *Contributions to Quaternary vertebrate paleontology: a volume in memorial to John E. Guilday*, (ed. H. H. Genoways and M. R. Dawson). Special Publication of Carnegie Museum of Natural History No. 8, pp. 370–83, Pittsburgh. {29}

Webb, S. D., MacFadden, B. J., and Baskin, J. A. (1981). Geology and paleontology of the Love Bone Bed from the late Miocene of Florida. *American Journal of Science*, **281**, 513–44. {15, ApC}

Webb, S. D., Morgan, G. S., Hulbert, R. C. Jr, Jones, D. S., MacFadden, B. J., and Mueller, P. A. (1989). Geochronology of a rich early Pleistocene vertebrate fauna, Leisey Shell Pit, Tampa Bay, Florida. *Quaternary Research*, **32**, 96–110. {27}

Webb, T., III and Bartlein, P. J. (1988). Late Quaternary climatic change in eastern North America: the role of modeling experiments and empirical studies. *Bulletin of the Buffalo Society of Natural Sciences*, **33**, 3–13. {30}

Weber, M. (1928). *Die Säugetiere: Einführung in die anatomie und Systematik der Recenten und Fossilen Mammalia*. Band II, Systematischer Teil (1928, unter mitwirkung von O. Abel). Gustav Fischer, Jena. {5}

Wegner, R. N. (1913). Tertiaer und Umgelagerte Kreide bei Oppeln (Oberschlesien). *Palaeontographica*, **60**, 175–274. [From Osborn 1936, p. 801] {ApC}

Weidmann, M. (1969). Le mammouth de Praz-Rodet (Le Brassus, Vaud). *Bulletin de la Société Vaudoise des Sciences Naturelles*, **70**, 229–40 + 3 plts. {19, 25}

Weinsheimer, O. (1883). Über *Dinotherium giganteum* Kaup. *Palaeontologische Abhandlungen*, **1** (3), 207–82 + 3 plts. {ApC}

Weiss, H. (ed.) (1985). *Ebla to Damascus: art and archeology of ancient Syria*. Smithsonian Institution Travelling Exhibition Service, Washington, DC. {21}

Weithofer, K. A. (1888). Einige Bemerkungen über der Proboscidier. *Morphologisches Jahrbuch*, **14**, 507–16. {12A}

Weithofer, K. A. (1891). Die fossilen Proboscidier des Arnothales in Toskana. *Beiträge zur Paläontologie Österreich-Ungarns und des Orients*, **8**, 107–240 + 21 plts. {9, 19}

Weithofer, K. A. (1893). Proboscidiani fossili di Val d'Arno in Toscana. Comitato geologico. *Memorie per servire alla descrizione della carta geologica d'Italia. Regno*, **4**, (2), 1–152. {9}

Welcomme, J. -L. (1994). Le plus ancien crâne de proboscidien d'Europe, *Gomphotherium hannibali* nov. sp. (Proboscidea, Mammalia), du Miocène inférieur du Languedoc (France). *Comptes Rendus de l'Académie des Sciences*, **319**, Série II(1), 135–40 [in French with English abstract and abridged English version]. {ApC}

Wells, N. A. and Gingerich, P. D. (1983). Review of Eocene Anthracobunidae (Mammalia, Proboscidea) with a new genus and species, *Jozaria palustris*, from the Kuldana Formation of Kohat (Pakistan). *Contributions from the Museum of Paleontology, The University of Michigan*, No. 26 (7), pp. 117–39. {4, 5, 6, 8, 34, ApA}

Wendorf, M. (1982). The fire areas of Santa Rosa Island: an interpretation. *North American Archaeologist*, **3**, 173–80. {24}

Wenner, A. M. and Johnson, D. L. (1980). Land vertebrates on the California Channel Islands: sweepstakes or bridges? In *The California Islands: proceedings of a multidisciplinary symposium*, (ed. D. M. Power), pp. 497–530. Santa Barbara Museum of Natural History. {24}

Wenner, A. M., Cushing, J., Noble, E., and Daily, M. (1991). Mammoth radiocarbon dates from the Northern Channel Islands, California. In *Proceedings of the Society for California Archaeology*, Vol. 4, (ed. M. D. Rosen, L. E. Christenson, and G. T. Gross), pp. 221–6. Society for California Archaeology, San Diego. {24}

West, R. M. (1980). Middle Eocene large mammal assemblage with Tethyan affinities, Ganda Kas Region, Pakistan. *Journal of Paleontology*, **54**, 508–33. {1, 6, 8}

West, R. M. (1983). South Asian middle Eocene moeritheres (Mammalia: Tethytheria). *Annals of Carnegie Museum*, **52**, 359–73. {4, 8, 34, ApA}

West, R. M. (1984). A review of the south Asian middle Eocene Moeritheriidae (Mammalia, Tethytheria). *Mémoires de la Société Géologique de France*, New Series, **147**, 183–90. {8, 34, ApA}

Western, D. (1989). The ecological value of elephants: a keystone role in African ecosystems. In *The ivory trade and the future of the African elephant*, (co-ordinator S. Cobb), Vol. 2, section 5.2. The Ivory Trade Review Group, Oxford. {34}

Western, D. (1991). When the forest falls silent. In *Elephants: the deciding decade*, (ed. R. Orenstein), pp. 83–95. Key Porter, Toronto. {34}

White, J. A. (1966). A new *Peromyscus* from the late Pleistocene of Anacapa Island, California, with notes on variation in *Peromyscus nesodytes* Wilson. *Contribution in Science, Natural History Museum of Los Angeles County*, **96**, 2–8. {24}

Whitehead, D. R., Jackson, S. T., Sheehan, M. C. and Leyden, B. W. (1982). Late-glacial vegetation associated with caribou and mastodon in central Indiana. *Quaternary Research*, **17**, 241–57. {27}

Whitmore, F. C. Jr, Emery, K. O., Cooke, H. B. S., and Swift, D. J. P. (1967). Elephant teeth from the Atlantic continental shelf. *Science*, **156**, 1477–81. {24, 27, 29}

Whittington, S. L., and Dyke, B. (1984). Simulating overkill: experiments with the Mosimann and Martin model. In *Quaternary extinctions: a prehistoric revolution*, (ed. P. S. Martin and R. G. Klein), pp. 451–65. University of Arizona Press, Tucson. {30}

Whybrow, P. J. (1987). Summary [on the Miocene geology and palaeontology of Ad Dabtiyah, Saudi Arabia]. *Bulletin of the British Museum (Natural History) Geology*, **41**, 367–9. {10}

Wight, A. W. R. (1980). Palaeogene vertebrate fauna and regressive sediments of Dur at Talhah, southern Sirt Basin, Libya. In *Geology of Libya*, Vol. 1, (ed. M. J. Salem and M. T. Busrewil), pp. 309–25. Academic Press, London. {8}

Williams, J. H. (1950). *Elephant Bill*. Longmans and Green, London. {29}

Williams, S. (1957). The Island 35 mastodon: its bearing on the age of Archaic cultures in the east. *American Antiquity*, **22**, 359–72. {30}

Wilson, R. W. (1936). A new Pleistocene deer-mouse from Santa Rosa Island, California. *Journal of Mammalogy*, **17**, 408–10. {24}

Windley, B. F. (1984). *The evolving continents*, (2nd edn). John Wiley & Sons, Chichester, England. {8}

Wittry, W. L. (1965). The Institute digs a mastodon. *Cranbrook Institute of Science, News Letter*, **35**, 14–19. {30}

Wolf, D. (1989). Report on skin remains from Monte Verde. Manuscript on file at the University of Kentucky, Lexington. {31}

Wolfe, J. A. (1978). A paleobotanical interpretation of Tertiary climates in the northern Hemisphere. *American Scientist*, **66**, 694–703. {30}

Wolkomir, R. (1991). New finds could rewrite the start of American history. *Smithsonian*, **21**, (12), 130–2, 134–44. {31}

Woodburne, M. O. (1961). Upper Pliocene geology and vertebrate paleontology of part of the Meade Basin, Kansas. *Papers of the Michigan Academy of Science, Arts, and Letters*, **46**, 61–101. {15, 27}

Woodburne, M. O. (ed.) (1987). *Cenozoic mammals of North America: geochronology and biostratigraphy*. University of California Press, Berkeley. {9, ApD}

Woodburne, M. O. and Swisher, C. C., III (1995). Land mammal high-resolution geochronology, intercontinental overland dispersals, sea level, climate, and vicariance. *Geochronology Time Scales and global stratigraphic correlation. Society of Economic Paleontologists and Mineralogists*. **Special Publication 54**, 336–64. {ApD}

Wright, V. C. and Leopold, E. B. (1982). An early Quaternary record of *Artemisia* steppe in the Snake River plains, Idaho. *Program and Abtracts, Seventh Biennial Conference American Quaternary Association*, p. 182. {27}

Wyss, A. R., Novacek, M. J., and McKenna, M. C. (1987). Amino acid sequence versus morphological data and the interordinal relationships of mammals. *Journal of Molecular Biology and Evolution*, **4**, 99–116. {5}

Wyss, A. R. *et al.* (1993). South America's earliest rodent and recognition of a new interval of mammalian evolution. *Nature*, **365**, 434–7. {ApD}

Xie, J. (1989). Late Tertiary strata and fossil mammals of Gansu, China. *Proceedings of International Symposium on pacific Neogene Continental and Marine Events*, **IGCP-246**, 119–27. {ApC}

Xue, X. (1981). An early Pleistocene mammalian fauna and its stratigraphy of the river You, Weinan, Shensi. *Vertebrata PalAsiatica*, **19** (1), 35–44 + 2 plts [in Chinese with English abstract]. {ApC}

Xue, X. and Zhang, Y. (1991). The Quaternary mammalian fossils and the fossil human beings in China. In *The Quaternary of China*, (ed. Z. Zhang and S. Shao), pp. 313–55. China Ocean Press, Beijing. {ApC}

Yabe, H. (1950). Three alleged occurrences of *Stegolophodon latidens* (Clift) in Japan. *Proceedings of the Japan Academy*, **26**, (9), 61–5. {12A, 17, ApA, ApC}

Yang, H., Golenberg, E. M., and Shoshani, J. (1996). Phylogenetic resolution within the Elephantidae using fossil DNA sequence from American mastodon (*Mammut americanum*) as a outgroup. *Proceedings of the National Academy of Sciences*, **93**, 1190–4. {34}

Ye, J. and Jia, H. (1986). *Platybelodon* (Proboscidea, Mammalia) from the middle Miocene of Tongxin, Ningxia. *Vertebrata PalAsiatica*, **24** (2), 139–51 + 2 plts [in Chinese with English summary]. {13, ApC}

Ye, J., Qiu, Z., and Chen, J. (1989). Comparative study of a juvenile skull of *Platybelodon tongxinensis*. *Vertebrata PalAsiatica*, **27** (4), 284–300 + 2 plts [text in Chinese and English]. {13}

Ye, J., Wu, W., and Jia, H. (1990). Reconstruction of the jaw-closing muscles of *Platybelodon tongxinensis* (Amebelodontidae, Proboscidea) and discussion of cranial evolution from long-jawed mastodont to short-jawed elephantid. *Vertebrata PalAsiatica*, **28** (4), 284–95 [in Chinese with English abstract]. {13}

You, Y. and Qi, G. (1973). New materials of Pleistocene mammals in Yuanmu, Yunnan. *Vertebrata PalAsiatica*, **11** (1), 66–85 [in Chinese]. {17}

You, Y., Liu, H. and Pan, Y. (1978). Plio-Pleistocene deposits and mammalian faunas of Yuanmou Basin, Yunnan. *Professional Papers of Stratigraphy and Palaeontology*, **7**, 40–67 (Geological Publishing House, Peking, China) [in Chinese with English abstract]. {17}

Young, C. C. (1938). A new *Stegodon* from Kwangsi. *Bulletin of the Geological Society of China*, **18**, 219–26. {17}

Young, C. C. and Liu, P. T. S. (1948). Notes on a mammalian collection probably from Yushe series (Pliocene). *Contributions Institute of Geology, Academia Sinica*, **8**, 273–91. {13}

Young, C. C. and Liu, P. T. (1950). On the mammalian fauna of Koloshan near Chungking, Szechuan. *Bulletin of the Geological Society of China*, **30**, 43–90. {17}

Yun, B. (1975). Two new species of *Stegodon* from Yunnan. *Vertebrata PalAsiatica*, **13** (4), 229–33 + 2 plts [in Chinese]. {17, ApC}

Zagwijn, W. H. (1960). Aspects of the Pliocene and early Pleistocene vegetation in The Netherlands. Unpublished D. Phil. thesis. University of Leiden. {20}

Zagwijn, W. H. (1963). Pollen-analytic investigations in the Tiglian of The Netherlands. *Mededelingen van de Geologische Stichting*, New Series, **16**, 49–71. {20}

Zagwijn, W. H. (1979). Early and middle Pleistocene coastlines in the southern North Sea basin. In *The Quaternary history of the North Sea*, (ed. E. Oele, R. T. E. Schuttenhelm, and A. J. Wiggers). (= *Acta Universitatis Upsaliensis, Symposia Universitatis Upsaliensis Annum Quingentesimum Celebrantis*, Vol. **2**. University of Upsala). {20}

Zagwijn, W. H. and Doppert, J. W. C. (1978). Upper Cenozoic of the southern North Sea basin: palaeoclimatic and palaeogeographic evolution. *Geologie en Mijnbouw*, **57**, 577–88. {20}

Zalasiewicz, J. A., *et al.* (1988). Stratigraphy and palaeoenvironments of the Red Crag and Norwich Crag formations between Aldburgh and Sizewell, Suffolk, England. *Philosophical Transactions of the Royal Society of London*, Series B, **322**, 221–72. {19}

Zalensky, V. V. (1903). Osteological and odontographic studies on the mammoth (*Elephas primigenius* Blum), and the elephants (*El. indicus* L. and *El. africanus* Blum). *Scientific results of the Imperial Academy of Sciences expedition to excavate the mammoth carcass found by the Beresovka River in 1901*, Vol. **1**, pp. 1–124. Imperial Academy of Sciences, St. Petersburg. {19, 25, 26}

Zapfe, H. (1954). Die Fauna der miozänen Spaltenfüllung von Neudorf a. d. March (CSR.): Proboscidea. *Sitzungsberichte-Ber. Österreichische Akademie der Wissenschaften Mathematisch-Naturwissenschaftliche Klasse*, Series I, (**163**), 71–87. {9}

Zhai, R. (=Chai, J.) (1959). Discovery of some Miocene mammals in Chin-an, Kansu. *Vertebrata PalAsiatica*, **1** (3), 139–40 + 1 photograph [in Chinese]. {13}

Zhai, R. (1961). On a collection of Neogene mammals from Ching-an, eastern Kansu. *Vertebrata PalAsiatica*, **3**, 262–8 [in Chinese with English summary]. {13}

Zhai, R. (1963). A new species of *Gomphotherium* from Changzhi, SE. Shansi. *Vertebrata PalAsiatica*, **7** (1), 59–61 + 1 plt [in Chinese with English summary]. {ApC}

Zhang, X. (1982). The fossils of Pliocene Elephantoidea from the basin of Lufeng, Yunnan. *Vertebrata PalAsiatica*, **20** (4), 359–65 + 2 plts [in Chinese with English abstract] {ApC}

Zhang, Y., Zong, G., and Liu, Y. (1983). A new species of *Palaeoloxodon* from Pingliang, Kansu. *Vertebrata PalAsiatica*, **21** (1), 65–8 + 1 plt [in Chinese with English abstract]. {ApC}

Zhao, Z. (1980). Fossil mammals from Jiaoshan Wuming, Guangxi. *Vertebrata PalAsiatica*, **18** (4), 299–303 + 1 plt. {ApC}

Zhao, Z., Lix, X., and Wang, L. (1981). Human fossils and associated fauna of Jiulengshan Hill, Guangxi. *Vertebrata PalAsiatica*, **19** (1), 45–54 + 2 plts [in Chinese with English abstract]. {17}

Zheng, S., Huang, W., Zong, G., Huang, X., Xie, J., and Gu, Z. (1975). *Huanghe Xiang*, pp. 1–46 + 20 plts. Science Press, Beijing [in Chinese]. {17, ApC}

Zherekhova, I. E. (1977). Description and measurement of Berelekh mammoth teeth. In *Mammoth fauna of the Russian plaine* [sic] *and eastern Siberia*, (ed. N. K. Vereshchagin). Proceedings of the Zoological Institute, Vol. 72, pp. 50–8 + 1 plt. Academy of Sciences of the USSR, Leningrad [in Russian]. {19, 26}

Zhou, M. (1978). See under Chow, M. (1978)

Zhou, M. and Zhang, Y. (1974). *Fossil elephants of China*. Science Press, Beijing. {6, 13, ApC}

Zhou, M. (=Chow, M.) and Zhang, Y. (1983). Occurrence of the proboscidean genus *Stegotetrabelodon* in China. *Vertebrata PalAsiatica*, **21** (1), 52–8 + 1 plt. [in Chinese with English abstract]. {12A, 17, ApA, ApC}

Zhou, X. (1988). The Pliocene micromammalian fauna from Jingle, Shanxi: a discussion of the age of Jingle red clay. *Vertebrata PalAsiatica*, **26** (3), 181–97 + 2 plts [in Chinese with English summary]. {17}

Zong, G. (1979). A new species of *Pentalophodon* from Pingliang, Kansu. *Vertebrata PalAsiatica*, **17** (1), 81–4 + 1 plt [in Chinese with English abstract]. {ApA. ApC}

Zong, G. (1987). Note on some mammalian fossils of Yanyuan, Sichuan. *Vertebrata PalAsiatica*, **25** (2), 137–45 + 2 plts [in Chinese with English abstract]. {13, ApC}

Zong, G. (1992). Occurrence of proboscidean genus *Stegolophodon* in China. *Vertebrata PalAsiatica*, **30** (4), 287–94 + 1 plt [in Chinese with English abstract]. {ApC}

Zong, G., Tang, Y., Lei, Y., and Li, S. (1989). *Hanjiang Zhongguoruchixiang* [= *The Hanjiang mastodont* (Sinomastodon hanjiangensis *nov. sp.*)]. Academic Press, Beijing [in Chinese with English summary]. {16, ApC}

AUTHOR INDEX

Citations for authorships of taxa are excluded from this index unless the authors are cited in references. Numbers in *italics* denote reference to illustrations. References to footnotes are designated with the letter 'n.' after the page number.

STRATIGRAPHIC, GEOGRAPHIC, AND LOCALITY INDEX

Numbers in *italics* denote reference to illustrations.
References to footnotes are designated with the letter 'n.' after the page number.

TAXA INDEX

- With the exception of listings in some appendices, this index includes proboscidean taxa mentioned in this volume (common names are in the Subject Index). For example, species and subspecies which were listed in Osborn (1936, 1942) and their senior synonyms as adopted in this volume are given in Appendix C.2; the majority are not listed in this index because they are easily accessible there (pp. 374–85). In addition, all taxonomic names below the ordinal level (e.g. family, genus, species and subspecies) which are listed in the tabulated sections of Appendices A, B, C.1, Addendum to C.1, and C.2 (pp. 349–85), may not be listed below unless they appear elsewhere in this volume.
- Following Article 31 (especially 31(b)) of the International Code of Zoological Nomenclature, we amended the suffixes of some species and subspecies to agree with the gender of the genus. These changes were made in the summary tabulated portion of Appendix C.1 (pp. 355–8; e.g. under *Gomphotherium*), in the right columns of pages 369–82, and elsewhere in the volume, unless the species is discussed in the original context or the name is within quotation marks.
- When the genus, species, or subspecies, is not considered valid in this volume, the symbol '=' appears inside parentheses followed by the valid taxon used in this volume [e.g., '*africana pharaohensis, Loxodonta* (=*Loxodonta africana*) 384'].
- Taxa designated as '*incertae sedis*' fit into two categories: those listed in the tabulated part of Appendix C.1 and those given elsewhere in the volume. The former is identified in parentheses without '=', e.g., '*barbouri, Morrillia* (*incertae sedis*) 357' and the latter is given either as '*falconeri, Mastodon* (=Proboscidea *incertae sedis*) 363 n.', or as '*pyrenaicus, Zygolophodon* (='gomphotheres' *incertae sedis*) 142'.
- Taxa with a complex structure such as subgeneric names or synonyms in parentheses are listed under the first entry (genus) followed by the second (subgenus or synonym or both). For example, for the entry '*willistoni, Trilophodon* (*Genomastodon*) (=*Gomphotherium productum*) 361 n.', *willistoni* is also listed under '*Trilophodon* (*Genomastodon*) *willistoni*'.
- Numbers in *italics* denote reference to illustrations. References to footnotes are designated with the letter 'n.' after the page number.

SUBJECT INDEX

For simplicity, most quotation marks have been deleted. Numbers in *italics* denote reference to illustrations. References to footnotes are designated by the letter 'n.' after the page number.

The manufacturer's authorised representative in the EU for product safety is Oxford University Press España S.A. of El Parque Empresarial San Fernando de Henares, Avenida de Castilla, 2 - 28830 Madrid (www.oup.es/en or product.safety@oup.com). OUP España S.A. also acts as importer into Spain of products made by the manufacturer.
Printed and bound by CPI Group (UK) Ltd, Croydon, CR0 4YY
14/09/2026
02216298-0001